Advances in
Human Factors, Ergonomics,
and Safety in Manufacturing
and Service Industries

Advances in Human Factors and Ergonomics Series

Series Editors

Gavriel Salvendy
Professor Emeritus
Purdue University
West Lafayette, Indiana

Chair Professor & Head
Tsinghua University
Beijing, People's Republic of China

Waldemar Karwowski
Professor & Chair
University of Central Florida
Orlando, Florida, U.S.A.

Advances in Human Factors, Ergonomics, and Safety in Manufacturing and Service Industries

Edited by
Waldemar Karwowski
Gavriel Salvendy

CRC Press
Taylor & Francis Group
Boca Raton London New York

CRC Press is an imprint of the
Taylor & Francis Group, an **informa** business

CRC Press
Taylor & Francis Group
6000 Broken Sound Parkway NW, Suite 300
Boca Raton, FL 33487-2742

First issued in paperback 2019

© 2011 by Taylor and Francis Group, LLC
CRC Press is an imprint of Taylor & Francis Group, an Informa business

No claim to original U.S. Government works

ISBN-13: 978-1-4398-3499-2 (hbk)
ISBN-13: 978-0-367-38386-2 (pbk)

Visit the Taylor & Francis Web site at
http://www.taylorandfrancis.com

and the CRC Press Web site at
http://www.crcpress.com

Table of Contents

Section 1: Enterprise Management

Section II: Human Factors in Manufacturing

Section III: Processes and Services

Section IV: Design of Work Systems

Section V. Working Environment

Section VI. Product and System Safety

Section VII. Safety Design Issues

Section VIII. Safety Management

Section IX. Hazard Communication

Section X. Occupational Risk Prevention

Preface

This book is concerned with the human factors, ergonomics, and safety issues related to the design of products, processes, and systems, as well as operation and management of business enterprises in both manufacturing and service sectors of contemporary industry.

The book is organized into ten sections that focus on the following subject matters:

I: Enterprise Management
II: Human Factors in Manufacturing
III: Processes and Services
IV: Design of Work Systems
V.   Working Environment
VI.   Product and System Safety
VII.   Safety Design Issues
VIII.   Safety Management
IX.   Hazard Communication
X.   Occupational Risk Prevention

Each of the chapters of this book were either reviewed by the members of the Editorial Board or contributed by them. We would like to thank the following Board Members for their invaluable contributions:

R. Badham, Australia
B. Das, Canada
P. Dawson, UK/Australia
E. Fallon, Ireland
E. Górska, Poland
A. Gramopadhye, USA
M. Grozdanovic, Serbia
W. Grudzewski, Poland
I. Hejduk, Poland
M. Helander, Singapore
H. Iridiastadi, Indonesia
W. Karwowski, USA
G. Kaucsek, Hungary
B. Kayis, Australia
K. Kogi, Japan
D. Koradecka, Poland
R. Koubek, USA
Y. Kwon, S. Korea
M. Lehto, USA
R. Lifshitz, Israel

H. Luczak, Germany
S. Maly, Czech Republic
G. M. Martínez de la Teja, Mexico
A. Matias, Philippines
P. Mondelo, Spain
I. Noy, USA
Y. Okada, Japan
H. Pacaiova, Slovak Republic
D. Podgorski, Poland
A. Polak-Sopinska, Poland
K. Saarela, Finland
D. Shinar, Israel
J. Sinay, Slovakia
M. Soares, Brazil
J. Stahre, Sweden
S. Trzcielinski, Poland
T. Van Der Schaaf, The Netherlands
M. Wang, PR China
M. S. Wogalter, USA
B. Zimolong, Germany

This book will be of special value to researchers and practitioners involved in the design of products, processes, systems and services which are marketed and utilized by a variety of organizations around the world.

April 2010

Waldemar Karwowski
University of Central Florida
Orlando, Florida, USA

Gavriel Salvendy
Purdue University
Sarasota, Florida, USA

Editors

Standardization as a Tool to Promote Improvement - Case Study of an Automobilist Factory

Leandro Wiemes [1], Giles Balbinotti [2],
Grazielle A. Coutinho [3], Leandro V. Vieira [4]

[1]Universidade Federal do Paraná – UFPR
Curitiba,PR, Brazil

[2]Universidade Federal de Santa Catarina – UFSC
Florianópolis, SC, Brazil

[3]Universidade Federal de São Carlos – UFSCar
São Carlos, SP, Brazil

[4]Universidade Tecnológica do Paraná - UTFPR
Curitiba,PR, Brazil

ABSTRACT

The quality structure of a production process is strongly supported on the standardization and therefore on how the activities and improvements in the job are recorded and documented. To achieve the best results in terms of quality dimensions and ensure customer satisfaction this paper presents a mechanism called Job Observation (structured method) to strengthen the standardization on a production line. Some definitions and relevant aspects are also discussed in a way to demonstrate the application of this method in a production line. Results and comments from the application of this methodology are also presented.

Keywords: Job observation, standardization, quality, production.

INTRODUCTION

The organization of the documentation in the manufacturing process can be broken down per job, which speeds and allows more versatility in the methods of updating information. However, the documents and records alone do not guarantee the quality levels as they were defined for the processes. Taking into account the philosophy of continuous improvement, provided in ISO 9001 requires the use of tools that contribute to the development, improvement and checking of manufacturing processes. The consolidation of standardizing on a production line allows achieving the best results in terms of quality dimensions (Quality, Cost, Time, Human Resource and Security) to ensure customer satisfaction for the product. A proposed mechanism for achieving this goal is the Job Observation, which culminates in one of the key elements to make viable the development cycle SDCA (Standard, Do, Check, Act) in the standardization process. With this method it is possible to identify compliance with the procedures or not during the performance of activities, aiming at a detailed observation to highlight any anomalies and find ways to progress. The knowledge by the supervisor of the manufacturing process to the reality of the documentary structure of the area to be noted is the basic premise for the Job Observation is well established. By using the observation tool in the workplace is possible to identify proposals for improvements related to: accident prevention, logistical problems, ergonomic conditions, changes in operational methods, and in some cases, even change the specifications of the product. To demonstrate the occurrence of certain situations in these industrial operations, mainly due to the speed of production and demand, causes often defective or even improvements are not easily detectable by the production coordinators. This paper presents a formal structure and consolidated the production process of the automobile industry, and demonstrates practical examples of implementation of the observation tool in the workplace through graphic examples and results of improvement.

According to Faye and Falzon 2009, Production in the automotive industry, based on assembly line work, is now characterized by lean manufacturing and customization. This results in greater flexibility and increased quality demands, including worker performance self-monitoring.

Perhaps, the most important dependent variable in industrial and organizational psychology is job performance. For all of the main applications of this branch of psychology, such as employee training and job redesigning, the focus is almost always on improving job performance. (Kahya apud Borman, 2007). However, a consensus seems to be developing that job boundaries are becoming blurred, as inter-job activities become the norm (Singh apud Drucker, 2007).

As (Singh apud Schneider and Konz, 2007) note, "...the implicit assumption has been that the specification of the tasks to be performed, and the knowledge, skills and abilities required for job performance are for a job as it currently exists, and/or has existed in the past. This assumption implies that the job is static..."

Sugimori et al, 1977, explain, nowadays it has become an international interest to respect humanity of workers in production shops. Toyota firmly believes that making up a system where the capable Japanese workers can actively participate in running and improving their workshops and be able to fully display their capabilities would be foundation of human respect environment of the highest order.

STANDARDIZATION

As contextualizes Wiemes (2009), the standardization of processes consists of activities to be performed in jobs, and should be held in the best possible way. For the application of standardization and control it from day to day, you can record smaller losses from rework or even scrapping of components and / or products, contributing significantly to increase the financial results of the organization. Another good contribution to the standardization provides is the provision of security of the person who performs the activities, as it also deals with working conditions and personal protective equipment (PPE) needed to perform the operation

The production supervisor should know the essence of all documentation on the work unit so that he can make a good observation on the job. The main documents that he should consider prior to its observation in the workplace are: Engineering Document, Operating Procedure (POP) and Operational Training Form.

Participation in these activities will also spread information in the organization and thereby improve the understanding of other stakeholders' view points. This may have the effect of creating understanding for not changing certain conditions as well as improving the readiness for change in other situations. The development work may create challenges to creativity and to joint group activities in problem solving. This may enhance team work and cooperation. Communication may thus improve between workers and between workers and management (Eklund apud Imai, 2000).

However, standardization alone is not an element which alone can ensure the quality of the production process. It is necessary that these activities are constantly monitored. Even as the production process is subject to the occurrence of interference, ranging from an abnormality in a specific operation, change from a kaizen, interference outside the process.

The standardization of an activity or process requires time and personnel training. The following documents were used to do standardization and to turn it structured and consistent: Engineering, Operating Procedure (POP), Document Line Balancing Form and Operational Training. These are the main basis of standardization and must be specified in the System of Quality Management in the organization.

These procedures provide insight into the requirements of a job and a wealth of information not readily obtainable by other means. In addition, they often provide the content for the development of scales and instruments used in the more structured procedures (Jones et al, 1991).

EXPLANATORY FLOWCHART

The methodology of Job Observation (JOB Observation - Automotive industry documentation, 2009) is to assess factors that impact a production process. To apply this methodology it will be necessary to define a schedule that includes the work station and the operator to be audited and the date of the Job Observation, scheduled for the month in progress. The planned evaluation lasts about one hour. In a comprehensive manner and on a unit with 15 workers, it should have provided in your schedule an audit every two days. This schedule can be put in numerical order or even random order, as is considered practical, thus, the work unit that provide more nonconformities may need attention.

The document used in the audit provides information about the job and the operator to be audited. It sets out criteria that must be assessed during the observation. To strengthen the audit is identified the reason why the observation is being performed. In this document can be identified the following items: Quality, Time and Security, and can also cover the other dimensions of quality (Cost and Human Resources). If no criterion presents negative conditions than the expected, the deviation is described and then is set an action plan containing the action to be taken to correct the problem identified, the responsible action and the execution time.

The main components of the Job Observation are Training, Standardization, Quality Analysis of the work unit, five senses (5S`s) kaizen and Environment, as described below:

Training - consists in identifying the operator in Table Versatility of the Work Unit. In this case, the employee must present situation in training or already trained and able to carry out the activities in (s) stand (s) of work. While in training in a job, the employee must be accompanied by a trained and able to perform the operation at the station (in case of failure of implementation carried out by the new operator). The training developer ends when he is able to perform the operations in the cycle time down to the line and there are no quality problems in operations.

Standardization - In this topic is rated the procedure performed by the operator (which is the order and content of each stage of the operations of the job) against the documentation that is being audited. Also identify any tools used during the execution of operation. Using as a background paper the POP (Operating Procedure) audits assessed the following key points and the reason for their existence, that is what is important to realize in a transaction and why. Exceptions and / or prohibitive situations can also be identified in the documents themselves and /or Engineering Documents. The continuous assessment based on the documentation of the station work to verify that the POP take into account the specifications recommended in the pieces that have the characteristics of safety and regulation, and the accompanying documentation is identified and available.

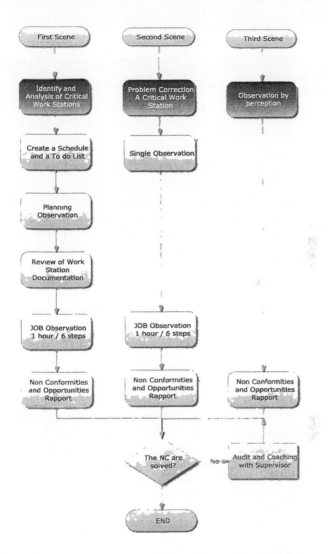

Figure 01 – Explanatory Flowchart of the Job Observation

Kaizen - according to the preconization of ISO 9000, the continuous improvement is a very important part in carrying out audits of certification and recertification. And it must be well presented and emphasized. Employees are encouraged to propose, make and evaluate the performance of a Kaizen. The completion of Job

Observation strengthens the implementation of the culture of continuous improvement, especially with the involvement of the person who will be directly impacted by the change. Some specifications can be identified by questioning the operator or, when the activities performed by the employee, performs the same movements and gestures at odds with good operating practices.

5S - the practice of using the senses, sorting, cleaning, health and self-discipline are also transmitted to employees in the form of training and subsequently identified the jobs on this situation.

Environment - also characterized as a management system, focused on the environment. It makes the employee aware of the possible impacts that he may cause to the environment in case of improper disposal of materials.

Items related to safety are strictly observed during the audit. They concern the safety rules of the company and proper use of PPE 's (Personal Protective Equipment). It carried out an evaluation in place to identify if there is proper identification of parts specified.

After performing the audit, the supervisor here called an auditor, has a responsibility to address nonconformities neatly placing them in a LUP (single list of problems), for a strict control of the shares.

The auditor, in the process must have full knowledge gained from the station to observe, that subtends the greater your knowledge the better your observation.

The Job Observation is the basis for a process improvement or a correction process. It is possible to say that it is the check in the SDCA cycle (Standard, Do, Check, Act), so it is a tool for obtaining performance. There may be have circumstances that have arisen an emergency post a comment, for example, when there is a problem related to security or the occurrence of an accident at work in unity, the hierarchy should specify to the auditors to pay more attention and greater focus on safety.

Three types of evaluations of the Job Observation can be consider: a planned - which is described in a schedule, the possibility - which are those that were not planned; observation and reflection - that is the cultural aspect, that every passage in the post its elements are observed. The observation is a reflection. The planned observation is the most important of the three by the fact that the process of prevention of birth is constantly evaluated.

The main contemplation of the process will be to finalize and develop a strategic objective of restructuring at work, showing strongly the technique of observation, as noted, is not to look, but use all five senses to learn the reality of a situation and to act. An experienced auditor will not leave their lanes without notice improvements and / or non-compliance.

However, the observation post is a technique that can be further explored at the time of observation. To transmit the feedback to the operator, which in turn may relay something they are interested to complete the work successfully.

For this, the supervisor should have developed a critical sense and a very accurate control of your manufacturing process.

A CASE STUDY

The case presented in this study was established in a production unit of the automobile industry located in the south of Brazil, whose production is doors cars with 48 vehicles for hour. The study was took place through an interview with the supervisor of the production line and an operator of the same work unit. The person responsible for carrying out the Job Observation is the supervisor of the unit and, as the rules of production system established by the company the supervisor must accomplish at least one Job Observation per day.

The estimated time to perform a Job Observation is 1 hour and 40 minutes in that 40 minutes is the period of time used to talk to the operator and the other part of time is used to audit the work station.

In the case presented by the supervisor and according to the identified evidences the supervisor said in his schedule of Job Observation it meets 100% of its operators with 100% of their posts within the period provided for the assessment.

For the audit, it is applied a standard form which has the main points to be observed during the Job Observation (Training, Standardization of the work station, Entertainment Unit, Records, Media Manufacturing, 5 senses, security, safety, environment and Kaizen). There are 18 specified criteria for the Job observation. Another field to be filled in the heading relates to the supervisor name, the hierarchical name, the unit and the workstation identification to be audited, the cycle time of the line and the name of the operator. There is a field where the supervisor must put the reason for the observation (Planned, Quality, Time, Security or Other) with the date of observation.

The supervisor interviewed uses a plan of action for the items identified as non-conformity in Job Observation. Through this action plan is constructed an indicator of the amount of non-compliance identified per month, as shown below in figure XY. This information considers also the totality of non conformities identified on the work unit.

For each non-compliance identified in this list there is an action plan associated with the name of the responsible, the deadlines for completion and their status.

To improve the methodology of the Job Observation, the Production Chief of the supervisor does a regular check of job observation, checking in the Gemba the changes in situations and demonstrates eventual more improvements. The principal orientation of the Production Chief is to disclose its board to the supervisors is that for each observation in the workplace should be doing it from the moment the focus of observation was established, is it quality, process compliance, and/or others. This is a very important guideline for when the supervisor does a job observation and it will be done all the time to identify improvements in the production process of the job observation.

Among the major non-conformities identified in the production line at the time of the job observation are related to how the operator performs the operation, i.e. e. differently than are specified in the statement of work pattern, the five senses and improvements to possible risks of accidents, such as incorrect posture of the operator

The comments emitted by the worker were: "The more I know the process more it can identify non-compliances and find clues for improvement"; "My boss asked to keep the focus on the Work Instruction, because we had many problems and as consequences, poor results. After that improved guidance" and "I like when my supervisor held a Job in my post, because we have a good conversation, if you have problems at work or at home, it helps me"

After an interview with the supervisor that uses the Job Observation, he related the following comments to express his opinion about this systematic: "The more I know the process more it can identify non-compliances and find clues for improvement" and "My boss asked to focus on the Work Instruction, because we had many problems on it and the consequences was poor results. After his guidance everything improved"

Depending on the type of trouble, it can be said that the Job Observation is considered a working tool that is beyond a simple observation. There is a good structure that follows this process, being a good tool to maintain standardization and especially the relationship between boss and subordinate.

DISCUSSION

In view of the case presented and as mentioned by Sugimori, et al (1977) compared the results obtained in this research, the Job Observation can be considered a tool of origin based on the Toyota System, and that has contributed significantly to the achievement of excellent results. As shown in the figure 02 below, it represents a reduction of non quality identified in the production line between the years 2007 and 2008. The identification of NPAD refers to the non quality identified in the production line and the unit refers over 1000 produced products.

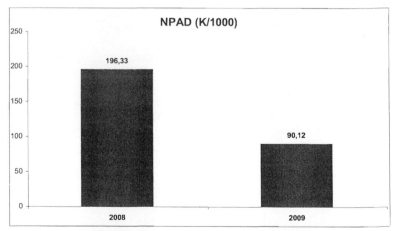

Figure 02 Quality indicator of Production Line.

The quality results were very significant, because the unit achieved a 46% reduction in quality output. In addition to the JOB Observation as the main tool for the result of quality improvement, there were other means, such as deployment of Poka-yoke (mistake-proof system) unit.

Importantly, the level of accidents in the unit where the supervisor who is using the tool Job Observation, contributed to reducing casualties over the years, as shown in the chart below, and to take prevention about these potential causes:

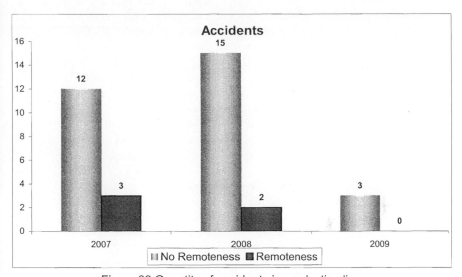

Figure 03 Quantity of accidents in production line.

The accident results has improved over the years, because it was been established means to reduce and avoid accidents, as showed in Figure 03. But according to comment of the unit supervisor, the primary factor for reducing accidents has been the use of the tool JOB Observation, through the identification of possible accidents, having as action plan the coaching with the operator in a way to prevent accidents.

According to Jones et al 1991, the importance of reviewing the procedures and work instructions in the daily life of a production system is that it has a key role in contributing to the success of the company.

CONCLUSION

According to the research presented a review of the documentation has good results for the growth of a company, making it more efficient and effective. A Job Observation shows significant results when applied with responsibility and accuracy.

The structure presented in the implementation of this tool is for large-scale, the existence of the SDCA cycle is of fundamental importance for the success of the tool, as a whole, indicating better quality.

Applying the tools together and using the Job Observation form, it will aid to point out problems and to diagnose the situations in a way to solve problems and avoid their recurrence, i.e. the problem will be completely eliminated. Obtaining quality revolves around the observation, noting that not only observe but also to analyze the situation where it is possible to identify the causes and improvements, not to mention also the prevention of the innumerable problems which presents the audit reports .

A JOB observation can be further explored as a tool to approach the supervisor with operator, getting a closer relationship and reliable, therefore bringing to the supervisor better results on the themes of his responsibility and quality in the unit production, reducing costs production, improving the operative mode, in addition to the search for better working conditions for operators.

In one of the realized JOB's was noted that the operator did not know the main activities of his job and thus was established as an action plan the supervisor, to train the operator in operative deficiencies, identified through the training on the job. "

Another important point was observed on employment security, because he did not use any PPE's (Personal Protective Equipment) as recommended. So, it was established that the supervisor would be monitored through daily coaching by the operator. This coaching was aimed to strengthening of the operator problems (work accidents) that may be generated by non-use of PPE's.

Therefore, based on the Job Observation that brings together key tools and levels of observation, it is summarized that it is the closest ally to the enterprise as a whole, because it will ultimately become possible to gather the main bases of the smooth operation of the company, supervisor and subordinate, which of which will work for the union is stable and self-criticism thus forming the backbone of the industry, or will be complete for the proper functioning of the production.

REFERENCES

Eklund, J. (2000). Development work for quality and ergonomics. Applied Ergonomics , pp. 641-648.

Faye, H., & Falzon, P. (2009). Strategies of performance self-monitoring in automotive production. Applied Ergonomics ,pp. 915-921.

Jones, J. W., Steffy, B. D., & Bray, D. W. (1991). Applying psychology in business: the handbook for managers and human resources professionals. NY: Lexinginton Books.

Kahya, E. (2007). The effects of job characteristics and working conditions on job performance. International Journal of Industrial Ergonomics 37 , pp.515–523.

Singh, P. (2007). Job analysis for a changing workplace. Human Resource Management Review , pp. 87-99.

Sugimori, Y., Kusunoki, K., Cho, F., & Uchikawa, S. (1977). Toyota production system and Kanban system materialization of just-in-time and respect-for-human system. International Journal of Production Research, pp. 553-564.

Wiemes, L. et al, (2009), Standardization in Production – The Experience in an Automobilist Factory, COBEM.

JOB Observation - Automotive industry documentation, (2009).

CHAPTER 2

Cognitive Engineering for Self-Optimizing Assembly Systems

Marcel Ph. Mayer, Barbara Odenthal, Marco Faber, Wolfgang Kabuß,
Nicole Jochems, Christopher M. Schlick

Institute of Industrial Engineering and Ergonomics
RWTH Aachen University
52064 Aachen, Germany

INTRODUCTION

Within the Cluster of Excellence "Integrative Production Technology for High-Wage Countries" at the Faculty of Mechanical Engineering of RWTH Aachen University, a research project has been established to study self-optimizing assembly cells. The concept of self-optimizing describes a concept that value stream orientated measures are monitored and effective interventions are anticipated, so that in spite of disturbances the quality of processes and products is continuously improved (Schmitt & Beaujean 2007). The application of existing knowledge to similar situations or new production cases as the core of self-optimization enables new perspectives of manufacturing and assembly systems.

A novel architecture of the cell's numerical control based on a cognitive architecture forms the technological basis of this approach. The cell's numerical control is termed a cognitive control unit (CCU) which is able to process procedural knowledge encoded in production rules and to control multiple robots. Thus, the CCU is able to take over tasks from the machining operator, which are repetitive, dull, dangerous and not too complex, and to cooperate with the human on a rule-based level of cognitive control. Clearly, knowledge-based behavior in the true sense of Rasmussen (1986) (and also skill-based behavior to a large extent) cannot be modeled and simulated by the CCU and therefore the experienced machining operator plays a key architectural role in the concept of self-optimization.

A numerical control that can autonomously plan on the basis of its own knowledge representation and based on the situation will have a significant impact on the operator or skilled workers scope of responsibilities (see Mayer et al. 2008, Schlick et al. 2009, compare Sheridan 2002).

In the following sections a laboratory experiment is presented, on the basis of which three rules for human assembly strategies are identified. Further these rules are validated by a second independent experiment. The empirically validated rules are implemented as production rules for the CCU and evaluated regarding performance in a simulation study.

EXPERIMENTAL ASSEMBLY CELL

In order to validate and to further develop the aforementioned CCU concept in a realistic production environment, a robot supported assembly cell is currently being built at RWTH Aachen University (see Kempf et al. 2008). The cell design is shown in Figure 1. The assembly cell is fed by a conveyor belt system. Two robots are integrated, of which in a first step only one is controlled by the CCU. The second robot takes over the separate task of feeding the assembly objects from a pallet onto the conveyor belt. An area is located in the middle of the assembly cell that is used as both a mounting surface and temporary storage area (buffer) for desired but not directly usable production parts. At present the workstation of the operator is separated by an optical safety barrier. The workstation's multimodal human-machine interface displays e.g. via an augmented vision system about the system status, and assists in problem solving. Detailed commentary regarding the multimodal interfaces can be found in Odenthal et al. (2009) and Schlick et al. (2009). It should be emphasized that the project's focus is not in optimizing the assembly of objects with arbitrary shape and surface, which involves very complex sensors and actuators, but in evaluating the concept and methodology of the novel automation approach.

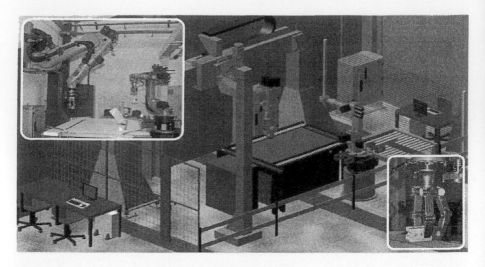

Figure 1: Layout of the Experimental Assembly Cell

Due to their design, many industrially manufactured workpieces are limited in that their assembly is only possible in a particular order, or that they only allow for a few procedural variations in their assembly (see Eversheim 1998). Therefore, these components do not seem to be suitable objects in order to demonstrate the full potential of a CCU. To study the concepts and methods witch scientific methods, mountable assemblies were chosen that allow for a high variation of assembly sequences. One of the requirements for the basic elements to be assembled is that they can be arbitrarily configured and are similar in their functionality. LEGO building bricks, from the Danish company of the same name, fulfill this requirement and were therefore used for the system design and evaluation. The bricks are also easy to describe mathematically because of their (semi-) symmetric geometry, much unlike complex free forming components (e.g. interior elements in an automobile). The (semi-) symmetric geometries of the basic elements allow for very complex assemblies to be constructed despite the simple basic structures, and even allow for manifold assembly sequences. This is easily shown by a simple example. Building a small pyramid with only 5 bricks and foundation of two by two bricks leads to 24 different assembly sequences.

SIMULATION OF A COGNITIVE CONTROL UNIT

In case of supervisory control it is crucial for the human operator to understand the assembly plan developed by the CCU. Basic assumption is that a robot control based on human decision making will lead to a better understanding regarding the intention of the technical system and to improve its conformity with operator's expectations. This can be referred to as cognitive compatibility.

To enhance cognitive compatibility, the question arises on how to design the symbolic representation of the knowledge base for the CCU to maximize the conformity with the operator's expectations. Proprietary programming languages that are used in conventional automation have to be learned case specific and do not necessarily match the mental model of the human operator. Focusing a human centered description to match the process knowledge to the mental model one promising approach in this special scenario is the use of motion description.

The cognitive architecture SOAR was chosen to simulate human cognitive functions, because its internal knowledge base is based on well-known production rules (if-then rules) (Leiden et al. 2001). A rule-based approach has the advantage of not needing to be preconditioned, as opposed to emergent systems for example artificial neural networks. SOAR is able to simulate rule-based human decision-making on the shop floor to a certain degree, and to take over repetitive, monotonous activities without showing signs of fatigue (Hauck et al. 2008). As aforementioned the simulation of true knowledge-based behavior (sensu Rassmussen 1986), cannot be modeled and simulated by the CCU.

Fundamental motions of the MTM-1 taxonomy were used for the description of the motion cycles of the assembly robots. The underlying hypothesis is that an execution sequence composed of empirically verified and expectation-conforming basic elements can be easily understood by humans and optimized, even when the controlled entity is a robotic gripper (see Gazzola et al. 2007, see Tai et al. 2004). The MTM fundamental motions implemented in production rules resp. SOAR operators are equal, and are therefore not defined in a process sequence. They correspond to the MTM-1-basic movements REACH, GRASP, MOVE (with parallel TURN), POSITION (with APPLY PREASSURE) and RELEASE. In addition, other rules are provided that include the physical constraints depending on the used basic elements (ex. joining-angle or conditions required for positioning an element), but also include information whether a supplied element can be installed directly or stored for a later assembly step in the buffer.

A simulation environment that closely follows the design of the aforementioned assembly cell was developed to evaluate the basic SOAR-MTM CCU. The conveyor belt has been simplified by substituting a tableau – much like a chess board – on which the bricks lie in the spaces. That way the random supply of a single brick as well as the supply of all required pieces can be varied for the deployment process. The areas of the workplace and the buffer have been incorporated into the simulation environment as independent areas.

Since the focus of the research is not initially concerned with process optimization, but instead with the evaluation of the concept, the time information in opposition to MTM-1 was not considered. This simplifying assumption is only acceptable if the bricks needed for assembly are present, or the component consists of identical bricks hence sum of distances and total time does not change for fixed start and end positions of the end-effectors in spite of different assembly sequences.

Repeated simulation runs of the basic SOAR-MTM CCU with differing part sets for feed (all necessary pieces were in place, random component feed including provision of unneeded bricks) show that the desired target assembly was assembled correctly and within the expected number of simulation cycles. However, it must be acknowledged that the variance of the observed assembly sequences is immense and despite the application of an anthropocentric taxonomy in the form of MTM-1, the question raised is whether the described approach is sufficient to ensure the conformity expectations of the system behavior for the user of the assembly cell.

The proposed basic model as seen from a techno-centric approach provides valid results in terms of complete, correct and target-oriented buildup strategies. Viewed from an anthropocentric perspective, the results appear to be inadequate, as they may not be compatible with human reasoning due to their high procedural variance. In other words, based on Marshall's Schema Model (Marshall 2008) the elaboration knowledge of the CCU is underrepresented (Mayer et al. 2009).

To investigate the effect of improved elaboration knowledge on conformity with expectations a first experiment with 16 persons (13 male, 3 female) was conducted to identify human assembly strategies. The subjects were given the task of assembling an entire LEGO-module based on a CAD drawing. In order to keep the results comparable with those of the assembly cell, restrictions were posed on how the task has to be processed. The assembly had to be done single handedly. It was also prohibited to assemble subgroups and grabbing multiple bricks. The target object was just a single-colored pyramid of 30 LEGO bricks.

The analysis of assembly strategies in the first experiment permitted the derivation of the following three generalized rules. These rules must be empirically validated for an expansion of the basic model:

1. The position of the first assembly brick lays in the left corner position of the subject's field of view (87.5% of the cases in the sample). Should this rule be applied to other assemblies such as those with a rectangular cross-section, then preference should be to have the bricks in the edge position (100%).
2. Bricks that can be positioned in the direct vicinity of other bricks during a certain assembly phase, are preferred (81%). This is hereafter referred to as the so-called 'adherence to the neighborhood relationship'.
3. The target object is assembled in layers, which are parallel to the mounting surface (81%).

VALIDATION OF SIMULATED COGNITIVE FUNCTIONS

EXPERIMENTAL PROCEDURE

A second experiment was carried out in order to validate the previously identified assembly rules. 25 subjects (14 male, 11 female) who are not normally occupied with manual assembly in their daily work took part. None of the individuals had participated in the first experiment. The mean age was 26.9 years old (SD = 3.4). The assembly tasks that were given the subjects were expanded when compared to the first experiment. Ten identical pyramids of 30 bricks were to be assembled in a timely manner so that despite the laboratory conditions, a training-state would be reached which is comparable to a small series production. The beginning and the end of each construction sequence were determined by double-clicking an installed push-button in the assembly area. The previously mentioned constraints regarding the structure, and namely one-handed assembly, as well as not being able to construct subgroups or grabbing of more than one brick at a time still held.

A detailed analysis of the time data based on an expanded group of participants can be found in Jeske et al. (2010).

CROSS VALIDATION OF THE IDENTIFIED RULES

If the empirically identified rules from the first experiment hold true for the second experiment, at least equal relative abundance should be found in the empirical data. Following this assumption, the following null-hypotheses can be formulated for statistical review:

- H_{01}: The relative frequency of the position of the first brick in the second experiment is equal to the data collected in the first experiment: a) left corner, b) edge position (attenuated form).
- H_{02}: The structure of the neighboring relations in the experiment occurs with the same relative frequency as in the first series of experiments.
- H_{03}: The level design in the second experiment occurs with the same relative frequency as in the first experiment.

To verify the null hypotheses, the χ^2-fit test was used on a significance level of $\alpha = 0.05$.

The results of the χ^2-fit test are shown in Table 1. According to Table 1 H_{01} has to be rejected. The observed frequency differs significantly from the data of the first series of experiments. Therefore, the null hypothesis is contemplated in the attenuated form, which is preferable in the edge position as the assembly is started. The requirements for the χ^2-test are not met for the attenuated form of H_{01} based on the observed frequency of the first series of experiments (Rasch et al. 2004). However, only 1.2% of the bricks were placed on internal positions, i.e. 98.8% of the bricks (100% in the first series of experiments) were placed on edge positions, meaning that the rule can be empirically confirmed.

According to Table 1 H_{02} must be rejected as well. Since the relative frequency of 93.8% is above the expected value from the first series of experiments (81%), it can be stated that the rule is adhered to more than expected.

Finally according to Table 1 H_{03} must be rejected. Since the observed relative frequency is 97.2% (expected value from the first series of experiments: 81%), it may also be noted that the rule is adhered to more than expected.

Table 1 Results of the χ^2-Fit Test

	OV$_{E2}$ (EV$_{E1}$)	df	χ^2	p
H$_{01}$ Rule 1 a)	80,4% (87,5%)	1	11.52	.00
H$_{01}$ Rule 1 b)	98,8% (100%)	---	---	---
H$_{02}$ Rule 2	91,2% (81%)	1	16.25	.00
H$_{03}$ Rule 3	97,2% (81%)	1	41.75	.00

α=0.05; OV$_{E2}$=Observed Value Experiment 2; EV$_{E1}$=Expected Value based on Experiment 1

INFLUENCE OF THE RULES ON PREDICTION ACCURACY OF HUMAN BEHAVIOUR

The cross validation of the investigated rules showed that despite Rule 1 in its original form all rules were even adhered to more strongly than anticipated. Hence the rules regarding the position of the first brick in the attenuated form (Rule 1), the adherence to the neighborhood relationship (Rule 2) as well as the buildup in layers (Rule 3) were formulated as SOAR production rules.

To investigate the influence of the aforementioned assembly rules, a simulation study was carried out with a self developed simulation environment. The reference simulation model was the basic SOAR-MTM model as described before, containing only the rules based on the fundamental motions of the MTM-1 taxonomy as well as the rules necessary to describe the assembly objects. Each additional simulation model was based on the reference model but additionally was enriched by one of the identified rules or combinations of those, e.g. SOAR-MTM CCU+ Rule 1 & 2.

Two facets are of primary interest when evaluating the identified rules: How accurate can each rule or combination of rules predict the next possible brick Assembled by the human? To which degree can the rules be generalized? To answer these questions the data of 250 assembly sequences of the second experiment were used. The dependent variable *LCP* and *RoG* are definrd to assess prediction accuracy and generalizability:

- *LCP* – Logarithmic Conditional Probability: *LCP* represents gives the logarithmic overall probability of an assembly sequence under study. It describes the conditional probability $p(x_t|x_{t-1})$ of a brick being placed by the human at a position under the assumption of a given state:

$$LCP = \sum_{t=1}^{n} \log_{10} p\left(x_t | x_{t-1}\right) \qquad (1)$$

- *RoG* – Ratio of Generality: In case a *LCP*-value is not defined for a given state in a sequence because the state was not reached by the human, the set of rules under consideration cannot simulate the following state and hence the complete sequence. The amount of assembly sequences that cannot be simulated is $n_{\notin LCP}$. A measure of the generality of a rule can be described as the following ratio with n_{total} describing all regarded sequences:

$$RoG = \frac{1 - n_{\notin LCP}}{n_{total}} \qquad (2)$$

For statistical analysis MATLAB R2009a was used. A Kruskal-Wallis test was carried out to test against differences of the *LCP*-values of different rules ($\alpha = 0.05$).

RESULTS

According to the simulation data, a significant effect for differences in the LCP-values were found (p=.00). To further determine which pairs are significantly different, a multiple comparison test was performed, using critical values from the t distribution, after a Bonferroni adjustment to compensate for multiple comparisons. The results of the multiple comparison test as well as the mean ranks and 95% confidence intervals of the rulesets can be seen Figure 2.

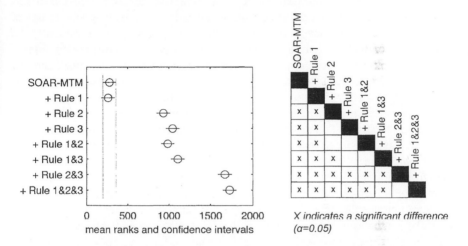

Figure 2: left: mean ranks and confidence intervals (Kruskal-Wallis) right: pairwise comparison (adjusted t-values acc. to Bonferroni)

Rule 1 does not have a significant effect on the prediction accuracy, whereas Rule 2 and Rule 3 have significant influences. The highest level of compatibility regarding human assembly, and thus the highest prediction accuracy ($LCP = -15.09$), occurs when all rules are combined. Overfitting-effects decrease the generalizability which results in poor overall prediction. Nevertheless the RoG is 0.916 in this case, i.e. only 8.4% of the assembly sequences cannot be simulated by this rule set. This result clearly shows that minor extensions to the knowledge base of a CCU can lead to a significant increase in the conformity of the assembly robot behavior with operator expectations.

Table 2 Simulation Results of the Prediction Accuracy and Generality

Rules	SOAR-MTM	SOAR-MTM + Rule 1	SOAR-MTM + Rule 2	SOAR-MTM + Rule 3	SOAR-MTM + Rule 1&2	SOAR-MTM + Rule 1&3	SOAR-MTM + Rule 2&3	SOAR-MTM + Rule 1&2&3
M_{LCP}	-25.53	-25.34	-20.63	-20.66	-20.38	-20.24	-15.23	-15.00
(SD_{LCP})	(1.01)	(1.06)	(1.82)	(1.15)	(1.90)	(1.35)	(1.41)	(1.47)
RoG	1	0.984	0.94	0.964	0.936	0.952	0.92	0.916

SUMMARY

In this paper, a novel approach for a "cognitive control" of a self-optimizing assembly cell was introduced. Based on simulated cognitive functions, parts of operator's original task spectrum, such as the scheduling of a process, can be transferred from the human to the robot in production systems. However, in order to be able to monitor the process, it is necessary to ensure that the decisions made by the CCU, i.e. establishing the assembly sequence, are understandable to the human operator. Additionally the assembly sequence should be planned so that it conforms to human expectations. A concept that is able to adapt system behavior to operator's expectations by using human-centered process knowledge based on the MTM-1 taxonomy was introduced and validated on the basis of an experimental assembly cell. The question of whether the subject given the assembly task simply executes generalized strategies was investigated using two independent experimental series. Three assembly strategies in the form of rules could be identified and validated: 1) The human begins the assembly at the edge. 2) The human preferable builds in the vicinity of neighboring objects. 3) The human prefers to assemble by layers.

Based on a simulation system it could be shown, how the consideration of the identified rules in the form of production rules in the knowledge base of the CCU has an effect on the predictability of an assembly robot. Further the generality of the rules was regarded.

AKNOWLEDGEMENT

The authors would like to thank the German Research Foundation DFG for the kind supportof the research on human-robot cooperation within the Cluster of Excellence "Integrative Production Technology for High-Wage Countries".

REFERENCES

Eversheim W (1998) Organisation in der Produktionstechnik, Bd. Konstruktion. Springer, Berlin

Gazzola V, Rizzolatti G, Wicker B, Keysers C (2007) The anthropomorphic brain: The mirror neuron system responds to human and robotic actions. NeuroImage 35:1674-1684

Hauck E Gramatke A, Henning K (2008) Cognitive Technical Systems in a Production Environment. In: Proceedings of the Fifth International Conference on Informatics in Control, Automation and Robotics. Hrsg. v. ICINCO: Madeira, Portugal

Jeske T, Mayer M, Odenthal B, Hasenau K, Schlick C (2010) Cultural influence on learning sensorimotor skills. In: Karowski W, Salvendy G (Eds) Proceedings of the 3rd International Conference on Applied Human Factors and Ergonomics (AHFE) 17.-20. July 2010. Miami, Florida, USA

Kempf T, Herfs W, Brecher C (2008) Cognitive Control Technology for a Self-Optimizing Robot Based Assembly Cell. In: Proceedings of the ASME 2008 International Design Engineering Technical Conferences & Computers and Information in Engineering Conference, America Society of Mechanical Engineers, U.S.

Leiden K, Laughery KR, Keller J, French J, Warwick W, Wood SD (2001) A Review of Human Performancer Models for the Prediction of Human Error. Prepared for: National Aeronautics and Space Administration System-Wide Accident Prevention Program. Ames Research Center, Moffet Filed CA

Marshall SP (2008) Cognitive Models of Tactical Decision Making. In: Karowski W, Salvendy G (Eds) Proceedings of the 2nd International Conference on Applied Human Factors and Ergonomic (AHFE) 14.-17. July 2008. Las Vegas, Nevada, USA

Mayer M, Odenthal B, Grandt M, Schlick C (2008) Task-Oriented Process Planning for Cognitive Production Systems using MTM. In: Karowski W, Salvendy G (Eds) Proceedings of the 2nd International Conference on Applied Human Factors and Ergonomics (AHFE) 14.-17. July 2008. Las Vegas, Nevada, USA

Mayer M, Odenthal B, Faber M, Kabuß W, Kausch B, Schlick C (2009) Simulation of Human Cognition in Self-Optimizing Assembly Systems. In: Proceedings of 17th World Congress on Ergonomics IEA 2009. Beijing

Odenthal B, Mayer M, Kabuß W, Kausch B, Schlick C (2009) Error Detection in an Assembly Object Using an Augmented Vision System. In: Proceedings of 17th World Congress on Ergonomics IEA 2009. Beijing

Rasch B, Friese M, Hofman W, Naumann E (2004) Quantitative Methoden. Springer, Berlin

Rasmussen J (1986) Information Processing and Human-Machine Interaction. An Approach to Cognitive Engineering, North-Holland, New York (NY)

Schlick C, Odenthal B, Mayer M, Neuhöfer J, Grandt M, Kausch B, Mütze-Niewöhner S (2009) Design and Evaluation of an Augmented Vision System for Self-Optimizing Assembly Cells. In: Schlick C (Hrsg) Industrial Engineering and Ergonomics. Springer, Berlin

Schmitt R, Beaujean P (2007) Selbstoptimierende Produktionssysteme. Zeitschrift für wirtschaftliche Fabrikation 9:520-524, Jahrg.102

Sheridan T (2002) Humans and Automation: System Design and Reserch Issues. John Wiley and Sons, New York

Tai YF, Scherfler C, Brooks DJ, Sawamoto N, Castiello U (2004) The Human Premotor Cortex is »Mirror« Only for Biological Actions. Current Biology 14:117-120

Chapter 3

Limits of Control in Advanced Technology and Consequences for Reassigning Accountability

Gudela Grote

Department of Management, Technology, and Economics
ETH Zurich
Kreuzplatz 5, 8032 Zurich, Switzerland

ABSTRACT

Regardless of a system's degree of automation, it is humans who are responsible for its functioning. The basic tenet of human-centred automation – that human operators need to be in control of technical systems – is derived from this responsibility. In view of human fallibility in general, as well as loss of control evidenced in accidents in particular, doubts are expressed about how achievable the goal of human control over technology really is. Reasons for lack of controllability can be found in the normative assumptions of those developing and implementing technology which may support a self-fulfilling prophecy of turning humans into risk factors. However, the ever increasing complexity of systems also has to be acknowledged as a limiting factor. It is therefore suggested to found system design on the premise of partial non-controllability of technology. This approach could help human operators to deal with system opaqueness and uncertainty better by providing systematic information on the limits of control and thereby also relieving them of some of their responsibility. At the same time, this approach would force system designers, the organizations operating the systems, and regulatory institutions to take on responsibility for the use of technical systems whose complexity can no longer be mastered entirely. Consequences for making decisions on system automation by regulators and companies are discussed within the larger realm of establishing a new politics of uncertainty on a societal level, which would be based on deliberately giving up the pretence of being in control always.

Keywords: automation, human control, accountability, system design

INTRODUCTION

The more automated systems are, the more the human operators have the role of supervisory controllers (Sheridan, 1987), implying that they monitor the correct functioning of the technical system and intervene when the technical system fails. At the same time, it may become impossible for the human to adequately fulfil this role due to increasing system complexity in combination with reduced opportunities for practising operational skills and maintaining adequate situation awareness during operation of the system (Bainbridge, 1983).

The ironies of automation described by Bainbridge can also be understood in terms of different approaches to managing uncertainty (Grote, 2009). A strong driver for technological development has always been the increase in control over production processes (Weitz & Shenhav, 2000). In this line of thinking, technology is regarded as a source of reliability and safety. However, high automation requires a very thorough understanding and high predictability of systems in order to define robust algorithms. This can best be achieved through minimizing uncertainties in the processes to be handled. Whatever tasks are left for human operators to carry out, have to be fulfilled within the tight bounds of technically prescribed operations. As a consequence, human operators are left with very little scope for action and insufficient control over the technical system in the case of disturbances, which may create new risks. Therefore, human-centred approaches to automation (*e.g.*, Hollnagel, 2003; Ulich, 1998) have stressed the importance of human control even of highly automated systems, especially with regard to the necessity of dealing with disturbances. In line with a coping with uncertainties approach, they argue that human operators need an understanding of the technical processes and adequate opportunities to influence those processes (Parasuraman & Riley, 1997). Adequate control furthers motivation and opportunity to develop adequate mental models of the technical system and its processes as well as the operator's readiness to intervene adequately.

In this paper, requirements for human control over automated systems are discussed as prerequisites for holding humans accountable for system functioning. Many methods have been developed to support system design that maintains human control over new technology (*e.g.*, Hollnagel, 2003). However, possible limits to human control in increasingly complex systems have to be acknowledged. It is suggested that better system design may result when partial non-control is explicitly taken into account. A new approach to design derived from this suggestion is outlined and consequences for reassigning accountability and handling technological risks more broadly are discussed.

CONTROL AND ACCOUNTABILITY IN AUTOMATED SYSTEMS

Many authors agree (*e.g.,* Boy, 1998; Grote *et al.*, 2000; Hauß & Timpe, 2002; Hollnagel & Woods, 2005; Ulich, 1998, Waterson *et al.*, 2002) that limits to automation are not only determined by technical feasibility and societal acceptance, but to a large part by the necessity to maintain human responsibility over system goals and their attainment, including all positive and negative (side) effects. Human control over technical systems, including tranparency, predictability, and sufficient means of influencing the systems, is considered to be the main prerequisite for taking on this responsibility. As Hollnagel and Woods (2005) point out, control not only concerns the ability to achieve desired outcomes, but also the ability to recover from disturbances and disruptions. The increasing complexity of automated systems, for instance, through increasing application of "autonomous" and "learning" systems, renders it more and more difficult to meet the conditions for human control, and thereby also raises the question of whether human operators can still be held accountable for the functioning of these systems.

In order to discuss issues of accountability properly, it is important to point out that every automated system is a socio-technical system, independent of its degree of automation, as the workerless factory, the driverless subway, or automated money transfer systems have been developed by humans for humans. Therefore, technical systems should never be looked at in isolation, but always as part of a socio-technical system, which includes humans operating the system and the formal and informal structures and processes within which they work. Furthermore, it is necessary to include in the system definition all those organizations and organizational units which are in charge of system design and maintenance, as well as those that are responsible for rules and regulations controlling system design and operation. This much broader framework helps to reframe questions on the distribution of control and accountability in complex socio-technical systems (Baram, 2007).

"HUMAN UNRELIABILITY": LACK OF HUMAN CONTROL OVER TECHNICAL SYSTEMS

There is more and more acceptance of the fact that accidents are most often caused by a complex combination of human, technical, organizational, social and environmental factors (*e.g.,* Reason, 1997). Frequently, within this complex interaction of many contributing factors, an unfortunate coupling of human and technology can be found. One example is the accident involving a Lufthansa Airbus A320 in Warsaw on the 14th of September, 1993 (Main Commission Aircraft Accident Investigation Warsaw, 1994), where the automatic algorithm (with no manual override) for braking the aircraft after touchdown on the runway was instigated too late. This algorithm is released only when there is a prespecified amount of pressure on both back wheels, which in this case did not happen immediately after touchdown due to a slight tilt of the aircraft and the resulting unequal pressure on the wheels. The delay in braking the aircraft resulted in the aircraft overrunning the runway and crashing into a mound of earth. The aircraft caught fire immediately and two people were killed. As a consequence of this accident, Lufthansa pilots were informed in more detail about the technical definition of the landing procedure and some technical improvements of the Lufthansa Airbus aircraft were implemented. The fully automatic control of reverse thrust and of the brakes during landing was left unchanged. Informing the pilots was meant to re-establish conditions for keeping them responsible, without actually changing the distribution of authority between human and technology.

The problems resulting from removing control from the human operator in this way have been described very well by Amalberti (1992, 1993). He assumes that human operators act on the basis of an "ecological risk management", which allows pilots to deal with their resource limitations by anticipating different courses of action, prioritization of actions, and active control forcing the actual situation to follow the anticipated one. He further argues that this way of dealing with risk is rendered more difficult by automation as transparency and flexibility are lost. Pilots react to this difficulty by trying either to outwit the technical system – for example they may enter non-existent wind into the computer in order for the computer to calculate a different, that is, the desired, approach angle – or by fully ceding responsibility to the technical system.

For technology to support the pilots' ecological risk management, system designers would have to consider human situative problem solving strategies more instead of assuming prescriptive optimal strategies. This would require acknowledging the human as being at least as much a safety factor as a risk factor. By viewing the human mainly as a risk factor and assigning the majority of functions to technology as the presumed safety factor, the human is turned into a risk factor instead. A self-fulfilling prophecy is created.

CONTROL AND ACCOUNTABILITY IN AUTOMATED SYSTEMS REVISITED

Beyond assumptions about the nature of humans and technology it is important to also question assumptions about organizations, that is, images of organization (Morgan, 1986), especially assumptions concerning possibilities and limitations of planning and control in organizations and the preferred ways of managing uncertainties (Grote, 2009). Abandoning the myth of full central control frees system designers to systematically support decentralized autonomy, constructive coping with the limits of planning, and deliberate choice between local and central control mechanisms. However, for such a change in perspectives on humans, technology, and organization to take effect, another even more fundamental assumption is needed. This assumption states that the most advanced technology would still be controllable by human operators if only system designers and buyers of systems would really want that and were willing to invest more resources in the development of such systems.

After decades of trying to develop theories and methods to affect such a change and as a consequence, design better systems, the recurring discussions on how successful these attempts have really been should be reason to seriously question this assumption itself (Dekker & Hollnagel, 2004; Dekker & Woods, 2002; Parasuraman *et al.*, 2008). What if the insufficient human control of technology is not caused by normative assumptions about humans, technology, and organizations, but by factual limitations of human control and, even more basically, human imagination, due to the ever increasing complexity of technology? Then, either technology development has to be stopped – which is no real option – or the (partial) lack of control has to be accepted. Technically, this acceptance is equivalent to the determination of unmanaged residual risks. On the human side, there is hesitance to admit to a lack of control due to problems of unmanaged accountability. The human operator is kept in the system as a backup where all problems come together and have to be dealt with. The fallibility of this approach and its abuse by system developers and the organizations operating the systems in order not to have to admit to the lack of control has been pointed out by Bainbridge (1983) with utmost clarity.

Polemically one could argue that the current interest in research on trust in technology at the level of human-machine interaction (*e.g.,* Lee & See, 2004; Muir, 1994; McGuirl & Sarter, 2006; Moray *et al.*, 2000; Rajaonah *et al.*, 2006) has its roots in the fact that – while still acknowledging that control would be better than trust – trust is all that is left to the human operator. Experiments have shown that technology is trusted most when trust in one's own competences is low (*e.g.,* Lee & Moray, 1992, 1994). From general psychology we know that self-confidence is strongly related to perceived personal control (Bandura, 2001). Trust may therefore be a consequence of lack of control. This agrees with sociological definitions of trust as a mechanism to cope with uncontrollable uncertainties (*e.g.,* Luhmann, 1979, 1988). In the system design literature, however, trust is often understood as a desirable user attitude based on familiarity with the system and confidence stemming from high transparency, predictability, and reliability of systems, thereby actually providing essential prerequisites for control (Lee & See, 2004; McGuirl & Sarter, 2006; Muir, 1994). To explore further the sociological notion of trust as substitute for control, might prove valuable in supporting system design based on the assumption of only partial controllability of technical systems..

DESIGNING PARTIALLY (NON-)CONTROLLABLE SYSTEMS

If one assumes that technology cannot be controlled fully by human operators, however hard system designers and operators try, the criteria for system design have to be changed. The main purpose of such new design guidelines would be to free human operators of their impossible role of trying to fulfil stop-gap and backup functions in complex socio-technical systems. Methods supporting adaptive system design indicate a move in a similar direction by allocating control fully and without human influence to the technical system in very stressful situations (*e.g.,* Inagaki, 2000; Moray *et al.,* 2000). However, the crucial issue of assigning responsibility and accountability is usually not dealt with in these design methods.

According to Kornwachs (1999) the main prerequisites for taking on responsibility are the freedom to make choices and to act in a chosen manner. If people are forced to act in a particular way they cannot be held accountable unless they have brought this situation upon themselves. Also, he argues that all necessary knowledge concerning the action, its purpose and its consequences, has to be available and attainable. He emphasizes that automation attempts to reduce complexity for the human operator in order to achieve these preconditions, but that at the same time, new complexities are created which may violate these conditions.

In order to provide the necessary preconditions for taking on responsibility, and thereby also control, the limits of control should be defined as clearly as possible. In those areas which are classified as outside the control of human operators, they cannot be held responsible. Taking the braking procedure in the Airbus A320 as an example, this would mean that the irreversible automation of the essential braking functions should be taught in pilot training and should also be indicated in the cockpit during landing. If mistakes happen in the execution of these functions, the system developer, or possibly the organization operating the system, should be held responsible, but not the pilots. Only if the pilots – in line with Kornwachs' definition – can be proven to have induced this situation deliberately or carelessly (*e.g.,* as a consequence of insufficient competencies) might they have to assume some of the responsibility. Even in such cases, the responsibility might lie more with the airline, especially if a particular pilot's lack of ability or knowledge has already surfaced earlier, for instance, during simulator training, and no action has been taken.

In a similar vein, Bellotti and Edwards (2001) have argued that in the design of context-aware systems, issues of intelligibility and accountability must be addressed. Context-aware systems are an important element of pervasive computing as they are able to identify different contexts and change their behaviour accordingly, for instance, allowing or prohibiting connections with other systems. Availability for personal contact in work teams, as an example, might be signalled on the computers or mobile phones of members of these teams based on each of them being in their office and not currently speaking with anybody. Bellotti and Edwards argue that context-aware systems need to be intelligible for users, which necessitates that these systems, before acting upon what they infer about the context must be able to represent to their users what they know, how they know it, and what they are about to do about it. Additionally, accountability must be ensured by providing explicit mechanisms that enforce user identification and action disclosure and by effective user control. The latter does not necessarily imply that the user is intimately involved in task execution but that the desired outcome is attained through an appropriate interaction between system and user.

In order to design socio-technical systems according to such guidelines, existing methods for describing and assessing technical, human, and organizational risks should be extended to clearly indicate zones of uncontrolled risks for both the organizations operating the system and the human operators at the sharp end. Once zones of uncontrollable risks have been identified, it would have to be decided whether or not functions in such zones could be fully automated and whether human operators can in any way be supported in acting appropriately despite their reduced level of control. Given the unpredictabilities in the defined zones of limited or no control, support for operators could only be in the form of heuristics. Process rules as suggested by Hale and Swuste (1998) could be such heuristics, that is, rules which do not specify a concrete action nor only specify the goal to be achieved, but provide guidance on what to do in order to find out what the right action is in a given situation. This could concern, for instance, information sources to be used, other actors to be involved, or definition of priorities.

At the same time, in these zones of limited control the responsibility for the safe operation of the system would remain with the system developer and potentially the organization operating the system, but not the human operator. The pressure to keep these zones small and thereby maximize control for the human operator would increase. Something similar is achieved already by US law allowing system operators to sue the system developer when his or her own erroneous action can be proven to be a consequence of bad system design (Baram, 2007).

A method for risk analysis that could be helpful for identifying zones of no or limited control is Hollnagel's (1998) CREAM (Cognitive Reliability and Error Analysis Method), which has recently been developed further into an Extended Control Model (Hollnagel, 2007). In Hollnagel's model four modes of control are distinguished: scrambled (choice of next action is close to random), opportunistic (choice of next action driven by narrow focus on current situation), tactical (choice of action based on following a known procedure) and strategic (choice of action determined by higher-level goals with longer time horizon). Control may concern anticipatory or compensatory action with each involving more specific activities of targeting, monitoring, regulating, and tracking. The four modes of control are assigned human error probabilities which are used to determine the overall probabilities of human error in event sequences potentially involving several switches between the different modes.

By analyzing possible event sequences for a particular human-machine system and the different control modes involved, zones of no or scrambled control can be identified and appropriate measures taken to handle processes in these zones. As an example of such an approach which already exists, one might take the 30 minute rule in nuclear power plants. This rule demands that after the occurrence of major non-routine events the process control systems in these plants are capable of keeping up a sufficiently safe level of operation for as long as 30 minutes without human intervention, thereby giving the human operators time to recover from a state of confusion and scrambled control and to regain tactical or strategic control. Also, different stages of action regulation have to be distinguished, that is, information seeking, decision-making, execution of decided actions, feedback on effects of action, and corrective actions. The lack of control may only concern some of these stages and also, may affect different stages differently. The outcome of a CREAM based analysis would help to make decisions on full or partial automation more systematically, aiming at a very deliberate match between control and responsibility. Also, instead of pretending that systems are safer than they are, in particular due to their increasing embeddedness in complex networks, it may be much better to regain overall control by admitting to areas of opaqueness and uncertainties in the system. Only then can the ability to cope with uncertainty and loss of control be trained and supported systematically.

TOWARDS A NEW POLITICS OF UNCERTAINTY IN SYSTEM DESIGN

Much of what has been said so far is not new, but the underlying attitude may be: Instead of lamenting the lack of human control over technology and of demanding over and over again that control be reinstated, the approach presented here states very explicitly that current and future technology, even with the best of system design, contains more or less substantial zones of no control. Any system design should build on this supposition and develop concepts for handling the lack of control in a way that does not delegate the responsibility to the human operator, but holds system developers, the organizations operating the systems, and societal actors accountable. This could happen much more effectively if uncertainties were made transparent and the human operator was relieved of his or her stop-gap and backup function. In order for such a change in perspective on system design to be successful, fundamental changes in regulatory policy and public attitudes have also to be effected. In line with Power's (2004) call for establishing a new politics of uncertainty, we all have to learn to accept that risks can, at best, be responsibly handled, but that they cannot be managed away.

To support this new thinking, there should be transparency of the criteria employed in system design decisions, the methods applied to determine measurements on these criteria, and the consequences of different decision alternatives for all stakeholders. In this paper, hopefully some first ideas have been provided on how decisions on the design of automated systems can be arrived at in ways that acknowledge limits of control without irresponsibly giving up control.

REFERENCES

Amalberti, R. (1992). Safety in process-control: An operator-centred point of view. Reliability Engineering and System Safety, 38, 99-108.

Amalberti, R. (1993). Safety in flight operations. In B. Wilpert & T. Qvale (Eds.), Reliability and safety in hazardous work systems (pp. 171-194). Hove: Lawrence Erlbaum.

Bainbridge, L. (1983). Ironies of automation. Automatica, 19, 775-779.

Bandura, A. (2001). Social cognitive theory: an agentic perspective. Annual Review of Psychology, 52, 1-26.

Baram, M. (2007). Liability and its influence on designing for product and process safety. Safety Science, 45, 11-30.

Bellotti, V. & Edwards, K. (2001). Intelligibility and accountability: human considerations in context-aware systems. Journal of Human-Computer Interaction, 16, 193-212.

Boy, G. (1998). Cognitive function analysis. London: Ablex.

Dekker, S. & Hollnagel, E. (2004). Human factors and folk models. Cognition, Technology & Work, 6, 79-86.

Dekker, S. & Woods, D.D. (2002). MABA-MABA or abacadabra? Progress on human-automation coordination. Cognition, Technology & Work, 4, 240-244.

Grote, G. (2009). Management of uncertainty - Theory and application in the design of systems and organizations. London: Springer.

Grote, G., Ryser. C., Wäfler, T., Windischer, A. & Weik, S. (2000). KOMPASS: a method for complementary function allocation in automated work systems. International Journal of Human-Computer Studies, 52, 267-287.

Hale, A.R. & Swuste, P. (1998). Safety rules: procedural freedom or action constraint? Safety Science, 29, 163–177.

Hauss, Y. & Time, K.-P. (2002). Automatisierung und Unterstützung im Mensch-Maschine-System. In K.-P- Timpe, T. Jürgensohn & H. Kolrep (Hrsg.), Mensch-Maschine-Systemtechnik - Konzepte, Modellierung, Gestaltung, Evaluation (S. 41-62).Düsseldorf: Symposion.

Hollnagel, E. (1998). CREAM - Cognitive Reliability and Error Analysis Method. Oxford: Elsevier.

Hollnagel, E. (Ed.) (2003). Handbook of Cognitive Task Design. Mahwah, NJ: Lawrence Erlbaum.

Hollnagel, E. (2007). Flight decks and free flight: where are the system boundaries? Applied Ergonimics, 38, 409-416.

Hollnagel, E. & Woods, D.D. (2005). Joint cognitive systems - foundations of cognitive systems engineering. London: Taylor & Francis.

Inagaki, T. (2000). Situation-adaptive autonomy for time-critical takeoff decisions. International Journal of Modelling and Simulation, 20, 175-180.

Kornwachs, K. (1999). Bedingungen verantwortlichen Handelns. In K.-P. Timpe und M. Rötting (Hrsg.), Verantwortung und Führung in Mensch-Maschine-Systemen (S. 51-79). Sinzheim: Pro Universitate.

Lee, J. & Moray, N. (1992). Trust, control strategies and allocation of function in human-machine systems. Ergonomics, 35, 1243-1270.

Lee, J. & Moray, N. (1994). Trust, self-confidence, and operators' adaptation to automation. International Journal of Human-Computer Studies, 40, 153-184.

Lee, J.D. & See, K.A. (2004). Trust in automation: designing for appropriate reliance. Human Factors, 46, 50-80.

Luhmann, N. (1979). Trust and power. Chichester. Wiley.

Luhmann, N. (1988). Familarity, confidence, trust: problems and alternatives. In D. Gambetta (Ed.) Trust making and breaking cooperative relations (pp. 94-107). New York: Blackwell.

Main Commission Aircraft Accident Investigation Warsaw, "Report on the Accident to Airbus A320-211 Aircraft in Warsaw on 14 September 1993,"

Warsaw, Poland, March 1994. Web version prepared by Peter Ladkin, URL: http://www.rvs.uni-bielefeld.de/publications/Incidents/DOCS/ComAndRep/Warsaw/warsaw-report.html

McGuirl, J.M. & Sarter, N.B. (2006). Supporting trust calibration and the effective use of decision aids by presenting dynamic system confidence information. Human Factors, 48, 656-665.

Moray, N., Inagaki, T. & Itoh, M. (2000). Adaptive automation, trust, and self-confidence in fault management of time-critical tasks. Jounral of Experimental Psychology: Applied, 6, 44-58.

Morgan, G. (1986). Images of organization. Beverly Hills, CA: Sage.

Muir, B. (1994). Trust in automation, part I: Theoretical issues in the study of trust and human intervention in automated systems. Ergonomics, 37, 1923-1941.

Parasuraman, R. & Riley, V. (1997). Humans and automation: use, misuse, disuse, abuse. Human Factors, 39, 230-253.

Parasuraman, R., Sheridan, T.B. & Wickens, C.D. (2008). Situation awareness, mental workload, and trust in automation: viable, empirically supported cognitive engineering constructs. Journal of Cognitive Engineering and Decision Making, 2, 140-160.

Power, M. (2004). The risk management of everything: Rethinking the politics of uncertainty. London: Demos.

Rajaonah, B., Anceaux, F. & Vienne, F. (2006). Study of driver trust during cooperation with adaptive cruise control. Travail Humain, 69, 99-127.

Reason, J. (1997). Managing the risks of organizational accidents. Aldershot, UK: Ashgate.

Sheridan, T.B. (1987). Supervisory control. In G. Salvendy (Ed.), Handbook of human factors (pp. 1243-1268). New York: Wiley.

Ulich, E. (1998). Arbeitspsychologie (4. Aufl.). Zürich: Verlag der Fachvereine; Stuttgart: Schäffer-Poeschel.

Waterson, P., Older Gray, M.T. & Clegg, C.W. (2002). A sociotechnical method for designing work systems. Human Factors, 44, 376-391.

Weitz, E. & Shenhav, Y. (2000). A longitudinal analysis of technical and organizational uncertainty in management theory. Organization Studies, 21, 243-265.

CHAPTER 4

A Procedure Approach for the Culture-Adequate Implementation of Production Systems

Ralph Hensel and Birgit Spanner-Ulmer

Professorship of Human Factors and Ergonomics
Chemnitz University of Technology
Chemnitz, 09107, Germany

ABSTRACT

Intercultural competence is a basic prerequisite for the success of enterprises activities in foreign countries. Due to that finding, ergonomics face the challenge to view the present ergonomic methods and concepts for the transferability to other cultures. The Professorship of Human Factors and Ergonomics at the Chemnitz University of Technology investigates that topic and examines the cultural dependence of scientific questions in the field of product and process ergonomics. In relation to that, an approach for a cultural concept, to structure and map cultural diversity, was introduced. Therefore, the focus of this contribution concentrates on the process ergonomic approach to solutions further challenged with the present approach regarding their cultural dependence. The main concern are questions in labor organization like culture adequate design of organizational structures and concepts as well as the transferability of elements in product systems and their implementation.

Keywords: Intercultural Management, Change Management, Production Systems, Intercultural Change Management

MOTIVATION

The development from a local oriented company towards a globally operating company is closely related to structural, functional, and personnel changes which have to be managed. However, reality proves that up to 75 percent of all initiated change processes fail or do not achieve their goals (Schirmer, 2000; Schreyögg, 2000). This trend worsens with change processes involving people of different cultures against the backdrop of intercultural diversity.

Particularly with regard to global acting enterprises, challenges can result from the changing conditions in the political-legal, the macroeconomic, the environmental, the technological, and especially, the socio-cultural business environment.

Seeing that, the internationalization of company activity affects not only the enterprise itself but also every single employee. On the one hand, it has an impact on the adaption and transferability of organizational and management concepts to foreign locations against the backdrop of fundamentally different society concepts of varying countries like management of organizational rationalization and structural measures or the implementation of production systems. On the other hand, intercultural differences can cause problems and conflicts in the direct, interpersonal interaction between members of different cultures, for instance, managers being sent as expatriate to foreign countries or the cooperation in multicultural work groups (Hensel and Spanner-Ulmer, 2008a).

To successfully implement production systems, knowledge of values and rules as wells as exploring the very cultural conditions are of crucial importance. Therefore, it is necessary to design, firstly, suitable methods and structures of production systems and, secondly, to create a change management, incorporating a culture-adequate strategy when implementing production systems.

INTERCULTURAL CHANGE MANAGEMENT

When having situations where cultures overlap, socio-economic environmental factors have an immense impact on the success of management processes. Variables of the socio-cultural environment are especially culturally determined values and attitudes, for example, religious philosophies, ethico-moral norms or traditional codes of conduct. Intercultural problems often result from assumptions of similarities towards foreign partners or the lack to understand other cultures. Hence, it is important to know the values and codes of the target country and to gain detailed information about the cultural conditions to avoid value conflict, misunderstandings, and misinterpretations (Hensel and Spanner-Ulmer, 2008b).

Fields of cultural comparative and intercultural management research explore different aspects of the intercultural management and provide general guidance for both the interaction between members of different cultures and the development of functional, structural, and personnel concepts (Keller, 1982). However, there is no integrative approach considering the cultural influence factors on the individual level as well as on company the level. Problems resulting from the design and implementation of changes in internationally operating companies cannot be solved with the help of the present research approaches.

According to Hofstedes (2006), culture as "collective programming of the mind" understands culture as the sum of norms, values, and the attitudes of a community and therefore, represents a collective orientation system for human behavior.

By introducing cultural dimensions, patterns of behavior, changing from culture to culture in its characteristics and distribution, can be described. At the Professorship of Human Factors and Ergonomics at the Chemnitz University of Technology a model to describe culture has been developed. On the basis of present cultural comparative approaches, which have been analyzed and evaluated, a comprehensive approach for cultural identification, uniting the 14 most important cultural dimensions, can be presented. Furthermore, this approach is based on the organizational culture model by Schein (1992) categorized into 5 basic assumptions representing the frame of reference: Concept of time, communication, truth and trust, interpersonal relations as well as power and performance (see Figure 1).

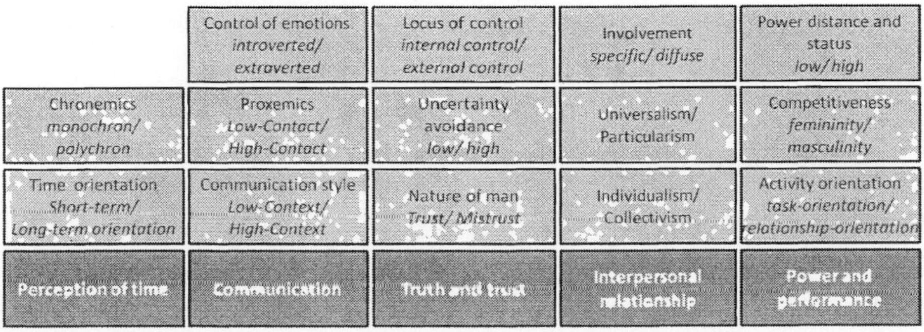

	Control of emotions introverted/ extraverted	Locus of control internal control/ external control	Involvement specific/ diffuse	Power distance and status low/ high
Chronemics monochron/ polychron	Proxemics Low-Contact/ High-Contact	Uncertainty avoidance low/ high	Universalism/ Particularism	Competitiveness femininity/ masculinity
Time orientation Short-term/ Long-term orientation	Communication style Low-Context/ High-Context	Nature of man Trust/ Mistrust	Individualism/ Collectivism	Activity orientation task-orientation/ relationship-orientation
Perception of time	Communication	Truth and trust	Interpersonal relationship	Power and performance

Figure 1. Model to describe culture (Hensel and Spanner-Ulmer, 2010)

The elaborated approach for cultural identification allows the structuring and illustration of cultural diversity. This further permits, with the help of the present secondary data, the description of different national cultures in empirical researches, and can finally be compared in terms of their culture-differentiating features (norms, values, attitudes) (Hensel and Spanner-Ulmer, 2008b). Providing such a basis, work organizational concepts can be analyzed according to their cultural dependence with the aim to, firstly, design methods and structures of production systems and, secondly, to design their implementation process by the means of a culture adequate change management with regard to cultural diversity.

At present, there is no solution approach in the research field of change management considering the intercultural diversity. According to the current state of science, theories and concepts of change managements, which prove to be valid depending on the national provenance of the scholars in certain countries and cultures, are regarded as universal and "culture-free". In contrast, soft elements of the management theory such as leadership style, motivation, conflict control or authority relations are identified to be "culture-bound" (Dülfer, 2006). Hence, change management concepts, predominantly developed in the German- and Anglo-Saxon-speaking area, cannot simply be applied to other countries and their cultures. Science tries to seek an answer to the implementation problem of change processes by approaching the topic from different perspectives. Thereby, the scientific field ranges from socio-psychological compliance research over to the empirical-socioscientific diffusion research and up to organizational research with economic research approaches (Frey, Stahlberg, and Gollwitzer, 1993; Rogers, 1995). Nevertheless, only partial aspects of the change management are considered without the change process being entirely examined throughout all phases and actor levels. Therefore, cultural factors of influence on the micro- and macro-level cannot be identified. Thus, the approach for cultural identification was further extended by a procedure model describing the compliance decision on the individual's level and the diffusion of the change in the company on the organizational level.

In Figure 2, the cyclic-reflexive phase model of the change management considers both approved and current literature recommending procedures for the change management. However, this model clearly differs from the present linear phase model because of its cyclic-reflexive procedure. To draw a conclusion from the various present phase concepts, a categorization into the phase "Initiation", as phase of problem recognition, problem definition and target setting, "Analysis" as phase of data collection and current state estimation, "Conception" as phase of concept development and planning, "Implementation" as phase of realization of the measures and "Stabilization" as phase of consolidation and measurement evaluation have been introduced.

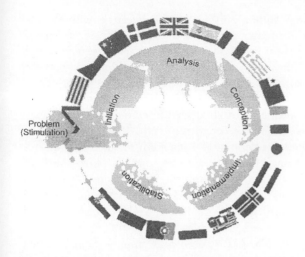

Figure 2. Cyclic-reflexive change process (Hensel & Spanner-Ulmer, 2008b)

PRODUCTION SYSTEMS IN THE AREA OF CONFLICT OF INTERCULTURAL DIVERSITY

Intercultural competence is a key prerequisite for the success of operative activities in foreign countries. Because of that, Ergonomics face the challenge to view the present ergonomic methods and concepts regarding their transferability to other cultures.

The Professorship of Human Factors and Ergonomics at the Chemnitz University of Technology explores this subject and examines the cultural dependence of ergonomic questions in the fields of product and process ergonomics. Thus, the approach for cultural identification, developed at the Professorship of Human Factors and Ergonomics, can help to question process-ergonomic concepts in terms of their cultural dependence. In the following, this will be illustrated by examples of work organizational questions. The central focus is the transferability of production systems, according to the Toyota Production System (TPS), at which single elements of production systems will be chosen by way of example.

With the aim to generate value-adding production processes, Japanese product principles, oriented on benchmark Toyota, are increasingly adopted in German companies. Nonetheless, the fact is often ignored that in Japan other social frameworks predominate and large intercultural diversities exist, when implementing production systems. Hence, elements of production systems cannot generally be applied to other countries and cultures as shall be seen in the case of Lean-Management, CIP, and team work.

The hierarchical structure of structural organization is determined by several cultural influences. In cultures like Germany, Scandinavia or the USA, in which the power distance, a dimension of cultural differences, between employees and employer is low, organizations are rather peripheral and even-leveled structured. Lean-Management-Concepts are more easily realized than in cultures with a high power distance like France or China where small hierarchical differences between the organization members have to be structurally indicated. But even the profile of the structural organization is culturally determined. Departmental Line Organizations accommodate the members of collectivistic cultures as they regard themselves more closely connected with the department and its line manager. The relationship network has a higher importance compared to the performance orientation. In contrast, matrix management can lead to conflicts on the basis of undefined competences for members, evolving from the personal loyalty towards the manager, of collectivistic cultures and high power distance. Matrix management is particularly adequate for cultures whose members work independent and responsible.

Concepts of personnel involvement like CIP (Continuous Improvement Process) can easier be realized in cultures with low power distance because the employees expect a participative and democratic leadership. Again, in contrast, in cultures with a high context dependency, difficulties may arise when identified problems are not openly discussed as it would represent criticism of the present situation. Therefore, especially in cultures with a high power distance, critique would be perceived as attack towards their manager and the management itself. In proactive and self-determined cultures from both the leader and the follower participation possibilities are expected. The personnel is aware of its responsibilities and regard them as command for action and participate intrinsically motivated within the continuous improvement process. In contrast, personnel in a fatalistic-formed, heteronomous culture expect guidance and behave rather passive under reference to the authority of the leader. This means that also members of a fatalistic culture can be involved through the concrete call to participate in the CIP, however, this procedure requires appropriate formal structures.

In this context, intercultural differences in personnel motivation shall also be addressed as it remains questionable how an inducement system can be designed to support, for instance, CIP-Activity. It was Hofstede (2006) who already proved that the question whether motivation is perceived intrinsically or extrinsically, depends on the culture. So, on the one hand, there are task-oriented cultures like the US, Italy or Japan which are rather concerned about money and material goods. On the other hand, there are relationship-oriented cultures like Scandinavian countries where people rather aim for quality of life and interpersonal relations. This has particularly been considered for the design of inducement systems because either performance-dependent payment and carrier as extrinsic motivational factors play an important role or the motivation can intrinsically be carried out via the work task itself. In individualistic societies like Germany or the US it is important to respect the individuality of the employee. That means, firstly, to honor individual performances and secondly, to attain, for example, through a Cafeteria-System an individualistic assimilation opportunity for the personal needs of the employee. However, in collectivistic cultures such as Japan or China personal needs mostly remain untouched. Therefore, work motivation is the idea of the "group's well-being" that means intrinsically gained through group membership. Hence, special attention has to be paid to the cohesion, which would interfere with personal benefits of individual achievement, within the group.

The concept of team work, which especially gained popularity through the MIT-Study, shall further be addressed. Nevertheless, this concept cannot be transferred one-to-one and has to be adjusted. The Japanese-Group-Work-Concept has a hierarchically higher positioned and more authorized group leader because of its high power distance what would not be accepted in German or American companies with regard to the low power distance of the personnel. Instead, in Germany there are only group speakers. In the USA individuals are far more individualistic than in Germany or even in Japan – group concepts face different problems, especially concerning group identity. In that context, the cultural dependence of motivation and payment, already mentioned above, shall be reemphasized, as according to individualistic or collectivistic values the group performance or the individual performance is supported. And additionally, according to the division into task- and relationship-oriented cultures an extrinsic motivation via payment or intrinsic motivation through group membership is more appropriate.

Cultural differences do not only have to be considered in the structural and methodical design of production systems but also in their implementation. As already indicated, reliable procedure concepts, successful in German or Anglo-Saxon language areas, cannot easily be applied to other countries or cultures. This problem is of special importance in international company activities when employees and leaders being sent as expatriate to foreign locations to realize rationalization plans, to transfer organizational concepts or production systems or to organize site relocations. Above all, the direct interpersonal interaction when cooperating in project teams or in the personnel management can lead to unexpected conflict situations.

Concrete conflict fields can arise when implementing production systems, for instance, the setup of an adequate project organization. Here, it is necessary to integrate the relevant power-, know-how-, and process-promoters into the change process to raise the acceptance on the side of the affected parties and to push the diffusion of the whole organization forward. In cultures with large power distance, power promoter can realize changes because the hierarchical low-positioned actors accept those who tell them what to do and who follow their decisions. Whereas, in cultures with small power distance personnel expects participation in the change process; a central diffusion system (Rogers, 1995) would instead block the diffusion of the change. Moreover, the open mindedness towards changes is varying according to the characteristic of the cultural dimension of the uncertainty avoidance. A situation of uncertainty results from the change process affecting the members of cultures with high uncertainty avoidance. This can lead to disorientation and loss of motivation. However, this uncertainty can be overcome by know-how-promoters helping with their competence to reduce the uncertainty. Closely related to that are complexity and level of intervention as well as the ability to plan the change. Past-oriented cultures perceive changes rather as threat making the compatibility towards present structures, norms, and values of great importance. Whereas present-oriented cultures face challenges and potential changes more neutral, however, they orientate themselves at the immediate short-term benefit of an action or change. Therefore, changes should take place over a longer period with incremental degrees of development and should pay attention to achieve Quick-Wins to ensure the continuity of the change process. Forward-looking planning is gaining importance in future-oriented societies. Readiness for risks, investment, and persistent target tracking support the realization of radical and drastic changes.

Additional potential for conflicts results of the direct interaction between members of different cultures when cooperating at project level and workshops. This concerns the communication behavior on a verbal and nonverbal level. Following that, interaction processes in low-context-cultures are marked by explicit, verbal communication, whereas communication in high-context-cultures is characterized by strong nonverbal communication. Moreover, conflicts and problems in high-context-cultures are not directly addressed as such. The working method differs concerning the systematic of procedures, the structuring, and persistence.

CONCLUSION: THE NEED FOR INTERCULTURAL INDUSTRIAL ENGINEERING

The presentation of the paper was aimed to emphasize the cultural dependence of ergonomic concepts and methods, especially of production systems. Those concepts and methods have not comprehensively been elaborated yet and sufficiently been considered in today's enterprises. Due to the approach at the Professorship of Human Factors and Ergonomics in the field of intercultural process ergonomics a first step could be made to describe culture and to identify cultural factors of influence on process ergonomic solution concepts. At present, these concepts are evaluated in operational businesses and are to be incorporated into globally operating companies in terms of a praxeological realization of ergonomic findings as intercultural Industrial Engineering (Spanner-Ulmer, 2009).

REFERENCES

Dülfer, E. (2001). *Internationales Management in unterschiedlichen Kulturbereichen.* München: Oldenbourg.

Frey, D.; Stahlberg, D.; Gollwitzer, P.M. (1993). *Einstellung und Verhalten: Die Theorie des überlegten Handelns und die Theorie des geplanten Verhaltens.* In: D. Frey, M. Irle (Hrsg). Theorien der Sozialpsychologie, Kognitive Theorien, Bd 1. Bern: Huber. S. 361–384.

Hensel, R.; Spanner-Ulmer, B. (2008a). *Vom lokal zum global agierenden Unternehmen - Strategien des interkulturellen Managements.* In: E. Müller, B. Spanner-Ulmer (Hrsg.). Wandlungsfähige Produktionssysteme. TBI'08 - 13. Tage des Betriebs- und

Systemingenieurs / II. Symposium Wissenschaft und Praxis. Tagungsband. Chemnitz: Institut für Betriebswissenschaften und Fabriksysteme.

Hensel, R.; Spanner-Ulmer, B. (2008b). *Change Management im Spannungsfeld interkultureller Unterschiede.* In: Tagungsband 54. GfA Frühjahrskongress. Dortmund: GfA Press. S. 833-836.

Hensel, R.; Spanner-Ulmer, B. (2010). *Interkulturelle Unterschiede im Arbeitsleben.* In: Tagungsband 56. GfA Frühjahrskongress. Dortmund: GfA Press.

Hofstede, G. H. (2006). *Culture's consequences. Comparing values, behaviors, institutions, and organizations across nations.* Thousands Oaks: Sage.

Keller, E. v. (1982). *Management in fremden Kulturen.Ziele, Ergebnisse und methodische Probleme der kulturvergleichenden Managementforschung.* Bern: Haupt.

Rogers, E. M. (1995). *Diffusion of Innovations.* New York: Free Press.

Schein, E. (1992). *Organizational culture and leadership. San Francisco: Jossey-Bass.*

Schreyögg, G. (2000). *Neuere Entwicklungen im Bereich des Organisatorischen Wandels.* In R. Busch (Hrsg.), Change Management und Unternehmenskultur (S. 24-66). München: Mering.

Schirmer, F. (2000). *Reorganisationsmanagement - Interessenkonflikte, Koalitionen des Wandels und Reorganisationserfolg.* Wiesbaden: Gabler.

Spanner-Ulmer, B. (2009). Industrial Engineering – ein interdisziplinärer Ansatz. In: Tagungsband 3. Symposium Wissenschaft und Praxis. 12.11.2009. Chemnitz.

Chapter 5

Shaping the Strategy of Knowledge Based Economy of Polish Enterprises

Hanna Włodarkiewicz-Klimek, Joanna Kałkowska

Institute of Management Engineering,
Poznan University of Technology
Poznan, Poland

ABSTRACT

The paper presents the factors influencing on Polish enterprises strategy shaping in conditions of knowledge-based economy. There will be pointed the area results from knowledge-based economy conception which are the direct and indirect chances for organization development. Pursuant to a carried out researches which the subject were the Polish enterprises strategy evaluation, there will be accomplished the relation analysis as well as the evaluation of environment influence degree for shaping the enterprises behavior. The ability enterprises to opportunities usage which are create by the knowledge-based economy development will be also evaluated.

Keywords: knowledge-based economy, knowledge potential, corporate strategy

INTRODUCTION

The basis resource of developing economy is a permanent ability of knowledge society potential proper usage. The modern development is connected with the knowledge and information permanent wining, transformation and usage. The European countries in 2000 accepted the common concept of knowledge-based economy. The development's postulate concerning the potential knowledge usage, were presented in Lisbon Strategy, which resolutions are still binding. The Lisbon Strategy focuses on four fundamental potentials, among which we can distinguish following ones:

- human resources, i.e. society of knowledge (which the part of knowledge is gathered in),
- innovation system (with the entrepreneurship, more concentrated for operations of companies, but also on the cooperation with science), it creates new knowledge in result of discoveries and innovations,
- information technologies facilitating the exchange of knowledge, also with foreign countries,
- institutional and legal environment, which creates conditions for development of presented domains; it constitutes from various institutions and regulation, etc. [Kałkowska, Włodarkiewicz-Klimek, 2009].

The accepted conception of knowledge-based economy development supported by capital came from European Union funds is a great background for enterprises development. The direct and indirect opportunities which are created by knowledge-development economy is a sign for adaptation the enterprises development strategy to new economy conditions.

DEVELOPMENT OF THE KNOWLEDGE BASED ECONOMY IN POLAND

The widely understood knowledge became one of the fundamental notions of contemporary economy and of the management theory and practice. From the one side the knowledge constitutes the basic development factor in the macroeconomic approach [Nahulis, 2003, Welfe, 2007], hence from the other side it is more and more often considered in microeconomic models, including the theory of enterprises [Maier, 2004]. Dominance of the knowledge in the social and economic life fruited in the nineties of the 20th century with introducing into the economic theory and practice is the concept of "knowledge based economy". Ambiguity in understanding the notion knowledge based economy causes that many sets of features with different degree of accuracy are being used in descriptions of this phenomenon. In the classic definition presented in 1996 by the OECD a knowledge-based economy is an economy which directly is based on the production, distribution and using the knowledge and the information [OECD, 1996]. In its report the World Bank is presenting another look at the idea of knowledge based economy claiming that the economy is becoming an „knowledge-based economy", when using and creating knowledge maintain permanently the centre of its processes of the economic development. A knowledge-based economy is an economy which uses knowledge as a motor of economic growth [World Bank, 2006]. Knowledge is treated here as a fundamental driving force of the economy, as a factor stimulating to progress. The entry of Poland to the European Union caused the necessity of the acceptance and the realization of the developmental conception common for all member states. The

Lisbon Strategy adopted by union countries assuming creating the most competitive and dynamic knowledge-based economy in the world, in his conceptual and executive shape is evolving up till today. On the level of individual member states national conceptions of the development of the economy are being created taking into account the common Lisbon program. In Poland the crucial document is the National Development Strategy 2007-2015, which is a superior document in a view of other strategies or programs. One of the most important element of the vision of Poland till 2015 is the construction of the knowledge based economy: "Poland have to develop its knowledge-based economy and apply a wide application of information and communication technologies in all areas..." [SRK, 2006 p. 26]. The development strategy is realized through numerous other documents: strategic as well as operational. National Strategic Reference Framework in 2007-2013 (NSRO) and operational programs related with it are crucial from the point of view of stimulation of the entrepreneurship and creating conditions for the enterprises development. The strategic objective of the National Strategic Reference Framework functioning has been determined as "creating conditions for the growth of competitiveness of the knowledge-based economy and entrepreneurship that can guarantee the increase of employment and social, economical and spatial coherence" [NSRO, 2007 p.40]. NSRO's goals are being realized through operation programs, which are based on redistribution of assets from EU founds. From the view point of development stimulation of the Polish knowledge-based economy by upgrading the number of enterprises, the most efficient instrument of such operation is the operation program "Innovative Economy 2007 – 2013". Its main objective is the "development of Polish economy on basis of innovative enterprises" [POIG, 2007 p. 61]. Its particular aims assume:

- increasing the innovation of enterprises,
- increasing the competitiveness of the polish science,
- increasing the role of science in the economic development,
- increasing the participation of innovative products of the Polish economy in the international market,
- creating durable and better places of employment,
- growth of using information and communication technologies in the economy.

Determined program has an amount of 9.7 bn. euro at its disposal, currently it is the most development-oriented operational program, responding to modern challenges of the economy and referred directly to enterprises, mainly to the sector of small and medium enterprises.

FACTORS AFFECTING THE SHAPING STRATEGIES OF POLISH ENTERPRISES IN CONDITIONS OF THE KNOWLEDGE-BASED ECONOMY

The strategy is a pattern and a plan integrating primary goals, politics and sequences of actions of the organization into a consistent system [Mintzberg et. al, 1998, p.5]. The strategy is settled in the context of the future, so it is shaping objectives of the enterprise, its internal potential through the identification of existing and predicted conditioning of the company created by its environment. A strategy is real or planned action, which have an essential importance for the being and the progress of the company. The strategy determines actions of the enterprise, which it is entertaining or it should take for achieving its goals and in this way perform its mission. [Kałkowska et. al, 2010].

The influence of the environment on the enterprise and in result, the necessity of adjusting its potential to changes occurring in particular segments of this environment, is determining for the company. The Lisbon Strategy and documents evaluating the efficiency of its implementation, as well as World Bank's publications Doing Business" emphasis that business plans and strategies of enterprises are in their 2/3 conditioned by company's strategic environment. The chances created by the knowledge-based economy can be considered through the prism of its basic pillars: human resource, innovation, information systems and institutional-legal environment.

In the area of human resource, i.e. creating the knowledge society, the most important chances creating the proper climate for development of enterprises in conditions of the knowledge based economy are following:

- development of the active employment policy aspiring to the balance between the flexibility and the security of employment – model flexicurity,
- investing in human resources through realization of the concept of life-long-learning,
- taking actions serving for forecasting, monitoring and qualification of abilities essential in the future,
- creating conditions to increase the mobility of workers.

The innovation system along with entrepreneurship and information technologies, create the key conditions for shaping strategies of enterprises in the knowledge-based economy; we are ranking following factors as most important ones:

- the functioning of European area of research enabling a fluent exchange of knowledge between the state, science and business,
- growth of national expenses for research and development,
- simplification of patent policy,
- creating conditions for development of information technologies,
- implementing information technologies in institutions of public

administration,

- redistribution of European founds for stimulation of growth of enterprises innovation.

Within the frames of the institutional and legal environment and amongst factors supporting the enterprise it is possible to enumerate following ones:

- maintaining the regime of macroeconomic indicators,
- realization of structural reforms,
- simplification of legal regulations,
- creating a dynamic business environment with a shorter red tape,
- creating favorable conditions for functioning of enterprises on the uniform market of European Union.

Enumerated factors are creating areas on which direct chances being stimuli to create strategies in conditions of knowledge-based economy arise.

IDENTIFICATION OF FACTORS SHAPING THE STRATEGY OF KNOWLEDGE BASED ECONOMY POLISH ENTERPRISES

The favourable climate for entrepreneurship development which is created by knowledge based is determining directly in long-distance behaviours and enterprises plans. On the above presented sources there were described the factors which illustrates the potential usage of knowledge based economy while creating the enterprises strategy. These factors are following:

- growing level of top managers education,
- investment areas,
- information technologies possession,
- own research and development area,
- patents,
- collaboration with research and development units,
- product and service export.

The presented factors are treated as an elements which appearance prove the enterprises tendency usage the knowledge based economy potential. The second research area concern the strategy concept identification focusing on strategic aims analysis and description of knowledge based economy enterprises further development.

RESEARCH METHODOLOGY

The research were carried out with random sample of twenty Polish enterprises which were Warsaw Stock Exchange listed in 2007-2009. The research sample consisted of 10% enterprises with no more than 10 years market tradition existence, 50% enterprises with the market existence between 10 and 20 years and 40% was a group of enterprises with more than 20 years market tradition existence. The biggest group among researched companies were trade-production ones (80%), than trade-service companies (10%) and trade only (10%). The research were carried out in January 2010. The subject of the research was the analysis of standard enterprises documentation as well as secondary information.

ENTERPRISES TENDENCIES IDENTIFICATION FOR KNOWLEDGE BASED ECONOMY POTENTIAL USAGE

The identification of the enterprises tendencies to use the knowledge based economy potential was the retrospective research. The analysis involved the last three years of enterprises activity and concerned the following issues: factors analysis which illustrate the knowledge based economy potential in creating enterprises strategy as well as strategic aims assessment in a context of description of knowledge based economy enterprises further development. The research results are presented below.

THE FACTORS ANALYSIS DETERMINING KNOWLEDGE BASED ECONOMY POTENTIAL USAGE IN CREATING THE ENTERPRISES STRATEGY

Identifying the factors illustrating the knowledge based economy potential usage in creating enterprises strategy the following described below results were received. First, analyzing the growing level of managers education being a favouring factor of increasing knowledge transfer in organizations it can be assessed as unused opportunity. Admittedly, 100% of top managers are high educated, however in the last two years only 40% of managers supplemented their knowledge by participation in post-graduate studies or other trainings (fig. 1). The lack of knowledge access can significantly limit both the opportunities identification in environment as well as enterprises strategic horizon.

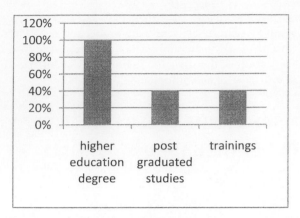

FIGURE 1. The growing level of top managers education (source: own study)

One of the important factors describing enterprises tendency to strategic development with knowledge potential usage are long-term investments. Identifying the types of investment, the following investment has dominated: real estate and other infrastructure than production equipment. Other investment concerned technology in general and Information Technologies infrastructure (fig. 2). Analyzing this research results it can be state that due to investment realization mainly concerning introducing new technologies including information technologies the new knowledge transfer into enterprise is observed. Thanks to it the new competencies leading to reach the strategic competitive advantage are created.

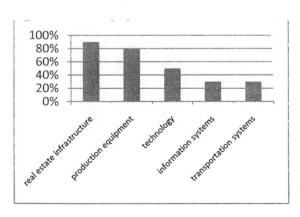

FIGURE 2. The investment areas (source: own study)

Analyzing the types of information technologies usage in management process it is worthwhile to notice that all researched organizations has been equipped with information technologies supporting management process. Half of those companies used integrated management systems, while the others applied dedicated information technologies software supporting finance and storage activity (fig. 3).

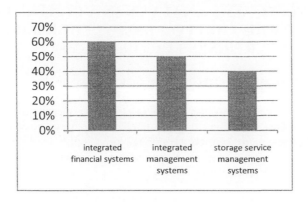

FIGURE 3. Information Technologies possession (source: own study)

The presence and accessibility of information systems in managing enterprises is the key importance concerning organization's knowledge increasing specially focusing on codified knowledge. The codified knowledge allows both to level the knowledge among the workers and to generate new ideas as well as new organizational knowledge creation. Moreover, the presence and development of information technologies are the one of most important aspects of organizations' internal structures adaptation to knowledge based economy.

Among important areas of strategic knowledge development in organizations, both the research and development activity as well as patents possessing should be distinguished. Just only 40% of researched organizations delivers the own research and development activity. Next 20% declare that they are the patent owners and 30% of researched enterprises make the collaboration with outside research and development units. The distinguished areas are the perfect sources of creating the competitive advantage, however, the research results shows that the organizations does not take these possibilities. Among factors influencing on that situation both the limitations concerning finance sources and lack of own research laboratories as well as proper human resource are listed. The last factor being a subject of the research was the export activity. The research presents that over than 80% of companies deliver export activity. The exchange of knowledge potential with foreign partners is a factor with very high development possibilities. According to trade and service exchange with foreign partners there is also knowledge transfer and exchange forcing the strategic advantage source.

THE STRATEGIC AIMS ANALYSIS IN A CONTEXT OF SOURCES DESCRIPTION FOR KNOWLEDGE BASED ECONOMY ENTERPRISES FURTHER DEVELOPMENT

The long-term enterprises development strategy was the subject of a second research area. The researches concerned the accepted strategic aims of organization. Among frequently occurring concept of organization development it can be distinguished following aims like strengthening the competitive position and branch consolidation. Further, there were also appointed aims mainly concentrated on sales networking development, export development, acquisitions and being a branch leader as well as costs minimalisation (fig. 4).

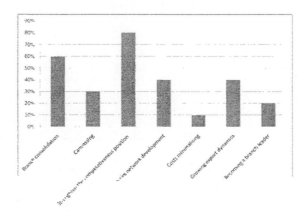

FIGURE 4. The areas of strategic aims concentration

Analyzing the organization's development conception taking into consideration the opportunities of knowledge based economy, the most developing aims are both branch consolidation, acquisitions and export dynamics increasing. Accomplishing of these aims will lead to fast and direct wining and knowledge introduction into the enterprises by the strategic partners potential usage. In the case of other aims conductions the knowledge increase rather gradually by the new competencies creation.

CONCLUSION

The development of knowledge based economy concept create of modern enterprises a lot of opportunities both for development and shaping internal enterprise's structure based on knowledge capital. The conducted research showed that in Polish enterprises it is possible to distinguish factors favouring opportunities usage created by knowledge based economy. These factors are following: investment development, information technologies implementation and usage as well as export activity. However, a lot of areas which could dynamics the organization's strategic development are unused and simply unperceived by the enterprises. According to this, both factors like research and development activity and collaboration with external research and development units can be listed. Such situation has direct correlation on strategic concepts declared by the enterprises. Only half of strategic aims created by the enterprises is connected with direct knowledge increase. Other aims are concentrated on enterprises' existing potential usage. The other important factor which should be also pointed out are the unused opportunities created by European Union funds. Only 20% among all researched enterprises were able to obtain with Union resources for their development. The changes dynamics of modern economy force the constant readiness and openness for the changing. The Polish enterprises which wants to be competing in creating their own strategies on the world markets have to be opened to use the appeared opportunities created by knowledge based economy development.

REFERENCES

Kałkowska J., Pawłowski E., Trzcielińska J., Trzcieliński S., Włodarkiewicz-Klimek H.(2010), *Zarządzanie strategiczne. Metody analizy strategicznej z przykładami,* Poznań: Wydawnictwo Politechniki Poznańskiej

Kałkowska J., Włodarkiewicz-Klimek H. (2009). (Eds.), *Managing Enterprises. Social Aspects*, Monograph, Publishing House of Poznan University of Technology: Poznań

Maier R. (2002, 2004) *Knowledge Management Systems: Information and Communication Technologis for Knowledge Management*, Berlin-Haidelberg: Springer

Mintzberg H., Quinn J.B., Ghoshal S. (1998), *"The Strategy Process"*. Prentice Hall: London,5

Nachulis R. (2003), Knowledge, Inequality and Growth In the New Economy, E.Elgar Cheltenham UK.

NSRO (2007), *„Narodowe Strategiczne Ramy Odniesienia 2007-2013 wspierające wzrost gospodarczy i zatrudnienie"*, Warszawa: Ministerstwo Rozwoju,

OECD (1996), The Knowledge-based Economy, Paris: OECD

POIG (2007), *Program Operacyjny Innowacyjna Gospodarka 2007-2013*, Warszawa: Ministerstwo Rozwoju Regionalnego

SKR (2006), *Strategia Rozwoju Kraju 2007-15,* Warszawa: Ministerstwo Rozwoju Regionalnego

World Bank (2006), Korea as Knowledge Economy. Evolutionary Process and Lessons Learned. Overview, Washington: World Bank

Chapter 6

Organizational Structure Designing and Knowledge Based Economy. The Research Framework for the Polish Enterprises

Edmund Pawlowski

Institute of Management Engineering,
Poznan University of Technology
Poznan, Strzelecka 11, 60-965, Poland

ABSTRACT

The paper is a part of the project "Adaptation of Enterprises Management Systems to Knowledge Based Economy". The essence of the research is a question if there are dependences between changes in enterprises' environment related to Lisbon Strategy, and internal changes in Polish enterprises focused on creating knowledge based organization. This paper is focused on organizational structure issues, and presents methodological approach to the empirical research. Empirical verification is planned in 2010 and 2011.

Keywords: Knowledge Based Economy, Organizational Structure, Organizational Design, Organizational Innovations

INTRODUCTION

In 2005, European Union members were obliged to implement " National Lisbon Strategies". In response, Poland Development Strategy for 2007-2015 has been prepared. The vision of Poland in 2015 includes development of knowledge based economy and wide range utilization of IT technology. Political declarations are not transferred to economy automatically, and that impact is searched rather rarely. Moreover, national and EU programs are only part of an enterprise's external conditions that influence innovative development process. In the above context, in 2009, Management Engineering Institute from Poznan University of Technology has started the project "Adaptation of Enterprises Management Systems to Knowledge Based Economy". The essence of the research is a question if there are dependences between changes in enterprises' environment related to Lisbon Strategy, and internal changes in Polish enterprises focused on creating knowledge based organization.

The aim of the project is:

- To work out some good practices referring to changes in enterprises (strategy, human capital, innovations, IT technologies, relationships with legal and administration environment) as a source of implementing the model of knowledge based organization.
- To find out the mechanisms of enterprises behavior that ignore or block the external impact and as a result the enterprise keeps an organization not adequate to coming opportunities and effective competition.
- To define barriers in environment and inside enterprises, that neutralize or negate the relationships between changes in environment related to knowledge based economy, and variables characterizing knowledge based organization.

The paper is a part of the research project focused on organizational structure issue. The main question is if and how Knowledge Based Economy Programs contribute to utilization of new knowledge and technology of organizational structure modeling and designing in Polish enterprises. The paper presents methodological approach to the empirical research. Empirical verification is planned in 2010 and 2011.

ORGANIZATIONAL STRUCTURE AS A SUBJECT OF EMPIRICAL RESEARCH

Empirical research on conditions of organizational structure have been carried on for decades. They were commenced in the 50's within a strand of sociological theory of an organization. J.Woodword proved that organization structure is influenced by production technology (Woodword, 1965). The research of Aston Group has partially confirmed this thesis and proved that the size of an organization is of a greater importance (Hicson et al. 1979). Searching for the influence of various environmental factors on the organizational structure the following were being examined: the influence of external controlling organizations on the level of centralization of the organization (Hinings, Lee, 1976), the influence of environment insecurity on the organizational structure (Lawrence and Lorsch, 1981), the influence of the level of environment changeability on a model of adopted organizational structure (Burns and Stalker 1981). The research of A. Chandler on the strategy of diversification of activity of larger enterprise showed strong influence of a new factor - strategy on accepted structural solution (Chandler, 1981). Mechanisms of influence of strategy on organizational structure are thoroughly described in literature (Thompson and Strickland, 1993, Mintzberg et al., 1998). A strand of sociological theory of organization has led to a distinction of three groups of context variables for organizational structures: organization's environment features, internal organization's features and company strategy. The structure itself has been considered from the perspective of five features called the dimensions of organizational structure: configuration, centralization of decision making, specialization, standardization, formalization.

The second research strand is the economic organizational approach. Changes of organizational structure are seen as one of the elements of organizational innovations. Research on organizational innovation focus on the role of organizational structures, learning process and adaptation of organizational changes (Oslo Manual, 2005, p.34). In the 90's the scope of statistical research was limited only to product and technology innovations. Later, the methodology prepared by the OECD extended the definition of innovation by distinguishing four classes: innovations in the area of products, innovations in the area of process, organizational innovations and marketing innovations (Oslo Manual, 2005). Unified research methodology OECD is a foundation of systematic comparative research used by OECD member countries. European activities on innovation are examined by two instruments: Community Innovations Surveys (CIS), and European Innovation Scoreboard (EIS). The CIS4 methodological recommendations focus on the following aspects of enterprises organizational innovations (CIS, 2006):

1. New knowledge management systems introduced for a better use of exchange information, knowledge and skills within an enterprise.
2. New management systems introduced for the production and / or supply operations.
3. Significant changes in the organization of work that increased or decreased employee decision making and their responsibility for their work.
4. Significant changes in the management structure of an enterprise, such as creating divisions or departments, integrating different departments or activities, adoption network structure etc.
5. New or significant changes in relations witch other firms or public institutions, such as through alliances, partnerships, outsourcing.

By comparing these aspects of organizational innovation to the dimensions of organizational structure in a social-organizational approach, we can find convergence of the subject of research in three dimensions:

1. configuration – described by the above aspect 4 and 5
2. specialization – described by aspect 4 and 5

3. centralization – described by aspect 3.

The characteristic is lack of dimension of standardization and formalization. I underline the lack of these dimensions because they are a part of methodology of organizational structure design.   The methodology also considers the environmental context and features of enterprise itself. The research also encompasses the influence of organizational changes on: time reduction to respond to customer or supplier needs, improvement of goods quality, cost reduction, employee satisfaction improvement, communication improvement, ability to develop new product or processes. Questionnaires are constructed based on a binary system of answers. Such methodology simplifies the statistical research but at the same time simplifies and shallows the conclusions. For instance, the assessment of importance of innovations, what is and what is not significant, remains in the area of subjective assessment of a respondent. Despite those simplifications, extensive statistical material is a valuable database for further analyses. As one of the examples can serve two Norwegian Community Innovation Surveys (CIS 3&4) to examine how the age and size of companies have different impacts on organizational innovation and on the effects of such innovation on firm performance (Sapprasert, 2008).

Third research strand is an attempt of combined usage of both previously mentioned strands: sociological and economic. The closest to the subject of this article is the research done in years 2006-2008 by the group led by A. Stabryla, regarding the organizational structures of enterprises in the context of their adjustment to knowledge based economy. The research encompassed 275 polish enterprises. Research questionnaire included 85 questions on various scales: binary, multiple choice and descriptive. Analysis regarded all dimensions of organizational structure in the context of its conditions and influence on enterprises' efficiency (50 questions), and the level of intellectual capital and organization's learning (35 questions). This research is characterized by methodological complexity (from the perspective of organizational structure theory), detaility (measured by a number of quantitative variables) and additional descriptive research which validate and supplement quantitative data.

This research is independent and despite many similarities cannot be directly compared with the results of OECD research. It is not a criticism but a postulate for own research, to integrate the methodology and increase the comparability of results.

While searching for literature prerequisites to personal research I have come to the following conclusions:

1.   The issue of methodology of organizational structure design has not been a subject of a separate empirical research. It appears indirectly in various surveys as a result of this design. In some aspects we can draw conclusions about used design approach based on identified organizational structure and in other aspects such reasoning is distorted by significant error resulting from lack of data.

2.   Literature on the other hand provides with extensive material regarding the methodology of conducting empirical research on organizational structure. Individual research approaches, including sociological and economic, can be

used together in one research scheme regardless to many changes between them.

CONCEPTUALIZATION OF RESEARCH

Each enterprise has an organizational structure. It is a result of informed or uninformed decisions made by management or a result of neglecting certain decisions. The influence of organizational structure on efficiency and effectiveness of reaching business goals depends not only on modernity of this structure and the level of its adjustment to the condition in which it operates. Efficiency and effectiveness depend also on the knowledge of managers about the organizational structure as a management tool. A very well designed organizational structure prepared by an outside consulting company is often poorly implemented and yields results opposite from intended as a result of low level of knowledge of company management and executive staff. Therefore an issue of organizational structure should be seen not only as a structure ex post but also as knowledge ex ante which has led to this structure.

Thus, a dependent variable occurring in research is a structure as such as well as codified and non-codified knowledge of organizational structure and methodology of its creation. This knowledge is contained in documents, information procedures and minds, abilities and experience of managers and executive staff of a company. The goal of the research is to define:

1. What is the level of modernity of this knowledge in enterprises?
2. Has this knowledge changed and how has it changed in recent years?
3. What are the major channels of diffusion of this knowledge from outside the company?
4. What are the major centers of this knowledge inside the enterprise?
5. What are the major factors blocking the diffusion of this knowledge from inside and outside of the company?
6. Do the realized programs of Lisbon Strategy have any influence on the actual state of this knowledge and what kind of influence is it?

The following independent variables will occur:

1. Changes in macro-environment (economic, political, legal, social, technological) and in particular, selected parameters monitoring the effects of economic crisis and legal changes in recent years.
2. A progress in realization of Lisbon Strategy in Poland and in particular the parameters monitored in Poland Development Strategy for 2007-2015.
3. Characteristics of company's sector and business environment (suppliers, consumers, competition, other institutions).
4. Characteristics of the organization itself (size, age, technological level, etc.)

OPERATIONALIZATION OF A RESEARCH MODEL

A research framework is presented on fig. 1. It includes further development of dependent and independent variables and definition of research hypotheses.

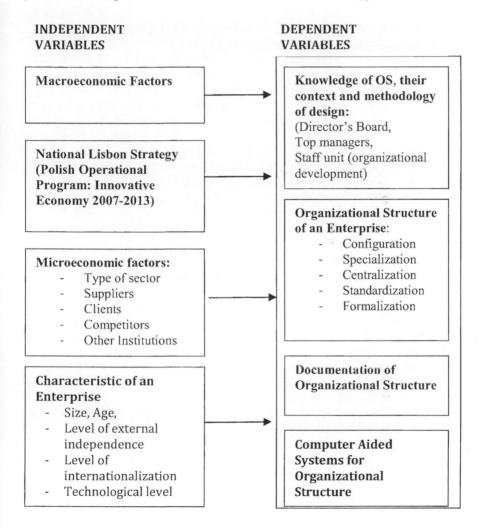

FIGURE 1. The framework of research

OPERATIONALIZATION OF DEPENDENT VARIABLE

Knowledge of organizational structures and methodology of its modeling
Design and implementation decisions regarding organizational structure are conditioned by various aspects of knowledge, experience and intuition. A chart of used knowledge can be arranged in five dimensions of design space (Pawlowski, 2009):

1. Knowledge of organizational structures and their practical interpretation (simple or multidimensional)
2. Methodology of organizational structure modeling (for example modeling of a structure of business processes, modeling of organizational charts, modeling of deployment of competencies, etc.)
3. Methodological approach to organizational design (diagnostic and prognostic approach, functional and process approach, free and standardized approaches)
4. Procedures of organizational structure design (there are several dozens of various procedures of organizational structure design of which between ten and twenty typical charts can be distinguished.)
5. Principles of organizational structure design, including:
   - Construction regularities and transformation of organizational structure
   - Strategy – structure correlation
   - Organizational structure conditionings and influence on effectiveness
   - Strategies and concepts of company management (like Agile, Lean, TQM, JIT ,…)

We can prove that systemized moving in this space increases the rationality of organizational changes. Accepting a defined objective of organizational changes, for example the increase of an Agile effect, a designer/decision maker makes choices in a systematized space of criteria, choosing solutions closest to the objective (Pawlowski, 2008).

An organizational structure ex post of a company created in the moment of research and additionally an assessment of changes within a period of 3 years. Methodological base of structure measurement will be the dimensions of a structure according to a concept of "Aston University": configuration, centralization of decision making, specialization, standardization, formalization.

Documentation of organizational structure should be the base for the evaluation of an actual, not just declared, level of modernity of used organizational knowledge. Documentation of a structure of business processes, process description charts, organizational structure charts, competencies deployment and decision making tables, workplace descriptions define the level of organizational knowledge.

Computer aided design and actualization of organizational documentation. It should demonstrate the level of usage of IT as a factor of flexibility of access to documentation and actualization of documentation of organizational structure and the speed of its designing.

OPERATIONALIZATION OF INDEPENDENT VARIABLES

Economic **macro-environment variables** will be registered based on national economic statistics. Data such as GNP, interest rates, unemployment ratio, structure of expenses on education, will be considered in a context of accordance of macro-economic trends with a trend of expenses on companies' innovations.

Variable monitoring the realization of Lisbon Strategy in Poland. The research conducted within an Operational Program Innovative Economy 2007-2013 will be used. 9,7 billion of Euro was appointed to this Program of which a majority is to be spent in enterprises (Piech, 2009, p.262). Program includes 9 priority lines: 1/research and development of modern technologies, 2/Infrastructure of R&D, 3/capital for innovations, 4/investment in innovative undertakings, 5/diffusion of innovation, 6/Polish economy on the international market, 7/information society - establishment of electronic administration, 8/information society - increase of innovativeness of economy, 9/technical assistance. Monitoring of this program is done with a use of a system of indexes, such as expenses on R&D in enterprises, own innovations in SME, innovations of SME with cooperation with others, venture capital, employment rate in the high tech industry. Some of those macroeconomic trends will also serve as a base for comparison with trends in examined enterprises.

Characteristics of a sector and business environment. Sector comparisons are to indicate if there are differences in the level of organizational knowledge in enterprises of different sectors (industry, commerce, services, finances). Research of business environment are to verify what are the directions of diffusion of knowledge from customers, suppliers and other institutions and if there are any.

Characteristics of examined organizations. Literature study showed that there is a relation between organizational innovations, including changes in organizational structure, and the features of the organization itself. We treat the following features as significant: size of an enterprise, level of external independence (legal and organizational), level of internationality of the enterprise, age of the enterprise, technological level, geographical location (region of a country)

A SCOPE AND METHOD OF RESEARCH

The research will include enterprises listed on Warsaw Stock Exchange. A primary research tools will be a questionnaire including both open and closed questions, supplemented with structured interviews and examination of enterprises source documentation.

HYPOTHESES

Considering the former empirical research regarding organizational structures as well as personal surveys in several enterprises I set the following main hypotheses.

1. The influence of Lisbon Strategy on the level of organizational knowledge regarding the modeling of organizational structure is of little significance, however it shows a growing tendency.
2. There is a significant relation between the level of enterprise's internationalization, sector of economy, size of an enterprise and the level of organizational knowledge in the area of organizational structure modeling.
3. Major channels of diffusion of knowledge to enterprise in the area of organizational structure modeling are:
   - international corporation headquarters forcing its own organizational strategy,
   - suppliers and customers demanding standards of quality in a delivery chain,
   - competitive enterprises in a business sector.

CONCLUSIONS

Organizational knowledge in enterprises regarding the methodology of organizational structures design has not yet been a subject of empirical study. Various research regarding organizational structure and its context variables may however be used in prepared methodology of empirical study of a structure design methodology. A research model being prepared is based on three pillars:
1. uses contemporary theoretical knowledge of organizational structure and its context variables,
2. uses contemporary knowledge from the area of organizational design methodology,
3. uses contemporary knowledge about knowledge based economy and theory of organizational innovation.

Research model was presented on the second level of operationalization, however it seems that it is sufficient for a substantial discussion.

REFERENCES

Burns T., Stalker G.M. (1981). Management Structures and Systems. in: Organizations by Design: Theory and Practice. Jelinek M., Litterer J.A., Miles R.E. (red.). Business Publications, Plano, Texas

Chandler A.D. Strategy and Structure. (1981). in: Organizations by Design. Theory and Practice. Jeli Jelinek M., Litterer J.A., Miles R.E. (red.). Business Publications, Plano, Texas

Hickson D.J., Pugh D.S., Pheysey D.C. (1979). Operational Technology and Organizational Structure: An Empirical Reappraisal.; in: Readings on Dimensions of Organizations: Environment, Context, Structure, Process, and

Performance. M. Zey-Ferrell (ed.). Goodyear Publishing Company, Santa Monica

Hinings C.R., Lee G.L. (1976). Dimension of Organizational Structure and their Context: A Replication; in: Organizational Structure: Extensions and Replications. Pugh D.S., Hinings C.R. (ed.). Saxon House, Westmead

CIS (2006). Community Innovation Statistics. From today's Community Innovation Surveys to better surveys tomorrow. AG, 6 September 2006

Lawrence P.R, Lorsh J.W. (1981). Environmental Demands and Organizational States. in: Organizations by Design: Theory and Practice. Jelinek M., Litterer J.A., Miles R.E. (ed.), Business Publications, Plano, Texas

Mintzberg H., Quinn J.B., Ghoshal S. (1998). The Strategy Process. Prentice Hall, London

Oslo Manual. (2005). Guidelines for Collecting and Interpreting Innovation Data, 3rd Edition, OECD/European Communities 2005, Polish Issue: Podrecznik Oslo. Zasady gromadzenia i interpretacji danych dotyczacych innowacji. Ministerstwo Nauki i Szkolnictwa Wyzszego, Departament Strategii i Rozwoju Nauki – Warszawa 2008

Pawlowski E., Pawlowski K. (2008). A Framework of Organizational Structure Designing for Agile Enterprises; in: Conference Proceedings: AHFE International Conference, jointly with 11th International Conference on Human Aspects of Advanced Manufacturing (HAAMAHA) 2008, edited by Waldemar Karwowski and Gavriel Salvendy, USA Publishing, Las Vegas

Pawlowski E. (2009). Designing the organizational structure of a company. A Concept of multidimensional design space. In: M. Csath, S. Trzcielinski. (ed.). Management Systems. Methods and Structures. Monograph. Publishing House of Poznan University of Technology, Poznan

Piech K. (2009). Wiedza i innowacje w rozwoju gospodarczym: w kierunku pomiaru i wspolczesnej roli panstwa. Instut Wiedzy i Innowacji, Warszawa

Sapprasert K. (2008). On Factors explaining Organizational Innovation and its Effects. In: TIK Working Papers on Innovation Studies. No. 20080601. University of Oslo

Stabryla A. (ed.) (2009). Doskonalenie struktur organizacyjnych przedsiebiorstw w gospodarce opartej na wiedzy. Wydawnictwo C.H. Beck, Warszawa

Thompson A.A., Strickland A.J. (1993). Strategic Management. Concept and Cases. Irwin, New York

Woodword J. (1965). Industrial Organization: Theory and Practice. London

Chapter 7

"The Operator Saves Our Day?" - Why Do We Need the Operator?

Kerstin Dencker, Lena Mårtensson, Åsa Fasth, Johan Stahre

ABSTRACT

The paper discusses possibilities for companies to become competitive by considering the levels of automation, information and competence. In a research project on assembly systems these three parameters are meant to create proactivity in the system. Proactivity is defined as *readiness for changes, occurrences and disturbances in production for the planning of a sustainable and reliable assembly system* The operator roles and corresponding tasks in the assembly system are being presented as well as their relation to automation. The importance of operator intervention in the system at disturbances is being stressed. Scenarios are given for what preconditions are needed for the operator to intervene correctly. The importance of social competence of the operators when putting together a team of operators is strengthened. The paper concludes that variable levels of automation, information and competence increase the action space for the operators.

Keywords: Task allocation, human-machine

INTRODUCTION

To foresee the future is a challenge of unknown proportions. We still need to do that on a competitive market. With the aim to run a smooth and highly efficient production process there is a need to foresee disturbances before they occur.

In a Swedish research project, an attempt has been made to find characteristics of a proactive assembly system, where levels of automation, information and competence are the key parameters. In the ProAct project, *proactivity* has been defined as a *readiness for changes, occurrences and disturbances in production for the planning of a sustainable and reliable assembly system*. The goal of the project has been to create an appropriate action space for the operator by means of variable levels of automation and variable levels of information. Five industrial case studies have been carried out, a model for increased action space for the operator has been designed as well as a web-based tool for development of proactive assembly systems.

A method for analysis of production logistics and production technology and related levels of automation has been developed, i.e. the DYNAMO++ method (Fasth, 2009). Further, a method for analyzing the information content and level has been used (Bruch et.al, 2008). This paper concentrates on a third critical parameter, the level of competence.

AUTOMATION

As a major issue, manufacturing automation, its role and application should be considered. The introduction and use of automation in human-machine systems raises many questions:

- How should technology be introduced?
- Which level of automation should be used and in what part of the system?
- Is a variable level of automation feasible and desirable? What are the trade-offs?
  What does increased level of automation mean to the understanding of the system?
- How complex can the system get before its operators loose
  their capability to understand and control the system?
- Monitoring, recovery mechanisms and graceful/intelligent degradation
  of the all systems elements (human and otherwise) require attention.

Automation is an indispensable part of manufacturing systems today. The level of automation is linked to the roles of operators in the system. Although there are many reasons to automate, it is questionable whether increasing levels of automation always result in a more cost-effective and productive system. Thus, identifying and implementing the right level of automation in a controlled way could be a way to maintain the effectiveness of a system. According to Frohm (2008), to be able to make the manufacturing system as robust, flexible and adaptable as possible, the system must be able to handle variations in the process, such as introduction of a new product, a tool change, a product disturbance etc. It is thus important to understand how to obtain a balanced manufacturing system that has the proper mix of operators and machines in order to e.g. obtain the highest profit possible without suffering any losses of product quality. One way to achieve this balanced manufacturing system is to separate the system into two basic classes of activities, i.e. information/control and physical work. To be able to measure and assess different levels of automation a methodology has been developed by Frohm, et. al. (2007) which is described as, "The relation between human and technology in terms of tasks and function allocation, which can be expressed as an index between 1 (total manual work) and 7 (total automate) of physical and cognitive support". The method has been further developed by Fasth (Fasth, 2007, 2008, 2009).

COMPETENCE

Increased flexibility, change-frequency, and levels of automation normally induce complexity in the manufacturing system. This complexity must be handled by increased competence among the operators and system-related personnel. The definition of competence adopted for the ProAct project was as follows: *"Theoretical knowledge and skills based on experience, attitude and ability for correct judgment and evaluation in order to use the knowledge in its proper context"*. Furthermore, sensory motoric factors to act from one's perceptions; cognitive factors to analyze, combine and see connections; personality factors like self confidence to do things; as well as social factors are also part of the competence of the individual.

For the production situation the competence level of the whole team is more crucial than that of a single individual. There is a need for one person in the team who has the sensory motoric skill to "fix things". There is also a need for a problem solver, who knows how to analyze a critical situation, as well as for a socially competent person to get the team together, so that the job is being carried out under the best circumstances.

ROLES AND TASKS OF THE OPERATOR

Operator competence levels in manufacturing should not be seen as fully generic but rather relative to the role in which the operator is acting at a specific moment. In the ProAct project an appropriate model for operators in automated or semi-automated manufacturing systems was adopted. The human supervisory control model was proposed by Sheridan (Sheridan, 1988, 1992) and can be used as a model when designing automated systems. The model has since been adapted for semi-automated manufacturing by Stahre (Stahre, 1995). The roles of the operators are planning, programming, monitoring, intervening and learning. These roles are relevant during normal operations as well as during disturbances and breakdowns. During planning the operator decides what should be done in the system. The plan is then transformed into a program and transferred to the automated sub-system. When programming is finalized, the process is started and monitored by the operator who continuously checks if the system is experiencing deviations from the plan and the program. If such deviations occur, the operator has to intervene, which is an important task to make the system run smoothly. Each intervention induces an operator learning process, where the valuable experience is memorized and the knowledge is used when planning of the next operation occurs.

Billings (1997) had a more general view on the role of the operator in complex systems. He suggests that the operator contributes with his creativity and flexibility. His human capabilities are used to scan all available sources for valuable information and to deal with incomplete knowledge and uncertainty.

These capabilities are usually not fully internalized but may be described using Rasmussen's three-level taxonomy for human behavior (Rasmussen, 1983), i.e. skill-, rule-, and knowledge-based behavior. The skill-based level is relevant when the operator is trained and has learnt the task by heart; it has become a well-integrated part of his knowledge. Skill-based behavior generally requires on-the-job training. Rule-based behavior is relevant when the operator uses established and internalized rules for how to carry out a task and relevant support can be given by rule-based decision support. Finally, the individual operator is often confronted with original and completely new tasks, whereby he needs to use all his knowledge education, and training to analyze and solve the problem. If his knowledge is not sufficient, he has to rely on external competence, e.g. references or colleagues to resolve the issue at hand.

RESULTS — AUTOMATION, INFORMATION, AND COMPETENCE

Five case studies were carried out in industrial assembly systems. Examples of the work environment are presented in figure 1 and figure 2. Observations of the work places were made with a detailed task analysis, Figure 3. Interviews were made with operators and production engineers. The results from the analysis were discussed in workshops, where possible improvements were discussed.

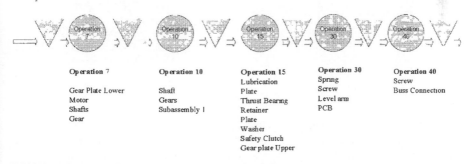

Operation 7	Operation 10	Operation 15	Operation 30	Operation 40
		Lubrication	Spring	Screw
Gear Plate Lower	Shaft	Plate	Screw	Buss Connection
Motor	Gears	Thrust Bearing	Level arm	
Shafts	Subassembly 1	Retainer	PCB	
Gear		Plate		
		Washer		
		Safety Clutch		
		Gear plate Upper		

FIGURE 1 Example of assembly tasks in the case studies

FIGURE 2 Different levels of automation in an assembly system

The operator roles described above can be developed into tasks, which has been done in Figure 1 with the example of assembly work. The detailed level of tasks is useful for the design of new systems. For example the operator has to *intervene* and handle disturbances due to lack of material or technical problems on a small or large scale. He also has to make quality checks of products and systems.

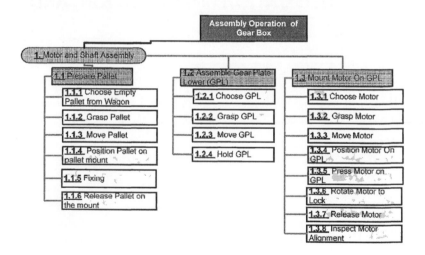

FIGURE 3 Hierarchical task analyze of an assembly system

From operator interviews it can be said that the challenges for the operators is to detect that a disturbance has occurred, to know whom to contact, to find the cause for the disturbance, to describe it to others, to correct the problematic component, to document the disturbance and perhaps the most difficult to find lasting solutions, so that you don't "have to solve the same problem 50 times!" To find lasting solutions requires time, which may be a scarce resource in a hectic production.

However, problem solving although it may be difficult is also considered a positive and intellectual challenge. The operator competence for handling disturbances has been analyzed by several researchers (Grothe, 2003; Erbe, 2003).

After every intervention in the assembly system something is *learnt,* which is valuable for future planning. In learning organizations the knowledge gained by one individual or by a team should be spread to other people in the organization for them to benefit from others' experiences. To teach your colleagues about your own experiences is a good way to create a proactive system. The concept of continuous improvements allows operators and other staff to come forward with their suggestions for improvements which leads to an efficient and creative organization.

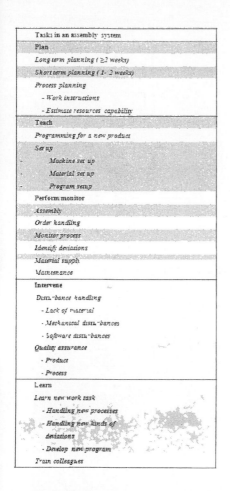

FIGURE 4 Operator roles and tasks in an assembly system

Having analyzed the roles of the operators in automated systems in general and the task for the operators in an assembly system in particular Figure 5 shows a spider web which includes a designation of knowledge-based level (K-6-7) for different roles as well as rule based (R-4-5) and skill based roles (S-1-3). It could be said that the material handling and order handling requires skills and rules to be carried out,

68

while planning is a rule based role and intervening and learning requires a knowledge based approach.

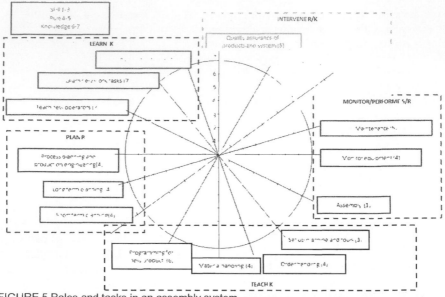

FIGURE 5 Roles and tasks in an assembly system

FIGURE 6 Example of short term planning performed by the operators.

DISCUSSION

The project related above has aimed at creating proactivity in assembly system. The three critical parameters have been level of automation, level of information and level of competence. Intervening has been pointed out as an important role for the operator. Dencker et. al (2009) presented an assessment model based on Endsley's concept of situation awareness. Table 1 shows the model, where different scenarios on the operator awareness of deviations range from not noticing to notice, interpret correctly and intervenes correctly. The scenarios are analysed from the point of view of perception, comprehension, projection, intervention and cause. Only at the final scenario "Operators/system notice the signals, are able to interpret and has the tools to act correctly" the operator has the correct competence and the correct information for acting in a proactive way, hopefully leading to efficiency in the system.

The table shows the importance of correct information and correct competence. The importance of a joint system wide information to be shared by everybody concerned could not be strengthened enough. In the interviews at the case study companies it was found that the teams had different routines for sharing information. Meetings were held every morning for planning of the day's work. Information was given on the production rate for every team. The operators themselves carried out the short term planning. Time was given to discuss improvements.

The competence level of the operators was a critical parameter. At one company a competence matrix was held showing the different courses taken by the individual to pursue different tasks. Some operators had a special competence for a particular task. Another company had an extra operator in each team to e.g. manage sick leaves. One company had a "super operator" in each team, a knowledgeable person who managed all tasks in the team. It was obvious that the competence of the team was equally important as that of the individual operator. The supervisors stated the importance of designing the teams appropriately. "You cannot have a team of only socially competent individuals, then the job will never get done" according to one of the supervisors. On the other hand, a team with no socially competent operator was equally bad.

It was said in the interviews that each department needs a strategy for competence development covering issues like

- the competence of the team as well as that for the individuals,
- competence development plan for the individual,
- experts vs novices,
- the number of novices in the team at the same time,
- the level of basic technical education of the team members,
- information/instructions at each work place.

Table 1. Scenarios for operator with different possibilities for correct actions.

Scenario	Perception	Comprehension	Projection	Intervention	Cause
Operators/system not able to notice signals from deviations	Not Ok	Don't exist	False	Wrong	Lack of knowledge No clear overview of the system
Operators/system notice the signals, are not able to interpret and therefore no ability to act or acts not correctly.	Ok	Not ok	False	Wrong	Lack of competence
Operators/system notice the signals, but don't think the interpretations are correct.	Ok	Not ok	False	Wrong	Unreliable information system Lack of experience
Operators/system notice the signals, are able to interpret but do not know what tool to use.	Ok	Ok	False	Wrong	No standardized work instruction No standardized work settings.
Operators/system notice the signals, are able to interpret but do not have the tools.	Ok	Ok	True	Wrong	work instruction Lack of standardized work settings
Operators/system notice the signals, are able to interpret and has the tools to act correctly.	Ok	Ok	True	Correct	Correct competence Correct information

The results from the five case studies showed that it is possible to design for a variable level of automation, that the information should be given at many different levels and that the level of the competence is most important on the team level. This gives a greater action space for the operator as shown in figure 7.

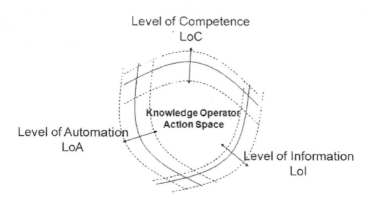

FIGURE 7 The operators action-space

Finally, an industrial manager once phrased his concern for the operator: "When things get out of hand in complex technical systems, you always expect the human operator to save our day".

ACKNOWLEDGEMENTS

The authors want to express their deep gratitude to the Swedish Governmental Agency for Innovation Systems (VINNOVA) for funding the ProAct project. We also want to express our gratitude to our ProAct project colleagues. Further, our thanks go to the participating companies: Parker Hannifin, Global Garden Products, Bosch Rexroth, Siemens Building Technologies, Stoneridge Electronics and Electrolux Home Products.

REFERENCES

Billings, C.E., (1997), *Aviation Automation: the search for a human-centered approach*. Lawrence Erlbaum Associates Publishers, New Jersey.

Endsley, M.R. and Kaber, D. (1999). Level of automation effects on performance, situation awareness and workload in a dynamic control task. Ergonomics, vol 42, pp 462-492.

Erbe, H.H., (2003), *Operating and understanding – intelligent interaction in human-machine systems*. Preprints of the 8th IFAC symposium on Automated systems based on human skill and knowledge, 2003, Chalmers University of Technology, Göteborg.

Fasth, Å. (2009) Measuring and Analysing Levels of Automation in assembly systems - For future proactive systems. *Product and production Development, Production systems*. Gothenburg, Chalmers University of Technology.

Fasth, Å., Bruch, J., Dencker, K., Stahre, J., Mårtensson, L. and Lundholm, T., (2010), Designing proactive assembly systems (ProAct) - Criteria and interaction between automation, information, and competence *Asian International Journal of Science and Technology in production and manufacturing engineering (AIJSTPME)*, 2/4.

Fasth, Å., Stahre, J. and Dencker, K. (2008) Measuring and analysing Levels of Automation in an assembly system. *The 41st CIRP conference on manufacturing systems* Tokyo, Japan.

From, J. (2008) Levels of Automation in production systems. *Department of production system*. Gothenburg, Chalmers University of technology.

From, J., Lindström, V., Winroth M. and Stahre, J. (2008) Levels of Automation in Manufacturing. *Ergonomia IJE&HF, 30:3*.

Grothe, G.,(2003), Uncertainty *management at the core of system design*. Preprints of the 8th IFAC symposium on Automated systems based on human skill and knowledge, 22-24 September 2003, Chalmers University of Technology, Göteborg

Rasmussen, J., (1983), *Skills, Rules, Knowledge, Signals, Signs and Symbols and other Distinctions Human Performance Models*, IEEE Transactions on Man, Systems and Cybernetics, SMC-13/3:257-266.

Sheridan, T.B., (1988) Task Allocation and Supervisory Control, in: Handbook of Human—Computer Interaction, Helander, M., (Ed.) , Elsevier Science Publishers, New York, pp. 159-173.

Sheridan, T.B., (1992), *Telerobotics, automation and human supervisory control*, MIT Press, Cambridge Massachusetts.

Stahre, J., (1995) Evaluating Human/Machine Interaction Problems in Advanced Manufacturing. *Computer Integrated Manuf. Systems,* Vol:8:2, pp.143-150.

Chapter 8

Some Metaphors of Agile Enterprise

Stefan Trzcielinski

Poznan University of Technology
60-965 Poznan
Poland

ABSTRACT

In Management Sciences some metaphors are quite often used to explain organizational and managerial phenomenon by use of well known phenomena from another area of life or science. One of managerial concepts which slowly becomes popular in industrial practice is Agile Enterprise. Such enterprise is oriented on undertaking short life time opportunities. To do so the enterprise has to be bright, intelligent, flexible, and shrewd.

In this paper an attempt is made to explain the brightness and intelligence of enterprise by metaphors appropriately of bright eyes of chameleon, and human intelligence. On this base some conclusions how to shape the both factors of enterprise agility are proposed. The research is a part of project run by the Institute of Management Engineering from Poznan University of Technology on "Adaptation of Enterprises Management Systems to Knowledge Based Economy".

Keywords: Agile enterprise, Intelligent enterprise, Metaphors

INTRODUCTION

The concept of agile manufacturing was introduced by Goldman et al. (1991) as that one that can increase competitiveness of company in changing and unpredictable business environment. More widely the concept was popularized in the book on Agile Competitors and Virtual Organizations (Goldman et al., 1955). When defining and describing the agility of company, the authors stresses on capability of operating profitably in a competitive environment of continually, and unpredictably, changing customer opportunities as well as on importance of knowledge about individual customers and the interactions with them. Such company is able to develop high customer-perceived value product through individualization, much quickly, at much lower cost, than it has done historically for its markets. This meaning of agility stresses on the enterprise ability to deliver a product fully adjusted to the customer needs and is frequently presented in the literature (Hormozi, 2001; McCullen and Towill, 2001; Toussaint and Cheng, 2002; Jin-Hai et al., 2003; Brown and Bessant, 2003).

Rick Dove (2001) defines agility as the "ability of an organization to thrive in a continuously changing, unpredictable business environment". He adds that being agile means being a master of change, and allows one to seize opportunity as well as initiate innovations. That can be interpreted that agile company not only is able to follow quickly the customers identified expectations but create his new needs as well. These two levels of agility refers to what is called by Dove (2005) as a reactive and proactive response ability when needs and opportunities are unpredictable, uncertain, and likely to change.

Another aspect of agility concerns the ability of the company to form alliances with other companies that already possess the resources required to cut the cost, shorten the customer respond, get access to expertise, facilities, or markets (Goldman, Nagel, 1995). Such level of agility expresses the enterprise capability to form the virtual organization, as it reconfigures its internal cross-unit network and external network with partners (Galbraiht, 1997). The external network is a supply chain and its agility is affected by its flexibility, complexity (Prater, et al., 2001; Weber, 2002), and capability of information technology (Swafford, et al., 2006). The network of the company with external partners represents an organizational form of agility which can be called "virtual production unit" (Trzcieliński, ed., 2007). The virtual production units are small and medium enterprises which are configured to undertake short life time opportunities perceived by the network creator.

The internal cross-unit network (internal virtuality) can be configured around seven dimensions which are: function, work place, competencies, work contract, equipment, functional interdependences and hierarchical interdependences (Trzcieliński, Wojtkowski, 2007). The increase of internal virtuality leads to structural decomposition of the enterprise and results in the increase of external virtuality as more work is teleworked, outsourced and contracted (Pawłowski, 2009). In the extreme stage the enterprise plays only the role of broker of the market and all the needed resources acquires from its industry environment. In result a virtual organization is created that is able to execute all the work of traditional company although it possesses a little if any human and tangible resources (Handy, 1997). The virtual enterprise that is "… rapidly configured, multi-disciplinary network of firms organized to meet a window of opportunity to design and produce a specific product" is called agile virtual enterprise and represents the higher level of agility (Cunha, Putnik, 2006).

Summarizing the above the following four levels of enterprise's agility can be distinguished:

- Agility as the ability to follow the customer needs as well as manufacture and deliver quickly customized product.
- Agility as the ability to create the needs of existing and new customers.
- Agility as ability to create and reconfigure quickly the virtual production units to undertake short life time opportunities.
- Agile virtual enterprise.

From the view point of shaping the agility of enterprise it is important to identify its features. It is worth to notice that the defining characteristic of agile enterprise is that it has to be short life time opportunities oriented (Goldman, et al., 1991; Goldman, Negal, 1995; Cho, et al., 1996; Dove, 2001). An opportunity is an event or system of events which happen in the macro and industry environment that favours achieving the enterprise goals. If so, the features are these which are necessary to use the opportunity. The logical analysis allows deducing the following features of agility of enterprise:

- Brightness that is the ability of the enterprise to perceive quickly the opportunities.
- Intelligence that is ability of the enterprise to learn, accumulate the knowledge and adapt to the changeable business environment.
- Flexibility is a feature of the disposable resources that extends the repertoire of tasks the enterprise is able to undertake.
- Shrewdness that is the ability of the enterprise to achieve its goals in very practical mode, even intuitively.

Although the above features seem to be morphological components of enterprise's agility, their understanding is not unambiguous. Simultaneously they are crucial to mould the organizational conditions that enable to convert the enterprise into the agile one. Below some metaphors are employed to clarify the understanding of two agility features – brightness and intelligence of enterprise.

BIOLOGICAL METAPHOR OF BRIGHTNESS - THE CHAMELEON BRIGHT EYES

The chameleon is an animal with good appetite and maybe therefore the nature equipped it in eyes which both distinguish it among other reptiles and let it perceive the flying insects (opportunities) in wide visual field. Its eyes can move independently over a large range of 180° horizontally and 90° vertically. It fixes the object of interest with quick eye movements. During binocular fixation of a prey, the visual fields overlap and both eyes converge. To prevent blurring of vision during its head movement, the gaze is continuously stabilized by the eyes movement in opposite direction to the movement of head, but with the same speed. In this way the image is stabilized in the eye's retinal (Haker, et al., 2003). To use the opportunity i.e. to capture the insect, the chameleon shuts with great precise its long tongue for the distance of the flying insect. Its very precise judgment of distance is achieved by coupled accommodation in both eyes (Bowmaker, et al., 2005). The accommodation occurs by shrinking the eye's retinal and it can be coupled or uncoupled in both eyes. To notice the prey the chameleon moves quickly its one eye knob and scan the environment when at the same time the second eye is turn on long perspective vision. The process is switched in short intervals between both eyes. When the insect is noticed, both eyes move forward to fixate the prey, while the head axis is aligned towards the target and coupled accommodation occurs shortly before the tongue shot. It happens to improve accommodation precision rather than to permit stereopsis (matching corresponding points) or triangulation (Ott, et al., 1998).

Visual perception of the insects by the chameleon is a good analogue of the enterprise's brightness. The insects fly quickly so the time they are in the reptile's field of strike is very short, what illustrates well the short life time opportunities. The chameleon eyes are the analogue of organizational units of enterprise which independently observe different segments of the environment. Usually the enterprise observes better and has more deep knowledge about industry than the macro-environment (Thompson, Strickland, 1993) that is an analogue to wider angle of visual field in horizontal than in vertical surface by the chameleon. The opportunities appear in different segments of the environment therefore they are indentified by different organizational units (chameleon's eyes). They can occur in the markets of: finance, labour, new technologies, row materials supply, customers, other producers or in the area of legal regulations, and therefore the segments are scanned by organizational units which "accommodation" is determined by their specialization. However when the opportunity is identified than it is necessary to assess if it is in the enterprise's "field of strike" regarding the disposable resources (chameleon's tongue) and what is the value of the opportunity and the associated risk (if the insect is in the chameleon's "menu" and if it fits to the anatomic constraints of the reptile). To get such assessment the opportunity has to be examined by different organizational units what can be treated as both chameleon eyes accommodation on the same object and what means a momentary overlapping of the visual fields of the organizational units. In results of the simultaneous accommodation the resource accessibility of the opportunity is assessed and the decision about using the opportunity can be taken (shut of chameleon's tongue).

METAPHOR OF ORGANIZATIONAL INTELLIGENCE - HUMAN BEING INTELLIGENCE

The intelligence is a complex category and therefore is defined in many ways. Sternberg and Detterman (1986) have grouped the different definitions in three groups (Strelau, ed. 1999). According to the authors the intelligence of human being is: (a) ability to learn on the base of own experiences, (b) ability to adapt to surrounding environment, (c) ability to recognize own learning processes and control them.

There are several reasons why people are interested in possessing higher level of intelligence but their synthesis is that higher intelligence allows being more effective and efficient in new situations. In case of enterprise we can talk about its better creativeness and innovativeness. In result such organization has more chance to extend its market share and keep its competitive position (Grudzewski, Hejduk, 2000). If so it is worth to use the knowledge about human intelligence and implement it as a metaphor to shape the intelligence of enterprise.

There can be distinguished the following groups of theories explaining the nature of human intelligence (Strelau, ed. 1999): factor, biological and cognitive theories (see Figure 1).

INTELLIGENCE OF THE ENTERPRISE IN THE LIGHT OF FACTOR CONCEPT OF HUMAN INTELLIGENCE

According to the factor theory elaborated by Spearman, the intelligence is mould by general and specific intellectual abilities (Strelau, ed. 1999). The general factor of intelligence is meant as a kind of "mental energy" that is a basic ability of nervous system to execute effectively the tasks. In enterprise the "mental energy" refers to intellectual capacity of staff and teams and their openness for challenges as well as to motivation system, organizational culture, and general rules of the organizational structure.

Among different concepts about the specific factors of intelligence there are particular two which delivers good analogues to organizational intelligence. The first elaborated by Cattel appoints on two groups of factors: fluid intelligence and crystallized intelligence (Sternberg, 1982; Strelau, ed. 1999). The former one is something like preliminary biological equipment that is "crystallized" in abilities acquired through learning, practice, and exposure to education. The crystallized intelligence manifests itself by knowledge and abilities important in given cultural context. In the enterprise these two kinds of intelligence represents respectively the general rules of functioning of the organization (corporate order) which in the process of retaining them become "crystallized" in organizational culture.

78

The second concept of specific factors of intelligence elaborated by Guliford, called "structure-of-intellect", classifies the factors according to three dimensions: Contents (of tests) i.e. tasks to be executed, mental Operations required in the tests, and types of Products which resulted from the operations. With 4 types of Contents, 5 types of Operations, and 6 types of Products Guilford has distinguished 120 specific factors of intelligence (Sternberg, 1982; Strelau, ed. 1999). This three dimensional "structure-of-intellect" can be a good inspiration for creation and studying in the enterprise a three dimensional structure of organizational intelligence. Than the Contents can represent Decision problems of enterprise (e.g. well structured, poorly structured, and unstructured), Operations represent Methods of management (e.g. operational research, TQM, JiT, Empowerment, etc., and heuristic methods), and Products represent types of Innovations (e.g. product, process, marketing, organizational, financial, etc. innovations). Such three dimensional space can be a useful tool for quick and effective adaptation of the enterprise to changing environment and the same can be treated as important factor of organizational intelligence.

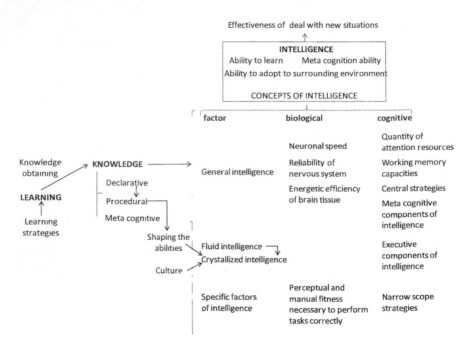

Figure 1. Concepts of human intelligence. (On the base of: Sternberg, ed., 1982; Strelau, ed. 1999).

INTELLIGENCE OF THE ENTERPRISE IN THE LIGHT OF BIOLOGICAL CONCEPT OF HUMAN INTELLIGENCE

Particular the highest level factor representing the general intelligence of human being can be partly explain from view point of biological concept of intelligence. The first aspect of this concept concerns foundation of intelligence that is the speed of transmission of impulses in the nervous system. People with speeder impulse transmission gain an advantage in resolving intellectual tasks. Additionally the time of reaction is shortened and therefore the effect of forgetting is reduced (Strelau, ed. 1999). The neuronal speed delivers a direct analogue of bandwidth of information channels in enterprise. Wider the bandwidth, more quick of information flow and shorter time of respond is possible. Also, if the information come when it is needed, the decision task can be undertaken immediately; if there is delay in information flow than not only the time of reaction is longer but the decision task can be postponed or even forgotten as well.

The second aspect of biological concept of intelligence concerns reliability of the nervous system. The increase of reliability makes shorten the time of reaction as fewer mistakes are done and therefore fewer corrections of the mistakes are needed. Nevertheless the most important is the influence of reliability on competent execution of tasks. If there are less mistakes in the impulses transmission then more actions are proper and this means improvement of the competence (Strelau, ed. 1999). Also in the enterprise the reliability of the "nervous" system, i.e. the information system, is crucial for both, the time of respond and the competent action. Less reliable information systems causes delays in obtaining information by the decision centre and allows information noise and the same decision taking on the base of wrong promises. Such actions cannot be competent.

The third aspect of biological theory of intelligence refers to the energetic efficiency of brain. It was proved that more "intelligent" brains use less energy (glucose) (Strelau, ed. 1999). The "intelligent brain" is a good analogue of enterprise's intelligent management system, which uses fewer resources (glucose) than other enterprises to achieve the same goals or results. Example of such intelligent system can be Lean management (Womack, Jones, 1996) or Agile enterprise (Trzcielinski, ed. 2007).

INTELLIGENCE OF THE ENTERPRISE IN THE LIGHT OF COGNITIVE CONCEPT OF HUMAN INTELLIGENCE

The levels of intelligence's factors can be also interpreted from the point of view of cognitive concept of intelligence. The first aspect of the concept refers to the quantity of resources of attention which are interpreted as amount of processing power of cognitive system. According to Lansman & Hunt, some activities depend on the knowledge when others depend on disposable resources of attention (Strelau,

1999). The resources determine the ability of the system to control all cognitive activities and processes. Because the fixed amount of resources of attention, and that different activities compete for them, therefore the level of execution of particular activity depends on the amount of resources it receives. More resources are disposable, more of them can be assigned to each of activities, and this leads to better execution of the all activities (Strelau, ed. 1999). This cognitive aspect of intelligence is a direct analogue to limited knowledge and resources which constrain some activities of enterprise. The resources are always limited but an intelligent enterprise, by creating virtual organization, is able to undertake some activities (opportunities) even if it does not possess all the required resources.

The second aspect of the cognitive concept of intelligence concerns the capacity of the working memory. The capacity is strongly limited. The working memory not only temporarily stores the information but participates in processing it as well. Lager capacity of working memory and slower speed of forgetting the information, then smaller the negative consequences of the deal between storing and processing the information. Because of this both factors are important promises of intelligence (Strelau, ed. 1999). In the enterprise the working memory is represented by specialists and functional organizational units which are expected to make the needed information available to the decision centres. The working memory is also represented by computer data-bases and data-warehouses as well as by different registers and records, either electronic or traditional.

The third aspect refers to cognitive strategies i.e. the qualitatively separate way in which the cognitive task is executed. Baron has distinguished two kinds of strategies: central and local (Strelau, ed. 1999). The central strategies are crucial to cognitive processes such as: searching for relations between information kept in the memory and information currently received, and also for checking the run and results of cognitive activities. The central strategies decide about general intelligence. The local strategies are important to narrow cognitive tasks (Strelau, ed. 1999). The central strategies can be considered as analogue of project management structure of enterprise. In such structure the teams which run the particular projects are responsible not only for storing and using all necessary date and information but also for observing the environment, obtaining new information, and adjusting the run of the project appropriately to changeable situations. Thus intelligent enterprise is able to modify its behaviour.

The forth aspect applies to meta cognitive components of intelligence, which have been distinguished by Sternberg (Strelau, ed. 1999). They are responsible for planning, supervising and controlling the activities which are important to execute the cognitive task as well as for analysis of feedback information and for accompanying learning process. Sternberg identified 10 meta cognitive components including e.g. perception of the problem, defining the problem or choosing a proper set of executive components (Strelau, ed. 1999). In the enterprise the "meta cognitive components" are obvious analogues of managers who are responsible for planning and organizing the work of the subordinators as well as for motivating, and controlling them.

CONCLUSIONS

To search for the opportunities, the brightness of enterprise can and should be supported by methods of "scanning" the environment and forecasting the events and situations which can be recognized as opportunities or threats. These methods are well popularized in the literature on strategic management and among others they include: PEST, SWOT, Delphi method, scenarios methods, or methods of white intelligence. Another method, which especially is suitable for agile enterprise, is so called "long tail". The long tail is a huge amount of unique orders (opportunities) extracted from the Data Base of Intentions (Trzcielinski, 2007). Such data base consists of queries inserted by the internet search engine.

The human intelligence metaphors seem to be very productive. Example can be the structure-of-intellect that can be a good inspiration for building the structure of enterprise intelligence. Also the metaphors which refer to neuronal speed, reliability of the nervous system, and the energetic efficiency of brain, enable to understand better how works the information system of enterprise and how it can be organized. Additionally the metaphor of energetic efficiency of brain creates new possibility for explaining why lean and agile enterprises can be categorized as intelligent organizations. The explanation lies in physical interpretation of "organizational energy". Because "energy" is an equivalent of "work" and "work" depends on "force" and "distance" (which in enterprise are represented appropriately by amount of resources and effectiveness of their use), than the same level of organizational energy depends on the relation between the two factors. Lean and agile enterprises reduce the amount of resources together with increasing the efficiency of their use.

To exploit the human intelligence metaphor in more complex way also the interrelations between learning, knowledge, and intelligence should be taken into consideration. This problem will be a subject of further research.

REFERENCES

Bowmaker, J.K., Loew, E.R., Ott, M. (2005). The cone photoreceptors and visual pigments of chameleons. Journal of Comparative Physiology A. Vol. 191 No. 10.

Brown, S., Bessant, J. (2003). The manufacturing strategy-capabilities links in mass customization and agile manufacturing – an exploratory study, Vol. 23 No. 7.

Cho, H., Jing, M., Kim, M. (1996). „Enabling technologies of agile manufacturing and its related activities in Korea", Computers and Industrial Engineering, Vol. 30 No. 3.

Cunha, M.M., Putnik, G.D. (2006). Agile virtual enterprises: Implementation and management support. Hershey: Idea Group Publishing.

Dove, R. (2001). Design Principles for Highly Adaptable Business Systems, With Tangible Manufacturing Examples. www.parshift.com/publications.htm

Dove, R. (2005). Fundamental Principles for Agile Systems Engineering, Proceedings of 2005 Conference on Systems Engineering Research (CSER), Stevens Institute of Technology, Hoboken, NJ.

Galbraith, J.R. (1997). "The reconfigurable organization", in: The organization of the future, Hesselbein, Goldsmith, Beckhard (Eds.). pp. 87-97.

Goldman, S.L. Preiss, K., ed.; Negal, R.N., Dove R., principal investigators, with 15 industry executives. (1991). 21st Century Manufacturing Enterprise Strategy: An Industry-Led View. Bethlehem: Iacocca Institute at Lehigh University.

Goldman, S.L., Nagel, R.N., Preiss, K. (1995). Agile competitors and virtual organizations. New York: Van Nostrand Reinhold.

Grudzewski, W.M., Hejduk, I. (2000). „Rozwoj i implementacja organizacji inteligentnej [Development and implementation of intelligence organization]", in: Przedsiebiorstwo pryszlosci [The enterprise of the future], Grudzewski, Hejduk (Eds.). Warszawa: Diffin, pp. 136-146.

Haker, H., Misslisch, H., Ott, M., Frens, M.A., Henn, V., Hess, K., Sandor, P.S. (2003). Three-dimensional vestibular eye and head reflexes of the chameleon: characteristics of gain and phase and effects of eye position on orientation of ocular rotation axes during stimulation in yaw direction. Journal of Comparative Physiology A. Vol. 189 No. 7.

Handy, C. (1997). "Unimagined future" in: The organization of the future, Hesselbein, Goldsmith, Beckhard (Eds.). pp. 377-383.

Hormozi, A.M. (2001). Agile manufacturing: the next logical step. Benchmarking: An International Journal, Vol. 8 No. 2.

Jin-Hai, L., Anderson, A.R., Harrison, R.T. (2003). The evolution of agile manufacturing. Business Process Management Journal, Vol. 9 No. 2.

McCullen, P., Towill, D. (2001). Achieving lean supply through agile manufacturing. Integrated Manufacturing Systems, Vol. 12 No. 7.

Pawlowski, E. (2009). "Designing the organizational structure of company. A concept of multidimensional design space", in: Management systems. Methods and structures, Csath M., Trzcielinski S. (Eds.). Publishing House of Poznan University of Technology, Poznan. pp. 107-122.

Prater, E., Biehl, M., Smith, M.A. (2001). International supply chain agility. Treadoffs between flexibility and uncertainty. International Journal of Operations and Production Management, Vol. 21 No. 5/6.

Spearman, C. (1927).

Sternberg, R.J., ed. (1982). Handbook of human intelligence. Cambridge University Press: Cambridge.

Sternberg, R.J., Detterman, D.K. (1986). What is intelligence? Contemporary viewpoints on its nature and definition. Norwood, NJ: Ablex.

Strelau, J., ed. (1999). Psychologia [Psychology]. Gdańsk: Gdańskie Wydawnictwo Psychologiczne.

Swafford, P.M., Ghosh, S., Murthy, N.N. (2006). A framework for assessing value chain agility. International Journal of Operations and Production Management, Vol. 26 No. 2.

Thompson, A.A., Strickland, A.J. (1993). Strategic management. Boston: Irwin.

Toussaint, J., Cheng, K. (2002). Designing agility and manufacturing responsiveness on the Web. Integrated Manufacturing Systems, Vol. 13 No. 5.

Trzcieliński, S., ed. (2007). Agile Enterprise. Concepts and Some Results of Research. Madison: IEA Press.

Trzcieliński, S. Wojtkowski, W. (2007). Toward the measure of organizational virtuality. Human Factors and Ergonomics in Manufacturing, Vol. 17 No. 6.

Weber, M.M. (2002). Measuring supply chain agility in virtual organization. International Journal of Physical Distribution & Logistic, Vol. 32 No. 7.

Womack, J.P., Jones, D.T. (1996). Lean thinking. Simon & Schuster: New York.

CHAPTER 9

Selling Human Factors and Ergonomics in a Successful Way: Creating Enthusiasm for Ergonomics

Peter Vink

TNO/Delft University of Technology
Landbergstraat 15, 2628 CE Delft
The Netherlands

ABSTRACT

In this tutorial the most crucial elements of a successful approach will be trained and also mistakes in previous projects will be shared with the participants. The literature indicates success factors like: setting a clear goal in terms of shareholder value and productivity, active participation of end-users and management and a process with a clear inventory of the problems, a structured approach, real tests and a steering group responsible for the guidance and end-users.

Keywords: Productivity, Participatory ergonomics, Success factors, Participatory Design.

INTRODUCTION

Ergonomics should be fun the human factors' specialist and their clients. Selling human factors and ergonomics is needed to support companies in being more productive and healthy. Creating enthusiasm for ergonomics could help in this process. However, many case studies show of course that it's not easy to develop ergonomic solutions that are really implemented and used (Koningsveld et al. 2005), but it is possible. In the tutorial the most crucial elements of a successful approach are discussed and also the mistakes in the projects are shared with the participants. In this paper important elements of successful ergonomics case studies are described.

Several studies mention success factors in ergonomic projects to arrive at better comfort, health and productivity (Koningsveld et al. 2005; Looze et al. 2001, Molen et al., 2005(1)). The literature indicates success factors like (1) setting a clear goal in terms of shareholder value and productivity, (2) active participation of end-users and management and (3) a process with a clear inventory of the problems, a structured approach, real tests and a steering group responsible for the guidance and end-users. These three points will be elaborated in this paper as it is the basics of the tutorial.

SETTING CLEAR GOALS

In a discussion with 19 ergonomic experts it was clear that it's important for all participants to know what the purpose of a study is (Vink et al. 1992). It is important not only to focus on health, comfort or well-being (Koningsveld et al. 2005). The most positive effects are shown when the goal is a combination of benefits for the company (e.g. financial benefits) and benefits for the workers (e.g. well-being). In many cases, financial benefits of a reduction of absenteeism by ergonomics interventions will not pay back the efforts connected to the implementation of the solutions and measures, at least not in a short time (Koningsveld et al. 2005). Productivity increase often pays off. For instance a case study focused on a better layout prevented the building of a new manufacturing facility, which is an enormous financial advantage (Rhijn et al. 2005). Therefore, a combination of health or comfort and productivity leads to better results and if possible should be made the goal for the project.

To begin a participatory ergonomics project an initiative should be taken somewhere, either within or outside the organization. There are different potential starting points. For example, workers, occupational physicians, sector organizations, or unions could be aware of a problem and initiate the approach. Of course, successful implementation is more likely when the workers and management take the initiative and set the goal or are in need of the improvement and thus desire improvements. A project were the sector organization took the initiative to reduce the physical workload among brick layers was not successful (Molen et al. 2005(2)). The innovations were hardly implemented.

ACTIVE PARTICIPATION

When the goals are set it is important to define the stakeholders or actors and their roles in the approach. The following groups are mentioned increasing the chance of a successful end result (Vink et al., 2008): top management, middle management, employees, ergonomist, HSE experts, designer, engineer, internal staff, organizational psychologist, purchasing etc. In one project the purchasing department was not involved (Vink, 2005, page 95-110). Initially this leaded to the fact that no new ergonomically correct furniture was purchased as the purchasing department had good deals with other suppliers. After convincing them of the use the purchasing was done as a consequence the implementation was 6 months delayed. The same study shows the importance of being involved as the 12 persons actively involved in testing the work stations had significantly more adapted in their work station than the other, who were also informed about the project and shared rooms with the persons doing the tests.

There is not much debate about the fact that workers should participate in a process towards more comfort and productivity. Already in 1988 A review of Cotton et al. (1988) argues that involving a group that is only representative of workers in a working group, rather than the workers themselves, will not lead to successful participation. It is also vitally important to have workers develop ideas for improvement in collaboration with managers (Vink et al., 2008).

Figure 1. Active involvement is an important aspect towards a successful end product.

Management involvement is also inevitable. In many cases, for example where decisions on budgets or organizational structures are involved, middle management or even top management should also participate, given their decision-making power. In an evaluation of more than 200 cases top management was mainly involved in the first stage of the process, where the boundaries and goals are set. It could be that management is not yet convinced of the need for ergonomic improvement. When management does not take the improvement initiative, they will need to be convinced as to the need for improvements, which may be difficult. A solution to this problem is to address the managers in their own language, as described by Kogi (1988). This can mean stressing an economic goal or benefit and giving concrete examples.

An essential part of the participatory ergonomics strategy is that direct participation of the stakeholders involved should be strived for in an improvement process (Vink et al. 2008). Not only an active role of management and employees is crucial, but also an active role by the ergonomist. A case of installation work improvement illustrates that a small role of the ergonomists led to a limited effect (Jong and Vink, 2000), but also other stake holders like facility management, HSE executives, suppliers, controllers etc should be taken into account (Vink et al., 2008).

STRUCTURED APPROACH WITH REAL TESTS

Sometimes having only goal leads to uncertainty among the participants. Their roles should therefore be defined and to stimulate active involvement roles should be defined in terms of their own interest (Noro and Imada, 1992). Another structure needed in the definition of the proves. A well-structured approach gives a lucid holdfast and clearness to all participants. The structure also makes it possible to monitor the process towards the end result in this process. Different stages can be discerned (Molen et al., 2005; Noro and Imada, 1992). But often the following stages are mentioned: Once the process is initiated, problems should be identified and solutions selected and developed. Then the solutions should be implemented and the effects, as well as the process itself, should be evaluated, with the results feeding back into the process.

It is important in product design or design of a working environment to measure the effects. Effects could be measured on dis)comfort, predictors of health and productivity during the design process (Vink, 2005). This can be done in several stages of the design process. (Dis)comfort can be measured in the first stage of design by questioning subjects on their experience with the now available products. It can be checked in showing various drawings of ideas. Or it can be checked by using virtual reality or mixed reality (Krassi et al, 2010). Of course in the last phase of the design process it can be checked by testing prototypes.

Part of the structured approach is also informing the participants in different stages. Also, focusing on the positive aspects of ergonomics and having a party when the next step of a project is achieved. A comparison of four projects (Vink et al., 2006) indicated that positive experiences with improvements, where people have wow experiences or are in a kind of flow has positive results. If end-users have positive experiences with the potential improvement (end-users feel or see the benefits) the chance of implementation will increase. The conclusion of this study is only based on four cases and further research is needed to support this theory.

CONCLUSION

It was the aim of this paper to support the content of the tutorial at the AHFE 2010 congress to create more enthusiasm for ergonomics. Many success factor are mentioned in the literature and some are reported in this paper. More focus on the positive side of ergonomics could be help. However, there is no universally agreed list of factors. Moreover It's all based on case studies. Based on these case studies there are indications that setting a clear goal, arrange active participation and having a structured process with celebrations of successes increase the change of success.

REFERENCES

Cotton JL, Vollrath DA, Froggatt KL, Lengnick-Hall ML, Jennings KR. Employee participation: Diverse forms and different outcomes. Academy of Management Review 13(1988): 8–22.

Jong AM de, Vink P. The adoption of technological innovations for glaziers; evaluation of a participatory ergonomics approach. Int. J. Ind. Erg 26 (2000) 939-46.

Kogi K. Improving working conditions in small enterprises in developing Asia, ILO, Geneva, 1988.

Koningsveld EAP, Dul J, Rhijn GW van, Vink P. Enhancing the impact of ergonomics interventions. Ergonomics 2005;48(5):559-580.

Krassi B, D'Cruz M, Vink P, 2010. ManVAR: a framework for imporving manual work through virtual and augmented reality (this book)

Looze, MP de, Urlings IJM, Vink P, Rhijn JW van, Miedema MC, Bronkhorst RE, Grinten MP van der. Towards successful physical stress reducing products: an evaluation of seven cases. Appl. Erg. 32 (2001), 525–534.

Molen HF van der, Sluiter JK, Hulshof CTJ, Vink P, Duivenbooden C van, Frings-Dresen MHW. Conceptual framework for the implementation of interventions in the construction industry. Scan J Work Environ Health 31,suppl.2 (2005) 96-103. (1)

Molen HF van der, Sluiter JK, Hulshof CTJ, Vink P, Duivenbooden C van, Holman R, Frings-Dresen MHW. Implementation of participatory ergonomics intervention in construction companies. Scand J Work Environ Health 2005;31(3):191-203 (2)

Noro K, Imada AS. Participatory Ergonomics. Taylor & Francis, London, 1992

Rhijn JW van, Looze MP de, Tuinzaad GH, Groenesteijn L, Groot MD de, Vink P. Changing from batch to flow assembly in the production of emergency lighting devices. International Journal for Production Research 43(2005) 3687-3701

Vink P (2005): comfort and design: principles and goos practice, CRC press, Boca Raton.

Vink P, Koningsveld EAP, Molenbroek JF. Positive outcomes of participatory ergonomics in terms of greater comfort and higher productivity. Applied Ergonomics 37 (2006): 537-546.

Vink P, Lourijsen, E., Wortel, E. and Dul, J. (1992) 'Experiences in participatory ergonomics: results of a round table session during the 11th IEA congress', Ergonomics, 35(2): 123-127.

Vink P, Nichols S, Davies RC. Participatory Ergonomics and Comfort. In: Vink P (Eds.) Comfort and Design: Principles and Good Practice. CRC Press, Boca Raton (2005) pp: 41-54.

Vink, P. Imada, A.S. Zink, K.J. (2008). Defining stakeholder involvement in participatory design processes. Applied Ergonomics 39, 519-526.

Chapter 10

National OSH Programme – the Polish Experience

Danuta Koradecka

Central Institute for Labour Protection – National Research Institute
Czerniakowska 16, 00-701 Warsaw, Poland

ABSTRACT

Occupational safety and health have a considerable value for the employee and the employer alike. Simultaneously, as work processes become more flexible, this knowledge grows in importance for society as a whole.

The substantial technical progress has not solved the problems of occupational safety and health. It has only shifted the core of the problems from chemical, physical and dust hazards to psychophysical and biological ones.

Labour protection – like art in the Renaissance – must now focus on people with their limited psychophysical abilities in the workplace.

Our research carried out among 10 000 industry employees in Poland has shown that there is a big difference between objective exposure to harmful conditions and subjective assessment of working conditions.

The parallel assessment of hazards, including objective methods and subjective estimation of exposed workers, constitutes the basis of human protection system in the working environment.

Many such solutions have been developed under the Polish national research programmes, carried out and coordinated by the Central Institute for Labour Protection – National Research Institute. These programmes concern OSH in line with the rapidly changing technologies (nano- and biotechnologies, human-robot interaction, active "intelligent" multifunctional personal protective equipment) and the labour market (flexible employment, telework).

The results of carrying out those programmes have made it possible to achieve, among others:

- a significant reduction in the number of people exposed to harmful factors and in the number of occupational diseases;
- the implementation of a system of differentiated social insurance premium (experience rating) related to the level of working conditions in an enterprise;
- a reduction in the number of employees entitled to early retirement (from ca. 1.2 to 0.2 million).

Thanks to the implementation of programmes, Poland as a Member State of the European Union since 2004 has harmonised its regulations and practices with the required standards of occupational safety and health. This process covered the entirety of working conditions with a view to preventing occupational accidents and diseases and satisfying the requirements of ergonomics.

Keywords: occupational risk assessment, OSH management, labour protection

INTRODUCTION

Occupational safety and health (OSH) in Central and Eastern European countries at the turn of the 1980s and 1990s was characterised by a considerable number of people exposed to exceeded levels of health standards on harmful chemical and physical agents (e.g., noise, vibration, hot microclimate), and fibrogenic dust (including carcinogenic asbestos).
The risk of injury was also high (lack of safeguarding of machinery working zone, lack of proper organisation of work and work space, etc.).

In 1980, approximately 16.5 million people were employed in the Polish economy. Of this number, studies of the Central Statistical Office (Główny Urząd Statystyczny, 1981) covered 12.7 million, including 1.9 million employed in conditions in which health standards were exceeded.

Currently, the number of persons employed in conditions where the hygiene standards are exceeded is 0.6 million (Central Statistical Office, 2008).

The data mentioned above could therefore suggest that the transformation in Poland did not have a negative impact on the basic indicators of the status of occupational health and safety.

Research carried out by the National Safety Foundation (USA), conducted in Poland together with the Central Institute for Labour Protection in the 1990s (Szejnwald et al., 2001; Broszkiewicz et al., 2002) confirmed this thesis, even though the initial prognosis of American scientists provided for a negative impact of the transformation upon the system of labour protection.

AIM

Adaptation of the OSH system in Poland to ensure its compliance with the international standards, in particular European Community standards (Council Directive 89/391/EEC of 12 June 1989).

METHODS

1. Analysis of the working conditions using national statistical data, objective field measurements and subjective assessment.
2. Launch of a social dialogue with a view to develop and realize a national research and implementation programme in the field of OSH.

RESULTS

Working conditions have been assessed so far by the national statistics, objective measurements (fieldwork) or subjective assessments (surveys) (Lindstrom et al., 2000; European Agency for Safety Health at Work, 2000; Rantanen 2001; European Foundation for the Improvement of Living and Working Conditions, 2000; European Foundation for the Improvement of Living and Working Conditions, 2007; Koradecka, editor, 2010). Our research involved simultaneously all the above mentioned methods. 10 000 workers took part in the research. As a result, the subjective assessment of working conditions has proven more negative than the national statistics and objectives measurements of occupational risk show (Koradecka et al., 2010).

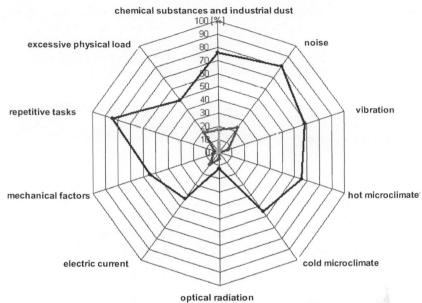

FIGURE 1. A comparison of the results of a questionnaire survey (▬▬▬),
field measurements (▬▬▬) and statistical data (▬▬▬).
Employees exposed to environmental factors (%) (Koradecka et al., 2010).

The costs of inappropriate working conditions incurred by the Social Insurance Institution (ZUS) have also turned out to be considerable. In 2008, the total cost of disability and family pensions as well as single benefits, paid out by the Social Insurance Fund, administered by ZUS, for occupational diseases, accidents at work and accidents on the way to and from work amounted to approximately EUR 1.1 billion, which constituted 3.6% of the total cost of compensations paid from the Fund. As the analyses conducted in the EU countries show, indirect costs (treatment, rehabilitation costs, losses incurred by employers) are 3-4 times higher than the costs of pensions and compensations. This would result in the costs due to improper working conditions amounting to EUR 3.3-4.5 billion annually, which in 2008 constituted ca. 1.1-1.5% of the total Polish GDP.

This situation has become the basis for the decision to set up government programmes concerning the improvement of safety and working conditions. Since 1995, they have been coordinated and carried out by the Central Institute for Labour Protection – National Research Institute in Warsaw. Ministries, relevant supervisions and control bodies as well as social partners cooperate in their implementation. The main objective of the current programme entitled "Improvement of safety and working conditions" (2008 – 2010) is to reduce the number of work-related accidents and occupational diseases, as well as the related economic and social losses, by developing and implementing innovative organisational and technical solutions aimed at the development of human resources, new products, technologies, methods and management systems.

This objective is being achieved through actions taken in the following areas:

- fulfilment of basic requirements included in legal regulations on occupational safety and health adapted to the provisions in force in the European Union,
- provision of special protection for persons working in high-risk sectors of the economy,
- development and implementation (e.g. in SMEs) of OSH management system methods with particular focus on occupational risk assessment and its reduction,
- increasing the awareness of employees and the society of the consequences of accidents at work and occupational diseases, and cost-effectiveness of preventive measures at the national and enterprise level,
- shaping a high level of safety culture among the employers and employees by developing a system of education and information on OSH requirements.

As a result of implementation of the OSH programmes, documentation has now been developed of more than 200 harmful agents in the work environment. The list of TLV-s (threshold limit values) currently in force in Poland contains 528 items (Regulation of the Minister of Labour and Social Policy of 29 November 2002; Koradecka, editor, 2010). In addition, original methods of measuring these agents in the working environment have been developed for approximately 100 harmful chemical agents. Approximately 300 OSH requirements have been standardised. Consequently, at present, the Institute supervises approximately 1 000 standards in this area (including ISO and EN standards).

In the dynamically expanding laboratories of the Institute, accredited in accordance with international standards, some 1 300 products as well as working environment parameters and personnel competencies (experts, lecturers) in the field of OSH are tested every year.

In addition, approximately 150 technical and organisational solutions in the area of human protection in the working environment have been developed, receiving 109 prizes at international and domestic exhibitions, with some covered by patent protection. These achievements have also enabled participation in tens of European research projects in such fields as nanotechnology or "intelligent" personal protective equipment solutions.

In order to systematise the OSH management programme, a package of standards compliant with European standards has been worked out. A computer programme making it easier to assess occupational risk and report its results to the competent supervision and audit agencies has been developed to facilitate the application of these standards by employers. This program is now used by more than 1 000 enterprises.

In order to create economic incentives, a system of differentiated insurance premiums advantageous to businesses that improve working conditions has been developed and implemented.

Enormous progress has been achieved in the development and implementation of a modern multimedia OSH education system (at university, school and nursery school level"). The Institute's website ranks third in terms of the number of visits in Europe (after Great Britain and France). The activities propagating the results of the programme cover some 5 million people annually.

Occupational safety and health is a fascinating, though complex discipline, as it covers the achievements of technical, biological and social sciences. All these disciplines have been developing very dynamically in the last decade.

Therefore, we expect further dynamic development of technology, and consequently, an increase in the need to be able to trust its safety. At the same time, a new generation of highly independent employees is entering the labour market. This may lead to revolutionary changes in the way occupational safety and health issues are perceived.

CONCLUSION

1. Research in the area of OSH is interdisciplinary and as such should be carried out and coordinated as national programmes.
2. Due to the production globalisation and markets penetration, OSH requirements should be comparable between various countries.
3. Poland, as a member of the European Community since 2004, has harmonised its OSH legal regulations and practices with those of the Community. This applies to all working conditions in order to prevent accidents and occupational diseases and also to meet the requirements of ergonomics.

REFERENCES

Broszkiewicz R., Szejnwald Brown H. (2002), Occupational and environmental status of Polish industry: a comparative study of the private and public sector, New Solution, vol. 12(3).

Central Statistical Office (GUS) (2001), Warunki pracy w 2001 r. [Working conditions in 2001]. Warsaw, Poland: GUS.

Central Statistical Office (GUS) (2008), Warunki pracy w 2008 r. [Working conditions in 2008]. Warsaw, Poland: GUS.

Council Directive 89/391/EEC of 12 June 1989 on the introduction of measures to encourage improvements in the safety and health of workers at work. OJ. 1989;L183:1–8. Retrieved February 10, 2010, from: http://eurlex.europa.eu/LexUriServ/Lex-UriServ.do?uri=CELEX:31989L0391:en:HTML.

European Agency for Safety Health at Work. (2000), Monitoring: the state of occupational safety and health in the European Union - pilot study. Luxembourg: Office for Official Publications of the European Union.

European Foundation for the Improvement of Living and Working Conditions. Third European survey on working conditions 2000. Luxembourg: Office for Official Publications of the European Communities; 2001. Retrieved February 10, 2010, from: http://www.eurofound.europa.eu/pubdocs/2001/21/en/1/ef0121en.pdf

European Foundation for the Improvement of Living and Working Conditions. Fourth European survey on working conditions. Luxembourg: Office for Official Publications of the European Communities; 2007. Retrieved February 10, 2010, from: http://www.eurofound.europa.eu/pubdocs/2006/98/en/2/ef0698en.pdf

Koradecka D., Pośniak M., Jankowska E., Skowroń J., Karpowicz J. (2006), Chemical, dust, biological, and electromagnetic radiation hazards. In: Salvendy G, editor. Handbook of human factors and ergonomics. 3rd ed. Hoboken, NJ, USA: Wiley.

Koradecka D., et al. (2010), A comparative study of objective and subjective assessment of occupational risk. International Journal of Occupational Safety and Ergonomics (JOSE), Vol. 16, No. 1.

Koradecka D., editor. (2010), Handbook of occupational safety and health. Boca Raton, FL, USA: CRC Press; In print.

Lindstrom K., Elo AL., Skogstad A., Dallner M., Gambarele F., Hottinen V., et al. (2000) Users guide for the QPSNordic, general Nordic questionnaire for psychological and social factors at work. TemaNord 2000:603. Copenhagen, Denmark: Nordic Council of Ministers.

Rantanen J., Kauppinen T., Toikkanen J., Kueppa K., Lehtinen S., Leino T. (2001), Work and health country profiles: country profiles and national surveillance indicators in occupational health and safety. Helsinki, Finland: Finnish Institute of Occupational Health.

Rozporządzenie Ministra Pracy i Polityki Społecznej z dnia 29 listopada 2002 w sprawie najwyższych dopuszczalnych stężeń czynników szkodliwych dla zdrowia w środowisku pracy [Regulation of the Minister of Labour and Social Policy of 29 November 2002 on the maximum admissible concentrations for agents harmful to heath in the working environment]. Dz. U. 2002; (217): item 1833.

Szejnwald Brown H., Angel D.P., Broszkiewicz R., Krzyśków B. (2001), Occupational safety and health in Poland in the 1990, in: A regulatory system adapting to societal transformation, Policy Sciences, vol. 34.

Chapter 11

Information Technologies in Concurrent Engineering in Selected Knowledge Based Economy Companies in Poland – Some Results of Research

Joanna Kałkowska, Hanna Włodarkiewicz-Klimek

Institute of Management Engineering,
Poznan University of Technology
Poznan, Poland

ABSTRACT

Application of advanced information technologies (IT) is one of the most important factors of implementing concurrent engineering into the production enterprise. The paper presents a brief overview of the various aspects of IT application in concurrent engineering in some polish manufacturing companies mainly focusing on the following issues: identification of information technologies supporting concurrent engineering in selected companies, what kind of information technologies are used in the area of communication and applied dedicated software, what is the information technology influence on different factors determining concurrent engineering, the assessment of information technologies development and determination of the actual state of their usage in selected enterprises.

98

Keywords: Information Technologies (IT), CAx Technologies, Concurrent Engineering (CE)

INTRODUCTION

At present, the knowledge-based economy companies[1] are searching for competitive advantage concerning quality of products, manufacturing costs and cycle of launching product. To achieve this goals, they implement modern concepts of management supporting by information technologies. One of such a method is concurrent engineering. Concurrent engineering successfully fulfilled with information technologies enable also efficient information management. The aim of this paper is to present some issues associated with applying information technologies in concurrent engineering in selected companies in Poland.

CONCURRENT ENGINEERING (CE)

Simply, Concurrent Engineering (CE) is the simultaneous product development process. Many individuals, managers and scientists use a lot definitions of concurrent engineering but the most important one was defined by the Institute for Defense Analysis in USA in 1986 and it is following: Concurrent Engineering is a "systematic approach to the integrated, concurrent design of products and their related processes, including manufacturing and support. This approach is intended to cause developers, from the outset, to consider all elements of the product life cycle cost, schedule and user requirements" (Ion, 1994; Syan and Menon, 1994; Tang et al., 1997). According to the above statements concurrent engineering is not just about "engineering" functions, in fact the process involves far more of the "product value stream" including suppliers, marketing personnel, service and support personnel and customer as well. Implementing CE in company impose

---

[1] The European countries in 2000 accepted the common concept of knowledge-based economy. The development postulate concerning the potential knowledge usage were presented in Lisbon Strategy, which focuses on four fundamental potentials, among which we can distinguish following ones: Human resources, i.e. society of knowledge (which the part of knowledge is gathered in), innovation system (with the entrepreneurship, more concentrated for operations of companies, but also on the cooperation with science), it creates new knowledge in result of discoveries and innovations, information technologies facilitating the exchange of knowledge, also with foreign countries, institutional and legal environment, which creates conditions for development of presented domains; it constitutes from various institutions and regulation, etc. (Kałkowska, Włodarkiewicz-Klimek, 2009). Taking into consideration the Lisbon Strategy it can be state that the companies which applied Information Technologies supporting concurrent engineering become a part of this strategy.

changes in management adopting to computer systems requirements. CE is also integrated set of management strategies guided to gain very high competitive product. The main aim of CE are:

- to reduce the time from "idea to market",
- to rise the quality of product,
- to rise the quality of manufacturing process.

Achieving the full benefits of implementing CE to the company is possible only when a significant number of changes at all levels of organization will be implement including application of advanced information technologies (IT). These changes can be summarized as follows:

- the introduction of multidisciplinary team-working involving personnel from all stages of the new product development process such as design, marketing, finance, manufacturing and purchasing (including subcontractors),
- simultaneous design of the product and manufacturing process,
- the use of concurrent engineering tools such as QFD (Quality Function Deployment), CCM (Controlled Convergence Matrix) and DFMA (Design for Manufacture and Assembly) and others, the use of appropriate project management tools against clearly defined and agreed cost, quality and delivery targets specified to achieve complete customer satisfaction and business profitability. A fast and efficient communication structure is essential (Bullinger, Warschat, 1996, Clausing, 1994, Ion, 1994; Hartley, 1992; Syan and Menon, 1994; Salomon., 1995).

SELECTED INFORMATION TECHNOLOGIES IN CONCURRENT ENGINEERING

One of the most important issue of concurrent engineering is application of information technologies (IT). This technologies enable to coordinate all tasks concerning new product development process as well as efficient information management. Information technologies are also recognized as the one of most important factors of development contemporary knowledge-based economy enterprises. Most of them already have automated support for various functions such use of computer-aided design, computer-aided manufacturing systems in design, engineering data management systems, etc. Achieving the full benefits of those systems is only possible by their integration. Such integration is guaranteed by CIM (Computer Integrated Manufacturing) which fundamental tasks is automatization and integration of manufacturing system elements. The CIM's structure is compound and it is changing along with the development and implementation of modern manufacturing technologies (Plichta, Plichta, 1999).

The literature points out a range of instruments constituting the CIM system, which are called CAx^2 technologies. Still, opinions concerning classification of particular CAx systems into the CIM structure are significantly diversified. There is an idea that CAx techniques represent an integral part of the CIM and can be distinguished as following:

- CAD (Computer Aided Design),
- CAP (Computer Aided Planning),
- PPC (Production Planning and Control,
- CAM (Computer Aided Manufacturing),
- CAQ (Computer Aided Quality Control) (Chlebus 2000, Salomon 1995).

Some researchers widens the above classification by implementing additional following systems:

- CAPP (Computer Aided Process Planning),
- TDM (Team Data Management and Total Data Management),
- EDM/PDM (Engineering Data Management/ Product Data Management) – systems serving for the engineering data management concerning the product,
- Workflow systems,
- MRP II/ERP systems (Brzeziński, 2002; Plichta, Plichta, 1999).

It is essential that these facilities are integrated into the CE approach taken by companies to gain optimal benefit. However, it is observed that the most popular systems used in researched companies are divided into two groups. First group are CAD/CAM systems. The CAD/CAM systems have the capabilities of three-dimensional shape modeling and the ability to derive physical properties like weight, center of gravity, etc. and to produce manufacturing data such as numerical control data and programme files. Among a lot of CAD/CAM system it is possible to distinguish the main following tools:

- 2D CAD – this level of CAD is entirely appropriate for non-complex designs and design documentation requirements. 2D CAD is a base for many 3D tools,
- 3D CAD – allows designers to see the outline of the parts in a three dimensional,
- Solids Modeling – most popular option for constructors and designers. It let to create shapes on the screen. Using this method, designers can see what they are actually designing and can correct issues as they work,

---

[2] CAx technologies are usually interpreted as computer methods and techniques applied in the integrated manufacturing, which integrate the area of technical functions and management. CAx is a specific mental shortcut from commonly known terms from the domain of computing science, such as CAD, CAM, CAE, etc. (Chlebus, 2000).

- NC/CNC programming and simulation – tools exist that automate file translation into NC programming language such the result can be used directly by NC machine. These tools simulate the machining of raw stock so that machine operations are know prior, avoiding machine errors, including scarp calculations.
- Rapid prototyping - this tool allows the automatic conversion of data for rapid prototyping equipment. Rapid prototyping is used by the design team to understand the product in its environment. Early prototypes can be simple models. Some are carved from wax or clay to show the shape, other models are made by machining out plastics on CNC machine (Salomon, 1995).

The next group constitutes Engineering Data Management (EDM) or EDM/PDM systems. Within the generic class of EDM/PDM systems there are many examples of them like Product Data Management (PDM) systems and Engineering Document Management Systems (EDMS). EDM/PDM systems manage engineering data - all the data related to the product and to the processes used to design, manufacture and support the product. Much of this data are created with computer-based systems such as CAD, CAM and CAE. EDM/PDM systems also manage the flow of work through those activities that create or use engineering data. They support techniques of Concurrent Engineering that aim to improve engineering workflow (Stark, 2000).

Among other systems supporting CE are communication systems which can be comprehended as a different forms and methods of information transfer. Contemporary communication systems not only enable transferring text messages or file with data, but also different type of media files very often containing complicated graphics, sounds or even a television picture. Communication technologies includes also stationary and mobile telephony (cell phones and satellites), fax and Internet technologies. Moreover, this group of instruments also embrace the communication method like "face to face" and traditional mail. The range of application communication systems has been partially shown in research carried out in enterprises of the automotive industry in Germany. This research has been carried out in 2002 by Uwe Lukas from the Centre of Computer Graphics ZDGV in Rostock. Its aim was identifying techniques applied in communication between members of virtual teams. In result of this research it occurred that the most common method of communication are still: traditional and mobile telephony, e-mail, fax and meetings "face to face" organized from time to time (Lukas, 2003).

At present, the communication is more and more often carried out with use of Internet technologies. This technology guarantees the fast flow of information and unlimited range. The most popular Internet technologies can be distinguished according to Curtis and Cobham and they are following: e-mail, World Wide Web, File transfer protocol, online telecommunications, discussion groups, IRC (Internet Relay Chat), Electronic conferences and project management tools (Curtis, Cobham, 2002).

INFORMATION TECHNOLOGIES IN RESEARCHED COMPANIES - THE SCOPE AND METHODOLOGY OF RESEARCH

The research were carried out in the beginning of 2010 with random sample of nineteen manufacturing companies of different branches. Only twelve of them identified the concurrent engineering and could assess their knowledge of it as well as the importance of information technologies. Seven of the researched companies represents electro–machine industry, two of them furniture branch, also cardboard box branch, food production branch and gas industry. In each of that companies there are construction, technology and production department, quality management, marketing as well as production units which manufactures the products. The researches were carried out with direct interview with management staff and with managers of individual divisions. The researches towards to get the answer for the following questions:

1. what is the knowledge of concurrent engineering concept,
2. what kind of advanced information technologies supporting concurrent engineering are applied,
3. what is the information technologies influence on selected factors determining concurrent engineering,
4. what is the actual stage of information technologies usage,
5. what is the level of information technologies development.

KNOWLEDGE OF CE IN RESEARCHED COMPANIES

The concept of CE is widely applied in world industry since many years. Because of that, it was interesting to identify knowledge about this concept through managers and specialists from different connected departments in selected companies. The researches embraced managers of following departments: construction, technology, production, quality management, marketing and sales. Synthesis of that research is shown in fig. 1. The point estimation is interpreted as following: 0-lack of knowledge, 1-very low knowledge, 2- low knowledge, 3-cursory knowledge, 4-good knowledge, 5-very good knowledge.

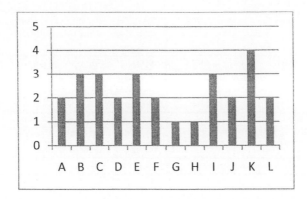

Figure 1. Knowledge of concurrent engineering (CE) concept in researched companies (source: own study)
(Description: A, B, D, E, J, K, L – electro-machine industry companies; C–food branch company, F –gas industry company, G –cardboard box branch company, H, I –furniture industry companies)

Analyzing the conception of concurrent engineering (CE) it can be state that the knowledge of it is relatively good and fluctuate between 1 and 4, which gives average results level near 2,5.

IDENTIFICATION OF IT SUPPORTING CE

The use of advanced IT is widely discussed in a literature (Bullinger, Warschat, 1996, Hartley, 1992; Syan and Menon, 1994; Salomon., 1995, Weiss, 1998). On that basis it was possible to differentiate IT tools into two groups. First group consist of technologies which support design process and the second one supports communication. In the mentioned companies, the most popular tools supporting design was all CAD/CAM systems (AUTOCAD, LogoCAD, Unigraphics, etc.) and MRP/ERP systems (SAP, IFS, Movex, Impuls, etc.). These technologies were also assigned to some phases of product development process (tab. 1). Concerning communication the following technologies were used: email, fax, phones, mobile phones, internet, intranet as well as internet communicators like Skype.

Table 1. Selected IT supporting some NPD phases (source: own study)

NPD Phase	Information Technologies
General product concept	SAP, IFS, Movex, SolidWorks, LogoCAD, AutoCAD, SolidEdge, ProENGINEER, Unigraphics
Study on construction	AION Configurator, Inventor, SolidEdge, IFS, Movex, SolidWorks, LogoCAD, AutoCAD, SolidEdge, ProENGINEER, Unigraphics
Study on technology	AION Configurator, IFS, Movex, Impuls, BAAN, PDM/ EDM
Prototype testing	Movex, FlowWorks, Unigraphics
Cost calculation and costs management	SAP, IFS, Movex,

IT INFLUENCE ON SELECTED FACTORS DETERMINING CE

Taking into consideration the importance of information technologies in concurrent engineering, the authors analyzed some factors which in smaller or higher degree determine the concurrent engineering. These factors are following: study on construction, study on technology, quality of products, quality and system of customer service, rising the quality and work efficiency, cost level reduction, competitive advantage, communication system, information access and usage, collaboration with cooperators, collaboration with deliverers. The point estimation of that influence is interpreted as following: 0-lack of influence, 1-very law influence, 2-low influence, 3-average influence, 4-essential influence, 5-substantial influence. Interpreting the above results it is necessary to take into consideration the pilot character of this research. It follows that the quite substantial influence of IT concern such a factors like study on construction, communication system, information access and usage (fig. 2). This situation seems to be rather rational taking into consideration the concurrent engineering issue as well the stage of IT development in particular enterprises.

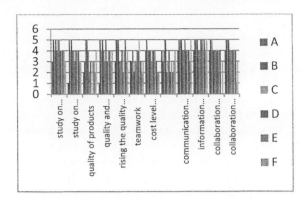

Figure 2. Information Technologies (IT) influence on particular factors determining CE (source: own study)

IT DEVELOPMENT AND ACTUAL STATE OF USAGE

The research shows that companies applied advanced information technologies but they do not used them effectively. Such situation seems to be dictated both with the limitations concerning knowledge of this technologies application and infrastructure. Moreover, the respondents rather positively assess IT resources but comparing in relation to their modernity level there is an observation that the IT are not fully used. It is also interesting that the arithmetical average of IT usage is exactly at the same level what technology development in researched companies (fig.3).

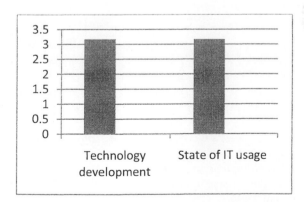

Figure 3. The IT development and state of usage (average assessment) (source: own study)

CLOSING REMARKS

In researched companies were carried out the analysis of applied IT in particular companies considering CE concept. Analyzing the research results it is important to remember that the companies were different area of activity. In general the results pointed out rather high importance of information technologies in concurrent engineering. Rationality of that research confirms researches carried out in Australia and Great Britain (Trzcielinski, 2003).

Concerning IT infrastructure the research included also questions concerning future investments and IT safety. Some respondents answered that they are going to invest in that technologies in the near future. Moreover, the companies are going to finance these investments partly with own capital, partly by loan bank and if possible with European Union support. The amount of this investment fluctuate from 100 000 till 400 000 Euro and will concern typical IT infrastructure and specialist software. Concerning safety the main point concerned the protection of all IT resources and transfer of data.

However the CE concept is widely applied in world industry, it is difficult to find it fully implemented in polish industry. This somehow depend on IT application as well as the specificity of enterprise. However, the IT are the factors which not only conditioning the successful CE. IT also influence on whole management system and constitutes one of the internal competitive advantage factors. What more, the respondents rather positively assessed the IT development level in researched companies perceiving them as dynamically developing regardless CE application. Anyway, economical forecasting's suggests that along with technology development, knowledge transfer and better access to modern IT, the CE seems to be better identified and used in polish industry taking into consideration the requirements knowledge-based economy.

REFERENCES

Brzeziński M., (ed.), *Organizacja i sterowanie produkcją*, Placet, Warszawa 2002.
Bullinger H. J., Warschat J., *Concurrent Engineering Simultaneous Engineering Systems*, Springer – Verlag, London 1996.
Chlebus E., *Zintegrowany rozwój produktu i procesów produkcyjnych*, Oficyna Wydawnicza Politechniki Wrocławskiej, Wrocław 2002.
Clausing D., *Total Quality Development*, ASME Press 1994.
Curtis G., Cobham D., *Business Information Systems; Analysis, Design and Practice*, Prentice Hall, 2002.
Hartley J. R., *Concurrent Engineering*, Productivity Press, Portland, Oregon 1992.
Ion W. J., *The key to Concurrent Engineering*, red. P. T. Kidd, W. Karwowski, IOS Press, Amsterdam, 1994.
Kałkowska J., Włodarkiewicz-Klimek H. (2009). (Eds.), *Managing Enterprises. Social Aspects*, Monograph, Publishing House of Poznan University of Technology: Poznań

Lukas U., *Synchronous tele-cooperation in virtual product creation*, Materiały warsztatów VIDA workshop, 2003.

Plichta J., Plichta S., *Komputerowo zintegrowane wytwarzanie*, Skrypt Wydziału Mechanicznego, Koszalin 1999.

Salomon T. A., *What every engineer should know about concurrent engineering*, Marcel Dekker, New York, 1995.

Stark J., *www.johnstark.com*

Syan Ch. S., Menon U., *Concurrent Engineering. Concepts, implementation and practice,* Chapman&Hall, Great Britain, 1994.

Tang N., Jones O., Forrester P.L. (1997). *Organizational groth demands concurrent engineering*. Integrated Manufacturing Systems, 8(1), pp. 29-34.

Trzcieliński S., *Lean management a wirtualność przedsiębiorstwa*, Prace naukowe Instytutu Organizacji i Zarządzania Politechniki Wrocławskiej, nr 73, 2003.

Weiss Z., *Projektowanie współbieżne – Concurrent Engineering*, Zakład Informatyzacji Systemów Produkcyjnych, Politechnika Poznańska, Poznań 1998.

Chapter 12

Total Ergonomics Management: A Modular Concept for the Integration of Ergonomics into Production Systems

Max Bierwirth, Ralph Bruder, Karlheinz Schaub

Institute of Ergonomics
Darmstadt University of Technology
Germany

ABSTRACT

In this paper a comprehensive management model is introduced that enables companies to identify and control physical stress at work. First experiences from the implementation in five different manufacturing companies are presented.

Keywords: Ergonomics, Management Model, Condition-Oriented Prevention, Work-related Musculoskeletal Disorders, Product-Realization-Process

INTRODUCTION

In recent years, efforts in health promotion programs have increased. Notwithstanding, work-related musculoskeletal disorders (wMSDs) remain a widespread and growing issue of concern in the manufacturing industry in Germany and other industrialized countries (HSE, 2009). In the future, wMSDs leading to absence and reduced employment ability along with an aging workforce with comparatively high wages will become an even greater challenge to these manufacturing companies facing worldwide competition.

The prevention of wMSDs is achieved through improvements in the design of working conditions and tasks as well as through influencing the health promoting behavior of individuals. So far, many workplace health promotion and prevention programs focus on behavior-oriented prevention such as fitness programs, control of alcohol and tobacco use and diet programs (Kramer et al., 2009). In general, these programs, however, affect only certain risk-groups and yield only effects on individual behavior. In addition, consequences and lasting success of these behavioral changes are seldom evaluated.

Actions of condition-oriented prevention, however, affect all individuals working in given respectively changed conditions and do not depend solely on sustaining behavioral changes of individuals and can thus be better controlled and sustained. Nevertheless, isolated condition-improvements always risk achieving only locally optimized conditions by shifting hazards to a different part within the production system. Condition-oriented approaches which are based on single solutions, therefore, cannot be effective in systematically reducing wMSDs. What is needed, therefore, is a systematic approach, that enables companies to identify and control physical stress at work that leads to wMSDs in a comprehensive manner. Such a comprehensive integration of ergonomics as condition-oriented prevention is described in the following.

THE MODULAR ERGONOMICS CONCEPT

The concept comprises four modules which describe different levels of the integration of ergonomics and a controlling module (see Figure 1). Each module defines process- and method-requirements that should be realized within the production system to integrate ergonomics. The modular structure allows to define necessary actions for each module based on the situation at hand in a given company.

Figure 1 modular ergonomics concept. (Adapted from Bruder et al., 2009)

MODULE 1: INTRODUCTION OF APPROPRIATE RISK ASSESSMENT TOOLS

Basis for the improvement of working-conditions is the analysis of the existing conditions and the identification of work system deficits and risks. For that purpose structured risk assessment methods are needed. Introduced risk assessment methods have to fit the existing load patterns and working conditions and should allow an assessment of all types of jobs in the production system. A two-stepped method-approach with so-called "screening tools" for a quick identification of possible risks and 2nd-level (expert) tools for detailed analyses of contributing factors and the deduction of purposeful improvements helps to evaluate a high quantity of workstations as well as to conduct detailed analyses. Training in these methods for selected staff members is needed to enable the company to conduct analyses without the involvement of external resources.

Without such assessment methods, potential risks and deficits cannot be detected and analyzed systematically. Likewise, contributing factors for existing risks may not be properly determined and consequently actions taken may not solve the effective problem. With appropriate assessment methods changes in the work system can easily be evaluated under ergonomic aspects. Standardized methods also help to communicate ergonomic issues in an objective and comprehensible way. Thereby, their application also raises the ergonomic awareness within a company.

Prerequisites for the introduction of risk assessment methods are a scientific foundation of all introduced methods and the acknowledgement of the assessment results within the company.

MODULE 2: DISSEMINATION AND INTEGRATED USE OF INTRODUCED TOOLS

Risk assessment methods provide information about risks, contributing risk factors and indications for improvements. The second module concerns the use of these tools and the findings they generate with the aim of a comprehensive optimization of working conditions. The methods and the resulting indications should be used and considered whenever working conditions are to be changed or at question.

After a risk has been identified, improvements are achieved only if changes in the work system are realized and a re-evaluation confirms the effectiveness of the actions taken. For this purpose, a problem-solving process such as a PDCA-Cycle with defined responsibilities and allocated resources should be established each time a risk has been identified (Deming, 1986). To avoid inconsistencies all introduced Screening and 2nd-level tools need to be aligned so that they do not produce contradicting results. Moreover, it should be defined which method is to be used in which context.

To disseminate the introduced methods and to gain acceptance, managers of various levels and departments should be enabled to understand the methods structure and their results. Furthermore, more staff members with different functions should be trained in conducting analyses. Analysis-findings and derived improvements should be evaluated and documented in terms of knowledge management (lessons-learned, best-practice) and used in future improvement cases.

Through the integrated use of assessment tools and their indications wherever ergonomics aspects are affected, an optimization of the work system in a comprehensive manner can be achieved. Within "Kaizen"-activities and for the definition of job rotation patterns, the systematic consideration of ergonomic aspects can contribute significantly to risk reduction.

The increasing use of assessment tools also facilitates the transparency of the ergonomic quality of the work systems. This also avoids the shifting of hazards from one work system to another. A structured and standardized form of the assessment results helps to include ergonomics aspects in decision making and benchmarking activities. The combination of the findings with other data can also help to optimize other issues, such as quality and productivity performance.

Prerequisite for the implementation of module 2 is the introduction of at least one appropriate and well-founded risk assessment method and the commitment of the management to support improvement activities with adequate resources.

MODULE 3: FORMALIZATION OVER THE ENTIRE PRODUCT REALIZATION PROCESS

Once appropriate tools are established and users are enabled to conduct analyses and to implement improvements, the usage of the different tools and their generated data over the entire product realization process (PRP) should be defined. This includes the definition of steps in the process when an analysis should be conducted, which method should be applied, and which consequences for consecutive process steps have to be drawn once a deficit or risk has been identified.

Usually, the formalization begins with the regulation of systematic analyses and improvements of existing work systems. Reactive measures, however, yield only limited improvement possibilities and entail high costs of changes (Sagot et al., 2003). If design solutions cannot be realized in an existing work system, organizational changes can at least reduce the risk exposure for each individual.

The most effective and less costly improvements, however, can be achieved in the design and development of a product as well as of production processes (Dul and Neumann, 2009). Ergonomic risk assessments are to be integrated into the development steps, and ergonomic requirements have to be formulated and checked within the quality gates over the entire PRP.

Although all specifications of the planned product and production process are available in full detail only very late in the PRP, many ergonomic aspects of the future work system can already be checked in earlier phases. For example, product weight and dimensions, targeted lot size, necessary assembly operations and forces are available quite early in the product development phase and can be assessed under ergonomic aspects to obtain indications of potential risks and possible solutions. For the assessment, additional methods may be needed (in addition to module 1). Furthermore, information about the ergonomic quality and potential risks resulting from the analyses of existing work systems (module 1+2) should be consulted.

With a standardized procedure and adequate resources stress in existing work systems can be controlled so that hazards and risks are, within the limitations of corrective ergonomics, reduced. Through the integration of ergonomics considerations into the quality gates of the PRP conceptual ergonomics are realized. In the phases of design and development the definition of almost all work system characteristics can be influenced. The assessment of potential risk factors permits to define or influence the work system characteristics so that they yield the lowest risk possible in respect to other given constraints. The use of information about the ergonomic quality and potential risks of existing work systems impedes the recurrence of existing design deficits. Thus, the ergonomic quality of future work systems is systematically improved.

Especially for the formalization, management support and controlling is the major prerequisite since responsibilities and process changes have to be assigned and to be followed. The availability of data from analyses and improvements at existing work systems (module 2) is another requirement to realize organizational learning in the work system design.

MODULE 4: ABILITY-ORIENTED PLANNING

Typically, physical abilities of workers are only considered at the assignment of jobs once an employment limitation has emerged. Module 4 takes the physical abilities of workers systematically into account to reduce the (individual) risk level, in addition to the comprehensive adaptation of work system design (module 1-3), even further.

To permit a systematic matching of worker abilities and job demands, standardized profiles of the physical abilities of individuals or categories of workers (e.g. elderly, tall, women) should be assessed. These profiles have to be compatible with the demand profiles obtained from the introduced risk assessment methods (module 1). Determined ability profiles can also be analyzed to reflect future planning standards. A forward-projection of the abilities of the actual workforce could reveal potential gaps between expected demands and abilities of the actual workforce in the future.

The systematic fit of worker abilities and workplace demands leads to an optimized worker-workplace (stress-strain) allocation in terms of ergonomic as well as productivity aspects. Through modules 1-3 risk levels, based on a normalized worker population, are comprehensively reduced. With module 4 the risk level can be reduced or adapted on a differentiated basis. Already, the differentiation of a few categories of workers can help to significantly reduce the risks for the "weakest" percentile of the workforce.

With the implementation of module 4 a comprehensive condition-oriented prevention is achieved that uses both ways of ergonomic work design: adapting the work system design and fitting the worker to the job.

Prerequisites for a comparison of ability profiles and demand profiles are the availability of compatible risk assessment data (modules 1-2) and a basic integration of ergonomics considerations into the planning process (module 3) should be realized.

TOTAL ERGONOMICS MANAGEMENT (TEM)

To integrate the modules described into a manufacturing company a management model is needed to control the implementation and the achievement of the desired results. An acknowledged management model for comprehensive organizational changes is the Total Quality Management (TQM) concept.

That is why; a similar model was developed for the management of the concept described above (see Figure 2). TQM focuses on the implementation and quality of processes in the entire organization which consequentially enable a company to achieve the desired results, i.e. a quality product, zero defects in the production processes, and, in the long run, satisfied customers (Tennant, 2001).

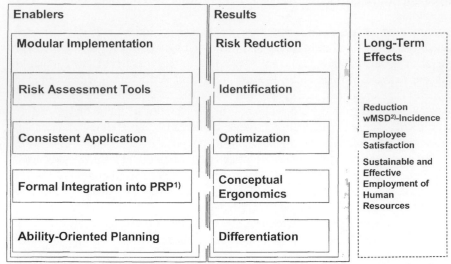

1) PRP = Product-Realization-Process 2) wMSD = work-related musculoskeletal disorders

Figure 2 Total Ergonomics Management Model

The desired results of the integration of ergonomics into a production system can be summarized as the reduction of wMSDs-incidence and resulting employment ability limitations, the sustainable and effective employment of human resources in the production system and at large a higher employee satisfaction.

Because of the extensive time gap between cause and effect in the case of condition-oriented prevention as well as the multi-causal influences on these result indicators, additional indicators are needed to manage this concept.

Although there are many contributing factors, it can be said that the characteristics of the work system represent certainly one important factor. Therefore, the improvement of the entire production system in terms of the reduction of risk can serve as an adequate intermediary result for long-term effects. This intermediary result can directly be influenced through the implementation of the modules and can be measured within a shorter period of time after implementation.

With the implementation of each module the level of integration of ergonomics in the production system is extended and the company is more and more able to reduce potential risks (enablers; vertical arrows). With the use of the implemented methods and processes, a partial reduction of existing or potential risks in the production system is realized (horizontal arrows). With the implementation and use of every following module a bigger part of the entire risk in the production is reduced (results; vertical arrows). With the full implementation of this modular concept, ergonomics are integrated into the production system in a holistic and comprehensive manner.

SUSTAINABLE IMPLEMENTATION THROUGH SELF-AUDITING

To support the implementation of organizational changes and the assessment of the quality of processes, auditing is a common method that is widely used in the context of the implementation and evaluation of quality management systems (ISO, 2008). Since the implementation of the Total Ergonomics Management system focuses also on organizational changes and improved process quality, a corresponding self-audit-program seems to be the most adequate management and implementation support. The audit-program measures progress in the implementation as well as improved process outcomes in terms of the expected risk reductions and helps to identify further actions.

All implementation activities of the model should start with a first audit to diagnose the actual situation in a company concerning the integration of ergonomics in the production system. Based on this initial analysis targets should be set and necessary actions should be determined. To support and sustain the targeted processes, the level of implementation should be evaluated regularly (as in any other audit) and the direct results concerning the risk reduction should be measured and communicated. Long-term effects should be monitored and documented to identify possible overlaying effects caused by other contributing factors other than the work system design.

FIRST EXPERIENCES OF IMPLEMENTATION

First experiences of implementation have been gathered from applying this approach to five different manufacturing companies in the automotive sector in Germany.

IMPLEMENTATION OF MODULE 1

The risk assessment methods in use vary widely from company to company. Officially released and approved screening tools such as the key-indicator method for the most obvious risks through manual material handling are the most wide-spread risk assessment methods (Steinberg et al., 2006). More sophisticated screening tools such as the Automotive-Assembly-Worksheet (AAWS) that consider more than one type of physical load, e.g. static postures, action forces and manual materials handling, are introduced only in bigger companies with a strong health and safety department due to the tool complexity and the required training efforts (Winter et al., 2006).

Changed patterns of work load in modern production systems, e.g. risks resulting from "Kanban"-activities, also call for tools that can analyze several risk factors and, at the same time, are of moderate complexity for the user. The key to success for introducing risk assessment tools seem to be their intuitive structure and the comprehensibility of their results. Specific 2nd-Level tools for detail analyses, e.g. for risks in repetitive tasks such as the Occupational-Repetitive-Actions-Method (OCRA), are rarely to be found as introduced assessment method (Occhipinti, 1998). Reasons given for this refer to the efforts in conducting such detailed analyses in exchange for the perceived limited value of their results.

IMPLEMENTATION OF MODULE 2

In bigger companies, empowered health and safety protagonists promote and support the training and the use of screening tools and help to realize effective improvements. In some cases, risk assessment results are transparently documented in so-called "ergonomics-maps" of the shop floor. This seems to be helpful to enhance management support and to avoid solely shifting hazards from one work system to another.

In smaller companies, physical risk factors are usually considered only in the context of the compulsory risk assessment. Here, physical stress and contributing work system deficits appear, if considered at all, in short checklists or at best in the application of a limited screening tool for manual materials handling.

Actions to harmonize and to clearly define the use of the introduced risk assessment methods have rarely been undertaken so far. Thus, in some companies, a variety of different risk assessment tools has been introduced over time and, depending on division, plant or department, different tools are used. Sometimes, different tools are even used for the analysis of the same situation but by different department representatives with a slightly different view point.

IMPLEMENTATION OF MODULE 3

The formalization is connected with the assignment of responsibilities and resources. In bigger companies, health and safety departments have designated ergonomics protagonists, while other companies charge industrial engineers to lead ergonomic improvements. In some cases this is done without allocating additional resources.

The integration of ergonomic checks and risk assessments into the product and production design and development in order to move from corrective ergonomic improvement to conceptual ergonomics is an easily accepted approach that, in turn, implies higher efforts during planning phases. Yet, analyses of several planning processes revealed that in most cases planning parameters that contribute to a potential risk are not evaluated under ergonomic aspects until the work system is finally realized. Typically, prescriptive limits for weights are given and, in some cases, for action forces. Aspects of repetition and exposure times are rarely considered. Another method used to realize better conceptual ergonomics are checklists that vaguely help the engineer to identify potential deficits in the design.

IMPLEMENTATION OF MODULE 4

For the disability management, usually a more or less standardized process is defined. Nevertheless, a connection between risk assessment results and the generated ability profiles for the disability management does not exist. In general, the examination of the impaired worker is done by an occupational health specialist using an impairment assessment tool. After that, work site visits are undertaken to find a workplace fitting this individual impairment. Systematic assessments of abilities of workers other than those with known impairments are not conducted.

DISCUSSION

In general, the described concept finds a high acceptance among the involved companies. Nevertheless, our experiences show that high efforts by different stakeholders are needed to achieve a total integration of ergonomics into the production system.

Often, ergonomics activities are not connected or remain isolated on-demand activities, as for example the introduction of risk assessment tools. Our experiences suggest that the introduction of new methods should be based on a formulated need which arises from specified situations in the work systems or which the future users of the tools have already encountered.

Although the dissemination of risk assessments may be easier to promote with a "one-for-all-tool", complex work situations in modern production systems require a set of different tools for appropriate risk assessments. Here, the harmonization between the tools is of paramount importance to avoid contradicting results that question the objectiveness of the risk assessment results. On the other hand, complex risk assessment tools which require intensive training and high assessment efforts may impede their dissemination and the integrated use in a formalized process. For the formalization, higher efforts are needed in bigger companies than in smaller companies with shorter decision cycles and hierarchies.

Even after a formalization and integration of defined ergonomic standards into the planning phases, strong management commitment is needed, in order not to compromise the compliance to ergonomic standards by short-term economic considerations.

To implement an ability-oriented planning as described above, high efforts are required and the available tools fulfil the necessary functions only to some extent.

Currently, this concept has been developed for the management of physical stress only. With adequate tools, however, also mental workload and resulting risks could be managed within this management model as well.

REFERENCES

Bruder, R., Rademacher, H., Schaub, K., and Geiss, C. (2009) "Modular Concepts for Integrating Ergonomics into Production Processes", in: *Industrial Engineering and Ergonomics: Visions, Concepts, Methods and Tools*, Schlick, C. (Ed.). pp. 385-396.

Deming, W.E. (1986), *Out of the Crisis*. Massachusetts Institute of Technology, Cambridge.

Dul, J., and Neumann, W.P. (2009), "Ergonomics contributions to company strategies." *Applied Ergonomics*, 40(4), pp. 745-752.

Health and Safety Executive (HSE) (2009), *Self-reported work-related illness and workplace injuries in 2007/08: Results from the Labour Force Survey*. Health and Safety Executive, UK

International Organization for Standardization (ISO) (2008), *ISO 9001:2008, Quality management systems—requirements*.

Kramer, I., Sockoll, I., and Bödeker, W. (2009), „Die Evidenzbasis für betriebliche Gesundheitsförderung und Prävention – Eine Synopse des wissenschaftlichen Kenntnisstandes", in: *Fehlzeiten-Report 2008,* Badura, B., Schröder, H., and Vetter, C. (Eds.). pp. 65-76.

Occhipinti E. (1998), "OCRA: a concise index for the assessment of exposure to repetitive movements of the upper limbs." *Ergonomics*, 41(9), pp. 1290-1311.

Sagot, J.-C., Gouin, V., and Gomes, S. (2003), "Ergonomics in product design: safety factor." *Safety Science*, 41, pp.137–154.

Steinberg, U., Caffier, G., Liebers, F. (2006), "Assessment of Manual Material Handling based on Key Indicators – German Guidelines", in: *Handbook of Standards in Ergonomics and Human Factors*, Karwowski, W. (Ed.). pp. 319-338.

Tennant, G. (2001), *Six sigma. SPC and TQM in manufacturing and services*. Gower, Aldershot.

Winter, G., Schaub, K., and Landau, K. (2006), "Stress screening procedure for the automotive industry: Development and application of screening procedures in assembly and quality control." *Occupational Ergonomics*, 6(2), pp. 107-120.

Chapter 13

Chaku-Chaku Assembly System – A New Trend in the Automotive Industry. First Results of the Ergonomic Evaluation on a Case Study

Enríquez-Díaz José-Alonso, Schiekirka Sarah, Arenius Markus,
Ekkehart Frieling and Sträter Oliver

Institute for Ergonomics and Process Management
University of Kassel
Heinrich-Plett-Straße Kassel
DE-34132, Germany

ABSTRACT

Currently, the international automotive industry is characterized by high competitive pressure concerning products and production processes. Consequently, companies attempt to increase competitiveness by adopting production concepts of the leading competitors like Toyota. Therefore, Toyota Production Systems (TPS) has been copied and adapted in many different ways. The method of implementation of TPS activities varies in the corporations. Due to the demographical change, a new discussion started in Europe, which addresses the recent challenge in creating work systems that maintain the working ability of older people. In the light of this demographical challenge, the present study examines the problems and benefits arising from the assembly line concepts favored by *Toyota* in a newly introduced Chaku-Chaku assembly system (N=41) and a conventional assembly system (N=64) in a large internationally active automotive company. The focus is on the ergonomic point of view and the consequences for an aging workforce.

Keywords: Chaku-Chaku assembly system, demographic change, automotive industry.

INTRODUCTION

Several German automotive companies have attempted to introduce or reproduce the concepts of the Toyota Production Systems (TPS) (Ohno, 2005) for about thirty years. This management philosophy as well as its derived concepts (e.g. lean production) take the expenditure of resources for improvement of production times, product quality and the reduction of costs by means of (re)designing the production work process into account (Womack, Jones & Roos, 1991). Thus, industrial-engineering and planning departments of automotive manufactures are today dealing with new production forms, which are based on TPS concept. Here, the Chaku-Chaku assembly system (Kono, 2004; Spanner-Ulmer, Frieling, Landau & Bruder, 2009) is a new trend of work organization, which has the principal goal of time reduction through waste elimination. Paralleling this trend, both the German and the population of other industrial nations are getting progressively older. This aging of the population is associated with an increasing number of older blue collar workers in industrial organizations. In order to address the physical and psychological changes introduced by this aging process, new production systems should be designed under the consideration of the age-related impairment of employees.

Therefore, taking these aspects into account, we need to clarify the extent to which these assembly systems affect task performance. The goal of this investigation is the analysis of the effects of the present Chaku-Chaku assembly system on employee performance with regard to age-based design. The data collection has been conducted within the framework of a larger project "Age-based work design in the automotive industry", which is financed by the DFG (German Research Foundation). The ergonomic evaluation included data describing features of current work systems. Additionally, an extensive questionnaire was used in order to explore the perception of the work characteristics and age-related differences on job performance and health. Workers of the gearbox assembly line in the same manufacturing plant served as control group.

Our sample consists of 51 blue collar workers from a large international automotive manufacturer. The employees work on the assembly of handbrake levers (average age: 32,1 years) and exhaust pipes (average age: 42,1 years). The workstations are arranged in the form of cells embedded in a Chaku-Chaku system. The control group consists of 64 workers with an average age of 38,5 years. The workplaces are arranged like a typical assembly line, with an average cycle time of 85 sec.

Subsequent to the data collection and the report of the results to the workers, a list of suggestions for improving the ergonomic environment was developed with the involvement of the workers, management and corresponding members of the industrial engineering department.

METHODOLOGY

SAMPLE

The studied population consisted of blue collar workers from three assembly lines from a globally active component factory for middle class cars in the automotive industry. The blue collar workers have a three-shift work system. The principal technical features of each assembly line are shown in Table1.

In the field of investigation A1, gearboxes are assembled in a typical assembly line manner. The work stations are arranged side by side and work pieces are indexed in linear motion from station to station. Each worker is solely responsible for one station and performs the assigned task after the assembled components are conveyed to the next station. At the time of the investigation, this assembly line was at the level of maximum production.

The employees of the fields of investigation A2 and A3 work on the assembly of handbrake levers and exhaust pipes. The product is assembled in assembly cells. Each cell consists of one to three interconnected workstations arranged in the form of a U-Shape with one worker being responsible for several work steps (multiple station work) (see Table 1). Therefore, the operator can start a machine while walking over to the next one (Baudin, 2007). The main assembly tasks are the loading of parts for the automatic assembly in the machines. The components are supplied by an operator (line-filler) through manual handling of boxes. This operator is exposed to diverse body postures during the work due to the variety of positions in which the material has to be handled. At the time of the investigation, the A3 assembly line was at the level of maximum production and the data was collected before the redesign of the work stations. The data collection in the A2 assembly line was conducted during the initial production phase. Altogether, data of 105 workers was collected in a cross-sectional design.

Table 1 Main characteristics of the studied samples.

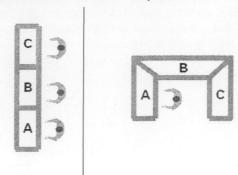

Field of investigation	A1 (N=64)	A2 (N=18)	A3 (N=23)
Product	Gearboxes	Handbrake lever	Exhaust pipe
Average cycle time (sec.)	85	25	55
Manufacturing method	Straight line	U-Shaped Cell (Chaku-Chaku)	U-Shaped Cell (Chaku-Chaku)

MEASURES

Workload Analysis

For the description and evaluation of the risk factors associated with work-related musculoskeletal disorders of the upper limbs (WMSDs), we used the Occupational Repetitive Actions method (OCRA), (Occhipinti, 1998). The OCRA-Method is based on a consensus document of the International Ergonomics Association (IEA) – a technical committee on musculoskeletal disorders (Occhipinti & Colombini, 2005).
For identifying and evaluating general poor working postures the Ovako Working postures Posture Analysing System (OWAS) was used (Karhu, Kansi & Kourinka 1977). The OWAS method assesses work postures for the back, arms, legs and the weight of the load handled.

Work Ability Index (WAI)

The WAI is a self-administered questionnaire for assessing the individual work ability. The concept of work ability is a very heterogeneous construct and can be defined as the ability of a worker to perform his or her job in relation to the specific work demands, the individual health condition and psychological and physical resources (Ilmarinen and Rantanen, 1999). In the present study we used the short version of the WAI (Nübling, Hasselhorn, Seitsamo & Ilmarinen, 2004).

Analysis Musculoskeletal symptoms

In the present study we solely used the standardized questionnaire which revolves around general information and physical disorders associated with nine anatomical regions (neck, shoulder, elbow, hands/wrist, upper back, low back, hips/thighs, knees, feet/ankles) (Kuorinka et al., 1987).

RESULTS

PERSONAL DATA

In the sample population A3 the mean age was 42.1 years (SD=8.9), 38.5 years (SD=11,1) in the sample population A1 and 32,1 years (SD=9,2) in the sample population A2. The majority (70 %) of the sample A1 and A2 is male, whereas the majority of blue collars (80 %) in the A2 sample population consists of females. The present assembly job is on average performed for 7.7 years in the age group "<45" and 16.0 years in the age group "≥45". Furthermore, the share of blue collar workers, which have completed technical apprenticeship, is 72 % of the total for the age group "<45" (missing data 13%) and 57% of the group "≥45" (missed data 18%).

WORKLOAD ANALYSIS

Representative work assembly activities during the operation of the Chaku Chaku line were collected by means of video recording equipment. Observational data was evaluated with pen and paper by the use of the OCRA-Index. This includes the evaluation of the muscle force and posture of the upper limbs (shoulder, elbow, hand wrist and fingers) when doing technical actions as well as the weight for "additional factors". Table 2 provides a summary of the evaluation parameter for two workplaces in the sample population A2. The workload analysis was made during the level of maximum production (approx. 800 units per shift).

Table 2 Workload on the upper limbs according to OCRA-Method (Chaku Chaku line A2).

Work-station	Limb	Action/min	Multiplier				OCRA index
			Force	Posture	Repetitive ness	Addi-tional	
A2-WS1	Right	40	0.99	0.70	0.70	0.85	3.0
A2-WS2	Left	52	1.0	1.0	0.70	0.85	2.6

The OCRA-Index is reported for the upper limb when carrying the heaviest load. In both cases the values identify a risk level that is quantified as "very low" (Colombini & Occhipinti, 2006). Advisable improvements to set up are recommended. It should be added that the handled weights are light (1 assembled handbrake lever = 650 gr.). However, the high movement frequency of the upper limbs plays a central role.

As mentioned above the line-filler carries out transport logistic tasks, which includes the manual lifting, carrying and holding of boxes containing components to be assembled and positioned on chute conveyors or flat belt conveyors when necessary. Compared to the operator, who works in the assembly cell, the provision of material has a large percentage of non- repetitive tasks.

An observational technique was used to evaluate working postures made by the line-filler of the A2 assembly line. The working postures were analyzed by means of direct observation of the activity. The observation was carried out in two representatives work intervals (each 45 min.). The observation was made on constant time intervals of 30sec.

The study revealed that 85% of the adopted postures are normal work postures and 15% of the adopted postures are slightly harmful. No harmful postures were reported by the OWAS method. However, it seems appropriate to consider the effect of the frequent handling of weights (boxes) lighter than 10 kg. on the total workload.

WORK ABILITY (WAI)

Our analysis revealed that employees working at the gear box assembly line had a mean work ability score of 35.6 (SD=6.7). 11% showed a poor (7-27 points), 27% a moderate (28-36 points), 31% a good (37-43) and 6% an excellent (44-49 points) work ability index. The mean work ability index of the handbrake lever assembly line workers was 36.8 (SD=4.7). A moderate and good index was obtained (44%). 6% had an excellent index while 6% of the persons did not give full particulars. In the exhaust pipe assembly line, the mean work ability score was 35.1 (SD=6.5).

13% of the employees had a poor work ability index. 31% had a moderate, 31% had a good and 4% had an excellent work ability index (Fig. 3).

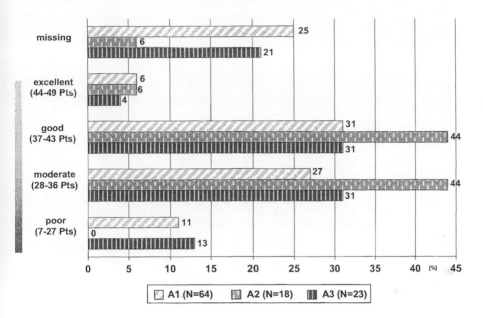

FIGURE 3: Frequency distribution for the work ability index.

Evidently, the workers of the handbrake lever assembly line that reached the highest mean work ability index have the least missing data and are the sole group without a poor category score. On the contrary, they have higher percentage rates in the moderate and good category score.

However, it is important to stress that the WAI index of handbrake lever assembly line workers is still low compared to other occupation groups. For example the mean work ability index of a group of nurses - working in hospitals, nursing homes, health centers and ambulant nursing services - with comparable age (30-34 years), is 38.1 (Hasselhorn, Tackenberg & Mueller, 2003).

MUSCULOSKELETAL SYMPTOMS

Fig. 4 shows the reported musculoskeletal symptoms collected with the Nordic Questionnaire. The workers of the gear box assembly line mainly reported problems concerning the neck (45%). At the second and third place are the lower back (41%) and the shoulders (33%). This should be seen in the light of the fact that 10,9% could not perform their work task during the last 12 month due to neck impairments, 14.1% due to shoulder problems and 7.8% due to back pain (in the following, these numbers are reported in brackets).

72% (38.9%) of the handbrake lever assembly line staff named lower back, 56% (16.7%) neck and 50% (22.2%) shoulder discomfort. 65% (26.1%) of the employees of the exhaust pipe assembly system of mentioned discomfort in the lower back, 57% (30.4%) discomfort in the neck and 52% discomfort in (4.3%) hands and shoulders (21.7%).

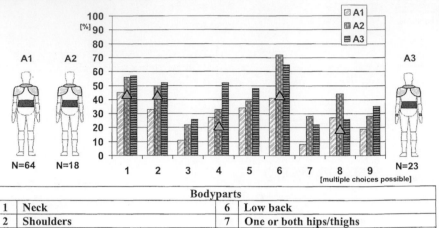

Bodyparts			
1	Neck	6	Low back
2	Shoulders	7	One or both hips/thighs
3	Elbows	8	One or both knees
4	Wrists/Hands	9	One or both ankles/feet
5	Upper back		Comparison values (BiBB/BAuA, 2006)

FIGURE 4 Percentage of musculoskeletal symptoms.

A representative survey (BiBB/BAuA) conducted in 2005/2006 in Germany, revealed that 43% of the sampled population (around 20.000 persons of the labor force) reported back problems, 44% neck/shoulder problems and 20% of hand problems. Thus, the reported symptoms of our sample are higher than these comparison values.

SUMMARY

The optimization of assembly manufacturing systems as well as the elimination of non-value-adding work activities are key strategies for the improvement of automotive industries. This strategy led to the development of new production systems under the Toyota management philosophy like the aforementioned Chaku-Chaku assembly system. The present study provides a first glance on the risks associated with the employment of age blue collar workers at such production systems.

From a biomechanical perspective, the assembly activities evaluated on the workplaces of the sample A2 (handbrake lever) show a very low risk level concerning the upper extremities (Table 2). However, additional analysis is needed in order to integrate the additional workload due to the walking in circles of the operator during the assembly work steps. With regard to the subjectively measurement of the ability to perform the assembly work in relation to the given demands (work ability), differences in the sample population were found. This makes sense as the different samples are characterized by different average worker age in the three lines. The mean age of the handbrake lever assembly line workers is 31.1 years. The mean age of the workers in other lines is distinctly higher (42.1 and 38.5 years). The reported musculoskeletal symptoms (Fig. 4) stand in contrast to the WAI results, where the workers of the handbrake lever assembly line reached the highest mean work ability index. However, the work in this line is characterized by high standardization, stereotypy and short cycles and could therefore lead to these high rates of complaints.

The next step in the project will consist of the ergonomic evaluation of the fields of investigations with regard to the given changes (output capacity, redesign of the workstations, etc.), as well as the combination of data from the questionnaires with objective and medical data of other samples of assembly line workers to explain and validate the results of the data obtained and especially for deducing age-based interventions for the studied areas.

REFERENCES

Baudin, M. (2007), Working with machines – The nuts and bolts of lean operation with Jikoda. Productivity Press.

BIBB/BAuA: Erwerbstätigenbefragung 2005/2006. Ausgewählte Ergebnisse., Online: www.arbeitsschutz.nrw.de [March. 2010].

Colombini D. and Occhipinti E. (2006), Preventing upper limbs work-related musculoskeletal disorders (UL-WMSDS): new approach in job (re)design and current trends in standardization. Appl. Ergon. 37 (4): 441-50.

Hasselhorn, H., Tackenberg, P. and Mueller, B. (Eds.) (2003), Work conditions and intent to leave the profession among nursing staff in Europe. Report no. 2003.

Ilmarinen, J. and Rantanen, J. (1999), Promotion of work ability during ageing. Am J Ind Med, 15(1), 21-23.

Karhu, O., Kansi, P. and Kuorinka, I. (1977), Correcting working postures in industry: a practical method for analysis. In: *Applied Ergonomics*, 8(4): 199-201.

Kono, H. (2004), Fundamental Principles and Side-Effects for Effective Manufacturing Revolution. In: Proceedings of the Fifth Asia Pacific Industrial Engineering and Management Systems – Conference 2004.

Kuorinka, I., Jonsson, B., Kilbom, A., Vinterberg, H., Biering-Sørensen, F. and Andersson G. (1987), Standardised Nordic questionnaires for the analysis of musculoskeletal symptoms. Applied Ergonomics 18, 233–237.

Nübling, M., Hasselhorn, H-M, Seitsamo, J. and Ilmarinen, J. (2004), Comparing the use of the short and the long disease list in the work ability Index questionnaire. In: Costa, G., Goedhard, J.A.W. and Ilmarinen, J. (Eds.). International Congress Series 1280. Amsterdam: Elsevier.

Occhipinti E. (1998), OCRA: a concise index for the assessment of exposure to repetitive movements of the upper limbs. *Ergonomics*, 41(9): 1290-331.

Occhipinti E. and Colombini D. (2005), The occupational repetitive action (OCRA) methods: OCRA Index and OCRA Checklist. In: Stanton, N., Brookhuis, K., Hedge, A., Salas, E. and Hendrick HW., (Eds.), Handbook of human factors and ergonomics methods. Boca Raton, Florida: CRC Press: 15:1-14.

Ohno, T. (1993), Das Toyota-Produktionssystem. Frankfurt/Main: Campus-Verl.

Spanner-Ulmer B., Frieling, E., Landau K. and Bruder, R. (2009), Produktivität und Alter. In: Landau, K. (Edt.): Produktivität im Betrieb, Tagungsband der GfA Herbstkonferenz, Millstatt: Ergonomia Verlag, 81-117.

Womack, J., Jones, D. and Roos, D. (1991), Die zweite Revolution in der Automobilindustrie. Frankfurt: Campus Verl.

CHAPTER 14

Digital Manual with Wearable Retinal Imaging Display for the Next Innovation in Manufacturing

Miwa Nakanishi[1], Tomohiro Sato[2]

[1]School of Engineering, Division of Design Science
Chiba University, Chiba 263-8522, Japan

[2]NID Research & Development Dept.
Brother Industries, Ltd., Nagoya 467-8561, Japan

ABSTRACT

This study discusses how digital manuals with retinal imaging display (RID) can be applied to conventional work scenarios, particularly focusing on their impact on workers' psychology. An RID is a type of wearable display, like glasses, that enables users to view an overlay of a digital image and the real world. We conducted an experiment in which the subjects assembled an object consisting of 10 parts, by referring to a manual presented using RID. We prepared three types of manual content: a picture manual in monochrome, a picture manual in full color, and a movie manual. The subjects were tested under three conditions in which each type of content was presented using RID. For comparison, the subjects were also tested under two additional conditions: the same picture manual in monochrome presented on a 17-inch normal display; and the same picture manual in monochrome presented on paper. Based on the experimental results, we make the following suggestions. When a paper manual is replaced by a digital manual, the workers' task performance improves. However, for highly skilled workers, differences in display type or content design do not always improve their performance further. On the other hand, differences in display type and content design can have a characteristic influence on the workers' psychology. In particular, 1) manuals with full-color pictures can work well in various situations; 2) manuals with monochrome pictures give workers a comfortable sense of tension and make them conscious of a professional atmosphere; and 3) movie manuals can add positive factors such as pleasure, high motivation, and strong emotional impressions.

Keywords: Retinal imaging display, digital manuals, users' psychology

INTRODUCTION

Progress in mechanization and automation of manufacturing is accelerating. However, manpower is still necessary, particularly when fine and complicated manipulation is required. In these cases, manuals stating instructions and procedures are typically used. Thus far, paper-based manuals have been common. However, paper manuals have recently begun to be replaced by digital manuals because of the following factors: compact display of digital manuals, development of GUI (graphical user interface) technologies, and a trend toward paperless workplaces.

Some workplaces have already introduced digital manuals with fixed or portable displays. However, these cases where workers' move or positions tend to be restrained are transitive or conditional measures. In this point, wearable displays that do not have any difficulty are greatly expected to be applied to digital manuals. Also, our previous research (Nakanishi, et al., 2007; Miura, et al., 2007; Nakanishi, et al., 2008; Kordic, (Ed.), 2010) has indicated that using digital manuals with see-through wearable displays improves workers' performance without adding a new physical load on them.

The original wearable displays have rarely been used beyond the experimental stage. In recent years, however, the functionality, usability, and design of wearable displays have remarkably improved to the point where this technology can be applied in practical situations. Therefore, this study focuses on the type of wearable displays with the strongest potential, RID (Retinal Imaging Display), and explores the situations in which digital manuals with RID can work effectively. (More details of RID are discussed in the following chapter.) Thus, this study aims to clarify the types of additional values that can be generated by applying digital manuals with RID in actual workplaces. In particular, we discuss what types of media and content influence the workers to experience positive psychological factors such as pleasure, motivation, and satisfaction. Our goal is to put digital manuals with RID to practical use, and thereby, make the next innovation in manufacturing.

METHOD

EXPERIMENTAL TASK

In the experiment, the subjects were asked to assemble an object consisting of 10 parts. Figure 1 shows the parts before they are assembled, and Figure 2 shows the object after the parts are assembled. The subjects were required to complete the assembly task correctly and quickly.

TYPES OF MANUAL PRESENTATIONS

In this experiment, an RID (developed by Brother Industries, Ltd.) was used as a media of manual presentation. The RID lets users recognize an image by directly projecting a small-power laser to their retina and scanning it at a high speed. The effective laser power is set at less than the safety standard. When the subjects wore the glass-like RID (Figure 3), they could observe both the real world and a 16-inch full-color image in the same field of view.

We prepared three types of manual content for the experiment: a picture manual in monochrome, a picture manual in full color, and a movie manual. The details of each type of manual are described below:

- *The manual with monochrome pictures (Figure 4)*
This manual consists of seven pages that illustrate the instructions and procedures of the assembly task. The procedures to complete the task are divided into seven steps and are explained with drawings and words.

- *The manual with full-color pictures (Figure 5)*
The content of this manual is the same as the manual with monochrome pictures. However, the drawings are colored to accurately represent the real objects.

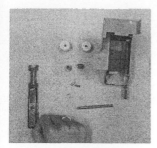
Figure 1. Parts before they are assembled

Figure 2. Object after the parts are assembled

- *The movie manual (Figure 6)*

This manual consists of five chapters that present the same procedures as the picture manuals, but with 3D animations rather than pictures. Each chapter is repeated automatically.

Figure 3. Retinal Imaging Display (RID)

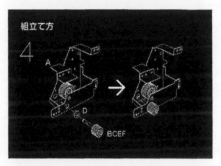

Figure 4. Part of the picture manual in monochrome

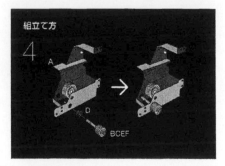

Figure 5. Part of the picture manual in full-color

EXPERIMENTAL CONDITIONS

The subjects were tested under three conditions in which each type of content was presented by RID. For comparison, the subjects were also tested under two additional conditions: the same picture manual in monochrome presented on a 17-inch normal display and the same picture manual in monochrome presented on paper.

In the conditions where the subjects used RID or normal display, they switched pages or chapters by key touch. In any condition, the subjects had to start from the first page or chapter and refer to it in turn according to the progress of the task.

Figure 6. Part of the movie manual

TRIALS

The subjects were 24 healthy students ranging from 22 to 24 years in age. Each subject practiced the task using the paper manual twice to learn the procedures. The subject then tested each of the five conditions once to record the data. In the first trial, each subject used the paper manual. In the subsequent trials, the subject used each of the other types of manuals according to the order assigned to each subject. The order was balanced to eliminate possible order effect.

DATA

We recorded the subjects' actions using a digital video camera. We also recorded the time of switching pages or chapters. In addition, the subjects answered a questionnaire after each of the following types of trials: using the monochrome picture manual with the normal display; using the monochrome picture manual with RID; using the full-color picture manual with RID; and using the movie manual with RID. This questionnaire included 21 questions, where each question consisted of a pair of words. The subjects evaluated each type of manual in comparison to the paper manual by giving scores of -3 to $+3$ to each question, assuming that all scores of the paper manual were set to zero. Table 2 summarizes all the questions.

Table 2. Questions
(The left column corresponds to the lower score, and the right column corresponds to the higher score.)

Question items (pair words)	
obscure	clear
complicated	simple
not relieving	relieving
dull	pleasant
old	new
tiring	not tiring
unfamiliar	familiar
tight	loose
boring	not boring
confusing	not confusing
tense	relax
not sharp	sharp
not easy to continue	easy to continue
uneasy to start	easy to start
not smooth	smooth
not exciting	exciting
boring	not boring
difficult to concentrate	easy to concentrate
not impressive	impressive
unfashionable	fashionable
not explicit	explicit
not progressive	progressive

RESULTS

TASK PERFORMANCE

Figure 7 shows the mean time to complete the task under each condition. The mean time was cut by 15% to 20% when the subjects used the digital manuals with the normal display or RID, compared with when they used the paper manual ($F = 2.46$, $p < 0.05$). However, there was no significant difference between the four conditions where the digital manuals were used. The authors had carried out another experiment where the subjects tried the same task without practicing it in advance and had found that the subjects could complete the task within a shorter time period when they used the movie manual than when they used one of the other types of manuals. Synthesizing these results, we can draw the following conclusion: when the subjects were highly dependent on the manual, a difference in patterns of the content significantly influenced the task efficiency. On the other hand, when the subjects were not highly dependent on the manual, a difference in the patterns of the content did not show a change in the task efficiency.

Errors were almost zero in each of the conditions. This may be because the subjects were highly skilled.

Summarizing the results concerning the task performance, the subjects who were accustomed to the task were hardly influenced by the difference in the content of the digital manual, although they were influenced by the difference between the paper manual and digital manual.

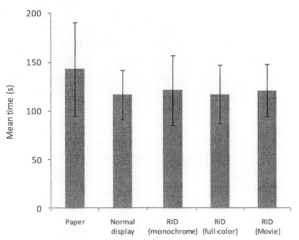

Figure 7. Mean time for a task

PSYCHOLOGICAL EVALUATION

Characteristics of each type of manual

We examined how each type of manual influenced the subjects' psychology by analyzing their answers to the questionnaire.

First, we categorized the questions by performing cluster analysis using Ward's method. Figure 8 shows a clustering tree of the questions. Based on the clustering, we categorized the 21 questions into four groups, where each group represents an aspect for evaluation of each type of manual. These four groups are task reliability, task efficiency, relax and easy, and motivation. Second, we calculated the mean scores of the questions included in each group, and defined them as the evaluation scores of each aspect. Figures 9-1, 9-2, 9-3, and 9-4 show the evaluation scores given to each condition for the following types of manuals: the manual on normal

136

display, manual with monochrome pictures on RID, manual with full-color pictures on RID, and the movie manual on RID.

These results show that every digital manual was evaluated higher than the paper manual in all aspects. On the aspect of "task reliability," the conditions of the manual on normal display, the manual with monochrome pictures on RID, and the manual with full-color pictures on RID, all scored around 0.9. However, the condition of the movie manual on RID scored higher than the three abovementioned conditions. The highly skilled subjects hardly committed errors, as described in the previous section, but they felt the movie manual more explicit and definite than other manuals.

On the aspect of "task efficiency," all conditions scored around 0.9. This result corresponds to actual task efficiency that was observed by the mean time (Figure 7), which means that the subjects felt more efficient when they used the digital manual, compared to when they used the paper manual.

On the aspect of "relax and easy," although the scores were around 0.3 in the conditions where the picture manual was presented by RID, the score was almost 0.0 in the condition where the movie manual was presented by RID. This means that the subjects found that the movie manual was not superior even to the paper manual from the viewpoint of "relax and easy." By contrast, the score was comparatively high in the condition where the normal display was used. We conclude that the normal display, which is very popular in our daily life, did not cause the subjects to experience any special atmosphere, while RID, which is one of the unprecedented human-machine interfaces, caused the subjects to experience an unusual atmosphere. Furthermore, the movie presented by RID seemed to make the subjects feel somewhat tense.

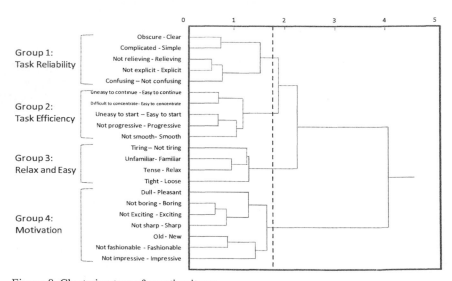

Figure 8. Clustering tree of question items

On the aspect of "motivation," the score increases in the order of the following conditions: the normal display, monochrome picture presented by RID, full-color picture presented by RID, and the movie presented by RID. This means that the subjects can feel much more pleased, excited, or impressed, as they use a colorful, moving, and wearable manual.

Summarizing the above results, a manual presented by a normal display is a well-balanced, acceptable manual. In comparison, a manual presented by RID can maintain task reliability and efficiency and can motivate workers, although it may make workers feel somewhat tense. In addition, if the movie manual is presented by RID, it can increase experiences of pleasure and emotional impressions.

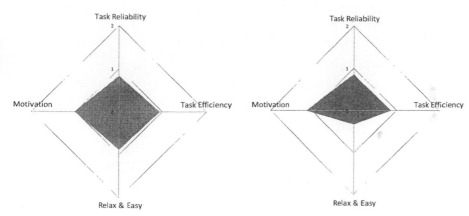

Figure 9-1. Evaluation from four perspectives in condition of "Normal display"

Figure 9-2. Evaluation from four perspectives in condition of "RID (monochrome picture)"

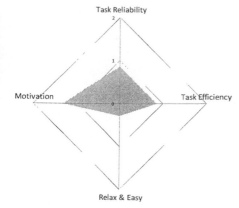
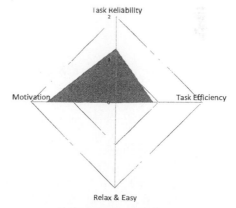

Figure 9-4. Evaluation from 4 perspectives in condition of 'RID (full-color picture)'

Figure 9-4. Evaluation from 4 perspectives in condition of 'RID (movie)'

Applicable Situations of Each Type of Manuals

Furthermore, we examined comparative positions of each type of manual and explored which situations each type of manual can be effectively applied to.

We analyzed the mean scores of each question using principal component analysis. The first two principal components were extracted. The first total variance was 50.4%, and the second total variance was 29.2%. Figure 10 shows the positions of each question given by the principal component loadings, where the y-axis corresponds to the first principal component, and the x-axis corresponds to the second principal component. We can find the items that have a professional and job-like image commonly in the negative y-direction. On the other hand, we can find the items that have a smooth and loose image commonly in the negative x-direction. Moreover, we can find the items that have a usual and familiar image commonly in the positive y-direction. And we can find the items that have an interesting and exciting image commonly in the positive x-direction.

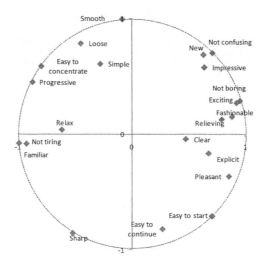

Figure 10. Positions of each items based on principal component loadings

Figure 11 shows the positions of each type of manual given by the principal component scores, where each type is plotted in the space spanned by the above-defined axes. Based on the relative locations of each type, we can make the following conclusions. The manual presented by the normal display enabled the subjects to relax and concentrate on the task. The monochrome picture manual presented by RID gave the subjects a serious, earnest, and job-like atmosphere. The movie manual presented by RID let the subjects experience pleasure, high motivation, and strong emotional impressions. And the full-color picture manual presented by RID, which has characteristics of other types of manuals, was accepted as an intermediate type between the other three types.

Summarizing the above examination, we suggest the following situations as an

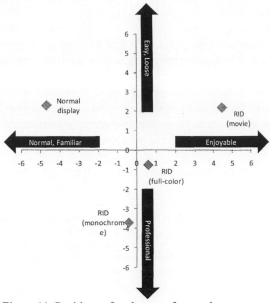

Figure 11. Positions of each type of manuals
based on principal component scores

effective use of each type of manual. First, if it is possible to fix displays, and if workers can continue to manipulate in a standing or sitting position, a normal display can be useful to present digital manuals. However, if this is not possible, or if it is desirable to allow workers to move freely, then using an RID is greatly expected. In this case, manuals with full-color pictures can work well in various situations. On the other hand, if it is necessary to give the workers a comfortable sense of tension and make them conscious of a professional atmosphere, using manuals with monochrome pictures is effective. Moreover, if it is necessary to add positive factors such as pleasure and high motivation to the workers, using movie manuals is recommended. Since movie manuals presented by RID give a strong emotional impression to workers, they should be effective for training or education purposes.

CONCLUSION

In this study, we discussed how digital manuals with RID are expected to change conventional manufacturing, particularly focusing on the impact on worker psychology. Based on the experimental results, we came to the following conclusions. When the paper manual is replaced by the digital manual, the workers' task performance improves. However, for highly skilled workers, differences in

140

display type or content design do not always improve their performance further. On the other hand, those differences can have a characteristic influence on the psychology of the workers.

In particular, 1) manuals with full-color pictures can work well in various situations; 2) manuals with monochrome pictures give workers a comfortable sense of tension and make them conscious of a professional atmosphere; and 3) movie manuals can add positive factors such as pleasure, high motivation, and strong emotional impressions. From these discussions, we can conclude that digital manuals with RID can add positive value to conventional manufacturing paradigms, with respect to not only task reliability and efficiency, but also worker satisfaction. Our next focus is on design processes of manual content presented by RID, to make workers more comfortable and to effectively motivate them.

ACKNOWLEDGMENTS

We thank Mr. Hirokuni Kato and Mr. Hideki Tani, who provided helpful comments and suggestions. And we received great help from Mr. Kenji Ota, Mr. Yusuke Ogura, Mr. Akihisa Senda, and Mr. Yusuke Higa.

REFERENCES

Nakanishi, M., Miura, T., and Okada, Y. (2007). How does the Digital Manual with Controllability and Perfect-Transparency Effect on Workers' Cognitive Processes? -Application of Augmented Reality Technology to Manufacturing Work-.: Proceedings of the 4th International Conference on CITSA(Cybernetics and Information Technologies, Systems and Applications), on CD-ROM.

Miura, T., Nakanishi, M., and Okada, Y. (2007). A Study on Human Interface for Control in Using AR Manual.: Proceedings of the 12th International Conference on HCI (Human Computer Interaction), on-CD-ROM.

Kordic. V(Ed.). (2010), Human factor guideline for applying AR-based manuals in industry, -in the book of "Augmented Reality," IN-TECH, 129–156.

Nakanishi, M., Ozeki, M., Akasaka, T., and Okada, Y. (2008). What Conditions are Required to Effectively Use Augmented Reality for Manuals in Actual Work., *Journal of Multimedia*, 3(3), 34–43.

Chapter 15

The Effect of Glove Fit on Task Performance

Swantje Zschernack, Jessica Stack

Department of Human Kinetics and Ergonomics
Rhodes University, Grahamstown, 6140, South Africa

ABSTRACT

The hand is the most complex of all of the anatomical structures in the human body. It has been found that hand injuries are among the most frequent injuries that occur to the body, predominantly occurring during industrial activities. It has therefore been concluded that more research is needed into protective factors, such as glove use. The design features of a glove emphasize either protection or performance. There is often a tradeoff between increased safety and performance capability when donning gloves. It has been determined that gloves which are fitted and comfortable for the worker may provide the best compromise between protective functions and decreased performance. This investigation aimed to assess the influence of glove fit on the performance attributes of industrial tasks. Glove fit was analyzed as 35 participants donned 3 different glove sizes during each test, including a best fitting glove, a glove one size smaller than best fitting, and a glove one size larger than best fitting. For each glove size, gloves of two differing materials were tested. A barehanded condition was also tested, totaling 7 gloved/barehanded conditions for each test. Significant differences ($p < 0.05$) were found between conditions for participants' dexterity, speed and accuracy and torque, indicating variance in performance due to glove fit.

Keywords: Glove fit, performance, electromyography

INTRODUCTION

Occupational injuries of the hand are common (Dias & Garcia-Elias, 2006), and are an economic burden for any industry (Skov, Jeune, Lauritsen, & Barfred, 1999). Despite a large number of published materials regarding upper extremity disorders, the hand has not been thoroughly considered from an ergonomics standpoint (Kumar, 2004). Due to the hand's complex anatomical structure, and intense usage (Muralidhar, Bishu, & Hallbeck, 1999) ergonomic interventions are vital (Kumar, 2001). The hand is the predominant medium for motor activity and the most important sensory and tactile organ (Napier, 1956), and therefore, perfect matching between hand devices, such as gloves, and hand characteristics is essential (Imrhan & Loo, 1989).

Many industries necessitate the use of industrial gloves with a pre-described level of protection, encompassing the thickness and the material of the glove. It is therefore impractical to investigate modification of these glove components. A glove component which can be altered, and which has been highlighted in the literature requiring further ergonomic investigation is glove fit. Ill-fitting gloves have been found to increase the likelihood that workers are exposed to physical strain, and result in them being less adept at completing industrial tasks efficiently and effectively. Bradley (1969) has shown that two glove factors, snugness of fit and flexibility, have a strong influence on the glove user. There are limited studies which have isolated glove fit and the effect of this hand-glove interface on components of worker performance. It is imperative that information and data be collected in order to highlight the degree to which the hand-glove interface inhibits performance and enhances muscular effort, therefore drawing attention to the potential of the simple intervention of addressing and adjusting glove fit for workers across all industries.

Therefore, the focus of this research project was the identification of the effect of a range of glove sizes on the performance attributes of industrial tasks. Further findings regarding participants' responses using electromyography and psychophysical rating scales have been omitted for reasons of practicality.

METHODOLOGY

Glove fit was analysed, using a one group design, as participants donned 3 different glove sizes for each test to be completed. This included a best fitting glove, a glove one size larger than best fitting, and a glove one size smaller than best fitting. For each glove size, gloves of two differing materials were tested: a neoprene glove and a nitrile glove. A barehanded condition was also tested. There were therefore 7 gloved/barehanded conditions for each test.

A battery of tests was designed for the study, each test being considered representative of the components of task performance in industry. The tests evaluated participants' maximum force-producing capabilities, in a pushing and pulling direction, torque, precision of force, speed and accuracy, and dexterity.

MAXIMUM FORCE TESTS

Push and pull

Participants' pushing and pulling force was measured. This was done using a uni-axal load cell (LSB-300 – 300lb). The load cell, fitted with a pushing and a pulling mechanism, was connected to the Biometrics Ltd DataLOG W4X8, by means of a single analogue channel, and calibrated accordingly. When participants exerted force, the changes in force production generated in the load cell were displayed on the laptop screen.

Participants were requested to maintain a seated posture; ensuring backs were against the back rest. The chair was positioned such that the participant was within optimal reaching distance. The handle, which the participants grasped, was attached to the load cell, and positioned at each participant's elbow height. Participants gripped the handle and generated maximal effort for 5 seconds. The effort was generated in either a pulling, or a pushing direction. This was repeated 3 times, with a rest interval of 30 seconds allocated between each exertion. This protocol was repeated twice, once in a pulling direction.

Torque production

Isokinetic testing was done on the CYBEX □ 6000, at a speed of 60°.s-1. This speed was used in order to best simulate the working environment, and because it is considered to be the optimal speed for strength measurements (Kumar, 2004). Participants were seated in front of the dynamometer, with the head assembly facing upwards, at an axis alignment of 90°. The height of the head assembly and distance from the participant were standardized, as the axis of the control was adjusted to be positioned in the transverse plane, aligned with the participants' xiphoid process. The participant was seated within optimal functional reach of the control.

The lever arm used was the universal tool adaptor. This provided the base for the force acceptance attachment (the control), which was a medium-sized, convoluted knob. This combination created the interface between the participant and the system, and moved radially about a fixed axis. The knob attachment was connected directly above the fixed axis, therefore isolating the required wrist movement for the test. The participant was instructed to grasp the knob attachment and produce maximum torque, maintaining a reasonably stationary forearm. This was achieved by abduction (external rotation) followed by adduction (internal rotation) at the radiocarpal joint. The participant performed the test by executing three maximal turns, in both directions. Participants' peak power, total work and average power data were recorded.

PRECISION OF FORCE

This test consisted of a 60 second static handgrip effort at 30% MVC. The 30% selected, was that of the maximum force that was produced during he maximum force tests (both pushing and pulling). This 30% MVC was chosen, as according to Monod's curve, and as seen in Figure 1, fatigue will not be incurred at this level, over the duration of a minute. In order to further guarantee that no fatigue would be sustained the 30% value is below that which is recommended by Monod's curve.

The participant was given the same instruction as for the maximum force tests, however, instead of exerting maximal effort, verbal feedback was given to the participant in order to initiate the required amount of force (30% MVC). The participant was then instructed to maintain this level throughout the 60 seconds, without visual or verbal feedback. This was done in both a pulling, and a pushing direction. The force trend, and deviation from the trend was calculated in both directions for each condition.

FIGURE 1 A graph adapted from Ma, Chablat, Bennis & Zhang (2009) illustrating Monod's curve, which indicates the relationship between duration of exertion and the percentage of maximum force exerted, as a function of fatigue. The red lines indicate the 40% MVC recommended for participants to exert for 60 seconds.

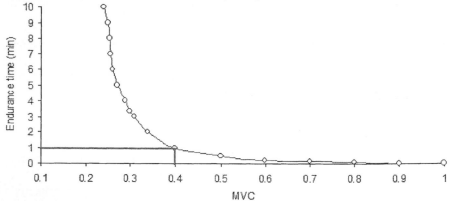

SPEED AND ACCURACY

A total of 25 yellow spherical targets appeared individually in a consecutive manner on a touch-screen. The participants were instructed to touch the target, as rapidly and accurately as possible using the second digit, as soon as it appeared on the touch-screen interface. After each target was selected, the same amount of time elapsed (1.5 seconds) before the next target appeared. Each task therefore lasted approximately 40 seconds (25 targets × 1.5 seconds). Targets jumped from point to point, and appeared 500 to 1000 ms sequentially.

The targets appeared in a random sequence to prevent participants from predicting target location before the targets appeared. The movement time and deviation from the target were the variables of interest during the test.

MINNESOTA DEXTERITY TEST

The one-hand turning and placing test was administered. A board was placed on the table, with a second board directly in front of the first board, 2 centimetres from the edge closest to the participant. The two boards were aligned and touching each other. One board contained 60 disks. All of the disks were placed in the holes, in the top board, with the black side facing up. The object of the test was to see how quickly the participant could pick up the disks from the top board (the board furthest away from the participant), turn the disks over, therefore showing the red side of the disks, and place the disks into the holes of the bottom board (the board closest to the participant). Participants began on the right, picking up the bottom disk, and inserting it into the top hole of the closest board, while turning the disk over to expose the other colour. Participants then picked up the next disk above the empty hole on the first board, and continued the pattern in the right column. This sequence was repeated in the second to the fifteenth columns, filling the entire bottom board. The participant's score was the total number of seconds taken to complete the trial.

RESULTS

Table 1. A comparison of the gloves sizes within each material, for each performance test.
(Significant differences (p<0.05) are indicated by p values.
n.s. specifies no significant differences (p<0.05).
Tests which generated no significant differences (p<0.05) whatsoever have been omitted.)

		Nitrile		Neoprene	
		Best fit	**Large**	**Best fit**	**Large**
Dexterity (s)	Small	n.s.	00.000590	n.s.	0.000256
	Best fit		0.001259		0.000186
Peak torque ext. rotation (Nm)	Small	n.s.	n.s.	n.s.	n.s.
	Best fit		n.s.		n.s.
Peak torque int. rotation (Nm)	Small	n.s.	n.s.	0.018712	n.s.
	Best fit		n.s.		n.s.
Average power ext. rotation (J)	Small	n.s.	n.s.	n.s.	n.s.
	Best fit		n.s.		n.s.
Average power int. rotation (J)	Small	0.037498	n.s.	n.s.	n.s.
	Best fit		n.s.		n.s.

As seen in Table 1 the dexterity results show significant differences ($p<0.05$) between the large glove, and the best-fitting glove and between the large glove and the small glove for both materials.

During the maximum torque test (internal rotation), participants exerted a significantly greater ($p<0.05$) peak torque with the neoprene best fitting glove than with the small glove size.

The same was found for average power (internal rotation) with the nitrile glove.

Table 2. A comparison of the barehanded condition with the best fitting condition of each material, for each performance test.
(Significant differences ($p<0.05$) are indicated by p values.
n.s. specifies no significant differences ($p<0.05$).
Tests which generated no significant differences ($p<0.05$) whatsoever have been omitted.)

		Best fit nitrile	Best fit neoprene
Dexterity (s)	barehanded	0.000112	0.000112
	best fit nitrile		n.s.
Reaction time (s)	barehanded	n.s.	0:041275
	best fit nitrile		0:032469
Target deviation (mm)	barehanded	n.s.	0:001321
	best fit nitrile		0:000946
Peak torque ext. rotation (Nm)	barehanded	0:000117	0:000117
	best fit nitrile		n.s.
Peak torque int. rotation (Nm)	barehanded	0:000117	0:000117
	best fit nitrile		n.s.
Total work ext. rotation (Nm)	barehanded	0:000117	0:000117
	best fit nitrile		n.s.
Total work int. rotation (Nm)	barehanded	n.s.	n.s:
	best fit nitrile		n.s.
Average power ext. rotation (J)	barehanded	0:000117	0:000117
	best fit nitrile		n.s.
Average power int. rotation (J)	barehanded	0:000117	0:000117
	best fit nitrile		n.s.

In Table 2 it can be seen that during the dexterity test, there was a significant difference ($p<0.05$) in the dexterity score between the barehanded condition, and each of the best fitted glove conditions.

For the speed and accuracy test, participants reaction time when donning the best fitting neoprene glove was significantly different (p<0.05) to that of the barehanded condition or the best fitting nitrile glove. The same was found for target deviation. During both internal and external rotation in the torque test, the barehanded condition generated significantly different (p<0.05) peak torque than the best fitting nitrile and best fitting neoprene gloves. The same was found for the total work (external rotation), and the average power (external and internal rotation) results

DISCUSSION

A hand performance parameter found to be adversely affected by glove use includes dexterity (Mital, Kuo & Faard, 1994; Muralidhar et al., 1999). This is in accordance with findings seen in Table 2, as the barehanded condition produced the most efficient results, with significantly less time to complete the task than with either best fitting gloves. It has been found that although fitting ease within a glove is necessary to a certain extent, and glove thickness is required for protection, excess ease and thickness cause a loss of dexterity, and feeling for the worker (Williams, 1979; Tremblay, 1989). Findings in Table 1 indicate that a larger glove sizes yields significantly greater dexterity performance times than both the best fitting and small glove conditions. Findings indicate that redesigning the glove to conform to the hand, wrist and forearm would be likely to improve performance with gloves. This is consistent with Tremblay's (1989) findings that a tight fitting glove will interfere the least with finger dexterity.

Gloves interfere with a person's hand movements (Mital et al., 1994), particularly precision of movement (Oborne, 1988). It has also been established that f excess glove ease and thickness are afforded by the glove, the time necessary to complete the task has been found to increase (Williams, 1979; Tremblay, 1989). Bensel (1993) tested various glove types and found that, relative to bare-handed performance, the gloves which were tested limited the speed at which the work could be accomplished. This was found to be true for the neoprene glove. Results in Table 2 agree with this literature, as the neoprene glove, although being a best fitting glove, generated significantly longer performance time, and deviation from the target than either the barehanded condition, or the nitrile glove condition. This may be attributed to the increased thickness of the neoprene glove, compared with the nitrile glove, therefore further increasing participants' hand circumference, and consequently impeding hand motion.

No significant differences (p<0.05) were recorded for the push and pull tasks, either maximal, or precision. This result is unexpected, as it has been established that the capability of hands to exert force is influenced by the use of gloves (Mital et al., 1994).

From this study, it can be established that pushing and pulling force is unaffected by glove size. However, force exerted during the torque test rendered significant results (p<0.05). There are conflicting findings on the effects of gloves on torque performance (Shih and Wang, 1997). According to Mital et al. (1994) torque exertions have been found to increase with gloves. Findings in Table 2 demonstrate that glove use significantly increases torque performance as compared with the barehanded condition. In Table 1 it can be seen that the best fitting glove condition generated significantly greater performance effects than the small glove conditions. This may be explained by the fact that the small glove condition was subjectively, verbally reported to limit participants' range of motion during the task. This finding concurs with Muralidhar et al. (1999), who found that the range of motion that the hand is capable of can be limited by glove dimensions.

CONCLUSIONS

Task performance has been found to differ with glove use, and glove fit. Particular task components which are affected are dexterity, speed and accuracy, and torque. It can be understood that glove use decreases participants' performance regarding dexterity, movement time and target accuracy; however, torque performance is enhanced with glove use.

Therefore, it is important to select the most appropriate glove for a particular application, and it is also essential to know the performance characteristics of gloves relative to the specific task to be completed. Before purchasing gloves, the employer should consider: the work activities of the employee, which should be studied in order to determine the degree of dexterity, speed, accuracy, hand movement and force requirements necessitated. Most importantly, the workers' anthropometrics and comfort should be considered, emphasizing glove fit; and therefore optimising worker performance, and compliance with safety regulations.

ACKNOWLEDGMENTS

The National Research Foundation South Africa is to be gratefully acknowledged for assisting with funding this research project.

REFERENCES

Bensel, C.K. (1993). The effects of various thicknesses of chemical protective gloves on manual dexterity. *Ergonomics, 36(6):* 687-696

Bishu, R.R., Bronkema, L.A., Garcia, D., Klute, G. & Rajulu, S. (1994). *Tactility as a function of grasp force: effects of glove, orientation, pressure, load, and handle.* Nasa Technical Paper NASA-TP-3474. Johnson Space Centre, May 2004. 1-23.

*Bradley, J.V. (1969a). Effect of gloves on control operation time. *Human Factors, 11(1):* 13-20. (see Mital et al., 1994)

Bellingar, T.A. &Slocum AC (1993). Effect of protective gloves on hand movement: an exploratory study. *Applied Ergonomics, 24(4):* 244-250

Dias, J.J. & Garcia-Elias, M. (2006). Hand injury costs. *International Journal of the Care of the Injured, 37:* 1071-1077

*Imhran, S.N. & Loo, C.H. (1989). Trends in finger pinch strength in children, adults and the elderly. *Human Factors, 31:* 689-701 (see Kumar, 2004)

Kumar, S. (2001). Theories of musculoskeletal injury causation. *Ergonomics, 44:* 17-47

Kumar, S. (2004). *Muscle Strength.* Canada: CRC Press

Krausman, A.S. & Nussbaum, M.A. (2007). Effects of wearing chemical protective clothing on text entry when using wearable input devices. *International Journal of Industrial Ergonomics, 37(6):* 525-530

Ma, L., Chablat, D., Bennis, F. & Zhang W. (2009). A new simple dynamic muscle fatigue model and its validation. *International Journal of Industrial Ergonomics, 39:* 211–220

Mital, A., Kuo, T. & Faard, H.F. (1994). A quantitative evaluation of gloves used with nonpowered hand tools in routine maintenance tasks. *Ergonomics, 37(2):* 333-343

Muralidhar, A., Bishu, R.R. & Hallbeck, M.S. (1999). The development and evaluation of an ergonomic glove. *Applied Ergonomics, 30:* 555-563.

*Napier, J.R. (1956). The prehensile movements of the human hand. *J. Bone Joint Surg. Br., 38B:* 902-913 (see Kumar, 2004)

*Oborne, D. (1988). *Ergonomics at work.* West Sussex: Wiley (see Krausman & Nussbaum, 2007)

Shih, Y. &Wang, M.J. (1997). The influence of gloves during maximum volitional torque exertion of supination. *Ergonomics, 40(4):* 465-475

Skov, O., Jeune, B., Lauritsen, J.M. & Barfred, T. (1999). Time off work after occupational hand injuries. *Journal of Hand Surgery, 24B(2):* 187-189

*Trembley, J.F. (1989). *Evaluation of functional fit and comfort of chemical protective gloves for agricultural workers.* Unpublished Master's thesis. Edmonton, Canada: University of Albert (see Bellingar & Slocum, 1989)

*Williams, J.R. (1979). Permeation of glove materials by physiologically harmful chemicals. *Am Ind Hyg Assoc J, 40: 877-882* (see Bellingar & Slocum, 1993)

Chapter 16

Ergonomic Evaluation of Workload from Deburring Action Done with the Vibrating Hand-Tool

Hossein Arabi*, Alexandre Morais*
Bronislaw Kapitaniak**, Christophe Gueydan*

* Service Ergonomie Industrielle PSA Peugeot Citroën, France

hossein.arabi@mpsa.com

**Unité d'Ergonomie UPMC Paris, France

bronislaw.kapitaniak@upmc.fr

ABSTRACT

In this study physical and physiological indicators were measured to define the level of hardness engendered by the action of deburring cast-iron parts with a vibrating tool. The measurements include: the forces applied to the tool, the muscular bioelectric activities (EMG) and the biomechanical parameters of the articular angulations. We also carried out these measurements on a simulated workstation in the laboratory environment. (Pre-study). The results show that the muscular forces used in working with the tool are high, considering the static character of the action. The risk of MSD is, according to the standard NF EN 1005-3, almost 4 times as high as the acceptable risk. The results of the measures of bioelectric activity of muscles engaged during the work with the tool also confirm an excessive effort. The repetitiveness of the task of deburring constitutes a very important risk factor. The calculated OCRA index is 6 times over the limit of unacceptable frequency. The task of deburring should be considered as implying a very important risk of MSD..

Keywords: physical workload, EMG, vibrating tools, forces measurement

INTRODUCTION

We realized the observations of professional activity and the measures of working strengths on the workplaces of foundry during the actions of deburring of cast-iron parts (crankcases and disks of brake) with a vibrating tool.

To refine the knowledge of hardness engendered by the action of deburring cast-iron parts with a vibrating tool, we realized in the laboratory physical and physiological measurements to define the level of hardness during the incriminated operations.

The realized measurements concerned:

- the forces applied to the tool,

- the muscular bioelectric activities (EMG),

- the biomechanical parameters of the articular angulations.

We carried out these measurements on a simulated workstation in the laboratory (prestudy) with a single subject (age 52, height 172 cm, weight 68 kg).

A comparison of the results from the laboratory with that from the workplaces allowed us to validate the utilize method on the field.

METHODOLOGY

The observations were realized on real workplaces on 4 operators during periods of 8 working hours. The operators exerted the task of deburring of crankcases and disks of brake (fig 1) with a vibrating tool (chisel). The action forces and the maximal forces (MVC) for each operator have been measured with a strain-gauge Belt (fig 2). The action forces have been explained in absolute value (daN) and in relative one (%MVC).

In the laboratory, the forces were measured with a strain gauge SML-100 MFG (302422) of a capacity about 50 daN. The forces exercised during the activity of deburring were measured by attaching the strain gauge to the arm of the subject on one end and to a fixed point on the wall on the other. Then the operator exercised the activity in a closest possible way of the real activity. These measurements were preceded by the measurements of maximal muscular forces (MVC) in order to establish the reference for calculating the relative forces in terms in percentage of the maximal force. MVC was measured in the axis of the professional gesture on a period of 3 s. The muscular bioelectric activity was measured by the EMG system Biometrics Ltd Datalog with electrodes EMG (Sx230 calibrated on 1000 Hz). It was measured on 4 muscular groups of the upper limb which holds the tool. The chosen muscles are: long supinator, biceps brachial, anterior deltoid and posterior deltoid. Electrodes were placed on the fleshy parts of these muscles; the reference electrode was placed on the wrist. The skin was cleaned with alcohol. EMG was recorded on the MVC exercise and on the professional activity.

Fig 1 Task of deburring the disk of brake

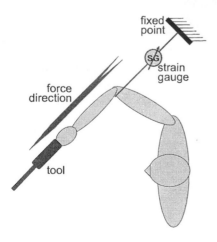

Fig 2 Schema of force measuring system on the workplace

The exercised activities were recorded in digital video (Sony Handycam DCR-TRV620E). The articular positions were estimated with the method BIOVECT. From the video recording realized during the exercise of deburring, articular positions were defined and freeze frames taken accordingly. The angular and vector analysis was realized on the chosen images. The results of the measurements were then sifted with the software CAPTIV (version 4.3.2.) allowing to synchronize all the measured parameters with the video and to obtain a quantitative treatment of the results. A part of the calculation of the muscular activity and the forces was realized with EXCEL.

RESULTS

The results of the direct observations of the activity show that the working time with the vibrating tool (chisel) is of the order of 150 min by post a day for the deburring of crankcase and 60 min for the disk of brake. In most of the cases the operators apply the rotation of posts every 30 minutes. The activity of deburring is exercised by cycles about 15 – 30 s for crankcases and 6 in 10 s for disks.

The results of measures of working strength realized on the workplaces show that the strengths of support on the tool vary between 3 and 7 daN it means between 8 and 18 % of MVC for crankcases and respectively between 2 and 5 daN it means between 5 and 13 % of MVC for disks.

The results of the measures realized in the laboratory on the simulated workplaces are the following ones. For our subject the maximal force to push the tool is approximately 25.4 daN. The forces applied to activate the tool vary between 3.6 and 7.1 daN with an average of 6 daN. They thus correspond to relative forces between 14 and 28% MVC with the average of 23.6 % MVC.

The working forces which reach on average 23.6% of the maximal force must be interpreted as very high, considering the static character of the operation, especially as the duration of this operation lasts for several dozens of seconds.

The results of the measurements of the bioelectric activity (EMG) are expressed in percentage of the registered maximal activity. Taking into account that the used equipment (Biometrics Datalog) does not allow to record the raw signal of the EMG, but only the already integrated and averaged signal (avg RMS), this method of calculation turns out to be the most useful and provides the most relevant information. We calculated two values of the integrated EMG. The first one is a mean value over the period of 2 minutes of activity of deburring and the second is a mean value of the continuous action during 15 s. The calculated values are presented in Table 1.

	mean 2 min	mean 15 s
Deltoid anterior	37.2%	46.0%
Deltoid posterior	12.5%	25.7%
Biceps brachial	8.1%	16.3%
Long supinator	7.4%	18.3%

Table 1. Results of iEMG over the period of activity of deburring

The results confirm the conclusion of the high workload according to the results of force measurements. Indeed, during a continuous static action, deltoids work between 25 and 50 % of their maximal activity, while the biceps and long supinator are solicited between 15 and 20 %. It is necessary to remember that the muscle can support the static load below 15 % of its maximal activity. The averages calculated on a complete cycle of deburring about 2 min indicate that deltoids are overloaded.

The vector biomechanical analysis was realized with the method BIOVECT on 2 photos (figure1) of most observed postures. The results of the angular analysis indicate that the working posture observed during the action with the tool is in most cases comfortable. Only the actions requiring more vertical holding of the tool engender the position of the right hand operating the tool in excessive extension at the level of the shoulder.

Figure 1. BIOVECT analysis of working postures

EVALUATION OF THE HARDNESS OF THE ACTION ON THE TOOL ACCORDING TO NF EN 1005 STANDARD

We applied NF EN 1005 standard: "Safety of machinery - Human physical performance, Part 3: Recommended force limits for machinery operation, Part 4: Evaluation of working postures and movements in relation to machinery and Part 5: Risk assessment for repetitive handling at high frequency".

As regards the limit of force applied to the tool, NF EN 1005-3 standard recommends an approach which consists at first in determining the specific action realized with the machine, the distribution of the applied forces and the characteristics of the population of users. The specific action is a push of the tool in the axis of the arm. The population of users can be considered as a male population aging from 18 to 55, having no medical limitations due to health. For these conditions, the standard recommends the acceptable maximal isometric force of the push of 10.9 daN. Because the gesture implies an evident movement, the standard recommends a coefficient of 0.8 (§4.2.2.1) and for the frequency included over 20/min a coefficient of 0.3. It is necessary to add a coefficient of duration of 0.8 (a duration between 1 and 2 hours). The recommended force (Fbr) thus becomes:

Fbr = 10.9 * 0.8 * 0.3 * 0.8 = 2.1 daN

The ratio FR/Fbr = 6/2.1 = 2.9

Thus, according to the standard, a ratio higher than 0.7 is to be avoided. Our ratio indicates a risk 4 times over the acceptable risk.

The working posture observed during the action with the tool can be considered as acceptable.

As regards the limit of the frequency of the action, NF EN 1005-5 standard recommends an approach which consists in work analysis. The task of deburring must be considered as repetitive acts with high frequency (§5.3.3.1). Because the average time of the cycle (15 s) is inferior to 30 s and the frequency of the technical actions of both upper limbs is over 40 technical actions per minute.

To calculate the OCRA indication we can calculate the value of RF (reference recommended frequency) according to the following equation:

$RF = CF \times PoM \times ReM \times AdM \times FoM \times (RcM \times DuM)$

where :

CF is the "constant of frequency" of technical actions per minute = 30

PoM, ReM, AdM, FoM are multipliers for the risk factors connected to the postures, to the repetitive acts, to the force and the additional factors.

RcM is the multiplier for the risk factor "lack of recovery".

DuM is the multiplier for the total duration of one or several repetitive tasks during a shift.

We can consider the following values of multipliers:

Pom = 0.7 (elbow posture over 45°)

ReM = 0.7 (time of cycle 15s)

AdM = 0.8 (vibration presents over 80% of the time of the cycle)

FoM = 0.35 (the applied forces are over 25 % of the maximal force)

(RcM × DuM) = 0.6

Consequently:

$$RF = 30 \times 0.7 \times 0.7 \times 0.8 \times 0.35 \times 0.6 = 2.5$$

FF (frequency of technical action recorded on the workplace) = 60/min

Thus the OCRA index = FF / RF = 60 / 2.5 = 24.3

This value of the OCRA index is considered as highly unacceptable. It indicates a high risk of MSD..

CONCLUSIONS

1. The results of the measures on the workplaces are confirmed by the results obtained in the laboratory. This result confirms the validity of the dynamometric measurements realized in the workplace in the field conditions.

2. The muscular forces used in working with the tool are high, considering the static character of the action. The risk of MSD is, according to the standard NF EN 1005-3, almost 4 times as high as the acceptable risk. The results of the measures of bioelectric activity of muscles engaged during the work with the tool also confirm an excessive effort (between 25 and 50 % of maximal efforts for the actions over 15 s, between 10 and 40 % for the averages measured over the periods of a cycle about 2 min).

3. The repetitiveness of the task of deburring constitutes a very important risk factor. The calculated OCRA index is 6 times over the limit of unacceptable frequency.

4. The task of deburring should be considered as implying a very important risk of MSD.

PROPOSITIONS

In order to decrease the risk, it would be preferable to reconsider the conception of the tool of deburring. Should it prove impossible, the organizational solutions would have to be envisaged. Among them, the rotation of the posts, the decrease of work pace and the ergonomic conception of the workstation should be privileged.

REFERENCES

AFNOR (2008) Recueil de normes françaises, Ergonomie des postes et lieux de travail, éd. AFNOR, CD-ROM.

INRS. (2000) Prévenir les Troubles MusculoSquelettiques. INRS, Paris.

Arabi H., Kapitaniak B., Morais A., (2009) Physiological evaluation of strain from deburring action done with the vibrating hand-tool. VII Conférence Internationale d'Ergonomie Karpacz

Arabi H., Vandewalle H., Kapitaniak B., Monod H., (2000) Evaluation of wheelchair users in the field and in the laboratory: Feasibility of progressive tests and critical velocity, Int.J. Ind.Ergonomics, 24, 483-491

Arabi H.,Vandewalle H., Kapitaniak B., Monod H., (1997) How to optimize the wheelchair for the sportive handicapped subjects. 13th Congress of the International Ergonomics Association, Tempere, Finland.

Bae S, Wei Zho, Armstrong T., (2009) Finger Motions in Reach and Grasp Work Elements, XVIII IEA Congress, Pékin

Colombini D., Grieco A., Occhipinti E. (1998) Occupational musculoskeletal disorders of the upper limbs due to mechanical overload, Ergonomics, vol.41, n°9

Forcier L., Kuorinka I., (2006) Work-related musculoskeletal disorders: overview, International Encyclopaedia of Ergonomics and Human Factors, 2ème edition, vol. 2, 1625-1632

Kapitaniak B., (2006) Static load, International Encyclopaedia of Ergonomics and Human Factors, 2ème edition, vol. 1, 580-583

Kapitaniak B., Peninou G., Heusch F. (1996) Software "BIOVECT" help for postural analysis, XI Annual International Occupational Ergonomics and Safety Conference, Zurich.

Malchaire J, Indesteege B, (1997) Troubles musculosquelettiques, analyse du risque INRCT, 1 vol, 122 p

Mariaux Ph., Demaret J-Ph., Freyens S., Vandoorne CH. (1998) Manutentions manuelles. Commissariat fédéral à la promotion du travail, Bruxelles.

Mathiassen S.E., Wulff Svendsen S., (2009) Systematic and random errors in posture percentiles assessed from limited exposure samples, XVIII IEA Congress, Pékin

Monod H., Kapitaniak B. (2003) Ergonomie. Masson, 2ème éd., 1 vol., 286 p.

Motmans R., Adriaensen T., Hermans V, (2009) Muscle activity during repetitive work, XVIII IEA Congress, Pékin

Occhipinti E., Colombini D., De Vito G., Molteni G., (2006) Exposure assessment of upper limb repetitive movements: criteria for health surveillance, International Encyclopaedia of Ergonomics and Human Factors, 2ème edition, Vol. 1, 1507-1509

Péninou G., Kapitaniak B. (2006) Vectorial analysis; ergonomics applied method. International Encyclopaedia of Ergonomics and Human Factors, 2ème edition, vol. 3, 3018-3022

Peninou G., Monod H., Kapitaniak B. (1994) Prévention et ergonomie. Masson, 1 vol., 120 p.

Roquelaure Y. et coll. (1994) Les troubles musculosquelettiques du membre supérieur liés au travail : physiopathologie et facteurs de risques., Rev. med. travail, tome 21, n° 3

Silverstein B., (2006) Work-related musculoskeletal disorders: general issues, International Encyclopaedia of Ergonomics and Human Factors, 2ème edition, vol. 2, 1621-1624

Young J., Armstrong T., Ashton-Miller J., (2009) Active and Passive Forces in Hand/Work-Object Coupling, XVIII IEA Congress, Pékin

Chapter 17

Effect of Grasp-/Contact-Characteristics of Snap Fasteners on Time Requirements and Electromyographic Activity for Snap-Fit Assembly

Hamed Salmanzadeh, Marianela Diaz-Meyer,
Verena Bopp, Kurt Landau, Ralph Bruder

Institute of Ergonomics
Darmstadt University of Technology
Petersenstrasse 30, 64287 Darmstadt, Germany

ABSTRACT

Snap-fit assembly has become more important in the automotive industry and within its subcontractors, as well as, in the consumer goods industry. The reason for this development is particularly related to the increased economic benefits that occur through the accelerated assembly task. However, the ergonomic aspects of snap fit assembly have not been sufficiently studied. Fasteners with sharp-edged head geometry can produce a displeasing perception on the finger pad during the insertion operation. Snap fasteners can also constrain the operators to use a pinch-grasp during the insertion. The high insertion forces during pinch-grasp, particularly due to the small friction coefficient in friction fit coupling, can lead to musculoskeletal disorders and repetitive strain injuries. This study investigates, in two different experiments, the influence of different conditions of grasp- and contact-characteristic of snap fasteners on assembly time. In the first experiment, the influence of the sharpness of the snap-fit on the insertion time and EMG was studied. The insertion force was set to 20N and 70N. The results showed a significant difference in insertion times of snap fasteners depending on the head geometry (i.e., sharpness). This effect was higher for higher predetermined insertion forces. The significant effect of sharpness on the electromyographic activity of the musculature of the thumb was also proved. According to these results, a second experiment was planned. In the second experiment, the influence of additional grasp and contact characteristics (i.e., sharpness, dimension and slipperiness) on the reach, move, insertion and assembly time was studied. It was found that the slipperiness and dimension had a significant effect on insertion time. Furthermore, the kinematic analysis pointed out that the move-time was significantly affected by only regrasping.

Keywords: delicate object, snap fastener, snap-fit assembly, insertion time, EMG

INTRODUCTION

The utilization of snap fasteners in the automotive industry is very common. It is assumed that more than 20% of all fasteners in a vehicle are snap fit components (Huebner, 2006). A snap-fit is an integral latching mechanism for attaching one part to another. Snap-fits differ from loose or chemical attachment methods in that it requires no additional pieces, materials or tools to carry out the attaching function (Bonenberger, 2005). Innovative manufacturers are adopting them, in order to reduce assembly times and the number of components to be assembled. Snap-fasteners, however, require a high degree of precision during assembly and optimization of the hand-finger coordination in order to allow sufficient grasp stability when handling, aligning and inserting. The ergonomic and micro-economic aspects of snap-fit assembly have not been sufficiently studied. The head geometry of snap fasteners, particularly in pipe and cable fastening, as well as, in profile- and strips fastening, may lead to complaints and cause impairments, if the workers mount these fasteners with a high frequency within a shift. For example, fasteners with sharp-edged head geometry can produce a displeasing perception on the finger pad during the insertion operation. On a continuing basis, it can lead to skin irritation. Snap fasteners can also constrain the operators to use a pinch-grasp during the insertion. The high insertion forces during pinch-grasp, particularly due to the small friction coefficient in friction fit coupling, can lead to musculoskeletal disorders and repetitive strain injuries.

The investigation of assembly times related to different conditions of grasp- and contact-characteristic of snap fasteners can help to compare various productivity levels. Furthermore, the results can be used as an indicator for the degree of difficulty within the snap-fit assembly process. The consideration of ergonomic aspects in an early stage of the design process of snap fasteners can contribute to an increase in long-term productivity.

There are research deficits in the product design and industrial engineering concerning snap-fits. Notably, the effects of snap-fit design on assembly time requirements were not studied. Boothroyd and Dewhurst (1983) and Poli et al. (1986) published estimations of the time required for handling and insertion using snap-fit systems. Handling and insertion times were estimated according to the orientation (rectangular or cylindrical), locking strength, number of steps in the locking operation, need for alignment, etc.

MTM-1 (DMTM, n.d.) is a highly defined method that provides designers and production planners with the tools for the calculation of handling and assembly times when the snap-fit is introduced into mass production operations (Landau et al., 2009). Concerning the insertion process, MTM-1 only accounts for factors such as handling, fit-class and symmetry. It is, however, not clear, whether these three factors can completely describe the snap-fit operation. For example, these three factors do not distinguish between different cases of snap-fit geometric characteristics (e.g., sharpness, slipperiness, flexibility, and dimension). For all these cases, MTM-1 proposes a global additional time, which should be added to the regular time. Furthermore, insert forces are neglected by the allocation of the parameters into the three fit-classes. Insertion forces are an important technical parameter for the assembly time of snap-fits.

MTM-UAS, Universal Analyzing System (DMTM, 2000), is a supplementary method, which contains a specific section for the analysis of snap-fit operations. The drawback of MTM-UAS lies on its high degree of aggregation, which makes the analyses unsuitable for the evaluation of design decisions (Landau et al., 2009).

Landau et al. (2009) classified the main time-related factors that affect the handling and assembly using snap-fits. These are the socket and the head geometry of the snap fasteners, as well as, the type of positioning (assembly motion). They calculated in a first step, the assembly time for several snap-fits using the MTM-1 procedure, without considering the above-mentioned systematic influences. This analytical method aimed at the comparison of the different snap-fits from the perspective of the productivity, with no consideration of ergonomic aspects. In this way, recommendations for snap-fasteners were derived. The main topic of this study is the exhaustive empirical investigation of selected grasp and contact characteristics of snap fasteners influencing the assembly time, in particular, the investigation of sharp-edged head, slipperiness and dimension of contact-surface.

Because handling snap-fits requires highly sensomotoric skills, higher muscular stress is expected (Goebel, 1996a, 2009). Goebel (1996b) describes this pre-stress in the muscles as a necessary physiological "basic load" during precision movements, which require additional stabilization. Diaz Meyer and Landau (2009) confirmed this finding in the case of handling delicate objects, namely, by moving an object filled with liquid. Based on the studies mentioned above, it is expected that muscular strain is affected by higher accuracy requirements when inserting sharp-edged snap-fits.

METHOD

In order to investigate different influence factors, the present study was divided into two major experiments. None of the participants reported any current musculoskeletal pain or injuries within the past six months. Due to the fact that the sensomotoric performance tends to be reduced starting in middle age people (i.e., 45 years) (Goebel, 1996b), subjects aged 20 to 30 years were chosen in both experiments. Furthermore, it was found in a preliminary test that the anthropometry of the thumb (thumb's width and thickness) affects the skin sensation. In order to limit the variability of skin sensation in subjects, the thumb's width and thickness around the 50th percentile ± standard deviation were used as additional selection criteria.

In the first experiment, 12 male students from the age of 19 to 30 participated in the study. The effect of sharpness (see Table 1) and insertion force was analyzed with respect to insertion time. All subjects in this study adopted a standardized posture in order to achieve reproducible results. The subjects were standing upright and facing the test apparatus on the table. An adjustable apparatus, fixed to a force plate, was used to insert two types of head geometry of snap fasteners (sharp-edged and even) (see Figure 1). The position of the subject was adjusted to the height of the experimental apparatus with a height-adjustable pedestal, so that the join patch could be reached with the tip of the thumb of his right hand.

Table 1 Grasp and contact characteristics of tested snap fasteners

grasp- and contact- characteristic	Snap-fastener		Experiment
sharpness (head geometry)	even	sharp-edged	first and second
dimension	large	small	second
slipperiness	slippery	non-slippery	second

The apparatus was adjusted for two predetermined insertion forces (20N and 70N). This resulted in the combination of four experimental variants. Eight independent insertion-positions in the test apparatus were used to repeat the insertion operation 16 times. A total of 64 measurements were carried out (4 experimental variants x 8 insertion position x 2 passes) for each subject. The experimental variants were randomly permuted. The insertion times and the actual measured forces of all participants for each of these four experimental conditions were recorded via a force plate (Kiag Swiss, Type 9261A). Simultaneously, the activity of six muscles (M. flexor digitorum superficialis, M. extensor digitorum, M. biceps brachii, M. triceps brachii, M. erector spinae and thenar muscles) in the right upper extremity and trunk was recorded with a surface electromyography.

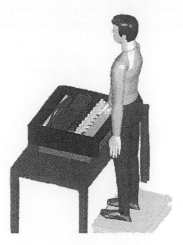

Figure 1. Standardized initial position of subjects.

Figure 2. Apparatus setting for the second experiment. 1, 3 and 5 indicate the three columms added for the supply of the snap fasteners. 2, 4 and 6 indicate the three insertion positions.

After analyzing the results of the first experiment, a second experiment was planned. In the second experiment, 30 male students from the age of 19 to 33 participated in the study. The purpose of this experiment was to investigate the influence of different grasp conditions on snap-fit assembly times such as: reach times, move times and insertion times. (In our experiment, we could not separate the grasping-time and actual reach-time, so reach-time consists of both elements).   In addition to the first experiment, some other factors, namely dimension, type of coupling (i.e., slippery) and regrasping was analyzed (see Table 1). The predetermined force was set only to 50 N. The experimental apparatus was the same as in the first experiment but the number of insertion positions was limited to three. An additional dispenser (three columns) attached to the experimental apparatus (see Figure 2) was used to supply the snap-fits. The distance between the centers of the insertion positions and the center of the columns associated with the supply of snap-fits was about the same. At each column a photoelectric sensor was mounted. By pulling a snap-fit a short current pulse was sent both to the LEDs as well as to the measuring channel of the analog-digital converter (HBM Spider8). This impulse-signal was utilized for the subsequent synchronization of the measuring instruments and for the separation into several motion time elements (reach, move, insertion and total assembly time). An average of three repetitions was used for the statistic analysis.

RESULTS AND DISCUSSION

FIRST EXPERIMENT

Although the force measuring apparatus was adjusted in the first experiment for equal predetermined insertion forces for both head geometries, the actual measured insertion forces were higher for sharp-edged snap-fits (see Table 2). It was found that the actual measured insertion force was significantly influenced by the sharpness of the head geometry of the snap fastener (see Table 3).

Both the mean and the standard deviation (SD) were higher for the sharp-edged snap-fits with larger predetermined insertion force. A linear relationship between the actual measured insertion force and the insertion time could not be proved.

Table 2 Descriptive statistics of actual measured insertion forces in Newtons (N= 12)

sharpness	predetermined insertion force	min	max	mean	SD
even	20 N	18.8	52.8	28.6	5.4
	70 N	63.1	137.7	85.6	13.6
sharp-edged	20 N	19.6	66.7	31.5	6.8
	70 N	55.9	159.9	90.0	18.0

Table 3 MANOVA for actual measured insertion force (N= 12)

source	df	MSE	F	p-value
predetermined force	1.00	636629.22	4175.55	0.000**
sharpness	1.00	2753.80	23.27	0.000**

*p<0.05. **p<0.01

Table 4 Descriptive statistics of insertion time in sec (N= 12)

predetermined insertion force	sharpness	mean	SD	mean time difference
20 N	even	0.96	0.29	0.07
	sharp-edged	1.03	0.32	
70 N	even	1.41	0.45	0.27
	sharp-edged	1.68	0.59	

Table 5 MANOVA for insertion time (N= 12)

source	df	MSE	F	p-value
predetermined force	1.00	58.74	398.22	0.000**
sharpness	1.00	5.61	38.61	0.000**

*p<0.05. **p<0.01

Table 6 MANOVA for normalized electromyographic activity (N= 12)

source	df	MSE	F	p-value
M. flexor digitorum superficialis				
predetermined force	1.00	20625.60	230.68	0.000**
sharpness	1.00	27.23	0.37	0.545
M. extensor digitorum				
predetermined force	1.00	43602.67	422.89	0.000**
sharpness	1.00	50.84	1.88	0.172
M. biceps brachii				
predetermined force	1.00	611.68	120.28	0.000**
sharpness	1.00	17.31	3.72	0.055
M. triceps brachii				
predetermined force	1.00	81782.83	288.42	0.000**

sharpness	1.00	223.24	6.10	0.014**
M. erector spinae				
predetermined force	1.00	2363.35	103.73	0.000**
sharpness	1.00	5.57	1.93	0.166
thenar muscles				
predetermined force	1.00	22447.37	131.26	0.000**
sharpness	1.00	14137.94	44.72	0.000**

*p<0.05. **p<0.01

Table 4 shows the descriptive statistics of insertion time in seconds for all examined variants. The statistical test (see Table 5) indicates that the predetermined force and sharpness have a significant influence on the insertion time. For a predetermined insertion force of 20 N, the subjects needed 0.07 seconds longer for the insertion of sharp-edged snap-fits (comparison of mean insertion time).

As expected, the predetermined insertion force has a significant influence on the electromyographic activity. It was noted that the sharpness only has a significant effect on the normalized electromyographic activity of the Triceps and Thenar muscles (see Table 6). In both cases, it was higher for the insertion of sharp-edged snap-fits.

The insertion of sharp-edged snap-fits is perceived by the subjects as rather painful. This produces a reaction affecting the form of the insertion process, performance, and actual insertion force of the subjects. It has also an influence on the muscular strain. Although the subjects were instructed to vertically insert all snap-fits, an oblique insertion of sharp-edged snap-fits was observed. This could result from the unpleasant subjective sensation in the thumb and fingers. This oblique insertion of the sharp-edged snap-fits required higher forces. The increased actual measured insertion forces are associated with the increased muscular strain (EMG). The insertion time was also longer for sharp-edged snap-fits than for even snap-fits.

SECOND EXPERIMENT

The results of statistical analysis are summarized in Table 7.

Table 7 P-values of the evaluation of time data (N= 30)

indep. var. / dep. var.	reach-time	move-time	insertion-time	total-assembly-time
sharpness	0.378	0.980	0.000*	0.505
dimension	0.951	0.001*	0.002*	0.000*
coupling / slipperiness	0.000*	0.211	0.000*	0.000*
regrasping	0.001*	0.000*	0.038*	0.000*

*p<0.05

The results show that dimension, slipperiness and regrasping significantly affected (p <0.01) the total assembly time. This is not the case for sharp-edged snap-fasteners, although the insertion time is affected significantly by sharp-edged snap-fasteners. The reason is that the other elements of snap-fit assembly time, such as, move-time and reach-time, are not influenced by sharpness. Table 7 also shows that the insertion time is affected significantly by all mentioned grasp conditions. In contrast, the move time was affected only by dimension and regrasping. This result was expected for regrasping but the significant prolonging effect of dimension on the move-time can be explained as follows: it can be assumed that the smaller snap-fits had to be gripped with caution and were therefore carefully moved unconsciously. But this is only one possible interpretation. It must be confirmed by a detailed analysis of the acceleration or velocity curves. The same applies to the significantly higher time requirement of the slippery snap-fits in reach-movements. But this can be shown only by further evaluation of the specific contexts.

CONCLUSION

The snap-fit assembly and snap-fit movements are affected by grasp-/contact-characteristics of snap fasteners. The insertion times and moving times for different conditions of grasp- and contact-characteristic of snap fasteners were investigated in two experiments. The consideration of these results in an early stage of the design process of snap fasteners can contribute to an increase in long-term productivity, as well as, a reduction of musculoskeletal disorders and repetitive strain injuries. The results of these experiments can only give qualitative information on the effects of limited grip and contact conditions. Nonetheless, quantitative information using a mathematical model is desirable. This would answer the question of which characteristic in comparison to the other has a larger effect on the assembly time. We can deduce the answer to this question from the results of this investigation.

ACKNOWLEDGEMENTS

The authors would like to thank all the participants for their valuable contribution to the completion of the study. This work was funded by the German Research Foundation (DFG) and supported with materials by Raymond Group .The Experiments took place in the motion laboratory of the Institute of Ergonomics (IAD), Darmstadt University of Technology.

REFERENCES

Bonenberger, P.R. (2005), *The first snap-fit-handbook*. Hanser, München.

Boothroyd, B., and Dewhurst, P. (1983), *Product design for assembly*. Boothroyd Dewhurst Inc., Wakefield.

Deutsche MTM-Vereinigung e.V. (DMTM) (2006), *MTM-UAS Training Manual*. Hamburg.

Deutsche MTM-Vereinigung e.V. (DMTM) (2008), *MTM-1, Training Manual*. Hamburg.

Diaz Meyer, M., and Landau, K. (2009), "Beurteilung muskulärer Beanspruchung bei Hand-Arm-Rumpf Bewegungen, die besondere Sorgfalt und Präzision erfordern." *Zeitschrift für Arbeitswissenschaft*, 3, 257–267.

Goebel, M. (1996a), "Electromyographical analysis of sensory feedback in movement control. " *Proceedings of the 1st International Conference on Psychophysiology in Ergonomics PIE 96*, Wuppertal, May 30 - June 1.

Goebel, M. (1996b), *[Analysis of human motor coordination and its fatigue by surface electromyography] Elektromyografische Methoden zur Beurteilung sensumotorischer Tätigkeiten*. Dokumentation Arbeitswissenschaft. Band 40, Schmidt Verlag, Köln.

Goebel, M. (2009), "Produktivität und sensumotorische Fertigkeiten." In: K. Landau (ed.) *Produktivität im Betrieb*, GfA Herbstkonferenz, Millstatt, Ergonomia Verlag, Stuttgart, 273–286.

Huebner, A., Irmer, W., Martinek, M., Pieschel, J., Zwickert, H., and Zinke, M. (2006), *Studienmaterial für Fertigungslehre*. Universität Magdeburg. Fakultät für Maschinenbau. Institut für Werkstoff und Fügetechnik.

Landau, K., Landau, U., and Salmanzadeh, H. (2009), "Productivity Improvement with Snap-Fit systems." In: *Industrial Engineering and Ergonomics: Vision, concepts, methods and tools*. Schlick, C.M. (Ed). Sprinter Verlag, Heidelberg-Berlin, 595–608.

Poli, C., Graves, R., and Groppetti, R. (1986), "Rating products for ease of assembly." *Machine Design*, 58 (19), 79–84.

Chapter 18

Ergonomic Design of the Viewing Angle in a Computer Numerically Controlled-Electro Discharge Machine Environment

Imtiaz A Khan, Mohammad Asghar
Mechanical Engineering Department,
Aligarh Muslim University, Aligarh-202002,INDIA

ABSTRACT

Dynamic situations are not fully controlled and affected by uncertain human factors. Anthropometric considerations are significant in the design of industrial systems. Present research work has considered the 'viewing angle', whose effect on operator's performance has been ergonomically evaluated in a CNC-EDM environment. In this work, the experimental data is analyzed through an ANOVA using SPSS statistical software. The result indicates that the viewing angle significantly affects the operator's performance in a CNC-EDM interaction environment. Further analysis revealed that a 21 degree viewing angle gives the optimal performance as far as a Human- CNC machine interaction environment is concerned.

Keywords: Visual Display, Viewing Angle, System Design, CNC-EDM Environment, Searching Task.

INTRODUCTION

With the high technology applications getting more widespread at the global level, the problems associated with the introduction of this hi-tech have also been generating more concern. Most part of such concern is reflected in occupational stresses in the form of poor job performance, waste leisure time, low level of job satisfaction, alcohol related problems and hence forth. One notable component of hi-tech era emerged in the shape of human-CNC machine interaction (HMI) that basically comprises of a CNC workstation and an operator. The use of CNC systems is increasing exponentially. This is accompanied with a proportionate increase in occupational stresses too in human operators. Previous studies pertaining to HMI by different researchers in the field revealed that all sorts of problems associated with the use of CNC machines could be traced in terms of physical characteristics of the CNC workstation, visual factors, psychological factors and postural factors. Present study relates to the visual factor that relates to the constrained postures of the CNC operators governed by the characteristics of the given workstation. The review of literature suggested that the original sources of postural stresses may be traced in terms of poor CNC workstation design. Nowadays, the major emphasis is on preventing musculoskeletal injuries in the workplace. These injuries create a significant cost for industry. One of the important considerations in the human-machine interaction environment is the viewing angle, which plays a key role in the   system design. Therefore, its effect on human performance in a CNC-EDM environment was explored in this study.

RELATED WORKS

Recently public and private organizations have engaged themselves in the enterprise of managing more and more complex and coupled systems by means of the automation. Modern machines not only process information but also act on the dynamic situations. These dynamic situations are affected by uncertain human factors. The viewing angle is considered frequently in the design of the systems like human-computer interaction, human-CNC machine interaction and so on. A review of the literature finds a relatively large number of studies on the viewing angle. Kong-King Shieh et al., (2007) determined the viewing distance and screen angle for electronic paper (E-Paper) displays under various light sources, ambient illuminations and character sizes. Findings of this study indicate that mean viewing distance and screen angle should be 495 mm and 123.7 degrees (in terms of viewing angle, 29.5 degrees below the horizontal eye level), respectively. Proper visualization of the background of surgical field is essential in the laparoscopic surgery. Another study for visual display unit (VDU) work environment was carried out by Svensson et al., (2001). In this study two viewing angles, namely 3 degrees above the horizontal and 20 degrees below the horizontal, were considered. The findings concluded that the load on the neck and shoulders was significantly lower at 3 degrees as compared to 20 degrees. Jan Seghers et al., (2003) explored that low VDU screen height increases the viewing angle and also affects the activity of the neck extensor muscles. Ayako Takata et al., (2002) determined the effects of the tilt angle of a notebook computer on posture and muscle activities. It was concluded in the study that at 100 degree tilt angle, the subjects had relatively less neck flexion. Visual display units are widely used in the industries. The optimization of their orientation is a critical aspect of the human-machine interaction and impacts on the worker health, satisfaction and performance. Due to increase in the visual and musculoskeletal disorders related to VDU use, a number of ergonomic recommendations have been proposed in order to combat this problem. Fraser et al., (1999) observed that, the monitor position, 18 degree below eye level had no significant effect on the position of the neck relative to the trunk while, the mean flexion of the head, relative to the neck increased 5 degrees. Burgess-Limerick et al., (2000) determined optimal location of the visual targets as 15 degrees below horizontal eye level. Recently computer-based consumer interactions are becoming common. Effective interactions are crucial for commercial success and aspects such as vision and control are important determinants. Adjustability effect of the touch screen displays in a food service industry was investigated by Batten et al., (1998). To determine the optimal viewing angle or range of a given touch-screen display, an anthropometric analysis was carried out. The results recommended the adjustable range of the touch-screen display as 30 to 55 degrees to the horizontal. Furthermore, the results of the study indicated that the subjects adjusted the touch-screen between the angles of 19 and 54.5 degrees to the horizontal. The study concluded that there is no optimal viewing angle of the touch-screen displays with a dynamic set of the user heights and a static workstation height. The displays should be adjustable through a range that accommodates multiple users and workstations and provides adjustment to compensate for other miscellaneous variables such as glare etc. Some Virtual Reality (VR) head-mounted displays (HMDs) can cause temporary deficits in binocular vision. On the other hand, the precise mechanism of the occurrence of visual stress is unclear. Mon-Williams et al., (1998) in their study pointed out that as vertical gaze angle is raised or lowered the 'effort' required to binocular system also changes. The results indicated that the heterophoria varies with vertical gaze angle and stress on the vergence system during the use of HMDs will depend, in part, on the vertical gaze angle. The results also illustrated that the angle at which the eyes have the smallest heterophoria measurement is somewhat below the ear-eye line (mean 34 degrees). As the gaze angle rises from this point the heterophoria becomes more divergent.

The reviewed researches have clearly indicated that the visual disorder is one of the major factors as far as human injuries in the computer controlled working environment are concerned. The above findings have been used to formulate the present study of the effect of viewing angle in a CNC-EDM interaction environment.

METHODOLOGY

Subjects

Experimental investigation was carried out with three groups of 18 subjects each. Groups were divided according to the variation in height of the subjects; i.e. (Group1) – Subjects of height 5' 9", (Group2) – Subjects of height 5' 6" and (Group3) – Subjects of height 5' 4". All subjects were male, age varied from 21-26 years with mean age of 23.72 years (S.D = 1.592).

Experimentation

In order to conduct the investigation, an experiment was designed in a controlled CNC-EDM (Computer Numerically Controlled-Electro Discharge Machine) wire cutting environment, at "The National Small Industries Corporation Ltd." (NSIC) Aligarh, India.

Three levels of Viewing Angle, namely 15, 21 and 28 degrees (see Figure 1) were considered on the basis of findings discussed in the related works and comprehensive surveys conducted at various EDM centers. Before actual start of the experiment, each of the subjects was asked to go through the instruction sheet served by the experimenter. Specific time interval was allowed to perform the actual error searching task for one set of the experimental condition. To start and stop the task instruction was given through prerecorded voice on a recorder. Error searching time constituted the index of the human performance. The performance of each subject at a pre-specified time was recorded through a specially designed error searching task (see Figure 2).

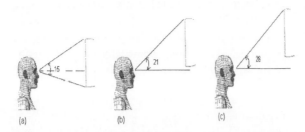

Figure 1. Showing the EDM monitor and considered viewing angles for (a) 5'9", (b) 5'6" and (c) 5'4" height subjects, respectively.

Figure 2.   Picture showing subject performing the error searching task.

Statistical analysis

The experimental data collected, in terms of subject's performance in a CNC-EDM environment, was investigated using statistical analysis with repeated measures. A method of comparison of the mean was used to determine the optimum level of viewing angle.

RESULTS

The analysis of variance pertaining to the single factor repeated measure type of statistical design was performed over the data collected. The result is shown in the analysis of variance (ANOVA) Table 1;

Table 1: Summary of Analysis of Variance, S-Source, VA- Viewing Angle, E-Error, T-Total, df-degree of freedom.

S	Type III Sum of Squares	df	Mean Square	F-value	P-value
VA	17.297	2	8.648	80.932	<0.0001
E	5.450	51	0.107		
T	858.501	54			

F-ratio was used for testing the statistical hypothesis, and the level of significance for the test was set to 0.01. It was concluded that;

(i) The null hypothesis, "Viewing Angle does not significantly affect the operator's performance in a CNC-EDM environment ", was rejected because of the aggregate's mean time difference (performance data in terms of error searching time). (ii) Null hypothesis rejected because the F-value$_{ov}$ = 80.932 (see Table 1) was greater than [F$_{0.01}$ (2, 51)]$_{cv}$ = 5.0472 obtained from the F-table using the values for degrees of freedom (2, 51). [*Where ov = observed value and cv = critical value*]. (iii) Null hypothesis rejected because the P-value for F-value = 80.932 was found to be less than 0.0001 i.e. (p<0.0001), which was less than the set significance level (α = 0.01).

Since the viewing angle had statistically significant effect so far as the error searching task was concerned, an attempt was made to develop a mathematical model to search for the relationship between human performance and the viewing level. Then linear and non-linear regression analyses were performed. For the case of non-linear, exponential, hyperbolic and power function models were examined. The criterion fixed for selecting the best model was the value of the co-efficient of determination, R^2 i.e., the best one would have the highest value of R^2. Proceeding this way the exponential model was found to have the maximum value (0.774) of the R^2. The best fit model had the following form:

$$Y = 0.025 * X^2 - 1.0724 * X + 15.067$$

Where, Y = Human performance in a CNC-EDM environment.

  X = Viewing Angle level.

For the above mathematical model, data were generated and a graph was drawn showing relationship between the human performance and viewing angle level (see Figure 3).

Statistical conclusion: The null hypothesis stated above was rejected since $F_{ov} = 80.932$ was greater than $F_{cv} = 5.0472$ (obtained from F-table). Furthermore, the computed probability value (p-value) i.e. [p<0.0001] meant that the test was strongly significant at 1%; hence Ho (null hypothesis) must be unequivocally rejected at the critical value of 1% because 0.0001 is << 0.01. Thus, the above result indicated that the null hypothesis was rejected and it was found that the viewing angle had a significant effect on human performance in a CNC-EDM environment. Variation in performance under different levels of viewing angle was shown graphically in Figure 3. To establish which one out of the three considered viewing angles was optimal, the data was further analyzed by the method of mean comparison proposed by Winer, (1971).

Table 2: Summary of the analysis

Where: 1: First treatment mean (at a viewing angle of 15 degrees), 2: Second treatment mean (at a viewing angle of 21 degrees), 3: Third treatment mean (at a viewing angle of 28 degrees).

Con-trast	Contrast sum of square	df	Mean square	F-value	P-value
1 vs 3	2.0736	1	2.0736	19.38	<0.0001
2 vs (1,3)	15.3228	1	15.3228	143.20	<0.0001

Analysis in Table 2 shows that all contrast were significant, because;

176

(i) F-value$_{ov}$ = 19.38 and F-value$_{ov}$ = 143.20, were greater than [F$_{0.01}$ (1, 51)]$_{cv}$ = 7.1595 (obtained from F-table). [*Where ov = observed value and cv = critical value*]. (ii) P-values for both F-value$_{ov}$ were found to be less than 0.0001 i.e. (p<0.0001), which was less than the set significance level i.e. α = 0.01.

Furthermore, analysis showed that there was a significant difference between aggregates and the contrast [1 vs 3] was marginally significant however, the F-value 143.20 for the contrast [2 vs (1, 3)] was more significant, so the second contrast hypothesis was rejected. This indicated that a 21 degree viewing angle level results in optimal operator performance (see Figure 3).

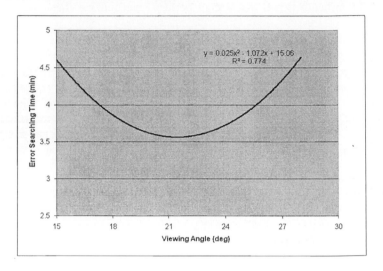

Figure 3. Graph showing the performance in terms of error searching time versus various levels of viewing angle.

DISCUSSION

World health organization (WHO) and Occupational Safety and Health Administration (OSHA) considers the cause of work related visual disorders and musculoskeletal diseases as multi-factorial. Factors such as work environment and the work performed are crucial from the ergonomic design point of view. Preferred term for conditions that are subjectively or objectively influenced or caused by the work is visual disorder. Many occupations are associated with a high risk of neck pain. Some risk factors can be identified, but the interaction between the factors is not much understood. It is important to recognize personal characteristics and other environmental and socio-cultural factors which usually play a key role in these disorders. Disorder and visual discomfort have been related to the VDU position and awkward posture. Factors which are considered to influence the activity of the eye muscles are horizontal distance between the worker and height of the VDU screen and the posture etc. (Westgaard et al., 1988). Present study was taken to develop a better understanding of the effect of viewing angle in a HMI environment. The CNC-EDM interaction system was targeted keeping in-view the exponential growth of the automation nowadays and the use of CNC machines in manufacturing and design. Therefore, the need of the moment is an efficient and effective ergonomic design of the CNC-workstations. Unorganized CNC machine working environment which does not meet the human capabilities is considered as a major source of stress and errors. Present work revealed that a 21 degree viewing angle gives the optimal performance as far as human-CNC machine interaction environment is concerned.

The above mentioned findings in some way or the other are similar to those obtained by some earlier investigators also. Kong-King Shieh et al., (2007), for example, found significant reductions in the eye muscle activity by modifying the workstation arrangement of an electronic paper displays. Dennerlein et al., (2006) based upon their study revealed that designing for the optimal configuration of a computer controlled workstation was necessary to eliminate the postural discomfort. In a VDU work environment, Svensson et al., (2001) found the optimum viewing angle which resulted lower load on the neck and shoulders. Also, Jan Seghers et al., (2003) explored that high viewing angle affects the activity of the neck extensor muscles. Hence it can be concluded that the visual factor play a key role in the effective and efficient ergonomic design of the human-CNC machine interaction environment.

CONCLUSIONS

Present study demonstrated that the viewing angle has a marked effect on the operator's performance.

On the basis of this study, the following concluding remarks are drawn:

(i) The level of viewing angle has a significant effect on the performance of CNC-EDM operators.

(ii) Findings of this work indicate that CNC-EDM systems should be re-designed so as to achieve a 21 degree viewing angle for optimal performance.

The finding of this work can be directly applied to the practical field which will improve the design of a CNC-EDM system. This work suggests that those responsible for the function and operation of CNC-EDM workstations would have to redesign the system to reduce injuries, as far as visual, musculoskeletal and other related problems are concerned.

The present result is very important for the system designers of tomorrow. It is expected that more studies would be undertaken in this regard in near future and the new human-CNC machine interaction systems would be designed accordingly.

REFERENCES

Ayako Takata, Hiroshi Jonaia, Maria Beatriz G. Villanueva, Midori Sotoyamaa and Susumu Saito. (2002), Effects of the liquid crystal display tilt angle of a notebook computer on posture, muscle activities and somatic complaints. International journal of industrial ergonomics, 29(4), 219-229.

Batten D.M., Schultz K.L. and Sluchak T.J. (1998), Optimal viewing angle for touch screen displays: Is there such a thing? International journal of industrial ergonomics, 22(4-5), 343-350.

Burgess-Limerick, Robin, Mon-Williams, Mark, Coppard and Vanessa L. (2000), Visual Display Height, The Journal of the Human Factors and Ergonomics Society, 42, 140-150.

Dennerlein J.K. and Johnson P.W. (2006), Changes in upper extremity biomechanics across different mouse positions in a computer workstation, Ergonomics, 49, 1456-1469.

Fraser K., R. Burgess-Limerick R., A. Plooy A. and D.R. Ankrum D.R. (1999), The influence of computer monitor height on head and neck posture, International journal of industrial ergonomics, 23(3), 171-179.

Jan Seghers, Arnaud Jochem and Arthur Spaepen (2003), Posture, muscle activity and muscle fatigue in prolonged VDT work at different screen height settings, Ergonomics, 46, 714- 730.

Kong-King Shieh and Der-Song Lee. (2007), Preferred viewing distance and screen angle of electronic paper displays, Applied Ergonomics, 38 (5), 601-608.

Mon-Williams M., Pooly A., Burgess-Limerick R. and Wann J. (1998), Gaze angle: a possible mechanism of visual stress in virtual, Ergonomics, 41(3), 280-285.

Svensson H.F. and Svensson O.K. (2001), The influence of the viewing angle on neck-load during work with video display units, Journal of Rehabilitation Medicine, 33, 133 – 136.

Westgaard R.H., Aaras A. and Stranden E. (1988), Postural angles as an indicator of postural load and muscular injury in occupational work situations, Ergonomics, 31(6), 915-933.

Winer B.J. (1971), Statistical principles in experimental design. 2nd edition, Tokyo: Mc Graw-Hill Kogakusha Ltd.

Chapter 19

User-Centered and Adaptive Support Systems for Production Systems

Jeronimo Dzaack, Bo Höge, Matthias Rötting

Chair of Human-Machine Systems
Department of Psychology and Ergonomics
Technische Universität Berlin
Franklinstr. 28-29, 10587 Berlin, Germany

ABSTRACT

Innovative production systems are complex and dynamic systems incorporating human operators and technical systems. Human operators are crucial parts and highly contribute to their resilience. Thus, it is necessary to integrate adaptive, contextual and user-centric interaction techniques into production systems. In this article an approach and a framework are presented that enable the predictive and automatic detection of human errors and malpractice and their contextual prevention. Therefore cognitive user models, actual knowledge of the systems and the operator and multimodal human-machine interacting mechanisms are used.

Keywords: Cognitive User Models, Ergonomics, Error Prevention, Interaction, Multimodality, Production Systems

INTRODUCTION

Innovative production systems are complex as well as dynamic environments incorporating human operators and technical components. In these socio-technical systems human operators are crucial parts and highly contribute to their resilience. Thus, it is important to provide methods to increase human reliability and performance. Both can be affected by many factors such as age, state of mind, emotions and propensity for common mistakes, errors and cognitive biases. A promising approach to overcome human errors and malpractice is to provide interaction methods that take into account human factors and that adapt to human errors or provide contextual support in crucial situations. This becomes more and more important because future production systems comprise not only products but also services (i.e. industrial product-service systems; Meier, Uhlmann & Kortmann, 2005). This development leads to more uncertainty of human operators within production systems due to changing working conditions, flexible action patterns and distributed knowledge within workflows and of processes.

In this article we present a theoretical approach leading to a practical framework that enables the predictive and automatic detection of human errors and malpractice (e.g. errors of emission or confusion) and their contextual prevention by either adapting the technical system or by providing multimodal support methods.

Approach

For user-centered and adaptive support of human operators within production systems we developed a theoretical approach that incorporates theoretical theories and practical applications. Cognitive user models (i.e. optimal user models) are applied to simulate operator behavior in a formal way that is compared with user behavior in real-time to detect irregularities between optimal and real behavior. Detection leads to the estimation of future system states. If a critical situation is predicted suitable actions are initiated (e.g. assistance functions, adaptation of the technical system). Additionally, our approach enables human operators to access online help and contextual support that is provided by external experts through state-of-the-art service devices (e.g. head-up displays, augmented reality components).

Three concepts are used within this approach: (1) cognitive user models, (2) condition-based regulations and (3) technical user support. In the following we give a short overview of these concepts.

COGNITIVE USER MODELS

Cognitive modeling attempts to provide symbol structures for selected cognitive processes and attempts to show that these symbol structures can generate the corresponding cognitive behavior (Tack, 1995). Modeling can be done within cognitive architectures that describe fixed mechanisms of cognition across uses and across tasks (Anderson, 2007) and are typically implemented in a computer program. Within cognitive architectures cognitive psychology concepts such as visual information processing and decision-making are used to simulate cognitive behavior. For an overview, see Pew & Mavor (1998).

Cognitive user models can be applied to predict the users' behavior and future needs by simulating observable user behavior. In the case of operator tasks cognitive user models are able to simulate operator behavior for specific tasks (Dzaack, 2008). Most cognitive user models are applied to predict user behavior offline. These models are used e.g. to evaluate the usability of designs (Jürgensohn. 2002). In the context of adaptive user interfaces in some cases cognitive user models are applied to predict emotions or workload of humans interacting with a technical system (Conati, 2002; Li & Ji, 2005).

Actual objectives in cognitive modeling are to simplify the model-building process, to provide concepts for sharing and reusing model components and to integrate emotions (Ritter, 2009).

CONDITION-BASED REGULATIONS

To err is human, so errors can never be totally avoided. But with efforts consequences of errors can be minimized, systems can be stopped into a failsafe mode and users can be assisted and advised before or close to the moment of making an error.

There exist already many successful concepts for avoiding errors while interacting with machines or computers, e.g. technical systems that control the user's inputs and proof them for correctness and appropriateness in the current context. This can be found in software applications to avoid false inputs for example in text forms or in a more complex way in fly-by-wire systems within aircrafts. In these systems steering inputs are controlled by control laws that limit the set of commands depending on the situation to prevent undesirable conditions of the aircraft (Favre, 1996).

Daily life condition-based warnings and notifications are known from all kind of systems, e.g. in the automotive sector. Navigation systems give warnings regarding the speed limited by comparing the movement of the GPS-sensor with the internal map data. To inform the user several mechanisms can be used such as a vibrating steering wheel, displays and spoken information. Assisting systems can already adapt to the driver's condition to reduce potential errors due to workload (Bachfischer, Bohnenberger, Hoffmann, Wäller & Wu, 2007). In the automotive sector also approaches exist for a context-sensitive error management during multimodal interaction with car infotainment and communication applications (McGlaun, Lang, Rigoll & Althoff, 2004).

Most of the systems to avoid errors or correct human behavior use condition-based regulations but do not take into account complex models of user behavior to adapt a system to the cognitive state or the demands of a user.

TECHNICAL USER SUPPORT

Especially in production systems user support plays a decisive role. The possibility to change aspects of the system (e.g. use models, personnel) over the full lifecycle increases the demand for adaptive and adaptable user support. Our understanding of support begins with printed manuals and ends with an interactive, individual, multimodal remote support. Between these extremes, digital versions of manuals, interactive video instructions and support by telecommunication (teleservice) can be found (Jaschinski et al., 1997). The usage of these support technologies leads to a closer relationship of customer and provider where the human being plays a decisive role (Massberg et al., 2000).

But until today, support by telephone seems to be the most used form of user support. Even experienced on-site service teams sometimes have to contact experts of the manufacturer. This was found in interviews with employees of 18 companies (e.g., Siemens, Eurocopter, ICH Merwede, Danisco) with the focus on the question, which kind of remote support they offer for their customers or which kind of remote support they request from machine manufacturers (Höge et al., 2009).

Adding a shared view to the verbal telecommunication, so that remote expert and on-site technician have a view onto the problem combined with eye tracking technology could possibly lead to a more intuitive, efficient and effective way of remote support.

Shared-Vision System

Based on the findings of Velichkovsky (1995), Brennan (2007) and Sadasivan et al. (2005) related to the supportive effects of gaze awareness in collaborative problem solving tasks, we have implemented full gaze awareness into a support system which we call shared-vision system. The system enables a novice and an expert to mutually recognize the visual focus represented by the gaze position. It also enables a shared view for both interacting communication partners onto the workspace of the novice.

FIGURE 1 The wearable part of the shared-vision system is based on a Liteye LE-750A-S head-mounted display and a SMI iView X HED eye tracker which are both mounted onto a headband. The head-worn eye tracking unit is placed in front of the user's left eye, the HMD is placed in front of the user's right eye. Above the HMD a digital scene camera is mounted.

The shared-vision system contains two parts, (1) a wearable device for the user and (2) a stationary device for the remote specialist. The wearable part of the shared-vision system is based on a head mounted display (HMD) and a head-worn eye tracking unit with a scene camera for the novice (see Figure 1). The objective of the HMD is to place information directly into the field of view of the user. The stationary part is based on a remote eye tracking unit which is mounted below a monitor and used by the expert. For verbal communication, expert and novice use headsets. Both parts of the shared-vision system are driven by personal computers which are connected by internet protocols.

For the application, the novice is equipped with the head-mounted system. He is connected by the internet with the expert, who is sitting in front of a PC monitor (see Figure 2). The scene camera of the novice shows the view onto the machine to the expert as well as the gaze information which is measured by the head-mounted eye tracker. The expert can see the view and the gaze position of the novice on his screen. For supporting the novice, the gaze information of the expert on his monitor are recorded by a remote eye tracker and transmitted to the novice's head-mounted optical see-through display. In this display the novice can see the gaze position of the expert augmented to his reality.

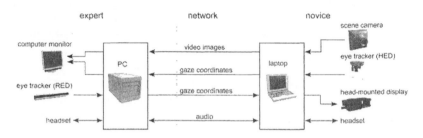

FIGURE 2 Schematic of the shared-vision system. Mobile and stationary parts are connected by internet protocols.

FRAMEWORK

The aim of the conceptual framework is to establish the fundamentals for contextual support of human operators within production systems taking into account human cognition and human factors. The contextual framework consists of two consecutive steps: (1) benchmarking the actual user behavior and (2) choosing an appropriate support mechanism for the human operator in a case of a potential future error. The conceptual framework is illustrated in Figure 3.

BENCHMARKING USER BEHAVIOR

The process of benchmarking user behavior is subdivided into three steps: (1) data gathering, (2) comparison of human and simulated data and finally (3) the benchmarking of differences in respect to the actual system state. In the following these three steps are described in more detail.

In the first step required data is gathered. Human data can be gained by using motion or gesture tracking systems. Technical data can be provided by sensors that are already integrated parts of technical systems or added for this purpose (e.g. sensors for process stability, target values and goal-orientation). For the simulation of operator behavior cognitive user models are used. These cognitive user models provide quantitative and theoretical data of ideal operators for specific tasks (e.g. eye movements, execution times). All data is processed and characteristics are extracted from both the online data and the simulated data (e.g. sequences, times, cornerstones) that are provided in a general-purpose format for ongoing analyses.

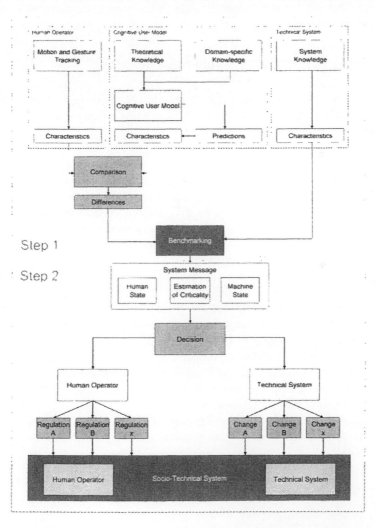

FIGURE 3   Conceptual framework for user-centered and adaptive support systems

within production systems.

The second step deals with the comparison of the simulated and the online operator data. Abstracted data of the cognitive user model is compared to online user data in real-time to reveal differences. These can occur for instance on the level of execution times or action sequences.

The last step comprises the benchmarking of the differences in context to the actual system state. By using machine learning algorithms pattern can be recognized in the data that allow forecasting critical situations or states of the production system. If a precondition for a critical situation is detected a message is generated that contains information on the machine state, the estimation about how critical the current situation is and the captured state of the human operator and is processed to the subsequent step.

CHOOSING AN APPROPRIATE SUPPORT MECHANISM

The second step of the framework deals with the choice of an appropriate support mechanism for the human operator. This consists of three steps: (1) deciding to assign the action to the human or the technical system, (2) definition of the granularity of the action and (3) the accomplishment of an action.

In the first step the incoming message of the pre-located step is analyzed by a rule-based decision making component. It has to be decided to assist the user, to change the machine state or both. For example if the situation is estimated critical, the machine could be stopped and the user informed about the reason.

On the second level the complexity and granularity of the future action should be made. On the human side decisions have to be made regarding the user's situation, knowledge, state and equipment. It has to be determined which modality is possible and optimal for the current state to inform or warn the user. If the user wears a head-mounted display, visual text warnings could be shown or augmented reality components could be displayed. On the machine side the processes regarding the machining can be adapted by changing the order, by re-organization the system state or stopping running processes and actions.

The third level facilitates the initiation of an action. Regulating notifications on one or even on both sides lead to a new state of the socio-technical system and thus affect both the human operator and the technical component that are computed within the first step of the framework.

Part of the research for the realization of the conceptual framework is also the analysis of possible dependencies between the human-operator and the technical components during the regulation. Another focus lies on the selection of corresponding support systems and in the design of the regulations for a multimodal support concept.

DISCUSSION

The presented approach enables to assist the human user or to change the machine state depending on the condition of the holistic socio-technical system. Prevention of and assistance in malpractice has several advantages for production systems: (1) the operator is protected and supported by a surrounding system and can be stopped by technical components before injuring himself or anybody else. (2) By rule of logic everything is possible after an error occurs, so the user can even do more mistakes and even impair the situation. By providing online benchmarking mechanisms all actions can be assessed and avoided by the technical system if necessary. And (3) interruptions due to human errors of planned processes, which lower the productivity and stability of processes, can be avoided by contextual and human-centric support. Considered as a whole, it allows integrating human cognition and human factors into production systems.

At the current state different parts of the theoretical framework are developed including cognitive user models, software and hardware frameworks as well as algorithms to gain, compare and benchmark data from different sources.

An open question is on which level of detail the comparisons between human and cognitive user model should be made. And it has to be shown that characteristics derived from theoretical knowledge are comparable to human data. First approaches for multilevel data analysis and comparison of cognitive user models can be found in Dzaack & Urbas (2009). A second problem is the huge variety of users performing a task (i.e. intra- and interpersonal differences). To overcome this problem, a solution could be to provide a set of cognitive user models for a task and to choose the one with the best fit for the comparison.

One very important challenge is the data comparison in real-time due to the large data sets and the complex structure of human cognition. A solution could be to use simple matching algorithms or to predict several user behaviors in advance. In latter, the predictions and its preconditions can be mapped to the current system and user state in order to decide an appropriate simulation as the case arises.

A weak point of our approach is that only notifications or support information can be provided in conditions which are already known or which are supposed to occur. To overcome this problem a learning mechanism could be integrated into the conceptual framework. Such a learning algorithm should feed backward the initiated action and its utility for the given context, the situation and the success to the cognitive user model and the benchmarking module. The cognitive user model could use the information to update its knowledge base (i.e. declarative and procedural memory). Analog the benchmarking module could update its knowledge base.

SUMMARY

The described approach enables a user-centric and contextual interaction within production systems. We provide a theoretical base for the benchmarking of operator behavior by cognitive user models and a mechanism to initiate appropriate support interventions whenever mismatches are revealed between the human and simulated behavior. The focus lies more on the assistance of the human operator than in supervision since controlling or even the feeling of it can have a negative impact on human operators.

All described concepts and theoretical accounts of this paper will be implemented, tested and evaluated in a scenario which is used in the collaborative research project SFB/TR29, i.e. manufacturing of components for watches by micro production.

The introduced approach cannot replace conventional and existing efforts to minimize malpractice of user interaction. But this approach can lead to a higher productivity of a socio-technical system and offers a high flexibility and mutability of human and technical parts. The approach also does not lead to error-free interactions of operators but will probably lead to more error-tolerant systems. We think that this approach clears the way for new methods and instruments for user-centered and user-involving design in production systems.

ACKNOWLEDGMENTS

We express our sincere thanks to the German Research Foundation (DFG, Deutsche Forschungsgemeinschaft) for funding this research within the Collaborative Research Project SFB/TR29 on Industrial Product-Service Systems – dynamic interdependency of products and services in the production area.

REFERENCES

Anderson, J. R. (2007), *How Can the Human Mind Occur in the Physical Universe?* New York: Oxford University Press.

Bachfischer, K., Bohnenberger, T., Hofmann, M., Wäller, C. & Wu, Y. (2007), Kontextadaptive Fahrerinformationssysteme am Beispiel eines Navigationssystems. *KI - Künstliche Intelligenz*, 3, 57-63.

Brennan, S.E., Chen, X., Dickinson, C.A., Neider, M.B. & Zelinsky, G.J. (2007). Coordinating cognition: The costs and benefits of shared gaze during collaborative search. In: Cognition, 106 (2008), (pp. 1465–1477)

Conati, C. (2002), Probabilistic assessment of user's emotions during the interaction with educational games. *Applied Artificial Intelligence*, 16, 555–575.

Dzaack, J. (2008), *Analyse kognitiver Benutzermodelle für die Evaluation von Mensch-Maschine-Systemen*. Dissertation. Technische Universität Berlin.

Dzaack, J. & Urabs, L. (2009), Multilevel Analysis of Human Performance Models in Safety-Critical Systems. *Proceedings of HCI International 2009 (DVD)*, San Diego.

Favre, C. (1996), Fly-by-wire for commercial aircraft: the Airbus experience. In:

Mark B. Tischler (Ed.) *Advances in Aircraft Flight Control* (pp 212-216). London: Taylor & Francis.

Höge, B., Schlatow, S., & Rötting, M. (2009). Geteilter Sichtbereich und Kommunikation durch Blickbewegungen als intuitive und effektive Unterstützung beim kollaborativen Problemlösen. In: M. Grandt & A. Bauch (Hrsg.) Kooperative Arbeitsprozesse. DGLR-Bericht 2009-02. (pp.349-364) DGLR e.V., Bonn.

Jaschinski , C., Spiess, M., May, S. (1997): TeleService - Eine Einführung, Forschungszentrum für Rationalisierung an der RWTII, Aachen, Sonderdruck 13/97

Jürgensohn, T. (2002), Bedienermodellierung. In: K.-P. Timpe, T. Jürgensohn & H. Kolrep (Eds.), *Mensch-Maschine-Systemtechnik – Konzepte, Modellierung, Gestaltung, Evaluation* (pp. 107–148). Düsseldorf: Symposion Publishing.

Li, X., Ji, Q. (2005), Active affective state detection and user assistance with dynamic bayesian networks. *IEEE Transactions on Systems, Man and Cybernetics*, 35, 93-105.

Massberg, W., Hermsen, M., Zuther, M., 2000, TELec – Multimedialer Teleservice: Technik – Organisation – Vermarktung – Erfahrungsberichte. Schriftenreihe des Lehrstuhls für Produktionssysteme, ISBN 3-8265-7408-7, Aachen: Shaker, 2000.

McGlaun, G., Lang, M., Rigoll, G. & Althoff, F. (2004), Kontextsensitives Fehlermanagement bei multimodaler Interaktion mit Infotainment- und Kommunikationseinrichtungen im Fahrzeug. *Useware 2004: Nutzergerechte Gestaltung Technischer Systeme*, Düsseldorf: VDI-Berichte Nr. 1837, 57-65.

Meier, H., Uhlmann, E., Kortmann, D. (2005), Hybride Leistungsbündel - Nutzenorientiertes Produktverständnis durch interferierende Sach- und Dienstleistungen. *wt Werkstatt-Technik online*, 95(7), 528-532.

Pew, R. W., Mavor, A. S. (1998), *Modeling Human and Organizational Behavior: Application to Military Simulations*. National Academic Press, Washington D.C.

Ritter, E. F. (2009). Two cognitive modeling frontiers. *Transactions of the Japanese Society for Artificial Intelligence*, 24(2), 241-249.

Sadasivan, S.; Greenstein, J.S.; Gramopadhye, A.K. & Duchowski, A.T. (2005). Use of Eye Movements as Feedforward Training for a Synthetic Aircraft Inspection Task. In: Proceedings of CHI 2005, (pp. 141-149), New York: ACM Press.

Tack, H. W. (1995). Wege zu einer differentiellen kognitiven Psychologie. In: *Bericht über den 39. Kongress, der Deutschen Gesellschaft für Psychologie in Hamburg* (pp. 172-185). Göttingen: Hogrefe.

Velichkovsky, B.M. (1995). Communicating attention: Gaze position transfer in cooperative problem solving. In: *Pragmatics & Cognition*, Vol. 3(2), (pp. 199-222).

CHAPTER 20

Design of a Knowledge Module Embedded in a Framework for a Cognitive System Using the Example of Assembly Tasks

Eckart Hauck, Daniel Ewert, Daniel Schilberg, Sabina Jeschke
ZLW - Center for Learning and Knowledge Management &
IMA - Institute of Information Management in Mechanical Engineering
RWTH Aachen University
Aachen, Germany

ABSTRACT

Highly automated manufacturing systems are currently limited in their capabilities to adapt to changes in a product. They rely heavily on pre-programmed execution cycles. The introduction of cognitive abilities into manufacturing systems is a promising approach for improving the flexibility of these systems. In this paper a framework is proposed with supports the design and implementation of cognitive systems in a production environment. Furthermore the design of a knowledge compiler and the structure of the knowledge base will be introduced and a description of the use case in the assembly domain is given.

INTRODUCTION

A manufacturing company in a high-wage country has to position itself in two dilemmata. The first dilemma lies between mass production with a very limited product range (scale) and the manufacturing of a large variety of products in small quantities (scope). The second dilemma involves the dichotomy between planning and value orientation. The vertices of this dilemma shape the so-called polylemma of production technology (FIGURE 1) (Brecher, 2007). In order to create a lasting competitive advantage over low-wage countries, it is necessary to develop methods to reduce or dissolve the polylemma of production. With the intent to satisfy customer's desires by offering a customized product at mass production costs, it is not sufficient to optimize just a singular vertex. Every change of one vertex affects also the corresponding vertex.

The Cluster of Excellence "Integrative Production Technology for High-Wage Countries" at the RWTH Aachen University is engaged in the reduction of the mentioned polylemma. In the integrative cluster domain (ICD) "Self-optimizing Production Systems", the issue of the second dilemma is discussed.

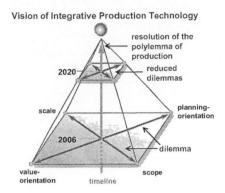

FIGURE 1 Polylemma of Production Technology

As a prerequisite for self optimizing behavior cognitive capabilities are essential (Hauck, 2008). Therefore the development of a system enacting cognitive behavior is one of the objectives in the ICD.

The paper is structured as follows: First the cognitive framework with the different software layers and the communication flows will be introduced. Then the knowledge compiler which is an important part of the framework will be described in more detail followed by the use case for such a system. Finally a conclusion and a perspective for future research will be given.

COGNITIVE FRAMEWORK

Cognitive systems which are suitable for production processes have to meet many functional and non-functional requirements (Hauck 2010) which have to be satisfied by the software architecture

The system has to work on different levels of abstraction. This means that the reasoning mechanism cannot work on the raw sensor data received from the production system. This demands a component which can aggregate the received information for the reasoning mechanism. To meet these requirements a multilayer approach for the software architecture of the system was chosen (Gat 1998).

FIGURE 2 shows the software architecture. The software architecture is separated into four layers which incorporate the different mechanisms required. The Presentation Layer incorporates the human machine interface and an interface for the modification of the knowledge base. The Planning Layer is the deliberative layer in which the actual decision for the next action is made. The Coordination Layer provides services to the Planning Layer which can be invoked by the latter to start action execution. The Reactive Layer is responsible for a low response time of the whole system in case of an emergency situation. The Knowledge Module contains the necessary domain knowledge of the system.

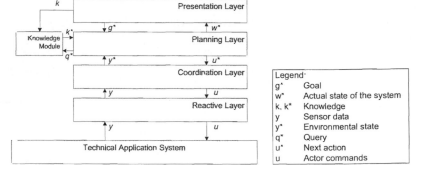

FIGURE 2 Software architecture of the cognitive system

At the beginning the human operator gives the desired goal g* to the Cognitive System via the Presentation Layer. This goal is then transferred to the Planning Layer where the reasoning component derives the next action u* based on the actual environmental state y* and the desired goal g*. The environmental state is based on the measured variables y from the sensors in the Technological Application System. In the Coordination Layer the raw sensor data y is aggregated to an environmental state y*. This means that the base on which all decisions in the Planning Layer are made is the environmental state y* at a given time. Therefore the decision process must not take too long, because the state of the Technological Application System can have changed significantly in the meantime.

The next best action u* derived in the Planning Layer is sent back to the Coordination Layer, where the abstract description of the next best action u* is translated into a sequence of actor commands u, which are sent to the Technological Application System. There, the sequence of commands is executed and the changed environmental state is measured again by the sensors. If the new measured variables y of the Technological Application System indicates an emergency situation the Reactive Layer processes the sensor data directly and sends the according actor commands to the Technological Application System.

The different layers of the software architectures are described in the following section. FIGURE 3 shows the software architecture and their components in more detail.

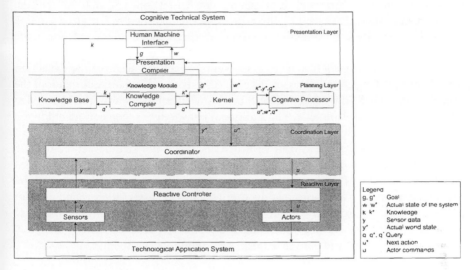

FIGURE 3 Software architecture with components

Presentation Layer

The Presentation Layer is responsible for the human machine interaction. It incorporates the human-machine-interface which is designed to meet the requirements of the interaction of such a system.

The operator specifies a goal description g, which is transferred to the Presentation Compiler. In case of an assembly task the description g can be the shape of parts to be assembled and the description of the location of the parts in the final assembly. This can be done, but is not restricted to, using a graphical representation, e. g. a CAD program. The Presentation Compiler has to translate this description g into g* which can be interpreted by the Cognitive Processor of the Planning Layer.

Due to constant changes in the environment the behavior of a cognitive system is not perfectly predictable in advance. Therefore, the actual state of the system should always be transparent to the user. The system state w* is given to the Presentation Compiler, where w* is aggregated to a human interpretable machine feedback w.

Planning Layer

The Planning Layer contains the core elements that are responsible for decision-finding and in the case of assembly planning in deriving an assembly sequence for a certain product. It contains the Kernel and the Cognitive Processor as components. The Kernel distributes the signal flows in the Planning Layer. The Cognitive Processor computes the next best action u^* based on g^* and the current environmental state y^*. If the Cognitive Processor cannot derive a next best action or cannot generate an assembly sequence it can send a query q^* for more information to the Knowledge Module.

The Kernel component then invokes the action execution according to the action returned by the Cognitive Processor. In case of a request for more information, the Kernel queries the Knowledge Module. The Knowledge Base returns the knowledge k^* via the Knowledge Compiler. The additional knowledge is then considered in the computation of the next best action or the next assembly step.

In the rare case that the Cognitive Processor could not find an action and the Knowledge Base could not return any knowledge, the Cognitive Processor queries the human operator. The user can then either give the next action or change the environmental state. This means that the user changes the assembly by hand without telling the system explicitly about this. The system then recognizes the new environmental state via the measured variables y, reasons about the new environmental state y^* and derives the next best action u^* based on y^*.

Coordination Layer

The Coordination Layer is the executing layer of the Cognitive System. It provides executable services to the Planning Layer. These services provide the actions the Cognitive Processor can use to generate the assembly sequence. The Coordinator has two responsibilities. It contains the different services and it processes the measured variables y received from the Reactive Controller in the Reactive Layer and aggregates this information to the current environmental state y^*.

Also, the Coordinator component receives the next action or a sequence of actions u^* to be executed by the Technological Application System. Every action u^* is a sequence of actor commands u. An example for an abstract action u* is a simple translational movement of a component and the subsequent assembly to another component. The action u* could be move(componentA, componentB). The service then sends a sequence of primitive movement commands to the actor, e. g. a robot. The sequence u is stored in the services in the Coordinator component and will be executed with the parameters given by u^*. u is then executed in the Technological Application System via the Reactive Layer. That way, the Planning Layer is exculpated from the details of the robot movements, e. g. the exact coordinates of the components-locations. The components providing the services can be agents based on cognitive architectures like Soar (Laird 1996). These agents can also handle non emergency exceptions like a misplacement of a component or other incidents.

Reactive Layer

The Reactive Layer and in it the component Reactive Controller is responsible for the low level control of the system. The vector of the measured variables y is observed for values which indicate a possible emergency situation. The Reactive Controller responds with the according actor commands u to either stop the whole system or to bring the actors in a safe position. This ensures low response times in case of an emergency.

Knowledge Module

The Knowledge Module contains the Knowledge Base with the necessary domain knowledge for the System to perform the assembly tasks. The domain knowledge k in the Knowledge Base has to be translated in a form which is interpretable by the Cognitive Processor. This is the responsibility of the Knowledge Compiler which is separated in two components. The first is the Reasoner component and the second is the Mediator (FIGURE 4).

The Reasoner queries the Knowledge Base and extracts the needed knowledge k out of the knowledge base. This knowledge is then translated into an intermediate format k' and transferred to the Mediator. The Mediator then compiles the knowledge k' into the syntax $k*$ which is then processed by the Cognitive Processor. FIGURE 4 shows the signal flows and the involved components. In case of an additional information request $q*$ by the Cognitive Processor the Mediator first translates $q*$ in q' and the Reasoner accesses the Knowledge Base to infer the requested information.

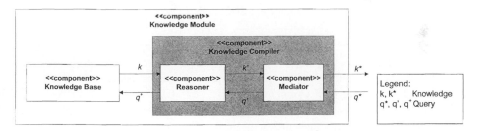

FIGURE 4 Knowledge Module with the Reasoner and Mediator component

DESIGN OF A KNOWLEDGE COMPILER

The Knowledge Compiler and its components work on the Knowledge Base which contains the domain knowledge for assembly tasks. The ontology of the assembly domain has four upper concepts. These are the concepts "Services", "Resources", "Objects" and "Assemblies". FIGURE 5 shows the structure and the

196

relations between the concepts. The concept "Services" has the two subconcepts "Operation" and "Action". An "Action" can be a primitive movement like a lateral transition of an "Object" or a transformation of an assembly state in another. multiple actions can be combined to an Operation. The "Resources" concept has one subconcept "Tools" which represent the different manipulators which can be used to enact different operations on "Objects". These manipulators can be "Handling_Tools" or "Screwing_Tools". The "Tools" described in these concepts have attributes describing their abilities. The "Objects" concept contains the "Composite" and "Parts" concepts. Both are connected by "Liaisons". All "Composites" consists of different Parts. The "Liaisons" contains the different relations which can exist between different "Parts" and "Composites". The last upper concept "Assemblies" contains whole assemblies and the assembly sequence which can be used in later processes. The formalism used for the domain knowledge is the Web Ontology Language (OWL) with the expressivity of a description language (Smith 2004).

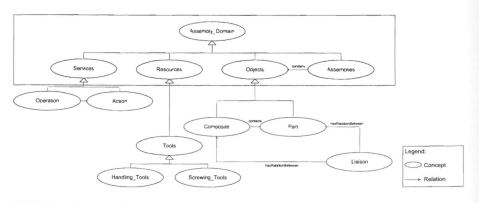

FIGURE 5 Structure of the ontology of assembly domain

The procedural knowledge which describes possible actions is stored in a description which ensures a direct representation of this knowledge in the ontology. The formalism supports the representation of rule based knowledge in the expressivity of full First Order Logic (FOL). FIGURE 6 shows the structure of the formalism embedded in the ontology by the example of a simple movement operation involving three different parts. The relations between the different individuals in the ontology are labeled with keywords. These are hasPosPrecondition, hasNegPrecondition, hasPosEffect, hasNegEffect, has1stParameter and has2ndParameter. With these keywords the algorithm can extract the description of the operation for the cognitive control elements. The predicates used in such a description, like onTop and clear are stored in a different concept which is not shown in the picture.

The algorithm used for the compilation of the knowledge searches the ontology for the keywords and builds the rules out of this representation. The compiler uses the OWL-API and the FACT++ Reasoner for queries of the knowledge base.

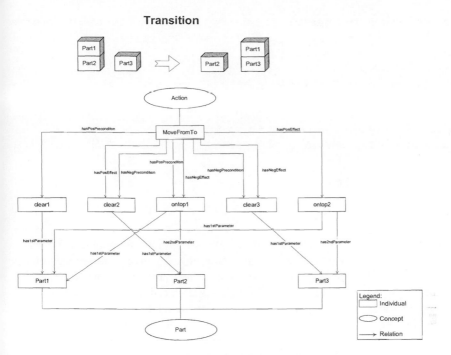

FIGURE 6 Structure for modeling procedural knowledge in the ontology

In this way the procedural knowledge can be used in a variety of agents, which work on rule based knowledge representation. The Soar architecture is one example. In case a new action has to be stored in the ontology the mediator translates the action in the described structure and stores it as a new action.

USE CASE ASSEMBLY PROCESS

The use case for the Cognitive System is the assembly of a product which is described by its CAD data. The system then plans and executes the assembly autonomously as far as possible with the robotic movements (Ewert et al. 2010). If some operations cannot be executed by the Cognitive System it cooperates with human operators. FIGURE 7 shows the schematic of the assembly cell. The cell consists of two robots and four conveyors belts whose velocity can be controlled separately by the cognitive system. In addition to that several light sensors and photo sensors for the color recognition are integrated in the cell.

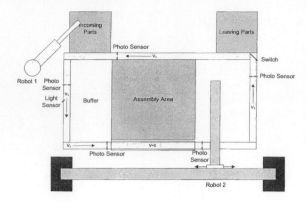

FIGURE 7 Schematic of the assembly cell controlled by the cognitive control unit

The workflow of the system is as follows. The first robot puts new parts randomly on the conveyor belt from the Incoming Parts area. The second robot, which is controlled by the cognitive system can either put the next part in the buffer or into the assembly area. In case the next part is not needed for the assembly it is transported to the Leaving Parts area. Because of the random feeding of the parts the assembly sequence has to be adapted continuously by the cognitive system. The next assembly step is chosen and executed according to the current assembly status. If the next action has to be executed by a human operator the system sends a signal to the operator. It waits then until the operation is done and continues with the sequence. Due to the monitoring of the assembly area the system can react to unexpected events like changes in the assembly.

RELATED WORK

In autonomous robots many architectural approaches for frameworks for technical systems enacting cognitive behavior are proposed (Karim 2006,, Gat 1998 et al.). These software architectures focus on the combination of a deliberative part for the actual planning process with a reactive part for motion control. Additional research for Cognitive Systems in production environments is done by Ding et al. 2008. Their research focuses on the implementation of cognitive abilities in safety controllers for plant control. In this context the human machine cooperation is the main evaluation scenario. All described approaches do not focus on the application of Cognitive Technical Systems in assembly operations.

In the context of the combination of procedural and declarative knowledge and the usage of this knowledge in an autonomous agent like Soar the Onto2Soar software is one approach. This approach differs from the one used here in that Onto2Soar translates the working memory of a Soar agent into an ontology to allow exchange of declarative knowledge between several agents, whereas we store all procedural knowledge of the system within an ontology.

CONCLUSIONS

The proposed framework and the embedded knowledge compiler are a suitable approach for the implementation of a cognitive technical system for assembly tasks. With this framework systems can be designed which can reduce the polylemma of production technology by decreasing the set up times and the needed amount of programming in manufacturing. The underlying concepts are not limited to assembly tasks but can be used in all production related tasks. Future work will focus on the implementation of different task descriptions in the assembly domain and the improvement of the capabilities of the cognitive processor. A later research topic is also the control of multiple assembly robots and their coordination in a safe manner.

The authors would like to thank the German Research Foundation DFG for the support of the depicted research within the Cluster of Excellence "Integrative Production Technology for High-Wage Countries".

REFERENCES

Brecher, C. et al. (2007), Excellence in Production, Apprimus Verlag, Aachen.

Ding, H. et al. (2008). A Control Architecture for Safe Cognitive Systems, 10. Fachtagung Entwurf komplexer Automatisierungsysteme, Magdeburg.

Ewert, D. et al. (2010), "Cognitive Assembly Planning using state graphs", *Proceedings of the 3rd international Conference on Human Factors and Ergonomics*, Miami.

Gat, E. (1998), "On Three-Layer Architectures" in *Artificial Intelligence and Mobile Robots*, Kortenkamp D. , Bonnasso R., Murphy R., (Ed.), pp. 195-211

Hauck, E.; Gramatke, A.; Henning, K. (2010), A Software Architecture for Cognitive Technical Systems Suitable for an Assembly Task in a Production Environment, in-tech web.

Hauck, E.; Gramatke, A.; Henning, K. (2008). Cognitive technical systems in a production environment, *Proceeding of the 5th international Conference on Informatics in Control, Automation and Robotics*, pp. 108-113.

Karim, S. et al. (2006). A Hybrid Architecture Combining Reactive, Plan Execution and Reactive Learning, Proceedings of the 9th Biennial Pacific Rim International Conference on Artificial Intelligence (PRICAI), China, August 2006

Laird, J.E.; Lehman, J.F.; Rosenbloom P. (1996). A gentle introduction to Soar, an architecture for human cognition, In: *Invitation to Cognitive Science*, MIT Press, Boston

Smith, M. et al. (2004). *OWL Web Ontology Language Guide,* http://www.w3.org/-TR/2004/REC-owl-guide-20040210/*OWL Web Ontology Language Guide,* http://www.w3.org/TR/2004/REC-owl-guide-20040210

Wray, R., Lisse, S., Beard, J. (2004) Ontology infrastructure for execution-oriented autonomous agents in *Robotics and Autonomous Systems,*pp. 113–122.

CHAPTER 21

Cognitive Engineering for Human-Robot Interaction - The Effect of Subassemblies on Assembly Strategies

Barbara Odenthal, Marcel Ph. Mayer, Nicole Jochems,
Christopher M. Schlick

Institute of Industrial Engineering and Ergonomics
RWTH Aachen University
52064 Aachen, Germany

INTRODUCTION

Future manufacturing systems will include human operators and robotic systems working together in a cooperative way because automated systems are often neither efficient enough for small lot production (ideally one piece) nor flexible enough to handle products to be produced in a large number of variants. In these human-(multi) robot systems, the robot could take over repetitive and dangerous tasks which are not too complex whereas the human operator could manufacture the product variants that require human knowledge or skills.

Concerning human-robot cooperation (HRC) in today's manufacturing systems the guideline 98/37/EG requires a strict separation between robotic systems and the human operator. This separation must be either a local separation or a temporal separation (Thiemermann 2005). To be precise the term HRC in this context is misleading, for there is no cooperation in the narrow sense of the word. The robotic system as well as the human operator is processing tasks sequentially regarding one workpiece to be produced. Suspending the local separation by maintaining the temporal separation, a synchronized processing of tasks in the same working area would be possible. Suspending the temporal separation by maintaining the local separation would lead to a concurrent work on different workpieces. To realize a cooperation of human operator and robot both separations have to be suspended. In the following this is referred to as direct human-robot cooperation (dHRC).

To implement dHRC on the shop floor the human operator in flow production systems still has to follow the "Takt" (cycle) of the technical system. From an ergonomic point of view a desirable scenario would be the opposite: The technical system following an ergonomic motion cycle of the human with a superimposed flexible cycle. Therefore the technical system has to be flexible regarding dHRC.

To expand flexibility to the human-machine system as a whole, the capabilities of the technical system have to be enhanced towards cognitive functions like assembly procedure planning and adaption which are not implemented in today's production facilities (cognitive automation, see Onken & Schulte 2010). Therefore a research project has been established within the Cluster of Excellence "Integrative Production Technology for High Wage Countries" at the Faculty of Mechanical Engineering of RWTH Aachen University that aims at the development of a self-optimizing assembly cell enabling flexible automation with the possibility of dHRC.

Following the paradigm that a numerical control of a technical system based on simulated cognitive functions will lead to a better understanding by the human operator regarding the intention of the technical system, empirical studies are presented in the following. The studies aim at the identification of human assembly strategies as a foundation for later implementation of product rules in the rule base of the cognitive architecture of the numerical control.

COGNITIVE CONTROL UNIT

A novel design of the cell's numerical control forms the basis of the automation approach. The cell's numerical control – here referred to as cognitive control unit (CCU) – accomplishes high-level information processing based on the cognitive architecture SOAR (see Leiden et al. 2001). SOAR belongs to the class of computational models which, compared to emergent systems like artificial neuronal networks, do not require initial training hence changes to the knowledge base can be performed quickly, which is of great interest especially for an application in an industrial environment. To a certain extent the CCU is able to simulate rule-based behavior of the human operator.

In such a cognitive controlled production cell the CCU is able to autonomously decide which part of a given task can be processed by a given system resource. Therefore, the role of the human operator changes:

- In classic supervisory control of automated manufacturing systems the role of the human operator includes planning of what needs to be done, teaching the system, monitoring the system state and if necessary intervening in the system (Sheridan 2002).
- In a cognitive controlled production cell on the other hand a so-called master assembly schedule has to be developed by the human operator first. This master assembly schedule primarily consists of a formal description of the desired final state of the product to be assembled and a set of useful

heuristics to reach the goal. Then, the assembly procedure planning and teaching (in terms of RC programming) are carried out autonomously on the basis of stored process knowledge that is encoded in rules (Schlick et al. 2009).

Since the CCU is able to resolve a certain class of assembly problems autonomously and therefore significantly reliefs the operator from repetitive and monotonic task processing, classic approaches for human-centered automation proclaiming a continuous involvement of the operator (e.g. Billings 1991) are not very meaningful in self-optimizing assembly cells. In case of not autonomously solvable manufacturing problems or in case of human triggered intervention, the human operator needs to be in the loop as quickly as possible (compare Bainbridge 1987) and must have detailed knowledge about the state of the technical system, the already executed tasks, and the systems objectives and constraints to make good or even close to optimal decisions. To cope with this inherent complexity, an important basic requirement for the ergonomic design of the CCU is the conformity with operator's expectations, especially in low level decision-making related to the product to be assembled, where the human operator is not continuously involved.

To enhance the human centered approach, the question arises on how to design the symbolic representation of the knowledge base for the CCU to ensure the conformity with the operator's expectations. Proprietary programming languages that are used in conventional automation have to be learned case specific and do not necessarily match the mental model of the human operator. Focusing a human centered description to match the process knowledge to the mental model one promising approach in this special scenario is the use of motion description. These motions are familiar to human operators from manually performed tasks hence those are easier to anticipate than complex programming code. Already established methods or taxonomies e.g. from process planning can be used. Since in production systems complex handling tasks have to be broken down into fundamental elements, a promising approach here is the use of the MTM system as a library of fundamental movements, ignoring the underlying time information in a first step (Mayer et al. 2008).

The MTM based description of the assembly process is seen to be a first step to ensure conformity with the operator's expectations. An additional step is the consideration of human assembly strategies. To identify these strategies laboratory studies were carried out.

EXPERIMENTAL ASSEMBLY CELL

In order to review the aforementioned conceptual CCU in a realistic production environment, a robot supported assembly cell is currently being built at RWTH Aachen University (see Kempf et al. 2008). The cell design is shown in Figure 1. The assembly cell is fed by a conveyor belt system. Two robots are available, of which in a first step only one is connected to the CCU. The second robot takes over the separate task of feeding the components from a pallet onto the conveyor belt. A surface is located in the middle of the assembly cell that is used as both a work surface and temporary storage area (buffer) for non-direct production parts. At present the workstation of the operator is separated by an optical safety barrier. The workstation's multimodal human-machine interface informs e.g. via an augmented vision system about the system status, and may possibly assist in problem solving (Odenthal et al. 2009, Schlick et al. 2009).

Figure 1: Layout of an Assembly Cell

Due to their design, many industrially relevant components are limited in that their construction is only possible in a particular order, or that they only allow for a few procedural variations in their assembly (see Eversheim 1998). Therefore, these components do not seem as suitable objects to observe in order to demonstrate the potential of a cognitive automated system. To fully illustrate the concepts and methods, mountable assemblies were chosen that allow for a high variation of assembly sequences. Therefore, one of the requirements for the fundamental elements needed for producing components is that they can be arbitrarily configured and are similar in their functionality. LEGO building bricks, from the Danish company of the same name, fulfill this requirement and were thus used for the evaluation. The (semi-) symmetric geometries of the basic elements can still allow for complex assemblies to be constructed despite the simple assembly structures, and even allow for many assembly sequences. This is easily clarified by a simple example. Building a small pyramid with only 5 bricks and foundation of two by two bricks leads to 24 different assembly sequences.

LABORATORY STUDY

In empirical studies Mayer et al. (2010) could identify human assembly strategies on the basis of a simple monochrome pyramid made of LEGO bricks. Based on a study with 16 subjects who were given the task to build up one pyramid (30 bricks), three hypotheses regarding „human" assembly strategies were formulated. In a second experiment with 25 subjects who had to build up ten identical pyramids each, the identified assembly strategies were evaluated and successfully validated and hence formulated as rules. In the following the subject group of this second particular experiment is called Group 0.

Based on the assembly sequences of the subjects the following three basic assembly rules were identified:

1) The position of the first assembly brick lies in the left corner position of the subject's field of view. Should this rule be applied to other assemblies such as those with a rectangular cross-section, then preference should be to have the bricks in the edge position.

2) Bricks that can be positioned in the direct vicinity of other bricks (in the same layer) during a certain construction phase are chosen with the highest preference. This is hereafter referred to as the so-called 'adherence to the neighborhood relationship'.

3) The target object is assembled in layers, where the planes are parallel to the mounting surface.

The intention of the empirical study is to investigate the influence of subassemblies in a structure on the assembly strategies. In other words, if there are subassemblies present, do the subjects tend to built up in subassemblies? If this holds true, a possible rule could be:

4) If subassemblies are visible, the target object is built up completely in subassemblies.

EXPERIMENTAL DESIGN

In the experiment, the influence of subassemblies on the assembly strategy of a human operator should be investigated. Therefore, two laboratory studies were carried out in which different assembly objects should be assembled which differ in color or structure or both color and structure (see Figure 2).

The assembly model (Model 0) of Mayer et al. (2010) was a monochrome solid pyramid with no obvious subassemblies. In the experiment described in this paper two different models are used. One model (Model 2) is also monochrome but consists of five subassemblies. So it differs from Model 0 regarding its structure. The other model (Model 1) has the same structure as Model 2 but the subassemblies are emphasized by the use of different colors for each subassembly.

Task: Two easy models consisting of 25 bricks resp. five subassemblies have to be built up five times (see Figure 2). All required parts to build the pyramids and an exemplary model of the assembly task are positioned on an area within reaching distance for the subject. The subject has to follow the following constraints:

- No preassembly of substructures is allowed
- Only one brick can be picked at a time

- Only one hand can be used to built
- The model has to be assembled on a defined area

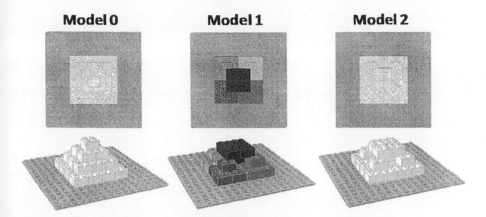

Model 0 **Model 1** **Model 2**

Figure 2: Model 0, Model 1 and Model 2 used in the assembly tasks

Procedure: The laboratory study was divided into two main phases. First, personal data were collected, e.g. age, profession as well as assembly skills relating to LEGO models. After the explanation of the assembly task and the assembly rules, the subject had to build the pyramid five times.

Subjects: A total of 18 subjects participated in the laboratory studies. The subjects were divided into two groups. The first group had to build the polychrome object (Model 1) and the second group had to build the monochrome object (Model 2).

- Group 1: The age of the subjects was between 20 and 32 years (mean 26.4, SD 3.5). The average experience in assembling LEGO models was 3.3 on a scale ranging from 0 (low) to 5 (high).
- Group 2: The age of the subjects was between 21 and 28 years (mean 23.8, SD 2.4). The average experience in assembling LEGO models was 1.5 on a scale ranging from 0 (low) to 5 (high).

HYPOTHESES

Based on the results described in Mayer et al. (2010), the following null-hypotheses can be formulated:

- H_{01}: (I) The relative frequency of the position of the first brick (edge position) in this series of experiments (Group 1 resp. Group 2) corresponds to the data collected in the test series distribution by Mayer et.al (2010) (Group 0).
  (II) The relative frequency of the position of the first brick (edge position) in Group 1 corresponds to the data collected in Group 2.
- H_{02}: (I) The structure of the neighboring relations in this series of

experiments (Group 1 resp. Group 2) occurs with the same relative frequency as in Group 0.

(II) The structure of the neighboring relations in Group 1 occurs with the same relative frequency as in Group 2.

- H_{03}: (I) The level design in this series of experiments (Group 1 resp. Group 2) occurs with the same relative frequency as in the Group 0.

(II) The level design in Group 1 occurs with the same relative frequency as in Group 2.

- H_{04}: The subassembly design observed in Group 1 occurs with the same relative frequency as in Group 2.

To verify the null hypotheses, the χ^2 test according to Pearson was applied with a significance level of $\alpha = 0.05$. The null hypothesis H_{02I} (Gr0-Gr2) and H_{03I} (Gr0-Gr2) cannot be tested by the χ^2-Test according to Pearson because the requirements for the test are not met according to Rasch et al. (2004). So that for testing H_{02I} (Gr0-Gr2) and H_{03I} (Gr0-Gr2) the Yates' χ 2-test was applied.

RESULTS

In Table 1 the relative frequencies of the empirical experiment are shown. The results regarding H_{04} are referred to as Rule 4 following the notation of the experiments in Mayer et al. (2010).

Table 1: Relative frequencies of rules in laboratory study

	Rule 1	Rule 2	Rule 3	Rule 4
Group 0	98.8% (edge)	91.2%	97.2%	---
Group 1	100% (edge)	12.5%	12.5%	66.7%
Group 2	100% (edge)	100%	100%	0%

The results of the χ 2-test according to Pearson resp. the Yates' χ 2-test are also given in Table 2.

For H_{01} (edge position of the first brick), one can see from the second column of Table 1 (Rule 1) that there is nearly the same value for the relative frequencies. Hence the rule can be empirically confirmed.

For H_{02} (in consideration of neighboring relations during construction), the null hypothesis (H_{02I} Gr0-Gr1) for Group 0 and Group 1 must be rejected. The degree of performance of Rule 2 is only 12.5% in Group 1 compared to 91.2% in Group 0. For Group 0 and Group 2 the null-hypothesis (H_{02I} Gr0-Gr2) cannot be rejected. Regarding the relative frequencies Rule 2 seems to be adhered to more for Group 2 than Group 0. The null-hypothesis (H_{02II} Gr1-Gr2) for Group 1 and Group 2 must be rejected. The degree of performance of Rule 2 is only 12.5% in Group 1 compared to 100% in Group 2.

For H_{03} (assembly by layers), the null hypothesis comparing Group 1 (polychrome model) with Group 0 (H_{03I} Gr0-Gr1, monochrome model) resp. Group

2 (H_{03II} Gr1-Gr2, monochrome model) must be rejected. Since the observed relative frequency is 12.5% as against 81% (Group 0) resp. 100% (Group 2). For Group 0 and Group 2 the null-hypothesis (H_{03I} Gr0-Gr2) cannot be rejected. Comparing the results of Group 0 and Group 2, the rule is adhered to more than expected, because each subject of Group 2 built the monochrome model layer by layer (100%).

For H_{04} (subassembly construction), the null hypothesis also must be rejected.

Table 2 Results of the χ^2-Fit Test

Rule	Hypothesis	F	df	p
2	H_{02I} Gr0-Gr1	154.089	1	0.000
	H_{02I} Gr0-Gr2	2.657*	1	> 0.05
	H_{02II} Gr1-Gr2	70.074	1	0.000
3	H_{03I} Gr0-Gr1	217.194	1	0.000
	H_{03I} Gr0-Gr2	0.267*	1	> 0.05
	H_{03II} Gr1-Gr2	70.074	1	0.000
4	H_{04} Gr1-Gr2	41.905	1	0.000

$\alpha = 0.05$; * Yates' χ^2-test.

According to the results given in table 1 and table 2, the following findings can be summarized regarding Rule 1 – 4:

1) Rule 1 (edge position) can be verified in the boundaries of this experiment for both Group 1 and Group 2.
2) In the case of the polychrome object where subassemblies are directly visible Rule 2 is not applicable, whereas for the monochrome object Rule 2 can be verified.
3) Due to the subassemblies incorporating multiple layers Rule 3 is not applicable to the polychrome object, whereas for the monochrome object Rule 3 can be verified.
4) Regarding Rule 4 a clearly visible separation of the subassemblies is necessary to have the human built up in subassemblies.

These findings lead to the conclusion, that the identified Rule 4 dominates Rule 2 and Rule 3 in the case of a clearly visible separation of the subassemblies.

SUMMARY

In this paper, a novel approach of a cognitive control for a self-optimizing production system was introduced. Based on the cognitive functions, parts of the original task spectrum, such as assembly procedure planning, can be transferred from the human to the machine in production systems. However, in order to be able to monitor the process, it may be necessary to ensure that the decisions made by the machine, i.e. establishing the assembly sequence, are understandable to the human operator. Additionally the assembly sequence should be planned so that it conforms to human expectations. A concept, being able to adapt system behavior to operator's expectations by using human-centered process knowledge based on the MTM-1 taxonomy was introduced and validated using the example of a robotized production cell. The question of whether the person given the assembly task simply executes generalized strategies was investigated using two independent experimental series.

The identified base of assembly rules of Mayer et al. (2010) could be extended: 1) The human begins the assembly at the edge. 2) The human preferable builds in the vicinity of neighboring objects. 3) The human prefers to assemble by layers. 4) The human tends to built up in subassemblies, if these are clearly separable by color.

The introduced experiment with 18 subjects shows a 66.7% buildup in subassemblies for Group 1 if the subassemblies are clearly separable, whereas no buildup in subassemblies was observed despite the same geometry for Group 2 if the subassemblies are not clearly separable.

The next step regarding the identified and implemented assembly rules is the investigation of the impact of the rules on the transparency of the system.

AKNOWLEDGEMENT

The authors would like to thank the German Research Foundation DFG for the kind support of the research on human-robot cooperation within the Cluster of Excellence "Integrative Production Technology for High-Wage Countries".

REFERENCES

Bainbridge L (1987) Ironies of Automation. In: Rasmussen J, Duncan K, Leplat J (Eds) New Technology and Human Error, Wiley, Chichester

Billings CE (1991) Human-centered aircraft automation: A concept and guidelines. In: NASA Technical Memorandum 103885, NASA-Ames Research Center, Moffet Field CA

Eversheim W (1998) Organisation in der Produktionstechnik, Bd. Konstruktion. Springer, Berlin

Kempf T, Herfs W, Brecher Ch (2008) Cognitive Control Technology for a Self-Optimizing Robot Based Assembly Cell. In: Proceedings of the ASME 2008 International Design Engineering Technical Conferences & Computers and Information in Engineering Conference, America Society of Mechanical Engineers, U.S.

Leiden K, Laughery KR, Keller J, French J, Warwick W, Wood SD (2001) A Review of Human Performancer Models for the Prediction of Human Error. Prepared for: National Aeronautics and Space Administration System-Wide Accident Prevention Program. Ames Research Center, Moffet Filed CA

Mayer M, Odenthal B, Grandt M, Schlick C (2008) Task-Oriented Process Planning for Cognitive Production Systems using MTM. In: Karowski W, Salvendy G (Eds) Proceedings of the 2nd International Conference on Applied Human Factors and Ergonomic (AHFE) 14.-17. July 2008. Las Vegas, Nevada, USA

Mayer M, Odenthal B, Faber M, Kabuß W, Jochems N, Schlick C (2010) Cognitive Engineering for Self-Optimizing Assembly Systems. In: Karowski W, Salvendy G (Eds) Proceedings of the 3rd International Conference on Applied Human Factors and Ergonomics (AHFE) 17.-20. July 2010. Miami, Florida, USA

Odenthal B, Mayer M, Kabuß W, Kausch B, Schlick C (2009) Error Detection in an Assembly Object Using an Augmented Vision System. In: Proceedings of 17th World Congress on Ergonomics IEA 2009. Beijing

Onken R, Schulte A (2010) System-Ergonomic Design of Cognitive Automation: Dual-Mode Cognitive Design of Vehicle Guidance and Control Work Systems. Springer, Berlin

Rasch B, Friese M, Hofman W, Naumann E (2004) Quantitative Methoden. Springer, Berlin

Schlick C, Odenthal B, Mayer M, Neuhöfer J, Grandt M, Kausch B, Mütze-Niewöhner S (2009) Design and Evaluation of an Augmented Vision System for Self-Optimizing Assembly Cells. In: Schlick C (Hrsg) Industrial Engineering and Ergonomics. Springer, Berlin

Sheridan TB (2002) Humans and Automation: System Design and Research Issues, John Wiley & Sons, Santa Monica

Thiemermann (2005) Direkte Mensch-Roboter-Kooperation in der Kleinteilemontage mit einem SCARA-Roboter. In: Westkämper E, Bullinger HJ (Eds) IPA – IAO Forschung und Praxis, Jost Jetter Verlag, Heimsheim

Directive 98/37/EC of the European Parliament and of the Council of 22 June 1998 on the approximation of the laws of the Member States relating to machinery

CHAPTER 22

Analysis of Lower Limb Measurements in Running Progress for High-Performance Slacks Design

Y. J. Wang, P. Y. Mok, Y. Li, Y.L. Kwok

Institute of Textiles and Clothing
The Hong Kong Polytechnic University
Hung Hom, Kowloon, Hong Kong, China

ABSTRACT

This study was undertaken to investigate the principal lower limb measurements in running progress and how these measurements influence slacks pattern design. A total of 10 male subjects, whose heights were among 170-175cm and BMI values are among 18-23, were recruited as volunteers in this study. Owing to the limitation of manual method and 3D body scanning, one new measuring instrument, body motion capturer with 8 cameras, was used in this study to collect body measurements in running progress. 36 landmarks for 13 body measurements were determined according to ISO 8559. Before formal measuring, corresponding markers for body measurements were fixed on subject's body with double-side tape. After the calibration of the instrument, each subject was required to keep natural standing posture and run on treadmill three times. The running progresses were recorded in body motion capturer system and body measurements were extracted.

The experimental results showed that each lower limb measurement in running progress has different changing values except ankle girth. Because of hip and knee joint movement, skin shapes around it were changed which result in hip and knee girth have the significant changing values (6.5 cm and 5.7cm, respectively). Meanwhile, back leg length also has significant changing values (6.5 cm). Moreover, analysis results of one-way ANOVA indicated that except waist girth, ankle girth, total crotch, crotch length and inside leg length as well as front leg length, effects of body motion on each measurement of lower limb were significant, which revealed different ease and cutting line should be designed in slack pattern to meet the comfort need of wearers on body movement. Based on experimental and analysis results, one new pattern construction method of slack was discussed and the dynamic block pattern was made in the final part of this research.

Keywords: Measuring method, Lower limb measurements, Wearing ease design, Dynamic block pattern

INTRODUCTION

Lower limb has capable of great subtlety in type and direction of movement, such as running and walking. In movements, positions of different joints in lower limb change significantly and body skin surfaces around joint are extended and contracted which result in corresponding body measurement changing. In order to provide necessary places for meet body measurement changes and provide un-restrictive movements of lower limb, suitable eases should be determined and incorporated into slack. Otherwise, body health and comfort are broken. Therefore, collecting lower limb measurements in motion state and finding the difference between body measurements in static and motion state are very necessary for slack design. Nevertheless, only a few research works have considered this aspect.

Huck et al. (1997) studied effects of different body postures on crotch (vertical circumference) ease design in overall and found that mean of added crotch ease in overall is 17.0 cm (6.7 inch). Choi and Ashown (2002) analyzed body movements and found more than wearing ease values should be added in working slack for pear farmer. Gu at al. (2007) conducted research works to determine suitable wearing ease values in protection apparel.

Although these experimental investigations provide a great number of information for designer to determine correctly wearing ease in slack design, these research works are still inadequate and no systematic. On the one hand, body dynamic postures studied in these research works are static. On the other hand, it has not been studied how to meet the requirements of lower limb movement by integrating the obtained wearing ease in patternmaking. For these reasons, an exploring experiment is carried out in this study to investigate lower limb body measurements in running progress with new anthropometrical technique. Based on collected measurements in state and motion state, the differences between measurements in different two states can be acquired which determine corresponding wearing ease values. Moreover, One-way ANOVA is used to analyze the effects of body motion on changes of lower limb measurements. Finally, how to design slack dynamic pattern according to measuring and analysis results is discussed.

REVIEW OF ANTHROPOEMTRIC METHODS

ANTHROPOMETRIC METHODS IN STATIC STATE

Many anthropometrical methods have been developed by researchers to collect body measurements. Bye, E. et al. (Bye, et al., 2006) indicated manual method with tape is the traditional measuring method in static state and also one effective method to collect body measurements when people make static body postures (Liu and Kennon, 2006). However, manual method with tape is an invasive method because measurer needs to touch subject. Moreover, measuring results are also inadequate which are influenced by measurers' operations.

3D body scanning technique has been used widely in apparel development because of its faster, less invasive and mass quantified information about body shape and proportion (Mickinnon and Istook, 2001; Loker, et al., 2005). In modern 3D body scanner, normal or laser light reflection is utilized to capture x, y, and z data points from body surface (Istook and Hwang, 2001). Based on these data, a 3D point cloud is formed in the scanning system. And then, reproducible data can be generated by processing, filtering and compressing raw information in file of 3D point cloud. Finally, critical anthropometric measurements can be extracted in 3D body scanning system. However, there are shortcomings in 3D body scanning: 3D scanner cannot capture special areas of human body such as armhole in static postures state and these missing places led that some key body measurements can not collected. For these reasons, one integrated anthropometrical method with 3D body scanning and manual method is regarded as effective method to obtain exact overall body measurements for product design. With it, lots of body size systems have been built (Zheng, et. al, 2007).

A lot of practical works in apparel industry have proved that traditional manual method and 3D body scanning are good way to collect body measurements in static state, but they are not suitable for body measurements in motion state. In static state, anatomical structures of human body provide the steady landmarks for body measurements, which is easy for designer to acquire body measurement with manual method or 3D body scanner. In motion state, landmarks become unsteadily because body skin surface extends or contracts. Without steadily landmarks, designer can not collect body measurement with traditional manual method and 3D body scanning technique.

ANTROPOMETRIC METHOD IN MOTION STATE

In order to record three dimensional movements and analyze characteristic of body movement based on the digital model, a great number of body motion analysis systems are developed and used in many areas (Dobrian and Bevilacqua, 2003). The overall process of body capturing and the corresponding analysis in the system involves these works: Firstly, markers covered with reflective tape (diameter is about 8-12 mm) are placed on reference points on different parts of the human body and the positions are determined by the research objectives. And then, the calibration of body motion analysis system is conducted. After the calibration, subject performs the designed movement and the coordinate of all fixed markers in each camera's view are stored in a data-station. At the same time, other files, such as Video and Sound file, can be acquired in the system. Based on extracting these stored data and processing these files, the system represents the paths of markers and create 3D digital model (Kapur, et. al, 2005). Finally, characteristics of human body in motion state can be analyzed.

Because body motion analysis system has the accuracy on capturing markers, it provide good platform to measure body in motion state. According to the definition of body measurement in ISO 8559-1989 Garment construction and anthropometric survey-body dimensions, landmarks on body skin are the foundations to acquire all body measurements. By measuring the distances between two or several landmarks, designer can obtain the corresponding body measurements. For example, Shoulder width is the horizontal distance between the acromion extremities. If the fixed markers in body motion analysis system have the same positions on body skin as landmarks for body measurements, body measurements in motion state can be obtained in motion analysis system by capturing the coordinates of all markers and calculating the obtained data.

EXPERIMENTAL WORKS

SUBJECTS

A total of 10 Chinese male subjects with ages 20-24 were recruited as volunteers in this study. Meanwhile, for ensuring that all subjects are in one size group, heights of all subjects are among 170-175 cm and values of BMI (Body Mass Index) are among 18-25. Before measuring, an introduction of the experiment was introduced to all subjects and their approvals were obtained.

LANDMARKS AND BODY MEASUREMENTS

Anatomical landmarks on human body were designed for obtaining 13 body measurements according to the definition of body measurements in ISO 8559: 1989. These 36 landmarks were distributed over three parts of human body: waist (8 markers), hip (8 markers) and leg (20 markers). Before formal measuring, markers covered with reflective tape were fixed on subject body with double-side tape, refer to Figure 1. Based on these landmarks, 16 body measurements were obtained which consists of following measurements: 1) girth: waist, hip, thigh, mid thigh, knee, calf as well as ankle. 2) Length: total crotch, crotch depth, and leg (outside, inside, front and back).

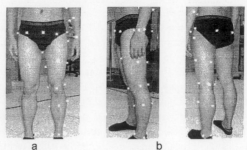

FIGURE 1 Positions of markers fixed on lower limb: (a) Front, (b) side and (c) back.

MEASURING METHOD AND INSTRUMENTS

In this study, one instrument is used to collect stature, weight and BMI value of subject in static state and VICON 8.0 body motion analysis system with 8 cameras is used to capture the coordinate of all markers fixed on body skin in static and running progress. Diameter of marker used in this experiment is 10 mm and the overall running progress of each subject is conducted on one treadmill.

MEASURING PROCEDURE

Firstly, stature, weight and BMI value of each subject were measured in a laboratory with the stable 24 ℃ temperature, 65% relative humidity and less than 1.0 km/h wind speed. And then, 36 markers were fixed on body skin of each subject according to measuring requirements and the coordinate of all markers in natural standing posture were captured with VICON 8.0 system. When the capturing works in natural standing posture are finished, each subject was asked to run on treadmill at the speed of 8 km/h. And the coordinate of all 36 markers in running progress were captured again with the same system. Finally, all captured data are processed in the system and exported as the files of Excel and data in Excel file are calculated to obtain body measurements in static and motion state.

EXPERIMENTAL RESULTS

MEASURING RESULTS IN STATIC STATE

Data in Table 1 show that mean of 10 subjects' stature values is 173.2 cm. Meanwhile, BMI values of all subjects (Maximum: 22.79/ Minimum 19.03) are among normal range: 18.5-25, which indicate they have normal body shapes (Kagawa et al, 2006). Maximum, minimum and mean of 30 body measurement of 10 subjects collected in natural posture are listed in Table 2.

Table 1 Measuring Results on Stature, Weight and BMI of Subjects

ITEMS	MIN.	MAX.	MEAN	STD. D
STATURE (cm)	170.00	175.00	173.20	2.04
WEIGHT (kg)	53.00	69.00	63.43	4.88
BMI	19.03	22.79	21.20	1.29

Table 2 Body Measurements of Subjects in Static State

BODY MEASUREMENTS		MIN. (CM)	MAX. (CM)	MEAN (CM)	STD. D
GIRTH	WAIST	70.00	78.00	72.30	2.36
	HIP	87.00	97.00	91.00	3.09
	THIGH	51.00	58.00	55.00	2.16
	MID THIGH	44.00	52.00	48.40	2.80
	KNEE	33.00	39.00	35.40	1.70
	CALF	34.00	40.00	37.40	1.79
	ANKLE	24.00	28.00	25.55	1.30
LENGTH	TOTAL CROTCH	70.00	75.00	71.70	1.49
	CROTCH DEPTH	24.00	27.00	25.90	1.37
	OUTSIDE LEG	101.00	104.00	102.50	0.97
	INSIDE LEG	72.00	74.00	73.10	0.88
	FRONT LEG	97.00	99.00	97.60	0.70
	BACK LEG	98.00	103.00	101.00	1.76

CHANGING VALUES OF LOWER LIMB MEASUREMENTS IN RUNNING PROGRESS

The greatest changing values of body measurements in running progress are shown in Table 3, when lower limb measurements of subjects taken from in static state have been subtracted from maximum values of the same measurements acquired in running progress.

Table 3 The Greatest Changing Values of Lower Limb Measurements in Running Progress

BODY MEASUREMENTS		MIN. (CM)	MAX. (CM)	MEAN (CM)	STD. D
GIRTH	WAIST	2.00	3.00	2.50	0.53
	HIP	3.00	9.00	6.50	2.27
	THIGH	1.00	4.00	2.20	1.14
	MID THIGH	2.00	4.00	2.70	0.82
	KNEE	3.00	9.00	5.70	2.00
	CALF	0.50	5.00	1.85	1.75
	ANKLE	0.00	0.00	0.00	0.00
LENGTH	TOTAL CROTCH	1.00	8.00	3.40	2.80
	CROTCH DEPTH	0.50	2.00	1.55	0.60
	OUTSIDE LEG	2.00	10.00	5.90	2.69
	INSIDE LEG	0.50	2.00	0.90	0.46
	FRONT LEG	1.00	5.00	2.90	1.79
	BACK LEG	5.00	8.00	6.50	1.08

From data in Table 3, it can find that except ankle girth, other 12 lower limb measurements have different changes in running progress. Among them, hip girth has maximum changing value (6.50 cm) in all girths and back leg length has the maximum changing values (6.50 cm) in all lengths. In additions, knee girth and outside leg also have significant changing values, 5.70 cm and 5.90 cm, respectively.

DETERMINATION OF WEARING EASE VALUES IN LOWER LIMB MEASUREMENTS

According to the means of lower limb measurements in Table 2 and mean of changing values in Table 3, wearing ease values in lower limber measurements and extending percentages of body skin surfaces can be acquired, as shown in Table 4. Because ankle girth has no changes in running progress, wearing ease value is 0 and it is not listed in Table 4.

Table 4 Wearing Ease Values and Extending Percentages of Skin Surface in Girths and Lengths of Lower Limb

GIRTHS (CM) / PERCENTAGE (%)					
WAIST	HIP	THIGH	MID THIGH	KNEE	CALF
2.50/	6.50/	2.20/	2.70/	5.70/	1.85/
3.45	7.14	4.00	5.58	16.10	4.95
LENGTH (CM) / PERCENTAGE (%)					
TOTAL CROTHCH	CROTCH DEPTH	OUTSIDE LEG	INSIDE LEG	FRONT LEG	BACK LEG
3.40/	1.75/	5.90/	0.90/	2.90/	6.50/
4.74	5.98	5.76	1.23	2.97	6.44

The percentages in Table 4 reveal that skin surface in knee girth has maximum extension (16.10%), the second is hip girth (7.14%) and the third is back leg length (6.44%). Therefore, knee girth, hip girth and back leg length are the important parts for wearing ease design. Skin surface in inside leg length has minimum extension (1.23%) among these lower limber measurements.

DISCUSSION AND APPLICATION

EFFECT OF BODY MOTION ON CHANGES OF LOWER LIMB MEASUREMENTS

Based on changing values of lower limb measurements in running progress, one-way ANOVA is used to analyze effects of body motion on changes of lower limb measurements. The analysis results of relationship between body motion and lower limb body measurements changes are listed in Table 5. From these data, it can find that effects of body motion on principal lower limb measurements are significant. But, body motions have no significant effects on five body measurements: waist girth ($p=0.78$), total crotch length ($p=0.49$), crotch depth ($p=0.55$), inside leg length ($p=0.98$) and front leg length ($p=0.73$), because p values of three measurements were more than 0 .05.

Table 5 Tests of Body Motion on Lower Limb Measurements

BODY MEASUREMENTS		DEGREE OF FREEDOM	F-VALUES	SIG.
GIRTH	WAIST	4	0.44	0.78
	HIP	4	15.69	0.00*
	THIGH	4	18.63	0.00*
	MID THIGH	4	13.10	0.00*
	KNEE	4	25.80	0.00*
	CALF	4	6.52	0.00*
	ANKLE	4	0.87	0.49
LENGTH	TOTAL CROTCH	4	0.78	0.55
	CROTCH DEPTH	4	0.05	0.98
	OUTSIDE LEG	4	14.39	0.00*
	INSIDE LEG	4	0.43	0.73
	FRONT LEG	4	25.46	0.00*

* $p \leq 0.05$: significance

From the analysis results in Table 5 and wearing ease values in Table 4, it can be summarized that measurements hip girth, thigh girth, mid thigh girth, knee girth, calf and outside leg length as well as back leg length are influenced by body motions and these measurements are principal parts when adding wearing ease values in slack pattern.

APPLICATION OF WEARING EASE VALUES IN SLACK DYNAMIC PATTERN MAKING

As shown in Figure 2-a, fit block pattern of slack are made based on the mean values of lower limb measurements in Table 2. And ten anatomical lines are marked in the pattern as reference line: Lines of waist, hip, thigh, mid thigh, knee, calf as well as ankle are the horizontal lines. Front and back line of slack pieces, outside line, inside line, centre front and back are the vertical lines.

Wearing Ease Design in Girths of Slack Block Pattern

Hip girth: Half of the ease value is added in two front pieces and another half in two back pieces. In front and back piece, wearing ease values are added in front and back line.

Thigh girth: Because ease value adding in hip girth is more than in thigh girth, work of ease design in thigh girth is finished when adding ease in hip girth, refer to Figure 2-b.

Knee girth: All ease values of elbow girth are added in front piece. As shown in Figure 2-a, two end points of thigh line in front piece are regarded as the fixed points and the thigh line is cut off. At the same time, front line from thigh line to cuff line is also cut off. Base on these two fixed points, right and left parts of front piece under thigh line are moved outward and wearing ease is added in elbow girth.

Mid high and calf girth: All wearing ease value in mid thigh, are added in pattern when adding ease in knee girth. In calf, half of total ease value is added in each side of corresponding girth lines.

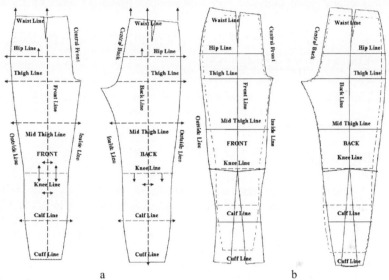

a   b

FIGURE 2 Ease designs in block pattern: (a) Fit block pattern, (b) dynamic block pattern.

Wearing Ease Design in Lengths of Slack Block Pattern

Back leg length: As shown in Figure 2-b, wearing eases value in back leg length is added in waist line, hip line, and cuff line

Outside leg length: Wearing ease value in outside leg length is added in waist line, hip line and knee line.

After adding in ease in girth and length of fit block pattern, new dynamic block pattern of slack with ease is made, as shown in Figure 2-b

CONCLUSIONS

In this research work, one new anthropometric method is used to collect lower limb measurements in running progress and static state. Based on collected lower limb measurements in static and running progress, the differences between measurement in two states and extension percentages are acquired which present that body measurements of lower limb have different changes in running progress except ankle girth. Among these measurements, hip girth and back leg length have maximum changing values. And skin surface in knee girth, hip girth and back leg length have significant extensions.

Results of one-way ANOVA reveal body motions have no significant influence on five measurements: waist, total crotch length, crotch length and inside leg length as well as front leg length. Only seven principal body measurements in lower limb, such as hip girth and back leg length, are influenced by body motions

Finally, application of wearing ease design in fit block pattern is discussed according to the determination of wearing ease values in body measurements influenced by body motions. By adding suitable wearing ease values in important parts, dynamic block pattern of slack is made which is the effective guide to develop different high-performance slack products.

ACKNOWLEDGEMENT

The authors would like to thank the Hong Kong Polytechnic University and HKRITA for financial support in this research project (project code: ITP/002/079TP, ITP/014/08TPand ITP/030/08TP).

REFERENCES

Bye, E., Labat, K. L., and Delong, M. R. (2006), "Analysis of body measurement systems for apparel." Clothing & Textile Research Journal, 24, 66-79.

Chang, I. C., and Huang, C. L. (2000), "The model-based human body motion analysis system." Image and Vision Computing, 18, 1067-1083.

Choi, M. S., Ashown S. P. (2002), "The design and testing of work clothing for female pear farmer." Clothing and Textile Research Journal, 20 (4): 253-263.

Dobrian, C., and Bevilacqua, F. A. (2003), Gestural controal of music using the VICON 8 motion capture system, available online at: http://music.arts.uci.edu/dobrian/motioncapture/

Huck, D. J., Maganga, O., and Kim, Y. (1997), "Protective overalls: evaluation of garment design and fit." International Journal of Clothing Science and Technology, 9(1), 45-61.

Istook, C., and Huang, S. J. (2001), "3D body scanning system with application of garment design and fit." International Journal of Clothing Science and Technology, 5 (2), 120-132.

Kagawa, M., Kerr, D., Uchida, H., and Binns, C. W. (2006), "Differences in the relationship between BMI and percentage body fat between Japanese and Australian-Caucasian young men." British Journal of Nutrition, 95, 1002-1007.

Kapur, A., Tzanetakis, G., Babul, N. V., Wang, G., and Cook., P. R. (2005). The 8th International Conference on Digital Audio Effects (DAFX-05), Madrid, Spain.

Liu, C., and Kennon, R. (2006), "Body scanning of dynamic posture." International Journal of Clothing Science and Technology, 18(3), 166-178.

Loker, S., Ashdown, S., and Schoenfelder, k. (2005), "Size-specific analysis of body scan data to improve apparel fit." Journal of Textile and Apparel, Technology and Management, 4 (3), 1-15.

Mickinnon, L., and Istook, C. (2001), "Comparative analysis of the image twin system and the 3T6 body scanner." Journal of Textile and Apparel, Technology and Management, 1 (2), 1-7.

Zheng, R., Yu, W., and Fan J. (2007), "Development of a new Chinese bra sizing system based on breast anthropometric measurements." International Journal of Industrial Ergonomics, 37, 697-705.

Chapter 23

The Impact of the Human Factor on Productivity in Advanced Production Systems

Matthias Göbel, Swantje Zschernack

Department of Human Kinetics and Ergonomics, Rhodes University
South Africa, email: m.goebel@ru.ac.za

ABSTRACT

Modern production systems involve highly specialised and complex machinery. Human activities thus shift towards control, maintenance and troubleshooting activities, as well as to manual tasks that are not automated due to flexibility requirements. In both cases the human factor plays an essential role for productivity, as labour costs of whole units and return of investment in machinery are affected.

The nature of those tasks comprises mostly of sensory motor activities, supervisory activities and problem solving activities. Whereas the both latter forms of work are researched extensively, the former one still lacks attention regarding modern forms of work.

This paper raises the question how to evaluate sensory motor tasks in terms of productivity. At a first glance one may consider precision and speed as performance indicators, however this does not allow evaluation of more complex activities, such as required in production systems. Further, sensory motor tasks are mostly framed by cognitive and a physical activity from whose it cannot be separated.

Drawing a line from the basics of movement organisation and performance evaluation to the impact of precision movement on productivity, this different contributing factors from different studies are linked to outline the conditions that are conducive or obstructive to sensory motor performance in a production environment. This is supplemented by the requirements for supervisory and problem solving tasks.

Keywords: Sensory motor tasks, Productivity, Production system

INTRODUCTION

Sensory motor abilities are an important performance feature for almost all work operations. They are central factors for assembly and control operations, however mechanical, motor and informative operations contain sensory motor components as well. This is valid for motor control during gross movements, e.g. shoveling, as well as for keyboard- and mouse movements during process control and management.

Apparently sensory motor abilities influence the productivity of the work process, depending on the operations characteristics possibly more than the physical demands or cognitive processes. However the influence of sensory motor abilities on productivity is not the same extent subjectively accessible like effort and exhaustion (and in many cases is also not directly measurable). For example for computer operations one is able to relatively simply observe how much time is spent for the sensory motor processes of the keyboard and mouse actions. However it is evident, that if information exchange between the human and the computer is performed during 10%-90% of the total working time the corresponding sensory motor processes must have a significant effect on the productivity, even if other cognitive processes are performed concurrently.

This is includes conscious as well as subconscious processing. Whereas sensory motor processes are mostly executed subconsciously (and can be controlled consciously only to a very limited extent), the central cognitive processes are consciously controlled. A focused analysis of motor control aspects is impaired by the fact that sensory motor control is framed by cognitive and physical subtasks, and most of those processes are performed unconsciously. In many cases even motor performance cannot be assessed straight (e.g. during vehicle control).

In the following a detailed analysis of motor control shall be outlined, here focusing on the effects of motor control on productivity.

PROFICIENCY OF MOTOR OPERATIONS

PRODUCTIVITY OF SENSORY MOTOR OPERATIONS

Productivity is the indicator for production efficiency initially defined as the relationship of output from input (onpulson 2009). As for *output* in this context only the essential performance is to be counted, this implies a further relationship to efficiency of the operations – even if not considered directly. In economic terms input and output are preferably expressed in monetary quantities, as these are the straight operational and control variables. Using the popular type of remuneration per time unit (for a given proficiency level) *input* can be expressed simply as time per unit or as unit labour cost. Output in this case is the generated surplus per activity unit. Any optimization of productivity on this initially simplified considerations level would thus be obtained through having per unit a reduction of input (less time required) and/or through an increased output (improved quality).

The previously employed immediate output/input consideration leaves many secondary, but nevertheless relevant variables initially unconsidered. In terms of an extended economic consideration, at least qualification and training costs (sensory motor skills demand significant training) as well as follow-up costs for human errors (which might sum up to considerable costs, e.g. in automotive or nuclear industry) would need to be considered.

PERFORMANCE CRITERIA OF SENSORY MOTOR TASKS

Human motor performance is mostly described as a feedback system with a limited information processing capacity. This is equally true for continuous movements and for discrete movements (Fitts 1954, Schmidt et al. 1979, Adams 1971, Oppelt & Vossius 1970). The limit in information processing capacity is linked to a processing delay (reaction time) that in a feedback system limits information pass through and thus control quality. This explains well the speed-accuracy tradeoff found with all types of targeted movements: Target precision and movement time show a reciprocal relationship, the slower a movement the higher the (relative) possible movement precision and vice versa. The only performance shaping variable is thus the human reaction delay. It can be shown that this type of performance limit is valid for all types of feedback systems, irrespective of the type of feedback and other feedback coefficients (Göbel 1996).

However typical reaction delays of 180 to 600 ms would not allow the level of performance typical humans may provide in daily operations. Reason for this is the learning process of anticipating environmental behavior and own reactions. If the movement behavior of a rolling or flying object can be anticipated by applying previous experience, responses can be initiated earlier in order to compensate most of the immanent reaction delay. This is, for example, the case when responding a tennis ball having a speed 10-20m/s or more. Göbel (2002) showed that anticipation allows reducing the effective reaction delay by about 90% to 20-25ms if the environment behaves reliable. As this anticipation process is trained during long term, short term changes may cause severe disturbances. This is the reason why even minor changes in reaction delay, for example as a consequence of alcohol consumption, may cause heavily impaired movement control.

PRODUCTIVITY OF SENSORY MOTOR TASKS

Applying the previously discussed aspects of motor behaviour and motor learning with regards to productivity criteria the following conclusions may be drawn:

The requirements on the sensory motor productivity are basically determined through the both parameters *precision* and *complexity of movement*. The complexity of movement is due to the frequency different movement sections, in which each sequence needs a new end point settlement and new motor programs for the following movement sequence. Derwort (1938) describes this effect as "rule of constant speed of shape": The circumnavigation of a large circle for example takes not significantly longer than that of a smaller one. The time related parameters are characteristic for the relative movement configuration, are but sparsely dependent on their size. The movement path with its visually anticipated kinks and reversal points determines movement time to a large extent, whereas the physical and anatomical performance conditions have only a minor effect.

The sensory motor ability is described through the information processing capacity or the effective reaction time. Depending on the problem, the dependent variable is *accuracy* or *speed*.

Changes in performance conditions (e. g. impaired visibility) will cause changes of the human reaction delay. Additional task components, that are not part of the sensory motor task, will require additional time (or cause a longer reaction delay), when performed simultaneously and using the same endogenous resources.

Such basic considerations end up, however, in a substantial complexity when applied for more varied movement sequences from real work tasks.

With regard to productivity, affecting variables on the one hand are those factors which describe the task complexity (movement amplitude, complexity/frequency, speed/precision, as well as performance conditions that influences the effective reaction time) and on the other hand the human factor, exclusively represented by the effective reaction delay. Human reaction delay thus is the one and only performance shaping factor on the human side. This raises the question which factors do influence the effective reaction time?

- Any change of performance conditions will affect the effective reaction delay, and thus disturb a most precise anticipation. This is true for task related as well as human related change.
- Human fatigue will lengthen the human reaction delay and thus impair the well trained anticipation. Thus performance drops down to a much greater extent than the reaction delay increases. Avoiding fatigue is thus essential to sustain performance (and thus productivity).
- Psychical strain causes a change of the parasympathetic activity level which again affects attention as well as physical tension (e.g. muscle tone). Changes of attention level affects again reaction delay as well as anticipation abilities, the physical tension affects the mechanical dynamics of the musculoskeletal system. Both effects might have a considerable impact on the effective reaction delay.

IMPACT ON PRODUCTIVITY

In the following it shall be demonstrated how the above named mechanisms will affect the achievement of sensory motor tasks.

THE DIFFERENT MOVEMENT PHASES

Movement studies with the methods of the system of Time and motion studies (e.g. Methods Time Measurement, MTM) are widely applied in industry. The calculated time serves as the basic planning criteria for work processes. In terms of a productivity consideration, the time allocation may allow for a capacity planning, which is crucial for optimization in terms of margins for improvement. However, this does less depend on time allocation, but much more on the sources of variability for the time needed to perform.

Figure 1 shows an example from manual assembly (Luczak & Göbel 1996) that might explain this effect. The upper part of the graph shows the time distribution for the different movement phases. The lower part of the graph in figure 1 shows the part of variance that each movement phase contributes to the variance of the whole cycle. The difference in shape is explains that specific movement cycles have much more impact on total time than in case of an equal relative variance (that would show a similar pattern that the time needed, see Figure 1, top).

FATIGUE

As mentioned before, fatigue has an important influence on the sensory motor activity. But how does sensory motor exhaustion originate and which factors promote or reduce their development? Local muscular fatigue normally does not occur during fine manipulative tasks, except for static muscular activity required for gravity compensation or co-contraction. Peripheral nerves and synapses in the spinal marrow are not known to fatigue during normal working periods. Fatigue can therefore most likely be caused by perceptual, cognitive or even central processes. It has a similar effect as alcohol, reducing working capacity and consequently an increased reaction time. Robertson & Zschernack (2009) found a 7% decreased motor performance when performing during a night shift (22h-6h) compared to performing the same task during a day shift (8h-16-h), jointly with a delayed visual reaction time (saccade latency).

Generally fatigue processes come into effect much earlier than it can be observed from an external perspective. Only once endogenous mechanisms through fatigue compensation (e.g. through load distribution) are depleted, fatigue phenomena become visible (Schmidtke 1965, Hockey 1997). This can be shown, for example for motor unit recruitment (DeLuca et al. 1982). For complex sensory motor processes proof of fatigue that is distinct from physical fatigue as well as from central fatigue is hardly achievable. However periodical fluctuations in activity and performance can be pointed out, which would indicate compensation reactions. Göbel (1996) analyzed the variation of performance and some other activity parameters during a manual assembly task by splitting into three different groups: (a) long and middle term trends, (b) cyclic fluctuations and (c) random variability. It was shown that cyclic fluctuations explained more than 25% of the total variability, which is more than the variance of long and middle term trends. Average period duration was about of 75 minutes throughout the parameters, slightly increasing with training level. However, further analyses might discover more detailed information about the compensatory and regulative functions that cause such fluctuations.

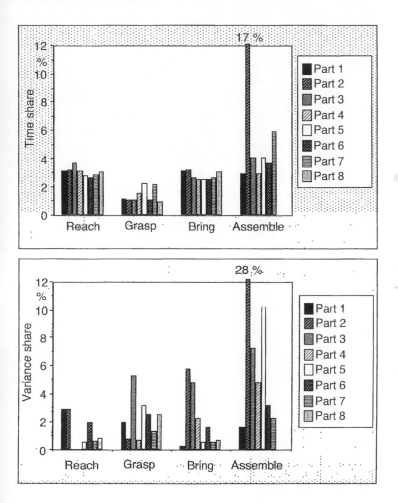

Figure 1: Distribution of assembly time (top) and assembly time variance (bottom) of a manual assembly task (10 workers, each 4 hours working time with average 650 repeated cycles).

INFLUENCE OF BODY POSTURE

Sensory motor tasks, that have to be performed in awkward body posture are common in industrial assembly, e.g. in the auto production industry. Initially any unbalanced body posture will induce an additional physical load to sustain an equilibrium, implying a risk of muscular fatigue and as a result an increased circulation activity. However, the segments of the human body build a biomechanical chain with flexible joints. Thus the stability of the trunk position will affect hand arm coordination and eventually target positioning. Although this is basically valid for any performance condition, positioning will be hampered with more unstable (awkward) body postures.

In spite of the clear arguments this effect is only rudimentary researched. Ngcamu et al. (2009) studied the effect of 8 different body postures on time to target and pointing precision for a visually stimulated pointing task. A supine position and overhead working result in a 8-10% reduced speed in comparison to upright sitting and standing as well as a light and strong forward bent or turned body. This is equally the case for low and high precision requirements. Pointing precision shows a similar characteristic with 10-15% lower performance in a supine position.

Compared to a force control task, requiring subjects to sustain a constant force between both hands and without any visual feedback, this task shows a decreased precision performance in a bent forward position and in overhead position, but no significant impairment when laying on the back. A performance variation of about 10% for different body postures is however significant in terms of productivity.

For a productivity consideration not only the effect of the body posture on the fine motor abilities remains, but also the reaction of the fine motor activity on the physical demands have to be taken into account. Precision movements might require an additional stabilization through co-contraction, which would then result in an additional muscular load. Göbel & Skelton (2009) however showed, that the superposition of awkward body postures and precision tasks have an indifferent or a compensatory interaction effect, depending on body posture and task setup. This means, that the fine motor movements tend to positively contribute to the dynamic stabilization of the body posture and consequently that no additional physical demand components through precision movements are resulted.

TRAINING AND PRACTICE

The effect of exercise and practice is standing matter of conceptual and empirical research since long due to the immense practical interest. Fundamentally the training process initially is focused on the composition of eligible motor programs, that then, over time, are more and more integrated for a more complex smooth movement. Further the integration of sensory feedback that becomes optimized. In doing so the coordination of the higher motor centres is shifted towards hierarchically lower, more autonomic but less flexible motor levels. As motor programs run subconsciously, training is defined fundamentally through repetition.

Thus the simpler the structure of a movement the faster it can be trained. In this context unconscious processing and settling of motor programs in passive phases plays an evenly important role as active training. Thus a suitable activity-rest ratio is essential for efficient motor learning.

In total there are only few options to optimize productivity in terms of training speed. Apart from avoiding disadvantageous conditions, this is about a suitable training-rest ratio and rather simple types of movement. The industrial practice to favor highly repetitive short-cyclic movements reflects (amongst other reasons) such advantages, albeit humanitarian aspects would speak for more varied tasks.

SENSORY FEEDBACK

An important factor in the movement control is, as shown, the sensory feedback. Here various modalities, mainly of a visual, haptic and kinesthetic nature play a role. For work design this raises the question to which extent the different types of sensory information do positively contribute to motor performance. An electromyographic signal analysis may reveal the degree of sensory feedback applied for different feedback loops according to the different process delay times (Göbel 1996, Göbel 2004). For a manual assembly activity, significant changes in the course of the learning process have been found: During the learning phase (in this case 20 hours) the impact of (monosynaptic and polysynaptic) reflex activity decreases whereas visual and haptic regularity control increases with time. Such a characteristic can be interpreted as a consolidation of movement pattern and therefore requiring less and less reflex control. With exercise the framing movement is increasingly controlled during execution by applying visual and haptic feedback.

From a productivity point of view it is of interest to which extent sensory feedback contributes positively to performance. Figure 2 shows the correlation of sensor activity with performance.

The effect of monosynaptic reflexes (the shortest feedback loop) as well as for polysynaptic reflexes is initially negative on performance, but the effect decreases with increasing practice. The use of haptic and visual feedback however takes place only with increased practice, and it has a positive effect on performance. The usage of haptic feedback seems to be effective only after reflex control has stabilized and takes even longer than for visual feedback and its onset requires a .

The graduation rate of 0.4 to 0.5 for the haptic and the visual modality shows, that an increase of sensory influences of for example of 20% to 25% on that level lead to an improved performance of an average of 10% for the entire assembly process. Hence the provision of suitable visual and haptic information that can be used for feedback is an essential feature for performance optimization.

Furthermore it can be shown, that haptic learning does occur exclusively in the work breaks, whereas through movement itself no learning effect has been proved. For the reflex level, however, the incorporation of feedback is practiced mostly during the movement performance.

230

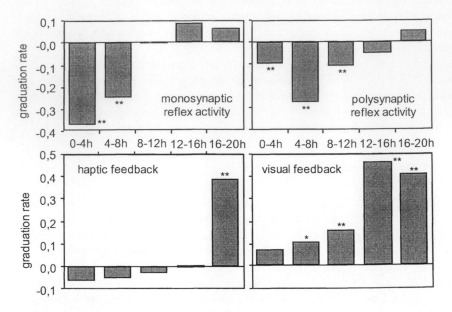

Figure 2: Correlation between feedback control and execution speed during a learning phase for a manual assembly task (10 subjects during 5 subsequent periods of each 4 hour duration; *: p<0.05; **: p<0.01; from Göbel 2004).

CONCLUSION

Sensory motor activities have a considerable influence on the productivity of work processes, as they are constituent of almost any work task. This is particularly the case in advanced environments where timeous and precise actions are essential, and often endurance capabilities are strained. Their effect is however not at all trivial to access and assess, as it is interlinked with many other performance shaping factors.

From a work design perspective many conditional variables may impair sensory motor performance if not suitably adjusted, but only few options do exist to improve motor performance from a properly designed level onwards. This basically cause by the facts that motor performance is mostly affected by the effective human reaction delay (including some compensation through anticipation). This can be lengthened through unfavourable circumstances, but through favourable circumstances it can hardly be shortened. Similarly the sensory motor performance cannot be improved by extra effort.

There are however numerous options to make use of the human performance potential to its maximum extent. This may be achieved through avoiding physical workload (e.g. awkward body posture), the provision of suitable visual and haptic feedback, and particularly through avoiding of fatigue and psychical stress. Otherwise a significant output loss is inevitable. Ergonomically design work is by far the most productive in terms of motor performance.

As such no ethical conflict exists between following humanitarian and following economical objectives, as any lack of human consideration will directly cause an impaired performance. Only for the preferred complexity of movements maximum productivity can be achieved for most simple movements, whereas aspects of humanisation will required a most varied task spectrum. In order to solve this conflict mediating measures like job enrichment, job enlargement and job rotation might be considered.

REFERENCES

Adams, J.A. 1971, A closed-loop theory of motor learning. *Journal of Motor Behaviour 3 (2)*, 111-149.

DeLuca, C.J., LeFever, R.S., McCue, M.P. & Xenakis, A.P. 1982, Behavior of human motor units in different muscles during linearly varying contraction. *J. Physiol.* 29, 113-128.

Derwort, A. 1938, Untersuchungen über den Zeitverlauf figurierter Bewegungen beim Menschen. *Pflügers Archiv ges. Physiol.* 236, 661.

Fitts, PM. 1954, The information capacity of the human motor system in controlling the amplitude of movement. *Journal of Experimental Psychology, 47:381-391.*

Göbel, M. 1996, *Elektromyografische Methoden zur Beurteilung sensumotorischer Tätigkeiten.* Dok. Arbeitswissenschaft, Bd. 40. Köln: Dr. Otto Schmidt Verlag.

Göbel, M. 2002, Laboratory estimation of driving safety during VDU device operation. In: H. Luczak, A.E. Cakir & G. Cakir (eds.), *WWDU 2002 - World Wide Work.* Berlin: Ergonomic Institut, p. 461-463.

Göbel, M. 2004, Electromyography. In N. Stanton et al. (eds.), *Handbook of Human Factors and Ergonomics Methods.* Taylor & Francis, 19-1 to 19-8.

Göbel, M. & Skelton, S.A. 2009, Interaction of awkward working posture and manupulation tasks on physical strain. Proceedings of the 17th Congress of the International Ergonomics Association IEA, 9-14 August in Beijing, China.

Hockey, G.R.J. (1997). Compensatory control in the regulation of human performacne under stress and high workload: A cognitive-energetical framework. *Biological Psychology 45,* p.73-93.

Lorenz, K.Z. & Tinbergen, N. 1938, Taxis und Instinktbewegung in der Eirollbewegung der Graugans. Zeitschrift für Tierpsychologie, 2, 1-29.

Luczak, H. & Göbel, M. 1996, Psychophysische Aspekte sensumotorischer Taetigkeiten. In K. Landau, H. Luczak & W. Laurig (Hrsg.), *Ergonomie der Sensumotorik*, Festschrift anlässlich der Emeritierung von Herrn Prof. Dr.-Ing. W. Rohmert. München, Wien: Hanser, 34-56.

Ngcamu N., Zschernack, S. & Göbel, M. 2009, The interaction between awkward working postures and precision performance as an example of the relationship between ergonomics and production quality. Proceedings of the 17th Congress of the International Ergonomics Association IEA, 9-14 August in Beijing,

Onpulson 2009. Definition Produktivität. http://www.onpulson.de/lexikon/produktivitaet.htm (visited 22 Juli 2009, 21:00).

Oppelt, W., Vossius, G. 1970, Der Mensch als Regler. Berlin: VEB Verlag Technik.

Robertson, J & Zschernack, S. (2009). The effects of fatigue on saccade latency and precision performance. A Thatcher, S Zschernack, A Todd & S Davies (Eds.). Proceedings of the 11th Conference of the Ergonomics Society of South Africa, ISBN: 978-0-620-45271-7, p 215-228.

Schmidt, R.A., Zelaznik, H.N., Hawkins, B., Frank, J.S., Quinn, J.T. (1979): Motor output variability: A theory for the accuracy of rapid motor acts. *Psychological Review* 86, 415-451.

Schmidtke, H. 1965, *Die Ermüdung*, Symptome-Theorien-Messversuche. Bern, Stuttgart: Verlag Hans Huber.

Chapter 24

Level of Automation Analysis in Manufacturing Systems

Åsa Fasth, Johan Stahre and Kerstin Dencker

Department of Product and Production development
Chalmers University of Technology
Gothenburg, SE-41296, Sweden

ABSTRACT

Manufacturing tends to be polarizing in terms of automation. Final assembly and other mass-customization operations in the production flow show an increase in allocation of functions to the operators. In contrast, early manufacturing phases, where high repeatability is possible and needed, show an increase in automation with high autonomy, adaptability, and robustness towards disturbances, with operators in a more supervision role. Manufacturing automation may benefit from a differentiated view; manufacturing automation has a strong physical component, i.e. the traditional physical ergonomic and task allocation view of operator work. On the other hand, increasing presence of information technology, manufacturing execution systems, and various operator support information have resulted in a radical change for the operator situation in terms of cognitive ergonomics.

This paper presents a methodology for analysis of semi-automated manufacturing systems, in terms of physical and cognitive levels of automation (LoA). The aim is to determine appropriate task allocation by generating a span of possible solutions for physical and cognitive automation levels. Results are based on ten industrial case studies from assembly operations in Swedish companies.

Keywords: Task allocation, levels of automation. Manufacturing systems, assembly

INTRODUCTION

In order to maintain sustainable production in an increasing globalization, current tradition for design and usage of automation in assembly systems may not be adoptable to the needs and future challenges that manufacturing companies are facing. Frequent changes in demands and requirements, both internal and external; trigger a plan for change in different manufacturing areas.

Smaller batches and shorter time limits for set-ups between products are some of the demands for the assembly systems caused by an increasing number of product variants i.e. mass customization. As a result, companies have to find increasingly flexible methods for assembling products and make the assembly system itself more proactive. Indentifying new strategies to reduce time parameters in a system e.g. cycle-time, set-up time and non value-adding tasks also becomes vital.

One solution is automation; which makes it important to determine the appropriate level of automation. A common industrial predisposition is to consider automation investments as a "binary" decision, even though choosing between humans or machines might actually be suboptimal. Several development trends towards highly automated production and shop floor workplaces were seen during the 1980's and early 1990's. At that time the predominant task allocation strategy was "left-over allocation". Since the late 1990's trends are changing, much due to the shortcomings of automation to fulfil cost and flexibility requirements. Another significant reason is the fundamental paradigm shift into "Lean Production". During a recent study tour in Japan, the authors noted significant automation potential for some of the factory lines visited. The response from company management was clear and immediate "We know, but robots cannot do kaizen" It appears that previous needs for **reactive** operator performance (e.g. disturbance handling) on the shop floor has been complemented and even overtaken by the need for **proactive** human behaviour in terms of continuous improvement (kaizen) and radical shop floor innovation (kaikau).

It is necessary to have different Levels of Automation depending on a range of requirements e.g. batch-size, flexibility, and competence. Further this measure of task allocation should preferably be separated into physical and cognitive levels of automation. This paper aims to present a methodology used for preparation of automation strategies, in order to determine task allocation with a span of possible solutions for cognitive and physical automation. Empirical data has been gathered from industrial assembly systems in Swedish companies.

TASK ALLOCATION

The classical task allocation strategy from 1951 (the so called MABA-MABA list) was proposed by Fitts (Fitts, 1951). It was an attempt to suggest allocation of tasks between humans and machines by treating them as system resources, each with different capabilities. Two examples; *M*achines *A*re *B*etter *A*t performing repetitive and routine tasks; while *M*en *A*re *B*etter *A*t improvising and using flexible procedures. At the time, this was a revolutionary thought causing a lot of debate.

Jordan (Jordan, 1963) argued whether you could actually compare men and machines; and that the two should be seen as complementary, rather than conflicting, resources when designing a man-machine system. Sheridan (Sheridan, 1995) suggested to "allocate to the human the tasks best suited to humans and allocate to the automation the task best suited to it. But, if tasks in which machines are better become automated and operators are still required to monitor the automation, maintaining full situation awareness (Endsley and Kiris, 1995), we might lose more than we gain.

Even fifty years after Fitts published his list, Hollnagel (Hollnagel, 2003b) argues that the machine (or automation) has been used for three main purposes over the years (which is in line with Fitts) i.e. to ensure more **precise** performance of a given function; to improve **stability** of performance by relieving people of repetitive and monotonous tasks; and to enable processes to be carried out **faster** and more efficiently. So, do Fitts' thoughts still prevail, or has research turned towards Jordan's argument?

The decision matrix suggested by Prince (Prince, 1985) was partly in line with Fitts in that some tasks were better performed by machines and some better by humans. But interestingly Prince also defined a set of tasks where the same task could and should be performed both by humans and by machines. Further, when there is no single allocation, the different resources need support from each other, which is in line with Jordan's argument. Hancock (Hancock and Chignell, 1992) argues that it is only when both human and machine can do the same task, the question of task allocation becomes an issue.

In line with Jordan, previous research (Hancock and Chignell, 1992, Kantowitz and Sorkin, 1987, Hou et al., 1993, Sheridan, 2000) agrees that the task allocation should been seen as a complementary between man and machine rather the divide the tasks solely to one. This complimentary of man and machine could be described as levels of automation.

LEVELS OF AUTOMATION (LOA)

Modern and traditional industry exhibits a broad spectrum of assembly systems, with varying degrees of automation, e.g. manual, semi-automated, or automated assembly (Rampersad, 1994). No simple way to make automation human-oriented exists that is applicable across all domains and types of work. Different processes and domains put different emphasis on flexibility, speed etc. requiring specific consideration of type appropriate automation. Also and how control can be enhanced and facilitated via proper automation design (Hollnagel, 2003b).

An extensive amount of research has been done in the area of levels of automation emphasizing different perspectives. A summary of some definitions and scales can be found in appendix 1. Automation research could be divided into three main groups;

- **Mechanical automation** (March and Mannari, 1981, Kern and Schumann, 1985, Groover, 2001, Duncheon, 2002)
- **Information and control** automation (Bright, 1958, Sheridan, 1992, Parasuraman and Wickens, 2008, Endsley, 1997, Parasuraman et al., 2000, Hollnagel, 2003a).
- **Combinations** of physical/mechanical and information/cognitive automation (Frohm, 2008).

Research on information and control has predominantly been used for military, aircraft (aerospace) and control room areas. In 2008 Frohm (Frohm et al., 2008) proposed a taxonomy and definition for levels of automation used in manufacturing systems, i.e. *"The allocation of physical and cognitive tasks between humans and technology, described as a continuum ranging from totally manual to totally automatic"*.

The taxonomy is a seven-step reference scale, for cognitive and physical LoA. Frohm (ibid) defined physical tasks as the level of automation for mechanical activities, *mechanical LoA,* while the level of cognitive tasks is called *information LoA*. Mechanical LoA is *WITH WHAT* to assemble, while Cognitive LoA is *HOW* to assemble on the lower levels (1-3) and *situation control* on the higher level (4-7).

A matrix integrating the two reference scales, as seen in figure 1, forms a 7x7 matrix, resulting in 49 possible solutions for task allocation, each including a physical LoA and a cognitive LoA. The figure also displays the division between human and machine assembling and monitoring the tasks.

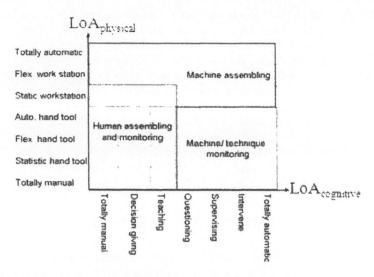

FIGURE 1 Joint matrix of physical and cognitive LoA (Fasth et al., 2009)

DYNAMO ++, A METHODOLOGY TO MEASURE AND ANALYZE LOA

To determine appropriate task allocation with a span of various levels of automation in assembly operations a methodology named DYNAMO++ was developed and validated (Fasth et al., 2010, Fasth et al., 2008 -b, Fasth, 2009). Initially, Value Stream Mapping (VSM) (Baudin, 2002) is performed in the current state of the system, in order to focus on the right tasks e.g. eliminating non-value adding tasks and improving necessary and value-adding tasks. Next, a Hierarchical Task Analysis (HTA) (Stanton et al., 2005) is executed by breaking down the value-adding assembly operations into tasks and sub-tasks, until it reaches a level, where only the human or the technology is responsible for *how* (cognitive) and *with what (physical)* the task is performed. The HTA is done to understand the structure of the assembly operations and to be able to discuss why different tasks are performed and why they are performed in that specific order. The different tasks in the HTA are subsequently compared to the LoA reference scales and given a value from 1-7 for the cognitive as well as the physical LoA and documented in the LoA matrix (fig. 1). The matrix provides a representation of the current LoA in the assembly system, but it also visualizes the company's general view of automation e.g. potential "binary" thinking regarding automation.

After the current state analysis, solutions for the future are proposed. A workshop is executed with the observers/researchers together with operators, production engineers and other stakeholders of the assembly system. The group jointly estimates relevant maximum and minimum LoA for each analyzed task. A minimum physical LoA can be related to ergonomics and heavy lifting, while cognitive LoA may be closely related to competence among operators or assembly instructions resulting from complex products with high variance. Further, the maximum LoA levels may propagate from feasible investment levels, levels of know-how, safety etc. Using respondents with good understanding of how observed tasks are performed, a good estimation of relevant maximum and minimum LoA levels can be achieved during the work-shop. Minimum and maximum levels are transformed into the LoA matrix, forming a Square of Possible Improvements (SoPI). The SoPI illustrates a span of possibilities, within the matrix, where companies believe their tasks and assembly operations could be successfully automated or performed manually. The SoPI is used for further analysis and system design specification. The DYNAMO++ methodology results in good data collection and insights into the current system functionality. It efficiently gathers possible solutions in a structured way, due to triggers for change delivering a well-substantiated demand specification list to the system designers.

CASE EXAMPLES USING THE DYNAMO ++ METHODOLOGY

During 2007-2009, the authors carried out ten case studies using DYNAMO++. Only real industrial assembly system case studies were made. The industries' reasons for redesigning their assembly systems were primarily: to increase quality, decrease time-parameters, and increase product and volume flexibility (Fasth et al., 2008 -a, Fasth and Stahre, 2008). A current state analysis was made using HTA and reference measurements of LoA. Measurements were gathered in the LoA matrix, as shown in figure 2. Data was gathered from six of the case studies. The tasks in each case are measured in one final assembly station.

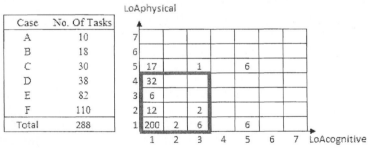

Case	No. Of Tasks
A	10
B	18
C	30
D	38
E	82
F	110
Total	288

FIGURE 2 Tasks observed from six case studies, documented in the LoA matrix

Approximately ninety percent of the tasks (fig 2) are assembled and monitored by humans, mostly done by hand and based on own experience (LoA= 1; 1), which is in line with Parasuraman and Wickens; "*humans are still vital after years of automation*" (Parasuraman and Wickens, 2008). The case question was whether the mechanical and cognitive level of automation should be changed. Results from the analysis (fig 3) showed that cognitive automation increased, although not extensively. Company goals towards increasing customization resulting in increasing information flow required new solutions for presenting information to the operators. Technical levels (i.e. automation) should be increased as well as the number of information modes. Human should still be in control of the task, but not merely in a monitoring role. In the LoA /SoPI matrix this means moving from

$LoA_{cog}= 1$ towards $LoA_{cog}= 3\text{-}5$. A majority of the companies did not prioritize an increase in mechanical LoA after the DYNAMO++ analysis.

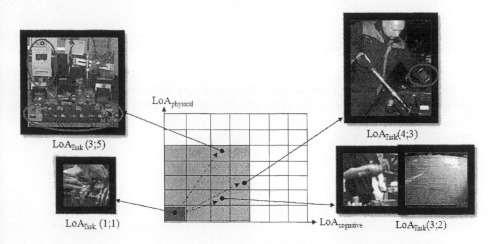

FIGURE 3 Example of a SoPI documented in the matrix for future state analysis

The SoPI shows a span of automation defining the performance boundaries of the task or operation solutions. It could be used for task allocation analysis and subsequently for a strategic step–by-step increase in automation.

In figure 4, the SoPI= (1-4; 1-5) which means that there are 20 different possible solutions including the current task that could be allocated. The figure also shows some examples of the different LoAs within the SoPI. From the LoA measurement (figure 3) and the SoPI analysis the effort, in terms of the number of tasks that needs changing when increasing automation, could also be estimated.

DISCUSSION AND CONCLUSION

The manufacturing complexity of increasing product customization in combination with cost awareness has forced companies to abandon previous strategies for full assembly automation. Task allocation in industrial manufacturing is therefore becoming increasingly important, specifically in assembly. The proposed method (DYNAMO++) for an analytical approach to "level of automation" analysis is the result of a broad review of reference scales addressing task allocation. The method emphasizes tasks that can be performed by humans as well as by automation, thus embracing the groundbreaking ideas of Fitts (Fitts, 1951), Jordan (Jordan, 1963) and Sheridan (Sheridan, 1995). The authors propose that the separation of cognitive and physical level of automation, while maintaining their relation in the LoA matrix, demonstrates an advance in the state of the art in task allocation. It should be emphasized that the exponential development of information technology could hardly have been foreseen by Jordan and Sheridan. Ten cases in real assembly system indicate that DYNAMO++ also provides industrially relevant results and increases quality of advanced, semi-automated manufacturing system analysis.

ACKNOWLEDGEMENT

The authors want to express their deep gratitude to VINNOVA (The Swedish Governmental Agency for Innovation Systems) for funding this research. Further, special thanks to the case companies and research colleagues in the ProAct project for the empirical data and interesting discussions.

REFERENCES

AMBER, G. H. & AMBER, P. S. (1962) Anatomy of Automation. *Prentice Hall.*

BAUDIN, M. (2002) *Lean Assembly: The Nuts and Bolts of Making Assembly Operations Flow,* Productivity Press.

BILLINGS, C. (1997) *Aviation Automation: The search for a Human-centered approach,* Mahwah, New Jersey, USA, Lawrence Erlbaum Associates.

BRIGHT, J. (1958) *Automation and Management,* Boston, USA.

DUNCHEON, C. (2002) Product miniaturization requires automation - but with a strategy. *Assembly Automation,* 22, 16-20.

ENDSLEY, M. & KIRIS, E. (1995) Out-of-the-loop performance problem and level of control in automation. *HUMAN FACTORS - Human Factors and Ergonomics Society.,* 37, 381-394.

ENDSLEY, M. R. (1997) Level of Automation: Integrating humans and automated systems. *Proceedings of the 1997 41st Annual Meeting of the Human Factors and Ergonomics Society. Part 1 (of 2), Albuquerque, NM, USA.* Santa Monica, CA, USA, Human Factors and Ergonomics Society, Inc.

FASTH, Å. (2009) Measuring and Analysing Levels of Automation in assembly systems - For future proactive systems. *Product and production Development, Production systems.* Gothenburg, Chalmers University of Technology.

FASTH, Å., BRUCH, J., DENCKER, K., STAHRE, J., MÅRTENSSON, L. & LUNDHOLM, T. (Approved for publication) Designing proactive assembly systems (ProAct) - Criteria and interaction between automation, information, and competence *Asian International Journal of Science and Technology in production and manufacturing engineering (AIJSTPME.*

FASTH, Å., LUNDHOLM, T., MÅRTENSSON, L., DENCKER, K., STAHRE, J. & BRUCH, J. (2009) Designing proactive assembly systems – Criteria and interaction between Automation, Information, and Competence. *The 42nd CIRP conference on manufacturing systems* Grenoble, France.

FASTH, Å. & STAHRE, J. (2008) Does Levels of Automation need to be changed in an assembly system? - A case study. *The 2nd Swedish Production Symposium (SPS).* Stockholm, Sweden.

FASTH, Å., STAHRE, J. & DENCKER, K. (2008 -a) Analysing changeability and time parameters due to levels of Automation in an assembly system. *The 18th*

conference on Flexible Automation and Intelligent Manufacturing - FAIM.
Skövde, Sweden.

FASTH, Å., STAHRE, J. & DENCKER, K. (2008 -b) Measuring and analysing Levels of Automation in an assembly system. *The 41st CIRP conference on manufacturing systems* Tokyo, Japan.

FITTS, P. (1951) Human engineering for an effective air navigation and traffic control system. Columbus,OH, Ohio state university.

FROHM, J. (2008) Levels of Automation in production systems. *Department of production system.* Gothenburg, Chalmers University of technology.

FROHM, J., LINDSTRÖM, V., WINROTH, M. & STAHRE, J. (2008) Levels of Automation in Manufacturing. *Ergonomia IJE&HF, 30:3.*

GROOVER, M. P. (2001) *Automation, production systems, and computer-integrated manufacturing,* Upper Saddle River, N.J., Prentice Hall.

HANCOCK, H. A. & CHIGNELL, M. H. (1992) Adaptive allocation by intellegent interfaces.

HOLLNAGEL, E. (2003a) *Handbook of Cognitive Task Design,* Mahwah, New Jersey, Routledge.

HOLLNAGEL, E. (2003b) The role of Automation in joint cognitive systems. *IFAC.*

HOU, T., LIN, L. & DRURY, C. G. (1993) An emperical studyof hybrid inspection system and allocation of inspection functions. *Internetional journal of human factors in manufacturing systems,* 351-367.

JORDAN, N. (1963) Allocation og functions between human and machine in automted systems. *Journal of applied psychology,* 47, 161-165.

KANTOWITZ, B. H. & SORKIN, R. D. (1987) *Handbook of human factors. Ch 3.3 Allocation of functions,* New York, Wiley.

KERN & SCHUMANN (1985) *Das Ende das Arbeitsteilung* Verlag Beck.

MARCH, R. & MANNARI, H. (1981) Technology and size as determinants of the organizational structure of japanese factories. *Administrative science quarterly,* 26, 33-57.

PARASURAMAN, R., SHERIDAN, T. B. & WICKENS, C. D. (2000) A model for types and levels of human interaction with automation. *IEEE transactions on system, man, and cybernetics - Part A: Systems and humans,* 30, 286-296.

PARASURAMAN, R. & WICKENS, C. D. (2008) Humans: Still Vital After All These Years of Automation. *HUMAN FACTORS - Human Factors and Ergonomics Society.,* Vol. 50, p. 511-520.

PRINCE, H. (1985) The allocation of function in systems. *Human Factors,* 27, 33-45.

RAMPERSAD, H. K. (1994) *Integrated and simultaneous design for robotic assembly,* Chichester, England.

SHERIDAN, T. B. (1980) Computer control and Human Alienation. *Technology Review,* 83, 60-73.

SHERIDAN, T. B. (1992) *Telerobotics, automation and human supervisory control,* Cambridge Massachussetts, MIT Press.

SHERIDAN, T. B. (1995) Human centred automation: oxymoron or common sense? *Intelligent Systems for the 21st Century, IEEE international.* Proceedings of: Systems,Man and Cybernetics.

SHERIDAN, T. B. (2000) Function allocation: algorithm, alchemy or apostasy? *International Journal of Human-Computer Studies,* 52, 203-216.

STANTON, N. A., SALMON, P. M., WALKER, G. H., BABER, C. & JENKINS, D. P. (2005) Human Factors Methods - A Practical Guide for Engineering and Design, Ashgate.

Author	Definition of Levels of Automation	Scale Mech	Scale Information and control
(Bright, 1958)	Divides the levels depending on who initiates the control, the human (1-4),the human together with automation (5-8) or the automation (9-17)	-	17
(Amber and Amber, 1962)	The extent to which human energy and control over the production process are replaced by machines	-	-
(Sheridan, 1980)	"The level of automation incorporates the issue of feedback, as well as relative sharing of functions in ten stages"	-	10
(March and Mannari, 1981)	Automaticity is defined in six levels from conducting the tasks manual, without any physical support, to fully automated cognition with computer control	6	-
(Kern and Schumann, 1985)	"Degree of mechanization is defined as the technical level in five different dimensions or work functions"	3 (9)	-
(Billings, 1997)	The level of automation goes from direct manual control to largely autonomous operation, where the human role is minimal	-	6
(Endsley, 1997)	The level of automation in the context of expert systems in most applicable to cognitive tasks such as ability to respond to, and make decisions based on, system information	-	10
(Satchell 1998)	The level of automation is defined as the sharing between the human and machines, with different degrees of human involvement	-	-
(Parasuraman et al., 2000, Parasuraman and Wickens, 2008)	"The interaction and task division between the human and the machine should instead be viewed as a changeable factor which can be called the level of automation"	-	10+ 4
(Groover, 2001)	"Level of mechanization can be defined as the manning level, with focus of the machines"	3	-
(Duncheon, 2002)	'Manual' tasks being those in which humans are responsible for conducting the task (e.g. application of epoxy). 'Semi-automatic' is a higher level of automation and involves automated alignment and application of epoxy by a robot. Material handling, on the other hand, is still conducted by humans, unlike 'automatic', where material handling is also automated.	3 (6)	-
(Frohm et al., 2008)	"The allocation of physical and cognitive tasks between humans and technology, described as a continuum ranging from totally manual to totally automatic"	7	7

CHAPTER 25

Cognitive Assembly Planning Using State Graphs

Daniel Ewert, Eckart Hauck, Daniel Schilberg, Sabina Jeschke

ZLW - Center for Learning and Knowledge Management &
IMA - Institute of Information Management in Mechanical Engineering &
IfU - Institute for Management Cybernetics
Faculty of Mechanical Engineering

RWTH Aachen University
Aachen, Germany

ABSTRACT

Assembly tasks are a big challenge for nowadays planning systems. Depending on the problem domain, a planner has to deal with a huge number of objects which can be combined in several ways. Uncertainty about the outcome of actions and the availability of parts to be assembled even worsens the problem. As a result, classical approaches have shown to be of little use for reactive (online) planning during an assembly, due to the huge computational complexity. The approach proposed in this paper bypasses this problem by calculating the complex planning problems prior to the actual assembly. During assembly the precalculated solutions are then used to provide fast decisions allowing an efficient execution of the assembly. ...

244

INTRODUCTION

THE POLYLEMMA OF PRODUCTION TECHNOLOGY

In the last years, production in low-wage countries became popular with many companies by reason of low production costs. To slow down the development of shifting production to low-wage countries, new concepts for the production in high-wage countries have to be created. The production industry in high-wage countries is confronted with two dichotomies. On the one hand there is the dilemma value orientation vs. planning orientation and on the other hand the dilemma scale vs. scope. These two dilemmas span the so called polylemma of production technology (Brecher et al., 2007) as shown in FIGURE 1. The cluster of excellence "Integrative Production Technology for High-Wage Countries" aims on reducing this polylemma. One approach to reduce the planning orientation by introducing automation in planning tasks is the project "Cognitive Control Systems for Production".

Vision of Integrative Production Technology

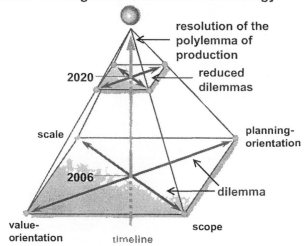

FIGURE 1 The polylemma of production

The project therefore develops a cognitive control unit (CCU) that aims to automate the assembly planning process, so that only a CAD description of the product to be fabricated is required as input to the system (Hauck et al., 2008).

TASK DESCRIPTION

The task of the CCU is to plan and control the assembly of a product described solely by its CAD data. An accordant description is entered by a human operator and the system then plans and executes the assembly autonomously. During the actual assembly the CCU cooperates with human operators: While most of the assembly actions are executed by the assembly robot, certain tasks can only be accomplished by the operator.

The presented approach is evaluated with the scenario depicted in FIGURE 2. The setup is a robot cell containing two robots (Kempf et al., 2009/Hauck et al., 2010), where only one robot (robot 2) is controlled by the CCU. The other robot (robot 1) delivers the separate parts for the final product in random sequence to a circulating conveyor belt. The CCU then can decide whether to pick up the delivered parts and put them into a buffer area or the assembly area for immediate use, or to refuse them.

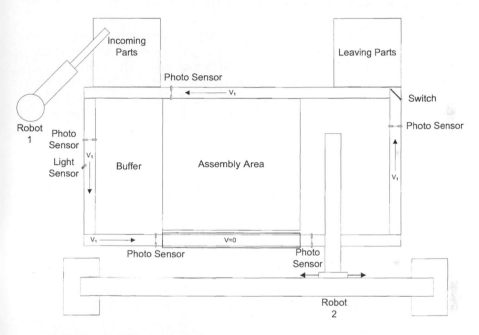

FIGURE 2 Diagram of the robot cell

Due to the random block delivery future states of the system cannot be predicted. The CCU is therefore facing a nondeterministic planning problem, requiring either an online re-planning during the assembly, whenever a not expected event occurs, or a plan in advance for all possible delivery sequences. Each of these strategies results in extensive computations, which lead either to slow responses during the assembly, or an unacceptable amount of preplanning. The approach proposed in this paper therefore follows a hybrid approach, based on state graphs as described in the remainder of the paper.

RELATED WORK

In the field of artificial intelligence planning is of great interest. There exist many different approaches to planning suitable for different applications. Hoffmann et al. (2001) developed the FF planner, which is suitable to derive action sequences for given problems in deterministic domains. Other planners are capable to deal with uncertainty (Hoffmann et al., 2005, Castellini et al., 2001). However, all these planners rely on a symbolic representation based on logic. The corresponding representations of geometric relations between objects and their transformations, which are needed for assembly planning, become very complex even for small tasks. As a result, these generic planners fail to compute any solution within acceptable time.

Other planners have been designed especially for assembly planning and work directly on geometric data to derive action sequences. A widely used approach is the Archimedes system by Kaufman et al (1996) that uses And/Or-Graphs and an "Assembly by Disassembly" strategy to find optimal plans. U. Thomas (2008) follows this strategy, too, but where the Archimedes system relies on additional operator-provided data to find feasible subassemblies, Thomas uses only the geometric information about the final product as input. However, both approaches are not capable of dealing with uncertainty.

STATE GRAPH BASED PLANNING

BASE IDEA

To allow for fast reaction times during the assembly the planning process is separated into an offline part, executed prior to the assembly, and an online part, executed in a loop during the assembly. The resulting system is drafted in FIGURE 3. While the Offline Planner is allowed computation times of up to several hours, the Online Planner's computation time must not exceed several seconds.

The task of the Offline Planner is the precalculation of all possible assembly sequences leading from the single parts to the desired product. These sequences are represented as a graph, which is transmitted to the Online Planner. During the assembly the Online Planner then maps the current system state to a state contained in the graph. Following, it extracts an assembly sequence that transforms this state into a goal state containing the finished product.

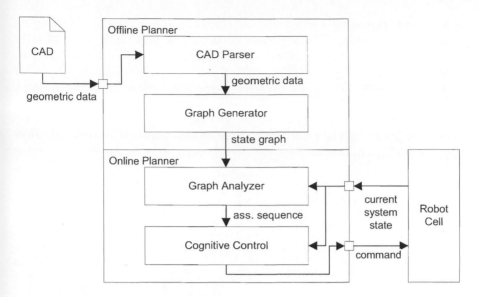

FIGURE 3 System Overview of the CCU

This assembly sequence is then handed to the Cognitive Control component which triggers the accordant robot commands, interacts with the human operator and reacts to unforeseen changes in the state. Thereby it is not bound to follow the given assembly, for example it can decide to invoke actions that move blocks from the conveyor belt to buffer and vice versa instead of continuing an assembly. The details of the described process are particularized in the following sections.

OFFLINE PLANNING

Graph generation

The CCU parses the geometric description of the desired product from a CAD file. To derive an assembly plan from this representation, the graph generator follows an "Assembly by Disassembly" strategy. Here, instead of planning how to assemble a construct from its parts, planning starts with the completed assembly. The algorithm then calculates all feasible segmentations of the assembly and recursively the segmentations of the resulting subassemblies. The algorithm ends when only single parts remain.

The separability of parts or subassemblies is computed by following the approach described in Thomas (2008). The algorithm computes for each pair of parts or subassemblies whether they can be separated without collisions with other parts. This information is then used to build a state graph as shown in FIGURE 4. The initial node for the generation contains a state description of the completed product. For all possible segmentations, successor nodes are created containing the state resulting out of the accordant segmentation. Single parts are not included in the states, so that every disassembly eventually ends within the node containing an empty state.

In the following step the translations between states are analyzed, now starting from the empty state. The algorithm analyzes, which additional parts are necessary and which action has to be executed to achieve the transformation between the states. This information is stored as meta data for each edge and the complete graph is sent to the Online Planner.

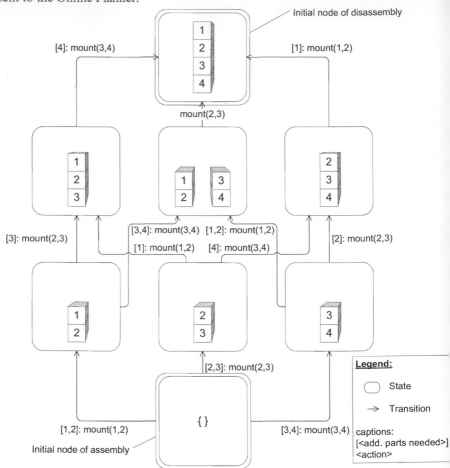

FIGURE 4 Example of a state graph representation

Graph pruning by mimicking human behaviour

Mayer et al (2009) found that assemblies executed by human individuals follow – in most cases – simple rules:

- Parts are placed next to each other. In other words, two separate subassemblies are not constructed in parallel
- Structures are build layer by layer from bottom to top
- Assembly begins at the boundary of the desired product

Incorporating these (or similar) rules into the graph generation results in two benefits: Pruning states that are inconsistent with the given rules, results in a smaller state graph and therefore in reduced computation times, both in the graph generation and also the graph analysis during the assembly.

As a second benefit, the human machine cooperation is improved: By following the given rules the systems behaviour meets the expectations of a human observer and therefore also increases the machine transparency.

However, by pruning the graph it no longer can be guaranteed that the later realized assembly sequence is optimal. As Kaufmann et al (1996) observed, "The tradeoff between computational time and optimality seems unavoidable". Therefore, the appliance of described rules has to be considered cautiously. Depending on the allowed computation time, the complexity of the product to be fabricated and the importance of machine transparency, the rigidity of appliance can be varied. This can be done by introducing weights for each rule, so that a state is only pruned from a graph when the sum of weights of violated rules exceeds a given threshold.

ONLINE PLANNING

Workflow of the Online Planner

During the actual assembly the Cognitive Control monitors the system's actual state. In the case of unforeseen events, for example the delivery of a new part to the buffer or the assembly area, the Online Planner is invoked to generate a new plan that takes the current situation into account. The Online Planner works in two phases: An update phase where the state graph is adapted to the current systems state, and the actual assembly sequence extraction.

Update phase

With this process, costs are assigned the state graphs edges. The cost of an edge depends on the assigned action and the availability of needed parts. In general, robotic actions receive lower costs than actions that must be executed by the human operator. Actions, that rely on parts, which are not accessible yet receive huge penalty costs. Due to this weighting of actions the Online Planner prefers currently realizable assemblies before assemblies that rely on parts, which are currently not present. However, reducing the penalty costs so that it becomes possible for a sum of realizable actions to outweigh the cost of one unrealizable action introduces speculation in the planning process: The algorithm could now chose, to prefer a short but yet not realizable sequence over a realizable that needs much more actions to fabricate the final product. This decision could turn out to be better, if the missing part is delivered to the assembly before the previously not realizable action is to be executed. This behaviour can be facilitated even more by assigning penalty costs depending on the distance of the action in question to the current state. Considering time: The further away the action to be executed, the lower are the penalty costs it retrieves.

Sequence detection

The algorithm now identifies the node of the updated graph which is matching the current state of the assembly. This node becomes the initial node from which an optimal path to the goal node has to be calculated. This is achieved using the A* search algorithm (Hart et al., 1968). A* chooses nodes for expansion with a minimal value for $f(x) = h(x) + g(x)$, where $h(x)$ is a heuristic for the distance from node x to the goal node, and $g(x)$ represents the present path costs. Here, $h(x)$ denotes the number of not correctly mounted parts and $g(x)$ is the sum of the costs of the traversed edges plus the costs for every necessary tool change along that path. The alteration between actions executed by the human operator and actions done by a robot also counts as a tool change, resulting in a preference for assembly sequences where human involvement is concentrated in single time slots.

COGNITIVE CONTROL

The Cognitive Control component is based on Soar, a cognitive framework for decision finding that aims on modelling the human decision process (Laird et al., 1987). It is responsible to invoke accordant actions based on the current world state received from the robot cell. It can request new assembly sequences from the Online Planner, invoke robot actions or communicate with an operator in order to proceed with an already received assembly plan. It also controls actions like placing parts, which cannot be mounted directly in the buffer area, discarding parts no longer used, etc.

OUTLOOK

Empiric studies need to show the capability of the approach to accomplish feasible assembly sequences for different products of varying complexity. It is to be determined, which weightings of realizable and not realizable actions lead to optimal results. Furthermore, research on how much graph pruning influences the average computation time is required, and if the gain in time is worth the risk of missing an optimal assembly. Additionally, the scalability and applicability of the approach to industrial use cases has to be examined.

CONCLUSION

The polylemma of production can be alleviated by incorporating automatic planning systems. Depending on the assembly setup, these systems have to face nondeterministic behaviour during the assembly, which increases the computational complexity by orders of magnitude. The approach presented follows a hybrid approach to deal with this problem. The actual planning problem is solved prior to the assembly by generating a state graph that describes all possible assemblies of the intended product. This graph is updated during the assembly and the at that time optimal assembly sequence is extracted. This process is by far less time consuming and allows a fluent assembly.

By taking into account how human proceed when assembling a product it is possible to speed up the graph generation. Situations during an assembly that would not occur during an assembly executed by humans are pruned from the graph. This results in reduced computation times and increased machine transparency by meeting human expectancy.

ACKNOWLEDGEMENT

The authors would like to thank the German Research Foundation DFG for supporting the research on human-robot cooperation within the Cluster of Excellence "Integrative Production Technology for High-Wage Countries".

REFERENCES

Brecher, C. e. a. (2007): *Excellence in Production*. Apprimus.

Laird, J. E.; Newell, A. and Rosenbloom, P. S. **(1987):** "SOAR: An Architecture for General Intelligence" *Artif. Intell. 33*, 1-64

Hauck, E.; Gramatke, A.; Isenhardt, I. (2008): „Einsatz Kognitiver Technischer Systeme im Produktionsumfeld." *AUTOMATION 2008 Lösungen für die Zukunft. VDI Berichte, Band 2032*, 305–308. VDI Verlag GmbH.

Hoffmann, J. (2001): "FF: The Fast-Forward Planning System" *The AI Magazine, vol 22*

Hoffmann, J., Brafman, R. (2005): "Contingent Planning via Heuristic Forward Search with Implicit Belief States" *Proceedings of the 15th International Conference on Automated Planning and Scheduling*, Monterey, CA, USA

Castellini C., Giunchiglia, E. and Tacchella A. (2001): "Improvements to sat-based conformant planning." *In Proc. of the 6th European Conf. on Planning*

Thomas, U (2008): *Automatisierte Programmierung von Robotern für Montageaufgaben*. Shaker

Hart, P. E., Nilsson, N. J., and Raphael, B. (1968): "A formal basis for the heuristic determination of minimum cost paths." *IEEE Transactions on Systems Science and Cybernetics, SSC-4(2)*, 100 - 107

Mayer, M.; Odenthal, B.; Faber, M.; Kabuß, W.; Kausch, B.; Schlick, C. (2009): "Simulation of Human Cognition in Self-Optimizing Assembly Systems" *Proceedings of the IEA2009 - 17th World Congress on Ergonomics* Beijing, China, 1-7

Kaufman, S., Wilson, R., Jones, R., Calton, T. & Ames, A. **(1996)**: "The Archimedes 2 mechanical assembly planning system" *1996 IEEE International Conference on Robotics and Automation Proceedings 4*, 3361-3368

Kempf, T., Herfs, W., Brecher, C. (2009): "SOAR-based Sequence Control for a Flexible Assembly Cell" *Proceedings of the 2009 IEEE Conference on Emerging Technologies and Factory Automation*, Palma de Mallorca

Hauck, E., Gramatke, A., Henning, K. (2010): *"A Software Architecture for Cognitive Technical Systems Suitable for an Assembly Task in a Production Environment"* Robotics, Automation and Control, InTech Education and Publishing

CHAPTER 26

Development of a Questionnaire for Prolonged Standing Jobs at Manufacturing Industry

Isa Halim[a], Abdul R. Omar[b], Alias M. Saman[b],
Ibrahim Othman[c], Mas A. Ali[b]

[a] Faculty of Manufacturing Engineering
Universiti Teknikal Malaysia Melaka
Locked Bag No. 1752, Pejabat Pos Durian Tunggal
76109 Durian Tunggal, Melaka, Malaysia

[b] Faculty of Mechanical Engineering
Universiti Teknologi MARA
40450 Shah Alam, Selangor, Malaysia

[c] Faculty of Art & Design
Universiti Teknologi MARA
40450 Shah Alam, Selangor, Malaysia

ABSTRACT

Prolonged standing has been identified as a potential threat for occupational health of workers in manufacturing industry. To minimize occupational injuries associated with prolonged standing, a reliable tool is required to investigate the root of the problems so that alternative solutions could be proposed. The purpose of the study is to develop a questionnaire that specifically used to assess difficulties in prolonged standing jobs. Literature survey, series of mini workshops, and pilot study have been performed to develop the questionnaire. Based on the carried out case study,

254

the questionnaire has shown a potential tool to manage prolonged standing jobs, however further work is required to establish the validity of the tool.

Keywords: Prolonged standing, Questionnaire, Manufacturing Industry

INTRODUCTION

Manufacturing industry can be associated with the challenging working environment, especially for metal-based productions. Usually, workers in the manufacturing industry perform jobs in prolonged standing as they need more frequent movement and handle heavy materials. Prolonged standing jobs may lead to discomfort and occupational injuries to the workers. In fact, current estimates show the prevalence of health risk in the United Kingdom (UK) to be more than 11 million workers or half the UK workforces caused by prolonged standing. In addition, health statistics estimated that hundreds of thousands of people in the UK have suffered from prolonged standing-related health problems and resulted over 2 million days sick leave a year (O'Neill, 2005). Usually, worker who performs job in prolonged standing particularly in an immobilization posture could lead to fatigue and pain in the lower limb muscles, lower back disorder, ache feet, legs swollen, even discomfort in the upper body parts such as neck and shoulder. As long term consequences, those symptoms promote severe injuries such as chronic venous disorders, circulatory problems, possibility of increase stroke risk, difficulty in pregnancy, and degenerative damage to the joints of the spine, hip, knees and feet. Even though it has been hypothesized as a threat to workers, job requirements such as repeated handling of heavy loads, mobility requirement for reaching and performing tasks, extended reaches and moves of substantial magnitude, manual downward forces of substantial magnitude, and mobility to monitor large areas, standing working position is the most practical choice.

HEALTH EFFECTS ASSOCIATED WITH PROLONGED STANDING JOBS

Prolonged standing contributes numerous disasters to the workers. It is well known that prolonged standing has been linked to the onset of work-related musculoskeletal disorders associated with lower back pain among industrial workers (Danik et al., 2009). A latest study reported that around 50% of healthy subjects reported lower back discomfort after exposed to 2 hours continuously standing (Gregory and Callaghan, 2008). Tomei (1999) in his study found that workers who exposed to more than 50% of working hours in standing position exhibited a higher frequency of chronic venous insufficiency than workers who spent less time in standing position. In addition to that, prolonged standing is also contributed to feet pain at the end of workday (Messing and Kilbom, 2001). As consequences of such work-related musculoskeletal disorders, productivity and job satisfaction of industrial workers could be affected (Uda, 1997).

Several studies agreed that prolonged standing could be a risk factor for preterm birth and spontaneous abortion among working pregnancy women (Eskenazi et al., 1994; Mozurkewich et al., 2000). Eskenazi (1994) reported that pregnant women who stood more than 8 hours in a working day have high chance for spontaneous abortion (provided by a previous history of spontaneous abortion).

ASSESSMENT METHODS FOR PROLONGED STANDING JOBS

Numerous studies have been carried out which focusing on prolonged standing at workplaces. The studies applied subjective method, objective method or a combination, and conducted either at a real workplace environment or a laboratory setting.

Studies related to prolonged standing that applied subjective method have been carried out to obtain psychological feedbacks of respondents, and normally they are carried out through a personal interview and questionnaire surveys. The common tools used for subjective method that has been reported are the Borg Scale or called Rating Perceived Exertion (RPE) and the Visual Analogue Scale (VAS). In sports and occupational ergonomics studies, Borg Scale is used to measure perceived exertion experienced by the tested subjects (Borg, 1970). The tool was originated by Gunnar Borg, whereby a scale of 6 to 20 was introduced. Later, the original scale is revised to 0 to 10 by many practitioners. Meanwhile, Visual Analogue Scale (VAS) commonly used to measure perceived exertion among tested subjects through questionnaires. VAS is a psychometric response scale to specify level of agreements of subjects to a statement by indicating a position along a continuous line between two end-points. Interestingly, VAS can be compared to Borg Scale as both tools have shown very similar results in terms of sensitivity and reproducibility (Grant et al., 1999).

Dissimilar to subjective method, objective method captures information on physiological and biomechanical responses of subjects to be tested. Usually, studies that apply objective method required scientific technical instruments which produce specific quantities such as frequency, distance, and temperature. Several common tools are optical leg volume meter, perometer, volumeter, and surface electromyography (sEMG). The optical leg volume meter has been recognized as an effective and reliable tool to evaluate leg volume measurement in detection of chronic venous insufficiency (Krijnen et al., 1997). The perometer is an electromechanical device that is used to measure limb cross section at multiple intervals for calculating limb volume. It has been applied to measure knee volume in subjects with impaired knee mobility (Man et al., 2003). The volumeter is a mechanical device that is utilized to measure volume of limb based on the amount of water displacement. In occupational ergonomics, sEMG has been considered a reliable tool to assess localized muscle fatigue (Kumar et al., 2004). Muscle fatigue due to prolonged standing exposures can be quantified by observing the changes in amplitude and frequency of electromyogram signals. When signal amplitude increases and power spectrum shifts to lower frequency, it indicates that the assessed muscles are in fatigue condition (Hostens and Ramon, 2005).

Based on thorough readings, the study found that a questionnaire that is specifically designed to capture information from the workers due to prolonged standing jobs becomes a necessary for ergonomics program in the workplace. Hence, the present study takes this opportunity to develop a questionnaire which can help ergonomics practitioners to improve occupational health of workers who are exposed to prolonged standing jobs, especially in manufacturing industry.

QUESTIONNAIRE DEVELOPMENT METHOD

Three methods were applied to develop the questionnaire: performing literature survey, organizing series of mini workshops, and carrying out a pilot study to finalize the design of the questionnaire. All methods will be explained in the next sections.

LITERATURE SURVEY

A literature survey related to prolonged standing exposures was carried out through the following hard-bound publications (articles, journals, and guidelines), on-line databases such as Science Direct, Scopus, and Springerlink. The selected articles were published from 1970 to 2009. The main purposes of the literature survey are to identify any relevant questions to be included in the questionnaire being developed, recognize health effects associated with prolonged standing, identify methods and tools that have been applied for assessing risks associated with prolonged standing, and explore alternative solutions to minimize health risk due to prolonged standing exposures. All information related to the study were compiled and discussed in a serial mini workshop.

SERIES OF MINI WORKSHOPS

When relevant information have been obtained from the literature survey, a mini workshop that involving a professor, lecturers, and research assistant was held to develop the questionnaire. During the workshop, the structure of the questionnaire is discussed by considering the background of respondents, time constraint, and functionality of the questionnaire. In addition, relevant questions, informative pictures were compiled and included to develop a questionnaire that is easy to understand, informative, yet reliable.

PILOT STUDY

A pilot study has been carried out to evaluate the usability of questionnaire. To do that, few postgraduate students and industrial workers have been assigned as respondents in the pilot study. The respondents evaluated the questionnaire by answering all questions and gave their comments on the following criteria: questions relevancy, easy to understand, time spending, and degree of difficulty of each question. After evaluation, the filled questionnaires were returned and all respondents' comments were considered to develop a new version of questionnaire.

THE QUESTIONNAIRE DESIGN

The final version of the questionnaire constituted four sections: (i) personal details and job activities, (ii) discomfort and pain that they experienced while performing jobs in prolonged standing, (iii) history of pain and treatment taken, and (iv) suggestion for improvement. Final design of the questionnaire can be found in Appendix I.

CASE STUDY

A case study was carried out in two metal stamping companies situated in Shah Alam, Malaysia to test the usability of the developed questionnaire. The main operation of the companies is metal stamping process. A set of developed questionnaire has been administered to the selected production workers to obtain their personal details and job activities, discomfort and pains that they experienced while performing jobs in prolonged standing, history of pain and treatment taken, and suggestion for improvement. The questionnaire form was filled by the workers during their morning and evening breaks. Furthermore, the researchers monitored the answering session so that the workers could ask any unclear questions. Each respondent took around 20 to 30 minutes to complete the questionnaire. Finally, all respondents were given a token of appreciation to appreciate their contribution. Before carrying out the study, the researchers have obtained approval from the Research Ethics Committee of Universiti Teknologi MARA.

RESULTS AND DISCUSSION

All collected data are keyed-in in a database and then interpreted through descriptive statistics and comparison analysis using Microsoft Excel 2007. Twenty male production workers participated in the questionnaire survey. In section 1, following personal details of worker were obtained:
- Age (mean = 25 years old, SD = 5.8 years)
- Work experience (mean = 6.4 years, SD = 5.6 years)
- Body mass (mean = 59 kg, SD = 11.8 kg)
- Experience of doing current job (mean = 2.4 years, SD = 3.4 years)
- Shoe size (mean = 7, SD = 1).

The survey involved 8 workers at stamping process production lines, 11workers performed jobs at maintenance department, and a worker worked at welding workstation. The companies run their production using three modes of working shift namely normal, morning, and evening. 10 workers worked in shift basis while another 10 workers worked from 8.00 am to 5.00 pm. Out of 20 workers, 19 of them performed their work in prolonged standing, 85% of workers involved in heavy lifting jobs, 14 workers manipulated jobs in body flexion, sitting on the vibrated area was very minimum (10%), and none of the worker worked in prolonged kneeling. Finally, both feet are open and parallel with each other, and one foot at front position and another one at the back were the most regular leg positions adopted by the workers during performing jobs in their usual workstations.

In section (2a) and (2b) of the questionnaire, the study observed that fatigue in the left wrist and left knee were the most common complaints as reported by 10 respondents for each problem. Besides that 9 respondents reported that they experienced fatigue in the right wrist. 8 different respondents claimed that they experienced fatigue in the lower back, right knee, left neck, left shoulder, and lower arms. Out of 20 workers, 8 and 7 of them experienced moderate fatigue and very fatigue respectively due to the mentioned complaints. In addition, 2 respondents reported that they experienced extreme fatigue. On the other hand, there is no respondent experienced no fatigue.

Section 2c investigates the level of fatigue in the lower extremities such as lower back (erectror spinae muscle), posterior leg (gastrocnemius muscle), and anterior leg (tibialis anterior muscle). Out of 20 respondents, 7 of them complained that they experienced moderate fatigue in the left and right lower back, while only one respondent reported no fatigue. In the left and right gastrocnemius muscles, majority of the respondents experienced very fatigue (7 respondents), however none was free from fatigue. Many respondents (8 workers) experiencing moderate fatigue in the left and right tibialis anterior muscles, in contrast 2 workers experienced no fatigue.

Section 2d examines the level of satisfaction of the workers with respect to their shoe, shoe-insole, shoe-base, and standing foundation. The survey found that almost half of the workers satisfied with their shoes and workplace condition. However, 25% respondents dissatisfied with the current that they have.

Section 2e reveals frequency of discomfort and pain experienced by the workers in the parts, muscles and tendons of both feet during performing jobs. Descriptive statistics found that more than half of the respondents reported that they never experienced discomfort and pain in the foot's part, in contrast few respondents (3 to 5 workers) always experienced discomfort and pain in their feet parts. In the muscles and tendons of foot, it was observed that majority of the workers reported that they never experienced discomfort and pain, however almost all respondents reported that they have experienced the symptoms in the ankle and under foot areas.

Comparison analysis have been performed to find any significant different in terms of fatigue in the general body between various categories such as workers' departments (workers who work at stamping process production lines and workers from handwork workstations), and normal working hour vs. shift basis. Based on t-Test (Two-Sample Assuming Equal Variances) ($\alpha = 0.05$) for workers' departments, it was found that there is no significant different between two groups of workers. In other words, either work at stamping process production lines or handwork workstations, they still experienced fatigue in the general body. Furthermore, t-Test observed that there is no significant different between workers who work in normal working hour and shift. In other words, both working modes have promoted fatigue to the workers.

In the other statistical analysis, correlation analysis was also carried out to determine any correlations between working in prolonged standing and fatigue in the general body and lower extremities such as lower back (erectror spinae muscles), posterior leg (gastrocnemius muscles), and anterior leg (tibialis anterior muscles). The carried out analysis found that there is a strong positive correlation between prolonged standing and fatigue in the general body as indicated by $r = 1$. Moreover, the analysis was also found that there is a positive strong association between prolonged standing and fatigue in the lower back, anterior and posterior legs.

Section 3 found that prolonged standing and condition of shoes were the main causes of discomfort and pain. Out of 20 workers, 7 of them took treatment to relief the pain, and pain killers and massage were found to be the most common treatments.

In the last section, more than half respondents suggested that standing with intermittent sitting, adequate work rest, and wearing proper shoes and shoe-insoles were the possible solutions to minimize pain and discomfort associated with prolonged standing jobs.

CONCLUSION

Significant findings pertinent to the development of questionnaire to capture workers' feedbacks due to prolonged standing jobs at manufacturing industry have been generated from this study. The carried out study concluded that:

- the developed questionnaire is functional and able to capture information from workers who are exposed to prolonged standing jobs in manufacturing industry,
- there is a correlation between prolonged standing exposure and fatigue in the general body and lower extremities, and
- the developed questionnaire has shown a potential tool to identify the main causes of discomfort and pain associated with prolonged standing jobs and possible counter measures to eliminate them.

ACKNOWLEDGMENT

The authors would like to acknowledge the Ministry of Science, Technology and Innovation (MOSTI) of Malaysia for funding this research under e-Science Research Grant, the Faculty of Mechanical Engineering of Universiti Teknologi MARA and Research Management Institute (RMI) of Universiti Teknologi MARA for providing facilities and assistance in carrying out this study. Special thank also goes to Miyazu (M) Sdn. Bhd. and Autokeen Sdn. Bhd. for the permission and ample opportunity to facilitate fruitful case study. Finally, the authors would like to thank Mr. Hadi for his helps in data collection.

REFERENCES

Borg, G. (1970), "Perceived exertion as an indicator of somatic stress". *Scandinavian Journal of Rehabilitation Medicine*, 2, 92-98.

Danik, L., Annick, C., Martin, D., Jean-Daniel, D., Janina, M. P., and Marcos, D. (2009) "Postural control during prolonged standing in persons with chronic low back pain." *Gait & Posture*, 29, 421-427.

Eskenazi, B., Fenster, L., Wight, S., English, P., Windham, G. C., Swan, S. H. (1994), "Physical exertion as a risk factor for spontaneous abortion". *Epidemiology*, 5, 6-13.

Grant, S., Aitchison, T., Henderson, E., Christie, J., Zare, S., McMurray, J., and Dargie, H. (1999), "A comparison of the reproducibility and the sensitivity to change of Visual Analogue Scales, Borg Scales, and Likert Scales in normal subjects during submaximal exercise". *Chest*, 116, 1208-1217.

Gregory, D. E., and Callaghan, J. P. (2008), "Prolonged standing as a precursor for the development of low back discomfort: an investigation of possible mechanism". *Gait & Posture*, 28, 86-92.

Hostens, I., and Ramon, H. (2005), "Assessment of muscle fatigue in low level monotonous task performance during car driving". *Journal of Electromyography and Kinesiology*, 15, 266-274.

Krijnen, R. M. A., de Boer, E. M., Ader, H. J., and Bruynzeel, D. P. (1997), "Venous insufficiency in male workers with a standing profession. Part 2: Diurnal volume changes of the lower legs". *Dermatology*, 194, 121-126.

Kumar, S., Amell, T., Narayan, Y., and Prasad, N. (2004), *Measurement of localized muscle fatigue in biceps brachii using objective and subjective measures.* CRC Press, New York.

Man, I. O. W., Elsabagh, S. M., and Morrissey, M. C. (2003), "The effects of different knee angles on knee volume measured with the Perometer(R) device in uninjured subjects". *Clinical Physiology & Functional Imaging*, 23, 114-119.

Messing, K. and Kilbom, A. (2001), "Standing and very slow walking: foot-pain pressure threshold, subjective pain experience and work activity". *Applied Ergonomics*, 32, 81-90.

Mozurkewich, E. L., Luke, B. Wolf, F. M. (2000), "Working conditions and adverse pregnancy outcome: a meta-analysis". *Obstetrics & Gynecology*, 95, 623-635.

O'Neill, R. (2005), *Standing problem.* Hazards 91.

Phyllis, M. K. (2002), "A comparison of the effects of floor mats and shoe in-soles on standing fatigue". *Applied Ergonomics*, 33, 477-484.

Tomei, F., Baccolo, T. P., Tomao, E., Palmi, S., Rosati, M. V. (1999), "Chronic venous disorders and occupation". *American Journal of Industrial Medicine,* 36, 653-665.

Uda, S. A., F., Seo, Yoshinga (1997), "Swell-preventing effect of intermittent exercise on lower leg during standing work". *Industrial Health*, 35, 36-40.

262

Appendix I

SECTION 1: PERSONAL DETAILS AND JOB ACTIVITIES

(1a) Employee number _____

(1b) Age _____

(1c) Gender ( ) Male ( ) Female

(1d) Working experience _____ year

(1e) Marital status ( ) Single ( ) Married

(1f) Body mass _____ kg

(1g) Height _____ m

(1h) Shoe size _____

(1i) Physical disability ( )Yes (specify) _____ ( ) No

(1j) Name of workstation _____

(1k) Job description: _____

(1l) Working experience of doing the current job: ____ year

(1m) Working mode .( ) shift ( ) normal working hour

(1n) Job activities and legs postures (please tick)

- ❏ Prolonged standing
- ❏ Standing with forward flexion
- ❏ Static standing
- ❏ Prolonged sitting
- ❏ Reaching goods
- ❏ Lifting heavy goods
- ❏ Prolonged bending
- ❏ Standing on the vibrated area
- ❏ Squatting
- ❏ Straight standing
- ❏ Standing with side bending
- ❏ Standing with body twisted
- ❏ Standing and walking frequently
- ❏ Standing and walking rarely
- ❏ Standing and sitting intermittently
- ❏ Prolonged kneeling
- ❏ Sitting on the vibrated area
- ❏ Handling imbalance goods.

(1o)   Please mark the empty box to indicate your legs position.

Diagram	Description	Your legs position
	both feet are open and parallel with each other	
	both feet are closed among themselves	
	one foot at front position and another one at the back	

SECTION 2: DISCOMFORT AND PAIN IN GENERAL BODY PARTS

(2a) Please specify types of discomfort and pain that you felt by marking the respective body parts.

1	Fatigue	2	Numbness	3	Tingling	4	Swollen
5	Strain	6	Hot/cold	7	Hamstring	8	Sprain

LEFT SIDE	BODY PARTS	HUMAN BODY DIAGRAM	BODY PARTS	LEFT SIDE
1 2 3 4 5 6 7 8	Neck		Neck	1 2 3 4 5 6 7 8
1 2 3 4 5 6 7 8	Shoulder		Shoulder	1 2 3 4 5 6 7 8
1 2 3 4 5 6 7 8	Upper back		Upper back	1 2 3 4 5 6 7 8
1 2 3 4 5 6 7 8	Forearm		Forearm	1 2 3 4 5 6 7 8
1 2 3 4 5 6 7 8	Elbow		Elbow	1 2 3 4 5 6 7 8
1 2 3 4 5 6 7 8	Lower arm		Lower arm	1 2 3 4 5 6 7 8
1 2 3 4 5 6 7 8	Wrist		Wrist	1 2 3 4 5 6 7 8
1 2 3 4 5 6 7 8	Lower back		Lower back	1 2 3 4 5 6 7 8
1 2 3 4 5 6 7 8	Thigh		Thigh	1 2 3 4 5 6 7 8
1 2 3 4 5 6 7 8	Knee		Knee	1 2 3 4 5 6 7 8
1 2 3 4 5 6 7 8	leg		leg	1 2 3 4 5 6 7 8
1 2 3 4 5 6 7 8	Foot ankle		Foot ankle	1 2 3 4 5 6 7 8

(2b) In general, please classify the severity of the reported symptoms by selecting the following scales:

0 · No fatigue   1 : Little fatigue   2 · Moderate fatigue

3 : Very fatigue   4| : Extreme fatigue

(2c) Do you feel any discomfort and pain in the following body parts during working hours, and/or after working hours? Please specify the severity of the discomfort and pain by selecting the provided scales.

Left side	Body parts	Right side
0 : No fatigue 1 : Little fatigue 2 : Moderate fatigue 3 . Fatigue 4 : Extreme fatigue	Lower back	0 : No fatigue 1 : Little fatigue 2 : Moderate fatigue 3 : Fatigue 4 : Extreme fatigue
0 : No fatigue 1 : Little fatigue 2 : Moderate fatigue 3 · Fatigue 4 : Extreme fatigue	Legs (back side)	0 . No fatigue 1 . Little fatigue 2 : Moderate fatigue 3 · Fatigue 4 : Extreme fatigue
0 : No fatigue 1 : Little fatigue 2 : Moderate fatigue 3 : Fatigue 4 : Extreme fatigue	Legs (front side)	0 · No fatigue 1 : Little fatigue 2 : Moderate fatigue 3 . Fatigue 4 : Extreme fatigue

(2d) Do you satisfy with your shoes and your workstation?

i. Shoes

☐ Very satisfy ☐ Satisfy ☐ Dissatisfy ☐ Very dissatisfy

Your comment ....

ii Shoe-insoles

☐ Very satisfy ☐ Satisfy ☐ Dissatisfy ☐ Very dissatisfy

Your comment ....

iii. Shoe-base

☐ Very satisfy ☐ Satisfy ☐ Dissatisfy ☐ Very dissatisfy

Your comment ....

iv. Standing foundation

☐ Very satisfy ☐ Satisfy ☐ Dissatisfy ☐ Very dissatisfy

Your comment ....

(2e) Please specify frequency of discomfort and pain in your feet while performing jobs in standing position (please refer to the scales).

0: Never   1: Rare (few in a month)

2: Frequent (few in a week)   3: Always (every day)

	Left side				Foot Picture		Right side			
A	B	C	D	E		A	B	C	D	E

MUSCLES AND TENDONS (LEFT SIDE) (please refer to pictures)																			
1	2	3	4	5	6	7	8	9	10	11	12	13	14	15	16	17	18	19	20

MUSCLES AND TENDONS (RIGHT SIDE) (please refer to pictures)																			
1	2	3	4	5	6	7	8	9	10	11	12	13	14	15	16	17	18	19	20

SECTION 3: HISTORY AND TREATMENTS

(3a)   When you noticed the discomfort and pain?

Month: ....   Year: ....

(3b)   How long the discomfort and pain occurred?

.... (days)   .... (week)   .... (month)   .... (year)

(3c)   During the last year, how often you experience the discomfort and pain?

...... times

(3d)   During the last week, do you still feel the discomfort and pain?

( ) Yes   ( ) No

(3e)   What are causes of the discomfort and pain?

....

(3f)   Have you take any treatments to relief the discomfort and pain?

( ) Yes   ( ) No

(3g)   If yes, what is treatment have been taken?

☐ Pain killer   ☐ X - ray   ☐ Massage   ☐ Injection   ☐ MRI

(3h)   Does the treatment able to relief the discomfort and pain?

( ) Yes   ( ) No

(3i)   How long medical leaves have been taken to relief the discomfort and pain?

.... (days)   .... (week)   .... (month)   .... (year)

(3j)   During the last 12 months, how long you have been prevented from doing regular job (e.g. home work)?

.... (days)   .... (week)   .... (month)   .... (year)

SECTION 4: SUGGESTION FOR IMPROVEMENT

(4a)   What are your suggestions to overcome discomfort and pain in your feet during performing jobs in prolonged standing? (you can tick more than one).

☐ Standing, alternate with sitting

☐ Sitting throughout the working hours

☐ Sufficient rest time for standing

☐ Do stretching in the feet/ easy exercise

☐ Use comfort shoe-insoles

☐ Wear appropriate shoes

☐ Install anti fatigue mats

☐ Other (please specify) .. ..

Chapter 27

Challenges of Beijing Road Transportation System: An Extended Application of TOC Think Processes

Guangyuan SHI[1], Sheng-Hung CHANG[2], Wei ZHANG[1]

[1] Department of Industrial Engineering, Tsinghua University, Beijing, China
[2] School of Management, Minghsin University of Science and Technology
Hsinchu, Taiwan

ABSTRACT

The thinking processes of the theory of constraints (TOC) have been successfully used in finding major constraints of production systems and improving productivity. In this study, we attempted to extend the thinking processes in finding major challenges of road transportation system, where it is difficult to draw a clear map of current reality tree (CRT). To extend the thinking processes, methods and procedures to create CRT was proposed and conducted, including focus group discussions, interview surveys, and questionnaires. The obtained CRT was validated through afterward questionnaires, which was used to improve the CRT. The methodology was applied to investigate the challenges of Beijing road transportation system. It is the preliminary application and the result could be used to provide suggestions for countermeasure design to improve the road transportation system.

Keywords: Transportation Safety, Theory of Constraints, Thinking Processes, Road Transportation System

INTRODUCTION

THEORY OF CONSTRAINTS AND ITS THINKING PROCESS

The Theory of Constraints (TOC), proposed by Dr. E. M. Goldratt in the 1980s, aimed at improving enterprise productivity and quality through continued systematic improvement (Reid & James, 2003). TOC adopts systematic approach to analyze a system, and assumes that any system must have a certain goal and also have certain constraints to achieve this goal. In other words, every system has at least one constraint and this constraint also means opportunity to improve such a system (Rahman, 1998). TOC treats constraints as positive factors and assumes that constraints determine the performance of a system and accordingly improving constraints will lead to improved system performance. TOC emphasizes that management effort and resources should be focused on system constraints to more effectively improve system performance (Rahman, 1998).

Three TOC paradigms have evolved in the 1980s and 1990s: logistics, global performance, and thinking processes. Originally, the logistics paradigm had managers looking for and elevating system constraints to increase production throughput. This included using drum-buffer-rope scheduling techniques that were implemented using the five focusing steps of TOC. Global performance measures were based on throughput, cost, and inventory. They allow managers to easily assess the impact of any given decision and help managers to focus on the institutional goal. Recently, thinking processes (logic tree, evaporating clouds, etc.) came into a more widespread use (Moss, 2002).

Thinking process is a method of problem measuring, analysis, and solving. Dr. Goldratt proposed this method in 1994 to find out underlying constraints within a system by common sense and logics. Thinking process emphasizes three main questions (Rahman, 1998): what to change? what to change to? and how to cause the change?. It uses five logic maps to answer these three questions: evaporating cloud (EC), the current reality tree (CRT), future reality tree (FRT), prerequisite tree (PRT), and transition tree (TT) (Kim et al., 2008).

CURRENT REALITY TREE (CRT)

Usually it is easy to identify the core problem of a simple system. However, when facing a complex system, it is often difficult to identify the existing problems and their reasons. And very often, these problems and reasons are interrelated. CRT is a powerful method for this kind of issue to analyze the current system reality through cause-effect logics. It mainly answers the question "what to change" and is the basis of thinking process. With CRT, it is possible to deeply analyze and identify the core problems that cause current system issues.

Currently, there are two methods to construct CRT: the traditional method and the evaporating cloud method (Kim et al., 2008). The latter one has many merits over the traditional one, such as easy to use and less time consuming. However, it is needed to make sure that the three undesirable effects (UDEs) are not included each other. Major steps of "3 evaporating clouds (3ECs)" include: (1) list the system's UDEs, (2) select 3 different UDEs and draw their ECs, (3) merge the 3 ECs into a core conflict cloud (CCC), and (4) construct CRT according to the CCC (Zhang & Zheng, 2009; Li & Chang, 2005).

OBJECTIVE OF THE STUDY

Road transportation system is a complex system for large cities. Beijing has a population of about 18 million. The road transportation system of Beijing is very complex with mixed traffics of automobiles, bicycles, and pedestrians. In the recent years, with the rapid increase of automobiles, traffic congestion and safety has become a major public problem. The aim of this study is to use TOC thinking processes to investigate the major challenges of Beijing road transportation system. Three ECs were used to construct Beijing road transportation system's CRT based on data gathered through interview surveys and focus group discussions.

However, it is needed to note that the approach we used to obtain the ECs may contain subjective bias. Therefore, we used specially designed questionnaires to validate the ECs and examine whether the ECs we obtained was consistent with afterward questionnaires result.

METHODS

GENERATING CRT USING CONVENTIONAL THINKING PROCESSES

The regular 3ECs were generated using conventional TOC Thinking Processes based on collected information, see logical thinking processes in Dettmer (2007). Specifically for this topic, interview surveys and focus group discussions were conducted to gather information from different kinds of road users in Beijing.

Figure 1 shows the obtained CRT based on the 3 ECs.

VALIDATING CRT

The obtained CRT may contain subjective bias from the researchers, accordingly validation was conducted to examine the consistency. The objectives of validation include: (1) Checking whether there were some "major causes" concluded by the researchers, but were considered "non-major causes" by the road users, (2) Checking whether there were any major causes ignored by the researchers, and (3) Quantifying the cause-effect relationship coefficients. The first two objectives are to make the cause-effect relationship complete and reasonable, while the third one is to quantifying the relationship for this kind of fuzzy relationships for some applications.

Regarding the validation process, each effect and its related causes in Figure 1 was considered as one unit or group, see Figure 2. Taking this unit for example, the following four questions were asked to participants to validate the cause-effect relationship:

(1) For the effect "Congestion during rush hours", how much percentage (please round to 10%) is due to "orderless urban traffic (such as UDE 5, 4, 9"?

(2) For the effect, how much percentage (please round to 10%) is due to "congestion at entrance/exit of ring roads"?

(3) For the effect, how much percentage (please round to 10%) is due to "too many vehicles on road"?

(4) For the effect, are there any other major causes in addition to the above? How much percentage (round to 10%)?

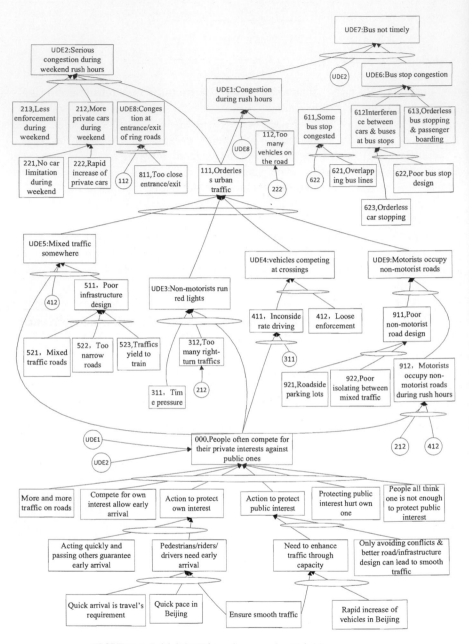

FIGURE 1. Initial CRT based on regular Thinking processes

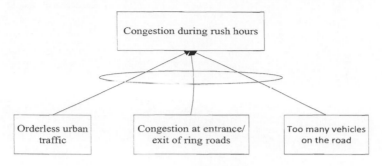

FIGURE 2. An example of validating unit

The first three questions are to validate the rationale and importance of the logical relationship between each cause and effect. If "0" is answered, then no logical relationship is considered for that specific cause-effect (in other words, the cause is not a real reason for the effect). If the number is between 1 to 10, then larger number means stronger cause-effect relationship. The last question is an open question and is used to check whether there are any major causes ignored by the researchers. The participants may add one or more causes and rate their importance. However, the final sum of the quantified numbers should be no larger than 10 (in other words, less than 100%).

Each effect and its corresponding causes in Figure 1 were taken as one unit and their corresponding questions were asked.

Dr. Goldratt proposed that usually a maximum of 3 major causes could explain 80% of the effect, and that eliminating these few major causes could almost eliminate the UDEs (Goldratt, 1998). Following this idea, the following rules were used to check the consistency.

Rule 1: For each of the cause-effect relationship in the CRT, if its coefficient is smaller than 1, then this relationship is considered not important and was eliminated.

Rule 2: For frequently raised opinions for open questions, it is needed to carefully analyze the problem and update the CRT if necessary. Re-validation is also needed for the updated CRT, see Figure 3 for the steps.

Rule 3: For a cause-effect logical unit, if the sum of all its relationship coefficients is smaller than 8, then some important causes might have been missed and accordingly it is needed to re-consider the logical unit.

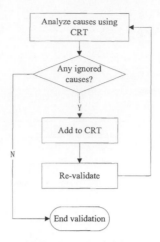

FIGURE 3. Validating processes for open questions

PARTICIPANTS

Participants were determined using the following criteria in order to achieve better result accuracy. Firstly, the participants should cover all major kinds of road users and stake-holders. In our study, the six kinds of road users and traffic administration policemen, including experienced private drivers, experienced professional drivers, novice drivers, bus passengers, bicycle riders, and pedestrians. Each kind has 5 participants. Secondly, the participants should have enough understanding of the studied system,. In this study, a majority of the 35 participants have been living in Beijing for over 10 years; see Figure 4 for their living period in Beijing. Finally, the participants should be able to conduct logic analysis.

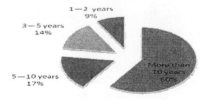

FIGURE 4. Living period in Beijing of the 35 particpants

DATA ANALYSIS

Experimental data analysis was conducted for each individual group. For non-open questions, the analysis was conducted in the following steps: (1) check the normality of cause-effect relationship coefficients obtained from the participants after outliers were removed (2) Calculate the average value of coefficients for each cause-effect relationship, (3) sum up the averages and obtain the total percentage of how much the causes can explain the effects.

Using the Rule 1 and Rule 3 above and follow these three steps to analyze Figure 2, the result is shown in Figure 5. Number near each arrow represents the importance of that cause for that effect. It can be seen that all the data satisfy Rule 1 and Rule 3. It is also noted that "too many vehicles on road" and "rush hour" are the two major causes of traffic congestion.

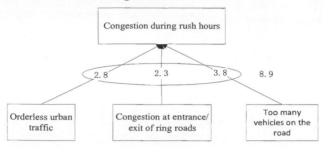

FIGURE 5. A validating unit of the improved CRT by questionnaires

For open questions, Rule 2 was used to analyze the experimental data. Take the effect "traffic congestion during rush hours" as an example, 9 of the 35 participants made replies. Table 1 shows the main causes to this effect as given by the participants. Among the causes, "Rapid increase of private cars" and "Too peaked rush hours" causes "too many vehicles on the road", and this accordingly causes "traffic congestion during rush hours". Therefore, they are indirect causes of this final effect. "Direction control not clear at some crossings" is the cause of "orderless urban traffic". We also think that "Traffic accidents " and "slow road development" might be major causes and accordingly we added them in our check unit and conduct another check using the same group of participants. The new result was shown in Figure 6. According to Rule 1, the coefficient of "slow road development" is 0.4 and accordingly was eliminated. The final result was shown in Figure 7.

Similarly, applying the above steps to each test unit will generate the improved CRT, see Figure 8.

Table 1. Opinions of causes of "traffic congestion during rush hours"

Causes	Proportion	Frequency
Traffic accidents in urban areas	1	3
Slow road development	1	1
Rapid increase of private cars	1	2
Direction control not clear at some crossings	1	2
Too peaked rush hours	1	1

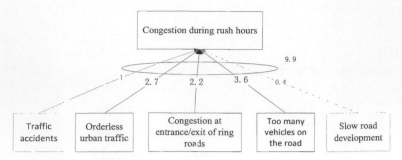

FIGURE 6. A validating unit of the improved CRT by open questions

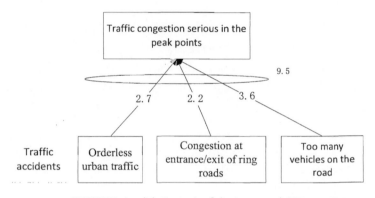

FIGURE 7. A validating unit of the improved CRT

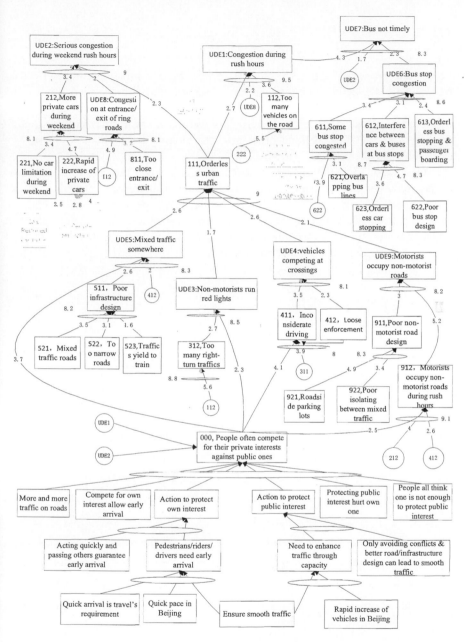

FIGURE 8. Finally improved CRT

DISCUSSIONS

Using the proposed approach, it is effective to minimize the bias effect of researcher's subjective opinions on the road transportation system study, and improve the understanding of the system. By quantifying each cause-effect relationship, it makes the CRT clearer. This preliminary work may provide a good way to extend TOC's thinking processes into looking for bottlenecks of service management systems.

One major limitation of the proposed approach is that the questionnaires used in this study contain too many questions and some of them were difficult to give a clear answer. To minimize the negative effect, a lot effort is used to find out suitable participants who can understand and give clear answers to the questions. It might be better to classify the questions for each kind of participants (pedestrians or drivers) to improve the reliability and validity.

REFERENCES

Dettmer, H.W. (2007), "The Logical Thinking Process-A Systems Approach to Complex Problem Solving", Milwaukee, Wisconsin ASQ Quality Press

Goldratt, E.M. (1998), "the constraints management handbook", The CRC Press, pp. 285-294

Kim, S., Victorial, J.M., John, D. (2008), "The theory of constraints thinking processes: retrospect and prospect", International Journal of Operations &Production Management, Vol. 28, No.2, pp. 155-184

Li, R.G., Chang, S.H. (2005), "TOC – from limited to unlimited", China Productivity Center (Taiwan)

Moss, H.K.(2002),"The application of the theory of constraints in service firms", PhD dissertation, Clemson University: United States -- South Carolina

Rahman, S.U. (1998), "Theory of constraints A review of the philosophy and its applications", International Journal of Operations & Production Management, Vol. 18, No. 4, pp. 336-353

Reid, R. A., James, R.C. (2003). "Applying the TOC TP: a case study in the service sector". Managing service quality , Vol. 13, No. 5, pp. 394-369

Zhang, Q.X., Zheng, Q.L, (2009), "Questions needs to be considered systematically – solving all questions in one time using TOC thinking processes", Taipei Times Culture

Chapter 28

An Extended Study on Human Performance in Control Rooms of Nuclear Power Plants

Huafei Liao, Jo-Ling Chang

Bechtel Power Corporation
5275 Westview Drive, Frederick, MD 21703
hliao, jjchang@bechtel.com

ABSTRACT

In an effort to investigate how human-system interface can impact human performance in nuclear power plant environment, 204 licensed operators from 10 commercially operating nuclear power plants in the US, 18 units in total, participated in a survey to examine 30 human errors due to human-system interfaces in nuclear power plant control rooms. The survey results were analyzed with factor analysis to derive a five factor structure: *Invisibility of System Status*, *Incorrect System Interface Design*, *Insufficient Support for System Diagnosis and Decision Making*, *Misoperations*, and *Manual Actions*. Corrective actions are suggested in the context of a decision-action model to prevent such errors in control rooms and/or mitigate their consequences.

Keywords: Human error, human performance, human factors, human reliability, human-system interface, factor analysis, nuclear power plants, control rooms

INTRODUCTION

Human performance and reliability are critical to the safety of nuclear power plants (NPP) because abnormal events in NPPs "have so often been the result of incorrect human actions" (IAEA 1988, p.19). On reflection of the causes of the Three Mile Island (TMI) accident, Kemeny et al. (1979) concluded that many incidents and accidents in NPP applications were due to insufficient recognition of the importance of the human beings who managed and operated the plants. Although issues related to human-system interfaces (HSI) in NPP environment have drawn more and more attention and many measures have been taken to prevent human errors since the TMI accident, incidents caused by human actions continue to occur.

In an attempt to understand the nature of HSI related human errors in NPP control rooms and then propose effective corrective actions to prevent the errors and/or mitigate their consequences, the present study took a statistical approach to the errors to identify their categories, occurrence patterns and trends, and reveal potential hidden interrelationships between them and their causal factors. The study commenced in November, 2008. A pilot study, which included responses from the first 138 participants, was conducted in March, 2009 (Chang et al., 2009). More participants were recruited between April, 2009 and June, 2009. Hence, compared to the pilot study, the analysis and conclusions presented in this paper were performed on and derived from more available data.

THEORETICAL BACKGROUND

In order to systematically analyze HSI related human errors, the present study analyzed 146 plant events listed under the categories of Control Room Operator Work Group and Man-Machine Interface Casual Factor in the Institute of Nuclear Power Operations (INPO) Operating Experience (OE) database between 12/3/1990 and 4/24/2008 to identify their contributing factors. The analysis results together with the categorization proposed by The United States Nuclear Regulatory Commission (U.S. NRC) in Human Factors Engineering (HFE) Guidelines (NUREG-0700) (NRC, 2007) were used to propose a hypothetical factor structure for HSI related human errors. As listed in Table 1, the factor structure first classifies the human errors into four categories based on the causes of the errors: *Operation Based*, *Controller Design Based*, *Deficient Indication Based*, and *Ambiguous Indication Based*. Each category is then broken down into individual contributing items. These items were identified in the analysis mentioned above and from the review of other literature available on control HSI (see Chang et al., 2009 for more detailed description for each item). Table 1 shows the 30 items with their primary reference sources listed. Items that do not have a reference source listed are derived from the literature review and cannot be pinpoint into a single source document.

Table 1 Hypothetical Factor Structure of Human-System Interface (HSI) Related Human Errors in Control Rooms of Nuclear Power Plants (NPP) (Chang et al., 2009)

Categories	Items	Sources
Operation Based	1. Operation movements.	INPO, 2004a
	2. Simultaneous operation.	Sheridan & Parasuraman, 2005
	3. Control room/simulator discrepancies.	INPO, 2007a
	4. Operate equipment incorrectly.	INPO, 2005a
	5. Inappropriate compensation.	INPO, 2000
	6. Over reliance.	Sheridan & Parasuraman, 2005
	7. Defeated safety features.	INPO, 2003a
	8. Inexperience.	INPO, 2002a
Controller Design Based	9. Operate on wrong equipment.	INPO, 2007b
	10. Controls too far apart.	Chen et al., 2005
	11. Controls too close together.	INPO, 2004b
	12. Incorrect function allocation – Manual actions designed to be automated.	Hugo & Engela, 2006
	13. Incorrect function allocation – Automated actions designed to be manual.	Hugo & Engela, 2006
	14. Equipment allowing failures.	INPO, 2006
	15. Work-around's.	
	16. Time limit to operation.	INPO, 1999
	17. No operator intervention allowed.	Naito et al., 1995
Deficient Indication Based	18. No alarm noting abnormal conditions and/or failures.	INPO, 2004c
	19. Insufficient plant information.	INPO, 2002b
	20. Boolean indication.	INPO, 2003b
	21. Unreliable indication.	INPO, 2003b
	22. No feedback.	Burns & Vicente, 2000
	23. No projection.	
	24. No trending.	Naito et al., 1995
Ambiguous Indication Based	25. **Control panel visually crowded.**	Grozdanovic, 2000
	26. **Color/Sound coordination.**	INPO, 2005b
	27. **Over-indication.**	
	28. **Non-intuitive control.**	
	29. **Display challenges.**	INPO, 2005c
	30. **Data searching.**	Naito et al., 1995

ANALYSIS AND RESULTS

PROCEDURE

A survey was developed based on the hypothetical factor structure shown in Table 1. Each survey question consisted of two parts. The first part asked power plant operators to indicate their opinions on whether an item was likely to cause an HSI error on a 7-point Likert scale ranging from *Strongly Disagree* (1) to *Strongly Agree* (7). The second part asked them whether the error was likely to be caused by *Operator Decision*, *Operator Action*, or *both* based on their past experience. The survey contained 32 questions, including two paired questions to estimate the internal consistency of participants' responses.

Ten commercially operating nuclear power plants, 18 units total, participated in the study. Several methods, e.g., telephone, email, and/or post, were used to contact the head of the Operations Department at each plant. Participants were directed to either finish the survey online or mail back completed hard copies of the surveys.

PROFILE OF PARTICIPANTS

204 licensed operators participated in the survey by June 20th, 2009. Out of the responses eight were completed on paper, while the remaining responses were collected online. Operations training instructors at various plants confirmed that a single operating plant had approximately 20 licensed operators to continuously staff its main control room. Therefore, the response rate of the survey was estimated to be 57%. The relatively low participation rate may be explained by the following three reasons. Firstly, since the invitation to the survey was distributed by the head of the Operations Department of each NPP, some plants sent the invitation to all operations personnel, while others selected small groups of individuals to participate in this study. Secondly, some licensed operators could not be reached due to training and/or plant outage. Thirdly, certain plants were not able to distribute the survey due to their policy restrictions, and some plants participated as one individual, which resulted only one survey response for the entire plant.

Of the 204 respondents, 14 were discarded due to low internal consistency (see discussion in the next section). The remaining 190 respondents had a mean age of 46.4 (SD = 7.32). 18.4% of them had 1 to 10 years of operations experience, 26.9% had 11 to 20 years, and 54.7% had over 20 years.

DESCRIPTIVE STATISTICS

The overall internal consistency of the survey responses as estimated by Cronbach's coefficient alpha is 0.64. This indicates that the survey has acceptable internal consistency.

The general characteristics of the survey results were examined. The mean scores for each of the items in the survey are between 3.5 and 6.2 and the standard deviations are between 0.88 and 1.65. The overall mean score of all responses is 5.1.

Items Q4 ("operate equipment incorrectly") and Q15 ("work around's") have the highest mean scores, 6.2 and 6.0, respectively, with the lowest standard deviations, 0.88 and 0.90, respectively. This indicates that the respondents tended to agree that incorrect operation of equipment and unnecessary corrective actions were the two leading contributors to human errors in control rooms. It should be noted that except for item Q12 ("manual actions designed to be automated"), all average responses on the rest survey items are above 4 (neutral), indicating that the respondents tended to agree that those items were contributors to human errors in control rooms. Both items Q12 and Q13 ("automated actions designed to be manual") involve function allocation between humans and machines; however respondents' attitudes toward these two items were in opposite directions (see Table 2). As such, it suggests that human errors can be more effectively prevented by improved automation level.

FACTOR ANALYSIS

Maximum likelihood factor analysis with varimax rotation was conducted to explore the hidden factor structure determined by the correlations among survey items. Five factors, whose eigenvalues are larger than 1.0, are retained. The five factors explain 73.7% of the total variance.

Factor loadings are presented in Table 2. The loading patterns are discussed below with reference to the hypothetical factor structure in Table 1 and the preliminary factor structure derived in our pilot study (Chang et al. 2009). As the preliminary factor structure was based on a smaller sample, we are more confident in the factor structure illustrated in Table 2 when there are discrepancies between the two.

- Factor 1 pertains to errors caused by inaccurate and insufficient information on the status of power plants. As such, Factor 1 is classified as "Invisibility of System Status". All Factor 1 items were originally categorized under "Deficient Indication Based Events" in the hypothetical factor structure and loaded on Factor 1 of the preliminary factor structure.
- Factor 2 relates to system design that violates basic human factors design principles by ignoring human physiological and psychological limitations. Therefore, Factor 2 is labeled "Incorrect System Interface Design". This factor contains a shuffling of items from the "Controller Design Based" and "Ambiguous Indication Based" categories. There is no clear match between Factor 2 and factors in the preliminary factor structure.

Table 2 Factor Analysis of Survey Items

Item No.[a]	Factor 1	Factor 2	Factor 3	Factor 4	Factor 5	Mean[b]	SD	Factor Mean	Decision-Action			χ^2_1
									Action (%)	Decision (%)	Both (%)	
Q21	**0.73**	0.06	0.19	0.24	0.16	**6.0**	1.03	5.7	14 (7.4)	131 (68.9)	45 (23.7)	94.4***
Q22	**0.59**	0.15	0.48	0.15	0.04	**5.3**	1.21		48 (25.3)	96 (50.5)	46 (24.2)	16.0***
Q19	**0.45**	0.30	0.26	0.38	0.01	**5.7**	0.98		17 (8.9)	131 (68.9)	42 (22.1)	87.8***
Q18	**0.44**	0.40	0.29	0.31	-0.06	**5.6**	1.18		61 (32.1)	94 (49.5)	35 (18.4)	7.0*
Q11	-0.04	**0.70**	0.06	0.13	0.24	4.8	1.37	5.1	139 (73.2)	31 (16.3)	20 (10.5)	68.6***
Q25	0.19	**0.55**	0.37	0.29	0.06	**5.1**	1.18		93 (48.9)	54 (28.4)	43 (22.6)	10.3**
Q13	0.26	**0.50**	0.14	0.15	0.10	**5.3**	1.13		87 (45.8)	50 (26.3)	53 (27.9)	10.0**
Q29	0.26	**0.42**	0.26	0.21	0.09	**5.1**	1.35		41 (21.6)	101 (53.2)	48 (25.3)	25.4***
Q20	0.21	0.24	**0.59**	0.23	0.12	4.8	1.31	4.7	24 (12.6)	141 (74.2)	25 (13.2)	83.0***
Q23	0.16	-0.05	**0.57**	0.12	0.09	4.2	1.38		44 (23.2)	111 (58.4)	35 (18.4)	29.0***
Q26	0.07	0.22	**0.46**	0.09	0.15	4.9	1.29		46 (24.2)	86 (45.3)	58 (30.5)	12.1**
Q24	0.22	0.26	**0.45**	0.28	0.04	4.9	1.27		24 (12.6)	132 (69.5)	34 (17.9)	74.8***
Q7	0.18	0.13	0.12	**0.62**	0.07	**5.3**	1.58	5.2	30 (15.8)	87 (45.8)	73 (38.4)	27.8***
Q6	0.01	-0.10	0.20	**0.46**	0.32	4.2	1.46		21 (11.1)	145 (76.3)	24 (12.6)	92.6***
Q4	0.29	0.22	-0.09	**0.41**	0.10	**6.2**	0.88		45 (23.7)	70 (36.8)	75 (39.5)	5.4*
Q12	0.00	0.20	0.11	0.16	**0.50**	3.5	1.36	4.0	69 (36.3)	87 (45.8)	34 (17.9)	2.1
Q10	0.30	0.42	0.13	0.19	**0.48**	4.5	1.52		142 (74.7)	30 (15.8)	18 (9.5)	72.9***
Eigenvalue	15.5	2.3	1.7	1.3	1.2							
Variance explained by each factor	5.7	5.1	4.8	4.1	2.4							
% Variance explained by each factor	19.0	17.0	16.0	13.7	8.0							
Cumulative % total variance explained	19.0	36.0	52.0	65.7	73.7							

*: $p < 0.05$; **: $p < 0.01$; ***: $p < 0.0001$

[a]: Reference Table 1 for Item Description.

[b]: Means above the overall mean (5.1) are in bold type.

Factor loadings (≥ 0.40) in bold type are considered to be significant. Five factors explained 73.7% of total variance.

- Factor 3 may be described as "Insufficient Support for System Diagnosis and Decision Making", since the items of this factor are related to information for operators to diagnose a system error and make a correct decision on how to respond to information from the system. Most of the items of Factor 3 are categorized under "Deficient Indication Based Events" in the hypothetical factor structure and loaded on Factor 2 of the preliminary factor structure.
- Factor 4 includes only items from the "Operation Based" category of the hypothetical factor structure and is named as "Misoperations". This factor maps closely to Factor 3 of the preliminary factor structure.
- Factor 5 relates to human errors caused by inappropriate design for manual actions and is labeled "Manual Actions" hence. This factor includes only items from the "Controller Design Based" category of the hypothetical factor structure and matches closely Factor 5 of the preliminary factor structure.

Group I (no incident) Correct Decision + Correct Action	Group II Incorrect Decision + Correct Action
	• Unreliable indication. • No feedback • Insufficient plant information • Display challenges • No trending • Color/Sound coordination • Boolean indication • Operate equipment incorrectly • Defeated safety features • Equipment allowing failures • Over reliance
Group III Correct Decision s + Incorrect Action	Group IV Incorrect Decision + Incorrect Action
• Control panel visually crowded • Controls too close together • Controls too far apart	• Non-intuitive control • No alarm noting abnormal conditions and/or failures • Time limit to operation • Incorrect function allocation – Manual actions designed to be automated

FIGURE 1. Populated Decision-Action Model

MODEL POPULATION

The survey items loaded on the factor structure was populated into a decision-action model proposed in Chang et al. (2009). Shown in Figure 1, the model classifies a survey item into one of the four groups based on whether a survey item is likely caused by a decision error or an action error. The four groups and their corrective action guidelines are briefly described below.

- Group I represents correct cognitive and manual action activities, hence no correction action is required.
- Group II represents situations where diagnosis is incorrect but the manual action correctly follows the misdiagnosis. This type of human error can be prevented by improvement to operations procedures, general guidelines, and pre-job briefings.
- Group III represents circumstances where manual actions are executed incorrectly with correct diagnosis. This type of human error can be prevented by additional operator training, peer checks, and management oversight.
- Group IV represents situations where diagnosis is incorrect and the subsequent manual actions are carried out incorrectly even under the wrong diagnosis. This type of human errors can be prevented by control room modifications with human factor re-evaluation to the extended condition.

DISCUSSION AND CONCLUSIONS

Human errors in control rooms have been identified to be a major contributing factor to various incidents that affect the safety and performance of entire plants. The present study took a statistical approach to investigate HSI related human error in NPP control rooms in an attempt to identify error categories, occurrence patterns and trends, and reveal potential hidden interrelationships between errors and their causal factors.

A hypothetical factor structure (see Table 1) of human errors in control rooms was first developed based on analysis from the INPO OE database and review of available literature on HSI in safety-critical industries. Then, a survey was conducted to collect opinions on the items of the hypothetical structure from 204 licensed operators in 18 NPP units in the US. Factor analysis on respondents' responses revealed a five-factor structure (see Table 2).

The fact that items Q4 and Q15 received the top scores indicates that manual task execution is the most significant contributor to human errors. This is reasonable considering that most operator actions in NPP control rooms are proceduralized; hence slips occur with the highest frequency. However, it does not mean that human factors issues related to manual actions are more important than those associated operators' cognitive activities. According to the decision-action model illustrated in Figure 1, most human errors are mistakes and can be attributed to operators' cognition. This implies that control room design should also focus on providing support for operators' rule- and knowledge-based behaviors. More research efforts are needed to examine human internal information flow and thinking process during disturbance and accidents that are not covered by operating procedures.

REFERENCES

Burns, C., & Vicente, K.J. (2000) A participant-observer study of ergonomics in engineering design: How constraints drive design process. *Applied Ergonomics*, 31, 73-82

Chang, J., Liao, H., & Zeng, L. (2009). Human-system interface challenges in nuclear power plant control rooms. *Proceedings of HCI (Human-Computer Interaction) International 2009*, July 2009, San Diego, CA, USA, pp. 729-737.

Chen, X., Zhou, Z., Gao, Z., G., Wu, W., Nakagawa, T., & Matsuo, S. (2005). Assessment of human-machine interface design for a Chinese nuclear power plant. *Reliability Engineering and System Safety*, 87, 37-44.

Hugo, J., & Engela, H. (2006) Function allocation for industrial human-system interfaces. Proceedings from: *The forth International Cyberspace Conference on Ergonomics*

IAEA (1988). *Basic safety principles for nuclear power plants* (Safety Series No. 75-INSAG-3). Vienna, Austria: International Atomic Energy Agency.

Institute of Nuclear Operations (1999). *Point Lepreau 1, 4/26/1999: Containment isolation system button-up during degassing of the degasser condenser* [908-990426-1]. Retrieved from http://www.INPO.org on August 14, 2008.

Institute of Nuclear Power Operations (2000). *Gentilly 2, 11/26/2000: Recirculated service water diesel motor pump 7131-P36 damaged* [851 001126-1]. Retrieved from http://www.INPO.org on August 14, 2008.

Institute of Nuclear Power Operations (2002a). *Cooper 1, 2/12/2002: Unintended increase in reactor power due to misoperation of reactor recirculation pump speed control* [298-020212-1]. Retrieved from http://www.INPO.org on August 14, 2008.

Institute of Nuclear Operations (2002b). *Gentilly 2, 2/12/2002: Local radiological alert due to a moderator leak in upgrading plants* [851-020212-1]. Retrieved from http://www.INPO.org on August 14, 2008.

Institute of Nuclear Power Operations (2003a). *Cernavoda 1, 6/12/2003: Inadvertent draining of in-service fire water tank 7140-TK1 results in start and damage of the diesel engine driven pump* [121-030612-1]. Retrieved from http://www.INPO.org on August 14, 2008.

Institute of Nuclear Operations (2003b). *Duane Arnold 1, 12/30/2003: High temperature in fuel pool because of procedure use problem* [331-000111-1]. Retrieved from http://www.INPO.org on August 14, 2008.

Institute of Nuclear Power Operations (2004a). *Millstone 2, 3/15/2004: Automatic reactor scram after a steam generator feed pump trip* [336-040315-1]. Retrieved from http://www.INPO.org on August 14, 2008.

Institute of Nuclear Power Operations (2004b). *Nine Mile Point 1, 5/4/2004: Two control rods scrammed during rod scram timing test* [220-040504-1]. Retrieved from http://www.INPO.org on August 14, 2008.

Institute of Nuclear Operations (2004c). *Point Lepreau 1, 6/6/2004: Primary heat transport (PHT) thermal transient* [908-040606-1]. Retrieved from http://www.INPO.org on August 14, 2008.

Institute of Nuclear Power Operations (2005a). *Darlington 2, 2/8/2005: Unit 2 turbine leading (normal) mode inadvertently entered* [932-041101-1]. Retrieved from http://www.INPO.org on August 14, 2008.

Institute of Nuclear Operation (2005b). *Susquehanna 2, 5/27/2005: B circulating water pump shutdown instead of B condensate pump* [388-050527-1]. Retrieved from http://www.INPO.org on August 14, 2008.

Institute of Nuclear Operation (2005c). *Gentilly 2, 7/12/2005: 3481-TK2 tank draining and 3481-P1 and P2 pump cavitations* [851-050530-1]. Retrieved from http://www.INPO.org on August 14, 2008.

Institute of Nuclear Operations (2006). *Perry 1, 7/9/2006: Reactor operation in unanalyzed region* [440-060709-1]. Retrieved from http://www.INPO.org on August 14, 2008.

Institute of Nuclear Power Operations (2007a). *Monticello, 3/14/2007: Half scram due to cold water transient during valve operation at Monticello Nuclear Generating Plant* [263-070314-1]. Retrieved from http://www.INPO.org on August 14, 2008.

Institute of Nuclear Power Operations (2007b). *Trillo 1, 11/28/2007: During routine tests, start-up of emergency diesel generator GY60 activated by reactor protection system* [715-071128-1]. Retrieved from http://www.INPO.org on August 14, 2008.

Kemeny, J. G. and other 11 authors (1979). *The need for change: The legacy of TMI -- Report of the president's commission on the accident at Three Mile Island*. New York, NY: Pergamon Press.

Naito, N., Itoh, J., Monta, K., & Makino, M. (1995) An intelligent human-machine system based on an ecological interface design. *Nuclear Engineering and Design*, 154, 97-108

Norman, D. A. (1981). Categorization of action slips. *Psychological Review*, 88, 1-15.

Rasmussen, J. (1983). Skills, rules, and knowledge; Signals, signs, and symbols, and other distinctions in human performance models. *IEEE transactions on systems, man, and cybernetics, SMC* 13, 257-266.

Reason, J. T. (1990). *Human error*. Cambridge: Cambridge University Press.

Sanders, M. S. and McCormick, E. J. (1993). *Human factors in engineering and design*. New York, NY: McGraw-Hill.

Sheridan, T. B., & Parasuraman, R. (2005). Human-automation interaction. *Reviews of Human Factors and Ergonomics*, 1, 89-129

Swain, A. D. and Guttman, H. E. (1983). *Handbook of human reliability analysis with emphasis on nuclear power plant applications* (NUREG/CR-1278, US Nuclear Regulatory Commission). Washington, DC.

CHAPTER 29

Design of a Collaborative Visualization Facility for Geoscience Research

Christopher D. White

Louisiana State University

ABSTRACT

The energy industry uses diverse data and complex earth and process models to design and manage energy extraction projects ranging from conventional oil and gas to novel enhanced recovery projects. These data seats models are often large, complex, and computationally expensive to maintain and use, motivating use of visualization to improve model quality and comprehension. A new reservoir geoscience visualization center is being created at LSU to address these needs in education and research. The geoscience visualization laboratory is funded by a gift of $1M for design, remodeling, equipment, software, and early-life maintenance and support. The selected room is approximately 800 feet, with adequate ceiling height and withnearby space to house noisy components. The design and construction are undertaken by a set of three committees: a steering committee controls funds and ensures the donor's intentions are honored, a user committee specifies desired capabilities, and the technical committee specifies components required to achieve the capabilities. The steering and technical committees work with the architect and contractor to ensure the capabilities are realized. The user committee has formulated an interactive, collaborative setting vis-à-vis an immersive or theater setting. The technical committee has specified a combination of high-performance workstations, large tiled display, projectors, and high performance compute, rendering, and file servers. The steering committee has provided further guidance on capabilities and esthetics. The laboratory is now in late design stages.

Keywords: Visualization, laboratory, collaboration, geosciences, energy

INTRODUCTION

The energy industry relies on geoscientists and engineers to locate, develop, and manage earth energy resources, whether for conventional sources such as coal, oil, and natural gas or for emerging sources such as shale gas, geothermal, and methane hydrates. These tasks require monitoring, fusion, and modeling of diverse data for a range of analyses and decisions.

The geoscience visualization laboratory centers on the subsurface elements of this challenge, which include geophysics, geology, geomechanics, reservoir engineering, seismology, and flow in porous media. The goal is to provide support for mapping complex geometries in three dimensions, correctly locating and estimating spatially varying properties, and viewing measurements and model results efficiently and informatively.

SCOPE

This project is in the design phase at the submission date of the abstract. For that reason, the emphasis is on capability and specification, with only a preliminary sketch included for orientation. We anticipate project completion in calendar year 2010.

Before proceeding to the laboratory per se, a quick discussion of geoscience tasks and visualization requirements is presented in the remainder of this introduction. Following the introduction, the organization of the design process is outlined. This is followed by a discussion of the current design, and we conclude with remarks on lessons and expectations.

GEOSCIENCE TASKS AND VISUALIZATION

Many geoscience and engineering tasks are especially visual, and will be emphasized in the design and implementation of this laboratory. For example, the correlation of seismic events (e.g., characteristic contrasts in seismic impendance evinced as consistent amplitude excursions) are picked and used to map horizons that bound intervals of interest. Although automated methods to pick events have long been available, these still must be checked "by eye." Because of the size (terabytes) of seismic data volumes, excellent visualization tools are needed. Other examples of strongly visual geoscience and engineering tasks include: mapping of transport properties; fault mapping (from pick anomalies); fracture identification from well logs; and identification of trends in measurements.

A second important area of application is in data quality control, especially when automated methods are used to model properties or pick horizons. Here, high performance is needed to rapidly slice or threshold models, to scroll between various properties, and to change viewpoints.

Finally, static views or animations of data and simulations can be used to support model comprehension. This is especially important in flow modeling, where the development plan – the number, type, and location of wells to be used to extract the resource – interacts strongly with the model parameters, and is subject to considerable uncertainty.

This laboratory will support computationally intense solutions such as seismic processing and flow simulation, but will concentrate on visualization sensu stricto.

REQUIREMENTS FOR GEOSCIENCE VISUALIZATION

Comprehension of geoscience problems poses a number of challenges. Geoscience data are of diverse types. For example, 3D seismic surveys my cover hundreds of square kilometers, with time series (traces) of reflected seismic amplitude recorded every few milliseconds at spatial locations only a tens of meters apart; the overall dataset size is commonly in the terabytes (Biondi 2006). Spatial models of reservoir geometry and properties (geomodels) may be geometrically complex and comprise millions of inferred properties (Oliver, Reynolds, and Liu 2008). Other data types include dense and accurate measurements of velocity, radiation, resistivity, etc., along wellbores (Ellis and Singer 2007); time series of pressures or rates measured at wells or groups of wells; and laboratory measurements of properties that must be associated with particular points in the reservoir.

Relevant geoscience data span vast scales in time, space. The simulation and interpretation of magnetic resonance used for rock property characterization is at a timescale of milliseconds and affected my approximately micron-scale geometry (Cohen and Mendelson 1982), whereas studies of petroleum origin and migration may span millions of years and hundreds of kilometers (Magoon and Dow 1994). Finally, critical parameters may vary widely: hydraulic permeability varies from nanodarcies for gas shales to hundreds of Darcies for hydraulic fractures approximately 12 decades.

Geoscience measurements are commonly indirect and/or intercorrelated. For example, electrical resistivity or magnetic resonance relaxation time may be used to estimate hydraulic connectivity. Construction of geomodels requires simultaneous visualization of such diverse data.

Geoscience data are uncertain. Not only are the normal observation errors incurred, but properties must often be estimated far from well.s This has given rise to stochastic and ensemble approaches. Not only do these models impose storage and computation burdens, but they are also difficult for geoscience teams to comprehend.

Energy recovery process models are nonlinear. Typically predictions and calibrations use numerical solutions of very large (currently up to 10^9 unknowns) systems of coupled equations. This is computationally expensive and can make it difficult to distill insights and strategies.

Jointly, these challenges – diverse data at many scales, complex interactions, uncertainty, and computational cost – motivate quality, collaborative visualization to support data exploration, model construction, and results interpretation.

THE DESIGN PROCESS

The sponsor gift specifies that the funds are to be used to support an interdisciplinary center for reservoir and geoscience visualization. This reflects the sponsor's conviction that visualization skills are a vital component in energy production, and the sponsor's commitment to support research and education in vital technology and education areas.

The implementation of the gift is being overseen by three committees with different roles:

- **The steering committee** ensures that the design is in accord with the donor's intent and college and university mission and needs.
- **The user committee** represents the needs and desires of university geoscience and engineering researchers and educators, to ensure that the design provides the capabilities they and their students need.
- **The technical committee** identifies the components and contractors needed to achieve the capabilities.

The roles and interactions of these committees are discussed in detail below, and interactions with the architect are discussed briefly.

STEERING COMMITTEE

This committee has several roles. First, it ensures that the technical plan is financially realistic within the scope of the gift. All expenditures must be approved by the steering committee. To date, it has approved expenditures for staff salary for design, refurbishing the designated room, travel to visit state-of-the-art visualization facilities in the region, and limited computer purchases to ensure interoperability.

Second, it ensures that the donor's intent is being honored. In particular, the visualization laboratory must be useful to a wide range of geoscience and engineering research and education, and it must be especially well suited for reservoir geoscience and modeling. The steering committee is also concerned with esthetics and image. That is, the laboratory should be as attractive as possible, while attaining required capabilities and staying within the gift budget. Moreover, the facilities capabilities and appearance must reflect favorably on the sponsor.

USER COMMITTEE

The user committee is drawn from a broad cross-section of LSU researchers with interests in high-performance visualization.

- Geoscientists specializing in hydrogeology, structural geology, seismology, high-pressure and –temperature reactive flows, geothermal energy, and thermohaline flows.
- Petroleum, chemical, and civil engineers engaged in pore-scale modeling, reservoir modeling, reservoir simulation, history matching, engineering hydrology, non-Darcy flow, and geothermal engineering.
- An applied mathematician with expertise in variational fracture, including application to geothermal systems.
- Computer scientists with interests in visualization and tangible interfaces.

This diverse group embraces the wide range of specialties needed to identify geoscience visualization needs, including expertise of interested "outsiders" such as mathematics and computer scientists. Over a dozen individuals have attended one or more of the preliminary design meetings, and several individuals have been involved in the entire process and interact extensively with the technical committee.

The user committee addressed several issues. First, what should the facility look like? There is a range of options, that can roughly be described as (a) immersive or cave-like, including a theater setting; (b) a multimedia classroom with very good visualization; (c) a high-end remote conferencing facility; (d) a workspace or laboratory. Of course, these particular roles are not exclusive, but some weighting must be selected.

Second, what rendering capabilities are needed? Practically, this means identifying a software stack, and to some degree the associated operating systems and possibly computing hardware that are needed. Here, the user committee tried to emphasize capability versus technology, but the line sometimes blurred.

Finally, what do the facility type (question 1) and rendering (question 2) imply about rendering and displays? That is, what sorts of displays (individual, shared, wall-mounted, projected, etc.) are needed to make the facility perform well?

TECHNICAL COMMITTEE

This group was kept small to ensure easier communication and decision-making. It consists of a representative from petroleum engineering (the author) to represent the user and steering committee input, and to coordinate. The coordinator is joined by a visualization specialist who has created a high-performance visualization services center in the main library at LSU, and a systems and network administrator from the college of engineering.

The technical committee usually attends user committee meetings to ensure good understanding and tight integration between technology and needed capability. At least one technical committee member normally attends the steering committee meetings to make progress reports and funding requests.

The technical committee meets with university facilities and services to plan the refurbishing of the room and ensure that utilities infrastructure is adequate. Finally, the technical committee coordinates with the architect on the design. For example, the technical committee conveys power, cooling, and weight data to the erchitect. The proposed design is also presented to the user committee by the technical committee and the architect.

Other user committee responsibilities include identification of vendors, soliciting bids, and ensuring that budget requests to the steering committee are reasonable and timely; soliciting visits from vendors who might supply products for the center; testing early implementations of hardware and software selected; and performing, or arranging for, installation of required components.

CURRENT DESIGN

In this section the general design decisions are summarized first. This is followed by more details about selected computer, display, and presentation capabilities. The design is not final.

GENERAL DESIGN DECISIONS

The chosen emphasis was on collaboration, by creating an environment where a group of several researchers and students could meet and share data and visualization resources. This led to a gently tiered seating arrangement that provides adequate sight lines for all participants while imposing minimal formality (Figure 1.1). The main displays are on the right side of this schematic.

The first two rows are dedicated to housing high-performance workstations with dual 24 in widescreen monitors. The monitors will be mounted on pivoting arms so that they can be tipped to improve sight lines from the rear when not in use. Each row accommodates 6 workers and 3 machines. The third row has full power and network ports for 6 laptops. A fourth row has informal, lounger seating (with power and network ports available). The accommodation is for 24, allowing as it to be used as a small classroom or moderate-sized meeting room.

The renovation of the room will include raising the ceiling for improved sight lines; installation of wallboard, ceiling panels, and carpet to improve acoustics; and efficient, dimmable lighting. The entryway (lower left of Figure 1.1) will be partially walled away to permit people to come and go without disturbing attendees. There is also storage or rack space in the cabinet at that location, and space to array literature or registration materials on the top of the cabinet. Signs describing the use and sponsorship of the room will be attached to the wall above the cabinet.

As detailed below, a variety of computing, storage, and display resources will be provided.

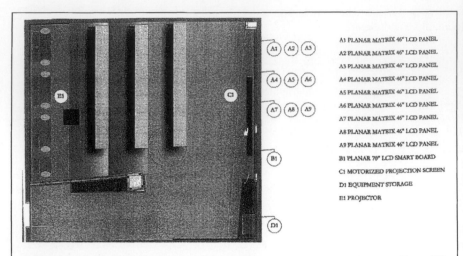

A1 PLANAR MATRIX 46" LCD PANEL
A2 PLANAR MATRIX 46" LCD PANEL
A3 PLANAR MATRIX 46" LCD PANEL
A4 PLANAR MATRIX 46" LCD PANEL
A5 PLANAR MATRIX 46" LCD PANEL
A6 PLANAR MATRIX 46" LCD PANEL
A7 PLANAR MATRIX 46" LCD PANEL
A8 PLANAR MATRIX 46" LCD PANEL
A9 PLANAR MATRIX 46" LCD PANEL
B1 PLANAR 70" LCD SMART BOARD
C1 MOTORIZED PROJECTION SCREEN
D1 EQUIPMENT STORAGE
E1 PROJECTOR

Figure 1.1 Preliminary design schematic; room is approximate 28 ft deep. From GD Architecture.

COMPUTING

The workstations will be near current desktop performance limitations, with at least 32 GB of RAM, multiple multicore processors, and large solid-state RAID drives for speed and low noise performance. Various users specified needs for Linux, Apple, and Microsoft operating systems. For example, some important reservoir modeling software is available only on Linux (for seismic processing) or Windows (some reservoir modeling applications). A significant minority of the users has OS X on their desktops, and it is important to support those users in the visualization laboratory. Therefore, the workstations will be a mix of dual-boot Windows-Linux and triple-boot Windows-OS X-Linux computers. Fortunately, we have experience in multiboot setups at the Visualization Services laboratory at LSU.

The compute server (a small Linux cluster) and graphics rendering cluster (GPUs for parallel rendering) are still being specified, and bids will not be solicited until the renovation is more nearly complete. The file server is an expandable network attached storage device. It will have dual network links: one directly to the Linux compute cluster, and one to a dedicated router for the laboratory. Compute, graphics, and file servers will be installed in a separate room with dedicated 10 Gbps connectivity.

To ensure a good user experience, we need to ensure good data access: movement of large data sets, or waiting for transfers, interferes with usefulness. We are working to connect our file server to other visualization centers with 10 Gbps or better speeds, and a local, dedicated router. We plan adequate server space for users to mirror data sets on that server so that they can be accessed in the visualization laboratory with our dedicated link. We will avoid creation of persistent user data on any machines within the visualization laboratory, to keep these machines available to all.

DISPLAYS, PRESENTATION, AND CONFERENCING

The main display in the room will consist of 9 thin-bezel (approximately 3 mm gap) modular HD LCD monitors arranged in a 3-by-3 tiled display (A1-A9, Figure 1.1). These flat panel displays were chosen for their simple, integrated, modular design; acceptable cost; lower power and cooling requirements; and much simpler, cheaper maintenance compared with projection solutions. Two HD projectors will be used, one to drive a smartboard (B1, Figure 1.1) and another presentation projector to project on a pretensioned screen (location to be decided).

Rather than using a traditional podium (as shown in Figure 1.1), a shallow, low table will be placed in front of the first row of workstations. Network, display, and power ports will be provided at several points along the table. This should simplify presentations. The smartboard will be driven from a computer in the casework at the front of the room (D1, Figure 1.1).

The main display and presentation projector will be accessible to rendering cluster, the 6 workstations in the first two rows, and the podium computer via a conference room controller and a matrix switch.

The room will be equipped with two cameras, one to broadcast the audience and the other to capture the speaker. The whiteboard or other projector could also be streamed out. The room will also be equipped with a display visible to the speaker to echo the outbound speaker camera. The moderate-sized room requires only a few speakers. Wired microphones will be placed in each rank of seats, and a wireless microphone will be available for a presenter.

In addition, several HD monitors may be mounted on the right wall (Figure 1.1) to allow presentations or slideshow displays. These monitors have not yet been specified, but video, power, and mounting connections are being installed as part of the room renovation.

LESSONS AND EXPECTATIONS

There have been a number of the normal delays in allocating space and starting design. In particular, creations of design and commencing construction have taken much longer than projected. If equipment had been ordered early, it would have been costlier for a similar level of performance, or lower performance. In fact, the thin-bezel tiled displays were not even available at the time of the first design specification. In that sense, delay has been beneficial (if frustrating).

Lesson learned, this delay is being exploited to get higher performance computing both on the desktop and in the servers. We have only ordered two workstations and the NAS file server, which will allow us to work out configuration issues for serving files with multiple OSs with various authorization methods. For the NAS file server, we have purchased a highly expandable but currently nearly empty unit. This will allow us to take advantage of anticipated improvements in disk performance, price, and capacity.

The users are anxious to have this facility online, and we plan to start physical renovation very soon. We expect to have moderately high demand for use by geoscience and engineering groups as a visualization workroom; researchers with geoscience and energy ties will be accorded first priority. After working meetings, we expect significant usage as a computer-equipped classroom, especially for smaller graduate classes. We will book other meetings and conferences as room scheduling allows.

ACKNOWLEDGEMENTS

The design and creation of this center is made possible by a gift from Chevron. Chevron's gift is specified to fund a reservoir visualization center at the Craft and Hawkins Department of Petroleum Engineering, Louisiana State University. Chevron professionals, especially Michael Geldmacher, have provided detailed technical and esthetic guidance. The LSU Foundation has provided guidance on gift management. Abundant technical guidance has been provided by Imtiaz Hossain, manager of the Visualization Services Center at Louisiana State University, and Chris Schwehm, Associate Director for System Administration and Network Management, LSU College of Engineering. Further guidance on computation and visualization has been provided by Blaise Bourdin, Associate Professor of Mathematics, LSU. Finally, a diverse cadre of potential users from many departments at LSU has aided in preparation of the capabilities specification.

The design schematic has been prepared by Fleming Ethridge of GD Architecture LLC, Baton Rouge.

REFERENCES

Biondi, B. (2006). *3D Seismic Imaging: Three Dimensional Seismic Imaging*. Society of Exploration Geophysicists.

Cohen, M. H., and Mendelson, K. S. (1982). "Nuclear magnetic relaxation and the internal geometry of sedimentary rocks." *Journal of Applied Physics*, 53(2), 1127–1135.

Ellis, D. V., and Singer, J. L. (2007) *Well Logging for Earth Scientists*. Springer.

Magoon, L. B., and Dow, W. G., eds. (1994). *The Petroleum System: From Source to Trap*. AAPG Memoir 60.

Oliver, D. S., Reynolds, A. C., and Liu, N. (2008). *Inverse Theory for Petroleum Reservoir Characterization and History Matching*. Cambridge University Press.

CHAPTER 30

Survey of Workload Factors for Technicians in Wind Turbine Maintenance Tasks: A Pilot Study

Guo-Feng Liang[1], Patrick Patterson[2], Sheue-Ling Hwang[3]

[1, 3] Department of Industrial Engineering and Engineering Management
National Tsing Hua University, 101, Section 2, Kuang-Fu Road
Hsinchu 300, Taiwan

[2] Department of Industrial Engineering, Texas Tech University
Box 43061, Lubbock, TX 79409-3061

ABSTRACT

The purpose of this study was to survey the perceived workload on technicians during maintenance tasks on wind turbines. Categories of condition influencing factors including human, task, system, environment, human-computer interface, and mental workload were designed into a subjective questionnaire survey. Factors were collected, analyzed and evaluated from previous studies or field interviews. To discover the critical items, fuzzy logic and fuzzy inference system were used to describe the degree of environment condition, degree of risk, and degree of effect on human body, for each item. Twenty-eight items were classified into three Influence Levels: Level 1 (low impact), Level 2 (intermediate impact), and Level 3 (critical impact). Result revealed that technicians ranked 'effort' as the number one contributing factor to workload with 'physical demand' and 'mental demand' as other important factors. In addition, contributing factors recognized by technicians working, classified as critical impact (Level 3) were (1) temperature and humidity variability in the wind tower (2) dangerous workplace, and (3) inconvenient workplace. The results not only provide reference data for wind turbine structures, but will also provide information to better prevent or reduce incidents or accidents on wind farms.

Keywords: Conditions influencing factors, safety and health, wind tower turbine, maintenance tasks.

INTRODUCTION

Effects of wind turbines on environment, human safety and health have been studied in recent years. Although the effects of wind farms appear to be less than nuclear power plants, it still has potential impacts on environment and on humans. Environmental issues include noise and vibration, soil erosion, threats to biodiversity that include habitat alteration and impacts to wildlife (International Finance Corporation, 2007). Compared with nuclear power's use of uranium mining, which destroys habitat and harms wildlife, wind power has impacts that are more modest on habitat wildlife. For wind turbines, noise can be problematic as they disturb people living near the wind farm (Western Australian Planning Commission, 2004; Berg, 2004). These noises are created from the turbine gearbox or generator, and aerodynamic noise (movement of the blades). In addition, occupational safety and health hazards specific to wind energy facilities and activities mostly include working at heights and working over water (International Finance Corporation, 2007). Therefore, wind farm developers need to consider all aspects of wind energy projects. Studies to date have focused mostly on the effects of wind turbine noise on the surrounding area. It is important to consider wind turbine impact on the technicians during maintenance tasks in the field.

In wind turbine maintenance tasks, on average, three to four corrective actions are required for each turbine. The mean down time per failure is two to four days with causes divided between mechanical and electrical problems (Leonardo Energy, 2007). Technicians who work inside the wind turbine have more safety and healthy problems than people who live in the surrounding area. They usually maintain the parts (i.e., generator component, rotor component or structural support component) of wind turbine, often climbing 200 to 300 feet (60 to 90 meters) in modern wind turbines. Technicians become fatigued from climb up and down the wind turbine. Therefore, physical and mental aptitudes for working at height are required in wind turbine. Confined space in wind turbine also limits technician activities, reduces work performance, and induces safety problems. According to the findings of Greef and Broek (2004), work conditions influence human performance. In addition, improved working conditions have positive effects on the company performance and add to the initial effects of the health and safety measures. Also, improvements in working conditions have a beneficial effect on productivity; productive and innovative companies generally have good working conditions (European

Foundation, 1998). Weather such as extreme cold or extreme heat is a challenge when technicians work in a wind farm. In cold weather, technicians may suffer intense shivering, feeling of cold and numbness, muscle tensing, fatigue, poor coordination, disorientation, blueness of skin, irregular pulse, slurred speech, retreat inward psychologically, dullness, and apathy. In hot weather, heat may result in problems of heat rash, heat cramps, heat exhaustion, and heatstroke.

In addition to the concerns of confined space, working conditions, and extreme weather, other factors relate to human safety and health as well as performance. In accident management situations, Kim (2003) developed 18 taxonomies into a new full-set of performance influencing factors (PIF) taxonomy, classifying them into four main groups:

- human (personal characteristics and working capabilities of the human operator)
- system (the human-machine interface, hardware system, and physical characteristics of the plant process
- task (procedures and task characteristics required of the operator)
- environment (team and organization factors, and physical working environment)

This study used the PIF taxonomy to consider how the human-computer interface in the wind turbine affects individual. Also technicians in the wind turbine suffer severe challenges such as working in tall and confined spaces, and in extreme weather. These present both mental and physical challenges. Therefore, the aim of this study was to survey how technicians conducting maintenance tasks in a wind tower were affected by the condition influencing factors (CIF). Further, to find out the criticality impact items of CIF, fuzzy logic (FL) and fuzzy inference system (FIS) were applied to describe the degree of influence with 'degree of environment condition (C)', 'degree of risk (R)', and 'degree of effect on human body (E)' in each item. Finally, important factors about the field maintenance tasks were investigated to provide designers or managers with data to establish standards and intervention methods.

METHOD

SUBJECTS

Questionnaires were given to 16 maintenance technicians (average age = 29.7 years; SD = 5.21 years) with an average of 1.82 years seniority (SD = 2.33 years seniority) at the Fluvanna Wind Farm, Fluvanna, Texas, U.S.A.

QUESTIONNAIRES OF DEMOGRAPHIC DATA

All participants filled out a questionnaire, consisting of demographic information about age, gender, height, weight, seniority, average working time per day, and injury history.

QUESTIONNAIRES OF SUBJECTIVE CIF

Factors that suffer from the CIF, and human safety and healthy in wind farm were investigated. Five categories, with 28 total items, were generated from Kim's taxonomies (2003), rated subjective measures of condition evaluation, effect evaluation, risk evaluation, and comments.

The 28 items were classified into five categories: Human (6 items), Task (5 items), System (4 items), Environment (7 items), and Human-Computer Interface (6 items). Working conditions, work risk, and condition effects on the wind farm were rated from 1 to 10. For the condition evaluation, a score of 1 meant working environment was not a problem and 10 meant working environment was poor. For the effect evaluation, the effect meant that when technicians conducted maintenance tasks in such a working condition, a score of 1 indicated the effect was low and 10 meant the effect was high. For the risk evaluation, risk meant that when technicians were affected by each item, a score of 1 indicated the risk was low and 10 meant the risk was high.

Studies have revealed that workload also affects human performance, health, and safety (Xie and Salvendy, 2000; DiDomenico and Nussbaum, 2005). Hart and Staveland's NASA-Task Load Index (NASA-TLX) method, using 10-point scales, was used to assess six variables: mental demand, physical demand, temporal demand, performance, effort, and frustration during maintenance tasks in wind farm. To investigate the critical factors from the condition evaluation, effect evaluation, and risk evaluation, fuzzy logic and fuzzy theory were used to quantify each item.

CLASSIFICATION OF INFLUENCE FACTORS WITH FUZZY LOGIC

Zadeh (1965) introduced the term fuzzy theory and fuzzy logic to mimic the human mind, which often deals with approximate rather than exact information. A rule-base system, the fuzzy inference system (FIS), is one of the most famous applications of fuzzy logic and fuzzy sets theory. A fuzzy inference system (FIS) uses fuzzy rules in a nonlinear mapping of input linguistic variables to a scalar output. A model of FIS contains four stages: (1) fuzzification, (2) inference engine and rule base, (3) aggregation, and (4) defuzzification.

To determine critical influence factors on wind farm job, this study applied fuzzy logic to map linguistic maintenance experts' opinions of working conditions, working risk, and working effect into crisp values. FIS was applied to assess the critical impact factors. This analysis uses linguistic variables to describe the 'degree of environment condition (C)', 'degree of risk (R)', and 'degree of effect on human body (E)' in each item. These linguistic variables were then "fuzzied" to determine the degree of membership, low, middle, and high. The inference engine determines the degree to which the antecedent is satisfied for each rule. The rule base is derived from combining fuzzy sets and the fuzzy numbers. The defuzzification procedure used the Center of Gravity (COG) method to calculate the crisp values. The crisp value then used in fuzzy set output to classify items into different groups. The fuzzy set output is defined as the linguistic term, Influence Level, and was described by three attributes: Level 1 (lower impact), Level 2 (intermediate impact), and Level 3 (critical impact). Let IL (Impact Level) with a range of [0, 10] represent the degree of total influence from environment condition, risk, and effect. For example, if IL is Level 3, it means the group of these factors impact mostly on maintenance technicians during wind farm tasks. From the Influence Levels, the critical items can easily be identified.

RESULTS AND DISCUSSIONS

SUBJECTIVE WORKLOAD ASSESSMENT

Results of workload survey from assessment of six subscales: (1) mental demand, (2) physical demand, (3) temporal demand, (4) performance, (5) effort, and (6) frustration during maintenance tasks in the wind farm are shown in Figure 1 (see Appendix).

Figure 1 reveals that technicians ranked 'effort' as the number one contributing factor to workload, followed by performance. Next, technicians ranked 'physical demand' and 'mental demand' as other important factors contributing to workload. Based on these results it can be seen that though effort, mental demand and physical demand were given a higher score by technicians, interestingly, performance was also given a higher score by technicians. This possibly suggests that though technicians ranked the effort that they put in their tasks at the wind tower turbine are high, and that these tasks require high mental and physical demand, these three factors do not have a negative impact on their performance.

FUZZY CRITICALITY ASSESSMENT

To obtain the critical factors, 28 items from five categories (Human, Task, System, Environment, Human-Computer Interface) were classified into three different influence levels. After the fuzzification process converts the environment condition, risk situation, and effect on human body into their fuzzy representations, matched with the premises in the rule base, crisp values were determined by defuzzification. Three Influence Levels were found. Group Level 1 represents items having minor effect on technicians. On the contrary, Group Level 3 represents items having a critical impact on the worker, which may induce health and safety problems. High degree of membership belonging to Level 2 also needs to be continuously monitored and analyzed to prevent the potential risk of hazard. It is obvious that (1) working time per day, (2) reliability of computer system, (3) noise from the wind farm, (4) labor numbers, (5) status of safety system or component, and (6) complexity of the procedure in the computer system are not primary interference factors for technicians. Among them noise from the wind farm was considered a minor effect during maintenance tasks. This finding was different from previous studies which focused on the mechanical noise and mechanical noise to bother people who live in the surrounding area (Western Australian Planning Commission, 2004; Harry, 2007; Berg, 2004).

The Level 2 degrees of membership, listed from high to low, were (1) complexity in wind tower tasks, (2) lighting in the wind tower, (3) training and experience, (4) communication & coordination, (5) availability and quality of information, (6) time pressure to address unusually events, (7) fatigue, (8) arrangement of work place proper, (9) graphic display in the monitor, (10) availability and quality of procedures, (11) indication & communication, (12) simultaneous goals/tasks, (13) status and trend of critical parameters, (14) physical vigor, (15) readable interface in the monitor, (16) feedback of work result, (17) text size in the monitor, (18) warning signal in the monitor, and (19) vibration in the wind turbine. These results showed that many problems still exist in wind tower tasks. Complex tasks, light, availability and quality of information, arrangement of workplace proper, procedures, graphic display and readable interface and warning signal in the monitor, and vibration in the wind turbine should be considered by a work designer. On the other hand, training and experience, communication and coordination, time pressure, fatigue, and physical vigor may arise from individual difference or job styles. These can be explained by Level 3, (1) inconvenient of workplace, (2) dangerous workplace, and (3) temperature and humidity variability. Therefore, more protection measures and tool design, keeping temperature and humidity for to provide suitable training can be considered to decrease time pressure, fatigue, and physical vigor as well as increase efficiency of communication and coordination.

CONCLUSIONS AND FUTURE STUDY

In this study, field interviews and a subjective questionnaire were used to investigate the technicians' workload and CIF during wind tower turbine tasks. From the result of workload survey, technicians spend more effort, mental demand, and physical demand to achieve acceptable job performance. Further, twenty-eight items were classified into three different influence levels (i.e., low impact, intermediate impact, critical impact) using fuzzy logic and a fuzzy inference system. Critical impact level consisted of inconvenient working place, very dangerous work place, and temperature and humidity variability.

Findings from this study can be used as a basis for identifying strategies to prevent or reduce their likelihood of injury from a maintenance tasks in the wind tower. Also, the 'risky' CIF can be identified and documented to develop intervention strategies for use in developing standard working policies in different areas of a wind farm. Further studies are needed to collect more data to address and compare the wind farms from onshore and offshore locations.

ACKNOWLEDGMENTS

This study was supported in part by National Science Council (NSC) of Taiwan Grant (NSC96 − 2221 − E007 − 083 − MY3).

REFERENCES

Berg, G.P. van den (2004), "Effects of the wind profile at night on wind turbine sound." *Journal of Sound and vibration*, 277, pp 955-970.

DiDomenico, A., Nussbaum, M.A. (2005), "Interactive effects of mental and postural demands on subjective assessment of mental workload and postural stability." *Safety Science* 43, 485–495.

European Foundation for the Improvement of Living and Working Conditions (1998), "The costs and benefits of occupational safety and health." Report.

International Finance Corporation (IFC) (2007), "Environmental, Health, and Safety Guidelines for Wind Energy." Report.

Kim, J.W., Jung, W. (2003), "A taxonomy of performance influencing factors for human reliability analysis of emergency tasks." *Journal of Loss Prevention in the Process Industries* 16, pp 479-495.

Leonardo Energy (2007), "Power Quality and Utilisation Guide." Report.

Greef, M.D., Broek, K.V.d. (2004), "Quality of the working environment and productivity-research finding and case studies." *European Agency for Safety and Health at Work,* p 56.

Western Australian Planning Commission (2004) "Guidelines for Wind Farm Development." *Planning Bulletin*, 67, pp 1-8.

Xie, B., Salvendy, G. (2000), "Review and reappraisal of modeling and predicting mental workload in single- and multi-task environments." *Work & Stress* 14(1), 74-99.

Zadeh, L. (1965), "Fuzzy sets." *Information and Control* 8(3), 338-353.

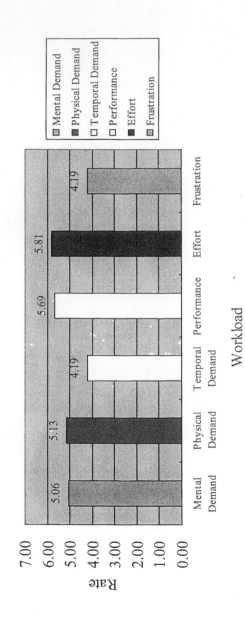

Figure 1 NASA-TLX workload assessment

Chapter 31

Human Factors in Oil Refinery Operation

David A. Strobhar, PE

Beville Engineering, Inc.
7087 Corporate Way
Dayton, OH 45459

ABSTRACT

Safe and efficient operation of oil refineries is critical to the US economy. However, fatalities and disruptions in production continue to occur as a result of human factors deficiencies. This paper outlines the history of human factors issues in refinery operation, including progress to date on improvement. Key issues and opportunities will be highlighted. The discussion is oriented toward the human factors professional that may not be aware of refinery operational issues.

Keywords: Oil refineries, process control, human factors, operators

INTRODUCTION

The need and benefits of considering human factors in the design and operation of complex systems has been appreciated by defense, airspace, medical, and other industries. One high-risk industry that has had limited use of systematic human factors analyses and interventions is oil refining. The lack of human factors use is not due to a lack of need, but in many cases an information gap (expanse). Industry professionals often do not know what human factors is or how it can help. Human factors professionals often don't understand the industry enough to offer suggestions. This paper is an attempt to bridge the gap, describing the industry and some of its human factors issues.

Oil refineries are a collection of 10-30 processing units which operate 24/7, 365 days a year, with shift staffing of three to seven operators per unit. The process units are interconnected via product streams and a reliance on common utilities (power, steam, hydrogen). Process units can handle 10,000 to 100,000+ barrels of product per day (42 gallons/barrel), often heating it to over 1000 degrees and over 2000 PSI pressure. There are over 100 major oil refineries in North America. The end products are gasoline, diesel, jet fuel, petroleum coke, butane, and propane. Some refineries have chemical units, further expanding their size and product slate. Operators work eight or twelve-hour shifts, with twelve-hour being more common, rotating every three to four days. Operators in the past were required to only have a high school diploma, but many companies now require an associate's degree from programs specialized for the refining industry. While long male dominated, US refineries have a high proportion of female operators.

Most process units at some point distill their products, using the difference in boiling points to separate different hydrocarbons. Some units are strictly distillation, while others use heat and catalysts to break or re-arrange carbon bonds.

The products must be treated to remove impurities, requiring their own processing units. A common impurity that is extracted is hydrogen sulfide, a highly toxic gas. Gasoline is not a single product; rather it's a blend of products from multiple units. Both incoming crude and outgoing product can be transported by ship, rail, pipeline, or truck. While all refineries have units carrying out similar functions, they vary in the details of their design and operation. This lack of commonality inhibits the ability to make changes that can be applied carte blanche across all units of given type.

Refineries have been operating for over 100 years, but they have been undergoing dramatic changes in technology over the last generation. The basic function of controlling the pressure, flow, level, and temperature of the hydrocarbon has not changed. However, what has changed is how it is done. At one time, operators manually opened/closed valves in the process unit, using local gauges to monitor process variables. The first change was indication of process variables being brought into a control room for monitoring purposes; and later, pneumatic control of many of the valves was added. An operator position was subsequently created to monitor and control the process, usually called the board or console operator. This position would be responsible for monitoring 500+ variables and controlling another 150-200 by manipulating valves. This latter number is described as the operator's span of control. Control of the variables was largely one at a time, with the operator entering the desired setpoint for each variable (e.g., flow, level, temperature, pressure) that was to be controlled. The control system changed the position of a valve to maintain the variable at setpoint. In the 1980's, the instrumentation was changed from panel mounted instruments to video display units using distributed control systems (DCS). With the advent of DCS, the control and monitoring no longer needed to be near the process unit as it had been previously. As a result, the control and monitoring was in many cases relocated to a remote control center, with the control of multiple units to the entire refinery from one location. In the last decade, advances in control system technology have dramatically reduced the need for operator intervention and the operator's span of control has doubled. Programs now control numerous variables at the same time.

Assisting the console operator are typically three or more field operators. While they still open/close valves and start/stop pumps that are not automated, their direct intervention in the process has been reduced. Their focus has shifted from operation to reliability – keep the equipment running. Visual inspections are performed during the course of the shift to detect malfunctions or potential malfunctions. Samples of the process material are routinely taken, with the laboratory tests either done by the operators themselves or by a central lab. Daily equipment must be prepared for maintenance work to be performed, and that work overseen to ensure it is done safely. All of this occurs outside on structures that can be over ten stories high. As a result, the work is done in all types of weather.

Documented cases of "operator error" related events have resulted in dozens of fatalities and millions of dollars in damage. Undocumented economic losses are likely equal to or greater than those documented. The errors have occurred both inside the control room and out on the unit. The control room errors tend to be misdiagnosis of events or missing the event entirely. The outside errors tend to be around maintenance that is being performed. Fatalities generally result from what is termed "a loss of containment," or basically that the hydrocarbon gets outside of the pipe. There are numerous ignition sources on a refinery unit when that happens, with fires and explosions the result.

While refining is a hazardous business, the risk and danger is often overemphasized by those new to the industry. Oil companies take great care to be as safe as possible. Safety is emphasized at all plants and safeguards for the process are continually evaluated and upgraded. Major fires and explosions, while tragic and news making, are infrequent events. While safety is paramount, these are still businesses. They need to keep costs down in order to turn a profit, so the units need to run reliably for long periods of time. Unplanned shutdowns and damage to critical equipment, while posing a safety issue, are also undesirable from an economic standpoint.

CONSOLE OPERATION

The console, or board operator, is often derisively referred to as the bored operator. Sitting in front of four to eight monitors, the console operator's primary function is to monitor process variables and make adjustments as needed. This is done through calling up on the monitors one of the 50-150 displays available. Operators rarely go longer than 10 minutes without looking at their monitors to check on the status of the process. The displays largely look like, or mimic, engineering drawings of the process, showing towers, pumps, and valves with the instrument data generally shown in digital format near its associated piece of equipment. Targets for the process variables are transmitted via written or email operating instructions/orders.

If one of a set of key variables deviates from its desired value too much, an audible alarm sounds and a description of the wayward variable shows on one of the screens. The set of possible alarms should be 500-1000, but it is often up to six times that amount. The average console operator responds to about six alarms per hour and makes an equivalent number of control changes. Information is obtained from and instructions given to field operators on changes that can only be made outside. The console operator is an information focal point on unit operation, communicating with engineers, management, and other console operators on operational issues. Administrative demands on the console operator include logging key changes that have occurred and reviewing procedures.

The console job radically changes when something malfunctions. Alarms can flood in, up to 50-75 per minute. The console operator must rapidly determine the cause and corrective course of action. If a significant change in unit operation is demanded, the operator must implement those changes on the DCS displays or give the appropriate instructions to the field or other console operators. If the process gets too close to pre-determined unsafe modes of operation, automatic shutdown systems actuate to remove energy from the process. The console operator must then react to and monitor the shutdown process.

The type of person performing the job has changed in the last decade. Originally the console job was a position on par with the outside. Anyone could and would work it after being qualified. Once an individual learned all the outside jobs, they would train and qualify for the console job. Individuals on a crew rotated through the console job over the course of a month. At some plants, the console job was associated with seniority. It was seen as requiring experience and a good job since it was indoors. As the span of control and level of automation increase, the job has become increasingly specialized. At first, once a person learned the console job, they would no longer rotate through the other field positions other than to maintain proficiency. Now, specialized selection tests are being used at many locations to identify those operators who would be good at the console job. Where the console operator once worked the field positions prior to handling the control duties, several locations have console operators controlling units in which they have never worked.

FIELD OPERATION

Field operators spend about 60% of their 12-hour shift on the unit. At the beginning of the shift, relief is made with the off-going operator, discussing any operational issues. The operator then reads the operating orders for the shift to determine if anything unusual needs to be done that day. A pattern of activities begins that is similar across units and refineries. The activities, which will repeat about every four hours, are –

1. Walk-down unit – Called a "round," the operator walks throughout the unit using their senses to determine if anything is out of the normal. Some readings of local instruments may be taken.

2. Collect Samples – Sample bottles are prepared and samples of the material taken.

3. Run Samples – Tests are performed on the samples at some locations and jobs, while most send the samples to a central lab.

4. Work with Maintenance – At any time maintenance can arrive on the unit, at which point the operator will need to ensure that the equipment is safe to work on through a process known as lock-out/tag-out (LOTO).

5. Carry out console operator requests – Also an unpredictable activity, the field operator may need to verify on local gauges what the console is seeing on the DCS, start/stop pumps and fans, or open/close valves.

Field operators face environmental issues not faced by the console operators. While refineries are never clean, different units have different levels of dirt and grime. Those units that process crude oil and heavier hydrocarbons (gas oil, asphalt) tend to be dirty. Those units that handle the lighter hydrocarbons (propane, butane) that come from the crude unit are relatively clean. Regardless, all units have towers that must be climbed, either via stairs or ladders. While elevators exist on the tallest units, they tend to be unreliable. Climbing ten stories in either 110 or -40 degree weather is not unusual. The heavy oil units tend to have large valves and the viscous nature of the material makes operation of those valves difficult. This often requires the assistance of a valve wrench, or persuader. As heat is essential to most processes, piping can often be very hot. Most plants now require field operators to carry portable gas monitors, generally checking for hydrogen sulfide gas.

Upsets for the field operators are a race against the clock. The faster the field operator can carry out their tasks, the less impact the upset will have on everyone. So it is in the field operators' self-interest to do what is needed as quickly as possible. While no two upsets are identical, there are general procedures for what must be done when something malfunctions. As many pumps cannot be started from the console (most plants allow the pumps to be shutdown from the console, but not all), power interruptions send the field operator to rapidly restart pumps. Product streams will need to be diverted so the potentially off-spec product does not contaminate good product already in tankage. A long time ago the philosophy was to try to "save the unit" and prevent it from shutting down. However, this proved to be not as safe as allowing the unit to shutdown and then restarting it under more controlled conditions with extra operators.

Automation has changed what the field operator needs to do in an upset. A generation ago, if a field operator did not rapidly close certain valves in the unit at the start of an upset, then major pieces of equipment could be damaged or destroyed, with ensuing fires. Those valves have now been automated. The field operator may close the same valve as before as added protection, but they are no longer the primary means of carrying out the function. Pumps and fans still need to be started/stopped in many cases, but most tasks that the operator did to prevent unsafe conditions from developing have been automated. Failure of the operator to quickly intervene may cost lost production, but with modern shutdown systems no one should get hurt or product released/spilled.

ISSUES AND OPPORTUNITIES

The original interest in human factors within the industry was around alarm systems. This stems from the ability to identify operator error in cases of poor design, "The alarm was present and you didn't respond." The cost of those errors could be readily identified. Alarm problems came with the advent of DCS, as now any point could have up to eight different alarms configured for it. The resulting flood of alarms resulted in operators missing many key indicators of problems, sometimes with disastrous results. Much of the initial effort has been on the selection of individual alarms. Expansion into how to best present alarm information has only recently begun.

As an extension of the alarm problem, presentation of other information, interface design, is growing in interest. The DCS allows the users to configure the interface, so there is no "standard" design. There are general patterns or trends, but almost every unit is different in some respect from every other. Changes in how color is used has slowly occurred, with it is generally understood that color is NOT to be used just to make the displays "pretty." The International Society for Automation is developing a standard on graphics design.

Interest in selection and training has increased with increasing automation. Higher levels of automation have resulted in less operator intervention in the process and fewer process disturbances. The skill set for the operator has changed, and the ability to practice those skills has diminished. As a result, interest has increased in how to ensure that the individual controlling process is proficient in their job.

Fatigue has become an issue in the past several years. How many days in a row can/should an operator work? Guidelines have been issued by the American Petroleum Institute, but like most guidelines to date, they rely on extrapolation of research from other domains. Solutions to the problems specific to the refining industry have been formulated, but neither tested nor proven.

Most human factors efforts to date have been by individual practitioners. Much of the work that the industry would consider "human factors" has been done by individuals with no formal training in the field. While intelligent and earnest, the lack of theoretical constructs is often apparent in the results. This changed in 2007 with the creation of the Center for Operator Performance, a collaboration of operating companies and DCS suppliers based at Wright State University. The COP funds research that is to meet academic standards, publishable in peer reviewed scientific journals. Research to date has occurred on alarms, displays, and training.

ABOUT THE AUTHOR

David Strobhar (PE) is the founder of Beville Engineering, a human factors engineering firm which has conducted since 1984 modernization, operator workload, and alarm analyses for over 40 oil/chemical companies (e.g., BP, ConocoPhillips, Chevron, Shell) at over 80 locations. Mr. Strobhar began the human factors department for General Public Utilities following the accident at

309

their Three Mile Island Nuclear Station. He has a BS in Human Factors Engineering from Wright State University in Dayton, Ohio.

CHAPTER 32

The Personas Layering Framework Applied to Consumer Services Design for Automotive Market

Alessandro Marcengo[1], Amon Rapp[2], Elena Guercio[1]

[1]Telecom Italia, Turin, 10148, Italy
[2]Telecom Italia, University of Turin, "Progetto Lagrange - Fondazione C.R.T.", Turin, 10148, Italy

ABSTRACT

This paper describes a research project dealing with the design of innovative internet based services for the Italian automotive market. The project has followed a twofold stream. The first one defined a relatively wide range of services with a potential interest, according to the technological, developmental trends and the specific needs of the Telco business sector. Meanwhile, there was a parallel stream that, through continuous confrontation with the users, and using a combination of user centred methodologies, allowed identifying, prioritizing and reshaping the services of real interest for the target users.

Keywords: Personas, Ethnography, Service Design, In-car Services, Qualitative Methods.

INTRODUCTION

This paper describes the user centric research project Telco@Car, dealing with the design of innovative internet based services for the Italian automotive market. The main objective of the project has been to prioritize the development of services well accepted by the end users in particular focusing on the driver. The project follows a twofold approach, one based on the *Technology* stream and one based on the *Users* stream (see Figure 1). The first one defined 12 services with a potential interest according to the technological and developmental trends, and the specific business needs of a telecommunication carrier. Meanwhile, the *Users* stream allowed selecting and reshaping the services of actual interest of the target users also considering the heavy constraints of the application environment. The user centric approach adopted within the project benefits from the *Personas Layering Framework,* an original work flow developed in our research area and already applied in several research project (Marcengo et al., 2009). The *Personas Layering Framework* maximizes the cost-benefits trade-off developing the set of Personas as composite entities, consisting of several layers, partly developed once and reusable in many contexts and partly developed on the basis of the application context of the different projects requiring a user centred perspective (Marcengo et al., 2009).

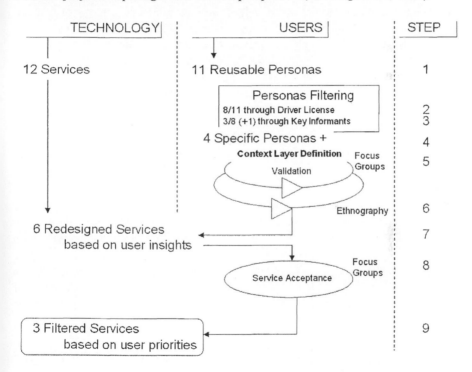

Figure 1. Technology stream, users streams, process steps

THE INITIAL SET OF SERVICES

In our *Technology* stream we identified 12 possible service concepts characterized by a technical feasibility (*technical criteria*) and a possible inclusion in the value chain of a typical Telco operator (*business criteria*). The following are the 12 service concept brought into the user research in order to verify their possible user acceptance; **In-car Internet Radio (A)** - The possibility to listen internet radio stations while travelling; **In-car Podcast (B)** - The possibility to listen specific podcast while travelling; **Home Music Repository (C)** - The possibility to access your own home media server while travelling; **Internet Browsing (D)** - The possibility to browse the "big Internet" from the dashboard; **In-car Walled Garden (E)** - The possibility to access an optimized web environment in terms of content and formats optimized for in-car usage; **Direct Webcam Access (F)** - The possibility to watch directly public or private Webcam from your in-car aggregator; **Video Broadcasting (G)** - the possibility to publish, in real time, you own on-board camera; **Multimedia Streaming (H)** - The possibility to access some media channels with visual component; **Augmented Reality (I)** - The possibility to access content, text, audio, video, directly related to one's own geographical position while travelling; **Mapped Friends (L)** - The possibility to be seen or to see friends of a social network in relation to one's own geographical position; **Contextual Community (M)** – The possibility to receive and produce content dynamically linked to a specific context (i.e. traffic info about the ring road) with a community having the same "interests" (i.e. same location) in relation to one's own geographical position; **Theme Community (N)** - The possibility to receive and produce content dynamically linked to a topic of stable interest (i.e. same bike brand, same food preferences, etc.) also in relation to one's own geographical position.

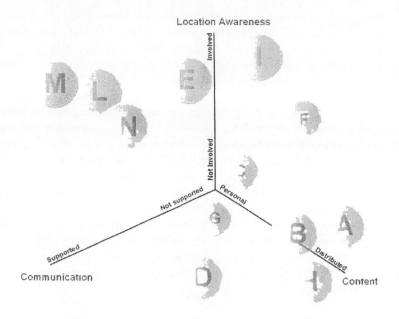

Figure 2. Services initial set

To better understand our arena and to identify the possible combination of features is useful to breakdown in three axes the entire in-car service landscape of interest for a Telco operator focused on a "consumer" perspective: Content (*personal vs. distributed*); Communication (*supported vs. not supported*); Location Awareness (*involved vs. not involved*). All the service concepts designed fall into the space delimited by our three axes depending by the characteristic components of each one (see Figure 2). Because the focus of this paper is on the human factors aspects we will not describe deeper the technical details of the services.

THE USER STREAM

In our *user stream* the core of our methodological choice has been the Personas methodology (Cooper, 1999; Grudin et al., 2002; Pruitt et al., 2006). This approach better represent long term needs and goals that are not subject to the rapid technological variations and trends. In this way it is possible to design medium/long term services by focusing on their practical and emotional aims and needs rather than merely on the actual technological framework. (Guercio et al., 2006) To maximize the cost benefits trade-off in developing the set of Personas within our research area it has been developed a *Personas Layering Framework*, which considers the single Persona as a composite entity, consisting of several layers, partly developed once and reusable in many context (the internal layer) and partly developed on the basis of the application context of the different projects requiring a user centred perspective (the *context layer*) (Marcengo et al., 2009).

In this way the process of Personas definition is articulated into two independent stages:

- *the first* targeted to the accurate definition of a basic set of Personas, reusable and valid for a whole research or business area. In this stage, through both quantitative and qualitative research methods, for each Persona that composes the set is created an *internal layer* that defines not only all the socio demographic characteristics, but also all the aspects that guide the Persona daily life that does not change through different   contexts   (i.e.   his   values   and   desires)
- *the second* is specific for each individual project that intends to reuse the basic set of Personas, and aims to define a *context layer* based on the usage context of the services under development. The context layer drives attitudes, needs and specific aims of that person for that context (e.g. home, car, etc.). This "layer" is built through ad hoc qualitative research techniques (key informants interviews, ethnography, scenarios, etc.) depending on the type of information crucial for the service (Marcengo et al., 2009).

This model can significantly lengthen the life span of the Personas set developed because the internal layer of all individual Personas can be reused within the same research or business area in different projects over the years while maintaining its validity and constituting a solid basis for guiding the design of different services but of course also provide techniques to filter the initial set according to the specific application context or even to add new Personas (Marcengo et al., 2009).

In our Research Area both qualitative and quantitative data were used to define the initial Personas set. The definition started with the analysis of the most relevant Italian research statistics institutes: CENSIS (Censis, 2005) for media consumption, ISTAT (Istat, 2007a) for time usage, ISTAT (Istat, 2007b, 2007c) for social trends and daily life conditions, ACI-CENSIS (Aci-Censis, 2008) for italian mobility. From these data were defined 11 different "stereotypes" for the Italian market of "consumers and early adopters of ICT and telecommunications services". These profiles were selected to reflect the middle class Italian population using ICT and TLC services. Great attention, in the definition of their internal layer, was given to the development of their personality traits, the definition of their relationship with media and technology, and their practical and emotional daily "goals". These "reusable" Personas have been, like in other research projects, also injected *(step1)* in the Telco@car service design process (see Figure 1) and then filtered according to the specific context of application.

From the initial set of 11 reusable Personas 3 have been discarded because not having the driving license *(step2)* and to better understand the personas suitability 10 semi-structured interviews with car dealers have been carried out. We chose the car dealers as key informants (Corbetta, 2003) because of their direct experience in customers needs, requirements and complaints about extra car equipment. This information helped us to understand if the users with described profiles could be really interested and willing to purchase services for their car, and also to identify any other user target to be more open to the adoption of new in-car services. The dealers were selected on the basis of the properties considered relevant on the research purposes (Mason, 1996): the completeness of the product range (from the budget car to the business car), the status and the diffusion of the brand in Italy, a mid range customer basis. The selected car dealers (Volkswagen, Opel, Citroen, Renault, Fiat/Lancia, Ford) were considered as perfectly representing these three properties because they are balanced and typical cases (Gobo, 2004). Data collected from car dealers have caused us to eliminate 5 more Personas of the remaining 8 *(step3)*, as being potentially too little interested in our service concepts and to their adoption in the future (i.e. Margherita, employee, 42 years old, married with two children). But on the other hand the car dealers suggested us the importance of a "new" Persona, Umberto, sales representative without children, who drives more than 50.000 km per year in his Audi A4 and therefore wants to make as comfortable and useful as possible his driver's cabin. The others members of the resulting set of Personas for the Telco@car project were: Alessio, 25 years, a university student in the faculty of economics, single, owns a Mini Cooper, which he uses every day to go to the university; Paola: 21 years, an university student of engineering, in a relationship, has a Fiat 500, that uses every day and sometimes for long trips during week ends; Giovanni: 47 years, graduate, manager in a telecommunications company, married with two children, uses a corporate X3 Bmw. So in the end we focused on 4 Specific Personas for Telco@car *(step4)*. Following our model a context layer for each Personas has been created for the car context (describing their patterns of usage of the car), while for Umberto also the internal layer has been developed. These layers were validated through 4 focus groups *(step5)*, one for each Persona. Each group consisted of 8 participants with the same profile of the Persona under study. Participants were asked to tell their lives and to compare it with the lives of the target Persona, illustrated through short clips, which were staged in the moments of everyday "typical" life (i.e. the working day).

Then we shifted our attention to their driving behaviors. Basically our need was that every Persona was accompanied by a description of his usual behavior, his objectives and needs while driving in order to further validate and enrich the context layer and to collect data about how the various devices inside the car (i.e. mobiles, car radios, etc.) were used during the activity of driving, and for what purposes, capturing the more possible user insights to filter or redesign our initial set of service concepts. To gather these elements we adopted an ethnographic approach observing drivers in their daily driving routine *(step6)*. In particular, we chose the rapid ethnography approach (Millen, 2000), consisting of a participant observation on the field that selects in advance and accurately the aspects that need to be observed, reducing costs and resources (Paay, 2008), and a series of contextual interviews. For each defined Persona it has been selected a real individual with substantially similar characteristics representing the typical and emblematic case (Gobo, 2004) and then was observed while driving his car on a normal day. Observers have always been 2 in order to have a double observation of the same patterns. All the sessions were video recorded. The observations were carried out on routes already decided by the participants basing that on their daily needs, and thus expressing their typical everyday life. The Personas "representatives" were observed and subsequently interviewed about the specific behavior and the main habits related to it. All participants were also asked to imagine or recall specific situations in which, with the help of some in-car service, they could have been improved their travel experience or resolved some difficulties.

The data collected shows the most used device inside the car by all Personas representatives (despite the availability on the automotive and CE Market of PND, Carputers, Media Player, etc.) is the dashboard radio. It is rated as the perfect source of information and media fruition, is not very intrusive and is not perceived as a distractive device. The two younger Personas use the radio for daily travel, but they prefer to listen their own music selection (mp3, cd, etc.) when travelling on longer trails. Alessio says *"Unfortunately I do not have the usb plug to connect my iPod to the car. It is the feature that I would like the most"*. Asked about his own personal music at home, Paola said: *"At home I have concentrated in one place all my collection of mp3s, and even if I had the iPod plug on this car, I should decide which music carrying with me on each journey ...doing not always the right choice"*. The data suggests that the car radio could be the right platform to place new connected services: the participants have been very interested to enhance the car radio platform enriching it with new features. Besides, the video screens are perceived as not optimal devices for in-car usage, too distracting and unsafe for the driver, and also with little interest for the passengers. Umberto says: *"Even if you could please the passengers providing them some video entertainment during the trip, I prefer to talk with them. I think the video even if it is used only by passengers, leads to no communication and make the trip boring, especially for the driver"*. The phone has a central role during the car journey, even if the actual technologies are not so safe while driving. The most problematic issue lies in the voice recognition technology that would be useful but is not currently perceived as being satisfactory. In fact Umberto says: *"Sometimes I say a name and the phone call another one, so you must always check that the system did not performed a wrong recognition ... It is easier to select the contact manually "*. So, it is indeed welcomed the opportunity of dialing and receiving hands free calls considering that part of the current Bluetooth pairing between personal devices and car systems still unsatisfactory. Some user raise the need of having the Internet on board but this issue receives conflicting opinions. None of the participants reported the usage of location awareness devices like PND (Personal Navigation Device). The investigation showed that the PND is seen mainly as a service suited for long journeys, and exclusively for orientation purposes: but a need aroused for an enhanced navigator that combines the location awareness with contextual information relevant to the subject, whether traffic or points of interest. Gianni said: *"Once on vacation I was looking for something interesting to visit in the vicinity of my hotel not having a guide or the possibility to access the Internet. Unfortunately my PND Navigator did not allow me to find anything! It would have been helpful if the device was connected to the network in order to receive information on where I was and advices on the best places in the area pinpointed by those who had been there before".*

Finally, in general, we have noticed from the observation, how other factors that we didn't take in account at first have influence in the use and adoption of new technological services in the car. The first relates to the perception of the own car that is reflected in the characterization (*customization*) of the driver's cabin by the participant. In the observed subjects we found two opposite attitudes: a low or almost absent customization of the driver's cabin, in the case of Umberto and Alessio and, on the contrary, a strong characterization of the space inside the car for Paola and Gianni. Umberto, for example, even using massively the car has a driver's cabin nearly empty, without any personal effects, showing little emotional investment in his car perceived as a simple mean of transportation and work environment. Totally different is the case of Gianni: his driver's cabin is very personalized, filled with objects, which are used for specific needs. During the interview the car is described by Gianni as an extension of his own home, even if not used for many kilometers/year: *"I would like to have the broadband on my car ... like home ... for instance ...if you have an appointment ... and is too early ... in those 15-20 minutes I could surf the web, or try something fun, or doing something that I did not had the time to do at home".* These different perceptions seem to affect the predisposition of the driver toward enriching the car with technological gadgets that could enhance the pleasantness and the comfort of that environment. The second factor relates instead with the general predisposition in the adoption of new technologies from the subjects observed. This issue was therefore investigated more deeply in the last phase of the research and it emerged, that the predisposition to the use and testing of new technologies in all areas of daily life is by far the most important factor for the emergence of any interest in Telco@car services.

FINAL RESULTS AND GENERAL FINDINGS

According to the user insights collected during the context layer definition of the 4 specific Personas we redesigned 6 over 12 services discarding the remaining 6 *(step7)* in order to keep only those applications that looked more close to the participants needs and not in opposition with their actual behaviour in the driver's cabin. Just to name a few *In-car Internet Radio (A)* is maintained as a service because it is consistent with the usage patterns already adopted by users at home or on the move via mobile device while *In-car Podcast (B)* was eliminated as a separate service but not as a feature. *Internet Browsing (D)* would seem to meet many needs especially during breaks and thus is maintained in contrast to *In-car Walled Garden (E)* that appears linked to a model of optimization and filtering of content not appreciated by the users. *Direct Access Webcam (F)* is removed along with *Multimedia Streaming (H)* because it is based primarily on a visual component so considered against the idea of total security that the service has to suggest to the driver. *Video Broadcasting (G)* is eliminated because we could not find any need, expressed or not, able to justify it. *Augmented Reality (I)* is maintained for the great interest that users expressed on the issues of user generated content and social networks, and because it can be also declined in non visual mode. In the last phase we carried out a service acceptance round of focus group *(step8)*. One by one, the 6 redesigned services were screened and commented in the focus room.

In the following we summarize the potential interest of each Persona as emerged from the research: **Paola** has proved to be one of the profiles most prone to the expansion in her car of the features that already use in everyday life. In fact, the ability to access her *music repository* or expanding the range of options for listening the car radio through Internet connection that already she use at home was a very welcomed opportunity. Even "totally new" services that will benefit from an increased location aware augmented reality were particularly appreciated. **Gianni** has the broader base of acceptance for the services offered, well 4 out of 6. This is probably due to the emotional meaning that he put in the car (among other things is the only Personas who has paid his own car and use it outside of daily trips). Particularly toward Paola he also appreciates the opportunity to freely access the Internet almost as he does at home; this is very interesting because of the margins of freedom in the use of web services that this concept take within. Gianni advocate a simple connection to the "big Internet" meanwhile raising some issues on ergonomics, ease of navigation and security that this approach could bring with. **Alessio** is interested primarily in two services: the *Augmented Reality*, which he considers very innovative and quite new in the panorama of UGC applications and the access to his *home music repository*. Alessio is in many cases an "expert", a "collector" and a "music organizer" so he is also keen to lift any deal that must be done nowadays through CD and USB keys to listen to personal music in the car. Rather less interested in the possibility of extending the range of available channels in his car through the introduction of internet radio. **Umberto** is the Persona with the most unexpected results. Despite of passing a very long time in the car, sometimes 8 hours a day, it seems less interested in the majority of services offered. This is partly due to the fact that his technological ability is lower than the other profiles, because it has no time to practice for studying or working in the office. The only service of interest appears to be widening the scope of its current car radio through the introduction of the internet radio.

In relation to these macro results we decided to give priority in our work to design and prototyping the following 3 services that have received the greatest number of consensus for all Personas *(step9)*. **In-car Internet Radio (A)**, as at home, user can listen to a channel not in the general manner, but in relation to specific themes or genres. Many users have also pointed out that many FM radios that they listen are also broadcasted via the Internet, thus giving the opportunity to listen them even outside the FM coverage areas. **Home Music repository (C)**. This service deals with the possibility to access your own home media server. The media server could be completely personal (located at home) or hosted (a personal space hosted by a service provider). No matter what time of content could be accessible (pictures, movies, songs, etc.) even if audio (music and podcast) looks a more suitable content for in-car fruition. **Augmented Reality (I).** The service is the intersection of distributed content partly generated by users (Flickr, Youtube, Wikipedia, etc.) with a dimension of location awareness that allows, in real time, while crossing an area, a neighbourhood, or a city a media streaming (according to the filter selected), providing so a semantic dimension of the territory generated by users. It is an opportunity to enrich the experience of travelling with a number of contextual media content. User really appreciated that, even suggesting the possibility of the simultaneous "generation" of this content (e.g. by writing a note or by sending a photo leaving then these contents on precise geographical coordinates, on the fly). Regarding the services on which we have now decided not to focus, we believe there are still issues to be investigated and the solutions to be proposed also have to take in account the "concerns" expressed by the participants involved in this research. For example, services that cross the social network area, variously shaped, with the location awareness dimension raise major concern about privacy issues. General findings indicate also that the greatest importance for the interest of innovative in-car services is the general attitude in the adoption of new technologies in various contexts of daily life. The 3 early adopters Personas have proved to be the more enthusiastic about new services. The emotional investment factor in personalizing their cars emerged as a result during the ethnographic phase, as an addition to individual predisposition to the adoption of new technologies in all areas of daily life and is important for the detection of highly predisposed target of the designed services. These results underline the importance for research, also industrial, to take into account factors not exclusively related to the effectiveness, utility and performance in the design of services: individual guidance, the different ways of perceiving the environments in which services will be placed and linked to different attitudes toward technologies, are equally crucial elements in predisposing the users toward a new service.

REFERENCES

Aci-Censis (2008), Rapporto Automobile 2008. L'auto libertà responsabile, Roma.

Censis (2005), Quinto Rapporto Censis-Ucsi sulla comunicazione in Italia 2005: 2001-2005 cinque anni di evoluzione e rivoluzione nell'uso dei media, Roma.

Cooper A. (1999), The Inmates Are Running the Asylum : Why High Tech Products Drive Us Crazy and How To Restore The Sanity. Sams, Indianapolis, Ind.

Corbetta, P. (2003) La ricerca sociale: metodologia e tecnica. III. Le tecniche qualitative. Il Mulino, Bologna.

Gobo, G. (2004), Generalizzare da un solo caso? Lineamenti di una teoria idiografica dei campioni. Rassegna Italiana di Sociologia XLV, 1.

Guercio, E., Marcengo, A. and Rapp, A. (2007), How to connect user research and not so forthcoming technology scenarios – The extended home environment case study. Journal of Social Science 2, 4, 203-208.

Grudin, J., Pruitt, J. (2002) Personas, Participatory design and product development: an infrastructure engagement. In Proc PDC 2002. CPSR, 144-161.

Istat (2007a), L'uso del tempo: Indagine multiscopo sulle famiglie "Uso del tempo" - Anni 2002-2003, Roma.

Istat (2007b), La vita quotidiana nel 2005: Indagine multiscopo sulle famiglie "Aspetti della vita quotidiana" Anno 2005, Roma.

Istat (2007c) Reddito e condizioni di vita: Indagine sulle condizioni di vita - Anno 2004, Roma.

Marcengo, A., Guercio, E., Rapp, A. (2009), Personas layering: a cost effective model for service design in medium-long term Telco research projects. In Proc. of HCII.

Marcengo, A., Guercio, E., Rapp, A. (2009), Telco@home: a seamless communication project with a user perspective. In Proc. of IE.

Mason, J. (1996) Qualitative Researching. Sage, Newbury park, CA 1996.

Millen, D.R. (2000), Rapid ethnography: Time depending strategies for HCI field research. In Proc. DIS 2000 ACM Press 280-286.

Paay J. (2008), From Ethnography to Interaction Design. Book Chapter in Handbook of Research on User Interface Design and Evaluation for Mobile Technology, Joanna. (Ed.), Lumsden,.

Pruitt, J., Adlin, T. (2006), The Persona Lifecycle : Keeping People in Mind Throughout Product Design. Morgan Kaufmann.

Chapter 33

The Process of Identifying and Implementing Ergonomic Controls in the Packaging of Motors

Ivana Wireman

Ohio Bureau of Workers' Compensation
3401 Park Center Drive, Suite 100
Dayton, Ohio 45414-2577

ABSTRACT

The goal of this case study is to address the process involved in identifying ergonomic deficiencies and implementing recommendations for packaging motors at one Ohio assembly plant. The weight and manual handling of motors and related effect on the safety and health of employees were of primary concern. The ergonomics initiative for redesign of the shipping area was prompted by the company's management as part of an overall facility improvement plan. Monies from the Ohio Bureau of Workers' Compensation (BWC) Safety Grant$ Program were utilized for the engineering controls implemented. The motor packaging area was analyzed for Cumulative Trauma Disorder (CTD) risk factors, material flow, equipment/process requirements, work practices, and historical data. The Ohio BWC CTD Risk Factor Assessment Form, along with other assessment tools was used to quantify risk. Nineteen (19) CTD risk factors were identified. Proposed recommendations focused on eliminating or substantially reducing the CTD risk. These recommendations were reviewed by company management. A team of experts, including employees of the Shipping Department, determined the final engineering controls implemented. Some of the final interventions differed from the

322

proposed recommendations by the ergonomics consultant. The economic cost analysis (payback method) yielded 5.35 years return-on-investment (ROI). Packaging productivity improved by 29%. Due to recent economic hardships the true impact of this redesign cannot be fully tested. The quality of the shipping box has improved in both the packaging integrity and visual appearance. Seventy-Four percent (74%) of CTD risk factors were eliminated because of the interventions. Packaging material costs decreased 8% to 26% and are dependent on box size and product configuration. The new packaging design has reduced the use of non-biodegradable liquid foam by 50% and promotes the company's policy on being green. The success of this project was due to thorough attention to detail and employee involvement. This project was initiated in October 2008 and implemented in October 2009 at a cost of $336,000.

Keywords: packaging, process, material flow, manual handling, Cumulative Trauma Disorders, green, engineering controls, macroergonomics

INTRODUCTION

SEW-Eurodrive, Inc. contacted the Ohio BWC for ergonomics assistance in their Shipping Department's high-volume United Parcel Service (UPS) line. SEW-Eurodrive, Inc. is recognized as a world leader in drive technology and a pioneer in drive-based automation. This Troy, Ohio facility employs a total of 122 workers with six (6) employees being dedicated to the Shipping Department--three (3) workers per shift. The company has a history of Cumulative Trauma Disorders (CTD) claims resulting in injuries to the back, upper extremities, and abdomen. With an average employee age of 43 years, the company elected to address proactive measures to prevent further injury to its greatest resource.

The company had four objectives for the ergonomics assessment. These include: (1) Prevent CTD injuries/illnesses; (2) Reduce costs associated with injuries/illnesses; (3) Improve material/process flow and be capable of integration with the new paint line project; and (4) Utilize packaging that was more environmentally friendly.

The purpose of this paper is to report the process involved in identifying the ergonomic deficiencies and implementing the appropriate recommendations.

METHOD OF ANALYSIS

The Ohio BWC Safety Grant$ Program uses the BWC CTD risk factor assessment form for identification of CTD risk as part of the application process. In addition, all Safety Grant$ applicants must document other variables which include loss experience, layout design, productivity, quality, and cost/benefit measures. Both dependent and independent variables are used to determine qualification for the Ohio BWC Safety Grant$ (Hamrick, 2001). These guidelines were followed in the ergonomics hazard assessment and documentation process for this case study.

DESCRIPTION OF THE PROBLEM

The Shipping Department is responsible for packing and shipping all products in a timely and efficient manner. The quality of the process is determined by how quickly the billing order is filled. The process starts when the motors arrive on a motorized overhead monorail from the paint department. The motors weigh between 10 and 80 pounds (with the average weight being 45 pounds). Motors are either manually or mechanically lifted on to a table or conveyor. Heavier motors can be lifted with an overhead hoist; this decision is left to the operator. Most operators did not choose to use the hoist. Next, the motors are manually lifted and placed inside pre-assembled cardboard boxes that are located on a cart. A fully loaded cart is manually pushed to the next work station. Liquid foam is added to the box. The box is then closed and sealed with several strips of tape applied to each box. The cart with sealed boxes is pushed back to a sorting pallet area. Each boxed motor is manually unloaded and sorted onto two pallets. The sorted UPS pallet is mechanically moved to the UPS weighing area, manually unloaded, weighed, and manually restacked onto an outgoing pallet. The outgoing pallet load is mechanically transferred to the warehouse. The following pictures show some of the job tasks:

Monorail   MMH motor to cart   Push cart   Sort boxes

There are 35 risk factors identified on the BWC CTD Risk Factor checklist. Nineteen (19) risk factors were identified within the duties of the Shipping Department employees and were reported as baseline data. The most significant risk factors are the weight of the motors, awkward postures required at various workstations, unnecessary re-handling (repetitiveness) of the product, forceful hand grips, and excessive push/pull forces required to move the loaded cart.

The NIOSH Lifting Guideline was utilized in the assessment of lifting tasks (NIOSH, 1981; Putz-Anderson and Waters, 1991). The 1991 NIOSH Lifting Guideline indicated that for all lifting tasks required for packaging motors, the Limiting Index (LI) and the Composite Lifting Index (CLI) exceeded 3.0 indicating engineering controls are recommended (Waters et al., 1993).

A digital force gauge was used to measure the initial force required to push/pull a fully loaded cart with boxed motors. The measured values varied from 50.1 to 94.5 pounds; the mean exceeded 65 pounds. The Liberty Mutual model, based on psychophysical data, was used to assess pushing of the loaded cart (Snook, S.H. and Ciriello, V.M., 1991). According to the parameters of the model, and based on 90% of the male population being capable of performing this task, the initial push force should not exceed 53 pounds. Additional guidelines from both General Motors (GM) and Chrysler offer lower limit values. GM's Design Ergonomics Worksheet (DEW) suggests a maximum initial whole body push/pull force of less than 30 pounds. The Chrysler Design Guidelines suggest a maximum two-handed push/pull force of 44 pounds. This author prefers to use the most conservative guideline. Therefore, for all pushing/pulling cart tasks, the DEW was exceeded and engineering controls are needed.

Material and process flow for the original configuration is shown in Figure 1. Material flow required significant re-handling of the product and long travel distances.

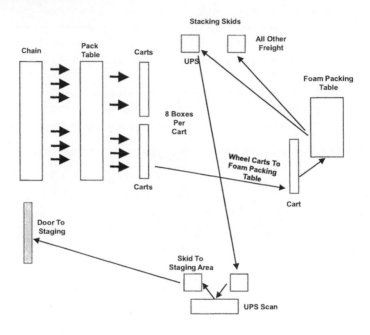

Figure 1. Material flow chart

DESCRIPTION OF THE PROCESS/SOLUTIONS

Initial recommendations were provided in a written report to the company by the ergonomics consultant. Solutions were engineering controls and did not address any administrative measures. The company determined that some of the proposed recommendations could not be implemented. Table 1 lists these recommendations and provides reasons for their denial:

Table 1: Consultant proposed recommendations

Consultant Proposed Recommendation:	Company's Decision		Reason for Denial:
	Approve	Deny	
Conveyor system	√		
Two (vacuum grippers) and gantry crane	√	√	• Based on volume of production only one gripper was needed
Automatic box former	√		
Automatic tape machine	√		
Scale in-line with conveyor	√		
Automatic shrink wrap machine	√		
Modify wall structure to connect packaging within the shipping department		√	• Cost prohibitive; • Damage integrity of structural wall; • Fire codes would be violated by altering fire wall; needed steel beam and sprinkler combination; • HVAC concerns; • Security issues due to door location.

Once the initial recommendations from the consultant were screened, the company allocated resources for a project team. The Engineering Department along with the Shipping Department determined how to implement the ergonomic recommendations. From the beginning, the company defined the Shipping Department as the 'customer'. The focus of the interventions was to reduce the quantified CTD risk factors and meet the criteria of the customer. The interventions were part of an overall plant evaluation and needed to integrate with future paint line modifications.

The next step of the implementation process involved the project team's decision to utilize and choose one expert in each field of conveyors and packaging. These consultants and the company's project team attended meetings, made assessments, and conducted simulations of the initial recommendations to determine feasibility and effect of material/equipment selection on the total process. The process is significantly influenced by paint curing time. The freshly painted motors could not be scratched and/or damaged by the packaging materials and/or process. During the three month simulation period, the team inadvertently modified both the packaging materials and the process. As the team changed one component, it had direct impact on another solution. The final solutions are summarized in Table 2:

Table 2: Final solutions

FINAL SOLUTION	UNFORSEEN MODIFICATIONS
Conveyor System (Power)	• Longer conveyor needed to allow for the 2 hour drying process time; • Layout/design needed to accommodate manual packaging option; in case of equipment breakdown/failure; • Conveyor heights needed to accommodate user anthropometrics and accommodate for drop-over distance of declination.
Conveyor slide	• Added to connect the monorail/scissor lift area to main conveyor and eliminate unnecessary manual lifting and/or requirement for a second manipulator/gripper; • Slide also used as a gate (aisle way); • Designing for angle of declination.
Gantry Crane	• Implementation of intervention was delayed due to oscillation concerns; braking mechanism of hoist failed and redesign of crane required welding of cross bars diagonally to crane.
Automatic box former	• Not needed due to process/box redesign; • Process does not lend itself to the packaging of uncured painted motors.
Adjustable drop-down unit	• Integrated into the conveyor system; • Provides a comfortable work height for employee to add sleeves/box over the motors.
Automatic tape machine	• New tape required a stronger adhesive and wider width; • Automatic sensing adjusts to two different box heights.
New packaging	• New sleeve packaging uses corrugated material with higher recyclable content; • New design eliminates box pre-assembly; • New design contains only one motor per sleeve.
Shrink fill and heat seal	• Automatically shrink wrap and heat seal each motor prior to sleeve packaging; • This option eliminated the use of liquid foam (a non-biodegradable product).
In-line scale	• None
Vacuum gripper	• Unit design and function needed to be intuitively operated; • Optimized label placement to prevent tear-off by vacuum gripper.
Automatic shrink wrap machine	• Not needed due to packaging redesign; • Cannot use due to paint curing properties.

INTERVENTION RESULTS

Due to the process/equipment modifications, the implementation process for installing the new equipment took six months to complete with a budget of $336,000. The new layout is shown in Figure 2 below.

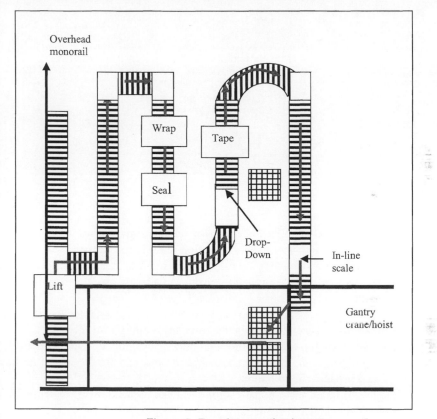

Figure 2. Post-intervention layout

Motors continue to come in on the overhead monorail. A pad/insert is placed on the scissor lift table and elevated to motor unload height; motors are unhooked from the monorail and positioned on the lift table. The scissor lift table is lowered to conveyor slide height and the motor is pushed to the in-feed conveyor. The power conveyor transports the motor to the shrink wrap station. The shrink wrap is applied mechanically to each motor and next heat sealed. The power conveyor is activated and the motor mechanically travels to the box/sleeve packaging station where the drop-down unit is located. The box and sleeve are placed over the motor and the shipping label is applied. The boxed motor is mechanically transferred to the automatic tape machine where it is closed, taped, and mechanically moved to the out-feed conveyor. The packaged motor is weighed at the in-line scale location and vacuum lifted onto an outgoing pallet. The outgoing pallet load is mechanically transferred to the warehouse for shipping. The following pictures show these operations:

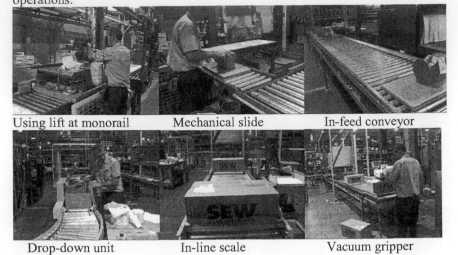

| Using lift at monorail | Mechanical slide | In-feed conveyor |
| Drop-down unit | In-line scale | Vacuum gripper |

SUMMARY

- The BWC CTD Risk Factor score for the post-intervention assessment is seven (7). This indicates a 74% reduction in CTD risk factors.
- The new process requires 'zero' lift (no manual lifting) of the motors, boxed motors, and eliminates push/pull forces associated with cart movement.
- The new layout and equipment selection is adaptable to all user populations and accommodates the company's aging workforce.
- Material savings resulted from the new packaging design/components. Packaging material costs decreased 8% to 26% and are dependent on box size and product configuration.
- The use of liquid foam in packaging has been reduced by 50%. This is an ecological goal that the company elected to meet.
- The new packaging uses recycled paper which is less porous, less costly, and environmentally friendly.
- Person-hours to package one motor improved by 29%. This improvement coincides with the company's upcoming goal of increasing productivity by 50%. Based on current production levels annual savings were calculated at $41,680. If production volume increases, the potential to package and ship 53 extra motors per two shifts is feasible. With packaging operating at full production an annual profit over $2.6 million may be realized.
- Economic cost justification using the payback method calculated a value of 5.35 years for the intervention to pay for itself.

- Customer feedback has been positive on the new packaging. The customer can more easily handle and store the single motor/box design.
- The quality of the shipping box has improved in both the packaging integrity and visual appearance. The company does not have a formal quantitative method to calculate quality of packaging.
- Before the intervention, two CTD claims were associated with the manual handling of the motors. No CTD claims in the Shipping Department have been documented since the intervention was implemented.
- No OSHA recordables in the Shipping Department have been reported since the intervention was implemented.
- Employees with the Shipping Department are extremely satisfied with the outcome of this project.
- The new material flow/process can be fully integrated into the new proposed paint line.

DISCUSSION

As an ergonomics consultant, it becomes apparent that a macroergonomics approach is needed when making recommendations that focus on engineering controls. One recommendation may impact many other facets of the operation that are either unknown and/or unforeseen by the consultant. Working with a team of experts and the employees affected by the intervention is critical in developing an economically feasible and workable solution.

REFERENCES

Christopher A. Hamrick (2001) "Proof that Ergonomics Works: Combined Results of Over 100 Independent Ergonomic Intervention Studies", *Proceedings of the Human Factors and Ergonomics Society 45th annual meeting*, p. 987-991.

Chrysler (2008). *Ergonomics Design Criteria*, "Manual Material Handling Design Criteria."

General Motors (2005), *The Design Ergonomics Worksheet* (DEW).

NIOSH (1981) *Work Practices Guide for Manual Lifting*. DHHS, Washington, DC: US Government Printing Office.

Mital, A., Nicholson, A.S. and Ayoub, M.M. (1997) *A Guide to Manual Materials Handling*, Second Edition, Washington: Taylor & Francis, pg. 75.

Snook, S. H. and V. M. Ciriello (1991). "The design of manual handling tasks: revised tables of maximum acceptable weights and forces." Ergonomics **34**: 1197-1213.

Thomas R. Waters, Ph.D., Vern Putz-Anderson, Ph.D., Arun Garg, Ph.D. (1994) Centers for Disease Control & Prevention, *Applications Manual For the Revised NIOSH Lifting Equation.*

Waters, T. R., V. Putz-Anderson, et al. (1993). "Revised NIOSH equation for the design and evaluation of manual lifting tasks." Ergonomics **36**(7): 749-76.

ACKNOWLEDGEMENT

Bill Tompkins, Engineering, SEW-Eurodrive, for his technical assistance and support with this project.

Mayme Larson, Human Resources, SEW-Eurodrive, for her assistance with this project. (Company website: www.seweurodrive.com)

CHAPTER 34

Product Maturity, Security and Software Engineering

Guy Boy[1], Richard Ford[1], Ronda Henning[2]

[1]Florida Institute of Technology

[2]Harris Corporation

ABSTRACT

The maturity of a product is defined by its capability to meet specified qualitative criteria at a given step in the design process. In this paper, we will focus on product maturity; a step beyond traditional process maturity as defined by Capability Maturity Model Integrated (CMMI)-like methods. Product maturity is measured through a series of high-level requirements, intensive formative evaluations and expert integration. A global philosophy is vital to guide complexity management throughout the design and development process. High-level requirements of the product must address the right function(s) as well as any security policy enforcement requirements. High-level requirements must be stable, correctly validated, clearly stated, well-written, and consistent. Scenario-based design should be promoted to appropriately test design solutions. The personnel resources assigned to the project are responsible for execution of the design. A knowledgeable technical manager must constantly enforce a global vision of the product and integrate possible partial visions throughout the design and development process. End-users should be included throughout the process to appropriately validate high-level requirements. Technical coordination of the various actors is required to insure effective and correct integration and operation reliability insurance. Management effectiveness must be integrated also and reduce the distance between actors. Continuity of the team organization throughout process must be preserved; internal personnel turnover should be carefully controlled. Specific care should be devoted to selecting appropriate of design and evaluation methods, especially for innovative products. Finally, high-level requirements should be evaluated from the start and used to define incremental integration checkpoints to support continuous evaluation, i.e., not at the end of the development process.

Keywords: Maturity, security, software engineering, scenario-based design.

INTRODUCTION

The last 30 years have seen an information revolution of incredible size. Information technology has become ubiquitous, and users clamor for more power, more functionality, and more speed. Hardware capabilities have increased at incredible rates, and long term trends, like Moore's Law, seem to indicate that the future will bring more of the same.

Amid this incredible rush of technology, the role of the software engineer has become critical to the safety and continuity of services ranging from Digital Television, food and power distribution, and industrial control systems. However, software often outlives its expected termination date, as was evidenced during the "Y2K" exercise. As such, the computer industry has begun exploring how to build software in ways that produce more predictable outcomes. In essence, the goal is to move from a "quick fix" model towards software that has well-documented properties and is built to support applications over a longer lifespan.

Aside from the rapid rate of hardware and software change, modern computer software is (typically) incredibly complex. As an example, the Windows XP operating system is built from approximately 40,000,000 lines of source code! In a project of this size and complexity, it is impractical (if not impossible) to prove that the code performs to a particular set of standards. Furthermore, it is not enough to demonstrate that a program produces certain output when given a particular input; one also needs to ensure that the program has no other undesirable side effects. These errors can range from benign (a miscolored rendering of a pixel) to catastrophic (a control system that incorrectly causes a car to accelerate). Additionally, software errors can also be exploited by an attacker, leading to security problems such as denial of service to authorized users. This is a matter of the maturity of the product and emerging practices associated with it (Boy, 2005).

As development organizations move toward more reliable software, interest has increased in how other disciplines have reduced product defect rates. In particular, we turn our attention to the aviation industry, which as a whole has an outstanding track record for low system failure rates. However, as the Software Engineering community follows the trajectory laid out by aircraft engineers, we should also take careful note of the ways in which the process-centric aviation approach has acted to the detriment of reliability. We aim to learn not only from aviation's successes, but also avoid some of the failures that are becoming apparent in the techniques used.

The remainder of this paper is laid out as follows. We describe the current state of security and the steps that software engineers have taken to address these deficits. We then examine the processes used within the aviation world and some of the lessons we can learn from them. Finally, we use these observations to describe a roadmap for the software engineering discipline that provides a way toward improved development processes and, ultimately, end product.

THE CURRENT STATE OF SECURITY

Based on empirical evidence, security has been handled poorly by existing software design methodologies. These methodologies have focused on product features and functionality at the expense of secondary requirements such as security. In part, this is due to the poor integration of secondary requirements into the development lifecycle, and the unique features of security that makes the attributes of a "known, secure state" rather hard to define. For most computer users, the sorry state of computer security will not be a surprise. However, when put in the context national security or financial impact terms, the issues raised by computer insecurity are staggering.

According to the 2008 CSI Computer Crime and Security Survey, almost 50% of participating organizations had suffered from a virus attack, and an astonishing 27% of respondents reported that they had suffered a targeted malware attack (Richardson, 2008). Financially, these attacks were expensive, with an average cost of $500,000 for computer security incidents involving financial fraud, and almost $350,000 for incidents involving "botnets" on the organization's network.

In addition to challenges faced by industry, information assets are a critical resource during times of war. As such, many of the world's nations have invested heavily in Computer Network Operations (CNO) in order to deny an adversary access to information during a battle. Notably, in a US-China Economic and Security Review Commission Report, the authors point out China's "sweeping military modernization program", that is "providing the impetus for the development of an advanced IW [Information Warfare] capability" (Krekel, 2009). As such, we must understand that computer security affects not only our day-to-day lives but also the geopolitical shape of the world.

Given these rather sobering observations it is important to consider the challenges developers have with the software development process in general and security in particular.

THE SECURE DEVELOPMENT PROCESS

When considering software engineering as a discipline, it is important to understand the different phases of the software life cycle. Typical models of this lifecycle include requirements analysis, system design, program design, coding, unit and integration testing, system testing, acceptance testing, and operation and maintenance. These steps are often organized into different models, such as the linear waterfall model, or the iterative spiral model. In addition, developers frequently make use of prototyping to obtain user feedback.

These development approaches are well known, and have been used for years in the industry. However, the rapidly-advancing nature of technology has lead to numerous issues with this idealized picture. Time to market, changing specifications, and system complexity have all negatively impacted software quality. To this end, the Software Engineering Institute (SEI) at Carnegie Mellon developed the "Capability Maturity Model" (SEI-CMM) to help formalize the often ad-hoc processes used by developers (Paulk et al., 1993).

The model breaks the process maturity state of an organization into five levels: Initial, Managed, Defined, Quantitatively Managed, and Optimizing. The idea is that within each level, the organization is moving toward higher quality, which equates to more predictable results. Note that CMM is ultimately not a purely process driven model; the most "advanced" organizations are permanently in the optimization process, striving to measure and improve production. However, the overall focus on CMMI is comprehension, measurement, and optimization of the processes by which products are built.

Despite this formalization, many programs have continued to be plagued by security problems. Perhaps the most widely known example of this is the Windows operating system by Microsoft. Several years ago, Microsoft began a massive campaign to improve its development practices with respect to security. This culminated in the "Windows security push" and the Security Development Lifecycle (SDL) (Howard & Lipner, 2003). The basic ideas of the push were based upon propagation of security practices through workforce education, code analysis and review, threat modeling, and the design change process. Based on Howard's reports, SDL has dramatically improved the security of Windows, and as such, is a step forward in large-scale software design. SDL is not revolutionary in that each step was known in the field at the time; however, SDL provided a comprehensive framework and tools that aided developers at every stage of the process. Overall, SDL is broken down into the following stages:

- Pre-SDL Requirements: Training
- Phase One: Requirements
- Phase Two: Design
- Phase Three: Implementation
- Phase Four: Verification
- Phase Five: Release
- Post-SDL Requirement: Response

During each stage of the process, everyone connected with the development of the product has guidelines that define best practices to be employed within that phase.

In addition to the SDL, there have been many other attempts to improve the security of program development. From formal methods to Agile Development techniques, software engineers have all driven toward improving the final output of a project with respect to security, and more broadly, safety. While these efforts to improve software engineering from a security perspective have borne fruit, there is considerably more work to be done (as ably demonstrated by the statistics we provided previously).

Part of this stems from the fundamentally different requirements of security versus functionality. That is, most functionality can be directly implemented;

requirements are typically of the form "the product must do x", where x can be measured. Conversely, security is often a negative requirement of the form "under no circumstance shall the product allow y". Logically, proving the non-existence of a behavior is difficult under many circumstances, and impossible under others.

Part of the challenge is also the relative immaturity of the science that underpins the software development process. Computer science is still rapidly advancing, and technological advancements in the underlying hardware are enabling previously impractical approaches for developers. Within this whirlwind of progress, the development team moves from project to project and technology to technology, never quite "finishing" a project, and never perfecting their process as the tools and methods can vary dramatically as a function of time.

This situation described above has created the perfect storm. As society becomes more focused on the use of information technology, attackers have become strongly incentivized by both financial and national gains. In addition, the rapid changes in the technology underpinning and the thirst for "new" and "better" products have led to an explosion in complexity even for products that we might consider to be rather simple (such as a document viewer).

COMPLEXITY AND MATURITY

Many people interact with a computer everyday without caring about its internal complexity... fortunately! This was not the case thirty years ago. Individuals who were using a computer needed to know about its internal complexity from both architectural and software points of view, in order to make it do the simplest tasks. Computer technology was not as mature as it is today. Computer users had to be programmers. Today, almost everyone can use a computer in some fashion, such as surfing the web or using a smart phone. However, internal complexity may become an issue in abnormal or emergency situations, such as when a fatal system error occurs in Windows, resulting in the dreaded "blue screen of death." Human-machine system complexity has been studied extensively in the aerospace sector (Boy & Grote, 2009; Donohue et al., 2008; Cummings & Tsonis, 2006).

A recent study on socio-technical complexity evaluation in the aviation world provided recommendations for human-centered design of controls, displays, system behavior, and system integration as well as design guidance for error management (Boy, 2007). This study was conducted in the framework of the regulatory context (CS 25.1302; AMC 25.1302; EASA NPA 15-2004). Perceived complexity criteria were developed to be usable and useful during certification, development and training. They include factors such as equipment novelty, the degree of integration of the equipment in the cockpit, and equipment complexity.

Internal complexity is about *technology maturity*. Maturity is a very complex matter that deals with the state of evolution of the technology involved, and especially reliability. Are both the finished product and underlying technology stabilized, that is, are they based upon established capabilities? Internal complexity of artifacts is not or barely perceived when these artifacts are mature and reliable, i.e., available with an extremely low probability of failure. In other words, when one can delegate with confidence and the work performed is successful, the complexity of the work delegated is not an issue. However, when the work is unsuccessful, one investigates possible causes for the failure; essentially, looking into the "black-box."

For that matter, the "black-box" should become more understandable, i.e., an appropriate level of complexity must be shown to the user, and the user should be able to interpret this complexity. Either the user is expert enough to address this complexity or needs to ask for additional support. For example, current sophisticated cars and trucks are so computerized that when something is suddenly going wrong, the driver may not be able to understand the situation and consequently comes to a decision that may cause inappropriate action. Particular kinds of purposeful information should be provided to avoid consequences that may cause even greater harm. Various levels of explanations should be available according to the situational context; this is a difficult thing to do when the context may be unpredictable.

Perceived complexity is about *practice maturity*. Is this technology adequate for its required use, and how and why do, or don't, users accommodate to and appropriate the technology? Answers to these questions contribute to a better understanding of perceived complexity, and further development of appropriate empirical criteria. Perceived complexity is a matter of the relationship between users and technology. Interaction with an artifact is perceived as complex when the user cannot do or has difficulty doing the perceived task. Note that users of a new artifact still have to deal with its reliability and availability, which are not only technological, but are also related to tasks, users, organizations and situations. This is why novelty complexity analysis and evaluation require a solid structuring into appropriate categories.

Users who interact with a complex system inevitably build expertise. They cannot interact efficiently with such a system as naïve users. There is an adaptation period for new complex tasks and systems because complex systems such as airplanes are prostheses. The use of such prostheses requires two major kinds of adaptation: mastering capacities that we did not have before using them (e.g., flying or interacting with anyone anywhere anytime); and measuring the possible outcomes of their use, mainly in social terms. All these elements are intertwined, involving responsibility, control and risk/life management. Therefore, provided evaluation criteria cannot be used by just anyone. They must be used by a team of human-centered designers who understand human adaptation. Unfortunately, in a finance driven industrial economy, human centered design is too costly to use for all but the most critical software due to the complexity and time involved.

MATURITY MANAGEMENT

It is expected that the complexity of systems varies during their life cycle. Therefore, both technology and practice maturities need to be taken into account along the life-cycle axis of a product by all appropriate actors; the "maturity axis" is a sequence of maturity checkpoints and redesign processes up to the entry of a software component into service. Even if the CMM was developed and extensively used to improve the efficiency of software production processes, it only partly contributes to the assurance of technology maturity, in the sense of software quality assurance. However, this method does not address directly either internal complexity or perceived complexity of the system being developed. This is why maturity checkpoints should strongly involve usability (Nielsen, 1993) tests during the whole life cycle, and more generally involve appropriately experienced people.

Complexity refers to the internal workings of the system, and perceived difficulty of the face provided to the user – the factors that affect ease of use. The history of technology demonstrates that the way to make simpler, less difficult user interfaces often requires more sophisticated, more intelligent, and more complex internal workings. Do we need intelligent interfaces if the intelligence and complexity should be internal to the system?

The interface is "the visible part of the system, where people need stability, predictability and a coherent system image that they can understand and thereby learn (Norman, 2002)." Norman's citation is very important today when there are layers and layers of software piled on top of each other, sometimes designed and developed to correct previous flaws of lower layers. Software engineers commonly talk about patches. This transient way of developing artifacts does not show obvious maturity. The technology maturity, and consequently internal complexity, of an artifact can be defined by its integration, reliability, robustness, resilience and availability. As previously stated, it is always crucial to start with good, high-level requirements that, of course, will be refined along the way. Problems arise when these requirements are weak and ambiguous.

A ROADMAP FOR SOFTWARE ENGINEERING...

This paper focuses on secure-software production and we claim that development maturity is a key impediment to this production process. However, current maturity methods are not fully appropriate to ensure secure software at product delivery time. These methods do not address product maturity directly and more importantly the maturity of the accompanying practices and system environment. We propose the following roadmap for software engineering to better support secure software production.

High-level requirements are key factors for the whole life cycle of a software product. Therefore they need to be carefully addressed. In the same way an architect is hired to build a house, an architect is needed to build software. These architects must be knowledgeable and skilled in human-centered design. They need to be able to involve users on the design and development process with appropriate participatory design techniques and tools, and a well-structured collaborative system engineering approach. Even if interaction design started as a human-computer interaction approach, it has now become a system engineering approach where interaction with the various mockups of the system being developed should be incrementally tested. In this case, interaction design means that all possible interactions between the various agents of the overall system are tested. By agents, we mean human and software agents.

The software architect should also be equipped with socio-technical analysis tools and methods. Since software is buried and integrated with multiple layers of software, internal complexity is a major concern and needs to be tested. In addition, the use of such systems induces the emergence of new properties that need to be discovered and appropriately tested. Various techniques could be used that are grounded in complexity theories and organization science (Schlindwein & Ray, 2004; Dooley, 2002; Hemingway, 1999; Hollnagel, Woods & Leveson, 2006; Leveson, 1995).

Finally, we cannot produce anything right without appropriate expertise, experience and hard work. This applies to secure systems!

REFERENCES

Boy, G.A. (2005). Knowledge Management for Product Maturity. *Proceedings of the International Conference on Knowledge Capture*. Banff, Canada. ACM Press, New York, ACM Digital Library.

Boy, G.A. (2007). Human Centered Development of Perceived Complexity Principles: Developed Criteria. Technical Report DGAC/EURISCO No. T 2007 201, France.

Boy, G.A. & Grote, G. (2009). Authority in Increasingly Complex Human and Machine Collaborative Systems: Application to the Future Air Traffic Management Construction. *Proceedings of the 2009 International Ergonomics Association World Congress*, Beijing, China.

Cummings, M. L. & Tsonis, C.G. (2006). Partitioning Complexity in Air Traffic Management Tasks. *International Journal of Aviation Psychology*, Volume 16, Issue 3 July 2006, pp. 277–295.

Donohue, G.L., Shaver, R.D. & Edwards, E. (2008). *Terminal Chaos: Why U.S. Air Travel is Brocken and How to Fix it*. Ned Allen, Editor in Chief, Library of Flight, AIAA, Reston, VA, USA.

Dooley, K. (2002), "Organizational Complexity," *International Encyclopedia of Business and Management*, M. Warner (ed.), London: Thompson Learning, p. 5013- 5022.

Hemingway, C.J. (1999). Toward a Socio-cognitive Theory of Information Systems: An Analysis of Key Philosophical and Conceptual Issues, *IFIP WG 8.2 and 8.6 Joint Working Conference on Information Systems: Current Issues and Future Changes*. Helsinki, Finland: IFIP. 275-286.

Hollnagel, E., Woods, D.D. & Leveson, N. (Eds.) (2006). *Resilience engineering: Concepts and precepts*. Ashgate, UK

Howard, M., & Lipner, S., (2003). Inside the Windows security push, in: *Security & Privacy*, IEEE, 1:1(57-61)

Krekel, Bryan, (2009). Capability of the People's Republic of China to Conduct

Cyber Warfare and Computer Network Operations, *The US-China Economic and Security Review Commission*

Leveson, N. (1995). *Safeware: System safety and computers*. Addison-Wesley, USA.

Nielsen, J. (1993). *Usability Engineering*. Academic Press, Boston, MA, USA.

Norman, D.A. (2002). Complexity versus Difficulty: Where should the Intelligence Be?, in *IUI'02 International Conference on Intelligent User Interfaces*. Miami, FL, USA.

Paulk, M., Curtis, B., Chrissis, M. & Weber, C. (1993). *Capability Maturity Model for Software (Version 1.1)*. Technical Report   CMU/SEI-93-TR-024.

Richardson, Robert, (2008). CSI Computer Crime & Security Survey, *Computer Security Institute*.

Schlindwein, S.L. & Ray, I. (2004). Human knowing and perceived complexity: Implications for systems practice, *E:CO,* vol. 6, pp. 27-32.

Chapter 35

Macro and Micro Ergonomics Application in a Medium Sized Company

Shahnavaz, H.[1], Naghib, A.[2], Samadi. S.[2]

[1]Center for Ergonomics of Developing Countries (CEDC)
Sweden

[2]Kaf Holding (Darougar)
Tehran, Iran

ABSTRACT

Ergonomics literature provides ample evident of many successful micro and macro ergonomics interventions and their positive impacts for both employees and employers of all sectors of the society. It is commonly accepted that the application of micro- and macro ergonomics is essential for improving working conditions, system efficiency and promotion of the working-life quality. While ergonomics has shown good potential for ensuring optimum technology utilization and proper technological development in the 'industrialized world', interest and attention paid to the subject is very low among organizations and industrial managers in the Industrially Developing Countries (IDC). Almost, two-thirds of the world's population in IDCs has little or no access to the vast knowledge base that makes ergonomics such an important tool for improving work environment and increase productivity.

The first phase of this project involved a macro ergonomics intervention process at Darougar Company, starting in spring of 2006. The main objective was to improve various systems and procedures of the company from a macro ergonomics point of view. The intervention process model introduced is based on three fundamentals, that is; management support, knowledge support and employees "participation. The process started with employee's training, team working and development of a feedback system. For the successful continuation of the process, it also was found necessary to develop a functional progress assessment system, recognition & reward system as well as a communication net-work system

Thirty three of the company's top and middle mangers took part in a workshop aimed to collectively formulate the company *vision* and develop the related *goals* for reaching the defined vision. The entire management capacity of the organization was directed towards a collective effort for this purpose. In a participatory process using the 'Future Workshop (FW)' technique, participants analysed various problems that they were experiencing for achieving the defined goals. Problem catalogues were developed for each goal and the problems were ranked in order of importance in different categories, under different tittles. Participants were divided into different working teams. 'Logical Frame work Approach (LFA)' was used to analyze problems, develop feasible and applicable solutions as well as implementing the proposed solutions for achieving the desired goals. Proposed solutions were later verified at 'Special Groups' and then approved by the Managing Director (MD) and some by the company board of directors for application. After four years of project progress, problems related to achieving 10 goals out of total 17 were analyzed and practical solutions are developed, many were approved and some are implemented or are in the process of implementation. The company is continuing successfully with the macro ergonomics intervention process. Currently problems related to two more goals are worked with at four different working groups.

In 2008 the company agreed to conduct a micro ergonomics intervention process at its largest factory. The aim was to improve the working conditions and productivity as the second phase of the overall ergonomics intervention process. The process started with training of thirty five factory male production and line managers, the deputy directors and the factory director (all males) in basic ergonomics and ergonomics intervention. The main objective was to create ergonomics awareness among factory managers and to teach them ergonomics methods and tools for improving working conditions, environmental conditions as well as productivity. The process started with introducing the 'Ergonomic Checkpoints' book (ILO, 1996). After basic ergonomics training, the factory managers were divided into several working teams. Teams were assigned to analyse all workplaces from ergonomics view point and provide improvement solutions. Solutions were then verified at the factory steering committee and later approval by the factory director. Approved solutions were implemented using factory resources.

During the past two years over 80 improvement projects have been implemented. Twenty five of the implemented projects resulted in productivity improvements and waste reduction. The total cost of the 25 implemented projects was $9430 US Dollars and the benefits according to the company calculation in form of improved productivity for just one year are was $386500 US Dollars, showing that the cost benefit after one year is more than 40 times the invested capital. The other 55 projects were mainly aiming at improving safety, health and satisfaction of the employees. The micro ergonomics process is continuing successfully at the factory.

Keywords: Ergonomics Intervention, Developing Country

INTRODUCTION

People in the Industrially Developing Countries (IDC) have little or no access to the vast knowledge base that makes ergonomics an important tool for work environment and productivity improvements. While ergonomics has shown good potential for ensuring optimum technology utilization and proper technological development in the 'Industrialized World', interest to the subject is very low among organizations and industrial managers in the IDC. Ergonomics literature of over half a century provides ample evident of many successful ergonomics interventions and their positive impacts for both employees and employers of all sectors of the society. It is commonly accepted that application of micro and macro ergonomics is essential for improving system efficiency, working condition, productivity, and promotion of quality of working life.

Ergonomics intervention and its potential to deliver benefits has been accepted and practiced in a small number of projects in IDC, as not many decision makers are conversant with the breadth and depth of ergonomics, O'Neill 2005. At present there are few practical examples to illustrate the benefits that can come from ergonomics intervention programs in IDC. It is therefore true to say that there is lack of awareness regarding the potential benefit of ergonomics in most IDC workplaces. Ergonomics as yet is not a global science and a major effort is required to make it known as a useful and practical discipline in that part of the world which is most in need of ergonomics input (Shahnavaz 2002). However, when applying the appropriate type of ergonomics, there would be improvements in quality, productivity, working conditions, occupational health and safety, reduction of rejects, and increases in profit (Yeow and Sen 2002).

The term intervention refers to efforts made to effect change and render such change stable and permanent (Westlander *et al.* 1995). The objective of ergonomics intervention is to design jobs that are possible for people to do, are worth doing and which give workers job satisfaction and a sense of identity with the company and protect and promote workers' health. An ergonomics intervention should therefore result in improving both the employees' well-being (health, safety and satisfaction) as well as the company's well-being (optimal performance, productivity and high work quality), Shahnavaz 2009.

According to Hägg 2000, different types of intervention programmes are identified with aims ranging from time-limited intervention to continuous processes. Ergonomics intervention, however, should be a well-planned and structured process and must be a local activity that meets the particular needs of local people. In IDC, it should focus on utilizing available potential and resources of the local enterprise for planned improvement. Ergonomics intervention can take different forms such as; Workstation improvement through physical changes, Operator training, Work organization change, Technology change and/or Work systems improvement. Factors affecting ergonomics intervention are: Management support and commitment, ergonomics awareness and know-how, employee participation in a democratic climate, motivation, organization and culture, legislation and inspection and cost effectiveness of the intervention programs. According to Burgess and Turner 2000, "commitment is particularly important for the successful implementation of projects and strategic programs". Worker participation is important for successful intervention. Participation of people in identifying ergonomics problems and developing solutions is shown to be an effective way of finding cost effective improvements. The success of low-cost improvements actually depends on the careful selection of feasible and local solutions acceptable by both management and employees.

PHASE 1- MACRO ERGONOMICS INTERVENTION AT DAROUGAR COMPANY

Macro-ergonomics is a socio-technical system approach to work system design (Hendrick and Kleiner 2001). It contains components at the larger level system such as organizational climate, organizational culture, organizational structure, technology, management practice and employee's participation. According to Hendrick 1995, an important outcome of macro ergonomics intervention is culture-change, where organizational culture is defined primarily by the organization's core values. Macro ergonomics' approach and participatory ergonomics process in IDCs help organizations, managers and workers for a better understanding of their problems at work as well as providing solutions for work environment problems and improvement of work system in any process, by finding common goal and effort. Ergonomics intervention is a process of change for improvement. As changes are happening very often, it is important to establish an intervention process that is able to deal with different kinds of problems which may occur as the result of change in technology, organization, environment, economy and politics.

METHOD

Darougar Company is a medium sized company in Iran with over 1000 employees producing detergent, washing powder, soaps and other hygiene products. The company was interested to conduct a macro ergonomics intervention program in order to improve its performance. After several meetings with the managing director, one of the authors of this paper (the ergonomist) was asked to propose a project for improving system efficiency and condition at the company. A theoretical model to guide the ergonomics intervention process has been proposed and accepted to be applied at the company. The intervention process is based on three fundamentals, that is to say; Management Support, Knowledge Support and Employee Participation. The process starts with employee training, team work and a feedback system designed and developed with the help of the three fundamentals. For the successful continuation of the intervention process, there was a need to design and develop a functional and acceptable progress assessment system, a recognition and reward system, as well as an appropriate communication net-work system for both bottom-up and top-down communication, which was developed and established at the early stage of the project. Training was an important part of a comprehensive intervention process, enabling knowledge access within the organization and implementation of the acquired knowledge in practice.

The ergonomics intervention starts with training which brings about awareness and action, aiding in the introduction of necessary ergonomics improvements. Training and transfer of ergonomics know-how is a dynamic process, where the external expertise and know-how provided by the *'knowledge support'*, i.e. the project manager (the ergonomics facilitator) meets the recipients' individual experiences and talents existing among participants of working teams in a fruitful interplay. The role of the ergonomist in the participatory ergonomics intervention process was as a change agent or a facilitator of change. The intervention process was built upon trust, competence, relevance and a strong tie to the practical reality, which was adjusted to the local needs and requirements and cultural reality of the enterprises in which the intervention was taking place.

In the proposed process, the employees learnt to analyse their work, working procedures and the working conditions, and propose feasible improvement solutions. The main objective of the ergonomics intervention project was to improve various systems and procedure of the company from macro ergonomics points of view. Systems and procedures such as; Human Resource Management, Customers' Relation, Productivity and Productivity Assessment, Quality Assurance, Employees' Motivation, Employee's Promotion, Health, Safety and Security Issues, Technology Improvement, Financial Resource Improvement, Welfare Facilities, etc. that the top and middle managers were mostly involved in and faced some problems.

The project manager (the ergonomist) informed the management that a company needs to have a well formulated vision with related goals, which considers the company's ambitions and resources and which is defined and accepted by all company managers and decision makers. In this way every manager would try its best to achieve the goals which he/she had contributed to in its formulation. It was assumed that as a result, the entire management capacity of the company would be directed towards a collective effort for accomplishment of the defined vision. The company needed to formulate a set of clear goals, achievable and measurable and should lead the organization to its defined vision within a certain period of time. Thereafter, for achieving each goal the company had to develop a plan and programme, identifying various activities that were required and the resources needed for reaching each individual goal. This was achieved in a participatory process in which all top and middle managers were actively involved.

The process started with a three day workshop, in a pleasant environment outside the company premises, in which 35 of the company's top and middle mangers took part. The workshop was aimed to collectively formulate the vision for the company and develop a set of goals for reaching the defined vision. Management was made aware of the fact that the company needed to have a well-formulated vision and related goals, accepted by all decision makers.

The workshop started with a written questionnaire asking each participant to write down what he/she assumed was the vision of the company and at least 3 goals for achieving the vision. The questionnaires were collected and the participants were divided randomly into 5 different groups. Each group received seven randomly selected questionnaires to work with in their group. The task of each group was to put together and combine the written vision and goals in the questionnaires, finding similarities and differences between the written answers. Thereafter, the groups came together and presented their findings for the whole group. It was noticed that it existed vast diversity and sometimes contradictory ideas about the general vision and goals of the company.

During the afternoon session of the first day, participants were trained in key features of a successful and prosperous organization. Core values and related issues such as productivity, participation and employees' commitment, motivation and motivating factors, importance of adapting a holistic system approach for productivity and quality, the employee value, customer value, environmental value, social image value, sustainability. etc. were discussed. The next day, participants were asked to work within their group and develop new scenarios for the company vision and related goals, based on the previously collected questionnaires information and the knowledge gained from the yesterday training session. They were also asked to evaluate their scenario with the SWOT (Strengths, Weaknesses, Opportunities, and Threats) analysis diagram. At the end of the second morning, the groups came together and presented their scenario and their SWOT analysis diagrams for the whole group. The afternoon of the second day was devoted to combining the work of the five groups and defining the final version of the company vision with related goals, approved and accepted by all managers.

On day three participants came together to prioritise the goals. They then selected the first priority goal to work with. The workshop was interested to find out what problems existed from a systems perspective to achieve this goal, and to find practical and acceptable solutions for solving the identified problems. This task is best achieved using the 'Future Workshop' (FW) technique (Junk and Müller 1987; Skoglind-Öhman and Shahnavaz 2004). The FW technique was described to the participants and the problem experienced by participants for achieving the first goal of '*high productivity*' was selected as the workshop theme. User's guides for workshop leaders and workshop participants prepared by Mohammed-Aminu Sanda (2006) were given as a reference to participants.

Participants managed to develop a problem catalogue related to achieving the first goal, which was 'high productivity' according to the FW procedure. After giving credits to the problems and prioritising them, they then divided the problems which received credits under four relevant titles (topics). Participants were asked to select one of the four titles which they had an interest in, and the knowledge to analyze and propose feasible solutions. Then the four "Working Teams" (WTs) began working on analysing each individual problem and proposing feasible and practical solution, using the techniques that they have learned from FW, Shahnavaz 2009.

At the end of the third day, the project facilitator presented a model for future activity of the WTs. The 35 top and middle managers were divided into four WTs. It was decided that the WTs should meet once a week for 3 hours continuing their activities for about 6 months.

Approximately every 6 months it will be a general meeting (Workshop) of WTs in a two day workshop. During the first day, the WTs will present the results of their last months' activity, followed by general discussion. Thereafter, two new goals will be selected for developing the problem catalogue according to FW procedure. Managers will then be divided into new WTs to work with the new goals. At the same time, one or two Expert Groups (EG) will be formed from the most knowledgeable and experienced manager in the related topics. The task of the EG is to finalize the output of the WTs, verify and make their proposed solution more practically applicable. The output of the EGs (final proposals for change) will then be sent to the Managing Director (MD) for approval and later will be implemented. During these 6 months, regular general workshops, activities of the WTs and EGs were reviewed and discussed.

In each of the workshop, the project supervisor conducted few hours of education, training company managers in relevant topics such as; Productivity measurement technique, Time management technique, Learning organization, Job satisfaction and Motivation, Project design and Project management (LFA).

After each workshop a complete report has been prepared with the help of a 'Project coordinator'. This report contained a summery of the workshop, the critical catalogue developed at the Future Workshop, the final report of each WTs, the findings and final solutions proposed by the EGs, and the plan for future activities. This report is distributed to all partners involved.

A female manager (one of the authors) was appointed as the 'project coordinator' for the whole project. She will be a contact person and in direct contact with the project supervisor. Project coordinator will make sure that working teams and expert groups meet regularly according to the planned schedule, keeping the dead lines, document their activities and send a standard report-form after each meeting to the project coordinator. Further, the coordinator is a support person for the WTs and EGs and a contact person between MD, project supervisor, the WTs and EGs.

It was also decided that team's activities and all decisions related to this project should be documented by the project coordinator and shall be available to all company managers involved in the project. The ergonomics intervention process was estimated to take more than 4 years to complete. Up to the end of 2009, ten out of 17 goals have been analyzed with regards to their practical and applicable solutions by the WTs and EGs. Further, two more goals are been worked with currently by the working teams.

RESULTS

The results of the first 3 days workshop were the development of company visions and set of 17 goals for achieving the defined vision by all company managers, which was unanimously approved. The 3 visions and related (goals) are as follows:

I- ESTABLISHING DAROUGAR COMPANY (DC) AS THE LEADING AND LARGEST DETERGENT AND COSMETIC PRODUCTS COMPANY IN IRAN.

The related goals for achieving this vision were identified unanimously by company managers at the workshop as:
1. Increasing market share,
2. Better marketing,
3. Customer centred company,
4. Better human resource management (recruiting, keeping and promotion of employees),
5. Offering products at reasonable prices,
6. Homogeneous distribution of product within the country,
7. Ensuring high productivity,
8. Increasing number of products to complete the product basket,
9. Improving the chain of production from raw material to finished product.

II- ENVISIONING DC AS BEING AT THE HEALTH AND HYGIENE SERVICE OF THE SOCIETY

Again the related goals for achieving this second vision were identified as follows:
1. Improving company image in the society,
2. Improving company culture,
3. Improving product quality.

II- SEEING DC AS MOVING TOWARD BECOMING A WORLD COMPANY

For achieving this vision the following goals were defined by company managers:
1. Better usage of opportunities for investment internally and externally,
2. More brands,
3. Cooperation with large international companies,
4. Improving export,
5. Innovation.

Macro ergonomics intervention process is continuing since the first workshop, which was conducted in 2006. During the past 3 years, by the end of 2009, six more two days workshops were conducted, in which all company managers have participated. During these workshops 28 Working Teams and 10 Expert Groups have been formed. Darougar manager worked for 3417 man hours in these teams/groups analyzing and finding solutions for achieving 10 goals. Two more goals are currently discussed by four working groups. The remaining five goals out of total 17 will be the work for future. Working teams have developed 28 proposals. These were then reduced to 10 final applicable proposals by the expert groups. These proposals are practical and applicable improvement solutions for the most common administrative and managerial problems facing the company management. It is believed that by applying these proposals the company will reach its defined goals for achieving it desired vision. However, out of these developed proposals 18 (35% of the total) are approved until the end of 2009 by the managing director and even some complex ones by the board of the company directors. Five improvement proposals are applied and 13 other proposals are in progress of implementation, See Figure 1 (see Appendix) and 2.

PHASE II- MICRO ERGONOMICS INTERVENTION

Darougar Company has several factories. The oldest and largest is located near Teheran. This factory produces many different detergents and cosmetic products and has over 450 employees. The company director agreed to the proposal to conduct a micro ergonomics intervention process at this factory for improving the working conditions and productivity as the second phase of the overall ergonomics intervention process.

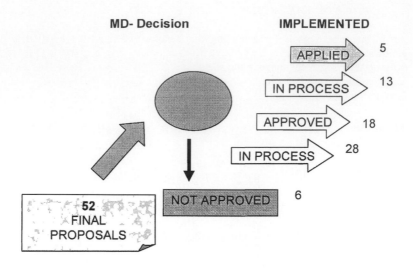

Figure 2- Final stage of implementation process: MD Decision

METHOD

The process started with a two-day workshop conducted in a pleasant environment outside the factory. Thirty five male production and line managers of the factory, together with the factory director and his deputy directors participated in this workshop. The aim of the workshop was to create ergonomics awareness among participants and to teach them ergonomics methods and tools for improving working conditions, environmental conditions as well as productivity. The workshop started with introducing the 'Ergonomic Checkpoints' booklet (ILO 1996). The project supervisor explained its content and showed participants how to use it at their work place.

In the afternoon of the first day, participants were divided into seven working teams (WT) to work with the ergonomics checkpoint material. This material includes 128 checkpoints related to nine different areas of basic ergonomics. Each team was assigned to work with all 128 checkpoints, trying to understand the content, 'why' and 'how' of each checkpoint. Participants were asked to discuss the material from their own work experiences, especially the experiences from their current workplace. They were also asked to write down their collective experiences in respect to each checkpoint, indicating if they proposed action at their current workplaces, and if so, its priority.

As the time of the workshop was very limited they were advised to select just a few checkpoints as training for this workshop and to continue their work after the workshop at the regular meetings planned for the future according to the introduced activity diagram. Thereafter, teams came together and each presented its work to the other teams, discussing their decisions regarding their proposed action and its priority, and presented their solutions for the identified problems. Despite it being the first time they were introduced to ergonomics and worked together as a team for identifying a shop floor problem and proposing solutions, they came up with very good local no-cost/low-cost solutions during this short time. This exercise greatly motivated the participants and they were all very excited to continue their work. Participants were informed that they should continue the work in their teams. They should try to find and analyze problems at their work-places and be vigilant about other health, safety and productivity problems that they may encounter at the factory in general. They should discuss their findings with their team members and try to develop applicable solutions for them. They should then write their proposed solutions in form of a project proposal, indicating activities required for solving the problems as well as the resources they needed and the time they think it would take to do the improvement. Proposals should be very precise, considering the possibility and acceptability of implementing their solutions from both management and employees perspective.

Participants were then trained in simple project design and presentation. Participants were informed that their proposals for change will be then discussed and assessed at the factory steering committee. If approved, it will be implemented using the company resources. After the workshop, the project supervisor, together with the project coordinator and the factory director, selected members for the Steering Committee (SC). Apart from the factory director, three of his deputies were selected for this committee. Two lists of activity, one for the working groups and one for the steering committee that was prepared by the project supervisor were discussed with participants and were approved unanimously.

A rewards system was designed to recognise the WTs' achievements and motivate workers for further engagement. All team members should receive an equal share of the reward. Four criteria were suggested and agreed to by the SC for assessing the effectiveness of the proposed project and deciding on the type of reward to be given to the WT. The maximum credit that each implemented project can receive is 100 points which will consist of: Increased productivity (35 credits), Economic savings due to resources saving and reduction in waste and reject products (35 credits), Increased safety and health (20 credits), and Increased worker satisfaction (10 credits).

Implemented projects will be evaluated by the SC on a regular basis according to defined schedule. Project evaluations will be conducted in the presence of the WT coordinator. The maximum financial reward for the proposed idea is approximately 25% of the total one year financial gain due to implemented project either because of increased productivity and/or economic saving. It was decided that a quarter of the one year gain should be given to the idea and three quarters to material, over-head and labour cost of the implemented project. If the project is implemented by the working team, they will receive another 25% for their labour.

The proposed financial reward categories and limits are a general guideline and are regarded as flexible. In cases of significant gain in productivity and or cost saving due to the implementation of an exceptional proposal, the SC together with MD can suggest higher financial rewards to the members of the working team.

RESULTS

The process, after two years has resulted in many benefits both for the factory management and employees. The benefits can be categorised under Direct benefits or Indirect benefits.

DIRECT BENEFITS

Performance of all units has been improved because of working teams' activities and implementation of the ergonomics checkpoints within factory workplaces. The working teams proposed and implemented over 80 no-cost and/ or low-cost solutions to small and medium problems, using mostly local resources. These changes included work-place and work-environment improvement; health and safety improvements; energy and resource saving; better house keeping; minimizing material handling; improving technology; and improving productivity. In short, all 128 checkpoints emphasized in the ILO Checkpoint book were checked, and improved if necessary. Out of the 80 implemented projects, 25 projects contributed to financial gains in form of increased productivity and/or resource saving. According to the Steering Committee's calculation, the total cost of implementing the 25 projects during the last 2 years were only $9430, and the benefits (Improved productivity and cost saving) for just one year is $366500, showing that the cost benefit after one year is more than 40 times, (see table 1). The low implementation costs of these projects are due to optimum utilization of factory resources. All projects were initiated, designed and implemented by the factory employees. They also used mostly scrap or unused material that were available in the factory. The other 55 projects were mainly aiming at improving safety, health and satisfaction of the employees. Projects such as improving lightings at workplaces and gate ways, improving machine safety and guarding and improving working postures and working conditions. Further, as a result of the working teams' activities during last 2 years, many working teams have received financial rewards and recognitions for their proposed and implemented projects. There are further, several projects in pipeline under implementation and under assessment by the SC. The micro ergonomics process is continuing successfully at the factory.

INDIRECT BENEFITS

The process initiated an active and conscious development environment at the factory. Employees are now more vigilant about their working conditions and are committed to cooperate and find solutions to all type of production and service problems. There is an atmosphere of competition between the working teams, each trying to come up with better cost saving or productivity improving proposals. According to the factory management the motivation among employees is now much higher compared to previous years. They are more committed to their work and use their experience and creativity to come with new ideas. There is better communication and the company is benefiting from utilizing its resources, especially its human resources, much more effectively than before. Employees are more engaged in company affairs and help in solving problems and contributing to company progress. As the result of the ergonomics project, they also benefited financially. Most of them received significant cash for their ideas and or helping to implement the ideas.

DISCUSSION AND CONCLUSION

Three years into the project many direct and indirect benefits for management and employees have been achieved. Performance of all units has been improved because of ergonomics implementation. The process has started an active and conscious development environment at the company.

Employees are more committed and use their knowledge and creativity to solve day to day problems, plan and develop policy procedures for future. It is now better communication and the company is fully utilizing its human resources. The participative culture is accepted and management is aware about the potential benefit of it.

The benefits can be categorised under 'direct' and 'indirect' benefits. The direct benefits are the various new ideas and solutions that are developed and implemented for various problems and obstacles that were identified for achieving the company goals. Several effective procedures have been developed; some are approved by MD and some by the board of directors. This process still is continuing until all the defined goals leading to the company vision are achieved. With regards to the indirect benefits, DC managers are talking about better coordination of various divisions towards achieving company goals, learning new methods for solving problems and better atmosphere for discussion, exchange of ideas and expressing own views. According to the company management, the moral and motivation among employees is now much higher compared to before. Managers are now much better informed about various problems that their colleagues from other divisions are facing. They are more familiar with their colleague's leadership style and the challenges they must confront to achieve the company goals. According to top management, it is now existing a much more relax and understanding environment for discussion and decision making, especially for complex issues, which requires involvement of several different divisions of the company.

This project started with macro ergonomics intervention, followed by micro ergonomics. The micro ergonomics intervention resulted much sooner in some tangible improvement. Working teams at the factory proposed several improvement changes that were quickly assessed by the steering committee and soon after implemented mostly by the team proposing the change. Most of the projects were no cost and/ or low cost, which were implemented using mostly the local resources. The micro ergonomic process resulted in significant cost saving and financial gain for the company. The benefit of the macro ergonomics process is more long term. More effective methods and procedures that are developed in the macro ergonomics process are helping the company to be more competitive in the national and global market. It can be concluded that changing administrative procedure and people's way of working from macro ergonomics point of view was not so easy among administrative personnel.

Table 1. The costs and benefits of implemented projects during two years at the factory

	PROJECT TITLE	Cost of Project in USD	Benefit after 1 year in USD
1	Improved emergency lighting	2 000	30 000
2	Improved liquid soap filling system	2 000	50 000
3	Re-charging fire fighting cylinders	0	600
4	Increased transport capacity	0	25 000
5	Installed alarm system for steam tanks	100	600
6	Changed cooling pump system	40	500
7	Improved air steering system	50	900
8	Re-used hot steam	150	4 200
9	Redesigned control system for liquid tank	800	3 500
10	Improved production capacity of shampoo line	200	3 000
11	Redesigned new soap cutting machine	100	38 000
12	Improved production capacity of line 1400	60	15 7000
13	Re-using cooling water	400	3 000
14	Increased capacity of storage room for packed material	300	7 000
15	Increased storage capacity for raw material	0	3 000
16	Designed automatic device for packing line	200	4 000
17	Changed cooling tanks	0	20 000
18	Re-using returned cooling water	150	7 000
19	Redesigned feeding system of line 2	250	4 000
20	Isolated steam pipes	200	1 600
21	Designed warning system for assembly line	150	7 000
22	Changed switches of mixing machine	30	3 000
23	Improved transport layout	150	2 000
24	Digitalization of press machine sensors	100	1 600
25	Redesigned layout of conveyer-belt	2 000	10 000
	TOTAL	**9430**	**386500**

REFERENCES

Burgess, R., and Turner, S. (2000), Seven key features for creating and sustaining commitment. International Journal of Project Management 18:225-233.

Hendrick, H. W. (1995), Future directions in macro ergonomics, Ergonomics 38: 1617-1624.

Hendrick, H. W., and Kleiner, B. M. (Eds.) (2001), Macro ergonomics, an introduction to work system design. Human Factors and Ergonomics Society, USA.

Hägg, G. (2000), Corporate initiatives in ergonomics – an introduction. Journal of Applied
Ergonomics 34: 3-15.

ILO. (1996), Ergonomic Checkpoints, Practical and easy-to-implement solutions for improving safety, health and working conditions. Geneva, International Labour Office.

Jungk, R., and Müller, N. (1987), Future workshops: How to create desirable futures. London, England. Institute for Social Interventions. ISBN 0948826398.

Mohammed-Aminu S. (2006), Four case studies on commercialisation of government Rand D agencies, Doctoral Thesis, Luleå University of Technology 33, Sweden.

O'Neill, D., (2005), The promotion of ergonomics in industrially developing countries. International Journal of Industrial Ergonomics 35: 163–168.

Shahnavaz, H. (2002), Ergonomics intervention in industrially developing countries, Key note speech, proceeding of the first national conference of the Iranian Ergonomics Society (IES), October 29-30.

Shahnavaz, H. (2009) Ergonomics Intervention in Industrially Developing Countries, Ergonomics in Developing Regions: Needs and Applications, 41-58. Taylor & Francis.

Skoglind-Öhman, I., and Shahnavaz, H. (2004), Assessment of future workshop's usefulness as an ergonomics tool. International Journal of Occupational Safety and Ergonomics (JOSE), Vol 10, No 2: 119-128.

Westlander, G., Viitasara, E., Johansson, A., and Shahnavaz, H. (1995), Evaluation of an ergonomics intervention programme in VDT workplaces. Applied ergonomics, Vol 26, No. 2: 83-92

Yeow, P., and Sen, R. (2002), 'The promoters of ergonomics in Industrially Developing Countries, their work and challenges', proceedings of the 3rd International Cyberspace Conference on Ergonomics, the CybErg 2002.

354

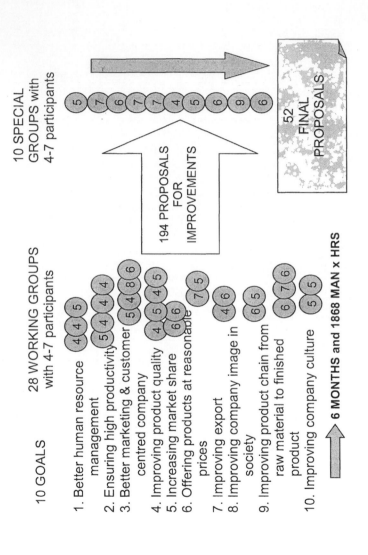

Figure 1- Working groups and special group activities during the last 3 years

CHAPTER 36

The Specification and Evaluation of Personalised Footwear for Additive Manufacturing

André S. Salles[1], Diane E. Gyi[2]

[1] Department of Design and Technology, Loughborough University,Loughborough, United Kingdom

[2] Department of Ergonomics (Human Sciences), Loughborough University,Loughborough, United Kingdom

ABSTRACT

Although additive manufacturing (AM) has potential in producing personalised items, such as footwear, it is unknown how to best capture and measure the foot in this context nor the short and medium term impact of such footwear on discomfort and injury risks. Therefore, research is currently being conducted to evaluate the short and medium term use of personalised footwear in terms of discomfort and injury risk; to identify measurement techniques for specifying and evaluating such footwear; and to determine the measurements required to specify personalised footwear suitable for additive manufacturing.

Participants had both feet scanned and 16 anthropometric measurements of the right foot taken. A single-blind paired samples experimental design was used and participants were paired according to: gender, age, body mass index and km ran per week. Thirty eight runners (19 pairs: 9 male pairing and 10 female pairing) were recruited. The two experimental conditions were: control and personalised. The personalised condition consisted of a pair of trainers fitted with personalised glove fit insoles that were designed and manufactured from the foot scans to match the exact plantar geometry of the individuals' foot. The control condition consisted of the same trainers, but fitted with a pair of insoles that were manufactured from the scans of the original insole shape, using the same material and thickness as the personalised condition.

Participants were allocated to one of the two experimental conditions and asked to wear the footwear for 3 months. Participants attended laboratory sessions at the start of the study (month 0), halfway (month 1.5) and at the end of the study (month 3) where the footwear was evaluated in terms of discomfort and injury risk. Discomfort was evaluated using a visual analogue scale. Injury risk was evaluated from plantar pressure distribution, rearfoot eversion, tibial internal rotation and peak vertical impact force of the right foot. The proposed paper will present the detailed methodology and approach of the study and discuss the early findings.

Keywords: Footwear, Additive Manufacturing, Anthropometry

INTRODUCTION

Additive manufacturing (AM), formerly known as rapid manufacturing, is potentially promising for producing personalised components, because of its geometric freedom and tool-less capability. In addition, AM can reduce unit costs, allowing production near the location they will be used, minimizing transportation and stock space (Hopkinson and Dickens, 2001).

The personalisation of footwear, in particular, can be advantageous for population groups, including older individuals, people with arthritis or diabetic foot problems. Personalised shoes can potentially provide a 'perfect fit' for the wearer. Studies indicated that 'fit' is the most important component of footwear not only because it is strongly correlated to comfort, but because it is speculated to be linked to injury and damage prevention (Cheng and Perng, 1999; Wunderlich and Cavanagh, 2001; Luximon et al., 2003). Too little or too much space in a shoe can be perceived as tight or loose respectively (Witana et al., 2004). Too tight a shoe will compress tissues leading to discomfort whereas too loose a shoe will lead to tissue friction because of the slippage between the foot and the shoe both causing blisters (Cheskin et al, 1987). In addition, poor shoe fit can cause undue pressure on the toes which can lead to deformities (Kouchi 1995; Kusumoto et al., 1996). In relation to specific population groups, a good fit can be even more important. For instance, recent reports indicate that the elderly population has wider feet than the shoes currently on the market, so they tend to develop forefoot pathologies (Chantelau and Gede, 2002; Menz and Morris, 2005). Also, individuals with diabetes have reduced pain sensation, so, unlike other population groups, they will not stop wearing footwear it is poorly fitted and this can start to damage the tissues (Chantelau and Gede, 2002).

The personalisation of footwear can also address personal preferences in terms of comfort. As comfort is influenced by an individual's foot characteristics, there is no comfortable shoe for everyone (Miller et al., 2000). Comfort is important because it is the main aspect that is considered when purchasing footwear (Cavanagh, 1980) and because it allows runners to maintain aerobic work for long periods of time (discomfort precedes pain).

In order to specify personalised footwear that is optimal to the individual, it is important to stress the importance of anthropometry. To provide a good fit, it has been speculated that at least 2 measurements in different dimensions in each region of the foot (forefoot, midfoot and rearfoot) are needed (Goonetilleke et al., 1997). The important measurements in determining individual preferences/needs, include instep girth, bottom width, heel height and toe box space (Goonetilleke et al., 1997; Cheng and Perng, 1999; Witana et al., 2004). Furthermore, foot shape plays an important role in the development of many types of injury (James et al., 1978; McKenzie et al., 1985; Cowan et al., 1993). Low arched (LA) individuals are likely to have more discomfort and greater rearfoot eversion because the lack of the arch and, consequently, the lack of shock absorbing capability leads to more foot and back injuries (Cheng and Perng, 1999; Williams III et al., 2001a). In addition, LA tend to prefer soft insoles in comparison to high arched (HA) runners that prefer harder ones (Mundermann et al., 2003). On the other hand, the HA foot is characterized by the longitudinal arch being more rigid, which makes it less efficient at absorbing impact shocks. They have more rearfoot inversion, resulting in higher lateral loadings and higher peak pressures (Cavanagh, 1980; McKenzie et al., 1985, Morag and Cavanagh, 1999; Williams III et al., 2001b). These are associated with a greater risk of injury, especially tibial shock, mechanical trauma and knee injuries (Cavanagh and Rodgers, 1987; Williams III et al., 2001a).

Although AM has much potential in the footwear field, it is not known how best to measure feet in this context nor even whether the short and medium term use of personalised footwear can affect discomfort and injury risk in comparison to the generic shoes currently available on the market. Hence, a 3 month study is being conducted at Loughborough University to investigate personalised footwear by providing participants with an AM glove fit insole. It is speculated that an insole which closely matches the foot of the individual will be more comfortable and reduce plantar pressure in comparison to a standard insert (Mundermann et al., 2003; Goske et al., 2006).

Therefore, the main objectives of this research are: (1) to evaluate the short and medium term use of the personalised footwear in terms of discomfort and injury risk; (2) to identify measurement techniques for specifying and evaluating such footwear; (3) and to determine the measurements required for specification for additive manufacturing.

METHODS

Recreational runners were recruited from gyms, running clubs, leisure centres and word by mouth in the Leicestershire area in the UK. Sampling criteria were: to run at least 5 km.wk^{-1}; have no reported musculoskeletal pain or injury for the last 12 months; 18-65 years old; and to have not used an orthosis for the last 12 months. A single-blind paired samples experimental design was utilized and participants were paired according to: gender, age, body mass index and km ran per week. The study was approved by Loughborough University's Ethical Committee. A total of 38 runners (19 pairs: 9 male pairings and 10 female pairings) were recruited. Participants were allocated to one of the two experimental conditions: control and personalised. Each experimental condition will be detailed in the next section. The footwear was evaluated in terms of discomfort and injury risk throughout a 3-month period. For that, participants were asked to attend to 4 laboratory sessions.

In laboratory session one, detailed anthropometric measurements were taken of the right foot following Hawes and Sovak (1994). The 16 measurements included girths, lengths, widths and heights.

Calculations enabled the classification of individuals according to the medial longitudinal arch:

- Arch ratio – height of the dorsum of the foot from the floor at 50% of the foot length divided by individual's truncated foot length (Williams and McClay, 2000);
- Arch index – calculated as the ratio of the navicular height to the foot length (Williams and McClay, 2000);
- Relative arch deformation (RAD) – calculated as:

$$RAD = \left(\frac{AHU - AH}{AHU} \right) \frac{10^4}{bodyweight}$$

where AHU is the measurement of the arch height taken in unloaded (i.e. 10% of weight bearing) position, AH is the arch height measurement taken in a full weight bearing (i.e. 90% of weight) position (Nigg et al., 1998).

Participants then had both feet scanned using a 4-camera 3 dimensional laser scanner (model: RealScan USB 200; 3D Digital Corporation, Newtown, CT, USA). Scans were taken with participants sitting on a chair, slightly resting their foot on the glass of the scanner (Figure 1).

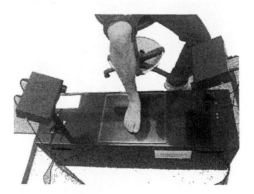

Figure 1. Participant having the foot scanned.

EXPERIMENTAL CONDITIONS

After the first session, the insoles (control and personalised) were designed and manufactured, as described below.

The personalised condition consisted of a pair of New Balance trainers (model: NB-757 Neutral Cushion) fitted with personalised 'glove fit' insoles. These insoles were designed and manufactured from foot scans to match the exact plantar geometry of the participants' feet from the heel to the base of the metatarsal heads. The foot scan data taken during the first session were manipulated (i.e. data were 'cleaned' to remove the 'noise' and unwanted data, smoothed, thickened to 2 mm and converted into a STL file) using a Geomagic Studio 10 software (version: 10; Geomagic, Inc, Durham, USA). Parts were manufactured from DuraForm PA (polyamide), using selective laser sintering, an AM process technology. The process was similar to the one described elsewhere (Salles and Gyi, 2009). The 2 mm thick insole made them relatively rigid, providing heel and arch support, but not correction of lower limb abnormalities.

The control condition consisted of the same trainers, but fitted with a pair of insoles that were manufactured from the scans of the original trainer insoles, using the same material and thickness as the personalised condition. Thus, the control condition had identical shape as the trainers' original insole, but was manufactured using additive manufacturing to have same hardness and material as the personalised insole.

After parts were manufactured, a microporous polyurethane foam was used to cover both insoles. This was to provide some comfort to the individuals and to make sure the insoles would fit the inside of the trainers. Hence, the only difference between the two conditions was their geometry: one was generic (control) and the other was personalised (personalised).

LABORATORY SESSIONS

The laboratory sessions are detailed in Table 1. Sessions 3 and 4 took place approximately 6 weeks (1.5 months) and 12 weeks (3 months) after session 2 respectively. The shoes were given to participants in session 2. Individuals were asked to only wear the pair of footwear trainers for jogging/running and were encouraged to contact the investigator if they had any concerns.

Table 1. Laboratory session schedule for the participants.

Week – Lab session	Data collected
Week 1 – Lab session 1	Anthropometric measurements of the foot; foot scans
Week 4 – Lab session 2	Discomfort and injury risk
Week 10 – Lab session 3	Diary, discomfort and injury risk
Week 16 – Lab session 4	Diary, discomfort and injury risk

Sessions 2, 3 and 4 followed the same protocols for the collection of discomfort and injury risk data. Discomfort was evaluated using a 150 mm visual analogue scale. Injury risk was evaluated measuring plantar pressure distribution, rearfoot eversion, tibial internal rotation and peak vertical impact force of the right foot. At the end of session 4, the pair of trainers (with the original insole) used in the experiment were given to the participants.

DISCOMFORT ASSESSMENT

At the end of the laboratory sessions, participants were given a 150 mm Visual Analog Scale (VAS) to measure self-perceived discomfort. The VAS was similar to one used by Mundermann *et al.* (2002), with the left of the scale indicating 'the most comfortable condition imaginable' and the right 'not comfortable at all'. Six aspects of the foot were covered: heel, midfoot, forefoot, fit, arch and overall (whole foot).

INJURY RISK ASSESSMENT

Injury risk was assessed from the biomechanical data collected: plantar pressure distribution, rearfoot eversion, tibial internal rotation and peak vertical impact force. The literature suggests that high values are positively related to increased injuries in runners (Nigg et al., 1998; Hreljac et al., 2000; Mundermann et al., 2004; Yung-Hui and Wei-Hsien, 2005).

To ensure that individuals ran at a speed of 2.78 m/s ($\pm$ 5%), electronic timing gates (model: SmartSpeed; Fusion Sport, Brisbane, Australia) were positioned in the middle of a 10-meter runway. Therefore, before starting the data collection, participants had 5 practice trials to run for 10 meters in order to familiarize themselves with the required speed. After that, an F-Scan Mobile (Tekscan Inc, South Boston, MA, USA) in-shoe plantar pressure distribution sensor (N/cm^2) was placed inside the shoe and recorded at 250Hz. Participants then ran 5 times under the same experimental condition for 10 meters whilst plantar pressure distribution was recorded. For the purpose of data analysis, the foot was divided in three regions: heel, midfoot and forefoot. Plantar pressure was captured for each region using a F-Scan Mobile Research software (version, 5.72; Tekscan Inc, USA). The mean of the peak pressure values were taken for each region of the foot during ground contact.

After 5 valid trials (i.e. speed was within the range accepted), the plantar sensor was removed from the shoe and 16 reflective markers (14 mm diameter) for tracking 3D movement were placed according to the Plug-In-Gait standard lower body modeling. Participants were then asked once again to run 5 times at the same speed range (2.78 m/s $\pm$ 5%) while kinematic data were collected with a 12 camera Vicon MX system (400Hz; Oxford Metrics, Oxford, UK). Ground reaction force was recorded at 800 Hz and the force plate (type: 9281; Kistler Instrumente AG, Winterhur, Switzerland) was synchronized with the kinematic data. For data analysis, tibial internal rotation, rearfoot eversion and vertical impact peak of the ground reaction force were captured for each participant under each experimental condition. Since the rearfoot was considered fixed on the ground for the majority of the stance, tibial internal rotation was defined as rearfoot adduction/abduction (i.e. transverse plane motion of the ankle joint), whilst rearfoot eversion was defined as the frontal plane motion of the ankle joint. Impact peak was defined as the first peak in the vertical ground reaction force data. The forces were normalized as times body weight (bw) and the ankle joint angles were normalized in relation to the data taken with the individuals wearing the trainers fitted with its original insole. Ground reaction force values were also used to determine the moments of heel strike and toe-off in stance phase.

ACTIVITY DIARY

At the end of laboratory sessions 2 and 3, participants were provided with a pedometer (model: NL-800; New-Lifestyles Inc, Lee's Summit, USA) and an activity diary. They were instructed to wear the trainers and the pedometer every time they went jogging/running for a 3-month period as well as complete the diary after each training session. The activity diary captured information such as: how long the running shoes were worn, pedometer reading of steps taken, any discomfort felt and any additional comments. Discomfort was once again measured using a 150 mm VAS. The diary was returned to the researcher in laboratory sessions 3 and 4.

DATA ANALYSIS

The paired samples Student's t-test was used to detect significant differences between the two experimental groups for the variables (discomfort ratings, plantar pressure distribution, rearfoot eversion, tibial internal rotation and vertical impact peak) in months 0, 1.5 and 3. The level of significance was accepted as $\alpha \leq 0.05$. Statistical Package for the Social Sciences (SPSS) software for Windows (Release 15.0, SPSS©, Inc., 2006) was used for all analyses.

The process of manipulating and manufacturing the insoles from the scans was evaluated according to the following: compatibility of the data taken from the foot scans (i.e. if the files worked in the Geomagic Studio 10 software), the software capability to manipulate the files, compatibility of the final data with the AM machines and durability of the material (DuraForm® PA).

RESULTS

The data collection started in June 2009. Thirty eight runners (19 pairings) have been recruited to take part in this research. To date, 4 pairings (8 participants) have completed the study. Six participants (2 from the control group and 4 from the personalised group) have discontinued the study. The paper will report on the data in relation to the 4 pairs of participants (Table 2).

362

Table 2. Descriptive (mean ± SD) statistics for the participants.

Group (n=8)	Age (yrs)	Height (cm)	Mass (kg)	Activity (km/wk)	Gender
Control (n=4)	39.25 ± 12.6	1.68 ± 0.07	61.5 ± 7.3	12.5 ± 8.7	2M and 2F
Personalised (n=4)	37.5 ± 11.7	1.66 ± 0.04	60.85 ± 3.95	11.25 ± 5.2	2M and 2F

The discomfort ratings taken in the laboratory sessions were generally low for both experimental conditions (Figure 2). The student's t-test indicated no significant differences between the two conditions throughout the 3-month period. However, participants reported less discomfort in the personalised condition, particularly in months 1.5 and 3.

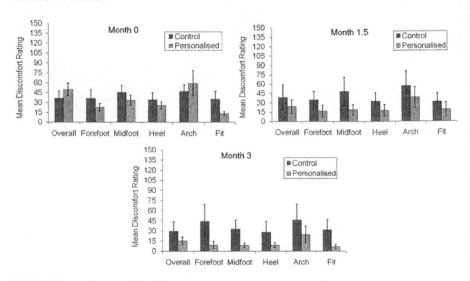

Figure 2. Mean discomfort ratings and standard error for the control and personalised conditions in months: 0, 1.5 and 3.

The mean values for rearfoot pronation and tibial internal rotation showed an increase in both conditions for the 3-month period, whereas vertical impact peak showed a reduction in the same period. On one hand, the pronation and tibial rotation values indicated an increase, while the vertical impact peak suggests a decrease in the risk of injury for the participants in the two groups. However, no significant differences were found between the control and personalised conditions for these injury risk variables over the 3 month period.

In addition, no significant differences were found between the two conditions for mean peak plantar pressure distribution (Table 3).

Table 3. Comparisons of mean peak pressure (unit: KPa) between the two conditions in months 0, 1.5 and 3. Data are presented as mean (SE).

Region	Condition	Month 0	Month 1.5	Month 3
Heel	Control	209.3 ± 69.2	219 ± 50.4	224.3 ± 71.4
	Personalised	211.6 ± 36.8	247.3 ± 12.8	208.4 ± 19.5
Midfoot	Control	99.9 ± 16.9	93 ± 14.7	103.3 ± 25.7
	Personalised	107.3 ± 20.6	103.9 ± 11.4	79.8 ± 6.4
Forefoot	Control	208.3 ± 52.5	190.7 ± 34.6	236.5 ± 67.7
	Personalised	245.5 ± 16.9	222.6 ± 17	167.9 ± 18.2

The activity diary supported the discomfort ratings taken during the laboratory sessions. The participants were instructed to report any discomfort (in any region of the body), during their running sessions. During the first 1.5 months of usage of the shoes, the participants in the control and personalised conditions experienced some discomfort in, on average, 63% (ranging from 40-75%) and 51% (30-85%) of their training sessions respectively. However, the analysis of the ratings indicated that this discomfort had a low mean rating. In the last 1.5 months of the study, on average, the runners in the control condition reported some discomfort in 38% (0-58%) of their training sessions, whereas the participants in the personalised group in 20% (0-80%). Both groups attributed their discomfort, in the majority of cases, to the insoles. Also, the arch region was the most cited area of discomfort by the two groups.

DISCUSSION

This study is currently being conducted and completion is expected in summer 2010. The data presented of four pairs of participants must therefore be interpreted carefully and in this context.

These preliminary results did not show any significant differences between the two experimental conditions for discomfort, but a trend can be noted. After a 1.5 month period, the personalised insoles had lower discomfort ratings for all the regions of the foot assessed. It is likely that the individuals need a period of accommodation with any footwear. The height and stiffness of the arch support, was found to be too intrusive in the beginning, but after 1.5 months they became used to it and their perception changed. The data from the activity diary supports this, especially with regards to the arch ratings. On the other hand, Mundermann et al. (2003) reports that custom made insoles are significantly more comfortable than a control flat insert for a period of 3 weeks only.

Yung-Hui and Wei-Hsien (2005) indicated that custom fabricated insoles can reduce plantar pressure, attenuate the impact force and are more comfortable than a shoe without such insoles. However, the injury risks variables did not show any significant differences or any patterns between the two conditions over the three month period. This is likely to be attributed to the small sample size.

The anthropometric measurements and scan data were used in the design of the personalised insoles. Two foot length measurements taken from the most posterior projecting point on the heel to the 1st and 5th metatarsal phalangeal joints indicated the length of the insoles. The navicular height was used to determine their height. As the discomfort ratings were low and the data were compatible with the software and hardware utilized, the process of the design and manufacture of the insoles was successful. The data taken from the scanner were compatible with Geomagic Studio. This software provided the appropriate tools to reduce the noise, delete unwanted data, fix the jagged edges on the boundary, smooth and thicken the parts in 2 mm. The final data file proved compatible with the AM machines. The material (polyamide) showed very good durability throughout the study. No signs of breaking were noted in the qualitative inspections during the laboratory sessions.

However, further analysis need to be carried out regarding possible correlations between the anthropometric data and the discomfort and injury risk variables. These will help with the identification of foot shapes that are more likely to develop discomfort and injuries.

In summary, the preliminary data set of four pairs showed no significant differences between the two conditions for the discomfort and injury risk variables for the 3 month period. The design and manufacture of the insoles were successful. Anthropometric measurements helped with the design, delimitating its length and height. The scan data were compatible with the software and AM machines and the material (polyamide) showed good durability.

ACKNOWLEDGEMENTS

We would like to acknowledge the IMCRC at Loughborough University who provided funding for this research.

REFERENCES

Cavanagh, P.R. (1980), *The Running Shoe Book*. Mountain View, Anderson World, California.

Cavanagh, P.R., and Rodgers, M.M. (1987), "The arch index: a useful measure from footprints." *Journal of Biomechanics*, 20(5), 547-551.

Chantelau, E., and Gede, A. (2002), "Foot dimensions of elderly people with and without diabetes mellitus – a data basis for shoe design." *Gerontology*, 48, 241-244.

Cheng, F.T., and Perng, D.B. (1999), "A systematic approach for developing a foot size information system for shoe last design." *International Journal of Industrial Ergonomics*, 25, 171-185.

Cheskin, M.P., Sherkin, K.L., and Bates, B.T. (1987), *The complete handbook of athletic footwear*. Fairchild Publications, New York.

Cowan, D.N., Jones, B.H., and Robinson, J.R. (1993), "Foot morphologic characteristics and risk of exercise-related injury." *Archives of Family Medicine*, 2, 773-777.

Goonetilleke, R.S., Ho, E.C.F., and So, R.H.Y. (1997), "Foot anthropometry in Hong Kong." *Proceedings of the ASEAN 97 Conference*, Kuala Lumpur, Malaysia, pp. 81-88.

Goske, S., Erdemir, A., Petre, M., Budhabhatti, S., and Cavanagh, P.R. (2006), "Reduction of plantar heel pressures: insole design using finite element analysis." *Journal of Biomechanics*, 39, 2363-2370.

Hawes, M.R., and Sovak, D. (1994), "Quantitative morphology of the human foot in a North American population." *Ergonomics*, 37(7), 1213-1226.

Hopkinson, N., and Dickens, P. (2001), "Rapid prototyping for direct manufacture." *Rapid Prototyping Journal*, 7(4), 197-202.

Hreljac, A., Marshall, R.N., and Hume, P.A. (2000), "Evaluation of lower extremity overuse injury potential in runners." *Medicine and Science in Sports and Exercise*, 32(9), 1635-1641.

James, S.L., Bates, B.T., and Osternig, L.R. (1978), "Injuries to runners." *The American Journal of Sports Medicine*, 6(2), 40-50.

Kouchi, M. (1995), "Analysis of foot shape variation based on the medial axis of foot outline." *Ergonomics*, 38(9), 1911-1920.

Kusumoto, A., Suzuki, T., Kumakura, C., and Ashizawa, K. (1996), "A comparative study of foot morphology between Filipino and Japanese women, with reference to the significance of a deformity like hallux valgus as normal variation." *Annals of Human Biology*, 23(5), 373.385.

Luximon, A., Goonetilleke, R.S., and Tsui, K.L. (2003), "Footwear fit categorization", in: *The Customer Centric Enterprise: Advances in Mass Customization and Personalization*, Tseng, M.M., and Piller, F.T. (Eds). Springer, New York, pp. 491-499.

McKenzie, D.C., Clement, D.B., and Tauton, J.E. (1985), "Running shoes, orthotics and injuries." *Sports Medicine*, 2, 334-347.

Menz, H.B., and Morris, M.E. (2005), "Footwear characteristics and foot problems in older people." *Gerontology*, 51, 346-351.

Miller, J.E., Nigg, B.M., Liu, W., Stefanyshyn, D.J., and Nurse, M.A. (2000), "Influence of foot, leg and shoe characteristics on subjective comfort." *Foot and Ankle International*, 21(9), 759-767.

Morag, E., and Cavanagh, P.R. (1999), "Structural and functional predictors of regional peak pressures under the foot during walking." *Journal of Biomechanics*, 32, 359-370.

Mundermann, A., Nigg, B.M., Stefanyshyn, D.J., and Humble, R.N. (2002), "Development of a reliable method to assess footwear comfort during running." *Gait & Posture*, 16, 38-45.

Mundermann, A., Nigg, B.M, Humble, R.N., and Stefanyshyn, D.J. (2003), "Orthotic comfort is related to kinematics, kinetics and EMG in recreational runners." *Medicine and Science in Sports and Exercise*, 35(10), 1710-1719.

Mundermann, A., Nigg, B.M., Humble, R.N., and Stefanyshyn, D.J. (2004), "Consistent immediate effects of foot orthoses on comfort and lower extremity kinematics, kinetics, and muscle activity." *Journal of Applied Biomechanics*, 20(1), 71-84.

Nigg, B.M., Khan, A., Fisher, V., and Stefanyshyn, D. (1998), "Effect of shoe insert construction on foot and leg movement." *Medicine and Science in Sports and Exercise,* 30(4), 550-555.

Salles, A.S., and Gyi, D.E. (2009), "The specification of personalised footwear for rapid manufacturing: a pilot study." *Proceedings of the fifth World Conference on Mass Customization & Personalization 2009*, Helsinki, Finland.

Williams, D.S., and McClay, I.S. (2000), "Measurements used to characterize the foot and the medial longitudinal arch: reliability and validity." *Physical Therapy,* 80(9), 864-871.

Williams III, D.S., McClay, I.S., Hamil, J., and Buchanan, T.S. (2001a), "Lower extremity kinematic and kinetic differences in runners with high and low arches." *Journal of Applied Biomechanics*, 17, 153-163.

Williams III, D.S., McClay, I.S., and Hamill, J. (2001b), "Arch structure and injury patterns in runners." *Clinical Biomechanics*, 16, 341-347.

Witana, C.P., Feng, J., and Goonetilleke, R.S. (2004), "Dimensional differences for evaluating the quality of footwear fit." *Ergonomics*, 47(12), 1301-1317.

Wunderlich, R.E., and Cavanagh, P.R. (2001), "Gender differences in adult foot shape: implications for shoe design." *Medicine and Science in Sports and Exercise*, 33(4), 605-611.

Yung-Hui, L., and Wei-Hsien, H. (2005), "Effects of shoe inserts and heel height on foot pressure, impact force and perceived comfort during walking." *Applied Ergonomics,* 36, 355-362.

Chapter 37

Tools of Occupational Biomechanics in Application to Reduction of MSDs

Danuta Roman-Liu

Central Institute for Labour Protection-National Research Institute
Czerniakowska 16, 00-701
Warsaw, Poland

ABSTRACT

Musculoskeletal disorders (MSDs) are widespread and occur in all kind of jobs. The most strenuous tasks, which are the cause of MSDs are those characterized by static load of muscles and those which involve repetitive tasks performance.

Data related to musculoskeletal load in work conditions as well as tools aimed at musculoskeletal load assessment provides occupational biomechanics. Methods, which serve to musculoskeletal load assessment can refer to external or internal load of musculoskeletal system. For internal load assessment well established and widespread method is electromyography. Fatigue assessment at low level force meets difficulties. An algorithm based on approximation function, which can be applied in fatigue analysis of EMG signal registered at low level of muscle contraction. Since there is a strong link between MSDs rate and occupational involvement in repetitive tasks of upper limbs, method dedicated especially to repetitive tasks, has been developed on the basis of seven degrees of freedom model of upper limb model and parameters, which characterize cycle load. In conclusion it can be stated that complex assessment of risk of MSDs development needs application of both internal and external load assessment methods.

Keywords: occupational biomechanics, electromyography, repetitive tasks, muscle forces

INTRODUCTION

Epidemiological studies seek to identify association between the development of MSDs and musculoskeletal load during performing work tasks. Numerous studies are conducted in Europe and worldwide to investigate potential interactions between physical and psychosocial risk factors in the workplace with symptoms of musculoskeletal disorder of the back, neck, lower and upper limbs. The main role in providing data related to musculoskeletal load in work conditions as well as tools aimed at musculoskeletal load assessment plays occupational biomechanics.

One of the most strenuous tasks, which are the cause of MSDs are those characterized by static load of muscles and those which involve repetitive tasks performance. Many studies proved that performing such tasks at even very low levels of load can be crucial factor in MSDs development. Solutions regarding optimum load of the human musculoskeletal system in occupational conditions are obtained as a result of research with application of various methods.

Musculoskeletal load can be referred to as external or internal load. External load is assessed as dependant on parameters related to body posture, exerted forces and time sequences. Internal load measures reflect reaction of human body on the exerted external load. It means that the later expresses both external load related to work place and individual characteristic of the worker including age, gender or health. External load is assessed using various methods which assess load based on the parameters describing positions of separate body elements, force exerted by a worker and time sequences of load. Internal load and worker's fatigue may be assessed using such methods as the analysis of blood pressure, energy expenditure or the analysis of an electric signal characterising muscle contraction, namely surface electromyography.

MSD

Musculoskeletal disorders (MSDs) is one of the expression used for pain in different body parts or diagnosed diseases of musculoskeletal system. It denotes health problems of the locomotors apparatus i.e. muscles, tendons, the skeleton, cartilage, the vascular system, ligaments and nerves. The symptoms of MSDs may vary from discomfort and pain to decreased body function and invalidity. The other terms which represent MSDs are cumulative trauma disorders (CTDs), repetitive strain injuries (RSI) or work-related upper limb disorders (WRULDs).

Work-related musculoskeletal disorders (WMSDs) can be defined as impairments of the body structures, which are caused or aggravated primarily by the performance of work and the effects of the immediate environment in which work is carried out. Most of work-related musculoskeletal disorders are cumulative disorders, resulting from repeated exposure to high or low intensity loads over a long period of time.

Cumulative WMSDs can stem from a wide range of factors, that together create an inadequate margin between people's work demands and the coping resources available to them. As risk factors in WMSDs development are considered physical and psychosocial factors present in a work environment. It is assumed that in work environment such physical factors as posture and movements, exertion of forces and time sequences of work tasks could cause damage to muscles and tendon tissues. It should be noticed, however, that not only strong muscle contraction could cause damage and pain, it can be also to low intensity of muscle contraction. There are also several personal factors that could have indirect effect on the occurrence of MSDs. Important role play individual characteristics such as gender, age, physical capacity, lifestyle, parenthood and household activities, inclination to musculoskeletal diseases, life style and personality traits. However, there are still difficulties in precisely conceptualizing and measuring mechanisms by which these factors influence work and health outcomes.

Musculoskeletal disorders (MSDs) are widespread and occur in all kinds of jobs. According to a European survey up to 25% of workers reported back pain and 23% muscular pain. Certain studies estimate the cost of work-related musculoskeletal disorders at between 0,5% and 2% of the Gross National Product.

National and European Union surveys identify musculoskeletal disorders (MSDs) as being responsible for a large proportion of the working days lost due to illness (Forth European Working Conditions Survey; TCRO; EUROSTAT).

About 0.8% of the European workers suffer from working conditions which cause 14 days or more of absence from work due to MSDs during the period of a year. Higher absenteeism occurs in women population than among men workers. The highest number of days out of work is among population between 40 to 54 years old (Figure 1).

However, WMSDs are not only health problems, they also are a financial burden to society. The costs are related to medical costs, decreased productivity, sick leave, and chronic disability.

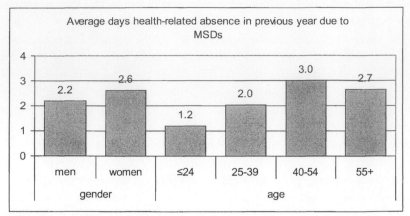

FIGURE 1. Number of days health related absence due to MSDs problems in groups of workers obtained from EU27 countries (own resources based on data from Forth European Working Conditions Survey; TCRO; EUROSTAT).

ASSESSMENT OF MSD RISK IN UPPER LIMBS

REPETITIVE TASK INDICATOR

Strong link between MSDs rate and occupational involvement in repetitive tasks performance indicates that assessment of musculoskeletal load during performing repetitive tasks plays significant role. There has been developed methods aimed at musculoskeletal load or risk of MSDs development assessment (Occhipinti, 1998; Moore and Garg, 1995).

Musculoskeletal load of upper limbs can, for example, be assessed on the basis of Repetitive Task Indicator (RTI), where upper limb posture is described by values of seven angles. RTI characterizes the upper limb load which results from performing work described by the means of parameters of repetitive task. Those parameters refer to duration of particular sequences of the work cycle and to its force, taking into account the posture of the upper limb. Time characteristic is described by the cycle duration (CT), number of phases in the cyckle (k) and duration of given cycle phase (DPi where i – number of work cycle periods, $1 \leq i \leq k$).

Force connected with performing a work task is expressed as a percentage of maximal force of a given type of force activity exerted in a given posture of the upper limb. In order to specify the maximal force, a predictive equation may be applied which will make it possible to assign maximum value of a given type as the function of the upper limb posture. Such predictive equations have been developed for the basic types of force activity such as handgrip, pinch grip, pushing, lifting, supination and pronation (Roman-Liu and Tokarski, 2005).

The repetitive task indicator (RTI) is a function of the load resulting from performing one work cycle of any duration (ICL), cycle duration (CT) and number of cycle phases (k) (Figure 2).

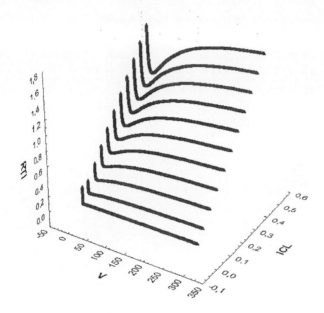

FIGURE 2. Repetitive Task Indicator as a function of ICL and V (CT/k) (own resources).

Value of the RTI can be related to one of three zones, which determine the risk as acceptable (low), conditionally acceptable (medium) or not acceptable (high). When RTI is below 0.5 the risk is acceptable. Value between 0.5 and 0.8 means that risk is conditionally acceptable. RTI value above 0.8 assess the risk as not acceptable.

RELATIONSHIP BETWEEN MUSCULOSKELETAL LOAD AND MSDS OCCURRENCE

The study on 15 types of work stands with application of RTI indicator has been conducted. RTI calculated for each work stand has been compared with occurrence of musculoskeletal disorders among workers working on the repetitive work stands of the same type. Signs of MSDs were detected on the basis of questionnaire. The questionnaire allowed for determination of indicator of musculoskeletal disorders in arm and forearm/hand region (DC). In the questionnaire there were also questions focused on carpal tuner syndrome diagnosis related to pain and numbness in hands. On the basis of those questions indicator of CTS has been assessed (CTS).

RTI for each work stand type was calculated on the basis of precise timing with specification of tasks, duration of cycle and determination of each cycle phases force. Left and right upper limb posture was determined. Figure 3 presents comparison of RTI with DC and CTS parameters for 15 analyzed types of work tasks. Generally for those types of work stands for which RTI was higher DC and CTS was higher too. Spearman correlation coefficent between RTI and indicator of arm and forearm/hand disorders equals 0,546 and correlation coeeficent between RTI and CTS indicator equals 0,806, which proved strong agreement between RTI value and MSDs occurrence.

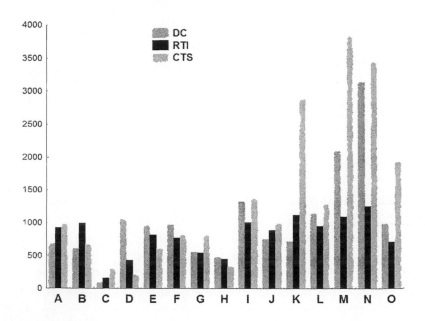

FIGURE 3. Indicators of musculoskeletal disorders (DC and CTS) and upper limb load for a specific type of work stand marked from A to O (for comparison reasons RTI values were multiplied by 40 and CTS by 100) (own resources).

SURFACE ELECTROMYOGRAPHY

Electromyography is based on registration of the electric activity from selected muscles involved in the performance of tasks. In research with electromyography application two types of electrodes can be used: needle electrodes and surface electrodes.

Indwelling electromyography is invasive methods and therefore is mostly applied in clinical research related to muscle diseases diagnosis.

Using surface electrodes makes it possible to register signals in a non-invasive and relatively easy way. Usually two active electrodes are used, however, lately also array electrodes (high density) are implemented (Figure 4).

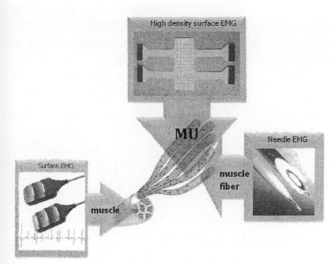

FIGURE 4. Scope of various EMG techniques (own resources).

Indwelling electromyography transfers data related to muscle fibers. EMG signal registered with application of surface EMG transfers information related to one muscle or even few muscles. Since array electrodes transfer information on a level of motor unit.

Analysis of specific attributes of EMG signal is one of the most often applied methods of assessing local muscular fatigue. Those data are obtained as a result of processing the EMG signal in time and frequency domains.

It has been shown that as a result of fatigue, EMG power spectrum tends to shift towards its low frequency. Physiologically, the frequency shift has been attributed to changes in conduction velocity, in intra-muscular pH and to modification in the recruitment and synchronization of the motor units and the fiber type. It is manifested, in changes in MF (Median Frequency) and MPF (Mean Power Frequency) parameter values. Muscle fatigue also leads to an increase in EMG signal amplitude. It is generally agreed that muscle fatigue is documented by the EMG signal only if there is both a decrease in a parameter analysed in the frequency domain and an increase in the parameter analyzed in the time domain (Merletti et al., 1997).

Muscle fatigue is not the exclusive reason for changes in the values of EMG parameters. Changes in the distribution of frequencies of the EMG power spectrum are also affected by changes in the force level, recruitment of new, not fatigued, motor units, proportion of muscle fibers, muscle type, muscle length (Lowery et al., 2002).

Global parameters of power spectrum such as mean and median frequency are well established measures of muscle fatigue. However, those parameters are sensitive not only to fatigue phenomena but also to variations in muscle force. This creates inaccuracy in measurements when muscle contraction is below 25% MVC (Maximum Voluntary Contraction). At low levels of muscle contraction muscle fibers can be switched off and exchanged by others, not fatigued and load can still be sustained. Such phenomena in parameters like mean and median frequency is demonstrated by decrease due to fatigue and then rapid increase in parameter value (Figure 4). Even if switching occurs to prevent exhaustion in motor units, it can indicate muscle fatigue and fatigue can be quantitively expressed with approximation of changes of EMG signal parameters by logarithmic curve and calculations of differences in the function values due to time of fatigue (Roman-Liu et al, 2004). Time of fatigue in case of unexpected increase is accepted up to the moment of that point.

On the basis of the equation of the regression function f(t) and ratio of fatigue time to time of load 'R' the fatigue index can be expressed as:

$$FI = \left| \frac{f(t_2) - f(t_1)}{f(t_1)} \right| R$$

Where FI is fatigue index; R- ratio of fatigue time to time of load; t_1 – initial time for a parameter analysed in frequency domain and final time for a parameter analysed in time domain; t_2 – final time for a parameter analysed in the frequency domain and initial time for a parameter analysed in time domain.

SUMMARY

Many studies proved that load sustained even at very low levels can be a factor in MSDs development. Despite the fact that there is widespread awareness of the problem and measures to limit development of MSDs are being undertaken, according to an European survey up to 25% of workers report back pain and 23% muscular pain. Musculoskeletal load can be referred to as external (dependant on parameters describing performed task related to body posture, exerted forces and time sequences) or internal load (expressed by reaction of musculoskeletal system to the external load). For the external load assessment Repetitive Task Indicator can be applied to assess upper limb load and risk of MSDs development. As internal load assessment method the most commonly EMG is applied. Difficulties in fatigue assessment on low levels of muscle contraction can be overcome by specific algorithms for analysis of global parameters. However, parameters sensitive to changes invoked by fatigue and not sensitive to change in muscle force would be more reliable measure of muscle fatigue than the global parameters of power spectrum. One of ways where solutions are being seek for is time-frequency analysis, the other is more profound analysis of power spectrum of EMG signal.

Since the internal load characterize both external load related to work place and individual characteristic of the worker including age, gender or health condition, it can be stated that complex assessment of risk of MSDs development needs application of both internal and external load assessment methods.

REFERENCES

Lowery, M., Nolan, P., and O'Malley, M. (2002), "Electromyogram median frequency, spectral compression and muscle fiber conduction velocity during sustained sub-maximal contraction of the brachioradialis muscle." *Journal of Electromyography and Kinesiology,* 12(2), 111-118.

Merletti, R., and Lo Conte, L.R. (1997), "Surface EMG signal processing during isometric contractions." *Journal of Electromyography and Kinesiology*, 7, 241-250.

Moore, J.S., and Garg, A. (1995), "The strain Index: A proposed method to analyze jobs for risk of distal upper extremity disorders." *American Industrial Hygiene Association Journal,* 56, 443-458.

Occhipinti, E. (1998), "OCRA: a concise index for the assessment of exposure to repetitive movements of the upper limbs." *Ergonomics,* 41 (9), 1290-1311.

Roman-Liu, D., Tokarski, T., and Wójcik, K. (2005), „Quantitative assessment of upper limb muscle fatigue depending on the conditions of repetitive task load." *The Journal of Electromyography and Kinesiology,* 14, 671-682.

Roman-Liu, D., and Tokarski, T. (2005), "Upper limb strength in relation to upper limb posture." *International Journal of Industrial Ergonomic,* 35, 19-31.

Roman-Liu, D. (2007), "Repetitive Task Factor as a tool for assmssent of upper limb musculoskeletal load induced by repetitive taks." *Ergonomics,* 50 (11), 1740-1760.

Forth European Working Conditions Survey. European Foundation for the Improvement of Living and Working Conditions. (www.eurofound.europa.eu)

Work-related musculoskeletal disorders – TCRO 2006. European Agency for Health and Safety at Work (Prevent, 2006)

EUROSTAT.http://epp.eurostat.cec.eu.int/portal/page?_pageid=1996,45323734&_
   dad=portal&_schema=PORTAL&screen=welcomeref&open=/popul/health/hs
   w/hsw_inj_pb/hsw_healthpb&language=de&product=EU_MAIN_TREE&roo
   t=EU_MAIN_TREE&scrollto=531

Chapter 38

Enhancement of MSD Problems in the Automobile Industry

Young-Guk Kwon[*], *Hyun-Joo Lee*[**], *Jung-Won Ryu*[*]

[*] Department of Safety Engineering,
College of Engineering
Seoul National University of Technology
Seoul, Korea, safeman@snut.ac.kr

[**] Department of Energy Safety Engineering,
The Graduate School of Energy & Environment
Seoul National University of Technology
Seoul, Korea, ilisoft@ilisoft.co.kr

ABSTRACT

Musculoskeletal disorder (MSD) is a health disorder caused by repetitive motion, inadequate working posture, excessive exertion of strength, body contact with a sharp surface, vibration, temperature, etc. MSD appears mainly in nerves, muscles and tissues around the neck, shoulders, waist, and upper and lower limbs. This study was conducted with an automobile part manufacturing company with around 800 workers.

In this study, we could save the loss of direct cost amounting to 140,800 US dollars (with 1,600 days as the minimum decrease in paid sick leave and 88 US dollars as basic direct labor cost per day) and indirect cost amounting to 4~18 times of the direct cost. Both the owner and the employees took interest in the prevention of musculoskeletal disorders continuously made a difference such that mutual trust was enhanced between the two parties, work environment was improved, job satisfaction was raised, and the incidence of musculoskeletal disorders and workers' stress decreased.

Keywords: MSD, job stress, workers, prevention cost effect, automobile part

INTRODUCTION

"Musculoskeletal burdened works" are works under Paragraph 5 of Subsection 1 of Section 24 of the Industrial Safety and Health Act and those specified by the minister of labor according to the load, speed and intensity of work and the structure of workplace. Work-related musculoskeletal disorders are health problems caused by repetitive motions, inadequate working posture, excessive use of strength, body contact with a sharp edge, vibration, temperature, etc. MSD appears in nerves, muscles and tissues around the neck, shoulders, waist, and upper and lower limbs.

Risk factors investigation and musculoskeletal disorders prevention management program required by law are for protecting workers' basic right of health. Industrial safety and health activities should be settled not as welfares but as a prerequisite of corporate activities and as companies' responsibility and culture. For these purposes, they are required a quantified research and an evaluation.

This study analyzed changes in the number of workers with musculoskeletal disorders and in lost work days for the last five years through investigating risk factors, medical management, and work environment improvement at automobile part manufacturing companies where many musculoskeletal burdened works exist.

METHODS

Subject and outline

The subject of this study was an automobile part manufacturing company with around 800 workers. In order to investigate risk factors, the labor and management formed a joint task force team. They conducted a complete survey of all executives and employees including contingent workers every three years. In addition, occasional surveys were conducted every year on changed work environment and processes where persons with medical problems take place. An early detection and early treatment system was established through ergonomic improvement activity once a month, and medical management once or twice a month by medical specialists (rehabilitation Medicine and oriental medicine) for musculoskeletal disorders.

PROCEDURE

Made a list of works

We selected a skilful worker at each department and appointed him as a person in charge of musculoskeletal disorders for the department, and educated the person two hours on the basics of musculoskeletal disorders prevention and management and how to make a list of works, and then the person in charge at each department analyzed and recorded all works in the department. Then, the health manager collected the lists, and if a list was inadequate, the manager visited the department and implemented the list.

Made a list of burdened works

The musculoskeletal TFT and specialists compared the list of works with 11 types of burdened works notified by the ministry of labor, selected burdened works from the list. Even though works that do not belong to the 11 types of burdened works, those them are expected to be a physical burden on the workers, we made precise evaluation to determine whether to include them as burdened works.

Surveyed symptoms

We distributed questionnaires to workers, ask them to fill the musculoskeletal disorder symptom survey sheet suggested by the Korean Occupational Safety and Health Agency, and collected the questionnaires. The health manager investigated symptoms, and those who need special medical attention and those with medical problems were examined first by a doctor with the rehabilitation medicine and sent immediately to cooperative hospitals and clinics for treatment and precise examination such as EMG and X-ray.

Investigation of the situation and condition at the workplace

For the investigation of the situation and condition at the workplace, the members of the musculoskeletal TFT and external specialists visited the workplace in person and interviewed the workers with regard to the selected burdened works and, at the same time, took photographs and made precise analysis.

Work analysis and evaluation

Work analysis and evaluation were performed in order to find problems in the selected burdened works through consultation with the labor and management and to make a decision on the priority of improvement. As we judged that problems in work times and working posture were serious, we developed an evaluation tool, which was software integrating RULA, REBA, QEC, etc. for assessing ergonomic risk and musculoskeletal disorders diagnosis as in Figure 1. Using this tool, we evaluated burdened works, identified high-risk works, and made and applied improvement plans. When the improvement plans were made, a decision tree was extracted using an expert system and priority was given to works that took less cost and time in improvement.

380

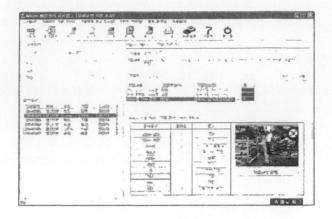

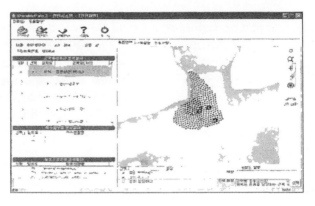

Figure 1: Ergonomic risk evaluation and musculoskeletal disorder diagnosis
system

RESULTS

Investigation of MSD risk factors by work and work environment improvement

Based on the results of evaluation, we applied many ergonomic improvement activities to the workplace including changing the way of moving Avante 1 automobile assemblies, solving musculoskeletal problems in the AV pad line control room, changing the method of loading anti pad products, improving the sequential process of click module, preventing difference in New Verna module MDPS and improving workability, improving the process of Santa Fe 1 assembly sub-process, installing fatigue prevention mats in Avante C/P, preventing musculoskeletal burdens in New Verna M/D, replacing the fabric of the Starex vacuum molding machine, and replacing New Avante M/D tools.

Medical management and the introduction of customized stretching

We examined those with musculoskeletal disorders identified in the symptom survey, and took measures as follows.

1) We managed those who complained with symptoms and those who were under long-term treatment at health and medical facilities inside the company.

2) The health manager informed the results of examination within 14 days to the workers.

3) The doctor with the rehabilitation medicine conducted examination twice semiannually, and diagnosed those with musculoskeletal problems, those who needed special medical attention, and those who wanted to be examined. For workers whose symptom was likely to develop into a disease, the doctor examined and took a proper medical measure, and explained the results to the workers. In addition, Oriental medicine doctors performed examination once a month in order to prevent development into a chronic disease, and activated the employees' health promotion activities. Table 1 shows cases of improvement through a medical management.

In the first complete survey in 2004, the numbers of treatment cases were increased due to the active early detection system but the total numbers of days lost in industrial accidents were decreased. Currently, the periodical medical management system has been settled, which includes 4 examinations by rehabilitation medicine doctors, 12 examinations by oriental medicine, and treatments by kinesitherapists each year.

4) We introduced stretching customized to each process and individual so that the workers can do exercise and manage their body continuously. In order to prevent musculoskeletal disorders, we provided information such as posters, videos, and pop-ups so that the workers do stretching before starting their work and in the middle of the afternoon. As the company recognized the effect, it opened a fitness center.

Changes resulting from musculoskeletal disorder prevention and management

Table 1 and Figure 2 show changes in lost work days with a result from the improvement in work environment, medical management, and the introduction of stretching.

Table 1: Work loss Unit: 1 US dollars, cases

	2004	2005	2006	2007	2008
Number of accidents	57	53	68	53	33
Number of musculoskeletal disorder patients	26	13	13	13	16
Work loss (number of sick leaves by industrial accidents)	2,039 (47)	1,629 (38)	1,070 (33)	621	439
Average days of sick leave per case	43.38	42.87	32.42	35.73	22.2

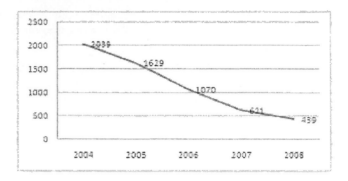

Figure 2: Lost work days by year

After the first investigation of risk factors in 2004, work environment has been improved steadily for five years, and through medical examination and stretching customized to individual workers' working pattern, the number of musculoskeletal disorder patients decreased by 40% and labor loss decreased by around 78% in 2008. The reason that the number of accidents increased was probably because workers' perception of musculoskeletal disorders was enhanced and this increased the number of workers subject to musculoskeletal disorders management and the number of patients counted as industrial accidents. On the contrary, lost work days caused by serious chronic diseases or intractable diseases decreased due to efficient stretching, consulting, examination, treatment, etc.

CONCLUSION

According to the results of this study with a workplace, musculoskeletal disorders can be minimized by prevention and management. Benefits from the prevention and management of musculoskeletal disorders show improvement of work environment, the relation between the labor and management, and productivity, and decrease in lost work days. From a long-term viewpoint, it can reduce financial losses and create the image of safe workplace.

In order to prevent and manage musculoskeletal disorders for a workplace, the labor and management formed a joint TFT, and promoted risk factor investigation and improvement through cooperation among the departments. When efforts were made steadily for five years to prevent and manage musculoskeletal disorders, they were decreased work loss considerably and prevented a financial loss of 140,800 US dollars (1,600 days of sick leaves × 88 US dollars of daily basic direct labor cost) and the loss of overhead cost 4~18 times larger than the reduction of financial loss. Most important results are that as the employer's will and cooperation between the labor and management work together, mutual trust is built up between the two parties and, at the same time, work environment is improved, job satisfaction is enhanced, and musculoskeletal disorders and workers' stress are reduced.

This study was conducted at an automobile part manufacturing factory, a representative workplace of musculoskeletal burdened works, so these results have a limitation with a study in a non-typical workplaces and service businesses.

REFERENCES

[1] Article 1 and 142 of Part 9 of Regulations on Industrial Health Standards

[2] Information Center of Government Legislation Agency, Article 24 of Industrial Safety and Health Act

[3] Korean Occupational Safety and Health Agency, Statistics of Industrial Accidents, 2000~2008

[4] Jeong Byeong-yong, Oh Sun-yeong, "Investigation of Risk Factors and Ergonomic Improvement in the Shipbuilding Industry," Ergonomics Society of Korea, Vol.24, No.1, pp27-35, 2005

[5] Mun Jae-dong, "Acknowledgement of and Compensation for Musculoskeletal Disorders as Occupational Diseases," 2005 Seminar on Musculoskeletal Disorders Prevention Technologies held by Korea Occupational Safety and Health Agency, 1st Theme, 2005

[6] Kim Yoo-chang, Lee Gwan-seok, Jang Seong-rok, Choi Eun-jin, "Economic Analysis of Musculoskeletal Disorders in Korea"

CHAPTER 39

The Investigation of Low Back Problems among Home Delivery Drivers

Chih-Long Lin[1], Yu-Ting Hung[2], Mao-Jiun Wang[2], Jiann-Perng Chen[3]

[1]Department of Crafts & Design,
National Taiwan University of Arts, Taipei, Taiwan (R.O.C.)

[2]Department of Industrial Engineering and Engineering Management,
National Tsing Hua University, Hsinchu, Taiwan (R.O.C.)

[3]Department of Rehabilitation Medicine,
Mackay Memorial Hospital, Hsinchu, Taiwan (R.O.C.)

ABSTRACT

Low back pain (LBP) and spine disorder are common problems for occupational drivers. For home delivery drivers, the Manual Materials Handling (MMH) task is the main risk factor that causes low back problems. The purpose of this study was to investigate the correlations between low back problems, and physical, biomechanical as well as psychosocial factors for the home delivery drivers. A total of 32 male home delivery drivers participated in the study. Two questionnaires including subjective musculoskeletal discomfort rating and job stress questionnaire were used. Besides, a rehabilitation doctor diagnosed the subjects' spine disorder (L5/S1 spine disc space narrowing degree) based on lumbar x-ray results. Furthermore, the biomechanical loadings of spine of twenty drivers while performing the MMH task were analyzed. The study results show that the prevalence of L5/S1 spine disc disorder among the home delivery drivers is 25%.

Over one third of the subjects experienced back/hip (56%), neck (47%), and shoulder (38%) discomfort. The scores of job stress scale for spine disc space narrowing group are greater than normal spine group, but the differences are not significant. The L5/S1 compression forces for normal spine group are significantly greater than spine disc space narrowing group (p<0.05). Moreover, the multivariable regression models were constructed to predict spine disorder from subjective, biomechanical and psychosocial factors. The percent of push goods with hands, the score of poor interpersonal relationship and job seniority factors are all significant in the regression models. The prediction models explained at least 75% of the variations.

Keywords: Home Delivery Driver, Low Back Pain, Spine Disorder, MMH, Psychosocial Factors.

INTRODUCTION

Working as an occupational driver is associated with a high prevalence of low back problems. Thirty years ago, Kelsey (1975) found that trunk drivers had higher prevalence of spine disorder than those of non-trunk drivers. In recent years, Raanaas and Anderson (2008) found that the prevalence of low back pain (LBP) in taxi drivers is almost 60%. Besides, over 80% of occupational trunk drivers reported LBP during the previous 12 months and there were 60% drivers still suffered LBP during the previous week (Robb and Mansfield, 2007).

The poor seated posture, whole body mechanical vibration and manual materials handling (MMH) performing are now widely recognized as causing low back problem amongst occupational drivers (Massaccesi et al., 2003; Mubarak et al., 2005; Okunribido et al., 2006). It is worth to note that the contribution of these factors was different for different occupational drivers. Okunribido et al. (2007) used validated questionnaire and observation method to investigate the prevalence and causing of LBP for city bus drivers. The results showed that bus drivers spent about 62% of the daily work time on driving with the torso straight and performed light MMH task. Thus, exposure to whole body vibration would be the most important factor of causing LBP for city bus drivers. Different from bus drivers who spent most time on driving, manual handling of loads is main task for delivery drivers. Okunribido et al. (2006) reported that the delivery drivers spent less than 30% of the workday on actually driving and no longer than 15 minutes at a time for continuous driving. However, the drivers spent about 38% of the workday performing MMH activities and the authors pointed out that high frequency of lifting and lifting immediately after driving are the potential risk factors for back injury.

For delivery drivers, various studies have found positive associations between van/truck driving and low-back pain (Robb and Mansfield, 2007; Okunribido et al., 2008). However, the researches have been focused on long-haul drivers and have often been investigated only in terms of subjective musculoskeletal discomfort ratings and whole-body vibration measures. Job-related psychosocial factors have been receiving increasing attention as potential significant etiological factors of musculoskeletal disorders (Hsu and Wang, 2003). The present study applied mixed methodologies, i.e., spine disorder diagnosis, subjective self-reported measures including musculoskeletal discomfort and psychosocial factors as well as biomechanical loading analysis for MMH task, to investigate the role of MMH and psychosocial factor for causing low back problems among short-haul home delivery drivers.

METHODS

SUBJECTS

A total of 32 male home delivery drivers completed the questionnaire and spine disorder diagnosis procedure. The average age was 38.0 (± 8.8) years, average body height was 1.71 (± 6.0) m and average body mass was 71.0 (± 12.6) kg. The mean job tenure was 101.2 (± 88.7) months. Furthermore, 20 out of 32 drivers were randomly selected and their biomechanical spine loadings were analyzed while performing MMH tasks.

SPINE DISORDER DIAGNOSIS

A rehabilitation doctor diagnosed the subjects' spine disorder based on lumbar x-ray results. The L5/S1 spine disc space narrowing degree were classified into 0, 1, 2, 3 levels: level 0 is normal, level 1 is slight disc space narrowing, level 2 is moderate disc space narrowing and level 3 is serious disc space narrowing.

SUBJECTIVE RATING MEASUREMENT

Two questionnaires, including subjective musculoskeletal discomfort rating (SMD) and Job Stress Questionnaire (JSQ) were used in the study. The SMD was modified from the Nordic Musculoskeletal Questionnaire (Kuorinka et al. 1987). The severity and frequency of discomfort were rated for 8 body areas (neck, upper back, low back/ hip, shoulder, elbow, wrist/hand, thigh/knee, and shank/ankle/foot). The JSQ was referred to the study of Hsu and Wang (2003) including 25-item job stress scale and 12-item job satisfaction scale.

BIOMECHANICAL LOADING ANALYSIS

First, the driver's task was classified into six categories, including lifting with two hands, lifting with one hand, manual cart using, paper work, walking, and pushing goods with hand. Further, the biomechanical loading analysis was focused on lifting with two hands and lifting with one hand tasks. The method of biomechanical loading analysis was based on the study of Neumann et al. (2001). The biomechanical analysis battery included digital video analysis, detailed biomechanical modeling, and a posture and load sampling technique. Twenty drivers were observed and videotaped twice when they loaded and unloaded heavy goods between platform and trailer in the morning and in the evening. For each driver, his trunk posture (9 categories), horizontal hand position (close, medium or far), hand force amplitude (8 categories), and hands used type (one hand or two hands) were recorded using a total of 432 categorical scales. Observations were made every ten seconds until the driver finished whole loaded or unloaded goods task. For each observation in the work sampling matrix (trunk posture, hand distance, load amplitude, hands used), a biomechanical analysis was performed by using a 3D Static Strength Prediction Program (3D SSPP) software.

STATISTICAL ANALYSIS

All variables were initially examined with an independent t-test for significant differences between spine disc degeneration group and normal spine group. Scores for the job stress scale and job satisfaction scale were analyzed by using principle component analysis with varimax rotation to assure the same dimensionality. The internal consistency was determined by using Cronbach's alpha coefficient. A regression analysis with a forward stepwise procedure was conducted to investigate the multivariable relationship to low back disorder. The significance level was set alpha = 0.05.

RESULTS AND DISCUSSION

SPINE DISORDER

The results of spine disorder diagnosis show that 24 subjects (75%) are classified into level 0, 5 subjects (16%) are level 1, 2 subjects (6%) are level 2, and 1 subject (3%) is level 3. In other words, 25% of subjects have spine disc space narrowing problem. Further, the mean age (38.9 years old) of spine disc space narrowing group (level 1, level 2 and level 3) is greater than normal spine group (37.7 years old). Beside, the body height of normal spine group (171.3 cm) is greater than spine disc degeneration group (169.0 cm). However, the independent T test result showed that the differences of mean age and body height between these two groups were not significant (p>0.05).

SUBJECTIVE MUSCULOSKELETAL DISCOMFORT RATING

As shown in Table 1, the prevalence rate for low back/hip discomfort (56%) was the highest comparing with the other body areas, followed by neck (47%) and shoulder (38%). On the other hand, the greatest severity was found in thigh/knee area (5.2 scores), followed by shoulder (5.0 scores) and upper back (4.4 scores). The reason was probably due to the drivers jumped up and down from cab and trailer frequently and resulted in home delivery drivers' thigh/knee area discomfort. Further, independent T test results show that the discomfort ratings for all areas were not significant. The average low back discomfort rating for spine disc space narrowing group and normal spine group was 4.3 and 3.8, respectively.

Table 1. The results of prevalence and severity of musculoskeletal discomfort.

Body area	N	Prevalence	Severity*
Low back/hip	18	56%	4.3
Neck	15	47%	4.0
Shoulder	12	38%	5.0
Wrist/hand	11	34 %	4.3
Upper back	9	28%	4.4
Elbow	9	28%	4.1
Shank/ankle/foot	8	25%	4.0
Thigh/knee	7	22%	5.2

*The score of severity is from 0 to 10.

JOB-RELATED PSYCHOSOCIAL FACTORS

From the results of factor analysis, four factors in job stress scale including lock of job control, lock of job encouragement, intensified work load, and unfriendly of interpersonal relationship were extracted with simple interpretation. For job satisfaction scale, only one factor, dissatisfaction about job content was identified. Table 2 shows the average scores of job-related psychosocial factors between spine disc space narrowing group and normal spine group. The scores of lack of job control, lack of job recognition, intensified work load, and unfriendly of interpersonal relationship for spine disc space narrowing group are greater than normal spine group. On the contrary, the score of dissatisfaction about job content for normal spine group is greater than spine disc space narrowing group. But, the results of independent T test show that the differences of job-related psychosocial factor between these two groups are not significant.

Table 2. The average scores of job-related psychosocial factors between spine disc space narrowing group and normal spine group

	Spine disc space narrowing group (N=8)	Normal spine group (N=24)
Job stress scale		
Lack of job control	2.49	2.31
Lack of job recognition	2.17	2.05
Intensified work load	3.34	2.82
Unfriendly of interpersonal relationship	2.18	1.72
Job satisfaction scale		
Dissatisfaction about job content	1.69	1.79

THE BIOMECHANICAL LOADING ANALYSIS

Over 6200 MMH samples were analyzed in the study. Table 3 shows the results of biomechanical loading analysis. Four drivers were classified into spine disc space narrowing group and sixteen drivers were classified into normal spine group. The differences in L5/S1 compression force between spine disc space narrowing group and normal spine group were significant on average compression force, the 1%ile compression force, the 50%ile compression force, and the maximum compression force ($p < 0.05$). It is interesting to find is that the L5/S1 compression forces for normal spine group are greater than spine disc space narrowing group. Moreover, the low back discomfort rating of the four drivers with spine disc space narrowing group are all positive. It suggests that the driver would use better posture, e.g. maintain trunk straight, to avoid discomfort feeling of low back when performing MMH and resulted in a decreased biomechanical spine loading.

Table 3. The results of L5/S1 compression force (Nt) between spine disc space narrowing group and normal spine group

	Spine disc space narrowing group (N=4)	Normal spine group(N=16)
Average*	2225.80	2706.58
Min	356.75	455.31
1%ile*	427.86	615.53
50%ile*	2184.50	2674.78
90%ile	3646.18	4096.72
99%ile	4667.92	5303.24
Max*	5153.00	6411.13

* $p<0.05$

THE LINEAR REGRESSION MODEL

For multiple regression analysis, the degree of spine disc space narrowing is used as dependent variables, and the subjective musculoskeletal discomfort ratings, job-related psychosocial factors (job stress scale and job satisfaction scale) and biomechanical loading data are used as independent variables. Result shows that the model is statistically significant ($p <0.001$) with coefficients of determination (R^2) greater than 0.75 (Table 4). The coefficient of determination 0.77 from regression analysis results is high enough to support the validating of the model, indicating the positive relationship between severity of L5/S1 spine disc space narrowing and these three factors. The percent of push goods with hands, the score of unfriendly of interpersonal relationship and job seniority factors were all significant in the regression models. Moreover, the standardized partial regression coefficient of the percent of push goods with hands is 0.3, greater than that of unfriendly of inter personal relationship (0.26) and job seniority ($2.17*10^{-3}$). The influence of push goods factor influence seems greater than other factors.

Table 4. Regression equation for the severity of L5/S1 spine disc space narrowing

Equation	R^2	Significance
Disc space narrowing degrees = -0.69+0.30*pushing goods with hands (%) +0.26*unfriendly of inter personal relationship (scores) +($2.17*10^{-3}$)*job seniority (months)	0.77	P< 0.001

CONCLUSION

The finding of this study indicates that 25% of subjects have spine disc space narrowing problem. The low back/hip, neck and shoulder area discomfort are rather common among the home delivery drivers. To avoid discomfort feeling of low back, the drivers with spine disc space narrowing problem adapt a better working posture to perform MMH task and results in a significantly lower biomechanical loading of spine than the drivers without spine problem. Moreover, the findings also show that the percent of push goods with hands, the interpersonal relationship and job seniority factors are more dominant than other factors for the severity of spine disc space narrowing. Therefore, an approach of integrating psychosocial support and ergonomics design improvement would be effective in reducing the prevalence of spine disc space narrowing problem.

REFERENCES

Hsu, W. H., Wang, M. J. (2003), "Physical discomfort among visual display terminal users in a semiconductor manufacturing company: A study of prevalence and relation to psychosocial and physical/ergonomic factors." *AIHA journal*, 64 (2), 276-282.

Kelsey, J. L. (1975), "An epidemiological study of acute herniated lumbar intervertebral discs." *Rheumatology,* 14 (3), 144-159.

Kuorinka, I., Jonsson, B., Kilbom, A., Vinterberg, H., Biering-Sørensen, F., Andersson, G., and Jørgensen, K. (1987), "Standardised nordic questionnaires for the analysis of musculoskeletal symptoms." *Applied Ergonomics*, 18 (3), 233-237.

Neumann, W.P., Wells, R.P., Norman, R.W., Frank, J., Shannon, H. and Kerr, M.S. (2001), "A posture and load sampling approach to determining low-back pain risk in occupational settings." *International Journal of Industrial Ergonomics,* 27 (2), 65-77

Okunribido, O. O., Magnusson, M., and Pope, M. (2006), "Delivery drivers and low-back pain: A study of the exposures to posture demands, manual materials handling and whole-body vibration." *International Journal of Industrial Ergonomics*, 36 (3), 265-273.

Okunribido, O. O., Shimbles, S. J., Magnusson, M., and Pope, M. (2007), "City bus driving and low back pain: A study of the exposures to posture demands, manual materials handling and whole-body vibration." *Applied Ergonomics*, 38 (1), 29-38.

Okunribido, O.O., Magnusson, M. and Pope, M.H. (2008), "The role of whole body vibration, posture and manual materials handling as risk factors for low back pain in occupational drivers." *Ergonomics,* 51 (3), 308-329.

Raanaas, R. K., and Anderson, D. (2008). A questionnaire survey of Norwegian taxi drivers' musculoskeletal health, and work-related risk factors. *International*

Journal of Industrial Ergonomics, 38, 280-290.

Robb, M.J.M. and Mansfield, N.J. (2007), "Self-reported musculoskeletal problems amongst professional truck drivers." *Ergonomics,* 50 (6), 814-827.

CHAPTER 40

Conception of a Task Analysis- and Screening-Method for Identifying Age-Critical Fields of Activity on the Basis of the Chemnitz Age Model

Mathias Keil and Birgit Spanner-Ulmer

Professorship of Human Factors and Ergonomics
Chemnitz University of Technology
Chemnitz, 09107, Germany

ABSTRACT

On the basis of the "Chemnitz Age Modell's" data research, an "age-differentiated task analysis- and screening-method (ATS)" is momentarily developed at the Professorship of Human Factors and Ergonomics at the Chemnitz University of Technology. With the help of ATS, task demands and standardized skills can be compared, and hence operation fields for the development and design of products and processes can transparently be presented on the basis of age-dependent performance factors.

Keywords: Chemnitz Age Model, Chemnitz Age-Database, age-differentiated task analysis- and screening-method

INTRODUCTION

Today and in future, the demographic development in Germany will be one of the most important operation fields for companies which need addressed this issue to maintain their competition and innovation ability. The increase of the mean age of German personnel will be more and more noticeable. Because of high-birthrates in the 1950s and 1960s (Statistisches Bundesamt, 2006) personnel will be increasingly old in Germany. Factors such as the gradual rise of the retirement age to 67 years (Bundesgesetzblatt, 2007), the discontinuation of state-subsidized part-time employment prior to retirement and the increase of the employment rate of the age-group of the 60 to 64 year olds (Eurostat – European Commission, 2008) will even enforce and fasten this process. Thus, this development will have far-reaching consequences for companies expressed by an increase of costs for the workforce, the increase of absence-rates due to illness, rise of personnel with limited performance, a lowered personnel flexibility, and the impending loss of innovation inability (Brandenburg and Domschke, 2007). Companies without an effective concept for the management with an ageing personnel risk a loss in performance ability and innovative strength (Buck, 2003; Baase, 2007).

Against the background of demographic changes, the present challenge for companies lies in the design of age-specific work systems. According to REFA (1993), a work system consists generally of the work task, the in- and output, the human as employee, the equipment, the work procedure, and the environmental influences. Due to age-dependent changes in performance ability, the human as system element becomes a central focus of examination. Referring to this, the elements of this work system should be designed in the manner that emerging deficits of the human performance ability are mostly compensated (Spanner-Ulmer et al., 2009). This is carried out by the principles of Human Factors and Ergonomics, and further the maxims of humanization and rationalization, that is: the design of the work system has to ensure that work demands are humane, effective, and efficient (Schlick et al., 2010). Apparently, there are only few ergonomic guidelines for age-differentiated concepts of work systems such as ISO TR 22411 (2008). Furthermore, merely any evaluation concepts exist which consider the variable "age" when evaluating the work station. At present, the "assistance system for age-differentiated work design and employee assignment" (Rademacher et al., 2009) developed by the Institute of Ergonomics, Darmstadt University of Technology in Germany, offers the potential to fill this gap, however the approach is limited to the evaluation of age-differentiated force- and motion sequences. An approach, relating to all the relevant age-dependent performance factors for the design of work systems, is continuously missing. Against the backdrop of age-related changes of performance conditions, the early cognition of age-differentiated design opportunities in product, process, and technology at the beginning process of product development and production, becomes the key component of success. The sooner age-relevant operation fields can be presented the greater is the creative leeway to obtain efficient solutions for human, technique, and organizations

(Spanner-Ulmer and Keil, 2009).

THE "CHEMNITZ AGE-MODEL" AND ITS SCIENTIFIC DATA BASIS

The foundation for an age-based design of products and processes focuses on the assimilation of the age-dependent changes of the individual's performance factors. Therefore, the "Chemnitz Age-Model" (see Figure 1) structures age-dependent performance factors based on 9 ability categories of humans.

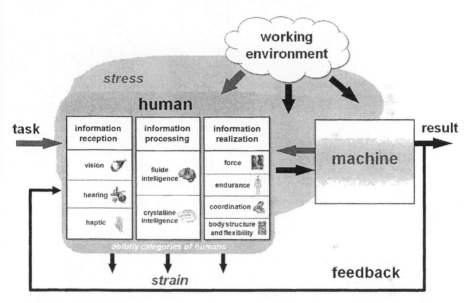

Figure 1. Chemnitz Age Model" on the basis of the structure scheme of human work (Schmidtke, 1993) with classes of human abilities.

Further, one ability category includes all age-dependent factors of the performance ability which positively or negatively influences the concrete ability in the ageing process (Keil and Spanner-Ulmer, 2009). In case of the ability category vision, the factors are, for instance, accommodative capacity, acuity, proprioception, and so on. In addition, with the term of work environment it becomes possible to summarize age-dependent physical (i.e. noise, vibrations, climate, lightning) and social influences (i.e. work motivation, shift work) during the task processing. The structure of the "Chemnitz Age-Model" was transferred in a database – the "Chemnitz Age-Database" (see Figure 2).

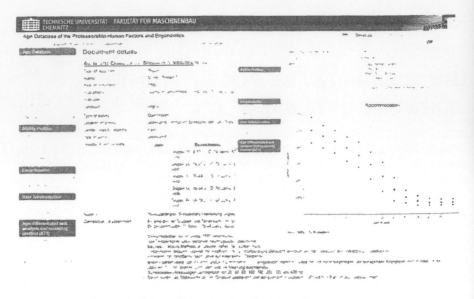

Figure 2. Screenshots of the "Chemnitz Age Database".

The goal is to gather a comprehensive status analysis from the present research data of medicine, biology, sports science, gerontology, ergonomics as part of engineering science, psychology, and sociology. To interpret and analyze this research data, the mental and physical changes, being typical in developments of "healthy ageing" male and female humans between 20 and 70 years of age, have to be identified. Additionally, work design possibilities for the influence of performance and trainability will be acquired for every ability category (Keil and Spanner-Ulmer, 2009). Changes of the performance factors during the ageing process can be sighted via ability profiles. Therefore, the research data is expressed in the value of the performance ability, and will further be related to the aging process. Through the consideration of the indicated standard deviation, interindividual scatter ranges of the specific performance feature can be shown. For this purpose, the "Chemnitz Age Database" already provides over 900 scientific-based studies on various age-differentiated performance features of the human. The "Chemnitz Age Model" successfully allows a scientific-based overview on all relevant age-dependent changes in physical and mental performance factors of the human which are essential for the design of work systems. On the basis of the present age-dependent reverence ranges, an "age-differentiated task analysis and screening method" is currently developed.

REALISATION OF THE "AGE-DIFFERENTIATED TASK ANALYSIS AND SCREENING METHOD" (ATS)

The classification of the subsystem human into information reception, information processing, and information realization, and the modularization into ability categories and relating performance factors allows the examination and evaluation of specific age-dependent abilities being necessary for a task. All task-relevant ability categories will initially be selected for the evaluation of the task demands. In the following, all age-relevant performance factors of every ability category will be evaluated, with the help of a questionnaire, according to their demand. To determine the demands, an 5-ary ordinal scale with the task demands ranging from 0 to 5 (0 – no relevance, 5 very high demands) is used. The increments of the level measurements are based on the physical parameters, such as luminance density or object seize, grounded on a valid data basis of selected studies. On that score, occurring correlations and dependences between the specific age-dependent performance factors are considered. The user is supported by guidelines for low, mediate, and high requests, and by examples of reference during the evaluation. Above that, it is possible to document the load duration through the measured time intervals (seconds or TMUs) of every task demand (see Figure 3).

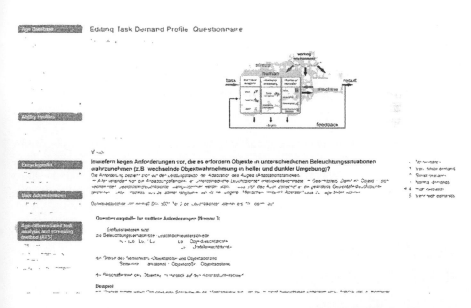

Figure 3. Evaluation of task demands with the help of the "age-differentiated task analysis- and screening-method (ATS)".

As a result, a request profile for the task on basis of the age-relevant performance factors can be generated. This can eventually be compared to the ability profile, in terms of standardized abilities of a chosen age group. The generation of the ability profile results from the scientific reference data of field research studies in the "Chemnitz Age Database" (see Figure 4).

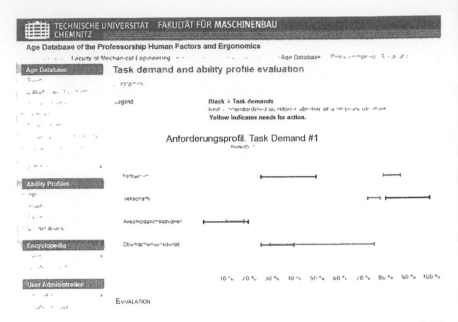

Figure 4. Comparison of the task demands with the standardized workforce abilities with the help of the "age-differentiated task analysis and screening method (ATS)".

The comparison of demands and standardized abilities allows age-critical demands, and hence operation fields for the product and process design are transparently represented. Need for action arises when task demands exceed the standardized reference value of the performance factor's ability. The stronger and longer this manner continues all the more will ergonomic design measurements be necessary.

RESULTS AND CONCLUSION / FUTURE PROSPECTS

In a first study of the Professorship of Human Factors and Ergonomics at the Chemnitz University of Technology, the analysis method for age-critical operation fields in the evaluation of an operational concept was tested (Jentsch et al., 2009).

Most hypotheses, derived from the comparison of demand and ability profile, could be validated. However, the analysis also showed that there are age-relevant influences which are not covered by this method. This lies, for instance, in the enhancement of age-relevant ability limitations due to the pathologic age effects. The further development of the method and its application in the industrial context are to allow the identification of age-critical operation fields in the early stages of product processes. This represents the prerequisite for the systematic integration of age-based standards in the process planning and product development.

REFERENCES

Baase, C.M. (2007). *Auswirkungen chronischer Krankheiten auf Arbeitsproduktivität und Absentismus und daraus resultierende Kosten für die Betriebe.* in: Fehlzeiten-Report 2006. Chronische Krankheiten - Zahlen, Daten, Analysen aus allen Branchen der Wirtschaft. Badura, B., Schellschmidt, H., and Vetter, C. (ed.). Berlin, pp. 45-59.

Brandenburg, U., and Domschke, J. P. (2007). *Die Zukunft sieht alt aus. Herausforderungen des demografischen Wandels für das Personalmanagement.* Wiesbaden.

Buck, H. (2003). *Alterung der Gesellschaft - Dilemma und Herausforderung.* in: Fehlzeiten-Report 2002 - Demographischer Wandel: Herausforderung für die betriebliche Personal- und Gesundheitspolitik. Badura, B., Schellschmidt, H., Vetter, C., and Astor, M. (ed.). Berlin, pp. 5-13.

Bundesgesetzblatt (2007). Teil I Nr. 16, *Gesetz zur Anpassung der Regelaltersgrenze an die demografische Entwicklung und zur Stärkung der Finanzierungsgrundlagen der gesetzlichen Rentenversicherung (RV-Altersgrenzenanpassungsgesetz).* Bonn, 30.04.2007; Bundesanzeiger Verlag.

Eurostat - European Commission (2008). *European Union Labour Force Survey-Annual results 2007* - Issue number 27/2008.

ISO TR 22411 (2008). *Ergonomics data for the application of ISO/IEC Guide 71 to products and services to address the needs of older persons and persons with disabilities.* Geneva.

Jentsch, M., Keil, M., Scherf, C., Kienast, H., and Spanner-Ulmer, B. (2009). *Bewertung und Evaluation von Bedienkonzepten für Infotainmentsysteme im Dualtask auf Basis des Chemnitzer Altersmodells.* in: 5. VDI-Tagung, Der Fahrer im 21. Jahrhundert – Fahrer, Fahrerunterstützung und Bedienbarkeit, 04.-05.11.2009, VDI-Berichte 2085, Düsseldorf. pp. 23-34.

Keil, M., and Spanner-Ulmer, B. (2009). *Chemnitz Age Model – an interdisciplinary research basic approach to characterize age critical performance factors.* 17th World Congress on Ergonomics – IEA 2009, 09.08.2009 - 14.08.2009, Peking/China.

Rademacher, H., Sinn-Behrendt, A., Landau, K., and Bruder. R. (2009). *Development of a tool for an integrative age-differentiated workload analysis.* 17th World Congress on Ergonomics – IEA 2009, 09.08.2009-14.08.2009. Peking/China.

REFA - Verband für Arbeitsstudien und Betriebsorganisation e.V. (1993). *Methodenlehre der Betriebsorganisation.* Teil 1, Grundlagen der Arbeitsgestaltung. 2. Auflage. München.

Schlick, C., Bruder, R., and Luczak, H. (2010). *Arbeitswissenschaft,* 3. vollständig überarbeitete und erweiterte Auflage. Heidelberg.

Spanner-Ulmer, B., Frieling, E., Landau, K., and Bruder, R. (2009). *Produktivität und Alter*. in: Produktivität im Betrieb. Landau, K. (ed.). Tagungsband der GfA Herbstkonferenz, 23.-25.09.2009. Millstatt, pp. 81-117.

Spanner-Ulmer, B., and Keil, M. (2009). Konsequenzen des demographischen Wandels für zukünftige Produktions- und Technologieabläufe am Beispiel der Automobilindustrie. in: *Zeitschrift Industrie Management, Technologiegetriebene Veränderungen der Arbeitswelt*. 2/2009, pp. 17-20.

Statistisches Bundesamt (2006). *Bevölkerung Deutschlands bis 2050, Annahmen und Ergebnisse der 11. koordinierten Bevölkerungsvorausberechnung*. Wiesbaden.

Chapter 41

Contemporary Aspects in Design of Work

Aleksandra Kawecka-Endler, Beata Mrugalska

Institute of Management Engineering
Poznań University of Technology
Poznań, Poland

ABSTRACT

The priority significance for quality improvement and productivity of work process has widely comprehended innovation (process innovation). The application of innovative solutions in this scope is crucial for each enterprise and it concerns both production and services. In the article the evolution of requirements in a process of shaping work is presented, the analysis of practical applications is shown and the main aspects and conditions of work processes are indicated. The examples of innovative solutions of work process, possible to apply in practice, are suggested.

Keywords: work process, human factors, work conditions

INTRODUCTION

Political, social, economical and technological transformations registered by both Polish and European industrial companies have caused changes in company organization. Thus, it has become necessary to adapt technique, technology and job organization to human psychophysical abilities. It is also vital to integrate operations aiming at reducing the negative effect of production processes on working conditions (in the company) and on the environment (in the company vicinity).

In the last twenty years Poland has experienced transformations which have brought about fundamental changes in the company structure and organization and in its surrounding. To meet various demands and to be competitive on national and international markets, Polish companies must cope with current tasks in a rational and methodical way. They must also analyze their execution and adjust it appropriately, taking into account organizational capabilities and surrounding hazards (Kawecka-Endler, 2007b).

First and foremost, the interest of human work relates to work effects by the assumption that the work is interesting and enables achievement of satisfactory salary. These two scopes of interests occur the most often separately in the existing work division in a society. They are described as exterior realization of work effects and are subjected to assessment by other employees. In the case of self-assessment, i.e. independent assessment of own work by the employee, which the basis is conscientiousness and reliability of performing work, both scopes can occur simultaneously. The basic criterion formulated for this scope is fulfillment of humanization rules which measure is specific work result also called productivity or work efficiency.

The analysis of effectiveness participation in so-called company's pure income (profit) enables to distinguish economic effectiveness or effectiveness of human work. The participation of human work in pure income can be calculated after deducting all expenditures borne in connection with assurance of specific (humanized) work conditions.

Only singular criteria of assessment, which allow for concrete work assessment, can be numbered among scope of interests or main criteria. They are given and result from ergonomics definition which is a science concerning human functional abilities in work processes in order to create such conditions, methods and work organizations that contribute to mental and physical development of human allowing gaining higher work efficiency.

Next, increase of human work efficiency is possible on the assumption that in productive and effective processes:

- employees will find working conditions free from threatens,
- appropriate standards concerning work content, work tasks, work environment and also salary and cooperation will be fulfilled,
- work space allowing for education and improvement of skills and development of employee personality will be organized (Bullinger et. al., 2003).

ERGONOMICS AND WORK PROCESS

According to the definition cited in many papers (introduced by the International Conference of Ergonomics in 1972 in Moscow):

Ergonomics/Human factors *is a science concerned with functional human possibilities in work processes in order to create such conditions, methods and work organization which contribute to intellectual and physical* *development of human and assure safety, health protection and employee's comfort allowing for getting better work efficiency.*

Ergonomics as applied science (recognized as the group of interdisciplinary sciences), applying practical rules and postulates formulated by such scientific disciplines as: physiology and work hygiene, anthropometrics, psychology and work sociology, praxeology, organization, is concerned with optimization of processes and functions executed by human in different systems defined as systems: man – machine – environment or more generally as man – technique system.

The problems concerning human work require detailed analysis in the following aspects:

1) adjustment of working means to human for the sake of his psychical and physical possibilities i.e. by correct choice and shaping of work stands, choice of ergonomically verified machines, tools and instrumentations,

2) creating system of motivational stimuli determining human development and his personality e.g. application of modern forms of work organization, building system of economical stimuli awarding a bonus for high quality and efficiency, guarantying systematic training, supplementation of education and possibilities of increasing qualification by work plant, appointing employees to employee's council or board of directors etc.,

3) organizational problems appearing mainly in labor division – correct solutions in this extent assure application of comprehensive job analysis which concerns the whole of issues related to work process realization e.g. correctly calculated time standards of work for individual works and operations; duration and frequency of breaks dependent on level of difficulty, load etc. (Handbuch der Arbeitsgestaltung…, 1980),

4) issues connected to safety and work hygiene and environmental protection which first and foremost involve application of so-called safety technologies and production processes that assure safety during their realization for direct executors (micro scale) and are natural for environment (macro scale) (Kawecka-Endler, 2007b).

CHANGE AS FUTURE CREATOR

Contemporary enterprises function in conditions of continuous and often turbulent transformations. The rules of competition analyzed within the space from industrial century (the 20th century) to informatization century (the 21st century) also underwent changes. In addition the requirements imposed on managers of today connected with managing changed dramatically. These transformations with reference to company management imply a number of new challenges which overcoming requires essential knowledge for their earlier expectation and understanding. The next step is the application of effective remedial measures. The ability of determining critical factors and their consideration in management process serves reinforcement of company competiveness (Bullinger et al., 2003).

The basics of future, progressive economic, technological, legal-political, cultural and social changes always depend on specific situation of the company, demanding precise description of factors presented at Figure 1.

NEW CHALLENGES OF WORK PROCESSES		
Changes of rules of competitiveness	Innovative potential	Change of values
• internationalization of markets • innovation dynamics of products and processes • sales markets • globalization • demographic structure • possessed means	• new products • process innovations • new forms of work organization and labor division • new forms of enterprises • free flow of employees • multiculturalism of teams	• responsibility for natural environment • age structure of employees • flexible working time • reduction of working time • conditions of sale • responsibility for quality on work stand

Figure 1. Changes of frame conditions in management zone of the enterprise. (Adapted from (Belllinger et al., 2003)

The characteristic trait of contemporary concepts of management used in practice is its variability. From one side, such a situation dynamizes development of theory and practice of company management, however, from the another one, it causes a number of doubts and ambiguities. They occur both on cognitive ground (content-related and methodological) and utilitarian ground.

As an effect of continuous changes in the surrounding new tendencies, concepts and strategies in enterprise management system such as: marketing, controlling, logistics, TQM (Total Quality Management), HRM (Human Resources Management), *lean management* or *reengineering* appear (Pacholski, 2003). Some of them complement one another, other are complementary or substitutive. The basis of the concept of HRM, TQM and *lean management* is human orientation. It is also a basis for TQM philosophy in which human is perceived as a client, employee or member of society (Kawecka-Endler, 2007a).

In the aspect of these conditions it is not difficult to notice what a huge role is fulfilled by knowledge which has to be systematically gathered, applied and verified.

WORK AND FORECAST OF ITS CHANGE

In the literature (national and international) directly or indirectly concerned with issues about work, its effects or factors influencing these effects (e.g. labor-consumption, productivity, efficiency etc.) a number of varied definitions, criteria and work aspects, which are basis of written classifications, can be found. In the authors' opinion two among work definitions formulated by H. Luczak (1991a; 1991b) precisely describe range and complexity of this problem in company activity:

1. *Work is a planned, aimed and managed activity, and it occurs in specific social conditions.*
2. *Work is a human activity occurring in cooperation with other people by application of technical means leading to a realization of specific aims which constitute production of goods and service.*

The following aspects of human activities in production process result from the second detailed definition:

- endogenous aspect (internal: effort, load),
- exogenous aspect (external: effects of goods and service production) (Laurig, 1992; Luczak, 1991a; Luczak, 1991b).

Both aspects are integrally connected with human work as well as all accompanying it factors and phenomena. The ordination of issues concerning human work process is the subject matter of praxeology[1]. Its tasks are an elaboration of recommendations and general solutions[2] adequate to varied fields of human activities in the society.

According to T. Kotarbiński (1973) *science of work includes issues of optimization of work conditions so that they favor health and employee's efficiency and assure specific productivity.*

Quoting the definition of work process it is necessary to mention the standard PN-EN 614-1+A1:2009 in which work process is understood as work course. According to this definition common activities of human, work means, materials, energy and information within the confines of work system follow each other in time and space. Work process concerns specific forms of work and may be evaluated by the application of suitable criteria. While discussing the matter of work process the following ergonomic rules should be also in mind i.e.: appropriate planning of breaks, work changeability, work enlargement, work enrichment (PN-EN 614-1+A1:2009).

A lot of authors, who work in this discipline, suggest different structures and methods of operation including in more broaden or narrower range issues of praxeology.

In Polish literature the classifications, in which science of work constitutes a part of human factors methodology and concentrates on principles of efficient

---

[1] Praxeology is a general methodology and concerns every human activities (Pszczołowski, 1978). According to T. Kotarbiński praxeology is a science more general than the science of work (Kotarbiński, 1973).
[2] Rationalization recommendations should be distinguished from directives about general and universal application.

working of human in different fields and human factors discipline analyses compound relations of man-technique system, are presented (Kotarbiński, 1973; Pszczołowski, 1978).

The science of work concerns a practical application of cognitive knowledge to realization of two basis aims:

- humanization of work - humanization aim,
- efficiency of work - economical aim.

The forecasts aiming at determination of direction changes of employees, organization and the same work elaborated for a new century in the 90th by Charles Handy were extremely aptness. This is the quotation (Handy, 1996):

"..a serious change has to take place in our way of thinking about organizations. They are organizers and employers and are minimalistic".

He also predicted that the number of workers employed in huge companies will decrease but the number of workers employed in smaller firms and self-employed will increase. Unfortunately, these changes will cause that much less employees will be generally without work for the sake of lack of specialist abilities needed both inside and outside enterprises. In Handy's opinion economic situation of a country (upward or downward) influences constant increase of number of the unemployed not to a large degree.

Example: in the years of British boom in 1985-90 industrial production increased by 19%, nevertheless employment of industrial workforce decreased by about 5%. Conclusion: employment decreases systematically but in the time of recession it falls down faster.

All these changes concern also the same work, its division, dimension and a manner of its realization and are presented at Figure 2.

WORK			
Classical paid work	Own work	Compulsory work	Social work

• full-time work
• part-time work
	• educational
	• additional
	  training
		• family
		• children
			• voluntary
			  organization
			• associations
			• honor
			  department

Figure 2. Types of work. (Adapted from (Materials of project, 2004))

SCOPE OF INTERACTION BETWEEN HUMAN AND WORKPLACE

The scope of interaction between human and his workplace can be analyzed in a number of aspects, among which the following are:

- manner of carried out tasks and demands made to their operators,
- adjustment of machines, equipment and tools for tasks on a given workplace,
- providing and using information,
- physical environment (O'Neill, 2007).

The requirements for individual work stands differ from each other similarly as do physical and intellectual skills of each person. In order to achieve optimal work results they should be matched to present abilities of human. The characteristics of operators should be also taken into account by choosing work means (Nelson, 1996). In the field of issues of human-workplace interaction send and received information at work stand are also numbered. Next, during the analysis of interaction of human with physical environment a type of environment (suitable for a given processes) and conditions assuring safety, should be defined. This scope should contain three fundamental groups of environmental factors such as physical factors, chemical factors and biological factors (Luczak et at., 2006).

In Figure 3 elements influencing on the relationship between human and his workplace taking into consideration direction of their interactions were presented.

HUMAN	
Psychology	
Anthropometry	
Psychophysical perception	Biomechanics Psychomotor skills
Instruments	Tools/Controls
Design of displays	Design of controls
WORKPLACE	

Figure 3. Interaction between human and workplace. (Adapted from (Zandin, 2001))

As it can be seen in Figure 3, interaction between human and workplace occurs by the application of right tools and/or controls. Adjusting them to the operator knowledge of biomechanics is used. It is a science about principles of build and functions of motor organ of human being in categories of mechanics. In practice each time the connection between the operator and the physical environment should be taken into account in an individual way (Będziński, et al., 2004). The application of biomechanical principles allows minimizing damage to muscles, joints and tissues (Zandin, 2001). At this stage of interaction information concerning psychomotor skills of human are also used. Next, by the application of tools and/or controls, controls, which allow human to influence on workplace, are designed. Feedback is provided from workplace to human. The way of its reception is determined by psychophysical perception which is an individual characteristic of each person. The common element, influencing the workplace as well as determining choice of the worker to the work stand, is anthropometrical data and psychological characteristics of the person.

The described above interaction between human and workplace is very crucial and vital from point of view of design of workplaces and choice of the worker to the specific work stand. Taking into consideration human factor allows to reduce occupational risk, minimize worker's sickness and accident costs, increase productivity, improve product quality and worker comfort on the job (Zandin, 2001).

INDUSTRIAL ACCIDENTS AT WORKPLACE

Increasing number of accidents registered during last years is one of the most important problems of Polish enterprises. Despite the development of technology, application of protection of work stands, control of the course and realization of processes and implementation of preventive actions, the number of accidents is constantly growing. Still in practice situations of belittling occupational risk and its assessment appear, or sometimes even lack of awareness of existence of potential accident risk from point of view of the workers but also from employers. Such state requires taking radical and efficient actions leading to upgrading work conditions and to raising the awareness and culture within the range of not following security. Statistical data indicate that industrial work accidents mostly result from not following principles of industrial security or generally speaking of bad work conditions (Handbuch der Arbeitsgestaltung…, 1980; Kawecka-Endler, 2003b).

The topic of working conditions arises partially from demands of economical practice but also from the necessity of taking into account legal regulation (domestic and European ones). According to the European Social Charter all employees have the right to safe and hygienic conditions of work. In Poland each employee has also this right (the Polish Constitution guarantees it) and the Labor Code obliges all employers to follow this regulation. Unfortunately, the realization of these regulations is far from perfect.

PRACTICE AND WORK PROCESSES

Models of solutions are a great simplification of real conditions because they are limited to the analysis of one or a few factors. However, it is known that there are connections and correlations between factors, and the main difficulty is their presentation in one general system which can be applied. In practice the most crucial feature of the work system is its openness to introduction of changes in order to shape working conditions the best. Each change requires specification of the aim, which is to achieve optimal effect (Kawecka-Endler, 2001; Kawecka-Endler, 2003a).

Shaping work should facilitate the employee its performing and minimize negative effects of its influence. The achievement of goal of shaping work is connected with getting unambiguous answer to a question: what conditions assure avoiding threatens for a human performing a given work in a specific time?
The specification of such working conditions requires:

1) selection of factors having an effect on a human being,
2) specification of acceptable values of parameters (Luczak, 1991a).

In practice each time it should start from a specification of objective factors because only such a description forms a basis to obtain intentionally shaped working conditions, well in an intended way. However, subjective influences should also be taken into consideration as additional variables (so-called intervention variables) (Kawecka-Endler, 2003b).

The elaborated model solutions present common points of reference characteristic for all working systems (also in Poland and other countries). A great complexity and diversity of models causes that not all relationships can be determined on the basis of cause-reason analysis, especially on planes of multigradual complexity.

CONCLUSIONS

Such functioning is characteristic of innovative processes, which guarantee effective building and introducing of a quality assurance system in a company. The essence of an innovative process is to plan and introduce changes in an enterprise. What is more, methodical and rational change introduction is the essential condition to the success of a modern company, which, as a result, should be able to gain:

- flexible organization, enabling quick adjustments to changing (and difficult to predict) market requirements,
- safe working conditions, determining the quality of work and production,
- chance to offer modern goods and services,
- good quality product at a competitive price,
- better productivity (total and partial),
- customer's satisfaction.

REFERENCES

Będziński, R., Kędzior, K., Kiwerski, J., Morecki, A., Skalski, K., Wall, A., and Wit A. (2004), „Biomechanika i inżynieria rehabilitacyjna". in: M. Nałęcz (Eds.), *Biocybernetyka i Inżynieria Biomedyczna*, Vol. 5, Polska Akademia Nauk, Warszawa.

410

Bullinger, H.J., Warnecke, H. J., Westkämper, E. (2003), *Neue Organisationsformen im Unternehmen*. 2nd Edition, Springer-Verlag, Berlin-Heidelberg.

Handbuch der Arbeitsgestaltung und Arbeitsorganisation (1980), Red. VDI-Gesellschaft, Düsseldorf, VDI-Verlag GmbH.

Handy, C. (1996), *Wiek paradoksu. W poszukiwaniu sensu przyszłości*. Dom Wydawniczy ABC, Warszawa.

Kawecka-Endler, A. (2001), "Innovative Aspects in Design and Manufacturing of Products." in: M. J. Smith, G. Salvendy (Eds.), *Systems, Social and Internalization Design Aspects of Human-Computer-Interaction*. Vol. 2, Lawrence Erlbaum Associates, Publishers Mahwah, New Jersey – London, 192-196.

Kawecka-Endler, A. (2003a), "Human factors as a determinant of quality of work." in: C. Stephanidis, J. Jacko (Eds.), *Human-Computer-Interaction: Theory and Practice*. Vol. 2, Lawrence Erlbaum Associates, Publishers Mahwah, New Jersey – London, 1371-1375.

Kawecka-Endler, A. (2003b), "Work conditions in assembly as a determinant of achieving a good quality of work," in: H. Strasser, K. Kluth, H. Bubb (Eds.), *Quality of Work and Products in Enterprises of the Future*. Ergonomia Verlag oHG, Stuttgart, 785-789.

Kawecka-Endler, A. (2007a), „Ergonomia w dydaktyce uczelni technicznej, a wymagania praktyki." *Zastosowania Ergonomii*, 65-66 (1-2), 175-184.

Kawecka-Endler, A. (2007b), "Safety and hygiene of work as a basis of the enterprises strategy". in: L. Pacholski, S. Trzcieliński (Eds.), *Ergonomics in Contemporary Enterprise*. IEA Press, Poland, 256-262.

Kotarbiński, T. (1973), *Traktat o dobrej robocie*. Ossolineum, Wrocław.

Laurig, W. (1992), *Grundzüge der Ergonomie*. Berlin, Köln, Beuth Verlag GmbH.

Luczak, H. (1991a), *Arbeitswissenschaft (Systematik, Arbeitsschutz)*. Vol. 1, Technische Universität Berlin, 3. Auflage, Berlin.

Luczak, H. (1991b), *Arbeitswissenschaft (Konzepte, Arbeitspersonen)*. Vol. 2, Technische Universität Berlin, 3. Auflage, Berlin.

Luczak, H., (2006), in: W. S. Marras, W. Karwowski (Eds.), *Fundamentals and assessment tools for occupational ergonomics*. 2nd Edition, Taylor&Francis, New York.

Materials of project (2004), *Netzwerk Arbeitsforschung in der neuen Arbeitswelt – FQMD*. Eigenverlag, TU Dresden.

Nelson, W. (1996), "Workstation Design". in: Q. Lee, A. Amundsen, W. Nelson, H. Tuttle (Eds.), *Engineering & Management Press*. Institute of Industrial Engineers, Norcross Georgia, USA.

O'Neill, M. J. (2007), *Measuring workplace performance*. 2nd Edition, Taylor&Francis, New York.

Pacholski, L. M. (2003), "Macroergonomic circumstances of the manufacturing company development". Proceedings of the XV Triennial Congress of the International Ergonomics Association and The VII Joint Conference of Ergonomics Society of Korea/Japan Ergonomics Society, *Ergonomics in the digital age: Safety and Health Miscellaneous Topics,* Vol. 6, Seoul, Korea, 556-559.

PN-EN 614-1+A1:2009 *Safety of machinery. Ergonomic design principles. Terminology and general principles.*

Pszczołowski, T. (1978), *Mała encyklopedia prakseologii i teorii organizacji.* Ossolineum, Wrocław.

Zandin, K. B. (2001), *Maynard's Industrial Engineering Handbook.* 5th Edition, McGraw-Hill.

Chapter 42

Teaching in Real Time – Development of a Virtual Immersive Manikin implemented in Real Time Systems

Frank Sulzmann[1], Katrin Meinken[2]

[1] Institute for Human Factors and Technology Management IAT
University of Stuttgart
Germany

[2] Fraunhofer Institute for Industrial Engineering IAO
Stuttgart, Germany

ABSTRACT

Work systems for manual assembling are a particular challenge in assembly planning. Next to the technical components of the workplace the component »human« has to be considered. Therefore not only the time requirement of the single process steps needs to be incorporated, also questions of ergonomic parameters become more important. Planning of work systems can be simplified and accelerated by applying immersive systems in combination with digital manikins. As existing manikins do often not meet the requirements of the user, Fraunhofer IAO is developing a flexible and modular built manikin for the planning and the evaluation of assembly work systems. The present paper will give an overview on the project status, the main features of the designed manikin and the ergonomic assessment method.

Keywords: digital manikins, virtual reality, real time systems, immersive assembly planning, ergonomic assessment

INTRODUCTION

Digital manikins are increasingly applied in real time systems, e.g. for ergonomic evaluation and analysis of work places, in systems using motion capturing as well as for virtual agents in architecture models. In the future the application of digital manikins for the evaluation of architecture models with dynamic populated human models or the use of virtual agents in order to simulate workplace scenarios as well as the dynamic analysis of physical exposures with digital manikins will become even more important.

Unfortunately state-of-the-art digital manikins do so far not provide all necessary functionalities in order to apply them for real time simulation and evaluation. Former projects, such as the development of the assembly planning system" iTeach" which combines the use of digital manikins and a simplified motion capturing to an interactive tool for workplace planning, show that currently available manikins are only limited applicable in immersive interactive real time systems as a simplified motion capturing does not suffice for displaying all necessary functionalities. (Hoffmann et al, 2007). Furthermore the source code of existing human models can often not be adapted or accessed, therefore a setup change or the programming of an add-on to the digital manikins with essential functionalities is not possible. At present the available interactively controllable human models are mainly intended and applied for the analysis of anthropometric parameters and the ergonomic assessment of static body postures such as seating positions in vehicles or airplanes. These models own functionalities that are not really necessary for the pure visualization of people but do consume a lot of computing time and therefore are slowing down the complete system. Controlling the manikin interactively via motion capturing is then only possible with limitations as the digital manikin will only be able to follow the actual performed movements in a much delayed way. Furthermore the available manikins do not offer reliable information on the dynamic musculo-sceletal exposure when conducting repetitive tasks as required for e.g. manual assembling. Functional limitations of state-of-the-art manikins are therefore given due to non-universal application possibilities and an insufficient dynamic strain and stress analysis.

Due to these facts Fraunhofer IAO develops a virtual immersive manikin which can be easily implemented in real time systems. The core idea of the research activity is to populate virtual realities with interactive people and to link them with analysis functions such as e.g. an ergonomic evaluation of workplaces. Main goal of the research activity is to develop a configurable digital manikin with a coupled motion capturing process for movement teaching and an ergonomic evaluation methodology for dynamic stress and strain analysis.

VIRTUAL HUMAN

Some of the currently available manikins like "Virtual Jack manikin" and "RAMSIS" can be used standalone in virtual reality applications. They have included a various number of modules for animation, automatic posture calculation, vision analysis, reach analysis and force analysis. Furthermore anthropometric databases are included to scale the body dimensions of the manikin correctly.

RAMSIS, JACK and Human builder are based on detailed skeletons with fixed numbers of joints and degrees of freedom. RAMSIS for instance has an internal skeleton with 53 joints and 104 degrees of freedom. There is no way to reduce the amount of joints and degrees. Furthermore it is not possible to turn off the analyzing functions for certain limbs. So in every frame of the simulation the whole skeleton is calculated and analyzed, even if the current studies are focused on certain limbs of the human body and these limbs are not affected at all. Therefore the possibilities are limited to adjust the manikins for special tasks and to get the needed performance for analysis in real-time. (Hoffmann et al, 2007)

Due to these facts Fraunhofer IAO develops a virtual immersive manikin which can be easily adjusted to the specific needs of current studies. The basis of the Fraunhofer manikin "Virtual Human" consist of a generic skeleton animation engine. The goal of these research activities is to get a lightweight, flexible tool in difference to commercial manikins like RAMSIS. Using a generic library enables to create manually manikins with the required level of detail regarding the amount of joints, bones and the resolution of the geometry.

CREATION OF MANIKINS

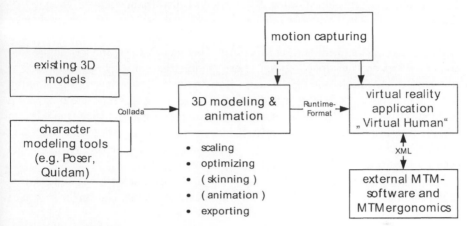

Figure 1: Toolchain

As shown in figure 1 the manikins can be created manually using special character modeling tools like Poser and Quidam. COLLADA is used to export the avatars. It defines an open standard XML schema for exchanging digital assets among various graphics software applications that might otherwise store their assets in incompatible file formats. Originally created by Sony Computer Entertainment, it has since become the property of the Khronos Group, a member-funded industry consortium, which now shares the copyright with Sony. COLLADA is able to store the geometry, complex materials, the skeleton and skinning data. (Arnaud 2006)

Once the manikin is imported into an advanced modeling and animation tool such as Autodesk Softimage it can be prepared for the use in a skeleton animation engine. Therefore it has to be scaled and optimized. Doing so the manikin can be adapted for the current needs.

As the last step the manikin has to be exported into the runtime format of the animation engine.

416

IMPLEMENTATION OF VIRTUAL HUMAN

The manikin is developed using the virtual reality framework "Lightning" of Fraunhofer IAO. The Lightning framework was developed for efficient creation of interactive high end virtual reality applications. It provides modules for description of the virtual environment and its functions. This includes 3-D-models, input- and output devices and functional blocks. These modules have a common data interface and can communicate through a data flow model. A two-layer programming concept facilitates configuration and programming using the run-time interpreted language Tcl/Tk, as well as implementation of performance-critical modules in C/C++. These modules are encapsulated in dynamic libraries which are loaded at runtime, to enable the development of application-specific extensions without the need for modifications of the core system. (Bues et al, 2001),

The skeleton animation engine and the analysis modules are implemented as Lightning modules. The architecture of the application "Virtual Human" is shown in Figure 2.:

Virtual Human

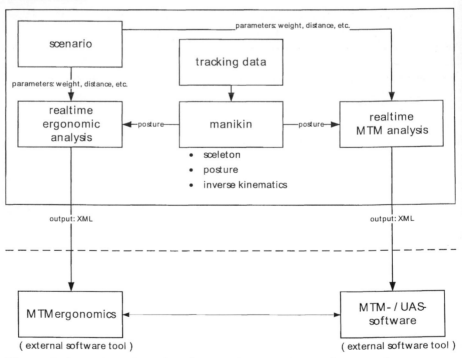

Figure 2: Overview virtual reality application "Virtual Human"

The manikin module consists of an integrated generic skeletal animation engine. It can be controlled using several ways. On the one hand the animation engine is able to handle various kinematic chains including inverse kinematics. For instance simple tracking targets are representing the end effectors of kinematic chains like arms or legs. On the other hand the poses can be modified directly either by using a motion capture suit for full body trackin. These features allow to operate in different modes such as analyzing the full body or to concentrate on discrete limbs only.

The scenario contains the working place in virtual reality. For instance a workbench with tools, sub-assemblies and so on. Certain parameters like distances and working height are implied in the geometry of the working place. Additional information is embedded into various items. In such a way action lists are added to sub-assemblies. Further input is provided by interactive 3D GUIs.

The real-time analysis modules are connected to the manikin and to the scenario's items. So they can collect the information which is needed for the analysis tasks. The results and additional data like postures can be saved using XML. External software tools like MTM Ticon and MTMergonomics will be integrated and used for further analysis of the stored data.

GENERAL APPROACH OF VIRTUAL HUMAN

Figure 3: Screenshot

The real-time analysis modules of "Virtual Human" are designed to analyze working processes of series production. The methods engineer teaches the manikin and performs the working processes in a intuitive way in the virtual reality. He is working at a virtual workbench and uses virtual craftsmen's tools to assembly virtual component parts.

A concept was developed to simplify the planning process as much as possible using metaphors to represent working processes in virtual reality. The Methods-Time Measurement (MTM) is used to structure the working process. MTM is a predetermined motion time system that is used primarily in industrial settings to analyze the methods used to perform any manual operation and to define the standard time in which a worker should complete that task.

Teaching the manikin defines certain actions with an starting point and an ending point. Implicit information like distances and information attached to virtual objects such the weight of a sub-assembly are the main input for the real-time MTM analysis tool of virtual human. Putting interactions of the manikin and embedded information together, the MTM module is able to break apart the process into its component actions, assign time values to each action and sum the times to calculate the total standard time.

At the beginning and the end of such a component action the postures of the skeleton are stored for further use in the ergonomics module. The analysis in the ergonomic module uses the structure determined by the MTM-module.

ERGONOMIC (DYNAMIC) EVALUATION

In the context of assembly planning the workers concerned have to be represented in an anthropometrical correct way. Also future work conditions need to be assessed and analyzed ergonomically. The overall goal is to predict the prospective exposure of the worker as well as the risk concerning occupational safety and health for taking corrective actions in advance (Zülch 2009). Digital manikins shall be used to visualize the work place conditions and to ergonomically assess the work processes in order to support an ergonomic design of future work places.

Surveys among constructing engineers using digital manikins show that essential functionalities of current applications could be improved: The manipulation of the body posture and the animation of the digital manikin are done with a high expenditure of time. Internal libraries as well as analysis and evaluation processes show strong deficiencies, leading to insufficient efficiency and acceptance of the applications. Also the effort of realizing a dynamic simulation and evaluation of ergonomic work places and conditions is very high with current digital manikins. The benefit in comparison is very little and is disproportionate to the effort needed to gain usable results. An advancement of analysis functions concerning force and posture is therefore a strong interest among users of digital manikins (Spanner-Ulmer et al 2009). Different methods for the statically evaluation of posture and static effort are well-known and applied in different state-of-the-art applications. Dynamic evaluation methods in contrast are so far not applied in the field of assembly planning. Dynamic aspects are usually covered by considering the frequency or the duration of motions in combination with external and torsional forces (Zülch 2009).

One goal of the Fraunhofer research activity was to couple the developed digital manikin with an existing dynamic ergonomic analysis tool. As research showed that none such method is yet developed, the dynamic ergonomic evaluation will be based on existing systems, which will be combined in order to achieve an approximation to a dynamic analysis. The methods used are RULA (Rapid Upper Limb Assessment) and the "Leitmerkmalmethode" (LMM), a method for analyzing the manual handling of loads. Input needed in order to apply both methods, such as load weight, body posture, angles or gender, can be retrieved from the digital manikin directly. Only little information, such as length of shift or frequency of movement has to be added manually. Therewith it will be able to analyze different body postures or work processes simple and effortlessly. (see above) The result will be a combined value of both methods indicating whether the ergonomic conditions need to be adapted in order to fulfill the standards or if a high ergonomic quality of the work place is already achieved. Results will be displayed in a descriptive and plain way, following the categorization of a traffic light: Color green indicating that all ergonomic standards are fulfilled and that the workplace does not need further revision. Color red indicating that the work place needs to be adapted to fulfill ergonomic standards. The demonstration of results will be orientated at the results display of the LMM as the method is well-known, approved and field-tested.

CURRENT STATUS OF VIRTUAL HUMAN

Currently the manikin will be equipped with a time determination tool and an additional analysis program for the ergonomic evaluation of the complete work cycle. This shall be realized by the coupling of already existing and approved evaluation methods in one self-contained dynamic strain and stress analysis which can be conducted in real time. Additionally the simulated work processes can be analyzed in detailed single steps using a downstream industrial standard software.

To verify the congruency of the real work cycle and the virtual simulation the system will be evaluated on the basis of different scenarios. Therefore defined work operations will be evaluated with conventional methods of analysis. Afterwards the workplaces as well as the work operations will be transferred to the virtual environment via motion capturing. The complete work cycle will then be analyzed with the developed manikin. Postures and movements will be evaluated in real time and time units for the single steps will be detected. The alignment of the received data from both the virtual world as well as the real workplaces will follow soon.

CONCLUSION

As shown above a flexible and modular built manikin has been implemented. Diverse options of configuration enhance the performance compared to commercial manikins like RAMSIS. With an integrated production process the manikins can be created highly use-oriented with an optimal level of detail and associated skeleton respectively. A flexible architecture does furthermore allow prospective enlargements as well as optimizations.

It can be assumed that in the future digital manikins will be increasingly used in real time systems for ergonomic analysis. An increased application is also expected in systems with virtual agents as well as in teaching systems. Also ergonomic questions will gain more importance in the process of assembly planning, especially regarding demographic change and the involved obsolescence of the employees.

REFERENCES

Spanner-Ulmer, B., and Mühlstedt, J. (2009), Virtuelle Ergonomie mittels digitaler Menschmodelle und anderer Softwarewerkzeuge. In Schenk, M. (Ed.) *„Digital Engineering - Herausforderung für die Arbeits- und Betriebsorganistion"* (pp. 151-170). Berlin: gito

Zülch, G. (2009), Perspektiven der Menschmodellierung als Werkzeug der digitalen Fabrikplanung. In Schenk, M. (Ed.) *„Digital Engineering - Herausforderung für die Arbeits- und Betriebsorganistion"* (pp. 107-124). Berlin: gito

Bues et al. (2001), Towards a Scalable High Performance Application Platform for Immersive Virtual Environments. In: *Proceedings of IPT/EGVE 2001* (pp. 44-52)

Hoffmann, H., Schirra, R., Westner, P., Meinken, K., Dangelmaier, M. (2007): iTeach: ergonomic evaluation using avatars in immersive environments. In: *Proceedings of International conference on human computer interaction.* Beijing

Arnaud, R., and Barnes, M. (2006). *COLLADA Sailing the gulf of 3D digital content creation.* Wellesley: Peters.

CHAPTER 43

Optimizing Humidity Level to Enhance Worker Performance in Automotive Industry

[1]Ahmad Rasdan Ismail, [2]Mohd Yusri Mohd Yusof, [2]Baba Md Deros,[3]Mat Rebi Abdul Rani, [1]Muhamad Mat Noor

[1]Faculty of Mechanical Engineering, Universiti Malaysia Pahang 26300 UMP, Kuantan, Pahang, Malaysia

[2]Department of Mechanical and Material Engineering Faculty of Engineering and Build Enviroment Universiti Kebangsaan Malaysia, 43600 Bangi, Selangor Malaysia

[3]Department of Manufacturing Industrial Engineering Faculty of Mechanical Engineering, Universiti Teknologi Malaysia 81310 UTM Skudai, Malaysia

ABSTRACT

The production of automotive parts is an important part of Malaysia's economy. The main work involved in producing automotive part is the manual assembly process which dependent on manpower capability. Thus the quality of the product heavily depends on the worker's comfort under the working conditions. Humidity is an environmental factor that has a significant effect on the worker's performance. Humidity level and productivity rate were observed in automotive factory. The data were analyzed using an Artificial Neural Network Analysis (ANN). The ANN analysis technique is a common analysis method used to determine the best linear relationship from the collected data. It is apparent from the linear relationship that the optimum value of production (value≈1) is attained when the Humidity is 54.86 %RH. At that level, the optimum value production rate (value≈1) for one manual production line in a particular company is successfully achieved. Through the ANN system, the optimum environmental factor can be predicted.

Keywords: Automotive, Environmental, Productivity, Humidity.

INTRODUCTION

The automotive industry in Malaysia represents large profit and investments. According to the automotive data released by Malaysian Automotive Association (MAA) in 2007, 441 678 vehicles were installed in manufacturing plants within Malaysia: 403 245 passenger vehicles and 38 433 commercial vehicles. This number increases every year. As a result, the demand of vehicle components (car / truck / bus) is rising. Therefore, the production of the components should be increased. More than 30 % of these components are fitted and produced manually. For components that require observation and human manipulation, such as the installation of an engine, the work must be done manually. Thus, the quality of the product heavily depends on the worker's comfort in the working environment. One major issue presented in the research of Shikdar, A.A. et al. (2003) relates to industries with improper workplace design, ill-structured jobs, mismatches between workers' abilities and job demands, adverse environments, poor human-machine system design and inappropriate management programs. This means that the comfort level of the workers in doing their job has not been previously considered. Tarcan, E. et al. (2004) indentifie irritated, sore eyes and throat, hoarseness, congesteion, excessive mental fatigue, headache and unusual tiredness as signs of negative workplace environmental conditions.

Light, noise, air quality and the thermal environment are considered factors that influence the acceptability and performance of workers (Olesen, B.W.,1995). Dua, J.K. et al. (1994) stated that reduced emotional health is manifested as psychological distress, depression and anxiety, whereas reduced physical health is manifested as heart disease, insomnia, headaches and infections. These factors can affect the comfort level of production operators in performing their jobs. When the comfort level at a workplace is increased, the productivity increases as well. In 2001, Ettner, S.L. and Grzywacz J.G. 2001 published a paper in which they stated that work environments a associated with the perceived effects of work on health. Workplace environmental conditions, such as humidity, indoor air quality and acoustics, have a significant influence on workers' satisfaction and performance (Atmaca, I. et al. 2007). These factors must be adjusted in order to increase comfort in the workplace. Temperature is the main parameter involved in this study. If the environmental temperature increases, the quality performance production will decrease (Bobko, N.A. et al. 2006). Thermal suitability is a factor related to workers' satisfaction and comfort in working in closed environments (Ismail, M.B., 2007). The relationship between thermal comfort and acceptability was investigated by Ismail M.B. (2007). Gagge, A.P. and Nevins, R.G. (1976) compared the effect of temperatures that deviate from those of optimum comfort, assessed by the percentage of comfort. Productivity declines by as much as 5 % to 7 % in high temperature rooms (Niemela, R. et al., 2002). Without ventilation, a building's occupants will first be troubled by odors, other possible contaminants and heat (Atmaca, I. et al., 2007). Ventilation systems are an important component in controlling the environmental temperature and play a key role in the comfort level of a working environment.

Productivity may be classified as the ratio of some output to some specified input (David Bain. 1982). Productivity is the relationship of an output in a particular organization to its input (John G.B. 1987). To measure productivity, Equation 1 can be used.

$$\text{productivity} = \frac{\text{output}}{\text{input}} = \frac{\text{results}}{\text{source}} = \frac{\text{efficiency}}{\text{effectiveness}} \tag{1}$$

Productivity is the relationship between goods output or service to employee input or non-human resources in production, such as total employee working hours and total machine operating hours (Vernon M.B. and Shetty, Y.K., 1981). Productivity basically consists of three components which are the production input, the production output and the effectiveness of product distribution from inputs to output (Bhattasali, B.N., Miss Bhattasali B.N., 1972). An Artificial Neural Network (ANN) is a calculation system for sets of data in a nonlinear experiment that uses system neurons with the ability of mapping a straight line (Wu H. et al., 2008). ANNs operate in a manner similar like the human brain to resolving problems that involve mathematical calculation (Zhou, C.C. et al. 2008). ANNs consist of input layer, hidden layer hiding and output layer. An ANN can build an accurate input-output experimental system (Lolas, S. and Olatunbosun O.A., 2008). In this study, Relative Humidity (RH) index is used to represent room humidity. RH is the percentage of water vapor pressure in the air with respect to the saturated water vapor pressure at a certain temperature. RH will be obtained by using Equation 2:

$$RH = \frac{\rho}{\rho^*} \times 100 \text{ \%} \tag{2}$$

where:

ρ is the water vapor pressure in air
ρ^* is the saturated water vapor in air at a certain temperature
RH is the relative humidity

MATERIALS AND METHODS

This study was carried out in a 200 m^2 area production set equipped with an air conditioner cooling system. During the study, one operator was chosen as the subject. The operator worked as usual, and the experimental measurement equipments did not restrict the operator in his work. The equipment was mounted near the operator, with a maximum range of 3 meters. All factors were recorded at 10 minute time intervals one working shift. This experiment was conducted for two working days (two shift days).

Before starting the data collection process, the equipments must be calibrated. The calibration process is vital to determining the accuracy of the data. This process was carried out by using a computerized method to ensure that calibration process followed the standard procedure. After calibration, these equipments was used to take readings at the study field. After completing the data observation, the equipments was calibrated again to verify the quality of the data and to facilitate data collection in the future.

After the data were measured and recorded, the data were analyzed using the ANN's process to obtain a linear regression of the production rate versus the relative humidity. This analysis was conducted repeatedly to produce the best reading. After obtaining the best data variation, the optimum value was determined.

In this study, productivity is determined by comparing the real output value with the target output. Based on Equation 1, productivity will be calculated as the ratio of real output (input) to target output (output). Below is the calculation used for productivity; the productivity will be calculated with the use of Equation 3.

Target output per shift   = 1400 units
Time for one shift   = 9 hours
Interval period   = 1.15 hours
Real time work   = (9-1.15) hours
                              = 7.45 hours
Target output for 10 minutes = 30.107 units

$$Productivity = \frac{real\ output}{30.107} \tag{3}$$

Relative humidity in this study was measured based on the percentage amount of water vapor existing in a gaseous mixture of air and water vapor in units of %RH. This relative humidity was measured using Quest-Thermal Environment Monitor equipment, as shown in Figure 1. This equipment uses Equation 2 to calculate the relative humidity in the study field. Every 10 minutes, relative humidity and production rate were recorded. To confirm whether the optimum value has achieved an acceptable range, the regression value (R) must be within the range of 0.5 to 1. Two factors were tested in this study. Firstly, the proportion factor between the production rate values after ANN analysis versus production rate before ANN analysis was evaluated and secondly, the proportion factor between the production rate values after ANN analysis versus the relative humidity values was investigated.

Figure 1: Quest-Thermal Environment Monitor

RESULTS AND DISCUSSION

After the relative humidity levels and production rates were obtained, the data were evaluated using ANN analysis to determine a linear relationship between the relative humidity level and the production rate. In the ANN analysis, the relative humidity values were set to be constant, while the dependent variable factor (production rate) was trained and altered. The production rate was manipulated to obtain a significant value.

Experiment day 1: During the experiment parameters such as relative humidity, real production and target production were measured and recorded. By using Equation 4, the productivity for the 1st day of the experiment can be obtained. The readings were recorded at 10 minute intervals, starting from 9:40 a.m. to 4:00 p.m. There was some idle time in the operation, due to the worker's break time and machine maintenance.

Figure 2 shows the original value of the production rate (shaped 'o') together with the production rate value obtained after ANN analysis (shaped '*') versus the relative humidity value for the first day of the experiment. The original value of the production rate is scattered randomly as a function of the relative humidity value. However, after ANN analysis, the production rate value shows a linearly decease pattern at relative humidity value of 47 – 50 % RH and 52 -54 % RH.

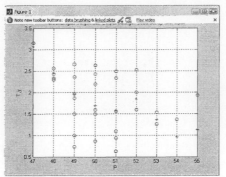

Figure 2: Production rate versus relative humidity

For the first day of the experiment, the R value is 0.52951 (Figure 3(d)) for the proportion factor between the production rate value after ANN analysis versus the production rate before ANN analysis. This shows that the relationship between the production rates after ANN analysis with the production rate before analysis is strong. Meanwhile, for the proportion factor between the production rate values after ANN analysis versus the relative humidity values, the R value is 0.5217. The strength of relationship between production rates after ANN analysis on relative humidity is fairly high.

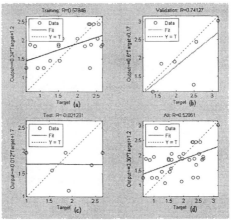

Figure 3: Production rate graph before ANN analysis versus production rate after ANN analysis for training department, validity, test and total.

Since both proportion factors have R value exceeding 0.5, can conclude that the relationship between the production rate after ANN analysis and the relative humidity is fairly linear. According to this relationship, the optimum production value (value≈1) will be achieved when the relative humidity value is 53.6582 % RH. This result is in agreement with the finding of Hitomi Tsutsumi et al. (2007) that the best relative humidity for work environmental is between 40 % RH to 50 % RH. This result is consistent with the ASHRAE Standard 62-2001 Ventilation for

Acceptable Indoor Air Quality, which states that a relative humidity range of 25 % RH to 60 % RH is optimal for human working conditions. According to an article provided by the Minnesota Blue Flame Gas Association, relative humidity ranges between 20 % RH and 60 % RH are comfortable for humans. The Engineering ToolBox has suggested that human comfort requires the relative humidity to be in the range of 25 % RH – 60 % RH. For relative humidity above 60 %RH, workers will feel uncomfortably damp.

Experiment day 2: The data were recorded at 10 minutes intervals from 9:40 a.m. to 4:50 p.m. There was some idle time in the operation, due to the worker's break time and machine maintance.

For the second day of the experiment, Figure 4 shows the original production rate value (shaped 'o') together with the production rate value after ANN analysis (shaped '*') versus the relative humidity value for 2nd day experiment. As can be seen in Figure 4, the production rate value before ANN analysis is randomly scattered with the relative humidity values. After ANN analysis (shaped '*'), the production rate value is still randomly scattered, but the data shows a decreasing pattern, although it does not decrease linearly.

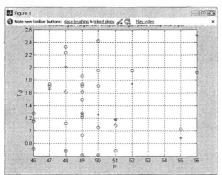

Figure 4: Production rate graph verses relative humidity value

For the 2nd day of the experiment, the proportion factor between the production rate values after ANN analysis versus the production rate before ANN analysis has an R value is 0.72165 (Figure 5(d)). This shows that the relationship between the production rate after ANN analysis and the production rate before analysis is strong. Meanwhile for the proportion factor between the production rate values after ANN analysis versus the relative humidity values, the R value is 0.3781. The strength of the relationship between the production rates after ANN analysis and the relative humidity is fairly weak. Although the R value is low, it represents the best value for the relative humidity data for that day.

428

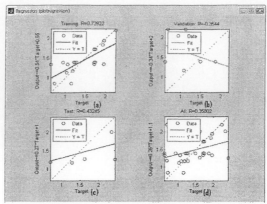

Figure 5: Production rate graph before ANN analysis versus production rate after ANN analysis: training , validity, test and total.

From the result that both proportion factor R values exceeded 0.5, can be concluded that the relationship between the production rate after ANN analysis and the relative humidity is quite linear. According to this weak relationship, the optimum value of production (value≈1) can be achieve when the relative humidity value is 54.86 % RH. This finding is in agreement with Hitomi Tsutsumi et al. (2007) findings that the best relative humidity for doing task while working is between 40 % RH to 50% RH. This result is consistent with ASHRAE Standard 62-2001 Ventilation for Acceptable Indoor Air Quality, which states that a relative humidity range of 25 %R H to 60 % RH is optimal for human working conditions. According to an article provided by the Minnesota Blue Flame Gas Association, relative humidity ranges between 20 % RH and 60 % RH are comfortable for humans. The Engineering ToolBox has suggested that human comfort requires the relative humidity to be in the range of 25 % RH – 60 % RH. For relative humidity above 60 % RH, workers will feel uncomfortably damp.

CONCLUSIONS

The objective of this study to determine the temperature level corresponding to the optimum value production rate (value≈1) for one manual production line in a particular company was successfully achieved. The optimum values were analyzed by an ANN's system with regression values, R, that were within an acceptable range and were strong. Through the ANN's system, the optimum environmental temperature level can be predicted. The relative humidity data for the first day of the experiment were used as the data for this experiment because it has a higher regression value. The optimum value of production can be obtained (value ≈1) when the relative humidity value is 54.86 % RH. Comfortable relative humidity values range between 40 % RH and 50 % RH. This result is consistent with ASHRAE Standard 62-2001 Ventilation for Acceptable Indoor Air Quality, which states that a relative humidity range of 25 % RH to 60 % RH is optimal for human working conditions. According to an article provided by the Minnesota Blue Flame

Gas Association, relative humidity ranges between 20 % RH and 60 % RH are comfortable for humans. The Engineering ToolBox has suggested that human comfort requires the relative humidity to be in the range of 25 % RH–60 % RH. For relative humidity above 60 % RH, workers will feel uncomfortably damp.

REFERENCES

B. N. Bhattasali, Miss G.Bhattasali.(1972). Productivity and Economic Development. Asian productivity organization. pp 1-8

Chang-Chun Zhou, Guo-Fu Yin, Xiao-Bing Hu. (2008). Multi-objective optimization of material selection for sustainable products: Artificial neural networks and genetic algorithm approach. Materials and Design 30: 1209-1215.DOI:10.1016/j.matdes.2008.06.006

David Bain. (1982). The productivity prescription. R.R Donnelley & sons Company. Pp 3-14

Dua, J.K., (1994). Job stressors and their effects on physical health, emotional health and job satisfaction in a University. J. Educ. Administ., 32: 59-78. DOI: 10.1108/09578239410051853

Ettner, S.L. and J.G. Grzywacz, (2001). Workers' perceptions of how jobs affect health: A social ecological perspective. J. Occupat. Health Psychol., 6: 101-131.

Gagge, A.P. and R.G. Nevins, (1976). Effect of energy conservation guideline, on comfort, acceptability and health. Final report of contract. http://www.osti.gov/energycitations/product.biblio.jsp?osti_id=5009419

Hitomi Tsutsumi, Shin-ichi Tanabe, Junkichi Harigaya, Yasuo Iguchi, Gen Nakamura. (2007). Effect of humidity on human comfort and productivity after step changes from warm and humid environment. Building and Environment 42, Pp : 4034-4042

Ibrahim Atmaca, Omer Kaynakli and Abdulvahap Yigit, (2007). Effects of radiant temperature on thermal comfort. Build. Environ., 42: 3210-3220. DOI:10.1016/j.buildenv.2006.08.009

Ibrahim Atmaca, Omer Kaynakli and Abdulvahap Yigit, (2007). Effects of radiant temperature on thermal comfort. Build. Environ., 42: 3210-3220. DOI: 10.1016/j.buildenv.2006.08.009

Ismail M. Budaiwi. (2007) An approach to investigate and remedy thermal-comfort problems in buildings. Building and Environment 42: 2124-2131. DOI:10.1016/j.buildenv.2006.03.010

John G.Belcher. (1987). Productivity plus. Gulf Publishing Company, Houston, Texas. Pp 1-10.

Minnesota Blue Flame Gas Association. Humidity and the Indoor Environment. http://www.engineeringtoolbox.com/relative-humidity-d_895.html. (19 January 2010).

N.A. Bobko, V.I. Chernyuk, Ye.Ye. Yavorskiy. (2006). Effects of time-of-day, work strain, noise and air temperature on human-operator performance under time pressure.Posters session 2 / International Journal of Psychophysiology 69: 242-275 .DOI:10.1016/j.ijpsycho.2008.05.142

Olesen, B.W., (1995). International standards and the ergonomics of the thermal environment. J. Applied Ergonom., 26: 293-302. DOI: 10.1016/0003-6870(95)00033-9

Raimo Niemela, Mika Hannula, Sari Rautio, Kari Reijula, Jorma Railio. (2002). The effect of air temperature on labour productivity in call centres-a case study. Energy and Buildings 34: 759-764. PII: S0 3 7 8 - 7 7 88 (0 2) 0 0 0 9 4 - 4

S. Lolas, O.A. Olatunbosun. (2008). Prediction of vehicle reliability performance using artificial neural networks. Expert Systems with Applications 34: 2360-2369. DOI:10.1016/j.eswa.2007.03.014

Shikdar, A.A. and N.M. Sawaqed, (2003). Worker productivity and occupational health and safety issues in selected industries. Comput. Ind. Eng., 45:

Summary of passenger vehicles, commercial vehicles and 4x4 vehicles assembled in Malaysia for the year 1980 to YTD march 2009. Malaysian Automotive Association (MAA). http://www.maa.org.my /info_summary.htmb3

Tarcan, E., E.S. Varol and M. Ates, (2004). A qualitative study of facilities and their environmental performance management of environmental quality. Int. J., 15: 154-173. DOI: 10.1108/14777830410523099

The American Society of Heating, Refrigerating, and Air Conditioning Engineers, Inc ASHRAE Standard 62-2001.Ventilation forAcceptable Indoor Air Quality.

The Engineering ToolBox. Recommended relative humidity level. http://www.engineeringtoolbox.com/relative-humidity-d_895.html. (19 January 2010)

Vernon M. Buehler, Y. Krishna Shetty. (1981). Productivity improvement. American management Associations. Pp 3-13

Wu Hao, Zhang Hongtao, Guo Qianjian, Wang Xiushan, Yang Jianguo ,(2008). Thermal error optimization modeling and real-time compensation on a CNC turning center. Journal of materials processing technology 207:172–179. DOI:10.1016/j.jmatprotec.2007.12.067

Chapter 44

Design for Manufacturing (DFM) – A Case Study of Repetitive Measures

Gunther Paul[1], Frank Wagner[2]

[1]Mawson Institute
University of South Australia, Mawson Lakes, SA 5095, Australia

[2]Fraunhofer IAO
70569 Stuttgart, Germany

ABSTRACT

Design for Manufacturing (DFM) is a highly integral methodology in product development, starting from the concept development phase, with the aim of improving manufacturing productivity and maintaining product quality. While Design for Assembly (DFA) is focusing on elimination or combination of parts with other components (Boothroyd, Dewhurst and Knight, 2002), which in most cases relates to performing a function and manufacture operation in a simpler way, DFM is following a more holistic approach. During DFM, the considerable background work required for the conceptual phase is compensated for by a shortening of later development phases. Current DFM projects normally apply an iterative step-by-step approach and eventually transfer to the developer team. Although DFM has been a well established methodology for about 30 years, a Fraunhofer IAO study from 2009 found that DFM was still one of the key challenges of the German Manufacturing Industry. A new, knowledge based approach to DFM, eliminating steps of DFM, was introduced in Paul and Al-Dirini (2009). The concept focuses on a concurrent engineering process between the manufacturing engineering and product development systems, while current product realization cycles depend on a rigorous back-and-forth examine-and-correct approach so as to ensure compatibility of any proposed design to the DFM rules and guidelines adopted by the company.

432

The key to achieving reductions is to incorporate DFM considerations into the early stages of the design process. A case study for DFM application in an automotive powertrain engineering environment is presented. It is argued that a DFM database needs to be interfaced to the CAD/CAM software, which will restrict designers to the DFM criteria. Consequently, a notable reduction of development cycles can be achieved. The case study is following the hypothesis that current DFM methods do not improve product design in a manner claimed by the DFM method. The critical case was to identify DFA/DFM recommendations or program actions with repeated appearance in different sources. Repetitive DFM measures are identified, analyzed and it is shown how a modified DFM process can mitigate a non-fully integrated DFM approach.

Keywords: Design for Manufacturing (DFM), Design for Assembly (DFA), Manufacturing Engineering

INTRODUCTION

An internet search for the term "Design for Manufactur*", in order to find publications and websites relating to the terms "Design for Manufacture", "Design for Manufacturing" and "Design for Manufacturability" which are used as synonyms, finds many of the 96 hits related to either Boothroyd Dewhurst's DFMA® software and services or other software and service providers. The review of the identified web pages as well as a literature database search resulted mainly in over 400 publications from the International Forums on DFMA in the years 1985-2009. Boothroyd & Dewhurst (2009) is a selection of the 20 best and most successful case studies presented during the International Forums on DFMA, and was evaluated instead of the complete 407 papers which were presented during the conferences. Other sources which give an overview of DFM are listed as references.

Design for Manufacturing (DFM) is the process of proactively designing products to: (a) optimize all the manufacturing functions: fabrication, assembly, test, procurement, shipping, service and repair; (b) assure the best cost, quality, reliability, regulatory compliance, safety, time-to-market, and customer satisfaction; and (c) ensure that lack of manufacturability doesn't compromise functionality, styling, new product introductions, product delivery, improvement programs, strategic initiatives and unexpected surges in product demand (Anderson, 2008).

Current product realization cycles depend on rigorous back-and-forth examine-and-correct approach so as to ensure compatibility of any proposed design to the DFM rules and guidelines adopted by the company. Compared to the more basic Design for Assembly (DFA) requirements (e.g. assembly in only one direction, reduction of manual assembly operations), DFM rules additionally consider aspects of standardization, modularity, process capability, tool wear, tool change reduction, product geometry for handling and packaging, part identification, Poka-Yoke and other factors influencing production time, complexity and cost (e.g. definition of minimum drill diameter; minimum radii; alloys to avoid chipping; unambiguous, self orienting assembly).

The DFM process itself is a step-by-step approach, starting with benchmarking, target setting, functional definition, evaluation criteria conceptual design and DFM evaluation. DFM projects are typically based on cross-functional team involvement, drawing upon expertise from manufacturing engineers, quality engineers, cost accountants, production personnel, product developers and program management. Used correctly, DFM can lead to a 25-30% reduction on the production cost without capital investment in new facilities.

Although this process ensures quality and ease of manufacturing in later stages of the production cycle, it is time consuming. However, if this rigorous back-and-forth examine-and-correct approach can be reduced to a minimum or even eliminated, the product development process and life cycle would become smoother. The key to achieving such reduction is to incorporate DFM consideration into the early stages of the design process.

The "R&D fit for future" study conducted at Fraunhofer IAO (Germany) in 2009 found that DFM is still a key challenge for the German manufacturing industry. According to the study, the most important trends in R&D over the next 3 years are

- Collaboration with other departments of the company, especially Manufacturing
- Digital support for all R&D processes and
- Development of customer specific products.

The case study is about application of DFM in an automotive powertrain engineering environment, based on a DFM project at a German automotive OEM started in 2007. The case study is following the hypothesis that current DFM methods do improve product design in a manner claimed by the DFM method. In order to gather data for the study, information oriented sampling from all available data was applied. These included the benchmarking database, program related DFA project reports, best of practice project reports, Harbour Reports®, part design workshop reports, production standard reviews, assembly prototype workshop minutes, program activity tracking and FMEA project reports. Overall 29 sources of data were analyzed. Furthermore, all key personnel involved in the product design, manufacturing engineering, production, customer service and quality control processes was invited to attend module specific DFM workshops. 9 topic specific workshops were conducted for production of the cylinder head, crankshaft, crankcase, piston rod and camshafts, as well as module specific assemblies and assembly of the electrical system, which together involved over 50 experienced and highly skilled cross-functional experts (see Ulrich and Eppinger, 2008) from 2 assembly plants. The workshops aimed at identifying design deficiencies and potentials relevant to the production and assembly processes, with a recognized capacity for cost reduction and quality improvement. The critical case for this study to disconfirm the hypothesis is to identify DFA/DFM recommendations or program actions with repeated appearance in different sources over time. All DFM workshop participants involved were OEM employees except the author.

THE DFM FRAMEWORK IS WELL ESTABLISHED

Design for Manufacturing and Assembly (DFMA®) efforts started with two separate thrusts: producibility engineering and design for assembly (DFA). Producibility engineering was an alternative to the older manufacturing reviews, which were typically held after the design was completed. Producibility guidelines focused on the producibility of individual parts rather than the total product with all parts assembled. Therefore, the major incentive was to produce simpler parts that were individually easier to manufacture. Producibility engineering guaranteed easily manufactured parts; as a result, the part count on the product often increased and caused a more complicated product structure. In many cases, the total product was more difficult and costly to manufacture and assemble (Rehg and Kraebber, 2005). On the other hand, the main basis for DFA was the reduction of the number of parts by eliminating or combining them to achieve a simplified product. It should be noted that DFA is covering not only manual assembly, but also automated assembly, high-speed automatic assembly and robot assembly (Boothroyd and Dewhurst, 1991). However, this approach on its own did not optimize the product development because using the best manufacturing process was not a high priority yet. In 1977, DFMA® started as an extension of the DFA process. The two methods, Design for Manufacturing (DFM) and DFA must interact because the key to successful DFM is product simplification through DFA. DFMA® is a holistic approach to design analysis because both the manufacture and the assembly of the finished product are considered simultaneously. This consideration of manufacturing and assembly during design results in lower total product costs (Rehg and Kraebber, 2005).

Current product development cycles involve three phases; (1) the concept phase, (2) the Digital Mockup (DMU) phase, and (3) the prototype phase. In the concept phase, the idea is generated, screened and tested (Molloy, Tilley and Warman, 1998). As a result a clearer concept for the design is obtained. At the end of the concept phase, a concept book is usually produced. Once the concept book is delivered, design engineers commence a digital design phase, usually referred to as the DMU. In this phase, designers would propose a solution that satisfies the criteria outlined in the concept book, which is then presented to the manufacturing team to assess its compatibility with the DFMA® rules and guidelines adopted in the company. Consequently, the manufacturing team would return to the design team with their suggested amendments and concerns. The design team will again try to accommodate for the amendments proposed by the manufacturing team; after which the design is again put forward for inspection by the manufacturing team. Once the design is approved by all parties, the physical prototyping phase commences. In this phase, the digital design is to be realized in a number of stages. First, prototypes for separate parts of the whole assembly are produced. Secondly, sub-assemblies undergo physical prototyping. Finally, a prototype for the whole assembly is produced. Figure 1 shows the various phases of a product development cycle, starting from the concept phase to the beginning of the production cycle.

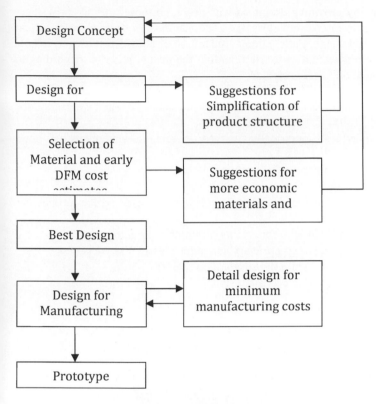

Figure 1: Typical steps taken in DFMA® (adapted from Anderson, 1990).

Once the whole assembly is prototyped, and the prototype justifies production, the production phase of the product life cycle takes over from the development phase. It can be noted that the above mentioned process is time consuming. As a matter of fact, it is even contradicting the DFM concept, which states (Anderson, 1990) that Design for Manufacturability is the practice of designing products so that they

- Be designed in the least time
- With the least development cost
- Providing the quickest and smoothest transition into production
- To be assembled and tested with the minimum cost and time
- Have a desired level of quality and reliability
- To satisfy customers requirements
- And compete well in the marketplace.

Although collaboration between the various teams involved in the development cycle ensures quality of the product, it also introduces various unnecessary loops to the cycle. A sensible solution to such problem is to enforce DFMA® rules and criteria on designs by providing designers with access to these rules at early stages of the development cycle. Nevertheless, the issue that has not been addressed so far is the generic method that will allow the consideration of DFMA® rules and criteria in early design stages without seriously retarding the design process. According to Allen (1990), typical DFM tools include

- Design axioms (use of fundamental principles of good design; see Lindbeck, 1995)
- DFM guidelines (empirically derived systematic statements of good design)
- DFA method (Boothroyd & Dewhurst, 1991)
- Taguchi method
- Manufacturing process design rules (design for a particular production process)
- Designer's toolkit (design for casting, design for stamping etc.)
- Computer-aided DFM (proprietary computer-aided DFM packages)
- Group technology (exploiting similarity of parts)
- FMEA (failure mode effects analysis)
- Value analysis

While the two first mentioned methods historically provide most advantages, manufacturing facility specific DFM is considered quickly advancing and gaining relevance and wide recognition for its great benefit.

It has been estimated that about 70 % of the life cycle cost of a product is determined by basic decisions made during product design (Nevins & Whitney, 1989). The product costs are fixed because important decisions about material, processes, and assembly requirements are settled early in the design. Any effort to reduce cost after the design stage can influence only the remaining 30% of product cost. In addition to cost issues, 80% of quality problems are often a result of poor design (Rehg & Kraebber, 2005).

TOWARDS A NEW DFM PARADIGM

While often a company's existing manufacturing processes are not capable of meeting the requirements of new designs, forcing manufacturing to develop new manufacturing processes concurrently with the design process (Priest & Sanchez, 2001), it could also be argued that the best use should be made of the existing manufacturing assets. Hence the business environment factors to be considered by the design team will include manufacturing strategy, schedule and resource constraints, production volume, product mix, outsourcing strategy and more. As DFM has been widely adopted across the industry, it is continually being developed to make guidelines more accessible for designers. The current design for X (DFX) approach is a knowledge based method, which attempts to incorporate all desirable characteristics, including quality, reliability, serviceability, safety, user and environmental friendliness, time-to-market, while minimizing lifetime and manufacturing cost (Bralla, 1999). Adding evaluations to each guideline that allows the designer to determine the cost gain achieved through its implementation (Ulrich and Eppinger, 2008), a taxonomy of possibly conflicting guidelines can be established. Boothroyd, Dewhurst and Knight (2002) have identified a persistent problem in DFMA® implementation, stating that "if the initial DFMA analysis is left until all or most of this data has been entered into the CAD system, it may be too late as there will then be a considerable reluctance to implement any substantial design changes in the product". To solve this issue, they suggest a link between CAD systems and DFMA® analysis, where the CAD application would be driven from the product structure in the DFA application. Vice versa, the DFA application would draw some of the required information for DFA analysis from the CAD system part data.

From the sources mentioned before, 1444 DFM requirements were clearly defined in a database and systematically attributed to parts in this study. An 11 digit numbering system based on the product data management structure was developed **(Error! Reference source not found.)**. The numbering system was required for mainly three reasons:

- Different programs use different part numbers
- Variants of the same part use different part numbers
- Numerous programs and engineers use non-uniform names for the same part.

Table 1: Part numbering system derived for DFM purposes

Main Module	Module	Sub-Module	Part Number
XX	YY	ZZ	abcde
(e.g. engine)	(e.g. air intake)	(e.g. air duct)	(e.g. air intake manifold)

While most DFM requirements applied to one specific part only, a series of DFM requirements needed to relate to a variety of parts. These were specified as "1:n" requirements. Additionally, the DFM requirements were classified into 13 categories and sub-categories, which were derived from literature and the underlying reports (Figure 2).

The approach adopted here requires relating particular DFM criteria to the appropriate entities, preferably at the earliest stage of the conceptual design. Furthermore, the modular labeling approach implemented allows for each semantic part to be assigned a unique part number across all programs. This database is interfaced to the CAD/CAM software, which requires an ontology and inference engine that restricts designers to the DFM criteria. Consequently, a notable reduction of examine-and-correct cycles can be achieved.

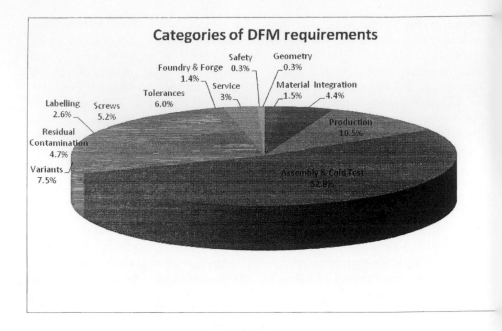

Figure 2: DFM requirement categories and their distribution

The sub-categories "Production", "Assembly & Cold Test" and "Foundry & Forge" would also form categories for allocating all other sub-categories. In the given structure, they comprise all items that could not be assigned to either of the other sub-categories. "Production" is encompassing the manufacturing entities where mainly cylinder heads, crankshafts, crankcases, piston rods and camshafts are machined. Based on the part relationship, all DFM requirements were then structured and reviewed for consistency (e.g. duplicates). The DFM requirements which were completely checked, categorized, numbered and illustrated went through a multi-stage approval process and were finally released into a thin client based DFM database. It was found that 675 or 47% of all listed requirements in the initial spreadsheet listing were either redundant or partially incorrect, indicating a contradiction with other DFM requirements. A small portion of DFM requirements in the category "Foundry & Forge" also conflicted with DFM requirements from the foundry& forge department. With all sources related to sequential studies, redundant DFM requirements document repetitive measures or improperly implemented DFM rules. One repetitive measure was the attempt to have elastomer washers supplied in an assembly, that are colored, include a securing device and presence indicator. This confirms the critical case and counters the hypothesis that previously applied DFM methods had improved product design. While DFM methods have proven to be reliable and effective in analyzing manufacturing processes and providing answers for process and product improvement, they have apparently failed to deliver a sustained improvement of product design.

CONCLUSIONS

As a result, the largest number of "TOP10 DFM requirements" in the case study - requirements that were deemed highest priority by senior management, following an ABC-Analysis - related to previously treated problems which repeatedly occurred (Figure 3).

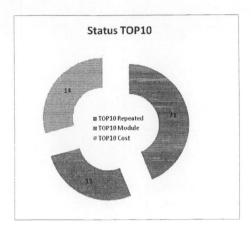

Figure 3: Origin of DFM requirements from TOP10 analysis

We therefore propose a new approach to integrate DFM with product design through a transparent interface. The DFM database is linked to the product data management system (PDM) and DFM requirements are integrated into the metadata of parts and assemblies as methods. The DFM requirements are then presented within the CAD/CAM/CAE system as a DFM label which is hyperlinked to a dataset within the DFM database. Drawing on the DFM methods of CAD objects, an instance which generates part/assembly templates automatically derives and calculates initial part features or constraints which provide a draft design to be further developed by the product designer.

It is obvious that DFM is lacking in efficiency, which may be partially attributed to extreme time and resource constraints common in product development environments. The situation is further aggravated as a generation of experienced product developers has been replaced by new staff in the last decade, in an attempt to optimize the use of CAD/CAM/CAE systems. It may also be assumed that many product developers are reluctant to use additional development systems (e.g. stand-alone DFM software) due to the complexity of their existing systems environment. This is consistent with findings of Boothroyd et al. (2002). The new process (

Figure 4) ensures DFM requirements to be reflected unanimously in product design across programs, and it supersedes any previously required control loops in the development process during the DFM stage (see **Error! Reference source not found.**).

440

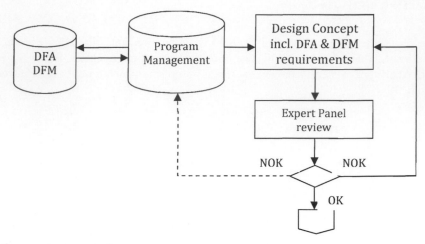

Figure 4: Revised DFM process layout

Effective application of DFM principles using this new process can provide a significant improvement over the engineering process (Anderson, 1990; Allen, 1990) in the areas of cost efficiency, quality, marketing, standardization, human resources, human factors, knowledge management, process management and innovation.

REFERENCES

Allen, C.W. (Ed.) (1990). *Simultaneous engineering – integrating manufacturing and design.* Society of Manufacturing Engineers, Dearborn.

Anderson, D.M. (1990). *Design for manufacturability: optimizing cost, quality, and time to market.* CIM Press, Cambria.

Boothroyd, G., and Dewhurst, P. (1991). *Product design for assembly.* Boothroyd Dewhurst.

Boothroyd, G., and Dewhurst, P. (2006). *How to Get Started on DFMA and Concurrent Engineering.* Boothroyd Dewhurst, Wakefield.

Boothroyd, G., Dewhurst, P., and Knight, W.A. (2002), *Product design for manufacture and assembly.* M. Dekker, New York.

Bralla, J.G. (Ed.) (1999). *Design for manufacturability handbook.* McGraw-Hill, New York.

Lindbeck, J.R. (1995). *Product Design and Manufacture.* Prentice Hall, New Jersey.

Molloy, O., Tilley, S., and Warman, E.A. (1998). *Design for manufacturing and assembly: concepts, architectures and implementation.* Chapman & Hall, London.

Nevins, J.L., and Whitney, D.E. (Eds.) (1989). *Concurrent design of products and processes: a strategy for the next generation in manufacturing.* McGraw-Hill, New York.

Paul, G., and Al-Dirini, R. (2009). "Design for Manufacturing – a review and case study." *Designing Futures, 45th Annual Conference Human Factors & Ergonomics Society of Australia, 22nd-25th November 2009, University of Melbourne.*

Priest, J.W., and Sanchez, J.M. (2001). *Product development and design for manufacturing. A collaborative approach to producibility and reliability.* Marcel Dekker, Basel.

Rehg, J.A., and Kraebber, H.W. (2005). *Computer-integrated manufacturing*. Pearson Prentice Hall, New Jersey.

Ulrich, K.T., and Eppinger, S.D. (2008). *Product Design and Development*. McGraw-Hill, New York.

CHAPTER 45

User Demands for New Mixed Reality Tools, First Results of the Manuvar Project

Peter Vink[1,2], Merle Blok[1], Gu van Rhijn[1], Tim Bosch[1], Hans Totté[3], Konstantinos Loupos[4]

[1]TNO, P.O.Box 718, 2130AS Hoofddorp, The Netherlands

[2]Delft University of Technology, fac Industrial Design Engineering
The Netherlands

[3]IPA Total Productivity, Kosterij 4
1721 PN Broek op Langedijk, The Netherlands

[4] Institute of Communication and Computer Systems (ICCS), 9 Iroon
Polytexneiou Street, 15773, Athens, Greece

ABSTRACT

This paper concerns the enhancement of the Mixed Reality system Ergomix, which is used to design assembly work stations. The evaluation was performed by 12 virtual reality experts in 5 cases, which was used as input for a new system. The new system should contain the key advantages of its predecessor, to be transparent (for good communication), easy to use (inputs from sketches and CAD) and be able to make the workers experience the new environment. Based on the results, the new requirements are: showing different views at the same time, having digital connections to other systems, making it portable for on site sessions and adding libraries with guidelines for acceptable musculoskeletal loadings. These adaptations will be developed in the ManuVAR project.

Keywords: Manual Handling, Participative Design, Mixed Reality, Production Improvement, Work Station Design, Human Factors, Participatory Design.

INTRODUCTION

Market forces have driven manufacturing enterprises and systems to continuously eliminate wasteful, non-value-added operations, as well as integrate the voice of the customer and worker (Genaidy and Karwowsi, 2007; Rhijn et al., 2005). Many companies have embraced the lean manufacturing philosophy in the pursuit of reducing wasteful activities and improving productivity and profits (Genaidy and Karwowsi, 2007). Automation, along with methods like lean manufacturing can be used to reduce manual handlings. However, automation is most suitable for moderate or large batch sizes. The currently growing level of customization and shorter product lifecycle results in smaller batch sizes, shorter delivery times and more variation in products. In order to produce these small batch sizes of varying products, production systems must be both flexible and efficient. The increased demands force manufacturers of complex assembled products to improve the flow of assembly orders together with a more efficient use of workers. The pressure on the organization is likely to increase the mental and physical stresses on the individual workers. To survive in the coming years, it is important for the companies to optimize their workstations. To support this workstation designs, TNO has developed an approach aimed both at involvement of work staff, at lead time reduction and at the improvement of the human assembly tasks from an ergonomic point of view (Rhijn & Tuinzaad, 1999). Participation of company representatives is crucial, for reasons that have been extensively indicated and discussed in previous papers on participatory ergonomics (Noro and Imada, 1992; Vink et al., 2008). Low cost activating tools for workstations design could help small and medium sized assembly enterprises to both optimize the assembly process and involve their workers. However, these tools are not always available in the form needed by the companies (Hallbeck et al., 2010).

This paper describes the envisioning technique or mixed reality system "Ergomix" as part of a design process of assembly workstations, and its application in assembly design. Additionally, requirements for a digital version of the ergomix are discussed based on five cases and the opinion of virtual reality specialists (participants in the the EuroVR network, see www.eurovr.com). Ergomix, a mixed reality simulation is a special form of Virtual Reality (VR) used to create environments wherein real world and virtual work objects are presented together in one environment (Hallbeck et al., 2010). Ergomix employs a chromakey technology, better known as blue-screen technology. This is similar to the weathercaster superimposed over a map and also allowing onscreen editing similar to football replay broadcasts and can be edited iteratively without pre-programmed configurations, all in real time. Ergomix simulation allows the focus to be on individual workplaces and workstations: the available space, the location of tools, equipment and components and the actual design of the workplace. In this 2-D technology, the workplace can be a photograph, hand drawing, computerized drawing, etc., while the user can actually use the tools or raw materials required in the manufacturing line during the simulation. A mix van be made from a side view, front view of view from above.

In the Ergomix, a real worker is the actor in his own "virtual" workstation and is asked to perform his usual assembly activities (Rhijn, et al., 2005).Actions (the virtual environment) are made directly visible to themselves and others (see Figure 1). The working height and, for instance, reach in the drawings can be immediately adjusted so that the right work posture is obtained. The drawings can be moved onscreen in real time and the whole scene can be scaled to estimate the representation of various sizes of workers. Together, the working group of participants, including employees, engineers, and management, human factor specialists as well as outside constructors of the workplace equipment or furniture decisions are made with respect to 'sitting, standing or semi-standing', working heights, orientation of machines and so on (Rhijn, et al., 2000) before it is built. Because the 'mix' is a video product it can be recorded as a movie and shown to others in the company

To define requirements for a new version of this mixed reality system five case studies were analyzed and a group of VR experts was interviewed after participating in an ergomix session.

CASE DESCRIPTIONS

To gather requirements for the new mixed reality systems five cases were chosen. Two cases concern the design of an interior of a vehicle, one case concerns a design of a part of an office and two cases are from the assembly work station design field.

FIGURE 1. The mix of a driver in a drawing in the wheel loader case/ Left the person in the ergomix. In the middle the mixed view. Right the wheel loader.

In case 1 the design of a wheel loader interior for Paus GmbH (SME wheel loader manufacturer) was optimized. The effects of the steering wheel position, pedal position and seat position on view and operator position are shown in the mixed reality and supported the manufacturer to choose the right design (see Figure 1). The lateral view was important in this case. A technical drawing was mixed with a real driver. Steering wheel position, pedal position and seat position were varied to find the optimum.

It appeared that the roof was sometimes a limiting factor and more space was created in this area, by lowering the seat. The manufacturer made the changes in prototypes and tested these again with real drivers in real work situations. After these tests with a real wheel-loader drivers were more satisfied and as a consequence the manufacturer as well.

In the 2nd case, a tram cabin interior was designed. In this case five tram cabin interior designs of different manufacturers were evaluated to optimize the fit between the driver and its environment. An experienced driver was mixed in all five tram interiors and in the mix "adjusted" into a 2.04 m tall person and a 1.64 m small driver, which was the range of the persons applying for the job. It was checked whether the person could do his work and see for instance obstacles of 1 m high 1 m close to the tram. This test influenced the choice of the tram company for the manufacturer and lateron supported the manufacturer in adjusting the design of the tram cabin interior. It appeared that a height adjustable pedal was needed to make shorter woman able to see satisfactorily the floor in front of the tram (i.e. increase their view space).

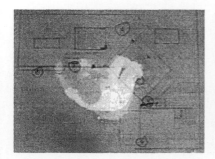 

FIGURE 2. Left: the mix of the person working at the drawing of the counter desk seen from above. Right: the counter desk in its final design.

Case 3 is an application of the ergomix in office interior design. The design of a new building of an area where 4 counter desks were positioned was evaluated and adjusted (see Figure 2). The counter desks were made by small companies and the municipality wanted to have a client oriented environment where employees could do their work pleasantly. The effects of various designs on safety and privacy of the counter desk worker as well as effects on productivity (speed of manual handling) and comfort (interviews) of the workers were evaluated. Based on the effects of the different designs adaptations were made. The manufacturer of the desk, the purchasing department of the municipality and the counter desk workers came to an optimal design, which was then implemented in the final version according to the results of this work. The applicability of the system in the various phases of the design is quite interesting since it was shown that that sketches of the designers could be used in an early design phase (see Figure 2) and CAD drawings in a later stage.. The top view was in this case important (see Figure 2 left side) as the counter desk worker could check reach ability of the various devices needed to help clients. The devices often used could be placed closer to the worker. All participants were satisfied, the only problem the manufacturing company mentioned is that the data (movements and sizes of humans) could not be stored digitally for use in other future designs.

In the 4th case, a coffee machine assembly work station was optimized for a manufacturer of professional-use coffee machines. The Ergomix demonstrated the value of having a maximum height for racks, for adjustability in assembly line height (to allow both standing and seated work and to handle machines with different dimensions and activities at different locations inside the machine). In addition, the workers have the pre-assembly directly connected to the final assembly, reported as a pleasant experience, with minimal handling, no storage in between with materials directly at hand and direct communication between pre assembly and final assembly. Use of the Ergomix raised the consciousness and awareness of the employees and process engineering on ergonomics. Since the equipment supplier was also partner in the Ergomix session, improvements could be easily and simultaneously adapted in the AutoCad drawings and the lines of communication among all the stakeholders were shortened.

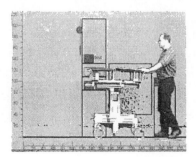

FIGURE 3. Left: the mix of the person showing the redesigned height-adjustable production cart for roof systems design and its realization in case 5 (right).

Case 5 concerned the assembly of roofs for cars. The project was initiated because of the need to set up an assembly cell for a new sliding roof. This company chose to base the assembly concept on existing concepts in the factory. During the assembly process, the product is situated on a moveable carrier while the personnel move along as the parts and sub-assemblies are supplied. The product is assembled while walking from one line zone to the next. In an Ergomix session, the position and dimensions of the racks and trolleys were determined. In the same Ergomix session, the optimal position was defined by trial and error. Thus the position of tools and assembly equipment and the walking and working spaces could be established, as shown in Figure 3. This resulted in improvements in the height at which the cover plate, mechanisms, sunshades and glass panel are supplied, as well as the distance needed to reach them.

EXPERT OPINION

Four VR experts participated in the ergomix session followed by the discussion on the technological improvements needed (Marcel Delahaye - COAT, Konstantinos Loupos- ICCS, Jeong Min Kim - Mitsubishi-Caterpillar, Ergon van der Broek - TUTwente). They experienced the system as they were mixed themselves in the work situations of the ergomix. Additional information was gathered by 8 other INTUITION partners (the EU-NoE INTUITION project was later continued in the EuroVR association) who received a video explaining the ergomix.

The participants in the session were questioned regarding the demands of small and medium sized enterprises on:
1. the explanation why this system is frequently used
2. the limitations and advantages of this system, and
3. improvement possibilities taking into account the user needs.

All experts agreed that there is now a gap between the technological more sophisticated VR systems and the needs of small and medium sized enterprises (SME). In several labs in Europe there are very sophisticated VR systems available that could support innovation in Europe. However, these are almost not used by SMEs. The ergomix is a very simple system that is frequently used, but on the other hand lacks precision and is not digitally connected. The challenge for Europe is to overcome this gap and find possibilities to grow to each other.

MAIN FINDINGS

1. Why used by SMEs? The frequent use of the ergomix system by SMEs was explained by the experts by the fact that it is real time, needs no preparation and input can be used from sketches, CAD and technical drawings. SMEs, who do not have their core business in designing for human operations, will not have their own VR system. It is therefore not their core business and their investment attitude is low. Often these companies do not have the time to wait for studies on the effects of design changes. That is why they want a fast design and quick results. Additionally, the ergomix does convince participants on the purpose of change and supports communication between various groups within the company (management, marketing, engineers, operators and designers). Another advantage of the ergomix, which fits with the SME demands, is that the system is transparent and is open to

use with real products, other software and hardware. The economic benefits (low price) of the system also increase its adaptability to different operational needs but also make it appropriate to a wide spectrum of budget figures.

2. Limitations: In Table 1 the main limitations are mentioned. The switch between views is mentioned as a problem. Now only one view can be seen, while for some actions it is important to see a top view (see Figure 2 left) simultaneously with a side view (see Figure 3 left). Other mentioned limitations are the 2D view (and not 3D) and the fact that the company should come to the TNO facility where the ergomix is running. A mobile portable system could easily solve this problem.

Table 1. Limitations and possibilities for improvement of the ergomix mentioned by the VR experts. (..) is the number of experts that mention the limitation.

Limitations	Improvement possibilities
Switch between views (top-side) needed(6)	3 Views/perspectives (at the same time
Not yet portable / not digital (5)	Make it digital and portable
2D view – not enough (2x 2D would be better) (4)	Try 2x 2D modeling
Cognitive point of view: system is purely physical not cognitive; at least color and contrast would be needed (3)	Add information about what the subject sees / eye tracking combined with 3D-modelling/simulation software
Not standard with rich visualization such as 3D modeling software (2)	Stereoscopic view: bring in depth on the screen for the viewer (w, or w/o glasses)
Not connect to ergonomic guidelines (2)	Add library of human postures with green, yellow and red postures
No library of objects available (2)	Make a library of objects
Long term effects cannot be tested (2)	Evaluated in a longitudinal study
No haptic feedback (2)	Add haptic feedback
	Add real objects / mock-up parts
	Incorporate EMG, EEG
	Add links to human modelling

3. Improvement possibilities: It is clear that the improvement possibilities are directly linked to the limitations. Three views visible at the same time (top view, side view and front view) is a solution for the limitation and a portable digital system can solve the location problem and the connection to other systems as well as the digital memory. The connection to ergonomic guidelines is also an improvement, which is valuable as participants see during the session whether for instance a bended neck is acceptable for the duration of the task or not.

DISCUSSION

The five cases demonstrate that the Ergomix simulation allows relatively fast and easy ergonomic intervention early in the design process. Other cases lead to the same conclusion (e.g. Vink and Rhijn 2006; Hallbeck et al. 2010). The workers and other participants experience their own workplace in practice, even when their workplace does not yet exist which increases acceptance of the workers and other stakeholders involved in the process. This experience of the work station before it is built is an important element in the adoption (Vink et al., 2008). Other achievements for the five case studies demonstrate that the designs could be optimized increasing productivity and reduction in musculoskeletal risk factors. After the redesign all the workplaces had the same or better performance sometimes due to better view and an improved working posture, sometimes due to the fact that devices and materials were within comfortable reaching zones. It is in this process important that the worker experiences the effect of the changes real time and that the environment can be adapted quickly with different ways of input: sketches, technical drawings and CAD files. These requirements should be kept in the new version of the system. In fact the VR experts come to the same conclusion: the system should be transparent (for good communication), easy to use (various inputs) and be able to let workers experience the new environment.

Additional requirements are shown in the right column of Table 1. The requirements (1) showing different views at the same time, (2) make it digital and portable and (3) adding a library of acceptable working postures in static and dynamic work will be taken very seriously as these adaptations will be made. To capture and register movements of body parts like arms and upperbody objectively the application of a motion capturing system could be explored. Apart from these improvements a connection with a digital human model will also be explored by connecting it to Jack (a digital human model). In this project assembly workstations will first be simulated and designed using improved Ergomix system. Physical load and productivity will be measured during the design phase. After that a real workstation (mock up) will be build using the Ergomix data and results. Again the physical load and productivity will be measured in the real world This research and development will be carried out in the ManuVAR project. ManuVAR is a European project that runs from 2009 through 2012 and comprises 18 partner organizations across 8 European countries. In this project existing virtual and augmented reality technology will be applied to improve systems and communications between people. A number of the main challenges (gaps) found in ManuVAR (Krassi et al., 2010) to apply VR in manual handling cases will certainly be solved in applying the new version of the Ergomix:

Problems with communication throughout life cycle

Poor user interfaces

Lack of technology accceptance

Inefficient knowledge management

Physical and cognitive stresses

Inflexible design process
Low productivity

CONCLUSIONS

The Ergomix is a simple mixed reality system filling the gap between available sophisticated VR systems and the VR needs of SMEs. Ergomix allows quick 2-D evaluations with easy on-site, real-time iterations. These 2x2D evaluations meet the needs of SME's for ergonomical workstation design. Ergomix costs are a fraction as much as the more traditional human modeling CAD system and similar evaluations take much less time than the traditional modeling, with no lengthy programming required (Hallbeck et al., 2010). The advantage of the Ergomix is the ability to change the parameters in real time while the end-users and other participants (including line supervisors and managers) are present. This allows for a more participatory approach in designing work stations.

In the future this system should be digital to connect it to other systems and portable to make on site sessions in the company possible. Connections to human motion capturing systems, human digital models and ergonomic guidelines could make the system more valuable to be able to show the effects on the whole body and to prevent musculoskeletal disorders. These connections is also useful to close the gap between low end and high end users without losing the flexibility and user acceptance of the already existing level of use, which opens supporting a wider variety of companies. By applying the system effects are expected on the seven gaps found in the ManuVAR project.

ACKNOWLEGDEMENT

The research leading to these results has received funding from the European Community's Seventh Framework Programme FP7/2007-2013 under grant agreement no. 211548 "ManuVAR".

REFERENCES

Genaidy, A. Karwowski, W. (2007). The Manufacturing Enterprise Diagnostic Instrument: A Tool for Assessment of Enterprise Systems Manufacturers. Human Factors and Ergonomics in Manufacturing, 17 (6), 521-574.

Hallbeck, S., Bosch, T, Van Rhijn, J.W., Krause, F., Vink, P. Looze M.P. de (2010). A tool for early workstation design for small and medium enterprises evaluated in five cases. Human Factors and Ergonomics in Manufacturing, in press

Krassi, B, DÇruz, M, Vink, P (2010). ManuVAR: a framework for improving manual work through virtual and augmented reality. This book

Noro, K. Imada, A. (1992). Participatory Ergonomics. London: Taylor and Francis.

Rhijn, J.W. van, Looze, M.P. de, Tuinzaad, G.H. Groenesteijn, L. Groot, M.D. de, Vink. P. (2005). Changing from batch to flow assembly in the production of emergency lighting devices. International Journal for Production Research 43, 3687-3701.

Rhijn, J.W. van, Tuinzaad, G.H. (1999). Design of efficient assembly flow and human centred workplaces in Dutch assembly companies. In: International Conference on TQM and Human Factors. Sweden, June 15-17.

Rhijn J.W. van, Looze M.P. de., Tuinzaad B. (2000). Design of efficient assembly flow and hu¬man centred workplaces in Dutch assembly companies. In: Zulch G, Rinn A, eds. Proceedings of the 5th Int. Workshop on simulation games in production manage¬ment. Karlsruhe, 163-72.

Vink, P. Imada, A.S. Zink, K.J. (2008). Defining stakeholder involvement in participatory design processes. Applied Ergonomics 39, 519-526.

Vink, P, Rhijn, J.W. van (2006). Mixed reality real time simulation to increase performance and involvement. In: Proceedings of the Workshop Virtual Reality: technology and applications. Braşov: Transilvania University of Braşov, Department of Product Design and Robotics, 59-64.

CHAPTER 46

An Analysis of the Impact of Trends in Automation on Roles in Radiotherapy Using Function Allocation

Enda F. Fallon[1], Liam Chadwick[1] and Wil J. van der Putten[2, 3]

[1] College of Engineering and Informatics and
[2] School of Physics
National University of Ireland Galway, Ireland

[3] Dept of Medical Physics and Bioengineering
Galway University Hospitals, Galway, Ireland

ABSTRACT

Currently, radiotherapy treatment is a complex process which utilizes advanced automated systems. However, until relatively recently the process was largely manual. Technologies under development, particularly in the areas of simulation modeling and treatment planning, offer the possibility of automating many of the functions that are currently carried out by radiotherapy staff. If adopted, they are likely to significantly change the roles of health care professionals. In this paper, an allocation of functions method, Levels of Automation (LOA) (Parasuraman et al., 2000) was used as the basis for an evaluation of the impact of adopting such new technologies on radiotherapists. The starting point for the analysis was an IDEFØ model of the radiotherapy treatment process previously developed by the authors. A number of functions which utilize technology extensively were selected for analysis. The LOA model was applied to the way in which each function was achieved in the past, the way they are currently achieved and the possibilities for achieving them in the future. The function allocation method proved to be useful in identifying trends in automation and also facilitated discussion among diverse disciplines with regard to the impacts of perceived future developments in the area. LOA models can provide a focal point for exploring the types and degrees of automation that are capable of being implemented in radiotherapy systems in the future.

Keywords: Allocation of Functions, Radiotherapy, Automation, Roles

INTRODUCTION

In recent years, radiotherapy has benefitted from several significant leaps in the equipment, technology and treatment types available for the delivery of care, e.g. the introduction of computer controlled accelerators and highly complicated software-based treatment planning models. New technology systems have been introduced that support critical treatment functionality across all aspects of the care process. Electronic Medical Records (EMR), Picture Archiving and Communication Systems (PACS), 3D treatment planning and simulation, and on-line and in-vivo dosimetry are now fundamental in supporting the delivery of new and advanced treatment types such as Intensity Modulated Radiation Therapy (IMRT) and Image Guided Radiation Therapy (IGRT) (Fallon et al., 2009a). These systems have also been introduced to support stakeholders in meeting the demands of increasing patient numbers, to facilitate more complex treatment plans and to reduce the potential risk of human errors which have been found to contribute to adverse events (Fallon et al., 2009b). The introduction of these technologies has impacted on the roles of professionals working in radiotherapy. For example in the department studied, radiotherapists currently sign-off on some of the radiotherapy treatment plans for breast cancer. Previously, this role was uniquely done by radiation oncologists. This change in role can be attributed to the increased capacity of the system due to the implementation of advanced technology and the need to effectively utilize this capacity. Future technological change is likely to have a greater impact on these roles, resulting in the re-allocation of functions between them. It is important that all system stakeholders are aware of these implications.

The main objective of the research reported in this paper is to analyze the impact of trends in automation on roles in radiotherapy systems using an allocation of functions methodology. The secondary objectives of the research are as follows;

- To trace the evolution of the professional staff roles in past, present and future radiotherapy systems using an allocation of functions approach
- To assess the impact of achievable automation on future practice in radiotherapy systems
- To comment on the usefulness of the LOA allocation of functions approach as a tool for post hoc evaluation of designs.

ALLOCATION OF FUNCTIONS

Allocation of functions is the term used to describe the activity of determining the functions to be performed by different components within a human-machine system. It is regarded by many human factors professionals as a fundamental component of the systems design process which can influence subsequent design thinking. The first tool to support the allocation of functions activity was the Men Are Better At / Machines Are Better At (MABA/MABA) lists developed by Fitts (1951) in which humans and machines were compared with respect to a set of performance criteria. In subsequent years, a number of more comprehensive, process oriented methods were developed which built-on the Fitts Lists approach and extended the comparative criteria, e.g. Meister (1985), Price (1985), Parasuraman et al. (2000). Jordan (1963) recognized that humans and machines are not comparable but complementary and therefore concluded that approaches to allocation of functions based on Fitts Lists were inappropriate. Clegg et al. (1989) and Waterson et al. (2002) developed methods which were grounded in socio-technical systems theory.

Critics of the concept of allocation of functions have questioned its usefulness and validity (Dekker and Hollnagel, 2004; Dekker and Woods, 2002; Fuld, 2000). Fuld (2000), maintained that design proceeded best by building on acceptable solutions and weeding out misfits rather than by top-down allocation of functions. Sheridan (1997) questioned whether functional allocations can be rationally determined, whether the process of doing so is an art or whether human factors practitioners should abandon their principles entirely with respect to the concept. At best, he advocates studied trial and error where the optimum designs will emerge through an evolutionary process. Dekker et al. (2004; 2002) castigated the usefulness of traditional engineering approaches involving functional decomposition and specification labeling them, MABA/MABA or Abracadabra? At a conference on the topic in Galway, Ireland, a number of other key concerns related to allocation of functions were raised, including; the study of the nature of work and work systems, i.e. work is transformed by new artefacts and therefore the a priori specification of work based on existing systems is futile, and the issue of emerging and articulating work in new or updated systems (Fallon et al., 1997).

The vast majority of applications of allocation of functions methods are in the hypothesis phase of design, (Fuld, 2000). In contrast, the work reported in this paper takes an alternative perspective and uses the Levels of Automation (LOA) method by Sheridan (1997) and Parasuraman et al. (2000) to assess the impact of automation decisions (allocations) on the various professional staff roles within radiotherapy systems. Effectively the method is being used in the testing phase of design (Fuld, 2000). Specifically, three radiotherapy systems are modelled using the LOA approach; a totally manual system, the existing paperless system in a public hospital in Ireland and a future system with potentially higher degrees of automation.

MODELLING LEVELS OF AUTOMATION (LOA)

Sheridan (1997) and Parasuraman et al. (2000) present a model of human interaction with automation based on two dimensions; Stage of Human Information Processing and Level of Automation. The four broad classes of functions in the model are; 1. Information Acquisition, 2. Information Analysis, 3. Decision and Action Selection and 4. Action Implementation. The degree of automation is divided into ten levels, with one indicating no automation and ten indicating completely autonomous automation, see Table 1. The model is not intended as a tool that will provide detailed solutions with respect to allocation of functions. Parasuraman et al. (2000) state that, "The model can be used as a starting point for considering what types and levels of automation should be implemented in a particular system. The model also provides a framework within which important issues relevant to automation design may be profitably explored". It is for the latter purpose that the model was chosen for the work reported in this paper.

OVERVIEW OF RADIOTHERAPY

The radiotherapy treatment process has the following generic stages, as shown in Figure 1 below. The process begins when a patient is referred, typically by their General Practitioner, to an outpatient clinic for assessment by a consultant radiation oncologist. If the patient requires treatment, the oncologist's clinical notes and the prescribed treatment protocol will be entered in the patient file and the patient will be added to the department's treatment schedule. If required the patient will be sent for positioning and immobilization preparation. He/she will undergo pre-treatment imaging, which includes CT scanning, simulation and patient marking for later treatment. A treatment plan will be developed based on the CT scan in conjunction with the prescribed treatment protocol. This plan will be used to treat the patient unless abnormalities are detected during the patient and treatment review, in which case changes to the treatment plan will be made under the supervision of the radiation oncologist.

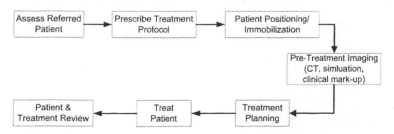

Figure 1. Generic Radiotherapy Treatment Flowchart

The research presented in this paper utilized an existing IDEFØ model of the radiotherapy treatment process, which has been developed by the authors (Fallon et al., 2009a). An example from the IDEFØ model is presented below in Figure 2. The diagram shown relates to the step 'Treat Patient'. The patient is positioned according to the specified set-up parameters (Patient Positioning/Immobilization), followed by checking of the patients CT scan and treatment plan images using image guidance, prior to dose administration. Differences in the images, resulting from physiological patient changes or shifts, can be compensated for through minor changes to the set-up parameters. Once treatment has been administered a chart review is completed and the treatment is recorded on the patients Quality Check List.

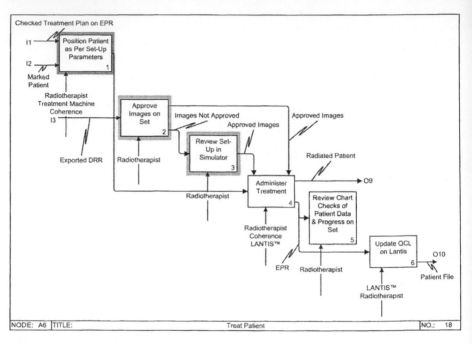

Figure 2. IDEFØ model for Radiotherapy Treatment Step 'Treat Patient'

MODELING RADIOTHERAPY USING LOA

In order to gain some insight into the trends in the development of automation in Radiotherapy, it was decided to model three phases of automation using LOA. The first phase began some years before 1990, prior to the introduction of significant computerization in radiotherapy. Most radiotherapy departments currently operate within the second phase which commenced around 2005. The level of automation associated with this phase is described in the introductory section of the paper. The third phase has yet to be realized, however potential levels of automation based on the authors' knowledge are presented. Due to space constraints, only the steps associated with pre-treatment activities such as simulation, treatment planning and image guidance during treatment are considered using the LOA model. Similarly, only the role of the radiotherapist is considered.

PRE-TREATMENT ACTIVITIES - SIMULATION (A)

Pre-treatment activities are those activities which prepare patient data for use in subsequent calculations and dose modeling. In radiotherapy this would now be called simulation. It involves, in terms of Parasuraman et al. (2000), information acquisition and analysis and as well as the planning of decisions and actions. Around 1990, automation would not have featured prominently here. The acquisition of patient specific information would have been done to a considerable degree through the use of mechanical devices such as, a pantograph. The patient (and the tumour target inside) would be considered a stationary object. Imaging of the patient to ensure correct positioning would have been performed using a few projection X-rays obtained using a conventional radiotherapy simulator. Only limited information on the size and location of the tumour would have been available using these methods. Data transfer from the contouring to the treatment planning system would have been done manually. Current pre-treatment activities are considerably more advanced and are now often called Virtual Simulation Systems in that they can simulate the real treatment process in a computer. Modern Computed Tomography and Magnetic Resonance Imaging devices produce 3D information of individual organs within the patient's body. The information obtained from these imaging devices can be used on its' own or in a merged fashion. In the latter case, the fusion of different imaging modalities is now a semi-automatic process with the computer suggesting best matches and the user (mostly medical practitioners) approving. The availability of this information allows for a manual, however accurate contouring of individual organ boundaries including the delineation of tumour as well as organs-at-risk. The latter are healthy organs which are not to exceed certain dose limits if significant morbidity is to be avoided. Data transfer between the scanning systems and the contouring systems is now fully automatic. Some automatic contouring of simple outlines (such as external skin) is now provided for and used routinely in the clinic. In a few years from now, automatic contouring will have advanced considerably and it can be expected that, based on atlases and libraries of patient anatomy, it will be possible to complete the contouring of normal anatomy automatically (Färber et al., 2005; Wu et al., 2009). In fact, the first commercial software package which does this has recently been introduced on the market, (Brainlab Oncology GmbH iPlan® RT). It can be expected that around 2015, motion (whether external, or internal organ motion within the patient) will also be incorporated into the pre-treatment, simulation process. This will be done in a way which is integrated with the actual radiotherapy treatment process. The respective degrees of automation described above are modelled in Figure 3 (A) using the LOA method.

TREATMENT PLANNING (B)

Treatment planning is the calculation of radiation doses inside the patient, based on knowledge of the patient anatomy, location of organs, placement of radiation beams around the patient and an accurate model of the radiation interactions of the photon (or electron) beams within the patient. Accurate treatment planning is essential in order to avoid incorrect dosage to the tumour and/or to normal tissue. In the 1990s treatment planning was often based on assumptions of a homogenous patient and the calculations were fairly simple. The treatment planner placed beams around the patient, applied weightings and then calculated the radiation dose at a point inside the patient. These calculations were often performed by hand and the use of computers was limited to automation of the manual calculations. This changed considerably in the later part of the 1990s with the introduction of novel and considerably more complex methods of delivering radiotherapy such as, e.g. Intensity Modulated Radiotherapy (IMRT) (Khan, 2003). The programs which calculate radiation dosage in IMRT automatically import patient data which was obtained during the simulation process. Also, instead of relying on the planner placing the beams and calculating the dose (often called forward planning) the systems allow a dose to be specified, including the dose to normal tissues, and then calculate optimum beam placement (called inverse planning). This is to a considerable degree an automated optimization process (using techniques such as simulated annealing) and the input of the planner is limited to that of a supervisor who ensures that the optimization is not stuck in a local minimum. The actual computer programs which calculate dose are still to a considerable degree analytical and although a considerable improvement over the earlier programs still have limitations, especially in areas with tissues of varying densities e.g. high density gradients. In the future, the analytical programs will be replaced by models which will trace individual photon or electron trajectories (so called "Monte-Carlo" trajectories). Increases in computer performance will make the use of these techniques routine. In addition, the use of inverse planning combined with so called "class solutions" (Khoo et al. 2003) will make treatment planning considerably more automated compared to today. The treatment planning process has already seen a marked shift from being performed by physicists (up to 1990) to being performed by radiotherapists and it may very well become a fully automatic part of the treatment process with skilled observers behaving more as system monitors than performing the actual planning. The respective degrees of automation described above are modelled in Figure 3 (B) using the LOA method.

IMAGE GUIDANCE DURING RADIOTHERAPY TREATMENT (C)

Image guidance during radiotherapy treatment is used as a method to verify the positioning of the patient during treatment and to ensure that the correct treatment volume is irradiated. In the early days of radiotherapy, it was only the latter which was used and it was for instance common to take what was known as a "port film"

(Khan, 2003). The primary purpose of this was not to affect the treatment process but it served mainly as a legal record to verify that the treatment had been performed as directed by the clinician. As treatments became more complex and especially with the capability of newer machines to shape the radiation field, the requirement to visualise the actual treatment became more acute. This led to the development of Electronic Portal Imaging Devices (EPIDs) which are now ubiquitous. Subsequent to EPIDs, other imaging devices such as ultrasound imaging systems are now used. In the near future, CT scanners entire and MRI systems will be tightly coupled to the actual treatment machines. These devices allow for easy and daily visualisation of the treatment volume. The software which controls them is also linked to the software and hardware which controls virtual simulation and this allows for daily verification of both patient intra-fraction movement and intra-fraction organ movement (inside the patient). Although technically possible, the link to allow for automatic patient registration has not yet been made. It can be expected that this link will be made in the near future and that an automatic patient registration link will be made between the software which performs the data acquisition during pre-treatment and the systems which perform image acquisition during treatment. In the 1990s, patient movements during treatment had to be authorised by radiation oncologists. Currently, radiation therapists carry out these movements which are suggested by the machine. In the future, it is probable that they will adopt a supervisory role monitoring the automatic machine execution of patient movements.

This small sample of three functions in radiotherapy demonstrates that evolving developments in automation have led to changes in allocation of functions among staff involved with the radiation therapy process. They have also caused a shift in responsibilities and changes in the skill sets required.

DISCUSSION

The application of the LOA model provided a useful framework for consideration of the impact of automation on professional roles in radiotherapy systems. It enabled the trends in automation from past implementations to future possibilities with respect to the three processes, patient simulation, treatment planning and image guidance during treatment, to be presented in a communicative manner.

The analysis encapsulated clearly the implications of achievable automation for functions in radiotherapy systems. However, it is clear that in order to answer detailed questions about operator roles the LOA model would need to be applicable at a task level. In order to achieve this, it may need to be combined with a model of human performance which supports direct interaction with automation.

It is clear that further work is required to tailor the model to radiotherapy systems. For example, Wandke (2005) extended the number of "action" stages of the model to include, (1) Motivation, activation and goal setting, (2) Action execution and (3) Processing feedback of action results. Also, Endsley and Kaber (1999) developed similar LOA models in the context of teleoperations. In essence all of these later models are variations on the concept proposed by Sheridan (1992).

REFERENCES

Clegg, C. W., Ravden, S., Corbett, M., and Johnson, G. (1989), "Allocating functions in computer integrated manufacturing: a review and a new method." *Behaviour and Information Technology,* 8(3), 175-190.

Dekker, S., and Hollnagel, E. (2004), "Human factors and folk models." *Cognition, Technology & Work,* 6(2), 79-86.

Dekker, S., and Woods, D. D. (2002), "MABA-MABA or Abracadabra? Progress on Human–Automation Co-ordination." *Cognition, Technology & Work,* 4(4), 240-244.

Endsley, M. R., and Kaber, D. B. (1999), "Level of automation effects on performance, situation awareness and workload in a dynamic control task." *Ergonomics,* 42, 462-492.

Fallon, E. F., Bannon, L., and McCarthy, J. (Eds.), (1997), *ALLFN'97, Revisiting the Allocation of Functions Issue: New Perspectives.* Proceedings of the First International Conference on Allocation of Functions. Galway, Ireland. IEA Press, Louisville, KY.

Fallon, E. F., Chadwick, L., and van der Putten, W. J. (2009a), Learning from Risk Assessment in Radiotherapy. In: Duffy, V. G. (Ed.), *Digital Human Modeling, LNCS 5620.* pp. 502-511. Springer, Heidelberg, Germany.

Fallon, E. F., Chadwick, L., and van der Putten, W. J. (2009b), "Risk assessment in Radiotherapy. Lessons from systems engineering." WC 2009, IFMBE Proceedings 25/XII.

Färber, M., Ehrhardt, J., and Handels, H. (2005), "Automatic atlas-based contour extraction of anatomical structures in medical images." *International Congress Series. CARS 2005: Computer Assisted Radiology and Surgery,* 1281, 272-7.

Fitts, P. M. (Ed.), (1951), *Human Engineering for an effective air navigation and traffic control system.* National Research Council, Washington.

Fuld, R. B. (2000), "The fiction of function allocation, revisited." *International Journal of Human-Computer Studies,* 52(2), 217-233.

Jordan, N. (1963), "Allocation of functions between man and machines in automated systems." *Journal of Applied Psychology,* 47(3), 161-165.

Khan, F. M. (2003), *The Physics of Radiation Therapy.* Lippincott Williams & Wilkins, Philadelphia, PA.

Meister, D. (1985), *Behavioral analysis and measurement methods.* Wiley, NY.

Parasuraman, R., Sheridan, T. B., and Wickens, C. D. (2000), "A model for types and levels of human interaction with automation." *Systems, Man and Cybernetics, Part A: Systems and Humans, IEEE Transactions on,* 30(3), 286.

Price, H. E. (1985), "The Allocation of Functions in Systems." *Human Factors: The Journal of the Human Factors and Ergonomics Society,* 27, 33-45.

Sheridan, T. B. (1992), *Telerobotics, Automation and Human Supervisory Control.* MIT Press, Cambridge, MA.

Sheridan, T. B. (1997), "Function allocation: algorithm, alchemy or apostasy?" ALLFN'97 Revisiting the Allocation of Functions Issue: New Perspectives,

Galway, Ireland.

Wandke, H. (2005), "Assistance in human-machine interaction: a conceptual framework and a proposal for a taxonomy." *theoretical Issues in Ergonomics Science,* 6(2), 129-155.

Waterson, P. E., Gray, M. T. O., and Clegg, C. W. (2002), "A Sociotechnical Method for Designing Work Systems." *Human Factors,* 44(3), 376-391.

Wu, X., Spencer, S. A., Shen, S., Fiveash, J. B., Duan, J., and Brezovich, I. A. (2009), "Development of an accelerated GVF semi-automatic contouring algorithm for radiotherapy treatment planning." *Computers in Biology and Medicine,* 39(7), 650-656.

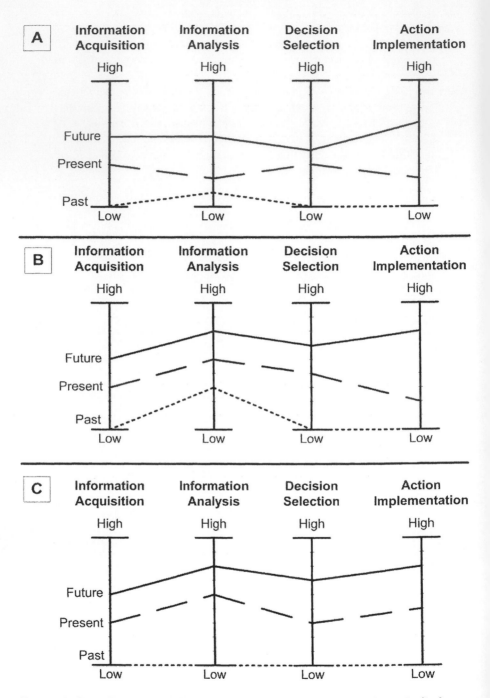

Figure 3. Past, Present and Future Levels of Automation for Three Radiotherapy Treatment Functions: (A) Patient Simulation, (B) Treatment Planning and (C) Imagine Guidance During Treatment (Shadowed Boxes in Figure 2 above)

CHAPTER 47

An Ergonomic Study on Measurement Operation Quality of an Inspection Laboratory

Hsiao-Ying Huang, Eric Min-yang Wang

Department of Industrial Engineering and Engineering Management
National Tsing Hua University, Hsinchu, Taiwan ROC

ABSTRACT

Keeping stability of operation or product quality is a required competency for quality control in manufacturing industry. For controlling the products quality, the relevant data must be measured and inspected to judge if the quality of products conforms to the specifications. Originally, the accuracy of the measured data varies from each other among different operators. Therefore, in this study the causes of measurement deviations among different operators were analyzed from ergonomics point of view. Improvements were proposed and implemented.

This study started with observing and recording the operations performed by the operators in the laboratory, followed by analyzing the mental process of gage reading judgment during the inspection, and ended with identifying the factors that might affect the operation quality and suggesting the improvement of the measurement method.

It was recommended that a standardized and easier feature identification technique should be helpful. The effects of the new method were verified by using

464

the Gage Repeatability and Reproducibility as an indicator. The results showed that that the data deviations caused by operators were reduced after the new method was implemented. The new measurement method could be disseminated to other departments for improving the measurement quality.

Keywords: Operation Quality, Task Analysis, Inspection Accuracy, Measurement System Analysis (MSA), Individual Difference

INTRODUCTION

Manufacturing quality monitoring depends on some control procedures, for example incoming quality control, in-process quality control, out going quality control etc. Qualified and unqualified products are judged by inspecting the measured data of these methods. Hence, the accuracy and reliability of the measured data is necessarily important. The inaccuracy of measured data not only affects the statistic meaning of quantitative analysis but may also lead to the failure of screening out unqualified products. Therefore, relevant research on improving the accuracy of the measurement system is considered an important goal for the industries.

This study took an inspection laboratory as a research platform. It was found that the inspection accuracy of measured data varies among different operators and has been annoying. This study aimed at finding out the causes to such deviations from the ergonomic point of view. It was expected that effective improvement could be suggested and implemented. Furthermore, the results could be extended to related industrial fields to minimize the measurement deviations.

The indicator of verifying the effects of the new method in this study was adopted from Automotive Industry Action Group's (AIAG) Measurement Systems Analysis (MSA) Manual in which "Gage Repeatability and Reproducibility (GRR)" is an important and significant measure. The main purpose of repeatability is to find out the deviation produced by inspection equipments, and the course of action was that the identical operator inspected the same product by the same inspection equipment for several times to gather the measured data and their deviations. The purpose of reproducibility is to find out the variability among different operators, and the course of action was that different operators inspected the same product by the same inspection equipment for several times to gather the deviations. To integrate estimated data of the GRR, verified whether the statistical characteristics are properly possessed in measurement system.

The indices of GRR are divided into the following three measurement levels representing different measurement performances:

- Level A: GRR<10% means the measurement is very reliable.

- Level B: 10%<GRR<25% means the measurement is good and acceptable.

- Level C: 25%<GRR means the measurement is inaccurate and unacceptable.

METHODS

Measurement operation of the factory's product, a liquid chemical solution, was chosen to be the experimental case in this study. The inspection laboratory in this factory is in charge of the quality inspection for all products. To carry out this important task, the inspection equipments are high precision electronic facilities.

SUBJECTS

Five inspection operators of the Quality Assurance Department were chosen as subjects. These subjects have known the inspection equipment and standard operating procedure very well. To avoid possible data distortion from over cautiousness caused by the mental stress of the operator, the subjects were re-ensured that their performance in the experiments would be merely for research purpose but not for personal performance evaluation.

EQUIPMENT AND MATERIAL MEASURED

Film thickness meter, optical microscope and viscometer were used for measurement in the experiment. The main products of the factory were chosen as the samples for measurement.

ENVIRONMENT

To avoid the measurement errors that might be influenced by the environment, the temperature and humidity of the laboratory were controlled with air conditioning. Before measurement, the sample temperature was kept constant for at least 15 minutes in the water tank associated with the viscometer. The brightness was invariable with all light sources on.

PROCEDURES

The laboratory in question is in charge of controlling the quality of the product, a liquid chemical solution, by measuring the diameter, the viscosity, and other characteristics of the liquid drop. In performing the measurement, operators have to judge and manually determine the measurement points and sections, which might produce data variations among measurements.

By recording the inspection procedure to analyze the mental process of using gage, main factors led to measurement deviations of different operators could be determined. Further, it was expected that new inspection method could contribute to easy measuring point identification and its standardization.

- To avoid disturbance to regular work and prevent inaccuracy measurement taken during busy and complicated working time, the experiment was conducted after the work hour.

- Experiment was carried out with one subject at a time while others were excluded to ensure all subjects were performing the experiments independently without affected by the impressions from observing other persons' experiments.

- The subjects were asked to "Think Aloud" and orally describe his/her own thinking, perception and judging process during performing the experimental tasks.

- The experimenter observed the inspection process and recorded relevant experimental data and information after each subject performed the task.

- The difference of thinking and performance among the subjects was analyzed to find out the connections between the accuracy of the performance data and the way of measurement.

DATA ANALYSIS

Data collected from the experiments were analyzed. The improvement and effects of measurement procedures were verified by using GRR as an indicator.

RESULTS

Table 1 and FIGURE 1 indicated that parts of operating steps might have easily produced the deviations with the different judgments of inspection operators. When the deviation occurred, the measurement methods would be modified. Then, the inspection operator preceded the inspection process again with the modified method. The GRR of the experimental data of the measurement system were gathered and analyzed.

For the verified equipments, the values of the GRR were obviously decreasing. This result indicated that the measurement system has been remarkably improved.

Table 1. Improvement of GRR

Equipment	Before	After
Film Thickness Meter	9.5	4.7
Optical Microscope	17.5	13.1
Viscometer	19.7	14.4

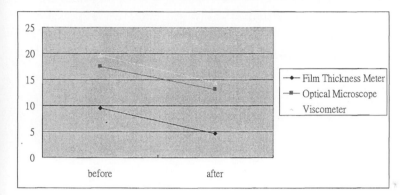

FIGURE 1. Improvement of GRR

DISCUSSION

In the process of operating measurement equipment, if a measurement needed to be done with an inspection operator's judgment, most probably, the operator would choose a number based on his/her mental thinking or personal preference. Moreover, the different personal mental thinking among these operators with different experiences had indicated that the main factor of the deviation of the measured data might be caused by individual difference. Therefore, to reduce the steps judged by the operators in the inspection process, or to modify the inspection methods to make the measurement points and sections easily identified and standardized, the deviation of measured data among inspection operators could be reduced substantially.

In this study, the experimental subjects were chosen in a small-scale inspection laboratory. The experimental results may not be general and comprehensive because of the limited participants. Therefore, an analysis method was suggested, which is to be used by different types of industry and the number of participants and participating departments should also be increased. This will be helpful to more significantly find out the main factors that makes the individual performance different.

CONCLUSIONS

This study discussed the deviation of measured data caused by different operators, and the deviation has been obviously decreasing by modifying the measurement method. The result indicated that the operation procedures can be redesigned and the measurement points can be identified easily will improve the measurement performance significantly. Therefore, the deviation of measurement system will be decreasing by considering the judgment in advance when the inspection methods are set up.

REFERENCES

AIAG Editing Group, (2002), Measurement System Analysis-Reference Manual (MSA), 3rd ed., Automotive Industries Action Group.

Jebaraj, D., Tyrrell, R. A., Gramopadhye, A. K., (1999), Industrial inspection performance depends on both viewing distance and oculomotor characteristics. Applied Ergonomics 30, 223 – 228.

Liu, J.-G., Frühauf, U., Schönecker, A., (1999). Accuracy improvement of impedance measurements by using the self-calibration. Measurement 25, 213–225.

Rozin P., (2003). Five potential principles for understanding cultural differences in relation to individual differences. Journal of Research in Personality 37, 273 – 283.

Garrett, S. K., Melloy, B.J., Gramopadhye, A. K, (2001). The effects of per-lot and per-item pacing on inspection performance. International Journal of Industrial Ergonomics 27, 291 – 302.

Chapter 48

A Cross-Cultural Study on the User Motivations and Behaviors of Online Knowledge Sharing

Eike Nohdurft[1], Qin Gao[2], Bin Zhu[2]

[1] RWTH-Aachen University
Germany

[2] Department of Industrial Engineering
Tsinghua University, Beijing, 100084 China

ABSTRACT

As the proliferation of the internet, the communication and collaboration among people are no longer hindered by space and time barriers. Cross-cultural communication and collaboration are taking place much more frequently than before. One popular example for internet-based services is online collaborative knowledge sharing, such as *Wikipedia*. The purpose of this study is to compare Chinese and German users on their motivations and behaviors in online collaborative knowledge sharing services. In the empirical part of this study, an interview and a questionnaire survey were conducted. From the results, we found that significant differences exist in the users' motivations and behaviors between Chinese and Germans. The cross cultural theories drawn from the physical world are also applicable to the virtual world.

Keywords: Knowledge Sharing, Cross-culture, Collaboration, Chinese, German

INTRODUCTION

The proliferation of internet technology changes people's life greatly. Many online knowledge sharing services, such as *Wikipedia*, provide people with an efficient platform to share and acquire information as well as knowledge. Thus, it becomes more and more convenient for people to acquire all kinds of information, such as economy, art, history and others. With the increasing popularity of online collaborative knowledge sharing services, the impacts of these online services rise in various domains.

In this study we investigated cross-cultural differences in Chinese and German users' motivations and behaviors. While the internet globalizes people's communication and collaboration, cultural conflicts may occur. Consequently, the cultural differences can influence the results and processes of online collaborative knowledge sharing. The study of users' motivations and behaviors could be meaningful in theory and practice.

THEORETICAL BACKGROUND

KNOWLEDGE SHARING

Lee (Lee, 2001) defined knowledge sharing as "activities of transferring or disseminating knowledge from one person, group or organization to another". Some researchers have studied users' motivations and behaviors during sharing knowledge. Hendriks (Hendriks, 1999) used the two-factor theory (Herzberg, Snyderman, & Mausner, 1993) to explain the motivations for knowledge sharing. He divided the influencing factors into hygiene factors and motivational factors. The hygiene factors include salary, status, company policy and interpersonal relations, while the motivational factors contain the hope for recognition, promotional opportunities, a sense of responsibility and the hope for reciprocity. Also the knowledge provider may be motivated by the challenge of knowledge work, operational autonomy and promotional opportunities. Hendriks (Hendriks, 1999) concluded that the hygiene factors can only stimulate the quantity of knowledge for sharing, instead of the quality of the knowledge, which is only influenced through the motivational factors. Chiu (Chiu, Hsu, & Wang, 2006) used the *social cognitive theory* (Albert, 1989) and *social capital theory* (Coleman, 1988) to explore the factors that influenced users' knowledge-sharing behaviors in virtual communities. Ryu (Ryu, Ho, & Han, 2003), Lin (Lin & Lee, 2004), and Hansen (Hansen & Avital, 2005) successfully applied the *theory of planned behavior* (TPB) (Ajzen, 1985) to explain the behavoir of knowledge sharing.

Based on the related theories and researches, in total of 11 factors which influence the motivations and behaviors of users during knowledge sharing were collected from the previous researches in this study. The 11 factors are listed below:

- *Ease of sharing knowledge* (Wu, 2003)

- *Reciprocity* (Wu, 2003; Hendriks, 1999; Chiu, Hsu, & Wang, 2006)
- *Sense of community* (Chiu, Hsu, & Wang, 2006; Sharratt & Usoro, 2003)
- *Challenge of work* (Hendriks, 1999; Wu, 2003)
- *Fear of losing face* (Huang, Davison, & Gu, 2008)
- *Keeping competiveness* (Ardichvili, Maurer, Li, Wentling, & Stuedemann, 2006)
- *Expectation of reward* for *shared knowledge* (Hall, 2001; Wu, 2003)
- *Expected relationship* through *knowledge sharing* (Wu, 2003)
- *Altruism* (Wu, 2003)
- *Recognition of job done* (Hendriks, 1999)
- *Fear of misleading* (Ardichvili, Page, & Wentling, 2003)

CULTURAL DIFFERENCES BETWEEN CHINESE AND GERMANS

A number of differences between Chinese and German culture have been identified in the existing cross-cultural research. According to Triandis's (Triandis, et al., 1988) cultural theory, the Chinese culture is a typical culture of collectivism and in-group orientation (Lockett, 1988), while the German culture tends towards individualism. Hofstede's (Hofstede & Bond, 1984) survey also confirmed this finding. In Hofstede's dimension of Individualism, the index of IDV (Individualism) for Chinese is 20, but German's is 67. Furthermore, Hall's (Hall, 1990) cultural theory (high-context and low-context) sheds light on the fact that, China is a high-context society, while Germany is a typical low-context society. In addition, Chinese are known for emphasizing on the social network (*guanxi* in Chinese) (Chen & Chen, 2004) and fear of losing face (Hu, 1944), while these aspects are not very important for Germans. These cultural differences may inevitably have impacts on the motivations and behaviors of online collaborative knowledge sharing.

The objective of this study is to examine the differences between Chinese and Germans in the motivations and behaviors of online knowledge sharing, with the aim of understanding the role that culture plays in knowledge sharing.

METHODOLOGY

The whole study was divided into two phases. In Phase 1, we conducted a series of semi-structured interviews in order to understand and verify factors influencing users' motivation and behaviors. In Phase 2, a cross-cultural survey was conducted to examine the difference between Chinese and German users.

PHASE1: INTERVIEW

A semi-structured interview was conducted to validate the 16 factors, which may have impacts on the users' motivations and behaviors. Based on the mentioned 11 factors extracted from the existing literature, we proposed another five factors based on the discussion results of focus groups. The rest five factors are: *knowledge quality, sharing knowledge just for fun, popularity of the service, anonymity,* and *response time.* From the pilot focus groups, we found that wiki services and BBS (Bulletin Board System) as the online collaborative knowledge sharing services were very popular among the Chinese and German participants. Thus, we employed wiki services and BBS as the samples in our interview and survey.

Participates

30 Chinese and 20 German users were interviewed. The 20 German participants were recruited out of RWTH Aachen University, while the 30 Chinese participants were from Tsinghua University. 66.7% of Chinese participants are male, and their average age is 23 years. 85 % of German participants are female, and their average age is 25 years.

Interview Results

14 factors were validated in the interviews, which were *ease of sharing knowledge, reciprocity, sense of community, challenge of work, altruism, recognition of job done, fear of misleading, knowledge quality, expectation of reward for shared knowledge, expected relationship through knowledge sharing, entertainment, keeping competiveness, fear of losing face, and social norm.* From the interviews, we found that German participants doubted the quality of knowledge from online knowledge sharing services more than Chinese did. 83.5 percents of Chinese participants reported that they shared knowledge just for fun, whereas only 50 percents of Germans reported such a motivation. Additionally, more Chinese participants (60%) expected rewards and recognition for knowledge sharing than Germans (20%) did. While 23.3 percents of Chinese participants mentioned that they had concerns of losing face due to sharing wrong knowledge, none of German participants worried about that. 30 percents of Chinese participants thought that it was an opportunity to enlarge one's personal social network, but none German participants mentioned that.

PHASE2: SURVEY

In order to examine the differences between Chinese and Germans, we conducted a questionnaire survey among Chinese and German users. The 14 factors obtained from interview results were 1) *ease of sharing knowledge,* 2) *reciprocity,* 3) *sense of community,* 4) *challenge of work,* 5) *altruism,* 6) *recognition of job done,* 7) *fear of misleading,* 8) *knowledge quality,* 9) *expectation of reward for shared knowledge,* 10) *expected relationship through knowledge sharing,* 11) *entertainment,* 12) *keeping competiveness,* 13) *fear of losing face,* and 14) *social norm.*

Questionnaire Construction

At the beginning of the questionnaire, participants were asked to provide demographic information. The rest items measured the 14 factors. The items of 11 factors were extracted from the existing literature (Wu, 2003; Ardichvili, et al., 2003; Hendriks,1999; et al.), whose reliabilities have already been validated. The items for the rest three factors (entertainment, knowledge quality and social norm) were self-designed according to the interview results. The primary items are listed in Table 1. A 5-point Likert scale was used to measure different levels of agreement from "1- totally disagree" to "5 - totally agree". After the English version was accomplished, three native Chinese speakers translated the original English questionnaire into Chinese version, and two native German speakers translated it into German version. In Chinese version, we used *Baidu Encyclopedia* instead of *Wikipedia*, because *Baidu Encyclopedia* is much more popular among Chinese, and similar to *Wikipedia*.

Table 1 Questionnaire items

Factor	Item
1) Ease of sharing knowledge	1.1 "it is very laborious to write down the knowledge"; 1.2 "the amount of the effort needed"; 1.3 "the bad user interface".
2) Reciprocity	2.1 "it is only fair to help others"; 2.2 "others will help me".
3) Sense of community	3.1 "I have a sense of belonging"; 3.2 "I have a strong feeling".
4) Challenge of work	4.1 "I am motivated to contribute a challenging topic".
5) Altruism	5.1 "If I see a topic needing rework, I will fix it"; 5.2 "giving is better than acquiring"; 5.3 "helping others is important".
6) Recognition of job done	6.1 "others will show gratefulness"; 6.2 "I like to receive compliments"; 6.3 "I expect my work to be valued".
7) Fear of misleading	7.1 "I will consider whether it is relevant"; 7.2 "I'm afraid of misleading others"; 7.3 "I don't contribute if it is not relevant".
8) Knowledge quality	8.1 "I'm afraid it is wrong"; 8.2 "I will use other services"; 8.3 "it is important for me it will be corrected".
9) Expectation of reward for shared knowledge	9.1 "I will receive status points"; 9.2 "I'm motivated by the status points"; 9.3 "I think I should receive reward".
10) Expected relationship through knowledge sharing	10.1 "the relationship is strengthened"; 10.2 "help to understand other"; 10.3 "my relationship network will be expanded".
11) Entertainment	11.1 "I might lose the control"; 11.2 ",knowledge loses its value";
12) Keeping competiveness	12.1 "Knowledge sharing is fun"; 12.2 "it is a leisure activity"; 12.3 "I share knowledge to kill time".
13) Fear of losing face	13.1 "I'm afraid other users will get a bad image"; 13.2 "other's

	thinking is important"; **13.3** "other's concern about the knowledge is important".
14) Social norm	**14.1** "I should contribute"; **14.2** "The service expects me to share knowledge"; **14.3** "Most of users contribute knowledge".

Participants

Chinese participants were recruited from Chinese universities, while Germans were from German universities. In total, we received 521 Chinese responses and 247 German responses. After eliminating incomplete and inconsistent answers, we got 71 valid responses of *Baidu Encyclopedia* and 83 valid responses of BBS for Chinese, while 62 valid responses of *Wikipedia* and 58 valid responses of BBS for Germans. These participants all had the experiences of contributing knowledge on *Wikipedia* or *Baidu Encyclopedia*. In this paper, we only present the results of *Wiki* services.

Procedure

The survey was administrated via internet. Both Chinese questionnaire and German questionnaire were web-based, and the questionnaire link was spread through email, blog, *facebook*, MSN (Microsoft service network), et al. Before filling the questionnaire, participants were asked to sign on the informed consent form. As an extrinsic reward, the German participants could win one of five 5€ "Amazon.de" coupons in a raffle, while the Chinese participants could acquire the research results if they were interested.

Results Analysis

The compared demographic data was shown in Table 2. The average computer and internet experience of Germans is more than that of Chinese. This may be caused by the degree of country development, with the more developed the country having the more advanced internet infrastructure.

Table 2 Demographic statistics of the survey participants

Demographics	Chinese	German
Male (%)	71.8%	61.3%
Average Age(years)	20.9	23.9
Education* (%)	95.8%	95.5%
Average Computer Experience	8.3	12.4

(years)		
Average Internet Experience	7.2	9.0
(years)		

* The index of education refers to the percentage of participants who got higher education than bachelor degree, including bachelor degree.

The Cronbach's Alpha of the 39 items which measured the 14 factors was 0.753 for Chinese sample and 0.826 for German sample, which showed that the internal consistency of the Chinese and German questionnaire was reliable.

The *Discrimination Index* of each item was calculated for two samples based on Truman Kelly's (Kelly, 1939) "27% of sample" group size. Except for Item 11.3, the *Discrimination Indices* of all items were above than 0.25. According to Ebel's (Ebel, 1954) criterion, the leaving 38 items were discriminative enough for two samples. Item 11.3 was deleted from the questionnaire.

We conducted Mann-Whitney U test to compare the means of two samples, as the normal assumption of T-test and ANOVA (Analysis of Variance) was not satisfied. The result of Mann-Whitney U test showed that except for Factor 8, significant differences existed in the leaving 13 factors. Furthermore, except for Factor 1, the value of the rest 12 factors is significantly higher for Chinese than German.

- **Factor 1: *Ease of sharing knowledge***
  (Chinese=2.69; German=3.65; $p < 0.001$)
- **Factor 2: *Reciprocity***
  (Chinese=3.68; German=3.00; $p < 0.001$)
- **Factor 3: *Sense of community***
  (Chinese=3.64; German=2.79; $p < 0.001$)
- **Factor 4: *Challenge of work***
  (Chinese=3.15; German=2.53; $p < 0.001$)
- **Factor 5: *Altruism***
  (Chinese=3.37; German=1.92; $p < 0.001$)
- **Factor 6: *Recognition of job done***
  (Chinese=3.74; German=2.31; $p < 0.001$)
- **Factor 7: *Fear of misleading***
  (Chinese=4.20; German=3.32; $p < 0.001$)
- **Factor 8: *Knowledge quality***
  (Chinese=3.62; German=3.62; $p = 0.609$)
- **Factor 9: *Expectation of reward for shared knowledge***
  (Chinese=2.39; German=1.90; $p < 0.001$)
- **Factor 10: *Expected relationship through knowledge sharing***
  (Chinese=3.21; German=1.74; $p < 0.001$)
- **Factor 11: *Keeping competiveness***
  (Chinese=2.42; German=1.62; $p < 0.001$)
- **Factor 12: *Entertainment***

(Chinese=2.96; German=1.97; p < 0.001)
- **Factor 13:** *Fear of losing face*
  (Chinese=3.11; German=1.97; p < 0.001)
- **Factor 14:** *Social norm*
  (Chinese=2.86; German=2.13; p < 0.001)

From the statistical analysis, we could get the following findings. In Chinese tradition, they respect reciprocity. Thus, Chinese users believe that the reciprocity (Factor 2) tradition should be retained in the online knowledge sharing communities, while German users pay less attention to that. The sense of community (Factor 3) during online knowledge sharing is higher for Chinese users. This is because the Chinese culture is group-orientated (Lockett, 1988). As a result, Chinese users are more concerned about the other users when sharing knowledge. They are more altruistic (Factor 5) and fear of misleading others (Factor 7). For Chinese users, sharing knowledge in an online collaborative community is not only a way of knowledge procurement but is also a possibility to establish new social relationships (Factor 10), while the need to develop new relationship for German users is significantly lower. This can be explained through the Chinese emphasize on the social network (*guanxi* in Chinese) (Chen & Chen, 2004). In addition, Chinese users pay more attention to receiving rewards in return for knowledge sharing (Factor 9), which is often given in the form of status points in a hierarchical reputation system. It shows that it is more important for the Chinese participants to gain a position in a hierarchical system. This is caused by the long feudal Chinese history. The position demand also leads to that Chinese want to keep competitiveness (Factor 11) in the community to promote their status. The Chinese face saving culture (Hu, 1944) also has a significant effect on users' motivation to share knowledge online (Factor 6 & 13).

With respect to Factor 1, the Chinese motivation is lower than the German. In the online knowledge sharing services, Chinese users usually copy and paste the information or knowledge from one website to another. This makes Chinese users pay less attention to the difficulty of knowledge sharing, for example, transforming the knowledge into a shareable form and handling human computer interface.

German culture is a typical low-context (Hall, 1990) and individual (Hofstede & Bond, 1984) culture, compared with Chinese culture. Consequently, the German users would like to provide clear and accurate knowledge during knowledge contributing. They consider more about ease of sharing knowledge than Chinese during knowledge sharing. The individualism of the German culture results in less concerns of social relationship, reciprocity, community, recognition of others and rewards.

DISCUSSIONS

The purpose of this study is to discover the cultural differences in user motivation and behavior of online knowledge sharing. The Chinese culture is a typical collective and high-context culture, while the German culture is an individual and low-context culture (Hofstede & Bond, 1984; Hall, 1990). Our study finds that this theory is also applicable to the virtual online activities, although it is based on the real world. Significant differences exist between Chinese and German users in the online collaboration knowledge sharing communities. Chinese users focus more on the social relationship and other users' opinions about them and their shared knowledge. Also, Chinese users are more concerned about rewards. The rewards could be other user's compliment or the points from the service developers. However, German users are less concerned about other users or the communities than Chinese do. German users pay more attention to the ease of sharing knowledge.

As the development of globalization, cross-cultural communication and collaboration take place more frequently. Our findings provide the service providers a strong reason that they need to design their services for cross-cultural communication and collaboration with a focus on cross-cultural differences. The research findings also provide sight to the design guidelines of online knowledge sharing services. For example, service providers should provide rewards to motivate Chinese users. They also should offer communication channels for Chinese, e.g. private messages, in order to satisfy their need of developing new relationships with other users. As for German users, they need to consider how to make the process of sharing knowledge easier.

One limitation for this study is the sample size, which is not big enough for factor analysis. As a result, the relationship between the 14 factors can not be dug out clearly in this study. Further research can be focused on the model development for these factors. Another limitation is the homogeneity of the samples. Almost all participants are students. Although student group occupies a large part of online knowledge sharing users, it is still not representative of the whole range of internet users.

REFERENCES

Ajzen, I. (1985), *From intentions to actions: A theory of planned behavior.* In J. Kuhl, & J. Beckman (Ed.), Action-control: From cognition to behavior (11–39). Heidelberg, Germany: Springer.

Albert, B. (1989), "Social Cognitive Theory." *Annals of Child Development*, 6, 1–60.

Ardichvili, A., Maurer, M., Li, W., Wentling, T., and Stuedemann, R. (2006), "Cultural influences on knowledge sharing through online communities of practice." *Journal of Knowledge Management*, 10(1), 94–107.

Ardichvili, A., Page V., and Wentling, T. (2003), "Motivation and barriers to participation in virtual knowledge-sharing services of practice." *Journal of Knowledge Management*, 7(1), 64–77.

Bock, G.W., and Kim, Y.G. (2002), "Breaking the myths of rewards: an exploratory study of attitudes about knowledge sharing." *Information Resources Management Journal*, 14, 14–21.

Chen, X.P. and Chen, C.C. (2004), "On the intricacies of the Chinese guanxi: a process model of guanxi development." *Asia Pacific Journal of Management*, 21(3), 305-324.

Chiu, C.M., Hsu, M.H., and Wang, E.T. (2006), "Understanding knowledge sharing in virtual communities: An integration of social capital and social cognitive theories." *Decision Support Systems*, 42(3), 1872–1888.

Coleman, J.S. (1988), "Social captial in the creation of human capital." *The American Journal of Sociology*, 94, 95–120.

Cronbach, L.J. (1951), "Coefficient alpha and the internal structure of tests." *Psychometrika* ,16(3), 297–334.

Davenport, T.H., & Prusak, L. (1998). *Working knowledge: How organizations manage what they know* (Vol. 1). Boston: Harvard Business School Press.

Ebel, R.T. (1954), "Procedures for the analysis of classroom tests." *Educational and Psychological Measurement*, 14, 352–364.

Hall, H. (2001). *Social exchange for knowledge exchange*. Managing knowledge: conversations and critiques .

Hall, E.T., and Hall, M.R. (1990). *Understanding cultural differences: Germans, French and Americans* (Vol. 1). Intercultural Press Inc.

Hansen, S., and Avital, M. (2005). *Share and share alike: The social and technological influences in knowledge sharing behavior*. Sprouts: Working Papers on information Systems , Vol.5 (13).

Hendriks, P. (1999). "Why share knowledge? The influence of ICT on the motivation for knowledge sharing." *Knowledge and Process Management*, 6(2), 91–100.

Herzberg, F., Snyderman, B.B., and Mausner, B. (1993). *The motivation to work* (Vol. 2). New Brunswick, USA: Transaction Publishers.

Hofstede, G., and Bond, M.H. (1984), "Hofstede's culture dimensions: an independent validation using Rokeach's value survey." *Journal of Cross-Cultural Psychology*, 15(4), 417–433.

Hu,H. C. (1944), "The Chinese concepts of "face"." *American Anthropologist*, 46, 45–64.

Huang, Q., Davison, R.M., and Gu, J. (2008). "Impact of personal and cultural factos on knowledge sharing in China." *Asia Pacific Journal of Management*, 25(3), 451–471.

Kelly, L.T. (1939), "The selection of upper and lower groups for the validation of test items." *Journal of Educational Psychology*, 30(1), 17–24.

Lee, J.-N. (2001), "The impact of knowledge sharing, organizational capability and partnership quality on IS outsourcing success. " *Information & Management*, 38, 323–335.

Lin, H.-F., and Lee, G.-G. (2004), "Perceptions of senior managers toward knowledge-sharing behavior." *Management Decision*, 42(1), 108–125.

Lockett, M. (1988), "Culture and the problems of Chinese management." *Organization Studies*, 9, 475–496.

Ryu, S., Ho, S.H., and Han, I. (2003). "Knowledge sharing behavior of physicians in hospitals." *Experts Systems with Applications*, 25(1), 113–122.

Sharratt, M., and Usoro, A. (2003), "Understanding knowledge-sharing in online communities of practice." *Electronic Journal on Knowledge Management*, 1(2), 187–196.

Triandis, H.C., Bontempo, R., and Villareal, M.J. (1988), "Individualism and collectivism: cross-cultural perspectives on self-ingroup relationships." *Journal of Personality and Social Psychology*, 54(2), 323–338.

Wu, S. (2003), *Exploring knowledge sharing behavior of IS personnel with theory of planned behavior*. Exploring Knowledge Sharing Behavior of IS Personnel with Theory of Planned Behavior, 104. Kaohsiung, Taiwan.

CHAPTER 49

Influences of Different Dynamic Office Chairs on Muscular Activation, Physical Activity and Posture

Rolf Ellegast[1], Kathrin Keller[1], Helmut Berger[2], Liesbeth Groenesteijn[3], Peter Vink[3]

[1]IFA – Institute for Occupational Health and Safety of the German Accident Insurance, Alte Heerstrasse 111, 53757 Sankt Augustin, Germany

[2]VBG – Accident Prevention & Insurance Association for Administration, Nikolaus Dürkopp-Str. 8, 33602 Bielefeld, Germany

[3]TNO Quality of Life, P.O. Box 718, 2130 AS Hoofddorp, The Netherlands & Delft University of Technology, The Netherlands

ABSTRACT

Prolonged and static sitting postures at VDU workplaces are discussed as risk factors for the musculoskeletal system. Manufacturers of office chairs have therefore created specific dynamic office chairs that contain structural elements to promote dynamic sitting and therefore to prevent musculoskeletal disorders. The aim of this study was to evaluate the effects of four specific dynamic chairs on erector spinae and trapezius EMG, postures/joint angles and physical activity intensity (PAI) compared to a conventional standard office chair. All chairs were instrumented with inertial sensors to measure chair parameters (backrest inclination, seat pan inclination forward and sideward) and tested in laboratory by 10 subjects performing 7 standardized office tasks. Muscular activation showed no significant differences between the specific dynamic chairs and the reference chair. The analysis of postures/joint angles and PAI revealed only few differences among the chairs, whereas the performed tasks strongly affected the measured muscle activation, postures and kinematics. The characteristic dynamic elements of each specific chair yielded significant differences of the measured chair parameters, but these characteristics did not seem to affect the sitting dynamics of the subjects performing their office tasks.

Keywords: Dynamic chairs, EMG, Posture, Physical Activity, office work

INTRODUCTION

Today more than 40 % of all employees in the EU are working at visual display units (VDUs) (Parent-Thirion et al., 2007). Several studies have shown associations between sedentary work and low back pain as well as degenerative changes of the intervertebral discs (Evans et al., 1989, Videman et al., 1990). Therefore design and sitting comfort aspects of office chairs have become an important issue to prevent musculoskeletal disorders at office workplaces (Looze et al., 2003). Manufactures of office chairs have further developed the concept of dynamic sitting by introducing structural elements that give the seat a dynamic mounting or active rotation of its own.

Some comparative studies have been conducted to analyse the effect of different seating concepts on muscular activity and posture. Wittig compared alternative seating concepts (pendulous chair, sitting ball, kneeling chair and raised chairs) to a standard office chair. No significant differences were found concerning erector spinae activation and lumbar spine flexion (Wittig, 2000). Moreover, the results of the subject's perception indicated that an alternative seat concept should not be used as a permanent seat instead of a standard office chair. Van Dieën et al. analysed the effect of three dynamic office chairs on trunk kinematics, trunk extensor EMG and spinal shrinkage (Van Dieën et al., 2001). As a result, trunk kinematics and erector spinae EMG were strongly affected by the different office tasks performed, but not by the chair type.

The aim of this study is to analyse the effect of four specific dynamic office chairs and a conventional dynamic chair on muscle activation, sitting postures/joint angles and physical activity in a comparative laboratory study. All specific dynamic chairs have features of a conventional dynamic office chair and in addition to that they all come with extraordinary dynamic features.

The research question of this study is:

Do these specific dynamic office chairs influence physical and muscular activity more than a conventional office chair?

METHOD

The study was carried out at a realistic VDU office workplace in laboratory. 10 experienced office workers (5 men and 5 women) volunteered as subjects in the laboratory tests. The mean ages were 35.4 years (SD 12.1 years) for the men and 34.8 (SD 12.7) for the women. Body heights ranged from 1.75 to 1.86 m for the men and from 1.62 to 1.68 m for the women. Body weights varied from 76 to 100 kg for the men and from 47 to 78 kg for the women. Each subject tested a total of five office chairs (4 particularly dynamic ones labelled A, B, C and E and 1 reference office chair – labelled D) during the performance of standardized office tasks, including reading and correcting text data, typing words in a word document with keyboard and mouse, an intensive tracking mouse task, sorting paper files and telephoning.

Body postures and movements were measured with the CUELA system (Ellegast et al., 2009). This person-centered measuring system consists of inertial sensors (3D accelerometers and gyroscopes) as well as a miniature data storage unit with a flash memory card, which can be attached to the subject. From the measured signals (sampling rate: 50 Hz), the following body/joint angles were calculated: Head inclination (sagittal and lateral), flexion/extension and lateral flexion of the spine in the thoracic (Th3) and lumbar spinal regions (L1 and transition to L5/S1), trunk inclination and the spatial position of the upper and lower legs (right and left).

From the kinematic measurements physical acitivity intensities (PAI) were determined by calculating a sliding standard deviation of the high-passed filtered vector magnitude of the 3D acceleration signals (time window: 1 s).

Surface electromyography (EMG) was used for measuring the muscle activity of the trapezius muscle (right/left) and erector spinae muscle (right/left) with the CUELA EMG signal processor for long-term analysis (Glitsch et al. 2006). To assess the EMG signals, the root mean square values (RMS values) were calculated from the raw EMG data over consecutive time windows (0.3 s). To normalize the RMS values, reference activities were performed at the beginning of all measurement so that all muscle activities stand in relation to a reference voluntary contraction (% RVC). The 100% RVC values for each muscle were defined as the median RMS values of a calibration interval at the beginning of each measurement (duration: 3 s), where the subjects were standing with their arms abducted at 90° holding a weight of 5 kg (females) or 10 kg (males). The 0% RVC values for each muscle were set as the minimum values of another calibration interval at the beginning of each measurement (duration 3 s), where the subjects were standing upright in a neutral relaxed posture.

To measure dynamically the adjustments of the office chairs, acceleration sensors (Analog Devices ADXL 103/203) were used for measurement of the backrest inclination and seat pan inclinations (forward and sideward). With FSR

(Force Sensing Resistors®, Interlink Electronics Inc. Luxembourg) pressure sensors (six each on the seat and backrest and three each on the armrests), the temporal and spatial extent of seat, armrest and backrest use were recorded.

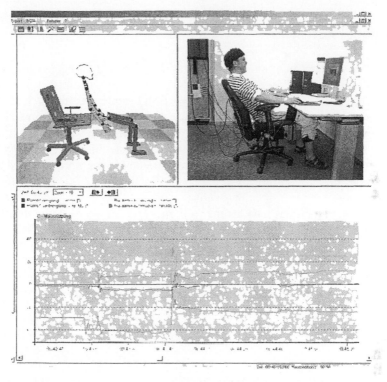

Figure 1 Data representation with the CUELA software by video (right, above), animated figure (left, above) and selected data time graph (below)

The measurement data can be depicted with the CUELA software together with the digitalized video recording of the workplace situation and a 3D animated figure (Figure 1). For statistical evaluation the software includes a calculation of the characteristic values of the frequency distributions of each joint/body angle, physical activity intensity (PAI), EMG recording and chair parameters.

With the measurement data ANOVA for repeated measures (General Linear Model, GLM) were performed to determine the significance ($p \leq 0.05$) of the main effects of the chairs (five chairs: A, B, C, E in comparison to reference chair D) and tasks (seven standardized tasks), as well as the chair x task interaction. The statistical assessment was done with the median values and 95th percentiles of the EMG RMS values, joint angles, PAI values and chair parameters.

RESULTS

MUSCLE ACTIVATION

The comparison of all specific dynamic chairs and the reference chair D shows no significant effect for all muscles.

For the m. erector spinae low median EMG amplitudes from 7 % RVC (chair C, erector spinae left) to 10.6 % RVC (chair B, erector spinae right) could be observed. Even the peak EMG amplitudes (95th percentiles) showed relatively low values varying from 19.8 % RVC (chair C, erector spinae left) to 27.8 % RVC (chair B, erector spinae left). For chair B high standard deviations were observed for the 95th percentiles (40.9 % RVC for m. erector spinae left and 27.0 % RVC for m. erector spinae right).

For both trapezius muscles the median EMG amplitudes were also low ranging from 4.2 % RVC (Chair C, m. trapezius left) to 15.3 % RVC (chair B, m. trapezius right). The peak EMG amplitudes (95th percentiles) varied from 25.8 % RVC (chair D, m. trapezius left) to 48.7 % RVC (chair B, m. trapezius right).

The statistical analysis showed significant effects between muscular activation and the performance of the standardized office tasks (exception: 95th percentiles, erector spinae left and right).

The analysis of the chair x task interaction revealed only for the m. trapezius right (mean and median values) significant results.

PHYSICAL ACTIVITY OF BODY PARTS

Overall the PAI values were very low for all chairs with median values not exceeding 0.8 % g.

The statistic comparison of chair A and the reference chair showed significant effects for the PAI values of L1 (mean and 95th percentile) and Th3 (95th percentile). This effect might be due to the specific dynamic feature of chair A, an electromotor that activates the seat pan.

All other comparisons of the specific dynamic chairs and the reference chair were not significant (exception: 95th percentile PAI head, chair C vs. D).

The highest peak values (95th percentiles) were measured for the lower leg ranging from 2.8 % g (chair A, lower leg left) to 4.6 % g (chair D, lower leg left).

The performance of the different standardized office tasks influenced the PAI values significantly (exceptions: median values of PAI lower leg left and right). The chair x task interaction showed only for the PAI thigh left significant results.

JOINT BODY ANGLES

The standardized tasks performed in laboratory had a significant impact on posture of the different joints, whereas no significant differences for the chairs and the comparison of the specific dynamic chairs to the reference chair could be found. The median values for the cervical spine flexion and trunk inclination angles were for all chairs in a neutral area (less than 5°) with high standard variations. During the performance of the standardized office tasks few asymmetric trunk postures were adopted, so that the trunk lateral flexion angles were close to the neutral posture varying from -2.3° (chair A) to -1.1° (chair B) for the 50th percentiles and from 0.4° (chair A) to 1.4° (chairs B and E) for the 95th percentiles.

The median values for the trunk flexion angles ranged from 28.4° (chair A) to 34.1° (chair E). The peak values (95th percentiles) differed not much from the 50th percentiles (minimum 33.8° for chair A and maximum 39.6° for chair C). According to the backward inclination of the pelvis during sitting, the L5 inclinations showed the expected negative angles with median values from -20.4° (chair E) to -13° (chair A). The 95th percentiles were still negative varying from -6.4° (chair A) to -12.6° (chair E).

The subjects were sitting in kyphotic lumbar spine postures on all chairs with median values ranging from 43.4 % (chair B) to 57 % (chair A) individual maximum kyphosis.

The characteristic values of the posture/joint angle frequency distributions differed not much between the different chair types with no significant effects. With the exception of the trunk lateral flexion the tasks had a highly significant effect on posture and joint angles. The analysis of the chair x task interaction showed only for the median values of the cervical spine flexion and trunk inclination angles significant effects.

CHAIR PARAMETERS

Concerning the chair parameters the comparison of chairs showed significant differences. The median values of the seat pan inclination angles varied from 0.2° (chair B) to -3.1° (chair E). The comparisons of the specific dynamic chairs B and C to the reference chair showed significant results for the median values and 95th percentiles. The lateral seat pan inclination angles showed nearly no lateral movements with angles close to the horizontal plane for chairs A, C and D. Only chair B and E showed noteworthy lateral peak angles (5th percentiles) of -1.1° (chair B) and -0.9° (chair E). Here, the seat pan inclination was particulary to the left. For that reason chairs B and E differed significantly from the reference chair D concerning the lateral seat pan inclination.

The median values of the backrest inclination angles were similar for all chairs ranging from -3.3° (chair B) and -8.2° (chair E). For the peak values of the backrest inclination angles chair E showed the highest backrest inclination angles (less than -

11° for the 5th percentile). This effect was not significant in comparison to the reference chair D (-8.1°, 95th percentile).

From the pressure distribution measurements on the seat pan and backrest the percentage of time spent in different sitting postures (front, central and back positions) was determined. As a result, no differences between the chairs were observed: The front and central sitting postures were preferred with mean percentages of approx. 40 % in front and 50 % in central sitting postures on all chairs. Rear sitting postures were only observed for approx. 10 % of the total measuring time for all chair types.

DISCUSSION

The aim of the present study was to analyse the effects of five different office chairs on body movements and muscular activation. Four specific dynamic chairs equipped with non-standard structural elements have been compared to a dynamic standard office chair as a reference in laboratory. Referring to the results of this study the research question (Do the specific dynamic office chairs influence physical and muscular activity more than the conventional office chair?) has to be negated for the here tested specific dynamic chair types.

For the m. erector spinae low activities with median values less than 20%RVC were measured for each standardized task. In average over all tasks the muscular activity did not exceed 11 % RVC. No significant differences of the m. erector spinae activity between the specific dynamic chairs (A, B, C, E) and the reference chair D could be determined. On the contrary the performance of the different office tasks had a significant effect on the m. erector spinae activity. These low level erector spinae activities were also reported in other studies.

Andersson et al. have performed several studies to examine erector spinae muscle activity during sitting tasks and observed that maintenance of sitting postures requires low-level muscle activity of the m. erector spinae (e. g. Andersson et al., 1975).

Wittig analysed alternative sitting concepts (e. g. swivel chair and pendulous chair) in comparison to a conventional office chair and found very low level activities of the m. erector spinae that were not significantly different for the different alternative chairs in comparison to the conventional chair (Wittig 2000).

Van Dieën et al. analysed different office tasks (word processing, reading and CAD work) in a comparative laboratory study with three different office chairs. The m. erector spinae activity was very low. The m. erector spinae activity was also strongly affected by the tasks performed, but not by the chair type (Van Dieën et al., 2001).

On average over all tasks the median activity levels for the m. trapezius were less than 16 % RVC. The statistical analysis lead to no significant differences between the specific dynamic chairs compared to the reference chair, but very significant differences for the m. trapezius activation levels for the task analysis were found.

The measurements of postures and joint angles yielded no significant differences between the specific dynamic chairs and the reference chair. On average over all tasks front and central sitting postures were adopted with more or less neutral trunk inclination angles and moderate trunk flexion angles. As expected, the kyphosis postures of the lumbar spine were dominant with median values ranging from approx. 43 % (chair B) to 57 % (Chair A) of max. individual kyphosis.

Many significant effects for the influence of performed tasks on postures and joint angles were found. This is inline with other studies.

Van Dieën et al. used a marker video system to analyse the effects of three different office chairs on trunk kinematics. The average position of the trunk at C7 was here not affected by the chair type, nor by the interaction of chair type and task, whereas the performance of the office task had a significant effect (Dieën et al., 2001).

In the study of Wittig no significant differences between alternative seat concepts and a conventional office chair were found concerning the lumbar spine flexion angles during sitting at VDU workplaces (Wittig, 2000).

The analysis of physical activity intensity (PAI) values resulted in low physical activities for all chairs and tasks. In average over all tasks the median PAI values did not exceed 0.8 % g for all sensors.

For the measurement of chair parameters (backrest inclination, seat pan inclination, seat pan inclination sideward) lots of significant differences between the specific dynamic chairs and the reference chair were observed, probably attributed to the variable possibilities of seat pan inclination of the five different chair types. Despite of these individual characteristics, the increase of mobility of the specific dynamic chairs, claimed by their manufacturers, could not be observed. However, the different dynamic characteristics of each specific dynamic chair did have no significant effects on muscular activation, almost no effects on postures, joint angles and physical activity in comparison to the reference chair.

CONCLUSION

The comparison of four specific dynamic office chairs to a conventional office chair in laboratory and field lead to almost no significant differences concerning muscular activation, postures/joint angles and physical activity during the performance of standard office tasks. On the contrary the performed tasks strongly affected the measured muscle activation, postures and kinematics. The characteristic dynamic elements of each specific chair yielded significant differences of the measured chair parameters, but these characteristics did not really affect the sitting dynamics. The results of the study emphasize that many aspects of workplace design, e. g. variability of tasks and work organisational factors, should be considered to avoid physical inactivity at VDU workplaces and therefore prevent musculoskeletal disorders.

ACKNOWLEDGEMENTS

This project was initiated by the VBG German Accident Insurance for Administration and funded by the German Social Accident Insurance (DGUV). The authors are grateful to Rene Hamburger, Ingo Hermanns and Merle Blok for their support. Special thanks also to Beverly Schlenther, Melanie Niessen and Daniel Annemaier for their help with data collection and analysis.

REFERENCES

Andersson G.B.J., Ortengren R, Nachemson A, et al. (1975), *The sitting posture: An elecromyographic and discometric study*. Orthop Clin North Am 6, 105 – 120.

Ellegast, R., Hermanns, I., Schiefer, C. (2009), *Workload assessment in field using the ambulatory CUELA system*. In: V. G. Duffy (Ed.): Digital Human Modelling, HCII 2009, LNCS 5620, Berlin: Springer, 221-226.

Evans, W., Jobe, W., Seibert, C. (1989), *A cross-sectional prevalence study of lumbar disc generation in a working populatio*. Spine 14, 60-64.

Glitsch, U., Keller, S., Kusserow, H., Hermanns, I., Ellegast, R.P., Hüdepohl, J. (2006), *Physical and physiological workload profiles of overhead line service technicians*. In: Pikaar, R.N., Koningsveld, E.A.P., Settels, P.J.M (eds.) Proceedings IEA 2006. Elsevier Ltd., Oxford.

Looze, M.P. de, Kuijt-Evers, L.F.M., Dieën, J.H. van (2003), *Sitting comfort and discomfort and the relationships with objective measures*. Ergonomics 46, 985-998.

Luttmann, A.; Kylian, H.; Schmidt, K.-H.; Jäger, M. (2003), *Long-term EMG study on muscular strain and fatigue at office work*. In: Quality of work and products in enterprises of the future: proceedings of the Annual Spring Conference of the GfA on the Occasion of the 50th Anniversary of the Foundation of the Gesellschaft für Arbeitswissenschaft e.V. (GfA) and the XVII Annual Conference of the International Society for Occupational Ergonomics & Safety (ISOES), Munich, Germany, May 07th - 09th.

Parent-Thirion A, Fernández ME, Hurley J, Vermeylen G. (2007) Fourth European Working Condi¬tions Survey. Dublin: European Foundation for the Improvement of Living and Working Conditions.

Van Dieën, J.H., De Looze M.P., Hermans, V. (2001), *Effects of dynamic office chairs on trunk kinematics, trunk extensor EMG and spinal shrinkage*. Ergonomics 44, 7, 739-750.

Videman, T., Nurminen, M., Troup, J. (1990), *Lumbar spinal pathology in cadaveric material in relation to history of back pain, occupation and physical loading*. Spine 15, 728-740.

Wittig, T. (2000), *Ergonomische Untersuchung alternativer Büro- und Bildschirm-arbeitsplatzkonzepte*. 1. Auflage. Bremerhaven: Wirtschaftsverlag NW Verlag für neue Wissenschaft GmbH 2000. (Schriftenreihe der Bundesanstalt für Arbeitsschutz und Arbeitsmedizin: Forschungsbericht, Fb 878).

Chapter 50

XDM - eXtensible Design Methods

André Neves, Fábio Campos, Sílvio Campello,
Leonardo Castillo, Leo Falcão

Department of Design
Universidade Federal de Pernambuco
Recife, PE, Brazil

ABSTRACT

This article discusses the results of a work developed in the last few years by a group of lectures and students of the Department of Design of the UFPE, Brazil. Our focus is the development of a design process consistent with the cyber culture of these current times. The major outcome is a process for artefact design named as eXtensible Design Methods [XDM]. The methods proposed have as a fundamental assumption the use of the cyberspace as a supporting environment. This strategy, we believe, confers to the activity of design a contemporary quality, bringing concepts and practices of the cyber culture into the context of the methodology of design.

Keywords: Exemplary Chapter, Human Systems Integration, Systems Engineering, Systems Modeling Language

INTRODUCTION

The cyber culture emerges from the social arrangements brought about by the communities of the cyberspace (Lemos, 2006). These communities expand and spread the use of the Internet, building up a new structure of worldwide relationships never before experienced.

On this contemporary scenario, the designer practice should also change and transform itself. It is in this sense that we propose a cyber culture perspective as a basis for the application of sharper, globally interlaced methods of design, capable to respond to different social contexts.

It is our focus the accomplishment of a design process that fits to the contemporary cyber culture. In this sense, a glance into the bigger picture of the development of design methods, points out to an evolving path. It is possible to identify a gradual move from hermetical geographic surroundings to the openness of environments with broader spatial limits as a cultural mark (Vries et al 1992).

As a major outcome of our investigation and our practice, we developed a process for artefact design named XDM (eXtensible Design Methods). XDM encompasses traditional methods of design, incorporating cyber culture habits in them during its application.

Aiming at the validation of XDM, we offered a group of courses in the undergraduate and graduate design programmes of UFPE. In those courses, the students designed digital games, costumes, graphic pieces, micro-electronic equipment and digital information systems using XDM.

At the present state of our research, two major points stand out. Firstly, the cyber cultural approach gives a great differential to XDM in relation to other design processes. Secondly, testing XDM with the students demonstrated that it is actually very easy to assimilate the use of the Internet as a platform for the methods proposed. This shows an important potential of XDM for pedagogical use.

The experiments point out that it is still necessary to improve the adaptation of certain methods when applying them to specific artefacts. It is also important to give some thought to the use of some Internet tools as a basis for the application of certain methods. Besides, it is required to apply XDM within a non-academic environment.

This document is divided in five more sections, besides this introduction. In the section 2, it is presented the design as a concept and as practice. The section 3 discusses the cyber culture, the time-space ambience within which we contextualize our research. Section 4 presents the XDM and its particularities. In section 5 it is raised some considerations about the experiment made. Section 6 draws on general conclusions about our work and points out possible developments of the research.

DESIGN

In this section the concept of design is presented, as it is understood within the research context.

Design as a concept

Within the scope of this work, Design is treated as an activity in which all kind of artefacts is conceived (Jones, 1991). The range of artefacts includes utensils, costume, graphic work, book, machine, environment, software, etc.

Specifying the definition, we adopt a current approach of the philosophy (Mora, 2001) that defines Design as related to an objective, a purpose. In this view, Design is placed between classical ideas of subject and object. Therefore, it is seen as opposite to arbitrary creation, without objective.

Design as practice

It is not our goal to present an ample and detailed picture of the evolution of processes and methods of design. However, this evolution will be briefly described in order to contextualize our proposal.

In the 1960s, researches investigating the processes and methods of design began to be built up. Encouraged by the increase in the complexity of problems, the amount of information involved in the search for a solution, and especially by the great and diverse demand of designs, the community began to look into the black box of the conception process. The aim was to make it clear and transparent the activity of design, allowing the replication of acts and procedures (Jones, 1992).

This movement engendered a first generation of design processes, structured in a linear form with each phase occurring after the previous being closed. In general terms, the process began by gathering information about the problem to be attacked, then was followed by phases of alternative designing, and finished by the selection of the best option (Burdek, 2006).

In that period, products were developed and their acceptance was evaluated only after launched in the market. The conception path was from the industry to the consumer. Among the methods used in this period it is possible to highlight the classic brainstorming, the morphologic box, and the semantic differential. These processes were mainly influenced by the aeronautic industry and have as most important authors, Rittel (1984), Asimov (1962), Munari (1998), Alexander (1964), Bonsiepe (1991), Lobach (2001) and Jones (1991).

The progress in methodology research and the consolidation of the activity of Design established a second generation of design processes. This second moment was largely noticeable by the fact that the phases did not occur in a linear form. On the contrary, it was depicted as a cycle, where each phase feeds back the other and the designer is able to go back to a previous phase at any time of the process.

This second generation is characterized by user-centred and argumentative methods. The solution of the problem is based on the satisfaction of users, who are seen as partners in the search for the best alternative. Among the methods developed in this period, definition of *personas* (Cooper, 1998), scenario planning (Fahey & Randall, 1998), and immersion techniques may be emphasized.

Therefore, a broad look into the development of design methodology points to an evolving path that comes out from the hermetic space inside the industry and gets to the openness of the exterior world. In this sense, we believe that a next generation of design methodology is about to emerge, adopting more dialogical structures with the real-world and incorporating concepts and habits of the cyber culture.

CYBER CULTURE

In this section, we present the cyber culture as a space-time environment where it is delimited the social-cultural universe of our work.

Cyber Culture as a concept

The term cyber culture has several meanings. However, it is possible to understand cyber culture as the social-cultural structure that emerges from the exchanges between society, culture and the new technologies based on microelectronics that started to appear in the 1970s, thanks to the convergence between telecommunication and computer systems.

Cyber Culture as a concern

Cyber culture emerges from the social dealing of communities in the virtual space. These communities are expanding and spreading the use of Internet. It is also spreading other communication technologies, making possible closer relations amid people from different parts of the world (Lemos, 2006). It relates directly to political, anthropological, social, economical, and philosophical dynamics of the individuals connected in the Web. The term is an attempt to encompass all the developments originated by the distinct behaviour that emerges from being connected world widely.

Cyber culture is not to be understood as driven by technology. Indeed, it is the establishment of a close relationship between the new social forms surfaced in the 1960s — a post-modern society (Farrell, 2006) — and the new digital technologies. In other words, cyber culture is the contemporary culture, which is strongly moulded by these new technologies. It is the present moment: home banking, intelligent cards, electronic vote, pages, palms, government taxes paid by Internet, Internet application forms, etc. These show that cyber culture is here and now impacting each human individual's life (Levy, 2001).

In this contemporary field, the design practice changes and re-shapes itself. That's why we propose a cyber culture approach as a basis for applying sharper and globally intertwined design methods.

XDM - EXTENSIBLE DESIGN METHODS

Process

Because of the conception process perspective of XDM, it may be placed closer to the second generation of design methods. The XDM approaches are user-centred and basically consider the industrial plant as a place where the conceived artefacts are made. However, although this is the major option made, XDM associates methods from both generations of design methodology and modifies them in order to incorporate aspects of contemporary cybernetics.

XDM builds up layers of information obtained from the application of design methods performed in cybernetic environment. That information is registered in "Weblogs", a type of journal published in the Internet which represents the stages of XDM process. A Weblog is made of a set of "posts" — messages inserted in chronological order by the author (or authors) of the Weblog. The posts of each Weblog may be commented by a third person through the Internet. In the case of XDM, we claim that the designer should invite a group of professionals and potential users to be part of the process. They may directly contribute to conceive the final design, commenting posts published during the process.

In XDM, design process is registered in a set of interconnected Weblogs: [1] *Weblog of problem exploring*, which aims registering information about the design to be made and to help the designer to increase her/his repertoire related to the artefact to be conceived; [2] *Weblog of alternative generation*, which has the goal of registering the possibilities of solution for the problem that has been worked out — this is built using methods that stimulate creativity; [3] *Weblog of alternative selection*, which has the objective of reducing the number of solutions, leading to an alternative close to the final design; [4] *Weblog of alternative evaluation*, which intends to recording critiques and reviews of the design proposed, using methods that points out problems and suggests necessary adjustments; and [5] *Weblog of description*, which has the aim of describing the final design, detailing aspects of production and use of the conceived artefact.

The creation of Weblogs is a cyclic process: new posts may be added to a Weblog adding information obtained during the elaboration of another Weblog.

Methods

As said above, the XDM conception process of artefacts involves the use of a set of design methods. However, the concept of extensibility of XDM is directly related to the fact that the designer may use a sub-set of the suggested methods. That depends on the type of artefact to be designed and/or the amount of time available for the project. Besides, new methods not listed here may be added to XDM.

The following list shows a set of methods we suggest as a basis for the XDM process.

Problem exploring methods

The main goal of such methods application is the elaboration of an open wide scenario about the artefact to be designed. It is from this information that the designer will conceive her/his solutions. To apply those methods, we suggest an intensive use of search engines and information recovering in the Internet. The paragraphs that follow describe each some of those methods and available Internet tools that may assist the designer in the task.

[1] **Definition of initial motivations**: in this method, the designer must describe the major aspects that encourage the design, making quick searches in databases and virtual social networks;

[2] **Market estimation**: this is an instrument to demonstrate the economic importance of the design. The designer must search for information in cyberspace databases in order to identify the promising market and potential users.

[3] **Similar design comparison**: this is an analysis of similar products in order to identify the main positive and negative characteristics. The designer must search for information through Internet search engines, using keywords related to the central theme of the project. Similar products must be gathered, their characteristics pointed out, and their images kept, when possible.

[4] **History evolution**: this is an investigative method that allows gathering a wider range of information about the evolution of a specific artefact along time. Using keywords for searching in the Internet, the designer registers the data about that evolution in a timeline.

[5] **Tendency analysis**: this is a research based on scrutinizing and identifying the tendencies of use and new technologies. The designer must search for information in the Internet, using core concepts and words related to the major theme of the project. The type of similar product that people seem to desire and the main innovative technologies related to the project must be registered.

[6] **Immersion research**: this is an investigation of positive and negative relations between users and similar artefacts. The designer must search for textual information and images using keywords within virtual communities and social networks in the Internet. The objective is to identify positive and negative aspects in the interaction between users and similar products, using the observations made by members of those communities or directly interviewing them.

[7] **Personas definition**: this is basically a specification of characters, which represent the user's profile of the artefact to be designed. The designer must define types of potential users based on all data analyzed during the problem-exploring phase. Then, characters encompassing the social-cultural characteristics of such users must be described.

Alternative generation methods

The main goal of such methods is to broaden the possibilities of resolution of the developing artefact. Within the application of the following methods, it is suggested the use of communication tools, both synchronous and asynchronous, available in the Internet. In the next paragraphs, we present the suggested methods for alternative generation:

[1] **Brainstorming**: a collaboration group aiming at the generation of possible solutions for a particular problem. Participants are encouraged to work out alternatives of each other's suggestions, building up new courses of action. The designer must establish a group of 4 to 12 participants directly involved in the developing product. The group should meet using a synchronous communication tool in the Internet — such as an instant messaging system. A coordinator must be defined, with the responsibility of exposing the problem, initiating the discussion, and organizing it, encouraging people to contribute with their own ideas. At the end of session, the coordinator must register and rank the most important ideas.

[2] **Brainwriting**: also a type of collaborative group, however the aim is to get to a level up using the ideas suggested during the brainstorming session. The designer must define another group of 4 to 12 people directly involved with the product being developed. A coordinator must also be chosen, with the same responsibilities of exposing the problem, organizing discussions, and ordering ideas. The coordinator must send by email the main ideas to be scrutinized by the group. Finally, he/she must register the evolution of the most important ideas.

[3] **Analogies and metaphors**: an activity of association between applied solutions to other fields and the artefact to be designed. The designer must search for artefacts in the Internet, not necessarily directly linked to the one being developed, but those that are somehow related to the possible solutions pointed out during the brainstorming or the brainwriting. The designer must register products that show characteristics possibly useful to the final design.

[4] **Morphologic box**: a matrix built in order to facilitate the development of new ideas and resolutions. It is a working principle to combine solutions for structural

and functional elements previously selected for the artefact. The designer must define a set of relevant features for the artefact configuration (the columns of the matrix) and build a matrix where each line represents possible variations of a specific attribute. Then, a characteristic of one cell may be combined with another's with the intention of creating an alternative configuration.

Alternative selection methods

The main purpose of this phase is to reduce the number of alternatives engendered.

[1] **Iterative evolution evaluation**: a process of alternative evolution where the designer develops her/his choices and checks how much each alternative gets closer or farther to an idealized option. Control is made using a focus group.

The designer defines based in her/his knowledge about the problem a set of expectations about the artefact. Upon those characteristics, contour line drawings picturing sketches of alternative designs are made. The sketches must be published in the Weblog of alternative selection and shown to online focus groups. The focus group must involve potential users and people from the industry and market sectors, willing to give their opinions about the alternative designs. The focus group opinion must be compared to the expectations defined by the designer. In the case of none of the alternatives be close enough to the expectations, the cycle starts again up to this point.

When at least one option stands up to the expectations, both from the designer and from the focus group, presentation drawings must be prepared, describing the option in detail to allow a more accurate understanding of the choice. The presentation drawing and a written description must then be published and shown to the focus groups once more, repeating the cycle. The opinions collected must be again compared to the expectations defined by the designer. In case of none of the alternatives are close enough to the expectations, the iterative evolution cycle must restart from the beginning. When one choice stands up to the expectations, both of the designer than of the focus groups, this is the solution that has to be adjusted and treated as the final choice.

Alternative evaluation methods

[1] **Heuristics evaluation**: analysis of the solution suggested for the problem, pointing out violations and proposing adjustments to minimize them. In this sense, we advocate a heuristic evaluation carried on the Internet.

The designer must define a collection of heuristics appropriate to the artefact designed, determining a rank of priorities and relevance. Then, those heuristics must be published in the Weblog of alternative evaluation. Images of the evaluated alternative must also be published. Experts and potential users may thus answer the heuristics and recommend adjustments to those they found infringed.

The designer must then verify each infringed heuristic, classify them according to the priority and implement the necessary correction in the alternative. After that, the design must once again be published in the Weblog to be evaluated by the experts and users. This cycle repeats itself until the design reach a satisfactory level in relation to the set of heuristics proposed.

Solution description methods

The aim of this phase is to present the final design to be implemented to a group of brief and understandable documents.

[1] **Use case**: a document that must have a kind of storyboard showing the artefact in use. The document must emphasize different use scenarios. The designer must list relevant cases of use to present the artefact and elaborate drawings showing different ways of using it, explaining each situation. The cases must be posted in the Weblog in order to be scrutinized by experts and potential users. After the critics, the designer must adjust the cases of use until reaching a satisfactory point for the group.

[2] **Technical specifications for production**: document which contains details of the artefact and its components. The designer must describe the core functionalities of the artefact and represent its components in a graphic manner. The description must be posted in the Weblog and presented to the production team. In case of doubts about the making of the artefact raise, the production team should be posted in the comments section of the Weblog with the purpose of allowing the designer to interact with the production team until reaching to a clear description that permits the production of the artefact.

CASE STUDIES

It was offered a group of courses in the undergraduate design programme of the UFPE, campus Recife, Brazil, intending to validate the XDM process. The courses were taught by different lectures, each of them with different skills and knowledge about specific types of artefacts. In those courses, students designed digital games, costumes, graphic jobs, microelectronic equipment and digital information systems.

Adaptations were made in the XDM for each course, adding, excluding or modifying details in the application of methods encompassed on each process document. These adaptations were presented to the students, which followed the methods proposed for the artefact design process.

At the end of the academic term it was obtained almost 50 projects developed using the XDM process.

The general results show that throughout the problem exploring phase the students exhaustively used the search engines and information recovery tools. These tools made a significant contribution to the quality of the final designs.

In the alternative generation phase, only a small part of the students used the synchronous communication tools in the Internet to perform brainstorming sessions. The vast majority performed face-to-face sessions with their course mates.

In general terms, the use of email messages was also widely employed on the application of the series of methods recommended. That was particularly true during the alternative selection and during the evaluation of alternatives to refine the final solution. The use of interaction tools by means of the proposed Weblogs was also small, confined to some students with a clear experience in using such tools.

FINAL CONCERNS

In this section, it is presented some of the major contributions foreseen on this work. It is also proposed some possible developments in our own research.

Contributions

Two core points draw our attention on this first stage of our research. Firstly, the conception of XDM itself, which brings a cyber culture approach into design methodology and offers a chief differential in contrast with other methods. Secondly, the experiment with the student demonstrated that they were not familiar to some design methods and showed great easiness in appropriating the use of Internet as a basis for applying XDM.

It is noticeable among students an important familiarity with the cybernetic environment. This general knowledge led to an trouble-free appropriation of some of the tools proposed for the application of methods, revealing the contemporary aspect of XDM.

Future Developments

It is possible to infer from the experiments made that we will have work hard on the adaptation of some methods when applied to specific types of artefacts. It is especially important to grow a better knowledge of the use of certain Internet tools for using them as a basis for some methods. It is also necessary to broaden the range of the experiments going out of the academic boundaries.

Besides, we initiated the definition of a tool for publishing specific Weblogs oriented to XDM. We expect to be able to apply this tool in the next experiments, hoping to reduce the initial resistance to open Weblogs in the Internet we detected throughout this first stage of the research.

REFERENCES

Alexander, Christopher (1964). Notes on the Synthesis of Form. Cambridge: Harvard.

Asimov, Morris. (1962) Introduction to Design. Prentice-Hall: New Jersey.

Bonsiepe, Gui (1991). Teoria e Prattica del disegno industriale. Feltrinelli: Milan.

Burdek, Bernhard E. (2006) Design: The History, Theory and Practice of Product Design. DuMont: Cologne.

Cooper, A. (1999). The Inmates are Running the Asylum: Why High-Tech Products Drive us Crazy and How to Restore the Sanity. Indianapolis: SAMS.

Fahey, Liam and Randall, Robert M. (1998) Learning from the Future. New Jersey: Wiley.

Farrell, John (2006). Paranoia and Postmodernism - the epilogue to Paranoia and Modernity: Cervantes to Rousseau. Cornell University Press: Cornell.

Jones, John Christopher. (1991) Designing Design. Architecture Design and Technology Press: London.

Jones, John Christopher. (1992) Design Methods. John Wiley & Sons: UK.

Lemos, André. (2006) Les trois lois de la cyberculture. Libération de l'émission, principe en réseaux e réconfiguration culturelle. Revue Société: Brussels.

Levy, Pierre. (2001) Cyberculture: Electronic mediation volume 4. University of Minnesota Press: Minneapolis.

Löbach, Bernd (2001). Design Industrial: bases para a configuração dos produtos. São Paulo: Edgard Blücher.

Mora, José Ferrater. (2006) Dicionário de Filosofia. São Paulo: Martins Fontes.

Munari, Bruno. (1998) Das coisas nascem coisas. Martins Fontes: Sao Paulo.

Rittel, H. (1984) Second-Generation Design Methods, in Developments in Design Methodology. John Wiley & Sons: UK.

Vries et al. (1992) Design Methodology and Relationships with Science: Proceedings of the NATO Advanced Research Workshop. Eindhoven, Sept. 29 - Oct. 2.

Chapter 51

Competence Profile of Top-Level Controllers

Tapio Salminen[1], Minna Kallio[2], Hannu Vanharanta[1], Barbro Back[2]

[1]Tampere University of Technology
Pori, Finland

[2]Åbo Akademi University
Turku, Finland

ABSTRACT

The paper aims at evaluating the profile of top-level controller using a new created self-evaluation computer application, called Strophoid. Strophoid consists of fifteen most important competencies of controllers. The collection of competencies was collected based on literature studies mainly from management accounting area as well as from interviews with some experienced controllers. Strophoid was used on a group of controllers, who worked in different international companies. They had to answer a number of statements with internet based Strophoid application about their current competencies and their proactive visions concerning the statements. The combined results showed that the controllers are very achievement oriented and master communication and self control well. The biggest improvement tension occurred between analytical thinking and conceptual thinking.

Keywords: Competence model, Controller, Strophoid

INTRODUCTION

The occupational work role of controllers or management accountants has changed during the last decades. The role is said to have changed from "bean counters" to business partners. Controllers are said to be "clear winners in the technical revolution that fostered the new information economy" (Russell et al 1999); they are now wanted experts in their organizations. This change now creates new kind of occupational work role demands. A great deal of attention has been paid to education i.e., does the university education really respond to the needs, which the forthcoming professionals will face or to the needs which the business will require (Bots et al. 2009, Kavanagh et al. 2008, Albrect et al. 2000). This paper will take another approach and we will study the competence profile the top level controllers have today and also the vision these controllers see in the future. Our approach is looking also in detail at the features that the top level controller must have, to be able to make most of their education and to be able to qualify for the controller's work role. Based on these features we construct first occupational work role ontology for controllers which is then translated to a computerized model, named Strophoid.

We will define the current and future profile using a self evaluation method. According to that method each respondent evaluates statements measuring competences according to his or her own personal current situation and future vision about his or her occupational work role, here the controller work. This construction method has been used to several times for other occupational work roles, i.e. competence models (cf. Kantola et al. 2007; Kantola et al. 2006; Chang et al. 2009 and www.evolutellc.com/).

The rest of the paper is organized as follows; we will first introduce the working roles of controllers. Then we will define the top level controller and presents the skills and abilities demanded to this specific occupational work role. Then we describe the created competence model, Strophoid. Finally, we will present the results, achieved by testing Strophoid with test subjects.

THE OCCUPATIONAL WORK ROLE OF CONTROLLERS

Partanen (2007) has used a set of metaphors to describe different working roles that controllers have. The roles are divided to three classes; information and controlling, interaction and leadership, and future oriented roles. The metaphor of future orientation is according to Partanen working like a co-driver. In this role the work is not a routine, but has a more free way to see the opportunities in business. The interaction and leadership roles have two metaphors; a bridge builder and a trusted man of the business leader. The bridge builder works connecting different parts of the global or international organization. The trusted man is needed supporting and participating in business decision making. The set of information and control has tasks described like an ambassador, a trainer, an interpreter, an information officer or even a social authority or supervisory police officer. The main thing is in different situations to pass information or work as an expert that either knows the solution of the problem or can process it further.

TOP-LEVEL CONTROLLER

As the work and tasks of controllers have become more demanding during the technical revolution of business we define the top level controllers as the ones, who have been or will be able to change form "bean counters" to business partners. Thus, we create the competence model of controllers/management accountants after the demands and changes described in literature. Burns and Scapens (2000) used the attribute "more proactive" as they talked about managements accountants, who are "part of the management team within a business process" Oxford English Dictionary describe the word "Proactive" as "creating or controlling a situation rather than just responding to it." We thought this is as an important part of being a top level controller, and measured it as a competence denominated "proactive thinking". Furthermore, Burns and Scapens consider the role of an accountant very important because s/he has knowledge of accounting and business, and is able to understand and advise on the impacts that actions could have on business.

Albrecht and Sack (2000) reported a study with an approach to develop accountant education to qualify the demands of business. They interviewed representatives of both groups. The four "top skills" were same for both groups: written communication, analytical/critical thinking, oral communication and computing technology, although the order was not exactly same for both groups. The fifth important skill according to the representatives of education was decision making, which was the sixth for the practitioners, and according to practitioners the fifth skill is teamwork.

Siegel et al. who have studied the competence and work of management accountants with a longer view (Siegel et al. 1996, Russell et al. 1999, Siegel et al. 2003), argue that becoming a top level controller, a business partner, is not a sudden happening but a process that is depending on controllers themselves. They have stated that "vision, hard work and perseverance are needed" (Siegel et al. 2003 p. 38). We have interpreted the before mentioned need as achievement orientation. In 1996, Siegel et al. introduced the Practice Analysis, a study to collect information of tasks and activities and as well of competencies and skills needed to do the work. According to the Practice Analysis the three most important knowledge skills and abilities (KSA) are: work ethics, analytical problem solving skills and interpersonal skills.

Russell, Siegel and Kulesza (1999) conducted a study and asked the management accountants to evaluate things they have learned in their job over the past five years. The most common answers were related to computers: skills, technology, networks and software. Others were mostly technical skills and facts of accounting and business, but 15% of the answerers mentioned communication and 9.5% other interpersonal skills. When they were asked to describe the most important knowledge, skill and abilities (KSA), they mentioned communication, ability to work as a team, analytical skills, solid understanding of accounting and an understanding of business functions. In our competence model we did not have the item of accounting, but all other important KSAs listed in the before mentioned study were included in our competence ontology. For example "understanding of business functions" is measured by two competencies; the *proactive thinking* and as well *comprehending strategy*. Bots et al. (2009) conducted a survey adapting the competency model of Birkett (2002). The model of Birkett consists of skills, which have been defined using five levels. The first level is divided to two skills; cognitive and behavioral. On level 2 the cognitive skills are divided to technical, analytical/design and appreciative skills, and the behavioral skills to personal, interpersonal and organizational skills. Those are then again divided to more detailed skills on next levels. On the last level of Birkett has 375 different skills called competency elements. Birkett as well defines five levels for management accountants regarding their experience; novice practitioner, assistant practitioner, competent practitioner, proficient practitioner and expert practitioner.

The research hypothesis was that there exists a rank sequence of competencies; during the first level of experience just a small set of competencies is demanded. Each higher level of experience is achieved as higher level competencies are yielded. Another research hypothesis, which they had, was that accountants perceive the most recently yielded competencies as the most relevant. Bots et al. have tested the model of Birkett by a survey and found that both hypotheses were supported.

THE COMPETENCE ONTOLOGY

Kantola (2005) lists various purposes of the competence model. It could be used in analyzing employees' performance, supporting organizational changes, training and development, recruitment and selection, career planning, flexibility analysis of workforce. The core of using the model is to improve the performance of the organization, which is dependent on employees. We have here created the competence ontology to be able to describe the profile of a top level controller, which competences s/he evaluates high and which low. Our model includes two groups of competencies; personal and social competencies (Table 1.). Personal competencies consist of self-knowledge, self control, cognitive ability and motivation. Self-knowledge corresponds to self-confidence i.e., the ability to value one's own opinions, to be able and willing to defend them and furthermore to be able to admit own limitations.

Self control corresponds to innovativeness, stress tolerance and self-control in different kinds of situations. Cognitive capability is a large and significant area of human's ability to perform in work. It corresponds to analytical thinking, conceptual thinking, professional and technical expertise, comprehending strategy, skills in information technology and proactive thinking. By analytical thinking we mean the ability to see the factors from which the task consists and furthermore to find the cause and consequence relationships of problems. The conceptual thinking is the ability to generalize the problem and capture the details of it.

Table 1. The Competence Ontology of the Strophoid Application.

Competence group	Competence class	Competence
Personal	Self-knowledge	Self-confidence
	Self-control	Innovativeness
		Stress tolerance
		Self-control
	Cognitive capability	Analytical thinking
		Conceptual thinking
		Professional and techical expertise
		Comprehending strategy
		Developing IT-skills
		Proactive thinking
	Motivating oneself	Achievement orientation
Social	Empathy	Organizational savvy
	Social skills	Communication
		Collaboration
		Teamwork and cooperation

With the professional expertise and IT skill we did not measure detailed facts, but the ability to recognize one's own level and receive new knowledge to develop it further. The comprehending strategy corresponds to how a person sees the characters of the organization s/he is working in, how well s/he knows it. The proactive thinking is like the analytical thinking in practice; the abilities to see the details in and effects on the whole. Motivation as a personal competence is measured as *achievement orientation* i.e., a person's attitude to benefit from the goals and targets and the ability to set rewarding and supporting levels for goals. Social competencies are further divided into two areas; empathy and social skills. The empathy is tested by measuring the organizational savvy, which means how well the person is conscious of and can use and understand the formal and informal relationships between organization members. By social skills we do here understand communication, collaboration and teamwork.

RESEARCH METHOD

This research was a pilot to test the competence measuring statements found in literature as well as together with the experienced controllers from business world. It was targeted to a group of controllers, who worked in different international companies. The controllers participated during the test time in a high level, advanced university business course. Thus we can assume that we reached the before mentioned top level controllers or at least a set of becoming top level management accounting professionals. They did the self evaluation during one of their education sessions. Two of the authors of the paper were available in case the controllers had any questions considering the method or questions. However, the participating controllers did not have any special problems to do the evaluation with the Strophoid application.

THE STROPHOID COMPUTER APPLICATION

Liikamaa (2006) has argued that each single competence has an influence on how the person performs. Competences are also synergetic factors and a group of competencies may together affect the performance. She states further that professional groups have competences in common, but that each group has also specific features. We have used this thought in selecting the competences of the Strophoid self evaluation method, i.e. the Strophoid model is based on literature and on other self evaluation competence models (cf. www.evolutellc.com/) as well as the information which we received from the experienced controllers.

RESULTS

In Table 2. the evaluations of the respondents regarding the current state have been ordered after the values of the Friedman test. The competences on the upper rows of the table, with the grey back color are estimated to belong to the highest level. The lowest ones are with the blue color and the white ones in the middle do not belong either to high or low values, they are considered neutral.

The highest evaluation the respondents made was achievement orientation. This is very understandable as the sample consisted of persons who participated in the advanced and demanding education program. As we also tested the model and statements we can assume the statements considering achievement orientation were not wrongly understood. Factors considering achievement orientation have been discovered by Siegel et al. as we mentioned before. The values of competence class Self Control are dispersed; self control has high values but innovativeness and stress tolerance, which belong to the same class, have low values.

Social competences; communication, collaboration and organizational savvy have high values only the current state of teamwork and co-operation is very low. But if we then go further and look at Table 4. we can see the Creative tension of teamwork is high, the need to develop on that area has been recognized. The evaluations of the respondents (table 3.) regarding the target state have been ordered after the values of the Friedman test. We recognize that the order is not the same in Tables 2. and 3. Communication, Analytical thinking and Organizational savvy received now the highest values.

Table 2. Current State.

	VALUE	n=11, α=0.05, min. diff.=3,35
1	11,6	Achievement orientation
2	10,6	Self-control
3	10,3	Communication
4	9,8	Comprehending strategy
5	9,5	Professional and techical expertise
6	8,9	Organizational savvy
7	8,8	Collaboration
8	8,6	Self-confidence
9	8,3	Analytical thinking
10	6,8	Stress tolerance
11	6,8	Proactive thinking
12	6,5	Conceptual thinking
13	5,6	Innovativeness
14	4,2	Developing IT-skills
15	3,7	Teamwork and cooperation
120	120	

Table 3. Target State.

	VALUE	n=11, α=0.05, min. diff.=3,13
1	11,6	Communication
2	10,4	Analytical thinking
3	10,3	Organizational savvy
4	10,3	Achievement orientation
5	10,0	Self-confidence
6	9,8	Self-control
7	9,3	Professional and techical expertise
8	9,1	Comprehending strategy
9	8,4	Collaboration
10	7,3	Conceptual thinking
11	5,7	Innovativeness
12	5,6	Proactive thinking
13	4,7	Developing IT-skills
14	4,0	Teamwork and cooperation
15	3,6	Stress tolerance
120	120	

In Table 4. the results have been ordered according to the value of Creative Tension, i.e. the difference between current and target situation. The recognized creative tension could work as a force to develop oneself further in the competence area.

Table 4. Creative Tension.

	VALUE	n=11, α=0.05, min. diff.=3,61
1	11,1	Analytical thinking
2	9,9	Conceptual thinking
3	9,8	Organizational savvy
4	8,9	Self-confidence
5	8,5	Communication
6	8,5	Collaboration
7	8,5	Teamwork and cooperation
8	8,3	Developing IT-skills
9	8,3	Innovativeness
10	8,0	Professional and techical expertise
11	7,1	Self-control
12	6,9	Comprehending strategy
13	6,4	Achievement orientation
14	5,9	Proactive thinking
15	3,9	Stress tolerance
120	120	

Table 4. depicts that the competence analytical thinking, which has been identified in several studies to be an important factor, has the highest value. For some reason the controllers have evaluated the current situation quite low and the target to be the second high and thus the difference is big. As this competence is identified very important and the group is achievement oriented, the result can be seen as consistent. Although the test group seems to have several competencies in which the 'driving force' that is creative tension is missing; there are some explanations to that. There are two competencies belonging to 'self control' and here we can assume that the test group already works on the satisfactory level of that. 'Comprehending strategy' has as well low values of creative tension, but both current and target values are high.

Here we can interpret the results so that they already have committed to their organization well. The current state of achievement orientation is perceived the highest competence, as it should be considering top level workers. The competence 'proactive thinking', which Burns and Scapens (2000) defined very important and which here regarded as a competence which is especially considering the controllers, has in this test had low values in all -current, target and creative tension scales. This may show a need to develop further the statements measuring this competence. Maybe they do not describe the core of top level controller working role precisely enough.

CONCLUSIONS

Albrecht and Sack (2000) listed the skills needed by controllers. They mentioned written communication, analytical/critical thinking, oral communication and computing technology, decision making and teamwork. Communication has high values both current and target estimation. As the creative tension is high too, the competence is regarded very important. Analytical thinking is not regarded high in current situation but the need to develop it is clearly recognized. Computing technology here developing IT-skill has quite low values but the creative tension is high.

Siegel et al. (2003) compressed the core of advanced controller work to "vision, hard work and perseverance is needed" and the achievement orientation was evaluated very high. Two thirds of competences have high values of creative tension, which is a force that drives onward. This shows that the controllers of our test group are top level. The research conducted did not bring out the competence of proactive thinking stated for example by Burns and Scapens (2000). As it is a very important competence according to literature, we should further develop the statements we have used.

ACKNOWLEDGEMENTS

The financial support of the Finnish Funding Agency for Technology and Innovation No. 33/31/08 is gratefully acknowledged.

REFERENCES

Albrecht, W.S., Sack, J. (2000), Accounting education: charting the course through a perilous, *Accounting education series* 16, 1-72.
Birkett,W.P. (2002), Competency profiles of management practice and practitioners, A report of AIB, Accountants in Business section in the International Federation of Accountants. New York: IFAC.
Burns, J. and Scapens, R. (2000), The Changing Nature of Management Accounting and the Emergence of "Hybrid" Accountants*, IFAC Press Center Articles library*, Nov 2000.
Bots, J. M., Groenland E., Swagerman D. M. (2009), An empirical test of Birketts competency model for management accountants: Survey evidence from Dutch practitioners, *Journal of Accounting Education*, 27, 1-13.

Chang, Y., Eklund, T., Kantola, J. & Vanharanta, H. 2009. International creative tension study of university students in South Korea and Finland. *Human Factors and Ergonomics in Manufacturing* 19 6, 528-543.

Kantola, J. I. (2005), *Ingenious Management*, Tampere University of Technology, Publication 568,Tampere, Finland.

Kantola, J., Czainska, K., Karwowski, W. and Vanharanta, H., (2007), International Competence Development Project for Emergency Response Personnel, Human Aspects of Advanced Manufacturing: HAAMAHA 2007, Poznan, Poland.

Kantola, J., Karwowski, W. and Vanharanta, H., (2006), Operators' creative tension and shift performance, *Paper and Timber, Paperi ja Puu*, 4/06, Helsinki, Finland

Kavanagh, M. H., and Drennan, L. (2008), What attributes does an accounting graduate need? Evidence from student perceptions and employer expectations, *Accounting and finance*, 48, 279-300.

Liikamaa, K. (2006), *Piilevä tieto ja projektipäällikön kompetenssit*, Tampere University of Technology, Publication 628,Tampere, Finland.

Partanen, V. (2007), *Talousviestintä johtamisen tukena*, Gummerus Kirjapaino Oy, Jyväskylä Finland, 2007.

Russell, K. A., Siegel G. H., and Kulesza C.S. (1999), Counting more counting less transformation in the management accounting profession, *Management accounting quarterly*, Fall 1999, 1-7.

Siegel, G., Sorensen, J., and Richtermeyer, S. (2003), Becoming a business partner, *Strategic Finance*, October 2003, 37-41

Siegel, G., and Kulesza, C.S. (1996), "Bud", The Practice Analysis of Management Accounting, *Management Accounting*, April 1996, 20-28

Chapter 52

Nonlinear Synchronization of Biceps and Triceps Muscles during Maximum Voluntary Contraction

David Rodrick[1], Santosh Erupaka[1],
Rohit Kumar Sasidharan[1], Waldemar Karwowsk[2]

[1]University of Michigan – Dearborn
Dearborn, MI 48128

[2]University of Central Florida
Orlando, FL 32816-2993

ABSTRACT

The objective of this pilot study was to explore nonlinear synchronization of biceps and triceps muscles during maximum voluntary contraction (MVC) protocol for biceps muscles. Generalized synchronization indexes (S-, H-, and N-index) were utilized to explore synchronization and interdependence of bivariate biceps and triceps time series data. It was found that the more robust H- and N-index were able to distinguish the dynamical changes in three stages (pre-MVC, MVC, and post-MVC) of MVC protocol. According to the H- and N-index values, synchronization between biceps and triceps muscles was more pronounced in pre and post-MVC stage than in MVC. ANOVA analyses confirmed the significant difference ($p < 0.01$) of those index values between pre-, MVC, and post-MVC stages. A linear synchronization measure, cross-correlation, was utilized that failed to capture any

512

synchronization patterns between biceps and triceps muscles. It is argued that that nonlinear synchronization indexes could provide useful insight to the synergistic muscle activities and their relationships with respect to the task, posture control and stability.

Keywords: Electromyogram (EMG), Maximum Voluntary Contraction (MVC), Nonlinear Synchronization

INTRODUCTION

Coactivation of an antagonist muscle (e.g., triceps or biceps brachii) occurs during a maximum voluntary contraction of an agonist muscle (e.g., biceps). This phenomenon can be treated as coupled oscillatory systems. In the last two decades, there have been increased interests in the study of synchronization between coupled systems, especially chaotic systems. In order to quantify the synchronization or interdependence between two systems, several generalized synchronization indexes were proposed (Arnhold, Grassberger, Lehnertz, and Elger 1999; Quian Quiroga, Arnhold, and Grassberger, 2000). Another recent study by Rosenblum, Pikovsky, and Kurths (2004) showed that nonlinear synchronization was more useful than linear techniques (e.g., cross-correlation) to explain biological systems.

This study is inspired by the fact that generally study of coactivation of agonist and antagonist muscles during concentric and/or eccentric conditions is overlooked and there is no such consensus on analysis of coactivations and their relationships. In recent years, studies found nonlinearity in surface EMG during isotonic contraction (Yang and Zhao, 1998; Yang et al., 1999), isokinetic contraction (Gupta et al., 1997, Lei et al., 2001), and isometric contraction (Anmuth et al., 1994; Gitter et al., 1991; Gitter and Czerniecki, 1995; Karwowski and Rodrick, 2003; Rodrick and Karwowski, 2004, 2006). Following the findings of studies, it was argued that the dynamic (in)stability of the muscle system could be explained by the chaotic invariant measures, such as, largest Lyapunov exponent, correlation dimensions, fractal dimensions, and Kaplan-Yorke dimensions. It was also argued that system instability occurs as a result of perturbation of internal and external constraints of the system during both static and dynamic conditions. In static condition, forceful exertion (external constraint) could cause more instability (higher largest Lyapunov exponent values) in the muscle system compared to non-loading condition as evidenced in Rodrick and Karwowski (2004, 2006). In light of these findings, it is justified to examine the nonlinear synchronization patterns of synergistic muscle activities.

Thus, the main objective of this study was to explore nonlinear synchronization patterns of the surface EMG of biceps brachii muscles of maximum voluntary contraction (MVC) and antagonist triceps muscles during MVC of biceps.

METHODS AND PROCEDURE

MVC POSTURE DESCRIPTION

In the MVC posture, the MVC of the biceps brachii muscle was used, and surface EMG of both biceps and triceps were recorded. The subject pulled on a chain upward in a sagittal plane, with the lower left arm flexed at 90 degrees (parallel to the ground), with upper arm parallel to the sagittal plane.

PARTICIPANTS

A total of six healthy male student volunteers took part in this study. The mean age (in years), height (in inches), and weight (in lbs) of the participants were 23.00 ($\pm$ 1.67), 68.83 ($\pm$ 2.56), and 137.67 ($\pm$ 16.35), respectively.

MVC MEASUREMENT PROTOCOL

To measure the MVC, participants were asked to exert their maximum effort (without hurting themselves) by pulling on a chain with a load cell in the above mentioned posture that isolates the muscle under consideration (Chaffin and Andersson, 1999). Each Participant went through 3 trials with a 5-minute rest between each trial. The participants were instructed to gradually increase force in the first 3 seconds (pre-MVC), maintain the maximal force for about 3 seconds (MVC), and gradually decrease force (post-MVC) in the last 3 seconds.

PROCEDURES

The subjects were briefed about the purpose of the study. Measurements of the anthropometric variables were also made. For EMG recordings, the standard EMG techniques (Marras et al., 1990) were used.

Data were collected at a sampling rate of 1000 Hz using MyoSystem 1400A of Noraxon USA, Inc. (Scottsdale, AZ). The EMG signals were pre-amplified, high-passed filtered at 10 Hz and low-passed filtered at 500 Hz. Raw EMG data were utilized for the data analysis.

RESULTS

EMG DATA

A total of 108 time series data were used for analysis. Each participant's biceps and triceps EMG data of each trial (3 trials) was divided into three MVC stages – pre-MVC, MVC, and post-MVC. Figure 1 shows an example EMG of biceps and triceps.

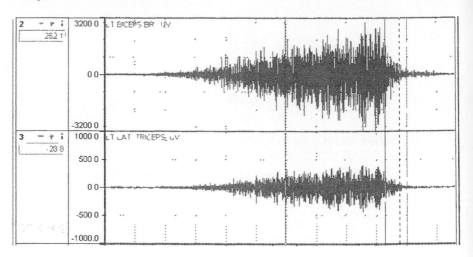

Figure 1. Illustration of biceps and triceps EMG signals

NONLINEAR SYNCHRONIZATION ANALYSIS

The followings steps were used to quantify the synchronization between biceps and triceps muscles in three stages of biceps MVC protocol.

Step 1. Determination of time delay
Time lag was determined by using average mutual information method. Time delay was selected where the first minimum information (in bits) was found. (Figure 2)

Step 2. Determination of minimum embedding dimensions
The minimum embedding dimensions were estimated using Cao's (1997) method. (Figure 3)

Step 3. Estimation of synchronization indexes
Timed delay and minimum embedding dimensions were used to estimate three indexes of synchronization: S-index (Arnhold et al., 1999), H- and N-index (Quian

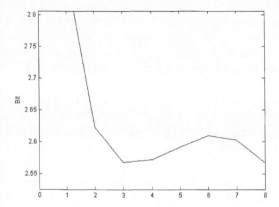

Figure 2. Time delay estimation by average mutual information

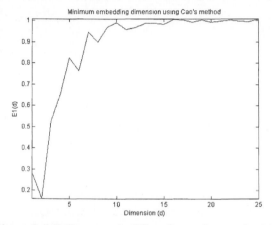

Figure 3. Minimum embedding dimensions estimated by Cao's method

Quoroga et al., 2002). Estimation of each index value required use of a pair of biceps and triceps EMG data. Thus, 54 synchronization values were estimated for each index. Each index computed two numerical values for a pair of biceps (x) and triceps (y) data under that notion that if x and y are oscillatory systems, lower values of these indexes would suggest independence and higher values would indicate synchronization of two systems or identical systems. For comparison purpose, cross-correlation between each biceps and triceps pair was calculated.

As mentioned in Quian Quoroga et al. (2000, 2002), the S-index is a normalized but less robust method. In S-index estimates so called normalized synchronization where a 0 value indicates independence of the systems while a value of 1 indicates complete synchrony. The authors showed that H-index was more robust method than S-index but this index would not estimate normalized values. In other words, a 0 value would indicate independence of the systems and positive values would indicate synchronization of the systems. Finally, the authors introduced a third index, N-index, which is normalized and in principle more robust than S-index. In N-index, as in S-, a 0 value would indicate independence of the systems whereas a value of 1 would indicate complete synchronization.

A 3x3 full factorial analysis of variance (ANOVA) was utilized to examine the effects of trials and MVC stages (pre-, MVC, and post-) on each of synchronization indexes. Tables 1 through 6 show the results of the main and interaction effects of trials and MVC stages on index values. Table 7 shows the results of main and interaction effect of trials and stages on cross-correlation values. Tables 1 and 2 show that there were no significant effects of trials and MVC stages on S x|y and S y|x indexes.

Table 1. ANOVA table of S x|y with respect to trial and MVC stages

Source	Sum of Squares	df	Mean Square	F	Sig.
Corrected Model	0.059	8	0.007	1.245	0.296
Intercept	4.073	1	4.073	693.034	0.01
stage	0.028	2	0.014	2.411	0.101
trial	0.003	2	0.001	0.223	0.801
stage * trial	0.028	4	0.007	1.173	0.336
Error	0.264	45	0.006		
Total	4.396	54			

Table 2. ANOVA table of S y|x with respect to trial and MVC stages

Source	Sum of Squares	df	Mean Square	F	Sig.
Corrected Model	0.064	8	0.008	1.201	0.32
Intercept	3.71	1	3.71	557.765	0.01
stage	0.025	2	0.012	1.873	0.165
trial	0.007	2	0.003	0.49	0.616
stage * trial	0.032	4	0.008	1.221	0.315
Error	0.299	45	0.007		
Total	4.073	54			

Table 3. ANOVA table of H x|y with respect to trial and MVC stages

Source	Sum of Squares	df	Mean Square	F	Sig.
Corrected Model	3.273	8	0.409	4.795	0.01
Intercept	13.786	1	13.786	161.611	0.01
stage	3.077	2	1.538	18.035	0.01
trial	0.069	2	0.035	0.407	0.668
stage * trial	0.126	4	0.032	0.37	0.829
Error	3.839	45	0.085		
Total	20.898	54			

Table 4. ANOVA table of H y|x with respect to trial and MVC stages

Source	Sum of Squares	df	Mean Square	F	Sig.
Corrected Model	2.887	8	0.361	4.792	0.01
Intercept	14.406	1	14.406	191.269	0.01
stage	2.823	2	1.412	18.742	0.01
trial	0.008	2	0.004	0.051	0.951
stage * trial	0.056	4	0.014	0.187	0.944
Error	3.389	45	0.075		
Total	20.682	54			

Tables 3 and 4 show that MVC stages have significant effects on H x|y and H y|x indexes. Post-hoc Tuckey HSD test shows significant difference between pre- and MVC and post- and MVC, while no significant difference between pre and post-MVC was found.

Table 5. ANOVA table of N x|y with respect to trial and MVC stages

Source	Sum of Squares	df	Mean Square	F	Sig.
Corrected Model	0.748	8	0.094	4.146	0.001
Intercept	4.639	1	4.639	205.554	0.01
stage	0.704	2	0.352	15.588	0.01
trial	0.012	2	0.006	0.266	0.768
stage * trial	0.033	4	0.008	0.364	0.833
Error	1.015	45	0.023		
Total	6.402	54			

Table 6. ANOVA table of N y|x with respect to trial and MVC stages

Source	Sum of Squares	df	Mean Square	F	Sig.
Corrected Model	0.699	8	0.087	4.278	0.001
Intercept	5.203	1	5.203	254.883	0.01
stage	0.688	2	0.344	16.846	0.01
trial	0.001	2	0	0.014	0.986
stage * trial	0.01	4	0.003	0.125	0.973
Error	0.919	45	0.02		
Total	6.82	54			

Tables 5 and 6 show that MVC stages have significant effects on N x|y and N y|x indexes. Post-hoc Tuckey HSD test shows significant difference between pre- and MVC and post- and MVC, while no significant difference between pre and post-MVC was found.

Table 7. ANOVA of cross-correlation with respect to trial and MVC stages

Source	Sum of Squares	df	Mean Square	F	Sig.
Corrected Model	.069a	8	0.009	0.202	0.989
Intercept	0.006	1	0.006	0.138	0.712
stage	0.001	2	0.001	0.017	0.983
trial	0.032	2	0.016	0.378	0.687
stage * trial	0.035	4	0.009	0.207	0.933
Error	1.911	45	0.042		
Total	1.986	54			

Finally, table 7 shows that there were no significant effects of trials and MVC stages on cross-correlation between biceps and triceps EMG data.

A paired sample t-test was used to examine any significant difference between S x|y and S y|x, H x|y and H y|x, and N x|y and N y|x. The tests found significant difference ($p < 0.05$) between S x|y and S y|x but there was no significant difference ($p < 0.05$) between H x|y and H y|x and N x|y and N y|x.

DISCUSSION

The main objective of this study was to explore the nonlinear synchronization between bivariate biceps and triceps EMG data during MVC protocol for biceps.

Analysis of variance of H- and N-index showed that synchronization is more pronounced in pre- and post-MVC stages than in MVC stage. Pre- and post-MVC stages could be considered as isotonic and eccentric while MVC is isometric contraction of the biceps. By definition of these indexes, synchronization is more distinct in isotonic and eccentric contraction while it is less obvious in isometric contraction of biceps muscles. The level of synchronization of pre-MVC was not significantly different from post-MVC conditions. This study also found that cross-correlation as a linear measure of synchronization (fundamentally coherence) fail to detect any synchronization pattern between biceps and triceps in any trials or any MVC stages suggesting that nonlinear measures are more sensitive compared to the linear measures. It can be argued from the findings that muscles behave more synchronously during dynamic contraction compared to static contraction.

From postural control and stability standpoint, if stability is defined as a tendency for an object/structure to return to an equilibrium state following a perturbation, then it can be conjectured that muscles behave in a more synchronous manner in the initial stage before going to a new state. For instance, biceps and triceps synchronization was more before the participants reached a steady state maximum voluntary contraction and resting condition.

The study found no significant difference between H x|y and H y|x, and N x|y and N y|x. Although by principle, H x|y and H y|x will not be exactly equal, statistical insignificant difference suggest leas asymmetry between biceps and triceps muscles. This suggests that during MVC even though force generation is greater for a prime mover, the synchronization is such that the prime mover does not drive the antagonist muscle(s) or vice versa. In other words, the focus is not on one-side.

REFERENCES

Anmuth, C. J., Goldberg, G., and Mayer, N. H. (1994), "Fractal dimension of EMG signals recorded with surface electrodes during isometric contractions is linearly correlated with muscle activation." *Muscle Nerve*, 17, 953-954.

Arnhold, J., Grassberger, P., Lehnertz, K. and Elger, C. E. (1999), "A robust method for detecting interdependences: application to intracranially recorded EEG." *Physica D*, 134, 419-430.

Cao, L. (1997), "Practical method for determining the minimum embedding dimension of a scalar time series." *Physica D*, 110, 43-50.

Chaffin, D. B., Andersson, G. B. J., and Martin, B. (1999), *Occupational Biomechanics*. John Wiley & Sons, New York.

Gitter, A. J. and Czerniecki, M. J. (1995), "Fractal analysis of the electromyographic interference pattern." *Journal of Neuroscience Methods*, 58, 103-108.

Gitter, A. J, Czerniecki, J. C. and DeGroot, D. (1991), "Muscle force and EMG activity relationship using fractal dimension analysis." *Muscle Nerve*, 14, 884.

Gupta, V., Suryanarayanan, S., and Reddy, N. P. (1997), "Fractal analysis of surface EMG signals from the biceps." *International Journal of Medical Informatics*, 45, 185-192.

Karwowski, W. and Rodrick, D. (2003), *Is chaos present in static postures observed at work: a nonlinear dynamics-based analysis of surface EMG signals.* Proceedings of the 47th Annual Meeting of the Human Factors and Ergonomics Society, Denver, CO.

Lei M., Wang, Z., and Feng, Z. (2001), "Detecting nonlinearity of action surface EMG signal." *Physics Letters A*, 290, 297-303.

Marras, W. S., Ferguson, S. A., and Simon, S. R. (1990), "Three Dimensional Dynamic Motor Performance of the Normal Trunk." *International Journal of Industrial Ergonomics*, 6, 211-224.

Quian Quiroga R, Arnhold, J., and Grassberger, P. (2000), "Learning driver-response relationships from synchronization patterns." *Physical Review E*, 61(5), 5142-5148.

Quian Quiroga, R., Kraskov, A., Kreuz, T., and Grassberger, P. (2002), "Performance of different synchronization measures in real data: a case study on electroencephalographic signals." *Physical Review E*, 65(4), 1903-1916.

Rodrick, D. and Karwowski, W. (2004), *Nonlinear behavior of muscle responses for four static postures observed at work.* Proceedings of the 48th Annual Meeting of the Human Factors Society, September 20-24, New Orleans, Louisiana, 1285-1289.

Rodrick, D. and Karwowski, W. (2006), "Nonlinear dynamical analysis of surface EMG signals of static postures observed at work." *Nonlinear Dynamics, Psychology, and Life Sciences*, 10(1), 21-35.

Rosenblum, M. G., Pikovsky, A. S., and Kurths, J. (2004), "Synchronization approach to analysis of biological systems." *Fluctuation and Noise Letters*, 4(1), L53-L62.

Yang, Z. and Zhao, G. (1998), "The phase space analysis of EMG signal." *Acta Biophys. Sinica*, 14(2), 257-261.

Yang, J., Liu, B., Peng, J., and Ma, Z. (1999), "The preliminary nonlinear dynamical analysis of surface electromyogram." *Space Med. Med. Engrg.*, 12(3), 185-187.

Chapter 53

Globe Stereothermometer- a New Instrument for Measurement and Evaluation of Irregular Radiant-Convective Load

Zuzana Oleksiaková[1], Šárka Bernatíková[2], Zdeněk Jirák[1,5], Miloslav V. Jokl[3], , Hana Tomášková[1,5], Stanislav Malý[4]

[1] University of Ostrava, Czech Republic

[2] Technical University of Ostrava, Czech Republic

[3] Czech Technical University in Prague, Czech Republic

[4] Occupational Safety Research Institute, Prague, Czech Republic

[5] Institute of Public Health in Ostrava, Czech Republic

ABSTRACT

According to EN ISO 7730 the estimation of operative temperature and RTA (t_{rA}-t_{rB}) is prescribed; it has been measured by Indoor Climate Analyzer of Bruel and Kjaer. It was proved that a new instrument, globe stereothermometer, can be applied for this purpose. It is a globe of 15cm in diameter divided into 6 segments;

the surface temperature of each segment is called stereotemperature (t_{stereo}). The mean value of all six stereotemperatures equals to globe temperature (t_g) for low velocities being equal to operative temperature. Radiant Temperature Asymmetry can be estimated from equations (6), (7) and (8) and from Figure 6. Stereothermometer is produced in Prague. The company also developed electronics and technology of the instrument.

Keywords: Operative temperature, Radiant temperature asymmetry, Local comfort criteria, Thermal comfort

INTRODUCTION

According to present standards (see e.g. ISO 7730) and directives (e.g. Czech Government Directive No.361/2007 Code) operative temperature is the basic criterion for thermal condition assessment of the environment. It is defined as "The temperature of the homogenous closed space in which a man could exchange the same amount of radiant and convective heat as in an investigated actual thermally non-homogenous space.

Radiant Temperature Asymmetry (RTA) is the recommended criterion for non-uniform thermal load on human body, e.g. from radiating window heated floor, heated or cooled ceiling, the workplaces at furnaces in iron and glass works, by the standard EN ISO 7730. It is valid for three categories A, B, C depending on various predicted percentage of dissatisfied people (PPD): A is the most comfortable, for the lowest PPD, C for the highest PPD. An example of prescribed values of RTA is presented in Table 1 (Table A4 EN ISO 7730) (Jokl 2002) (Petráš et al. 2004).

Table 1. RTA by EN ISO 7730:2005

Category	Radiant Temperature Asymmetry - RTA			
	Warm ceiling	**Cool wall**	**Cool ceiling**	**Warm wall**
A	< 5	< 10	< 14	< 23
B	< 5	< 10	< 14	< 23
C	< 7	< 13	< 18	< 35

THE ESTIMATION OF THE OPERATIVE TEMPERATURE AND OF RTA UP TO NOW

Operative temperature can be measured directly by a globe thermometer whose relationship between heat transfer coefficients by convection and radiation(ratio $h_c/(h_c+h_r)$ and $h_r/(h_c+h_r)$) would be the same as with the human body-this is valid for the globe of 10 to 15 cm in diameter and for low air velocities 0.1 to 0.2 m/s. The higher the globe diameter the higher is the impact of mean radiant temperature on the measured globe temperature value (Jokl 2002, Hemzal 2008).

RTA is measured up to now by Indoor Climate Analyzer type 1213, Bruel and Kjaer, Denmark. There is a special sensor for this purpose called RTA (Radiant Temperature Asymmetry) transducer MM 0036 (Figure 1), By means of PT100 radiant heat is estimated coming from two opposite sides A and B: surface temperatures t_{rA} and t_{rB} of a small plane are measured and their difference (t_{rA} - t_{rB}) is the RTA.

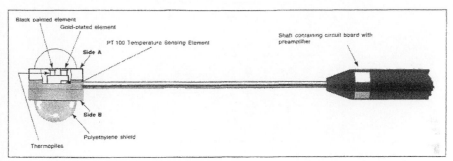

Figure 1. RTA transducer (MM 0036) of Indoor Climate Analyzer from Bruel and Kjaer.

A NEW WAY OF OPERATIVE TEMPERATURE AND RTA ESTIMATION

It is based on a new instrument application, the so called globe stereothermometer.

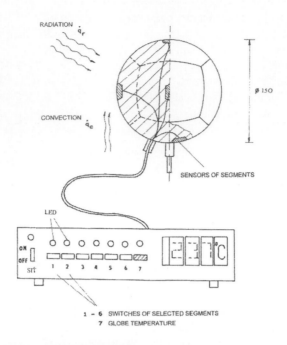

Figure 2. Globe stereothermometer scheme.

Figure 3. Globe stereothermometer photo.

PRINCIPLE OF STEREOTHERMOMETER

First have a look at an ordinary globe thermometer (GT). There is a big difference in success of GT and other instruments, developed for the same purpose: GT is used from the year 1923 continuously as a part of many national standards, government directives and hygienic prescriptions. There is a simple reason for it: – GT does not try to be a model of human body, it is only a part of human body thermal equilibrium equation, operative temperature can be substituted by globe temperature. Of course, globe temperature includes exactly only air and radiant temperature thus as a criterion of human comfort the relationship on other factors must be taken into account, on air velocity and humidity, clothing, activity, adaptation and exposure time.

What is valid for the whole GT can be written approximately also for a part of it, for its segment. For GT of 15cm in diameter six segments have been proved quite satisfactory (originally 18 segments were tested). The surface temperature of each segment is called stereotemperature, i.e. the mean value of all six stereotemperatures equals globe temperature.

For the scheme of instrument see Figure 2 and Figure 3 there is a photo. It is produced in Prague. The company also developed electronics and technology of the instrument (dipl. Ing. L. Vajner).

Supposing a) the equality of heat transfer coefficients for irradiated and non-irradiated segment, b) the stereotemperature of non-irradiated segment equals globe temperature, the simple equation based on segments heat balance can be derived

$$t_{rA} - t_{rB} = (t_{stereo} - t_g)(1 + \frac{h_c}{h_r}) \quad [°C] \qquad (1)$$

or

$$t_{stereo} - t_g = (t_{rA} - t_{rB})(\frac{h_c}{h_c + h_r}) \quad [°C] \qquad (2)$$

where $t_{rA} - t_{rB}$...   RTA [°C]

t_{stereo}... stereotemperature, i.e. temperature of exposed segment [°C]

t_g...   globe temperature [°C]

h_c...   heat transfer coefficient by convection [$W^{-1}.m^2.K$]

h_r...   heat transfer coefficient by radiation [$W^{-1}.m^2.K$]

These equations cannot be applied in practice because the calculation of heat transfer coefficients is a difficult problem. Therefore the experimental estimation of equations is necessary.

EXPERIMENTAL ESTIMATION OF THE RELATIONSHIP BETWEN RTA AND THE DIFFERENCE STEREOTEMPERATURE MINUS GLOBE TEMPERATURE

The impact of vertical and horizontal surface has been tested in the climatic chamber(dimensions 3x2x2 m) (Figure 4). Keeping globe temperature constant 24 °C the temperatures of the vertical surfaces was chosen 14, 19, 24, 29, 34 a 44 °C ,the temperatures of the horizontal surfaces 45, 53 a 58 °C (Table 2).

Table. 2 Conditions in climatic chamber

Vertical radiant surface							Horizontal radiant surface				
Exp. No.	t_g (°C)	t_{rA} (°C)	Δ t_{rA}-t_g (°C)	Exp. No.	t_g (°C)	t_{rA} (°C)	Δ t_{rA}-t_g (°C)	Exp. No.	t_g (°C)	t_{rA} (°C)	Δ t_{rA}-t_g (°C)	W/m²
1	24	24	0	4	24	29	+5	7	24	45	21	100
2	24	14	-10	5	24	34	+10	8	24	53	29	150
3	24	19	-5	6	24	44	+20	9	24	58	34	200

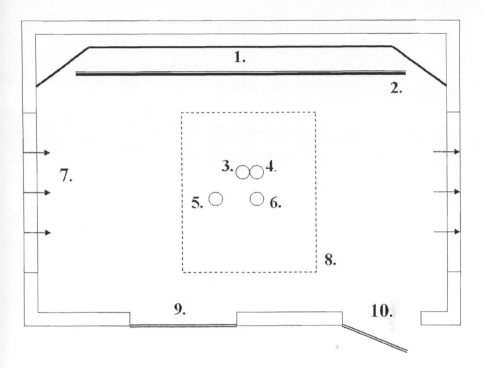

Figure 4. Measurements in climatic chamber

Where 1. vertical warm radiant panel, 2. vertical cool radiant panel, 3. radiant temperature, 4. stereothermometer, 5. air velocity, 6. globe thermometer Vernon-Jokl, 7. direction of air flow, 8. ceiling radiant panel, 9. control room window, 10. entrance.

RTA, t_{rA}, t_{rB}, air flow velocities(chosen 0.25 m/s, 0.5 m/s a 1.0 m/s) air temperatures, relative air humidities were measured by Indoor Climate Analyzer Type 1213(Bruel and Kjaer) respecting EN ISO 7726.

The stereotemperatures t_{stereo} (the temperature of the exposed segment) and globe temperature were measured by stereothermometer.

The results are presented in Figure 5 and 6.

The temperature t_{rA} (temperature of the irradiated plane side) against the stereotemperature (temperature of the irradiated segment) for three velocities (0.25 m/s, 0.5 m/s and 1.0 m/s) is presented in Figure 5. Correlation coefficients are high ($R^2=0.9797$ up to 0.9908), thus the graph can be applied to practice for the estimation of t_{rA} on the basis of measured stereotemperature t_{stereo}:

$$t_{rA}= 3.4153 \, t_{stereo}-57.004 \quad [°C] \text{ for } v=0.25 \text{ m/s} \quad (3)$$

$$t_{rA}= 4.14 \, t_{stereo}-74.971 \quad [°C] \text{ for } v=0.5 \text{ m/s} \quad (4)$$

528

$$t_{rA} = 5.46\ t_{stereo} - 108.33 \quad [°C] \ \text{for} \ v=1.0 \ \text{m/s} \qquad (5)$$

RTA ($=t_{rA} - t_{rB}$) depends on the difference stereotemperature minus globe temperature (t_{stereo} minus t_g), see Fig. 7.and following equations:

$$t_{rA}-t_{rB} = 3.6242(t_{stereo} - t_g) - 0.5098 \quad [°C] \ \text{for} \ v=0,25 \ \text{m/s} \quad (6)$$

$$t_{rA}-t_{rB} = 4.3807(t_{stereo} - t_g) - 1.3907 \quad [°C] \ \text{for} \ v=0.5 \ \text{m/s} \quad (7)$$

$$t_{rA}-t_{rB} = 5.1507(t_{stereo} - t_g) - 0.2010 \quad [°C] \ \text{for} \ v=1.0 \ \text{m/s} \quad (8)$$

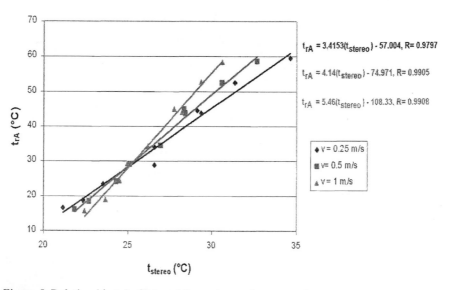

Figure 5. Relationship trA=f(tstereo) for various v from experiments

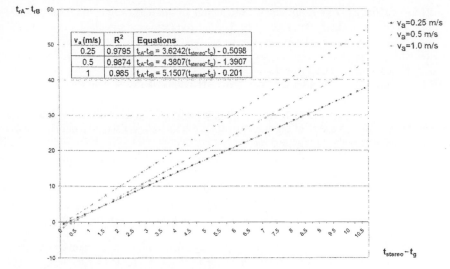

Figure 6. Relationship RTA = trA-trB= f(tstereo-tg) from experiments

DISCUSSION

The estimation of RTA has been possible only by Indoor Climate Analyzer (Bruel and Kjaer) up to now. The instrument is highly sophisticated and thus expensive. A new instrument, globe stereothermometer, can be also used for this purpose being much more simple and thus cheaper. Electronics and technology of this instrument was also a problem; being solved enough is now produced professionally.

CONCLUSIONS

The new instrument - globe stereothermometer – is a globe of 15cm in diameter divided into 6 segments; the segments temperature is called stereotemperature. Based on measurements in climatic chamber it has been proved that instrument allows estimation of RTA from the difference t_{stereo} minus t_g (see Figure 6) and equations (6), (7) and (8).

Acknowledment: The work was carried out within the research project of MLSA , identifier: MPS0002595001.

REFERENCES

EN ISO. 1998. *EN ISO 7726: 1998,* Ergonomics of the thermal environment-Instruments for measurements of physical quantities.

EN ISO. 2005. *EN ISO 7730: 2005,* Ergonomics of the thermal environment – Analytical determination and interpretation of thermal comfort using calculation of the PMV and PPD indices and local comfort criteria.

Hemzal,K. 2008:Operative temperature as a criterion of human body equilibrium.In Czech.VVI 17,2008,1:16-23.

Jirák, Z., Jokl, M. V., Šebasta D., Tomášková, H., Bernatíková, Š. and Malý S.2007:Use of Globe Stereo Thermometer for Evaluation of Irregular Radiation Load. Central European Journal of Public Health. Supplement, vol. 15 (JHEMI vol. 51) 2007, S24

Jokl, M. V. 1966. The way of infrared heat load estimation over the human body surface.In Cech, Patent No. 117894, Prague.

Jokl, M. V.and Tůma, V. 1988. Directional globe thermometer. In Czech.Patent No.236203.Prague

Jokl, M. V. 1990. The stereothermometer: A new instrument for hygrothermal constituent nonuniformity evaluation. ASHRAE Transactions 96, 1990, No. 3435, p. 13-18.

Jokl, M. V. 1991. Feuchtemessung nach "Kunstkopf"- Prinzip. Heizung-Luftung-Haustechnik 42,1991,1:27-32.

Jokl, M. V. 1991. Stereoteploměr – nový přístroj pro hodnocení nerovnoměrnosti tepelně-vlhkostní konstituenty prostředí. Čs. Hygiena 36, 1991, No. 1: 14-23.

Jokl, M. V. 2002.Health Residental and Working Environment.In Czech. Academia, Prague 2002.

Jokl, M. V. and Vajner, L. 2003.Globe Thermometer. Useable Pattern No. 13547. Written down 1.8.2003. Owner VÚBP Prague.

Petráš, D., Koudelková and D., Kabele, K. 2004.Hydronic and Electrical floor Heating.In Czech. Jaga, Bratislava 2004.

Chapter 54

The Human Body Reaction on Non-Uniform Radiant-Convective Load

Sarka Bernatikova[1], Zuzana Oleksiakova[2], Zdenak Jirak[2], Miloslav Jokl[3], Hana Tomaskova[2,5],Stanislav Maly[4]

[1] VŠB-TU Ostrava, Faculty of Safety Engineering,Ostrava, CZ

[2] University of Ostrava, Faculty of Health studies, CZ

[3] ČVUT Praha, Faculty of Civil Engineering, CZ

[4] Occupational Safety Research Institute, Prague, CZ

[5] Institute of Public Health, Ostrava, CZ

ABSTRACT

Thermal - humidity conditions are important factors that effect the subjective feeling of human comfort and work productivity as well. The thermal comfort is defined as a state of mind expressing thermal environment satisfaction. Dissatisfaction may be caused by overall discomfort of the organism or discomfort of each part of the human body.

The investigation is focused on physiological response on non-uniform radiant – convective load in experimental conditions of climatic chamber. The terms range is within the limits of values of microclimatic conditions set by legislative for a calendar year.

532

Experimental measurements are realized on the basis of physical factors of thermal environment, physiological response of the organism, subjective feelings and mental performance in a group of 24 experimental subjects.

Keywords: Non-uniform Radiant-Convective Load, Climatic Chamber

INTRODUCTION

The problem of irregular radiant-convective load is encountered in industrial, but also office buildings and municipal sector. The basic document in the assessment of the irregular heat loads in the EU is EN ISO 7730:2005 [1]. This standard is based on forecasts of a man feeling the heat by calculating predicted mean thermal sensation (PMV) and predicted percentage of dissatisfied people (PPD), which will feel discomfort in the environment in terms of excessive heat or cold. The aim of the study was to determine response to physiological indicators (skin temperature, body temperature, heart rate, sweat production and mental performance) and subjective feelings of people on the erratic experimental radiant-convective load in experimental conditions in a climatic chamber.

METHODS

The investigation was done under experimental conditions in a climatic chamber and includes measurements of physical factors on the thermal environment, the measurement of physiological responses of subjects, subjective feelings of heat and cold and the effect on mental performance of subjects. Part of the effect on mental performance is the subject of a separate work [8]. Climatic chamber (fig. 1) has an internal space 2 x 3 m.

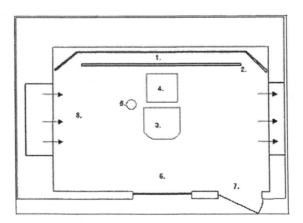

Figure 1. Chart climatic chamber

1. Vertical warm radiant panel, 2. Vertical cool radiant panel, 3. Experimental person, 4. Table with PC, 5. Measuring point, 6. Control room window, 7. Entrance, 8. Direction of air flow

The modified air is fed into the chamber from the engine room by a hole sized 1 x 1 m in the direction of the long axis of the chamber and is conducted as a large hole in the wall opposite chamber. Radiant panel forms the entire back wall of the chamber. A panel with negative radiation might be prepended before the thermal panel, if necessary. Chamber allows the setting of laminar airflow of 0.2 $m.s^{-1}$ to 2 $m.s^{-1}$, the air temperature in the range of 0 to 60°C, the intensity of unilateral radiation on a large scale with a maximum exceeding 200 $W.m^{-2}$ and relative humidity, depending on temperature of the range of 30 to 90%. An automatic regulation is controlled from the control room of the chamber. Details of heat-humidity conditions in the chamber and the values of physiological parameters sensed from each experimental persons are fed into the control room. Physical and physiological data are recorded continuously throughout the shift and stored in computer memory.

Experiments have been held on a group of 24 women (students, University of Ostrava), aged 20-24 years. Experiments were divided according to the final ball thermometer temperature (t_g) into 3 stages. t_g was ranged just about 22°C in all experiments in the first stage, in the second stage t_g was 19°C and in the third stage t_g was 25°C. Thermal resistance of clothing was chosen to match thermal neutral conditions in all experiments. Thermal resistance of clothing was 0.6, 1.2 and 0.5 clo in various stages. Each phase included 8 trials during which the intensity of radiation from the back wall of the chamber ranged from -97 to 153 $W.m^{-2}$. Air flow ranged between 0.25 to 0.3 $m.s^{-1}$ (average value at each stage was 0.28 $m.s^{-1}$ in all experiments. The relative humidity ranged in individual experiments in a physiological range of 30 to 60%.

The subjects were exposed to various conditions for 1 hour. The subjects sat at the computer facing the glowing back wall of the chamber during the experiments and playing computer games to address demands for focussing attention and short-term memory. Own experiments were preceded by instruction, during which the subjects became familiar with experimental conditions and undergone training in computer games to achieve steady performance .The order of conditions within each stage were randomly changed so that the subjects were not informed of conditions in the chamber. Heat-humidity conditions measured in the level of 110 cm from the floor near the head of an experimental person were continuously measured and recorded in the computer memory. Measured conditions were as follows: the resulting ball thermometer temperature (t_g) and stereotemperature (t_{stereo}) measured by a ball streothermometer developed by us [6, 7]. Dry air temperature (t_a), airflow (v_a), relative humidity (RH) and temperature of the radiation in the direction of radiation (t_{rA}) and from the opposite wall (t_{rB}). Moreover, values of tg and ta in the level of 60 and 10 cm from the floor were continuously measured [2].

The physiological parameters were continuously measured in half-minute intervals. The value of heart rate was recorded in computer memory, skin temperature was continuously measured on 6 sites of body surface area (forehead, chest, back, forearm, thigh and calf). The average temperature of the skin (t_{sk} average) was calculated as a weighted average according to equation [3]:

$$t_{sk} \text{ average} = t_{sk} \text{ forehead} .0,07 + t_{sk} \text{ chest}.0,15 + t_{sk} \text{ back}.0,19 + t_{sk} \text{ forearm}.0,19 + t_{sk} \text{ tight}.0,19 + t_{sk} \text{ calf}.0,21.$$

Core body temperature was measured at the ear drum and under the tongue at the beginning and at the end of exposure. Water loss in sweat and respiration was measured by weighing people (clad only in underwear) before and after the exit from the chamber. At the end of the stay in the chamber the subjects completed a questionnaire in which they assess the overall and local thermal sensations. Thermal sensations were expressed in a seven-point scale in accordance with EN ISO 7730 [1] in the range from -3 to 3 from the optimum conditions (table 1). Correlation and regression analysis was used for statistical analysis, processing was carried out in programs of the Stata v.9 [9]

Table 1 Seven-point thermal sensation scale (EN ISO 7730)

+ 3	Hot
+ 2	Warm
+ 1	Slightly warm
0	Neutral
- 1	Slightly cool
- 2	Cool
- 3	Cold

RESULTS

THERMAL CONDITIONS

Thermal conditions in the various stages and tests are given in table 2. The difference between the radiant temperature in the direction of radiation and the resulting temperature ball thermometer $\Delta (t_{rA}-t_g)$ varied within each stage from -9 to + 34°C. Vertical thermal imbalance, expressed by differences in the values of resulting temperature of thermometer measured at the level of 110 and 10 cm from the floor $\Delta (t_g110-t_g10)$ was highest in trials with the highest radiation irregularity.

Table 2 Heat –humidity conditions in the individual stages and experiments

Stage	$t_{g\ (110)}$ °C	t_{rA} (°C)	t_{rB} (°C)	$t_{a\ (110)}$ °C	$t_{g\ (60)}$ °C	$t_{g(10)}$ °C	RH %	v_a m/s
I.	22,06	16,19	22,84	22,50	21,39	20,83	28,33	0,25
	22,10	19,08	22,63	21,89	21,51	21,40	29,41	0,27
	22,02	22,57	22,24	20,79	21,18	21,59	47,41	0,27
	21,96	24,96	22,40	20,39	19,71	19,80	59,27	0,25
	22,01	31,99	22,84	19,84	18,61	17,58	54,36	0,25
	22,30	40,04	22,78	18,84	17,00	15,91	60,36	0,28
	22,11	45,34	22,72	17,86	14,47	14,38	51,68	0,30
	21,98	52,75	23,75	18,78	19,40	16,97	39,23	0,36
II.	18,68	15,95	19,89	18,46	18,40	18,82	29,70	0,24
	19,02	20,49	19,71	17,85	18,33	18,92	35,48	0,26
	18,89	22,17	20,03	17,37	18,15	18,69	44,11	0,26
	19,06	24,56	20,40	17,79	17,93	18,01	47,47	0,28
	19,11	32,16	21,06	17,93	16,39	16,14	43,51	0,28
	19,20	40,36	21,33	16,75	14,70	13,92	51,59	0,30
	19,17	46,07	22,10	16,60	14,67	14,27	61,80	0,33
	19,11	52,24	22,67	16,76	14,42	14,50	52,50	0,32
III.	25,04	16,39	25,74	26,28	23,35	20,47	37,98	0,29
	25,02	19,21	25,40	25,22	23,97	21,73	36,86	0,27
	25,14	21,91	25,33	24,90	24,31	22,94	36,47	0,26
	25,07	24,78	24,83	24,47	24,24	23,46	32,61	0,28
	25,17	31,64	25,03	23,84	23,56	21,17	63,32	0,28
	25,06	40,15	25,32	24,66	22,90	16,77	57,35	0,30
	25,16	45,98	24,33	22,93	22,14	16,69	53,82	0,28
	25,61	50,37	24,90	24,51	23,30	16,45	44,72	0,31

t_g – globe temperature (measured at the level of 110, 60 and 10 cm from the floor)
t_{rA} – radiation temperature of the direction of radiation
t_{rB} – radiation temperature in the opposite direction
t_a – air temperature
RH – relative humidity
v_a - air velocity

PHYSIOLOGICAL RESPONSE OF THE ORGANISM

The difference in body core temperature (measured at eardrum) before and at the end of exposure varied depending on the t_{rA} in the range $\pm 0.2°C$. The average water loss in sweat and breathing reached depending on t_{rA} from 30 to 70 grams / hour, which corresponds to the loss of water from 720 to 1680 g/24 hours. The highest values were recorded around the $t_{rA} = 22°C$, the second peak was reached in experiments with the highest value $t_{rA} = 53°C$. The average temperature of the skin varied in all stages depending on the value of $\Delta (t_{rA}-t_g)$ in the physiological range of 31.4 to 33.2°C, with all values t_{sk} average of stage III ($t_g = 25°C$) were in the upper range of temperature zone, while the value of stage II ($t_g = 19°C$) were the lowest (fig. 2). Forehead skin temperature at all stages increased depending on increasing $\Delta (t_{rA}-t_g)$ value (fig. 3), while calf skin temperature decreased (fig. 4). The difference between the temperature of the skin of the head and calf at all stages depended on the difference $\Delta (t_{rA}-t_g)$ increases (fig. 5).

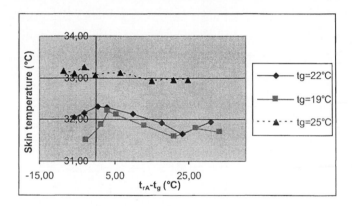

Figure 2. The relationship between skin temperature and temperature differences

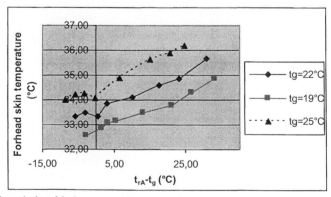

Figure 3. The relationship between skin temperature and temperature differences

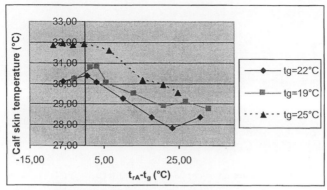

Figure 4. The relationship between calf skin temperature and temperature differences

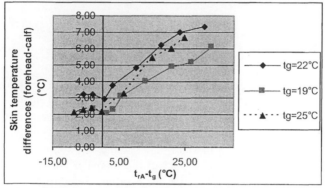

Figure 5. The relationship of skin temperature differences (forehead - calf) on the temperature differences

SUBJECTIVE SENSATION OF HEAT

The overall feeling of heat and local warmth to the head and legs shows a similar dependence on the value of Δ $(t_{rA}-t_g)$ as the temperature of the skin. This relationship is expressed by curves, however, unlike the skin temperature of more fluctuating. The overall feeling of warmth does not correlate with an average temperature of the skin, but is depended mainly on the local feeling of heat on the trunk (fig. 6).

538

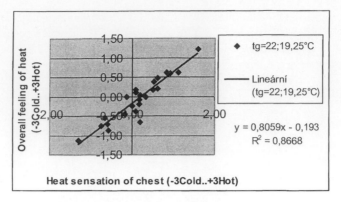

Figure 6. The relationship between overall heat sensation and local heat sensation of chest

DISCUSSION

Physiological characteristics, especially skin temperature and temperature of the body have proved to be very sensitive indicators of both overall and uneven radiant-convective loads. Forehead skin temperature and calf temperature are very sensitive to the horizontal and vertical non-uniform thermal stress. Level curves of the sitting occupant are depended on the thermal state of the environment, e.g. operative, respectively resulting temperature of a ball thermometer. According to skin, we found similar temperature dependence as the skin, we found of value Δ $(t_{rA}-t_g)$ and subjective evaluation of both the local and the overall feeling of heat or cold. The results are in contrast to the physiological parameters both by considerable inter- and intra-individual variability according the heat of emotions. The differences in the perception of thermal comfort for men and women are applied in the case of the interindividual as we could show in our previous work (4), where we watched a group of men and women at the same age. Whereas men, women reacted more sensitively to changes in thermal state of the environment and their thermal optimum is shifted to higher levels of thermal state of environment. Therefore, we chose only women for our research. The results of our work shows that the local warmth and cold in the head and legs well correlate with the temperature of the skin on the forehead and calf, while the overall feeling correlates primarily with the local sense of trunk heat.

CONCLUSIONS

The temperature of the skin is a very good indicator of uneven radiant-convective load. The temperature of the skin in the calf, depending on the size of Δ $(t_{rA}-t_g)$

decreases, while the temperature of the skin on the forehead is rising. Level curves are depended on the thermal environment, in terms of the resulting of temperature of thermometer. Local subjective feelings of heat and cold on the head and legs have the same dependence on the value of Δ $(t_{rA}\text{-}t_g)$ as the temperature of the skin. There are considerable interindividual differences between the experimental individuals.

The work was carried out within the research project of MLSA, identifier: MPS0002595001.

REFERENCES

[1] EN ISO 7730:2005 Ergonomics of the thermal environment – Analytical determination and interpretation of thermal comfort using calculation of the PMV and PPD indices and local thermal comfort criteria

[2] EN ISO 7726 Ergonomics of the thermal environment – Instruments for measuring physical quantities

[3] Jirák Z., Jokl M., Štverák J., Pechlát R., Coufalová H.: Correction factors in skin temperature measurement. J. Appl. Physiol., 38, 1975, 4, s. 752-755

[4] BERNATÍKOVÁ Š., JIRÁK Z., JOKL M.V., TOMÁŠKOVÁ H., Malý S., KILIÁN V. : Irregular thermal load – Results of subjective evaluation (czech). České pracovní lékařství, 9, 2008, 2-3, s.60-66.

[5] JIRÁK Z., TOMÁŠKOVÁ H., JOKL M.V , BERNATÍKOVÁ Š. ŠEBESTA D., Malý S., KILIÁN V.: Response of physiological indices to irregular radiation load in experimental conditions of climatic chamber (czech). České pracovní lékařství, 9, 2008, 4, s 125-130

[6] JOKL, M.V., VAJNER L.: Ball thermometer. Used model No. 13547. Registered 1.8.2003

[7] JOKL, M.V., Jirák Z., Šebesta D.: A new way of the uneven heat load evaluation of man. Heating, ventilation, installation, 5, 2005, s.223-224

[8] Malý S.: Effect of parallel radiant and convective heat action in conditions of moderate thermal environment on the reliability of human performance. Doctoral thesis VŠB-TU Ostrava, Faculty of Safety Engineering, 2007, 101 s.

[9] STATA Statistical software: Release 9.0 College Station,TX: Stata Corporation 2005

Chapter 55

A Study on the Evaluation of Management Plan for Whole-body Vibration by ISO 2631-1

Su-Hee Kim[1], Tae-Gu Kim[1,*], Young-Sig Kang[2], Kwan-Hyung Yi[3]

[1]Inje University
[2]Semyung University
[3]KOSHA

ABSTRACT

Actually, exposure of repeated whole-body vibration is directly related to musculoskeletal system disorder. Professional exposure to whole-body vibration is reported to relate to epidemic of degenerative change to vertebra organization including the lumbago disorder, sciatica, lumbar HIVD (Herniated Intervertebral Disc). Therefore, this paper describes the effect of vibration to the human body from rapid-transit train in KTX(Korea Train eXpress) and Sincansen based on ISO 2631-1. When these compare with measurement values for the x,y,z axes, the z (top and bottom) axis is higher than the x (front and rear), y (left and right) axes by 2 to 3 times. As a result, it is expected that the horizontal vibration has less effect than vertical vibration on the driver's body. The vector sum value for KTX is higher at 0.191m/s^2 than for Sincansen at 0.151m/s^2. The KTX exposure limit time is 143.27 hours, while Sincansen is 229.23 hours. The values for KTX and Sincansen are evaluated as "Health effects not documented/observed" for the "Health guidance caution zone" in ISO 2631-1.

Keywords: Whole-body vibration, ISO 2631-1, Health guidance caution zone, Exposure limit time.

INTRODUCTION

According to the Construction and Transportation Department, the number of passengers on KTX in 2007 was about 102,000 a day, an increase of 2.2% from 2006 and 14.6% from 2005. On Sincansen the number is also increasing. As industries develop, people are interested with not only material wealth but improving the quality of life and also begin to pay attention to effects of body shock from transportation.
The body shock means impacts made by rotating machinery and structure effects on through body through the buttocks and feet, and causes stability degradation, interruption of activities, health deterioration, irritability and nausea. These effects are mainly concerned with musculoskeletal disorders and an exposure to the body through shock in the work place is related to increasing back pain disorder, sciatica, and lumbar spine, including disc or degenerative changes in the organization.
  Therefore this research evaluates the effect of vibrations from KTX of Korea and Sincansen of Japan which impacts on the body using international body shock measurements from ISO 2631-1 to evaluate the strength of whole body vibration exposure, the exposure limit time and the human impact. Fig. 1 shows the evaluation process for the measurement of whole-body vibration.

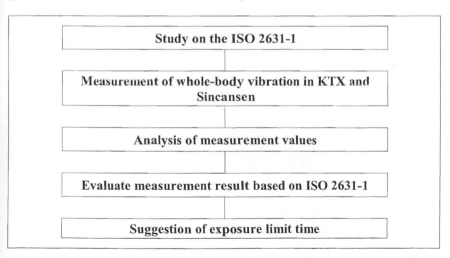

FIGURE. 1. Evaluation process for measurement of whole-body vibration

DESCRIPTION OF METHODS

ISO 2631-1 is the international human vibration standard for periodical, periodic, and temporary whole-body vibration measurement. The frequency bands range from 1Hz to 80Hz. The fundamental purpose is to measure human health and comfort, the probability of vibration perception and the incidence of motion sickness with the basic vibration values written as m/s². The other values include crest factor and vibration dose value by the fourth power vibration dose method. The body positions include sitting, standing and lying posture, and measurement directions include x (front, back), y (left, right), z (up, down). Fig. 2 shows body positions in ISO 2631-1. The method of whole-body vibration is given below.

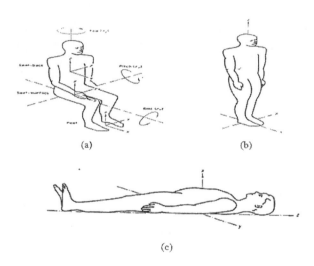

FIGURE. 2 Body positions in ISO 2631-1

R.M.S (ROOT-MEAN-SQUARE)

ISO 2631-1 is for mechanical vibration on the human body and measures the amount of vibration by evaluation of radiated forces along the x-axis (left, right), y-axis (before and after), and z-axis (up, down) of the translational 3-axis. A 3-axis accelerometer multiplied by a frequency weighting factor (x-, y-axis = 1.4, z-axis = 1) obtained by applying the measures for the effective r.m.s. (root-mean-square) is evaluated by the following equation.

$$a_\omega = (\frac{1}{T} \int_0^T [a_\omega(t)]^2 dt)^{\frac{1}{2}}$$

-------------------------------(1)

$a_\omega(t)$ is virtual value of the radiated vibration (m/s², rad/s²) and T is the measured time.

CREST FACTOR

Crest Factor is defined as the weighted r.m.s. value of the frequency signals to the maximum instantaneous peak of the signal for the weighted mean value.

$$CF = \frac{\max(a_\omega(t)) : (peakvalue)}{r.m.s(a_\omega)} \quad \text{------------------(2)}$$

VDV (VIBRATION DOES VALUE)

VDV (Vibration Dose Value) is the amount of vibration calculated by the radiation power using the average acceleration time.

$$VDV = \left\{ \int_0^T [a_\omega(t)]^4 dt \right\}^{\frac{1}{4}} \quad \text{-------------------------(3)}$$

eVDV(ESTIMATED VIBRATION DOSE VALUE)

eVDV (estimated Vibration Dose Value).

$$eVDV = 1.4a_\omega T^{\frac{1}{4}} \quad \text{-------------------------------(4)}$$

MTVV (MAXIMUM TRANSIENT VIBRATION VALUE)

MTVV (Maximum Transient Vibration Value).

$$MTVV = \max[a_\omega(t_0)] \quad \text{----------------------------(5)}$$

The acceleration values are evaluated through Equation (5) by weighting the frequency assignment criteria for the three kinds of relations.

$$a_{vs} = \sqrt{k^2{}_x a^2{}_{\omega x} + k^2{}_y a^2{}_{\omega y} + k^2{}_z a^2{}_{\omega q}} \quad \text{-----------------------(6)}$$

avs (vector sum) : The actual value used to evaluate the frequency weighted acceleration.

kx = ky = 1.4, kz = 1 indicates the frequency weighting.

THE STANDARD OF A DECISION OF THE AMOUNT OF VIBRATION

Fig. 3 is the evaluation methodology of ISO 2631-1. From Fig.3, select the axis to measure first, set weighting factors on the default frequencies in the selected axis, and then measure the r.m.s. Using the r.m.s. value, calculate the crest factor value. If the value is less than 9 apply to r.m.s. rating line in the "Health guidance zone". If the value is equal to or greater than 9, VDV and MTVV must be applied to evaluate[6] the value. If MTVV / r.m.s.> 1.5 or VDV / eVDV> 1.75, apply to the rating line of VDV. If MTVV / r.m.s. ≤1.5 or VDV / eVDV <1.25, apply to r.m.s. rating line. In both cases of r.m.s peak rate greater or smaller than 9, the vector sum is calculated by using equation (6), and applying this to the rating line.

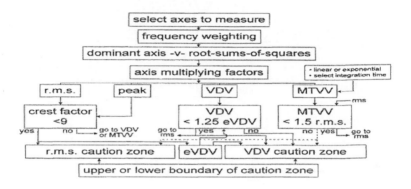

FIGURE. 3 Methodology of evaluation indicated in ISO 2631-1(1997) [7]

Fig. 4 is the health guidance caution zone quoted in ISO 2631-1 and the dotted line is the r.m.s. rating line that defines the phase of the exposure level of health warnings from 2.8 m/s^2 to 5.6 m/s^2. Solid lines represent the VDV assessment and define health warning-level phase between 17 m/s$^{1.75}$ and 8.5m/s$^{1.75}$.

Above the top line is "Health risks are likely" and between the lines is "Caution with respect to health risks". Below the bottom line is "Health effects not documented / observed". These are for evaluating the effect on humans and depend on the duration and impact. In ISO 2631-1, the maximum exposure limits is more than 10 minutes as defined in equation (7).

$$a_{vs} = 5.6 \left(t_0/t \right)^{\frac{1}{2}} \quad -------------------------------(7)$$

a_{vs} : Vibration level at the time of radiation tolerance (m/s^2), t_0 = 10min, t = Radiation exposure time (min).

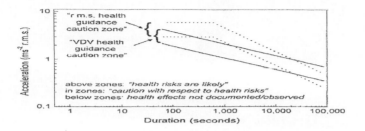

FIGURE. 4 Health guidance caution zone in ISO2631-1[7]

RESULT

The measurement target of the study is the impact on the human body from KTX train in Korea and Sincansen train in Japan. The measurements were conducted on these high-speed lines. Therefore, using speeds from 250 km/h to 300 km/h, the experiment was conducted from DongDaegu station to GuangMyuong station of KTX, and Fukuoka station to Hiroshima station of Sincansen. Measurement time is about 60 to 90 minutes. The same researcher measured the same section. Measurement conditions are shown in Table 1.

(a) KTX in Korea   (b) Sincansen in Japan

FIGURE 5. The type of measurement

Table 1. Condition of measurement

Rapid-Transit Railway type	Sampling Time (s)	Number (N)	Speed (km/h)	Interval
KTX	5025			E.Daegu ~ Gwangmyeong
Sincansen	3635	2	250~300	Fukuhoka~ Hirosima

Measurement position was sitting position from ISO 2631-1 as shown in Fig. 6. The installed acceleration sensor was able to measure the x,y,z axes between the seat and driver's buttocks. The measuring equipment consisting of acceleration sensors and analyzers is shown in Fig. 7

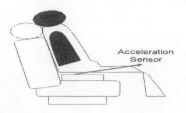

FIGURE. 6 Measuring position

(a) KTX   (b) Sincansen

FIGURE. 7 Installation of measuring device in Rapid-Transit train

RESULTS OF THE EXPERIMENT, KTX

The measurements on the KTX high-speed strain from Dongdaegu to Guangmyung are in Table.2. Comparing the 3-axis result values, the z-axis value represents the vibration up and down which is approximately 2-3 times higher than the x-axis values that represents the vibration from side to side and the y-axis that represents vibration left and right. Calculating CF using ISO 2631-1, results are in excess of 9. Therefore the MTVV / r.m.s value should be calculated to determine the assessment.

Table 2. Measurement result for KTX

Number	Axis	Sampling time (s)	r.m.s (m/s^2)	VDV ($m/s^{1.75}$)	MTVV (m/s^2)	Peak (m/s^2)	CF
	x		0.063	0.828	0.875	1.275	20.24
	y		0.064	1.41	1.890	2.713	42.39
1	z	3635	0.144	1.63	1.086	1.875	13.02
	Vector sum		0.191	-	-	-	-
	x		0.06	0.857	0.993	2.221	37.02
	y		0.053	1.54	2.087	2.833	53.45
2	z	3852	0.157	2.11	2.194	1.911	16.06
	Vector sum		0.193	-	-	-	-

As shown in Table. 3, the MTVV / r.m.s values are greater than 1.5 therefore the VDV assessment should be used. Using the VDV vector sum of the two values for the values of radiation exposure time, following VDV assessment on fig.4, the first and second measure both are the "Health effects not documented / observed" areas and are not found in any health effects. In the case of the maximum limit of exposure time, measured Vector sum value is $0.191 m/s^2$ and $0.193 m/s^2$, therefore according to equation (7), the values are 143.27and 140.32 hours respectively.

Table 3. Value of decision "Evaluating line"

Number	Axis	MTVV/r.m.s	
1	x	13.89	
	y	29.53	
	z	7.54	Higher than 1.5 → Evaluate using the "VDV evaluating line"
2	x	16.55	
	y	39.38	
	z	13.97	

EXPERIMENTAL RESULTS OF SINCANSEN

The analysis of Measurements on Sincansen train are shown in Table. 4. This study found that on the human body the up and down(z-axis) vibration effects were more than the left and right(y-axis) effects and the front and back (z-axis)effects. The Crest Factor is greater than 9.

Table 4. Measurement results for Sincansen

Number	Axis	Sampling time (s)	r.m.s (m/s^2)	VDV $(m/s1.75)$	MTVV (m/s^2)	Peak (m/s^2)	CF
1	x	5025	0.049	0.979	1.184	1.673	34.14
	y		0.059	0.840	0.624	0.755	12.80
	z		0.106	1.40	0.959	1.413	13.33
	Vector sum		0.151	-	-	-	-
2	x	5360	0.047	0.640	0.455	0.997	21.21
	y		0.029	0.446	0.382	0.865	29.83
	z		0.133	1.830	0.752	2.185	16.43
	Vector sum		0.154	-	-	-	-

As shown in Table. 5, the MTVV/r.m.s value exceeds 1.5, so we must use the VDV evaluation line. This table shows Sincansen has no effects on the human body as the results fall in the "Health effects not documented/observed" section. The exposure limit time for Sincansen is 229.23 hours and 220.39 hours, which is about 30 hours longer than KTX.

Table 5. Value for decision "Evaluating line"

Number	Axis	MTVV/r.m.s	
1	x	24.16	
	y	10.58	Higher than 1.5
	z	9.05	→ Evaluate using the
2	x	9.68	"VDV evaluating line"
	y	13.17	
	z	5.65	

An analysis of values for KTX and Sincansen shows values for both are two to three times higher in the vertical vibration(z-axis) than in the horizontal vibration(x,y-axis). The Vector sum, value for Sincansen is $0.151 m/s^2$, which is lower than the $0.191 m/s^2$ for KTX. Also the exposure limit times calculated average 225 hours at Sincansen and 143 hours at KTX, a gap of about 100 hours. The estimated Crest Factors are greater than 9 for the Health guidance caution zone so we use the VDV evaluation. Both trains have no effects on the human body.

CONCLUSIONS

To evaluate the effects of vibration on the human body from the rapid-transit trains in KTX and Sincansen based on ISO2631-1, we conducted an experiment with the obtained results are summarized as follows.

First, according to the results of the analysis on the values from KTX and Sincansen, both values shows the z-axis vertical vibration to be 2to3times higher than the x,y-axis horizontal vibration.

Second, the vector sum, vibration value at Sincansen is $0.151 m/s^2$, which is 79 percent of the KTX, value of $0.151 m/s^2$. Also the exposure limit times average 225 hours in Sincansen and 143 hours in KTX, with a gap of about 100 hours.

Third, the estimated Crest Factors are greater than 9 for the Health guidance caution zone that we use the VDV evaluation. Both trains show "Health effects not documented/observed" that has no effects on human body.

A more detailed analysis of whole-body vibration is asked by expanding the number of trains later.

REFERENCES

Korean Railroad, Annals of railway statistics, *Annual report*, 2007.

Griffin, M.J, Handbook of Human Vibration, Academic Press, London, U.K, 1990.

Bovnezi, M. and C. T. J. Hulshof, "An Updated Review of Epidemiologic Studies on the Relationship between Exposure to Whole-body Vibration and Low Back Pain", *Journal of Sound and Vibration*, 215(4), pp. 595 ~ 611, 1998.

International Organization for Standardization, Mechanical Vibration and Shock - Evaluation of Human Exposure to Whole-body Vibration, Part I : General Requirements, ISO 2631-1, 1997.

J.WanSub, Comprehensive understanding about Whole-Body vibration, *Korea Society of Noise and Vibration*, Volume 14, Part 5, pp. 6 ~ 14, 2004.

Korean Industrial Safety, "Exposure Character study on Whole-Body vibration(II) - intra-city bus drivers", *papers*, 2006.

Micheal J. GRIFFIN, Predicting the Hazards of Whole-Body Vibration Considerations of a Standard, pp. 83 ~ 91, 1998.

KangYoun, K. HoChul, K. SangYoung, L. JinHun, "Results on Noise and Vibration of KTX", *Korea Society of Noise and Vibration*, pp. 120 ~ 125, 2003.

WanSub, P. YongHwa, Micheal J. GRIFFIN, "The Analysis on Standard Method of Risk Prediction about whole-body Vibration and Repeated Shock" *Korea Society of Noise and Vibration*, Volume 10, Part 1, pp. 160 ~ 167, 2000.

Chapter 56

Optical radiation (Visible, Ultraviolet and Infrared) at the Workplaces

Agnieszka Wolska

Central Institute for Labour Protection – National Research Institute
Czerniakowska 16, 00-701 Warsaw, Poland

ABSTRACT

Some amounts of optical radiation are beneficial for humans but excessive exposure can cause many negative health effects to the skin and eyes and also can affect the immune system. Biological effects can be induced only by absorbed radiation. We could distinguish two types of reactions in biological tissues induced by optical radiation: photochemical (in UV and VIS range) and thermal (in VIS and IR range). Exposure limit values (ELVs) represent conditions under which it is expected that nearly all individuals may be repeatedly exposed without acute adverse effects and based upon best available evidence, without noticeable risk of delayed effects. New individual Directive 2006/25/EC of the European Parliament and the Council on the minimum health and safety requirements regarding the exposure of workers to risks arising from physical agents (artificial optical radiation) concerns measures protecting workers from the risks associated with artificial optical radiation. The examples of workstations where results of the exposure measurements indicate high occupational risk arising from optical radiation are: electric and gas welding. The employer is obliged to provide protection from exposure via combination of engineering controls, administrative measures and personal protective equipment (PPE).

Keywords: optical radiation, adverse health effects, occupational exposure

INTRODUCTION

Optical radiation is a portion of the non-ionizing part of electromagnetic spectrum in the wavelength range between 100 nm and 1 mm. The spectrum of optical radiation is divided into ultraviolet radiation (UV: 100-400 nm), visible radiation (VIS: 380-780 nm) and infrared radiation (IR: 780 nm - 1 mm). Usually ranges of ultraviolet and infrared radiation are divided into bands: A (near), B (medium) and C (far). According to CIE No. 17.4 (1987) Publication these bands cover the following wavelengths ranges of ultraviolet:

- UV-A: 315-400 nm
- UV-B: 280-315 nm
- UV-C: 100-280 nm

and of infrared:

- IR-A: 780-1400 nm
- IR-B: 1400–3000 nm
- IR-C: 3000 nm–1 mm.

Optical radiation exists as natural component of solar radiation which is essential to normal human development and activity. Both its insufficiency and excessive exposure can cause negative biological effects. Optical radiation is also produced artificially by electrical and technological sources. Electrical sources of this radiation are utilized in many industrial processes (e.g.: drying, curing, photpolymerization, engraving, photochemical reactions, disinfection), medical treatments (phototheraphy) or cosmetics (solaria). This radiation appears also as a by-product during some operations like welding, plasma cutting or emitted by hot objects like blast-furnaces.

HEALTH HAZARDS ARISING FROM OPTICAL RADIATION

Biological effects can be induced only by absorbed radiation. There are two types of reactions in biological tissues induced by optical radiation: photochemical (in UV and VIS range) and thermal (in VIS and IR range). The effects of exposure depend on physical parameters of radiation (wavelength, spectral power distribution of radiation), the amount of absorbed dose of radiation, optical and biological features of exposed tissue (skin, eye). The effect of optical radiation on skin tissue depends on the irradiance of the incident radiation, absorption of tissues at the incident wavelength, duration of exposition and the effects of blood circulation and heat conduction in the affected area. There are acute and chronic effects of exposure to optical radiation. Acute effects symptoms develop rapidly, usually during 24 hours after exposure. Chronic effects symptoms develop slowly due to long and continuous exposure. The acute effects of skin exposure to optical radiation are burns and erythema. Usually it is thermal injury following temperature elevation in skin tissues or photochemical injury from excessive levels of UV. Depending on the wavelength of optical radiation the adverse effects on eye can be located in its different parts. It is related to the spectral absorption in the particular parts of the eye i.e. cornea, aqueous, lens, retina. Ultraviolet radiation with wavelengths of 200–215 nm and infrared with wavelengths of 1400 nm or greater are absorbed in the cornea, so the radiation from these ranges can cause damages in cornea. Near ultraviolet is absorbed in the lens and that is why it is mainly responsible for lens injuries. Visible radiation and near infrared radiation are transmitted to the retina and they are responsible for possible retina damages. Relationship between eye and skin injuries and the range of optical radiation is presented in table 1.

Table 1. Eye and skin injuries and the range of optical radiation

Range of radiation	Eye injury	Skin injury
UV	*Acute:* Keratoconjuctivitis (welder's flash) Photokeratitis Conjunctivitis *Chronic:* Cortical cataract Pterygium Squamous cell carcinoma of the conjunctiva Squamous cell carcinoma of the cornea	*Acute:* Erythema Sunburn photodermatoses *Chronic:* Skin aging Malignant melanoma Squamous cell carcinoma Basal cell carcinoma
VIS and IR-A	*Acute:* Photoretinitis Retinal burns	*Acute:* burns
IR-B and IR-C	*Acute:* Corneal burns and lesions *Chronic*: Cataract (glass blower's cataract)	

PROTECTION OF WORKERS EXPOSED TO OPTICAL RADIATION

Since optical radiation can cause adverse effects on the health and many workers are exposed to optical radiation during performing their work tasks, it is necessary to ensure safe work conditions and protect workers in order to avoid possible negative health effects. First of all there is a need to assess the occupational risk which exists on particular workstation and then introduce the adequate measures to protect workers. Studies of the spectral effectiveness of optical radiation for a specified harmful health effects were the base for determination the criteria for health hazard evaluation and exposure limit values by different international organizations (ICNIRP, CIE, ACGIH).

The base for hazard evaluation is comparison between the level of exposure to optical radiation and the adequate exposure limit values (ELVs). Exposure limit values are limits of exposure to artificial optical radiation which are safe for workers. They are established on available medical experience and scientific findings and state the intensity of radiation below which acute or chronic adverse effects on health are not generally expected. ELVs reflect the wavelength of the radiation, the potential type of damage and the overall period of exposure. Physical variables that are reported in the ELVs include those such as radiation strength, radiation range, type and density. In some cases these variables are weighted using spectral weighting function specified for particular biological response (like: actinic action spectra for UV, photochemical retinal hazard, retinal thermal hazard). It is meant that limiting the occupational exposure to UVR for artificial sources in the workplace to values below ELVs provide sufficient protection of workers.

International recommendations for ELVs are established by International Commission on Non-Ionizing Radiation Protection of the International Radiation Protection Association (ICNIRP/IRPA). In 2006 the European Community established Directive 2006/25/EC of the European Parliament and of the Council on the minimum health and safety requirements regarding the exposure of workers to risks arising from physical agents (artificial optical radiation). The establishment of the Directive 2006/25/EC was an attempt to standardize the criteria of evaluating hazards related to optical radiation (Wolska, Dybczyński, 2010). The attachments to that directive contain criteria and exposure limit values for non-coherent and coherent (laser) radiation, which were mostly based on ICNIRP recommendations and CEN standards. The directive includes obligations for employers concerning: determination of exposure and assessment of risks, provisions aimed at avoiding or reducing risks, workers information and training and provisions concerning health surveillance. All these provisions have to be implemented in all countries of the European Community by April 27, 2010. Provisions of that directive are obligatory and risk assessment shall be done on all workstations where workers can be affected by artificial optical radiation. It requires identification and evaluation of the exposure to optical radiation. If the exposure limit values at the workplaces are exceeded, the risk is assessed and high and suitable protective measures must be applied in order to reduce exposure to a safe level.

Criteria and exposure limit values for non-coherent optical radiation are the same as these recommended by ICNIRP. The separate criteria and limit values are established for ultraviolet, visible and infrared radiation both for eye and skin.

The fundamental criteria for evaluating health hazards arising from optical radiation depend on its range:
- for ultraviolet radiation it is to prevent cornea and conjunctiva inflammation, cataract, erythema, skin photoaging and skin cancers;
- for visible radiation it is to prevent photochemical and thermal injuries of retina. Intensive visual radiation, especially in the range of blue light between 400 and 500 nm can produce retinal damage or diseases. Exposure time below 10 s causes mainly thermal injuries, whereas longer exposure – photochemical injuries;
- for infrared radiation it is to prevent thermal injuries of the cornea, lens, retina and skin. Hazards are determined separately for the retina, the cornea and the lens, and for the skin.

According to the provisions of the Directive 2006/25/EC, the employer is obliged to perform risk assessment, additionally taking into account any effects concerning health of workers belonging to particularly sensitive risk groups. That concerns also workers who suffer from illness which exclude their exposure to ultraviolet radiation (like different type of photodermatoses) or workers who take the medicines which could increase their sensitivity to UV, which are called photosensitizers. The risk assessment should also give particular attention on any possible effects on workers health resulting from workplace interactions between optical radiation and photosensitizing chemical substances. Such chemical substances, which could be present in work environment, include: benzocaine, benzoyl peroxide, carbamazepine, chlordiazepoxide , coal tar, etc.

OCCUPATIONAL EXPOSURE

Excessive exposure to optical radiation can cause harmful effects on human health; therefore, it is necessary to assess occupational risk related to that environmental factor at workstations with sources of optical radiation. That is why it is often necessary to measure relevant parameters of optical radiation and compare them with exposure limit values. Measurement methods and commonly used radiation measurement devices are presented in the European standards EN 14255-1 and EN 14255-2. The results of measurements of ultraviolet, visible and infrared radiation parameters which are carried out at electric welding (MAG welding of steel St3 protected with $ArCO_2$; rutile electrode ER 146 BAILDON, dimension: 3.25 mm, and welding current: 123 A) and gas welding (welding of steel St3; acetylene: oxygen - 4:1) workstations are presented in table 2 and 3. Values higher than exposure limit values indicate high occupational risk (i.e., a health hazard occurred).

Table 2. Sample results of measurements of UV radiation for electric and gas welding (adapted from Wolska, Dybczyński, 2010)

Workstation	Exposed part of the body	Exposure time (total)	Irradiance (effective)	Radiant exposure (effective)	Exposure limit value (Directive 2006/25/EC)
		s	W/m^2	J/m^2	J/m^2
Electric welding	eyes	10,080	0.75	7,560	30 (effective)
	face (skin)		1.09	10,987	
	hands (skin)		29.3	295,344	
Gas welding	eyes	10,080	0.0014	14.1	
	face (skin)		0.0022	22.1	
	hands (skin)		0.0033	33.3	

Table 3. Sample results of measurements for eyes exposed to visible and infrared radiation during electric and gas welding (adapted from Wolska, Dybczyński, 2010)

Workstation	Range of radiation	Exposure time	Angular subtence	Irradiance	Exposure limit values (Directive 2006/25/EC)
		s	mrad	W/m^2	W/m^2
Electric welding	VIS	10,080 (total)	10	0.64 (effective)	0.01 (effective)
	IR	19 (single)	-	22.7	1987
Gas welding	VIS	10,080 (total)	10,8	0.0051 (effective)	0.01 (effective)
	IR	35 (single)	-	191	1251

Comparison of results for electric and gas welding indicates that much higher hazard arising from ultraviolet and visible radiation. occurred during electric welding. However, infrared radiation also exists during welding process. its intensity doesn't constitute a hazard both for electric and gas welding. Results show that eyes and skin must be adequately protected against UV and VIS radiation

556

during the whole time of welding. There is a group of welders who start process of welding with uncovered eyes because they cannot see the object of welding through the welding shield properly. There is a strong evidence that starting electric welding with uncovered eyes even for a few seconds can be enough to exceed ELVs for UV and VIS radiation. The right choice of personal protective equipment (PPE) for face and eyes for welders is not equal with low risk until it is properly used. Periodical check-ups of PPEs and observance of the rules concerning their proper use by welders is still very important.

PROTECTION AGAINST OPTICAL RADIATION

The general rule of protection against overexposure to optical radiation is a combination of three preventive measures (Wolska, Dybczyński, 2010):
- time of exposure should be minimized
- distance to the source should be maximized
- shielding against radiation

The hierarchy of control measures depends on the source and performed risk assessment. Where the source of optical radiation constitutes a hazard, according to directive 2006/25/EC, the employer is obliged to provide protection from exposure via a combination of engineering controls, administrative measures and personal protective equipment (PPE). Engineering controls include for example installing shielding, enclosing the source, use of interlocks, built-in radiation detectors and alarms (visual and/or auditory). The administrative measures are: information and training of workers about hazards and safety aspects, limiting the time of exposure and increasing the distance of an authorized user from the source, monitoring the hazard by periodic measurements of optical radiation at workstations, according to current legislation. PPE is used if engineering and administrative controls do not provide appropriate protection of workers (Wolska, Dybczyński, 2010).

There are many different sources of optical radiation which could be identified at the workplaces but not all of them can affect the worker health. Sometimes the knowledge of the source radiation could be enough to exclude it as a hazardous. But in many situation measurements or calculations which are the base of risk assessment must be done by competent personnel.

REFERENCES

Wolska A., Dybczyński W. (2010). "Non-coherent optical radiation", in Handbook of Occupational Safety and Health, D. Koradecka ed, CRC Press, Boca Raton, 271-292

Directive 2006/25/EC of the European parliament and of the Council of 5 April 2006 on the minimum health and safety requirements regarding the exposure of workers to risks arising from physical agents (artificial optical radiation) (19 individual Directive within the meaning of Article 16(1) of directive 89/391/EEC), Official Journal of the European Union L 114/38

EN 14255-1:2005: Measurement and Assessment of Personal Exposures to Incoherent Optical Radiation - Part 1: Ultraviolet Radiation Emitted by Artificial Sources in the Workplace, European Committee for Standarization (CEN), Brussels

EN 14255-2:2005: Measurement and Assessment of Personal Exposures to Incoherent Optical Radiation - Part 2: visible and infrared Radiation Emitted by Artificial Sources in the Workplace, European Committee for Standarization (CEN), Brussels

Guidelines on Limits of Exposure to Broad- Band Incoherent Optical Radiation (0. 38 to 3μm) (1997) International Commission on Non-Ionizing Radiation Protection (ICNIRP). Health Physics, vol. 73(3), s. 539-554.

Guidelines on Limits of Exposure to ultraviolet radiation of wavelengths between 180 nm and 400 nm (incoherent optical radiation) (2004) International Commission on Non-Ionizing Radiation Protection (ICNIRP). Health Physics, vol. 87(2), s. 171-186.

CHAPTER 57

Human Factors Considerations in the Detection of Gas Leaks

Michael S. Wogalter[1], Kenneth R. Laughery[2]

[1]Psychology Department,
North Carolina State University
640 Poe Hall
Raleigh, NC 27695-7650, USA

[2]Psychology Department
Rice University
MS-25, P.O. Box 1892
Houston, Texas 77251-1892, USA

ABSTRACT

A scenario of a residential propane fire and explosion case is described. The apparent cause was a corroded pipe in the basement which leaked propane and was ignited by the spark generated by the start of a clothes dryer. The victim who was severely burned reported not smelling gas despite an odorant, ethyl mercaptan, being added to the gas. Reasons why the gas was not detected by smell are described. Electronic gas detectors are available but the gas supplier never communicated this fact to many of its customers. The added value of electronic gas detectors is described.

Keywords: propane, olfaction, smell, sensors, detectors, warnings

It was a balmy August day in Lincolnville, Arkansas. Jack Simpson was working on a project in the old barn built in the early 1900s. The previous house on their land had burned down about 10 years before their current 1950s ranch-type home was constructed.

Jack's wife, Mary Lee, was in the basement just starting to take the first load of laundry out of the washer. Their two kids were in their bedrooms; Sam was playing a 3-D video game and Jon was watching something on TV. The time was approximately 4:15 pm, nearly time to start thinking about dinner. Suddenly, right next to Mary Lee, there was a large flash of light, explosion, and fire. Later Mary Lee reported that she was in the basement of her home doing routine things related to laundry. She was putting wet clothes in her dryer when the area exploded around her. She did not remember closing the dryer door and pushing the start button, but she might have forgotten because the flash fire happened at about that time. When she first realized what had occurred, she saw she was on fire and started screaming as she ran up the stairs and out of the house.

Hospital records describe severe burns on both legs, the top of her feet, and her right arm. Medical treatment involved multiple skin grafts over the course of a year. The extensive scarring will require more surgeries and pain. She's describes the pain as intense and unbearable. She is reminded of disfigurement every time she dresses and moves her legs. The scar tissue is not so elastic as her other skin.

Investigation into the cause of the fire points to a corroded, cracked copper pipe supplying propane gas (also called liquid petroleum or LP) to the dryer. Apparently there was water moisture on the pipe at some point and combined with vibrations from the dryer caused the pipe to degrade and crack. That was the explanation given by an engineering expert hired by the Plaintiffs, Mr. Bertrand Rakeholder. He based it on electron micrography (showing discoloration, corrosion, and fracture marks), as well as other situational factors, such as the location of the pipe. He concluded that the flash fire was caused by a propane leak from a fractured corroded copper pipe connected to the dryer. The spark from the electric dryer being started by Ms. Simpson was probably the ignition source. The Defendant's expert had a somewhat different theory about the cause. Dr. Craig Seymour, a metalurgist from California, concluded the pipe had a manufacturing defect that was exacerbated by an installation that failed to include flexible pipe at the dryer connection.

The family had been getting gas from Northern Amalgamated Propane ("Amalgamated") since they purchased the house from the original owner about 15 years ago. Northern owns the tank in the backyard and delivers propane on a regular basis, about every month, but less frequently in the summer compared to the winter. The Simpsons never had a problem with propane before this event. Mary Lee reports that she did not smell gas before the explosion. She knows what propane smells like because she has sometimes smelt it when she starts the range/stove in the kitchen.

Propane by itself is an odorless gas. Since people cannot smell it, an extra chemical (ethyl mercaptan) is usually added to the gas to give it its characteristic odor to help with leak detection. Sometimes the smell is described as being like sewer gas or the stench of dead rodents. Some gas companies periodically send a sheet to customers with scratch and sniff patches that contain the chemical odorant so that people will learn what propane gas with odorant smells like. Other companies, like Amalgamated, only send these sheets to new customers. Amalgamated started this practice about three years ago when it was purchased by a Little Rock-based conglomerate. Existing customers would not have received this material. None of the Simpsons state that they have seen or smelled a scratch and sniff patch.

Adding the odorant to the gas to help alert users of a gas leak is clearly a good idea. It would aid in detection compared to without it. When gas is smelled, even in small amounts, gas companies advise that people vacate the premises quickly and not to use any electrical switches or anything else that could create a spark, including telephones, until after having exited the premises. However, people commonly report smelling gas when starting their stove, just like Ms. Simpson does, and yet they do not vacate each time they smell it.

However, this odorant warning system is not perfect. Leaks and resulting fires and/or explosions still occur even with the added odorant. Industry associations such as the National Fire Protection Association (NFPA) and the National Propane Gas Association (NPGA) publish materials (e.g., pamphlets/brochures) that state unequivocally that odor detection as a method of hazard protection cannot be counted on as being 100% reliable.

The Simpsons' filed a lawsuit against the propane retailer. As Plaintiffs, they alleged that the Defendant Amalgamated was at fault in causing severe injury to Mary Lee Simpson. The lawsuit documents submitted to Arkansas State Court in the County of Lincoln by the Plaintiffs claim that the seller had superior knowledge about the characteristics and hazards associated with propane but did not communicate them to less knowledgeable consumers like the Simpsons. Amalgamated supplied and retained ownership of the propane tank in the Simpson's yard and delivered propane on a regular basis, but it had not done any kind of inspection or check whether the system had any leaks or potential for leaks, i.e., it did not do any inside-the-home inspections or do any leak tests during the 15 years it supplied gas to the Simpson house. The Simpsons as consumers did not know how to recognize problem pipes that had corroded over time.

The Plaintiffs contend that the LP seller knew or should have known that propane odor is not always detected. In fact, materials found in the Defendant's possession included industry/trade publications that addressed this detection issue. Indeed, some of this information that the Defendant sends to new customers includes this information, although the Simpsons as existing customers would not have received it. (No records have ever been produced by the Defendant confirming the mailing of any such materials.)

The only printed warning information relevant to this case were on invoices that Jack Simpson received to pay the bills. On the back of them was some warning text in small type in light pink ink and embedded in other print material. This particular warning text concerned egress procedures when the odor is detected: Evacuate immediately and do not to turn on any light switch or use a phone, and call the gas company once outside the home. Mr. Simpson, who pays the bills, says he never noticed anything on the back of the invoices, and the other family members never dealt with paying the gas company bills.

Interestingly, even if one were to assume that all materials that Amalgamated alleges to have provided to its customers (regardless of whether they are actually true or not), the warnings were deficient. This was the opinion of the Human Factors expert that was retained by the Plaintiffs, Dr. Richard Raster, a psychology professor from a major southern university. According to his analysis, the warnings were defective with respect to manner, method, and content. According to him, the point of effective warnings is to alert people about hazards and to motivate them to carry out safety-appropriate actions to avoid harm to themselves and others. The warning information failed to communicate some fundamental aspects of the LP gas detection.

Amalgamated had two major components of their LP leak warning system: (a) ethyl-mercaptan odorant added to the gas to provide an olfactory cue, and (b) printed safety information. To be effective as a warning, the odorant must first be detected by the olfactory receptors before it can alert users to the presence of propane. Olfactory sensitivity can be reduced by several factors. One major factor is that some people chronically lack some or all of the ability to smell the odor. Genetics and illness can cause reduction in the ability to smell. Also chronological age reduces olfactory sensitivity (Doty et al., 1984; Gilbert & Wysocki, 1987; Stevens & Cain, 1985). Conditions that swell or clog the nasal passages and extra mucus will limit or prevent odors from reaching the olfactory receptors (Murphy & Cain, 1980). Colds and other respiratory difficulties can affect olfaction and hinder an individual's ability to detect odorized propane. Moreover there are still other factors that can decrease olfactory function (e.g., Katotomichelakis et al., 2007). Competing odors in the environment could interfere with detection of odorized propane (by disguising or masking the smell). Tobacco smoke, cooking smells, chlorine bleach, and musty damp odors can reduce the likelihood of odor detection (Fang, Clausen, & Fanger, 1998; Gunnarsen & Fanger, 1992; Stone & Bosley, 1965). It is possible that in the basement there were other odors present that could have masked the odor of escaping gas. The washing machine was adjacent to the dryer, so damp odors, detergent, and bleach from the laundry could have interfered with propane detection. Propane gas is heavier than air. The gas tends to settle in lower levels and follows the floor downwards. Basements can collect gas particularly when there is very limited airflow. Ms. Simpson said that she was not using bleach, but the basement sometimes smelled musty and damp. She does not remember any smell that day before the fire, including the laundry detergent's "fresh" smell.

Another potential shortcoming associated with reliance on odor as a warning mechanism is a phenomenon sometimes called "odor fade." This condition is a loss of detectability of propane that has previously been odorized due to physical/chemical reactions. Another condition is "odor fatigue." Over time, the sensory systems habituate or adapt to the smell, and this adaptation results in reduced awareness of odorized propane. Detecting a momentary scent of propane odor is normal when initiating use of appliances that run on propane. That is how Ms. Simpson says she learned what propane smelled like. Because of adaptation/habituation, people might not detect it even though the gas and odor is present (Dalton, 2004). This effect is important in situations in which the gas odor is present while sleeping. Upon awakening from sleep, individuals may not notice the smell due to adaption/habituation while asleep. Turning on light switches at night is an all too common reaction, unfortunately.

Thus, there are several known factors that result in persons may not smelling propane, and a leak that could go undetected exposing people to a risk about which they are unaware. Clearly, a different, separate way to detect the presence of gas is necessary. Fortunately, electronic LP gas detectors are now currently available. These systems would be beneficial, because they provide an effective additional measure to detect LP gas leaks. These devices could provide detection when humans may not be able to, such as when there are competing smells, odor fade or fatigue, conditions of anosmia, colds/clogging of nasal passages, and during times they are asleep or immediately after awakening, etc. It is a backup warning system for persons who are not adequately sensitive to smell of the odorant due to one or more of the reasons already mentioned.

Mary Lee and Jack Simpson testified that they did not know there were electronic LP gas detectors on the market. They also did not know there were numerous reasons for not smelling a propane gas leak. They were emphatic in saying that if they had been made aware of the need for electronic gas detectors, they would have purchased them. The human factors expert said if an electronic gas detection warning been sounded, then Ms. Simpson would been adequately warned of the existence of leaking propane gas in their basement. The Simpsons characterized themselves as safety conscious. At the time of the incident, they had two working smoke detectors and a CO detector in their home which was documented in the Lincoln County Fire Department's fire investigation report.

While it is not clear what safety materials the Simpsons received, none of the materials adequately emphasized the availability and need for electronic LP gas detectors. Some LP retailers sell LP gas detectors. Amalgamated did not sell them, although it knew that some LP retailers do. The Plaintiffs argued that information and warnings about electronic LP gas detectors needed to be conveyed by the Defendant sellers of propane. If the company was not going to sell the detectors directly to consumers then the company should have told them where they could purchase them.

Instead of a loud screeching alarm, currently available technology allows digitally recorded voice warnings to be given. This technology would help distinguish between sounds of other sensor alarms such as smoke and carbon monoxide detectors (e.g., see Haas and Edworthy, 1996). Voice warning and instructions would aid in identifying the hazard, instructing what to do, and providing information about consequences of not complying.

These electronic devices are not perfect, however. For example, they may give false alarms (i.e., alert people when there is no propane hazard) or miss detection of LP gas, such as when the batteries are dead or removed. People may over-rely on them. However, if working properly it would provide a benefit by supplementing and extending odor detection. Thus, a working electronic LP gas detector would serve as backup or redundant cue to the total gas detection system and it could also extend detection by placing them in areas where residents are not.

Amalgamated employees overly relied on people's ability to smell a propane leak to avoid fire and explosion events, even though they should have known that detection via smell can fail. Amalgamated's employees testified that they believed the presence of odorant to be an effective method of detection, and that there is little extra value of any added electronic equipment such an gas detector. Amalgamated had access to information that said electronic detectors had value, but it failed to pass on this relevant safety information to many of their customers including the Simpsons.

In briefs filed with the Court, Plaintiffs also argued that Amalgamated should have performed leak tests and examination of the gas plumbing on some regularly scheduled basis. That no such tests and no examinations were made inside the Simpson house in 15 years constituted negligence by the Defendant.

Ms. Simpson stated that she knew that propane was potentially dangerous, but she did not realize the extent of the danger it posed. Consumers partly base their decision to purchase and use consumer products like propane based on an assumption that companies would not sell (and that government would not allow) truly dangerous products to be sold. People expect an adequate warning for dangerous products, especially those capable of causing severe injury or death. Consequences, such as their house exploding/burning down and risk of occupant injuries of deaths were not effectively warned about. Further, Mary Lee Simpson and family were not made aware of the deficiencies of odor dection and the existence of, the need for, and where to obtain electronic gas detectors.

Authors' Note: Names of entities and details have been changed to protect privacy and confidentiality rights.   The scenario and description are based on several prototypical LP gas fire and exposion cases.

REFERENCES

Dalton, P. (2004). Olfaction and anosmia in Rhinosinusitis. *Current Allergy and Asthma Reports, 4,* 230-236.

Doty, R. L., , (1984). Smell identification ability: Changes with age. *Science, 226,* 1441-1443.

Fang, L., Clausen, G. & Fanger, P.O. (1998). Impact of temperature and humidity on the perception of indoor air quality. *Indoor Air, 8,* 80–90.

Gilbert, A. N., & Wysocki, C. J. (1987). The smell survey results. *National Geographic, 122,* 514- 525.

Gunnarsen, L. & Fanger, P.O. (1992). Adaptation to indoor air pollution, *Environment International, 18,* 43–54.

Katotomichelakis, M., Balatsouras, D., Tripsianis, G., Davris, S., Maroudias, N., Danielides, V, & Simopoulos, C. (2007). *Rhinology, 45,* 257-258.

Murphy, C., & Cain, W. S. (1980). Taste and olfaction: independence vs. interaction. *Physiology & Behavior, 24,* 601–605.

Stevens, J. C., & Cain, W. S. (1985). Age-related deficiency in the perceived strength of six odorants. *Chemical Senses, 10,* 517-529.

Stone, H., & Bosley, J.J. (1965). Olfactory discrimination and Weber's Law. *Perceptual and Motor Skills, 20,* 657-665.

Chapter 58

Impulse Noise

Jan Zera

Central Institute for Labour Protection – National Research Institute
Czerniakowska 16, 00-701 Warsaw, Poland

ABSTRACT

This paper presents a brief overview of requirements on noise measurements, impulse noise measurements and hearing damage risk criteria for noise exposure. First presented are the requirements referring to general assessment of noise, including those set by the European Council Directive 2003/10/EC. In following sections discussed are specific issues concerning the effect of impulse noise on hearing, including the usage of hearing protectors to reduce the exposure to continuous noise and impulse noise.

Keywords: Noise, noise regulations, impulse noise, impulse noise damage risk criteria, hearing protectors

INTRODUCTION

Noise is a common pollutant at workplaces and in community life. Since the beginnings of noise measurements and investigations of the effect of noise on humans a clear distinction has been made between continuous (steady-state) noise and impulse noise. The effect of those two kinds of noise on hearing is different therefore different hearing damage risk criteria are used for continuous and impulse noise. In this paper procedures for noise measurement and criteria for the assessment of hearing damage risk caused by noise exposure are presented in reference to continuous and impulse noise.

CONTINUOUS NOISE

As a measure for evaluation of continuous noise, A-weighted sound level is used. A-weighting reduces the level of low-frequency components of sound to reflect the variations in sensitivity of the human ear at different tone frequencies. The A-weighting filter corresponds to the attenuation of low-frequency sound transmission in the middle ear and reflects a relatively mild damaging effect of low-frequency sounds on hearing. Decades of research have demonstrated that the assessment of noise exposure with the use of A-weighting in the frequency domain well agrees with the data on hearing damage risk caused by noise. For continuous noise, a time vs. exposure level trade-off holds, leading to an equal-energy rule which assumes that hearing damage is related to the total A-weighted energy accumulated by the auditory system. In Europe, the noise exposure limit during an eight-hour work day is 85 dB(A). To maintain a constant amount of accumulated energy, each 3-dB increase in sound level requires halving of exposure time. The 3-dB rule is used in many countries, for instance, in Europe and in Canada. The OSHA regulations set a 90 dB(A) limit for 8-hour working day and a 5-dB exchange rate with doubling of exposure time (Earshen, 2000). In the US, the OSHA regulations are considered sufficient to minimize hearing damage risk, but they do not follow the equal energy rule.

EUROPEAN REGULATIONS

European Directive 2003/10/EC (2003) lays down minimum requirements for the protection of workers from hearing risks and health risks arising or likely to arise from exposure to noise. These minimum measures have to be maintained by European States, but countries can use more stringent protective measures. A novelty in the Directive, as compared to previous regulations, is that in addition to exposure limit values, introduced are the so-called lower and upper exposure action values. The meaning of the lower and the upper action values can be explained on an example of the use of hearing protective devices. When noise exposure exceeds the lower exposure action value, the employer is obliged to make individual hearing protectors available to the workers. When noise exposure equals or exceeds the upper exposure action value, individual hearing protectors must be worn, and such conditions have to be indicated by special warning signs.

Directive 2003/10/EC defines three parameters to characterize noise: the daily noise exposure level, $L_{EX,8h}$, in dB(A), the weekly noise exposure level, $L_{EX,8h}$, in dB(A), and the C-frequency weighted peak sound pressure level, $L_{C,peak}$, in dB (C). The two first parameters differ in the averaging periods which are an eight-hour work day and a five-day eight-hour work week. These parameters are applicable to all kinds of noise in the workplace, including impulsive noise. The exposure averaging periods follow the recommendations of ISO standard 1999:1990.

Exposure limit values defined by the Directive are $L_{EX,8h}$ = 87 dB(A) and L_{Cpeak} = 140 dB (which corresponds to peak pressure of 200 Pa). The upper and lower action values are defined respectively by $L_{EX,8h}/L_{Cpeak}$ of 85 dB(A)/137 dB (C) and 80 dB(A)/135 dB(C). In many European countries, e.g., in Germany, Poland, Estonia, Sweden or Denmark, the daily noise exposure level $L_{EX,8h}$ is set to 85 dB(A), and $L_{C,peak}$ is set to 135 dB. These limits, established before the introduction of Directive 2003/10/EC, are by 5 dB lower than the respective levels required by the Directive so leaving them unchanged fulfills the Directive requirements.

According to Directive 2003/10/EC, reduction of hearing damage risk should be obtained by the following means: using work methods that produce less exposure to noise, the choice of work equipment emitting the least possible noise levels, proper design and layout of workplaces and workstations, providing workers with adequate information and instruction, noise reduction by technical means (reducing airborne noise by shields, enclosures, sound-absorbent coverings as well as reducing structure-borne noise by damping or isolation), appropriate maintenance programs for work equipment, the workplace and workplace systems, proper organization of work to reduce noise, limitation of duration and intensity of noise exposure, and appropriate work schedules with sufficient rest periods.

IMPULSE NOISE

Impulse noise is most often generated by a sudden release of energy into the air, such as a gun shot or an explosion, or by collisions of objects, such as those often encountered in the industry, for example in drop-forge workshops. Due to short duration and high amplitude of impulse noise, the rules used for the assessment of hearing damage risk caused by exposure to steady-state noise do not apply to impulse noise. Even a single occurrence of a high-amplitude acoustic impulse may cause irreversible damage to hearing. At peak sound pressure levels of above 140 dB, the duration vs. exposure level trade-off does not hold. All exposures to peak sound pressure levels to higher than 160-dB should be considered extremely damaging to hearing.

Regulations that comply with the Directive 2003/10/EC are solely based upon limitations imposed on the level of impulse noise. The most important parameter is the L_{Cpeak} designed to measure the effect of impulse noise on hearing. The limit value for this quantity is reached earlier (for lower noise levels) than the impulse daily noise exposure level $L_{EX,8h}$.

The differences in the effects of continuous and impulse noise on hearing led to the development of a number of impulse noise hearing damage risk criteria. These criteria has been mostly developed for military environment, to protect soldiers from hearing loss at their daily training. Three major impulse noise hearing damage risk criteria were developed over the years: criterion by the Committee on Hearing, Bioacoustics, and Biomechanics (CHABA; Ward, 1968) in the US, then later by Pfander (Pfander et al., 1980) and Smoorenburg (1982) criteria used in Germany and in the Netherlands. All these criteria use peak sound pressure level and some measure of effective impulse duration, called A, B, C or D- duration (ANSI S12.7-1986, ISO 10843:1997). Effective duration is a measure of time interval, during which the impulse produces impact on the hearing system. Regardless of the details of specific criteria, it is assumed that impulses of both high peak level and long effective durations A, B, C, D are more damaging to hearing. Therefore, all criteria require that longer-duration impulses must be associated with lower peak sound pressure levels. The differences between definitions of A-, B-, C-, and D-duration come from either a desire to represent a large variety of time waveforms of acoustic impulses, or from an assumed difference between the effect of various kind of impulses on hearing.

A-duration is the duration of a shock wave of compressed air that arrives after a single explosion, such as produced by a weapon. This kind of an impulse, often called A-type impulse or a Friedlander wave (Hamernik nad Hsueh, 1991), has been described in a detailed mathematical model. It is assumed that both the duration of increased pressure and the peak pressure value contribute to the hearing damage. The value of A-duration is a proper description for a single isolated impulse. For a series of repetitive impulses B-duration is used. B-duration is a sum of periods in a quasi oscillating waveform in which the pressure envelope exceeds 10% of the peak pressure value. Both A- and B- durations are used in the CHABA criterion (Ward, 1968).

The measure of C-duration, introduced by Pfander (Pfander et al., 1980), is the total duration in which the signal exceeds one-third (-10 dB) of the peak sound pressure. D-duration (Smoorenburg, 1982) is similar to B-duration as it takes into account the total time in which the pressure envelope exceeds one-third (-10 dB) of the peak sound pressure. All criteria that use peak pressure vs. effective impulse duration are based upon epidemiological data which indicate that symptoms of temporary threshold shift are developed in only a small fraction of the population, when peak pressure and cumulated A-D durations are kept within certain limits.

Implementation of criteria based on A, B, C, or D duration is often difficult as standard sound level meters are not suitable for the measurement of impulse durations. For this reason criteria developed for military environment are not commonly used for the measurement of impulse noise at workplaces. Using a sound level meter it is easily to determine instead the values of peak sound pressure level and C-weighted sound pressure level.

Measurements of impulse noise are difficult to perform in industrial environment as impulse and continuous noise usually occur simultaneously, therefore the effects of both types of noise on hearing should be considered together. For this purpose a hearing damage risk impulse noise criterion developed by Dancer (Dancer and Franke, 1995) is convenient. Dancer's criterion uses an equal energy principle. It is assumed that the total energy accumulated by the ear due to impulse noise exposure cannot exceed the $L_{EX,8h}$ value. This criterion is limited to lower peak sound pressure levels, as at very high levels the equal energy principle does not hold. Nevertheless, this measurement method is similar to that used for continuous noise which makes it easier to measure continuous and impulse noise occurring simultaneously.

There is evidence that continuous and impulse noise occurring simultaneously are more damaging to hearing than each of those types of noise separately. Even for peak levels not exceeding the limit values, the presence of impulsive components in noise of otherwise identical A-weighted exposure level causes more severe physiological effects upon hearing, such as larger temporary threshold shift and longer recovery time after the exposure (Strasser, et al., 1999).

HEARING PROTECTORS

If the risks arising from exposure to noise cannot be prevented by other means, appropriate, properly fitted individual hearing protectors should be made available to workers and used. In setting exposure limit values, the determination of the worker's effective exposure takes into account the attenuation provided by the worn hearing protectors. The sound attenuation by hearing is typically assessed by the REAT (Real Ear at Threshold) method (ANSI/ASA S12.6-2008, EN 24869-1:1992, ISO 4869-1:1990). Regardless of the differences between specific procedures, the REAT method is based upon audiometric measurements in which the sound attenuation provided by a hearing protector is determined as the difference in hearing threshold measured in the field when the hearing protector is donned and doffed. This method is most suitable for determining protection provided by an earmuff or earplug during exposure to continuous noise. In impulse noise, peak sound level attenuation by a hearing protector varies when the impulse duration changes. It has been showed that peak level attenuation of a single earmuff can be as large as 30 dB for short A-duration impulses (gunshots) and as low as a few decibels for long A-duration impulses (large cannon firings, Ylikoski et al. 1987; laboratory generated impulses, Zera and Mlynski, 2007). This difference can be attributed to the frequency content of various impulses. Long A-duration impulses are associated with maximum energy localized in low frequency range in which earmuffs do not attenuate sound very well. In contrast, short A-duration is associated with impulses of high frequency energy content being well attenuated by earmuffs, attenuation of which reaches 40 dB in this frequency range. Thus, standardized measurements of hearing protectors do not provide adequate information on their performance in impulse noise. For this reason, a classification of types of impulses is used, and accordingly suggested corrections in hearing protection attenuation to account for impulse noise (EN 458:2004, ISO 9612:2009). Correction values used in normative documents for describing the effectiveness of hearing protectors in conditions of exposure to impulsive noise are complex and often not quite clear.

This work was supported by the Polish Ministry of Education and Science, Grant No. 3.R.01 P.W.

REFERENCES

ANSI/ASA S12.6-2008. Methods for Measuring the Real-Ear Attenuation of Hearing Protectors.

ANSI S12.7-1986 (ASA 62-1986). American National Standard. Methods for measurement of impulse noise,

Dancer A. and Franke R. (1995) Hearing hazard from impulse noise: a comparative study of two classical criteria for weapon noises (Pfander criterion and Smoorenburg criterion) and the L_{Aeq8} method, Acustica - Acta Acustica 3, pp. 539-547.

Directive 2003/10/EC (2003). Directive of the European Parliament and of the Council of 6 February 2003 on the minimum health and safety requirements regarding the exposure of workers to the risks arising from physical agents (noise).

Earshen, J. J. (2000) Sound measurement: Instrumentation and Noise descriptors", in: "The noise manual", Berger, E. H. at al. (Ed.). pp. 41-100.

EN 24869-1:1992. Acoustics – Hearing protectors – Subjective method for the measurement of sound attenuation .

EN 458:2004. Hearing protectors – Recommendations for selection, use, care and maintenance – Guidance document.

Hamernik R. P. and Hsueh, K. D. (1991) Impulse noise: Some definitions, physical acoustics and other considerations, J. Acoust. Soc. Am. 90, pp. 189-196.

ISO 10843:1997. Acoustics – Methods for the description and physical measurement of single impulses or series of impulses.

ISO 1999:1990. Acoustics -- Determination of occupational noise exposure and estimation of noise-induced hearing impairment.

ISO 4869-1:1990 Acoustics – Hearing protectors – Part 1: Subjective method for the measurement of sound attenuation.

ISO 9612:2009 Acoustics – Determination of occupational noise exposure – Engineering method.

Pfander, F., Bongartz, H., Brinkmann, H. and Kietz, H. (1980) Danger from auditory impairment from impulse noise: a comparative study of the CHABA damage-risk criteria and those of the Federal Republic of Germany, J. Acoust. Soc. Am. 67, pp. 628-633.

Smoorenburg, G. F. (1982) "Damage risk criteria for impulse noise," in: New perspectives on noise, edited by R. P. Hamernik, D. Henderson, and R. Salvi (Raven Press, New York), pp. 471-490.

Strasser, H., Irle, H., Scholz, R. (1999). Physiological cost of energy-equivalent exposures to white noise, industrial noise, heavy metal music, and classical music, Noise Control Eng. J. 47, pp. 187-192.

Ward, W. D. (1968) Proposed damage-risk criterion for impulse noise (gunfire), Working Group 57, NAS-NRC Committee on Hearing, Bioacoustics, and Biomechanics, NAS-NRC Committee on Hearing, pp. 1-10.

Ylikoski, J., Pekkarinen, J, and Starck, J. (1987) The efficiency of earmuffs against impulse noise from firearms, Scand. Audiol. 16, pp. 85-88.

Zera J., and Mlynski, R. (2007) Attenuation of high-level impulses by earmuffs, J. Acoust. Soc. Am. 122, pp. 2082-2096.

Chapter 59

The Smell of Danger: Natural and LP Gas Odorization

William S. Cain

Chemosensory Perception Laboratory
Dept. of Surgery (Otolaryngology)
University of California, San Diego
La Jolla, CA 92093-0957, USA

ABSTRACT

Natural gas and LP-gas initiated fires and explosions take a serious toll on human beings and on property. To avert such accidents, suppliers of gas have long relied upon odor to warn of impending danger. The system, rooted in measurements in the early 20th century, has changed little in modern times. The odorants themselves have limitations, as does the recipient of this "warning." Under conditions of an explosion, the warning odorant should behave like a fire alarm and drive the person with a normal sense of smell from the space. Even when effective, the odorants do not seem to have that specific effect. The situation only magnifies vulnerability for those many people who have impaired olfaction. Some accidents occur because some consumers miss the odor and some because consumers "negotiate" with themselves about what the odor means. Both the nature of the accidents that occur and field studies indicate that the gas odor leaves much work to be done by other means of communication in order to diminish the threat of gas fires and explosions.

Keywords: Natural gas, LP-gas, Propane, Fire, Odor, Warning

INTRODUCTION

According to the National Fire Protection Association (www.nfpa.org), the annual number of fires in the U.S. first ignited by natural or LP-gas equals about 5,000. The number of civilians who die in such events equals about 80 and the number injured equals about 500.

Although most injured persons in society would choose life over death, for some victims of burns, the choice would seem to test anyone's sense of resolve to live. Some victims undergo numerous surgeries to achieve even minor increases in functioning, such as the ability to close an eye, straighten fingers, or have a semblance of an ear or nose. Some live perpetually in Jobst garments and some become recluses because their injuries evoke horror or because they cannot thermoregulate (e.g., perspire). Medical costs often reach hundreds of thousands, even millions of dollars. Needless to say, medical bankruptcy abounds.

This societal burden, in addition to the approximate $150 million property loss, would seem to merit considerable attention, but receives surprisingly little. Central to this entire state of affairs lies the most primitive of warning signals, an odor. Society has afforded surprising deference to that odor insofar as it has not crowded the environment with competing signals to what we might think of as the "olfactory channel." Occasionally, odorization of gases has become a prospect in occupational safety, but only for specialized use, e.g., prevention of asphyxiation from inert gases in enclosed spaces (Cain et al., 1987).

WARNING AGENTS

Non-combusted gas has most commonly had a sulfur (aka mercaptan) smell, generally described as skunk-like, though usually discriminable from that. The odorant in LP-gas, ethyl mercaptan, may smell like scallions, like a skunk, or like rotten eggs, depending upon its level. The odorants in natural gas, commonly blends that contain t-butyl mercaptan at their core, have a generally skunk-like character, though with more than trivial variation. The widespread addition of such agents to gas began in the decade before World War II and has changed little. The variety of blends has changed to meet certain technical needs, but remain a variation on a theme.

When a gas explosion occurs, the relevant personnel, e.g., accident investigators, fire inspectors, normally raise the question of whether the gas was odorized, meaning, "Did someone put odorant into the gas as required by law?" Although essentially an unavoidable question, the answer is almost always "yes," and therefore not especially informative. One can accordingly assume that under some circumstances, the odorized, non-combusted gas would have warned of its presence. The odor warning should, however, behave like a fire alarm. False alarms aside, a fire alarm will rarely fail to communicate impending danger. The hearing impaired may fail to hear an acoustic alarm, but an annoying alarm should make a space essentially uninhabitable to the hearing and should drive them out. A gas-warning agent should do the same for those who can smell when danger from a fire or explosion looms. If the warning agent does not, then in one categorical sense, it has failed.

INDETERMINATE FAILURE RATE

How should one calculate the failure rate of gas systems? The statistics above from the NFPA imply a non-negligible problem, with 5,000 gas-ignited fires per year. Nevertheless, to get a sense of rate, one needs to know that about 75,000,000 consumers use gas. Hence, less than one in ten thousand "fails" annually, or, stated positively, success rate equals better than 99.99%. This number undoubtedly underestimates success, perhaps egregiously, since one cannot expect every gas-ignited fire to represent a failure of the gas system, i.e., escape of non-combusted gas. Such a high success rate deserves some degree of praise, for it represents competent design and engineering. From the standpoint of human factors, it deserves an indeterminate degree of praise for it masks the statistic of how often gas escapes and the action of a person averts an accident. The rate of failure may climb by orders of magnitude in that case. Most likely, no one computes it for no one knows how often a consumer stops flow that could have led to an accident.

Up to this point in this report, the distinction between natural gas and LP-gas installations has required no attention, but that will change below. Natural gas comes from a utility that may supply hundreds of thousands of customers and may have good records of complaints. If a natural gas leak occurs, the customer may shut off gas at an appliance, but may not know how to shut it off to the house. He may, however, open windows and doors and call for help. Depending upon where the leak occurs, e.g., a basement vs. a ground floor, those actions may avert an accident.

An LP-gas supplier may have a few hundred customers and limited records about complaints. If an LP-gas leak occurs, the customer may know how to shut off the gas at its source, a tank in the yard. Because each installation exists as a complete unit, stem to stern, it invites more independent fiddling. Tanks run out of gas, so that pilots go out, and the rural user of LP-gas often considers himself capable of maintenance of the system, a practice that may end in an explosion. In the meantime, he may have had many occasions to detect leaked gas, whereas the city customer whose gas comes from a central pipeline may never interact with his system and may never have smelled actual leaked gas. (Some utilities actually invite customers to call for a scratch-and-sniff card to experience the gas odor!)

So, at best, only large utilities that supply natural gas may have any "data" about how often the odor of gas has actually aborted an accident. Whether they share the information is another matter. Unfortunately, what serves to avert a natural gas accident will not necessarily do the same for LP-gas, the more dangerous of the two fuels (see below).

PERFORMANCE OF THE ODORANT

The choices of odorants underwent some empirical investigation in the early part of the 20th century and in such comparisons certain mercaptans proved adequate (Fieldner et al., 1931). Interestingly, these sulfur compounds and various sulfides often gave back to natural gas the odor it had when it came out of the ground as so called "sour" gas. The addition of the compounds to purified methane or propane, the principal LP-gas, required decisions about the level to achieve. The codified standards that translated into industry practice had to meet a performance criterion, not a physical criterion. The normal standard in widespread practice, and generally written into law, says that odorized gas must have a distinct odor to a person with a normal sense of smell at a level of one-fifth the lower limit of flammability. Simple on the surface, the standard leaves unspecified what constitutes a normal sense of smell. One could quibble about that, but if one grants that at the point of design most odorants at the rate of injection recommended by the odorant manufacturer will meet the standard, one can ask whether the odorant always performs according to design. Here, one gets to examine the underside of gas warnings.

Two questions immediately come to mind: 1) Does the odorant-gas mixture behave physically and chemically as designed? 2) Do more than a vanishingly small number of people have enough impairment of the sense of smell to fail to perceive the threat of escaped gas?

The term "odor fade" has become somewhat fashionable in arguments about how well gas odorants behave (Lemoff, 1992). All odorants in gas can diminish in concentration through a mechanism of sorption. Sites for such include soil, piping, construction materials, and condensed vapor (Kneibes, 1971; Parlman and William, 1976; Williams, 1976). The odorant ethyl mercaptan may also react relatively quickly with rust. The consequences of these processes may explain why some accidents occur. Nevertheless, the field of gas odorization generally lacks validation that the "design" level routinely occurs. That is, for any single odorant or blend, one can calculate the vapor concentration expected at any criterion level of warning, such as one-fifth the lower limit of flammability, and compare it to what happens in actual buildings.

Why might one suspect that the actual concentration may lie below the design concentration? Often the term stenching agent occurs with respect to gas odorization. A stench at five times a "distinct" level should motivate action much like a fire alarm (Cain and Turk, 1985). No one should think twice about the danger, yet anecdotal data give a different impression, one where the person may ponder the danger and may say, "I wonder what that smell is?" or "I wonder what that smell means?"

PERFORMANCE OF THE HUMAN

At any given time, around 2 % of the population has an impaired sense of smell, and the percentage climbs dramatically among the elderly (Murphy et al, 2002). Some people are born without a sense of smell, some lose smell from head trauma, such as falling off a bike, some lose smell permanently from an upper respiratory infection, some have chronically stuffy noses (rhinitis), and so on. Depending upon the depth of their loss, these people may smell little or no odor in the presence of leaking gas. This group does not qualify as "vanishingly small." So, whereas one might answer "yes and no, with a large component of maybe" to the question about whether the gas-odor mixture behaves as designed, one must answer "yes" to the question of whether "enough" people have impaired olfaction.

As one digs deeper, one realizes magnitudes of the challenges of human factors with respect to gas. How does one protect a consumer without a sense of smell? Historically, warnings brochures would tell customers that some of them might not register the odorant in gas and they should establish whether they have the ability to smell it. They do not say, "If you do not have the ability you may unknowingly become victim of an explosion and should probably switch from gas to another fuel." They might mention other ways to detect gas leakage, such as hissing, a cloud of vapor, or an apparent outage, but mostly they have implied: Roll the dice. It illustrates how the suppliers can see a conflict of interest regarding safety and loss of customers.

By its very nature, gas is a hazardous substance. It exists as a consumer product because it provides a societal benefit despite its hazards. Product manufacturers, such as those who make control-valves, appliances, and appurtenances have improved design of ignition and delivery, but gas will still escape some of the time. No one dictates to a company how it must communicate warnings. Some show a gun-shy approach by avoiding the term explosion in favor of the milder term fire, when indeed leaked gas may and often does explode. The image of a fire allows for people to escape unharmed. The image of an explosion allows for instant death. The word explosion might make a consumer respect gas more, but some must think that it might make the consumer choose another fuel.

In addition to the factors that impair odor detection entirely, others may impair it more circumstantially:

- The odorants used in gas do not awaken people from sleep, a condition that affects everyone, not only during a night's sleep, but also during naps (Fieldner et al., 1931; Arzi et al., 2010). A person asleep has essentially no sense of smell.

- Other sources of odor, such as birdcages, cat litter boxes, sewer drains, mustiness, paints, household chemicals, may interfere with a person's ability to register an odor as gas. Masking odors may not only diminish the perceived odor magnitude of the gas, which would delay reaction, but may actually assimilate the odor perceptually so that a consumer may never realize that a clearly perceived odor comes from gas and not from an uncapped drain (Cain, 1988). (With a little stretching, this category can include imagined sources of odors, such as dead animals, in a wall or crawl space.) Brochures may warn consumers to remove any odoriferous material, such as household chemicals, from the vicinity of appliances. This conjures up images of cluttered basements, not of homes built on slabs, where the absence of basements puts furnaces, water heaters, and dryers, right next to the most odoriferous room in the house, the kitchen.

- Everyone knows from experience that olfactory sensitivity wanes during exposure to an odorant. Older people adapt faster than younger people (Stevens et al., 1989). It appears that the sleeping person adapts, just as does the non-sleeping person. Hence, a person who lights a cigarette upon awakening may blow up the house with no hint of a surrounding explosive level of gas.

- Aging deserves special mention with respect to impaired olfaction. Not only do aged persons have a higher frequency of clinical loss, even normal aged persons have lower sensitivity and will perceive odors as weaker than younger people (Stevens and Cain, 1985, 1987a). They have even been shown to have higher thresholds and lower perceived magnitude for gas odor in particular (Stevens and Cain, 1987b; Stevens, et al., 1987). As noted, aged persons will also adapt faster. Because of a more sedentary existence, aged persons may make themselves more vulnerable to a slowly increasing level of gas. They may also nap in place more frequently.

LP-gas poses some special problems:

- Propane (butane) and its odorant are heavier than air and may collect in low places. This means that a leak a person might smell at ankle level may prove undetectable at nose level. Basements often have windows at head height and the air infiltration they provide might fail to vent the gas.

- In some cases, floors may have areas (e.g., step-down alcoves) that could serve for local collection of vapor, so that even a very small amount of leakage could achieve a flammable level in that local spot. Most people do not know that LP gas can burn at just over 2% concentration. They may not have much intuitive feel for that amount, but it does suggest that danger may "accumulate" suddenly. A fire that sets one cuffs alight can envelope the person and cause lethal burns.

- The odorant in LP gas has more inherent physical and chemical vulnerability than that natural gas.

- Rural culture in some parts of the US seems to encourage a do-it-yourself attitude regarding maintenance and even installation of gas appliances. This means that untrained people may work on gas systems. LP suppliers can encourage such behavior by having a policy that their personnel will not work inside a dwelling.

-Because the LP tank can go empty, the gas system will need to be treated as if new and tested for leaks, something few homeowners do properly or even know must be done.

Field Results

Field results of how well people know or would respond to the odor of gas suggest that the public may not respond the way a gas supplier might assume. In a particularly important study, Whisman et al. (1978) looked into how well people in a rural area would relate ethyl mercaptan to gas in various attentional states. The investigators had participants visit a normally decorated house trailer that contained ethyl mercaptan at levels as high as 250-fold above the average threshold. When they left, participants answered a questionnaire about whether they detected an odor they could relate to gas. Whereas 100% of non-elderly persons noticed a distinct gas odor at one-fifth the lower limit of flammability, when specifically directed to look for it, 90% did so when generally directed to note various kinds of sensory stimuli, 80% did so when given no directions, and 70% did so when given mild distraction. The relationship between level of attention and performance gave serious cause for concern for it implied that the gas odor warning needs more than just its presence to signal danger. It needs a consumer ready to attend to it.

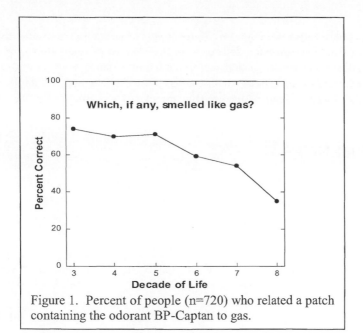

Figure 1. Percent of people (n=720) who related a patch containing the odorant BP-Captan to gas.

In the National Geographic Smell Survey performed in 1986, one of six scratch-and-sniff patches sampled by readers in a pull-out booklet contained BP-Captan, a t-butyl mercaptan blend. The authors considered the patch identified if a respondent chose "foul" for its identity (Wysocki and Gilbert, 1989). Among young adults, just fewer than two-thirds did so. In a follow up study, Cain et al. (1992) found about 70% performance to the question (Fig. 1), "Which, if any patch, smelled like gas?" Approximately 70% of adults under 50 answered correctly. In view of how the other patches smelled of banana, cloves, rose, musk, and a material called androstenone (musky, uriny to those who could smell it at all). The failure to relate the odor to gas again provided a sober result. Performance did not differ between those who did and those who did not have gas service. As expected from much other research, performance deteriorated progressively with age after middle age.

CONCLUDING REMARK

This report has focused more on the human factors challenge to warn consumers about gas leaks than on formulation of solutions. All suppliers of gas support safety and deliver information about safety to customers. Most provide print material, some deliver radio or TV spots. Some provide outreach via health fairs and go into schools. Some authors, e.g., Cain et al. (1992), have offered suggestions about how to improve communication of safety. Since this report will appear in a volume and be presented at a meeting devoted to applied ergonomics and safety, the author looks forward to discussion of the matter so that realistic solutions come forward. One development of considerable recent interest is the presence of methane/propane detectors on the shelves of retail stores. Although the gas industry has treated such detectors with ambivalence, fearful of false alarms and the expense of unnecessary house calls, these devices will inevitably become part of the solution against the horrible effects of gas explosions and fires on humans.

REFERENCES

Arzi, A., Sela, L., Green, A., Givaty, G., Sobel, N. (2010). "The influence of odorants on respiratory patterns in sleep." Chemical Senses, 35, 31-40.

Cain, W. S. (1988). "Olfaction". In R. C. Atkinson, R. J. Herrnstein, G. Lindzey, and R. D. Luce (Eds.), Stevens' Handbook of Experimental Psychology, Vol. 1: Perception and Motivation, rev. ed. New York: Wiley. Pp. 409-459.

Cain, W. S., Turk, A. (1985). "Smell of danger: An analysis of LP-gas odorization." American Industrial Hygiene Association Journal, 46, 115-126.

Cain, W. S., Leaderer, B. P., Cannon, L., Tosun, T., Ismail, H. (1987). "Odorization of inert gas for occupational safety: Psychophysical considerations." American Industrial Hygiene Association Journal, 48, 47-55.

Cain, W. S., Cometto-Muñiz, J. E., Gent, J. F., Cox, E. P., III. (1992). "Odorization of Natural Gas: Perception of Odor and Communication of Safety." Report to Stone and Webster Engineering, New York, NY.

Fieldner, A. C. et al., (1931). "Warning Agetns for Fuel Gases," Monograph No. 4. Washington, D.C.: Bureau of Mines, U.S. Department of Interior.

Kniebes, D. V. (1971). "Soil adsorption of odorant compounds." Institute of Gas Technology Odorization Symposium. Chicago: IGT. Pp. 261-269.

Lemoff, T. C. (1992). "Fuel gas odorization," Supplement 3, in T.C. Lemoff (Ed.), National Furel Gas Code Handbook. Quincy, MA: National Fire Protection Association and American Gas Association. Pp. 495-504.

Murphy, C., Schubert, C.R., Cruikshanks, K. J., et al., (2002). "Prevalence of olfactory impairment in older adults." Journal of the American Medical Association, 288, 2307-2312.

Parlman, R. M., Willliams, R. P. (1979). "Penetrabilities of gas odorant compounbds in natural soils." American Gas Association Operating Proceedings, D-294-D-299.

Stevens, J. C., Cain, W. S. (1985). "Age-related deficiency in the perceived strength of six odorants. Chemical Senses, 10, 517-529.

Stevens, J. C., Cain, W. S. (1987). "Detecting gas odor in old age." Annals of the New York Academy of Sciences, 510, 644-646.

Stevens, J. C., Cain, W. S. (1987). Old-age deficits in the sense of smell gauged by thresholds, magnitude matching, and odor identification. Psychology and Aging, 2, 36-42.

Stevens, J. C., Cain, W. S., Weinstein, D. E. (1987). "Aging impairs the ability to detect gas odor." Fire Technology, 23, 198-204.

Stevens, J. C., Cain, W. S., Schiet, F. T., & Oatley, M. W. (1989). "Olfactory adaptation and recovery in old age." Perception, 18, 265-276.

Whisman, M. L., Goetzinger, J. W., Cotton, F. O., Brinkman, D. W. (1978). "Odorant evaluation: A study of ethanethiol and tetrahydrothiophene as warning agents in propane." Environmental Science and Technology, 12, 1285-1288.

Williams, R. P. (1976). "Soil penetrabilities of natural gas odorants." Pipeline and Gas Journal, 203(3), 32-46.

Wysocki, C. J. and Gilbert, A. N. (1989). "National Geographic Smell Survey: Effects of age are heterogenous." Annals of the New York Academy of Sciences, 561,12-28.

Chapter 60

The Effect of Local Illumination on Light-on Test Inspection Performance in TFT-LCD Manufacturing

Li-Jen Twu, Chih-Long Lin, Mao-Jiun Wang

Department of Industrial Engineering
and Engineering Management
National Tsing Hua University
101, Section 2, Kuang-Fu Road, Hsinchu, Taiwan, ROC

ABSTRACT

TFT-LCD (thin film transistor liquid crystal display) industry is one of the most important industries in Taiwan. Light-on test is one of visual inspection tasks in TFT-LCD panels manufacturing process. The operators of light-on test inspection are working in a high luminance contrast ratio (low ambient illumination and high local illumination) environment. The purpose of this study was to examine the effect of different local illumination and work-rest schedule on inspectors' visual fatigue and task performance. A total of 9 well-trained female operators participated in this experiment. The experiment involved four levels of local illumination: (1) 2600 lx (with original lamp); (2) 6000 lx (with new lamp); (3) 2600 lx (with new lamp plus polarizer panel), and (4) 1900 lx (with new lamp plus polarizer panel and lamp half-covered). The work-rest schedule had two conditions: (1) working 29 minutes and resting 1 minute, then repeat once, and (2) working 58 minutes and resting 2 minutes. At the beginning and the end of each experiment combination, the subjects took critical flicker fusion (CFF) tests to measure their visual fatigue. The results showed that the local illumination effect was significant on task performance. Using new lamp (6000 lx) and new lamp plus polarizer panel (2600 lx) had lower miss rate than that of the original lamp (2600 lx). It seems that increasing local illumination to 6000 lx, and using polarizer panel to reduce high luminance contrast can enhance the inspection performance. The results also showed that the local illumination and work-rest schedule did not show significant influence on visual fatigue. This might be due to the operators were adapted to the work environment. A further study is necessary by involving more environmental and task factors in evaluating the light-on test inspection performance.

Keywords: TFT-LCD industry, inspection, illumination, critical flicker fusion frequency, miss rate.

INTRODUCTION

The manufacturing processes of TFT-LCD (thin film transistor liquid crystal display) include array process (Array), cell process (Cell) and assembly process (Module) sequentially. In order to make sure high quality TFT-LCD panels output, different inspection tasks are included such as optical inspection, electrical inspection and human eyes inspection in the manufacturing process (Su et al., 2000). In cell process, the TFT-LCD panels need to be inspected in a high luminance contrast ratio (low ambient illumination and highly local illumination) environment, and this task is so called light-on test inspection for examining the panel defects. The light-on test is performed by human eyes to check panels outside under high local illumination (2600 lx) first, and then to place panels on a test machine turning on light from panels backlight for further defect inspection. These tasks are performed in a low ambient illumination (98 lx in average). The inspectors working in this high luminance contrast ratio environment may feel eye strain, poor eyesight due to the limited resting time (Wang and Wu., 1999). Because of this extreme luminance contrast environment, it is necessary to evaluate the effect of different local illumination and work-rest schedule on inspectors' visual fatigue and task performance.

METHOD

SUBJECTS

A total of nine well-trained female operators who worked in a TFT-LCD plant participated in this experiment. Their working experience was 1.8±0.5 years. All subjects had good eyesight and had a clear understanding about the experiment procedure.

APPARATUS AND MATERIAL

Four levels of local illumination were involved, including (1) 2600 lx (with original lamp); (2) 6000 lx (with new lamp); (3) 2600 lx (with new lamp plus polarizer panel), and (4) 1900 lx (with new lamp plus polarizer panel and lamp half-covered). The purpose of polarizer was to reduce high contrast effect to inspectors' eyes. The four different illumination conditions are shown in Figure 1 to Figure 4. Figure 5 illustrates the light on test inspection station.

Figure 1 2600 lx (with original lamp)

Figure 2 6000 lx (with new lamp)

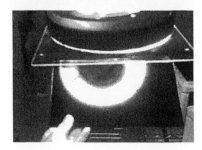

Figure 3 2600 lx (with new lamp plus polarizer panel)

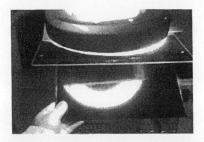

Figure 4 1900 lx (with new lamp plus polarizer panel and lamp half-covered)

Figure 5 the wok environment of light-on test inspection

The two work-rest schedules designed were: (1) working 29 minutes and resting 1 minute, then repeat once, and (2) working 58 minutes and resting 2 minutes. Critical flicker fusion (CFF) test (Model No. 502, Takei Kiki Kogyo Co., Japan) was used to evaluate inspectors' visual fatigue. There were 70 experimental panels used for this study, including 40 good panels and 30 no-good panels. The inspectors' task performance were classified into four categories: (1) judging good panels as good (OK/OK); (2) judging no-good panels as no-good(NG/NG); (3) judging good panels as no-good(OK/NG), and (4) judging no-good panels as good(NG/OK).

EXPERIMENTAL PROCEDURE

At the beginning of the experiment, all subjects took CFF test four times including starting from CFF 10 Hz upward and from CFF 60 Hz downward with two times each, then calculating the average of the four trials as the pre-CFF value. After the experiment procedure, the same CFF test procedure was repeated, and the average value was taken as the post-CFF value. The difference value between the post-CFF and pre-CFF value was taken as the indicator of visual fatigue.

Each subject had to finish a total of 8 experiment combinations (four illumination levels by two work-rest schedules). The panels were randomized each time to make sure that the subjects would not able to predict the defect conditions of the panels. Subjects should keep the same inspection pace as they usually did at work. If the subject did not inspect more than 40 panels in an experiment run, she should continue the experiment until reaching the specified number of panels. After each experiment run, the task performance in terms of OK/OK, NG/NG, OK/NG and NG/OK percentage were obtained.

STATISTICAL ANALYSIS

One way analysis of variance (ANOVA) was performed to evaluate the effects of different illumination levels and work-rest schedules, on the visual fatigue and task performance. Then, Duncan's multiple range test was conducted as a post-hoc testing.

RESULTS AND DISCUSSION

Table 1 shows that mean values of CFF test and the inspection performance under different experiment conditions. Except for inspecting under 2600 lx (with original lamp), the inspectors' post-CFF value was lower than the pre-CFF value.

Table 1 Mean value of CFF difference and task performance

CFF difference value (Hz)		Illumination level[1]				Work-rest schedule	
		2600 lx (with original lamp)	6000 lx	2600 lx (with new lamp plus polarizer panel)	1900 lx	W29R1	W58R2
		0.32	-0.28	-1.02	-0.28	-0.33	-0.3
Task performance (%)	OK/OK	59	60	60	58	60	59
	NG/NG	32	33	34	34	33	34
	OK/NG	1	2	1	1	1	1
	NG/OK	8*	5*	5*	7*	6	6

*Significant at $p < 0.05$

In TFT-LCD industry, the ratio of judging no-good panels as good products is called miss rate or leakage rate. In here, the effect of different illumination level was significant on NG/OK (miss rate) performance. The results of Duncan's post-hoc testing showed that inspecting under illumination level 2 (6000 lx, with new lamp) and level 3 (2600 lx, with new lamp plus polarizer panel) had significantly lower miss rate (5%) than the miss rate under illumination level 1 (2600 lx, with original lamp)

(8%). But there was no significant difference between illumination level 2 and level 3. Miss rate is a serious problem to panel makers for both cost and quality considerations. The results showed that increasing local illumination to 6000 lx, and using polarizer panel to reduce high luminance contrast can enhance the inspection performance. Although the results showed that the local illumination and work rest schedule did not show significant influence on visual fatigue. This might be due to the operators were adapted to the work environment.

CONCLUSION

The finding of this study indicates that increasing local illumination and using polarizer panel to reduce high contrast could enhance the inspection performance. This information is useful for the TFT-LCD panel manufactures for increasing quality performance. A further study was suggested by involving more illumination design and task factors in evaluating TFT-LCD light-on test inspection performance.

REFERENCES

Lu, C., Sheen, J., Su, S., Kuo, S., Yang, Y., and Kuo, C. (2007). Work environment and health effects of operators at light-on test process in TFT-LCD plants. Ergonomics and Health Aspects, HCII 2007, LNCS 4566, pp 113-117.
Su, C., and Huang, C. (2000). LCD light-on test instrument. J Mechanical Ind. 207, pp 146-151.
Wang, M., and Wu, S., (1999). Application of ergonomics in semiconductor industry. Ind Safety Technical, 30, pp 33-41.

Chapter 61

Predicting Image Quality on a Mobile Display Under Different Illuminations

Po-Hung Lin[1], Wen-Hung Kuo[2], Sheue-Ling Hwang[3]

[1]Department of Industrial Engineering & Management Information
Huafan University

[2]Department of Information Management
National Formosa University

[3]Department of Industrial Engineering & Engineering Management
National Tsing Hua University

ABSTRACT

This study was to construct the image quality models under different illumination levels and images through Group Method and Data Handling (GMDH). The models built the relationship between physical measurements and human perception. The results indicated that the effects of luminance, contrast, correlated color temperature (CCT), and resolution were significant on perceived image quality under 1500 lux. However, color temperature was not a significant physical characteristic and the interaction between luminance and resolution was found under 7000 lux. From the results of the experiment, it is considered that outdoor environment (7000 lux) may not be suitable for using mobile displays. Finally, once the image quality model is built, the human subjective ratings can be obtained if we imported the measurements of significant physical characteristics. The subjective ratings can be provided for mobile display manufacturers to improve the quality so that their products can really meet the customer requirements.

Keywords: Luminance, Contrast, CCT, Resolution, Mobile display, Image quality, GMDH

INTRODUCTION

Because of the convenience and portability, mobile display is more and more popular in the market and has been widely used and served as multifunctional purposes, such as multimedia, digital camera, and personal schedule. Thus, the importance of image quality on mobile displays has been receiving great attention recently. It is known that image quality evaluation from a user's viewpoint is essential and image quality assessment from subjective human perception is a critical issue in related literature (Bech et al., 1996; Besuijen and Spenkelink 1998). Several physical characteristics, such as luminance, contrast, CCT, color gamut, and resolution, are used as the index for the users to evaluate the image quality on different electronic devices. For TVs, Rajae-Joordens & Heynderickx (2004) investigated the effect of resolution, sharpness, contrast and luminance on image quality. With regard to mobile displays, Kim et al. (2006a) examined the relationship between individual physical characteristic and image quality, such as resolution, CCT, peak-white luminance, respectively. Most of the previous studies investigated the effect of individual physical characteristic on image quality. In fact, the physical characteristic such as resolution, CCT, luminance, and contrast can be considered simultaneously as multiple physical characteristics to investigate their co-effect as the function of image quality. Thus, more information such as the interactions between these physical characteristics might be found. Based on above arguments, the proposed image quality models were constructed using above physical characteristics.

It is worthy to investigate the effect of ambient illumination since people usually use mobile displays in different lighting conditions. For instance, a mobile display can be used in daylight, night, indoor office, and outdoor environments. Kubota (2006) investigated the measurement of light incident on a mobile display in 112 environments including indoor and outdoor environments, such as living room, library, university campus, bus stop, etc. In addition, Kim et al. (2007) developed the image-color-quality models under overcast, bright, and very bright conditions. In general, it is expected the image quality of mobile displays may be varying with ambient illumination since the ordinary users may use them in both indoor and outdoor environments. The category F (the range of illumination level is 1000-1500-2000 lux) in IES Lighting Handbook (Kaufman and Christensen, 1984) is suitable for very small size visual task in indoor environment. Thus, it is considered that 1500 lux is the appropriate illumination level considered in this study for indoor environment. On the other hand, too much illumination may result in a more harsh reflection or glare and degrade legibility (Sanders and McCormick, 1993). For outdoor environment, excess illumination may affect the image quality of mobile displays. Since the outdoor illumination was simulated from a color assessment cabinet used in this study, 7000 lux is the illumination level we set. In this study, 1500 lux (which typically represents the indoor office) and 7000 lux (which simulates outdoor environment) were the two illumination levels considered.

Recently, neural network is a well known methodology to construct the models. In general, neural network can be constructed without any assumptions between predictors and response and is widely applied into different industries and academic fields. Among different neural networks, Group Method and Data Handling (GMDH) was one of the better-know neural network methods originated by Ivakhnenko (1968). It is a practical methodology of data analysis for non-linear models and can solve the problems when only few data are available, the architecture of the function is not clear, or too many input variables are used. Hence, GMDH was applied in this study to construct the model.

To sum up, the purpose of this study was to construct the image quality models using luminance, contrast, CCT, and resolution under 1500 and 7000 lux through GMDH. The models represent the relationship between physical measurements and human perception. Once the image quality models were constructed, the human subjective ratings can be obtained to provide the mobile display manufacturers for understanding the requirements of their products from users' viewpoints.

METHODOLOGY

SUBJECTS

Thirty three college students (M = 24.21, SD = 3.09) took part in this experiment. All subjects have normal or corrected to normal vision.

APPARATUS

An OPTEC 2000 vision tester and Standard Pseudo Isochromatic charts were used to test the visual acuity and color vision of the subjects before the experiment. A Nokia N73 (display resolution, 320×240) mobile phone was used to test the subjective image quality of the subjects and a color assessment cabinet (VeriVide CAC 120-5) with a diffuse light source of TL84 was used to set the illumination levels. The color measurement was measured by the Minolta Spectrometer CS-100.

CONDITION OF WORKPLACE

Fig. 1 shows the experiment workplace arrangement. The mobile display was positioned on a table with 73 cm in height. The front edge of the table was 20 cm from the display center. The inclination angle of the mobile display was 105° for the vertical axis (Turville et al., 1998). Since the length of diagonal line (d) of mobile display screen approximates 6 cm, the distance between VDU and spectrometer should be at least 4d (VESA, 2001). Thus, the viewing distance was 25 cm and the subject's head was restrained by a chinrest 15 cm above the table. Before conducting the experiment, the height of seat could be adjusted by the subjects to make them comfortable.

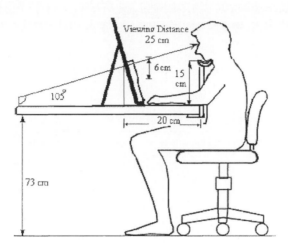

Figure 1.The arrangement of the workspace in the experiment

EXPERIMENTAL VARIABLES AND IMAGE MANIPULATION

Subjective image quality rating was the dependent variable in the experiment. Independent variables were four physical characteristics: luminance, contrast, CCT, and resolution. In the experiment, we set up different levels of each physical characteristic to be compared. For the level settings, please refer to Table 1. As to the image manipulation, the portrait image was selected from ISO 12640. In addition, the levels of luminance, contrast, and resolution for the image were adjusted by Photoshop software. On the other hand, the levels of CCT were adjusted by the program we developed for this experiment.

For the luminance, three increasing levels and three decreasing levels were set from the original level. In addition, 1, 0.991, 0.98, and 0.636 were set as the four levels for contrast. In order to let the subjects distinguish the differences of CCT, 9300, 8500, 7500, 6500, and 5000 were the sequent levels from higher to lower CCTs. Finally, due to the maximum resolution on Nokia N73 is 320×240, the descending levels from 320×240 to 80×64 with the same scale interval (i.e., 320×240, 260×208, 200×160, 140×112, and 80×64) were adjusted by Photoshop software.

A program was developed to measure luminance, contrast and CCT. Since the red (R), green (G), and blue (B) of the mobile display is the device dependent colors, they should be converted to device independent colors (i.e., XYZ). Before calculating the luminance, contrast, and CCT from images, a color conversion matrix was the important element for mobile display characterization. For the detailed calculation, first of all, the three primary colors RGB and the gray levels were displayed in Nokia N73 individually. Then we used the spectrometer to measure their trichromatic coefficients (x and y) and a tristimulus Y. After obtaining x, y and Y, we can easily calculate the other tristimulus values (X and Z). Thus, the X, Y, Z values of the above ten colors were obtained. Through normalization, we got the 3×3 color conversion matrix. Once obtaining the color conversion matrix, we used it to convert RGB to XYZ.

Table 1. The levels of physical characteristics

No.	Physical characteristics	Levels
X1	Luminance (lux)	3.46, 5.741, 8.697, 13.033, 18.217, 24.247, 31.512
X2	Contrast	1, 0.991, 0.98, 0.636
X3	CCT (k)	5000, 6500, 7500, 8500, 9300
X4	Resolution (total pixels)	80×64 (5120), 140×112 (15680), 200×160 (32000), 260×208 (54080), 320×240 (76800)

As to physical measurement of luminance, we can get the RGB of each pixel in an image in advance. The RBG was converted to XYZ through color conversion matrix and then the Y value of each pixel was obtained. After averaging each Y value of an image, the average luminance ($\bar{Y}$) was obtained, such as 3.46 lux. For contrast, $C = (L_{white} - L_{black})/(L_{white} + L_{black})$, the definition of the Swedish Confederation of Professional Employees (TCO) (2003), was used in this study. As mentioned before, we could get the XYZ of each pixel. The pixels which have the biggest Y (Y_{max}) and smallest Y (Y_{min}) were recorded and then Y_{max} and Y_{min} were used as L_{white} and L_{black} to get the value of contrast. For CCT, it represents the correlated color temperature of the white point in an image. After getting the XYZ of the white point, the x, y can be calculated through the XYZ. We calculated the CCT according to the formula of McCamy (1992) that $T = -473n^3 + 3601n^2 - 6861n + 5514.31$, where $n = (x - 0.3320)/(y - 0.1858)$.

EXPERIMENTAL TASKS AND PROCEDURE

The experiment consisted of two stages. Subjects were asked to evaluate subjective image quality under 1500 lux and 7000 lux in stage one and stage two, respectively. In each stage, the Likert's nine point scale was utilized for the subjects to assess the subjective image quality in terms of the different images (levels). In order to avoid the memory effects of subjects, the sequence of the images on each physical characteristic was displayed randomly.

The procedure of the experiment was as follows.

- The visual acuity of the subjects was obtained from the vision tester. If the score was lower than 0.8, the subject was not allowed to join the experiment.
- The subject was asked to review all the images on each physical characteristic before evaluating them.
- The subject was asked to evaluate the images on each physical characteristic under 1500 lux
- Ten minute break was given after the subject finished the experiment in the first stage.
- As in step (3), the subject was asked to evaluate the images on each physical characteristic under 7000 lux.

Finally, the subjects were asked to express their subjective perception and opinions for evaluating the images under 1500 lux and 7000 lux.

MODEL CONSTRUCTION AND DATA ANALYSIS

In order to predict the human perception on a mobile display, the relationship between the physical characteristics (Xi) and subjective image quality rating (Y) were investigated. For developing an image quality model of a mobile display, four physical characteristics, including luminance (X1), contrast (X2), CCT (X3), resolution (X4) were considered as input variables. On the other hand, subjective image quality rating (Y) was the dependent variable. Due to two illumination levels, two models were presented in this study.

For data analysis, the thirty-three raw data from the subjects were averaged to form one record for each level of each physical characteristic. There are seven levels in luminance, four levels in contrast, five levels in CCT, and five levels in resolution. To sum up, twenty-one sets of data were obtained. In this study, seventeen sets of the data as a training set were used to build the model through NeuroShell software 2. In addition, four sets of the data as a checking set were used to verify the model.

RESULTS

For investigating the differences of the models under indoor and outdoor environments, 1500 lux and 7000 lux were the two levels to be discussed as follows.

Model construction under 1500 lux

Under 1500 lux, seventeen sets of the data were used to develop an image quality model. The model is built as follows.

$$Y = -31 + 1.6X_1 + 8.8X_2 + 0.0045X_3 + 0.000071X_4 - 0.083X_1^2 - 0.00000035X_3^2 + 0.0012X_1^3 \qquad (1)$$

Equation 1 indicated that luminance (X1), contrast (X2), CCT (X3), and resolution (X4) were the significant variables. Mean Square Error (MSE) of the model was 0.25 and R square (R^2) of the model was 0.935. After constructing this model, it is necessary to be validated. Four sets of the data were used to validate the model (see Table 2). In this model, the four estimated values were very close to the real values and the real values were all in the 95% confidence intervals. Thus, through the validation from 95% confident intervals, this model is appropriate for predicting image quality.

Model construction under 7000 lux

Under 7000 lux, the model is built as follows.

$$Y = -68 + 5.8X_1 + 5.7X_2 + 0.00081X_4 - 0.016X_1^2 + 0.000067X_1X_4 \qquad (2)$$

Equation 2 also indicated that luminance (X1), contrast (X2) and resolution (X4) were the significant variables. However, CCT (X3) was not significant. MSE of the model was 0.162 and R^2 of the model was 0.939. Four sets of the data were used to validate the model (see Table 3). From Table 3, we can find that the four estimated values were very close to the real values and the real values were all in the 95% confidence intervals. Thus, we can conclude that this model is appropriate for prediction.

Table 2. Model validation under 1500 lux

Luminance	Contrast	CCT	Resolution	Real value	Estimated value	Low bound of 95% C.I.	Upper bound of 95% C
6.935	0.991	5884	76800	4.55	5.03	4.01	6.05
9.16	1	7500	76800	6.24	5.92	4.90	6.94
16.696	0.98	6597	76800	7.18	6.69	5.67	7.71
11.239	1	6513	5120	1.73	1.82	0.8	2.84

Table 3. Model validation under 7000 lux

inance	Contrast	CCT	Resolution	Real value	Estimated value	Low bound of 95% C.I.	Upper bound of 95% C.I.
.142	0.997	5000	76800	4.67	4.15	3.34	4.96
.721	1	6597	32000	4.36	4.27	3.46	5.08
.935	0.991	5884	76800	3.45	3.62	2.81	4.43
.561	1	6597	15680	3.30	3.17	2.36	3.98

DISCUSSION

This study implemented GMDH to construct the models to predict image quality on a mobile display. Through the calculation of the neural network, the results indicated that all the predicting values fell in the 95% confidence intervals. In addition, R^2 of the proposed models approximate 0.94. It means that these models can be used to predict the perceived quality accurately. In addition, the importance of the two proposed models was the acquisitions of human subjective ratings (from 1 to 9) based on the measurements of significant physical characteristics of a mobile display. Then the manufacturer can produce a mobile display with high image quality according to the physical characteristics. The manufacturers should emphasize these physical characteristics in production so that the customer satisfactions could be raised.

The two proposed models also provide the different information for different conditions. The results of this study indicated that luminance, contrast, CCT and resolution were the four significant physical characteristics for the mobile display under indoor environment (1500 lux). Previous studies considered luminance, CCT and resolution as critical physical characteristics in mobile display image quality under a dark room (Kim et al. 2006a; Kim et al. 2006b). Some studies (Rajae-Joordens and Heynderickx 2004; Kuo, Lin et al. 2007) investigated the effects of luminance and contrast in TV image quality in the dark rooms. Above studies were conducted under the dark environments. However, in general, the mobile displays may be used more often under the lighting conditions. Thus, this study provided the information that luminance, contrast, CCT and resolution were the important physical characteristics under indoor environment.

However, CCT was not significant under 7000 lux. In terms of the survey in the experiment, 91% subjects reported that it is more difficult for them to distinguish the levels of the physical characteristics under 7000 lux, especially for CCT. In the experiment, some of the subjects preferred high CCT images (bluish) but the others chose low CCT ones (reddish). Under the brighter environment, the difference of bluish images and reddish images are less apparent than that under a darker environment. Thus, it is reasonable to infer that CCT was not significant under a brighter illumination. Based upon above arguments, we may suggest that the mobile display manufacturers should utilize the limited capitals to manufacture a high quality mobile display which possesses high quality luminance, contrast, and resolution.

Actually, from the subjects' subjective perception and opinions, 84.8% subjects reported that they may give lower ratings under higher illumination levels due to lower visual acuity during the experiment. Vogel et al. (2007) pointed out that a dark environment is more suitable to view mobile displays than a 126 lux lighting environment. Under outdoor environments, the mobile display image quality on a lower illumination level (7000 lux) is also better than that on the brighter levels (35000 and 70000 lux) (Yoo et al., 2007). In this study, 1500 lux and 7000 lux were the two illumination levels we set. According to the results of this study and above research, we could conclude that it is not suitable for people to view the mobile displays under too bright conditions, such as 7000 lux in this study.

In addition, it is shown that there is no product term in equation (1). It means that the image quality of a mobile display can be judged separately by each individual physical characteristic in an indoor environment. However, as shown in equation (2), the product term between luminance (X1) and resolution (X4) was demonstrated. It implied that subjects' subjective ratings are affected by different combinations of the two physical characteristics under a brighter illumination situation. Therefore, it is necessary to find the proper combinations of different physical characteristics to improve image quality of a mobile display under a brighter environment. From the results, we can infer that it may be easier to manufacture a mobile display which is used under indoor environment since mobile display manufacturers just need to consider the individual physical characteristic. On the other hand, it is more difficult to consider the combination of different physical characteristics under a brighter environment, such as luminance and resolution. However, the results of this study provided the information for the combination of different physical characteristics and it is worthy to be considered in mobile display design.

CONCLUSION

The main contribution of this study was to develop the image quality models under different illumination levels. The two proposed models provide different information for two illumination levels and it is the useful information for mobile display design. The results indicated that luminance, contrast, CCT, and resolution were the significant physical characteristics under 1500 lux. However, color temperature was not significant and the interaction between luminance and resolution was found under 7000 lux. From the results of the experiment, it revealed that outdoor environment (7000 lux) was not suitable for the users to use mobile displays. After constructing image quality models, we can predict the image quality from the mobile displays if the measurements of significant physical characteristics were imported. This information is very essential for mobile display manufacturers since the ratings were obtained from users' viewpoints.

REFERENCES

Bech, S., Hamberg, R., Nijenhuis, M., Teunissen, C., Jong, H. L. d., Houben, P., & Prainanik, S. K. (1996), "The RaPID Perceptual Image Description Method (RaPID). " *SPIE*, 2657.

Besuijen, K., & Spenkelink, G. P. J. (1998), "Standardizing visual display quality. " *Displays*, 19, 67-76.

Heynderickx, I., & Langendijk, E. H. A. (2005), "Image Quality Comparison of PDP, LCD, CRT and LCoS Projection." *SID 05 DIGEST*, 36, 1502-1505.

Ivakhenko, A. G. (1968), "The group method of data handling; a rival of the method of stochastic approximation." *Soviet Automatic Control*, 13(3), 43-55.

Kaufman, J., & Christensen, J. (1984), *IES lighting handbook*. New York: Illuminating Engineering Society of North America.

Kim, Y.-J., Luo, M. R., Rhodes, P., Choe, W.-H., Lee, S.-D., Lee, S.-S., Kwak, Y.-S., Park, D.-S., & Kim, C.-Y. (2007), "Image-color-quality modeling under various surround conditions for a 2-in. mobile transmissive LCD." *Journal of the SID*, 15(9), 691-698.

Kim, Y. J., Luo, M. R., Rhodes, P., Choe, W., Lee, S., & Kim, C. (2006a), "Affective Attributes in Image Quality of a Mobile LCD." *Society for Imaging Science and Technology*, 494-498.

Kim, Y. J., Yoo, J., Luo, M. R., Rhodes, P., Cheung, V., Westland, S., Choe, W., Lee, S., & Kim, C. (2006b), "Image Colour-Quality Modelling for Mobile LCDs." *Society for Imaging Science and Technology*, 159-164.

Kuo, W.-H., Lin, P.-H., & Hwang, S.-L. (2007), "A framework of perceptual quality assessment on LCD-TV." *Displays*, 28(1), 35-43.

McCamy, C. S. (1992), "Correlated Color Temperature as an Explicit Function of Chromaticity Coordinates." *Color Research & Application*, 17, 142-144.

Rajae-Joordens, R., & Heynderickx, I. (2004),"Effect of Resolution, Sharpness, Contrast and Luminance on Depth Impression and Overall Image Quality for a 2D TV." *SID 04 DIGEST*, 35, 1370-1373.

Sanders, M. S., & McCormick, E. J. (1993), *Human Factors in Engineering and Design*. Singapore: McGarw-Hill.

TCO'03. (2003), *Flat Panel Displays (1.1 ed.): The Swedish Confederation of Professional Employees*.

Turville, K. L., Psihogios, J. P., Ulmer, T. R., & Mirka, G. A. (1998), "The effect of video display terminal height on the operator: a comparison of 15 and 40 recommendations." *Applied Ergonomics*, 29(4), 239-246.

VESA. (2001), *Flat panel display measurements standard (second ed.)*. San Jose, CA: Video Electronics Standards Association.

Vogel, R., Saha, A., Chakrabarti, K., & Badano, A. (2007), "Evaluation of high-resolution and mobile display systems for digital radiology in dark and bright environments using human and computational observers." *Journal of the SID*, 15(6), 357-365.

Yoo, J. J., Kim, Y.-J., Luo, M. R., Choe, W., Lee, S., Lee, S., Park, D. S., & Kim, C. Y. (2007), "Image quality difference modelling of a mobile display." *SPIE*, 6494(1-10), 41-49.

CHAPTER 62

The Elements of Work Environment in the Improvement Process of Quality Management System Structure

Adam Górny

Institute of Management Engineering
Poznan University of Technology
11 Strzelecka St., PL – 60-965 Poznan, Poland

ABSTRACT

The work environment is a central component of a business organization's management system and plays a pivotal role in enabling it to achieve desired economic benefits. The above also applies to managing quality.

One reliable measure of a company's ability to attain such benefits is the satisfaction of both its internal clients (the buyers of its goods and services) as well as the external ones (the employees tasked with carrying out successive interlinked processes). As of late, process-based approach lies at the heart of advanced management systems. The structure of such systems must incorporate those components of the work environment recognized as process distortions. The intensity of such distortion bears on process effectiveness and, as a consequence, on the company's competitive position.

It is therefore in the best interest of any systemically managed business organization to recognize such distortions and find ways to mitigate their impact.

Keywords: Human Factors, Management systems, Quality, Work environment, Continuous Improvement

INTRODUCTION

At a time of ongoing globalization and mounting competition, business managers must meet market demands while reaching for new effective solutions that allow them to take lead of the market. In such an effort, a key role is attributed to ensuring client satisfaction which is most commonly achieved by resorting to systemic quality management, often described in ISO 9000 standards. Client satisfaction with goods or services is a key prerequisite for establishing a functional quality-oriented corporate management system. The operation of a company to systemic management guidelines has been based on eight quality management principles linked with a number of critical management areas some of the most important of which are to ensure a safe and friendly work environment.

The work environment needs to be recognized as critical for the satisfaction of all parties involved in economic cooperation. Client satisfaction is a function of clients' perception of the extent to which their needs and expectations are met.

An equally important precondition for achieving the desired quality is to adopt the process approach wherein key systemic concerns are addressed by reference to a network of interlinked processes and mutual relationships between inputs and outputs.

IMPROVEMENT OF QUALITY MANAGEMENT SYSTEM STRUCTURE

Quality management systems rely on properly-defined continuous improvement guidelines based on quality engineering requirements which cover all issues essential for the operation of an organization. To achieve continuous improvement, which constitutes the cornerstone of the systemic approach, the organization needs to develop a set of cultural criteria which in turn are essential for ensuring management effectiveness.

Management through organization culture boils down to establishing a multifaceted value system to create desirable standards of conduct and high performance standards (Stańczyk, 2006; Górny, 2008). The development of a proper culture integrated with the management process is seen as an opportunity to achieve economic success which inspires one to employ intangible resources to create competitive advantage. The identified aspects of organizational culture seen in the context of safety play a pivotal role in defining and recognizing central areas of the work environment in the process of their development.

Quality cannot be discussed in isolation from working conditions. This is particularly true for any area of management. In order to manage their organizations efficiently, managers need to diagnose their work environment and then continually improve it by setting the stage for installing system elements which are essential for ensuring management efficiency and favorably affect the work environment.

To develop a system by the process-oriented method, it is essential to recognize client needs and demands as well as analyze costs and benefits to identify processes and areas for improvement. The prerogative in doing so is to identify distortions that undermine process effectiveness, i.e. factors that foil efforts or render ineffective processes defined with the use of indicators. The desired outcome of any process is to achieve a proper degree of cooperation which involves an understanding of mutual needs and expectations of all process participants.

A key to ensuring proper cooperation with the clients by enabling them to receive products and services that meet their needs is to take a broad view of the development of a quality management structure. One must bear in mind that clients are the pivotal element of any process which translates directly into reputation and loyalty. Clients are described as key to corporate operations. In this context, working conditions are a central criterion for client satisfaction. Seen in this manner, it is not only the external client but also the internal one that is of particular importance for an organization's commercial success.

In keeping with the nature of the process approach, as employees perform their work, they deliver results to others down the line acting as links in a chain of relationships that span all across the organization (Hamrol, 2005). Hence, by identifying the scope of an organization's business (in terms of its significance for the completion of tasks and client satisfaction) with proper account taken of the role of the internal client, one can define client satisfaction prerequisites and their significance for operating efficiency so as to set an organization's short- and long-term growth objectives.

WORK ENVIRONMENT IN THE IMPROVEMENT PROCESS

In order to benefit from the deployment of a management system in any field of an organization's operations, a number of primary as well as secondary factors need to be considered. This guarantees the achievement of the company's desired outcomes. An analysis of preconditions for the effectiveness of a quality management system shows that it is next to impossible to adopt a proper approach to client satisfaction if work environment issues are left out of the equation (Hamrol, 2005). The issue that remains up to the company to sort out is to recognize and fulfill occupational health and safety requirements. Defining such requirements appears to be a major challenge.

The work environment is to be seen as critical for business outcomes (the outcomes of manufacturing or service provision operations). Organizations which fail to recognize work environment needs often end up having to pay the price as they lose market positions and their competitive advantage (Górny, 2006; Łech, Górny, 2004). Further, a competitiveness analysis calls for consideration of the availability of production factors. The key criteria for competitiveness which constitute means of production include the human factor, described as the productive value of employees, and the social capital defined by the role a business organization plays in its social environment.

To ensure its competitiveness, a company must utilize its full potential – doing so is frequently essential for its social development in a market economy. In the context of developing the work environment, the competitive advantage is defined as a set of features perceived by the market and particularly valued by clients who recognize companies that actively seek to improve their working conditions. To formulate a growth strategy that fits a given organization, one needs to define client needs and investor expectations. A crucial part of such an effort is to develop a way to pursue the company's mission which is a function of its objectives formulated by the principles of systemic approach.

The result is a simple relationship which demonstrates that continuous improvement is a cornerstone of systemic business management which should extend to working conditions. The most essential among various working conditions are the environmental factors found in the working space.

The work environment is defined by factors that exert a specific impact on the processes carried out in a given organization. These include:

- deleterious factors whose impact results and may result in injuries,
- harmful factors which commonly lead to health conditions linked with working conditions,
- strain which causes employees to experience discomfort.

The impact of environmental factors is a key criterion in assessing whether the work environment meets the optimal requirements as defined for each occupational category. An assessment of the effects of strenuous, harmful and deleterious factors on the efficiency of systemic processes is shown in Table 1.

By incorporating work environment factors in the overall quality management system an organization is forced to establish links which support work environment management. As a consequence, every process pursued in a company will, to a specified extent, be defined as a risk and strenuousness factor seen as a distortion. The structure of such links is shown in Figure 1.

The process structure given above allows an organization to include work environment issues in the existing quality management system while additionally serving as a starting point for the development of a system for the management of quality and work environment at various levels.

TABLE 1: Assessment of the impact of distortions on systemic process efficiency

Factor type	Factor description	Impact on systemic process
Strenuous	significantly compromises working comfort impacts upon occupational health and safety with an intensity which depends on time of exposure	limits opportunities to achieve desired process outcomes
Harmful	generates risks to employee health should be seen as a potential cause of work-related health conditions	constitutes a significant obstacle to the achievement of desired system outcomes
Deleterious	generates risks to employee safety should be seen as a potential cause of work related accidents	prevents the achievement of desired system outcomes

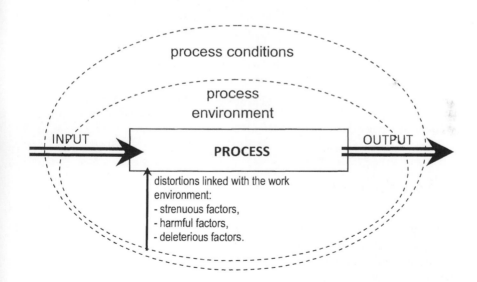

FIGURE 1. Process structure incorporating work environment
(Source: author's work)

A critical prerequisite for market success is for a business organization to define explicitly those components of the corporate quality management system that may underpin the work environment factors it has recognized as essential – doing so will greatly improve its edge over the competition. This in turn, will increase the extent to which the work environment contributes to the bottom line of any manufacturing or service operation. To adopt systemic solutions allowing for the effective management of work process safety, an organization should, among others (based on: Górny, 2009; Karczewski, 2000; Lewandowski, 1998):

- formulate a safety policy statement or at least incorporate occupational safety issues in all of its policy areas,
- disseminate such a policy statement among its labor force (by means of training, posters, etc.)
- select a Management Board member to take charge of occupational health and safety,
- identify areas within the organization that are critical for occupational safety and define occupational health and safety procedures for such areas,
- regularly review safety and use review conclusions to formulate corrective measures,
- allocate adequate resources to the achievement of the desired safety level.

By ensuring public approval of company growth with the help of a proper approach to occupational health and safety, an organization will boost its competitive advantage. For that to happen, it needs to send a clear message to its prospective clients, investors, lenders and the public at large. It also needs to establish its image as a caring organization by means of a marketing campaign targeted at the public and entities critical for its commercial and financial success. By putting in place a management system that promotes proper working conditions, the company will gain a substantial competitive advantage. An equally critical growth factor to be achieved by relying on the company's intangible resources is its reputation and image. If properly employed, the two will improve an organization's market recognition and, consequently, attract top employees which will strengthen the company's negotiating position vis-à-vis business partners.

CONCLUSIONS

The recognition of the need for public approval of company growth may and should be viewed as a competitiveness factor.

Business organizations increasingly face the need to incorporate work environment factors in their management systems to lay the groundwork for the deployment of systems relying on total quality management. This allows them to ensure continuous improvement and growing effectiveness in satisfying the needs and expectations of all parties involved in a given process. Goodwill is central for promoting growth by employing an organization's intangible resources. This area also includes care for working conditions, responsibility for the health and safety of all employees and recognition of the work environment as a factor for effectiveness and efficiency of production and service provision operations. By making proper use of such means of boosting a company's market position, an organization will gain better market recognition and will be perceived as one that proactively upholds the principles of social responsibility, which in turn will improve its negotiating position vis-à-vis business partners. The human factor on which such responsibility is focused lies at the heart of developing management conditions in keeping with the organization culture (Nowak, Pacholski, 2009). It is the organization culture, therefore, that defines the required scope of innovation.

An organization's commitment to pursue social objectives needs to be communicated by marketing means to the general public and other concerned parties critical for its commercial success and the bottom line (Górny, 2006; Radziszewska, 2008). Any company which has incorporated occupational health and safety concerns into its safety management system (many such companies obtain proper certificates) will be able to demonstrate its commitment to achieving the desired effects and continually improve its management system.

REFERENCES

Górny A. (2008). Ergonomics in occupational safety formation – ergonomic requirements in system managements of industrial safety, Foundations of Control and Management Sciences, No 11, pp. 127 – 137.

Górny A. (2008). The application of occupational risk assessment in process of forming a proper level of occupational safety – in aspect of Polish law, in: P. R. Mondelo, W. Karwowski, K. L. Saarela, A. Hale, E. Occipinti (Eds.). A Coruna: Proceedings of the 6th International Conference on Occupational Risk Prevention (ORP'2008: Commitment towards prevention: a corporate responsibility).

Górny A. (2006). Zarządzanie bezpieczeństwem pracy w budowaniu przewagi konkurencyjnej przedsiębiorstwa, Zeszyty Naukowe Uniwersytetu Szczecińskicgo, seria: Ekonomiczne Problemy Usług No 34 (540), pp. 295 – 302.

Górny A. (2009). Zarządzanie bezpieczeństwem pracy jako czynnik przewagi konkurencyjnej przedsiębiorstwa, Zeszyty Naukowe Politechniki Poznańskiej, seria: Organizacja i Zarządzanie, no 54, pp. 15 – 25.

Górny A., Dahlke G. (2005). The OHS management trough using of the TQM strategy elements, in: L.M. Pacholski, J.S. Marcinkowski, W. Horst (Eds.). Ergonomics and work safety in information community. Education and researches, Poznań: University of Technology, Institute of Management Engineering.

Górny A., Łech S. (2003). The Deming's Cycle in Improvement Process of Work Condition Quality, in: L. M. P acholski, J. S. Marcinkowski, W. Horst (Eds.). Work Quality Conditions in Researches and Education in Ergonomics, Work Safety and Labour Protection, Poznań: University of Technology, Institute of Management Engineering.

Hamrol A. (2005). Zarządzanie jakością z przykładami, Warszawa: Wydawnictwo Naukowe PWN.

Karczewski J.T. (2000). System zarządzania bezpieczeństwem pracy, Gdańsk: Ośrodek Doradztwa i Doskonalenia Kadr.

606

Lewandowski J. (1998). Zarządzanie jakością. Jakość, ergonomia, bezpieczeństwo pracy, ochrona środowiska, Łódź: Marcus.

Łech S., Górny A. (2004). Ocena kosztów wdrożeń rozwiązań techniczno-organizacyjnych doskonalących środowisko pracy w przedsiębiorstwie budowy maszyn, in: Gospodarka w obliczu eurotransformacji, J. Stankiewicz (Ed.). Zielona Góra: Uniwersytet Zielonogórski.

Mazur A. (2009). Shaping quality of work conditions, in: Health protection and ergonomics for human live quality formation, G. Dahlke, A. Górny (Eds.). Poznań: Publishing House of Poznań University of Technology.

Nowak W., Pacholski L. (2009). Human factors in improving process of the company organizational culture, in: Macroergonomics vs. social ergonomics, L. Pacholski (Ed.). Poznań: Publishing House of Poznań University of Technology.

Radziszewska A., Borkowski S. (2008). Społeczna odpowiedzialność przedsiębiorstwa jako istotny warunek jego sukcesu rynkowego, in: Organizacja w warunkach nasilającej się konkurencji, J. Stankiewicz (red.). Zielona Góra: Uniwersytet Zielonogórski.

Stańczyk S. (2006). Problemy zarządzania przez kulturę organizacyjną w małych przedsiębiorstwach, Zeszyty Naukowe Uniwersytetu Szczecińskiego, seria: Ekonomiczne Problemy Usług No 2 (427), pp. 239 – 245.

CHAPTER 63

Cultural Influence on Learning Sensorimotor Skills

Tim Jeske, Marcel Ph. Mayer, Barbara Odenthal,
Katharina Hasenau, Sven Tackenberg, Christopher M. Schlick

Institute of Industrial Engineering and Ergonomics
RWTH Aachen University
52062 Aachen, Germany

ABSTRACT

Manual assembly and manufacturing operations require sensory and motor skills – so called sensorimotor skills. Since skills must be acquired, the first time sensorimotor performance might be insufficient contrary to multiple repetitions. The period of time from a tasks' first execution until reaching a predefined level of performance, referred to as *learning time*, is investigated in the paper. The goal is to support process planners as well as personnel dispatchers to predict learning time. For this purpose, inter alia, cultural influences have been examined in laboratory experiments with European and Asian subjects (n=60). The results prove the occurrence of a learning effect and show no significant cultural influence.

Keywords: training, learning time, cognition, culture

INTRODUCTION

Future manufacturing systems will include human operators and robotic devices jointly working together in a cooperative way. Highly automated systems are often neither efficient enough for small lot production (ideally one piece) nor flexible enough to handle products to be produced in a large number of variants. Here a robot could take over repetitive and dangerous tasks which are not too complex whereas the human operator could handle the variants that are hard to automate. Hence, the human operator might have to adapt to constantly changing tasks.

This holds true already for todays manufacturing systems: Initial operation as well as restructuring of manual production and assembly areas go along with new as well as changed tasks.

The majority of the aforementioned manual tasks require sensory and motor skills – so called sensorimotor skills. Since skills must be acquired and in case of changing tasks must be adapted to the changed task, workers cannot perform any of these tasks productively from the very beginning (Rohmert et al. 1974). Productive workmanship, however, is defined in this context as achievement of reference performance in terms of target time. Required target times can be determined by using predetermined motion time systems (PMTS) like Methods-Time Measurement (MTM) or Work Factor (WF).

BACKGROUND

Learning time – referred to as the period of time a worker needs to practice the fundamental skills and to reach the predetermined target time – can be described only ex-post on the basis of learning curves. Even though already Smith (1776) and Babbage (1832) observed a relation between a task's repetition and an increasing performance of its execution, the first theory of this relation including a mathematical learning curve description for industrial use was developed by Wright in 1936 (Laarmann 2005). Wright's learning curve formula is:

$$t_n = n^{-k} \cdot t_1 \quad \text{with} \quad \begin{array}{l} t_1 \text{ - time of first execution} \\ n \text{ - number of repetition} \\ t_n \text{ - time of nth execution} \\ k \text{ - proportionality} \end{array} \quad (1)$$

As Wright's theory was criticized regarding the possibility of an infinite increasable performance, Levy developed a theory as well as a mathematical description which takes this criticism into consideration and leads to a limit (Hieber 1991). Levy's learning curve formula is:

$$t_n = (t_1 - c) \cdot e^{-k(n-1)} + c \quad \text{with} \quad \begin{array}{l} t_1 \text{ - time of first execution} \\ n \text{ - number of repetition} \\ t_n \text{ - time of nth execution} \\ k \text{ - proportionality} \\ c \text{ - limit} \end{array} \quad (2)$$

A prognosis about the expected learning time can be provided neither by means of PMTS nor by any other generally accepted method (Bokranz & Landau 2006). This results in considerable uncertain scheduling in production and assembly, e.g. low adherence on delivery dates for internal and external customers.

APPROACH

To support process planners as well as personnel dispatchers in their daily work a novel method for forecasting learning times of simple manual operations is under development (Jeske et al. 2009). The development of this method is part of the research project FlexPro, which is public funded by the German Federal Ministry of Education and Research (grant no. 01FH09019). The project focuses on further development of flexible production systems. Flexibility regarding production systems is a broad term that covers a range of definition of different subcategories, such as machine flexibility, process flexibility, performance flexibility, to mention only a few (e.g. Hofman 1990). Flexibility in production/manufacturing is a term mostly related to the technical equipment. Even in more up to date literature flexibility is only used in the boundaries of the technical system (e.g. Milberg 2003, Sheridan 2002, Chryssolouris 2006).

To expand flexibility to the human-machine system as a whole, abilities of the technical system have to be enhanced towards cognitive abilities like decision-making or problem-solving, which are not implemented in todays production facilities. Therefore, an additional research project has been established within the Cluster of Excellence *Integrative Production Technology for High Wage Countries* at the Faculty of Mechanical Engineering of RWTH Aachen University. This project aims at the development of a self-optimizing assembly cell enabling flexible automation with the possibility of flexible direct human robot cooperation (Mayer et al. 2009). An innovative design of the cell's numerical control forms the basis of this novel automation approach that accomplishes high-level information processing based on the cognitive architecture SOAR (see Leiden et al. 2001). Hence to a certain extent this cognitive control unit is able to simulate rule-based behavior of the human operator (Mayer et al. 2009).

Holistic production system shall allow the modification of their manufacturing principle depending on order backlog, e.g. from autonomous production or one-piece-flow to series production, which goes along with the aforementioned restructuring measures.

Previous studies have identified various factors that influence learning times, and can be separated into two areas: influencing factors which are characteristic of the work task, for example cycle time of a single work task (analyzed in PMTS) and its repetition frequency, and influencing factors that describe aspects of training. The latter include the methods of work instructions as well as additional training measures (Jeske et al. 2009).

In this paper the individual influence on learning sensorimotor tasks is analyzed. For that purpose it is examined to what extent the worker's cultural background influences his learning time. The study of cultural influences shall enable the development of a certain procedure with the ability to forecast the learning time for workers from different cultural background.

Preliminary results of this study have already been published in Jeske et al. 2010.

METHOD

The study was based on empirical laboratory experiments which have been conducted by the second and third author of this contribution. The goal to carry out these laboratory experiments was to identify cultural influences on learning times as well as on human assembly strategies of simple geometric objects.

PARTICIPANTS

Altogether sixty subjects, who were half from European and half from Asian cultures, participated in the study. The European participants were German and assigned to sample A while the Asian participants were Chinese and assigned to sample B. Gender was balanced in each sample. The average age in sample A (mean: 27.10 years; SD: 4.163; Min: 21; Max: 41) is higher than that in sample B (mean: 22.87 years; SD: 2.224; Min: 19; Max: 29). The participants majority is right-handed (sample A: right: 23; left: 2; sample B: right: 25; left: 0).

TASK

The investigation required the volunteers to perform multiple executions of a simple manual assembly operation. Specifically, the task was to build a pyramid from a predetermined number of 30 building blocks which are identically regarding their size and color. For this purpose a standardized sitting workplace has been set up. This workplace layout was symmetric to have identical conditions for right-handed as well as for left-handed subjects.

EXPERIMENTAL VARIABLES

The participants' cultural background was designated as independent variable. The time consumptions for assembling pyramids (execution times in seconds) were the dependent variables.

PROCEDURE

Before starting the main trial the participants were asked to fill out a questionnaire regarding demographic data such as age, gender etc. Additionally, the educational level, the experience with assembly and the experience with the used building blocks have been queried.

The task was explained by means of guidelines in the respective language (German or Chinese). While explaining all participants were told to have unlimited time for completing their tasks but were asked to build up in a timely manner. Furthermore, all subjects were informed about their hands to be filmed during the trials.

To achieve a certain degree of training and to stimulate a learning effect the task had to be executed ten times by each subject.

STATISTICAL ANALYSIS

The statistical analysis was calculated with the help of the statistical software package SPSS Version 18.0. The subjects' performance have been analyzed with the help of a multi-factor analysis of variance with repeated-measures. Thereby, the cultural backgrounds as well as the repetition of the building task execution were analyzed as main effects. For significant results the effect size ω^2 was calculated (see Field 2005 p. 452). Rank correlations have been analyzed with a two-tailed test according to Spearman. The chosen level of significance for each analysis was $\alpha=0.05$.

Regression models were computed with Matlab R2009a.

RESULTS AND DISCUSSION

According to figure 1 already the average execution times and their 95% confidence intervals clearly indicate an increasing participants' performance in both samples: Starting with comparatively high values for the first execution, time consumption initially decreases fast and reaches a nearly stable level from which performance can be increased only marginally.

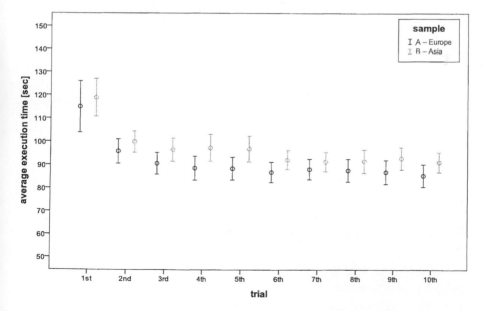

Figure 1. Average execution times and 95% confidence intervals of both samples

The statistical analysis shows a significant main effect of the repetition of the building task execution on execution times (F=51.034; df=2.539; p=0.000). This proves the aforementioned appearance of increasing performance with a large effect size of $\omega^2=0.352$.

The finding is confirmed by results of a multi-level comparison of means using a Bonferroni post-hoc test (see table 1). Especially the first execution differs significantly from all further executions (each p=0.000). Moreover, the second execution and partially the third differ from all further executions significantly (p<0.05).

Thus, the analysis of data proves the existence of learning time in terms of the fact that sensorimotor skills can be learned and trained. This was found without creating huge time pressure.

Table 1: Mean differences of execution times [sec] (multi-level comparison)

	1	2	3	4	5	6	7	8	9	10
1		19 062*	23 438*	24 008*	24.378*	27 553*	27 353*	27 444*	27 224*	28 809*
2	19 062*		4.376*	4.946*	5.316*	8.492*	8 291*	8 382*	8 163*	9 747*
3	23.438*	4 376*		57	94	4.116*	3 915*	4 006	3.787	5 371*
4	24 008*	4 946*	.57		37	3 545	3 345	3 436	3 216	4 801
5	24 378*	5.316*	94	.37		3 175	2 975	3 066	2 846	4 431
6	27 553*	8 492*	4 116*	3 545	3 175		- 2	- 109	-.329	1 256
7	27 353*	8 291*	3 915*	3 345	2 975	- 2		091	.129	1 456
8	27 444*	8 382*	4 006	3 436	3 066	.109	091		-.219	1 365
9	27 224*	8.163*	3 787	3 216	2 846	.329	.129	- 219		1 584
10	28 809*	9 747*	5 371*	4.801	4 431	1 256	1 456	1 365	1 584	

*The mean difference is significant at the 0.05 level

The subjects cultural backgrounds have no significant main effect on execution times (F=3.259; df=1; p=0.076). Thus, the results do not indicate significant differences between European or Asian cultural backgrounds regarding the acquisition of skills while executing simple sensorimotor tasks. Based on the conducted study there is no cultural influence on learning simple sensorimotor tasks.

Furthermore, no significant interaction between the cultural background and the repetition of the building task execution was discovered (F=0.609; df=2.539; p=0.583).

INFLUENCES OF AGE AND EXPERIENCES

All subjects were entitled to university entrance, however, they differ in terms of assembly and building blocks experience. According to the average values, European participants (mean: 2.02) got higher experience than the Asian participants (mean: 1.23). The same holds true for the experience with building blocks: European participants (mean: 2.85) got higher experience than Asian participants (mean: 0.97).

The participants' characteristics have been analyzed for both samples regarding their relation to the execution times as well as their interrelation (see table 2). The handedness could not be analyzed due to a small number of left-handed cases (two of sixty).

Significant correlations between the execution times and the participants' characteristics could not be proven. Especially, there is no significant correlation between the execution times and participants' age in any sample.

Table 2: Non-parametric two-tailed analysis of correlation (Spearman's Rho)

Sample		Age	Assembly exp.	Building blocks exp.
A – Europe (Germany)	Age	1.000	-.014	.029
	Assembly exp.	-.014	1.000	.446*
	Building blocks exp.	.029	.446*	1.000
B – Asia (PR China)	Age	1.000	-.384*	.179
	Assembly exp.	-.384*	1.000	.154
	Building blocks exp.	.179	.154	1.000
	*Correlation is significant at the 0.05 level (two-tailed)			

The analysis shows a significant correlation (sample A: $r=0.446$; $p=0.014$) respectively no significant correlation (sample B: $r=0.154$; $p=0.417$) between the experience with assembly and the experience with building blocks. Since the subjects of sample A have higher experience in both areas this is possibly due to a preference assembly affine people have for those building blocks. Since there is no significant influence on execution times neither from assembly experience nor from building blocks experience this correlation is not taken into consideration for further investigations.

Additionally, sample B demonstrates a significant negative correlation between the participants' age and their experience with assembly ($r=-0.384$; $p=0.036$). This correlation cannot be satisfactorily explained by now. Since there is no significant influence on execution times neither from age nor from assembly experience this correlation is not subject of further investigations.

MODEL FITTING

After proving a significant learning effect on the basis of an ANOVA, the predictive models are necessary to take into account. To analyze the fit of data to Wright's model (see equation 1), a best fitting power function has been approximated (see figure 2). The power function fits not well since the explanation of variance R^2 reaches values of -0.1809 (sample A) respectively 0.075552 percent (sample B).

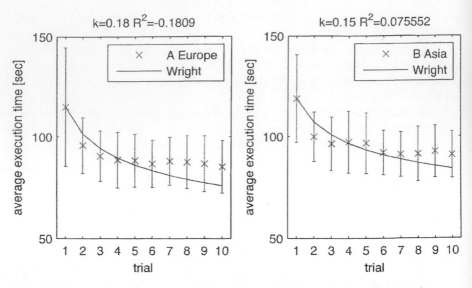

Figure 2. Fitting of average execution times to a power function for both samples

The fit of the acquired data to Levy's model (see equation 2) was analyzed, too (see figure 3). The exponential function fits much better than the aforementioned power function since the explanation of variance R^2 reaches values of 0.56864 (sample A) respectively 0.47439 (sample B). The calculated limits c differ marginally (difference=4.6sec).

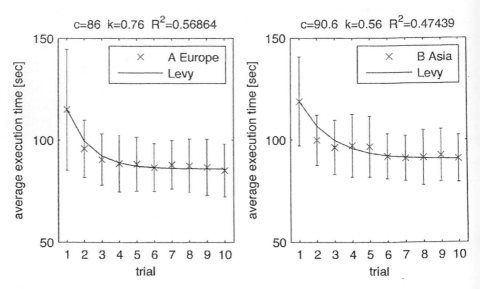

Figure 3. Fitting of average execution times to an exponential function for both samples

The existence of a limit in Levy's model leads to a comparison with the theoretical amount of time needed according to a predetermined motion time system (PMTS), specifically Methods-Time Measurement (MTM). Using the basic system (MTM-1) – the most precise system of MTM – a required time of 75sec can be predetermined. This value differs 14.5 percent (sample A, difference=11.0sec) respectively 20 percent (sample B, difference=15.6sec) from the observed limits. This fit can be regarded as good. Since MTM assumes middle-good trained persons the occurred deviations can be explained partially with a comparatively low degree of training the subjects reached. Furthermore, the instruction at the trials beginning set no restriction in time and offered unlimited time in order to avoid time pressure.

Moreover, the conducted laboratory experiments were analyzed regarding cultural influences on the development strategy of simple geometric objects. The interested reader can find details in Mayer et al. 2010.

CONCLUSION AND OUTLOOK

According to the results of the laboratory study presented in this paper, no cultural difference on learning simple sensorimotor tasks was found.

Further investigations are planned to focus on additional characteristics of employees such as formal qualifications, competencies and general motor skills regarding their influences on learning sensorimotor skills. For this purpose, empirical laboratory experiments have to be designed, conducted and analyzed. Intended subjects are students as well as industrial assembly operators who have to execute an industrial relevant task which appears in daily work of assembly operators.

Additionally to personal influencing factors those of the individual task as well as those of the chosen (vocational) training method will be taken into account when developing and evaluating the forecast model of learning time by means of case studies.

REFERENCES

Babbage, C. (1832). On the economy of machinery and manufactures. Philadelphia: Carey & Lea.
Bokranz, R. Landau, K. (2006). Produktivitätsmanagement von Arbeitssystemen. Stuttgart: Schäffer-Poeschel.
Chryssolouris (2006). Manufacturing Systems – Theory and Praxis. New York: Springer.
Field, A. (2005). Discovering Statistics Using SPSS. London: Sage Publications.
Hieber, W. L. (1991). Lern- und Erfahrungskurveneffekte und ihre Bestimmung in der flexibel automatisierten Produktion. München: Vahlen.
Hofman (1990). Fehlerbehandlung in flexiblen Fertigungssystemen. München: Oldenburg Verlag.

Jeske, T. Hinrichsen, S. Tackenberg, S. Duckwitz, S. Schlick, C.M. (2009). Entwicklung einer Methode zur Prognose von Anlernzeiten. In: Gesellschaft für Arbeitswissenschaft e.V. (ed.). Arbeit, Beschäftigungsfähigkeit und Produktivität im 21. Jahrhundert, Bericht zum 55. Kongress der Gesellschaft für Arbeitswissenschaft vom 4. - 6. März 2009. Dortmund: GfA-Press.

Jeske, T. Mayer, M.Ph. Odenthal, B. Hasenau, K. Schlick, C.M. (2010). Kultureller Einfluss auf das Erlernen sensumotorischer Fertigkeiten. In: Gesellschaft für Arbeitswissenschaft e.V (ed). Neue Arbeits- und Lebenswelten gestalten, Bericht zum 56. Kongress der Gesellschaft für Arbeitswissenschaft vom 24. - 26. März 2010. Dortmund: GfA-Press.

Laarmann, A. (2005). Lerneffekte in der Produktion. Wiesbaden: Deutscher Universitäts-Verlag.

Leiden, K. Laughery, K.R. Keller, J. French, J. Warwick, W. Wood, S.D. (2001). A Review of Human Performancer Models for the Prediction of Human Error. Prepared for: National Aeronautics and Space Administration System-Wide Accident Prevention Program. Ames Research Center, Moffet Filed CA.

Mayer, M.Ph. Odenthal, B. Faber, M. Kabuß, W. Kausch, B. Schlick, C.M. (2009). Simulation of Human Cognition in Self-Optimizing Assembly Systems, In: Proceedings of the IEA2009 - 17th World Congress on Ergonomics, 09. - 14. August, Beijing (CD-ROM), Beijing, China.

Mayer, M.Ph. Odenthal, B. Faber, M. Kabuß, W. Schlick, C.M. (2010). User centered cognitive engineering for self-optimizing assembly systems. In: Karowski W, Salvendy G (Eds). Proceedings of the 3rd International Conference on Applied Human Factors and Ergonomics (AHFE) 17.-20. July 2010. Miami, Florida, USA.

Milberg (2003). Die agile Produktion. In: Klocke, F. Pritschow, G. (eds) (2004). Autonome Produktion. Berlin: Springer.

Rohmert, W. Rutenfranz, J. Ulich, E. (1974). Das Anlernen sensumotorischer Fertigkeiten. Frankfurt a. M.: Europäische Verlagsanstalt.

Sheridan, T.B. (2002). Humans and Automation: System Design and Research Issues. Santa Monica: John Wiley & Sons.

Smith, A. (1776). An Inquiry into the Nature and Causes of the Wealth of Nations. London: printed for W. Strahan and T. Cadell.

Wright, T.P. (1936). Factors Affecting the Cost of Airplanes. Journal of the Aeronautical Sciences, 3(4): 122–128.

Chapter 64

Proactive Vision for the Safety Culture in a Finnish Chemical Plant

Pasi L. Porkka, Marjo Salo-Pihlajamäki, Hannu Vanharanta

Department of Industrial management and Engineering
Tampere University of Technology at Pori
Finland

ABSTRACT

Safety culture has been defined as one of the key factors of an organization's success. In this paper we define ontology of the safety culture. In our ontology we define features of the safety culture that we find to be the most important ones. The selection of the features is based on literature review and discussions with the management of the chemical plant, where the research was done. 39 respondents answered survey, which consisted of 59 statements concerned with 17 safety culture related variables.
Keywords: Safety, safety culture, organization behavior, safety consciousness, safety evaluation

INTRODUCTION

Safety has been defined as one of the key factors of an organization's success. Alone in Finland, in year 2007, there was 143 500 cases of compensations by insurance companies due to industrial accidents. Accidents bring increase in payments, delays to production and negative publicity.

Safety was formerly defined through the lack of accidents. However, this kind of definition does not help an organization to identify and anticipate potential risks. To be able to manage safety, one needs a proactive way to view this subject. An organization's manager has an advantage if he can foresee organization's behavior and potential actions in all situations, even when all seems to go on well.

Organizational safety culture has many different definitions. The one thing common to all definitions is that safety culture is a complex entity, which usually includes the trinity of an organization, work itself (job) and a worker, a person. The organization takes care of the environment and safety management. Work itself includes safe behavior when working, also known as safety at work. A worker has his own values and attitudes, which show in the context of safety. Also the atmosphere in the working environment consists of individuals' behavior. We have combined features from Coopper's (1998) model and from European Agency fo Safety and Health art work' s (EU-OSHA) (2008) models, into our work. The combined model is shown in Figure 1.

Figure 1. Safety culture model modified from Coopper(1998) and EU-OSHA (2008).

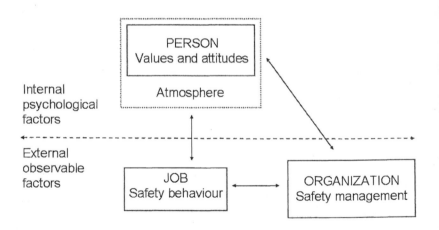

RESEARCH METHODOLOGY

We have made an ontology, which includes features from all three factors presented in Figure 1. In our model there are 17 characteristic measuring different features of safety culture. According to Cooper's model, safety culture is an entity formed through externally observable and internally psychological factors, as well as the interaction between work, organization and people. According to the model, safety culture may be analyzed by examining three sub-groups: the management system of safety, safety atmosphere / climate and attitudes as well as behavior. Also, the characteristics are seen to hold the issues of creating new knowledge and learning by doing.

There are two main categories in Serpentine: the learning environment and organisational knowledge creation activities, which are further divided into eight competence groups. The four first competence groups shown in Figure 2., are partly based on the model of organisational learning environment by Scott I. Tannenbaum (Tannenbaum 1997). The four latter sub-groups are based on the theory of organisational creation of knowledge by Nonaka and Takeuchi (Nonaka 1995).

Figure 2. Ontology of safety culture

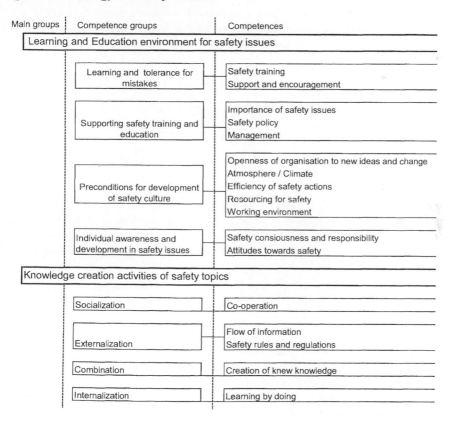

The contents of the competences are subsequently given a brief examination.

Safety training aims at canvas the needs for training, sufficiency, regularity and quality

Supporting and encouraging refer to how an organisation sees the behaviour supporting safety, f. ex. whether safety-conscious behaviour is rewarded and whether employees are encouraged to make suggestions to improve safety.

Importance of safety issues refers to those view point that reflect the prevailing attitudes toward bringing up safety issues in the organisation, f. ex. whether safety issues are given sufficient attention.

Safety policy aims at focusing at those practises, attitudes and methods, prevailing in the background, that organisation holds toward safety and development of safety culture, f. ex. how organisation appreciates health and safety.

Management reflects the attitude with which the management meets safety, how accessible the employees find their supervisors and the possibility to participate in safety issues.

Organisation's openness to new ideas and change canvasses, among other things, whether the organisation enables the questioning working practises and methods and whether the organisation appreciates all opinions in finding improvements to safety issues.

Atmosphere reflects, among other things, how the organisation finds dissenting opinions and overload of personnel caused by stress.

Efficiency of safety actions aims at charting how the resources invested in safety are returned, f. ex. how regularly safety implementation is monitored and evaluated, whether all accidents are examined, the learning of safety consciousness is monitored and whether the organisation follows safety rules and regulations.

Resourcing for safety reflects how the organisation has enabled safe working, f. ex. whether assignments and instructions are clear enough, there is enough time to complete the assignment safely.

Working environment reflects whether organisation takes into account environmental safety points in planning work, the condition of the environment is monitored and necessary improvements are implemented.

Safety consciousness and responsibility refers to how employees have taken in the dangers relating to their work and the working environment as well as safety-related goals of the organization, whether employees know their own contribution to safety at work and how to responsibly report danger situation and safety deficiencies to their supervisors.

Safety attitudes reflect the attitudes, employees hold on maintaining safety, f. ex. whether safety instruction are breached should it facilitate working methods or is safety taken into account in every day operations.

Co-operation refers, among other this, to how the various levels of the organization interact with each other and whether ill relations have significance toward safety.

Flow of information reflects whether work related issues are sufficiently informed, whether knew knowledge is easily accessible and the topic of safety risks are openly discussed in the organization.

Safety rules and regulations chart, among other things. the clarity, understandability and accessibility of safety regulations and whether their compliance contradicts common working methods and practices.

Creation of new knowledge aims at examining whether employees are also given sufficiently information outside their working task and whether accessing information relating to a new task is easily and rapidly accessible.

Learning by doing examines, f. ex., how rapidly well-found methods are disseminated in the organization and how rapidly the organization is able to manage erroneous situations and is able to come through them.

To study these competences with self-evaluation, 2-5 statements were specified to each competence. We came up to 59 statements to be presented in the questionnaire. For each statement the respondent should estimate the current and the target values for that statement. From the values of statements, with fuzzy logic deduction, each competence is given a single value.

The values to statements are given with clicking the mice. The current is the value of the statement at the moment. The target value represents the respondent's wish of the statement's value in the future. The difference of target and current states can be considered as the creative tension or proactive vision of the statement (Senge, 1990). The difference is called creative tension, when it is focused to some human factor, and proactive vision, when it is focused to some system variable.

We use the proactive vision as the proof of motivation. If there is big proactive vision for some competence, then people are motivated to improve that competence. Therefore it is also wise, from employer's point of view, to arrange education to improve those competences.

CASE STUDY

The research was done in a Finnish chemical plan. The respondents were from two different departments with 28 and 11 people answering the survey. Their demographics are shown in Table 1.

Table 1. Demographics data of respondents

Department	Sex	Position	Age	
	Female	Worker	<31:	2
	2	22	31-40:	3
A	Male	Manager	41-50:	1
28	24	6	>50:	2
	N/A	N/A	N/A	20
	2	-		
	Female	Worker	<31:	4
	2	5	31-40:	2
B	Male	Manager	41-50:	4
11	9	6	>50:	-
	N/A	N/A	N/A	1
	-	-		
	Female	Worker	<30:	6
	4	27	31-40:	5
Total	Male	Manager	41-50:	5
39	33	12	>50:	2
	N/A	N/A	N/A	21
	2	-		

To give respondents the possibility to freely speak out and say their opinion about the employer and facilities at the working place, answering to the survey was made anonymously. There was no way employer could find out who was giving and what kind of answers. The amount of missing values can be due to this anonymity. The respondents were afraid that the demographic data would reveal their identity. The reason, why there is none missing values in position is, that we know the range of user accounts given to both workers and managers. Even though a respondent gave no demographic data, we know his position.

ANALYSIS AND RESULTS

From the statistics point of view the survey data is of nominal scale. The nominal scale means that numbers (the values) are used merely as a means of separating the answers into different classes or categories (Conover, 1999). The different answers with only single respondent are comparable. From the data we can conclude only the ranks of single person. Due to the data's feature of being of nominal scale, the traditional statistic functions (sums, means, etc.) are not applicable when analyzing group results.

There are several valid and suitable statistical methods for this kind of nonparametric data. All these methods use rankings of the data instead of the actual input values. The values to competences of a single person can be ranked (from 1 to 17). The rankings of several respondents can then be summed together, gaining valid group results.

In our study case there are several related samples in the data. The most powerful test for several related samples, where the number of different variables is more than six, is the Friedman test (Conover, 1999).

The Friedman test gives also one extra benefit compared to the sums and averages. With Friedman test one can calculate the minimum statistical difference (MSD), the sums must have, after which they are regarded unequal. This difference may be calculated with different significance levels. We have used the significance level $\alpha=0.05$. The dotted lines in Table two are groupings gained with the use of MSD. Atmosphere is at the top with a value 13,1. All features, whose value is within the MSD, are statistically equal with the atmosphere. So the line is drawn to (13,1 – MSD = 13,1 – 1,73 = 11,37). All features, whose value is over 11,37 belong to the upper, or the highest group. The line for the bottom group is gained equally by adding the MSD to the minimum value, (5,1 + 1,73 = 6,83).

Table 2. Proactive vision of all respondents, n=39, α=0.05, MSD=1.73

SUM of Rankings	Competence
13,1	Atmosphere
11,7	Resourcing for safety
11,4	Flow of information
10,1	Co-operation
10,1	Doing by learning
9,8	Support and Encouragement
9,8	Safety attitudes
9,7	Safety directions and regulations
9,7	Working environment
9,3	Efficacy of the safety actions
8,9	Leadership
7,8	Creating new knowledge
7,4	Safety training
6,6	Safety awareness and Responsibility
6,5	Organization's openness to new ideas
6,0	Prioritizing safety issues
5,1	Safety policy

Atmosphere, Resourcing for safety and Flow of information are those two features that are found to be in bad shape, since there is the highest need to be improved. However, the organization has done certain things correctly. Safety policy, Prioritizing safety issues, Organization's openness to new ideas and Safety awareness and responsibility need the least enhancement.

When the two departments were compared, there were no significant differences with their answers. However the comparison of management and workers showed quite anticipated differences as showed in table 3.

Table 3. Comparison of creative tensions of management and workers

Competence	Manage-ment	Workers	Difference
Safety attitudes	9,8	2,8	6,9
Doing by learning	10,1	7,9	2,2
Resourcing for safety	11,7	10,1	1,7
Safety training	7,4	6,3	1,1
Flow of information	11,4	10,7	0,7
Prioritizing safety issues	6,0	5,5	0,4
Leadership	8,9	8,7	0,2
Support and Encouragement	9,8	9,6	0,1
Efficacy of the safety actions	9,3	9,5	-0,2
Atmosphere	13,1	14,0	-0,9
Safety awareness and Responsibility	6,6	7,6	-1,0
Co-operation	10,1	11,4	-1,3
Organization's openness to new ideas	6,5	8,0	-1,5
Safety policy	5,1	6,8	-1,7
Safety directions and regulations	9,7	11,7	-2,0
Creating new knowledge	7,8	9,9	-2,1
Working environment	9,7	12,4	-2,7

The competences are ordered in the difference between the values given by workers and management. The more positive is the difference value, higher it was within the management compared to workers. Consequently more negative the difference is, higher was the ranking within workers.

CONCLUSIONS

One must keep in mind, that when talking about proactive vision, higher the value means more change required. In all different analysis Atmosphere was always on the top. It gained the highest proactive vision as seen in Table 2., but also in Table 3. it has the highest value in both management's and workers' opinion. So it was clearly the most significant single feature that needed change.

Also Resourcing for safety and Flow of information were on the upmost part. Respondents wished that those two were also in better shape. Good news for the organization is that respondents found that the organization was certainly interested in workers safety. The respondents found that there was very little to be done with Safety policy, Prioritizing safety issues and Safety awareness and Responsibility. This result shows that the organization truly cares about its workers. Also good news is that the Organization is open to new ideas.

The comparison of workers and management revealed somewhat expected result. The management is more concerned about the safety attitudes. The workers found that there wasn't that much to improve within their attitude. However, the current state of safety attitudes, according to workers themselves, was rather poor. The workers did not see the importance of their own attitudes when trying to make their working place safer.

REFERENCES

Conover, W. J., 1999. *Practical nonparametric statistics.*, John Wiley & Sons, New York

Coopper, D. (1998), *Improving safety culture: a practical guide*. John Wiley & Sons Ltd. Chichester.

EU-OSHA (2008) European Agency for Safety and Health at Work *Yrityksen turvallisuuskulttuuri*, (http://osha.europa.eu/fop/finland/fi/good_practice/ turvallisuus/kulttuuri.stm) (28.02.2010).

Nonaka, I &Takeuchi, H. 1995. *The knowledge-creating company: How Japanese companies create the dynamics of innovation.* New York, Oxford University Press. 284 p.

Senge, P. 1990. *The Fifth Discipline*, Doubleday

Tannenbaum, S. I. 1997. *Enhancing Continuous Learning: Diagnostic Findings From Multiple Companies.* Human Resource Management, Vol 36 (4). pp. 437 – 452.

CHAPTER 65

Psychosocial Aspects of Work

Dorota Zolnierczyk-Zreda

Central Institute for Labour Protection – National Research Institute
Czerniakowska 16, 00-701 Warsaw, Poland

ABSTRACT

Some psychosocial aspects of work, such as job insecurity, work precariousness, work intensity, and workplace bullying that has been found to be the main European emerging psychosocial risks related to occupational safety and health have been presented in a paper. The overview containing some main figures as well as measures to prevent those risks at the European level has pointed for a serious concern these risks obtained among stakeholders. The Polish case in relation to these risks has also been shown, as well as the recent relevant studies aimed to better understand and to tackle these contemporary psychosocial risks.

Keywords: psychosocial aspects of work, job insecurity, work precariousness, work intensity, workplace bullying

INTRODUCTION

Psychosocial factors of work are those aspects of the design and management of work and its social and organizational contexts that may have the potential for causing psychological or physical harm (Cox & Griffits, 2005). They may also, like high levels of influence, meaning, predictability, social support, high rewards or suitable demands, be the road to a higher quality of job and workers' well-being.

Although nowadays, there is a strong tendency in occupational psychology to look for the zest of work and for its sources in psychological aspects of work, the economic crisis shows us how fragile these aspects are and how easily they can deteriorate due to an external situation putting workers' health at risk.

EMERGING PSYCHOSOCIAL RISKS - THE EUROPEAN BACKGROUND

As reported by the European Foundation for Improving Living and Working Conditions in Dublin in its Fourth European Survey on Working Conditions, work-related stress is among the most commonly reported causes of illness in workers affecting more than 40 million individuals across the EU (European Foundation for Improving Living and Working Conditions, 2007). The economic costs of work-related stress are widely recognized as accidents and sick leave absence, low performance and decreased productivity.

Job insecurity, precarious work, work intensification, workplace violence and bullying are those psychosocial factors that have been only recently identified as the emerging psychosocial risks (*Expert forecast on emerging psychosocial risks related to OSH, 2007*).

Globalization leads to higher competition with its pressure on cutting costs, outsourcing, temporary and precarious employment, high work intensity and unstable labour markets. Additionally, **job insecurity** is increasing due to economic crisis. Job insecurity has been found to be linked with significant both physical health (ex. coronary heart disease, and MSD) and mental health (ed. depression, anxiety, sleeping problems) impairment (Sverke, et all, 2002; Ferrie et al., 2001). Therefore, ensuring career and employment security has become one of the main European policy objectives. There is also a great political debate: "employment versus work", which signifies the dilemma whether to combat unemployment through ensuring workers employment or through maintaining their job security. Developing skills and competencies through life-long learning and flexible work arrangements are possible ways to secure employment.

Over the period of 4 years 1991-2005 the systematic rise in non-standard employment in the EU has been observed, half of the "new" jobs have not been permanent, with regard to a type of contract. Although flexible work arrangements are highly recommended as a measure to tackle job insecurity and to reconcile work and family life, they have usually a feature of **precariousness** that is:
-   low level of certainty over continuity of employment;
-   low control over such aspects of work, like working conditions, income, working hours;
-   low level of protection (mainly social protection);
-   insufficient income or economic vulnerability (Rodgers & Rodgers, 1989).

Precarious work can also be defined as employment with low quality and little opportunity for training and career progression. Like job insecurity, precariousness has been found to be related to an increased risk of illness and injury (Tucker, 2002). The research shows that whether working time arrangements can be judged as good or bad for employees depends on the manner in which flexibility are implemented and on the degree of choice that employees have to regulate their working life (Janssen & Nachreiner, 2004).

Another measure to ensure the secure and decent employment is to encourage lifelong learning. The access to this training is especially important for older and less qualified workers, particularly nowadays with increasing use of information technology. Thus it requires measures for all groups of workers, irrespective of age, sex or contract status.

The data shows that, as it has been mentioned earlier, **work intensity** is clearly on the increase, with rising numbers of workers performing their work at high speed and to tight deadlines. Twenty five percent of all workers have to work at very high speed all of their worktime (ESWC, 2007). According to this data, high-skilled blue-collar workers are more exposed to temporal constraints than their white-collar counterparts. Long working hours are commonly regarded as one of the work intensification indicator. Regarding their negative impact on both physical and mental health, and life quality, including work-family imbalance (der Hulst, 2003), a vivid debate has taken place at European level on how to prevent extended working hours practices at enterprise level. The European Directive on Working Time was launched in 1993 forbidding employers to extend the weekly working time above 48 hours and to ensure employees a minimum rest period of 11 hours in each 24 hours and a minimum number of paid leave days per annum. As a result of this Directive, since the early nineties, a steady reduction in the length of the working time in the EU has bee observed, but a trend changed in 2005 following accession of the New Member States, where average working hours have been longer.

Preventing extended working hours is a necessary condition for reconciling working and non-working life which is relevant mostly to people with caring responsibilities, that is women. It has been found that considering their double workload – women still work the longest. Since gender equality has received special attention in the EU policy, it will hopefully result in developing some strategic measures to tackle barriers to the integration of women and men in the labour market and to effective reconciling work and family for both men and women.

Workplace violence has been found as another emerging psychosocial in Europe. The term refers to "incidents where persons are abused, threatened or assaulted in circumstances related to their work, involving an explicit or implicit challenge to their safety, well-being, and health. Bullying and harassment occurs when one or more workers or managers are abused, humiliated or assaulted by colleagues or superiors. The third party violence refers to violence from clients, customers, patients and pupils and the like" (Leka, Vartia, Hassard, Pahkin, Sautela, & Cox, 2008, p. 136). According to the 4th EWCS (2007), 6% of the workforce had been exposed to threats of physical violence, 4% to violence by other people and 5% to bullying and/or harassment at work. Moreover, women more than men have suffered from all forms of violence, bullying has been mostly found in services:

14% of workers from the education and health sector, public administration and defense sector, and hotels and restaurants, as well as by 12% of workers from the transport and communication sector and 9% of workers from the wholesale and retail trade sector. About 4% of all workers have been a subject of external physical violence (coming from clients, students, patients, etc), and 1,.5% of all workers were a subject of violence coming form colleagues. Physical violence has been experienced most in the education and health sector (11% of employees), followed by the public administration and defense sector (95%) and hotels and restaurants (7%).

There is a large body of findings showing that workplace bullying and violence negatively affects physical and mental health, leading to a variety of symptoms, like depression, irritability, problems with concentration, insomnia, chronic fatigue, back pain or headache (Vartia, 1996; Kudielka & Kern, 2004; Hansen et all., 2006). Moreover, it has been found that the greater the frequency and the more serious form of violence, the more severe the effect is (Di Martino, Hoel & Cooper, 2003). Regarding the damaging outcomes of violence and harassment at work in people who are both directly and indirectly concerned by this issue and in companies, the social partners signed the European Framework Agreement on Violence and Harassment at Work. The general recommendation to tackle violence and bullying in the workplace is to develop written policies and procedures, like codes of conduct, to implement preventive measures including quality leadership and employees training, and to seriously treat workers' complaints related to violence and bullying.

EMERGING PSYCHOSOCIAL RISKS AT WORK - THE POLISH CASE

This short overview of the emerging psychosocial risks at work in Europe, could be a background for outlining the Polish case relating to these risks. Generally speaking, the picture concerning psychosocial work conditions in Poland, as well as in other New Member States of the EU is worse from the picture of the "old Europe". This is mainly due to economic transition these countries have recently experienced. The data of the latest European Survey on Working Conditions conducted by the Dublin's Foundation shows that it is particularly the case with regard to work intensification (extended working time), stress and fatigue caused by work. It has been found that 35% of Polish workers experience stress at work and more than 65% of them state that work negatively influence their health. Moreover, Poland has been found to be the fourth European country with the largest population working more than 48 hours weekly, and the fourth one with the largest number of workers perceiving work-life imbalance. Almost one third of Polish workers complain about job insecurity (the fear of job loss within 5 months) what places

Poland on the third least secure labour markets in Europe, behind the Czech Republic and Slovenia.

National data confirms the findings of the latest ESWC. As far as work intensity is concerned, it has been found that in 2009 60% of Polish workers worked more than 40 hours weekly, and more than 24% complained about overload and constant time pressure (GUS, 2009), in the same 2009 year the unemployment rate in Poland was 11,.9%. However, the strong tendency for flexible work arrangements as a measure for job insecurity is being observed in Poland.

It has been found in the research carried out in CIOP-PIB, that flexible working time is associated with lower levels of experienced work-related stress and higher levels of mental health. That is particularly the case for elderly and female workers, where the traditional double workloads are still in place.

The considerably better situation relates to workplace violence and bullying in Poland, which has the second smallest percentage of workers complaining on being exposed to this problem at work. National data shows that 1,.6% of Polish workers is exposed to bullying at work, and 1,.9% experience external violence. Inconsistently to the European findings, the highest level of bullying in Poland is observed in such sectors like transport and public administration.

Additionally, Poland has some specific legal regulations that oblige employers to prevent workplace violence and harassment and to ensure the victims of such violence the appropriate compensation.

The European Framework of Psychosocial Risk Management which has been developed with the contribution of experts from CIOP-PIB has pointed for the necessity of dealing with these risks both at the enterprise and at the macro levels (Leka & Cox, 2008). Psychosocial risk management at the macro level should include various types of actions like: development of policy and legislation, the specification of best practice, and standards at national or stakeholder's levels. These should finally lead to the signing of stakeholders' agreements towards a common strategy, the signing of declarations at the European or international levels, and the promotion of social dialogue and social corporate responsibility in relation to the issues in question. According to the Framework, the management of psychosocial risks on the enterprise level should be based on risk assessment, development of action plan and on introducing the particular intervention to improve the situation. The examples of such interventions are recognized both on the European and on the international levels. Since psychological well-being is of increasing importance, and is regarded as a part of workplace health promotion, there have been conducted studies in CIOP-PIB aimed at developing and evaluating the effectiveness of psychosocial risks management for general working population, for particular sectors (ex: education, health care, which are additionally female dominant sectors) and for specific work-related stress.

An example of such intervention directed at employees performing work which can trigger high psychosocial risk (emotional demands, violence, etc) was the cognitive behavioral intervention for teachers. The two-day burnout intervention programme focused at enhancing coping with the stresses observed in teachers' work. Karasek's job stress model was used as the theoretical framework. The aim of the intervention was to teach participants to better deal with high job demands and low job control. Some cognitive-behavioural methods overcoming workload and enhancing a sense of self-mastery and relations with students were introduced in the workshop. Fifty nine teachers were randomly assigned to an experimental or to a control group. Results showed that emotional exhaustion, perceived workload and somatic complaints decreased significantly in the intervention group (Tab. 1).

Table 1: MANOVA involving *Time x Intervention* Effects of the intervention

Variable	F	p	μ
Intellectual demands	F(1,56)=4.34	.042	.072
Psychological demands	F(1,56)=1.72	.19	.030
Overload and role conflict	F(1,56)=6.47	.014	.104
Behavioural control	F(1,56)=28.88	.000	.324
Cognitive control	F(1,56)=1.13	.29	.021
Support from supervisors	F(1,56)=.419	.52	.008
Support from co-workers	F(1,56)=.98	.32	.018
Emotional exhaustion	F(1,56)=9.47	.003	.152
Personal accomplishment	F(1,56)=.262	.616	.005
Depersonalization	F(1,56)=.006	.937	.000
Intensity of somatic complaints	F(1,56)=4.34	.042	.073

The greatest effect of the intervention was observed with regard to an increased behavioural job control. It was concluded that teaching participants how to better manage their work environment (e.g. avoiding workload, improving relations with students) could help them in changing their perception of stressful job characteristics and reducing burnout (emotional exhaustion) and somatic complaints.

Another stress-management intervention that has been developed in CIOP-PIB for an occupational group struggling with high psychosocial risk was mindfulness-based stress management programme. The programme was designed to help managers to enhance their coping with stress related to work intensification, a need for flexibility and running at full capacity almost around the clock. This study used a randomized controlled study design consisting of eight 2-hour sessions. The goals of this intervention were the following:

- decrease the perceived work-related stress,
- enhance coping skills,
- increase emotional and physical well-being.

The outcomes of this study have shown that generally their goals have been obtained, that is managers participating in the programme perceived significantly lower level of stress than those who did not participated in the intervention. Likewise, the significant improvement in emotional well-being has been observed in the experimental group, specifically less negative and more positive affect than in the control group, lower level of anxiety and anger. An increase in self-esteem has been another effect of the intervention, as well as the change from emotional coping into task-oriented, active coping.

CONCLUSIONS

A number of initiatives both at the policy and organizational level have been implemented in the recent past to effectively tackle the most prevalent psychosocial risks. However, the emphasis must be placed at conducting careful analysis of these interventions and to evaluate them in order to promote societal learning process.

REFERENCES

Cox, T. & Griffits, A. (2005) Monitoring the changing organization of work: A commentary. *Sozial-und- Praventivmedizin, 47,* 354-355.

European Foundation for Improving Living and Working Conditions. (2007). *The fourth European Survey on Working.* Office for Official Publications of the European Communities, Luxemburg.

European Agency for Safety and Health at Work. (2007). *Expert forecast on emerging psychosocial risks related to OSH*: Office for Official Publications of the European Communities, Luxemburg.

Ferrie, J. E.; Shipley, M. J., Marmot, M. G., Martikainen, P., Stansfeld, S., & Smith, G. D. (2001). Job insecurity in white-collar workers: toward an explanation of associations with health. *Journal of Occupational Health Psychology,* 6, 26-42.

Harrington, J. M. (2001). Health effects of shift work and extended hours of work. *Occupational and Environmental Medicine, 58,* 68-72.

Janssen, D. & Nachreiner, F. (2004). Health and psychosocial effects of flexible working hours. Revista Saude Publica, 38, 11-18

Kudielka, B. M. & Kern, S. (2004). *Cortisol day profiles in victims of mobbing (bullying at the work place): preliminary results of a first psychobiological field study.* Journal of Psychosomatic Research, 56, 149-150.

Leka, S. Vartia, M., Hassard, J., Pahkin, K., Sautela, A., & Cox, T. (2008). Best practice in interventions for the prevention and management of work-related stress and workplace violence and bullying In. S. Leka & T. Cox (Eds.) *The European Framework for Psychosocial Risk Management: PRIMA-EF* (pp. 136-173). Institute for Work, Health and Organizations, University of Nottingham.

Leka, S., Cox, T. & Zwetsloot, G. (2008). The European Framework for Psychosocial Risk Management (PRIMA-EF) In. S. Leka & T. Cox (Eds.) *The European Framework for Psychosocial Risk Management: PRIMA-EF* (pp 1-16). Institute for Work, Health and Organizations, University of Nottingham.

Rodgers, G. & Rodgers, J. (1989). *Precarious jobs in labour market regulation: the growth of atypical employment in western Europe.* International Institute for Labour Studies, Free University of Brussels, Brussels.

Sverke, M., Hellgren, J., & Naswall (2002). No-security: a meta-analysis and review of job insecurity and its consequences. *Journal of Occupational Health Psychology, 7,* 242-265.

Tucker, D. (2002*). "Precarious" non-standard employment – A review of the literature.* Departement of Labour, New Zeland, http:wwwpsa.org.nz?library/other/dol%20non-standard%20employment

Vartia, M. (1996). Consequences of workplace bullying with respect to the well-being of its targets and the observers of bullying. *Scandinavian Journal of Work, Environment and Health*, 27(1), 63-69/ .

Di Martino, H., Hoel, H., & Cooper, C. (2003). *Preventing violence and harassment in the workplace.* European Foundation for Improving Living and Working Conditions, Office for Official Publications of the European Communities, Luxembourg.

634

van der Hulst, M. (2003). Long workhours and health, *Scandinavian Journal of Work, Environment and Health, 29(3),* 171-188.

Główny Urząd Statystyczny (2009). *Wypadki przy pracy i problemy zdrowotne związane z pracą (Accidents at work and work-related health problems).* Główny Urząd Statystyczny, Warszawa.

Żołnierczyk-Zreda, D. (2005). An intervention to reduce work-related burnout for teachers. *Journal of Occupational Safety and Ergonomics,* 11, 423-430.

Chapter 66

New Tools for Analyzing Work Activities in the Process Industry

Petr Skrehot

Occupational Safety Research Institute
Prague, Czech Republic

ABSTRACT

Analysis of work activities is one of the pillars of system safety in the process industry. Their purpose is to determine the current status of implementation of individual work operations and verify compliance with the prescribed working pattern and work instructions. It turns out that deficiencies on the level of control documentation, especially in the description of implementation of subtasks still endure even in companies with a high level of system safety and safety culture. Many inconsistencies of prescribed requirements and real process of work activities can be discovered for a detailed view. Mentioned inconsistencies can be revealed by using Hierarchical Task Analysis (HTA), which is well known from perform of human reliability analysis. Although this method is simple, it is also time-consuming. Accordingly to simplify its use in operational practice, we developed a software tool that simplifies its application. The purpose of this article is to introduce the instrument „HTA module" and to demonstrate the possibilities of its practical use.

Keywords: Work System, Hierarchical Task Analysis, Predictive Human Error Analysis, Software Application, Human Error, Systems Safety

INTRODUCTION

Occupational Safety Research Institute completed the development of new tools for human factors (HF) reliability analysis. In 2006 was developed a modified version of the HTA method (Hierarchical Task Analysis). In 2007 it was transferred to software form and progressively tested in industry. Quality of outputs and working with the program was evaluated as excellent by users. So it was possible to proceed to the coupling method HTA with the method PHEA. PHEA method is designed for the human errors analysis using the outputs from the HTA for the next stage of analysis. In conjunction with the HTA, PHEA allows to perform complex, and while quite detailed analysis of the HF reliability. The specific human errors database and its HEP values and also a performance influencing factors (PIF) database were gradually built into HTA-PHEA method for maximum width of HF reliability and mistakes analysis. This complex methodology was then transferred to the software form and so developed an instrument "HTA-PHEA Analysis", which is undoubtedly one of the most advanced instruments for performing this type of analysis.

MODULE HTA 1.5

HTA has been used for long time primarily in the European chemical process industry in the context of training (Embrey, 2000). It is a systematic method for identifying the various objectives which must be achieved during required task performance, and the way in which these objectives are combined in the task scheme (Harris et al., 2005). At each level of analysis, starting with the highest levels sub-objectives, sub-tasks are defined, which are required to meet the overriding objective (along with the plan, which establishes their order). This procedure is successively repeated with each of the sub-operation level, where all operations (necessary to achieve the target) are targets for lower level of analysis. Once all subtasks (and operations to be done to achieve them) are described (at each subordinate level) the careful description of the task can be obtained.

The method is described in detail by Stanton (Stanton, online) and the current form of HTA method, which was developed by progressive improvements of traditional task analysis approaches, was designed by Patrick, Spurgeon and Shepherd in 1986. The method provides the task hierarchy by dividing of the task analyzed (Skrehot, 2008a). The procedure followed in the HTA method reflects the quality of the organizational aspects affecting the HF reliability and the potential protective barriers disruptions in the working system, which may be associated with the HF effects, but it is not further analyzed by this procedure.

Within HTA three types of analysis are carried out:

- requirements and objectives analysis of specific tasks (e.g. find out what is required of the employee during working operations),
- analysis of activities carried out within the task (e.g. identifying weaknesses in

the current way of performing work activities) and

- analysis of protection or barriers relating to each task (e.g. in the form of control systems redundancy, appropriateness of feedback informing about the correct / incorrect specific task performance, etc.).

The task diagram is HTA output, which reflects in an understandable form the performed task structure - from the simplest sub-tasks to meet the final (top) objective. The text output is associated to the diagram in the form of a table summarizing qualitative information about sub-tasks analyzed, including identified weaknesses and proposed measures for their elimination. The table describes the feedback information about requested action fulfillment or failure.

For the purpose of HF reliability analysis treatment according to the European Seveso directive requirements (in the Czech Republic it is Act No. 59/2006 Coll. about the prevention of major accidents), our Institute has developed a modified version of this method. This version was published in the HTA work independently (Skřehot, 2008b).

PREDICTIVE HUMAN ERROR ANALYSIS

PHEA analysis is aimed at predicting specific HF errors during specific activities performance. The method is part of a complex methodology SHERPA, but it can be used individually or as being in connection with HTA method. Modeling of error types that may occur in man-machine system is probably the most important aspect of assessing and reducing the HF contribution on the accident risk (Sandom and Harvey, 2004). As part of this process it is also considered how these estimated errors may be eliminated even before the manifestation of their negative consequences. This approach is based on cognitive psychology (Harris et al., 2005). Inputs for the analysis consist of information about the tasks structure and plans, which is extracted from the HTA, and also PIF impact assessment (Embrey, 2000). These data can be partly obtained deductively from the HTA; in part it is necessary to collect additional information needed.

The principle of error analysis is based on the fact that the relevant human errors are identified to any sub-task by using predefined taxonomy, where the errors are classified into 6 error modes (Action Errors, Checking Errors, Retrieval Errors, Communication Errors, Selection Errors, and Plan Errors). Then the reliable errors types for each sub-task are selected from this taxonomy by analyst (Stanton and Young, 1999) and from these the specific relevant errors are considered, i.e. errors which may realistically rise from working system - whose emergence can be expected. Because our modified PHEA, (differently from the original method), already contains a predefined error database, it is possible to proceed systematically within its use during analysis, as well as to identify such errors, which would not be considered without the use of this database.

SOFTWARE "HTA-PHEA ANALYSIS"

Software "HTA-PHEA analysis" is based on the modified PHEA method, which w
developed. This method differs from the classical PHEA that it allows for each potentia
vulnerability to evaluate its possible consequences, its occurrence probability (HEP) an
the HEP value correction under the current level of barriers preventing the erro
occurrence (or the emergency development), and also the PIF effect on the huma
reliability. Then remedial measures or measures for eliminating the risk are suggeste
according to the subjective analyst's opinion. Using the outputs from HTA and by thei
analysis using the modified PHEA allows performing a comprehensive HF reliabilit
assessment in the working system, including a quantitative HEP estimate.

The analysis required to perform 7 basic system steps:

1. Problem determination;
2. Task Analysis;
3. Risk subtasks selection;
4. Human error analysis;
5. Human error probability estimation;
6. Performance influencing factors analysis;
7. Proposal of measures to reduce errors.

Steps 1, 2 and 3 are carried out in the framework of HTA analysis and the other steps ar
then included in the PHEA analysis and they are described below.

STEP 4: HUMAN ERROR ANALYSIS

The most commonly encountered error mode in carrying out of the assigned tasks i
"Action Errors" which are made during the actual implementation of sub-workin
operations when the system status is changed (Stanton and Young, 1999) (e.g. controlle
value on production equipment was set wrong by operator). The error mode "Checkin
Errors" (e.g. poorly implemented control of subordinate employee) usually includes th
obtaining data process such as the level or status verification through visual inspectio
Error mode "Communication Errors" is related to obtaining information either from a
external source (e.g. transmitters) or from memory. The error mode "Communicatio
Errors" includes direct communication between two individuals and also indirect (vi
computer, writing, etc.). These errors are particularly relevant when there is a need fc
coordinated activities of several people in the team. Error mode "Selection Errors" i
related to the realization of the wrong selection between alternative operations, where i
the need to realize the explicit choice between two alternatives, such as manual instead c
automatic (and conversely). It can be physical objects or parts of technical equipment (e.g
valves, buttons, etc.) or process activities. The specific error modes structure briefl

presented above is described in detail in the literature (ICHE, 1994, Stanton and Young, 1999).

Identifying of specific relevant HF errors is necessary to realize for each sub-task (i.e. operation), which is found in HTA task diagram (expressed by hierarchy) usually on the lowest position and is no longer divided into the other sub-tasks (it is usually 2 to 4 level in HTA) (Stanton and Young, 1999). At least one of the error modes above (see Table 1) is assigned to this sub-task. Next those error types, which can be made by operator during the sub-task performance, are chosen from these error modes. Then it is possible to define specific relevant errors - e.g. by selection from the error database from modified PHEA or adding entirely specific errors which are not listed in the database (but they are eligible for the concrete conditions of the task analyzed).

STEP 5: HUMAN ERROR PROBABILITY ESTIMATION

The aim of this part of analysis is the relevant HF errors probability evaluation. The relevant errors are the errors which may occur during the working activities. This phase of PHEA analysis is not easy and it should be emphasized, that although it is leading to a specific numerical values of the probability that the error will be made by operator, this value is encumbered by the considerable uncertainty. HEP database has been created by the data collection from various expert sources, which give generic or statistical data from various types of process industry sectors and obtained in different time periods. This is therefore a mean value and for analysis it is necessary to correct them in a real working system according to local conditions, such as the existence of protective barriers, according the level of the material-technical support, staffing, training quality of the operators, control activities, etc. An important role in this correction is also the possibility that the correction will be made before the error shows its adverse consequences (e.g., the error is registered by the operator or by another worker or hardware). Against to statistically "average" error occurrence probability (HEP), which is the mean value obtained from various expert sources, it is possible that the real error probability under consideration will be more or less contrast. This correction of HEP value in the modified PHEA is carried out by selecting from the various severity categories:

- low (L) –the error occurrence at the current level of security is not expected;
- medium (M) – the error has been previously reported, but the current security level limits its repetition enough;
- high (H) – the error has already occurred several times (at the various members of the working team) or repeatedly by the same operator and it has to be counted with its occurrence at the current level of system security.

HEP values at M level corresponding to the mean values obtained from the literature (Kirwan, 1997; Grozanovic and Stojilkovic, 2006; Sharit, 2006; Vincek and Haight, 2007) and were derived from observation and analysis of emergency situations. Then the values for L and H levels were carefully derived using expert estimates.

STEP 6: PERFORMANCE INFLUENCING FACTORS ANALYSIS

Because of the fact, that human behavior and therefore his propensity to errors are significantly influenced by external factors, it is necessary incorporate assessment of their impact on human within the HF error analysis. It is particularly the assessment of the work organization impact, differentiation of tasks, human-machine interaction, human-environment interaction and also interaction between human beings themselves, i.e., human-human interaction inside the working system (social factors). In summary, all these factors are called performance influencing factors (PIF). PIF, as it is reported by a number of scientific papers, can be divided into 4 main groups (Working environment, Workplace and performed task characteristics, Organizational and social factors and Worker characteristics).

For characterization of the working system level using the PIF it is necessary to introduce a uniform evaluation system of their significance (Embrey, 2000; Skřehot, 2008a). This means in practice the implementation of their relative valuations. For this purpose a qualitative variable called "critical PIF" was incorporated into the modified PHEA method. It has three levels, which relate to whether HF reliability may be:
- increased –category I (Improve);
- unaffected – category N (Normal);
- reduced – category W (Worse).

Because it is only a qualitative assessment, this fact does not affect the HEP value of considered HF errors. During design of the preventive measures the output of PHEA gives information, whether it is also necessary to count the effect of specific PIF and decide if it is necessary to improve their quality or conversely to maintain the current level

STEP 7: PROPOSAL OF MEASURES TO ERROR REDUCTION

The last phase of PHEA is creating of possible strategies to error reduction or their prevention. Methods to increase the correction/recovery likelihood may be also used (Sandom and Harvey, online). Except potentially critical performed operations in which a person can make a mistake, it is also important to identify the PIF, which may be considered to have the greatest impact on errors occurrence. At this stage, the analyst (or analyst team) requires a brainstorming session. The goal should be to find a mechanism which can effectively prevent the errors or minimize its consequences. Reduction strategy is usually connected to one of the following three areas:
- change the environment, including the man-machine system;
- operators training and increasing their skills;
- improvement of working procedures, work organization, communication, etc.

Design of software "Analysis HTA-PHEA 1.0"

It is clear from the description above that the full analysis performance using HTA-PHEA methods is very time consuming. Therefore we developed a computer program for greater efficiency and simplification the whole analytical process. This program includes both, a section dedicated to the HTA analysis and another section to the PHEA analysis. These sections are interconnected. PHEA section contains an error database (includes 113 kinds of human errors), PIF database (contains 54 factors) and quantitative assessment of specific human error occurrence probability. Understandable interface then allows sufficient clarity of conducted analysis and the user can use the tool intuitively without special training. The software operating is simple and it is done through the dialog interface with windows. The user must first construct a task diagram and gradually add the necessary data to the analysis. First of all it is necessary to perform HTA analysis and subsequently it is possible to proceed to the PHEA analysis by the lowest task forces. This includes error analysis, HEP correction, PIF analysis and proposal of measures to human error reduction (see Figures 1 to 4). The conducted analysis is possible to export to jpg or pdf format, both in form of scheme or table output.

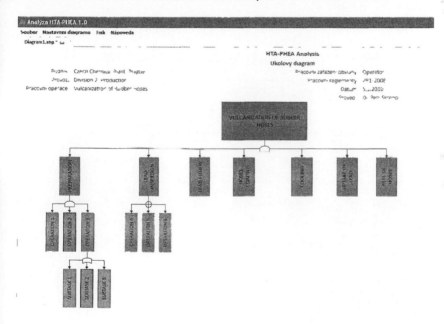

FIGURE 1: Step 1 a 2: Task diagram structure and sub-tasks analysis.

642

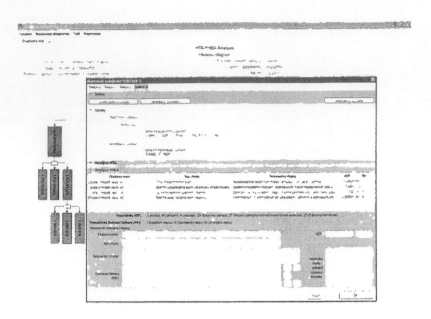

FIGURE 2: Step 3: PHEA - Information analysis about performed sub-tasks.

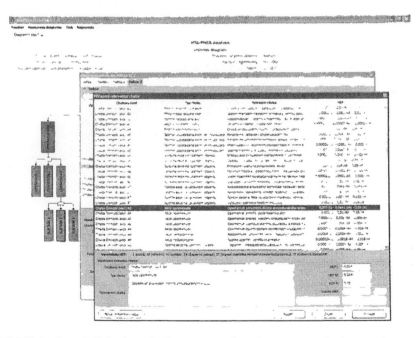

FIGURE 3: Step 4: Selection of relevant errors from human error database and correction of HEP values.

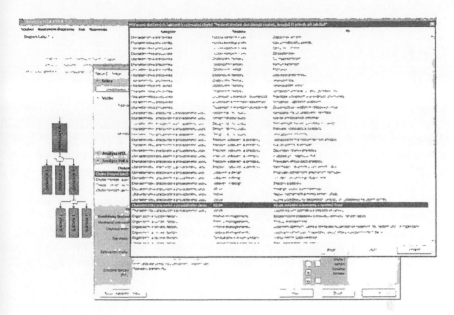

FIGURE 4: Step 5: PHEA - PIF analysis.

CONCLUSION

The HTA and PHEA methods were briefly introduced in the article. These methods have been proved very useful for the HF reliability and mistakes assessment, and especially the integrated HTA-PHEA method, which includes a quantitative assessment of the specific error occurrence probability. It was found that this method is usually considered like a useful tool for HF evaluation by HF experts, because it offers a comprehensive HF reliability analysis. Because the accidents are always caused by the negative influence of external factors, the modified HTA-PHEA method allows their integration into the process of HF reliability analysis., It is possible to identify the most likely errors which a man can commit by using knowledge of the performed task and individual work operations. It is also possible to propose specific measures for the occurrence error reduction. For this approach, which we proved in practice, we have developed the software tool "The HTA-PHEA analysis 1.0", which will help in applying the method into practice.

644

REFERENCES

Embrey, D. (2000). *Performance Influencing Factors (PIFs)* [online]. Human Reliabilit
Associates Ltd., 2000 [cit. 2008-07-30]. WWW
<http://www.humanreliability.com/articles/Introduction%20to%20Performance%20′
nfluencing%20Factors.pdf>.

Grozanovic, M.; Stojilkovic, E. (2006). Framework for human error quantification : UDC
331.468. *Facta Universitatis : Philosophy, Sociology and Psychology*, 2006, Vol. 5
No. 1, pp. 131-144.

Guidelines for preventing human error in process safety (1994). American Institute o
Chemical Engineers, 1994. P. 192. ISBN 0-8169-0461-8.

Harris, D.; Stanton, N. A.; Marshall, A.; Young, M. S., Demagalskij, J.; Salmon, P
(2005). Using SHERPA to predict design-induced error on the flight deck
Aerospace Science and Technology, 2005, Vol. 9, Issue 6, September 2005, pp. 525
532.

Kirwan, B. et al. (1997) The validation of three Human Reliability Quantificatior
techniques - THERP, HEART and JHEDI : Part II : results of validation exercise
Applied Ergonomics, 1997, Vol. 28. No.1, pp. 17-25.

Sandom, C.; Harvey, R. S. (2004). *Human factors for engineers* [online]. IET, 2004 [cit
2009-09-24]. 361 p. ISBN 0863413293. WWW
<http://books.google.cz/books?id=F8JXoln8FYQC&pg=PA162&lpg=PA162&dq=%22Predictive+Hum
n+Error+Analysis%22&source=bl&ots=grBbg5X9IY&sig=-
N1YHSSSV71WtXXt6OtWJBBO4Ak&hl=cs&ei=C067Sv6PI8j4_AbXtb2lDQ&sa=X&oi=book_result&
ct=result&resnum=2#v=onepage&q=%22Predictive%20Human%20Error%20Analysis%22&f=false>.

Sharit, J. (2006). Human Error : chapter 27. In Salvendy, G. Handbook of Human Factor
and Ergonomics. 3rd ed. Hoboken : John Wiley & Sons, 2006. ISBN 0-471-44917-2
P. 708-760.

Skrehot, P. (2008a). Využití faktorů ovlivňujících výkonnost obsluhy při hodnocen
spolehlivosti lidského činitele a kultury bezpečnosti. *Spektrum*, 2008, Vol. 8., No. 1
pp. 41-45. ISSN 1211-6920.

Skrehot, P. (2008b). *Posuzování spolehlivosti člověka v pracovním systému pomocí analý
úkolů.*, 1. vyd. Praha : Výzkumný ústav bezpečnosti práce, 2008. 28 p. (Bezpečn:
podnik). ISBN 978-80-86973-22-7.

Stanton, N. A. (2006). *Error Taxonomies*. In Karwowski, W. International Encyclopedi
of Ergonomics and human Factors. 2nd ed. Boca Raton : CRC Press, 2006. P. 706
709. ISBN 0-415-30430-X.

Stanton, N. A.; Young, M. (1999). *Guide to Methodology in Ergonomics : Designing fo
Human Use.* [s.l.] : [s.n.], 1999. P. 29-39. ISBN 0-7484-0703-0.

Stanton, N. A. *Hierarchical task analysis : developments, applications, and extensions* [online]. Uxbridge : Brunel University, School of Engineering [cit. 2009-09-25]. WWW: <http://www.hfidtc.com/pdf/reports/HTA%20Literature%20Review.pdf>.
Wincek, J. C.; Haight, J. M. (2007). Realistic human error rates for process hazard analyses. *Wiley InterScience*, Vol. 26, Issue 2, June 2007, pp 95-100.

Chapter 67

An Empirical Assessment of Driver Motivation, Emotional Response and Driving Conditions on Driver Risk Decisions

Yu Zhang, David B. Kaber

Edward P. Fitts Department of Industrial and Systems Engineering
North Carolina State University
Raleigh, NC 27695-7906, USA

ABSTRACT

Problem: Limited empirical evidence has been developed for models of motivation in driving. **Approach:** We used a high-fidelity driving simulator. Driver conformance with social norms was manipulated by varying traffic patterns. Time and performance-based payment systems were used to assess the effect of incentive on driver motivation. The complexity of the driving environment was investigated as a mediating factor. **Findings:** Risky behavior, specifically higher driving speed, was observed with the performance- versus time-based payment system. Drivers conformed with social norms associated with specific traffic patterns. Higher roadway complexity interacted with the motivational factor manipulations and produced lower driving speeds. **Conclusion:** Motivational models can be supported by empirical evidence on driver performance and are promising for explaining and understanding driver risk-taking behavior.

Keywords: Driver motivation model, perceived safety, traffic patterns, incentive factors, driving environment complexity

INTRODUCTION

Understanding how drivers adapt to dynamic driving environments is considered to be critical in the development of new technologies to support driver information needs and overall roadway safety (Michon 1985). Driver "behavior adaption" has been extensively investigated in terms of perceptual and motor skills under different road conditions. However, driving, as a self-paced task, is not limited to "skills" (Williams and O'Neill 1974). Drivers determine their task demands to a large extent based on emotional factors and motivation to a goal (Näätänen and Summala 1974). Motivations, such as time goals, conservation of effort, maintenance of speed and progress, pleasure in driving, need to be taken into account in order to accurately model driver behavior. Hence, there has been a recent increase in research devoted to exploring the influence of motivation on driving (Näätänen and Summala 1976).

Two approaches have been taken to develop models of motivation in driver behavior (Summala 2007), including: (1) the motivation measurement approach, which attempts to identify measures to explain driver behavior; and (2) the proxemics approach, which explores driver behavior in space and time dimensions. The motivation measurement approach includes uni-dimensional and multi-dimensional models. Among these, Näätänen and Summala (1976) multi-dimensional threshold model of driver action has been used to interpret risky decisions under various conditions. In their model, driver behavior is modified not only according to changes in the degree of complexity of traffic situations, but it is based on perceptions of risk. Driver goals also play an important role in the model, as they dictate the level of task difficulty. This model also makes three major assumptions: 1) monitoring for present or anticipated risks is the major inhibitory mechanism in driver behavior; 2) drivers' goals and motives push them towards performance limits; and 3) safety margins are key indicators of driver control. The third assumption of the model is related to a key concept of the proxemics approach, i.e., safety margins reflect driver time and spatial judgments.

Motivational factors influence driver behavior by triggering emotions. Emotions can be closely related to states of driving. As suggested by Russell and Barrett (1999), emotion can be broken-down into two major dimensions, pleasure and intensity. Comfort in driving is considered as a pleasant experience without any strong emotions being triggered. Drivers also expect normal daily driving to be a "fluid experience". It has been demonstrated that these motives, comfort and fluidity of experience, play an important role in determining daily driving behavior (Näätänen and Summala 1976). However, if these motives expand based on driver pursuit of strong emotions, such as thrill-seeking, this can easily lead to hazardous behavior, including speeding. Other motivations in driving, include being in a hurry, coping with social pressure (caused by other driver behaviors) and competition, which may cause driver anxiety.

One example of how motivations work in the driving domain was Summala's (2007) observation of social norms. Because rule following is a major motivation for safe driving and avoiding fines, compliance with rules makes drivers feel "comfortable" or "pleasant". Summala (2007) discussed that compliance with rules (e.g., speed limits) is based not only on the law, but also on social norms. For example, drivers tend to maintain closer headways in high density traffic (within 1.7s headway time of a lead vehicle). Many drivers are uncomfortable with large headways in this situation due to the distance being outside the social norm to which they are accustomed (Ohta 1993). In addition, Haglund and Åberg (2000) found that driver behavior is closely linked to the behavior of other drivers; that is, a driver who perceives others to drive at excessive speeds is also more likely to drive faster than a driver who perceives others to comply with limits (also see Ulfarsson, Shankar et al. 2001). However, the above studies all involved observations of real-life driving environment conditions. Thus, the motivations of drivers were not controlled manipulations under specific traffic environment.

In addition to social norms, roadway complexity may also affect driver risk assessments. For instance, it has been documented that visual clutter due to complexity of the environment may affect driver visual attention and other cognitive capabilities, leading to different perceived safe distances to surrounding vehicles (Jin and Kaber 2009; Zhang, Jin et al. 2009). Driving under complex conditions, drivers may perceive the need to be more cautious in order to compensate for demands on cognitive resources. Interestingly, it has been observed that although interstates are designed for smooth flow, which may lead to perceptions of less complexity, lane changes are performed less often and with lower urgency, as compared to local highways (Olsen 2003). This may be explained by driver motivation to conform with social norms. For example, high-traffic density in urban environments may cause feelings of frustration leading to lane changing and reduced headways (Ohta 1993). The visual clutter due to roadway complexity and the need for following social norms may conflict with each other in determining driver perceived safety. There is a need to more fully explore the potential interaction effect of these variables on driver behavior.

In order to extend the current knowledge of driver adaptation in terms of motivation and emotion factors, this study aimed at: (1) providing empirical evidence of the influence of extreme emotions (urgency, anxiety and tension) on driver risk-taking decisions by triggering emotions with incentives, i.e., drivers received more rewards if they completed a goal within a certain time; (2) investigating the effects of social norms on driver behavior by simulating different traffic patterns (other drivers' behavior); and (3) further exploring the effects of environment complexity on driver behavior.

METHOD

APPARATUS

A STISIM Drive™ M400 driving simulator (System Technology Inc., Hawthorne, California.), was used to present subjects with different driving environments and to assess their risk-taking decisions on speed under various motivation and traffic conditions. The simulator modeled a Ford Taurus with a drag coefficient of 0.32 and maximum braking rate of 1.4g. The simulator provided a 135 degree field of view of the driving environment through three 37 in. HDTV monitors. The simulator included a modular steering unit with a full-size steering wheel, turn indicators and a horn, as well as a modular accelerator and brake pedal unit, and audio system.

PARTICIPANTS

Based on pilot study results (mean responses and variations in driver performance under different test conditions), a total of 10 participants were recruited for the study to ensure a test power $(1-\beta)$ greater than 0.80 for each of the fixed experiment manipulations. All participants were required to have a valid driver's license with no restrictions and 20/20 vision or to wear corrective glasses or lenses. The recruitment of participants was balanced for gender. The mean age of the participants was 23.3 yrs. (SD=1.34). The average driving experience was 4 yrs. (SD=1.56).

DRIVING TASKS

Eleven different driving tasks, including three training and eight experiment tasks, were used in the research. Each of the tasks lasted for 12~15mins. Subjects were asked to do their best in exhibiting normal, daily driving behaviors. Although there were speed limit and crossing signs along the road, no simulated enforcement for speeding was imposed on the subjects. Because of the lack of a direct penalty for speeding, it is possible that driver behavior may have been more liberal, including higher than normal speeding rates. However, any such effect was expected to occur across the test conditions. Drivers were also aware that speeding might increase their chance of being in a crash and that they would not receive additional incentives for performance, if they collided with another car.

The first training trial required subjects to follow a lead vehicle on a two-lane road with horizontal curves to ensure they had sufficient skill for maintaining a lane. The second training trial included a two-lane road on which a lead vehicle varied its speed according to a sinusoidal speed profile. This trial was intended to familiarize subjects with distance judgments in the simulation. The driver was asked to maintain a comfortable distance from the lead vehicle. The third training trial was aimed at familiarizing drivers with passing maneuvers. Subjects were asked to follow a car at a comfortable distance and then, after the lead vehicle slowed down, overtake it.

The experiment trials exposed subjects to four traffic patterns (described in next section), with two replications under a city and a country environment. Subjects were required to respond to the traffic patterns and the environments according to

650

their own experiences. They were also required to respond to the incentive factor (described in next section), i.e., act as if they were in a hurry when paid for performance vs. time. Subjects were required to follow a predefined driving route, including three turn maneuvers. The route was the same throughout all experiment tasks for all subjects.

INDEPENDENT VARIABLES

Roadway Complexity: Under the complex condition, participants were presented with an inner-city environment, including large commercial buildings along the road and many pedestrians. There were two lanes on each side (left and right) of the road, divided by double yellow lines (solid or dashed; see Figure 1). There were 12 four-way intersections in the environment that divided the driving route into 13 blocks. Speed limits changed between intersections in all trials following the sequence 40, 20 and 40 mph, according to school zones, road configuration, etc. The speeds limits were used to encourage social norms under different traffic patterns and to influence driver behavior. Traffic lights at the intersections all had the same durations: yellow time was 4~5 sec; red time was 15~45 sec; and green time was equal to red. The simple scenario presented subjects with a country driving environment, which included a few houses, water tanks and barns along the road (see Figure 1). The number of pedestrians was substantially less than in the city environment. Similar to the city environment, there were 12 four-way intersections that divided the driving route into 13 blocks. The roadway characteristics were the same as in the city environment.

FIGURE 1. Image of complex and simple driving environment.

Traffic Pattern: Traffic patterns were used to assess the influence of behavior of other drivers, i.e., social norms and pressure, on subject driver speed decisions. Traffic patterns were simulated by manipulating traffic volume and traffic speed, as shown in Table 1. Four out of nine combinations of traffic speed and volume, representing typical driving conditions, were examined in the current study. The four traffic patterns included "traffic jam", smooth traffic flow at the speed limit ("normal" driving condition), "speeding", and "school zone". During the smooth traffic flow section, slow vehicles (traveling 10 mph slower than the speed limit) occasionally appeared in each lane, forcing the driver to perform overtaking. Each of the four test patterns was presented in two street blocks (each 2500 ft. long) in each test trial. Based on prior observational studies (Ohta, 1993), it was expected that this distance would allow subjects sufficient time to experience social norms according to the traffic patterns and possibly conform.

Incentive Factors: In half the test trials, drivers were motivated through a performance-based payment system. Drivers arriving at a predefined destination on-time and without accident received compensation more than double that received by those arriving late or having suffered any accident. The exact pay rate for drivers was $5/trial. In this way, driver motivation and emotional state (arousal) were manipulated. A regular time-based payment system was offered for the remaining four trials with a rate of $5/hr.

Table 1. Control factors in traffic patterns.

Volume vs. Speed	Slow (v<20mph)	Normal (35~40mph)	High (>45mph)
High (>20 cars in view/min)	Traffic Jam	---	---
Normal (10~12 cars in view/min)	School Zone	Normal	---
Low (5~7 cars in view/min)	---	---	Speeding

Note: The speed limit adopted in this study for "normal" driving was 40 mph.

EXPERIMENTAL DESIGN AND DEPENDENT VARIABLES

A split-plot design was used for the experiment (Montgomery 2006). The whole-plot level of the design contained two factors, including driving environment complexity and states of driver motivation. Driver exposure to each combination of whole-plot factors was repeated twice. In order to allow sufficient time for the incentive factor to influence driver behavior, participants were randomly assigned to start with either the time- or performance based payment system and to complete four trials under that system. They continued with the other payment system for the remaining trials. Each trial presented one of the two levels of complexity, i.e., either rural or city. Environment complexity was balanced within payment system. The order of driver exposure to the levels of complexity was assigned in such a manner that half of the subjects began with the high complexity scenario and the other half with the simple scenarios.

The subplot in the design, a single trial, contained four types of traffic patterns, including the traffic jam, school zone, smooth traffic flow at the speed limit, or speeding. There were 10 blocks in each simulated driving environment. As previously mentioned, each traffic pattern was featured in each of two blocks. The first and last blocks of a trial always involved normal driving. The remaining eight blocks of a trial were randomly assigned to one of the four traffic patterns. Data was only collected from the 2nd block until the 9th block of a trial. For each traffic pattern segment, data collection did not begin until the subject's vehicle was 300ft into the segment. The frequency of occurrence of specific traffic pattern sequences in the design was balanced to address carryover effects from one pattern to another under a particular incentive system and complexity condition combination.

The STISIM Drive™ software logged data for each experiment trial and the logs were used to derive speed measures, including maximum speed, average speed, and percentage of time in violation of the speed limit.

RESULTS

Maximum Speed: A logarithmic transformation was applied to this response. The transformed data conformed with the assumptions of the ANOVA. ANOVA results revealed that payment system ($F(1,67)=15.73$, $p=0.0002$), environment complexity ($F(1,67)=6.92$, $p=0.0106$) and traffic pattern ($F(3,219)=367.7$, $p=<.0001$) all had significant effects on maximum speed. The performance-based payment system produced an increase in maximum speed (42.0 ± 0.3 mph) compared to the time-based payment system (44.7 ± 0.3 mph). In addition to this, drivers maintained higher maximum speed under the rural (42.2 ± 0.3 mph)versus city (44.3 ± 0.3 mph) driving environment. Tukey-Kramer's test revealed that all four traffic patterns were significantly different from each other. The maximum speed occurred when other vehicles were speeding (55.2 ± 1.1 mph), followed by normal driving (47.4 ± 1.11 mph) and the school zone (40.7 ± 1.1 mph). The minimum speed occurred when driving in a traffic jam (30.2 ± 1.1 mph).

Speeding percentage: Non-parametric test results based on the Wittkowski procedure (1988) revealed that traffic pattern ($\chi(3)=386.2$, $p=<.001$) had a significant effect on the percentage of time drivers committed speeding violations. Post-hoc test results for traffic pattern revealed that driving in the school zone produced the highest speeding rate ($82\pm2\%$), followed by zones in which other vehicles were speeding ($64\pm2\%$) and when other vehicles maintained the speed limit ($27\pm2\%$). In the traffic jam, no speeding was observed. It was suspected that speeding violations observed in the school zone might be attributable to carryover speed by drivers from earlier high-speed segments. The traffic jam pattern provided no opportunity for drivers to speed because of the high-traffic density.

In general, the speed limit was much lower in the school zone, as compared to other three traffic patterns (25 mph vs. 40 mph). In order to determine whether drivers drove faster at the beginning of the school zone blocks and committed more speeding violations as a result of higher speeds under other, earlier traffic patterns, the two blocks of a school zone were divided into four segments, including Block 1-Early, Block 1-Late, Block 2-Early, and Bock 2-Late (1100 ft/segment). The percentage of speeding across these segments in the school zones was then analyzed by using the Wittkowski procedure (1988). The non-parametric test revealed that the four segments had no significant difference in speeding violation rates ($\chi(3)=1.9302$, $p=0.1587$). This finding indicates that speeding in the school zones was more likely due to the low limit and lower traffic density allowing drivers the opportunity to speed.

DISCUSSION

Motivation Factor (Payment System): The results of the experiment suggested that the performance-based payment system motivated drivers to perceive urgency and to hurry in many situations. Drivers commented that they felt tension in test trials (cf., Russell and Barrett 1999). They increased their maximum speed (41.99±1.04mph vs. 44.73±1.04mph) and exhibited more risky driving behavior under the performance payment system. Therefore, driving incentive or motive is a critical factor to include in any motivational driving model.

Complexity: Drivers produced higher maximum speed under the rural scenario compared to city (42.41±1.04 mph vs. 44.31±1.04mph). It may be possible that visual clutter, due to the increased environment complexity, led to a more conservative choice of speed.

Traffic Pattern: The current study findings were consistent with previous research. Driver vehicle speed and speed deviations were affected by the speed of vehicles in adjacent lanes and lane speed deviations (Ulfarsson, Shankar et al. 2001). The largest value for maximum speed appeared when other vehicles were speeding, followed by driving in traffic at the speed limit, in a school zone and in traffic jam. Contrary to expectation, drivers speed more in the school zone compared to the other three traffic patterns. As previously mentioned, this may have been due to the low speed limit and normal traffic volume in the zone leading to more speed violations. However, the speed in the school zone was much lower compared to driving in traffic either at the speed limit or above the speed limit. This suggests that although drivers had a tendency for frequent speed violations, they still behaved more conservatively when driving in the school zone.

CONCLUSION

The contributions of the present research include:1) empirical justification for factors in Näätänen and Summala's model of driver action (1976) - variables that were statistically significant in determining driver safety decisions were identified; 2) identification of additional variables, specifically environment complexity, that mediate the influence of emotions on driver behavior; and 3) demonstration of an effective experimental methodology for manipulating driver emotional states. There was strong empirical evidence of the effects of social norms on driver behavior, induced through different traffic patterns (other drivers' behavior), as well as levels of roadway complexity. This was tested in terms of speed performance measures. Drivers appear to be influenced by surrounding vehicles, e.g. they maintain high speeds in speeding segments, and they exhibit caution in speed control based on visual clutter in driving environments. With respect to assessing the influence of extreme emotions (urgency, hurry and tension) on driver risk-taking decisions, emotions were triggered in this study through motivational factors. The performance payment system caused drivers to hurry and they tended to take more risks by increasing their speed.

In general, any new motivation model based on Näätänen and Summala model should incorporate the factors of: (1) social norms; (2) task difficulty; and (3) motivational/emotional factors.

Applications: In addition to these theoretical contributions, the current study indicates that new in-vehicle technologies should be developed to provide adaptable warnings or assistant drivers based on emotional states. Changes in driver risk-taking decisions occur, in part, due to emotion state changes. Driver emotions can be assessed by various real-time monitoring systems, such as heart rate monitors (for valence (happiness) level), and galvanic skin response (for arousal). Such monitoring systems could provide real-time measures to predict speed changes or other risk-taking behaviors associated with extreme emotions. In this way, in-vehicle warning or assistant systems could be programmed to make corresponding modifications, such as increased warning intensity or sensitivity when sensors on drivers indicate tension, etc.

Limitations: The main limitation of this study is that only one motivational factor was assessed, i.e., an incentive system for creating driver tension. It is possible that other emotions may also play significant roles in driver behavior, e.g., excitement, depression. In addition to this, although the current study successfully manipulated a motivational factor, it was limited in terms of quantifying the exact extent to which driver perceived safety criteria changed relative to specific levels of the factor. Additional research is needed to assess a finer-grain manipulation of driver incentive and its impact on speed control or other measures. Furthermore, the driver population investigated in current research was limited (the mean age of the participants was 23.3 yrs). It is possible that middle-age drivers and elderly drivers may have different driving behaviors due perceptions of different social norms and different sensitivities to motivational factors. Future studies should consider using an even larger sample of drivers in repeated testing.

It should also be noted that, due to limitations in the fidelity of the driving simulator, drivers were aware that their safety would not be compromised in the experiment trials. Hence, they may not have experienced the same level of stress as would occur in real-world driving under similar circumstances and, consequently, they may have behaved more aggressively. Thus the results observed in this study, should also be validated with real world data under similar conditions.

Future Research: Future studies should be conducted to examine motivational factors that may cause other emotions (e.g. depression and excitement) in modeling driving behavior. In addition, there is a need to investigate the behavior of drivers with different backgrounds in terms of exposure to traffic. Drivers who mainly drive in city environments may be accustomed to high-density traffic, and hence, perceive less environment complexity compared to drivers who mainly drive in rural environments, when each type of driver is exposed to the same level of roadway complexity. Thus, there is a need to study individual differences in terms of driver background on perceptions of visual clutter and any corresponding effect on safety decisions.

REFERENCES

Haglund, M. and L. Aberg (2000), "Speed choice in relation to speed limit and influences from other drivers." *Transportation Research Part F*, 3(1), 39-51.

Jin, S. and D. B. Kaber (2009), "The role of driver cognitive abilities and distractions in situation awareness and performance under hazard conditions", proceedings of the IEA 2009 17th World Congress on Ergonomics, Beijing,China.

Michon, J. A. (1985), "A critical view of driver behavior models: What do we know, what should we do." *Human behavior and traffic safety*, 485-520.

Montgomery, D. C. (2006), *Design and analysis of experiments*, John Wiley & Sons, New Jersey.

Näätänen, R. and H. Summala (1974), *Road-user behavior and traffic accidents*, North Holland/American Elsevier, Emsterdam/New York.

Näätänen, R. and H. Summala (1976), "A model for the role fo motivational factors in driver's decision-making." *Accident Analysis and Prevention*, 6, 243-261.

Näätänen, R. and H. Summala (1976). Road-user behavior and traffic accidents, North-Holland.

Ohta, H. (1993), "Individual Differences in Driving Distance Headway." *Vision in Vehicles IV*, 91-100.

Olsen, E. C. B. (2003), *Modeling slow lead vehicle lane changing*. Industrial and Systems Engineering. Blacksburg, Virginia, Virginia Polytechnic Institute and State University. Ph.D.

Russell, J. A. and L. F. Barrett (1999), "Core affect, prototypical emotional episodes, and other things called emotion: Dissecting the elephant." *Journal of Personality and Social Psychology,* 76(5), 805-819.

Slovic, P., M. L. Finucane, et al. (2004). "Risk as Analysis and Risk as Feelings: Some Thoughts about Affect, Reason, Risk, and Rationality." *Risk analysis*, 24(2), 311-322.

Summala, H. (2007), "Towards Understanding Motivational and Emotional Factors in Driver Behaviour: Comfort Through Satisficing", in: Modelling Driver Behaviour in Automoitve Enviornment. C. Cacciabue (ED), pp. 189-207.

Ulfarsson, G. F., V. N. Shankar, et al. (2001), *Travel aid*. Washington State Transportation Center (TRAC), Washington.

Williams, A. F. and B. O'Neill (1974). "On-the-road driving records of licensed race drivers." *Accident Analysis and* Prevention 6: 263-270.

Wittkowski, K. M. (1988). "Friedman-type statistics and consistent multiple comparisons for unbalanced designs with missing data." *Journal of the American Statistical Association*, 83(404), 1163-1170.

Zhang, Y., S. Jin, et al. (2009), "The effects of aging and cognitive stress disposition on driver situation awareness and performance in hazardous conditions", proceedings of the IEA 2009 17th World Congress on Ergonomics, Beijing,China.

Chapter 68

Human Systems Integration Risk Management at the U.S. Department of Homeland Security

Darren P. Wilson[1], Thomas B. Malone[2],
Larry Avery[3], Janae Lockett-Reynolds[1]

[1]U.S. Department of Homeland Security
Washington, DC, USA

[2]Carlow International Incorporated
Potomac Falls, VA, USA

[3]BMT Designers and Planners
Arlington, VA, USA

ABSTRACT

This paper describes the process developed within the U.S. Department of Homeland Security (DHS) for mitigating human systems integration (HSI) risks in homeland security technology and systems. The significance of the paper is that it describes how DHS is mitigating risks to human performance and safety in emerging technology and systems where the concern is not only for the typical systems users such as the operator and maintainer, but also for the general public (involved and non-involved personnel in the area).

When human systems integration risks are mitigated, the product or system is more usable (intuitive, easy to use), more reliable (reduced potential for human error), safer (reduced hazards to safety and occupational health), and more affordable (e.g., lower ownership costs due to reduced design/redesign costs, human support costs, training costs, human error and accident costs, and maintenance costs).

The application of human systems integration risk management in system acquisition and science and technology development involves identifying, prioritizing and mitigating risks to human performance, including human capability, workload, health and safety and well being. Human systems integration risk management at DHS differs from risk management in other organizations (e.g. DoD) in that, in weapon systems the concern is for primarily for operator, maintainer, and team performance and safety, while in homeland security systems the concern for human performance and safety also includes all other involved personnel in the area (e.g., first responders, personnel associated with aligned systems), and the public (e.g., pedestrians or travelers being screened, bystanders, passers-by etc.) who might be affected by the operation of the technology or system.

In this approach, potential human systems integration risks are identified through a variety of methods including, but not limited to, human performance requirements analysis, lessons learned from legacy systems, observations of human walkthroughs, subjective reports, accident reports, modeling and simulation (M&S) exercises, and test and evaluation (T&E) events. A potential risk is identified as an expected problem for human performance, workload, safety, or occupational health. Risks have three components: (1) the root cause or hazard which creates the potential for the risk, (2) the consequence (or effect) of the occurrence of the potential risk, (3) probability (or likelihood) of occurrence.

Human system integration risks are driven by factors which are either directly causal (i.e., root causes) or which have the potential to increase likelihood or the severity of consequences (i.e., contributory or aggravating factors). These include aspects of system operation both internal (within the human) and external (attributed to system design and/or the operational setting/environment). Internal factors include fatigue, stress, attention/memory lapses, and inadequate capability, such as lack of appropriate training or skills, misunderstanding of the situation, and inadequate fitness for duty (the extent to which the human is fully rested, capable, motivated, vigilant, attentive, healthy, and ready). External factors include faulty design, complex or incomplete procedures, unavailable or confusing information, inadequate implementation and/or installation of the design, excessive task difficulty, tight time constraints for task performance, poor communications, inappropriate policies and procedures, excessive workload, erroneous expectations, inadequate environments, and inappropriate interaction with automation.

Potential human systems integration risks are analyzed in a three sub-step process consisting of first estimating the impact, or the consequences, should the risk occur, then estimating the likelihood that the risk will occur, and finally calculating a risk priority index. The risk priority index is a calculation based on likelihood and severity. The risk index, when informed by technical and economic feasibility, provides an estimate of the "true risk". These risks are tracked in a HSI risk database created for each technology or system development program.

The Human Systems Research and Engineering (HSRE) Program in the DHS S&T Human Factors/Behavioral Sciences Division has been developing processes, tools and metrics for managing risks in emerging systems and technology and to facilitate risk informed decision making. This paper describes the processes, tools and metrics that are being used today, and that are inherent in this HSI risk management process and practice developed for DHS technology and systems. The paper provides a description of how the risk management process is being applied to specific classes of DHS technology and systems, and the results realized so far in this application.

Keywords: HSI, Risk Management, Homeland Security

INTRODUCTION

Mr. R.D. Jamison, former Under Secretary, National Protection and Programs Directorate Department of Homeland Security, in testimony before Congress (Jamison, 2008) stated that in the context of homeland security, estimating risk includes characterizing three key factors: threats, vulnerabilities, and consequences. Terrorist threats can change rapidly and adapt to new security measures, making the estimation of threat extremely challenging. Vulnerabilities are usually identifiable through subject matter expert judgment and "red team" exercises that probe for weaknesses, but they vary widely for different scenarios or types of threats. The direct consequences of an attack are fairly straightforward to calculate, but it is very difficult to quantify indirect consequences, potential cascading effects, and the impact on the public psyche. In addition, integrating terrorism risk assessments with other hazard risk assessments, such as natural disasters, is difficult. For these reasons, and many others, risk management in homeland security remains a complex and arduous undertaking.

One type of risk that is more manageable within DHS is human systems integration risk. A major step in ensuring a given system or product design maximizes human performance and therefore over all system performance, is to minimize human systems integration risk so a given system or product design is usable, reliable, safe and affordable),and eliminates risks to end-users that will adversely impact human performance. A human systems integration risk is the potential for an unwanted outcome resulting from an incident, event, or occurrence based on difficulties achieving human performance objectives within defined cost, schedule, and performance constraints. Human performance objectives include required levels of human capability, proficiency, workload, health and safety. A human systems integration risk analysis can be performed to identify, assess, and mitigate human factors risks that can be expected to adversely affect human performance, in all of its dimensions.

At the U.S, Department of Homeland Security, human factors is encompassed in human systems integration (Wilson et al, 2009). HSI in system acquisition and science and technology development is concerned with identifying, prioritizing and mitigating risks to human performance. HSI risks include safety and health hazards, situations where the human cannot reliably perform tasks as required, excessive workload, and environmental conditions that degrade performance capability of health and safety. The integration of HSI risks extends beyond the tracking of HSI risks or risks with HSI implications in the program's risk database, but also addresses planning of HSI mitigation efforts. This planning includes the cooperation of other program elements, and application of HSI effort to assist in the mitigation of non-HSI risks which nevertheless have HSI implications.

HSI RISK MANAGEMENT PROCESS

This process for identifying and mitigating DHS HSI risks is illustrated in Figure 1. This HSI risk management process is based in part on a methodology developed by the DoD's Defense Safety Oversight Council (DSOC) to facilitate the identification, analysis, and mitigation of design induced human injury risk, designated the Human Engineering and Ergonomics Risk Analysis Process (HEERAP) (Avery et al, 2008, 2009). The major steps in this process include risk analysis, risk prioritization, and risk mitigation.

HSI RISK ANALYSIS

HSI risk analysis embraces the initial three steps of Figure 1, identifying and characterizing the risk and defining the root cause. Potential HSI risks are identified through a variety of methods including, but not limited to, human performance requirements analysis (HPRA), IISI lessons learned, observations of end user (operator, maintainer and public) walkthroughs of task sequences, end user subjective reports, accident or incident reports, casualty reports, modeling and simulation (M&S) exercises, or test and evaluation (T&E). A potential HSI risk is defined by three components: (1) the root cause or hazard, (2) the consequence of the hazard, should it occur, and (3) likelihood of the hazard occurring. Hazard, per the DHS Risk Lexicon (2008), is defined as a natural or man-made source or cause of harm or difficulty.

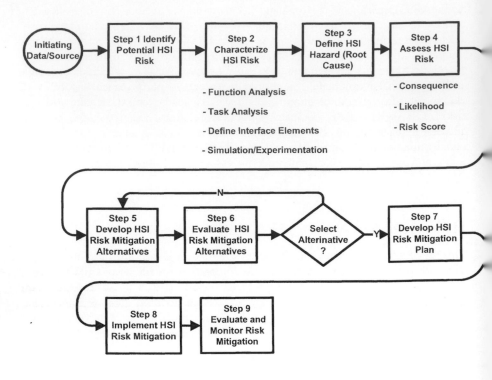

FIGURE 1. The HSI process for identifying and mitigating risks

Risk Characterization

To fully develop an understanding of HSI hazards and analyze HSI risks, the HSI risk needs to be characterized by how the human interacts with the technology or system and what factors may impinge on human performance. This characterization focuses on defining how functions have been allocated between humans and machines or automation, describing the roles of humans and machines in the performance of functions, developing an analysis of human task performance requirements, and, where possible, performing simulation/modeling or experimentation to further understand the HSI risk and the conditions under which it exists. To the extent possible, information already available from the program should be used, such as the results of function analysis, task analysis, or lessons learned from past experience. Where the program is part of a DHS Science and Technology (S&T) project, information that is available to use to characterize the risk may be limited, as programmatic documentation may not yet have been created. Therefore, the HSI risk should be characterized based on past experience, discussions with other subject matter experts (SMEs), or other standard HSI tools. The tools listed below are typically used to help characterize HSI risks.

Function Analysis

Function analysis focuses on identifying functions for selected mission scenarios and allocating functions between human-only, human with machine aiding, or machine-only, though in many cases, machine-only has a human in a supervisory role. The function allocation is concerned with defining the roles of humans and machines in the performance of functions. The machine component includes mechanical, electrical, and computer-based technology including automation. The function analysis and allocation, in most cases, should have already been done as part of the front end analysis for the technology or system.

Task Analysis

Task analysis focuses on characterizing the tasks the human will perform for each function. This characterization includes identifying the task, steps or activities in the task (sub-tasks), information required to perform the task, actions required by the human to perform the task, communication links to other humans as well as machines, performance requirements such as time and accuracy, and the user interface elements such as the operational environment (e.g., noise, lighting, time of day, operational tempo, etc.). As with the function analysis, the task analysis should have already been performed as a part of the front-end analysis and should be made available to the HSI Specialist.

Modeling/Simulation and Experimentation

The identified risks can be explored using modeling/simulation or some form of experimentation. This will allow for an empirical validation of the risk as well as provide a better understanding of the elements of the risk. Being able to perform either modeling/simulation or experimentation assumes that there is appropriate data and/or an actual or virtual prototype representation of the technology user interfaces available to use which may depend on the level of development and/or the stage of the development life cycle.

Define HSI Hazard

HSI hazards (root causes) are factors that directly cause or significantly contribute to the occurrence of an HSI risk. These include aspects of the system application both internal (within the human) and external (attributed to system design and/or the operational setting). In some cases, it may seem that the hazard is the same thing as the risk. In reality, however, the hazard differs from the risk. The hazard represents the causal or contributing factors for the risk, whereas the risk represents to potential negative outcome. "For example, poor user interface design may seem like the risk but in reality the risk is design induced human error due to the hazard of the poor design.

HSI RISK PRIORITIZATION

HSI risks are analyzed in a three sub-step process consisting of first estimating the consequence of the hazard if it occurs, second, estimating the likelihood that the risk will occur, and finally calculating a risk priority score. The estimates of likelihood and severity of consequences should be made using existing information from lessons learned, experimentation, or other data as well as with input from subject matter experts (SMEs). Using the consequence and likelihood ratings, HSI risks are classified and assigned a risk score using a table as illustrated in Table 1. This classification and score establishes the overall risk for an identified HSI issue or concern. This score is then used to rank different risks, judge the acceptability of a risk, and establish risk mitigation priorities. Once the risk scores have been established, the priority should be on reducing the highest risks with respect to human performance, injury, costs and program risk.

Table 1. Human System Integration Risk Index

	Impact		
Probability of Occurrence	**Low**	**Medium**	**High**
High	M	H	H
Medium	L	M	H
Low	L	L	M

HSI RISK MITIGATION

Risk mitigation addresses activities to be implemented to reduce the likelihood of occurrence and/or the consequences of the risks. Activities directed at reducing the level of risk involve analysis (including modeling, simulation and experimentation) addressing applications of technology, redesign, or other actions to reduce likelihood of occurrence and/or impact of the consequences, and experimentation to assess the impact of proposed changes.

Risk mitigation strategies focus on identifying mechanisms for reducing the risk to a minimal acceptable level. Risk mitigation typically falls into the following four broad categories:

- Avoiding risk by removing exposure to the hazard and/or the consequence.

- Controlling the hazard or consequence to reduce the potential for reduced human performance or safety.

- Transferring the risk by shifting some or all of it to another entity, asset, system, network, or geographic area.

- Accepting the level of risk and continuing on the current program plan.

To develop risk mitigation strategies, the HSI Specialist should work with system/technology SMEs to develop a thorough understanding of the risk (performance, injury, cost, and programmatic), the root cause, and the system design. All potential alternatives for reducing the risk should be identified and listed. These potential alternatives should range from engineering solutions that require redesign of hardware, software and workspace; redesign of tasks; use of procedures and job aides to reduce risk; and training. The preferred mitigation strategy will always be to design the risk out, but when that is not feasible the likelihood of the risk should be reduced.

The most acceptable mitigation alternative, in terms of cost traded off against risk reduction, should be selected for each HSI risk. As each mitigation strategy is developed, metrics that can be used to measure how well the mitigation strategy has reduced the risk should be defined. In some cases, the metrics may be applied as soon as the mitigation strategy is implemented. In other cases, the metrics require a more longitudinal data collection effort.

Risks are tracked to determine the extent to which mitigation efforts are successful, and that the level of risk is being reduced as scheduled. When a risk mitigation approach has been selected and implemented in accordance with the risk mitigation plan, an HSI evaluation must be conducted to verify that the risk has been mitigated. This evaluation could take the form of M&S, experimentation, or conduct of T&E exercises. This HSI evaluation will rely on reliable and valid HSI risk mitigation metrics. After evaluating that the risk has been mitigated to the appropriate level, any residual risk must be accepted by the appropriate approval authority. After the risk has been mitigated, it should still be tracked throughout system operation to ensure that residual risk does not become greater during actual operational conditions than has been anticipated during its evaluation.

EXAMPLES OF HSI RISK MITIGATION AT DHS

HSI risk analysis is new to DHS and is currently being piloted on multiple programs in 2010 and baseline metrics are being collected. Some emerging examples of where HSI risk mitigation process should be applied in DHS are described below.

AIRCRAFT PASSENGER & BAGGAGE SCREENING

The Risk

One of the more challenging areas for DHS risk management more is the area of aircraft passenger screening. Here the risk is the danger caused to an aircraft, its passengers, and persons on the ground resulting from one or more adversaries smuggling on-board an aircraft an explosive device or a weapon to be used in achieving control of the aircraft when aloft. The HSI risk is that a passenger screener will fail to detect that a passenger is carrying, on his or her person or in carry-on baggage, an explosive device or weapon.

Mitigating the Risk

Inherent in the previously identified risk, are the errors that result as screeners attempt to detect and classify information. Human sensory perception and cognition are fundamental to the ability to detect the presence/absence of targets in the surrounding environment, and to accurately classify and respond to those targets. More often than not, decisions take place in the presence of time pressure and uncertainty as in the case of the passenger screening environment. There are numerous theories that provide insight into these human behavioral characteristics and should be considered as mitigation strategies are developed for the passenger screening risk.

Attempts to mitigate the risk are described by Drury (2008) who states that the three aspects of aviation security inspection performance where humans have a large impact are: missed threats (failure to stop a threat), false alarms (stopping a person that is not a threat) and excessive time taken to process each passenger. All translate into two system performance measures: risk and delay. Human factors engineering (a component of HSI) applied to aviation security inspection has addressed each of these aspects and system performance measures. A good example is the Threat Image Projection (TIP) system which presents images of guns, knives and IEDs to screeners performing an X-ray screening task. This counteracts the known human tendency to detect fewer threats when there is a low probability that any single item contains a threat. TIP has the added benefit of providing embedded training and performance measurement for screeners. Projecting an artificial threat image during a screener's actual shift acts as a motivator, as well as reduces monotony, but it must be technically well executed to prevent non-threat-related artifacts from cuing the screener that a TIP image is being displayed or about to be displayed. The DHS S&T HFD/HSRE Program will continue to investigate those processes fundamental to human information processing (e.g., perception, attention, memory) to determine how best to support the human-machine interaction, thus reducing errors in perception and threat classification by providing design solutions that account for this particular human limitations and capabilities.

FIRST RESPONDER COMMUNICATIONS

The Risk

The DHS National Emergency Communications Plan (2008) identifies the risk of first responders being unable to communicate, and describes goals and objectives to enhance the operability of emergency communications systems. The plan, however, does not address the challenges to human performance in these systems, specifically the requirements for usability of user interfaces, and reliability (reduced potential for human error) in system design. A recent GAO report (2009) noted that interoperability, the ability to communicate across different organizations and jurisdictions as needed and authorized, is still a major need and that first responders had identified technological and human factors issues as the primary deficiencies that limit interoperability. The major HSI concern in emergency communications as well as situation awareness of first responders is human error. As an example, in December, 2008 the *Philadelphia Inquirer* reported that a series of human errors caused a 40-minute failure in the city's police radio system. In addition, the 9/11 Commission Report indicated that a transceiver hadn't been turned on properly, impeding radio transmissions, an error that should have been avoided and whose likelihood could have been reduced with the appropriate HSI analysis, design and test activities.

Mitigating the Risk

Too often the recommendation to combat high human error rates is more training, rather that addressing the problem at its core: the design of user interfaces to improve usability and reliability and reduce the likelihood of human error. The DHS S&T HFD HSRE Program has developed standard processes for enhancing technology design for usability and reliability (reducing the incidence and impact of human error), and reducing human performance risks in complex systems. In addressing human error reduction in emergency communications as well as first responder situation awareness, the approach is to develop a set of scenarios that are challenging to first responder performance, and define requirements for communications, collaboration and maintaining situation awareness. The requirements process will be supported with assessments of human performance issues in legacy systems and structured data collection interviews with first responders involved in these systems. Reviews will identify lessons learned and human performance high drivers in the legacy applications. Products will be user interface design criteria and evaluation criteria to assess the usability and reliability aspects of proposed technology or in the design of new technology.

PORT AND HARBOR SECURITY

The Risk

The GAO (2006) reported that ports play an important role in the nation's economy and security. Ports are used to import and export cargo worth hundreds of billions of dollars generating jobs, both directly and indirectly, for Americans and our trading partners. Ports are also important to national security by hosting naval bases and vessels and facilitating the movement of military equipment and supplying troops deployed overseas. Since the terrorist attacks of September 11, 2001, the nation's seaports have been increasingly viewed as potential targets for future terrorist attacks. Ports are vulnerable because they are sprawling, interwoven with complex transportation networks, close to crowded metropolitan areas, and easily accessible. Ports contain a number of specific facilities that could be targeted by terrorists, including military vessels and bases, cruise ships, passenger ferries, terminals, dams and locks, factories, office buildings, power plants, refineries, sports complexes, and other critical infrastructure (GAO, 2006).

In 2002 the GAO also noted that a terrorist act involving chemical, biological, radiological, or nuclear weapons at a large seaport could result in extensive loss of lives, property, and business, a disruption of the operations of harbors and the transportation infrastructure (bridges, railroads, and highways) within the port limits, extensive environmental damage and a disruption of the free flow of trade

According to Ho et al (2009), HSI in port/harbor security systems is directed at optimizing the performance of security personnel in identifying threats, avoiding the threat, defending assets, establishing and monitoring barriers, repelling an attack and minimizing damage. The goal of HSI in this context is to optimize performance of these personnel by providing them with efficient and effective technology, weapons, information, knowledge, decision aiding, and procedures to defeat the threat.

Mitigating the Risk

There are challenges, however, in applying HSI to port/harbor security system design. Some challenges include: (1) designing human machine interfaces to support surveillance vigilance and situation awareness, (2) reduction of the incidence and impact of human error, (3) reduction of human workload, (4) support of decision making and provision of decision support, (5) designing displays that are intuitive and integrated, and (6) reduction of risks associated with denying access or engaging an adversary. This paper provides a rationale for the importance of considering these challenges when developing system performance requirements.

CONCLUSION

The risk analysis and mitigation process is a key component to mission success. However, it is clear that this not only pertains to system elements such as hardware and software, but to the human component as well. This HSI Risk Management process developed for DHS technology and systems is based on existing practices for the identification and classification of risk factors for the purpose of developing successful mitigation strategies that are both schedule and cost effective. As evidenced by the examples provided, there are ample opportunities within DHS to apply this process with the goal of reducing the risk of human induced error that may lead to lost productivity, efficiency, effectiveness, or even injury and fatalities. Not only does this approach provide a means of supporting the successful design and development of DHS systems and technologies across all security domain areas, it also supports the department's goal of ensuring the safety of the nation's most valuable asset—the human (operator, maintainer, and the general public).

REFERENCES

Avery, L., Malone, T., Parker, C., Ruttenberg, A. (2008). Human Engineering and Ergonomics Risk Analysis Process (HEERAP) Parts 1 and 2. Prepared by BMT Designers & Planners, Inc. and Carlow International for Concurrent Technology Corporation, Washington, DC.

Avery, L., Malone, T., Geiger, M. (2009). Human Engineering and Ergonomics Risk Analysis Process, Enhancing the Ability to Identify and Reduce the Risk of Design Induced Human Injury. Paper presented at the Human Systems Integration Symposium (HSIS) 2009, Annapolis, Maryland.

Department of Homeland Security (2008), National Emergency Communications Plan

Department of Homeland Security, (2008), DHS Risk Lexicon, Risk Steering Committee, Department of Homeland Security, Washington DC.

Drury, C. G. (2008) Testimony to the Subcommittee on Technology and Innovation of the House Committee on Science and Technology, April, 2008.

GAO Report to Congress, (2009) Emergency Communications: Vulnerabilities Remain. June, 2009

GAO Report to Congress, (2006), Maritime Security: Information Sharing Efforts are Improving, GAO-06-933T.

GAO Report to Congress, (2002), Port Security: The Nation faces Formidable Challenges, GAO-02-993T.

Ho, M., Wilson, D.P., Lockett-Reynolds, J., Malone, T.B., "Human Performance Challenges in Port/Harbor Security Systems" NATO Advanced Research Workshop Human Systems Integration to Enhance maritime Domain Awareness for Port/Harbour Security Systems, Opatija, Croatia, December 8-12, 2008.

Jamison, R.D. (2008) Testimony before the Subcommittee on Transportation Security and Infrastructure Protection, Homeland Security Committee, United States House of Representatives by the Under Secretary, National Protection and Programs Directorate, Department of Homeland Security.

Wilson, D.P., Malone, T.B., Lockett-Reynolds, J., and Wilson, E.L., (in press, 2009). A Vision for Human Systems Integration in the U.S. Department of Homeland Security. Proceedings of the Annual Meeting of the Human Factors and Ergonomics Society, San Antonio, Texas, 2009.

Chapter 69

The Impact of Repeated Cognitive Tasks on Driving Performance and Visual Attention

Ying Wang, Bryan Reimer, Bruce Mehler, Jun Zhang, Alea Mehler, Joseph F. Coughlin

MIT AgeLab & New England University Transportation Center
Cambridge, MA 02139-4301, USA

ABSTRACT

This study assesses the degree to which three demand levels of an auditory delayed digit recall task impact visual attention, pupil diameter and simulated driving performance. Changes in horizontal gaze dispersion and reduced lateral variation during the dual task periods indicate a more centralized allocation of visual attention during periods of heightened cognitive load. This pattern was consistent across the first and second presentations of each task. At the highest demand level, pupil diameter increased significantly over other dual task periods and single task driving. Pupil diameter was moderately impacted by task repetition, suggesting some habituation to the novelty of the task. The overall results indicate that pupil diameter was a more sensitive measure of changes in workload with repeated task exposure than visual attention and driving performance measures in the context of the simulation. In an on-road study with the same secondary task, gaze became significantly more centralized as the level of cognitive workload increased. Although this level of discrimination was not replicated here, taken together, the results further highlight the usefulness of gaze dispersion as an indicator of cognitive workload.

Keywords: Driver Distraction, Cognitive Workload, Visual Attention, Pupil Diameter, Physiology, Driving Simulation, Validity.

INTRODUCTION

The impact of distraction on driver safety continues to receive broad attention. Visual demands associated with text messaging and in-vehicle display usage clearly distract a driver's visual attention from the roadway. Less obvious are the changes in attention associated with cognitive demands. In situations of increased cognitive load, a driver's eyes are more likely to remain fixated on the center of the roadway (Reimer, 2009; Harbluk et al., 2007; Victor et al., 2005; Sodhi et al., 2002); however, his or her ability to react to central stimuli may still be impaired (Strayer & Drews, 2004; Lamble et al., 1999). These findings, taken together with data presented by Recaret and Nunes (2003), suggest that cognitive distraction may generally interfere with visual processing and awareness of the operating environment. With added cognitive demand, more general changes in lateral and longitudinal control have also been observed (Reimer, 2009; Engström et al., 2005). Given concerns over cognitive overload, the development of advanced workload managers have been proposed (Reimer, Coughlin & Mehler, 2009) and measures of visual behavior have been suggested as potentially key attributes (Green, 2004).

The impact of cognitive distraction on eye behavior has been characterized in a number of different ways. Identifying fixation patterns, Harbluck et al. (2007) show that drivers reduce their frequency of gaze to peripheral area on roadway, instruments, and mirrors. Using an approach based upon raw gaze processing, Sodhi et al. (2002) and Reimer (2009) show a more centralized allocation of gaze points during heightened periods of cognitive workload. Mehler et al. (2009) suggest that physiological based measurements may be more sensitive measures of task demand than driving performance at certain load levels. Pupil diameter is a physiological signal, that in controlled environments, can also be captured using an eye tracking system. Pupil diameter has been shown to increase with added cognitive workload (Tsai et al., 2007).

Studies in our laboratory have used three increasingly challenging levels of an auditory delayed digit recall task to explore the impact of cognitive distraction (Mehler et al., 2009; Reimer, 2009; Reimer et al., 2009). Results presented in Reimer (2009) show that drivers alter their allocation of visual attention to the roadway prior to the appearance of changes in vehicle performance. In this work, participants were familiarized with the task just prior to the assessment periods and little is known about the extent to which learning or the novelty of the task impacts workload or outcome measures. In situ, secondary tasks are often repetitive and familiar to the operator, e.g. individuals engage in multiple repetitions of a task or group of tasks with common features. In this paper, we report on a driving simulator experiment conducted to assess the impact of repeated exposure to the task on gaze dispersion and driving performance. In addition, pupil diameter was assessed as a potentially more sensitive measure of workload (Tsai et al., 2007; Van Orden et al., 2001). Finally, a comparison with on-road trials (Reimer, 2009) is provided to consider the relative sensitivity and validity of gaze dispersion measures generated during driving simulation studies.

METHODS

PARTICIPANTS

A total of 47 individuals were enrolled in the study. Data from 12 participants was not considered in the analyses due to poor eye tracking quality, equipment or protocol failures. The remaining sample of 35 consisted of 18 males and 17 females with an average age of 24.3 (SD = 3.1) years. Participants were required to be active and experienced drivers. This was defined as driving three or more times a week and having had a valid driver's license for three or more years. Participants reporting an accident in the past year were excluded. To enhance eye tracking signal quality, individuals who wear eye glasses were excluded (contacts acceptable). All participants were compensated a minimum of $50 for participation.

APPARATUS

The study was conducted in the MIT AgeLab driving simulator, "Miss Daisy". This medium fidelity, fixed-base simulator, is constructed of a Volkswagen New Beetle situated in front of a projection screen. A virtual environment is generated based upon the drivers' interactions with the roadway using STISIM Drive™ (System Technologies, Inc., Hawthorne, California). Driving performance data generated by the simulation system was logged at 10 Hz while a faceLAB® eye tracker with software release 4.6 (Seeing Machines, Canberra, Australia) logged data at 60Hz. The simulation environment was designed to limit the impact of contextual changes on driver workload; see Mehler et al. (2009) for details.

An auditory delayed digit recall task (n-back) with three levels of difficulty was presented that increased cognitive workload without directly impeding manual operation of the vehicle (Mehler et al., 2009). At the lowest level of demand (0-back), participants were asked to respond to a series of auditory stimuli (single digits, 0-9) by immediately repeating out loud the last stimulus heard. At the more moderate demand level, participants were required to respond with the next-to-last stimuli (1-back) and, at the most challenging, the second-to-last stimulus (2-back).

PROCEDURE

Participants received a brief introduction to the experimental procedure and asked to review and sign an informed consent. Participants were provided written instructions on how to complete the three levels of the secondary task and asked to follow along as a research associate read them aloud. Training included n+1 practice trails for each of the demand levels. Additional repetitions of the instructions and practice trails were presented if a participant committed more than two errors on the last trial of the 1-back and more than four errors on the final trial of the 2-back.

Subjects were then seated in the driving simulator. Full head models were created using the standard faceLAB® protocol. A review of instructions for the secondary tasks as well as another n+1 practice trials at each difficulty level were played from an audio recording. Instructions on operation of the simulator and 8 minutes of driving for habituation to the simulator followed. After a 5minute rest period, the experimental protocol (figure 1) began. During each dual task block (A and B), the three levels of the secondary task were presented. Each level of the task started with 20 seconds of instructions followed by four 30 second trials. Each trial comprised of 10 stimuli (single digits 0-9) with an inter stimulus spacing of 2.25 seconds. Two minutes of single task driving occurred between the first and second tasks and second and third tasks.

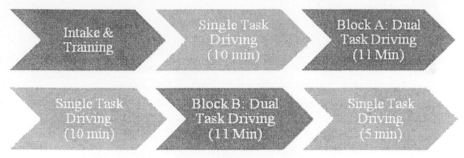

FIGURE 1 Overview of the experimental procedure

DATA ANALYSIS

Output measures were computed over ten 2 minute periods. Six were during dual task driving; the other four were drawn starting 2 minutes and 30 seconds before the first task in the dual task blocks and 30 seconds after the last task. Visual attention measures were computed based on raw gaze positions (see Reimer, 2009 for details). Outliers were identified and converted to missing values based on a box plot analysis and then replaced by the series mean. Statistical comparisons were computed in SPSS® (version 13.0) using a repeated measures general linear model (GLM) procedure with a Greenhouse-Geisser correction. Differences among significant main effects were assessed using pairwise t-tests. Significant results are reported when $p < .05$.

RESULTS

SECONDARY TASK PERFORMANCE

Consistent with Mehler et al. (2009) and Reimer (2009), the number of errors on the secondary task increased with task difficulty (F(1.53, 53.49) = 23.14, p<.001) although overall performance remained quite high. Across both repetitions, participants responded correctly 99.3% (SD = 3.7), 99.0% (SD = 5.2), and 91.7% (SD = 8.8) of the time for the 0, 1, and 2-back respectively. Task performance did not differ significantly across repetitions (F(1,35) = 0.49, p =.49). This lack of an experience effect with the second repletion is likely a result of proficiency developed during the extensive pre-training prior to assessment in the simulation.

VEHICLE CONTROL

Table 1 displays the mean and standard deviation values for the driving performance measures. Velocity did not vary significantly by demand level (F(4, 136) = 1.99, p =.10), but a marginal effect of repeated presentation appears (F(1, 34) = 3.99, p =.05) with velocity across the first set of tasks a modest 1.12 km/h faster than the second. A significant difference in the variation in velocity appears across demand levels, F(4, 136) = 2.71, p<.05, but not repetitions, F(1, 34) = 0.15, p =.70. Post-hoc comparisons show that this effect is limited to variation in velocity during the 2-back task where it was larger than all of the other periods.

Standard deviation of lane position varied significantly by period, F(4, 136) = 9.71, p<.001, but not repetition, F(1, 34) = 0.41, p = .53. Post-hoc comparisons show that, compared to single task driving, variation in lateral position decreased during all of the dual task periods. Decreases in lateral variation with cognitive demand were also observed in Reimer (2009) and Engström et al. (2005). In summary, the added demand of the secondary tasks had a marginal impact on the measures of simulated driving performance considered. Repetition of tasks did not result in a substantive shift in driving performance.

Table 1 Mean and standard deviation of vehicle control measures

Cognitive task period	Mean forward velocity (km/h) (n=35)		Forward velocity variation (km/h) (n=35)		Lateral position variation (cm) (n=35)	
	1st Trial	2nd Trial	1st Trial	2nd Trial	1st Trial	2nd Trial
Pre-baseline	109.3 (7.1)	108.8 (6.8)	2.81 (1.94)	2.96 (1.78)	21.48 (7.74)	20.95 (5.90)
0-back	108.6 (7.3)	107.7 (6.1)	2.90 (2.09)	2.88 (1.83)	19.17 (6.78)	20.26 (7.77)
1-back	108.8 (7.4)	107.0 (6.5)	3.09 (1.67)	2.69 (1.43)	18.26 (6.54)	17.47 (5.19)
2-back	109.0 (7.6)	107.6 (6.1)	3.52 (2.39)	3.51 (1.67)	17.49 (6.79)	18.13 (7.00)
Post-baseline	108.3 (6.7)	107.2 (6.4)	2.96 (1.62)	2.95 (1.73)	21.48 (7.37)	22.64 (7.61)

Note: standard deviations in parenthesis

VISUAL ATTENTION

Data from 29 subjects was available for the assessment of gaze dispersion. (See Reimer (2009) for details on data filtering.) As illustrated in Figure 2, there was a significant effect of demand level on horizontal gaze dispersion, $F(4, 112) = 3.58$, p<.01, but no effect of repetition, $F(1, 28) = 0.22$, p = .64. Overall, gaze was more centralized during the higher cognitive workload of the dual task periods. Post-hoc tests show that all three of the dual task periods trend lower than baseline (p<.1) and differ significantly from the recovery period (p<.05). The baseline and recovery periods are not statistically different (p = .55).

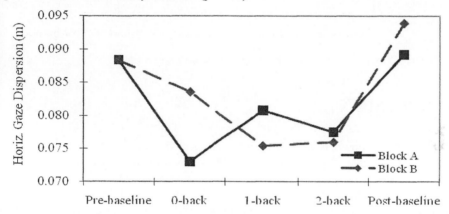

FIGURE 2 Drivers' horizontal gaze dispersion by period and repetition

Pupil Size

Figure 3 illustrates mean pupil diameter by period and repetition. A significant effect of demand level appears on pupil diameter, $F(2.10, 37.76) = 6.23$, p<.01. Post-hoc comparisons show that pupil diameter during the 2-back task was significantly larger than all the other periods (p<.01) and marginally greater during the 1-back as compared to the 0-back (p=.098). Pupil diameter was significantly smaller during the second repetition of the tasks, $F(1, 18) = 7.23$, p<.05.

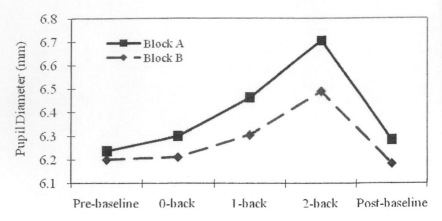

FIGURE 3 Drivers' mean pupil diameter by period and repetition

DISCUSSION

Consistent with previous work (Engström et al. 2005; Harbluk et al., 2007; Mehler et al., 2009; Reimer, 2009; Victor et al., 2005), this study provides further evidence that drivers' vehicle control and visual attention change with increased levels of cognitive workload. Greater longitudinal velocity variation, less lateral movement, narrowed horizontal gaze, and larger pupil size were all observed during the secondary task periods. However, post-hoc analyses show that only the most demanding level, the 2-back task, produced statistically significant changes compared to single task driving on the majority of measures. This suggests that the workload associated with the first 2 levels of the task did not yet have an easily observable impact on basic vehicle control; however, the overall centralization of gaze observed is consistent with a potential impact on visual attention. None of the pairwise comparisons showed differences between pre-task and post-task periods, indicating that drivers showed good recovery from the added cognitive demand periods in the return to single task driving.

As we have suggested elsewhere (Wang et al. 2010), the validity of driving simulation for modeling real-world behavior will vary depending on the task and measures of interest. Compared to two previous field studies using the same n-back task (Reimer, 2009; Reimer et al., 2010), the impact of this secondary task was less pronounced in the simulator and less sensitive at discriminating between the three levels of demand. Under actual driving conditions, Reimer (2009) observed distinct effects of the level of cognitive secondary demand on horizontal gaze dispersion. The effect was replicated and extended across three age groups in Reimer et al. (2010). The difference between the size of the effect seen in simulation and in these on-road studies may be due in part to the actual demand of the primary task of driving being greater on a real road and in actual traffic than in simulation.

One might logically expect that physical characteristics of simulation influence the extent and range of horizontal gaze dispersion. This might be particularly the case in single screen simulators where there are no objective reasons for a participant to scan left or right beyond the boundaries of the projection screen. As a consequence, there would be less overall range in the horizontal scan path and, hence, less sensitivity for detecting a concentration of gaze. Interestingly, Victor et al. (2005) present results comparing field data with results obtained in simulation with 150° and 135° screens, yet report a similar reduction in gaze dispersion in simulation compared to the field. Thus, even in a 150° simulation, with its larger field of view, the overall percentage of attention to peripheral regions appears to be less than in actual driving. In summary, while the overall pattern of gaze constriction during the dual tasks versus single task driving was consistent in our work with the n-back task on-road, the magnitude and discrimination sensitivity was less under simulation. In contrast, in Wang et al. (2010), where the primary visual measure of interest was the percentage of time a participant directed his or her gaze on or off the roadway during interaction with a navigation device, simulation was found to provide a highly accurate model of behavior observed under real driving.

A substantive body of research supports the position that pupil diameter changes with level of attention and effort and that such change can be treated as a useful indicator of changes in mental workload (Recarte and Nunes, 2003, Van Orden, 2001; De Waard, 1996). In agreement with this work, in the first presentation trial of this study, participants' mean pupil diameter increased significantly from 6.2mm pre-task to 6.7mm during the 2-back task (a change of 8.1%). With the second presentation trial, a change from 6.2mm pre-task to 6.5mm during the 2-back task (a 4.8% increase) was observed. Converting the data presented by Recarte and Nunes (2003) and Van Orden (2001) into percentage values, the changes observed in our participants fall in the mid to high range of pupil dilation responses to workload reported by these authors. Compared to the other measures presented here, pupil diameter appears to better represent the incremental changes in demand presented by the 3 levels of the n-back task.

In the study of the impact of cognitive workload on driving, little published research is available detailing the effect of multiple presentations of the same cognitive challenge. This leaves open a question as to the extent to which the patterns of behavior observed may vary with learning or with the reduction of novelty. The second repetition of the series of varied levels of task demand in this study was designed to begin to look at this issue. Performance on the secondary task its self did not vary significantly between the first and second presentations. This indicates that the significant pre-training in the task prior to the active simulator phase of the study was successful in removing a leaning component from this aspect of the task. No overall significant effect of repetition was observed on the standard deviation of velocity, lateral position, or horizontal or vertical gaze dispersion. These findings are generally supportive of the position that some generalizability can be assumed from the pattern seen in the first presentation.

For pupil diameter, the pattern of dilation increase with level of workload was quite similar across the two presentations. However, there was a significant effect on absolute levels. During the second presentation there was essentially no change from the pre-task baseline to the low demand condition of the 0-back. Dilation then increased in an incremental fashion from the 0-back to the 1-back to the 2-back, but the mean dilation value during the second presentation was lower for each corresponding level of the secondary task relative to the first presentation. Since the objective task is the same in both presentations, the basic workload is the same. Two other factors that can effect dilation are novelty and fatigue (Heitmann, 2001; Chi, 1998; Van Orden, 1998). It is possible that there was some fatigue component with the second presentation, but a decrease in the novelty effect on engaging in the secondary task while driving the simulator seems a more likely explanation.

CONCLUSION

These results provide additional insight on the use of driving performance and visual attention measures as a component of cognitive workload assessment. Changes in the standard deviation of velocity and lane position, horizontal gaze centralization, and pupil diameter occurred as expected during the heightened demand of the secondary cognitive task. However, at the levels of demand studied, the sensitivity of the driving performance measures and horizontal gaze concentration for discriminating between the 3 levels of the n-back task was limited. As we have presented elsewhere (Mehler et al., 2009), for the type of demand studied here, physiological measures such as heart rate and skin conductance are more sensitive than driving performance measures for differentiating discrete changes in workload prior to the point at which the driver's available resources begin to become saturated. It seems likely that the reduced sensitivity of horizontal gaze concentration seen here compared to field studies (Reimer, 2009; Reimer et al., 2010) is due to limitations in the simulation environment; it is important to recognize that the sensitivity and relative validity of simulation for modeling real-world behavior will vary depending on the nature of the simulation and the variable of interest (Wang et al., 2010). Perhaps most interesting is the consistency of the overall patterns observed across the first and second presentations of each task for each measure, increasing our confidence in being able to generalize the findings across repetitions.

Acknowledgements

The authors gratefully acknowledge the support of the United States Department of Transportation's Region I New England University Transportation Center at the Massachusetts Institute of Technology and the Santos Family Foundation.

REFERENCES

Chi, C.F., and Lin, F.T. (1998). "A comparison of seven visual fatigue assessment techniques in three data-acquisition VDT tasks." *Human Factors*, 40 (4), 577-

590.

De Waard, D. (1996). *The Measurement of Drivers' Mental Workload*. Master thesis, I ISBN 90-6807-308-7. University of Groningen, Haren, The Netherlands.

Engström, J.A., Johansson, E. and Öslund, J. (2005). "Effects of visual and cognitive load in real and simulated motorway driving." *Transportation Research Part F*, 8 (2), 97-120.

Green, P. (2004). *Driver distraction, telematics design, and workload managers: safety issues and solutions*. Research Report No. 2004-21-0022, University of Michigan Transportation Research Institute, Ann Arbor, MI.

Harbluk, J.L., Noy, Y.I., Trbovich, P.L. and Eizenman, M. (2007). "An on-road assessment of cognitive distraction: Impacts on drivers' visual behavior and braking performance." *Accident Analysis and Prevention*, 39 (2), 372-379.

Heitmann, A., Uttkuhn, R., Aguirre, A., Trutschel, U. and Moore-Ede, M. (2001). "Technologies for the monitoring and prevention of driver." *Proceedings of the First International Driving Symposium on Human Factors in Driver Assessment, Training and Vehicle Design*, Aspen, CO.

Lamble D., Kauranen, T., Laakso, M. and Summala, H. (1999). "Cognitive load and detection thresholds in car following situations: safety implications for using mobile (cellular) telephones while driving." *Accident Analysis and Prevention*, 31 (6), 617-623.

Mehler, B., Reimer, B., Coughlin, J.F. and Dusek, J.A. (2009). "The impact of incremental increases in cognitive workload on physiological arousal and performance in young adult drivers." *Transportation Research Record*, 2138, 6-12.

Recarte, M.A. and Nunes, L.M. (2003). "Mental workload while driving: Effects on visual search, discrimination and decision making." *Journal of Experimental Psychology: Applied*, 9 (2), 119-137.

Reimer, B. (2009). "Cognitive Task Complexity and the Impact on Drivers' Visual Tunneling." *Transportation Research Record*, 2138, 13-19.

Reimer, B., Coughlin, J.F. and Mehler, B. (2009). "Development of a Driver Aware Vehicle for Monitoring, Managing & Motivating Older Operator Behavior." *Proceedings of the ITS-America*, Washington, DC.

Reimer, B., Mehler, B., Coughlin, J, Godfrey, K., and Tan, C. (2009). "An on-road assessment of the impact of cognitive workload on physiological arousal in young adult drivers." *Proceedings of the First International Conference on Automotive User Interfaces and Interactive Vehicular Applications (AutomotiveUI 2009)* (ACM Digital Library), Essen, Germany, 115-118.

Reimer, B., Mehler, B., Wang, Y. and Coughlin, J.F. (2010). "The impact of systematic variation of cognitive demand on drivers' visual attention across multiple age groups." Manuscript under review.

Strayer, D.L. and Drews, F.A. (2004). "Profiles in driver distraction: Effect of cell phone conversation on younger and older drivers." *Human Factors, 46*(4), 640-649.

Sodhi, M. and Reimer, B. (2002). "Glance analysis of driver eye movements to evaluate distraction." *Behavior Research Methods, Instruments, & Computers*, 34 (4), 529-538.

Tsai, Y.F, Viirre, E., Strychacz, C., Chase, B. and Jung, T.P. (2007). "Task performance and eye activity: predicting behavior relating to cognitive workload." *Aviation, Space, and Environmental Medicine*, 78 (5), 176-185.

Van Orden, K.F., Limbert, W. and Makeig S. (2001). "Eye activity correlates of workload during a visuospatial memory task." *Human Factor*, 43 (1), 111-121.

Van Orden, K.F., Jung, T.P., Makeig, S. (1998). "Eye activity correlates of fatigue during a visual tracking task." *Proceedings of the Human Factors and Ergonomics Society 42nd Annual Meeting*, 1122-1126, Santa Monica, CA.

Victor, T.W., Harbluk, J.L. and Engström, J.A. (2005). "Sensitivity of eye-movement measures to in-task difficulty." *Transportation Research Part F*, 8 (2), 167-190.

Wang, Y., Mehler, B., Reimer, B., Lammers, V., D'Ambrosio, L., and Coughlin, J.F. (2010). "The validity of driving simulation for assessing differences between in-vehicle informational interfaces: a comparison with field testing." *Ergonomics*, 53(3), 404-420.

Chapter 70

HRA in China: Model and Data

Dai Licao*[1,2], Zhang Li[2,3], Li Pengcheng[2]

1 School of Info-physics and Geomatics Engineering
Central South University, Changsha 410083, China

2 Human Factor Institute, University of South China
Hengyang , 421001,China

3 Hunan Institute of Technology, Hengyang, 421003, China

ABSTRACT

Human reliability analysis (HRA) is generally viewed as quite an important part in probabilistic safety assessment (PSA). For the past decade, all Chinese operating nuclear power plants have conducted PSA and meanwhile HRA was carried forward. In this paper, THERP+HCR model is presented that describes how the operators' behavior is structured. A major problem of HRA is the lack of particular data which comply with the actual state of operators. Hence, HCR parameters are modified by experiment in the context of a Chinese NPP. In addition, the corresponding data of THERP are extrapolated. A case study is presented to show how the model and data are used.

Keywords: Human reliability analysis, THERP+HCR, modification and extrapolation, case study

INTRODUCTION

Up to now, there are 11 operating nuclear reactors which produce an annual 9078MW electricity power on Mainland China. 20 reactors are under construction and a couple of more are planned. To ensure safety and make the operating units meet the requirement of regulatory bodies of China, all the operating units have finished Probabilistic Safety Assessment (PSA) to assess their operating state.

The relative importance of HRA is increasing while the technical reliability is increasing (Oliver and Heiner 1999). As a part of PSA, qualitative and quantitative criteria of HRA to assure that the models and data are used in a correct and consistent way and thus to provide a realistic representation of the plant safety are material. However, HRA techniques, including models and data, mostly originated in the counties which are very experienced in the operation of nuclear power plants (NPPs). For the past decade, Chinese analysts did HRA in a careful way to establish acceptable models and made efforts to collect data agreeable with the actual conditions in Chinese NPPs.

TRADITIONAL HRA MODELS

To date, there have been several tens of methods to analyze human reliability. It requires a significant amount of experience to be able to select, among a wide range, the model(s) and correlations that are the most appropriate, and that will pass the scrutiny of a peer review process (Lydell, 1992). THERP (Swain and Guttmann, 1983) and HCR (Hannaman etal, 1984) are nowadays the prevailing methods.

TYPICAL ANALYSIS MODEL: TECHNIQUE FOR HUMAN ERROR RATE PREDICTION (THERP) AND HUMAN COGNITIVE RELIABILITY MODEL (HCR)

Technique for Human Error Rate Prediction (THERP)

THERP assumes that operators' response is divided into a sequence of behavior units. Analysis is based on a human event tree (as shown in Figure 1).

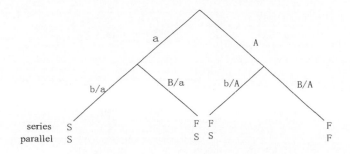

FIGURE 1 HRA event tree for series or parallel system
(adapted from NUREG/CR-1278)

The failure probability of a certain sub-task is expressed by basic human error probability (BHEP). It can be found in relative THERP tables according to the patterns of the sub-tasks. But HEP varies greatly due to different human quality and scenario, so in order to acquire the actual HEP in HRA event tree it must be modified by performance shaping factors (PSF). The modification can be expressed by an equation: HEP=BHEP • (PSF)1(PSF)2...Typical PSFs are radiation and training etc..

Human Cognitive Reliability Model (HCR)

HCR Model (Hannaman etal, 1984) is mainly used in human cognitive processing reliability. Two hypotheses are made in it. The first is that all the behavior types of human actions can be classified into skill type, rule type and knowledge type according to the Rasmussen behavior identification tree (Figure 2). The second is that the probability of every behavior error is only related to the proportion of permitted time to execution time ($t/T_{1/2}$) and complies with Weibull distribution:

$$p = e^{-\left\{\frac{t/T_{1/2}-\gamma}{\alpha}\right\}^{\beta}}$$

$T_{1/2}=T_{1/2}, \, n \times (1+k_1) \times (1+k_2) \times (1+k_3)$

In the equation: t: the allowable response time for the operator
$T_{1/2}$: the execution time of the operator
$T_{1/2},n$: the execution time in general condition
k_1: operator experience
k_2: stress
k_3: man-machine interface
α, β, γ: behavior type parameter of operator
Concerning parameter k_1, k_2, k_3 and α, β, γ, refer to Table 1 and Table 2.

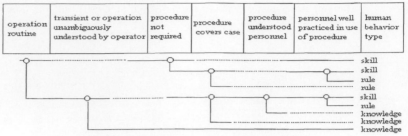

FIGURE 2. HCR behavior type identification tree

Table 1. Behavior shaping factor of HCR model and its value

(Hannaman etal,1984)

Behavior Type	α	β	γ
Skill	0.407	1.2	0.7
Rule	0.601	0.9	0.6
Knowledge	0.791	0.8	0.5

Table 2. Parameters α β γ (Hannaman etal,1984)

Operator experience K_1		Stress level K_2		Quality of operator/plant interface K_3	
expert, well trained	-0.22	situation of grave emergency	0.44	excellent	-0.22
average knowledge training	0.00	situation of potential emergency	0.28	good	0.00
novice, minimum training	0.44	active, no emergency	0.00	fair	0.44
		low activity	0.28	poor	0.78

THERP + HCR HRA TECHNIQUE

THERP and HCR techniques have different focuses in HRA. THERP mainly deals with sequence actions with no consideration of time pressure, and HCR focuses on time-related cognitive behaviors. In a NPP, when an accident happens, the operators need firstly detect the abnormality of the system. And then they are directed by alarms and annunciations to enter into the relative procedures to make diagnosis. Action is subsequently executed. That is, in response to an accident, operators' behaviors are basically made of three parts: detection, diagnosis and operation (Figure 3).

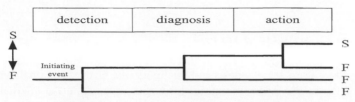

FIGURE 3   Operator's post-accident behavior model

In such event, if THERP is solely used, it may lead to a rough measurement of "diagnosis"; if HCR is solely used, the action errors may not be identified and considered. Therefore THERP and HCR are combined to form a complete chain of post-accident operators' response: during diagnosis period, HCR is used to evaluate its failure probability; for the possible action errors THERP is used to make an evaluation.

The model is developed according to the time progress of the operators' response after an initiating event (Figure 4). T0 is the start point of an accident. Thermo hydraulic time is generally calculated on the basis of the consumption time of water in steam generator (SG).It is assumed that reactor immediately trips when accident happens. Therefore T0 starts when reactor trips. T0 represents two time points, one is the time when accident begins, and another is the time when reactor trip happens. Point E represents the end of operator's response. It is determined by how much time is needed for operators handling the accident. In practices, sometimes it stands outside E0, indicating that in case of such accidents human definitely fail to manage the accident.

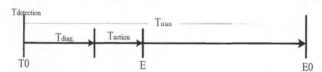

FIGURE 4. Time progress of the operator's response after an initiating event

In PSA, maximum thermo hydraulic available time is calculated to present HRA analysts a time duration t in HCR model. Figure 4 shows that the response time for operators consists of three parts, that is:

$$T_{response} = t_{detect} + t_{diag} + t_{action}$$

As in a main control room (MCR), alarms and annunciations flood when accident happens, operators would immediately detect them, t_{detect} is very short. $t_{detect} \approx 0$ is assumed. t_{action} is time needed for critical actions of operators , which are represented by HRA event tree (see section 3.2). t_{diag} and t_{action} can be obtained through interviews with operators and simulator. More details of the models are illustrated in section 4.

MODIFICATION OF HCR DATA

HCR is used to assess the diagnosis and treatment of a plant's abnormal events. It is closely related to the available time to terminate or mitigate the aftermath of an accident. Development of the Human Cognitive Reliability (HCR) model was funded by the Electric Power Research Institute (EPRI) in around 1983-4 to bridge a gap in the HRA as far as cognitive models of NPP operators in accident situations were concerned. A draft report was issued by EPRI's contractor NUS in 1984 (Hannaman et al.,1984). In order to acquire particular data acceptable in a Chinese NPP, an experimental study was conducted in a Chinese reference NPP on a full-size simulator (Zhang et al., 2007). The experiment involves 23 abnormal events and 55 human interactions covering 3 cognitive types: skill-base, rule-base and knowledge-base. The response time and conditions of 38 operators were recorded and the data were analyzed and processed to develop human cognitive reliability (HCR) model parameters which match with the features of reference NPP and its staff.

HCR model parameters of operators in a Chinese reference NPP (Table 3) and the Weibull Distribution fitting curves (Figure 5) of the response error probability of operators and standardized time are listed below.

Table 3. Weibull Distribution parameters in HCR Model in a Chinese reference NPP

Applied in	α	β	γ	σ
Skill-based	0.87	1.79	0.29	0.45
Rule-based	0.88	1.63	0.30	0.50
Knowledge-based	1.18	0.94	0.20	1.28

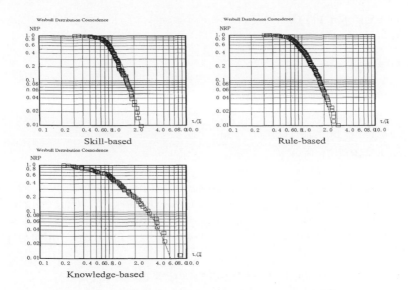

FIGURE 5. Response probability curve of skill-based, rule-based and knowledge-based operators in HCR model in a Chinese reference NPP

HRA EVENT TREE

Typical THERP HRA event tree considers errors of omission (EOM) .In THERP + HCR, only the execution of critical actions is included in THERP HRA tree. The failure mode of the execution of critical actions is erroneous selection. Figure 6 shows it.

DEPENDENCE

THERP emphasizes on the dependence between human failure events. However, the dependence between operators in main control room (MCR) might need more attention due to the organizational arrangement and task allocation in Chinese NPPs in which the main duty of SRO is to supervise ROs. Thus the dependence between SRO and RO is considered. Its calculation is based on THERP table 20-4 and 20-17. As SROs all have excellent education and are well trained, based on the description in Chapter 16 and table 20-6, supervision failure probability is assumed as 3×10^{-3}.

CASE STUDY

EVENT

OBPCC – OPERATORS FAIL TO INITIATE BPCC TO COOL DOWN STEAM GENERATOR WITHIN 60 MINUTES

SCENARIO

End shield pipe breaks, cooling storage could be exhausted within 60 seconds. 5 seconds after the initiating event, reactor stepbacks to 2% of the full power. While the reactor starts stepback, temperature and pressure should be reduced in main heat transport system. Operators should initiate SGPC to cool down within 60 minutes.

THERMO HYDRAULIC AND INTERVIEW PREREQUISITE

1) Thermo hydraulic calculation shows that within 60minutes, operators have to initiate BPCC to cool down steam generator (SG).

2) After reactor trips, operators confirm and memory-check the plant status. Interview with 3 different groups and watching the video record of operators' training on full-size simulator show that operators need an average $t\frac{1}{2}$, n=5 minutes to make diagnosis and make the decision to initiate BPCC.

3) Interview with 3 different groups of operators and watching the video record of operators' training on full-size simulator show that the execution time of initiating BPCC is approximately 2 minutes (ta).

4) Investigation and interview show that the behavior involved in managing the accident strictly follow the procedures. The behavior is rule-based.

5) Conservative assumption is made that operators have average training.

6) Interview with 6 operators in three different shifts show that under such condition stress level is situation of potential emergency.

7) Quality of operator/plant interface is good.

8) Written document shows that in the course of dealing with the event, a reactor operator is in charge of execution while a senior reactor operator supervises.

ANALYSIS AND ASSUMPTION

1) The probability of event failure is approximately the sum of the detection failure, diagnosis failure and failure of execution.

$$P_r \approx p_1 + p_2 + p_3$$

Where:

P_r is the total failure probability of the event;

P_1 is the detection failure;

P_2 is the diagnosis failure;

P_3 is the failure of execution.

2) Alarms and annunciations indicate obviously the happening of accident. Operators would detect it in a great probability. Therefore the detection failure is very small and assumed: $P_1 = 1 \times 10^{-4}$;

3) The diagnosis failure of P_2 is calculated by HCR model;

4) The execution failure of P_3 is computed by THERP on the basis of HRA event tree of the operators' execution steps.

CALCULATION

1) $p_1 = 1 \times 10^{-4}$

2) $p_2 = e^{-\left\{\frac{td/t_{1/2}^{-\gamma}}{\alpha}\right\}^{\beta}}$

Where:

$k_1 = 0$ (average training level)

$k_2 = 0.28$ (situation of potential emergency)

$k_3 = 0$ (good operator/plant interface)

So:

$t_{disg} = T_M - t_{action} \times (1 + k_1)(1 + k_2)(1 + k_3) = 57.44 \, \text{min}$

$t_{1/2} = t_{1/2n} \times (1 + k_1)(1 + k_2)(1 + k_3) = 6.4 \, \text{min}$

The value of α, β, γ is from Table 3, therefore:

$p_2 = 2.24 \times 10^{-5}$

3) HRA event tree is as follows (Figure 6 and Table 4):

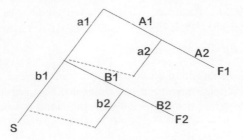

FIGURE 6   HRA event tree for OPBCC

Table 4. Event tree calculation

Failure limbs	HEP	EF	Source (THERP table)
A$_1$ Ro fails to select correctly SGPC mode switch 63614-HS-12	0.0005 ×2	10	20-12(5) 20-16(4)
A$_2$ SRO fails to correct RO	0.003×2 0.056		Assumption in section 3.3 20-16(4) 20-17(10-15)
B$_1$ Ro fails to select the temperature adjuster	0.0005×2	10	20-12(4) 20-16(4)
B$_2$ SRO fails to correct RO	0.003×2 0.056		Assumption in section 3.3 20-16(4) 20-17(10-15)

$$p_3 = F_1 + a_1 F_2 = A_1 A_2 + a_1 (B_1 B_2) \approx 1.12 \times 10^{-4}$$

4)
$$P_r \approx p_1 + p_2 + p_3 \approx 1 \times 10^{-4} + 2.24 \times 10^{-5} + 1.12 \times 10^{-4} \approx 2.34 \times 10^{-4}$$

DISCUSSION

THERP+HCR model is based on the assumption that detection, diagnosis and action are clearly cut off from each other, e.g. operators use a certain period time to detect, to diagnose and act. As in an actual accident situation, operators make diagnosis and simultaneously take plenty of actions, it is hard to judge how much time are respectively devoted to them. However, it appears that this segmentation would result in very conservative estimates of the total failure probabilities of coping successfully with abnormal events. In the absence of data which would permit full consideration of time dependencies, this technique is considered acceptable (Swain, 1987).

Modification of HCR parameters is based on a full-size simulator in a reference plant in China. There are two concerns needing close attention. One is the difference between simulator experiments and real life abnormal events. Another is small size sample acquired in a reference plant could not possibly represent the overall situation of Chinese NPPs and more samples should be studied to obtain more realistic data and more factual representation of post-accident human behavior.

ACKNOWLEDGMENT

The financial support by National Natural Science Foundation of China (No.70873040) is gratefully acknowledged.

REFERENCES

Hannaman, G.W., Spurgin, A.J. and Lukic, Y.D.(1984),*"Human Cognitive Reliability Model for PRA Analysis"*, NUS-4531,1984, NUS Corporation, San Diego, CA

Lydell. B.O.Y.(1992), *"Human reliability methodology, A discussion of the state of the art"*, Reliability Engineering and System Safety, Vol. 36 , pp.15-21

Oliver Strater, Heiner Bubb (1999), *"Assessment of human reliability of plant experience: requirements and implementation"*, Reliability Engineering and System Safety, Vol. 63, pp.199-219

Swain A.D. (1987), *"Accident Sequence Evaluation Program Human Reliability Analysis Procedure"*, NUREG/CR-4772,1987, Sandia National Laboratories, California

Swain A.D., Guttmann H.E. (1983), *"Handbook of Human Reliability with Emphasis on Nuclear Power Plant Applications"*, NUREG/CR-1278, United States Nuclear Regulatory Commission, Washington DC

Zhang Li, He Xuhong, Dai Licao, Huang Xiangrui (2007), *"The Simulator Experimental Study on the Operator Reliability of Qinshan Nuclear Power Plant"*, Reliability Engineering & System Safety, Vol. 92, No.2, pp. 252-259

Analysis of Human Errors in Emergency Situation in a Nuclear Power Plant Digital Control Room

Zhang Li[1,2], Dai Licao[2], Li Pengcheng[1], Cheng Jinghua[1]

[1]Human Factors Institute, University of South China
Hengyang 421001, China

[2]Hunan Institute of Technology, Hengyang 421003, China

[3]School of Info-physics and Geomatics Engineering
Central South University, Changsha 410083, China

ABSTRACT

In nuclear power plants, man-machine interfaces (MMI) are being transformed from conventional panels to computer workstations. Therefore, operators' behavior in modern main control room differs from what they do in a conventional one. This paper makes a comparison between digital instrumentation & control system and analog one in the perspective of system characteristics, man-machine interface and operators themselves. Main cognitive tasks and possible failure modes of operators' monitoring and detection, situation assessment, response planning and response implementation in emergency situation are analyzed. In addition, based on the study on the operators who are receiving a training program on a reference experimental Chinese digital simulator, main influencing factors which would lead to possible human errors are presented.

Kewords: Human errors, digital instrumentation & control system, emergency situation

INTRODUCTION

With the development of computer and control technology, instrumentation and control (I&C) system is becoming modernized from analog to digital control. Man-machine interfaces (MMI) are transformed from conventional panels to computer workstations. Hence, the situation factors affecting human reliability change , including information display, processing control, input and output, decision support, organization structure, communication and working place environment etc.. The cognitive process, behavior and task characteristics of operators change accordingly (Wang, 2006; Liu, 1997; Xu et al, 1997; Liu, 2008; O'Hara and Brown, 2002). In case of emergencies, these new features might become the direct /indirect causes initiating human errors. In this paper, the possible human errors and the main factors initiating them in emergency situation in a digital control room are analyzed to make efforts to prevent and decrease the happenings of human errors.

COMPARISON BETWEEN DIGITAL CONTROL

SYSTEM AND CONVENTIONAL ANALOG SYSTEM

Though conventional analog control system is comparatively reliable and safe, in comparison with modernized digital control system, analog control has its inherent deficiencies, such as functional inadequacy, lower economic competitiveness and difficult technological innovation progress.

Digital control system has following advantages: (1) the automation degree and integrity of control increase significantly. Simple and single parameter and target control is developed into multi-parameter and multi-target one. Adjustment and adaptation can be made under consideration of various disturbances. No setpoint drifting exists. Performances are greatly improved, such as error tolerance, self-detection and automatic adjustment. (2) Soft control, multi-hierarchy distribution and segmented function realization enhance the system reliability. (3) Decreased manual manipulation and interferences, concentrated working place and detailed information assistance make operating more reliable. Operators could fulfill monitoring and manipulation sitting before the central control panel instead of getting around the main control room on analog control system. Table 1 compares analog system with digital control system (Dusic, 1997; O'Hara and Higgins, 2002).

Table 1. Comparison between a conventional analog system and a digital control system

dimensions	system	analog control system	digital control system
Characteristics of system	display	gauges, analog display	digital display, large-screen overview display, much detailed information display, no setpoint drifting
	controller	knobs, buttons, switches, and push-buttons controls et al	computer workstation-based control screen interfaces and pointing devices, interface management tasks
	function	analog circuits, independent of the logical consequence	digital circuits, computer-based procedures, high degree of integration and automation, complex system structure, automatic calibration, various improvement capabilities, such as fault tolerance, self-testing, signal validation and process system diagnostics
characteristics of man-machine interface	types of man-machine interface	authorization	authorization
	access to information	getting around the control room	sitting before control terminates
	response implementation	getting around the control room	sitting before control terminates
	task characteristics	relatively complicated	extra interface management tasks, increased cognitive load
	characteristics of emergency procedures	event oriented, paper-based emergency procedures	state-oriented, computer-based emergency procedures,
The characteristics of team	level of information sharing	partly shared	information is shared
	team's commu-nication and cooperation	less	more

Though advantages are obvious in comparison with conventional analog control system, in the perspective of human factor, the new characteristics of digital system would possibly affecting operators in an unknown manner in emergency situation.

HUMAN ERROR MODE ANALYSIS OF

OPERATORS' BEHAVIOR

In a digital control system, in case of emergencies operators acquire information through plant display system (PDS) and conduct a series of cognitive behaviors to accomplish the post-accident tasks guided by the computer-based procedure (CBP). The main cognitive tasks (O'Hara and Higgins, 2002), are shown in Figure 1. Various human errors would appear in the course of management of the accidents.

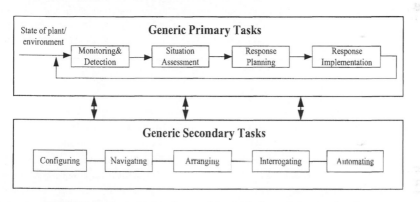

FIGURE 1. Operators' main cognitive tasks in emergency situation

POSSIBLE HUMAN ERRORS INVOLVED IN PRIMARY TASKS

(1) Monitoring and detection
Monitoring and detection are acquiring information from system, including monitoring the parameters and detecting plant abnormal conditions. Since information search is affected not only by the design of the control panel (data-driven search) but also by the perception of likely system failures and candidate goals (model driven search), it is useful to examine a number of information search errors which would lead to cognitive errors. Table 2 shows a taxonomy of information search errors based on the application of a HAZOP- style analysis on the available information, data, and oral instructions (Kontogiannis,

694

1997).

Table 2. Errors in gathering information cues, data, and instructions
(Kontogiannis, 1997)

Information search errors	Description
Missed (not done)	Failed to detect or identify information cues or data from procedures
Delayed (later than)	Cues or data are not identified in time (delayed detection)
Skipped (not accepted)	Cues or data are identified but ignored because they are not seen as relevant
Misread (more/less than)	Error in reading instruments, reading data from procedures, or receiving oral instructions
Mistrusted	Failure to verify unreliable cues or instructions. Mis-calibrated instruments can give rise to errors in trusting unreliable cues
Irrelevant (other than)	Operators gather cues which are not pertinent to the situation
Insufficient (part of)	Cues or data gathered by operators are not sufficient to understand situation or make appropriate decision
Redundant (as well as)	Operators spend a lot of time gathering cues or data in excess of what is required to understand the problem. Redundant cues may give rise to delays in interpretation and decision making

(2) Situation assessment

When operators detect abnormal events in a plant, they would identify and assess the situation to form a reasonable logic explanation on the plant condition. This process is situation assessment. It is covered by two relative models, e.g. situation model and mental model (O'Hara and Higgins, 2002), which are in consistent with Endsley's situation assessment model (Endsley, 1995)which consists of three parts: perception of elements in current situation, comprehension of current situation (situation model) and projection of future status (mental model). The first step to situation assessment is to gather the information in relation to analysis objectives, such as location and meaning of parameters and physical nature (color, size) etc. Then the comprehension of current situation is based on the gathering of basic elements and their identification and correlation analysis etc. to provide a comparatively complete and reasonable description of plant condition. Finally, in the light of the preliminary analysis operator's mental model is formed to project the future plant state. The particular possible human errors in situation assessment are shown in Table 3.

(3) Response planning

When situation assessment is completed, actions and response thereafter should be planned. In emergency situation, in order to get through the target, the options should be evaluated to select an optimized plan (see Table 3).

(4) Response implementation

Response implementation covers behavior errors. Behavior errors occur in four-dimensioned time and space and can be expressed by external error modes. This paper classifies response implementation according to Hollnagel's categorization of "time, type, object and place" (see Table 3) (Hollnagel,1998).

Table 3. Classification of cognitive and action errors

Situation Assessment			Response Planning	Response Implementation
Level 1: Perception of Elements in Current Situation	Level 2: Comprehension of Current Situation	Level 3: Projection of Future Status		
No perception	No interpretation	No anticipation	No special plan	Action at the wrong time
Inadequate Perception	Inadequate interpretation	Inadequate anticipation	Inadequate plan, only achieves part of goal	Action of wrong type
False Perception	False interpretation	False anticipation	False plan	Action at wrong object
Untimely Perception	Untimely interpretation	Untimely anticipation	Untimely plan and plan generates side-effects	Action in wrong place

POSSIBLE HUMAN ERRORS INVOLVED IN SECONDARY TASKS

In digital control system, in order to accomplish primary tasks, operators need to execute secondary tasks, e.g. interface management task. It includes display configuration, navigation, display adjustment, examine and shortcut setting etc.. Human error mode in connection with secondary tasks could be categorized from dimensions of time, space, target and nature of the tasks. More particular categorization could be described by key words such as, none, too early, too latter and incorrect etc.

MAIN INFLUENCING FACTORS OF HUMAN ERROR IN DIGITAL CONTROL SYSTEM

Kim and Jung (Kim and Jung, 2003) made a comparison between two categories of human error influencing factors and developed a full set of influencing factors.

Though there exists plenty of influencing factors, this paper only focuses on the influences of those in digital control system.

INFLUENCES OF DIGITAL CONTROL SYSTEM UPON HUMAN ERRORS

The digital control system is designed on the basis of the operating experiences. But digital control system possesses changes which have seldom been characterized in factual plant control. On the digital control simulator in a reference Chinese plant, tens of operators are receiving relative training program. The program shows that the main influences of digitalization on human errors are as follows:

(1) Uncertain parameter display location
In a digital control system, post-accident actions are based on CBP, e.g. state-oriented procedure (SOP). When under abnormal conditions, operators should acquire information, make diagnosis and manipulate on the screen. In order to get through SOPs reliably and timely, many affiliated displays and pictures are added to assist operators. Sometimes same parameter is shown on different places with the progress of operation. In emergency situation, it would possibly lead to misreading of the parameters.

(2) Information collision and overloading
Alarms and annunciations are displayed in a total different way than in a conventional plant control room. Conventional displays are on different panels and locations. In digital control system, information may collide and operators may be overloaded by the roll, overlapping, hierarchical configuration of information and large amount of pictures.

(3) Keyhole effect
The limited viewing area of PDS and computer display terminates brings about a new issue which is referred to as the "keyhole effect"(Woods, 1990). Operators are required to navigate repeatedly and get focused on a small area of the interface without recognizing the overall status of the plant, just like the view from outside of a door through a keyhole. The keyhole effect interferes with operator situation awareness. In emergency situation, operators may lose part of situation awareness due to view limitation. For example, operators may not confirm the equipment state because of the overcastting picture(s).

INFLUENCES OF TASKS ON HUMAN ERRORS

(1) Influences of interface management task on human errors
In emergency situation, in order to execute primary tasks to alleviate accident, operators should perform interface management task, e.g. the secondary task. Brookhaven National Laboratory made a lot of studies on the influences of

interface management task on human performance and plant safety. Result shows two kinds of undesirable effects exist (O'Hara etal, 2002):(a) interface management task takes some of the memory resource and undermines operators' cognition upon the primary tasks; (b) when in the event of high workload, in order to decrease the operational work of interface management task, operators have a tendency to give up visiting and searching for the important information in regard to primary tasks.

(2) More frequent operations add to cognitive workload and time pressure of operators
In digital control system symptom-oriented procedures (SOP) are used. SOP requires more operation and confirmation. An operational task includes a series of manipulations, such as searching for pictures, opening a certain window, clicks (switch on/off, start, stop etc.), confirming instructions (safety equipment), executing instructions and close current display window etc. These frequent manipulations add to cognitive workload and time pressures of operators and tend to lead to human errors.

(3) Execution of SOP makes it difficult for operators to detect the critical plant abnormalities
When in the execution of SOP, operators are loaded by plenty of tasks. Unless otherwise agreed or instructed by plant organization, operators should execute procedures repeatedly. In event of accidents, plant status change greatly and output great amount information. Operators could not acquire information directly as they do in conventional control system. There exists great possibility that key deficiency alarms and annunciations are not detected. Judgment and manipulation may be delayed. In some cases, operators may ignore equipment failure information.

EFFECTS OF SOP UPON HUMAN ERRORS

Procedures are critical to plant safety. The complexity of CBP increases the risk of overlooking certain steps in SOP or mistaking execution. SOP includes main menu and operation menu. When in the execution of main menu, operation menu is required to perform a particular action. Main menu is separated from worksheet. The times and levels of calling procedures increase.

EFFECTS OF TEAM COMMUNICATION

In emergency situation, team cooperation and decision are needed. Digital control system increases the amount of operators sharing plant information and communicating. It changes the team structure and information exchange stereotype. In digital control system, information sharing and communication are among operators and between operators and oncall personals. In addition, SOP increases information record and information connecting points. All these change the conventional dependence between operators and tend to increase human errors.

DISCUSSION

In digital control system of a nuclear power plant, human errors in emergency situation have become potential factors affecting plant safety. It is very important to study the relationship between human errors and main influencing factors, the important ratings and probabilities of influencing factors and how to quantify them and integrate them into probabilistic safety analysis model. All these need further study.

ACKNOWLEDGMENT

The financial support by National Natural Science Foundation of China (70873040) is gratefully acknowledged.

REFERENCES

Dusic, M. (1997), *"Safety Issues for Digital Upgrades in Operating NPPs"*, IEEE Sixth Annual Human Factors Meeting, Vol. 4, pp. 1-6

Endsley, M.R. (1995), *"Toward a theory of situation awareness in dynamic systems"*, Human Factors, Vol. 37, No. 1, pp. 32-64

Hollnagel, E. (1998), *"Cognitive Reliability and Error Analysis Method"*, Oxford, Elsevier Science Ltd.

Kim, J.W., Jung, W. (2003), *"A taxonomy of performance influencing factors for human reliability analysis of emergency tasks"*, Journal of Loss Prevention in the Process Industries ,Vol. 16, pp. 479-495

Kontogiannis, T. (1997), *"A framework for the analysis of cognitive reliability in complex systems: a recovery centred approach"*, Reliability Engineering and System Safety, Vol. 58, pp. 233-248

Liu, W.R. (1997), *"The development of instrumentation & control system in a nuclear power plant"*, Automatic Instrumentation, Vol. 18, No. 9, pp. 1-5

Liu, S.J. (2008), *"Influences of interface management task upon operators in main control room"*, Nuclear Safety, Vol. 1, No. 2, pp. 162-167

O'Hara, J.M., Brown, W.S.,. Lewis, P. M (2002), *"The effects of interface management tasks on crew performance and safety in complex, computer-based systems: overview and main findings"*, U.S. NRC, NUREG/CR-6690, Vol. I & II

O'Hara, J.M., Higgins, J.C., Stubler, W.F. (2002), *"Computer-based procedure systems: Technical basis and human factors review guidance"*, U.S. NRC, NUREG/CR-6634,Vol.1

Wang, H.J., Ma, J.Q. (2006), *"Digital Instrumentation & Control in a Nuclear Power Plant"*, Automation Exhibition, Vol. 10, pp. 16-18

Woods, D.D. (1990), *"Navigating through large display networks in dynamic control applications"*, Proceedings of the Human Factors Society 34th Annual Meeting, Orlando, Florida, USA

Xu, X.L., Guo, R.J. (1997), *"Digital instrumentation & control system and its application in a nuclear power plant"*, High technology telecommunication, Vol. 1, pp. 59-63

Chapter 72

Measuring the Performance of a Triple-Target Visual Search Task

Ruifeng YU, Zhexin KONG

Department of Industrial Engineering
Tsinghua University
Beijing, China

ABSTRACT

Visual search tasks are widely applied in the fields of industry, military, medicine and daily life. The paper describes the present research results in the field of visual search tasks of multiple targets. An experiment measuring the performance of a triple-target visual search task was conducted here. According to the result of the experiment, it was confirmed that the search strategy is random not systematic. We also find that the search time for the second and third targets are conspicuous shorter than the theoretical ones. The target independence on random visual search tasks of multiple targets is discussed.

Keywords: multiple targets, independence, random visual search, search model

INTRODUCTION

Visual search has long been a research topic in human factors and ergonomics areas. Visual search tasks are applied in various industrial filed as part of industrial inspection tasks, e.g. microchip inspection, pipeline inspection, industrial quality control, aircraft fuselage inspection (Drury 1990), etc. It is also common in military environments and many medical situations.

Until now, there are researches about every aspect of visual search field, e.g., two stage process modeling of visual search (Drury 1975, 1990), optimal stopping time model, visual search strategies from systematic to random (Arani et al. 1984), training and feedback about visual search, etc. In all of these, search performance is one of the most important theme in visual search research, which is often been presented as percentage of faults detected against search time. Krendel and Wodinsky (1960) derived a mathematical model for random search expressing the cumulative probability of detection P(t) for a single-target search task as a function of time t as in Equation (1) below.

$$P(t) = 1 - (1 - Psg)n \qquad \text{or} \qquad P(t) = 1 - e(-mt) \tag{1}$$

where m = -ln (1 - Psg) / T; Psg = the probability for target detection in a single glimpse, T= the sum of fixation time and movement time of the eyes, and n = the number of independent glimpses.

By introducing factor of visual lobe, Morawski et al. (1980) expanded the random search model as follows:

$$F(t) = 1 - e^{-(\frac{P_l a}{AT})t} \tag{2}$$

where A = stimulus field area, a = visual lobe area, Pl = the probability that a target will be detected if it is fixed, and T = average time for one fixation.

When generalizing the random search model for single target search to multiple (Morawski et al.1980, Dury and Hong 2000), one important premised assumption is that the search time of each target are mutually independent which means the search time of former targets would not affect that of latter ones. However, the obtained results of two experiments about double-targets revealed that there is a memory searching effect in the second target search from past fixations in prior searches (Chan and Yu 2009, Chan and Chan 2000). So we conduct the following experiment to validate the random visual search model with triple targets.

EXPERIMENT

SUBJECT

Ten Tsinghua University students participated in this experiment whose ages ranged from 23 to 26. Each of them has the normal near foveal acuity and none of them has any former experience in such kinds of experiments and visual search tasks.

STIMULUS MATERIAL

The experiment used stimulus images of 160×103mm shown on a 15-inch high resolution (1280×1024) display. Both target and background characters were of 2.0×2.0mm. Each image contained three different targets, either three out of the four arrows: $<$, $>$, $\wedge$, $\vee$ (these four targets had been proved to be no significant difference in conspicuity (Chan and Courtney 1993)). And the background were homogeneous characters of regularly spaced X. The targets' type and position were randomly chosen by a random generator which also determined that no target would be placed on edges or close to the centerline separating the upper and down quadrants (after subject detected each target, he was asked to decide the quadrant of the target- upper or down). Figure 1 is one of the examples.

FIGURE1. An example of the stimulus materials used in the experiment

PROCEDURE

Before the formal experiment, the subjects were asked to do the following things: write down their personal information, read the Experiment Instruction to have basic knowledge of the experiment, and do practices until they were familiar with the experiment software. After these preparations, the formal experiment was gone by these following steps: firstly, subject was required to sit 600mm from the display, press the TEST button, put in their name and ID number and fixate on the centre point of the first screen before they start.

704

FIGURE 2. Get-ready screen in the experiment

Then, they press Start button, one of the stimulus materials like figure 3 was shown. The subjects were required to search targets as quickly as possible. Once they found the first target, they clicked the left button of the mouse and determined the target type and location. After this, they went back to the original stimulus material to find the second and third targets just like the first one. After all three targets were found, the stimulus material would be changed to the next one. Each subject was tested 10 randomized stimulus materials.

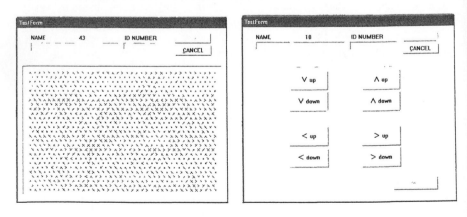

FIGURE3. One cycle of a target detection

RESULTS

DESCRIPTIVE STATISTICS

When in a trial all of the three targets were correctly located, we regard this trial 'correct', otherwise 'incorrect'. 96 correct responses out of 100 (96%) were recorded in this experiment. The statistics of search times per stimulus are shown in the following table 1.

Table 1 Descriptive statistics for search times (seconds) of the three targets

Search Time Statistics	First Target	Second Target	Third Target
Mean	1.62	3.04	4.45
SD	0.78	1.65	2.74
Minimum	0.9	1.7	2.5
Maximum	4.7	5.7	9.0
Median	1.5	2.9	4.1

From the descriptive statistics we can know that the search time of second target is less than twice of the first one, and the search time of third is less than three times of the first one.

RANDOM SEARCH MODEL

The proportion of stimuli searched within a certain time was taken as the cumulative percentage of detection for that time. Figure 4 shows the cumulative percentages of detection $P(t)$ against search times for the three targets. All of the three curves increased exponentially with time which means a good fit of the random search model not the systematic one (in this case the curves would be linear). The regression coefficient square of the first, second and third target line was 0.98, 0.94, 0.97 separately ($p<0.001$).

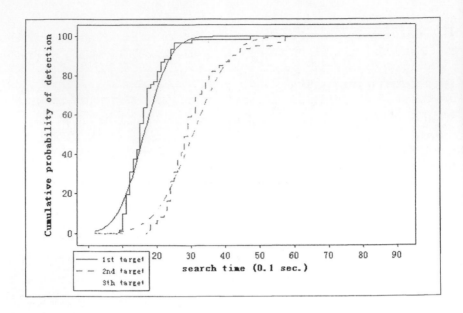

Figure 4. Cumulative percentage detection

The number of m in equation 1 for the three target are 0.208, 0.124 and 0.09 respectively (2.3 :1.4 :1), which is not the theoretical value of 3 :2 :1. So we can guess that due to the memory effect, the detection time of the latter target would be shorter than the theoretical value which assumes the detection time of every target would be independent.

DISCUSSION

According to the result of the experiment, the very high correlation coefficients between the linear regression of ln[1-P(t)] and search time for the three target confirm the search strategy is random not systematic. It is true that not all the subjects would adopted an exact random search strategy, but in our experiment circumstance where number of targets is absolutely small to the background, subject intend to search targets more random than systematic which is well described by the exponential search models.

In our experiment, we also find that the search time for the second and third targets are clearly shorter than the theoretical ones. During search for the second and third targets, subject might avoid the regions which were scanned during first target search time. Also in the experiment, we find 8 readings were shorter than 300 milliseconds (Chan and Courtney 1993 proved typical saccadic movement and fixation duration was 290 milliseconds). It is quite likely that subject had already detected more than 1 target during the first target detection duration.

Compared to equation 2, we can guess that in multiple-target search task, not only the number of target remaining change during the searching, but also the stimulus field area (parameter A in equation 2) change due to the memory effect. When search for the latter targets, the stimulus field area actually become smaller than these of former, so the search time of latter targets would also be shorter than the theoretical ones.

REFERENCES

Arani, T., Karwan, M.H., Drury, C.G. (1984), "A variable-memory model of visual search", *Human Factors*, Vol. 26, pp. 631-639

Chan, A.H.S., Courtney, A.J. (1993), "Inter-relationship between visual lobe dimensions, search times and eye movement parameters for a competition search task", *The Annals of Physiological Anthropology*, Vol. 12, pp. 219-227

Chan, A.H.S., Chan, C.Y. (2000), "Validating the random search model for a double-target search task", *Theoretical Issues in Ergonomic Science,* Vol. 1, No. 2, pp. 157-167

Chan, A.H.S., Yu, R.F. (2009), "Validating the random search model for two targets of different difficulty", *Perceptual and Motor Skills*.

Drury, C.G. (1975), "Inspection of sheet materials: model and data", *Human Factors*, Vol. 17, pp. 257-265

Drury, C.G. (1990), "*Visual search in industrial inspection*", In: Brogan, D. (Ed.), Visual Search. Taylor and Francis, London, pp. 263–276.

Drury, C.G., Hong, S.K. (2000), "Generalizing from single target search to multiple target search", *Theoretical Issues in Ergonomic Science*, Vol. 1, No.4, pp. 303-314

Krendel, E.S., Wodinsky, J. (1960), "Search in an unstructured visual field", *Journal of the Optical Society of America*, Vol. 50, pp. 562-568

Morawski, T.B., Drury, C.G., Karwan, M.H. (1980), "Predicting search performance for multiple targets", *Human Factors*, Vol. 22, pp. 707-719

Chapter 73

Efficacy of Central Alarm System on Nurses' Line of Care in a Surgical Intensive Care Unit in Singapore

Ranieri Yung Ing Koh[1], Taezoon Park[1], Fun Gee Chen[2]

[1]School of Mechanical and Aerospace Engineering
Nanyang Technological University, Singapore

[2]National University Hospital, Singapore

ABSTRACT

Multiple alarms in the ICUs are installed with the main aim of alerting the medical staff of potentially critical situations when the caregiver to patient ratio is high and attention on each individual patient may be compromised. When the number of individual alarms per patient gets too numerous, a central alarm system is put in place to help monitor all patients' status with one alarm system and a single computer monitor. This study aims to identify the possible factors that may affect the efficacy of the central alarm system put in place in the surgical ICU of a local hospital in Singapore, and seeks to evaluate the impact of the system on the line of care of patients by the nurses in the ICU. Preliminary observations and interview data collected during a 2-hour session in the surgical ICU have shown that the physical setup, rate of false alarms and perceived cues are factors affecting the efficacy of the central alarm system. Desensitization of the ICU nurses due to high false alarm rates also suggest that other cues are used to alert the ICU nurses of changes in patient status. Other factors such as the differences in day and night shifts are also identified as possible contributors to the impact of the central alarm system on the line of care of the ICU nurses.

Keywords: central alarm system, intensive care unit, alarms, nurses, patient care

INTRODUCTION

Extensive research has been done on the design of alarms within the context of healthcare, to determine the threshold of signals for alarms or smart alarm systems, and means through which alarms should be designed to best suit human behaviour (Charbonnier and Gentil 2007; Imhoff, Kuhls et al. 2009; Kindsmuller, Haar et al. 2009). Together with the development of new technologies, the number of audio and/or visual alarms in the Intensive Care Units (ICUs) has seen a dramatic increase over the years (Chambrin, Ravaux et al. 1999). In 2003, the Joint Commission on the Accreditation of Healthcare Organizations made clinical alarm safety one of its patient safety goals, further emphasizing the need to pay attention to alarm issues in health and patient care (Edworthy and Hellier 2006).

Multiple alarms in the ICUs are installed with the main aim of alerting the medical staff of abnormal changes in the patients' condition or equipment malfunction. It is especially useful when the caregiver to patient ratio is higher than 1:1 and attention on each individual patient may be compromised. Devices such as patient vital signs monitors and therapeutic devices have their own limit alarm systems that generate visual and acoustic alarms to alert the healthcare staff on duties when a need arises (Charbonnier and Gentil 2007; Imhoff, Kuhls et al. 2009). One therefore assumes, a priori, that alarms are imperative to the aiding of nurses in their line of patient care, and are required to warn the nurses when a patient requires attention or medical assistance.

However, the number of alarms in the ICU wards can total to about 40 for a single patient undergoing mechanical ventilation (Chambrin, Ravaux et al. 1999), and intensive care nurses have a high nursing workload in addition to monitoring alarms. ICU nurses spend a large amount of time moving around and performing patient and non-patient care activities, and must continuously respond to the needs of the patients and their families, routinely interacting with the most intensive emotional aspects of life (Gurses and Carayon 2009). In a study done by Gurses and Carayon (2009), it was reported that too many alarms was one of the factors that attributed to the noisy work environment which was identified as one of the performance obstacles of the intensive care nurses. In a bid to mitigate the additional load caused by the numerous alarms, an alternative alarm implementation by means of a central alarm system has been commonly put in place parallel with the other alarms. The central alarm system makes use of a few general alarm sounds that direct the ICU nurses to a "master caution" panel (a computer monitor or visual display) that indicates the situation being signaled (Edworthy and Hellier 2006). After referring to the master panel, the ICU nurses can then locate the problem location and head to the respective beds.

Many alarms, however, as they now exist in most monitoring systems, are usually perceived as unhelpful by medical staff because of the high rate of false alarms; that is, alarms with no clinical or technical significance (Chambrin 2001). In a study of alarm occurrences done by O'Carroll, (Ocarroll 1986), of 1455 alarm soundings, only 8 represented critical and potentially life-threatening risk to the patient. Another study done by Leung et al (Leung et al 1997) over 50 operations also suggested that alarms only indicated patient risk for 3% of the time that they went off. In fact, multiple studies showed that a large proportion of all threshold alarms in the ICU do not have a real clinical impact on the care of the critically ill (Ocarroll 1986; Lawless 1994; Tsien and Fackler 1997).

False alarms have been classified by Imhoff et al (2009) as either: technically false alarms, clinically false alarms, or technically-clinically false alarms. Technically false alarms are situations where alarms respond to preset thresholds that may not be applicable to the variable thresholds in reality; clinically false alarms are situations where the alarms respond to preset thresholds that may not be relevant clinically; technically-clinically false alarms are situations caused by interventions of the medical staff (Imhoff, Kuhls et al. 2009).

The large majority of false alarms pave the way to a possible desensitization of the ICU nurses towards true alarms (Meredith and Edworthy 1995). In one study, it was observed that an average of 390 alarms were activated a day for a 10-bed ICU, out of which only a quarter were followed by a nurse's or physician's action; 72% of them did not result in any medical action (Chambrin, Ravaux et al. 1999). Moreover, the alarm limits, which are manually set, might be adjusted to dangerously wide thresholds or are even completely disabled to just to reduce the nuisance from false alarms (Koski, Makivirta et al. 1990).

Separate studies have dealt with noise levels in the ICU (Gurses and Carayon 2007; Ramesh, Rao et al. 2009; Siebig, Kuhls et al. 2009), optimal design of alarms to appropriately reflect the patient's status (Charbonnier and Gentil 2007), impacts on the patients, and work performance obstacles of the nurses in the ICU (Gurses and Carayon 2009). However, the efficacy of the central alarm system in relation to the line of care of the patients by the nurses has not been widely researched or justified. It is important to better understand the central alarm system to evaluate its efficacy such that it does not become an add on to the nuisance caused by alarms and their false cries, leading to further mistrust of the ICU nurses in the alarms and ultimately defeating the purpose of the its implementation.

This study therefore aims to examine and justify the impact of the implemented central alarm system in an ICU on the points of decision making in patient care by the ICU nurses, therefore providing substantial evidence to support or deny the importance of a central alarm system in an intensive care unit.

METHODS

This study will make use of naturalistic observations and interviews to carry out the data collection. The study will involve a total of 22 ICU nurses working during either the day or night shift in the surgical ICU of a major hospital in Singapore.

The data collection is broken down into 3 phases.

The first phase involves a 2 hour observation of the work processes in the ICU and a 30 minute interview with one of the senior ICU nurses on duty. This phase serves to help better understand the work flow and responsibilities within the ICU and more importantly, of the ICU nurses. A quick sampling session is also carried out by observing the central alarm display together with the doctor on rounds for a total of 22 minutes and understanding the alarms that go off and the corrective actions that should be taken.

The second phase of the study involves a pilot study which will be carried out before the main data collection. It will include a 2-hour observation session during the day shift, and a 2-hour session during the night shift. The collection of data from 2 different shifts is to elicit data that can yield the comparison between the efficacies of the central alarm system during the different shifts. Staffing during different shifts may vary, and therefore the impact of the central alarm system may likewise change.

During each observation period, at least 2 observers will be located at the central alarm monitor. Both observers will only monitor the alarms that go off for a single bed out of the thirteen beds in the ICU. One observer will record the alarms that sound off on the central alarm system relating to the bed, while the other observer will follow up with the corrective action taken by the ICU nurse following the alarm (if any). A quick and short interview will be carried out with the ICU nurse following the action of care taken to investigate the motivation behind the action and the actual corrective action following the alarm.

From the pilot study, possible problems will be acted on and resolved, and the full study will be carried out thereafter. The full study will include 10 2-hour observation sessions for the day shifts and 10 2-hour observation sessions for the night shift. The full study will also be based on naturalistic observation and subsequent short interviews with the ICU nurses on duty.

The data collected from the 20 sessions will then be discussed with an experienced medical doctor to determine: the nature of the alarms that sounded off (true or false), the urgency of the alarms, the correct meditative action to be taken, and the appropriateness of the corrective action taken by the ICU nurses. A qualitative and quantitative analysis will then be carried out on the data collected to evaluate the efficacy of the central alarm system on the line of care of patients by the ICU nurses.

Institutional Review Board (IRB) and Human Use Committee and Patient Agreement approvals are obtained prior to the second phase of study initiation.

RESULTS

This research is still in progress, with only the first phase of the data collection completed. Results from the observation studies have been documented and used to aid the planning and execution of the experimental procedures. The first phase of data collection has been conducted for the morning shift of the surgical ICU. The surgical ICU has 13 beds in total and 11 of them were occupied by patients.

QUALITATIVE RESULTS

Preliminary observational studies have been carried out to draw a hypothesis regarding the efficacy of the central alarm system.

The preliminary observations and interviews identified the ICU nurses' decision making on the proceeding of patient care independent of the central alarm system. The interview also revealed that the ICU nurses during the day shifts are usually put in charge of 2 patients per ICU nurse. Each ICU bed is located within an enclosed room that is negatively pressured to avoid the spread of infection through air, therefore hindering the propagation of the sound. The beds are built with transparent glass walls, allowing the ICU nurses to keep watch on the patients. The ICU nurses have to toggle their responsibilities between the 2 patients, usually side by side, and have to record the vitals of the patient hourly and sometimes half hourly. The ICU nurses are usually seated in front of the computer stations located just outside the bed-rooms, with one computer station per room. The ICU nurses face both the computer screen and the patient in the room at the same time, keeping track of their patients' status by watching the patient monitor located inside the ICU bed-rooms. Alarms generated by the patient monitors and therapeutic devices cannot be heard from outside; only visual alarms will alert the ICU nurses. As an ICU nurse cannot watch 2 monitors in 2 different rooms at the same time, there is a common practice to put the screen displays of both patients' vitals together on one monitor only.

The central alarm system for the surgical ICU is located in the middle of the ICU. The central alarm system reflects a concise representation of the vitals for all the 13 beds in the ICU. When a certain vital reading goes beyond the thresholds that are set by the ICU nurses, an alarm will sound off. The signal is delivered through both auditory and visual channels. The visual alarm is indicated either by an alert in a red box, or a yellow blinking perimeter of the vitals statistics. The auditory alarms are categorized into 3 types: a low pitch, single beep; a high pitch, single beep; a blinking, double beep.

Alarms generated by the central alarm system are periodically going off every few minutes. However, there is no one attending to the alarms that are going off, and occasionally, the alarms from the central alarm system are treated as annoyance and muted or turned off. The ICU nurses are not located anywhere in the direct visual range of the central monitor. The alarms can also be turned off from within the ICU bed-rooms, after which there will be an alert on the central monitor that reads "alarm paused".

QUANTITATIVE RESULTS

Two short sampling sessions of the central alarm system for 8 minutes and 12 minutes respectively indicated a high number of false alarm rates arising from technical issues that were frequently ignored. Only the frequency of the alarms was recorded. In the total 20 minutes of observation, the alarms went off 72 times. There were 11 beds being monitored by the central alarm system, out of which 4 of them showed a red alert for the entire of the first 8 minute sampling session. No action was taken to mitigate 3 of the alarms, and 1 alarm was turned off. In the second 12 minute sampling session, no corrective action was taken to mitigate the alarms as well. Of the 41 alarm counts that went off, there were only 8 where the ICU nurses reacted to them – either by taking a glance or by shutting off the alarm.

None of the alarms required immediate attention or urgent corrective action. A few observed causes of the false alarms were motion artefacts, intermittently increased heart rate in patients with absolute arrhythmia, the moving or repositioning of the patient causing the disconnection of some of the devices. However, among the multiple false alarms, there were also a few true technical alarms which required the nurses to take preventive actions.

DISCUSSION

The high frequency of the alarms going off on the central alarm system with no immediate attention reflects desensitization of the ICU nurses towards the system. The noise contributed by the central alarm system has seemingly become a background noise for the ICU nurses due to several hypothesized reasons.

PHYSICAL SETUP OF THE ICU

The physical set up of the ICU and the location of the central alarm system does not seem to encourage vigilant patient care and monitoring by the ICU nurses. The central alarm system is located behind all the ICU nurses while they are seated facing their patients. Since sound is not transmittable through the walls of the ICU bed-rooms, the nurses should rely on auditory alarm from the central system for any alerts of changes in patients' status. However, the causes of the alarms on the central alarm system – the bed number, the vital that has exceeded the threshold values, and the reflected situation, cannot be identified from where the ICU nurse is seated. Once an alarm sounds, the ICU nurse has to walk away from her patient and over to the central alarm system, to check first whether the alarm was due to her own patient or another patient from another bed. Only then can she identify what the problem is before heading back to her workspace to help the patient.

The redundancy in this process of identifying and reacting to an alarm is considerable, and sometimes unnecessary. Should the alarm be a false alarm, the attention off the patient would then be uncalled for. This would then encourage the ICU nurses to stay put at their computer stations in front of the patients, and pick out visual alarms from the patient monitors located within the ICU bed-room instead.

RATE OF FALSE ALARMS

With 3.6 alarms going off per minute, and only a couple of true technical alarms within the 20 minutes, the ICU nurses would have been desensitized towards the central alarm system. The ICU nurses do not take the alarms from the central alarm system seriously, because they cannot trust that the system is truly alerting them of an emergency situation. In fact, the ICU nurse admitted that they do not pay much attention to the central alarm system when interviewed.

The ICU nurses are usually more interested in alarms that can warn them of emergency cases, or alarms that they can differentiate between clinical issues and technical issues. The nurses would also place more trust in a central alarm system that does not generate more than 50% false alarms.

From this simple observation, it can be hypothesized that if the number of false alarms decrease, the trust in the system by the ICU nurses would increase, leading to a higher dependency of the system put in place in the ICU. With the existing state of the central alarm system, the ICU nurses would naturally depend more on the cues from the patient rather than from the alarm system.

PERCEIVED CUES

Another important question of this study which will be answered in the second and third phases of data collection is whether the ICU nurses carry out their line of patient care and perform corrective actions due to the central alarm system alerts or due to other perceived cues which they pick out while watching their patients.

From the rate of false alarms, the observed layout and the work processes of the ICU nurses, it may be possible that the ICU nurses depend highly on other perceived cues from the patient and the patient monitors rather than the alarms from the central alarm system. The efficacy of the central alarm system does not seem to aid the decision making process of the ICU nurses regarding patient care. The interview revealed that the nurse focuses more on reading the vital signs hourly or half hourly to understand the patient's situation, suggesting that the central alarm system is not a main source of information. If that is the case, the central auditory alarm does not provide useful information, but increases ambient noise which may influence the focused attention of the nurses. However, during the night shift when the manpower of caregiver drops, the reliance on the central alarm may be higher because the number of patient per nurse increases.

This information will be obtained during the second and third phases of data collection when the nurses will be interviewed following any corrective action performed after an alarm sounds off. This will elicit substantial information that can confirm or reject the results from the observation and interview conducted in the initial phase.

CONCLUSION

The implementation of alarms in patient monitors and therapeutic devices in the ICU is with the intention of alerting the medical staff on duty so that they are able to proceed with their routine activities without having to keep a permanent watch on the patient. However, the countless alarms and the high rate of false alarms (Blum and Tremper; Siebig, Kuhls et al.; Yathish, Sheppard et al.; Koski, Makivirta et al. 1990; Lawless 1994; Meredith and Edworthy 1995; Chambrin 2001; Imhoff and Fried 2009) have resulted in desensitization of the medical staff and reduced their trust in the credibility of these device alarms.

To mitigate the problem of the numerous alarms, a central alarm system has been commonly put in place in the ICUs to give an overview of the beds and the patients that require attention. The specific alarms that are sounding are also reflected on the central display. However, the rate of false alarms does not seem to be improving; other factors such as the physical layout and the design of the central alarm system seem to discourage the effective use of the system.

The efficacy of the central alarm system has not been widely studied, and its usefulness not fully exploited. The preliminary observations in the surgical ICU of a major hospital have shown that the location of the system, the rate of false alarms and the design of the central alarm system interface may not promote the extensive and effective use of the system. With the initial phase completed, the rest of the study will seek to justify the three identified factors as indicators of the efficacy of the central alarm system. It is hypothesized that the alarms arising from the central alarm system does not aid the ICU nurses in their line of care for the patients during the day shifts, but this circumstance may differ during the night shift when each ICU nurse has a higher number of patients to watch.

The completion of the study will therefore provide an objective evaluation of the central alarm system in the surgical ICU for both the day and night shifts, thereafter providing recommendations in the areas in which the system can be improved to put it to good use.

REFERENCES

Blum, J. M. and K. K. Tremper "Alarms in the intensive care unit: Too much of a good thing is dangerous: Is it time to add some intelligence to alarms?" Critical Care Medicine **38**(2): 702-703.

Chambrin, M. C. (2001). "Alarms in the intensive care unit: how can the number of false alarms be reduced?" Critical Care **5**(4): 184-188.

Chambrin, M. C., P. Ravaux, et al. (1999). "Multicentric study of monitoring alarms in the adult intensive care unit (ICU): a descriptive analysis." Intensive Care Medicine **25**(12): 1360-1366.

Charbonnier, S. and S. Gentil (2007). "A trend-based alarm system to improve patient monitoring in intensive care units." Control Engineering Practice **15**(9): 1039-1050.

Edworthy, J. and E. Hellier (2006). "Alarms and human behaviour: implications for medical alarms." British Journal of Anaesthesia **97**(1): 12-17.

Gurses, A. P. and P. Carayon (2007). "Performance obstacles of intensive care nurses." Nursing Research **56**(3): 185-194.

716

Gurses, A. P. and P. Carayon (2009). "Exploring performance obstacles of intensive care nurses." Applied Ergonomics **40**(3): 509-518.

Imhoff, M. and R. Fried (2009). "The Crying Wolf: Still Crying?" Anesthesia and Analgesia **108**(5): 1382-1383.

Imhoff, M., S. Kuhls, et al. (2009). "Smart alarms from medical devices in the OR and ICU." Best Practice & Research Clinical Anaesthesiology **23**(1): 39-50.

Kindsmuller, M. C., M. Haar, et al. (2009). Designing User Interfaces for Smart-Applications for Operating Rooms and Intensive Care Units. Human-Computer Interaction - Interact 2009, Pt Ii, Proceedings. T. Gross, J. Gulliksen, P. Kotzeet al. **5727**: 684-695.

Koski, E. M. J., A. Makivirta, et al. (1990). "FREQUENCY AND RELIABILITY OF ALARMS IN THE MONITORING OF CARDIAC POSTOPERATIVE-PATIENTS." International Journal of Clinical Monitoring and Computing **7**(2): 129-133.

Lawless, S. T. (1994). "CRYING WOLF - FALSE ALARMS IN A PEDIATRIC INTENSIVE-CARE UNIT." Critical Care Medicine **22**(6): 981-985.

Meredith, C. and J. Edworthy (1995). "ARE THERE TOO MANY ALARMS IN THE INTENSIVE-CARE UNIT - AN OVERVIEW OF THE PROBLEMS." Journal of Advanced Nursing **21**(1): 15-20.

Ocarroll, T. M. (1986). "SURVEY OF ALARMS IN AN INTENSIVE THERAPY UNIT." Anaesthesia **41**(7): 742-744.

Ramesh, A., P. N. S. Rao, et al. (2009). "Efficacy of a low cost protocol in reducing noise levels in the neonatal intensive care unit." Indian Journal of Pediatrics **76**(5): 475-478.

Siebig, S., S. Kuhls, et al. (2009). "Noise in intensive care units. Do the alarms for subspecialties differ?" Anaesthesist **58**(3): 240-+.

Siebig, S., S. Kuhls, et al. "Intensive care unit alarms-How many do we need?" Critical Care Medicine **38**(2): 451-456.

Tsien, C. L. and J. C. Fackler (1997). "Poor prognosis for existing monitors in the intensive care unit." Critical Care Medicine **25**(4): 614-619.

Yathish, K., B. Sheppard, et al. "Alarms in the intensive care unit - a snapshot study." Anaesthesia **65**(3): 318-319.

Chapter 74

The Effect of Backpack Weight on Gait Parameters

Chao-Yin Wu [1,2]   *Chih-Long Lin*[3] *Meng-Jung Chung*[1] *Mao-Jiun Wang*[1]

[1] Department of Industrial Engineering and Engineering Management,
National Tsing Hua University

[2] Department of Rehabilitation Medicine,
Mackay Memorial Hospital, Hsinchu

[3] Department of Craft and Design
National Taiwan University of Arts

ABSTRACT

The purpose of this study was to examine the effects of backpack load, speed and gender on gait parameters. Twenty subjects (10 male, 10 female, aged: 21-28 years) were ask to carry a backpack loaded with 0%, 10%, 15% and 20% of their bodyweight (BW) and walked on their preferred walking speed (PWS), 80%PWS and 120%PWS. The changes of lower extremity joint motion, ground reaction force (GRF) and perceived exertion during walking were investigated. The results indicated that the effects of load were more dominant than speed and gender. PWS decreased with increasing backpack weight. With backpack weight increased, gait cycle and double support time increased while stride length remained the same. Decrease of pelvis anterior tilt, hip extension and increase of pelvis posterior tilt were found while backpack load increased. For GRF, the vertical component increased with increasing backpack. Post hoc analysis showed greater differences between backpack weight of 0% BW, 10% BW and 15% BW, but little differences between backpack weight of 15% BW and 20% BW were found. In conclusion, people tend to walk slower with heavier carriage. Gait pattern changed were adapted by proximal joints. It was recommended that backpack weight should be limited to the maximum of 15%BW. Subject's age or activity level can be included in future investigations to gain more insights about backpack injury prevention as well as human fitness with backpacking.

Keywords: backpack, gait, electromyography, ground reaction force

INTRODUCTION

The effects of backpacking on physiological responses have been evaluated over the years. The early backpacking studies were conducted with military purpose where soldiers often need to carry heavy supplies while marching from one place to another. It was reported that up to 68 kg weight was carried by British infantryman in the Falklands operation. Haisman (Haisman, 1988) pointed out that the ability of carrying a heavy weight was not only influenced by the environment, but also the carrier's body type (ie. Weight, back lengh, or waist girth). Thus, it is reasonable to study the optimal backpack carrying weight being represented as percentage of body-weight. Moreover, because of the increasing prevalence of back pain (Brackley and Stevenson, 2004; Skaggs et al., 2006), the load carrying of school children has drawn high attention. Although it is still uncertain that whether heavy backpack is the cause of children's back pain, American Academy of Pediatrics recommended "the backpack should never weigh more than 10 to 20 percent of the student's body weight" in preventing back injuries among school children.

Load carrying is commonly seen among hikers. People carry different weight depends on the distance of walking and the pace to accommodate their energy level. Many factors such as age, sex, height and weight may affect walking speed. People may need to adjust their walking speed according to the walking environment. It is important to investigate the change of gait parameters with different load carriage and walking speed.The purpose of this study is to investigate the effect of backpack weight and walking speed on kinematic and kinetic change of human gait.

METHOD

SUBJECTS

Twenty healthy subjects from the university community volunteered to participate in the study. The inclusion criteria included no current or history of musculoskeletal pain or pathology and no range of motion limitation of four extremities or spine. Each subject was informed about the purpose and procedure the study prior to the experiment. General anthropometric information were collected and they are listed in Table 1.

EXPERIMENTAL DESIGN

The experiment was a three-factor factorial design. The independent variables included gender (male and female), backpack load (0, 10%, 15% and 20% body weight) and walking speed (80%, 100% and 120% preferred walking speed). The dependent variables included lower extremity joint range of motion, ground reaction force, and perceived exertion.

Table 1 The means and standard deviations of the anthropometric data

Sex	No.	Age (yr.)	Height (cm)	Weight (kg)	Leg length (cm)	BMI (Kg/m2)
male	10	23.8 (2.4)	174.0 (5.3)	63.8 (7.0)	81.7 (3.8)	21.0 (1.7)
female	10	23.4 (3.1)	164.0 (5.4)	56.2 (6.7)	78.6 (5.0)	20.9 (1.9)

*$p<0.05$; **$p<0.01$

INSTRUMENTATION

Measurement of kinematics

The motion capture system (Proreflex MCU 240, Qualisys, Sweden) with six digital cameras were used to measure the movement of the lower extremity joints. The sampling rate was set at 60 Hz with low-pass filtering at 6 Hz. A total of 37 reflective markers were used to build the Visual 3D hybrid model developed by C-motion Inc. The data of pelvic, hip, knee and ankle joint range of motion were processed by using the Visual 3D software (C-motion Inc., USA).

Measurement of kinetics

Two force plates (Kistler 9281 B, Switzerland; Bertec 4060-08, USA) were used to collect the vertical ground reaction force (VGRF). The sampling rate was set at 600 Hz. Three data points within a gait cycle were collected for analysis including first maximum VGRF (Fz1), minimum VGRF (Fz2), and second maximum VGRF (Fz3). The VGRF data were normalized by each subject's body weight and were expressed in %BW.

Measurement of perceived exertion

The Borg rating of perceived exertion scale (Borg CR-10) was used to measure the subjective physical load of 11 different body parts, including: neck, shoulder, upper back, mid-back, low back, anterior thigh, posterior thigh, knee, anterior shin, posterior shin and ankle.

EXPERIMENTAL PROCEDURE

The experimental procedure was divided into two parts and was executed in two separate days to avoid fatiguing of the subjects. All measurements were performed by one investigator to eliminate inter-rater variations. In the first day, after the experiment procedure was explained to the subject, basic anthropometric data were measured. The preferred walking speed (PWS) while carrying 0, 10%, 15% or 20%BW backpack was then measured by asking the subject to walk on a treadmill. The sequence of weight carrying while testing the PWS was randomized. The PWS was determined by using the method described in Dingwell and Marin (Dingwell and Marin, 2006). After the PWS was determined, a metronome was used to match the cadence of walking. Its tempo was recorded and used as a control of walking speed at the second day of the experiment. This method was used in the determination of 80% and 120% of PWS as well. Between each testing of treadmill walking, subject was asked to rest at least 5 min to avoid fatigue effect.

In the second day, first was the set-up of static Visual 3D hybrid model with 37 reflected markers attached to the subject. Then, a total of 12 walking combinations (4 loads x 3 speed) were performed by each subject. The sequence of walking combinations was randomized. Before each actual walking combination, subject was asked to practice the tempo recorded at the first day of the experiment. Ten successful trials of each walking combination were recorded. At the end of each walking condition, the subject was asked to fill out the Borg CR-10 form. Between each walking condition, the subject was asked to rest at least 5 min to avoid fatigue.

DATA ANALYSIS

Analysis of variance was performed to evaluate the effect of walking speed, backpack load, and gender on the response measures. Duncan's multiple range test ($\alpha < 0.05$) was conducted as a post-hoc testing. Statistical analyses were performed using the statistical analysis software SPSS v.16.0.

RESULTS AND DISCUSSION

THE TEMPORAL-SPATIAL PARAMETERS (TSPS)

The effects of load and walking speed on TSPs are summarized in Table 2. The load effect on PWS, gait cycle and double support time was significant. The PWS decreased as the load was increased. However, in Duncan's test, there was no difference between 15% and 20%BW load. The gait cycle increased as the load was increased especially at 20%BW load condition. Double support time showed greater increase as load reached to 15%BW. This finding was similar to those reported by Martin (Martin, 1986).

able 2 The mean and Duncan grouping of TSPs

TSPs	Load (L) (%BW)				Sig.	Speed (S) (%PWS)			Sig.	Gender (G)		Sig.
	0	10	15	20		80	100	120		F	M	
PWS (m/s)	1.05^a	1.01^b	0.98^c	0.97^c	**	0.91^a	1.01^b	1.09^c	*	1.01	1.00	
Stride length (m)	1.19	1.19	1.19	1.19		1.19	1.19	1.19		1.17	1.20	
Gait cycle (s)	1.15^a	1.19^{ab}	1.19^{ab}	1.23^b	*	1.31^a	1.19^b	1.07^c	*	1.16	1.22	
ouble Support Time (s)	0.11^a	0.11^a	0.13^b	0.13^b	**	0.14^a	0.12^b	0.11^c	*	0.12	0.13	GSL

< 0.05, **$p < 0.01$, GSL=interaction between load, speed and gender

JOINT ROMS

Table 3 shows the effect of load, walking speed and gender on the mean of the maximum, minimum and total joint motion in pelvis, hip, knee and ankle. The pelvic tilt to posterior with load increase. When carrying 15% and 20%BW load, the pelvis moved from anterior tilt to posterior tilt and the total joint motion increased as well. The load effect can also be seen in hip motion. Hip flexion angle and total range increased as load increase. Chow (Chow et al., 2005) also found an increased hip ROM with increasing load. Knee and ankle joint motions were not affected by load which indicated that the load effects were adopted by the proximal joints (pelvis and hip).

The effect of walking speed was seen on pelvis, hip, knee and ankle. The pelvis moved more posterior with larger range at slower walking speedAs walking speed increase, an increased knee flexion and decreased ankle dorsi-flexion were also found. Joint motions differences between female and male were seen in pelvic

ROM and ankle ROM. Female showed less motion in pelvis but greater motions in ankle. There was no significant interaction between load, speed and gender.

GRF

Table 4 shows the effect of load, speed and gender on GRFs. With load increase, $Fz1$, $Fz2$, and $Fz3$ all increased significantly. On the other hand, $Fz2$ decreased significantly with increasing speed. Holt (Holt et al., 2005) reported an increase of GRF while carrying 40%BW load. In this study as the load reached to 20%BW, $Fz3$ increased about 16%.

Table 3 The mean angle and Duncan grouping of Pelvis, Hip, Knee and Ankle in sagittal plane

	Load (L) (%BW)				Sig.	Speed (S) (%PWS)			Sig.	Gender (G)	
	0	10	15	20		80	100	120		F	M
Pelvic Max (ant. tilt)	8.0^a	7.1^b	7.0^b	5.8^c	**	7.0	7.1	6.8		8.2	5.8
Pelvic Min (post. tilt)	2.9^a	0.3^b	-0.9^c	-1.3^c	**	-0.3^a	0.5^b	0.7^b	*	2.3	-1.8
Pelvic ROM	5.1^a	6.8^{ab}	7.9^{bc}	7.1^c	**	7.3^a	6.7^{ab}	6.2^b	**	5.8	7.7
Hip Max (flexion)	22.7^a	24.7^b	25.5^b	26.7^c	**	24.1	25.3	25.3		24.7	25.1
Hip Min (extension)	-16.8	-15.8	-16.1	-15.2		-15.9	-16.1	-16.0		-17.4	-14.5
Hip ROM	39.5^a	40.5^{ab}	41.6^bc	41.9^b	**	40.0^a	41.4^b	41.2^b	*	42.1	39.7
Knee Max (flexion)	69.5	69.2	68.9	70.2		68.3^a	69.7^b	70.4^b	**	71.5	67.4
Knee Min (flexion)	4.7	5.3	4.9	6.0		4.9	5.0	5.8		7.9	2.6
Knee ROM	64.8	63.9	64.0	64.2		63.44	64.65	64.58		63.6	64.8
Ankle Max (dorsi-flex.)	16.6	16.9	17.1	17.5		17.7^a	16.9^b	16.5^b	**	17.6	16.5
Ankle Min (plantar-flex)	-11.5	-11.6	-11.9	-11.7		-11.1	-12.0	-11.9		-14.4	-9.0
Ankle ROM	28.1	28.5	29.0	29.2		28.8	28.9	28.4		32.0	25.4

$^*p < 0.05$, $^{**}p < 0.01$

Table 4 The mean and Duncan grouping of GRF (%BW)

	Load (L) (%BW)				Sig.	Speed (S) (%PWS)			Sig.	Gender (G)		Sig.
	0	10	15	20		80	100	120		F	M	
Fz1	94^a	97^a	103^b	106^b	**	100	98	101		100	100	
Fz2	68^a	72^a	80^b	82^b	**	82^a	73^b	72^b	**	77	73	
Fz3	98^a	104^b	111^c	114^c	**	108	105	108		108	106	

*$p<0.05$; **$p<0.01$

BORG CR-10

Figure 1 shows the effect of load, speed and gender on the mean value of Borg CR-10 scale. Load had significant effect on the perceived exertion in every part of body, especially in shoulder area. In general, the perceived exertion of body at 10%BW reached to the same level of perceived exertion of lower extremities at 20%BW. This showed the backpack weight had greater effect on the body than the legs.

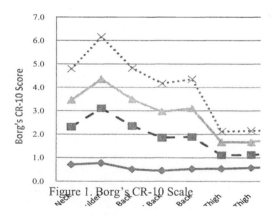

Figure 1. Borg's CR-10 Scale

CONCLUSION

This study investigated the effect of load, speed, and gender on gait kinematics, kinetics, GRF and perceived exertion. The results showed that the influence of load was greater than that of walking speed or gender. The effect of load was found to be higher in the proximal joints (pelvis and hip) than the distal joints. With increasing load, pelvis tilt posterior and hip flexion increased. The three main GRF investigated all increased with increasing load. The load effects on perceived exertions were all significant, especially in shoulder area. This information provided some insights about the change in physiological, kinematic and kinetic responses with increasing load. It can be useful for recommending the load limits

724

for backpackers.

REFERENCE

Brackley HM, Stevenson JM (2004), Are children's backpack weight limits enough? A critical review of the relevant literature. *Spine* 29: 2184-2190

Chow DHK, Kwok MLY, Au-Yang ACK, Holmes AD, Cheng JCY, Yao FYD, Wong MS (2005), The effect of backpack load on the gait of normal adolescent girls. *Ergonomics* 48: 642-656

Dingwell JB, Marin LC (2006), Kinematic variability and local dynamic stability of upper body motions when walking at different speeds. *Journal of Biomechanics* 39: 444-452

Haisman MF (1988), Determinants of load carrying ability. *Applied Ergonomics* 19: 111-121

Holt KG, Wagenaar RC, Kubo M, LaFiandra ME, Obusek JP (2005), Modulation of force transmission to the head while carrying a backpack load at different walking speeds. *Journal of Biomechanics* 38: 1621-1628

Martin PE (1986), The effect of carried loads on the walking patterns of men and women. *Ergonomics* 29: 1191-1202

Skaggs DL, Early SD, D'Ambra P, Tolo VT, Kay RM (2006), Back Pain and Backpacks in School Children. *Journal of Pediatric Orthopaedics* 26: 6

Chapter 75

DHM for Ergonomical Assessment of Home Appliances

Marco Mazzola, Ezio Preatoni, Nicola Emmanuele, Maximiliano Romero, Fiammetta Costa, Giuseppe Andreoni

INDACO Department
Politecnico di Milano
Milano, Italy

ABSTRACT

The ergonomic evaluation of the human-product interaction is still an open issue in Physical Ergonomics. The Proactive Ergonomic paradigm requires the introduction of new methods and indexes for the virtual comfort rating of motion, simulating the daily practice. Recently a new Index, *Method for Movement and Gesture Assessment* (MMGA), has been proposed by the authors for studying both ergonomic occupational effort and perception while performing dynamic tasks, and a theoretical assessment has been presented for the construction of reaching comfort maps .

The aim of this work is to evaluate the accuracy of the MMGA Index, comparing the score of different users while executing practical tasks interacting with three different commercial products. Traditional Ergonomics Indexes are defined mostly for a postural static evaluation of human motion and the new Index evaluates the kinematics of the entire task refining the measure. The MMGA has been tested evaluating human motion data acquired with an optoelectronic system.

The MMGA is in this work tested to point out significant differences among the different products and to be proposed as a relevant method for design practice in physical ergonomics.

Keywords: Ergonomic Index, MMGA; discomfort rating

INTRODUCTION

The design of the human-product interaction is traditionally based on tools proper of the Physical Ergonomic evaluation.

Anthropometry and Biomechanics are the main methods studied and improved through the generation of new digital human models and evaluation criteria based on the evidence of the parametric analysis obtained with the spread diffusion of optoelecronic technology in motion capture.

The effort of the research in Physical Ergonomics has been addressed to the introduction of quantitative or semi-quantitative indexes as a tool for the postural and gestural analysis of the interaction between the man and the environment at first, and of the product user interaction in the industrial design practice later.

Recently, a new method for quantifying the ergonomics of working tasks based on the kinematics of the executed movement have been proposed by the authors. This method, named Method for Movement and Gesture Assessment (MMGA) , is based on the measurement of the joint motion, and, consequently, on the availability of suitable technologies, such as: optoelectronic systems for motion analysis; and a dedicated software for further data processing. The starting point for the ergonomic index computation is the body kinematics, which is expressed as joint angles through a biomechanical model (as in the case of motion analysis systems or video-recording techniques), or through direct measure

The MMGA Index has been compared with the LUBA Index in the generation of comfort reaching maps while executing simple tasks interacting with a neutral reaching surface, but, according to the nature of this evaluation criterion, it is necessary to test the work in a more specific action research.

In this paper the comparison of the ergonomics of three different models of dishwashers with the MMGA methods is presented. In particular, the new ergonomic index is analyzed in a statistical global description of the comfortness of the three different home appliances, and then in the more specific motor strategies analysis of the most critical tasks.

MATERIALS AND METHODS

The Method for Movement and Gesture Assessment (MMGA) index is composed of three factors: a) the joints kinematics, b) an articular coefficient of discomfort for each joint that we will define as Discomfort score, c) a body normalization coefficient estimating the "weight" of the ergonomic contribution of each joint to the movement.

For the lower limb we applied an upper-limb corresponding scale, weighted on the mass of the lower-limb portion involved.

Eleven healthy adults (5 men and 6 women) participated to this study. Every participant was informed about experimental procedures and signed a written informed consent before participation. Total-body kinematics was recorded through a six-cameras (TVCs) optoelectronic system (Vicon M460, Vicon Motion System Ltd, Oxford Metrics, Oxford, UK) working at 120 Hz. TVCs were placed so that a volume of about 3 x 2 x 2 m was covered. Calibration procedures were carried out before each experimental session. The following variables were considered for this study: wrist, elbow, knee and ankle flex-extension; shoulder and hip flex-extension, intra-extra rotation and abdo-adduction; trunk flex-extension, rotation and lateral bending.

The acquisition protocol is based on a biomechanical model implemented through 33 passive and reflective markers placed on the body surface, according to the Vicon Plug-In Gait marker models (Vicon Motion System Ltd, Oxford Metrics, Oxford, UK), except for the markers on the head that are non relevant in this analysis and therefore excluded. For each subjects anagraphical and anthropometrical data have been collected.

Kinematic Data have been acquired while subjects interacted with three different dishwashers; named respectively M33, L46 and H11. Model M33 and H11 are similar models, with a dropping down front panel a single internal space for glasses and dishes. The L46 model is an industrial model, and present a double front door, each one of that opening as a cassette. The two doors give access to two different space. The three products have been place into the acquisition space (fig.1a).The objects to be placed in the different spaces have been prepare on a working surface upon the dishwashers, and the stating points have been fixed for a better repeatability of the tasks analysis (fig. 1b)

Figure 1. a) Experimental set-up; the three different model placed in the working space. b) fixed position for the different objects to be used.

The tasks selected for the analysis concern the product accessibility i.e opening the dishwasher door and trays, and usability i.e placing objects in the corresponding place. In particular:
- (T1) drop down of the front panel;
- (T2) placement of one glass from the working table to the specific place on the upper silverware basket, the closest to the user;

- (T3) placement of one glass from the working table to the specific place on the upper silverware basket, the farthest to the user;
- (T4) placement of one glass from the working table to the specific place on the lower silverware basket, the closest to the user
- (T5) placement of one glass from the working table to the specific place on the lower silverware basket, the farthest to the user;
- (T6) detergent filling.

Before the acquisition, each subject familiarized with the different products and tasks; after that the starting points has been fixed (fig.2). The user started and concluded the single task assuming the orthostatic reference position; the motor tasks sequence has been kept for the three models; the objects used for the usability tasks have been placed in fixed points on the workplace; three repetition for each task have been recorded for each subject and dishwasher model.

Figure 2: A subject selecting the most comfortable starting position before the acquisition protocol.

The MMGA index was calculated for each tasks for the comparison of the results.For each selected task statistical relevant differences between the different dishwasher models have been verified through the application, of the Friedman's Test with a confidence level of 5%. Post-hoc analysis with Bonferroni's Method have been carried when applicable.

The most critical tasks have been analyzed evaluating the different joints contributions and evidenciating the most relevant one for the index score. Maximum and minimum value of the tasks analysis have been compared.

RESULTS AND DISCUSSION

Results are presented in Figure 3, 4 and in Tab 1 and are expressed in % and reported as median and interquartile range (<IQR>). High percent represents high discomfort rating. 100% is the maximum discomfort rating defined for MMGA.

The Statistical significant differences have been revealed mostly for the accessibility tasks (P<.001) and for one of the usability tasks, the soap filling (P<.001). Examining in details the different movements, the comfort Index for T1 resulted lower in the opening of the upper cassette of the L46 model (median <IQR>: 18.9 <8.9> %) comparing with the opening of the lower cassette of the same dishwasher (45.2 <24.9> %) and with the dropping down of the front panel door of the M33 model (36.9 <15.5> %). No significant differences merged between the L46-up panel and H11that presents median values similar to the M33, but a different distribution of the comfort values (36.9 <35.2> %). Considering T6, the comfort Index related to the upper cassette of the L46 model (27.3 <12.0> %), it has been revealed lower than the one of the M33 (62.6 <27.8> %).

The Trunk movement has been revealed the most impacting factor for the increasing/decreasing of the discomfort index. In Task 1, the high values of trunk flex-extension (fig 4.a) induced the increase of the MMGA index, and consequently the reduction of the comfort perception in the task execution.

The opening of the upper cassette of L46 requires lower angles values in the whole task. On the opposite, the maximum values have been recorded using the lower cassette. M33 e H11 presented similar patterns, with low values at the beginning of the task, and a final increase due to the need of guiding the panel until the end of the task. A trunk flexion reduction could be obtained with the flexion of the lower limb, even if the presence of the door make this strategy difficult to adopt. Concerning the lateral bending and the rotation of the trunk, individual behaviours have been characterized by a similar medium pattern and a strong inter-subjective variability, as shown in fig 4b and 4c.

Task 6 determined results similar to task 1, while the others (T2 – T5) did not revealed statistical differences in the households.

TASK	M33	L46-up	L46-lw	H11
T1	36.9% (15.5%)	18.9% (8.9%)	45.2% (24.9%)	36.9% (35.2%)
T2	20.2% (9.9%)	20.1% (9.3 %)	\	18.6% (11.2%)
T3	31.0% (20.3)	29.9% (6.7%)	\	32.2% (16.8%)
T4	34.8% (26.1%)	\	40.5% (14.8%)	33.4% (11.6%)
T5	55.1% (28.2%)	\	54.7% (28.1%)	44.3% (7.4%)
T6	62.6% (27.8%)	27.3% (12.0%)	64.3% (32.5%)	78.6% (18.0%)

Table 1. Comfort Index (MMGA) measured while performing the 6 motor tasks interacting with the three products; results are expressed in % and reported as median (IQR). High percent represents high discomfort.

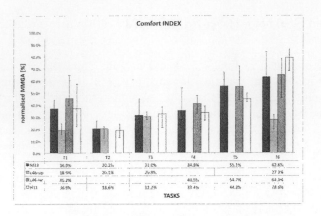

Figure 3. Comfort Index MMGA calculated on the analysed population. -up and .low rapresent the upper and the lower cassette of the L46 dishwasher model. Results are reported as median and interquartile range. (*)= statistical relevance (Friedman + Bonferroni, P<0.05

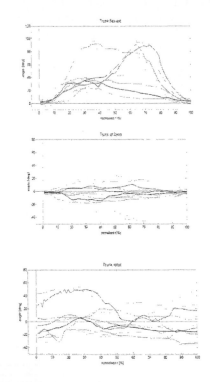

Figure 4. Trunk Flex-Extension (a), lateral bending (b), rotation (c) while executing task T1.

Concerning the motor strategy analysis, Task 6 has been evaluated for model M33. The aim of this comparisonwas to investigate the causes of the high difference in the limits scoring of MMGA for this task. In figure 5. is represented the subjects who scored the maximum value and in figure 6 the one who scored the minimum one. The trunk flexion is represented for the whole trial, and the motion in frontal and sagittal plane is figured at the starting time, in the reaching time (i.e the instant where the subject bring the detergent) and the filling time. It is possible to evidenciate that the trunk flexion values are higher in the most discomfortable trial, according to the MMGA measure. It is also possible to detect the relevant differences in motor strategies, for whom the worse condition is determined by a higher trunk flexion and a lateral bending in reaching the detergent, while the most comfortable task adopt a strategy that compensate the lower trunk flexion with the knee flexion, minimizing the MMGA scoring. These results demonstrate how the MMGA in ergonomical assessment s a sensitive tools to discriminate different postural situations.

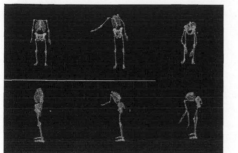 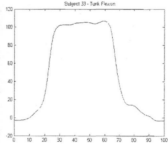

Figure 5: left) a view of the motor strategy adopted by a subject on frontal and sagittal plane. right) Trunk flex-extension during the complete trial. The higher values of the trunk flexion is determined by a static knee posture.

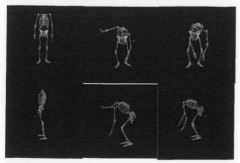 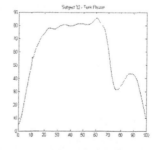

Figure 6: left) a view of the motor strategy adopted by a subject on frontal and sagittal plane. right) Trunk flex-extension during the entire trial. The lower values of the trunk flexion is balanced by a knee flexion.

732

The MMGA index has proved to differentiate the comfort level of easy tasks providing a coherent ergonomic ranking of movements; e.g. in the ergonomic assessment of tasks related to accessibility it presented the most MMGA index relevant differences between the models. The difference are strictly related to the motor strategy assumed by the subjects, and some design evidence are inferable to that. This can definitely be translated in design issues for the product development. Data from the MMGA index currently relate to a quantitative computation of the joints motion captured on real subjects but it might be integrated into a human motion simulation software for implementing proactive ergonomic analysis in the virtual prototyping process.

REFERENCES

Kee, D., Karwowski, W.: *LUBA: an assessment technique for postural loading on the upper body based on joint motion discomfort and maximum holding time*. *Applied Ergonomics*. 32(4), 357 –366 (2001).

Zatsiorsky,V., Seluyanov, V.: *The mass and inertia characteristics of the main segments of the human body*. In: H. Matsui, K. Kobayashi, eds. Biomechanics VIII-B, International series of Biomechanics, 4B: pp. 1152--1159. Human Kinetics Publishers, Champaign, Illinois, USA (1983).

Andreoni G., M. Mazzola, O. Ciani, M. Zambetti, M. Romero, F. Costa, E. Preatoni, Method for Movement and Gesture Assessment (MMGA) in Ergonomics, in Proceedings of HCI 2009

M. Romero M., M. Mazzola, F. Costa, and G. Andreoni, An Integrated method for a qualitative and quantitative analysis for an ergonomic evaluation of home appliances, in Proceedings of Measuring Behavior 2008

Davis, RB III, Ounpuu, S, Tyburski, D, and Gage, JR (1991). A gait data collection and reduction technique. Human Movement Sciences 10, 575-587

Kadaba MP, Ramakristan HK, Wooten ME (1990) "Measurements of lower extremity kinematics during level walking". Journal of Orthopaedic research, 8:383-392

Grood ES, Suntay WJ (1983) " A joint coordinate system for the clinical description of three-dimensional motions: Application to the knee", ASME Journal of Biomechanical Engineering, 105: 136-144

Chapter 76

A Comparison of Different Shoe Insoles in Gait Performance

Meng-Jung Chung[a], Chih-Long Lin[b],
Che-Huai Chang[a], Mao-Jiun Wang[a]

[a]Department of Industrial Engineering and Engineering Management
National Tsing Hua University
101, Section 2, Kuang-Fu Road, Hsinchu, Taiwan, ROC

[b]Department of Craft and Design
National Taiwan Universityof Arts
59, Section 1, Ta-kuan Road, Panchiao, Taipei, Taiwan, ROC

ABSTRACT

This study aims to evaluate the insole design effect on gait performance. A total of 15 healthy males participated in this study. Six different types of insoles were evaluated, including memory-foam insole, hard-metatarsal-pad insole, soft-metatarsal-pad insole, adjustable-arch-support insole, ergo-design insole, and barefoot condition. The response measurements include plantar pressure, joint motion, ground reaction force, and rating of perceived exertion.

Significant differences among different insoles were found in most of the response measures. The joint range of motion in pelvic tilt was the greatest in the memory-foam insole than the other selected insoles. When wearing both types of insoles with metatarsal pad, the plantar pressure under metatarsal areas would transfer into midfoot area, and resulted in a higher subjective discomfort in the midfoot area. Further, the insoles with hard arch support design (i.e. adjustable-arch-support insole and hard-metatarsal-pad insole) caused a smaller range of motion in pelvic tilt and a higher discomfort in arch area. Moreover, the barefoot condition showed the worst gait performance in most of the response measurements.

In summary, this study demonstrates that different insole designs and materials have different influence on gait performance. The findings can provide very important information for ergonomic insole design.

Keywords: insole, foot pressure, joint range of motion, ground reaction force.

INTRODUCTION

Various insoles made of different materials with different function and appearance tend to influence the gait performance. Goske et al. (2006) investigated 27 insole designs of the combination of three insole conformity (flat, half conforming, full conforming), three insole thickness (6.3, 9.5 and 12.7 mm) and three insole materials (Poron Cushioning, Microcel Puff Lite and Microcel Puff) on plantar pressure distribution. They found that the conformity of the insole is the most important design variable, whereas peak pressure is relatively insensitive to insole material selection (Goske et al., 2006). Hinz et al. (2008) evaluated the effectiveness of cushioning insoles (made of EVA foam or neoprene) on load reduction in comparison with conventional insoles. They reported that the neoprene insole has the lowest peak pressure with significant   load reduction under metatarsal area (Hinz et al., 2008).

Recently, memory-foam is one of the innovative insole materials available in the market. However, there are only a few reports on evaluating the memory-foam insole effect on gait performance (Ruano et al., 2009). Besides, metatarsal pad has been used as an effective orthotic device for redistributing forefoot plantar pressures (Chang et al., 1994; Hsi et al., 2005). Nowadays, the metatarsal pad design is added on to the conventional insoles. The functions of the alternative insole designs (e.g. heel cup, arch support, metatarsal cushion, and heel cushion) have been reported in many previous studies as well (Goske et al., 2006; Chiu and Wang, 2006; Hinz et al., 2008). It is interesting to evaluate the effects of the insole with the combination of different insole designs on gait performance. Thus, five different insoles available in the market as shown in Table 1 were selected for evaluating their effects on gait performance. The response measures include plantar pressure, joint motion, GRF, and perceived exertion.

METHOD

SUBJECTS

A total of 15 males who were students and staffs in a university participated in this study. They were 25.1 ± 3.1 years old, 173.8 ± 3.2 cm tall, and weighted

67.7 ± 5.3 kg. All subjects were free from musculoskeletal or lower extremity problems.

APPARATUS AND MATERIAL

Six different types of insoles were evaluated, including memory-foam insole (MF), ergo-design insole (ED), adjustable-arch-support insole (AAS), soft-metatarsal-pad insole (SMP), hard-metatarsal-pad insole (HMP), and barefoot condition. The characteristics of each insole are shown in Table 1.

Motion capture system (Proreflex MCU 240, Qualisys, Sweden) with six digital cameras was used to measure the movement of the lower extremity joints. The joint motions measured here include the joint angles of hip, knee, and ankle in sagittal and coronal planes. The vertical ground reaction force (VGRF) was recorded using two force plates (Kistler 9281 B, Switzerland; Bertec 4060-08, USA). Three measurements following the definition of White et al. (1996) were taken, including heel-strike, foot-flat, and toe-off stages. Plantar foot pressure data were recorded and analyzed by using a plantar pressure measurement device (Footscan® system, RSscan, Belgium). Foot pressure data were measured in seven zones including the hallux, 2-5th phalanges, 1st metatarsal, 2-3th metatarsal, 4-5th metatarsal, arch, and heel zones. The peak pressure for each plantar zone for a series of steps was calculated. Borg CR-10 rating scale was used to assess the perceived exertion of foot and plantar areas, including ankle, hindfoot, instep, forefoot, hallux, 2-5th phalanges, metatarsal, midfoot, and heel. Borg CR-10 is a 10-point psychophysical assessment scale, while 0 represents "nothing at all" and 10 for "maximal".

Table 1 The specifications of the experimental insoles (US size 9.5)

Item	Memory-foam insole (MF)	Ergo-design insole (ED)	Adjustable-arch -support insole (AAS)	Soft-metatarsal -pad insole (SMP)	Hard-metatarsal -pad insole (HMP)
Material	Memory foam and EVA	PU	PU	PU	PU
Weight	14.5g	74.5g	66.0g	52.5g	80.5g
Length	28cm	28cm	28cm	28cm	28cm
Breadth [a]	9.0cm	9.4cm	9.2cm	9.3cm	9.9cm
Thickness [b]	MP: 0.8cm AH: 0.8cm HL: 0.8cm	MP: 0.6cm AH: 2.6cm HL: 1.5cm	MP: 0.3cm AH: 2.0cm HL: 0.4cm	MP: 0.6cm AH: 2.6cm HL: 1.4cm	MP: 0.6cm AH: 2.5cm HL: 1.2cm
Arch support	No	Yes (soft)	Yes (adjustable; hard)	Yes (soft)	Yes (hard)
Forefoot cushion	No	Yes	Yes	Yes	Yes

Metatarsal pad	No	No	No	Yes (soft)	Yes (hard)
Heel cup	No	Yes	Yes	Yes	Yes
Heel cushion	No	Yes	Yes	Yes	Yes

[a.] The forefoot breadth in the shoe.

[b.] MP means the vertical distance from first metatarsal to insole bottom; AH means the vertical distance from arch to insole bottom; HL means the vertical distance from heel to insole bottom.

EXPERIMENTAL PROCEDURE

At the beginning of the experiment, all subjects were given information about the purpose and procedure of the experiment. The basic information and the relevant anthropometric data were collected. Thirty-seven markers were placed on the important anatomical landmarks on each side of the body such as: the heel, second metatarsal, ankle, knee, hip and pelvis. During the experiment, each subject was asked to take a warm up walking (3 km/hr) on the treadmill for 10 min, and then was instructed to walk on a 8-m walkway (1.2 m width path) with free walking speed. Joint motion, GRF, and plantar pressure were collected when the subjects walk through the center of walkway. After walking task, each subject had to give subjective assessment using the Borg CR-10 scale. The subjects rested at least 10 min prior to the next walking speed session to avoid fatigue effect.

STATISTICAL ANALYSIS

One way analysis of variance was performed to evaluate the effect of insole types on the response measurements. Duncan's multiple range test was conducted as a post-hoc testing. The response measurements include plantar pressure, joint range of motion, GRF, and subjective discomfort.

RESULTS AND DISCUSSION

JOINT MOTION

Table 2 shows that the insole effect on lower extremity joint motion of sagittal plane and coronal plane was not significant, except for max pelvic down obliquity ($p < 0.05$). The results of Duncan's post-hoc testings show that the range of motion on pelvic down obliquity while walking with MF insole (3.4°), SMP insole (3.1°) and barefoot (3.0°) was greater than that of walking with ED insole (2.8°), AAS insole (2.8°), and HMP insole (2.7°). Walking with MF insole and barefoot showed the greater pelvic obliquity was probably due to the lack of arch support and results in a greater forefoot pronation. A greater forefoot pronation may cause a greater pelvic obliquity. Although the SMP insole has the arch support feature, it is too soft to provide enough support during single-limb-stance period. Besides, the addition of metatarsal pad may cause a unstable gait that results in a greater pelvic obliquity as well. On the other hand, both HMP and AAS have hard arch support design, and thus the pelvic obliquity was smaller than the insole without arch support.

Table 2 The ANOVA results of joint motion

	Item	df	F-ratio	Sign.
Sagittal plane	Max pelvic tilt	5	0.807	
	Min pelvic tilt	5	0.441	
	Max hip flexion	5	0.059	
	Max hip extension	5	0.944	
	Max knee flexion	5	1.688	
	Max knee extension	5	1.318	
	Max ankle flexion	5	2.223	
	Max ankle extension	5	0.724	
Coronal plane	Max pelvic up obliquity	5	0.585	
	Max pelvic down obliquity	5	2.397	*
	Max hip abduction	5	1.085	
	Max hip adduction	5	0.596	
	Max knee varus	5	1.452	
	Max knee valgus	5	0.503	

*Significant at $p < 0.05$, **Significant at $p < 0.01$

VERTICAL ROUND REACTION FORCE

The insole effect was significant on VGRF in foot-flat stage ($p < 0.05$) (see Table 3). The results of Duncan's post-hoc testings show that walking with barefoot (85.4% BW) and HMP insole (85.2% BW) was significantly higher than using the other insoles. Walking with barefoot had the highest VGRF, while walking with MF insole (83.0% BW) had the lowest VGRF. These results reveal that walking with barefoot or with a hard metatarsal pad could not effectively attenuate shock during single-limb-stance phase and resulted in a higher VGRF. On the other hand, walking with MF insole had the lowest VGRF during the stance phases than using other insoles made of PU. It seems that the insole made of memory foam can have better absorption of ground impact than the insole made of PU.

Table 3 The ANOVA results of VGRF

Item	df	F-ratio	Sign.
Heel-strike stage	5	1.540	
Foot-flat stage	5	2.573	*
Toe-off stage	5	1.966	

*Significant at $p < 0.05$, **Significant at $p < 0.01$

PLANTAR PRESSURE

The insole effect was significant on peak pressure in 2-3 th metatarsal, 4-5th metatarsal, arch, and heel zones ($p < 0.01$) (see Table 4). Walking with barefoot showed the highest plantar pressure in the above mentioned zones than walking with any of the selected insoles. This phenomenon reveal the fact that walking with insole can successfully decrease the plantar pressure. It is interesting to find that the insole with hard metatarsal pad (i.e. HMP) showed the lowest pressure in metatarsal areas but had the highest pressure in arch area than the other insoles. This finding indicates that hard metatarsal pad can release the pressure in metatarsal area but may cause an increased pressure in arch area.

Table 4 The ANOVA results of plantar pressure

Item	df	F-ratio	Sign.
Hallux	5	1.768	
2-5th phalanges	5	2.292	
1st metatarsal	5	1.668	
2-3th metatarsal	5	5.781	**
4-5th metatarsal	5	9.588	**
Arch	5	5.268	**
Heel	5	17.279	**

*Significant at $p < 0.05$, **Significant at $p < 0.01$

RATING OF PERCEIVED EXERTION

Table 5 shows that the insole effect was significant on the RPE in hindfoot, metatarsal, midfoot and heel areas ($p < 0.05$). Generally speaking, barefoot walking would have a greater plantar pressure in metatarsal and heel areas and thus the highest RPE scores were found in metatarsal (3.5) and heel areas (3.1). The results also indicate that walking with metatarsal pad insoles (i.e. HMP and SMP) showed a higher RPE scores in midfoot area. It was probably due to that the subjects were asked to walk with insole with metatarsal pad, which is different from walking with conventional insoles in daily life.

Table 5 The ANOVA results of RPE scores

Item	df	F-ratio	Sign.
Ankle	5	2.263	
Hindfoot	5	3.680	**
Instep	5	0.889	
Forefoot	5	0.749	
Hallux	5	1.952	
2-5th phalanges	5	1.501	
Metatarsal	5	9.703	**
Midfoot	5	4.815	**
Heel	5	5.909	**

*Significant at $p < 0.05$, **Significant at $p < 0.01$

CONCLUSION

This study investigated the gait performance in five types of insoles with respect to different insole designs and materials. Our findings indicate that insoles with metatarsal pad can release the pressure in metatarsal area but caused a higher discomfort in midfoot area. Insoles made of memory foam can have better absorption of ground impact force than the insole made of PU. Moreover, using the insoles with arch support design showed the smaller pelvic obliquity than walking with insoles without arch support. These findings provide very useful information for insole design.

REFERENCE

Goske, S., Erdemir, A., Petre, M., Budhabhatti, S., and Cavanagh, P. (2006), "Reduction of plantar heel pressures: Insole design using finite element analysis." Journal of Biomechanics, 39(13), 2363–2370.

Hinz, P., Henningsen, A., Matthes, G., Jäger, B., Ekkernkamp, A., and Rosenbaum, D. (2008), "Analysis of pressure distribution below the metatarsals with different insoles in combat boots of the German Army for prevention of march fractures." Gait and Posture, 27(3), 535–538.

Ruano, C., Powell, D., Renshaw, D., Chalambaga, E., and Bice, M. (2009), "The effects of insoles on loading rate in level running." International Journal of Exercise Science, 2(1), S9.

Chang, A.H., Abu-Faraj, Z.U., Harris, G.F., Nery, J., and Shereff, M.J. (1994), "Multistep measurement of plantar pressure alterations using metatarsal pads." Foot & Ankle International, 15(12), 654–660.

His, W.L., Kang, J.H., and Lee, X.X. (2005), "Optimum position of metatarsal pad in metatarsalgia for pressure relief." American Journal of Physical Medicine and Rehabilitation, 84(7), 514–520.

Chiu, M.C., and Wang, M.J. (2006), "Professional footwear evaluation for clinical nurses." Applied Ergonomics, 38 (2), 133–141.

White, S.C., Tucker, C.A., Brangaccio, J.A., and Lin, H.Y. (1996), "Relation of vertical ground reaction forces to walking speed." Gait and Posture, 4, 167–208.

Chapter 77

The Evaluation of Double-Layers Clothing on Clothing Microclimate, Physiological Responses and Subjective Comfort

Te-Hung Chen, Chih-Long Lin, Mao-Jiun J. Wang

Department of Industrial Engineering and Engineering Management
National Tsing Hua University, Hsinchu, Taiwan 30013, R.O.C.

ABSTRACT

This study aims to investigate the influence of the double-layers clothing comfort in cleanroom environment. For the experiment design, twenty subjects including ten males and ten females completed all phases of the experiment. Each subject was asked to participate four treatment combinations with four underwear clothing levels (100% Cotton, 70% Cotton+30% Polyester, 65% Polyester+35% Cotton, 100% Polyester) × one semiconductor task (Dynamic task) × one cleanroom clothing (Class-100, Moisture permeability : 380 $g/m^2 \cdot h$, Air permeability : 2.5 $cm^3/cm^2 \cdot sec$). The microclimates in three body regions, the moisture absorption, skin temperature, and subjective responses were measured.

The results indicate that wearing 100% Polyester underwear caused significant increase in inner microclimate relative humidity (p<0.01). Wearing 100% Cotton underwear not only caused significant increase in inner microclimate relative humidity (p<0.01), but also in underwear's absorption. The major reason is that cotton fiber tend to trap water molecules and allow less water vapor to pass through as comparing to other fiber materials. Further, wearing the blending fiber underwear even caused lower relative humidity in inner and outer clothing microclimate (p<0.05). The clothing comfort of wearing 65% Polyester+35% Cotton underwear was greater than that of wearing 70% Cotton+30% Polyester. The findings suggest that wearing 65% Polyester+35% Cotton underwear caused a significant increase in subjective comfort feeling. Overall, the fiber material absorption and water vapor transportation of underwear are the major factors affecting the comfort of wearing double-layers clothings.

Keywords: Clothing Comfortable, Double-layers Clothing, Cleanroom Clothing, Clothing Microclimate, Semiconductor Fabrication.

INTRODUCTION

The semiconductor industry has placed considerable emphasis on minimizing the particles dispersed from cleanroom clothing. Moreover, people are the major contaminant source of most particles introduced into a controlled environment. In order to prevent personnel from releasing contamination, worker must wear a cleanroom clothing before entering the cleanroom. The critical consideration in the cleanroom is to control to a minimum the free particles in the atmosphere. Traditionally, the fabrics used to control contamination are very tightly woven. Prolonged wearing of clean room suit often causes some problems that workers feel warm, wet and uncomfortable. Thus, it is necessary to improve air permeability and moisture permeability of cleanroom clothing to increase comfort.

Clothing is used to assist human body to maintain a thermal balance between the metabolic heat which it generated and lost to the environment through the heat transfer by conduction, convection, radiation and evaporation (Ingram and Mount, 1975; Spencer, 1976; Kwon et al., 1998). Thus, the hygroscopicity and moisture permeability may be of crucial importance for double-layers clothing. Some researcher have found that the water absorptive property of the underwear in the protective clothing system for firefighters may alter the heat loss and further influence the heat and the cold stress (Lin et al., 2005). Other researchers found that the moisture transport through a garment will have influence on the microclimate between the garment and the body beneath, the thermal contact feeling of the wearer, and the thermoregulatory response of the body (Dai et al., 2008).

However, there have been few attempts to establish a direct relationship between cleanroom clothing and underwear. To understand the water absorption and moisture transfer in clothing systems, we need to measure water absorptive and moisture transport from skin to environment through double-layer fabrics. Hence, this study aims to investigate the influence of the double-layer clothing comfort in cleanroom environment.

METHOD

Twenty subjects (10 males and 10 females) with age ranging from 21 to 27 years volunteered to participate in this study. The subjects were fully informed of the purpose, procedures and potential risk of this experiment. The descriptive characteristics of the subjects (mean±SD) were age (23±2 years), height (167±8 cm), weight (59±10 kg) and BMI (21±2).

Four underwears (coded as A, B, C and D) were evaluated in this study as (shown in Table 1). The skin temperature was measured from chest, shoulder, and back by three thermistors (NX-TMP1A, Mind Media B.V., Nederland). All temperatures were recorded at 64 Hz and were continuously recorded and saved into data storage equipment (NeXus-10, Mind Media B.V., Nederland). All logged data were wireless transferred to a computer and the results were displayed every second on the computer screen to have continuous information about the skin temperature while the experiment was running (software: BioTrace+). Microclimate inside the clothing was measured using data logger (TK 500, Dickson, USA) located in the chest, shoulder, and back regions. Microclimate temperature and relative humidity (RH) were recorded at 0.1 Hz. The absorption property was evaluated by calculate the weight difference between beginning and end of each treatment condition.

Table 1. Materials and composition of the underwear.

NO.		A	B	C	D
Underwear					
Fiber composition		100% Cotton	70%Cotton+ 30%Polyester	35%Cotton+ 65%Polyeste	100% Polyester
Moisture Regain Rate (%)		8.5	6.1	3.2	0.4
weight (g)	M	108.0	124.0	198.0	142.0
	L	122.0	132.0	212.0	142.0
	XL	118.0			140.0

*Moisture Regain Rate was according to Chinese National Standards 2339-L3050 (1987).

744

The experimental sessions were done at ambient temperature of 22± 0.5 °C and relative humidity of 43± 3 % in a climatic chamber (HT-9745A, Hung Ta Instrument Co., LTD., Taiwan). The temperature and relative humidity level of chamber was simulated a semiconductor clean room environment. The experimental procedure involved preparation and operation phases. In the preparation phase, every participant wore the same clean room suit and slacks that prepared for the study as controlled clothing. Then, three skin temperature probes were attached to the chest, shoulder and back with surgical tape. Three microclimate data loggers were located in the same places where were between the inner side of clean room suit and the outer side of underwear. There were another three microclimate data loggers were located between the inner side of underwear and the skin. The operation phase involved walking, pod lifting and resting tasks, shown in Figure 1. A participant first walked on a treadmill for 7 minutes at a speed of 3.1 km/h, then lifted a 5.6 kg load every 10 s from 72 cm to 112 cm for 3 minutes and rested 2 minute. Each experimental session took 12 minutes and each participant had to complete 5 sessions. Thus, a total of 60 minutes was involved in each treatment condition. The subjective thermal questionnaire was used at 60 minutes for each treatment condition.

ANOVA and Duncan multiple range tests wre performed using STATISTICA[TM] version 6 (StatSoft, Inc., USA). The significance level of 0.05 was used throughout this analysis.

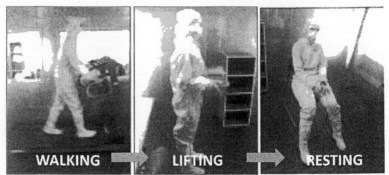

Figure 1. Cleanroom dynamic task

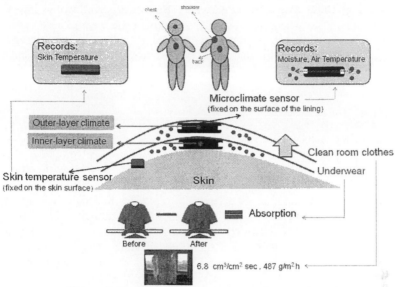

Figure 2. Experimental measures and sensor layout.

RESULTS AND DICUSSION

Absorption

The ANOVA results in table 2 show a significant gender and underwear effect on absorption ($p < 0.01$). For gander effect, it reveals that male's absorption (4.36 g) was higher than that of females (2.22 g). The results show that male subject tends to be easy to sweat than that of female subject in this task condition. For underwear effect, this clearly shows that 100% Cotton underwear absorption (5.2 g) was higher than that of other materials underwear. The 100% Cotton underwear had higher hygroscopicity capability and water survive on fiber because of hydrophilic fabric. The results indicate moist condition will be accrued in the double-layer system. On the contrary, since 100% Polyester underwear was hydrophobic fabric, it had lower hygroscopicity and water survive on fiber.

Skin temperature

Table 3 shows that the female's chest skin temperature (33.67 ± 0.20 °C) was significantly higher than that of the male's chest skin temperature (32.25 ± 0.25 °C) ($p < 0.01$). In addition, the female's skin temperature on the average was higher than that of the male's skin temperature, because the female's blood vessel dilation responses was significantly higher than that of the male (Kenny and Jay, 2006). A significant interaction between gender and underwear on shoulder skin temperature was observed ($p < 0.05$). Further, the results of Duncan's multiple range tests ($\alpha = 0.05$) had significantly higher shoulder skin temperature (34.3 °C) indicate the female's when wearing 100% Polyester underwear. But, had significantly lower skin temperature (32.5 °C) when wearing 100% Cotton underwear. These results are consistent with those of absorption results. Accordingly, the cerebrum nerve center must adjust to skin temperature, that a decreased caused skin temperature and sweat reduce.

Microclimate

The results of the ANOVA indicate that wearing 100% Polyester underwear caused significant increase in inner microclimate relative humidity (p<0.01). Wearing 100% Cotton underwear not only caused significant increase in inner microclimate relative humidity (p<0.01), but also an increase in underwear's absorption (Table 4). The major reason is that cotton fiber trend to trap water molecules and allow less water vapor to pass through as comparing to other fiber materials. The result shows polyester fabrics will be transported to the largest amount of heat and moisture (2.64 >1.51, 1.71, -0.44 %R.H.), and therefore, a human body would feel more comfortable with Polyester fabrics. This is in complete agreement with Kyunghoon's (2007) results. There is no statistically significant difference on microclimate relative humidity (p>0.05). In addition, wearing 100% Cotton underwear were increased about 7°C in inner and outer microclimate-temperature.

Although the polyester fabrics transports the largest amount of heat and moisture, the significant increase of outer layer humidity will cause invalid moisture transfer by cleanroom clothing (the cleanroom clothing moisture permeability rate < 100% Polyester underwear moisture permeability rate). Thus, both humidity increased will bring discomfort feeling. It is suggested that 100% polyester clothing was used alone. The air circulation can cause single clothes to effectively move H_2O, but the double-layers clothings is not without limitations.

Table 2. The moisture absorption (g) results.

Absorption		Gender**		AVG.
		Male	Female	
Underwear**	A	6.13	4.23	5.18
	B	5.38	3.13	4.25
	C	3.55	1.38	2.47
	D	2.38	0.13	1.25
AVG.		4.36	2.22	3.29

※**p<0.01

Table 3. Skin temperature results.

Gender**	Underwear*	Skin temperature (℃)		
		chest	back	shoulder
Female	A	33.49	33.80	32.49[a]
	B	33.53	33.51	33.36[b]
	C	33.76	33.85	33.49[b]
	D	33.91	33.93	34.34[c]
AVG.		33.67**	33.77	33.42
Male	A	32.51	32.97	33.47[b]
	B	32.02	33.60	32.58[b]
	C	32.41	33.56	33.65[b]
	D	32.06	33.08	32.77[b]
AVG.		32.25**	33.30	33.12

※*$p<0.05$, **$p<0.01$

Table 4. Microclimate-humidity

relative humidity (%R.H.)	Inner-layer climate**			Outer-layer climate		
	0 min(1)	1h avg.(2)	(2)-(1)	0 min(1)	1h avg.(2)	(2)-(1)
A	51.74	60.29	8.55	46.34	47.85	1.51
B	51.95	56.89	4.94	46.80	46.36	-0.44
C	48.66	51.83	3.17	44.30	46.03	1.73
D	52.25	60.28	8.02	44.26	46.90	2.64

※ 1h avg.= (chest+shoulder+back) microclimate-humidity/3
　　AVG means the average measurements of the groups.

Table 5. Microclimate-temperature

Temperature (℃)	Inner-layer climate			Outer-layer climate		
	0 min(1)	1h avg.(2)	(2)-(1)	0 min(1)	1h avg.(2)	(2)-(1)
A	29.79	30.55	0.77	26.81	27.54	0.73
B	30.59	30.67	0.09	27.50	27.58	0.08
C	30.12	30.33	0.21	27.34	27.52	0.18
D	31.04	30.96	-0.09	28.19	27.96	-0.23

※1h avg.= (chest+shoulder+back) microclimate-temperature/3

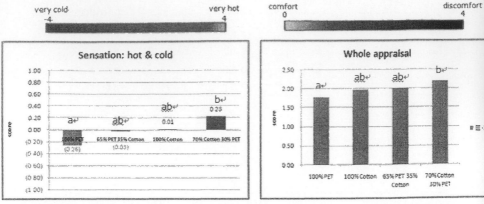

Figure 3. subjective responses results: sensation and whole appraisal (a, b: Duncan grouping code.)

Subjective responses

The underwear effect on sensation was significant ($p<0.05$)(see Figure 3). Different materials underwear would affect hot and cold sensation. A significant increase in hot sensation was found when wearing cotton fiber underwear, while a significant in cold sensation was found when wearing polyester fiber underwear.

CONCLUSIONS

The effectiveness of fiber material absorption and water vapor transportation of underwear are the major factors affecting the comfort of wearing double-layers clothings. Our results indicate that the inner clothing made of synthetic fabrics (such as a 70% Cotton+ 30% Polyester shirts and 65% Polyester+35% Cotton underwear) has higher clothing comfort. It is not recommended to wear 100% Cotton or 100% Polyester clothing as the inner clothing of double-layer clothing since it can increase inner microclimate humidity. Moreover, the findings suggest that wearing 65% Polyester+35% Cotton underwear caused a significant increase in subjective comfort feeling.

REFERENCES

CNS 2339-L3050:1987 Method of Test for Mixing Ration of Fibers Mixtures.
Imamura, R. (2000). The effect of sportswear material construction on body temperatures during exercise in windy conditions. In : Werner J, Hexamer M (eds) (2000) 9th International Conference on Environmental Ergonomics.

Shaker Verlag, Aachen, pp.323-326.

Imamura, R. (2002). Sweat accumulation in a kendo ensemble during indoor summer training. In : Tochihara T, Ohnaka T (eds) 10th International Conference on Environmental Ergonomics. Elsevier, Fukuoka, Japan, pp.535-538.

Ingram, D.L. and Mount, L.E. (1975). Man and Animals in Hot Environments (K. E. Schaefer Ed.), pp.5-23, Springer, Berlin, Heidelberg, New York.

Kenny, G.P. and Jay O. (2006). Gender differences in postexercise esophageal and muscle tissue temperature response. American Journal of Physiology, 292(4), pp.1632-1640.

Kwon, A., Kato, M., Kawamura, H., Yanai, Y. and Tokura, H. (1998). Physiological significance of hydrophilic and hydrophobic textile materials during intermittent exercise in humans under the influence of warm ambient temperature with and without wind. European Journal of Applied Physiology, 78(6), pp.487-493.

Lin, Y.W., Jou, G.T., Camenzind, M., Bruggmann, G, Bolli, W. and Rossi, R. (2005). Effects of heat and moisture transfer in firefighter : a study of the effect of underwear on physiological property and thermal protection in firefighter's clothing assembly. In : 11th International Conference on Environmental Ergonomics. Ystad, Sweden, pp.462-466.

Kyunghoon, M., Yangsoo, S., Chongyoup, K. and Kyunghi, H. (2007). Heat and moisture transfer from skin to environment through fabrics : A mathematical model. International Journal of Heat and Mass Transfer, 50, pp.5292-5304.

Spencer-Smith, J.L. (1976). The physical basis of clothing comfort. Clothing Research Journal, 4, pp.126-138.

Dai, X.Q., Imamura, A.R., Liu, G.L. and Zhou, F.P. (2008). Effect of moisture transport on microclimate under T-shirts, European Journal of Applied Physiology , 104(2), pp.337-340.

CHAPTER 78

Display/Control Quality in Small Touch Based Screen

Ying-Lien Lee

Department of Industrial Engineering and Management
Chaoyang University of Technology
Taichung County, 41349, Taiwan

ABSTRACT

Although intuitive in its nature, touch-based interface has the problem of poor performance time and accuracy. It also suffers from the finger occlusion problem where the objects underneath the contact area are obscured. Several techniques have been proposed to solve these problems, In this research, a technique called Fingerhit is proposed. Fingerhit uses a pseudo finger-occluded area to perform hit-testing and is lift-point determinant when activating an object on the screen. An experiment is conducted to compare the proposed technique against two other techniques called Cursor and Touch pointer. The results show that Fingerhit is better in terms of learning, performance time, and accuracy. Subjective responses also show that Fingerhit is more favorable compared to the other two techniques. We also discuss the implications and limitations of this research.

Keywords: Touch-base interface, Finger occlusion, Interaction technique

INTRODUCTION

Touch-based interface has the benefit of directness, robustness, and intuitiveness. It allows quick learning, rapid performance, and good hand-eye coordination. It also removes the needs of positional feedback (hence no cursor is needed), mechanical moving parts, and extra working space. Users are arguably more satisfied with touch-enabled design (Potter, Weldon, & Shneiderman, 1988; Sears & Shneiderman, 1991). Touch screen has been widely adopted in all kinds of devices, ranging from industrial equipments, control rooms or decks, Point-of-Sale (POS) systems, Personal Didital Assistant (PDA), mobile phones, and so on. Overall, touch-based interface is a very promising way of enriching Human-Computer Interaction (HCI).

However, touch-based interface still has its share of problems. Users suffer from high error rate and low precision while using touch-based interface (Potter et al., 1988; Sears & Shneiderman, 1991). The screen content is also obscured by user's finger (Albinsson & Zhai, 2003; Benko, Wilson, & Baudisch, 2006; Shneiderman, 1991). Researchers are trying to deal with these problems with innovative techniques, such as Shift (Vogel & Baudisch, 2007) and various other techniques (Albinsson & Zhai, 2003; Olwal & Feiner, 2003; Ren & Moriya, 2000).

To solve the aforementioned problems, a technique called Fingerhit is proposed and evaluated in this research. Fingerhit uses the finger-occluded area to perform hit-testing (the process of determining whether the cursor is on an object or not). To our knowledge, most touch-based interface still uses the mouse-based paradigm where a finger-controlled point cursor is used to perform hit-testing. In addition, the selection mechanism is different. In Fingerhit, activation of an object is made by lifting finger on the object regardless of the location of the initial contact point. In the mouse-based paradigm, however, a pair of mouse button down and up on the same object is required to activate the object. Most touch-based interface adopts this paradigm. So, to activate an object, users have to lift their fingers on the same object where the initial contact is made.

We conduct an experiment to study how well the technique works. In the Fingerhit experiment, a task similar to Fitts' Task is used. The technique is compared against two other techniques called Cursor and Touch pointer. Cursor is the most common form of touch screen control technique where the contact point on the screen is registered as a point and the on-screen cursor will be moved to that point. Hit-testing is performed by determining whether the point and the object in question have intersection when the finger is lifted. Touch pointer, a built-in feature in Microsoft Windows Vista, is an implementation of Offset Cursor (Sears & Shneiderman, 1991). Essentially, it is still a cursor. But whenever you contact the screen and then lift the finger, a cursor and a transparent virtual mouse are shown. The virtual mouse is offset so that you can drag the virtual mouse around to control the movement of the cursor.

THE CONSIDERED AND PROPOSED TECHNIQUES

In this section, we describe the design of the technique included in the comparison of the experiments, as well as the technique proposed in this research.

FINGERHIT

Fingerhit is a technique that uses finger-occluded area to perform hit-testing which is a process of determining if one shape is overlapping another shape. For computer mouse, hit-testing determines whether the mouse cursor is over an object. Most touch-based interface conveniently borrows the hit-testing of mouse and use the contact point to perform hit-testing. However, the contact point is not actually a point. When our fingers touch the screen, an area is formed and the portion under this area is occluded. In fact, the obscured area is also affected by the relative position of the eyes and the finger when a contact is made on the screen. Instead of just using one point to perform hit-testing, Fingerhit uses the finger-occluded area to do that.

Using finger-occluded area to perform hit-testing has some advantages. First, it can help solving the ambiguity in the process of hit-testing. Consider the case in Figure 1. The four rectangular objects are buttons, while the gray oval shape is the contact area formed by the finger which is depicted by the darker sketch line. Traditional hit-testing will only consider the center point of the occluded area and do the hit-testing with that point. In Fingerhit, the object that is hit is the object with largest portion being occluded. So, in this case, the upper left button is hit when the finger is lifted. When there are more than one object with the same portion being occluded, Fingerhit judges it as a tie and there is no hit. Alternatively, some kind of additional processing could be applied to break the tie, such as by comparing the center-to-center distance between the finger-occluded area and the objects in question.

In addition, Fingerhit is lift-point determinant when it comes to activating an object. In most touch-based systems, the contact point and the lift point has to be on the same object in order to activate the object. So, when users contact an object only to find that they choose the neighboring one, they cannot just glide their fingers to the right object and lift to activate the right one. This is not the case for Fingerhit. No matter which object is underneath the initial contact point, it is the object underneath the lift point that will be activated.

In the original design, the finger-occluded area would dynamically change its area. The rationale behind such design is that people have different sizes of index fingers. We sometimes operate touch-based interface with our thumbs, too. In addition, when different level of pressure is applied, the resulting finger-occluded area is different from one time to another. However, at the time of this writing, the Software Development Kit (SDK) for our hardware has not been released and no time table has been given. As a compromise, Fingerhit is designed so that the diameter of a pseudo finger-occluded area is adjustable. It's a per-user setting in our prototype system.

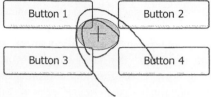

Figure 1. Using finger-occluded area to determine which object to highlight.

CURSOR

In this paper, Cursor refers to the technique where the finger contact point translates into a point and directly controls the coordinates of an on-screen cursor, which is much like the way mouse works. However, we modify its behavior so that it is also lift-point determinant.

TOUCH POINTER

In Microsoft Windows Vista, a new touch technique called Touch pointer is available for touch enabled systems. When users touch the screen, the cursor is positioned where the contact is made. In addition to the cursor, a virtual mouse is displayed in the proximity, too. The location of the virtual mouse varies so that it won't be obscured by the edge of the screen. When users want to move the cursor, they can simply tap a new location or drag the body of the virtual mouse and the cursor will move in parallel. Users can activate an object by tapping the left button of the virtual mouse twice, or select an object by tapping the left button once, just like using a real mouse. Likewise, tapping the right button of the virtual mouse will reveal the context menu of the object selected. In this research, Touch pointer is also made lift-point determinant in the experiments.

EXPERIMENTS

An experiment is conducted to investigate various hypotheses concerning Fingerhit. Participants receive financial compensation for their participation. To motivate our participants, an additional amount of monetary reward will be given to the participant with best performance in terms of time or accuracy.

FINGERHIT EXPERIMENT

This experiment is to study the Fingerhit technique and how it performs comparing to other two techniques under different conditions. Each session of the experiment is administered to one participant by two experimenters. Before the experiment begins, a consent form is presented to the participant. When the consent form is signed, one experimenter will set up the camcorder and the hardware platform, while the other will aid the participant filling out a background questionnaire collecting information such as age, gender, handiness, experience related to computer and touch screen, and the width of their index finger which is measured by the experimenter with a precision ruler.

After the questionnaire is filled out, the Fingerhit prototype system is launched to start a session. The width of the participant's index finger is entered in a dialog box, which in turn determines the diameter of the pseudo finger-occluded area. Then, the session begins with a tutorial and practice period for the first of the three techniques. After the participant is comfortable with the technique, he or she starts to do the formal experiment tasks identical to those they just practice. After the first technique is finished, the second and the third follow. A rest period is inserted between the presentations of each technique.

Participants

Twelve participants were recruited from the graduate school of the Department of Industrial Engineering and Management, Chaoyang University of Technology. Among them, there are seven male and five female students, whose average age is 23.75 years old. All of them are right-handed and have normal or corrected to normal vision. All participants reported that they use computer at least once per day. As to touch screen experience, eleven participants reported that they use touch screen once or twice per week. Only one participant reported frequent use (twice per day) of touch screen. Their experience of Touch pointer is also inquired, and none of them reported any.

The width of the index finger of our participants ranges from 1.00 cm (49 pixels on our hardware platform) to 1.45 cm (71 pixels) with a mean of 1.25 cm (61 pixels) and a standard deviation of 0.156. By applying t-test, a significant difference of the width of their index finger is present ($p < 0.05$).

Apparatus

The hardware platform used in this research is a Dell Latitude XT convertible Tablet PC running Microsoft Windows Vista Business. Its dual mode touch screen, developed by N-Trig Corporation, features electrostatic pen computing and capacitive touch technology. The operating system of the Tablet PC also provides two other touch-based techniques, Cursor and Touch pointer, which we are going to compare against our design. On the software side, a prototype system implementing the design of Fingerhit is developed using Java and Piccolo SDK (Bederson, Grosjean, & Meyer, 2004).

Experimental design

A 3*3*3 within-subject repeated measurement factorial design is used, where each factor is balanced to counter the order effect. A participant has to do the specified task under each treatment combination six times, yielding at least 162 data entries per participants. The three factors are as follows:

- Technique: three levels (Cursor, Fingerhit, Touch pointer, or C, F, T respectively)
- Width: three levels (W1, W2, and W3, or 10, 49, and 100 pixels respectively)
- Distance: three levels (D1, D2, and D3, or 260, 520, and 810 pixels respectively)

Given the levels of Width and Distance, the Indices of Difficulty in this experiment are 1.85, 2.63, 2.66, 3.19, 3.54, 4.13, 4.75, 5.73, and 6.36, with a range of 4.51.

Tasks

The task a participant has to perform is similar to Fitts' Task. When a task starts, a "ding" sound is played and a rectangular start button shown in the left portion of the screen. A participant has to activate the start button to start the task. Once the start button is activated, the button itself disappears and the target button is shown to the right of the start button. If the target button is activated correctly, the "click" sound is played and the participant can advance to the next trial. If the participant lifts his or her finger outside the target, which is recorded as an error, he or she has to do the same trial again. When an error occurs, a "chord" sound is played. The three factors vary as the experiment moves on. For Fingerhit, hit-testing is performed with the pseudo finger-occluded area.

Dependent variables

Several data are logged during the experiment. First, the time and coordinates when a participant contacts the start button is recorded. Since the activation is achieved via lifting the finger, the time and coordinates when the participant lifts his or her finger (and activate the button) is recorded, too. These pieces of information are also recorded for the target button, along with a field indicating whether it is a miss or hit.

Several pieces of information are derived from the logged data. First, the error rate is defined as the ratio between the numbers of error and total trials. Second, the

756

performance time is defined as the time span from the contact on the start button to the lift off the target button. Other important metrics such as Index of Difficulty and the distance between the target center and the points of contact and lift are also derived.

RESULTS

The learning effect plays a role during the experiment. For all three techniques, the performance time of the first repetition is significant longer than that of other repetitions that entail. The 95% confidence intervals are shown in Figure 2. To have stable results, the data from the first repetition is removed from the analysis.

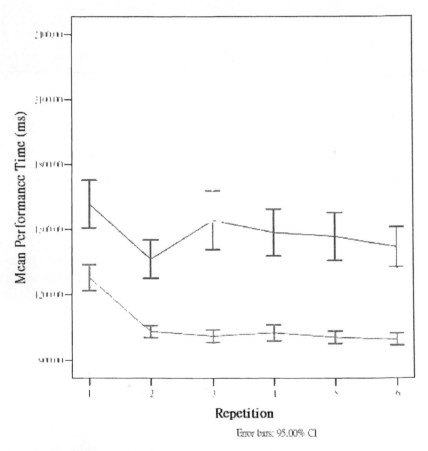

Figure 2. The graph of 95% confidence intervals of the performance time in each repetition for the three techniques in the Fingerhit experiment.

All main factors, as well as two of the interaction terms, have significant effect on the performance time, as Table 1 shows. The Technique factor is significant ($F(2, 22)=27.108$, $p<0.05$), where participants seem to perform well when using the Fingerhit technique. Two other main factors Width and Distance are also found to be significant, with F-test values of $F(2, 22)=27.711$, $p<0.05$ and $F(2, 22)=10.468$, $p<0.05$, respectively. The interaction term Technique*Width is significant, with F-test value of $F(4, 44)=20.539$, $p<0.05$.

Table 1. The ANOVA table of the performance time of the Fingerhit experiment.

	df	F	Sig.
Technique	2	27.108	0.000
Width	2	27.711	0.000
Distance	2	10.468	0.001
Technique*Width	4	20.539	0.000
Technique*Distance	4	0.696	0.599
Width*Distance	4	4.790	0.030
Technique*Width*Distance	8	2.659	0.012

Although the error rate is not very high (ranging from 0 to 0.19) throughout the experiment, it is found to be significantly affected by Technique ($F(2, 22)=8.735$, $p<0.05$), Width ($F(2, 22)=14.388$, $p<0.05$), and Technique*Width ($F(4, 44)=3.685$, $p<0.05$), as shown in Table 2.

Table 2. The ANOVA table of the error rate of the Fingerhit experiment.

	df	F	Sig.
Technique	2	8.735	0.002
Width	2	14.388	0.000
Distance	2	2.334	0.120
Technique*Width	4	3.685	0.011
Technique*Distance	4	1.962	0.177
Width*Distance	4	1.562	0.201
Technique*Width*Distance	8	0.834	0.575

DISCUSSION AND CONCLUSIONS

In the Fingerhit experiment, learning effect is quite evident for three techniques. Fingerhit has a learning rate as good as other two techniques, and remains stable throughout each repetition while Cursor and Touch pointer have less stable performance. It can be observed that during the experiment, our participants are confronted with the problem of obscured mouse cursor (of the Cursor technique) or jumpy virtual mouse (of the Touch pointer Technique).

As target width becomes larger, our participants respond almost as quickly when using Cursor and Touch pointer. For the Cursor technique, it is obvious that bigger target means easier activation of targets. For the Touch pointer technique, our participants also learn that they don't have to drag the virtual mouse along if they can activate a target by simply touching and lifting their fingers on the target. This is exactly the same way how they use the Cursor technique, which explains why the response time converges for these two techniques as the target width grows larger.

Performance times are found to be significant different in terms of three main factors and some interaction terms. Fingerhit outperforms other two techniques and has stable performance as target size changes. Only at the W2 and W3 levels can other two techniques be on a par with Fingerhit. Our participants tend to make less error when using the Fingerhit technique and when the target size is large.

In general, our hypotheses of Fingerhit are supported by the data we collected from our participants. With Fingerhit, our participants perform the task faster and make less error. As to user acceptance, it is arguable that our participants are more comfortable with Fingerhit since they respond and perform faster and make less error. An error-prone or hard-to-use user interface leads to user frustration.

REFERENCES

Albinsson, P. A., & Zhai, S. (2003). High precision touch screen interaction. Paper presented at the Proceedings of the SIGCHI conference on Human factors in computing systems.

Bederson, B. B., Grosjean, J., & Meyer, J. (2004). Toolkit Design for Interactive Structured Graphics. IEEE Transactions on Software Engineering, 30(8), 535-546.

Benko, H., Wilson, A. D., & Baudisch, P. (2006). Precise selection techniques for multi-touch screens. Paper presented at the Proceedings of the SIGCHI conference on Human factors in computing systems.

Olwal, A., & Feiner, S. (2003). Rubbing the Fisheye: Precise Touch-Screen Interaction with Gestures and Fisheye Views. Paper presented at the ACM Symposium on User Interface Software and Technology.

Potter, R. L., Weldon, L. J., & Shneiderman, B. (1988). Improving the accuracy of touch screens: an experimental evaluation of three strategies. Paper presented at the Proceedings of the SIGCHI conference on Human factors in computing systems.

Ren, X., & Moriya, S. (2000). Improving selection performance on pen-based systems: a study of pen-based interaction for selection tasks. ACM Transactions on Computer-Human Interaction (TOCHI), 7(3), 384-416.

Sears, A., & Shneiderman, B. (1991). High Precision Touchscreens: Design Strategies and Comparisons with a Mouse. International Journal of Man-Machine Studies, 34(4), 593-613.

Shneiderman, B. (1991). Touch screens now offer compelling uses. Software, IEEE, 8(2), 93-94.
Vogel, D., & Baudisch, P. (2007). Shift: a technique for operating pen-based interfaces using touch. Proceedings of the SIGCHI conference on Human factors in computing systems, 657-666.

CHAPTER 79

The Factors of Image Quality in Autostereoscopic Display

Yu-Ting Ding*, Sheue-Ling Hwang*, Kai-Chieh Chang**, Samper Wang***, Yueh-Yi Lai****, Wen-Hung Liao*****

*Department of Industrial Engineering and Engineering Management National Tsing Hua University 101 Section 2, Kuang-Fu Road, Hsinchu Taiwan
**Characterization & e-CAD Div., AU Optronics Corporation No. 1 Li-Hsin Rd., 2 Hsinchu Science Park Hsinchu, Taiwan

***Display Technology Center, Industrial Technology Research Institute Rm. 257, Bldg.17, 195, Sec. 4, Chung Hsing Rd., Chutung, Hsinchu Taiwan
****Measurement Standards Center, Industrial Technology Research Institute Rm.300, Bldg. 16, 321, Sec. 2, Kuang Fu Rd., Hsinchu, Taiwan

*****Measurement Technology Dept., AU Optronics Corporation No. 23 Li-Hsin Rd. , Hsinchu Science Park Hsinchu, Taiwan

ABSTRACT

"Crosstalk" is probably one of the most annoying distortions in 3D displays and the main resource of visual fatigue. In order to understand the level of "crosstalk" in different factors, the research will analyze the factors related to crosstalk through ergonomics evaluations in autostereoscopic displays, including subjective evaluation for questionnaire and the innovative method - Simulation of Crosstalk.

The result of this research revealed that two types of the evaluation methods had great explanatory ability. It is demonstrated that the database for the stereoscopic image quality will be build, and may provide the industries of 3D display with critical references.

Keywords: 3D display, image quality, crosstalk

INTRODUCTION

3D DISPLAY TECHNOLOGIES

The 3D display technologies can be classified into two types, stereoscopic and autosteroscopic displays.

Stereoscopic displays require users to wear a device or helmet, such as anaglyph glasses and head mount display, that ensure the left and right views are seen by the correct eye (Holliman, 2005).

Autostereoscopic displays can present the 3D stereoscpic images to the viewer without wearing glasses, goggles or any device. It can separate the left and right views and transmit the strereoscopic images directly to the correct eye (Halle, 1997).

CROSSTALK

In this study, we focus on the autostereoscopic dispalys. It has been more than a decade from the birth of 3D display products, and all the display designers follow eagerly, making an ideal autostereoscopic display to correspond with the market needs. For the ideal display, the width of the the window has to make equal to the separation of eyes which for the viewers are at variance with the range 55-70 mm. But variations in the different autostereoscopic display slit width will result in degradation in brightness and imperfect separation image between the left eye and right eye (Yuanqing, 2005).

In a perfect stereoscopic display, the image of right eye should be seen by the right eye only and should be invisible to the left eye completely and vice versa. In any case, complete separation of left and right eye image is often impossible in autostereoscopic display. The term is called crosstalk (the left eye view leaks through to right eye view and vice versa) (Kooi and A. Toet, 2004).

HUMAN FACTORS ASSESSMENT

Visual comfort is contrary to visual fatigue. Visual fatigue refers to a lower performance of the human vision system (HVS), which could be objectively measured, while visual comfort is its subjective counterpart (Lambooij, et al 2007; Uetake, et al 2000). In this review we will introduce the evaluation methods of human factors, comprising objective and subjective evaluation.

Objective evaluation methods of visual fatigue are pupillary diameter, near and light pupillary reactions, critical fusion frequency (CFF), visual acuity, near and far point, refractionability, visual field, stereo acuity AC/C ratio, heterophoria, convergency, spatial contarst sensitivity, color vision, light sense, blink rate, tear film breaking time, pulse rate and respiration time (Lambooij, et al 2007; Uetake, et al 2000).

Standards of subjective measures, defining plenty assessment methods, such as recommendation 500 of the International Telecommunication Union/Radio Communication (ITU-R), describing subjective assessment methods are used to establish the performance of television system using measurements that more directly anticipate the reactions of those who might view the systems tested (ITU, ,2000).

METHOD

The study was designed to evaluate the effects of crosstalk during autostereoscopic display experiment. The factors contained such as screen's brightness, viewing angle, viewing distance and different types of autostereoscopic displays.

The experiment was divided into two parts: effect of subjective evaluation questionnaire and effect of Simulation of Crosstalk.

The result of pretest indicated that the best viewing angle was not in the middle of the disaplay, and thus we took angle calibration. Four experts of Taiwan TFT LCD Association helped to find the best range of horizontal viewing angles, and got the average values to redefine the viewing angles.

PARTICIPANTS

Thirty-three students (23-28 years old) participated in this experiment. All subjects had a visual acuity >0.9 (as tested with the Landolt C test), good stereo depth perception and color discrimination (tested with "OptecTM 2000" vision tester). Twenty subjects had no experience of watching 3D displa. However, there were thirteen subjects who had the experience of watching 3D display for fewer than three hours. We randomly deployed the participants with three-level (High, middle and low) brightness groups and assigned them to the experimental conditions.

STIMULI

Subjective evaluation

The stimulus material used in the subjective evaluation consisting of two scene images (Figure 1) and varying in the screen's brightness levels (high, middle and low) as shown in Table 1, the calibrated viewing angle (0°, 5°, 10° and 15°) and viewing distance (60 cm and 120cm). Other independent variables were different types of autostereoscopic displays (lenticular advanced type - 2D plus depth, and parallax barrier type). The dependent variables were the subjective evaluation of autostereoscopic display questionnaire.

Table 1 The value of Brightness level

Level	Barrier type	Lenticular (2D plus depth)
H	$55.15(cd/m^2)$	$213.45(cd/m^2)$
M	$39.26(cd/m^2)$	$167.93(cd/m^2)$
L	$23.36(cd/m^2)$	$122.4(cd/m^2)$

The statistical analyses applied on the subjective data were scale ranking and the univariate analysis of variance (UNIANOVA). A categorical scale ranging from 1 to 7 was used. Where "1" represents bad image quality and "7" represents excellent image quality.

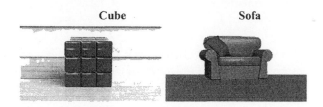

Cube   Sofa

FIGURE 1 Two images for subjective evaluation

Simulation program of Crosstalk

Simulation program of Crosstalk is a novel method for this study, and was fit into the standard of OpenGL (Open Graphics Library).

In this study, the appearance of Crosstalk is not the electrical or optical mixing of left and right eye as we know in literature. Instead, it means that the subjects see the stereoscopic image, and the perception of image imperfect separation from the view of left and right eyes. Besides, the subjects perceive different levels of Crosstalk when they watched the stereoscopic image with impact factors.

The statistical analyses were using Simulation program of Crosstalk and Multivariate analysis of variance (MANOVA). The independent variables were display type, brightness, viewing angle and viewing distance and dependent variables were Left Crosstalk % and Right Crosstalk %. The test pattern was a geometric figure (the object color, R: 220, G: 0, B: 0; the background color, R: 220, G: 220, B: 220) in this experiment (Figure 2). The display image was as same as the test pattern in Simulation program of Crosstalk.

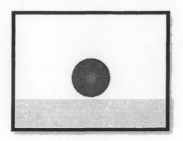

FIGURE 2 Test image in Simulation program of Crosstalk

EXPERIMENTAL ENVIRONMENT AND APPARATUS

Experimental environment

In the experiment, the experimental environment was dark, and the wall was painted with black ink, preventing from the reflection of the wall.

Experimental procedure

The purpose of the experiment was to analyze the factors related to Crosstalk, and compared the effects of image quality in different types of autostereoscopic displays.

The experiment consisted of the following steps:

1) Before the formal experiment, all subjects had to take the basic visual exams, including visual acuity, color discrimination, and pass the test of stereopsis visual ability. If the subjects fail the test of visual exam, they were unable to adapt to the experiment.

2) We illustrated the procedure of the experiment to make sure every subject could realize the whole task.

3) All the subjects would be divided into three groups with brightness levels (High, middle and low) randomly.

4) The formal task could break up into four parts, and the entire task had 8 runs (4 viewing angle * 2 viewing distance) at random. First, subjects were asked to observe the specific stereoscopic images and estimate with the subjective questionnaire with the original pictures. Second, they saw the geometric figure and then asked to manipulate the Simulation of Crosstalk to simulate what they see in

display.

5) Subjects had to take a rest (5-10 minutes approximately), avoiding the visual fatigue which may disturb the result of task.

6) Every subject had to repeat the 4) with different autostereoscopic displays to finish the entire task.

RESULTS

SUBJECTIVE EVALUATION

Effect of Type, Brightness, Viewing angle and Viewing distance level

The stereo image quality was impacted by all main effects, including type (F=16.245, p<.001), brightness (F=19.626, p<.001), viewing angle (F=20.009, p<.001) and viewing distance (F=185.973, p<.001) level.

The interaction between type and brightness (F=8.601, p<.001), brightness and viewing angle (F=5.319, p<.001), type and viewing angle (F=31.951, p<.001), type and distance (F=6.22, p=.011), and distance and viewing angle (F=6.877, p<.001) were significant for the subjective evaluation. On the contrary, the interaction between brightness and distance (F=.244, p=.828) and type, brightness and distance (F=.561, p=.526) were not significant.

The result showed that interactions among type, brightness and viewing angle (F=5.363, p<.001), brightness, distance and viewing angle (F=7.777, P<.001) and type, distance and viewing angle (F=5.059, p<.01) were significant for the subjective evaluation. However, interactions among type, brightness and distance (F=.561, p=.526) were not significant. And type, brightness, distance and viewing angle interaction was found significant (F=11.013, p<.001).

Interactions among the factors

In medium range of brightness, the image quality of Lenticular advanced type was obviously higher than that of Barrier type. However, there is no difference between these two types of display when it comes to low or high brightness

SIMULATION PROGRAM OF CROSSTALK

Effects of Type, Brightness, Viewing angle and Viewing distance level

The left Crosstalk % and right Crosstalk % was all influenced by type (MANOVA; Wilks's λ=.478, p<.05), brightness (MANOVA; Wilks's λ=.979, p<.05), viewing angle (MANOVA; Wilks's λ=.961, p<.05) and viewing distance (MANOVA; Wilks's λ=.961, p<.05) level.

The interaction between type and brightness (MANOVA; Wilks's λ=.936, p<.05), type and distance (MANOVA; Wilks's λ=.968, p<.05), type and Viewing angle (MANOVA; Wilks's λ=.942, p<.05), brightness and viewing angle (MANOVA; Wilks's λ=.935, p<.05), distance and viewing angle (MANOVA; Wilks's λ=.970, p<.05) were significant. However, interaction between brightness and distance (MANOVA; Wilks's λ=.995, p>.05) was not significant for the left and right Crosstalk %.

Interactions among type, brightness and viewing angle (MANOVA; Wilks's λ=.909, p<.05), type, distance and viewing angle (MANOVA; Wilks's λ=.918, p<.05), brightness, distance and viewing angle (MANOVA; Wilks's λ=.793, p<.05) were significant influence in Crosstalk%. On the contrary, interactions among type, brightness and distance (MANOVA; Wilks's λ=.992, p>.05) was not significant.

Interactions among factors

When subjects watched 3D image in Barrier type display, they perceived higher Crosstalk percentage was as the screen brightness was high than that as brightness was middle or low. On the contrary, in Lenticular advanced type, low brightness resulted in the highest Crosstalk % than middle and high brightness level did

DISCUSSIONS

TWO TYPES OF EVALUATION

From the result of 3.1 and 3.2, one can see that significant main and intercation effects on Subjective evaluation and Simulation of Crosstalk are same.

The Crosstalk percentage of left and right side was not completely symmetrical. For lenticular advanced type, subjects watched the 3D images from 120 cm, Crosstalk % was the highest at viewing angles 15 on the left. However, it had the highest Crosstalk % at viewing angles 10 on the right.When subjects watched the

3D images in Barrier type autostereoscopic display from 60 cm away, Crosstalk % was the highest at viewing angle 0 on the left. Nevertheless, it had the highest Crosstalk % at viewing angles 10 on the right.

EFFECT OF VIEWING DISTANCE OF 60 CM

Viewing distance was one of the independent variables, and the ideal viewing distance of IOSH (Institute of Occupational Safety and Health) and VESA (Video Electronics Standards Association) recommendation was 50-60 cm in 20-inch LCD. But does it really fit in autosteroscopic displays when users are watching?

From subjective evaluation questionnaire, we found that most subjects felt more comfortable at viewing distance of 120 cm than 60 cm

CONCLUSIONS

In this study, we compared subjective evaluation and a new appraisal method— Simulation program of Crosstalk for understanding the effects of stereoscopic image quality. From the result of data analysis, we could find 1) the appropriate range of viewing angles with calibration for evading deadzone 2) the suitable brightness level in two types of autostereoscopic display 3) the better viewing distance.

ACKNOWLEDGMENT

This study was supported by grant from Taiwan TFT LCD Association (TTLA) under project number A663TT4000-S11.

REFERENCES

Halle,M,(1997) *Autostereoscopic displays and computer graphics.* ACM SIGGRAPH, vol. 31, pp. 58–62.

Holliman, N.,(2005) *3D Display Systems," in Handbook of Optoelectronics.* vol. II, J. P. Dakin, T. R.G.W. Brown, and Francis, Eds.

ITU,(2000) *ITU-R BT.1438 Subjective Assessment of Stereoscopic Television Picture..*

Kooi, F. L. and Toet, A.(2004) *Visual comfort of binocular and 3D displays.* Displays, vol. 25, pp. 99-108.

Lambooij,M.T.M., IJsselsteijn,W.A. and Heynderickx, I (2007) *Visual Discomfort in Stereoscopic Displays: A Review.* in 2007 SPIE-IS&T, pp. 64900I-1-64900I-13.

Uetake, A. Murata,A., Otsuka, M. and Takasawa,Y.(2000) *Evaluation of Visual*

Fatigue during VDT Task. IEEE International Conference on system, machines and cybernetics, vol. 2, pp. 1277-1282.

Yuanqing,W..(2005) *High brightness auto-stereoscopic LCD with parallax backlight.* Proc. of SPIE, pp. 155-161.

Chaper 80

Comfort and Discomfort in Gripping Exertion

Yong-Ku Kong [a], Dae-Min Kim [a], Kyung-Sun Lee [b], Myung-Chul Jung [b]

[a] Department of Industrial Engineering,
Sungkyunkwan University
Suwon, 440-746, South Korea

[b] Department of Industrial and Information Systems Engineering,
Ajou University
Suwon, 443-749, South Korea

ABSTRACT

This study examined the relationships between gripping exertion and comfort and discomfort by measuring at various force exertion levels (%MVCs). Each of twenty-four male participants exerted ten different gripping forces (10% ~ 100% MVC at intervals of 10%) with a multi-finger force measurement (MFFM) system, and rated his subjective ratings by using of visual analog scales (-50= very discomfort and 50= very comfort). Analysis of variance showed that it significantly changed from comfort to discomfort as %MVC increased, especially around 65% of MVC. These results would be useful for hand tool design and further research including other physical factors is needed to find their effects on comfort and discomfort.

Keywords: Gripping, Force Level, Comfort, Discomfort

INTRODUCTION

Comfort and discomfort are controversial issues in ergonomics for product and workplace designs. Some understand comfort and discomfort as one opposite feeling but others understand them as two different feelings. There are also divergent opinions about factors causing comfort and discomfort. De Looze et al. (2003) suggested that the physical factors such as human physical capability, physical product feature, and task performance were more related with discomfort, whereas the subjective factors such as human emotion, product aesthetics, and psychosocial task environment were more related with comfort in their 'comfort and discomfort model for sitting'.

Among physical factors, a griping exertion level is probably a considerable factor for comfort and discomfort evaluation in product use, especially hand tools. Thus, the aim of this study was to identify the effects of different levels of gripping exertions (submaximal voluntary contractions; %MVCs) on subjective comfort/discomfort by using visual analog scales.

METHODS

PARTICIPANTS

Twenty-four male students participated in the experiment. The means and standard deviations of their age, height and weight were 24.3 ± 2.92 years, 178.1 ± 5.88 cm, 75.8 ± 13.34 kg, respectively.

APPARATUS

For measuring gripping force, a multi-finger force measurement (MFFM) system was made of synthetic resin by computer-aided design (SolidWorks 2008, Dassault) and rapid prototyping (3D Printers-SST 1200, Strarasys). Four miniature load cells (Model 13, Honeywell) installed in the middle of the handle were wired to an AD board and software (USB-6259 DAQ and LabVIEW, National Instrument) to obtain and analyze all gripping force from four load cells (Figures 1 and 2). The MFFM system was also designed to adjust grips spans (ranged from 45 to 65mm, with 5 mm increments) by replacing custom-made slip-on adapters (Kim and Kong, 2008).

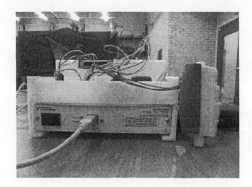

Figure 1. Multi-finger force measurement (MFFM) system

PROCEDURES

After informed consent, the participant sat on a chair, hanging the shoulder down, flexing the elbow at 90°, and remaining the wrist neutral. The participant squeezed the handle of the MFFM system three times for calculating average maximum voluntary contraction (MVC) of gripping. The experimenter then randomly set a certain level of submaximal voluntary contraction (%MVC) in the experimenter's display (Figure 2a). Without notified of the level of %MVC, the participant squeezed and maintained the handle by reaching the movable red pointer to the fixed blue pointer on the participant's display (Figure 2b). All the force measurements followed the Caldwell regimen (Caldwell et al., 1974). The visual analog scale (VAS) of 100 mm (-50 = very discomfort, 50 = very comfort) was provided after each measurement for comfort and discomfort evaluation of gripping exertion.

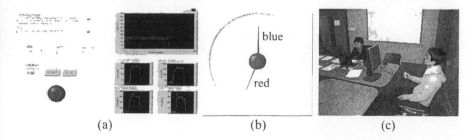

(a)   (b)   (c)

Figure 2. Experimental setup: (a) display for the experimenter only, (b) display for the participant only, and (c) experimental scene

EXPERIMENTAL DESIGN

The independent variable for analysis of variance (ANOVA) was ten levels of %MVC (10% ~ 100% MVC at intervals of 10%), and the dependent variable was subjective rating of VAS. The Tukey's Studentized Ranged (HSD) test followed ANOVA at a significance level of 0.05.

RESULTS

The results of ANOVA revealed that there was a significant effect of %MVC on the subjective ratings of comfort and discomfort ($F(9, 182) = 47.57$, $p < 0.0001$). The Tukey test showed 8 different %MVC ranges of 10% ~ 20%, 20% ~ 30%, 30% ~ 40%, 40% ~ 50%, 60%, 70% ~ 80%, 80% ~ 90%, and 90% ~ 100% (Table 1). Generally, the low levels of %MVC were rated as comfort and the high levels were rated as discomfort. Interestingly, the ratings changed from comfort to discomfort around 65% MVC (Figure 3).

Table 1. Tukey test results of %MVC

%MVC	Gripping force (N)		Subjective rating		
	Mean	SD	Mean	SD	Tukey
10	38.6	6.06	37.1	13.57	A
20	74.4	10.36	30.4	13.88	AB
30	110.8	15.16	24.8	14.43	BC
40	146.0	20.34	17.1	16.60	CD
50	182.2	24.30	12.4	14.72	D
60	220.2	33.44	3.4	19.74	E
70	256.0	34.09	-6.0	20.31	F
80	295.0	40.39	-12.4	24.24	FG
90	330.3	48.79	-19.7	22.21	GH
100	370.6	53.06	-24.3	21.11	H

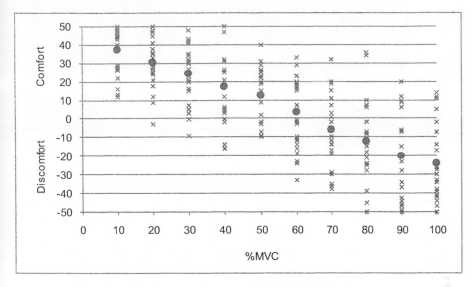

Figure 3. Subjective ratings of comfort and discomfort by %MVC (The circled dot represents a mean for each %MVC)

DISCUSSION

Many studies utilized comfort (Li, 2002; Groenesteijn et al., 2004; You et al., 2005; Kim and Kong, 2008; Lee et al., 2009) or discomfort (Freund et al., 2000; Spielholz et al., 2001; Kuijt-Evers et al., 2005; Kong et al., 2008) to evaluate products or workplaces. Zhang et al. (1996) showed that subjective factors were associated with comfort, while physical factors were associated with discomfort; however, De Looze et al. (2003) introduced that discomfort could influence comfort at the certain levels of human physical capability through a comfort and discomfort model.

The results of this study revealed that a negative effect of force levels of gripping exertion on comfort and discomfort; the lower force exerted, the more comfort (or the higher force exerted, the more discomfort). In addition, there was a certain level of submaximal gripping force exertion (about 65% MVC in this study) at which the participants changed their feeling from comfort to discomfort in a gripping exertion.

In conclusion, this study examined comfort and discomfort of the hand associated with gripping exertion levels. The results found in this study would be useful for hand tool design, and further research would be necessary to clearly identify the relationship between physical factors and comfort and discomfort.

REFERENCES

Caldwell, L.S., Chaffin, D.B., Dukes-Dobos, F.N., Kroemer, K.H.E., Laubach, L.L., Snook, S.H., Wasserman, D.E. (1974), "A proposed standard procedure for static muscle strength testing." *American Industrial Hygiene Association Journal*, 35, 201-206.

De Looze, M.P., Kuijt-Evers, L.F.M., and Van Dieën, J. (2003), "Sitting comfort and discomfort and the relationships with objective measures." *Ergonomics*, 46, 985–997.

Freund, J., Takala, E.P., and Toivonen, R. (2000), "Effects of two ergonomic aids on the usability of an in-line screwdriver." *Applied Ergonomics*, 31, 371-376.

Groenesteijn, L., Eikhout, S.M., and Vink, P. (2004), "One set of pliers for more tasks in installation work: The effects on (dis)comfort and productivity." *Applied Ergonomics*, 35, 485-492.

Kim, D.M. and Kong, Y.K. (2008), "Development of the "A Development of the "Adjustable multi-finger force measurement (MFFM) systems" for research of handtool-related musculoskeletal disorders, *Applied Human Factors and Ergonomics, 2nd International Conference*, Caesars Palace, Las Vegas, USA, 14-17 July.

Kong, Y.K., Lowe, B.D., Lee, S.J., and Krieg, E.F. (2008), "Evaluation of handle shapes for screwdriving, *Applied Ergonomics*, 39 (2), 191-198.

Kuijt-Evers, L.F.M., Twisk, J., Groenesteijn, L., De Looze, M.P., and Vink, P. (2005), "Identifying predictors of comfort and discomfort in using hand tools." *Ergonomics*, 48, 692-702.

Lee, S.J., Kong, Y.K., Lowe, B.D., and Song, S. (2009), "Handle grip span for optimizing finger specific force capability as a function of hand size", *Ergonomics, 52 (5), 601-608.*

Li, K.W. (2002), "Ergonomic design and evaluation of wire-tying hand tools." *International Journal of Industrial Ergonomics*, 30, 149-161.

Pheasant, S.T., and Scriven, J.G. (1983), "Sex differences in strength: Some implications of the design of handtools." *Proceeding of the Ergonomics Society's Annual Conference*, 9-13.

Spielholz, P., Bao, S., and Howard, N. (2001), "A practical method for ergonomic and usability evaluation of hand tools: A comparison of three random orbital sander configurations." *Applied Occupational and Environmental Hygiene*, 16, 1043-1048.

You, H., Kumar, A., Young, R., Veluswamy, P., and Malzahn, D.E. (2005), "An ergonomic evaluation of manual Cleco plier designs: Effects of rubber grip, spring recoil, and worksurface angle." *Applied Ergonomics*, 36, 575-583.

Zhang, L., Helander, M.G., and Drury, C.G. (1996). "Identifying factors of comfort and discomfort in seating." *Human Factors*, 38, 377-389.

CHAPTER 81

The Measurement of Grip-Strength in Automobiles: A New Approach to Detect Driver's Emotions

Felix W. Siebert, Michael Oehl, Hans-Ruediger Pfister

Institute of Experimental Industrial Psychology
Leuphana University of Lueneburg
Germany

ABSTRACT

This experimental study investigated grip-strength as a new non-invasive method to detect emotion. A positive emotion (happiness) and a negative emotion (anger) were examined regarding their influence on grip-strength applied to the steering wheel. Results confirmed and extended preliminary findings of LaMont (2000): Subjects showed increased grip-strength while driving a car when strong emotions like anger or happiness were induced. Implications for further research as well as for praxis will be outlined.

Keywords: Traffic Psychology, Detection of Driver's Emotions, Grip-Strength

INTRODUCTION

Human factors are the main reason of traffic accidents in the European Union (European Union, 2001). To counter this, politics focused on improving safety measures that are implemented in the traffic surroundings, e.g., guard railing, while car manufacturers improved passenger protection. Albeit these improvements, research suggests that emotions are a main cause for maladjusted driving that leads to accidents (e.g., Deffenbacher, Oetting, and Lynch, 1994; Mesken, Hagenzieker, Rothengatter, and de Waard, 2007; Nesbit, Conger, and Conger, 2007). Therefore the focus of this experimental study lies on the *detection of emotions in automobile drivers*.

DETECTION OF DRIVERS' EMOTIONS

Since the methods used to detect emotions needed to be applicable in an automobile, we have only chosen *non-invasive* measurements that do not interfere with the drivers' ability to control the car.

Earlier studies tried to detect emotions in general with a number of non-invasive measurements, e.g., speech (Scherer, Johnstone, and Klasmeyer, 2003) and facial expression (Cohn, and Ekman, 2005). To detect drivers' emotions, a new additional measure – *grip-strength* – was applied and evaluated in this current experimental study. Since there is no data to support the claim that only a certain valence (positive or negative) of emotions leads to maladjusted driving, we have chosen happiness as a prototypical positive emotion, and anger as a prototypical negative emotion to be induced in our experiment. Moreover, these two emotions are experienced very frequently by drivers in traffic (Levelt, 2003; Mesken et al., 2007). We classified emotions following the approach of Scherer's multi-dimensional emotion-space (Scherer, 2005). Scherer suggests that emotions can be classified by two dimensions, i.e., valence and arousal. Anger and happiness rate high on arousal. While anger ranks negative, and happiness ranks positive on the valence dimension. Emotions were induced in participants on the basis of the appraisal-theory (Lazarus, 1991).

EMOTIONS

As explained in the previous paragraph, the goal of this study was to find non-invasive measurements that can differentiate between emotions of different valence. This paragraph links emotions to physiological changes in the human body, thereby laying the ground for the use of physiological measurements to detect emotions.

Research hasn't found an unambiguous definition for the term emotion (Janke, Debus and Schmidt-Daffy, 2008; Scherer, 2005; Schmidt-Atzert, 1996). Most definitions have in common that they take into account different components of emotions. In this study we used the definition of Scherer (2001), as the basis for our research, in which emotion is defined as *"an episode of interrelated, synchronized changes in the states of all or most of the five organismic subsystems in response to the evaluation of an external or internal stimulus event as relevant to major concerns of the organism"*. The five subsystems described by Scherer are presented in table 1 (adopted from Scherer, 2005). They include a cognitive, a neurophysiological, a motivational as well as a motor expression component. Each of these components is linked to a part of the nervous system. It is debated if every subsystem has its own unique reaction-pattern for different emotions. At least for the facial expression a consensus seems to have been reached leading to a development from the facial action coding system (Ekman, and Friesen, 1978) to modern day automatic emotion recognition by facial expression. This study focused on the detection of the valence of emotion by measuring the neurophysiological component of emotion.

Table 1 Relationships between organismic subsystems and the functions and components of emotion (according to Scherer, 2005)

Emotion function	Organismic subsystem and major substrata	Emotion component
Evaluation of objects and events	Information processing (CNS)	Cognitive component (appraisal)
System regulation	Support (CNS, NES, ANS)	Neurophysiological component (bodily symptoms)
Preparation and direction of action	Executive (CNS)	Motivational component (action tendencies)
Communication of reaction and behavioral intention	Action (SNS)	Motor expression component (facial and vocal expression)
Monitoring of internal state and organism-environment interaction	Monitor (CNS)	Subjective feeling component (emotional experience)

Note: CNS = central nervous system; NES = neuro-endocrine system; ANS = autonomic nervous system; SNS = somatic nervous system.

While research suggests that emotions affect physiological activities of the human body, no consensus has been reached on whether there are discrete variables that accompany every specific emotion. Kreibig, Wilhelm, Roth, and Gross (2007) found that fear and sadness can be distinguished from a neutral emotional state by using cardiovascular, electrodermal and respiratory response patterns. Kreibig et al. achieved a detection rate between 69.00 % and 84.50 % analyzing fear and sadness.

In our research we pursue a multi-methodic approach, while only using non-invasive measures that could be applied in an automobile. In addition, the primary goal of this study is to detect different emotional states of the driver by measuring the driver's grip-strength applied to the steering wheel. This study follows the definition of Scherer (2001, 2005). Emotions take effect on different parts of the nervous system.

The two central research questions addressed to our current experimental study are:

1) Is it possible to detect emotions by driver's grip-strength applied to the

steering wheel as a new additional non-invasive measure to detect emotions in an automobile?

2) Is it possible to differentiate emotions that rank different on the dimension of valence but have an equal arousal level?

METHOD

This experiment was based on a one-factorial design with repeated measurements. The *independent variable* was the *induced emotion in the driver*. The induction was between-subjects three-tiered in *neutral vs. anger* vs. *happiness*. The *dependent variable* was driver's *grip-strength applied to the steering wheel*.

In the first part of this paragraph, we will explain the procedure of our study. Then the method to induce emotions is presented in further detail. At the end of this paragraph, the measurement of grip-strength is outlined.

Participants ($N = 22$), i.e., 10 male and 12 female at the age of $M = 26.55$ years ($SD = 9.77$), were asked to drive predefined routes in a driving-simulator STISIM DRIVE™ 2.0. The routes consisted of country roads as well as city roads and there was sparse oncoming traffic. The design of the routes was modeled after real life traffic-routes, to ensure high validity. Every participant drove at least two practice routes to get used to the driving simulator. After these practice routes, the two experimental routes followed in repeated measurements for the dependent variable. The first experimental road was without induced emotions, i.e., the *neutral route*. The second experimental route was with induced emotions, i.e., the *emotional route*.

After participants finished the neutral route they were asked to do a short test of their theoretic knowledge about driving. This test was used to induce a specific emotion in participants. The method of induction is explained later in this paragraph. Following this short test, participants drove the second experimental route, i.e., the emotional route. The first neutral and the second emotional experimental route were identical, but this time the surroundings of the road and the appearance of the cars on the road were changed in order to leave participants unaware that they were driving on the same route as before.

In this experiment we induced emotions as the independent variable. Participants were randomly assigned to three different experimental groups, labeled according to the induced emotions *anger, happiness,* and *neutral*. The neutral group without induced emotion served as a baseline for the dependent variable grip-strength as well as for the control of sequence effects due to repeated measurements. Anger and happiness were chosen due to their high occurrence in traffic (Levelt, 2003; Mesken et al., 2007). *Emotions were induced through a real-life situation*. Participants were asked to take a computerized test consisting of 15 questions, measuring their theoretical knowledge about driving. The test was similar to the test used in the German driver license test. Participants in the neutral group received no feedback concerning their performance in this test. Participants in the happiness group received the following feedback message (translated from German): "Thank you for your participation. You answered 15 out of 15 questions right." This message was given regardless of the true performance, to induce happiness in the participants. In the anger group the test consisted of questions of high difficulty. Also, the handling of the mouse was hindered, since its sensitivity was turned down. Furthermore, the test-program crashed at the last question and participants were asked to take the test again. The second time they took the test, it crashed again. The second crashing was followed by standardized remarks by the test supervisor, blaming the participant for the crash of the program. This was done to turn the frustration of participants into anger, following the appraisal theory (Lazarus, 1991). We evaluated the induction of emotions in pre-tests.

As *dependent variable* served the *driver's grip-strength* applied to the steering wheel as a new measure to detect driver's emotions in an automobile environment. To measure grip-strength in this experiment, we used a unique steering wheel provided by the KOSTAL Group. The steering wheel had a built in fiber that measured deformation caused by force applied on the steering wheel by participants' hands. Forces were measured all around the steering wheel, giving a unit less indication for the applied grip-strength. Figure 1 shows the schematic structure of the steering wheel. On the upper left a cross section of the steering wheel is shown while no pressure is applied. On the bottom left the same structure is shown with applied pressure. The upper right shows a single fiber that is used to measure the grip-strength. On the bottom right, the alignment of multiple fibers is shown within the steering wheel.

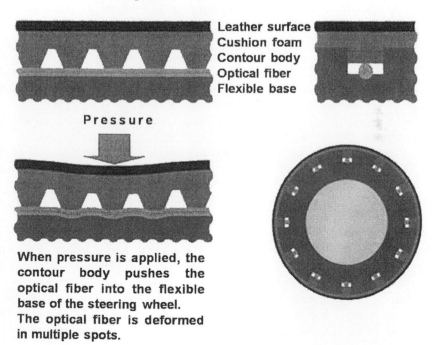

Leather surface
Cushion foam
Contour body
Optical fiber
Flexible base

Pressure

When pressure is applied, the contour body pushes the optical fiber into the flexible base of the steering wheel. The optical fiber is deformed in multiple spots.

FIGURE 1. Schematic structure of the grip-strength sensitive steering wheel (KOSTAL Group).

RESULTS

Taking into account the small sample size ($N = 22$) as well as the explorative approach of this study, the results are only analyzed descriptively. During the session of one participant technical difficulties occurred. Consequently the data of this participant were not included into further analyses. Since the steering-wheel *measuring of grip-strength was unit less*, it is important to see the results in the context of the possible maximum of applied grip-strength. Pre-tests showed that grip-strength could go up to approximately 12,000. Overall, participants showed an average grip-strength of approximately $M = 1,903$ ($SD = 775.12$) while driving. Due to the unit-less measurement of the grip-strength in our experimental study and additionally due to the large differences in applied grip-strength between participants, especially the relative change of grip-strength within subjects (neutral vs. emotional route) is significant (table 2).

Participants in the *neutral group* ($n = 5$) showed little change in grip-strength, comparing the neutral route to the emotional route. During the neutral route, the average grip-strength was $M = 2,000.71$ while during the emotional route it was $M = 2,066.97$. This is an increase of only 1.66%.

To find possible effects of sequence due to repeated measurements, it is additionally important to analyze the grip-strength of the neutral group (neutral vs. emotional route) in this regard. If there would have been signs of fatigue for example, they would have shown in the data as a decrease in grip-strength. Results suggest that there is no effect of sequence, since there is a minimal increase in grip-strength in the neutral group.

The *happiness group* ($n = 9$) showed an increase in grip-strength. In the neutral route the average grip-strength was $M = 1,911.16$ and in the emotional route it was $M = 2,053.54$. This is an increase of approximately 8.19% in grip-strength.

In the *anger group* ($n = 7$) there was also an increase in average grip-strength within the participants. During the neutral route their average grip-strength was $M = 1,656.51$ and during the emotional route they applied a grip-strength of $M = 1,758.36$. This equals an increase in grip-strength of approximately 9.40%. These results, including the standard deviation are shown in table 2.

Analysis of the data suggests that there are differences in the applied grip-strength between men and women. Men ($n = 10$) had an average grip-strength of $M = 2,247.81$ ($SD = 754.66$) in the neutral route, while women ($n = 11$) had an average grip-strength of $M = 1,522.23$ ($SD = 488.53$). Men also had a higher minimum (1039.76) and maximum (3248.28) grip-strength while driving the neutral route compared to women's minimum (975.08) and maximum (2,628.87) grip-strength. These findings are similar to the research of Eksioglu and Kizilaslan (2008), i.e., male drivers apply a higher grip-strength than female drivers. This might be an explanation for the high overall grip-strength in the neutral group compared to the other emotional groups, since it consisted of four men and one woman. There were three men and six women in the happiness group, and three men and four women in the anger group.

Table 2 Grip-strength mean, standard deviation and relative change (%) of the three experimental groups (neutral vs. emotional route)

	Neutral route	Emotional route	Change of grip-strength
Neutral group (n = 5)	M = 2,000.71 (SD = 816.49)	M = 2,066.97 (SD = 1036.66)	+1.66%
Happiness group (n = 9)	M = 1,911.16 (SD = 743.36)	M = 2,053.54 (SD = 834.45)	+8.19%
Anger group (n = 7)	M = 1,656.51 (SD = 720.69)	M = 1,758.36 (SD = 774.72)	+9.40%

CONCLUSIONS

Our first research question asked if it was possible to detect emotions by driver's grip-strength applied to the steering wheel as a new additional non-invasive measure to detect emotions in an automobile. Our results suggest that grip-strength can contribute to the detection of emotion. Grip-strength measures were increased in the 16 participants of the anger and happiness group, compared to participants of the neutral group. Participants that felt emotions had a higher grip-strength than participants that drove without emotion. The increase was slightly higher in the anger group (9.40%), compared to the happiness group (8.19%).

Our second research question asked if it was possible to differentiate emotions that rank different on the dimension of valence but have an equal arousal level. It is still not clear if a differentiation between anger and happiness by using grip-strength is possible. There was only a 1.21% difference comparing the change of grip-strength in the anger and happiness group. Considering the small sample-size as well as the explorative approach of our study, we cannot rule out the possibility that grip-strength can contribute to the differentiation of high arousal emotions. Our results for differences of grip-strength between men and women are similar to the results of Eksioglua and Kizilaslan (2008). Men applied a higher grip-strength while driving than women. Future research should take into account these differences, especially in the formation of the different experimental groups. These groups should consist of an equal number of men and women.

For future research it is important to refine the measurement of grip-strength. The steering-wheel provided to us, was a unique prototype. Therefore fine-tuning in the algorithm of the calculation of grip-strength may be possible.

Our results suggest that grip-strength can help to differentiate between high- and low-arousal emotions or at least between a non-emotional state of the driver and a critical emotional state like anger. Since critical emotional states lead to maladjusted driving, this differentiation could help in preventing accidents.

In its application this research could lead to adaptive advanced driver assistance systems (ADAS) depending on the emotional state of the driver and thereby increasing safety of traffic, e.g., an alerted automatic brake-system when the driver is detected to be angry and thus probably will show maladjusted traffic behavior. The ADAS may be realized within a multi-methodic system, taking into account additional non-invasive measures, e.g., facial expressions or speech, thereby increasing the validity of the emotion detection.

ACKNOWLEDGEMENTS

This research was funded by grant FS-Nr. 2006.63 from the 'Arbeitsgruppe Innovative Projekte' (AGIP) of the Ministry of Science and Culture, Lower Saxony, Germany. We would like to thank the Volkswagen Group (Group Research Electronics and Vehicle Research / HMI and Human Factors) as well as the KOSTAL Group for technical support. Many thanks go to Janina Suhr and Jennifer Hohmann for their help in conducting the experiment as well as to Claus Hunsen and Wolfgang Kozian for their help in analyzing the experiment and their support in applied computer science.

REFERENCES

Cohn, J.F., and Ekman, P. (2005), Measuring facial action. In J. A. Harrigan, R. Rosenthal, and K. R. Scherer (Eds.), *The new handbook of methods in nonverbal behavior research* (pp. 9-64). Oxford University Press, New York, USA.

Deffenbacher, J.L., Oetting, E.R., and Lynch, R.S. (1994), Development of a driving anger scale. *Psychological Reports, 74*, 83–91.

Ekman, P., and Friesen, W.V. (1978), *Facial action coding system. A technique for the measurement of facial movement*. Consulting Psychologists Press, Palo Alto, USA.

Eksioglua, M., and Kizilaslan, K. (2008), Steering-wheel grip force characteristics of drivers as a function of gender, speed, and road condition. *International Journal of Industrial Ergonomics, 38*, 354-361.

European Commission (2001), *White paper-European transport policy for 2010: time to decide*. Office for Official Publications of the Europe Communities, Luxemburg.

Janke, W., Debus, G., and Schmidt-Daffy, M. (2008), *Experimentelle*

Emotionspsychologie – Methodische Ansätze, Probleme, Ergebnisse [Experimental Emotional Psychology]. Pabst Science Publishers, Lengerich, Germany.

Keil, A., Smith, J.C., Wangelin, B.C., Sabatinelli, D., Bradley, M.M., and Lang, P.J. (2008), Electrocortical and electrodermal responses covary as a function of emotional arousal: A single-trial analysis. *Psychophysiology, 45*(4), 516-523.

Kreibig, S.D., Wilhelm, F.H., Roth, W.T., and Gross, J.J. (2007), Cardiovascular, electrodermal, and respiratory response patterns to fear- and sadness-inducing films. *Psychophysiology, 44*, 787–806.

LaMont, K. (2000), Effect of Induced Emotional State on Physical Strength. *In Perspectives in Psychology*. University of Pennsylvania, PA, USA.

Lazarus, R.S. (1991), *Emotion and Adaptation*. Oxford University Press, Oxford, USA.

Levelt, P.B.M. (2003), *Praktijkstudie naar emoties in het verkeer* [Emotions in Traffic]. *SWOV Report R-2003-08*. SWOV, Leidschendam, Netherlands.

Mesken, J., Hagenzieker, M.P., Rothengatter, T., and de Waard, D. (2007), Frequency, determinants, and consequences of different drivers' emotions: An on-the-road study using self-reports, (observed) behaviour, and physiology. *Transportation Research Part F: Traffic Psychology and Behaviour, 10*(6), 458-475.

Min, Y., Chung, S., and Min, B. (2005), Physiological evaluation on emotional change induced by imagination. *Applied Psychophysiology and Biofeedback, 30*, 137–150.

Nesbit, S. M., Conger, J. C., and Conger, A. J. (2007), A quantitative review of the relationship between anger and aggressive driving. *Aggression and Violent Behavior, 12*(2), 156-176.

Scherer, K.R. (2005), What are emotions? And how can they be measured?. *Social Science Information, 44*(4), 695-729.

Scherer, K.R., Johnstone, T., and Klasmeyer, G., (2003), Vocal expression of emotion. In R.J. Davidson, H. Goldshmith, and K.R. Scherer (Eds.), *Handbook of affective sciences* (pp. 433-456). Oxford University Press, Oxford, GB.

Scherer, K.R. (2001), Appraisal Considered as a Process of Multi-Level Sequential Checking. In K.R. Scherer, A. Schorr, and T. Johnstone (Eds.), *Appraisal Processes in Emotion: Theory, Methods, Research* (pp. 92–120). Oxford University Press, New York, USA.

Schmidt-Atzert, L. (1996), *Lehrbuch der Emotionspsychologie* [Emotional Psychology]. Kohlhammer, Stuttgart, Germany.

Chapter 82

On Quantification Method of Step Visibility: Basic Investigation on Influence of Observation Conditions

Kang Yonghak, Mikami Takamasa

Department of Mechanical and Environmental Informatics
Graduate School of Information Science and Engineering
Tokyo Institute of Technology
2-12-1, O-okayama, Meguro-ku, Tokyo, 152-8552, JAPAN

ABSTRACT

Steps and stairs are common features in our day-to-day living and working environment—found outside, in public parks, busy urban streetscapes, individual building sites, and within every home. Unfortunately, the occurrence of accidents on or around steps is thus inevitable. A failed visibility to see or recognize a step remains the most common attribute in these accidents. Surprisingly, appropriate methods for examining and quantifying step visibility have yet to be established. In this research, sensory scales of step visibility were created in order to examine the effects of three factors on step visibility in sensory testing methods; visual characteristics of steps, observation distances from steps and observation actions while viewing step. Results revealed that step visibility tends to decline as observation distances increase, and the quality of step visibility varies based on the unique and individual visual characteristics of a each step. Moreover, we assumed that step visibility has a correlation with horizontal deviation, vertical deviation, and surface roughness of steps, and relative pattern size on both upper and lower landings, confirming that relative rating of step visibility changes and are based heavily on observation distances.

Keywords: steps, visibility, observation condition, visual characteristic, safety

INTRODUCTION

Steps, as shown in Fig. 1, exist widely in all living environments. Some steps lack clear visibility, resulting in unforeseen falls of consequential stumbling, and are potential causes for life-threatening accidents.

Official accident reports are regularly recorded concerning falling accidents on and from stairs and steps, as shown in Table 1. However, from such statistics, it remains difficult to identify data related specifically to step visibility—the actual situations of the accidents remain vague.

Past research on the safety of steps has been carried out focusing on— the shock absorbance of landing surfaces during descending motion from the viewpoint of load (Ono, 1992), or foot clearance during stair descending from the viewpoint of effects of age and illumination (Hamel, 2005). However, concerning step visibility, limited research has succeeded in proposing comprehensive and practical measures for quantifying step visibility (Ono, 1993).

This research investigates developing a method for quantifying step visibility from the viewpoint of safety. Sensory tests were carried out based on the visual easiness of steps. Assessors, who have normal eyesight, observed steps from a descending direction. An additional aim was to examine the effects of three factors on step visibility with sensory testing methods; visual characteristics of steps, observation distances and observation actions.

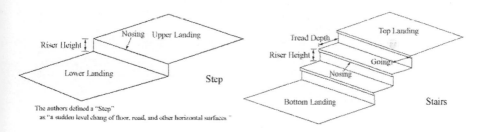

FIGURE 1 Schematic comparison of a step and stairs

Table 1 Statistics of accidental deaths concerning stairs and steps in Japan
(ICD-10, W10)

Code No	Year Content	1995	2000	2004	2005	2006	2007
W10	Fall on and from stairs and steps	720	653	671	687	666	652

METHODS

OUTLINE OF THE TEST

As shown in Table 2, two sensory scales were formulated, applying the method of successive categories. The scales are visual easiness of steps in standing and walking action. (Visibility scales).

STEP SAMPLES

Thirty-two step samples, shown in Table 3, were designed so as to provide a wide range of visual variety. Each sample was created with unique materials that can be differentiated by texture, gloss, and pattern, for its upper and lower landing areas and the nosing part. A riser height level of 20 cm was set for sample tests.

In addition, it will be our future works to examine the influence of other environmental factors such as handrails and side walls to the visibility of steps.

OBSERVATION CONDITION

Four observation distances were set for the test, 0.7m, 3.5m, 7m and 14m, determined by eyesight and footstep of persons.

Two observation actions for assessors were set during testing. The first was to remain still during observation (standing action). The second was to walk one or two steps back and forth during observation (walking action).

During the test, we did not limit the observation time for assessors, and instead, asked them to watch the samples carefully.

In the test, 32 samples were designed, four observation distances, and two observation actions. Therefore, a total of 256 answers were received per person.

LIGHTING CONDITION

Sensory tests were conducted in a darkroom where lighting conditions were controlled. The darkroom, as shown in Fig. 2, consists of a box room where step samples were set. A 14m long pathway, directly adjacent to the box room, is where observations by assessors were made.

The dimensions of the box room are 2.4m in height, 1.1m in width and 3.0m in depth. The box room has four D65 fluorescent lights installed on the ceiling. The position and direction of lights were set maintaining an almost uniform illuminance level on both upper landing surface areas.

In addition, lighting conditions were limited to only one condition. It will be our future work to examine the influence of various lighting conditions.

The pathway, as shown in Fig. 2, allows the observation frame to be adjusted according to observation distance. The observation hole, within the observation frame, is located 1.4m above the pathway floor.

ASSESSORS OF THE TEST

Eleven physically unimpaired persons were tested; nine males and two females. Assessors age and eyesight range are shown in Table 2.

It will be our future work to investigate cases of sight restricted persons.

Table 2 Outline of test

Test Method	Method of successive categories (Absolute judgement)		
Observation Actions	Standing action		Walking action
Scales	Step visibility scales		
Question and Answers	How well can you recognize the existence of this step?		
	(6) extremely well	(5) very well	(4) well
	(3) slightly well	(2) faintly well	(1) I cannot recognize it at all
Assessors	11 adults; nine males and two females (Age : 19~21, Eyesight : 0.8~1.5)		
Samples	Sample 1~32 (32 types) × Observation distance 0.7, 3.5, 7, 14m (4 levels) : Total 128 types Riser height : 20cm (1 level)		
Observation condition	Observation position : Horizontal distance from nosing – 0.7, 3.5, 7, 14m (4 levels), Vertical distance from floor – 140cm Observation hole : Height 4cm × Width 20cm, Observation time : No limit		
Lighting condition	Source of light : D65 fluorescent lamp × 4		

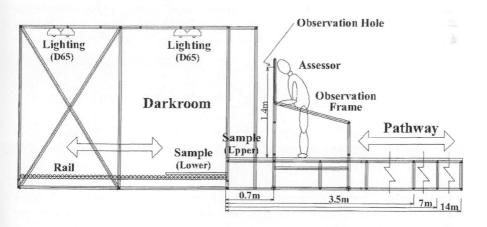

FIGURE 2 Layout of darkroom and pathway

Table 3 Photos of samples

Sample 1	Sample 2	Sample 3	Sample 4	Sample 5	Sample 6	Sample 7	Sample 8
Sample 9	Sample 10	Sample 11	Sample 12	Sample 13	Sample 14	Sample 15	Sample 16
Sample 17	Sample 18	Sample 19	Sample 20	Sample 21	Sample 22	Sample 23	Sample 24
Sample 25	Sample 26	Sample 27	Sample 28	Sample 29	Sample 30	Sample 31	Sample 32

RESULTS AND DISCUSSION

RESULTS ON SENSORY TEST

Raw test data was obtained by method of successive categories. This raw data was analyzed under both standing and walking actions to produce two sensory scales: visibility scale under standing action and visibility scale under walking action. Each sensory scale produced 128 individual values. (32 samples by 4 observation distances)

Coincidentally, the raw test data was examined for statistical significance by variance analysis.

THE EFFECT OF OBSERVATION DISTANCE ON STEP VISIBILITY

Fig. 3 and Fig. 4 show an example of the relation of the two scales of step visibility with varying observation distances on both the standing and walking actions.

From Fig. 3 and Fig. 4, on the case of each observation action, we understand that a farther observation distance equates to a reduction in step visibility evaluation for all samples. Moreover, we understand that the relative rating of step visibility for samples varies based on altering observation distances. For example, we understand that an easily recognized step at a shorter distance has no assurance of easy recognition at a further distance.

In a following section, we examined the relationship of visual characteristics of each step sample with the above-mentioned observation distances.

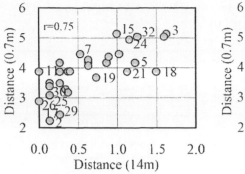

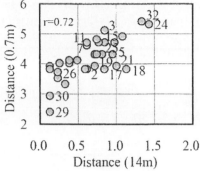

FIGURE 3 Correlation of two scales of step visibility under varying observation distances (Standing action)

FIGURE 4 Correlation of two scales of step visibility under varying observation distances (Walking action)

THE EFFECT OF OBSERVATION ACTION ON STEP VISIBILITY

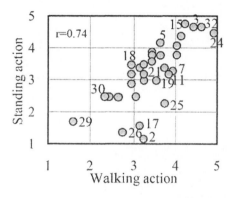

FIGURE 5 Correlation of the two visibility scales with both observation actions (3.5m)

Fig. 5 reveals the relationship of the two visibility scales under both observation actions.

From Fig. 5, in regards to step visibility, we understand that most samples show similar tendencies at both standing and walking actions. Moreover, we understand that certain samples reveal beneficial qualities in step visibility under walking action, but reveal ineffective qualities under standing action. In other word, the samples are unable to correctly predict step visibility under walking action based on standing action evaluations, because the relative ratings are different and independent of each sample.

In a following section, we examine the relationship of visual characteristics of step samples and the above-mentioned observation actions.

TRIAL EXTRACTION OF THE VISUAL CHARACTERISTICS OF STEPS THAT RELATE TO STEP VISIBILITY

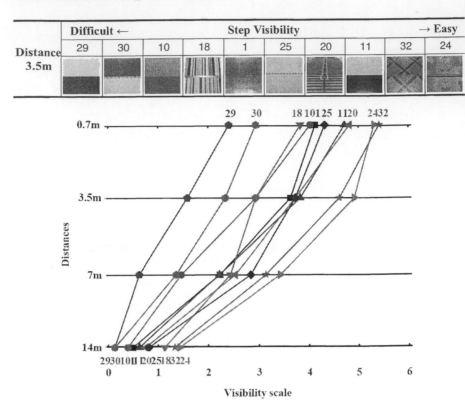

FIGURE 6 The relationship between visibility scale of steps with changes in observation distances (Walking action, 10 typical samples)

Fig. 6 is an axial expression graph exhibiting step visibility of 10 typical samples with varying observation distances, under walking action. The graph exhibits more difficult step visibility samples towards the left and easier step visibility samples further to the right. In addition, the series of photos within the graph represent samples at an observation distance of 3.5m. Similarly, step photos are organized according to step visibility evaluations with smaller scale value samples towards the left and larger scale value samples further to the right.

From results of Fig.6, we expected four characteristic visual factors of steps that relate to the existence of steps on the following.

The first factor is the easiness in recognizing the horizontal deviations of a step, based on isolated left and right eye observations. The second factor is the easiness in recognizing the vertical deviations while in motion. The third factor is the easiness in recognizing the surface roughness of steps. The fourth factor is the easiness in recognizing the relative pattern size on both upper and lower step landings.

We considered characteristics of each sample concerning these four factors on the following.

Samples that performed well due to horizontal deviation are samples 1, 11, 18, 20, 24, 32.

Samples that performed well due to vertical deviation are samples 20, 24, 25, 32.

Samples that performed well due to surface roughness of steps are samples 1, 10, 24.

Samples that performed well due to relative step size are samples 1, 24, 32.

Next, we comprehensibly examined potential causes of two specific samples, 29, 30, that exhibited very low visibility evaluations.

Concerning sample 29, the surface is smooth, and has a plain color, thus, assessors exhibited strong difficulty in recognizing any roughness or pattern on the surface. In addition, the recognition of any horizontal or vertical deviation was difficult due to the relative pattern size on both upper and lower landings, as well as the in ability to have a sharp, visual focus on the step.

Concerning sample 30, the upper surface landing has a very fine, longitudinal stripe pattern and the lower surface landing has a plain color with no pattern, thus making it very difficult to recognize the horizontal deviations. In addition, the upper surface landing's roughness differs from the lower landing's smooth surface, resulting in an inability to focus on the sample. Furthermore, the relative pattern sizes on both landings make it very difficult to recognize.

THE RELATIONSHIP OF A CHANGE IN STEP VISIBILITY DUE TO OBSERVATION DISTANCES AND VISUAL CHARACTERISTICS OF STEPS

From Fig. 3 and Fig. 4, we examined and grouped the changing tendency of step visibility due to observation distances in table 3 and table 4. Those concerned the 16 samples that exhibited this characteristic tendency.

Table 4 Changing tendency of step visibility based on varying observation distances (Standing action, 16 samples) (○ : Good △ : Normal × : Bad)

Sample	2	3	5	7	11	15	17	18	19	21	24	25	26	29	30	32
Short distance	×	○	△	○	△	○	×	△	△	△	○	×	×	×	△	○
Long distance	×	○	○	×	×	△	×	○	△	○	○	×	×	×	×	○

From table 4, samples that exhibit easy recognition evaluations due specifically to relative pattern size, horizontal deviation, surface roughness, and display large patterns on surfaces, are easy to recognize at both shorter and longer distances. Samples that are composed of minute surface roughness or reveal similar patterns, colors, or other visual characteristics, on both upper and lower landings, are easy to recognize step existence at the shorter distance. However, as distances increase, these samples become more difficult to recognize. Samples that exhibit no visual characteristics on either upper or lower landings are difficult to recognize step existence in both shorter and longer distances.

Table 5 Changing tendency of step visibility based on varying observation distances (Walking action, 16 samples) (○ : Good △ : Normal × : Bad)

Sample	2	3	5	7	11	15	17	18	19	21	24	25	26	29	30	32
Short distance	△	○	△	○	○	○	△	△	△	△	○	△	△	×	×	○
Long distance	△	△	△	△	△	△	△	○	△	○	○	△	×	×	×	○

From table 5, samples that are easy to recognize in terms of relative pattern size, horizontal deviation, vertical deviation and surface roughness, and displays large surface patterns, are easy to recognize step existence at both shorter distance and longer distances. However, samples are to have easier recognition of vertical deviation of lower landing, to be easier recognition of the existence of steps. Samples that are composed of minute surface roughness or reveal similar patterns, colors, or other visual characteristics, on both upper and lower landings, are easy to recognize step existence at the shorter distance. However, as distances increase, these samples become more difficult to recognize. Samples that exhibit no visual characteristics on either upper or lower landings are difficult to recognize step existence in both shorter and longer distances.

THE RELATIONSHIP OF DIFFERENCES IN STEP VISIBILITY BASED ON OBSERVATION ACTIONS AND VARYING VISUAL CHARACTERISTICS OF STEPS

From Fig. 5, the following concerns the effects of step visibility due to observation actions.

Samples that are easily recognizable with both standing and walking actions are samples 3, 15, 18, 19, 21, 24, 32. Samples those are difficult to recognize relatively during walking action compared with standing action are samples 5, 30. Samples that are easily recognizable relatively during walking action compared with standing action are samples 2, 7, 11, 17, 25, 26. Sample 29 was difficult to recognize under both standing and walking actions.

Samples that are easy to recognize—in terms of relative pattern size, horizontal deviation, vertical deviation, and surface roughness—and display large surface patterns render themselves easy to recognize under both observation actions.

Inversely, samples that are difficult to recognize—in terms of relative pattern size, horizontal deviation, vertical deviation and surface roughness—render themselves difficult to recognize under both observation actions.

Samples that are difficult to recognize based on vertical deviations, even when these samples exhibit easily recognizable horizontal deviations, remain difficult to recognize relatively under walking action.

Inversely, samples that are easily recognizable based on vertical deviations, even when these samples exhibit difficulty in recognition of horizontal deviations, are easily recognizable relatively under walking action.

CONCLUSION

In this research, visibility scale of steps was composed applying the sensory test method. Physically unimpaired person of eyesight observed steps from a descending direction. Moreover, we examined the effect of 3 factors on the composition of visibility scale; the visual characteristics of steps, observation distances and observation actions.

We obtained the following findings.

- Reduction in step visibility is tendencies of farther observation distances.
- The quality of step visibility differs with the unique visual characteristics of steps. Under even a short observation distance with similar, controlled lighting, certain samples revealed poor step visibility.
- We assumed that the easiness in recognizing steps is strongly correlated with horizontal deviation, vertical deviation, and surface roughness of steps and the relative pattern size on both upper and lower landings.
- The relative rating of the visibility of steps is changed by observation distance. Observation distance even at long distance, there are relatively small steps and big steps of reducing the visibility of steps.
- From the viewpoint of visual safety, it was desirable to exhibit qualities of good step visibility from both shorter and longer distances.

Henceforth, considerations in regards to the effects of step visibility due to various lighting conditions, we plan to develop and capture this necessary knowledge.

REFERENCES

Ono, H., Mikami, T., Nekomoto, Y., Yokoyama, Y., and Takahashi, H. (1992), "Study on evaluation of safety of difference in level." *Journal of Struct. Const. Eng.*, AIJ, (432), 19-27.

Hamel, K. A., Okita, N., Higginson, J. S., and Cavanagh, P. R. (2005), "Foot clearance during stair descent: effects of age and illumination" *Gait and Posture*, (21), 135-140.

Ono, H., Yokoyama, Y., Mikami, T., and Kusumoto, J. (1993), "Confirmation of Existence of imperceptible differences-in-level and investigations on the influence of extent of field of view and illuminance on perceptibility of difference-in-level from viewpoints of colors and patterns on the surface." *Journal of Struct. Const. Eng.*, AIJ, (443), 25-33.

Chapter 83

A Decision Support System for Unusual Behavior Detection in Nuclear Power Plants

*Yuan-He Wang, Pei-Chia Wang, Sheue-Ling Hwang, Rong-Huei Hou**

Industrial Engineering and Engineering Management
National Tsing Hua University
Hsinchu, Taiwan

**Atomic Energy Council
Taipei County, Taiwan

ABSTRACT

In highly secured areas such as nuclear power plant, alarm of the surveillance system indicates the nuclear safety is under threat, and which might be an extremely serious situation. In this study, an unusual behavior detection support system (UBDSS) is constructed to provide the nuclear power plant monitoring center an automatic mechanism to detect if a person in monitor screen shows unusual behavior, such as destroying facilities, moving abnormally, etc. Furthermore, it can also detect worker who has sudden heart attack symptom. The goal was to promote safety of an NPP.

Keywords: Unusual Behavior, Surveillance System, Nuclear Power Plant

INTRODUCTION

Nuclear power brings welfare and convenience to humanity. However, once radiation release, it may endanger life and cause pollution of the environment. For this reason, operation safety is the first priority of Nuclear Power Plant (NPP).The Safety protection system of NPP depends on software and hardware facilities to be not invaded. In addition, workers for daily supervision and maintenance may miss operation for some reasons and bring to an accident. In general, most accidents are caused by human error, and the human error is committed unintentionally. If somebody does improper operation intentionally, he will be the "insider", who possesses professional knowledge and ability, and destroys the safety system of NPP because of his aberration. Such a case has not happened in NPP, but it is not impossible (Wang, 2005). Once this case comes about, it will damage property and life seriously.

To prevent hazards the insiders intend to rise, Surveillance system is installed in NPP and there is considerable quantity of cameras around the range of plant to convey monitoring video data to the security personnel. Since there are a large number of video cameras monitored by few surveillance personnel, it is impracticable to visualize simultaneously the behavior of all the observed objects with such an amount of data provided to the user, in order to quickly and correctly detect danger situations. This reality provokes an overcharge of responsibility and loss of efficiency in the surveillance personnel (Duque et al., 2007).

Through constructing an intelligent video monitoring system, safety unit of NPP can automatically detect moving object and retrieve in database to indentify if dangerous event occurs, and further send out warning to staff. Nevertheless, researches at present mostly focus on the technology of object tracking and feature recognition, and very few study mention the information processing of such pattern as well as how it transfers to warning signal.

SITUATION AND SCENARIO OF UNUSUAL BEHAVIOR DETECTION IN NPP

Generally speaking, entrance of NPP is installed X-ray security inspection equipment, and thus any person attempt to carry dangerous items will be detected as passing through the inspection gate. Besides, there exist many highly control areas in NPP, and only the person with permission can enter and exit freely. These areas are highly secured and the access is controlled by entrance code or ID scan. However, if an insider holding permission to the areas attempt to destroy some facilities, the security equipment will be ineffective. Moreover, the two-person rule is performed to prohibit unauthorized procedures and the access by an individual to control material as preceding some specific task under safety concern. Therefore, the object this study focuses on is an unarmed individual who is preceding a task. Certainly, the follow-up simulated surveillance video was shot to record behavior of a single person. The unusual behavior was defined from three points of view; first is movement, which includes: (1) force majeure such as falling down and passing out; (2) destruction action such as destroying equipment. The Second one is moving trajectory that movement is in unusual trajectory such as zigzagging and moving back and forth. The third is moving speed including: (1) velocity changes continuously; (2) stop and stay. For the purpose of simplification, only the "stop and stay" moving speed were simulated here.

CONSTRUCTING THE UNUSUAL BEHAVIOR DETECTION SUPPORT SYSTEM (UBDSS)

In order to obtain video for behavior movement analysis purpose, the simulated surveillance video was shot with a DV camera. The scene was set to be a path with a door near the camera side. Persons were required to pass through the path a few times and each time in a different way.

Utilizing the image process algorithm, one can detect if there is a person showing up in the screen, and further proceeds the function of tracking and shape extraction; the location of target can be locked on and the coordinate can be calculated by tracking task, in addition, the contour shape of target can be obtained by shape extraction task. To begin the classification analysis of data obtained from previous step and meanwhile, to describe human behavior in a human comprehensible manner, the variable were transformed properly according to the three unusual behavior points of view this study defined.

The contour shape of human will be encoded, and the concept was based on OWAS system which describes the human gesture in a numerical way (Karhu et al., 1977). With an adjusted OWAS system to fit the requirement of this study, the human gesture can be represented by 5 parts, which are Head, Back, Upper limb, Lower limb and Axis. Therefore, the human shape data can be transformed into variable, say X, and express as (X_1, X_2, X_3, X_4, X_5) to describe the observed movement.

Tacking task can lock on the location of target and calculate its coordinate, and two following variables can be obtained by calculation of coordinate: trajectory and moving speed. In order to quantify trajectory, the human body in surveillance screen was regarded as a point called center point, and connected line of center points of three successive frames compose a vector angle which can be the trajectory clue. Hence, angle was regarded as the variable to describe trajectory. On the other hand, dividing the displacement of two successive center points by time, moving speed of human in screen can be obtained.

The previous data of gesture, angle and speed were then input into decision tree algorithm separately to construct decision tree classifiers. Next, the testing data was used to validate the classifiers; if the result was acceptable, the decision tree model would be used to predict class of new data, in this case, usual or unusual. In this process, the testing data was acquired by randomly draw from all samples, and remain data was the training data. Training data was input into decision tree software to execute the decision tree algorithm. The software used here was WEKA 3.6.1 (Witten and Frank, 2005). Testing data was used to validate the classified result of decision tree.

Finally, the UBDSS was constructed on the basis of trained decision tree. New video data was imported in UBDSS to judge if unusual behavior happened. If so, UBDSS would continuously produce alarm signal to inform security personnel detected unusual behavior until security personnel take any action.

RESULT OF CLASSIFICATION

The classification result of the variables, which are gesture, moving angle and moving speed, are as follow. In 170 training data of gesture, there are 151 data that can be correctly classified, which indicated the correctly classified rate was 0.888; focusing on the data belong to unusual behavior, the precision rate was 0.966, the recall rate was 0.884, and the false negative rate was 0.116, which means there is about 11% unusual behavior failed to be classified. In 119 training data of moving angle, there were 88 data that can be correctly classified, which indicated the correctly classified rate was 0.739; focusing on the data belong to unusual behavior, the precision rate was 0.982, the recall rate were 0.651, and the false negative is 0.349. In 102 training data of moving speed, there were 88 that can be correctly classified, which indicated the correctly classified rate was 0.862; focusing on the data belong to unusual behavior, the precision rate was 1, the recall rate were 0.837, and the false negative is 0.163.

The developed decision tree model was used to predict the value of the testing data, and validation process was applied by means of comparing the true value and the predicted value of testing data. In 33 testing data of gesture, there were 28 data to be correctly predicted, and the correctly predicted rate was 0.848; when it comes to data that belong to unusual behavior, the precision rate was 0.821, the recall rate is 1, and the false negative is 1, which implied that all the unusual behavior can be correctly detected from the classification process. In 25 testing data of moving angle, there were 19 data can be correctly predicted, and the correctly predicted rate was 0.76; when it comes to data that were belong to unusual behavior, the precision rate was 1, the recall rate is 0.697, and the false negative is 0.303, which meant there were 30% unusual behavior failed to be detected. In 39 testing data of moving speed, there were 35 data to be correctly predicted, and the correctly predicted rate was 0.897; as to data that were belong to unusual behavior, the precision rate was 1, the recall rate is 0.875, and the false negative is 0.125, which indicated there were 12% unusual behavior can't be detected.

To sum up, when new data were input, the correctly predicted rate of unusual behavior of the three variables were 0.848 for gesture, 0.76 for moving angle and 0.897 for moving speed. The reason why the correctly predicted rate of moving angle was too low is that unusual behavior may happen in a way that the trajectory seems to be normal. A real example is moving forward slowly but straight. However, because the variables are dependent to each others, the unusual behavior may be detected through different classifier in the same time. Accordingly, the system was set to activate the alarm when unusual behavior of any types of these three variables was detected, which provided a better detection result of unusual behavior.

CONCLUSIONS

Since the topic involves nuclear safety issues and related literature is confidential, there is little information available. In addition, this study was to discuss the NPP workers' unusual behavior; nevertheless, the surveillance video data of NPP is not opened to public owing to nuclear safety and workers' privacy. As a result, the simulated video data may not properly describe the real situation. Once there is unusual behavior that does not exist in the previous categories, the system may wrongly classify it. Despite of the limitations, the UBDSS that automatically detect unusual behavior by analyzes surveillance video data can be used to assist NPP safety personnel to practice surveillance task. Since the video screens to be monitored are many but the number of workers are usually no more than two, the system can lower the mental workload of workers, and thus decrease the occurrences of human error. In this way, the safety of NPP can be strengthened.

REFERENCES

Wang, L.H. (2005) "Report of the Eighteenth International Training Course", Physical Protection of Nuclear Material and Nuclear Facilities, Atomy Energy Council, New Mexico.

Duque, D., Santos, H. and Cortez, P. (2007) "Prediction of Abnormal Behaviors for Intelligent Video Surveillance Systems", Proceedings of the 2007 IEEE Symposium on Computational Intelligence and Data Mining.

Karhu, O., Kansi. P. and Kuorinka, I. (1977) "Correcting working postures in industry: A practical method for analysis", Applied Ergonomics, Volume 8, pp. 199-201.

Witten, I. H. and Frank, E. (2005) "Data mining: practical machine learning tools and techniques", Morgan Kaufman publisher, MA.

Chapter 84

A Proposal of Fall Risk Evaluation Method on Patient's Characteristics

S. Fukaya, Y. Okada

Fac. of Science & Technology, Dept. of Administration Engineering
Keio University
3-14-1, Hiyoshi, Kohoku-ku, Yokohama-shi, 223-8522, Japan

ABSTRACT

In medical scene, the fall accident of patients is one of the accidents that occur frequently. So, to evaluate the fall risk of patients is realistic to reduce fall accidents in medical scene. But, the majority of hospitals don't define the procedure to evaluate patient's fall risk considering a characteristic of each hospital. Therefore we propose the advanced method of evaluating fall risk using FRAS table to restrain fall accidents. By using this method, each hospital can lead patient's fall risk. So, we discuss the realistic how to use the information of fall risk with risk managers. And, we propose the method of concentrating the conscious for the patients classifying as high fall risk, and contribute to reducing the rate of fall accidents.

Keywords: Patient's Fall Accidents, Risk Evaluation, Medical Scene

INTRODUCTION

In a medical scene, the fall accident of patients is one of the accidents that occur frequently. At a certain hospital, the rate of fall accidents occupies about 64% of total accidents (FIGURE 1).

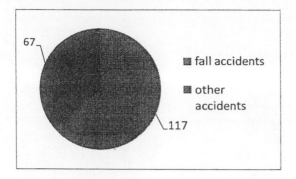

FIGURE 1 Rate of total accidents in *hospital X*

Fall accidents affect the patient's quality of life, such as delay of the recovery and decline in activities of daily living. Therefore, risk managers and nurses consider the method for preventing the fall accidents of patients.

TO REDUCE FALL ACCIDENTS

To prevent the fall accidents of patients, the measures of mechanical against fall accidents are carried out in medical scene. There is the method of attaching fence and handrail beside bed, for example. Then, to restrict patient's action gets patients not to fall when the patient rolls over, stands up or walks. But, to restrict patient's action causes the problems of patient's human right and privacy. Also, fence interrupts nurse's other services frequently. Therefore, it is difficult to prevent fall accidents substantially by mechanical measures. To prevent fall accidents completely, there is the method that sensor senses patient's action and fence automatically comes out around patient's bed. However, this method is impossible practically to cost a great deal.

So, medical risk managers consider showing nurses the information of patient's fall action by sensors. But, if medical risk managers use sensors about all patients, the sensors show nurses much information except patient's fall action, and interrupt the nurse's service. Therefore, to evaluate the fall risk of patients is realistic in medical scene. By attaching sensors and fence toward the patients evaluated as possibility of fall accident is high, we restrain fall accidents. We define *possibility of fall accident* as" fall risk". In addition, by taking care of the patients evaluated as high fall risk by nurses, we restrain fall accidents. The care of nurses performs communication, restroom instruction and walk assistance, for example. The care of nurses reacts to the patient's action evaluated as high fall risk, too.

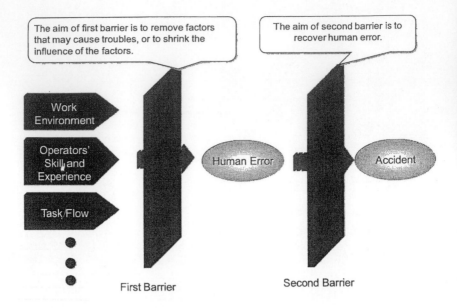

FIGURE 2 Fundamental policy on human error prevention strategy

By evaluating the fall risk of patients, we wish to restrain fall accidents using the measures of mechanical and the care of nurses, like a FIGURE 2.

THE METHOD TO EVALUATE FALL RISK

As one of patient's fall risk estimation methods, there is the Fall Risk Assessment Score table, called as FRAS table. FRAS table is recommended by Japanese Nursing Association. This table is often used as reference to evaluate the fall risk of patients in Japanese hospitals and constructed with about 30 check-items about patient's characteristics.

Table1 shows FRAS table using fall risk evaluation in a certain hospital. Nurses select each check-item of this table by watching the characteristic of patient, and add up the evaluation score of each check-item. And, nurses compare the total score of the patient with the criterion of fall risk. The criterion of fall risk is divided into three levels as "high level", "middle level" and "low level". High level shows the patient causing fall accidents frequently. Middle level shows the patient being apt to cause fall accidents. Low level shows the patient having possibilities of fall accidents. The manual using FRAS table is written as "It is necessary for you to review evaluation score and fall risk level regularly in each hospital." But, to review evaluation score and fall risk level regularly is difficult for the hospitals.

Table 1 Example of FRAS table using fall risk evaluation

	Classification	Characteristic (check-item)	Evaluation score	Check
A	Age	More than 70 years old or less than 9 years old	2	
B	Sex	Man	1	
C	Past history	Fall experience	2	
		Faint experience		
D	Feeling	Visual difficulty	1	
		Hearing difficulty		
E	Function difficulty	Paralysis	3	
		Feel numbness		
		Joint difficulty		
F	Activity territory	Weekness in leg muscles	3	
		Using wheel chair or a stick		
		Moving with assistance and wobble		
		Bedridden		
G	Recognation	Confusion	4	
		Dementia		
		Decline in comprehension power		
		Uneasy action		
H	Medicine	Pain-relief drug	1 (each item)	
		Medicine of Parkinson's disease		
		Narcotics agent		
		Antihypertensive diuretic		
		Hypnotic agent		
		Laxative enema		
		Chemical cure		
I	Evacuation	Inability to control urine flow	2 (each item)	
		Need assistance in bath room		
		Frequent micturition		
		Indwelling urethral catheter		
		Get to bath room at night		
		Long distance to bathroom		

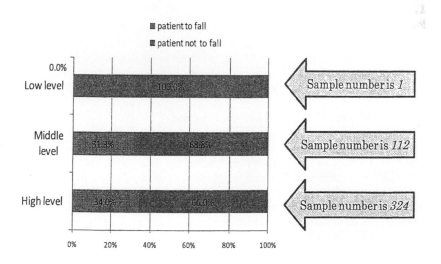

FIGURE 3 result of using FRAS table in *hospital A*

FIGURE 3 shows the result of using FRAS table in *hospital A*. From this result, the nurses express opinions.

· Nurses don't feel that the patient evaluated as "high level" tends to cause fall accident.
· The patient evaluated as "middle level" falls frequently.

Therefore, FRAS table isn't used effectively by the risk managers for the restraint of fall accidents. There are some problems to improve to using FRAS table.

· Nurses gather the number of fall accidents as personal history.
· There are not definite procedure to gather patient's data.
· The importance of the fall factors is not decided.

In this study, we propose the advanced method of evaluating fall risk using FRAS table to restrain fall accidents.

IMPROVEMENT OF USING FALL RISK EVALUATION

We improve the problems of the existing FRAS table.

· Nurses gather the number of fall accidents as personal history
The check-item of FRAS table is made as "Fall experience". Therefore, we confuse the fall experience whether the patient falls in the hospital or in the patient's home. So we decide on gathering the data of patient's fall accidents from entering hospitals.

· There are not definite procedure of each check-item to gather patient's data
There are no definite procedures of each check-item, and the criterions of each nurse are different. To solve problems, we discuss the common criterions of gathering patient's data with risk managers in each hospital. As a result, we propose the procedure that common criterions are specified.

· The importance of the fall factors is not decided.
The importance of all fall factors is one, and there are the cases that risk managers set the evaluation score by risk manager's experience and intuition. But, the importance of all factors isn't same. So, using Multiple Regression Analysis, we determine the importance of each fall factors. We use patient's fall experience as criterion variable. And we use the check-items of FRAS table as explanation variables. Therefore, Multiple Regression Analysis leads the coefficients of explanation variables. Also, as a result of Multiple Regression Analysis, we reduce the explanation variables that F statistic is under 2. So, we determine the coefficients of leaved explanation variables as the importance of fall factors.

Based on the improvement of the problem, we propose the procedure to evaluate patient's fall risk. The method of analysis is made up of three steps.

Step 1:
Prepare the method of gathering
patient's data for using FRAS table

Step 2:
Introduce the equation that can estimate
the fall risk in the target hospital

Step 3:
Determine the fall risk threshold

FIGURE 4 Patient's fall risk evaluation method

FIGURE 4 shows easily three steps of the procedure to evaluate patient's fall risk.

[Step 1] Prepare the method of gathering patient's data for using FRAS table
We decide on gathering the data of patient's fall accidents from entering hospitals. And, we discuss the common criterions of gathering patient's data with risk managers in each hospital, and we propose the procedure that common criterions are specified. As a result, risk managers can gather the data of the patient's condition for using FRAS table. If conditions of patients applied to fall factor in FRAS table, we filled in 2. If conditions of patients didn't apply to fall factor in FRAS table, we filled in 1.

[Step 2] Introduce the equation that can estimate the fall risk in the target hospital
On gathering the patient's data of fall risk for using FRAS table, we analyze fall risk evaluation equation. We use patient's fall experience as criterion variable and the check-items of FRAS table as explanation variables. Then, Multiple Regression Analysis leads the coefficients of explanation variables. Also, as a result of Multiple Regression Analysis, we reduce the explanation variables that F statistic is under 2. So, we determine the coefficients of leaved explanation variables as the importance of fall factors. As a result, we lead fall risk evaluation equation for analyzing fall risk.

[Step 3] Determine the fall risk threshold
Using fall risk evaluation equation, we determine the threshold of fall risk for each hospital to classify the each patient into whether fall risk of each patient is high or low.

By performing three steps, we can select the patients tending to fall in each hospital.

CASE STUDY

We apply the procedure to evaluate patient's fall risk to hospital and show the result of analysis. We take data from ten hospitals and analyze each hospital. We gather data in *hospital A* during past two months, for example. The patients of analyzing in *hospital A* consist of *1170* patients. To gather data, we decide on gathering the data of patient's fall accidents from entering *hospital A*. And, the risk managers of discussing common criterions gather the data of the patient's condition for using FRAS table. Table 2 shows the extract of the check list utilizing FRAS table for gathering patient's fall risk.

Table 2 Fall risk evaluation check list

				FALL RISK EVALUATION CHECK LIST.							
	A. Age	B. Sex	C. Past history	D. Feeling		E. Action difficult			...	Criterion variable
Room number	Age	Man 2 Woman 1	Faint experience	Visual difficulty	Hearling difficulty	Paralysis	Feel numbness	Joint difficulty	....	Fall experience
200	73	2	1	2	1	2	2	1	...	2
201	87	1	1	1	1	1	1	1	...	1
202	86	1	1	1	1	1	1	1	...	1
...	...	...	...	...	...	...	...	...	...	...

Like a Table 2, if conditions of patients applied to fall factor in FRAS table, we filled in *2*, and if conditions of patients didn't apply to fall factor in FRAS table, we filled in *1*. We use patient's fall experience as criterion variable and the check-items of FRAS table as explanation variables. Then, Multiple Regression Analysis leads the coefficients of explanation variables. Also, as a result of Multiple Regression Analysis, we reduce the explanation variables for F statistic of 2 and less. So, we determine the coefficients of leaved explanation variables as the importance of fall factors. As a result, we lead fall risk evaluation equation for analyzing fall risk.

Fall risk (hospital A) = 1.00*(using wheel chare or a stick) + 0.90*(moving with a wobble) + 0.75*(decline in comprehension power) + 0.75*(hypnotic)

$$(1)$$

Equation (1) is the derived function for *hospital A*. Using equation (1), we set the criterion for *hospital A* to classify the each patient into whether fall risk of each patient is high or low.

In this process, we set the criterion as discussing fall risk evaluation equation with risk managers in *hospital A*, and considering error of the first kind and the second kind and being able to expect amount each error to minimize. As a result, we set the criterion as *4.5*. If the risk evaluation value of each patient is over *4.5*, we classify a patient as high fall risk. If the risk evaluation value isn't over *4.5*, we classify a patient as low fall risk.

Table 3 Expected classifying data based on past data

		fall accident possibility		
		low	high	
past	didn't fall	396	392	50.3%
data	fell	50	332	86.9%
		88.8%	45.9%	

Table 3 shows expected classifying data based on past data. Rows show whether patients actually occurred fall accidents in past data, line show whether the patient's fall risk is high or low as a result of analysis. From Table 3, only *45.9* percent of the patients classifying as high fall risk actually tend to cause fall accidents. But, we find that *88.8* percent of the patients classifying as low fall risk don't tend to cause fall accidents. So, nurses can concentrate the conscious for the patients classifying as high fall risk. As a result, we consider the application for the restraint of fall accidents in each hospital.

PRESENTATION OF INFORMATION TO MEDICAL SCENE

In this study, we propose the advanced fall risk evaluation method using FRAS table. As a result, we select the patients of high fall risk in each hospital. But, risk managers don't use the information of fall risk really. So, we discuss the realistic how to use the information of fall risk with risk managers. Then, in this study, we propose the method that the risk managers of *hospital A* attach the star mark nearby patient's information board at nurse station that shows the patients classifying as high fall risk (FIGURE 5).

Room Number	Patient's Name	the star mark
101	Y.A	☆
102	A.Y	
103	R.T	
104	A.F	
105	Y.K	☆
106	S.T	

The star mark means "the fall accidents possibirilty of this patient is high."

FIGURE 5 Image of information board with the star mark

To test the effect of star mark, we compare in hospital ward using the star mark with in hospital ward not using the star mark.

In FIGURE 6, we compare the percentage of fall accidents in hospital ward using the star mark with in hospital ward not using the star mark during four months.

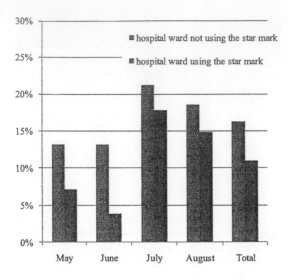

FIGURE 6 Process of patient's fall accidents percentage

From FIGURE 6, we find out that patient's fall accidents percentage decrease using the star mark during four months, especially in May and June. Using the star mark based on classifying data according to the procedure for the fall risk evaluation, nurses can concentrate the conscious for the patient that frequently occur the fall accidents and then fall accidents reduce. But, patient's fall accidents percentage using the star mark is similar to patient's fall accidents percentage not using the star mark in July and August. Because, we consider that many patients get out of shape and nurses are very busy to take care of the patients in July and August.

THE DISCUSSION OF THE FALL RISK METHOD

In this study, we propose the advanced method of evaluating fall risk using FRAS table to restrain fall accidents. But, to use fall risk evaluation method generally, there is the objective problem about the check-item of FRAS table.

Ex. The criterion of "Visual difficulty" by nurses
 · The goodness of the patient's eyesight
 · The illness of eye
 · Wearing glasses

Like the example, each nurse sets the different criterions to the ambiguous check-items. Therefore, it is necessary to define each check-item checking the characteristic of the patients in common criterion. So, we define nurses, associate nurses and care workers as the person of the criterion checking the characteristic of patients. And we consider doing each check-item objectively.

CONCLUSION

In this study, we propose the advanced method of evaluating fall risk using FRAS table to restrain fall accidents. Therefore, we make the definite procedure to gather patient's data through discussions with risk managers in each hospital, and we propose the procedure to estimate the patient's fall risk using FRAS table. Also, we propose the method of concentrating the conscious for the patients classifying as high fall risk, and contribute to reducing the rate of fall accidents.

REFERENCES

[1] Masaki SHIDA "A Study on the Establishment and Application of the Model of Falling", 2006, in Japan
[2] Erika WARASHINA "A Study on the decreasing Incidents of Falling among Patients in Hospitals", 2004, in Japan
[3] Hiroko ISHIBASHI, Etsuro YAMAGUCHI, Mihoko TAKAGI, Naomi YANI, Tomoko KIMURA, Tokiko KISHIMOTO, Naoko YUHARA and Eiichi TERASAWA ,"Evaluation of validity and improvement of a tumble assessment", 2007, in Japan

Chapter 85

Direct-to-Consumer (DTC) Prescription Drug Advertising: Exploring Self-Reports of Media Exposure and Associated Behaviors

Richard C. Goldsworthy[1], Christopher B. Mayhorn[2]

[1]The Academic Edge, Inc.
Bloomington, IN 47407-5307,USA

[2]North Carolina State University
Raleigh, NC 27695-7650,USA

ABSTRACT

The present research explored several hazard/risk communication issues in direct-to-consumer (DTC) advertising of prescription drugs. In a one-on-one approach interview, 2773 participants were asked: (a) whether DTC advertisements were encountered in the past thirty days, (b) what media was used to communicate the DTC advertisements, and (c) whether participants engaged in a number of associated behaviors as a result of their exposure to DTC advertisements. A large percentage of participants (85.4%) reported encountering DTC ads during the last thirty days, most commonly via television, magazines, and the internet. Examination of health-related behaviors associated with DTC exposure demonstrated the multidimensional impact of the advertisements. Participants felt comfortable asking a physician about a symptom or drug they had seen in a DTC ad during regular doctor visit; however, 20% felt too rushed to do so. More than twenty-one percent of participants attributed exposure to DTC ads as a reason for scheduling appointments with their doctors in the first place. Unfortunately, a significant percentage (17.7%) also indicated that they were more likely to borrow a prescription medication from a friend if they had seen a DTC ad. Human factors/ergonomics professionals can play an important role in further understanding, and addressing, how DTC advertisements affect primary and secondary health-related consumer behaviors.

Keywords: warnings, hazards, safety, pharmaceuticals, healthcare.

INTRODUCTION

Direct-to-consumer (DTC) advertising refers to the mass media promotion of prescription medications to the public. Guideline changes initiated by the US Food and Drug Administration (FDA) in 1997 effectively made it easier for manufacturers to advertise their drugs directly to the public (Marinac, Godfrey, Buchinger, Sun, Wooten, & Willsie, 2004). As a result of the tremendous growth of this kind of advertising in the last decade, it is likely that most of the U.S. public frequently sees many DTC prescription drug advertisements in a variety of broadcast media formats. However, exact estimates of exposure and media usage are scarce within the literature.

U.S. Federal regulations require that DTC drug advertisements convey both risk and benefit information. However, it is not clear that drug manufacturers' advertisements give adequate coverage of the risks relative to the benefits. DTC has been previously identified as a key area in which human factors professionals can play a significant role (Wogalter, Mills, Paine, & Smith-Jackson, 1999), yet our involvement to date has been limited. For example, in a study of DTC advertisements on the internet, risk information was found to be much more difficult to access because of its separated and "distant" presentation from the benefits information (Vigilante & Wogalter, 2001). Although patients may gain an incomplete understanding of the risks associated with taking a particular prescription medication, prior research suggests that patients actively engage in information seeking as a result of exposure to DTC ads (Vigilante, Mayhorn, & Wogalter, 2007). Unfortunately, it is unknown how information seeking associated with DTC advertisements translates into behaviors that have public health consequences.

Specifically, it has become increasingly more apparent that DTC influences interactions between doctors and their patients. For example, Findlay (2002) estimated that in 2001 between 8.5 to 12.6 million people received a prescription from a physician as a direct result of DTC drug advertising, and the numbers have undoubtedly increased since then. Whether or not DTC exposure influences a patient's willingness to schedule doctor's visits or to inquire about advertised drugs is unclear.

A chief concern is that people may be making assumptions about drug safety that might lead to a number of associated behaviors that have public health consequences when they encounter DTC advertisements yet very little research has investigated this aspect. One such associated behavior is medication borrowing. Goldsworthy, Schwartz, and Mayhorn reported the results of a survey in which as many as 36% of the participants admitted to loaning or borrowing prescription medications (2008). How this potentially dangerous behavior is tied to erroneous beliefs about prescription medication properties gleaned from DTC ads is unknown.

The present research investigates several questions related to DTC prescription drug advertising and associated behaviors. Specifically, these concerned: (a) whether participants had encountered a DTC ad in the past 30 days, (b) what media formats were encountered, and (c) whether participants reported engaging in behavior such as medication borrowing and information seeking as a result of DTC exposure.

METHOD

INTERVIEW

As part of a related research project, 2773 participants were queried about their experiences with DTC ads in a nationally distributed survey that utilized a one-on-one approach interview conducted in malls and other public places. Interview sessions for each participant approximated 25 minutes. Efforts were made to recruit diverse participants by using a stratification quota for adolescents, males, and Hispanics of 20%. Inclusion targets for other racial and ethnic groups mirrored the overall 2000 U.S. Census levels.

Participants ranged in age from 12 to 45 years with 21.4% of the sample categorized as adolescents aged 12-17 and the remainder evenly distributed across the age range. Forty two percent of the sample was male. Self reports of racial identity indicated that the sample was diverse: 49.6% Caucasian, 21.6% African American, 21.4% Hispanic, and 5.4% Asian.

EXPOSURE TO DTC ADVERTISEMENTS

The interview contained a series of questions concerning the participants' experiences with respect to DTC drug advertisements (commercials).

Participants were first asked to respond "yes" or "no" to whether they had seen or heard prescription drug advertisements in the past 30 days. Participants were then asked to respond "yes" or "no" to whether they encountered these advertisements in the following media formats:

- Radio
- Television
- Newspaper
- Magazine
- Billboard
- Direct Mail
- Email
- Internet

SELF REPORTS OF ENGAGING IN ASSOCIATED BEHAVIOR

A second set of questions concerned the participants' associated behaviors that results from DTC advertisement exposure.

Participants were asked to respond "yes" or "no" to the following questions:
- Has an advertisement prompted you to seek more information about a drug?
- Has seeing an advertisement made you think you had a particular illness?
- Has seeing an advertisement led you to schedule a visit with a doctor?
- Has seeing an advertisement led you to ask about symptoms or medicines during a regular doctor visit?
- Have you wanted to ask your doctor about symptoms or medicines that you saw in advertisements but felt rushed?
- Would you feel comfortable asking a doctor about a drug if you saw an advertisement for it?
- Would you be more likely to borrow a prescription medication from a friend if you had seen it advertised?

RESULTS

All "yes" and "no" responses were coded as 1 and 0, respectively. Statistical analyses were conducted in SPSS (SPSS Inc, Chicago, Ill).

EXPOSURE TO DTC ADVERTISEMENTS

As Table 1 illustrates, the vast majority (85.4%) of participants reported encountering a DTC advertisement in the last 30 days. The most frequently reported media formats for DTC ads were television, magazines, and the internet. The least frequently reported media outlets for DTC ads were direct mail and email.

SELF REPORTS OF ENGAGING IN ASSOCIATED BEHAVIOR

A significant portion of participants reported engaging in a number of health-related behaviors as a direct result of exposure to DTC ads. For instance, information seeking behaviors were apparent from the responses such that 53.9% reported feeling comfortable asking doctors about drugs seen in an advertisement and 32.6% reported asking about symptoms or medicines during a regular doctor visit. Interestingly, 21.5% of the sample reported scheduling a doctor's visit after seeing an advertisement. Half feel comfortable asking their doctor about medications they learn about through DTC; however, 1 in 5 indicated they wanted to ask their provider but felt too rushed to do so. 17.7% of the sample (n = 489) reported that they would be more likely to borrow a prescription medication from a friend if they had seen it advertised.

DISCUSSION

These results describe some specific aspects of how DTC prescription drug advertisements affect some consumers. There were several notable results with respect to risk perception and hazard communication researchers.

A large percentage of people are being exposed to DTC drug advertisements in a variety of media such as television, magazines, and the internet. Second, significant percentages of the sample reported an increased likelihood of engaging in further information seeking as a direct result of exposure to a DTC advertisement. Participants reported feeling comfortable asking doctors about drugs seen in an advertisement and asking about symptoms or medicines during a regular doctor visit Third, participants attributed exposure to DTC ads as a reason for scheduling appointments with their doctors. Finally, a sizeable percentage of the sample (17.7%) reported that they were more likely to borrow prescription medications from a friend if they had seen a DTC advertisement.

These findings are consistent with current trends in healthcare where the patient is taking a more active role in making decisions. Unfortunately, the quality of the information that the public is drawing from DTC pharmaceutical advertisements is questionable. A recent study indicated that the content of many advertisements is confusing, difficult to understand, and that side effect information is not sufficient. Moreover, errors occur, including conflating over-the-counter (OTC) and prescription medications, and the outcomes of such active engagement are not always be positive, as in the case of medication sharing. Online drug purchases may exacerbate this issue by making it easier for people to self-medicate without a doctor in the process, thereby bypassing gatekeeping.

DTC drug advertisements appear to be prompting some people to discuss advertised prescription drugs as potential treatment options with their doctors (Marinac et al., 2004). Surprisingly, previous research also indicates that physicians are unlikely to deny DTC-driven requests for medication prescriptions (Vigilante, Mayhorn, & Wogalter, 2007). So, even when consulting a provider, if they ask, they get it; and DTCA ads do prompt them to ask.

Given the diverse and multidimensional effects DTC advertising may engender, HF/E researchers and practitioners can contribute their expertise in a number of areas in this domain. First, HF/E knowledge regarding cognition, behavior, and message design principles can be applied to DTC advertisements. Second, HF/E professionals might actively assist federal regulatory agencies such as the FDA and the Department of Justice to design interventions that protect the public from unintended risks. This involvement should extend beyond the immediately apparent, primary, health behaviors, such as requesting medication and asking about disease, to a broader perspective that involves secondary and tertiary health-related behaviors, such as scheduling appointments and, further removed from the message intended by DTC, prescription medication sharing. A significant part of our potential involvement is precisely to help maintain this 'bigger picture' perspective on the ways messaging affects consumer behavior and how to capitalize on the positive effects—including increased information seeking and health awareness—while minimizing negative ones—such as unnecessary prescriptions and non-recreational prescription sharing.

An important limitation of the current study was the lack of data from seniors. As an age group, seniors may be more susceptible to the marketing messages employed in the advertisements. On average, seniors also tend to take more medications, both prescription and over-the-counter, raising the risks associated with drug interactions. However, the information presented in DTC advertisements may not always be sufficient to fully educate the seniors as to risks associated with drug interactions. Future surveys should incorporate senior populations to determine the effects of DTC drug advertisements on their purchasing and use habits including online purchasing.

REFERENCES

Findlay, S. D. (2002). *DTC advertising: Is it helping or hurting?* (Statement before the Federal Trade Commission Health Care Workshop). National Institute for Health Care Management. Washington, DC: US Government Printing Office.

Goldsworthy, R. C., Schwartz, N. C., & Mayhorn, C. B. (2008). Beyond Abuse and Exposure: Framing the Impact of Prescription Medication Sharing. *American Journal of Public Health, 98*(6), 1115-1121.

Marinac, J. S., Godfrey, L. A., Buchinger, C., Sun, C., Wooten, J., & Willsie, S. K. (2004). Attitudes of older Americans toward direct-to-consumer advertising: Predictors of impact. *Drug Information Journal, 38*, 301-311.

Mayhorn, C. B., & Goldsworthy, R. C. (2009). Borrowing prescription medication: Implications for healthcare warnings and communications. In *Proceedings of* *the Human Factors and Ergonomics Society 53rd Annual Meeting* (pp. 1608-1611). Santa Monica, CA: Human Factors and Ergonomics Society.

Vigilante, W. J., Jr. Mayhorn, C. B., & Wogalter, M. S. (2007). Direct-to-consumer (DTC) drug advertising on television and online purchases of medications. In *Proceedings of the Human Factors and Ergonomics Society 51st Annual Meeting* (pp. 1272-1276). Santa Monica, CA: Human Factors and Ergonomics Society.

Vigilante, W. J., Jr. & Wogalter, M. S. (2001). Direct-to-consumer (DTC) advertising of prescription medications on the world wide web: Assessing the communication of risks. In *Proceedings of the Human Factors and Ergonomics Society 45th Annual Meeting* (pp. 1279-1283). Santa Monica, CA: Human Factors and Ergonomics Society.

Wogalter, M. S., Mills, B. J., Paine, C. S., & Smith-Jackson, T. L. (1999). Application of cognitive principles to the design of direct-to-consumer advertising of prescription medications. In *Proceedings of the Human Factors and Ergonomics Society 43rd Annual Meeting* (pp. 515-519). Santa Monica, CA: Human Factors and Ergonomics Society.

Table 1:
Self-reported exposure to drug advertising and associated behaviors.

	Yes No.	Yes (%)	No No.	No (%)	Valid N (#)
Drug Advertising					
Considering the past 30 days, have you seen or heard prescription medication advertisements?	2362	85.4	405	14.6	2767
Have you seen advertisements with the following media?					
Radio	1456	61.6	906	38.4	2362
Television	2173	92.0	189	8.0	2362
Newspaper	1218	51.6	1144	48.4	2362
Magazine	1738	73.6	624	26.4	2362
Billboard	1211	51.3	1151	48.7	2362
Direct Mail	908	38.4	1454	61.6	2362
Email	998	42.3	1364	57.8	2362
Internet	1532	64.9	830	35.1	2362
Associated Behaviors					
Has an advertisement prompted you to seek more information about a drug?	691	24.9	2080	75.1	2771
Seeing an advertisement made me think I had a particular illness.	777	28.1	1987	71.9	2764
Seeing an advertisement led me to schedule a visit with a doctor.	596	21.5	2167	78.5	2763
Seeing an advertisement led me to about symptoms or medicines during a regular doctor visit.	900	32.6	1863	67.4	2763
I wanted to ask my doctor about symptoms or medicines I saw in advertisements but I felt rushed.	554	20.1	2207	79.9	2761
I would feel comfortable asking a doctor about a drug if I saw an advertisement for it.	1488	53.9	1271	46.1	2759
I would be more likely to borrow a prescription medication from a friend if I had seen it advertised.	489	17.7	2268	82.3	2757

Chapter 86

Causal Relation of Negative Event Occurrence – Injury and/or Failure

Juraj Sinay, Anna Nagyová

Technical University of Kosice
Faculty of Mechanical Engineering
Department of Safety and Quality production
042 00 Košice, Slovakia

ABSTRACT

One of the important parts of Risk Management System is dangerous identification and also Hazard Situation identification. It includes activities, which together with risk minimizing or elimination are included in successful prevention, the most effective method in Risk Management System. For effective algorithm selection in the frame of negative event prevention is necessary to recognize its occurrence process, which is marked as a causal relation occurrence. The most specific definition of this relation might help to determinate appropriate method for its disconnection as a tool for effective prevention of injury and/or failure occurrence.

Keywords: Hazard situation, danger, injury

INTRODUCTION

Failure, occupational disease or injury is a sudden, undesirable and unexpected event which can damage persons but also technical units, thus leading to a break or a cut in the planned operational state.

To prevent or minimise these failure or accidents it is neccessary to systematically identify, analyse but mainly describe their causes as well as the course of their formation. The causes, when not identified, minimised and removed on time, can result in the future negative events occurence.

First chart depicts relation called causal dependency of failure, occupational injury and/or accident occurrence, including five, time independent periods:

- danger,
- hazard situation,
- initiation,
- damage,
- loss.

This can be outlined on the basis of a fact that such a functional dependency applies to all sorts of failures and accidents thus their occurrence is not by chance but rather a kind of regularity. In practice knowing the course of this causal dependency is inevitable to create a system for its discontinuance which means prevention of failure, occupational injury and/or accident.

Mutual dependence of the individual phases of a negative event occurrence is depicted in Fig. 1, the process is irreversible one, always happening in one direction. The phases depend on each other mutually (Sinay, 1990).

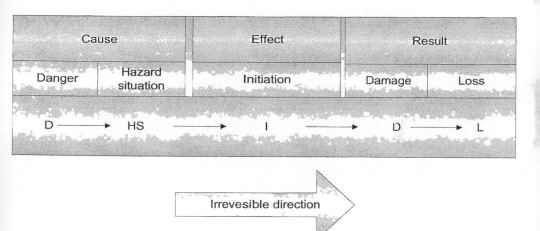

Fig.1 Mutual dependence (Sinay, 1990)

DEFINING THE PHASES OF FAILURE, OCCUPATIONAL INJURY AND/OR ACCIDENT OCCURRENCE CAUSAL DEPENDENCY

The individual connections within the causal dependency of failure, occupational injury and/or accident occurrence will be made clear by using a lifting machine, the one of frequently used within different types of logistic systems.

Danger

It is a machine, a subject, a technology, an act or even a man **attribute/quality** to inflict damage leading into the loss – negative phenomenon.

The danger, if not activated, is of no interest for scientific analysis as then there is no damage or loss.

Example 1:
Tower crane – its particular properties can lead to its fall, causing human or technical loss. However, when not used for lifting the loads, in wind absent weather or in the absence of „force majored" (e.g: an earthquake), there is no damage and subsequent loss – Figure 2.

Example 2:
Dynamic qualities of an overhead crane as a flexible construction, induce oscillations which are then transferred to the operator point – into the cabin space. These oscillations are result of non-stationary phenomenon's occurring during overhead crane operating.
These include:

- inertia effect when increasing the speed of work movements,
- stochastic geometric deviations of the crane's track,
- operating wheels wear,
- swinging of the load on the hanging unit,
- bumper impact.

Hazard Situation

It is a state when an object – a person and/or a machine in a defined space and time is able to activate imperil. It arises when the object becomes active and the person or the thing occurs in the subject's working area.
        A person or some material gets into the working area of the tower crane after its activation, the work movements then can, for example, imply swinging of the load on the hanging unit – Figure 3. There, **the Hazard situation area**, can be defined.
        During the overhead crane operation, as the result of non- stationary phenomenons, its oscillations are transferred to all parts of construction. They also influence the operating staff – human factor - which can have negative impact on

their health in case this activity is a long-term one.

In the professional literature available, or in the normative such as STN EN ISO 14121-article 3.4, the terms „dangerous event" and „dangerous situation" are used which can be identified with the Hazard Situation.

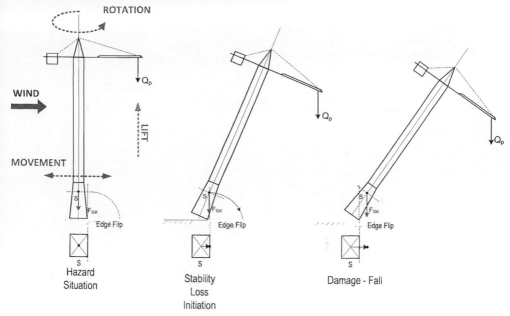

Fig. 2 Phases of causal dependency (Sinay, 1990)

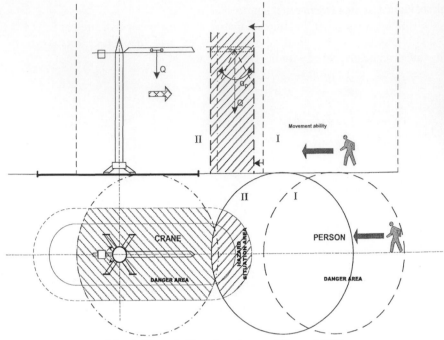

Fig. 3 Hazard Situation Areas (Sinay, 1990)

Risk

The term risk is used more or less in all languages to, in the same way, express the extent (potential) of the danger and its denotation as the relation between
- a negative event – loss, injury, accident occurrence probability and
- the results implied by these.

The risk expresses the danger extent; both parameters depend on underlying factors, which is manifested in various methods used for its quantitative consideration (Sinay, 1990).

Damage

It is an active stage of a negative phenomenon occurrence causal dependency. In this stage - bearing in mind the effective prevention, the exact definition of the process course – if applicable – is the main task. This can be formulated as the time function, eventually the usage period function. The course of the damage process can be defined as the continuous one (stochastic or deterministic) – Figure 4a and 4b and (metal material fatigue, material wear, corrosion, vibrations cummulation and their effect on people) as the step one – Figure 4c (damage to fragile materials –

glass, violent break of wire rope – cut).

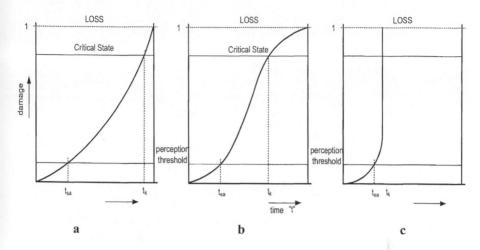

Fig. 4 The course of the damage process (Sinay, 1990)

The process of damage as the stage in which precautions (discontinuation of the negative phenomenon dependency) can be done must be defined as time function and the knowledge of it determines the effective precautions choice. The following must be met for these:

- the occurrence, the course and the duration of the damage must be definable (e.g: the course of the steel construction fatigue fracture, defining the oscillation intensity and its transfer function within the whole overhead crane construction),
- it can not be the step change of the state,
- during the evaluation process the object considered must be fulfilling its functions,
- the damage process must be, with the use of measuring chains - preferably effective on-line methods, controlled quantitatively.

Knowledge of the damage stage course is a vital part of knowing the technical subjects' risks and leads to a proper choice of maintenance strategy, inspections planned, defining the intervals for medical preventive check ups and controls - the precautions for its minimalization.

The damage processes can be of two kinds:

- progressive,
- degressive

for which time of reaching the negative phenomenon perception state „t sa" and the time up to damage occurrence - critical state „t k" is the longest possible, that is the progressive course of the damage – Figure 5.

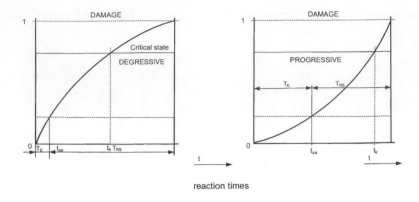

reaction times

Fig. 5 Reaction times

With regard to effective precautions these measures must be taken as the prior ones:
- the damage process should be defined as progressive ones already in the stage of project or construction preparation,
- suggest the measures for damage intensity decrease - modern methods of technical equipment maintenance, on-line monitoring of operating status,
- make redundant systems (airplanes, mechatronic systems),
- define the time of service unit being exposed to oscillation (e.g:on an overhead crane).

Loss

The term means (including other languages):

- physical injury, damage to health and/or
- machine break down or an accident resulting in the loss of its functioning (e.g.: machines or machinery complex systems).

Negative phenomen initiation

It is the crucial stage of negative phenomenon causal dependency. The impulse for system balance failure origins here. It can be activated by a man (improper work behaviour), technology (safety equipment activity absence, improper choice of crane seat damping – the man itself again) and the environment (undefined non-stationary phenomenons during the overhead crane operating, wind during the lifting machine operating, seismic effect, geological compound of the soil) In most cases defining the initiation beginning is accidental one thus can be least effectively used as a preventive measure. The damage initiation in case of oscillation influence on man in the crane cabin can be defined by border rates of oscillation frequency and real acceleration. Their mathematical relation implies the time during which the servicing person can be exposed to cabin oscillation. The time when the rates are exceeded can be taken as the moment of negative phenomenon initiation, oscillation related injury being the case.

THEORY OF MEASURES PREVENTING FAILURE AND/OR ACCIDENT AS THE PRECAUTIONS FUNDAMENT

The main role of all precautions as the part of effective risk and health safety management is to analyse all the stages of negative phenomenon causal dependency during the machines and machinery technological lifetime. This means considering all the aspects of their technical life – machine history, technologies, materials, working environments and environments – so that the procedures for causal dependency discontinuance can be suggested in its initial stages. Every shift of discontinuance towards the damage, into the actual stage of damage, implies the limitations of effectiveness and is financially demanding – e.g. returns of new TOYOTA cars at the beginning of the year 2010. This process is depicted in Fig.6.

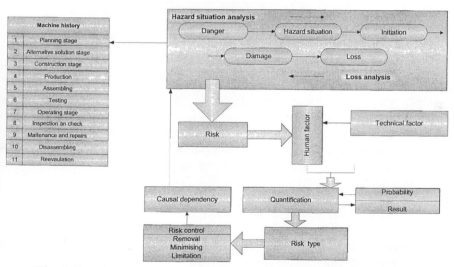

Fig. 6 Causal Dependency Discontinuation in the frame of Machine History

Danger The example of causal dependency of risk - oscillation related injury of an operator on an overhead crane

When dealing with the causal dependency discontinuation in case of occupational

826

injury the following must be considered: the power flow spreads from its source around the whole crane construction (see Fig. 7). This causes the oscillation process spreading into the servicing point – the cabin. Traditionally, the cabin is stuck to the steel construction by a fixed connection. That results oscillate in a direct transfer not only into the cabin but also all the parts of the human body.

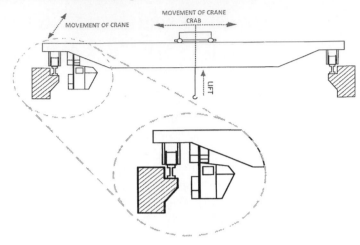

Fig. 7 Overhead crane scheme

a.   *Causal dependence discontinuation in the damage stage* – this can only be realized when after the first signs of negative oscillation impact the crane operator stops working. In this phase, the effective health diagnosis of the crane operator must be done to set the optimal deadlines of preventive check ups.

b.   *Initiation period use for causal dependence discontinuation –* This is possible in case of an on-line monitoring system installed in the overhead crane. The system registers actual operating time, the crane load, oscillation frequency and its characteristic values. When evaluating these parameters in form of multi criteria function, the time periods of crane operating are set not to be longer than the acceptable values.

c.   *Causal dependency discontinuation in the imperil stage –* the optimal solution to be used in the period of machine construction. This includes:

-   active flexible hinge of the cabin – minimizing the oscillation effect on the operator,
-   crane remote control – the operator is out of the power flow, placed outside the crane construction but still in its operating area,
-   use of automatic overhead cranes – the operator is outside the crane construction as well as its operating area (Sinay, Bugar, Bigos, 1988).

These measures tend to be the most effective ones, but, during their application, the economic parameters have to be taken into account, too.

CONCLUSIONS

Accident, failure or occupational injury occurrence is not an accidental process. It has its rules which can be defined on the basis of their causal dependency. The aim of all the „higher safety" measures or risk minimizing is to know the course of causal dependency negative phenomenon's occurrence. The analysis has to include the knowledge of machine and machinery operating state and particular operating process technologies. The team approach is needed to integrate professional knowledge to reach the outputs - safe, ergonomically suitable machines and machinery systems. This approach gains its importance with regard to new technologies such as mechatronic systems, nanotechnologies, biotechnologies, new kinds of renewable energy sources and the like. This seems to be one of the challenges for future generation of scientists in accordance with saying „Safety first!

The paper was written as the contribution to the project VEGA č. 1/0240/09 Research of Integrated Risk Management System methods of the technical equipment and industrial technologies.

REFERENCES

Sinay, J.: Beitrag zur Qualifizierung und Quantifizierung von Risiko-Faktoren in der Fördertechnik – dargestellt am Beispiel von Hebezeugen, VDI Verlag Düsseldorf/SRN, Reihe 13, Fördertechnik Nr. 36, 1990, ISBN 3-18-143613.

Sinay,J.: Bugar,T.: Bigoš,P.: Dynamische Analyse des Systems Kabine-Brückenkran als Mittel zur Vewirklichung humaner Arbeitsbedingungen. Hebezeuge und Fördermittel, Berlin, 28 (1988)12, pg.:366-369

Making it Right: The Critical Performance Influence Factors for Offshore Drilling and Wireline Operations

Birgit Vignes[1], Jayantha P. Liyanage[2]

[1]Petroleum Safety Authority
Norway

[2]University of Stavanger
Norway

ABSTRACT

Human factors and human performance have gained considerable attention in the petroleum sector in Norway over the last few years. This has been driven, to a large extent, by aims to improve control room design and developments related to integrated operations and their complex impact at the work performance level. The principal argument by major oil producers and the authorities is that the actual work performance has to be better understood in the Human – Technology - Organization (HTO) context, to reduce system vulnerability and to improve performance from a safety and environmental point of view.

The Petroleum Safety Authority, Norway (PSA), has performed a number of projects and audits on human factors, particularly related to driller and wireline operators' work situation. The driller and the wireline operator actively engage in the control of the well barrier status and the management of well integrity, which are considered critical for HSE, for instance in terms of limiting integrity incidents, leaks and well control situations. Updated and relevant information on the well barrier status enables the companies to manage operational integrity in a more proactive manner and thereby to ensure "healthy" and safe wells, and prevent or reduce potential risks.

The purpose of this paper is to present the critical performance influencing factors in drilling and wireline operations based on the experience from projects and audits. Performance influencing factors (PIFs) affect the human performance in terms of human abilities to perform actions in a safe and efficient manner, as well as our ability to improvise and improve work performance. Lack of identification and classification of PIFs is generally seen as a major causal factor for human errors, poor decision making, and reduced task performance. The understanding of interaction between HTO and the knowledge about PIFs is critical to achieving the goals of safe and effective drilling and well operations.

INTRODUCTION

Human factors focuses on human beings and their interactions with products, equipment, procedures, environments and facilities used in work and everyday living. Human factors seeks to improve the things people use and the environment in which they use these things to better match the capabilities, limitations and the people's needs. Human factor has to be included in the engineering and the design of a workplace or a working environment because it influences the work performance.

Human factors today has developed to a stage that allows the performance analysts to have a detailed view of complex systems operations involving dynamic interactions between human, technology and organization (HTO). The human factor methods and knowledge acquired through years of experience and R&D work can be used to assess and improve the interaction between human, technology and organization[11,15,16,21,24] of complex systems such as nuclear, oil & gas production, chemical processing, etc. This allows efficient and safe operations of those settings to be realized, taking into account people's opportunities and limitations, as well as needs[25,38,39,40]. This discipline has evolved over the years, and has today ramified into various industrial and public sectors.

The human errors are not often predictable, and may be caused by a number of factors outside the control of the individual. For instance, once a 30" casing weighing 8.5 tones fell down on the drilling cabin at West Epsilon in 2007 [20]. The casing crushed the roof of the drilling cabin, and the end of the casing ended up in the driller's chair. The persons in the drilling cabin received minor injuries. The direct cause of the accident was that the lifting collar was not properly closed and locked when the casing was lifted. However, there were other underlying causes such as deficiencies in the design of the drill floor, construction of the lifting collar, operating instructions, follow-up of safety notices, competence, planning and implementation, deficient management, and breach of procedure. Interestingly, years of studies and investigation in the petroleum and the petro-chemical sector have shown that there is a complex connection between various factors in the good working environment and operational safety. A fatal explosion occurred at the Texas City Refinery in 2005[23], and other major incidents such as Bhopal, Chernobyl, Piper Alpha, etc. provide a range of examples highlighting how different factors are connected under a given circumstance leading to catastrophic consequences. These historical events have also shown that human errors are also a precondition for learning and the opportunity to develop and improve the systems[37]. The key to such a learning process is to take human experiences into account when designing equipment and technologies and make these systems able to handle errors. In principle, it is important to maintain a comprehensive focus and control interactive factors that have a notable influence on human performance.

PERFORMANCE INFLUENCE FACTORS

People are bombarded by all sorts of stimuli from within their own bodies and from their environment at work, at home, while driving a boat or just sitting on a bench in a park. These stimuli are received by their sense organs: eyes, ears, nose, taste, and sensory receptors in the skin. All these stimuli are processed by the human brain and can affect human performance[43].

There are various factors that influence human performance in a given operational setting. Such factors can be explicit as well as implicit in making an impact on the human's ability to perform actions in a correct and efficient manner. A given situation can easily lead to heavy losses when various technical, work, personal, organizational, and even operational environment factors are combined together. Under such conditions, the chances for erroneous performance by the operators of complex systems are seemingly quite high, for instance, the gas subsea blow-out at Snorre A (2004)[49]. A gas blow-out on the seabed, resulting in gas under the facility occurred during the work of pulling pipes out of the well in preparation for drilling a sidetrack. The PSA characterizes this incident as *one of the most serious to occur on the Norwegian continental shelf,* because of the potential of the incident and the failure of the barriers in the planning, implementation and follow-up of the operation. There were no physical injuries in connection with the incident.

Performance influencing factors (PIFs) are often the reason for human errors. Knowledge about these factors for a given operational setting gives valuable information, so that specific measures can be taken to reduce negative effects on human performance. In this context, systems view - or macro ergonomics - has gained much attention lately, where various combinations of complex sets of influence factors are subjected to study in different operational settings. Recently, Azadeh (2005)[15] studied human performance in the control room in a thermal power plant. The results of the study showed that macro ergonomic factors such as safety and organizational procedures, teamwork, self-organization, job design, and information exchange influenced human performance in the power plant. As the traditional control rooms today are gradually transformed into remote operation centers, much attention has also been paid to human performance in collaborative environments. In 2008, Bayerl[17] performed an empirical study during implementation of Collaborative Environments (CEs), covering the factors that influence personnel acceptance, satisfaction and attitudes regarding CEs. Bayerl identified eight factors that were contributing to the success of the CEs, inclusive of an engagement process including the offshore personnel and management, staffing, preparation and training, etc. Brannigan (2008) [16] maintains that human factors is a critical discipline that has to be considered in the planning, design and operation of the modern remote operation centres to ensure operational safety. Such remote operational environments may have specific challenges of their own with respect to the human and how he/she interacts with tasks, machines and the environment, in comparison to traditional control room settings. The operators who are controlling and monitoring tasks from a remote operation centre for an offshore oil & gas production facility, for instance, can get engaged in the direct control and the decision making processes that rely greatly on different performance characteristics. Much of the discussion here, at least on the Norwegian continental shelf, is about for instance, cognitive abilities, knowledge management, training, etc. Notably, as much as the design of the physical environment and technology were important aspects, the current experience from the Norwegian shelf is such that the post implementation analysis and performance follow-up process are also critical in improving human performance under complex operational settings.

PIFs affect the human and the ability to perform actions correctly and efficiently. PIFs can be divided into various dimensions, but can primarily be classified as work-related and human-related factors[1,12,13]. According to Exprobase (2008), work-related factors can include the economy, environment, equipment, interaction, layout, routines, personnel policy, etc., while human-related factors can take the form of physiological- and psychological-factors. Different researchers may tend to classify the factors differently, and Figure 1 shows how Redmill (1997) views the PIFs.

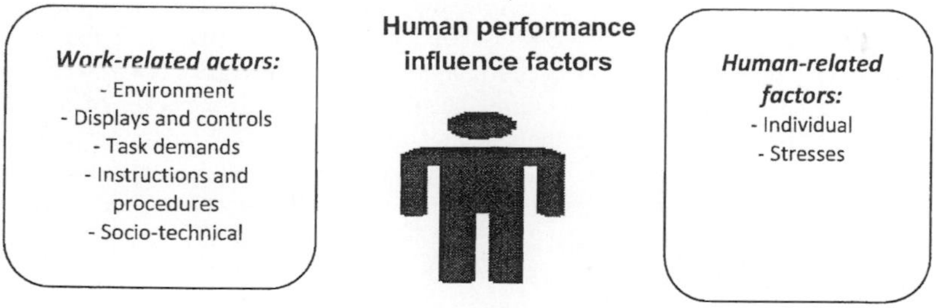

Figure 1: Human performance influence factors in principal include work-related and human-related factors.

The environment is one aspect of work-related elements that has implications for our health and level of performance. The environment typically includes such factors as temperature, humidity, illumination (including glare and reflections), noise, lighting, air quality and vibrations[1,43,44]. Each of these environmental factors has some defined requirements in guidelines and regulations. Recently, the RNNP report (2008)[18] showed an alarming number of noise-related injuries in the petroleum industry. This means that major employee groups are exposed to noise and thus are at risk of developing hearing injuries.

With respect to the displays and controls, in principle they need to be readable and easy to operate and to give feedback when something is wrong. The displays and controls have to be identified and comparable with the total system when designing a new system or when performing modifications. It is important for the system designer is to be aware how many displays and controls are needed where to locate them, colors and placement, etc. According to Sanders (1992), there are several aspects to discuss when selecting different displays and controls such as the physical arrangements of the displays and the controls (Ray Ray (1979)), the selection of rotary or stick types of controls, the placement of the controls, and whether the displays should be positioned in different planes; it is also important to include the reaction time or sensitivity of the controls.

The task demands and characteristics can involve frequency, workload or interaction with other tasks. The characteristics can typically be physical, regarding memory, vigilance, duration, attention, or critical nature. The human perception of task demands and the attention with which a task is performed have often been a discussion point during various complex operations. For instance, the aircraft disaster of Eastern Air Lines Flight 401 near Miami (1972) [30, 31] occurred because of complexities related to perception and attention. The pilot, co-pilot, and the flight engineer had become fixated on a faulty landing gear light and had failed to realize that the autopilot had disconnected. Consequently, the flight crew did not recognize the plane's slow descent and the aircraft eventually struck the ground. The crash occurred because of the flight crew's failure to monitor the flight instruments during a malfunction of the landing gear position indicator system. The disaster killed 98 passengers.

The instructions and procedures also have a critical role on the human performance characteristics. In principle, they need to be readable, meaningful, easily accessible, and to have a user-friendly format. It is important that the user understands the intention in the instructions, and the procedures need be clear. The instructions and procedures should be related to the level of detail and they should always define the selection and location. The documentation should frequently be updated, and applicable revisions should always be easy accessible. An example illustrating how the procedures can affect human performance comes from a fatal accident on the Gyda platform in 2002[19]. A man was killed because he was crushed between two containers during a lifting operation. The Norwegian Petroleum Directorate (NPD) performed an investigation of the accident and the important causes were found to be many and collective violations of the applicable procedures. Serious defects were identified in the operator's management system; relevant personnel did not have sufficient knowledge of the lifting procedures, and breach of procedures was a problem. The petroleum industry in particular has a number of such examples.

The socio-technical aspects that can affect the human performance, can include, for instance the team structure, roles and responsibilities, attitude to safety, communication, rewards and benefits, etc. The resources available, the work hours and breaks, the social pressures and conflicts are also other aspects that can be included in the socio-technical performance influence factors[1]. With respect to the current development in the offshore industry, the integrated operations (IO) concept is expected to bring major changes to the traditional socio-technical aspects. This is due to principal changes in terms of offshore-onshore roles and responsibility delegation, changes in organizational structure, staffing, management systems, collaborative technology, and the form of operational interactions expected. The activities on land and offshore are being merged into a single operations unit; work is controlled and organized in real time and in different parts of the world. Some offshore activities will be run from land-based organizations independent of distance, sea, and shore. IO may lead to new technological opportunities and changes in responsibilities and roles related to the transfer of decision-making authority across time zones and cultural or national boundaries. IO may thereby create new risks as much as generating further opportunities to become better at productivity and HSE.

As aforementioned, in addition to pure work-related factors, performance can also be influenced by inherent human-related factors that often relate to the current state of a person at any given point in time. This implies that the state of the human varies largely, from one given instance to another, resulting in different psychological and physiological status. Psychological conditions may affect the way human beings behave. The mental state of humans, and errors related to the emotional state of mind, can occur because of various personal events, even divorce or separation, death in the family or other work-related complaint or protests. The "I don't care" attitude can easily lead to lack of concentration and motivation, and may even oversee the expected behavior and precautions on work performance as specified in rules and regulations. Human error related to the emotional state of mind can never be totally eliminated, but it is important to implement measures and practices that allow the operators to have a better control of psychological status during work performance [35].

There are a number of personal factors that can affect human performance because each individual has different capacities, training, experience, skills, knowledge, attitudes, motivation, etc. In addition, there are other individual factors that can affect performance, such as personality, physical conditions, and risk perception. Humans have emotions that can influence the behavioral patterns at work. People respond to the work culture and, for, example new employees will follow the examples of other employees, even if they are given training and thereby know how to perform. A false sense of security can also be developed when nothing ever goes wrong, and thereby influence how to perform the work in a safe and secure manner. It is important to overcome unsafe and risk-taking attitudes and behaviors and to have the attention on developing a safety culture that includes training and education related to work performance[35].

Stresses come into play for various reasons, for instance lack of experience, time pressure, many distractions, heavy workload, and monotony in work or working in a high-risk environment. Stresses can also occur when the workers experience fatigue or they feel isolated. Shift work incentives are also a performance influence factor that could lead to stresses.

Most of those factors are under the control of the company and are shaped by the organizational policies and practices, as well as human qualities and characteristics. However, the experience from various investigations reveals that quite often the management does not necessarily always have a complete overview or practical knowledge about the influence of working conditions and organizational provisions on the human performance. The management may be often seeing intervening in handling specific situations of concern mostly as follow-up work after audits, investigations, or incidents. However, prior knowledge on various performance-shaping factors applicable to complex operational situations is a prime managerial requirement for the promotion of safe and good working organization. Such knowledge in principle aims at deploying early organizational interventions prior to serious events and incidents. Importantly, it also has a great potential to give confidence and motivation to the operators who particularly take up challenging, complex, and safety-critical tasks. In complex work settings, particularly where safety risks are high, this is an absolute necessity to achieve outstanding results, through better awareness of performance influence factors.

The Petroleum Safety Authority in Norway (PSA) has performed a number of projects and audits on human factors related to various challenging work settings. Recent developments related to integrated operations (IO), and its potential impact on traditional work practices, has triggered much of the ongoing projects and programs. Over the last few years, drilling and well intervention in particular has gained much attention due to the impact of ongoing change processes on the drill floor[4,5,6]. The early work performed clearly showed the lack of knowledge on human factors as well as understanding on complex PIFs. The projects underlined that, with respect to the ongoing change processes, the knowledge about PIFs and the understanding of interaction between HTO is critical to achieving the goals of safe and effective drilling and well operations. The next sections discuss the drilling and wireline operations work setting, and the critical PIFs based on the outcomes of the projects and the audits performed.

DRILLING AND WIRELINE OPERATIONS

The driller and the wireline operator lead the operation and manage well control and well integrity. Well integrity is defined in NORSOK D-010 as *an application of technical, operational and organizational solutions to reduce risk of uncontrolled release of formation fluids throughout the life cycle of the well*[8]. The well is designed with well barriers which are installed to prevent unintended influx (kick), cross-flow and outflow to the external environment. If the well barriers fails this could lead to loss of well control, kick, or blowout.

Human factors and the human-organization-technology context in drilling and wireline operations can be presented as puzzle pieces with elements like layout, management, manning, training, procedures, operation/work tasks, planning and equipment, time to conduct tasks, organization, communication/cooperation and the human[4]. The puzzle pieces illustrate how changes in one condition can influence the others. It is therefore important to evaluate how changes in one area may affect another when performing changes in one or several of the elements. Due to the complex connection and interaction between these elements, even a slight change in one condition or element can easily influence the others. For example, the driller or wireline operator has to use a new type of equipment. This situation may lead to new and updated procedures, more training related to the new equipment, changes in manning because the operators need a supervisor the first time they use the new equipment, and maybe it will take longer to conduct the task because the operator is not used to handling this type of equipment and needs more time to plan the job. Both drilling and wireline operations have a large tendency for errors during work performance due to the complexity and the dynamic nature of the work setting. There are many interactions in this process in technical, organizational, and individual terms. The connections have to be functional and fail-proof to ensure the safety integrity of the work setting. This implies that the chances for even a minor error need to be identified and properly evaluated in order to have a clear understanding of potential risk.

DRILLER

Drilling is an example of an area with great challenges in the interaction between human, technology and organization. For example, the driller must maintain control of the well, lead the work on the drill floor, and deal with technically advanced, screen-based solutions in the drilling cabin. It may thus be challenging to understand, operate, and maintain an overview of all the incoming data - and simultaneously maintain control and an overview of what is physically taking place on the drill floor.

The drillers' work situation:
- work area is offshore in the driller cabin
- work tasks are drilling and tripping operations
- leading the operation of the drilling crew at drill floor and managing well control
- one driller and one assistant driller at each shift (12-hour shifts)
- operating equipment with severe technical complexity in the drilling cabin together with the totality of the drilling module
- cooperating with deck personnel, crane operators, service personnel and contractors
- cooperating and teaching the assistant driller
- dependent on extensive use of cameras, telephone and radio communication due to poor visibility
- operating with minimum "down time"
- some administrative work

WIRELINE OPERATORS

Well intervention by means of wireline is a method of reaching operational objectives. The tools and equipment are conveyed into wells either through an "open hole" without surface pressure, or through special pressure retaining equipment which allows the tool string to be conveyed into live wells with full production pressure. Wireline services encompass slick, braided and electric line, plus specialized services such as H_2S service and heavy-duty units for fishing. Operational objectives are mechanical operations (setting plugs), well clean up (removal of sand or debris in the well), explosive services (punching or perforation) and data acquisition (production logging)[8].
The wireline operators' work situation:
- work area is offshore and outdoor
- the crew includes three or four operators each shift (12-hour shifts)
- the wireline operator has a crew operation leader

- maintaining control of the well
- carrying out well operations and well interventions
- dealing with technically advanced equipment
- working outdoors and at different deck levels: pipe deck, hatch deck, and weather deck
- communicating within the wireline operation crew, the driller, the crane operators, the pumping crew, the deck personnel and contractors
- operating with minimum "down time"

Drilling and wireline operations represent a very specific and a complex work setting in the offshore environment. The extent of use of technology for advanced drilling and well operations, the complex nature of the task, the level of psychological and physiological engagement demanded of drillers and wireline crews, the specific hazardous character of the work station, etc., contributes greatly to the need of better understanding of the dynamics of this complex setting. This provided the basis for the Norwegian Petroleum safety authority, to launch a number of projects looking into the interesting puzzle of this work situation.

CRITICAL PERFORMANCE INFLUENCE FACTORS IN DRILLING AND WIRELINE OPERATIONS

Embrey (2000)[41] describes the PIFs as *those factors which determine the likelihood of error or effective human performance*. Basic human error and PIFs are factors that together can create critical operational situations. Error likelihood can be minimized and the performance can be improved when the relevant PIFs are identified and optimized. The elements covered in the PSA study were the challenges in the driller and wireline operators' work situation offshore. The drillers and the wireline operators operate equipment with severe technical complexity, and the work situation includes continually monitoring, analyzing and problem solving based on a number of operational factors. The driller and the wireline operator actively engage in the control of the well barrier status and the management of well integrity, which are considered critical in terms of limiting integrity incidents, leaks, and well control situations. They are cooperating with deck personnel, crane operators, service personnel and contractors. The driller is also dependent on extensive use of cameras, telephone and radio communication due to poor visibility. The critical performance factors in drilling and wireline operations highlighted here are identified according to Redmill (1997), and are based on what was learnt from those practical studies performed by PSA[4,5,6].

ENVIRONMENT

The wireline winch was old and had several challenges related to the ergonomic, noise, lighting, visibility, vibrations, and climate (temperature and humidity). The winch was badly isolated, and the operators were exposed to a high level of noise and vibrations when performing wireline operations. The design of the wireline winch could have negative consequences for muscle stresses, concentration, and the wireline operators' attention when performing well operations. This working situation may have an impact on the operators' performance, and thereby increase the risk of well incidents and accidents[6,45].

Noise and oil vapor
The surveys showed that the drillers experienced irritating noise and bothersome oil vapor odor in the drill cabin. Unfortunately such irritating noise and odor may contribute actively to the drillers' feeling of tiredness, fatigue, and even stress during the work shifts[4].

Climate - Windwalls
The wireline operators are working outdoors in all kinds of weather conditions and they reported the lack of "wind walls" (wind protection) in their working areas at the facility. The use of wind walls can extend the tolerance of the operators for cold conditions. According to Enander (1989), a reduction in core temperature and/or muscle temperature reduces the physical functions and results in a decrease in muscle strength and endurance. The wireline operators' working situation without "windwalls" in a cold environment can have an impact on their performance[6, 43, 45].

Lighting
The wireline operators reported that there was unsatisfactory lighting at the wellhead deck, hatch deck and weather deck at the facility. Unsatisfactory lighting can have an impact on the performance of the wireline operators', since lighting can make details easier to see and colors easier to discriminate without producing discomfort or distraction [6,43].

DISPLAYS AND CONTROLS
Alarms and drilling systems[4]
Several drillers had experienced the alarms or drilling systems not functioning in an optimal manner, i.e. they experienced a large number of unnecessary alarms. The alarms did not provide guidance related to the actions the driller should take. Also the drilling systems did not provide support in critical situations, and the drilling systems rarely provided an early alarm when something was going wrong. It is important that the drilling system and the alarms provide support to the drillers and help them to prioritize the most important task at the right time, particularly in critical situations. This example shows that the alarms and the drilling systems may not function optimally in a given critical situation and this may lead to mistrust of the drilling support system by the operational crew, which may eventually influence the driller's performance.

Design and equipment layout[5]
It is seen that more and more equipment is installed in the drilling areas to meet the new requirements, and new technology is widely adopted related to drilling and well operations. Older installations were designed for manual drilling operations and the space for new and advanced drilling and well operation equipment on the drill deck was not considered when the rig was built. Modifications of older constructions to make space for new and more modern equipment have become relatively complicated with notable challenges for the operation. If the new equipment is to be utilized as intended, there is often

limited space where the equipment can be installed. The consequence of increasing equipment on the drill deck can contribute toward a lack of space, smaller work rooms, reduced accessibility to equipment, reduced visibility, increased use of cameras, and increased workload.

TASK DEMANDS AND CHARACTERISTICS

Workload[4]

The results in the surveys showed that one in four drillers sometimes work so hard that they are pushing the limit of what is prudent. This is a situation that shows the central challenge in the driller's workday. About half of the drillers felt that they have to perform many tasks during their work performance, resulting in an excessive workload. The drillers' perception is that a substantial amount of time is also spent on other assigned work tasks, such as administrative tasks including logging of the work performed. The drillers also comment that a lot of time is spent on daily maintenance work within their responsibility[4].

The wireline operators have reported that they experience challenges related to the workload. The wireline operators must maintain control of the well; perform the wireline operations at the pipe deck, hatch deck, and weather deck; and deal with technically advanced equipment. There are challenges related to workload, and stresses, when the operators work on different decks, to maintain control and overview of what is physically taking place in the well, especially when some of the operators in the team have less experience, and when the operators have a lot of work to perform[6, 45].

Work order permits system[6]

The control room operators felt that the installation has many work order permits and several simultaneous activities. It was reported that it was difficult to get an overview of acceptable risk and when to take necessary actions. The control room operator suggested a monitoring tool that ranged the work order permits related to risk and that illustrated the number of orders in each working area.

INSTRUCTIONS AND PROCEDURES

Procedure[6]

The wireline operators felt that the procedures and the management documentation were too comprehensive. The procedures in wireline operations included the operational procedures and about 26 installation procedures. The operators also considered that the documentation was not easily accessible.

Procedures and administrative systems[4]

A majority of the drillers reported that there were too many administrative systems and procedures that they must deal with. This gave rise to situations where some drillers indicated that they took shortcuts. One third of the drillers responded that they did not know whether the procedures existed for some work operations, and several of the drillers reported that they did not always have the time to read the applicable procedures. In total, seven of the 187 drillers stated that they often took shortcuts in relation to the procedures.

Work programs[4]

Nearly one half of the drillers felt that the work programs were difficult to understand. This may indicate that some drillers considered that they had too little time to quality control the given programs. Some drillers also stated that there were times when the daily work programs were not signed by the tool pusher/drilling supervisor. This could also indicate the lack of quality control of the existing work programs, and the shortfalls in the work routines.

SOCIO-TECHNICAL

Manning and competence

The wireline team includes three or four operators depending on the operational situation. This work setting can be affected and challenged, particularly when a team includes one or several new operators without adequate experience. The experienced operators have to quality check the job done by the less experienced ones, and at the same time perform their own task. In such situations, the wireline personnel may have less time to carry out the activities in accordance with the legislation, and in addition are handling hazardous situations and critical operations that have a large accident potential.[6].

The wireline team formally includes one team leader and, in the survey, the operators reported that the team leaders did not have proper leadership training. This shows that the wireline crew leader may lack the necessary competencies to lead the operational crew, which eventually can directly affect the social interaction patterns as well as the work performance of the team[6].

Planning, cooperation and change management[5]

Planning is a prerequisite for successful drilling and well operations. A number of areas for improvement were revealed, inclusive of involvement of operational personnel and planning, particularly due to major changes in the operational practices. A drilling entrepreneur has reported in the survey that: *"Planning for when something unforeseen takes place during operations needs to be improved. The work can be well planned from the start, but changes can take place. Then there is a challenge to get the organization to stop and design a new plan where the changes are considered in the planning for the ongoing work".*

Management[5]

A professional practice of drilling management was seldom reported in the drilling area during operations. Operative presence is greatly reduced due to the changes in administrative tasks and routines. The drilling management's limited presence and participation in the operative work may contribute toward negative consequences, i.e. lack of understanding of drilling conditions, reduced overall view over drilling operation, and reduced familiarity to daily challenges. This has large implications for the operational decisions taken by the management, which eventually can influence performance characteristics for the offshore operational crews.

INDIVIDUAL

Physical conditions[4]

Half of the drillers experienced muscle pain and eye fatigue when operating the drilling systems. The drillers' physical ailments can be directly connected to the design of the technical systems. The physical conditions can lead to unilateral strain on the eyes, neck, back and arms. The drillers' physical conditions can also be linked to their work routines with long working days (12-hour shifts) and the reported high work pressure.

STRESSES
Stresses due to time pressure[5]
There are challenges related to the operational personnel's (i.e. wireline operators) work situation when the operator companies call for personnel to board the installation immediately prior to carrying out well operations. The operators who enter the rig on short notice often get limited time for necessary preparations related to the job, limited time to familiarize and orientate themselves at the installation, restricted HSE information related to the installation and related to the relevant procedures and work description. This shows that the operators who have to board the installation on short notice to perform complex well operations may have psychological challenges that may easily enhance stress levels and reduce performance abilities due to the high-risk nature of offshore work settings.

DISCUSSION

The identified critical performance influencing factors within drilling and wireline operations are presented in the earlier chapters and the data is based on the experience from PSA projects and audits. The performance influencing factors can be divided into various dimensions, but can primarily be classified as workplace-related and human-related factors[1,12,13]. Environment is one of the workplace-related factors which include elements such as lighting, temperature, and noise. The element itself can not produce work output, but lighting can make details easier to see and colours easier to discriminate without producing discomfort or distraction. High noise levels pose serious threats to our hearing. There are several methods to reduce the noise levels at the source and along the path to the receiver, as well as to reduce the exposure levels of people through the use of hearing protection devices. It is important to have an active awareness of the problem and a concerted effort to deal with the problem constructively Effort should be made to optimize people's comfort, protect their health and maximize the performance when they are exposed to a cold environment during work. Proper clothing is very important for reducing the adverse effects of cold related to comfort, health and performance. The use of warm clothes, gloves and wind walls can extend the tolerance of people for cold conditions. The use of warming room can be useful when the tolerance levels to cold are exceeded, to warm the body and hands[43]The regulations related to noise, outdoor work, and lighting at an offshore facility are defined in the facilities' regulations.

The survey presented findings related to the drillers' and the wireline operators' task demands and characteristics. The interaction with many tasks, workload, manning, and stresses were the factors that influenced the drillers' and the wireline operators' performance. The drilling and wireline companies should therefore analyze how to improve the drillers' and the wireline operators' working situation related to the manning, responsibility, and working tasks. The surveys showed that *one in four drillers sometimes work so hard that they are pushing the limit of what is prudent*. The survey also presents challenges related to the alarm and drilling systems. The drillers reported that the alarms or drilling systems did not function in an optimal manner, i.e. they experienced a large number of unnecessary alarms and the drilling systems did not provide support in critical situations. This situation can influence the driller's performance because of mistrust of the drilling support system. This situation indicates that the driller's working situation has to be analyzed and improved before incidents and accidents occur.

There are work order permits systems like monitoring tools, on the market, that range the work order permits related to risk and illustrate the number of orders in each working session. This kind of tool can improve the overview of the many work order permits and the several simultaneous activities at the installation and thereby improve the working situation and the performance of the control room operators.

The survey indicates that the drillers' and the wireline operators' management documentation was too comprehensive; there were many administrative systems, instructions, and procedures. The work programs were difficult to understand and the documentation was not easily accessible. This situation led to a working environment in which the operators and drillers took shortcuts and the operators had little time to quality control the given work programs. According to the Activity regulations, Section 22 Procedure, *It shall be ensured that procedures are established and used in such a way as to fulfill their intended functions*. According to the guidelines to the activity regulations, *the procedures should be unambiguous, user-friendly and adapted to the users' competence. The users of the procedures should take part in the formulation and revision of such procedures.*

CONCLUSION

The critical performance influencing factors (PIFs) in drilling and wireline operations are presented based on the experience and the data from projects and audits. The PIFs affect the human performance, the human abilities to perform actions in a safe and efficient manner and the human ability to improvise and improve work performance. The lack of identification and classification of PIFs is generally seen as a major causal factor for human errors, poor decision making, and reduced task performance. The understanding of interaction between HTO and the knowledge about PIFs is critical to achieving the goals of safe and effective drilling and well operations. There is a wide range of factors that affect human performance under different working conditions. Such factors introduce different effects on the task performance by humans. In general, the PIFs can either be work -related, or inherently human-related. A good overview of these factors is very critical in constant organizational efforts to reduce work performance risks.

The offshore petroleum industry represents a high-risk complex operational environment. Drilling and wireline operations in particular are a sensitive work setting, especially from a safety risk perspective. Over the last few years, there

842

have been some growing concerns on the risk and integrity aspects of drilling and well operations, due to notable changes in the operational environment. In this paper, PIFs were used to elaborate on the several challenges within drilling and wireline operations. The challenges were highlighted under such elements as environment, displays and controls, task demands and characteristics, instructions and procedures, socio-technical, individual and stresses. In order to improve human performance within drilling and wireline operations, an understanding of factors that have a critical effect on daily operations performed by drillers and wireline operators is required. This is the very basis for a safety-conscious and risk-aversive work environment. By providing this information, it is assumed that organizational practices and managerial attitudes towards a safe and productive work place are reviewed so that proactive measures can be implemented to achieve best performance. The management of relevant organizations has a critical role and responsibility in this context.

REFERENCES

1. Redmill, F. and Rajan, J. (1997) *'Human Factors in safety critical systems'*, Oxford, Butterworth-Heinemann. Chapter; *'The causes of human error'*, pp 37-65.
2. Bailey, R.W. (1996) *'Human performance engineering'*, Prentice Hall. Chapter: *'Human engineering acceptable performance'*.
3. NORSOK standard D-010 Rev. 3, August 2004; *Well integrity in drilling and well operations*. NORSOK standard is developed with broad petroleum industry participation by interested parties in the Norwegian petroleum industry and is owned by the Norwegian petroleum industry represented by the Oil Industry Association (OLF) and Federation of Norwegian Manufacturing Industries (TBL). Standards Norway is responsible for the administration and publication of this NORSOK standard.
4. Åsland, J.E., Det Norske Veritas AS and Heber, H., Petroleum Safety Authority Norway (2007) *'Human factors in drilling and well operations. The Driller's work situation'*, [Online], Available www.ptil.no
5. Åsland, J.E., Det Norske Veritas AS and Heber, H., Petroleum Safety Authority Norway (2005) *'Human Factors in Drill and Well Operations; Challenges, projects and activities'* [Online], Available www.ptil.no
6. Petroleum Safety Authority Norway (2008) *'Audit of the wireline operators' work situation'* [Online], Available www.ptil.no
7. Petroleum Safety Authority Norway (2009a) *'The HSE regulations for the petroleum activities'*, [Online], Available www.ptil.no
8. Aker Solutions (2009) *'Wireline Services'*, [Online], Available http://www.akersolutions.com/Internet/IndustriesAndServices/OilAndGas/Wellinte rvention/Wirelineservices/default.htm
9. Petroleum Safety Authority Norway (2009b) *'HTO / Human factors'*, [Online], Available http://www.ptil.no/hto-human-factors/category140.html
10. Bento, J.P. (2000) *'Menneske – Teknologi – Organisasjon. MTO- analyser av hendelsesrapporter'*. Published by Oljedirektoratet.
11. [Beznosov, K., Fels, S., Iverson, L., and Fisher, B. (2006) *'HTO Admin: Human, Organization, and Technology Centered Improvement of IT Security Administration'*. [Online], Available: http://lersse-dl.ece.ubc.ca/record/107
12. Expro base (2008) *'Performance influencing factors'*. [Online], Available: http://www.exprobase.com/Default.aspx?page=16
13. Reason, J. (1997) *'Managing the risks of organizational accidents'*, Ashgate Publishing company, Burlington. Chapter; Maintenance can seriously Damage your system, pp 146-148.

14. Bjerkebaek, E., and Eskedal, T.S. (2004) SPE Paper 86597 *'Safety Assessment of Alarm Systems on Offshore Oil and Gas Production Installations in Norway'*. Published by the Petroleum Safety Authority Norway at the seventh SPE International Conference on Health, Safety and Environment in Oil and Gas Exploration and Production held in Calgary, Canada.

15. Azadeh, A., Nouri, J., and Fam, M. (2005) *'The Impact of Macroergonomics on Environmental Protection and Human Performance in Power Plants'*. Published at the Iranian J Env Health Sci Eng, Vol. 2, No. 1, pp 60-66.

16. Brannigan, J., Veeningen, D., Wiliamson, M., and Gang, Z. (2008) SPE paper 112219 *'Human Factor in Remote Operation Centers'*. Published by SPE at an Intelligent Energy Conference and Exhibition held in Amsterdam, Netherlands.

17. Bayerl, P.S., Lauche, K., Badke-Schaub, P., and Sawaryn, S. (2008) SPE paper 112104 *'Successful Implementation of Collaborative Environments; Human Factors and Implications'*. Published by SPE at an Intelligent Energy Conference and Exhibition held in Amsterdam, Netherlands.

18. Petroleum Safety Authority Norway (2009c) *'Risk level 2008: Expects visible results from HSE projects'* [Online], Available: http://www.ptil.no/news/risk-level-2008-expects-visible-results-from-hse-projects-article5497-79.html

19. Petroleum Safety Authority Norway (2009d) *'Many and collective violations of procedures caused the fatal accident on the Gyda platform on 1 November 2002'* [Online], Available: http://www.ptil.no/news/many-and-collective-violations-of-procedures-caused-the-fatal-accident-on-the-gyda-platform-on-1-november-2002-article231-79.html

20. Petroleum Safety Authority Norway (2009e) *'Investigation of incident on West Epsilon 14 September 2007 with notification of order''* [Online], Available: http://www.ptil.no/news/investigation-of-incident-on-west-epsilon-14-september-2007-with-notification-of-order-article3579-79.html

21. Dekker, G.F. and Bergen, E.A.V.D. (1996) SPE paper 35792 *'Human Factors in E&P facility design, a participatory approach'*. Published at the International Conference on Health, Safety and Environment held in New Orleans, Louisiana.

22. Noyes, J., and Bransby, M. (2001) *'People in control. Human factors in control room design'*. London, The Institute of Electrical Engineers. Chapter; Integrated platform management system design for future naval warship, pp 293-306.

23. Hamilton, W.I., Cullen, L., and Reeves, G. (2007) SPE paper 108267 *'Integrating people, plant and process in the design and operation of process industry assets'*. Published at the SPE Asia Pacific Health, Safety, Security and Environment Conference and Exhibition held in Bangkok, Thailand.

24. Baby, R. (2008) SPE paper 117830 *'Integrity Management during Design stage'*. Published at the International Petroleum Exhibition and Conference held in Abu Dhabi.

25. Høivik, D., Throndsen, T.I. (2005) SPE paper 96455; *'Human factors – Health and safety as a priority in design experience from the Norwegian Petroleum Industry'*. Published at the SPE Asia Pacific Health, Safety and Environment Conference and Exhibition held in Kuala Lumpur.

26. Pipitsangchand, S., and Somata,P. (2002) P. SPE paper 73990 *'Improving HSE Performance by management of human factor'*. Published at the SPE International Conference on Health, Safety and Environment in Oil and Gas Exploration and Production held in Kuala Lumpur, Malaysia.

27. Swinstead, N. (2006) SPE paper 98633 *'Human Factors program to reduce lost-time Injuries in Oilfield Operations'*. Published at the SPE International Conference on Health, Safety and Environment in Oil and Gas Exploration and Production held in Abu Dhabi.

28. Swinstead, N. (2004) SPE paper 86758 *'Human Factors Program Eliminates Safety Hazards of Marine Seismic Streamer – Handling Operations'*. Published at the seventh SPE International Conference on Health, Safety and Environment in Oil and Gas Exploration and Production held in Calgary, Canada.

844

29. Sethi, M.G. (2008) IPTC paper 12848 *'Lip Control and Protective System Upgrade – Proving a Case for Consideration of Human Factors in Design'*. Published at the International Petroleum Technology Conference held in Kuala Lumpur, Malaysia.

30. Kilroy, C. (2008) *'Special Report: Eastern Air Lines Flight 401'* [Online], Available: http://www.airdisaster.com/special/special-ea401.shtml

31. Sekicho (2002) *'Eastern Airlines flight 401'* [Online], Available: http://everything2.com/title/Eastern%2520Airlines%2520flight%2520401

32. National Transportation Safety Board (2004) *'NTSB says pilots excessive rudder pedal input led to crash of American flight 587; Airbus rudder system design & elements of airline pilots training program contributed'* [Online], Available: http://www.ntsb.gov/Pressrel/2004/041026.htm

33. Petroleum Safety Authority Norway (2009f) *'The chemical and physical working environment'* [Online], Available: http://www.ptil.no/working-environment/the-chemical-and-physical-working-environment-article4146-148.html

34. Petroleum Safety Authority Norway (2009g) *'Integrated operation'* [Online], Available: http://www.ptil.no/integrated-operation/category143.html

35. Wong, W. (2002) *'How did that happen? Engineering safety and reliability'*, London and Bury St Edmunds, UK Professional Engineering Publishing. Chapter; Human factors, pp 61-84.

36. Burgos, R., Allcorn, M., Mallalieu, R., and Vicens, J. (2007) SPE paper 100164 *'Eliminating Human Error during Coiled Tubing Operations'*. Published at the SPE/ICoTA Coiled Tubing and Well Interventions Conference and Exhibition held in Woodlands, Texas.

37. The Department of Explosives (2009) *'Chapter four – Statistical information of the department of Explosives'* [Online], Available: http://explosives.nic.in/wmchap4wo.pdf

38. Spiller, R. (1998) SPE paper 46767 *'Human Factors: The Trade Union Response'*. Published at the SPE International Conference on Health, Safety and Environment in Oil and Gas Exploration and Production held in Caracas, Venezuela.

39. Grant, P.J., Senior, L.M., and Lake, J.H. (2005) SPE paper *'Step Change towards a zero Incident Work Environment – Managing the Human Factor'*. Published at the SPE Asia Pacific Health, Safety and Environment Conference and Exhibition held in Kuala Lumpur, Malaysia.

40. Delve, K. (2006) SPE paper 101378 *'Training – Maximising the Human Asset'*. Published at the Abu Dhabi International Petroleum Exhibition and Conference held in Abu Dhabi.

41. Embrey, D. (2000) *'Performance influencing factors (PIFs)'* [Online], Available: http://www.humanreliability.com/articles/Introduction%20to%20Performance%20Influencing%20Factors.pdf

42. Exprobase (2008) *'Performance influencing factors'* [Online], Available: http://www.exprobase.com/Default.aspx?page=16

43. Sanders, M.S. and McCormick, E.J. (1992) *'Human factors in engineering and design'*, Singapore, McGraw-Hill Book Co-Singapore, Seventh edition. Chapter; 'Introduction' and 'Environmental conditions'

44. Hendrick, H.W., Kelner, B.K. (1999) *'Macro ergonomics – Theory, methods and applications'*, Lawrence Erlbaum Associates, Mahwah, New Jersey, London. Chapter; Part I 'Introduction to macroergonomics'.

45. Vignes, B. and Liyanage, J.P. (2010) *'Integrity in drilling and wireline operations in offshore assets: A case on the potential error causation paradigms'*.

46. Ray, R.D., and Ray, W.D. (1979) *'An analysis of domestic cooker control design'*. Ergonomics, 22, 1243-1248.

47. Enander, A. (1989) *'Effects of thermal stress on human performance'*. Scandinavian journal of work and environmental health, 15 (suppl. 1), 27-33.

48. Petroleum Safety Authority Norway (2005) *'The PSA's investigation of the gas-blowout on Snorre'* [Online], Available: http://www.ptil.no/news/the-psa-s-investigation-of-the-gas-blowout-on-snorre-article1850-79.html

Chapter 88

Surface Pressure Analysis in Prosthetics and Orthotics

J. Zivcak, R. Hudak, M. Kristof, L. Bednarcikova

Institute of Technical Sciences
Technical university of Kosice
Letna 9, 042 00, SVK

ABSTRACT

Measurement processes in prosthetic and orthotic practice play critical role in efficiency of the treatment. Pressure is one of the most important parameters that indicate the condition and tendency of the technological process, testing and wearing of the new prosthetic and orthotic aid. The pressure distribution between the human body and trunk orthosis was analyzed using thin pressure indicating film. The main goal of the research was to design a new pressure measurement methodology for optimization of the treatment from the ergonomic, correction and efficiency point of view.

Keywords: Adolescent Idiopathic Scoliosis (AIS), Back Brace, Orthoses, Pressure Indicating Film,

INTRODUCTION

Adolescent idiopathic scoliosis (AIS) affects 1–3% of children in the at-risk population of those aged 10–16 years. The aetiopathogensis of this disorder remains unknown (idiopathic), with misinformation about its natural history. Non-surgical treatments are aimed to reduce the number of operations by preventing curve progression. Although bracing and physiotherapy are common treatments in much of the world, their effectiveness has never been rigorously assessed.

Van den Hout (2002) [5] performed pressure measurements which are applicable and of value for research on the working mechanism of brace treatment. Therefore, they performed pressure measurements on new THORACO LUMBO SACRAL ORTHOSES (TLSO) with a more or less dynamic thoracic brace pad.

Delphine Périé (2003) [6] analyzed Boston brace biomechanics, pressure measurements and finite element simulations were done on 12 adolescent idiopathic scoliosis patients. The results showed that high thoracic pads reduced both thoracic and lumbar scoliotic curves more effectively than lumbar pads only. The study suggests that mechanisms other than brace pads produce correction and contribute to the force equilibrium within the brace.

Stefano Negrini and colleagues (2009) [13] carried out the third study published with respect to SRS criteria, and it is the first one that has also fulfilled the SOSORT criteria for bracing studies. The former criteria provide the methodological framework while the latter give the clinical framework so as to gather the best possible data on this kind of treatment. The number of patients is low, but the population is comprehensively selected and cohesive. Results: Median reported compliance during the 4.2±1.4 treatment years was 90%(range 5-106%). No patient progressed beyond 45°, nor was any patient fused, and this remained true at the two-year follow-up for the 85% that reached it. Only two patients (4%) worsened, both with single thoracic curve, 25-30° Cobb and Risser 0 at the start. They found statistically significant reductions of the scoliosis curvatures (-7.1°): thoracic (-7.3°), thoracolumbar (-8.4°) and lumbar (-7.8°), but not double major. Statistically significant improvements have also been found for aesthetics and ATR. Conclusion: According to our results, in patients at risk it is possible to avoid surgery, provided the patients follow their prescriptions and adhere to the regime of treatment. By respecting the SOSORT criteria and focusing on compliance, a complete conservative treatment based on bracing and exercises will produce much better results according to the SRS criteria than have been reported previously. These results should be verified in the future with a prospective paper that will also include drop-outs, which are failures of treatment. This paper demonstrates the importance of the human approach together with the technical aspects of the treatment.

Technological advances have much improved the ability of surgeons to safely correct the deformity while maintaining sagittal and coronal balance. Much has yet to be learned about the general health, quality of life, and self-image of both treated and untreated patients with AIS.[1] The pressure range at orthotics treatment in adolescent idiopathic scoliosis was analyzed to manage or even reduce spinal curvatures while waiting for skeletal maturation. By applying specific pressure points on the torso, the brace treatment attempts to modify mechanically the scoliotic spine shape and control progression of the spinal curvatures. . [4] Many clinical studies used standard radiographs to assess the brace effectiveness. To analyze spinal brace biomechanics, pressure measurements at orthosis testing in four adolescent idiopathic scoliosis patients were done.

To analyze spinal brace biomechanics, pressure measurements and finite practice exam were performed on 4 adolescent idiopathic scoliosis patients. The aim was to analyze the spinal brace effectiveness using a finite practice exam and experimental measurements. All the measurements and our research contribute new knowledge to this field. Each piece of knowledge might be useful for orthopedists, orthopedic specialists and for all those who also work to improve the properties of the back brace for idiopathic scoliosis. Biomechanical pressure between the body and orthosis is an important parameter during the treatment. Its results will also be helpful for innovations in the design. Specific prosthetic and biomedical experiments were carried out by pressurex sensitive films (SENSOR PRODUCTS INC., USA) in COP Company of Kosice and at department of Biomedical Engineering, Automation and Measurement, Technical University of Kosice, Slovakia.

MATERIALS AND METHODS

There are few biomechanical pressure studies of spinal brace effectiveness and proper function. Biomechanical pressure action is not completely understood. For our research we used pressure sensitive films (Figure 1). Pressurex sensitive film is an affordable and easy to use tool that reveals the distribution and magnitude of pressure between any two contacting, mating or impacting surfaces. Pressure indicating sensor film is thin (0, 1016 to 0, 2032 mm) which enables it to conform to curved surfaces. It is ideal for invasive intolerant environments and tight spaces not accessible to conventional electronic transducers. [2]

Pressurex is a mylar based film that contains a layer of tiny microcapsules. The application of force upon the film causes microcapsules to rupture, producing an instantaneous and permanent high resolution "topographical" image of pressure variation across the contact area.

Apply pressure, remove it and immediately the film reveals the pressure distribution profile that occurred between the two surfaces. Like Litmus paper, the color intensity of the film is directly related to the amount of pressure applied to it. The greater the pressure, the more intense is the color. Pressurex has a wide array of uses. Our pressure indicating film acts as a force sensing resistor between patient's skin and orthoses. It can measure surface pressure distribution whether it is used as an impact force sensor, seat pressure sensor, as a strain gauge or even as nip impression paper.

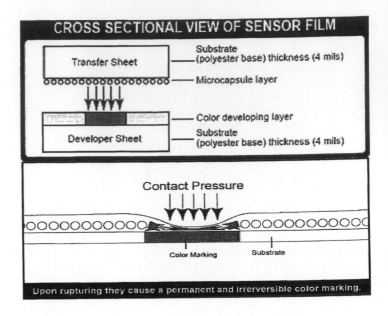

FIGURE 1: Pressure sensitive film technology (PRESSUREX, SENSOR PRODUCTS INC., USA)

Table I Types of pressure sensitive films

Type	Pressure RANGE
MICRO	0,14 – 1,4 kg/ cm2
ZERO	0,5 – 2 kg/ cm2
ULTRA LOW (A)	**2 – 6 kg/ cm2**
SUPER LOW (B)	**5 – 25 kg/ cm2**
LOW (C)	**25 – 100 kg/ cm2**
MEDIUM	100 – 500 kg/ cm2
HIGH	500 – 1300 kg/ cm2
SUPER HIGH	1300 - 3000 kg/ cm2

The procedure of Presurex application includes: applying the pressure, removing it and the film reveals the pressure distribution profile that occurred between the two surfaces. Like Litmus paper, the color intensity of the film is directly related to the amount of pressure applied to it. The greater the pressure, the more intense is the color. (Table I) [2]

Three different ranges of pressure indicating films were applied between the trunk orthosis and the human body (Figure 2.).

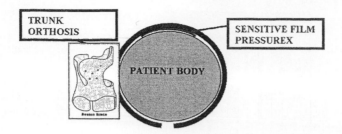

FIGURE 2: Scheme of measurement system between the body and trunk orthosis.

Figure 3 shows the methodics of pressure indicating film (PIF) tapes from application of the different pressure ranges tapes to pelots like films to the application of real-time sensoring system.

RESULTS

An initial study was performed in 3 female and 1 male patients (children) with scoliosis treated with the spinal orthosis BOSTON brace system. An experimental protocol was composed by the acquisition of two sets of pressurex sensitive films (Figure 4).

The Topaq software system is designed to be used in conjunction with Pressurex sensitive films (Figure 5). Utilizing a specially adapted flatbed scanner, Topaq scans and interprets these films and determines the exact pressure applied at every point across the film surface at resolutions up to 1000 DPI. The maximum film dimension that can be scanned at one time is 12" x 17".

The software has many analysis tools that give flexibility in scanning the entire images or just small areas of an image. Statistics can be generated on local regions or the entire scanned area.

Topaq software displays the image in 3 distinct formats; as the actual scanned image looks, as a pseudo color representation of the scanned image with different colors corresponding to different ranges of pressure, and as a three-dimensional image.

Additionally, Topaq allows you to plot a histogram and log all the data collected for conversion to formats accepted by other popular spreadsheet programs for further analysis. [2]

Finally, all generated images can be saved to disk and/or printed out on a Laser or color printer. (Figure 5)

850

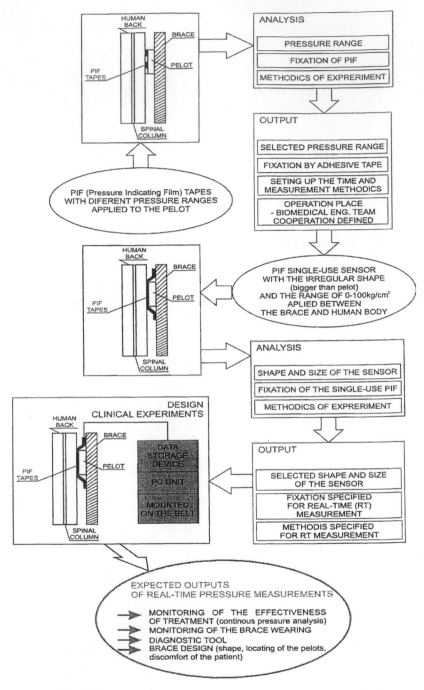

FIGURE 3: Scheme of the pressure indicating films application between the human body and brace.

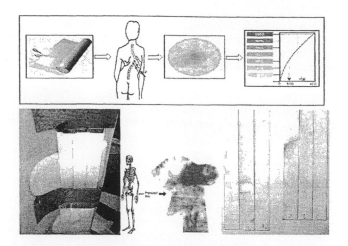

FIGURE 4: Sensitive films during the measurement process.

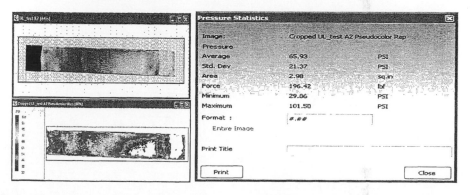

FIGURE 5: Topaq output of pressure sensor films after the measurement.

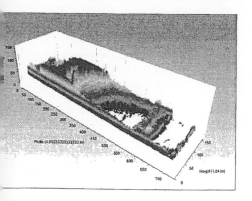

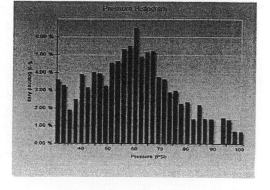

FIGURE 6: Topaq 3D reconstruction and the histogram.

A personalized finite element modeling of the pressure was generated from the 3D reconstruction and the histogram of the patient's geometry.

The brace treatment was simulated by the application of equivalent forces calculated from the pressure measurements.

The results of every measurement confirm our prognosis about pressure extant in back brace during the measurement. After analysis of our measurements, we have got the results of pressure range between the patient's body and orthosis. The values are given in the summary sheet. (Table 2)

In the summary table are 3 selected types of pressure sensor films: A – ultra low, B – super low, C – low. Numbers are equal to the number of patients.

Table 2 : Pressure average from the measurement

PRESSURE AVERAGE [kg/ cm2]		
Measurement	I	II
A1	3,56	3,44
A2	3,20	3,32
A3	3,69	3,33
A4	3,27	3,39
B1	7,99	8,36
B2	9,76	8,37
B3	8,11	7,51
B4	8,12	7,89
C1	31,41	32,94
C2	34,20	32,43
C3	35,45	35,04
C4	34,10	35,39

CONCLUSIONS

Our initial research opened a new view on the construction of new surface measurement system. This is a new challenge for the research, orthopedists and technicians. Additional research is necessary to get the clear outputs for the field of the diagnostics of scoliosis treatment. Research is moving into the new direction with new methods, technologies and materials. Nowadays, the new cooperation has started with the Center of Orthopedics and Prosthetics, Kosice. The plan is to measure the contact pressure between residual limb and prosthetic socket by pressure sensors.

We wish to contribute to more profound research of pressure between the body and orthosis. Our second study will be aimed to dynamic measurements using electronic pressure sensors in orthosis during the treatment. A new real-time sensor system with the implementation of Sensor Product Company sensors will be designed. The methodology for new clinical measurements and evaluation of the results will be proposed for the effective treatment, ergonomics, correction features and efficiency for spinal orthoses. The results will be then entered into the computer and compared.

ACKNOWLEDGMENT

"The presented paper was prepared by the support of the Slovak Grant Agency of Ministry of Education and Slovak Academy of Sciences – VEGA 1/0829/08 and VEGA 1/4263/07"

REFERENCES

[1] Stuart L Weinstein, Lori A Dolan, Jack CY Cheng, Aina Danielsson, Jose A Morcuende: Adolescent idiopathic scoliosis, The Lancet, Volume 371, Issue 9623, 3 May 2008-9 May 2008

[2] Sensor products INC (2009). TACTILE PRESSURE EXPERTS-[online] (May 11,.2009) The Sensor products Website : http://www.sensorprod.com/static/lamination.php

[3] Weis, H.-R. (2007) „Is there a body of evidence for the treatment of patients with Adolescent Idiopathic Scoliosis (AIS)." Research report No. 10.1186/1748-7161-2-19, BioMed Central Ltd., This article is available from: http://www.scoliosisjournal.com/content/2/1/19

[4] Noel, J.: Advantages of CT in 3D Scanning of Industrial Parts, 3D Scanning Technologies Magazine., Vol. 1, No. 3, 18 — December 2008.

[5] Van den Hout JA, van Rhijn LW, van den Munckhof RJ, van Ooy A.Interface corrective force measurements in Boston brace treatment. *Eur Spine J* 2002.
This abstract is available from:
    http://www.ncbi.nlm.nih.gov/pubmed/12193994,

[6] Périé, D., Aubin, C.-E., Petit, Y., Beauséjour, M., Dansereau, J., Labelle, H. : Boston Brace Correction in Idiopathic Scoliosis: A Biomechanical Study This article is available from: http://www.medscape.com/viewarticle/461584

[7] Hutníková, L., Hudák, R., Živčák, J.: Scoliosis severity diagnostics: the new application for automatic X-ray image analysis, In: Infusing Research and Knowledge in South East Europe: 3rd Annual South-East European Doctoral Student Conference : Thessaloniky, 26-27 June, 2008, ISBN 978-960-89629-7-2, ISSN 1791-3578

[8] Živčák, J., Petrík, M., Hudák, R. et al.: Embedded Tensile Strength Test Machine FM1000 – An Upgrade of Measurement and Control, Trans Tech Publications, Switzerland, Solid State Phenomena, Vols. 147-149 (2009) pp 657-662

[9] Toporcer, T., Grendel, T., Vidinský, B., Gál, P., Sabo, J., Hudák, R.: Mechanical properties of skin wounds after atropa belladonna application in rats. In: Journal of Metals, Materials and Minerals. vol. 16, no. 1 (2006), p. 25-29. ISSN 0857-6149.

[10] Gal, P., Kilik, R., Spakova, T., Pataky, S., Sabo, J., Pomfy, M., Longauer, F., Hudak, R.: He-Ne laser irradiation accelerates inflammatory phase and epithelization of skin wound healing in rats. In: Biologia. roč. 60, č. 6 (2005), s. 691-696. ISSN 0006-3088.

[11] Weinstein, S. L., Dolan, L. A., Cheng, J. CY, Danielsson A., Morcuende J. A (May 2008) "Adolescent idiopathic scoliosis.", The Lancet, Volume 371, Issue 9623, 1527-1537.

[12] Weis, H.-R., Werkmann, M., and Stephan, C. (2007) „Brace related stress in scoliosis patients – Comparison of different concepts of bracing " Research report No. 10.1186/1748-7161-2-10 BioMed Central Ltd.,

854

This article is available from: http://www.scoliosisjournal.com/content/2/1/10

[13] Negrini S., Atanasio, S., Fusco C. and Zaina F., (september 2009) „Effectiveness of complete conservative treatment for adolescent idiopathic scoliosis (bracing and exercises) based on SOSORT management criteria: results according to the SRS criteria for bracing studies" SOSORT Award 2009 Winner ISICO (Italian Scientific Spine Institute), Via Roberto Bellarmino 13/1, 20141 Milan, Italy . Research report No. 10.1186/1748-7161-4-19 *Scoliosis* 2009, This article is available from: http://www.scoliosisjournal.com/content/4/1/19

Chapter 89

Adaptability of Operators when Exposed to Hand-Transmitted Vibration at Work with Electric Hand-Held Tools

Světla Fišerová

Czech Republic

ABSTRACT

The article deals with the measurement and evaluation of hand-transmitted vibration exposure in the case of persons having various durations of working adaptation. From the point of view of occurrence of unfavourable health effects, what matters most is hand vibration transmitted from a vibrating handle or another held subject.

Keywords: Hand-transmitted vibration, repetitive measurements, electric hand-held tools, adaptability of operators

INTRODUCTION

Hand-transmitted vibration occurs at work with electric, pneumatic and hydraulic tools, i.e. hand-held drills, grinders, saws, cutters, planes, spanner wrenches, rammers, hammer drills, hammers, and other similar tools. In professional practice, electric hand-held tools are widely used. Vibration is transmitted to the hands of the operator from the grasped handle of a hand-held tool and from the hand-held material being machined or from another grip on the tool being used. Depending upon the type of tool, the working position of the operator, and the location of the workplace, vibration may enter only one arm or both arms simultaneously and may be transmitted by the hand and the arm to the shoulder. Hand-transmitted vibration is a source of discomfort and possible decrease in labour efficiency. The long-term use of vibrating mechanized tools is connected with the occurrence of symptoms of diseases affecting blood vessels, nerves, bones, tendons, joints, muscles and fibrous tissues of hand and forearm. The level of exposure of the person who is subject to hand-transmitted vibration is determined by the intensity of vibration, which is defined as the total weighted level of vibration acceleration and the duration of exposure. Other factors influencing the effects of exposure of operator's hands to vibration are working methods and operator's skill, quality of made temporal analysis of work, working posture of hand and arm, type, condition and weight of hand-held tool being used, properties of material being machined, microclimatic conditions and thermal situation in the working place, noise, health condition and use of medicines and/or habit forming substances influencing the human circulatory system. In view of the above-mentioned factors, the total weighted level of hand-transmitted vibration acceleration declared by the manufacturer usually differs from a resultant value (in accordance with relevant international standards) measured during a specific activity under real conditions.

Research being conducted at the teaching laboratory for labour safety and ergonomics at the Faculty of Safety Engineering of VŠB – Technical University of Ostrava focuses on the verification of influence of operator's skill depending upon the duration of adaptation on the magnitude of vibration transmitted to hands at work with electric hand-held tools. Research is directed towards the confirmation of the assumption that during the same activity and under the same conditions of measurement of hand-transmitted vibration, an unadapted person is exposed to a higher exposure than an adapted person.

METHODS

In the Czech Republic, methods of vibration exposure evaluation even in the area of hand-transmitted vibration are in accordance with valid legal rules of European Union. The latest directive concerning vibration affecting humans, the Human Vibration Directive 2002/44/EC, was adopted on the 25th of June 2002. It determines the minimum health and safety requirements regarding the exposure of workers to the risks arising from physical agents – vibration. Vibration is measured and evaluated according to standard methods, which are understood as methods contained in the Czech technical standard, under conditions determined in the Government Decree No. 148/2006 Coll., on health protection against unfavourable noise and vibration effects. Several ten valid international standards dealing with vibration exist. For hygiene measurement and vibration evaluation the standard ČSN EN ISO 5349:2002 Measurement and evaluation of human exposure to hand-transmitted vibration, Part 1 – General requirements, Part 2 – Practical guidance for measurement at the workplace is used. Before the hand-transmitted vibration measurement itself, the workplace must be assessed and the activity of workers must be analysed. The analysis concentrates on the position of the workplace, the description of performed activity of workers, including the description of working position, used tools and their technical conditions, used materials and subjects being machined, the time image of real exposure to hand-transmitted vibration.

CHARACTERISTICS OF OCCUPATIONAL EXPOSURE EVALUATION

For the needs of hygiene evaluation of hand-transmitted vibration, the vibration acceleration level L_a and the RMS value of vibration acceleration a_e are taken as determining quantities.

The vibration acceleration level L_a is given by the following relation

$L_a = 20 \log (a/a_0)$ [dB]

where
a   is the instantaneous vibration acceleration in $m.s^{-2}$
a_0 is the reference vibration acceleration level in $m.s^{-2}$
a_0 is 10^{-6} $m.s^{-2}$

The RMS value of vibration acceleration a_e is given by the following relation

$$a_e = \sqrt{\frac{1}{T} \int_0^T a^2(t)dt} \qquad [m.s^{-2}]$$

where
a(t)   is the instantaneous acceleration in $m.s^{-2}$
T   is the time for which the RMS value of acceleration in s is to be determined.

Vibration acceleration levels and RMS values of vibration acceleration are vibration quantities that are substitutable by each other. The weighted level of vibration acceleration L_{aeq} is such vibration acceleration level that corresponds to the frequency correction for the given way and conditions of transmission and the direction of vibration. It is expressed in dB. The total weighted level of vibration acceleration is given by the vector sum of weighted RMS values of acceleration along three orthogonal axes. In the case of hand-transmitted vibration, the vibration should be evaluated with regard to its effects on humans in bands of frequency of 8 – 1000 Hz. Hand-transmitted vibration is measured and stated in the three directions of an orthogonal coordinate system. The total weighted level of vibration acceleration is after calculation stated as RMS VTV (Vibration Total Value).

MEASUREMENT OF HAND-TRANSMITTED VIBRATION

Vibration is standardly measured in the point of transmission of vibration to a human body. This means in the point where vibrating equipment is in contact with a human organism. When evaluating the unfavourable effects of vibration transmitted to humans, the way of transmission, the dominant direction and the frequency of vibration are crucial.

For the evaluation of directional effects of vibration, a human body coordinate system and a hand coordinate system were determined, in which measurement is carried out. The subject of evaluation is translational vibration. The frequency of vibration is observed especially with a view to reduction in unfavourable effects of vibration at the resonant frequencies of human organism. The basicentric system of coordinates corresponds to the biodynamic system being adjusted commonly in the plane y-z so that the direction y is parallel to the axis of the handle. Local hand-transmitted vibration is measured in all three directions of the orthogonal system; the directions correspond to the axes x, y and z of the biodynamic system.

Biological effects of vibration depend on the connection between a hand and a source of vibration. This connection can considerably influence the size of measured values (Smetana). That is why the measurement must be made in the course of action of forces that are representative of connection between a hand and a vibrating mechanized tool, a handle or subject being machined during some typical activity of the tool or typical working practice.

STRATEGY AND MEASUREMENT METHOD

To confirm the formulated assumption, repeated measurements of hand-transmitted vibration at machining the same material were made under standard microclimatic conditions at the workplace of maintenance and administration of buildings and laboratories of Faculty of Safety Engineering (henceforth referred to as FSE), VŠB – Technical University of Ostrava.

FIGURE 1 An example of measured work activity (sensor mounted on the handle of the tool)

The measurements were taken in the period March – April 2008 (Action 1) and March – April 2009 (Action 2). In both the cases, the material being machined was spruce of prismatic shape with the width of prism of 60 mm.

The electric hand-held tool that was a source of hand-transmitted vibration was in the first case an electric belt sander BOSCH PBS 75AE and in the other case am electric plane NAREX EDH 82.

The measurements were carried out as follows. Hand-transmitted vibration was recorded at activity performed by an adapted and an unadapted person; the values of exposure to vibration transmitted to the right hand grasping the handle of the tool and to the left hand handling the frontal grip of the electric tool were recorded. One comprehensive measurement includes altogether six measurements of exposure along three orthogonal axes of action of vibration transmitted to both operator's hands. The duration of individual repeated partial measurements (20 sec) is in accordance with the requirement of relevant international standard.

Operators who performed the measured activity were selected persons adapted and unadapted. In either Action (1, 2) two different persons always participated.

An adapted person is the operator who executes the given activity for more than one year, and this activity is included in a common specification of job of his /her profession. In both the cases of selected adapted persons, the persons were workers of workplace of maintenance and administration of buildings and laboratories of FSE; the common activity of them being the use of electric hand-held tools in the preparation of samples for teaching laboratories and the maintenance of laboratories and buildings.

An unadapted person is the operator who is acquainted with the working practice and the principles of safe use of the tools; however, the person carries out the activity without previous practical experience in the use of electric hand-held tools at any similar activity whatsoever. The selected unadapted persons participating in the measurement were FES students in the field of study, Safety Engineering.

In the framework of Action 1, one comprehensive measurement was made in the case of adapted and one comprehensive measurement in the case of unadapted person. In the framework of Action 2, four comprehensive measurements were carried out in the case of both the persons. Before the commencement of both the actions, test measurements had also been made and evaluated.

For the implementation of measurements in Action 1, vibration measuring equipment B&K, namely a ⅓ octave signal analyzer 2260 ObserverTM and Front-end 1700, i.e. equipment for vibration recording in three axes of action, and a three-axial vibration sensor, type 4524 were used. As for Action 2, a B&K device for vibration measurement and analysis VA 4447 with a possibility of connection of an accelerometer of type 4520 were used. The device VA 4447 is equipped with its own software Vibration Explorer. In all cases the measuring devices were calibrated using a calibrator B&K of the type 4294 before and after each measurement.

Instrument equipment fulfils requirements for ⅓ octave band measurement that correspond to requirements for detailed measurement in Accuracy Class I with accuracy of $\pm$ 0.8 dB.

The vibration acceleration sensor – accelerometer must be attached in accordance with the requirements of relevant international standards to avoid influencing measurement uncertainty. The attachment of sensors used for all realized measurements was performed uniformly. Detailed pictures of attachment of the sensor to the handle and also the frontal grip of used electric hand-held tool are there in Figures given below.

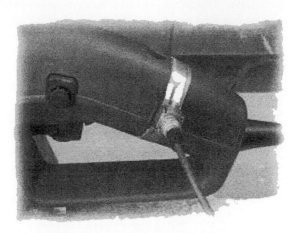

FIGURE 2 A detailed picture of attachment of the accelerometer to the handle of used electric hand-held tool

FIGURE 3 A detailed picture of attachment of the accelerometer to the frontal grip of used electric hand-held tool

RESULTS

By carrying out altogether five comprehensive measurements of vibration transmitted to the hands of adapted and unadapted operators, partial results were acquired. Those were evaluated by means of software products forming part of measuring instrument equipment. All partial results for each partial measurement are provided in both numeric and graphic form. In the following figures there are examples of graphical output for the measured quantities from one partial measurement for axis x of adapted as well as unadapted person.

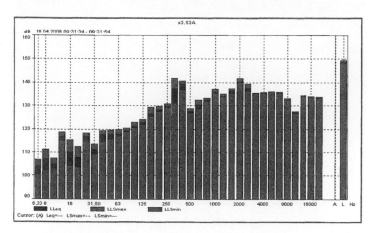

FIGURE 4 A spectrum of vibration along the axis x acting on the right hand of adapted operator (partial measurement)

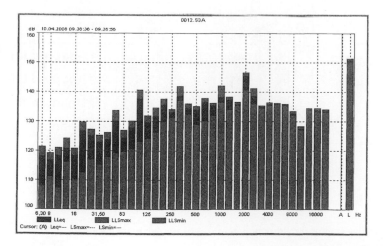

FIGURE 5 A spectrum of vibration along the axis x acting on the right hand of unadapted operator (partial measurement)

Resultant values of partial measurements were calculated using SW to RMS Vibration Total Values and processed to a resultant summary table and a graph for the right (RH) and the left (LH) hand of adapted as well as unadapted person of operator. The results concern the calculation for five comprehensive measurements designated as Action 1 and Action 2.

Table 1 Calculated RMS VTV results

Operators RMS VTV [dB]	Action 1	Action 2.1	Action 2.2	Action 2.3	Action 2.4
Adapted RH	131.7	128.8	128.7	128.6	128.8
Adapted LH	128.6	126.9	127.2	127.6	127.4
Unadapted RH	132.8	129.5	129.5	129.3	131
Unadapted LH	131.8	128.1	129.4	129.8	130

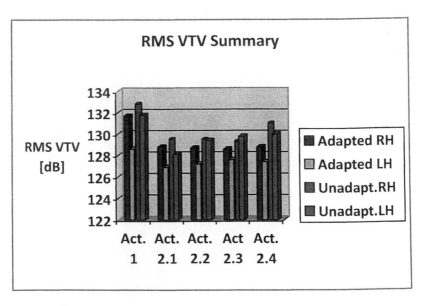

FIGURE 6 A graphical representation of measurement results recalculated to RMS Vibration Total value.

DISCUSSION

Results of repeated measurements of hand-transmitted vibration have confirmed the assumption that persons unadapted to work with electric hand-held tools are exposed to a higher exposure to vibration than adapted persons. In all the measured cases, the obtained values were always higher with unadapted persons (the left as well as the right hand) and also along all three orthogonal axes. Only in a third of the cases, an increased exposure was found in the interval of declared measurement uncertainty.

Results of measurement of exposure to hand-transmitted vibration also show that the exposure of the left hand differs from that of the right hand of both the adapted and the unadapted person.

For the left hand of the unadapted operator, values were in comparison with the right hand of the adapted operator higher and moved in the range of 1.2 – 3.2 dB. For the right hand, a vibration exposure in the case of the unadapted operator was higher and moved in the range of 0.7 – 1.2 dB. This also proves that the vibration exposure of the right hand of the operator differs from that of the left hand of the operator in the course of activity being performed.

In the case of the adapted person, the load is divided equally to both the hands and confirms that the pressure is applied mainly to the handle of the tool used.

Results of measurements in the unadapted person show that the pressure acts alternately on both the hands, which confirms the assumption of insufficient skill and experience in work with electric hand-held tools.

In the case of measurement in the framework of Action 2, which took place in the year 2009, all measurements were made with the same unadapted person. From measurement results it is obvious that the short-term use of tools for the purposes of experimental research did not lead even to a moderate gradual decrease in exposure at repeated measurement.

In real practice, it is other factors, such as the selection of a suitable type of hand-held tool and its proper maintenance that influence the level of exposure to hand-transmitted vibration.

In the Czech Republic, the limit for hand-transmitted vibration is stricter than that recommended by the relevant EU directive. In spite of this, occupational diseases due to hand-transmitted vibration occur permanently over and over again. Of the total number of occupational diseases in the Czech Republic, the proportion of these is about 17 %.

CONCLUSION

Repeated measurements of vibration transmitted to the hands of operators at work with electric hand-held tools have confirmed that unadapted persons and persons with short-term training are exposed to higher exposure to hand-transmitted vibration than adapted persons. The determination of a suitable time of training,

864

when the pace of work of the operator will be lower, will lead to the acquiring of required skill and experience, and thus to the mitigation of influence of hand-transmitted vibration on the health condition of the operator.

REFERENCES

Hovorka,V., and Fišerová,S. (2008), *Vibrace v pracovním prostředí.* Bakalářská práce oboru Bezpečnost práce a procesů, Faculty of Safety Engineering, VŠB – Technical University of Ostrava, Czech Republic.
Michálek,M., and Fišerová,S. (2008), *Rizikový faktor vibrace.* Diplomová práce oboru Bezpečnostní inženýrství, Faculty of Safety Engineering, VŠB – Technical University of Ostrava, Czech Republic.
Smetana,C. (1998), *Hluk a vibrace.* Sdělovací technika, Praha, Czech Republic
Valášek,P., and Fišerová,S. (2009), *Vibrace přenášené na ruce při práci s ručním elektrickým nářadím.* Diplomová práce oboru Bezpečnostní inženýrství, Faculty of Safety Engineering, VŠB – Technical University of Ostrava, Czech Republic.

Chapter 90

Effects of Cell Phone Conversations on Driving Performance in Japan

Mi Kyong Park, Takayoshi Machida,
Kimihiro Yamanaka, Mitsuyuki Kawakami

Division of Management Systems Engineering
Tokyo Metropolitan University
Tokyo, 191-0065, Japan

ABSTRACT

The use of a hand-held cell phone by a driver while driving in Japan has been prohibited since 1999. However, the number of traffic violations related to the using of hand-held cell phones in 2008 was about one million two hundred thousand, and it has been increasing each year. The aim of this study is to examine impacts of cell phone conversations on the driving performance of fatigued drivers. Ten male students performed No-task (NT) and conversation (cognitive tasks) with hands-free (HF) and hand-held (HH) cell phones while driving in city traffic in which the maximum speed limit is 40km/h. Each participant was given a practice run on a monotonous road for more than 50 min in order to become familiar with the driving simulator and to induce fatigue while driving. Reaction time to step on brake in response to pop-up event was significantly slower in HH conversation than in NT, but there was not a significant difference between HF conversation and NT in reaction time. In addition, NT showed relaxed braking deceleration compared with both HF and HH conversations at red traffic lights. The present study suggests that talking on the phone while driving has negative impacts on performances of fatigued drivers.

Keywords: Cell Phone Conversation, Fatigue, Driving Performance, NASA-TLX, Heart Rate

INTRODUCTION

Since the passing of the Road Traffic Act that banned the use of a hand-held cell phone by a driver while driving in 1999 in Japan and its revision in 2004, the number of accidents caused by using hand-held phones while driving has substantially decreased. When using a hand-held cell phone while driving, drivers must remove one hand from the steering wheel to hold and operate the phone, glance away from the road at least momentarily, maintain the phone conversation, and respond to the constantly changing road and traffic conditions. This means that its use includes a high potential risk for traffic accidents. Nevertheless, according to the National Police Agency of Japan, the number of traffic violations related to the using of hand-held cell phones in 2008 was about 1,200,000, and it has been increasing each year. This may reflect the past survey's result that most drivers who use a cell phone while driving use a hand-held phone, although 75% of them acknowledged that it is very often extremely dangerous (RAC Motoring Services, 2001).

In addition, numerous studies have reported that using a hands-free cell phone while driving was not any safer than hand-held phone or driving without a cell phone (Strayer and Drews, 2004; McEvoy et al., 2005; Ishigami and Klein, 2009), and cell phone conversation caused negative impacts on driving performance, which include missing or reacting slower to critical signals and changing stop lights (Consiglio et al., 2003; Hancock et al., 2003; Horberry et al., 2006). However, most of these studies have not considered the effects of cell phone conversation on driving performance when drivers become tired. Indeed, driver's fatigue could be a critical potential threat to road traffic safety. Therefore, this study examined the impacts of cell phone conversations on the driving performance of fatigued drivers.

METHOD

PARTICIPANTS

Ten male students participated in the study. They were healthy, with a mean age of 23.1 ± 1.1 years. All participants possessed a valid driver's license, and written informed consent was obtained from all participants after a full explanation of the experimental purpose and protocol.

CELL PHONE CONDITIONS

There are two different cell phone conditions (Hands-free and Hand-held cell phones) and each phone condition has two in-vehicle distraction conditions (Conversation and No-task), resulting in four experimental conditions.

CONVERSATION TASKS

Participants received a telephone call in two cell phone conditions and engaged in a conversation with math problems as the cognitive task. They were instructed to give the first answer after subtracting a single-digit number from a double-digit number and then continuously give the second answer after multiplying each number of the first answer (e.g., $52 - 6 = \underline{46} \rightarrow 4 \times 6 = \underline{24}$). A research assistant located at a remote location conversed with the driver using the cell phone, asking the questions and recording the drivers' responses.

MEASUREMENTS AND DATA ANALYSIS

Participants' driving performance was measured using SFZ-NU-FD-01 software by HONDA R&D and assessed in three categories: driving maintenance, attention lapses, and reaction time. Driving maintenance was assessed via the recording of standard deviation of lane position (m), which is the position of the vehicle center with respect to the road's central dividing line. In this study, attention lapses were defined as (a) driver failed to scan the intersection at a stop sign; (b) driver stopped in the absence of a stop sign; (c) driver failed to notice that pedestrians were crossing when turning into a side street from a main street. Attention lapses were assessed as recorded data using an eye tracking device (nac EMR-8B). Response times were taken for various driving events. These events included the mean time to step on the brake in response to pop-up events and the brake reaction time from initiation of slowing to the stop line at the red traffic lights. The latter means that the longer the time, the greater relaxed braking deceleration is made by drivers.

In addition, heart rate was measured to examine physiological response during cell phone conversation while driving. From the Electrocardiogram (WEB-5000, NIHON KOHDEN, Japan) data with a sampling frequency of 1 KHz, heart rates were derived. Perceived workload and fatigue feeling were assessed using the NASA Task Load Index (TLX) and a 100-mm visual analog scale, respectively. The NASA-TLX is a multi-dimensional rating scale (0–100) that provides an overall workload score based on six subscales: mental demand, physical demand, temporal demand, performance, effort, and frustration. In this study, the Weighted Workload (WWL) was used for the overall evaluation points.

PROCEDURE

Participants were introduced to the driving simulator (HONDA R&D) and shown the relevant functions of the vehicle. Previous studies reported that sixty percent of drivers become tired within the first hour of driving (Galinsky et al., 1993; Skipper and Wierwille, 1986), and monotonous road condition (Thiffault and Bergeron, 2003) and long-time driving (Hakkane and Summala, 2001) causes driving fatigue. Thus, the participants in this study wore the eye tracker and drove a monotonous practice route which causes boredom for over 50 min to familiarize themselves with the simulator vehicle and the eye tracker or to induce driving fatigue, followed by 5 min of each driving (with cell phone conversation or No-task) in city traffic in which the posted speed limit was 40km/h. After the first driving condition, the participants drove the next driving condition after 5 min of rest. The experiments of two different cell phone conditions were performed on different days to control similar fatigue level induced by driving. In other words, participants drove driving route 1 with Hand-held phone conversation and No-task conditions on one day and driving route 2 with Hands-free phone conversation and No-task conditions on the other day. The presentation order of these was counterbalanced across participants to control for order effects across conditions. Each driving route was approximately 1.6 km in length in city traffic, with the same number of hazards and traffic conditions. Participants were instructed to drive as they normally would and obey the road rules. In addition, they were required to take any action that they judged necessary to avoid collisions with other vehicles or pedestrians. Heart rate was continuously measured while driving. After each condition, participants were presented with the perceived workload using NASA-TLX and fatigue feeling for the previous drive.

STATISTICAL ANALYSIS

All data were expressed as the mean value ± standard deviation (SD) and were analyzed statistically using the Student's t-test between No-task and conversation distraction. $p < 0.05$ was considered significant.

RESULTS

DRIVING PERFORMANCE

For standard deviation of lane position, there were no significant differences between conversation distraction and No-task in both Hands-free (0.167 vs. 0.119) and Hand-held (0.128 vs. 0.101) cell phones. For attention lapses, there were no significant differences between conversation and No-task in both Hands-free (0.56 ± 0.88 vs. 0.22 ± 0.44) and Hand-held (1.38 ± 1.06 vs. 0.75 ± 0.89) cell phones.

Reaction time to step on brake in response to pop-up event was significantly slower in Hand-held conversation than in No-Task, but there was no significant difference between Hands-free conversation and No-Task (see Figure 1). Especially, a collision happened in Hand-held conversation condition.

In addition, No-Task showed relaxed braking deceleration compared with both Hands-free and Hand-held conversation in the brake reaction time from initiation of slowing to the stop line at red traffic lights (see Figure 2).

HEART RATE

For heart rate, conversation distraction was higher than No-task in both Hands-free and Hand-held cell phone conditions (see Figure 3).

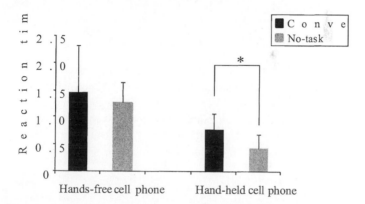

Figure 1. Comparison of reaction time to step on brake in response to pop-up event between conversation and No-Task in each cell phone condition (mean value ± SD, * $p<0.05$).

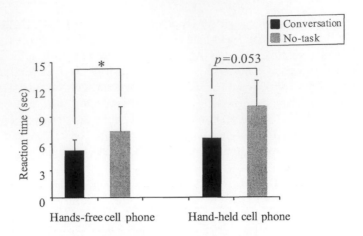

Figure 2. Comparison of the brake reaction time from initiation of slowing to the stop line at the red traffic lights between conversation and No-Task in each cell phone condition (mean value ± SD, * $p<0.05$).

NASA-TLX SCORES AND FATIGUE FEELING

For NASA-TLX WWL scores, the participants perceived remarkably higher workload in conversation distraction than in No-task in both Hands-free and Hand-held cell phone conditions (see Figure 4). In addition, they had significantly higher complaint rates of fatigue in conversation distraction than in No-task in both Hands-free (69.00 ± 17.13 vs. 45.50 ± 25.98, $p<0.05$) and Hand-held (76.50 ± 17.65 vs. 56.00 ± 20.39, $p<0.01$) cell phone conditions.

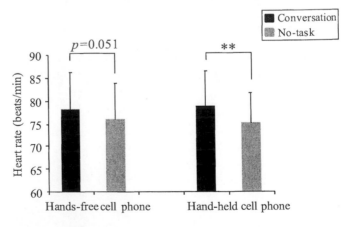

Figure 3. Comparison of heart rate between conversation and No-Task in each cell phone condition (mean value ± SD, ** $p<0.01$).

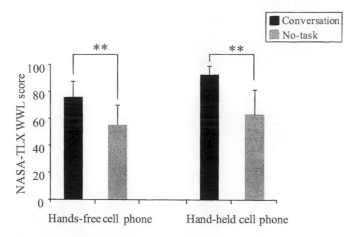

Figure 4. Comparison of NASA-TLX WWL scores between conversation and No-Task in each cell phone condition (mean value ± SD, ** $p<0.01$).

DISCUSSION

Phone conversation of fatigued drivers did not impair performance aspects of standard deviation of lane position and attention such as failing to scan peripheral environments. This indicates that cognitive conversation task might lead to increased arousal level of fatigued drivers, but it was very limited because drivers showed a quite slow braking reaction time in response to pop-up event in Hand-held cell phone conversation in which one hand must be removed from the steering wheel to hold the phone. Phone conversation was also significantly shorter than No-task in the brake reaction time from initiation of slowing to the stop line at traffic signals regardless of cell phone style. It suggests that fatigued drivers might be distracted with deceleration control and estimation of distance by phone conversation or drivers' distance judgment ability might decrease with fatigue. The results of subjective fatigue scores and perceived mental workload provide support for this. Drivers felt more fatigued and perceived high mental workload under phone conversation than in no-distraction task while driving regardless of cell phone style.

Our study is consistent with the previous studies, in which using a hands-free phone while driving was not as safe compared with using a hand-held phone or driving without a cell phone (Strayer and Drews, 2004; McEvoy et al., 2005; Ishigami and Klein, 2009). This phenomenon was also observed in heart rate. Generally, drivers consistently get tense while driving because they must respond to the changing road and traffic conditions, and cautiously check for obstacles and hidden hazard; in addition, in this study, all phone conversation regardless of cell phone style while driving led to drivers' sympathetic nervous system activation. Drivers' increased heart rate enhanced drivers' stress; as a result, drivers had a high

872

complaint rate of driving fatigue and perceived high mental workload. These results indicate that phone conversation itself regardless of cell phone style or fatigued drivers' excess stress may produce potential risk for dangerous traffic accidents. Therefore, the present study suggests that talking on the phone while driving has negative impacts on the performances of fatigued drivers.

REFERENCES

Consiglio, W., Driscoll, P., White, M., Berg, W.P. (2003) Effect of cellular telephone conversation and other potential interference on reaction time in a braking respone. Accident Analysis and Prevention Volume 35, 495-500.

Galinsky, T.L., Rosa, R.R., Warm, J.S., Dember W.L. (1993) Psychophysical determinants of stress in sustained attention. Human Factors Volume 35 No. 4, 603-614.

Hakkanen, H., Summala, H. (2001) Fatal traffic accidents among trailer truck drivers and accident causes as viewed by other truck drivers. Accident Analysis and Prevention Volume 33 No. 2, 187–196.

Hancock, P.A., Lesch, M., Simmons, L. (2003) The distraction effects of phone use during a crucial driving menuver. Accident Analysis and Prevention Volume 35, 501-514.

Horberry, T., Anderson, J., Regan, M.A., Triggs, T.J., Brown, J. (2006) Driver distraction: The effects of concurrent in-vehicle tasks, road environment complexity and age on driving performance. Accident Analysis and Prevention Volume 38, 185-191.

Ishigami, Y., Klein, R.M. (2009) Is a hands-free phone safer than a handheld phone? Journal of Safety Research Volume 40, 157-164.

McEvoy, S.P., Stevenson, M.R., McCartt, A.T., Woodward, M., Haworth, C., Palamara, P., Cercarelli, R. (2005) Role of mobile phones in motor vehicle crashes resulting in hospital attendance: a case-crossover study. British Medical Journal Volume 331 (7514), 428.

National Police Agency. (2008) Annual Statistics 2008, Japan .

RAC (Royal Automobile Club) Report on Motoring 2001. Feltham, RAC Motoring Services, UK.

Skipper, J.H., Wierwille, W.W. (1986) Drowsy driver detection using discriminate analysis. Human Factors Volume 28, 527-540.

Strayer, D.L., Drews, F.A. (2004) Profiles in driver distraction: effects of cell phone conversations on younger and older drivers. Human Factors Volume 46 No. 4, 640-649.

Thiffault, P., Bergeron, J. (2003) Fatigue and individual differences in monotonous simulated driving. Personality and Individual Differences Volume 34 No. 1, 159–176.

CHAPTER 91

Ergonomics Study on the Visual Environment at Urban Uncontrolled Intersections Based on Visibility Simulation Approach Applying 3-Dimensional Computer Graphics Software

Midori Mori, Sadao Horino, Noboru Kubo
Faculty of Engineering,
Kanagawa University
3-27-1 Rokkakubashi, Kanagawa-ku, Yokohama 221-8686 JAPAN

ABSTRACT

In Japan, a quarter million crossing collisions occur at intersections, accounting for one-fourth of the total accidents in 2008, which is the second most frequent pattern among traffic accidents. Three-fourths of crossing collisions occurred at urban intersections. Our previous studies revealed that the risk of collision depended on the quality of the visual environment at intersections, including direct and indirect visibility provided by traffic convex mirrors. Ergonomics guidelines for installing mirrors were proposed by authors to assure their performance in ensuring crossing safety. This study aimed to assess the quality of the visual environment at

874

intersections and discuss effective countermeasures for preventing collisions, especially safety enhancements related to direct and indirect visibility. Visibility simulation studies, applying 3-dimensional computer graphics software, were conducted at actual and hypothesized uncontrolled intersections. The visibility was assessed and optimal solutions were derived, which provided mirror images that ensure the maximum visible distance from the crossing road as well as minimal blind areas. We observed that the fine adjustment of the depression/horizontal angle of the mirror plate, horizontal position of the mirror plate, and width of major and minor roads were critical factors affecting indirect visibility. The visibility simulation approach was efficient and effective in demonstrating that direct and indirect visibility could be improved by installing mirrors or adjusting the angle and position of the mirror plate and pole, based on the ergonomics guidelines. Minimum ergonomics requirements for intersection visibility were proposed to prevent crossing collisions at urban intersections.

Keywords: Visibility Simulation, Traffic Convex Mirror, Crossing Collision.

INTRODUCTION

Nearly a quarter million crossing collisions occur at intersections (208,290 cases, 27.2%) representing the second most frequent pattern among traffic accidents, following rear-end collisions (239,126 cases, 31.2%) and accounting for one-fourth of the total accidents (766,147 cases) in Japan in 2008. The frequency of crossing collisions has doubled during the last 20 years, with 75% occurring at urban intersections and 70% at uncontrolled intersections (National Police Agency, 2007).

A large number of investigations and measures designed for the prevention of traffic accidents have been conducted from the traffic/road management authority viewpoint, focusing mainly on human factors. As a result, safety enforcement and education, rather than engineering redesign or control of road traffic environment, have been emphasized as preventive measures (ITARDA, 2005; ITARDA, 2007). Ergonomic studies on preventing crossing collisions, however, should focus on what really happened and why accidents occur frequently at uncontrolled intersections (Horino et al., 2003; Mori et al., 2008a, b).

The authors revealed that the risk of a crossing collision depended on the quality of the visual environment at the intersection, including both direct and indirect visibility, provided by traffic convex mirrors at intersections (Mori et al., 2008a, b). Our ultra wide angle photograph analysis revealed that most of the visible ranges in the right/left directions at 19 medium or smaller sized intersections in an urban community area were insufficient to confirm safety and the quality of direct visibility was associated with the collision frequency. In addition, it became clear that more than half of the 23 mirrors installed at the intersections had marked shortcomings as a device for ensuring crossing safety; thus, appropriate installation conditions for ensuring crossing safety were discussed. Few studies are available concerning the availability of traffic convex mirrors (Moukhwas, 1987) and the present official guidelines regarding mirror installation (Japan Road Association,1980, 2006) were considered to be insufficient in terms of their visibility criteria. The authors, therefore, proposed the following ergonomic guidelines for installing mirrors (three ergonomic requirements): (1) Road lane shall be located in a central part of the mirror image; (2) Blind area shall be deleted in the mirror image; and (3) Road surface markings shall be explicit in the mirror image (Figure 1) (Mori et al., 2008a, b).

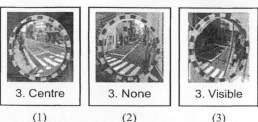

(1)   (2)   (3)
(1) Road lane shall be located in a central part of mirror image
(2) Blind area shall be deleted in mirror image
(3) Road surface markings shall be explicit in mirror image

Figure 1 Ergonomics guidelines for installing traffic convex mirrors and example mirror images satisfying the three visibility requirements.

This study discusses effective countermeasures for preventing crossing collisions, especially safety enhancements from the perspective of direct and indirect visibility. However, there are no available desk methods in civil engineering for examining conditions for mirror installation or facilitating improvements to supplement direct visibility. Therefore, this study develops a highly accurate visibility simulation approach, applying three-dimensional computer graphics (3DCG) software, and discusses the appropriate installation conditions that ensure crossing safety, in terms of the direct/indirect visibility requirements.

METHOD

DEVELOPMENT OF VISIBILITY SIMULATION APPROACH APPLYING THREE-DIMENSIONAL COMPUTER GRAPHICS SOFTWARE

A visibility simulation approach, applying 3DCG software (Shade), was developed to efficiently and effectively demonstrate that direct and indirect visibility can be assessed and improved by installing traffic convex mirrors or adjusting the angle and position of the mirror plate and pole.

FIELD CASE STUDY: VISIBILITY SIMULATION STUDY OF THE VISUAL ENVIRONMENT AT AN ACTUAL INTERSECTION

This study was conducted through two complementary approaches: an actual assessment and problem solving of the visual environment at an uncontrolled intersection; and theoretical analysis to clarify the optimal or appropriate conditions (ergonomic requirements).

A visibility simulation study and a field experiment were performed to measure and assess the direct/indirect visibility from the driver's viewpoint of an uncontrolled intersection in an urban community area; the residents recognized the danger of this intersection and requested that the road administrator install mirrors. The Road Traffic Law in Japan requires drivers to drive on the left side of the road, and drivers with lower priority must stop momentarily at a legal stop line in crossing uncontrolled intersections. The road configurations, various items (public/private facilities, etc.), and visual environment from the driver's viewpoint in a passenger car (1.2 m high) at a legal stop line and at the boundary line between an intersection and the crossing road were simulated in all four directions (Figure 2).

The direct/indirect visibilities were assessed whether a driver with lower priority could view the entire safe distance [D1 > D0] (more than 50-60 m from the boundary line) or not [D1 < D0] (Mori et al., 2008a, b).

D0: Minimum right/left directional ranges of view for safety confirmation [the value calculated as the travel distance of an approaching car with higher priority while a car with lower priority crosses the intersection].

D1: Visible distance at the crossing road [the measured value].

Optimal solutions that met the condition of the mirror image reflecting the maximum visible distance from the crossing road as well as minimum blind areas were analyzed by controlling the following dimensions of the mirror plate: horizontal position, ground height, depression angle, and horizontal angle.

As a field experiment, a mirror (800 mm in diameter, with a 3000 mm radius of curvature) was installed and a simulation plan for the optimal solutions was reproduced to verify consistency with the actual mirror image.

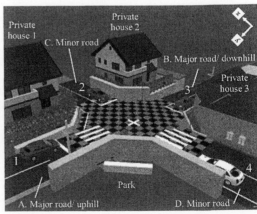

Figure 2 An uncontrolled intersection located in a residential area, used for the visibility simulation. Left: Photo taken from the perspective of the driver in a passenger car No.3 (Yokohama, Kanagawa, Japan). Right: Diagram of the actual intersection for analyzing the optimal mirror installation.

THEORETICAL STUDY: SYSTEMATIC VISIBILITY SIMULATION ASSESSMENT OF THE VISUAL ENVIRONMENT AT TYPICAL INTERSECTIONS IN COMMUNITY ROADS

A systematic visibility simulation study was performed to measure and assess the direct/indirect visibility from the driver's viewpoint in a passenger car, in the same manner as the field case study described above, at hypothetical uncontrolled intersections in an urban community road. The optimal solutions were analyzed to clarify appropriate conditions (ergonomic visibility requirements) by controlling the following parameters (Figure 3):

1. Association between mirror image and road configurations (width of road)
    (1) Independent variables:
        -width of minor road (three conditions: 4, 5, and 6 m)
        -width of major road (five conditions: 6, 7, 8, 9, and 10 m)
    (2) Dependent variables: depression angle, horizontal angle
    (3) Fixed conditions: horizontal position, ground height of mirror plate
2. Association between mirror image and horizontal positions of mirror plate
    (1) Independent variables:
        -horizontal position of mirror plate (12 conditions: 12 locations surrounding the diagonal spot (M) at the left corner with reference to the X- and Y-axes)
    (2) Dependent variables: depression angle, horizontal angle
    (3) Fixed condition: width of minor road (4 m), width of major road (6 m), ground height of the mirror plate

These conditions were determined based on the fact that the view ranges to the right was crucial for drivers on the left-hand side, and our research findings that setting mirrors at the left corner for the right view was recommended for installing mirrors and ensuring indirect visibility.

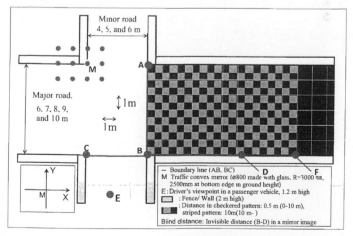

Figure 3 Diagram of the hypothetical intersection for analyzing optimal traffic convex mirror installation based on the visibility simulation approach.

RESULTS AND DISCUSSION

DEVELOPMENT OF VISIBILITY SIMULATION APPROACH APPLYING THREE-DIMENSIONAL COMPUTER GRAPHICS SOFTWARE

The visibility simulation approach using 3DCG software was developed and the demonstration test showed high consistency between the simulation study and an experiment in a test course in terms of mirror images (Figure 4). It was considered that visibility could be assessed and an optimal solution could be achieved.

Figure 4 An example of visibility simulation applying 3DCG software: Mirror image experimentally installed in a test course (left: actual image; right: image from simulation).

FIELD CASE STUDY: VISIBILITY SIMULATION STUDY OF THE VISUAL ENVIRONMENT AT AN ACTUAL INTERSECTION

Indirect visibility simulation and ergonomic field experiment of the visual environment from the driver viewpoint at an uncontrolled intersection

An optimal solution was simulated for indirect visibility provided by the mirrors in four directions from the perspective of the driver. It was confirmed that the driver was guaranteed indirect visibility (visible distance from the crossing road: e.g., direction C: 60 m and D: 55 m in vehicle No. 3) adequate for safety (Figure 5).

Figure 5 Consistency of mirror images between the field experiment (left) and the simulated optimal solution for the left/right views provided by mirror from passenger car No. 3 (right) (mirror at corner AC and AD)

The demonstration test was performed in a field experiment to verify the optimal solutions for installing mirrors based on ergonomic requirements. A high consistency was found between the mirror images in the simulation and the field experiment, in terms of conditions such as the size and position of subject images, blind areas, and the visible distance at the crossing road (Figure 5).

Variation of the mirror image affected by fine adjustment of mirror angle

The visible distance in the mirror image installed at corner AD was varied by approximately 10 m by making fine adjustments to the horizontal angle of the mirror plate in 1 degree increments (Figure 6), and similar effects were confirmed for the depression angle. These findings suggested that fine adjustment of the depression/horizontal angle of the mirror plate were critical factors for ensuring the indirect visibility (the safe visible distance).

(1)   (2)   (3)
                                        Optimal solution

Figure 6 Remarkable change in the visible distance in mirror images affected by fine adjustment of the mirror angle (corner AD) (horizontal angle/visible distance: left −41°/40-50 m, center −42°/50-60 m, right −43°/60-70 m (optimal solution), depression angle: 7°, distance per black and white markings: 10 m)

THEORETICAL STUDY: SYSTEMATIC VISIBILITY SIMULATION ASSESSMENT OF THE VISUAL ENVIRONMENT AT TYPICAL INTERSECTIONS IN COMMUNITY ROADS

Association between the mirror image and road configuration (width of road)

Indirect visibility simulations were performed systematically for 15 intersection patterns: combinations of widths of minor roads (4, 5, and 6 m) and major roads (6, 7, 8, 9, and 10 m), which were hypothesized as typical intersections in community roads. The optimal depression angle/horizontal angle of mirrors (ø800 made with glass, R=3000 mm) installed at the left corner M were adjusted to meet the following preconditions: a minimum blind distance existed from the crossing road, and a visible distance greater than 50-60 m in a mirror image, the blind distance was defined as the invisible distance between B and D in a mirror image (Figure 3).

We observed that there was no condition with a mirror image with no blind areas, that ensured a visible distance greater than 50 m; the blind distance was minimum (BD: 2.5m) for the combination of a minor road (6 m) and major road (6 m). The blind distance was inversely proportional to the width of the minor road; however, it was proportional to the width of major road (Figure 7 and Figure 8).

The theoretical visibility simulation study revealed that the indirect visibility provided by mirrors was insufficient for drivers entering the intersections to confirm the presence of a crossing vehicle and avoid a crossing collision.

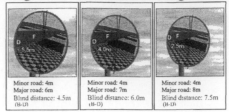

Figure 7 Association between the mirror image and width of road—Variations of a simulated mirror image.

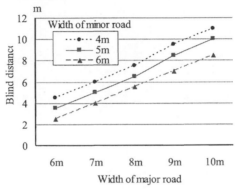

Figure 8 Association between the mirror image and width of road—the distribution of blind distances in simulated mirror images.

Association between the mirror image and the horizontal position of the mirror plate

To assess how the blind distance in a mirror image is affected by the position of the left corner of intersection (12 locations surrounding spot M, four locations on the x-axis: +1, 0, −1, and −2 m; and three locations on the y-axis: +1, 0, and −1 m), the optimal depression angle/horizontal angle of mirrors were adjusted to meet the preconditions, given the smallest intersections (a 4 m wide minor road and 6 m wide major road). We observed that the blind distance in the mirror image decreased by displacing the mirror in the negative direction on both the x- and y-axis (Figure 9). The best position was the leftmost spot in the nearest row (x: −2 m, y; −1 m), from the driver's viewpoint in a passenger vehicle stopping at the boundary line.

Tradeoffs between optimal solutions from the aspect of the blind area in the mirror image and direct visibility

Sometimes drivers had to confirm the mirror at the position distant from the boundary line, about 10 m at maximum, because of the pedestrian crossings and the legal stop line. Under these circumstances, drivers could not directly confirm the mirror installed in the best position, because direct visibility was obstructed by items such as walls, fences, and utility poles. The tradeoff problem occurred in intersections where the appropriate position was restricted despite of the optimal solution from the aspect of blind area in a mirror image. It was necessary to solve these problems systematically so as to ensure compatibility with direct visibility.

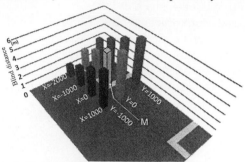

Figure 9 Association between the mirror image and the horizontal position of the mirror plate—the distribution of blind distances in simulated mirror images.

CONCLUSIONS

A visibility simulation approach, applying 3-dimensional computer graphics software was developed by the authors. Visibility was assessed and an optimal solution was obtained that ensures the maximum visible distance at a crossing road with an image having minimum blind areas by installing mirrors or adjusting the angle and position of the mirror plate and pole based on ergonomic guidelines (three ergonomic requirements). The visibility simulation demonstrated that a slight change in the angle and horizontal position of the mirror plate significantly influenced the visible distance. These findings suggested that fine adjustment of the depression/horizontal angle and horizontal position of mirror plate were critical factors for adequate indirect visibility for safety.

In conclusion, the visibility simulation approach using 3DCG software was efficient and effective in examining the conditions for good direct and indirect visibility. Crossing collisions are highly associated with the quality of the visual environment for drivers entering intersections, and improvements to indirect visibility by installing mirrors appropriately and removing obstructing items should be given a high priority. Minimum ergonomic requirements for intersection visibility were proposed to prevent crossing collisions at urban intersections.

ACKNOWLEDGEMENTS

This research was partially supported by a Grant-in-Aid for Scientific Research (C) from the Japan Society for the Promotion of Science (#18510149: 2006-2008,

#21510184: 2009) and a Grant for Encouragement of Joint Research from Institute of Technological Research, Kanagawa University (2007-2009). The authors received generous cooperation with this research project through their graduate theses from Mr. H. Sugiyama, Mr. M. Ono, Mr. Y. Fukunaga, Mr. S. Watanabe, Mr. H. Nishimura, and Ms. E. Noguchi. In addition, the authors received great advice and cooperation from the Kanagawa Prefectural Police, the Road Bureau and Road Administration Office of Yokohama City, the Local Road/Environment section of the Ministry of Land, Infrastructure and Transport Road Bureau, the Japan Construction Association of Traffic Signs and Road markings, the Traffic Convex Mirror Association, the Japan Automobile Research Institute, the National Institute for Land and Infrastructure Management, and the Residents' Association of Hino Residential Area (Konan-ward, Yokohama City). The authors would like to express their heartfelt thanks for their great contributions.

REFERENCES

Horino S., Ueyama M., Mori M., & Kitajima S. (2003). "Ergonomic assessment and improvement for accident dangerous intersections without traffic signal light." *Proceedings of the 15th Triennial Congress of the International Ergonomics Association*, 5, 387-390.

Institute for Traffic Accident Research and Data Analysis (2005), *Analysis of Human Factors in Crossing Collision*. ITARDA Information, No. 56.

Institute for Traffic Accident Research and Data Analysis (2007), *Crossing Collisions in View of Road Environment*. ITARDA Information, No. 69.

Japan Road Association (1980), *Guideline for Traffic Convex Mirror Installation*. Japan Road Association (Maruzen) (in Japanese).

Japan Society of Traffic Engineers (2007), *Design and Plan of Level Crossing-Basic Edition: the 3rd Edition (Revised Version)*, Japan Society of Traffic Engineers (Maruzen) , Tokyo (in Japanese).

Mori, M., Horino, S., Kubo, N., and Kitajima, S. (2008a), "Ergonomics proposal for visibility requirements at urban intersections in Japan for preventing frequent crossing collisions." *Proceedings of the Applied Human Factors and Ergonomics 2nd International Conference*, CD-ROM.

Mori, M., Horino, S., Kitajima, S., Ueyama, M., Ebara, T., Itani, T. (2008b), "Ergonomics Solution for Crossing Collisions based on a Field Assessment of the Visual Environment at Urban Intersections in Japan." *Applied Ergonomics*, 39(6), 697-709.

Moukhwas, D. (1987). "Road junction convex mirrors." *Applied Ergonomics*, 18, 133–136.

National Police Agency. (2007). *Statistics 2006 Road Accidents Japan*, International Association of Traffic and Safety Sciences, Tokyo.

Chapter 92

The Major Quality Factors on Two-view Autostereoscopic Displays

Pei-Chia Wang, Hsin-Ying Huang, Chih-Fei Chuang, Sheue-Ling Hwang,
Kuen Lee**, and Jinn-Cherng Yang**

National Tsing Hua University
Hsinchu, Taiwan 300

** Industrial Technology Research Institute
Hsinchu, Taiwan 310

ABSTRACT

This study was carried out via mirror-type autostereoscopic display, and aimed at investigating the relation between 3D cues and perspective image quality. The independent variables included shade and brightness contrast ratio. The dependent variables were subjective evaluation toward image quality. The result of the experiment showed that shade significantly affected viewers' perspective image quality, but their interaction was not statistically significant. The brightness contrast of the object and the background and the interaction of brightness and shade influence the sense of depth. In addition, the experience of viewing autostereoscopic displays does not affect the judgment of subjective stereoscopic for viewers. In order to promote the 3D display in the market, industry will need to drive them to support the content creators and provide the features and ease of use that end users require. It is expected to acquire the optimal combination of 3D cues and build up a guideline for well-performed autostereoscopic display.

Keywords: Autostereoscopic displays, Quality, 3D Cue, Shade, Brightness

INTRODUCTION

Since the introduction of television, one of the most interesting developments is the creation of stereoscopic display. Basing on the principal of human vision system (HVS), each eye seeing slightly different images to create a sensation of depth by a process known as stereosis (Mansson, 1998), stereoscopic displays use left/right separation to present convincing 3D effect. Stereoscopic TV provides people a whole new experience, where the dominant factor is no longer image quality but the fundamental change in image features, enabling them to watch their content in three dimensions.

Research on stereoscopic and auto-stereoscopic display has made it possible to provide viewers images and videos with a high-quality perception of depth and with a greater sense of realism. A distinction is made between stereoscopic displays and autostereoscopic displays: Stereoscopic displays entail an optical device to direct the left and right eye images to the appropriate eye, while autostereoscopic displays have the built-in separation technique in the display screen (Meesters et al., 2004; Lyndon el al., 2006; Seuntiëns et al., 2005, 2006, 2007; Wang, 2009).

Viewers don't need to wear the special glasses to watch autostereoscopic displays. But there are some issues in viewing autostereoscopic displays. 3D cues are the major factors which affect the visual quality on viewers, and they may increase viewer's mental and physical loading. This research aims at investigate the impact of 3D cues on the 3D viewing experience by a means of human factors assessment.

METHOD

APPARATUS AND STIMULI

A stereo photo includes several 3D cues, such as shade, occlusions, relative size, depth and so forth (Lyndon Hill, 2006). The pictures with the most 3D cues were chosen in the experiment, which were Hanging Decoration, Table & Chairs, Flowers, and Cup.

The experiment was carried out in the dark room. The temperature was about 24 to 27 ℃. All the experimental pictures were appeared in a two-view mirror-type display. It contained two LCDs in the left and right sides, and two mirrors in the center to create a stereo picture. The apparatus included a head fixer, a personal computer, a keyboard, a mouse, a laptop and a stereo acuity tester. The experimental environment was shown in Figure 1.

Figure 1. Experimental environment

EXPERIMENTAL PARAMETERS

The independent variables were shade and brightness. In the pictures of Flowers, Hanging Decoration, and Cup, the backgrounds were brighter than the object in the foreground. There were three different levels, 1%, 10%, and 30%, for the brightness ratio of the object and the background. Table & Chairs was the picture, where the object was brighter than the background, and the brightness ratio of background and object was 30%.

The dependent variables were subjective image quality questionnaire by seven points of Likert's scale, which were perspective evaluation towarding the degree of the sense of light and shade, the sense of depth, stereoscopic feeling, viewer crosstalk, and harmonization.

PARTICIPANTS

25 participants were recruited in the study. The anticipants included 10 engineers who worked on the relevant issues of autostereoscopic display, and 15 graduate students who have never experienced autostereoscopic display.

EXPERIMENTAL PROCEDURE

The participant who passed the stereo acuity test could participate in the formal experiment. At first, the purpose and the procedure of the study was described to the participant. There was a frame fixed the head of participants. Human eyes were fixed in the center of two mirrors in order to watch the 3D pictures properly. Before the formal experiment, four experimental pictures were used to explain the questions in the subjective image quality questionnaire. The experimental pictures had different combination of two main factors and each one was randomly displayed in the mirror-type display. After the picture was shown for 5 seconds, the participant had to answer orally five questions associated with subjective image quality to the experimenter. The participants used the mouse to display the next picture. An all-black photo was displayed during the 20 seconds interval when participants were taking a rest. The number of experimental picture was totally 40.

RESULTS

3D CUES

Based on the MANOVA (Multiple Analysis of Variance) as shown in Table 1, shade statistically affected all subjective image quality (F=14.079, p-value<0.001), but brightness (F=1.336, p-value>0.05) and the interaction between shade and brightness (F=1.245, p-value>0.05) did not affect the viewer's stereoscopic judgment.

Table 1 MANOVA for 3D Cues

Multivariate Tests[c]

Effect		Value	F	Hypothesis df	Error df	Sig.
Intercept	Pillai's Trace	.974	5483.148[a]	5.000	740.000	.000
	Wilks' Lambda	.026	5483.148[a]	5.000	740.000	.000
	Hotelling's Trace	37.048	5483.148[a]	5.000	740.000	.000
	Roy's Largest Root	37.048	5483.148[a]	5.000	740.000	.000
Shade	Pillai's Trace	.087	14.079[a]	5.000	740.000	.000
	Wilks' Lambda	.913	14.079[a]	5.000	740.000	.000
	Hotelling's Trace	.095	14.079[a]	5.000	740.000	.000
	Roy's Largest Root	.095	14.079[a]	5.000	740.000	.000
Brightness	Pillai's Trace	.018	1.336	10.000	1482.000	.206
	Wilks' Lambda	.982	1.338[a]	10.000	1480.000	.205
	Hotelling's Trace	.018	1.339	10.000	1478.000	.204
	Roy's Largest Root	.016	2.378[b]	5.000	741.000	.037
Shade * Brightness	Pillai's Trace	.017	1.245	10.000	1482.000	.257
	Wilks' Lambda	.983	1.245[a]	10.000	1480.000	.257
	Hotelling's Trace	.017	1.245	10.000	1478.000	.257
	Roy's Largest Root	.013	1.961[b]	5.000	741.000	.082

a. Exact statistic

b. The statistic is an upper bound on F that yields a lower bound on the significance level.

c. Design: Intercept+Shade+Brightness+Shade * Brightness

According to the ANOVA (Analysis of Variance) as shown in Table 2, the picture with and without shade significantly influenced the sense of light and shade, the sense of depth, stereoscopic feeling, and harmonization (p-value<0.001), and brightness only affected the sense of light and shade (p-value<0.001), and the interaction between shade and brightness (p-value<0.001) statistically affected the viewer's perception of depth.

Table 2 ANOVA for 3D cues

Tests of Between-Subjects Effects

Source	Dependent Variable	Type III Sum of Squares	df	Mean Square	F	Sig.
Corrected Model	Light_and_shade	82.983[a]	5	16.597	10.382	.000
	Depth	63.319[b]	5	12.664	9.458	.000
	Stereoscopic	71.735[c]	5	14.347	10.745	.000
	Harmonization	67.872[d]	5	13.574	8.740	.000
	Viewer_crosstalk	14.187[e]	5	2.837	.914	.471
Intercept	Light_and_shade	18046.721	1	18046.721	11289.671	.000
	Depth	17501.505	1	17501.505	13071.104	.000
	Stereoscopic	18381.825	1	18381.825	13766.386	.000
	Harmonization	15980.592	1	15980.592	10289.217	.000
	Viewer_crosstalk	5891.205	1	5891.205	1896.928	.000
Shade	Light_and_shade	60.492	1	60.492	37.843	.000
	Depth	52.801	1	52.801	39.435	.000
	Stereoscopic	65.121	1	65.121	48.770	.000
	Harmonization	64.533	1	64.533	41.550	.000
	Viewer_crosstalk	.432	1	.432	.139	.709
Brightness	Light_and_shade	13.683	2	6.841	4.280	.014
	Depth	1.875	2	.937	.700	.497
	Stereoscopic	2.315	2	1.157	.867	.421
	Harmonization	2.144	2	1.072	.690	.502
	Viewer_crosstalk	10.867	2	5.433	1.749	.175
Shade * Brightness	Light_and_shade	8.808	2	4.404	2.755	.064
	Depth	8.643	2	4.321	3.227	.040
	Stereoscopic	4.299	2	2.149	1.610	.201
	Harmonization	1.195	2	.597	.385	.681
	Viewer_crosstalk	2.888	2	1.444	.465	.628
Error	Light_and_shade	1189.296	744	1.599		
	Depth	996.176	744	1.339		
	Stereoscopic	993.440	744	1.335		
	Harmonization	1155.536	744	1.553		
	Viewer_crosstalk	2310.608	744	3.106		
Total	Light_and_shade	19319.000	750			
	Depth	18561.000	750			
	Stereoscopic	19447.000	750			
	Harmonization	17204.000	750			
	Viewer_crosstalk	8216.000	750			
Corrected Total	Light_and_shade	1272.279	749			
	Depth	1059.495	749			
	Stereoscopic	1065.175	749			
	Harmonization	1223.408	749			
	Viewer_crosstalk	2324.795	749			

a. R Squared = .065 (Adjusted R Squared = .059)

b. R Squared = .060 (Adjusted R Squared = .053)

c. R Squared = .067 (Adjusted R Squared = .061)

d. R Squared = .055 (Adjusted R Squared = .049)

e. R Squared = .006 (Adjusted R Squared = -.001)

BRIGHTER OBJECT AND BRIGHTER BACKGROUND

To Compare 30% brightness contrast ratio of object and background (brighter object) and 30% brightness contrast ratio background of and object (brighter background) on the five subjective stereoscopic questionnaires. The main factor of brightness contract ratio significantly affected viewer's stereoscopic feeling via statistically analysis in Table 3 (F=4.219, p-value<0.05). Furthermore, as shown in Table 4, the result of ANOVA indicated only the sense of depth affected the brightness contract ratio, the sense of light and shade, stereoscopic feeling, harmonization, but viewer crosstalk didn't influence the brightness contract ratio.

Table 3 MANOVA for brighter background and brighter object

Multivariate Tests[b]

Effect		Value	F	Hypothesis df	Error df	Sig.
Intercept	Pillai's Trace	.974	3630.277[a]	5.000	494.000	.000
	Wilks' Lambda	.026	3630.277[a]	5.000	494.000	.000
	Hotelling's Trace	36.744	3630.277[a]	5.000	494.000	.000
	Roy's Largest Root	36.744	3630.277[a]	5.000	494.000	.000
Brightness	Pillai's Trace	.041	4.219[a]	5.000	494.000	.001
	Wilks' Lambda	.959	4.219[a]	5.000	494.000	.001
	Hotelling's Trace	.043	4.219[a]	5.000	494.000	.001
	Roy's Largest Root	.043	4.219[a]	5.000	494.000	.001

a. Exact statistic

b. Design: Intercept+Brightness

Table 4 ANOVA for brighter background and brighter object

Tests of Between-Subjects Effects

Source	Dependent Variable	Type III Sum of Squares	df	Mean Square	F	Sig.
Corrected Model	Light_and_shade	4.418[a]	1	4.418	2.622	.106
	Depth	7.688[b]	1	7.688	5.548	.019
	Stereoscopic	1.568[c]	1	1.568	1.180	.278
	Harmonization	.072[d]	1	.072	.051	.821
	Viewer_crosstalk	9.248[e]	1	9.248	2.898	.089
Intercept	Light_and_shade	11975.618	1	11975.618	7108.598	.000
	Depth	10746.248	1	10746.248	7755.268	.000
	Stereoscopic	11654.792	1	11654.792	8772.273	.000
	Harmonization	10269.512	1	10269.512	7343.624	.000
	Viewer_crosstalk	4657.352	1	4657.352	1459.268	.000
Brightness	Light_and_shade	4.418	1	4.418	2.622	.106
	Depth	7.688	1	7.688	5.548	.019
	Stereoscopic	1.568	1	1.568	1.180	.278
	Harmonization	.072	1	.072	.051	.821
	Viewer_crosstalk	9.248	1	9.248	2.898	.089
Error	Light_and_shade	838.964	498	1.685		
	Depth	690.064	498	1.386		
	Stereoscopic	661.640	498	1.329		
	Harmonization	696.416	498	1.398		
	Viewer_crosstalk	1589.400	498	3.192		
Total	Light_and_shade	12819.000	500			
	Depth	11444.000	500			
	Stereoscopic	12318.000	500			
	Harmonization	10966.000	500			
	Viewer_crosstalk	6256.000	500			
Corrected Total	Light_and_shade	843.382	499			
	Depth	697.752	499			
	Stereoscopic	663.208	499			
	Harmonization	696.488	499			
	Viewer_crosstalk	1598.648	499			

a. R Squared = .005 (Adjusted R Squared = .003)

b. R Squared = .011 (Adjusted R Squared = .009)

c. R Squared = .002 (Adjusted R Squared = .000)

d. R Squared = .000 (Adjusted R Squared = -.002)

e. R Squared = .006 (Adjusted R Squared = .004)

3.3 NOVICES AND EXPERTS

To compare the difference of perspective evaluation between the novices and the experts on the five subjective image quality questionnaires, the outcome of MANOVA (Table 5) and ANOVA (Table 6) indicated the experience on the autostereoscopic displays didn't affect the viewer's subjective evaluation toward all questions in the subjective image quality questionnaire (p-value>0.05).

Table 5 MANOVA for experience

Multivariate Tests[b]

Effect		Value	F	Hypothesis df	Error df	Sig.
Intercept	Pillai's Trace	.968	881.121[a]	5.000	144.000	.000
	Wilks' Lambda	.032	881.121[a]	5.000	144.000	.000
	Hotelling's Trace	30.594	881.121[a]	5.000	144.000	.000
	Roy's Largest Root	30.594	881.121[a]	5.000	144.000	.000
Experience	Pillai's Trace	.015	.425[a]	5.000	144.000	.831
	Wilks' Lambda	.985	.425[a]	5.000	144.000	.831
	Hotelling's Trace	.015	.425[a]	5.000	144.000	.831
	Roy's Largest Root	.015	.425[a]	5.000	144.000	.831

a. Exact statistic

b. Design: Intercept+Experience

Table 6 ANOVA for experience

Tests of Between-Subjects Effects

Source	Dependent Variable	Type III Sum of Squares	df	Mean Square	F	Sig.
Corrected Model	Light_and_shade	.010[a]	1	.010	.005	.942
	Depth	.444[b]	1	.444	.253	.616
	Stereoscopic	1.361[c]	1	1.361	.861	.355
	Harmonization	2.668[d]	1	2.668	1.500	.223
	Viewer_crosstalk	.588[e]	1	.588	.563	.454
Intercept	Light_and_shade	3660.250	1	3660.250	1964.522	.000
	Depth	3664.284	1	3664.284	2082.708	.000
	Stereoscopic	4023.788	1	4023.788	2545.262	.000
	Harmonization	3620.028	1	3620.028	2034.762	.000
	Viewer_crosstalk	375.068	1	375.068	359.275	.000
Experience	Light_and_shade	.010	1	.010	.005	.942
	Depth	.444	1	.444	.253	.616
	Stereoscopic	1.361	1	1.361	.861	.355
	Harmonization	2.668	1	2.668	1.500	.223
	Viewer_crosstalk	.588	1	.588	.563	.454
Error	Light_and_shade	275.750	148	1.863		
	Depth	260.389	148	1.759		
	Stereoscopic	233.972	148	1.581		
	Harmonization	263.306	148	1.779		
	Viewer_crosstalk	154.506	148	1.044		
Total	Light_and_shade	4086.000	150			
	Depth	4061.000	150			
	Stereoscopic	4396.000	150			
	Harmonization	3996.000	150			
	Viewer_crosstalk	552.000	150			
Corrected Total	Light_and_shade	275.760	149			
	Depth	260.833	149			
	Stereoscopic	235.333	149			
	Harmonization	265.973	149			
	Viewer_crosstalk	155.093	149			

a. R Squared = .000 (Adjusted R Squared = -.007)

b. R Squared = .002 (Adjusted R Squared = -.005)

c. R Squared = .006 (Adjusted R Squared = -.001)

d. R Squared = .010 (Adjusted R Squared = .003)

e. R Squared = .004 (Adjusted R Squared = -.003)

CONCLUSIONS

3D cues are the major indicator for viewer's subjective stereoscopic evaluation. In terms of the result of this study, shade significantly influenced on the four stereo indicators (stereoscopic feeling, the sense of depth, harmonization, and the sense of light and shade) besides viewer's crosstalk. Brightness affects the perception of the feeling of light and shade. The brightness contrast of the object and the background and the interaction between brightness and shade influence the sense of depth. Therefore, shade is the vital factor to the image quality of the 3D picture.

There is no difference between the subjective stereoscopic evaluation for experts and novices. As a result, the experience of viewing autostereoscopic displays does not affect the evaluation of subjective stereoscopic for viewers.

There have been many challenges to create suitable images for 3D displays. In order to promote the 3D display in the mass market, industry will need to tackle all of them to support the content creators and provide the features and ease of use that end users require. It is expected to obtain the optimal combination of 3D cues and build up a guideline for well-performed autostereoscopic display.

ACKNOWLEDGE

This study is supported by ITRI, Taiwan.

REFERENCES

Lyndon Hill (2006), "3D Liquid Crystal Displays and Their Applications".

Mansson, J. (1998), "Stereovision: A model of human stereopsis," Lund Univ. Cognitive Science, Tech. Rep.

Meesters, L. M. J., IJsselsteijn, W. A., and Seuntiens, P. J. H. (2004), "A survey of perceptual evaluations and requirements of three-dimensional TV", IEEE Transactions on Circuits and Systems for Video Technology, 14(3), 381–391.

Seuntiëns, P. J. H., Meesters, L. M. J., & IJsselsteijn, W. A. (2005), "Perceptual attributes of crosstalk in 3D images", Displays, 26, 177-183.

Seuntiëns, P. J. H. (2006), "Visual Experience of 3D TV", Eindhoven University of Technology and Philips Research Eindhoven, the Netherlands.

Seuntiëns, P., Vogels, I., & van Keersop, A. (2007), "Visual Experience of 3D-TV with pixelated Ambilight", Proceedings of the 10th Annual International Workshop on Presence, October 25 - 27, Barcelona, Spain.

Wang, X., Yu, M., Yang, Y. & Jiang, G. (2009), "Research on subjective stereoscopic image quality assessment", Proceedings of SPIE, 7255.

Chapter 93

Impact of Occupational Health & Safety Activities and Safety Climate on Industry Accident Rates in South Korea

K.H Yi[1] , H.S Jung[2]

[1] Occupational Safety & Health Research Institute, KOSHA
Incheon 403-711, South Korea

[2] Department o f Preventive Medicine, College of Medicine
The Catholic University of Korea
Seoul 137-701, South Korea

ABSTRACT

The intent of this study in particular was to assess the safety climate of the workplace in South Korea at national level and how occupational injury rate was affected as consequence. A study conducted in 2005 by Occupational Safety and Health Research Institute (OSHRI) to evaluate the National Survey for Occupational Safety and Health Tendency (NSOSHT) reported 595 cases of workplace injury. As a tool of assess the safety climate, data were categorized into safety attitude, commitment to safety, safety policy, and safety administration and also looked at how safety climate, health and safety committee, and health and safety code affected the rate of occupational injury. When adjusted model was applied, occupational injury showed strong relationship to the number of employees, followed by health and safety code, and safety attitude. From these findings, it is essential to establish health and safety codes in work places that do not currently have it in place or appears to be inadequate standards, strengthen health and safety education, actively engage in health and safety administration, promote employee participation, enact health and safety regulations to enforce all workplaces would effective aid in decrease the rate of occupational injury in South Korea.

Keywords: Safety climate; Occupational injury rate

INTRODUCTION

Workplace injuries not only have profound effects on the personal well-being of employees, with fatal consequences in some instances, they also affect the business activity in various ways such as through a loss of work hours, economic loss, increase in business insurance premiums, and even, indirectly, corporate public image (Larsson & Björnstig, 1995; Jovanović et al., 2004). In order to reduce the rate of occupational injury, the South Korean government revised its Industrial Safety and Health Acts in 1991 and 2001 so as to reinforce the role of safety and health officers who directly monitor health and safety issues in the workplace. It also formed a health and safety committee to discuss health and safety issues with employees and workers who represented both laborers and managers, and mandated businesses to establish health and safety codes within their organizations (Ministry of Labor, 2008).

Despite the above efforts, South Korea still has relatively high rates of occupational injury. In 2006 the average incidence of occupational injury in South Korea was 0.77%, which led to 72 million lost worker-days and an economic loss of US$15 billion. Moreover, the economic loss had increased by 4.56% and the number of lost worker days had increased by 2.85% compared to the previous year (Ministry of Labor, 2007).

Ahn et al. (2008) found that theaverage crude rates of occupational fatal injury between 1998 and 2001 were 13.8 and 3.6 workers per 100,000 in South Korean and the United States, respectively. In addition, Won et al. (2007) reported that the rates of occupational injury were 1.79 and 3.13 workers per 100,000 in Japan and South Korea, respectively. In order to reduce the rate of occupational injury, it is critical to identify the contributing factors and attempt to correct them. There are two possible categories of factors that affect the rate of occupational injury: (1) job-level factors within the work organization, which include the characteristics of individual work stations in terms of the job performed and their format, and other individual-level factors; and (2) company-level factors, which encompass the structure, safety climate, and practices within the organization, the philosophy and state of labor-management relations, workplace health promotion activities, and the presence and effectiveness of health and safety committees (Shannon et al., 2001).

Thus, the present study aimed to clarify the state of the safety climate and its effect on occupational injury rate in South Korea using data from the National Survey for Occupational Safety and Health Tendency (NSOSHT) conducted in 2005 by the Occupational Safety and Health Research Institute (OSHRI).

SUBJECTS AND METHODS

Sample

The NSOSHT was performed to support safety and health policy, to promote occupational injury prevention, and to prioritize safety and health activities (Oh, 2008). The OSHRI conducted annual studies from 2002 to 2005 at the national level to evaluate the state of workplaces, and will continue such studies every 3 years thereafter (OSHRI, 2008). The NSOSHT performed in 2005 included manufacturing, nonmanufacturing, and construction companies with more than 5 employees, with the 2,633 included manufacturing companies being stratified by region, employment size, and industrial category, and data being collected from April to July in 2005 (OSHRI, 2005). A trained data collector visited the companies and obtained self-reported data from one person in each company, such as a health manager, safety manager, supervisor, or CEO. The NSOSHT consisted of nine items: (1) general characteristics, (2) injury-related data including occupational injuries, accidents, and estimated absenteeism, (3) costs related to health and safety, (4) management status related to health and safety,including the presence of a health and safety committee and the adoption of a health and safety code, (5) education related to health and safety, (6) management of chemical products, (7) collaborations on health and safety, (8) perspectives about health and safety, and (9) the perceived safety.

Data related to health examinations, sales, injuries, accidents, absenteeism, costs, education, and management were derived from the records kept by each companies, and data related to perspectives of the safety activities and safety culture were collected from one person in each company.

A total of 619 of the 2,633 manufacturing enterprises had experienced organizational injuries in 2005, of which 24 unclear injury reports were eliminated. Because this study was related to the safety activities and safety culture and the perspective might differ between CEOs and employees, data from 31 obtained via CEO reports were also excluded. Therefore, data from 564 companies were finally analyzed in this study.

Measurement

The rate of occupational injury was represented as a percentage of the number of injured workers relative to the average number of workers employed during 1 year. There were 15 items related to surveying safety activities and the safety climate in the NSOSHT. Each question was answered on the following grading scale related to the degree of agreement: "Not at all", "Do not agree", "Neither agree or disagree", "Agree", and "Agree strongly" these answers were assigned scores of 1–5 points, respectively.

Although the survey items were not originally designed for surveying the safety climate, they were considered appropriate for this purpose, and hence items were categorized into two factors based on the results of exploratory factor analysis and a review of a previous study (Dedobbeleer & Beland, 1991). Dedobbeleer and Beland (1999) categorized safety-climate issues into two factors: (1) the management commitment to safety, which relates to the degree to which workers are aware of and conform with safety procedures, and (2) worker involvement in safety activities, which refers to the active participation of those involved in management of the workplace.

The safety climate can influence the rate of occupational injuries, and hence the established health and safety committee and health and safety code were also included. A health and safety committee in a company deals with various issues related to health and safety in order to obtain consensus with both the employer and employees, and a safety code is a document related to the mutual responsibility for problems related to health and safety.

Statistical Analysis

The safety-climate items used in this study had not been validated previously, and hence cross-validation was performed using exploratory factor analysis with the two randomly assigned samples (Hair et al., 1998). Using a principal-axis analysis with a Varimax rotation according to the two factors, items related to the safety climate were selected in Group I. The generalization of a safety climate was cross-validated in Group II using the same extraction and rotation technique. In addition, internal consistency was quantified using Cronbach's alpha coefficient.

We used t-tests or ANOVAs to assess differences in the occupational injury rate according to the characteristics of the manufacturing industries. Multiple regression analysis was used to explore the significance of correlations between the occupational injury rate and the presence of a health and safety committee, the adoption of a health and safety code, and the safety climate. Because in almost every company the occupational injury rate was lower than 0.05 and the distribution of injury rate was skewed to the left, we applied a logarithmic transformation to the injury rate as the dependent variable in the multiple regression analysis.

RESULTS

The characteristic of this study's target is such as Table 1. Male account for 87.6% and Female account for 12.4%. Based on the age, 30-39 make up 45.5%, the top rate of the study, 40-49, 20-29 continue and over 50 are the least. Based on occupation, supervisor make up the top rate of 46.6% and next is safety manager with 44.5%. Though it was expected that a business proprietor would take different stand against other occupation's respondent about Workplace's safety climate, there is no significant difference in workplace's safety climate score based on occupational classification. So this research includes all occupation.

Table 1. Characteristics of person who answer the questionnaire

Variable	Category	N	%
Gender	Male	521	87.6
	Female	74	12.4
Age	20~29	77	12.9
	30~39	271	45.5
	40~49	188	31.6
	50 and older	59	9.9
Type of position	President	31	5.2
	Safety manager	265	44.5
	Health manager	22	3.7
	Supervisor	277	46.6

Cronbach's alpha reliability coefficients for the safety climate. Cronbach's alpha was 0.882 for the management commitment to safety, 0.818 for worker involvement in safety activities, and 0.854 for the safety climate in this study.

The job descriptions quoted by employees were safety manager, health manager, and supervisor. The rate of occupational injury was higher in workplaces with 5–49 employees than in those with 50 or more employees (F=174.174, p=0.000), and higher in workplaces with 41 or more working hours per week (F=5.597, p=0.001).

There were 213 workplaces with a health and safety committee, and their reported rate of occupational injury was significantly lower than that in workplaces with no oversight committee (t=11.928, p=0.000). A health and safety code was established in 418 workplaces, and their rate of occupational injury was also lower than that in workplaces with no such code (t=9.994, p=0.000). The occupational injury rate was also lower in workplaces with better worker involvement (t=6.973, p=0.000) and safety climates (t=5.463, p=0.000).

The factors affecting the occupational injury rate concerning the safety climate of the workplace as analyzed whilst accounting for the number of workers and work hours are listed in Table 3. According to the model, the factors that significantly decreased the rate of occupational injurywere the presence of a health and safety committee (p=0.000), the adoption of a health and safety code (p=0.000), and a better safety climate (p=0.008). The variance was most affected by the presence of a health and safety committee (β=–0.954) followed by the adoption of a health and safety code (β=–0.825) and a better safety climate (β=–0.211).However, the number of employees and the working hours per week did not significantly affect the occupational injury rate in the model.

Table 2. Occupational injury rates according to the characteristics of the manufacturing industries

Variable	Category	N(%)	M±SD	t/F	P	Duncan Grouping
Job description of employee	Safety manager	265(47.0)	2.03±3.12[b]	53.940	0.000	a>b
	Health manager	22(3.9)	1.28±1.00[b]			
	Supervisor	277(49.1)	5.87±5.59[a]			
Number of Employees	5-49	235(41.7)	7.45±5.54[a]	174.174	0.000	a>b
	50-299	167(29.6)	1.50±0.92[b]			
	300 and more	162(28.7)	1.18±2.38[b]			
Working hours per week	40 and less	107(19.0)	2.27±4.16[b]	5.597	0.001	a>b
	41-45	150(26.6)	4.18 ±4.48[a]			
	46-50	143(25.4)	4.71±5.66[a]			
	51 and more	164(29.1)	3.95±4.69[a]			
Health and safety committee	Exist	351(62.2)	5.36 ±5.43	11.928	0.000	
	Does not exist	213(37.8)	1.46±2.21			
Health and safety code	Exist	146(25.9)	7.34±5.07	9.994	0.000	
	Does not exist	418(74.1)	2.68±4.18			
Management commitment to Safety	Poor and average	288(51.1)	3.98±4.54	0.456	0.648	
	Good	276(48.9)	3.79±5.20			
Worker involvement in safety activities	Poor and average	345(61.2)	4.89±5.38			
	Good	219(38.8)	2.31±3.40	6.973	0.000	

| Safety Climate | Poor and average | 348(61.7) | 4.70±5.18 | 5.463 | 0.000 |
| | Good | 216(38.3) | 2.57±4.01 | | |

Table 3. Affecting factors on occupational injury rate by multiple regression analysis

Variables	Parameter estimate	Standardized b	Pr>|t|
Intercept	2.140	.393	.000
Number of employees	-2.669E-05	.000	.305
Working hours per week	.005	.006	.340
Health and safety committee	-.954	.113	.000
Health and safety code	-.825	.123	.000
Safety Climate	-.211	.079	.008
F	78.922		
Adj R-sq	0.356		

DISCUSSION

The standard definition of safety climate can vary between studies performed in different countries due to differences in social environments and conditions. For example, in the United States, Brown and Holmes (1986) considered that the following three factors defined the safety climate: (1) employee perception of management's concern with their well-being, (2) employee perception of management's response to the concern, and (3) employee's perception of the existing physical risk. In contrast, Varonen and Mattila (2000) defined the safety climate in Finland in terms of the organizational responsibility, safety attitude of workers, safety supervision, and company safety precautions. Based on Dedobbeleer and Beland (1991), the present study identified two dimensions of the safety climate: the management commitment to safety and worker involvement in safety activities. There were several similarities in the items included in previous study and the NSOSHT, such as "How much do supervisors and other top management seem to care about your safety?" (Dedobbeleer & Beland, 1991) and the "Supervisor considers safety in the workplace seriously" (in the present study). In addition, there were items related to the management commitment to safety, such as "Is the proper equipment for your tasks available at your job site?" (Dedobbeleer & Beland, 1991) and "Are you encouraged to use personal safety equipment?" (in the present study).

Although validation of safety-climate measurements was not an objective of this study, exploratory factor analysis and reliability tests were performed because the items used for the safety climate in this study had not been validated previously. In order to identify a two-factor solution in accordance with the study of Dedobbeleer and Beland (1991), three items in the NSOSHT were deleted based on item selection criteria. The reliabilities of each dimension and the total safety climate were identified for the remaining 12 items, and the high internal consistency for the total safety climate and high interitem correlations represent evidence that the safety-climate measurement tool would produce consistent results. The NSOSHT consistsof comprehensive questions related to general safety and health trends, and hence there was the limitation that the items related to the safety climate were not specific to the safety climate. Another limitation of this study is that acquiring data from one person per company does not necessarily provide a full perspective of the safety climate. However, the employees who answered the questionnaires had job descriptions of safety managers, health managers, and supervisors, and hence they should all have been aware of the various issues related to health and safety in their company, meaning that their opinions can be considered to be representative.

The factors that significantly decreased the rate of occupational injury were the presence of a health and safety committee ($p=0.000$) and the adoption of a health and safety code ($p=0.000$). Existing regulatory laws that mandated the creation of health and safety committees consisting of representatives from both management and employees and the created health and safety codes need to be considered from the perspective of the safety climate in South Korea. Renner(2004) also indicated that the most important factor for preventing workplace accidents is promoting employee participation. The presence of a health and safety committee with employee representation is highly regarded in South Korea, and Shannon et al. (1997) claimed that an active relationship between the management and the health and safety committee is important to reducing the incidence of occupational injuries. A health and safety committee can also have positive effects on employee's job satisfaction, organizational unity, and administrative work, by both reducing injuries in the workplace and having a far-reaching influence in positively changing attitudes and actions toward safety (Michael et al., 2005). The present study also indicates the significance of a health and safety committee in reducing the occupational injury rate and suggests the need to continuously promote active engagement in the workplace.

In addition, the rate of occupational injury is lower in workplaces that have a health and safety code than in those that do not. Many South Korean workplaces still do not recognize the importance of health and safety issues (Kwon & Chung, 2001; Lee et al., 2006) and do not execute any administrative health and safety activities with autonomy within the workplace organization.

The health and safety code was the most important factor in reducing the occupational injury rate, and interest in health and safety issues will ultimately be expressed as a regulatory code. Therefore, establishing health and safety codes in workplaces that do not currently have them would be greatly beneficial to decreasing the rate of workplace injury in South Korea.

Since the presence of a health and safety committee and the adoption of a health and safety code were the significant factors for reducing the rate of occupational injury, appropriate regulations and institutional policies should be effective at preventing injury in South Korea.

We also examined the safety climate and occupational injury rate at the national level from the perspective of employees. The key finding was that better safety climates were associated with lower occupational injury rates. This result was consistent with the those of a meta-analysis of 35 studies by Clarke (2006), who found that the safety climate was positively correlated with occupational accidents and injuries. Huang et al. (2006) also found that the safety climate is a critical factor predicting the history of self-reported occupational injuries. In contrast, in a study using objective injury data, Smith et al. (2006) found that the association between the safety climate and injury risk was no longer significant when the data were adjusted for injury hazard among industries.

The present study investigated the relationship between the safety climate and the injury rate in South Korea by stratified surveyed sampling. Future studies should employ safety-climate questionnaires that reflect the specific social environment and conditions in South Korea when examining the relationship between safety climates and occupational injury rates.

REFERENCES

Ahn YS, Bena JF, Bailer AJ (2008) Comparison of unintentional fatal occupational injuries in the Republicof Korea and the United States.InjPrev10,199-205.

Bhattacherjee A, Bertrand JP, Meyer JP, Benamghar L, Otero Sierra C, Michaely JP, Ghosh AK, d'Houtaud A, Mur JM, Chau N, Lorhandicap Group (2007) Relationships of physical job tasks and living conditions with occupational injuries in coal miners. Ind Health 45, 352-358.

Chau N, Bourgkard E, Bhattacherjee A, Ravaud JF, Choquet M, Mur JM; Lorhandicap Group (2008) Associations of job, living conditions and lifestyle with occupational injury in working population: a population-based study. Int Arch Occup Environ Health 81, 379-389.

Gershon RR, Karkashian CD, Grosch JW, Murphy LR, Escamilla-Cejudo A, Flanagan PA, Bernacki E, Kasting C, Martin L (2000) Hospital safety climate and its relationship with safe work practices and workplace exposure incidents. Am J Infect Control 28, 211-221.

Ghosh AK, Bhattacherjee A, Chau N (2004) Relationships of working conditions and individual characteristics to occupational injuries: A case-control study in coal miners. J Occup Health 46, 470-478.

Gyekye SA (2005) Workers' perceptions of workplace safety and job satisfaction. Int J Occup Saf Ergon 11, 291-302.

Jovanović J, Aranđelović M, Jovanović M (2004) Multidisciplinary aspects of occupational accidents and injuries. Working and Living Environmental Protection 2, 325 – 333.

Korea Occupational Safety & Health Agency (2006). The National Survey for Occupational Safety and Health Tendency(NSOSHT).

Kwon OS, Chung CK (2001). The awareness, knowledge and attitude of the employers and employees regarding occupational health among industry with less than 5 workers. Korean J Occup Health 40(3), 87-98.

Larsson TJ, Björnstig U (1995) Persistent medical problems and permanent impairment five years after occupational injury. Scand J Soc Med 23, 121-128.

Michael JH, Evans DD, Jansen KJ, Haight JM (2005) Management commitment to safety as organizational support: relationships with non-safety outcomes in wood manufacturing employees. J Safety Res 36(2), 171-179.

Ministry of Labor (2007) 2006 Statistics of Industrial Accidents, Seoul, Korea.

Ministry of Labor (2007) High-Five Movement, Seoul, Korea.

Ministry of Labor (2008). Industrial Health and Safety Act, Seoul, Korea.

Nakata A, Ikeda T, Takahashi M, Haratani T, Hojou M, Swanson NG, Fujioka Y, Araki S (2006) The prevalence and correlates of occupational injuries in small scale manufacturing enterprises. J Occup Health 48, 366-376.

Renner P (2004) Systems of safety and active worker-participation strategies for a safe workplace: the philosophical and structural underpinnings of the labor institute, and the paper, allied-industrial, chemical and energy workers international union, accident prevention programs. New Solut 14, 125-137.

Shannon HS, Mayr J, Haines T (1997) Overview of the relationship between organizational and workplace factors and injury rates. Safety Sci 26, 201-217.

Shannon HS, Robson LS, Sale JEM (2001) Creating safer and healthier workplaces: Role of   o rganizational factors and job characteristics. Am J Ind Med40,319-334.

Siu O, Phillips DR, LeungT(2004) Safety climate and safety performance among construction workers in HongKong: The role of psychological strains as mediators. Accid Anal Prev 36,359-366.

Smith GS, Huang YH, Ho M, Chen PY (2006) The relationship between safety climate and injury rates across industries: The need to adjust for injury hazards. Accid Anal Prev 38, 556-562.

Varonen U, Mattila M (2000) The safety climate and its relationship to safety practices, safety of the work environment and occupational accidents in eight wood-processing companies. Accid Anal Prev 32, 761-769.

Won JG, Ahn YS, Song JS, Koh DH, Roh JH (2007) Occupational Injuries in Korea :A Comparison of blue-collar and white-collar workers' rates and under reporting. J Occup Health49,53-60.

Chapter 94

OHSMS Based on Ergonomics and BBS

Sonia Marino [1], Maria Donisi [2], Marco Testasecca [1]

[1] Integronomia (ergonomics and sustainability research association)
Rome, ITALY

[2] ISPESL (Istituto Superiore Prevenzione e Sicurezza sul Lavoro)
Monte Porzio Catone, Rome, ITALY

ABSTRACT

An OHSMS uses a cyclic sequence of phases, planning, implementation, monitoring and reassessment of the system. This is a dynamic process undergoing continuous development and improvement. The system has to reach the intended aims in terms of commitment and involvement of all corporate functions. The systems management presents characteristics fully in accordance with the methods used in ergonomics but also in the BBS. For example, it must adapt to the specific characteristics of the organisation, improving the ability to adjust to the evolution of the regulations and norms of good techniques and including the workers and their representatives and the continuous improvement of the cycle. Once the potential of both approaches (ergonomics and BBS) has been assessed, the differences which characterise them, as well as the possible common points which can make them complementary and collaborative in attaining the same goal, it is possible to integrate them and systematise them in a process of the formulation of a specific OHSMS.

Keywords: Ergonomic, Behaviour-Based Safety, BBS, Occupational Health and Safety Management System, OHSMS

INTRODUCTION

The study seeks to design an Occupational Health and Safety Management System (OHSMS) that is based on a melding of ergonomic and Behaviour-Based Safety (BBS) methodologies. To begin with, we shall consider how each of the two methodological approaches, Ergonomics and BBS, analyses and assesses the factors that can contribute to the occurrence of a damaging event. The assessments they make are intended to enhance accident prevention and safety in the workplace in a manner that is compatible with the structural and organizational constraints that naturally subsist in the working environments of any productive enterprise. As regards the ergonomic aspects of the workplace, the emphasis is very much on the need for information and training. Our comparative study begins with a synthesis of the specific theories of the two disciplines, then goes on to consider what each achieves when applied to companies differing in nature and size. The results are described in some of the texts referenced in the bibliography.

RESULTS

According to Standard BS OHSAS 18001, an Occupational Health and Safety Management System forms part of the overall system of management that an organization uses to develop and implement a policy for the control of occupational risk. A system of health and safety, moreover, uses a cyclical sequence of phases consisting of: planning, implementation, monitoring and re-examination. It is a dynamic process that is continuously being put into effect and improved.

The Occupational Health and Safety Management System has to attain a set of objectives, to which end it calls on the commitment and involvement of all the departments of the entity in which it is being implemented. A closer look reveals that an OHSMS is characterized by a number of elements that accord completely with the objectives of both Ergonomics and BBS. For example, it adapts to the specific characteristics of an entity; it improves the entity's own ability to embrace regulatory changes and adopt good practices; it involves workers and their representatives; and it requires a process of continual improvement of the cycle.

With its anthropocentric vocation, Ergonomics sees the human person as central to the generative impulse that drives forward planning, the organization of space and the creation of products that are suitable for the user. Ergonomics seeks to maximize the level of observance of and responsiveness to the requirements of the workplace, and reflect the psychophysical characteristics both of the operator who produces and of the person who uses the products. The creation of conditions and situations that are conducive to good levels of comfort is therefore the primary, most valid and most effective preventative measure that can be taken.

Ergonomics is characterized by two fundamental aspects. The first relates to its field of application. Only in so far as it can be made operational and be concretely applied does Ergonomics acquire a meaning and a right to exist. Similarly, only through its effective application can its principles and methods assign central importance to the human person in the process of configuring spaces, objects, processes.

The worker is the fulcrum of a system around which ergonomic correlations and processes can be constructed. The building blocks themselves consist of: the place of work, the tasks of the job, the production processes, the organization of production and the management of safety. Experience obtained from several production environments indicates that conformity to ergonomic principles is not only a valid strategy for avoiding discomfort, sickness, physical and mental fatigue, negative stress and accidents, but is also an excellent way of increasing productivity, improving results, enhancing efficiency and capitalizing on energies and resources.

BBS, on the other hand, focuses on the adaptive modulation of behaviour by the worker himself or herself, and seeks to achieve complete congruity of content and context between the worker and his or her task, the specific working conditions and expected performance. BBS succeeds in bringing about statistically significant decreases in the incidence of human error, and has proved itself capable of enhancing the performance of the worker/operator in situations of emergency in relation to the handling of actual, probable or possible risks. BBS is a discipline that seeks to improve safety by enforcing the adoption of a set of shared values that become a cardinal part of the safety culture. In order bring about a significant improvement in individual and group safety, BBS first needs to instil values, espoused and shared by workers, relating to the culture of safety and behaviour modification. To do so, it makes use of established scientific techniques borrowed from the discipline of behaviour psychology, modulated to suit the character and size of the working environment in question.

The human factor is absolutely central to both disciplines, but they have markedly different perspectives. One (Ergonomics) sees the human factor as the generating power, the other (BBS) as an element that can be improved through modification. If, however, the two were made to work in conjunction with each other and combined within the one system, then it should be possible to get them to work as complementary factors and to maximize the benefits of both, and thus move beyond their differing visions of the human factor. In this way, the differences between them could be turned into a creative force capable of forging new operational tools.

Indeed, both Ergonomics and BBS emphasize the importance of posture and related issues relating to the use of machinery and the risk present in the workplace and in work processes. Similarly, a common point of departure for both disciplines is the close preliminary observation of occupational activities and the identification of the related risks. In the case of BBS, the act of observation is aimed almost entirely at detecting incorrect and risk-laden behaviour by workers as a preliminary to the training phase, which is overwhelmingly concentrated on modifying unsafe behaviour. Oversimplifying, we can say that BBS leverages the power of peer group pressure on the individual in order to re-programme and improve working practices and establish a virtuous cycle of behaviour.

In Ergonomics, errors of behaviour – once planning failure is discounted as a cause – are treated as a natural consequence of poor adaptation. That is to say, cognitive dysfunctions are created by the occupational context, and to remedy them we must go to the environmental, psychological and behavioural source. An important part of the remedial intervention is to instruct workers, who can thereby expand their experience and scope of competence, not simply by addressing their weaknesses, but also by dwelling on their strengths.

Further, Ergonomics is particularly interested – with a view to mitigation and/or elimination –in the assessment of negative factors that may be present in the entity. It is similarly concerned with the structure of personal relations, and seeks to pinpoint the root causes of occupational stress.

The results from the case studies cited below suggest that BBS can greatly reduce the number of accidents in the workplace and tends to create a cooperative environment. As for Ergonomics, in addition to reducing the number of accidents, it is successful also at reducing the incidence of work-related sicknesses and in improving the psychophysical well-being of workers. This is no small matter in light of the obvious truth that a person who feels well works well.

The strength of BBS, which consists in concentrating almost all its efforts on the human behaviour, may also be its weakness, because it neglects that broad-ranging and multifaceted vision that typifies Ergonomics whose aim is to bring about a substantial improvement in the quality of work and hence the psychophysical well-being of the workers, with quantifiable benefits for the company. This is precisely the point of convergence that we are considering here. Ergonomics can be seen as countervailing force to the drastic approach of BBS.

The application of an OHSMS embodying both disciplines points the way towards a more fluid and more efficient systemic approach to safety, one that can lead to a virtuous circle and therefore genuine and continuous (as opposed to merely regulatory or bureaucratic) improvements.

In short, an Occupational Health and Safety Management System of this sort, while remaining within the parameters of current legislation, could bring about a transformation of safety and make it an integral part of the organization of an entity. Safety would therefore become the fountainhead of the entity's new organizational culture, a culture that naturally seeks to effect improvements and is based a dynamic of cyclical retroaction and strengthening. Thanks to its multidisciplinary aspects, Ergonomics seeks to synthesize well-being in the workplace and increase productivity, and, along with BBS, reduce accidents and illness.

Plan – Do

Assess risk by means of a careful analysis of its sources within the processes of the entity (Ergonomics) and within the patterns of behaviour of workers themselves (BBS).

Determine, prepare and apply the necessary technical and organizational safety measures to ensure occupational safety. The measures must not, however, be limited to the setting up of a safety and accident prevention system, but must also involve all the personnel in an entity, including workers (Ergonomics is fundamental for participation and BBS is fundamental for the institutional of a mechanism for behaviour oversight and instruction). Involvement of contractors (Ergonomics)
Selection of personal protective equipment (PPE) and collective protective equipment (CPE) (Ergonomics and BBS).
Planning of emergency procedures.

Check – Act

Setting out the procedures in writing is a necessary but not sufficient first step. It is also of fundamental importance to engage all managers and personnel in the periodic monitoring of implementation. The approach to ensuring the system is functioning is proactive. Whereas strategies shall be selected by the directors of the company/entity and objectives set by managers, their achievement is up to the workers. For the system to work, therefore, it is absolutely essential to secure the collaboration and involvement of the workers (from an ergonomic perspective) both during the monitoring phase and during the successive re-examination of the system. This process may even lead to an overhaul of the entity's organization. Ergonomics may turn out to be fundamentally important for the reorganization/redesign of the organization in order to secure worker well-being and, at the same time, higher productivity and an acceptance of change and innovation.

CONCLUSIONS

The potential of both disciplines (Ergonomics and BBS) and the differences between them are evident, but it remains possible to combine them as complementary elements in the pursuit of a common goal. It is possible to integrate the two within the framework of an Occupational Health and Safety Management System.

Both conceptions have the same part of departure: the human person; and both have the same objective: the prevention or reduction of risk. It is in their drive to achieve this objective, however, that their methods and actions diverge. Ergonomics is multidisciplinary and interdisciplinary. Its scope ranges from the assessment and resolution of failures of design and processes to problems arising from stress and social issues. BBS, on the other hand, is a specialized discipline that focuses on the behaviour of the worker. At first glance, these differences might seem to suggest that the two disciplines cannot be applied at the same time, that communication between them is impossible and they cannot be made to work together for the sake of occupational risk prevention. Furthermore, the methods adopted by BBS to construct safe behaviour contain an important element of psychological persuasion techniques that have very little in common with Ergonomics. Yet these techniques, if used appropriately and with balance, have an important potential to be used as complementary elements in the application of ergonomic methods. This can be achieved if their use is based on a solid sense of commitment to the single-minded pursuit of the goal of healthy, safe and continuously improving conditions in the workplace.

A systemic vision of occupational health and safety will use all available instruments without preconceptions and point the way towards new and innovative approaches. Safety is a theme of great complexity and sensitivity that requires vigilant and rigorous management. The quest for perfection is a moral undertaking because of the countless human and social implications inherent in the theme, which is also closely entwined with practical questions of feasibility, service delivery, the characteristics of the sector, efficiency of production and market competitiveness.

REFERENCES

Bandini Buti L., (2008) Ergonomia olistica, Franco Angeli

McSween T.E., (2008) Scienza & Sicurezza sul lavoro: Costruire comportamenti per ottenere risultati (ed. italiana), A. A. R. B. A.

BS OHSAS 18001:2007

Di Martino V., Corlett N., (2005) Organizzazione del lavoro ed ergonomia. Come è possibile migliorare le condizioni operative, la qualità e la performance e affrontare con successo le sfide del terzo millennio, Franco Angeli

Cenni P., (2003) Applicare l'ergonomia, FrancoAngeli

Andreani A., Baldacconi A., Raveggi F., (2003) I Sistemi di Gestione della Salute e della Sicurezza sul Lavoro, ISL 8/2003

Hendrik H.W., Ergonomics: an international perspective, in Karwowski W.E., Marras W,S., (1999) The occupational ergonomics handbook, CRC, Boca Raton, Florida

Osborne D.J., Leal F., Saran R., Shipley P., (1993) Person centered ergonomics, Taylor & Francis, Londra

Marino S., Donisi M., Papale A., (2009) Behavior Based Safety ed Ergonomia. Due metodologie, uno stesso obiettivo: la diminuzione degli infortuni sul lavoro, atti 72° Congresso Nazionale della SIMLII, Firenze, 25/28 novembre 2009

Chapter 95

Safety and Risk Philosophy in Maintenance Management

Hana Pacaiova

Safety and Quality Department, Mechanical Faculty
Technical University of Kosice
Letna 9,042 00 Kosice, Slovakia

ABSTRACT

Traditional, maintenance management is accepted as a service for production line, but not as an adequate partner. There are a lot of problems in safety area. Hazard prevention of maintenance staff by the standard maintenance activities, hazard prevention comes from the failure with possible serious consequences. Risk analysis as a base tool for good management decision, hast to be standard tool how to planned maintenance activities, how to organize reliable and safe maintenance processes and how to effective mange overall maintenance process. In the present time there are so many supports how it can be achieve. But the main role is to change the behavior not only of maintenance management and maintenance staff but also an enterprises management approach. This article tries to describe base Risk analysis products applied in Maintenance Management and assess their advantages and disadvantages.

Keywords: Risk analysis, RBI, RCM, FMEA, Maintenance

INTRODUCTION

European Agency for Safety and Health at Work (EU – OSHA) in Bilbao (Spain), organises every year campaign to support of safety improving in specific areas. Last two years were these campaigns oriented to support the development of applicable tools and procedures for risk assessment processes with orientation to small and medium enterprises. New campaigns for the period from 2010 to 2011 is orientated for monitoring the improvement of awareness about maintenance importance for health and safety work place, as well as employees protection, who performed them.

The goal of campaign is underline these societies and enterprises, where well organise Maintenance Management supported specified OH&S society's aims, but also aims in social safety protection linked to major accident related to industry hazard activities and environmental and quality production area.

European Federation of National Maintenance Societies (EFMS) generate activities for this campaign as a support through all its partners, i.e. individual National Maintenance Societies (NMS), to monitor and follow injuries and health damages as effect of maintenance activities jobs and find the tools for decreasing this reality in the range 1% of maintenance costs. For this purpose EFNMS created European Health and Safety Committee – EHSC, which closely cooperates with National Maintenance Societies, but also with European Agency for Safety and Health at Work.

According to particular results from NMS analyses, the percentage number of significant occupational accidents is connected with maintenance activities and equipment repairs, which are range from 20 to 47%.

Some analysis in 2007 and 2008 done by National Labour Inspectorate of Slovakia, estimated that this parentage in Slovak's enterprises is 25% (Gecelovska, 2009).

Most frequently causes of occupational fatality injuries with maintenance activities are pressurized container explosion (approx. 50%).

Hence, it is clear that OH&S in maintenance staff activities is very important aspect which influences level of Safety Management in societies or industrial enterprises.

RISK FACTORS IN MAINTENANCE ACTIVITIES

Inadequacies in OH&S area are sources and causes of work injuries, occupational diseases and other healthy harms according to different work's factors. Impropriate work conditions de-motivate workers and increase their physical and psychical loading and could be sources of stress and that way weaken the workers performance and their concentration on work duties. Effects are lower work effectivity and productivity, lower production quality and quality of provided services. This facts, degraded image of organisation on the business market and also ability of implementation of its products and services activities. Overall result of this is negative economical and social effects not only on employer, but in finally effect also on employees and global society economy.

From Slovak Labour inspectorate check records, a lot off employers keep only formal checking of work place hazards and risk assessment.

Safety Management as well as working-out Safety program, is often understanding as only safety techniques duties, and only in small measure are cover also responsible management activities. Table 1 shows summary of safety non conformity (investigation) statistics in particular checking areas and activities in chosen enterprises in years 2006 and 2007.

Table 1: Non conformity with OH&S legislative requirement (NLI, 2007 2008)

Assessing areas	Number of safety non-conformity	
	year 2007	year 2006
Working conditions	1 673	2 323
Allocation and state of PPE	1 414	1 719
Safety Management	7 301	7 197
Labour Organisation	635	711
Working environment	691	790
Operational buildings and objects	7 083	8 134
Working activities	881	1 194
Collective contracts	17	17
Labour directives and labour regulations	14 784	15 160
Pressured, electrical, lifting and gas Equipments	12 307	15 383
Others machinery and equipments	1 033	1 070
Special machinery and equipments	670	841
Summary of machinery	14 010	17 294
Overall S U M A R Y	48 489	54 539

910

Present safety lacks in the technical machineries and equipments state mainly are related with non-adequate maintenance and repair activities (Olejnik, K., 2009), also with not doing prescribes controls, inspections and tests, with using of physical and morally wears out machinery and technical equipments. Often missing complete operation, maintenance and line up documentations, which aren't functional protective and blockage machinery equipments, guard covers are missing and control elements are without functionality marks.

It is clear, that by safety factors investigation in the term of maintenance management it is necessary to oriented into tree main areas:

- R_{MA}: risks related to **maintenance activities** providing, which can influence overall OH&S enterprise level (hazards of maintenance staff),
- R_{FO}: risks related to bad provided or omitted maintenance activities (machinery **failures**), which consequences can have effect on harm of employees/**operators**,
- R_{FS}: risks related to bad provided or omitted maintenance activities (machinery **failures**), which consequences can have effect on health and safety of society - major accidents.

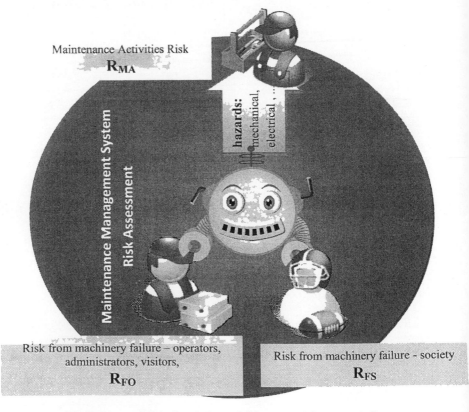

Fig.1. Types of Risks in relation to Maintenance Management System

Risks related to **maintenance activities** providing **(R_{MA})**
The injuries causes, which became as a result of maintenance activities is possible to specify as follow:

- doing activities by person without qualification or not enough qualified /trained person/,
- bad work organization, using danger techniques,
- individual factors – indisposed at the moment, abilities, crafts, work under the stress of maintenance staff,
- impacts of work environment (e.g. dust, explosion hazard, failure hazard).

Risks **related to bad provided or omitted maintenance activities** (machinery **failures**), which consequences can have effect on **harm of employees/operators (R_{FO})**

Theirs causes can be:
- corroded machinery or its parts,
- machinery with inadequate or not-functional or disassembly protective and blockage machinery equipments,
- not enough reliable equipment for safe operation,
- hidden machinery failures.

Risks **related to bad provided or omitted maintenance activities** (machinery **failures**), which consequences can have effect on health and safety of society - major **accident (R_{FS})**
The causes can be the same as previous, however often are relate to not adequate analysis in the plane phase of maintenance strategy and with underestimate hazard related to specific equipment operation (e.g. dangerous substances tank, pipelines).

MAINTENANCE MANAGEMENT CONCEPTIONS

Maintenance management can be characterize as "all management activities, that determine the maintenance objectives, strategies and responsibilities and which are realized by maintenance planning, managing, controlling and improvement of organizational methods included economical aspects.

Maintenance is defined (according to EN 13306: Maintenance Terminology) as the combination of all technical, administrative and managerial actions during the life cycle of an item intended to retain it, or restore it, a state in which could be perform the required function.

But maintenance management is in praxis always underestimated, mainly from the purpose of supervising maintenance costs without aspects of indirect costs related to possible failure consequences, where just the effective maintenance management is already important thing influencing effectivity of overall enterprise's management.

Therefore maintenance costs can be more closely describe as (Bentley, J,P.,1999):
- ✑ costs of *preventive maintenance*:
  - **non recurring costs**, e.g. cost for testing and controlling of equipments, tools, costs for initial training of maintenance staff, creating and executing of documentations, initial spare parts costs,
  - **recurring costs**, e.g. labor costs, spare parts costs, consumption items, on going training of maintenance staff,
  - regular changes of parts, e.g. oil changes, in demanding time period.

- ✑ costs of *corrective maintenance:*
  - **non recurring costs,** e.g. costs for test equipment, tools, same as by preventive maintenance,
  - **recurring costs**, e.g. cost for labor, spare parts, consumables costs, on-going training costs of maintenance staff,
  - **consequential costs,** due to loss of production or capacity including costs for compensation and loss of income (e.g. safety, environmental, quality losses).

Dependability of product / machinery (according to EN 60300-3-3: Life Cycle Costing) is the collective terms used to describe the level of product availability by given condition i.e. reliability performance, maintainability and maintenance support performance.

Costs related to level of define dependability's elements then describes overall life cycle cost of item (see Figure 2).

Higher acquisition costs can lead to improvement of reliability and/or maintainability, so can improve availability with consequence of lower operational and maintenance costs.

Maintenance Costs is possible to define as follow:
- costs for object renewal and cots for its corrective maintenance,
- costs for preventive maintenance,
- consequential costs.

Consequential costs (indirect costs) are in principle indirect failure effects losses, is concerned:
- warranty costs,
- liability costs,
- costs due to losses of revenue,
- costs for providing an alternative (external) service.

It is necessary in phase of planning maintenance activities to apply the risk analysis (assessment), which detect another group of possible consequential costs, e.g.:
- enterprises image (reputation),
- losses as result of safety or environmental breach of legislative,

- costs due losses of status on the market.

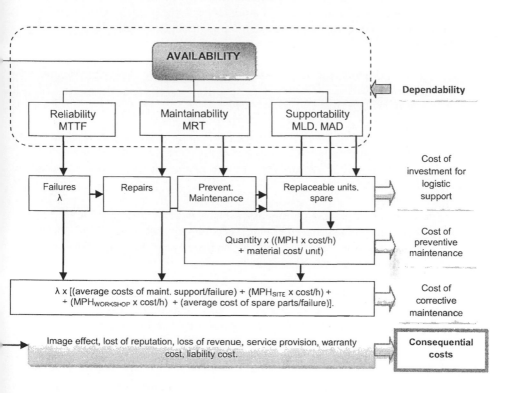

Fig. 2 Relationship between dependability and its elements and LCC
(according to standard EN 60300-3-3: Life Cycle Costing)

MTTF – mean time to failure, MRT- mean repair time, MLD – mean logistic delay, MAD – mean
administrative delay, MPH – maintenance person hours.

Key factor in effective enterprises management is on the one side the status of
maintenance as a maintenance management system with specific management, main
and supporting processes, however as integrated part of other management systems
(quality, environmental, safety) on the other side. For maintenance staff it is not
problem to understand and define failure as negative event and its consequence
understand as hazard of management objectives defined in the integrated politics of
enterprises management. To understand the failure only as problem of downtime
and number of spare parts is in present maintenance only "homesickness history".

From this reason the optimalization of maintenance activities, their operative and
effective management, demand implementation lot of methods and procedures of
risk assessment into the maintenance management tools, which were applied
privilege by safety technician and economical experts.

Recently, in the maintenance conceptions exist which are applied in maintenance management according to the type of equipments in the plant, or which type of risk is possible by useful maintenance conception to manage or eliminate.

There are:

⤷ **RCM** - *Reliability Centered Maintenance*

⤷ **RBI** - *Risk Based Inspection*, **SMIP** - *Pipeline Integrity Management System*,

⤷ **TPM** - *Total Productive Maintenance*.

Key element of **RCM** is FMEA analysis (Failure mode and Effect Analysis), through its application is possible to optimize maintenance activities based on "Risk Priority Number - RPN" of analysed failure mode and its effects (according to RCMII using Risk Matrix).

RBI is define as a process of risk assessment and risk management of integrity losses of pressure equipment (tank, pipe), according material wear- out. This risk is primary possible to manage by checking and analysing the state of equipment – build the effective inspection programme.

According o standard CEN/TS 15173 is **PIMS defined as**: management safety system, oriented on pipeline integrity. The objectives of pipeline integrity management system is to use inspection, monitoring and maintenance as prevention for "constriction" pipes integrity, i.e. integrity which treat the societal safety, business objectives or environment.

TPM is a complex of activities oriented on improvement of overall equipment effectiveness by optimalization working conditions in "Man-Machine-Environment" relation.

CONCLUSIONS

All these conception in maintenance management are lead to achieve optimalization of maintenance activities with effective cost and by minimizing prospective losses.

Theirs background is mostly from USA (RCM, RBI, PIMS) or from Japan (TPM).

Last years showed, by enterprises effort to keep legislation requirements and minimalization of losses, that it is necessary to built maintenance management on the systematic base. The exploitation of tools and methods for risk assessment in the planning phase and choosing adequate maintenance strategy (predictive, preventive, corrective) in the relation to identified failure mode and its effects, it seems as most effective and important step.

Present European standards in the maintenance management and dependability management support some elements of maintenance managements describing, e.g. how to execute the maintenance documentation (EN 13460), base requirement for qualification of maintenance personnel (TNI CEN/TR 15628), definition of Key performance indicator in maintenance (EN 15341) and so on. In case to build of maintenance management some help can be describe in European standard EN 60300-3-14: *Dependability management: Maintenance and maintenance support* which purpose is „to outline, in a generic manner, management, processes and techniques relate to maintenance and maintenance support that are necessary to achieve adequate dependability to meet the operational needs of the customer".

The research which led to these results, obtained financial fund from Seventh Frame program iNTegRisk based on Grand Agreement Nr. .CP-IP 213345-2 and along with Slovak Research and Development Agency based on Agreement Nr. DO7RP-0019-08.

REFERENCES

Bentley, J,P.(1999) Reliability & Quality Engineering.Addison Wesley Longman, ISBN 0-201-33132-2.s 137-139.

Haarman, M.- Delahay, G.(2004) Value Driven Maintennace. Mainnovation, p.20-39. ISBN 90-808270-2-9.

Legát, V. (2006) The world´s trends in Maintenance Management/Světové trendy v managementu údržby. Údržba/Maintenance, roč. V, č.3/2006, s.6-8. ISSN 1336-2763.

Moubray, J. (1997) RCMII, Industrial Press Inc., New York, USA, p.348, ISBN: 978-0-8311-3164-3.

Olejnik, K. (2009) Safety system of operating of building vehicles related of lack of visual transfer aspect. Napedy i sterovanie. miesiecznik naukowo-techniczny, Nr.4, 1/2009, p. 106-109.

Pačaiová, H. (2005) Failure mode and effect analysis. In: Practical Handbook for safety technician. Nové Zámky: Verlag Dashofer, p. 5, ISSN 1336-7668.

Pačaiová, H. (2006) Maintenance strategy as a basic tool for major hazard prevention. In: Sharing knowledge and success for the future : Maintenance dialogue between practice and science: Congress reports of 18. Euromaintenance 2006, 3. Worldcongress of maintenance 2006,, 20.-22. June 2006, Basle, Switzerland. Bern: Guildo Walt, H.-Joachim Behred, p. 503-508., ISBN 978-3-9523151-01.

Pačaiová, H - Grenčík, Juraj (2008) The advantages of safety integration in Maintenance Management/ Výhody integrácie bezpečnosti do riadenia údržby. In: ÚDRŽBA 2008 : Sborník 5. mezinárodní odborné konference : 6.-7. listopad 2008, konferenční centrum AV ČR - zámek Liblice. Praha: Česká společnost pro údržbu, 2008. p. 33-40, ISBN 978-80-254-2500-8.

Wintle, J.B.- Kenzie, B.W.(2001) Best practice for risk based inspection as a part of plant integrity management. HSE 2001. Crown copyright, 363p., ISBN 0 7176 2090 5.

Chapter 96

Safety Culture - An Effective Tool for Major Hazard Prevention

Milos Palecek

Occupational Safety Research Institute
Prague
Czech Republic

ABSTRACT

Analyses of disasters and unwanted events prove that human factor failure is the most frequent reason for them. It is obvious that the only implementation of safety management system in an enterprise will not bring about the sought-after results. The evidence shows that in many cases systems of safety management are often embodied by a set of out-of-practice manuals and handbooks with which workers do not identify themselves. Implementation and reinforcement of safety culture as a set of values, approaches, competence and behavioural patterns that determine engagement, style and competence of an organisation's programmes in Occupational Safety and Health. The paper focuses on the opportunities of assessment of safety culture level in an enterprise as well as up-grading possibilities.

Keywords: Safety, Safety Culture, Safety Climate, Human Failure, Human Error

NTRODUCTION

afety and risk prevention have been developing mostly in the terms of quality erformance for several decades. A standpoint to solve a problem with organisational nd/or technical measures was prevalent until the 70s, now the attention focuses rather n the behaviour and conduct of employees, i.e. human factor.

In the seventies and eighties of the 20th century the procedures, practice and rogress were regarded as a predominant factor influencing safety. This view had hanged due to the disaster at Chernobyl as well as the growing interest of the business vorld in the development of a strong organisational culture. The definition of safety ulture reconciled the influence of attitudes and also the influence of structure and elated safety culture to personal attitudes, intellectual habits and the style of an rganisation. Even though attitudes were generally intangible, not grasped, it has been ecognised that they may lead to major tangible results. At the end of the nineties of the 0th century people were aware of culture complexity and its multilevel character.

hase One: Safety is based on rules and regulations

1 this phase, an organization perceives safety as an external requirement rather than a ehavioral aspect which helps reach a success. Safety is perceived as a technical problem 1at can be solved in congruence with rules and regulations.

he characteristic traits of this phase include particularly:

) Problems have not been foreseen and an organization will not solve them until they have emerged.

) Communication between organization departments and individual functions is insufficient.

) Cooperation and shared decision-making process are restricted.

) People who make mistakes are blamed for being incapable of meeting requirements and regulations.

) Management role is understood as a task to enforce the regulations.

) Little attention is paid to monitoring, comprehension or studies inside and outside the organization which generally leads to a defensive attitude to criticism.

People are considered as elements, parts of a system (mechanical perception).

Unfriendly relation realms between the management and workforce.

People are awarded for being obedient and the outcomes regardless long-term results.

Phase Two: Safety is regarded as a significant goal of an organization

In this phase, an organization regards safety as a significant goal happening without external requirements. The management mainly concentrates on the technical and process solutions. Here, safety is understood as a set of goals and tasks with responsibility to have fulfilled the tasks given. In this phase, an organization often realizes that no further improvement would happen after a certain time when safety trends had improved.

The following traits are particularly characteristic of this phase:

1) Raising awareness of the influence of culture at the workplace takes place. Yet it is not understood why more severe control and schooling sessions do not bring about an expected improvement in safety.

2) The company management enhances a communication between the departments and organization functions.

3) Reaction of the management to the faults consists in the implementation of other controls, procedures and the providing of further education.

4) The management role is to secure the fulfillment of tasks and laying down explicit working tasks to the employees.

5) Organization is eager for such a learning that is based on the experience of external groups, particularly concerning a new technology and the best practice.

6) The relations between the management and the employees are hostile, although a number of opportunities for the discussion on the common goals can be found.

7) People are awarded for task excess regardless long-term results.

8) Interaction between people and technology is considered rather from a view of boosting the effectiveness of technology.

9) An organization supports team work.

10) The dealing with the problem remains reactive in an organization, yet it is possible to have more anticipation in relation to the issues that come into planning phase.

Phase Three: Safety can be always improved

In this phase, an organization has accepted its idea of permanent improvement and has applied this concept in safety. The organization emphasizes communication, schooling/training, management style, improvement in efficiency and efficacy enhancement. People in the organization perceive an impact of the overall culture on safety.

The following traits are characteristic of this phase:

1) The problems are foreseen and solved prior to their origin.

2) Cooperation between departments and the functions in the organisation is good.

3) No clash between the meeting of production tasks and safety.

4) Nearly every fault is viewed from the standpoint of process variability with a highlight on understanding of what had happened rather than the search of the person to blame and a subsequent accusation.

5) The management task is perceived as coaching towards an improvement in safety.

6) The process of learning from other persons is appreciated within an organization both from inside and outside.

7) People are respected and awarded for their contribution.

8) The management and employees support themselves.

9) People are aware of the impacts of culture problems and these impacts are taken into consideration in the decision-making process.

10) People are awarded for the improving of the attitudes as well as for the results.

11) People are regarded as a significant part of the system of a company and attention is paid to the satisfying of their needs, not only to reaching technical efficiency.

Nowadays, on the one hand, most workers are aware that it is insufficient to comply with legal regulations, yet on the other hand it is necessary to implement a proactive approach and to prevent emergencies of unwanted events in the fields of operational safety and OSH.

Practical experience mostly shows that the sole implementation of OSH management system will not bring desirable results to an enterprise. Analyses of incidents and accidents at work have proven that the reasons are due to a failure in the management system. It has often shown that the so-called *virtual* system of safety management, which represents handbooks and manuals, does not reflect the existing practice in a number of cases. It would be a mistake to maintain that a safety management system in an enterprise is unreasonable. It bears evidence that one has to do something more to attain the required functionality. This entails the implementation of safety culture and its sustainable growth.

Safety culture does not supplement OSH management system but is complementary and reinforces the outcome of the latter.

It seems obvious that safety culture may not be tentatively defined as "a special field of culture, wherein and when safety is of special corporate interest". It is necessary to remember that culture is a trait of a group rather than pertaining to an individual. There may be various different cultures, they may overlap or divide into subcultures in an enterprise. Culture will always concern the groups of workers and never individuals. Nevertheless, company management perceive culture rather as a phenomenon of an individual and not of an enterprise.

The required level of safety may be attained by means of an irreversible relation and cooperation of company management at all levels. The safety level is aimed at reaching safety culture which is really accepted not only by the company employees and management but also by suppliers as one of the key priorities and values.

The present situation in the Czech Republic

The obligation to secure operational and occupational safety is vested into an employer *inter alia* by the Labour Code and Act on Major Hazard Prevention.

The Labour Code focuses on securing safety and health protection at the workplace. The Act on Major Hazard Prevention is aimed at prevention of accidents in an enterprise and machineries dealing with excessive dangerous chemicals. Both legal regulations necessitate identification and risk assessment and implementation of preventive measures. Moreover, the Act on Major Hazard Prevention demands laying down principles of safety policy and implementing safety management system – a safety programme.

As most small and medium-sized companies have their management systems implemented under ISO 9000 – quality and ISO 14000 – environment, the supplementation with a system of safety management and health protection under OHSAS 18000 does not present a major problem. A safety programme in major hazard prevention structurally pegs to the systems. These matters create a good point departure for implementation and safety culture up-grade.

Safety Climate vs. Safety Culture

Good safety culture reflects the values that are shared at an organisation's each level and are based on the belief that safety is important and everyone is liable for it within an organisation. A safe culture may be only where the role of interpersonal relations and communication at the workplace are not undermined. An absolutely essential change in thinking and superiors' attitude, this it to say not to see only from the standpoint of one's own position in the hierarchy but to respect the needs of the whole.

The quality of work climate has got significant importance to occupational safety and to the traits of attitude and relation to obeying by regulations. A refusal or underestimation may be the reason for the failure and poor management of a situation. This may lead up to emergencies, industrial calamities or occupational accidents. A good safety culture is not endangering anyone within or beyond the company. It is an outcome of positive attitudes at the workplace and has to concern everyone – from the president to a new person working on the inferior post. A stress is important on the mutual significant and measurable safety and health protection, politics and action which serve a pattern rather than a regulation. It is also characterised by personal refinement (education) on all organisation levels and also a capability to be liable for one's behaviour. Care for an overall positive atmosphere, friendly working environment, safe working procedures, harmonic and friendly interpersonal relations, good health of the employees, their safety and personal development as well as the care for a health friendly working environment and harmless impact on the overall land ecology. Most factors are at the same time stress and violence prevention at the workplace, therefore i purposeful to combine efforts in occupational safety with the efforts in the creating of ιe safety culture in a company, namely in connection with stress prevention, excessive ɔdy strain etc.

The organisations possessing a positive safety culture are characterised by ɔmmunication based on the mutual confidence, shared perception of importance and ɔnfidencc in the effectiveness of the prevention measures undertaken. And redominantly, they continuously assess safety. In practice they use the mechanisms ith the help of which they gather safety information, assess occupational safety and rganise the workers' meetings in order to ascertain how one could work more safely. hese mechanisms are used not only for the solving of problems concerning ɔcupational safety but also for securing better problem identification and solution on a ιy-to-day basis.

Regular inspection of its procedure / processes relating to operational safety ave to be encrypted into organisation as a standard part of its safety strategy. Among ιese processes ranks personnel instruction and recruitment, feed-back from operational ׀perience, management of project changes, inspection of operational regulations and so ι. The self-regulation is aimed at a new view and to let the competent individuals or rgans that are not a part of a hierarchic structure of production management, to provide ιdependent safety assessments and drafts of new approaches and solutions. Assessment f workers' competency is a part of the programme of their orientation or education. The ιles and principles of safe behaviour are more important than technical skilfulness or ׃tailed knowledge of operational rules or procedures. A broader instruction is necessary ɔ ascertain that the workers will understand the importance of their obligations correctly ; well as the outcomes of the mistakes they may make. It could happen that the staff ould neither pay thorough attention to the emerging safety problems with no ɔmprehensible understanding of safety and nor would perform appropriate principles ιe to the misunderstanding of possible risks.

It is necessary to reach and maintain a high level of safety and safety of ɔeration. This would keep the operation requiring new technologies being employed by ι appropriate number of highly qualified workers who are aware of technical and Iministrative safety requirements. These workers have to be appropriately motivated so ιat they could follow a positive attitude towards any safety requirement. A positive titude is one of the basic prerequisites of safety culture. It is necessary for the workers ɔ participate in a number of educational and training programmes (which are subject to ɔntinual revisions that are to secure their up-date and correctness).

Superiors realise that individual workers understand their responsibilities, the sponsibilities of their closest co-workers and department management correctly as well ɔ the way of supplementing responsibilities by the ones pertaining to other groups. The ιders should secure that all work related to nuclear safety is conducted in line with the ·ictly laid down processes and developments. Monitoring and controlling systems will ׃ introduced and checked upon their effectiveness to work to effect this inspection. The se for laying down and controlling of the development of safety is to build up a ׃rarchy of papers – from the strategic instructions to detailed operational regulations d work commands. All paperwork should be checked, verified and revised pursuant to ׃ valid quality enhancement system in a given organisation. The management should

provide for full qualifications and competency of the staff subordinated to fulfil t
duties. The superiors have to encourage, assess and search for the ways of concre
awarding for particularly recommendable attitudes towards safety. It is necessary that a
awarding system does not support increase in production at the expense of safe
lowering. Workers' mistakes should serve a source of valuable experience and t
workers should be encouraged to the thorough determination, notification a
improvement of the imperfections in their own and others' work in order to ave
possible future problems either on one's own or others' side. The management shou
enforce their own responsibility for disciplinary corrective measures in case of recurrin
shortcomings at work or a severe dereliction, otherwise safety would be endangere
They have to perform it in a very cautious way as any recourse may lead up to t
concealment of the mistakes. An overall positive attitude towards occupational safety
the envisaged outcome.

In the ultimate result a sound operation depends on individuals' behaviours wh
are influenced by motivation and attitude both the overall staff and a group. Safe
culture concerns personal engagement and responsibility of all individuals involved i
any activity related to the safety of operation. A universal thinking concentrated c
safety should be regarded as the key element which enables to accept a self-critic
attitude, to prevent from the feeling of self-contentment, to engage in favour of a hig
level of safety and to foster personal responsibility and self-control in significant safe
activities. An individual's responsibility should be defined and documented in
sufficiently detailed way so that any ambiguities could be excluded. Individual
approaches are to a big extent influenced by their work environment. Organisation
development and operational rules that form the work environment and support positi
attitudes towards safety are the key elements to an effective safety culture.

The workers are actively interested in information and regulations which mig
improve the safety of their behaviour In the companies characterised by a "sound" safe
culture. They have positive attitudes to safety requirements, are attentive towar
emergencies and not foreseen circumstances and should they encounter an unknow
hazard they would approach experts for assistance. Cooperation should play a cruci
role. All working groups participate in defining the aspects of occupational safety a
practical applications cooperatively. They are not eager to pin the blame one to anoth
or to enforce safety-related questions on another group as a sort of penalisation but th
collaterally cooperate in the solutions to safety problems.

The following features are characteristic of a sound safety culture:

-   Awareness of the safety importance pertaining to everyone involved

-   Knowledge and competency raised through instruction, schooling and se
    education

-   Motivation through leadership, management through indicated goals, award a
    recourse system and by means of accepted and created attitudes

-   Supervision including audits and revisions with on prompt reaction on critic
    attitudes of individuals

Responsibility stemming not only from a formal authorisation, a list of responsibilities and their comprehension by an individual but making a worker be aware of them.

Basic traits of an elevated safety culture

The following traits of an elevated safety culture have been enlisted:

1) manager behaviour: top managers support safety by means of serving a model, pattern for other persons, showing their commitment in their behaviour, attitudes towards safety, source placement, including time to solve safety problems and particularly the time spent on safety improvement, their behaviour in the field of safety has to be obvious to their colleagues and other workers.

2) Self-assessment: the aim of self-assessment is to enhance increased safety efficiency by means of a direct engagement of people in critical testing and improving their own work and work results. Possible flaws can often be detected in time rather than lower the reserve in safe operation. A strong commitment into the process of self-assessment may motivate a worker in the efforts of the permanent improvement in safety.

3) It is always possible to improve safety: organisations are incessantly dissatisfied with the level of safety they have attained and are permanently looking for an improvement. Such a recognised value reflects the fact of a self-assessment that is widely applied within an organisation.

4) Proactive and long-term perspective: the plans should contain short-term, medium-term and long-term goals proving that an organisation is actively prepared for the future and also the measures to enable the changes.

5) The quality of documentation and processes: the documentation should be detailed and pregnant and easy to understand. Responsibility for preparation and verification should be laid down clearly. The documentation should be used both for work and for instruction. It is important for the documentation to be available for the workers in the version for work use.

6) Conformity with regulations and processes: the importance of this feature is obvious for safety. The processes should determine what to do in case of emergency in case it is not covered by the existing regulations or work progresses. Violation of the regulations and progresses is a clear mark of a weak safety culture.

7) Clear tasks, authorities and responsibilities: it is essential not to have unclear tasks or responsibilities in the field of safety. Job descriptions are to define these roles and responsibilities clearly. Responsibility means that the aims are clearly settled, a progress in fulfilling is regularly assessed and consequences for fulfilment and non-fulfilment are determined.

8) Motivation and contentment at work: workers' behaviour is being influenced by their motivation to work and contentment at work. Leadership and attention paid to job description is reflected in an organisation.

(9) Involvement of all workers: the workers should not have a feeling that safety is no their business unless they are involved in the detection and solution of the problems Safety is an area in an organisation to which everyone can contribute.

(10) Fair working conditions: safety may be impaired when the workers are occupationally overburdened. Their morale and attention to the safety-related questions will lower. It is essential that the managers anticipate the impact o unusual situations or the restructuring of an organisation on an employee to aver possible problems.

(11) Performance measurement in safety: it is important for an organisation to measure its performance in safety and communicated the results of the measurement and the trends discovered to all employees. Measurement of activities is determined for the improvement in safety. Here, an organisation should not only rely on acciden statistics. Such an approach is too reactive.

(12) Cooperation and team work: it is important to the workers to work effectively in a team, particularly when conflicts are to be averted and at the same tome provide that the team activity goes across various levels of competency. Successful team should be awarded within an award system existing in an organisation.

(13) Conflict mediation: an organisation should have corresponding processes fo conflict mediation, otherwise conflict escalation or its concealment and re-emergin; elsewhere are at risk. An ease with which the workers may draw the attention o other colleagues is a trait of such a workforce that contains self-confidence. Safety related conflicts are particularly harmful as they may degenerate to mutua accusations and mutual distrust.

(14) Relations between the management and the workers: a relation of mutual respec and openness between the workers and managers. Both parties should hav confidence in the capacity of mutual approach as well as problem solution. Som organisations may be built more hierarchically than other ones but a hierarch should not prevent from mutual respect.

(15) Top safety priority: a great number of organisations declare that safety is their top priority. Still, activities and behaviour not always prove this recognised value Trustworthiness of an organisation is lower if the reality does not comply with recognised value.

(16) Openness and communication: a good communication is needed if the workers ar to work effectively in an organisation. The employees have to be confident withi their qualifications and have to have an opportunity to share their concerns betwee themselves, either being a group or individual. An organisation may apply a grea deal of information channel, communication modes. Therefore an organisation ha to make use of a number of information channels to communicate with employee If an organisation declares itself to this value it will steadily support opennes between its employees afterwards.

(17) A will to learn: this value may be mostly viewed as the philosophy of a organisation wherein an approach to any problem is considered as an opportunity t learn something. It is a will to learn from the others and to share one's ow

experience with the others. An organisation revises the surrounding environment and adapts to the environment changes incessantly. An effort of the organisation to learn is very important for learning from the safety problems to determine their real primary causes.

(18) Time focus: it is important to keep equilibrium between the past, the presence and the future. An excessive focus on one period and exclusion of the other ones may bring about problems. Equilibrium in activity planning should be found. It is essential that the equilibrium is in employees' work. Time focus will be influenced by broader social and national cultures.

(19) Standpoint relating to the mistakes: mistakes may be regarded either as an opportunity to learn from or an opportunity for punishment. Organizations may influence employees' opinion on mistakes in a broader social culture. It is very important that the employees could draw attention to the safety-related mistakes in a fearless manner of being punished; otherwise the knowledge of a mistake will be suppressed. Another employee may be stricken with this very mistake in the future.

(20) Opinion on workforce: opinion on workforce may have an important influence on how people are being treated in an organization. On the one hand, if people are regarded as undisciplined and interested only in themselves, they will be incessantly subject to controls and their activities will be monitored. On the other hand, should people strive at the enforcing of their capabilities by means of personal development and trustworthiness, they might be managed in a more flexible way enabling them to take on responsibility. Both ways of assessing the workforce may have their tasks while improving safety, yet the latter is a prevalent opinion that will have bigger outcomes in the long run.

Safety climate

Safety climate differs from safety culture particularly in the following way:

1. Safety climate is a psychological phenomenon which is usually defined as perception of the safety status in time,

2. Safety climate deals with indefinite, intangible questions (problems) like situation factors, environmental factors.

3. Safety climate is a temporary (transitory) phenomenon, „a snapshot" of safety culture which is relatively variable and change prone.

Safety climate is perceived as an attribute which is composed of two factors: commitment of company management to safety and responsibility of the employees. This is a superficial trait of safety culture distinctive from attitudes and perceptions of the workforce in time. Factors of safety climate should reflect perception of safety policy, operation rules, and the awarding process. Other factors of safety climate should reflect the depth in which the employees believe that safety is a value internally pertinent to the organisation.

Safety climate may be described and assessed by means of *hard* facts of adopting various measures; safety culture is reflected only in *soft* data on subjective perception of safety ascertained mainly by means of in-depth interviews and questionnaires. Influencing and improving of safety climate may lead to improvement in safety culture. Positive safety climate results in higher effectiveness and safety results. If we wish safety culture was enhanced, it is necessary to start with an assessment and strengthening of safety climate. Hence, it is helpful to define the appropriate indicators.

The advantages of indicators are the following:

- its use makes it possible to observe the trends,
- managers pay higher attention to what is measured and use of the indicators will raise the interest in this concept.

The indicators identify to which extent an organisation shows the following traits of safety culture:
- Safety is a clearly reconciled value,
- Safety-led leadership is clearly expressed
- Safety is integrated into all activities
- Safety responsibility is clearly defined
- Safety is enforced by learning

The most significant emerging obstacle in implementing safety culture is not to appreciate the depth and force of culture. Culture is deep, ample, and stable. We are not able to manage culture; culture will manage us without knowing its extent in which it happened. The problems should be investigated and reverted to at an early stage so that safety could not be depleted imperatively. A thorough study of events and findings received from the faced problems in safety culture and safety management will be another important step concerning this effort.

In conclusion: **One should be aware that it is easier to change culture rather than create a new one.**

CHAPTER 97

What Constitutes Typical Adolescent Behavior and How Does It Differ from Adult Conduct?

Michael J. Vredenburgh, Ilene B. Zackowitz, Daniel R. Spencer,
Michael R. DeTaboada, Alison G. Vredenburgh

Vredenburgh & Associates, Inc.
Carlsbad, CA 92008, USA

ABSTRACT

Adolescence is a period that includes children ranging in age from 12 to 18 years. At this time, children are beginning to develop their own identities and are greatly concerned about peer approval. Adolescents are at great risk for being involved in injury incidents due to developmental characteristics unique to this age group. As human factors and safety consultants, we analyze multiple incidents every year that involve adolescents. In order to gain perspective on adolescent behavior, we developed a survey that was administered to the adolescents themselves. This differs from much past research on the topic that relied on hospital data and parent perceptions. This paper discusses common adolescent risk-taking behavior and their limited self-protective behavior.

Keywords: Adolescent, Safety, Behavior, Risk-taking, Injuries

INTRODUCTION

Unintentional injury is the leading cause of death and disability for adolescents (Agran, Winn, Anderson, Trent, & Walton-Haynes, 2001). Overall, unintentional injury rates are highest among adolescents ages 15 to 18 (Grossman, 2000). Using National Health Interview Survey data, Danesco, Miller, and Spicer (2000) determined an injury rate of 25/1000 for children 13 through 21 years of age, or 20.6 million injuries a year. The estimated cost of unintentional injuries to this population was $347 billion annually. These statistics reflect the fact that more adolescents engage in high-risk behavior than their younger peers or their elders. In fact, risk-taking behavior is normal for this age range (Steinberg, 2004).

In the field of forensic human factors, a relevant question when performing an analysis of an injury incident is: Was the person involved in the incident acting reasonably? Attorneys often seek human factors experts to explain human behavior and evaluate "reasonableness of conduct." Reasonableness of conduct refers to behaviors consistent with societal norms for a given population (Zackowitz & Vredenburgh, 2005). Age is an important consideration when evaluating reasonableness of conduct because behaviors that are perceived as unreasonable for an adult may be common for adolescents.

ADOLESCENT DEVELOPMENT AND PSYCHOLOGY

Adolescence is defined as the period between 12 and 18 years of age. This period of development is usually characterized by increasing independence, autonomy from the family, greater peer affiliation and importance, identity formation and physiological and cognitive maturation (Igra & Irwin, 1996). These children exhibit an intense need to be approved by peers, to be independent from parents and other adults, and to develop their own unique identity. These children desperately want the approval of their peers and will often engage in dangerous behaviors to gain the coveted "cool" status (Ellickson, Lara, Sherbourne, & Zima, 1997).

At this age level, children are capable of advanced cognitive processes including abstract thought, hypothetical reasoning and speculation about future events. Physically, these children are well coordinated, have improved endurance and are better able to judge distances than younger children. Socially, they are extremely concerned with peer acceptance and become more daring in their choices of activities and behaviors (Brown & Beran, 2008).

Despite the fact that adolescents often have the cognitive ability to make sound decisions and use abstract thought, they have many outside pressures that must be evaluated and considered. Peer pressure, egocentrism and personal fables are natural parts of adolescence that can greatly impact behavior. Similarly, invulnerability is common among this age group. Adolescents tend to believe that others are more vulnerable to injury, illness and other negative outcomes than they are themselves

(Quadrell, Fischoff & Davis, 1993). These types of age-related issues may be considered when analyzing why an adolescent has engaged in risky or rule-breaking behavior and whether that conduct could be expected given the facts of the case coupled with the adolescent's developmental stage.

ADOLESCENTS AND RISK-TAKING BEHAVIOR

Risk taking behaviors are the most serious threats to adolescent health and well being. Negative potential consequences of these behaviors include severe disability and death (Igra & Irwin, 1996). Adolescents often have a false sense of security: they feel immune to injuries and don't think bad things will happen to them. Therefore, they frequently take part in risky behaviors like playing with fire and risky recreational activities like snowboarding and riding ATVs.

Human factors consultants are frequently asked to investigate civil cases that involve injured adolescents. Typical cases fall into several categories including sports and water sports-related incidents, driving, incidents involving products, falls, fire, guns and incidents that occur at school. Typical accidents involving adolescents include being struck by an automobile while crossing the street, burns, and falling from an unintended height (off a roof, for example).

Children in this age range often play with products that have the potential to injure despite being commonly thought of as "playthings." For example, the air guns associated with paintball, airsoft pellets and BBs all have severe eye injury potential. While adolescents may be aware of this hazard, a study regarding these "toy" guns found that even when shooting at each other, 18% chose not to wear eye protection at least half of the time and females were significantly *less* likely to wear eye protection when participating in these activities, a finding quite different from adult females who typically are more like to practice self-protective behavior (Vredenburgh, Vredenburgh & Kalsher, 2006). Similarly, sports such as football, and ride-on athletics like cycling, skateboarding and snowboarding often result in injury.

Driving also poses serious threats to adolescent safety. Adolescents of driving age are more likely to be involved in motor-vehicle incidents due to minimal experience and high potential for distraction. To help address the peer influence issue, the law has changed to require a one-year period after receiving their license before new drivers can transport other teens.

In order to assess self-reported exposure and occurrence of injuries, we have developed a survey to address actual risk-taking behavior in the common adolescent injury categories described above as perceived by the adolescents themselves. This methodology differs from previous studies in the field as we have surveyed the adolescents themselves, not their parent or guardian, in determining the level of risk taking behaviors in high school students. Most previous research regarding adolescent injuries is based on parent report or analysis of hospital data banks.

METHOD

A survey was developed to determine typical adolescent risk-taking and self-protective behaviors regarding common accident scenarios involving teenagers. Injury data were also collected regarding these activities. Survey items include questions about sports, driving, use of consumer products, and school-related injuries.

The survey was pilot-tested and revised to ensure understanding of the assessment items. Surveys were administered to students at two high schools by three high school student interns (during classes with consent from teachers). Participants were told that their participation was anonymous and voluntary.

Participants were 144 adolescents ranging in age from 13 to 18 with a mean of 14.7 years old and a SD=1.04. There were 69 males and 70 females (5 did not report their sex) in the study. Their GPA ranged from 1.0-4.8 with a mean GPA of 3.23 and a SD=0.88.

RESULTS

Overall, 48% of the participants reported experiencing injuries resulting from an accident that required medical treatment.

RECREATIONAL ACTIVITIES

Many of the participants have participated in recreational activities that resulted in injury-producing accidents. Table 1 indicates the overall participation in recreational activities and injury rates to participants.

TABLE 1. Sporting Injury Rates

Activity	% participate	% injured
Bike, scooter, skateboard	73.2	60.4
Snowboard/ski	31.1	11.4
All sports	98.1	47.3
ATV	27.8	9.2

Of the almost 75% of the participants who ride bikes, scooters or skateboards, 60% have experienced injuries serious enough to require medical treatment.

Participants were asked to what extent they protect themselves by wearing personal protective equipment (PPE) when riding bikes, skateboards and scooters.

Table 2 depicts the frequencies and percent of participants who use protective gear. 95 (71.97%) of the participants of these activities report that they wear PPE *half of the time or less*. These findings are consistent with the Vredenburgh, et al (2006) study findings regarding low eye protection usage while playing with "toy" guns.

TABLE 2. Use of Protective Equipment

Frequency of use	Frequency	Percent
Almost Never	65	49%
50% of time	30	23%
Most of the time	27	20%
100% of time	10	8%

Of the people who reported wearing protective equipment, almost all reported wearing a helmet. Two people reported full dirt bike gear and two reported knee and elbow protection.

Participants were asked if they have access to a gun. 19% of the participants reported that they have access to a gun, and of those, 25% reported that they have used the guns without supervision.

Many accidents involving teenagers involve water sports. Therefore, participants were asked if they have ever been injured while diving head first into water without knowing how deep it was; 48% of the respondents reported that they had. They were also asked if they ever ignored posted warnings indicating that it is not safe to swim or dive; 48% reported that they ignored warnings.

DRIVING SAFETY

Of the 25 (18.4%) participants who drive, 65% drive with a learner's permit and 35% with a license. New drivers and those with a permit are not allowed to drive non-family under 18 years of age. When asked if they have driven other kids when their license did not allow it, 14% of the drivers reported that they did. 13% of the drivers reported that they had been in an accident while driving.

FALLS FROM HEIGHT

Participants were asked if they were injured from accidentally falling from a height or while jumping down from a height. 46 (34%) reported injuries after accidentally falling and 31 (23%) were injured after intentionally jumping.

INJURIES FROM FIRE

Participants were asked if they ever played with fire and had it get out of their control. 40 (29.4%) respondents lost control of fire and 20 (15%) were burned enough to seek medical care. 89 (65%) have been with other adolescents who were playing with fire.

CONSUMER PRODUCTS

Participants were asked if they ever needed medical treatment after being injured while using a consumer product; 45 (39%) reported an injury. Of those injured, most have been injured while using sporting equipment. Table 3 depicts the injury categories.

TABLE 3. Product-related injuries

Product	Frequency	Percent
Tools/Knives	25	48.1
Sporting equipment	20	38.5
Chemicals	4	7.7
Toys	1	1.9
Other	2	3.8
Total	52[*]	100

*The injuries total more than 45 because some participants had multiple injuries.

Participants were then asked how often, if ever, they would read instructions or warnings on products. They were also asked under what types of conditions they would read warnings. Table 4a provides the frequencies that they read instructions and warnings. For the participants who chose to read warnings, Table 4b provides situations when participants would read warnings.

Of the 121 participants who reported that they read warnings at least occasionally, the most common type of product that they read warnings for was for medication (24; 20%). Eight (6.6%) respondents said they read for chemicals, 4

(3.3%) for toys, 3 (2.5%) for flammable items and 3 (2.5%) for guns. Other responses were tampons, trampoline, tools, ATV and food.

TABLE 4a. Frequency That Participants Read Instructions And Warnings

	Always	Sometimes	Rarely	Never
Instructions				
Freq.	29	72	24	14
%	20.9	51.8	17.3	10.1
Warnings				
Freq.	25	64	32	17
%	18.1	46.4	23.2	12.3

TABLE 4b. Situations When Warnings Were Read

Situation	Frequency	Percent
New product	60	44.8
Looks dangerous	63	47.0
Know someone injured	11	8.2
Total	134*	100

*Some participants read warnings for multiple reasons

SCHOOL SAFETY

Participants were asked if they had been injured in school. 65 (47%) reported that they were injured enough to require medical treatment. While most of these injuries occurred in PE, sports or lunch, there were two injuries in science class and one that occurred in English class. 9 (6.8%) were caused by fighting. 28 (20%) reported that they injured other students while fighting. Throwing of objects at students is very common; 113 (82%) of the respondents reported seeing other students throwing objects. 76 (56%) admitted to throwing objects themselves and 31 (23%) said that they had been injured by an object thrown at them.

Of the 158 objects reported to have been thrown by the participant or observed being thrown, the most common objects thrown were pencils and pens (47; 30%), food (26; 16%), paper (24; 15%), balls (18; 11%), rocks (18; 11%), erasers (8; 5%), books (4; 3%), chairs (4; 3%) and paper clips (3; 2%) and scissors (2; 1%). Also thrown were a trashcan, knife, bullet, and a shoe.

ADOLESCENTS' SEX, SELF-PROTECTIVE BEHAVIOR AND INJURY RATES

Overall, there was no significant difference in accident rates between male and female participants. T-tests were performed to compare males and females on behavior and injuries resulting from the various activity types. Females were found to be more likely to ski and snowboard than males $(t(133)=3.12, p<.01$; they were also injured significantly more than males while skiing and snowboarding $(t(77) = 2.16, p<.05)$. Males were significantly more likely to have been burned while playing with fire $(t(129)=2.51, p<.05)$. Males were also injured more in fights $(t(135) = 2.18, p<.05)$ and threw objects at other students more often than females $(t(134)=2.45, p<.05)$.

Another significant finding was that adolescent males reported that they read instructions $(t(137)=2.49, p<.05)$ and warnings $t(136)=1.99, p<.05)$ significantly more frequently than females did.

INFLUENCE OF ACADEMIC PERFORMANCE AND AGE

Overall, the academic performance of the participants was not related to their behavior or injuries. A one-way ANOVA was used to determine whether behavior or injuries resulting from the various activity types differed as a function of academic performance. The only significant finding relating to academic performance was the likelihood of injuries at school, which differed as a function of grades $(F(3,127)=4.66, p<.01)$. Specifically, post-hoc analyses using the Bonferonni procedure revealed that the A-students were significantly *more* likely to be injured than the students with a GPA below a 2.0 (D or lower). No other comparisons were significant (all p's >.05). The "A" students, however, were no more or less likely to read warnings or instructions than the other students (all p's >.05).

There was no significant difference among the different ages on any injury or behavioral measure (all p's >.05).

CONCLUSIONS

The results of this study are consistent with previous research regarding adolescent behavior: adolescents engage in risky activities and are injured as a result of them. Namely, of the 75% of participants who use bikes, scooters and ATVs, the majority of them (60%) have been injured while riding. Almost half of the students indicated they had been injured while participating in other sports.

The reported use of PPE is also consistent with what is known about children at this age: they feel that negative consequences are unlikely to occur to them. Only 7% of respondents indicated that they use PPE 100% of the time. Interestingly, despite the fact that the majority of respondents have been injured in one of the investigated categories, half of them almost never wear PPE. This indicates adherence to the personal fable that bad things only happen to other people.

Similarly, the reported frequency of reading warnings among the survey sample was low. Only about 20% indicated they always read warnings, with medication warnings being read most often. Again, the false belief that bad things only happen to other people may explain the low prevalence of reading warnings.

An interesting finding is that the students with the highest grades were injured at school significantly more often than the participants with the lowest grades. Further research to determine why this occurred would be important. Another interesting finding is that boys read instructions and warnings more often that the girls; research regarding adults almost always finds that women are more likely to read and comply with warnings.

These results make it clear that during a forensic investigation, adolescents involved in an injury incident, should not necessarily be held to the same standards as adults are. Other factors such as peer group influence, supervision, safety rule enforcement, activity location and equipment design may be relevant to accident causation (Cheng, Fields, Brenner, Wright, Lomax, & Scheidt, 2000).

Individual physical fitness is another factor that may influence whether and how severe an adolescent may be injured. Evidence suggests that poor physical fitness in some youths may result in more severe and frequent injuries (Emery, 2003). Prevention strategies for reducing the severity of injury incidents may involve increasing the physical fitness of adolescent populations.

These results reinforce the notion that what is reasonable behavior for an adult may not necessarily be the same for adolescents. Although many adolescents are at the cognitive developmental stage to be able to make well-informed decisions, it is the norm for this age group to be strongly influenced by outside forces. Therefore, decisions based on some of these outside forces may not be atypical for this cohort and may constitute common adolescent behavior.

An example of a case where the reasonableness of conduct of an adolescent is inconsistent with that of an adult may be a pedestrian versus auto collision case. A middle school student may rush across a street to catch a bus and is hit when violating a car's right of way. On the surface, the child's behavior seems unreasonable. However, upon further analysis, it may be determined that the child is rushing to catch a bus because he knows the later bus will hold older kids who bully and tease him. This information is very important to the analysis because it caused the unsafe behavior. An adult would likely not have a similar outside influence affecting his behavior. Therefore, in some situations, behavior that is unreasonable for an adult may not be for an adolescent.

REFERENCES

Agran, P.F., Winn, D., Anderson, C., Trent, R. & Walton-Haynes, L. (2001). Rates of pediatric and adolescent injuries by year of age. *Pediatrics*, 108(3).

Brown, T. & Beran, M. (2008). Developmental stages of children. In R. Lueder & V. Rice (Eds.). *Ergonomics for Children: Designing products and places for toddlers to teens*. Chapter 2. Boca Raton, FL: Taylor & Francis.

Cheng, T.L., Fields, C.B., Brenner, R.A., Wright, J.L., Lomax, T. & Scheidt, P.C. (2000). Sports injuries: An important cause of morbidity in urban youth. *Pediatrics, 105(3)*, E32.

Danesco, E.R., Miller, T.R., & Spicer, R.S. (2000). Incidence and costs of 1987-1994 childhood injuries: Demographic breakdowns. *Pediatrics*, 105(2).

Ellickson, P.L., Lara, M.E., Sherbourne, C.D. & Zima, B. (1997). Adolescence: Forgotten age, forgotten problems. *RAND Research Review, 21(1)*.

Emery, C.A. (2003). Risk factors for injury in child and adolescent sport: A systematic review of the literature. *Clinical Journal of Sport Medicine, 13(4)*, 256-268.

Grossman, D.C. (2000). The history of injury control and the epidemiology of child and adolescent injuries. *The Future of Children*, 10(1), 23-52.

Igra, V. & Irwin, C.E. (1996). Theories of adolescent risk-taking behavior. In R. DiClemente, W. Hansen & L. Ponton (Eds). *Handbook of Adolescent Health Risk Behavior*. New York: Springer.

Quadrel, M.J., Fischoff, B. & Davis, W. (1993). Adolescent in (vulnerability). *The American Psychologist*, 42(2), 102-116.

Steinberg, L. (2004). Risk taking in adolescence: What changes and why. *Adolescent Brain Development: Vulnerabilities & Opportunities*, 1021, 51-58.

Vredenburgh, A.G., Vredenburgh, M.J., & Kalsher, M.J. (2006). Adolescent risk perception and self-protective behavior regarding airsoft and paintball Guns. *IEA2006: 16th World Congress on Ergonomics*. Elsevier Ltd.

Zackowitz, I.B. & Vredenburgh, A.G. (2005). Preschoolers, adolescents and seniors: Age-related factors that pertain to forensic human factors analyses. In Y.I. Noy & W. Karwowski (Eds.), *Handbook of Human Factors in Litigation*. Chapter 35. Boca Raton, FL: CRC Press.

Chapter 98

Developing a Method to Evaluate the Employee Satisfaction on Safety Management

Masahiro Tomita, Yusaku Okada

Graduate School of Science & Technology
Keio University
3-14-1, Hiyoshi, Kohoku-ku, Yokohama-shi, 223-8522 Japan

ABSTRACT

In a past study, we had developed HEMAS (Human error management assessment system) that can evaluate organizational safety consciousness. HEMAS clarifies problems of safety management in organization but it cannot suggest solution of these problems. In this study, we developed a method to evaluate employee satisfaction on safety management. Developing the method to evaluate the employee satisfaction constructs a first-step of the guideline for planning the effective strategies that can promote safety management.

Keywords: Employee Satisfaction, Safety Management, Human Error

INTRODUCTION

In order to be a trusted organization in society, accident prevention activities are necessary. However, accident prevention activities are difficult tasks for many safety managers, and often they feel hard to activate their measures to improve the condition. This happens because they often fail to figure out employees' compliments and lack of understandings about the current accident prevention activities. In other words, the problem is gaps of a sense of values for the safety activities between managers and employees. As a result, accident prevention activities often treated from employees as a meaningless and troublesome act, and fail to gain their agreement and cooperation.

In a past study, we had developed HEMAS (Human Error Management Assessment System) that can evaluate organizational safety consciousness. We had grasped gaps of the sense of values for the safety activities between managers and employees by combining questions of the questionnaire of HEMAS into five groups. However, we cannot extract factors of the gaps. To decide directionality of accident prevention activities, safety managers discuss with human error consultant about the result of HEMAS.

Therefore, we aimed to develop a method to evaluate employee satisfaction on safety management. Developing the method to evaluate the employee satisfaction supports the safety managers to execute accident prevention activities without the consultant. Through developing the method to evaluate the employee satisfaction, we will be able to construct the guideline for planning the effective strategies that can promote accident prevention activities.

The flow of this study is shown below, and FIGURE.1 shows the position of this study.

1. HEMAS (Human Error Management Assessment System)
2. Discussion of the gaps of the sense of values for the safety activities
3. Developing a method to evaluate the employee satisfaction

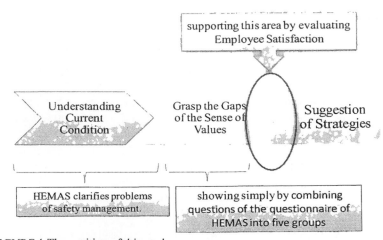

FIGURE 1 The position of this study

HEMAS

HEMAS consists of 55 question items. There are 20 questions of basic concept to ask the fundamental views against human error, 5 questions to ask an atmosphere whether reports of an error are easy to do, 10 questions to ask the inspection system including checklist and 5 questions to ask whether reports of troubles are easy to do, and so on. We marked a point to the questions so that a total of each question group becomes 100. And we made the low consciousness group score, because the low consciousness group is apt to be the obstacle of the accident prevention activities. The low consciousness group score = *(percentage of 40-60 pt group)*0.2 + (percentage of 20-40 pt group)*0.6 + (percentage of 0-20 pt group)*1.0.*

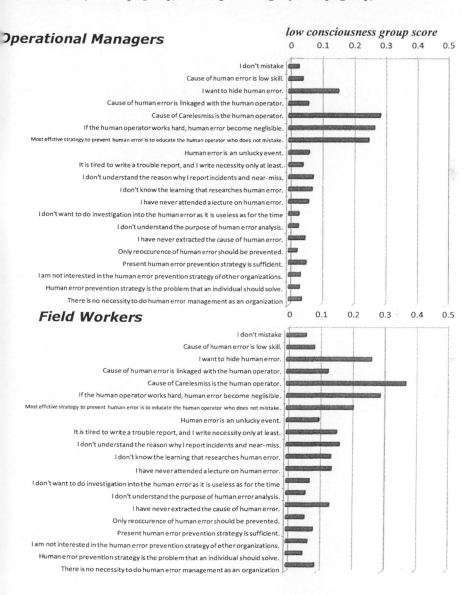

FIGURE 2 An example of a gap of sense of values by HEMAS

Many Japanese enterprises (76 enterprises, 37604 persons) have executed this method to decide a policy of the safety activities (see appendix). We can grasp the gaps by difference between the managers' result and employees' result.

DISCUSSION OF THE GAPS OF THE SENSE OF VALUES FOR THE SAFETY ACTIVITIES

We can obtain the gaps of sense of values for the safety activities by comparing bar graph of each question between managers and employees. However, only quantitative indices cannot show the detailed problems of the organization. So the safety manager should collect lots of data about human factors in the organizations. We had applied PCA (Primal Component Analysis) to scores of 55 questions of HEMAS, and had found several sets of questions which are relevant to different kinds of problems. From the result of PCA, we combined questions of the questionnaire into five groups (FIGURE.3). FIGURE.3 shows especially large gaps of a sense of values for the safety activities in "only personal character factors are chosen as causes of incidents" and "Only personal skill and safety mind are taught as safety education".

From the result of HEMAS, we can grasp the gaps of the sense of values for the safety activities between managers and employees. However, we cannot extract factors of the gaps and suggest directionality that shrinks the gaps. To decide directionality of accident prevention activities, safety managers discuss with human error consultant about the result of HEMAS.

Therefore, we aimed to develop a method to evaluate employee satisfaction on safety management. To maintain safe condition, employee satisfaction and motivation are very important. For this reason, we thought that evaluating the employee satisfaction is effective to extract factors of the gaps. We introduce an example of "confirmation by pointing finger". For the highly-motivated employees, it is an effective activity. They may find troubles. But, for lowly-motivated employees, they only point their own finger to a physical object while uttering ascertainment items. Such activity means very little.

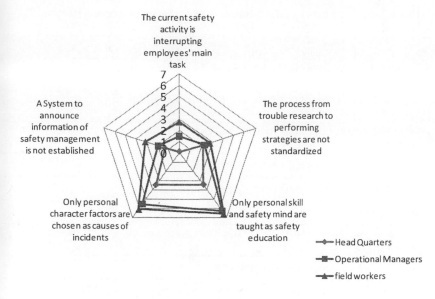

FIGURE 3 A radar chart showing gaps of sense of values simply

DEVELOPING A METHOD TO EVALUATE THE EMPLOYEE SATISFACTION

Through discussion with 12 safety managers in general hospitals, the method to evaluate the employee satisfaction was developed. And we made questions by focusing on features of staffs eager in safety activities and features of staffs feeling job satisfaction. The evaluation method of the employee satisfaction is comprised of all 23 questions consisting of four categories of "learning", "communication", "morals" and "environment" (Table.1). Managers evaluate the employee of own workplace by selecting answer form two or three choices of each question. For example, about question No.12 "In your division, does staff prepare to begin work on time?", there are three answers; "Some staff comes after the start of the business day.", "All staff is ready to begin work on time.", "All staff is ready to begin work 5 minutes before starting." For managers lacking in knowledge of the safety management and field site, evaluating their own division by this method raises the level of their knowledge.

We can grasp feature of each division by evaluating the employee satisfaction. In evaluation results of the employee satisfaction, improving question items marked low score is effective to shrink the gaps.

Table 1 question items of the method to evaluate employee satisfaction

Study	
1	In your division, does staff take doeswn a note when they attend a training workshop?
2	In your division, does staff instruct operators according to each operator's ability?
3	In your division, does staff inquire and examine information that they don't know?
4	In your division, does staff hold study sessions?
5	In your division, does staff participate in study sessions?
Communication	
6	In your division, does stuff obtain the information of the patient from a patient's record and so on?
7	In your division, does staff say hello?
8	In your division, does staff nurse a patient after saying something?
9	In your division, does staff say their thought clearly?
10	In your division, does staff say something to associates depending on their situation?
11	In your division, does staff point out wrong things?
Moral	
12	In your division, does your division prepare to begin work on time?
13	In your division, does staff know the rules decided by medical safety?
14	In your division, does staff keeps their work place clean and tidy?
15	In your division, do staff's clothes conform to office regulations?
16	In your division, does staff answer the nurse call actively?
Work	
17	In your division, does staff react to what was ordered?
18	In your division, does the staff work in cooperation with each other?
19	In your division, does staff understand cleanliness / dirtiness?
20	In your division, does staff change the work schedule for reason of personal circumstances?
21	In your division, does staff write reports?
22	In your division, does staff bring up problems of the workplace?
23	In your division, does staff act depending on the situation?

CONCLUSIONS

In this study, we discussed the gaps discussion of the gaps of the sense of values for the safety activities by the results of HEMAS, and developed a method to evaluate employee satisfaction on safety management. Based upon the method to evaluate the employee satisfaction, we wish to constructs the guideline for planning the effective strategies that can promote safety management. For example, by studying relation between the evaluation results of HEMAS and employee satisfaction, we will be able to feed back comments depending on the both evaluation results. Through these studies, the guideline for planning the effective strategies will be constructed. It supports the safety managers to execute accident prevention activities without the consultant.

REFERENCES

Okada, Y., Human Error and Performance Shaping Factors, the 3rd International Forum on Safety Engineering and Science, 2002

Okada, Y., Human Error Management of Performance Shaping Factors, International Conference on Computer-Aided Ergonomics, Human Factors and Safety, 2005

Nakayama, Y.; Okada, Y., Development of Human Error Management Support Program, the 16th Congress of the IEA, 2006

Mori, Y.; Yokomizo, A.; Okada, Y., Assessment of Human Error Management in Organizations, AHFEI the 2nd International Conference, 2008

Yokomizo, A.; Mori, Y.; Okada, Y., An Analysis of Human Error Management Data on Organizations -Tendencies and Features of Organizational Human Error Management-, AHFEI the 2nd International Conference, 2008

Yokomizo, A.; Tomita, M; Mori, Y.; Okada, Y., A Study on Improvement of Safety Management in Organizations, Nordic Ergonomic Society, 2009

Tomita, M; Yokomizo, A.; Mori, Y.; Okada, Y., A study of method to enhance safety management activities based upon consciousness data about human error prevention strategies in an organization, Nordic Ergonomic Society, 2009

APPENDIX

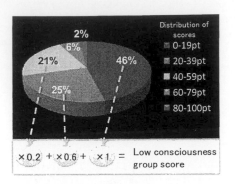

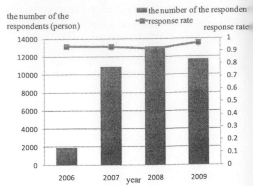

FIGURE 4 About the low consciousness
group score

FIGURE 5 The change of the number of the
answerers and the answer rate

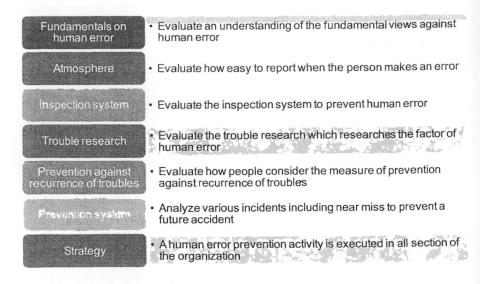

FIGURE 6 Seven categories of HEMAS

Table 2 question items of HEMAS

Fundamentals on human error	
1	I don't mistake
2	Cause of human error is low skill.
3	I want to hide human error.
4	Cause of human error is linkaged with the human operator.
5	Cause of Carelesmiss is the human operator.
6	If the human operator works hard, human error become negligible.
7	Most effective strategy to prevent human error is to educate the human operator who does not mistake.
8	Human error is an unlucky event.
9	It is tired to write a trouble report, and I write necessity only at least.
10	I don't understand the reason why I report incidents and near-miss.
11	I don't know the learning that researches human error.
12	I have never attended a lecture on human error.
13	I don't want to do investigation into the human error as it is useless as for the time
14	I don't understand the purpose of human error analysis.
15	I have never extracted the cause of human error.
16	Only reoccurrence of human error should be prevented.
17	Present human error prevention strategy is sufficient.
18	I am not interested in the human error prevention strategy of other organizations.
19	Human error prevention strategy is the problem that an individual should solve.
20	There is no necessity to do human error management as an organization
Atmosphere	
21	At his post of yours, he is not scolding the person who did human error (a reason is not asked).
22	Punishment (a reason is not asked) is not given to having caused the error at its post of yours.
23	At its post of yours, there is no trend to scorn the person who caused the error.
24	There is atmosphere which is easy to report an error.
25	Neither an error nor cause pursuit of a trouble is replaced with clarification of responsibility.

Inspection system	
26	It is not the check which merely attaches a RE point.
27	The importance and difference of that are attached to the check item.
28	The check list has managed periodically whether it has the suitable amount of work.
29	It has improved periodically about the contents of inspection, and an item.
30	The inspection person has full knowledge of the present spot about the contents of work, a worker's characteristic, etc.
31	Communication with the inspection division spot is achieved.
32	The inspection mistake and the error in the spot are divided.
33	The matter discovered by inspection is fed back to the spot in detail.
34	An inspector's training period is fully taken.
35	There is almost no difference of an inspector's skill.
Trouble research	
36	Analysis of a trouble example has not made only the report of a person concerned with a trouble a basis.
37	The administrator has visited the trouble generating spot himself.
38	There is nothing only at the trouble generating time and the flow of all work until it results in a trouble is investigated.
39	It is analyzed to the factor (PSF) of human error.
40	It can be investigating to not only a direct factor but an indirect factor, and a potential factor.
Prevention of recurrence	
41	The measure is not reflection of a person concerned with a trouble.
42	The measure is not only consciousness reform of its post including the error persons concerned.
43	Measure planning according to PSF is made.
44	Whenever a trouble occurs, a measure is not formed, but the measure based on the total of an analysis result is worked on.
45	It is referring to the others' measure.
Prevention system	
46	There is post which performs mainly trouble measures inter the office.
47	The incident case and the near miss case are collected.
48	A class of human error is performed regularly in the office.
49	The human error education in which even not only an administrator but the on-site worker spread is performed.

50	Trouble cases, incident cases, near miss cases and the analysis result of those are developed in the office.
Strategy	
51	All administrators have the right recognition about human error prevention activities.
52	The information on human error activity of other types of industry is positively collected as well as the other company in the same trade.
53	There is relation with external man and company which can hear an opinion about human error management.
54	The risk about human error is evaluated exactly.
55	The view of human error management has been common sense within a station.

Chapter 99

A Proposal of a Training Program of Risk Managers Applying Incident Reports in Medical Organizations

Tatsuya Fukuda, Y. Okada

Fac. of Science & Technology, Dept. of Administration Engineering
Keio University
3-14-1, Hiyoshi, Kohoku, Yokohama 223-8522, Japan

INTRODUCTION

In the hospitals, it is important to execute the sustainable safety activity. One of major safety activity is to analyze the incident cases. The incident is a slight case with the possibility of causing an accident though not influenced patient's life and body. Here, it is so important to analyze which information about the incident case.

In the safety activity, there are a lot of taking measures for the recurrence prevention based on detailed hearing and the analysis for the accident that actually happens. The approach on the recurrence prevention of preventing that a similar accident happens again by working on such a past accident is important. However, when the factor of the accident is seen, only the clear factor such as "Mistake" and "Carelessness" tend to be enumerated. Such an activity is insufficient as the factor that leads to the prevention of the accident. It is important to extract not only the clear factor but also a lot of factors that lurk in the back side of the clear factor from a past accident. Decreasing these latent factors leads to the prevention of a severe accident. In a word, it means that taking measures based on analyzing incidents will consequentially prevent a severe accident. This idea is based on the Heinrich' hierarchy "There are accident of 29 medium degrees and 300 slight cases in the back side of one severe accident ". When an accident occurs, the possibility that a lot of incidents have been lurking till then is high.

Therefore, it is important to take measures based on the analysis of not only an accident but also an incident case. However, such a safety activity isn't actually functioning in a lot of hospitals. Therefore, in this research, we investigated an actual situation of the safety activity by using an incident case for ten hospitals. And, we made the support tool which encourages the safety activity and growth of the risk manager, which is a person who is related to the business of a safe medical treatment, in the hospital.

SURVEY

First of all, we investigated how incident cases are collected and analyzed on the medical field. As a result of investigation, we obtained that incident cases are collected by the report form in most medical institutions. When the incident case occurs, the person concerned examines about the content, the factor and the measure of the incident etc. at an individual level, and writes the incident report

Thus, incident cases are collected. However, the essential item of the generation content and the generation factor is a free description form. Therefore, because the hospital's staff doesn't understand what they should write, the content such as "It was my attention shortage." and "It was my confirmation shortage." is often written in the point of the generation factor. Moreover, similarly as for the improvement idea, the content like the spirit theory such as "Working carefully." and "Confirming it without fail." were often written in the point of the measure . And because the staff has a negative image easily in the report of the failure, the report like the reflection sentence is often submitted.

Table 1. Frequencies of factors in incident reports

Insufficient verification	40
Not notice or forget	17
Mistake	14
Patient (e.g. The patient who is easy to fall down)	14
Inappropriate of communication	8

Table 1 is showing of the ratio of the described factor based on 93 incident reports actually submitted. Most these ratios are the self's factors such as insufficient confirmation, not notice and mistake. These self's factor is related only to measures like the spirit theories, and not connected with enough measures. Therefore, it is preferable to extract factors other than the self's factor and the patient's factor.

Moreover, the staff of the safety management of each section arranges the written report. Some of these are talked about at the conference in the section. Measures are often taken based on it. However, measures are taken to such an individual case but measures to consider all cases cannot usually be taken based on analyzing all cases and managing the tendency to the factor, because there is a lot of incident number of cases and a lot of reports like the reflection sentence. Figure 1 is a questionnaire whether the collection and the analysis of an incident case can be performed at each section. The object person is a 46 staffs of the medical institution such as the risk managers that actually works on safety management.

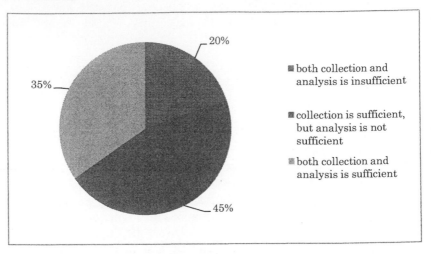

Figure 1.   Survey about collection of incidents in hospital

As above, the factors extracted from the report are almost self's factors and the patient's factors. Moreover, it is difficult to analyze all incident cases because there is a lot of incident number of cases and a lot of reports like the reflection sentence. Therefore, it is a current state that the efficient strategy is not performed enough on the medical field.

PROPOSAL

To improve this state, we proposed the supporting tool to which the safety activity with the incident was able to be performed effectively. It is important to make the staff notice the existence of the factor that lurks on the back side of the factor or the factor except the self-factor. Moreover, it is important to extract not a detailed analysis that covers all factors but the factor frequently shown or the factor that should be extracted at least. And, it is important that arranging information of the case, and analyzing from the viewpoint of statistics and the tendency and so on. Then, we made the PSF list that put a typical factor group that should be extracted at least. We chose 17 PSFs, which are frequently seen in the work of hospital based on past case, from 98 PSFs. We classified these PSF into four categories (Gestalt, Affordance, Preview, and Workload). We referred to GAP-W type PSF when these factor groups was classified (Yukimachi and Nagata, 1999-2005). This list is used as a support tool when the factor is analyzed in addition to an incident report.

Table 2: The list item

Gestalt	
Lack of skill	A skill about the work is short.
Uncertain work	The work is depending on personal judgement.
	There is not a manual, manual is vague.
Lack of knowledge	Knowledge about the work is short.
	A basic medical knowledge is short.
Affordance	
Difficultly of verification	The work is hard to verifyy impossible to verify.
	A verification item is uncertain.
Difficultly of discrimination	Indication of the apparatus is bad.
	There are plural similar objects.
Imperfection of apparatus	it is hard to use an apparatus.
	How to use operation appliance is incomprehensible.
Lack of criterion for judgement	There is not a criterion is lacking in concreteness.
Preview	
Difficultly of prediction	It is hard to get information to predict it.
	It is hard to predict latent danger.
Inappropriate communication	Instructions, communication and guidance from the boss is not appropriate.
	Cooperation with the other post is not good.
Lack of preparation	Begin work with incomplete preparation.
	Daily lack of preparation.
Unplanned work	The work has interruption or changes.
Workload	
Excess of information and direction	A lot of information is given from plural people or places at the same time.
Complicated work	There is much incidental work such as preparations, a record, and the transaction.
	The work needs processing at the same time.
Psychological burden	Independent work.
	Fear of the failure is severe.
Bad work environment	There is not enough area about the work.
	The duty of the long time.
	Illumination is dark.
Physical burden	The work needs hard posture or movement.
	Long time work.
Distraction of attention	Cannot distribute mind to surroundings in a focalization.

EXAMINATION

The hospital staff analyzed the factor with the PSF list against the incident report actually submitted in a hospital. The object person is the administrator, who is like the highest medical safety manager, and the manager who is taking charge of the safety management in each section. As a procedure, managers extracted the factor for an incident case which was submitted at each section. Next, administrators extracted the same one again. The object data used 93 incident reports actually collected by ten sections of the hospital. Table 3 showed those sections of detailed incident reports. We verified the data collected by this method by using the statistical hypothesis. We call the method that extracts the factors of incidents from reports of free description as 'Free Description Based', and the method that extracts the factors of incidents by the PSF list as 'Check List Based'.

Table 3: Sections of detailed incident reports

section	supplement	the number of collection
A3	acute period internal medicine(respiratory, emergent medicine, artificial dialysis section)	10
A4	neurosurgery, urology	5
A5	orthopedics, ophthalmology, otolaryngologist	11
B3	chronic phase internal medicine, the section where 70% or more patient who has physical disability certificate	10
B4	chronic phase internal medicine, the section where 70% or more patient who has physical disability certificate	8
B5	health facilities for recuperation	12
C3	cardiovascular section	12
C4	surgery, gastroenterological medicine	3
C5	orthopedics	6
ICU	intensive-care unit	11
OPE	Operating room	5

INVESTIGATION RESULT

We investigated the difference between 'Free Description Based' and 'Check List Based'. For 93 incidents same medical risk managers in a hospital tries to extract the incident factors by above two methods. As the result we obtained that 'Check List Based' method can support to extract many latent factors, shown as Fig.2 (see Appendix).

Table 4: Ratio of extracted factor

	The ratio of *Free Description Based* to entire extracted factor	The ratio of *Check List Based* to entire extracted factor
Insufficient verification	29.63%	0.00%
Not notice or forget	12.59%	0.00%
Mistake	10.37%	0.00%
Patient	10.37%	0.00%
Lack of skill	0.00%	6.47%
Uncertain work	2.22%	11.33%
Lack of knowledge	3.70%	7.44%
Difficultly of verification	0.00%	7.12%
Difficultly of discrimination	0.74%	1.62%
Imperfection of apparatus	4.44%	0.65%
Lack of criterion for judgement	0.00%	2.59%
Difficultly of prediction	4.44%	13.59%
Inappropriate communication	5.93%	7.12%
Lack of preparation	0.00%	6.15%
Unplanned work	0.00%	3.88%
Excess of information and direction	2.22%	3.56%
Complicated work	1.48%	2.27%
Psychological burden	2.22%	6.80%
Bad work environment	4.44%	5.50%
Physical burden	2.22%	1.94%
Distraction of attention	2.96%	11.97%

In Fig.2, we observed that a lot of factors such as the self's factor and the patient's factor were described in the report of the free description. The ratio of self's factors and patient's factors in the examination of '*Free Description Based*' is about 63%. In '*Check List Based*', such factors were not appeared, and conversely background factors of self factors can be extracted. In addition, the average of extracted factors in '*Free Description Based*' is 1.45, but the average number in '*Check List Based*' is 3.32. Thus, we confirmed that '*Check List Based*' can extract more factors than '*Free Description Based*'.

Because writing in the report is a person concerned of an incident case, the person writing the report of a free description and the person using the PSF list are different people. Therefore, we can't compare simply but it is sufficiently appreciable that factors were extracted for all cases. It is the reason that we thought that it is impossible to analyze all incident cases in a free description. Moreover, it will be able to be said that the PSF list makes the staff notice the existence of various factors.

Next, the factor of the report was able to be arranged to the settlement by the factor extraction of this method. Then, the feature that the section and the manager individual had was examined by using the technique of testing of statistical hypothesis for this data. We could proposer useful knowledge for safety management.

At first, we focused a feature of each section. We statistically examined a lot of factors, which is a factor that seems that a lot of these factors lurk in the section, extracted in the section. We can take measures that are appropriate for feature of each section by feeding back the tendency to these factors. We verified with the test of the difference of the ratio. The ratio is a ratio of the factor extracted in the entire hospital and a ratio of the factor extracted in each section in the data extracted by medical safety manager. We assumed the hypothesis that there was no difference in two ratios extracted in the entire hospital and each section for a certain item, and verified it. The reason intended for administrator's data is that the result is not consistent because the person extracting in each section is different for manager's data

Table 5 (see Appendix) shows the value which is statically valuable. The vacant cell means that there is no difference in the ratio of factors extracted in the entire hospital and each section. In addition, we set parameters as following

x_i means PSF. In Table 5, x_1 is "Lack of skill".

y_j is Section. In Table 5, y_1 is "A3".

Q_{ij} is One side P value which is statically valuable on x_i in y_j.

And it means that the smaller Q_{ij} is, the higher possibility that a lot of factors lurk in the section is. Next, we defined R_{ij} and σ_j to evaluate state of each section.

$$\begin{cases} if \quad Q_{ij} \leq 0.05 \quad R_{ij} = 1 \\ \quad Q_{ij} > 0.05 \quad R_{ij} = 0 \end{cases}$$

$$\sigma_j = \sum_j R_{ij}$$

The section which is $\sigma_j > 5$ has problem and were observed with A4 (Neurosurgery, Urology), B5 (Recuperation type ward), ICU (Intensive-Care Unit), OPE (operating room). This problem is that a lot of factors exist. We could obtain the tendency of the factor that lurks in each section based on these.

Another point is individual's characteristic of an analyst. Here, we focused on two kinds of role; one is medical risk managers in each section (RM), the other is the supervisor on medical risk management in a hospital (SRM). RM works in a section, and so RM has many information about actual workplace of the section, atmosphere in the section, personality of workers in the section, and environment in the section. In addition, he/she can observe the latest condition of incident occurrence. On the other hand, SRM has been studied the risk management, and has lots of knowledge about prevention strategies. The experiences and work condition makes effects of factor extraction. Therefore, we investigated the difference between RM's extraction results and SRN's extraction results.

As a result, in 309 extraction factors, RM extracted 160 factors. In addition, we verified the difference between RM and SRM by the McNemar's test. From the result of the test, we obtained the Table.6.

When the result of the factor that the manager and the administrator extracted is totaled, the number of extracted factors is 309. The manager extracted 160 factors and the ratio of these factors is 52%. In a word, these factors are that the administrator didn't extract these factors or the administrator judged that the influence of other factors is larger.

Then, we verified with the McNemar's test, which is the test of the difference of the ratio when there is correspondence, and compared the result of the extract of the manager with the result of the extract of the manager. We picked up the factor that either the extracting person didn't especially noticed easily based on the result. The ratio is the ratio only the manager extracted and the ratio only the manager extracted to each item. We verified the hypothesis that there is no difference in the ratio of extracts of both sides by comparing these two ratios.

As a result, there were five factors that the hypothesis was refuted by significance level 0.01 in 17 factors That is, these five factors are factors considered that there is a difference in the ratio of extracts of both sides. Two factors which are "Distraction of attention" and "Difficultly of prediction" are factors extracted a lot of factors by the manager, that is, factors that the manager doesn't notice easily. Three factors which are "Complicated work", "Lack of preparation" and "imperfection of apparatus" are factors extracted a lot of factors by the manager, that is, factors that the manager notice easily but the manager doesn't easily.

Table 6:   Factors extracted a lot by one analyst

		Manager	
		Checked	Not checked
Administrator	Checked		Distraction of attention Difficulty of prediction
	Not checked	Complicated work Lack of preparation Imperfection of apparatus	

Generally, there is neither the correct answer nor the incorrect answer in the factor analysis. It is difficult to conclude that a certain factor doesn't exist at certain section. Therefore, according to the factor that the manager extracted and the manager didn't extract, we don't consider that the manager is wrong. It might be preferable to understand the difference in the aspect and the idea by feeding back various aspects like this result.

It came to be able to do the factor analysis simply by using the PSF list, and various factors came to be extracted. Moreover, because the report of each factor was able to be arranged, we were able to know the tendency to the aspect that the person who had extracted doesn't notice easily the factor and the factor a lot of lurking in the section by testing of statistical hypothesis. We were able to extract profitable knowledge the safety management was advanced by these.

However, the people who had actually extracted the factor enumerated some problems as following,

- The explanation of the factor of the list is understood when staffs extracted the factor and the meaning of the factor is not understood easily.
- The staff hesitates to judge whether the factor is extracted.

It is still understood that the factor extract is not easy from these opinions. The difficulty of the factor analysis is not canceled completely still. The following problems are caused from these opinions. As following,

- If the meaning of the factor is not understood easily, the extract on the factor of an incident case and the factor of the PSF list is not related well.
- The person who extracts and the person who doesn't extract show up to the same factor.
- The staff cannot get hold of a common image to the factor.

We are working to improve this difficulty now. Actually, six staff in five hospitals extracted the factor by using the PSF list for 15 incident cases. And, the staff examined about the reason that extracted or didn't. We want the staff to be able to extract the factor more easily by improving the meaning and the vagueness of the factor based on this reason. And, because it comes to be able to extract the factor easily, the following can be expected. As following,

- The time of the factor extract is shortened.
- The sense of resistance of the factor extract with the PSF list is lost.
- The interest of an incident case increases.
- The number of reports increases.

However, there is a problem with insufficient information in an incident report in the other. Therefore, the following opinions were asked. As following,

- The extract of the factor becomes vague because staffs don't understand the situation of the occurrence.
- There are items of the factor can't be extract because staffs don't understand the situation of the occurrence.

As previously noted, the essential item of content and factor of generation are free descriptions. Therefore, it is necessary to review information and the form, etc described in an incident report. Additionally, it is important that we set what kind of information should be written, which range should be written and how the report should be written and educate them to the registrant.

CONCLUSION

We obtained that the safety activity that staffs in the hospital refer to an incident case doesn't function enough. In this study, to improve those problems, we proposed the PSF list as the supporting tool. An incident collection and the analysis brought various values to the medical safety in a wide point. However, various points that should be improved remain in an incident report and the PSF list. As future tasks, we want to solve the problem caused when staffs extracted factors with the PSF list. And, we aim at the proposal of making of a new PSF list.

Moreover, there is the problem that effective measures cannot be derived from the result of the analysis though the collection and the analysis are performed. Accordingly, in the future, we shall overlap the PSF list and the measures list which can derive measures from the analyzed factor, and to aim at the proposal of the design of an incident report of the site support type.

REFERENCES

[1] Swain, A. D., & Guttman, H. E. (1983). Handbook of human reliability analysis with emphasis on nuclear power plant applications.. NUREG/CR-1278 (Washington D.C.).

[2] Yukimachi, T. , Nagata, M. : Reference List on Basis of GAP-W Concept and a Case Study, Human Factors in Japan, Vol.9, No.1, 2004.7, pp46-62

[3] Yukimachi, T. , Nagata, M. : Study of Performance Shaping Factors in Industrial Plant Operation and GAP-W Concept, Human Factors in Japan, Vol.9, No.1 2004.7, pp7-14

[4] Sagawa, N. , Fujitsuka, R. , Furukawa, A. ,Okada, Y. :A Study on Usability of a Reference List in Nursing Duties, Proceedings of the 38th Annual Meeting of Kanto-Branch, Japan Ergonomics Society, 2008, pp47-48

[5] Akihisa FURUKAWA, Yusaku OKADA: A Proposal of Collection and Analysis System of Near Miss Incident in Nursing Duties. HCI (9) pp497-502 2009

[6] Akihisa FURUKAWA, Yusaku OKADA, Nami SAGAWA, and Ryuko FUJITSUKA: A Proposal of a Training Program of Risk Managers Applying Incident Reports in Medical Organizations. Proceedings of Asia Pacific Symposium on Safety 2009 pp352-355

Table 5: One side P value

PSF \ Section	A3	A4	A5	B3	B4	B5	C4	C5	ICU	OPE
Lack of skill	0.110			0.291	0.429	0.415			0.354	
Uncertain work		0.487		0.291	0.429	0.415	0.399		0.354	
Lack of knowledge	0.050	0.438		0.198				0.164	0.471	0.103
Difficulty of verification			0.106				0.407	0.372		0.058
Difficulty of discrimination		0.377	0.385		0.225	0.434	0.192	0.451		0.377
Imperfection of apparatus	0.063			0.455		0.298			0.246	
Lack of criterion for judgement		0.125							0.123	0.019
Difficulty of prediction				0.367		0.316	0.127		0.031	
Inappropriate communication		0.183				0.109		0.306		0.491
Lack of preparation		0.110	0.433		0.105	0.066	0.307			0.021
Unplanned work		0.021			0.407	0.065				0.467
Excess of information and direction		0.013	0.084	0.455	0.173	0.298				0.112
Complicated work				0.149						0.008
Psychological burden	0.057			0.224			0.426	0.398		0.007
Bad work environment		0.085	0.178				0.215			
Physical burden	0.000									0.081
Distraction of attention					0.127	0.289				0.001

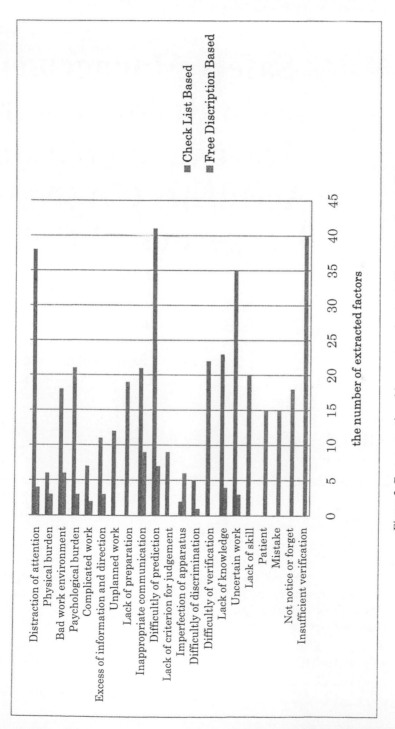

Figure 2. Factors mentioned in reports and extracted by using reference list.

Chapter 100

Analysis of the Process Safety Management Systems in Relation to the Prevention of Major Accidents Caused by Dangerous Chemical Substances

Pavel Forint

Ministry of the Environment
100 10 Prague
Czech Republic

ABSTRACT

The respective European Union Council Directive 96/82/EC, so-called Seveso II Directive, laid down the obligation to set up the safety management system in establishments or installations where the dangerous chemical substances are present in the specific amount. Major accident prevention policy and safety management system is the document prepared by the operator and the purpose of that is to decrease risk by the systemic approach to the process safety management. The systemic approach is highly effective preventive tool for the decreasing of the undesired event probability and minimization of the undesired impacts that could arise. This analysis focuses on the process safety management systems and their possible use in the prevention of the major accidents in the integrated way.

Keywords: management, systems, prevention, chemical substances, accidents

INTRODUCTION

For the purpose of implementing the operator's major-accident prevention policy and safety management system account shall be taken of the following elements:

(a)　　the major accident prevention policy should be established in writing and should include the operator's overall aims and principles of action with respect to the control of major-accident hazards;

(b)　　the safety management system should include the part of the general management system which includes the organizational structure, responsibilities, practices, procedures, processes and resources for determining and implementing the major-accident prevention policy;

(c)　　the following issues shall be addressed by the safety management system:

- organization and personnel — the roles and responsibilities of personnel involved in the management of major hazards at all levels in the organization; the identification of training needs of such personnel and the provision of the training so identified; the involvement of employees and, where appropriate, subcontractors;
- identification and evaluation of major hazards — adoption and implementation of procedures for systematically identifying major hazards arising from normal and abnormal operation and the assessment of their likelihood and severity;
- operational control — adoption and implementation of procedures and instructions for safe operation, including maintenance, of plant, processes, equipment and temporary stoppages;
- management of change — adoption and implementation of procedures for planning modifications to, or the design of new installations, processes or storage facilities;
- planning for emergencies — adoption and implementation of procedures to identify foreseeable emergencies by systematic analysis and to prepare, test and review emergency plans to respond to such emergencies;
- monitoring performance — adoption and implementation of procedures for the ongoing assessment of compliance with the objectives set by the operator's major-accident prevention policy and safety management system, and the mechanisms for investigation and taking corrective action in case of non-compliance; the procedures should cover the operator's system for reporting major accidents of near misses, particularly those involving failure of protective measures, and their investigation and follow-up on the basis of lessons learnt;
- audit and review — adoption and implementation of procedures for periodic systematic assessment of the major-accident prevention policy and the effectiveness and suitability of the safety management system; the documented review of performance of the policy and safety management system and its updating by senior management.

DEVELOPING OF SAFETY PROGRAM

Under Czech legislation the safety program is a document in which the operator demonstrates the implementation of a functional accident prevention management system (safety management system) in the required structural and material completeness or submits a specific plan for the implementation of the system in real time periods with a reasonable level of risk, together with introducing functional obligation for the fulfillment of the plan in the set deadlines. The requirements laid down in the document should be proportionate to the major-accident hazards presented by the establishment.

An operator of an establishment or installation is obliged

- to prepare a safety program based on the results of risk assessment and includes within principles for prevention of major accidents, the structure and system of safety management to protect lives and health of human beings, domestic animals, the environment, and property;
- to submit a draft safety program or its updating for approval to the Regional Authority; on the basis of a Decision of the Regional Authority, an operator shall be obliged to include in the program preventive safety measures related to potential domino effects;
- to proceed according to the safety program for the protection of the lives and health of human beings, domestic animals, the environment and property, to inform the employees and other natural persons present in the establishment or at the installation of the risks of a major accident, of preventative safety measures and of desirable behavior in case of the occurrence of a major accident.

THE INTEGRATION OF THE PARTICULAR MANAGEMENT SYSTEMS IN SINGLE ESTABLISHMENTS

An integrated management system is a management system that integrates all of an organization's systems and processes in to one complete framework, enabling an organization to work as a single unit with unified objectives.

With an integrated system, organization becomes a unified whole, with each function aligned behind a single goal: improving the performance of the entire organization. Instead of "silos", you have a genuinely coordinated system: one that's greater than the sum of its parts, and can achieve more than ever before. An integrated system provides a clear, holistic picture of all aspects of organization, how they affect each other, and their associated risks. There is less duplication, and it becomes easier to adopt new systems in future. An integrated management system allows a management team to create one structure that can help to effectively and efficiently deliver an organization's objectives. From managing employees' needs, to monitoring competitors' activities, from encouraging best practice to minimizing risks and maximizing resources, an integrated approach can help an organization achieve their objectives.

Integrated management is relevant to any organization, regardless of size or sector, looking to integrate two or more of their management systems into one cohesive system with a holistic set of documentation, policies, procedures and processes. Typically, organizations most receptive to this product will be those who have maturing management systems and who wish to introduce other management systems to their organization with the benefits that those bring.

A combined audit occurs when management system audits related to different audit criteria (standards) are conducted simultaneously. An integrated audit is the assessment of an integrated management system, which is a single management system that fully or partially integrates the requirements of different audit criteria.

By using a combined or integrated approach across all audit criteria, efficiencies can be realized that may not be achieved by auditing individual management systems, including:

- Lower certification costs, through a reduction of auditor-days and/or expenses;
- Fewer interruptions for your organization – you can meet all requirements in a single audit;
- Streamlined processes and more consistent objectives across multiple systems;
- Reduced documentation, including a single audit report.

Many organizations choose to combine audits to ISO 9001 and ISO 14001, or ISO 14001 and OHSAS 18001. Many other combinations are possible.

In enterprises handling dangerous substances the management of risks related to chemical accidents has a relationship with these management systems for occupational health and safety (OH&S), environment (E) as well as quality (Q). Responding to customer's requirements and societal needs, and for the purpose of minimizing risks related to OH&S, E and Q, many enterprises have established and maintained multiple management systems in parallel to meet the requirements of all these MSS. ISO Guide 72 provides the common principles and framework for management system standards, which means that most management systems will have some elements in common. Core elements that could be examined are:

- Clearly defined management responsibilities and accountabilities for Safety & Health, Environment and Quality management systems;
- Hazards, associated risks and the way in which these are managed;
- Procedures associated with the management of hazardous activities;
- Safety & Health, Environment and Quality inspections of plants and processes;
- Training activities and methods.

INTEGRATION IN INDUSTRIAL PARKS AND MULTI-OPERATOR SITES

A special case for the management systems application is the implementation of the Seveso II Directive requirements in industrial parks and multi-operator sites particularly challenges associated with implementation and tools and approaches that have been developed. Some questions arise in that context:

- Who holds operational responsibility for site risk management?
- How should the specific risks of chemical parks be managed, for example, emergency planning for the entire site, potential domino effects, etc.?

Two different models for managing functions common to all installations are on the table:

- The "major user" concept in which the dominant installation takes responsibility for overall site management, or
- The "off site operating company" in which an independent company takes responsibility for overall site management.

The hazard potential of an industrial site does not change when there are multiple operators, as long as the processes and structures are unchanged. However, the relationship of each installation to the site is changed and site management responsibilities are also changed accordingly. In the light of these facts seems to be very effective to use the common elements from the particular management systems of the individual operators and establish the common management system in the integrated way. Principles for maintaining effective co-ordination between installations of common concerns, and in particular maximize safety efforts have to be based on establishing of common rules for the site, information exchange among installations on a regular basis and sharing specialized services among installations.

The location of Seveso II establishments in the Czech Republic [Figure 1.] and the types of industry running there [Figure 2.] is partly organized in several industrial chemical parks [see Table 1.].

Table 1: Parameters of the Seveso establishments groups in the Czech Republic 2003 - 2008

Parameter	2003	2004	2005	2006	2007	2008
Number of groups of establishments	8	6	6	7	3	4
Average number of establishments per group	3	3	3	2	2	2
Number of establishments in the smallest group	2	2	2	2	2	2
Number of establishments in the biggest group	5	5	6	6	6	6

These data have also been effective in determining responsibilities for managing risks associated with domino effects in individual industrial parks. The over time experience has led to a convergence in regard to understanding the problem of domino effects as well a convergence in approaches to managing the safety.

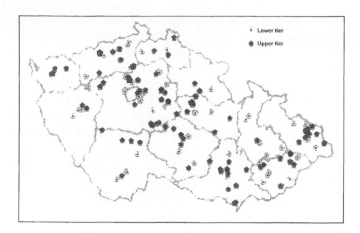

Figure 1. Seveso establishments [lower and upper tiers] location in the Czech Republic

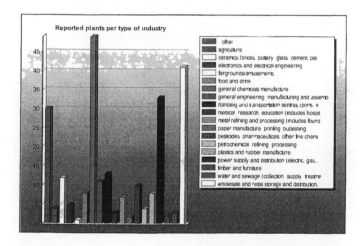

Figure 2. Reported plants in the Czech Republic per type of industry

Some recommendation based on the experience of these parks operation could be formulated in response on the requirements of the respective legislation regarding major accident prevention policy [MAPP].

MANAGEMENT SYSTEMS IN MAPP

The major accident prevention policy is the most important document in the field of major accident prevention and it is a part of the single safety policies which are the combined control of areas of major accident prevention, fire protection, protection of safety, health protection, emergency planning, security and human, property and information protection.

The objective of the single safety policies is to achieve the maximum level of protection to humans, the environment, property and information.

The most important component of the single safety policies is prevention in all areas and in particular in the area of the possible occurrence of major accidents.

The company management based on a detailed analysis of their sources of risk and in cooperation with specialized contractors establishes the major accident prevention policy.

The top management performs regular evaluation of all major accident prevention policies.

All employees are regularly trained in individual areas of the safety system and the principles thereof so they can be applied in their daily work.

The administrative authorities and also the public on request provide the major accident prevention policy.

Cohesive rules are drawn up for the formation, control and updating of the site's major accident prevention documentation. Documentation is accessible to all site operators and combined processes and knowledge are introduced into the system thanks to the use of specialized external contractors.

Regular controls performed by the state administration on the individual companies facilitate the rapid transfer of the increasing legislative requirements. This leads to a further increase in the quality of the whole major accident prevention management system because managers and technicians of the individual plants are actively involved. Example: in order to improve the quality of the major accident prevention system a HAZOP study is performed by managers, technicians and expert mechanics whereby improving their know-how and at the same time increasing the quality of the system.

An inseparable part of the company's major accident prevention system is the safety management system that is made up of the following parts:

- Major accident risk assessment
- Domino effects at the site
- Organization and personnel (training, risk familiarization)
- Control of operations
- Control, record keeping, commenting and approval of modifications
- Emergency planning
- Monitoring the fulfillment of the Program
- Audits, controls, group training
- Common emergency investigation.

CONCLUSIONS

Risk analysis and safety management is an inherent part of any management of change as well as environment and quality considerations. Basic principles of the integrated approach in the prevention of the major accidents:

- „One area – one safety" approach is used, confirmed by mutual agreement in written form.
- Internal computer network employed, nevertheless the necessity of face-to-face meetings appeared (enforced by insurance companies)..
- Agreed by local authority and inspection.
- Daily reporting of changes, possible dangers etc.
- Regular common meetings of management.
- Common communication with public.
- Non-standard situations and time tested.

Awareness of common approach to safety must be strongly supported by managements of enterprises. Participation of top management is extremely useful.

Evaluation: no punishment but „lessons learned".

REFERENCES

BS 8800:1996 Guide to occupational health and safety management systems

Council Directive 96/82/EC on the control of major-accident hazards

Directive 2003/105/EC of the European Parliament and of the Council of 16 December 2003 amending Council Directive 96/82/EC

IEC 31010 Risk management – Risk assessment techniques

ILO Occupational safety and health convention, 1981 (No. 155)

ILO-OSH:2001 Guidelines on occupational safety and health management systems

IMS-BSI Implementing and operating of IMS

ISO 14001:2004 Environmental management systems – Requirement with guidance for use

ISO 31000 Risk Management – Principles and guidelines on implementation

ISO 9000:2000 Quality management system – Fundamental and vocabulary

ISO 9001:2000 Quality management systems – Requirements

ISO Guide 72 Guidelines for the justification and development of management system standards

ISO/IEC Guide 73 Risk Management Vocabulary Guideline For Use in Standards

NTA 8620: Netherlands Technical Agreement, Requirements for a safety management system

OHSAS 18001:1999 Occupational health and safety management systems

Responsible Care Management Systems Guidance

SFK-GS-31 Aid for integration of a safety management system pursuant to Annex III of the Hazardous incident ordinance 2000 within existing management systems

Program "Safe Enterprise" as a Supporting Tool for Effective Occupational Safety and Health Management System

Sarka Vlkova

Occupational Safety Research Institute
Prague, Czech Republic
Jeruzalemska 9, 116 52 Prague 1

ABSTRACT

Health and safety at work is now one of the most important and most highly developed aspects of EU policy on employment and social affairs. Thanks to the adoption and application in recent decades of a large body of EU laws, it has been possible to improve working conditions in the EU member states and make considerable progress in reducing the incidence of work-related accidents and occupational diseases. The new strategy for 2007-2012 proposes to step up our ambition and to aim for a 25 % reduction in the total incidence rate of accidents at work by 2012 by improving health and safety protection for workers and as one major contribution to the success of the growth and employment strategy. By one of approaches in the Czech Republic how to meet the target is implementation of occupational safety and health management system with the help of free available program "Safe Enterprise", which is introduced by following text.

Keywords: occupational safety and health (OSH), management system

INTRODUCTION

The boom of civilization and the expansion of knowledge are reflected in a number of human activities, in the technologies and the technical equipment used, in the existence and use of chemicals etc. Often then unconsciously, there are adverse events with various acute and chronic consequences and impacts. To reduce these negative effects are used in addition to technological measures also various organizational arrangements and economic tools. Among these tools include, inter alia, management systems. Their essence is in general terms control and direction of the activities of organizations to achieve the expected effects on the basis of predefined set of rules and procedures.

The management techniques are today applied to practically all areas of organization's management, such as product quality, employee safety and business, information security, affecting the environment, financial decision-making, planning, allocation of resources etc. All of it is done mostly to strengthen the position of the organization market, strengthen the competitiveness of its products, increasing the prestige and profits. Whereas the purpose of these systems is the way of management of organization at all levels, which ensure the required level of product quality, level of safety or the level of impacts to the environment.

Any employer would be aware that the wealth of companies consists mainly of its employees. Therefore, they should create such conditions that ensure their safety and well-being and thus lead to their identification with the philosophy and values of the company. This is in the Czech Republic significantly supported by the national program "Safe Enterprise", which defines the requirements for OSH management system and was developed mainly to support the systematic safety management and thus to raise its level and overall level of corporate culture.

PROGRAM "SAFE ENTERPRISE"

Ministry of Labour and Social Affairs of the Czech Republic and the State Labour Inspection Office announce the Program "Safe Enterprise" as a voluntary activity for organizations and entrepreneurs. This program is a specification of requirements for OSH management system to support the gradual increase both of the company's safety level and the overall corporate culture.

THE AIM OF THE PROGRAM

The essence of the program "Safe Enterprise" is the interconnection of several sub-areas within the overall management. The program covers issues of occupational safety and health, environmental protection, safe purchasing, fire protection and the major accident prevention.

The program aims to:
- increase the level of companies' safety,
- support effective management system of OSH,
- help meet national and European legal requirements.

The program "Safe Enterprise" is based on:
- integration OSH management with other company control structure and implementation of effective management system,
- cooperation between employees and management in raising the level of safety,
- methodical support of professional supervision organs of state enterprises registered in the program.

The result should help with the practical implementation of the following principles:
- to give equal priority to safety, health protection and the environment, such as economic aspects,
- to manage the business to increase the level of workers and the public health and the level of environmental protection,
- to strengthen the awareness of responsibility to protect workers own health and their cooperation in enhancing safety,
- to include safety, health and environmental aspects into the design of all existing or new operations (activities, products, processes or workplaces),
- to inform the local and district offices, customers and the public about the risks caused by hazardous substances used and the safety measures taken,
- to provide information about how to safely dispose of the products supplied to customers,
- to cooperate with public administration authorities and local authorities in the prevention of accidents and improving the protection of human health and the environment in the region.

Requirements for OSH management system defined by this program are based on requirements specified by the handbook of the International Labour Organization standards and OHSAS 18001. Thus management systems implemented by all those documents are corresponding. Nevertheless the program "Safe Enterprise" presents slightly more complex approach. And that is especially given by the fact that it covers except the area of OSH in a defined range also the requirements for environmental protection, fire protection and quality. The result is that rules and procedures of this program enable enterprises establish a functional, transparent, efficient and cost-effective safety management system corresponding to important EU directives e.g. 96/82/EC on the major accidents prevention or 96/61/EC concerning integrated prevention and pollution control and the resulting legislation of the Czech Republic.

BASIC ELEMENTS

The OSH management system is based on five key elements as shown in Figure 1.

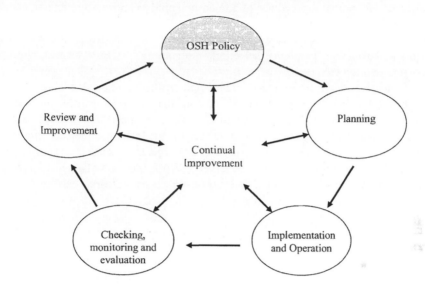

FIGURE 1. Key elements of OSH management system according to program "Safe Enterprise" (translated form Safe Enterprise, State Labour Inspection Office, 2009)

Management strategy and OSH Policy

Top management should demonstrate the leadership and commitment necessary for the OSH management system to be successful and to achieve improved OSH performance. The organization's OSH Policy has to be appropriate to the nature and scale of its identified risks and should guide the setting of specific objectives.

The policy is, as a minimum, required to include statements about the commitment of an organization to:

- the injury and illness prevention,
- implementation of this policy in cooperation with employees,
- the compliance with legal requirements both by employees and by other interested parties (e.g. suppliers),
- continual improvement in OSH management including environment protection, support of workers´ well-being and increasing the level of safety culture.

The OSH Policy should be reviewed periodically to ensure that it remains relevant and appropriate for the organization.

Organization and Planning

The management system has to enable the enterprise meet its goals leading to effective operation, rational use of human potential as well as minimizing economic costs and losses in terms of safety. This is in addition to establishing responsibilities and obligations of workers also the addressing adequate working conditions and work environment, employee selection, motivation systems, ensuring appropriate communication etc.

Objectives have to be clearly identified, realistic, understandable for all and controllable. Planning must be based on the hazard identification and risk assessment. An important requirement is to develop specific indicators, parameters and criteria necessary for the evaluation of the effectiveness of implemented measures. The procedure for achieving the objectives must be determined and planned with regard to available resources. Work must be planned in a way that creates rational decision-making and responsibilities which are clear and reasonable.

Implementation and operation

Top management should identify roles of all employees in organization with respect to the OSH management and make sure they are aware of their responsibilities and what they are accountable for.

Within the frame of implementation of OSH management system organization has to ensure the followings:

- any person performing tasks that can impact safety has to be competent on the basis of appropriate education, training or experience,
- establish specific procedures for necessary communication (internal communication among the various levels and functions of the organization, communication with contractors and other visitors at the workplace, other communication),
- employee participation,
- emergency preparedness and response,
- keeping essential documentation.

The overall objective of operational controls is to manage risks to fulfill the OSH Policy. It should be established and implemented as necessary to manage the risks to an acceptable level, for operational areas and activities, e.g. purchasing, research and development, sales, services, offices, off-site work, home based working, manufacturing, transportation and maintenance. Operational controls can use a variety of different methods, e.g. physical devices (such as barriers, access controls), procedures, work instructions, pictograms, alarms and signage. The operational controls need to be implemented, evaluated on an ongoing basis to verify their effectiveness, and integrated into the overall OSH management system.

Checking, monitoring and evaluation

An organization should have a systematic approach for measuring and monitoring its performance on a regular basis, as an integral part of its overall management system. Monitoring involves collecting information, such as measurements or observations, over time, using equipment or techniques that have been confirmed as being fit-for-purpose. Measurements can be either quantitative or qualitative.

Monitoring and measurements can serve many purposes such as:
- tracking progress on meeting policy commitments, achieving objectives and targets, and continual improvement,
- monitoring exposures to determine whether legal (and other) requirements have been met,
- providing data to evaluate the effectiveness of operational controls, or to evaluate the need to modify or implement new way of controls,
- providing data to proactively and reactively measure the organization's performance,
- providing data to evaluate the performance of the OSH management system,
- providing data for the evaluation of competence.

The organization has to conduct internal audits of the OSH management system in planned intervals to ensure that OSH management system:
- conforms to planned arrangements,
- has been properly implemented and is maintained in the appropriate way,
- is effective in meeting the organization's policy and objectives.

Review and improvement

Top management shall review the organization's OSH management system, in planned intervals and these reviews should focus on the overall performance of the OSH management system with regard to:
- continuing suitability of the management system (Is the system appropriate to the organization, dependent on its size, the nature of its risks etc.?),
- adequacy (Is the system fully addressing the organization's OSH policy and objectives?),
- effectiveness (Is it accomplishing the desired results?).

Reviews shall include assessing opportunities for improvement and the need for changes to the OSH management system, including policy and objectives.

Obligations of Participants of the Program

There are several basic obligations for program participants which they have to meet to obtain the certificate "Safe Enterprise":
- to implement OSH management system according to program requirements,
- ask for inspection by the all interested public authorities (the Czech Environment Inspection, Fire Brigade, Regional Hygiene Station, Mining office board),
- ask for audit by the State Labour Inspection Office.

After certification there are new obligations for program participants. They have to regularly at the end of year provide to regional Labour Inspection Office information about the performance of their OHS management system. There is the need to define specific safety performance indicators, which will demonstrate positive actions in safety and the continual improvement of OSH management system.

These indicators are follows:
- Technical equipment,
- Work organization,
- Personal protective equipment,
- Near-misses,
- Accidents, incidents,
- Working environment and conditions,
- Health care,
- Environment,
- Fire protection,
- Tenders,
- Statistics of accidents at work.

Another obligation is to evaluate the performance, progress in safety and effectiveness of OSH management system with the outcome of other legal entities.

CONCLUSION

Program "Safe Enterprise" has been implemented in enterprises since 1996 and until today it was modified with many changes and improvements in connection with the evolving priorities of the European Union and legislation.

The strong point of this program is methodical support and consultation of the State Labour Inspection Office which is free. This represents in the current times of economic crisis and related cost savings significant advantage compared to financially demanding services of commercial consultancy and auditing organizations.

Program "Safe Enterprise" is the set of requirements for OHS management system in connection with environment protection, major accident prevention and fire protection. Such a broad scope of requirements indicates that obtaining the certificate "Safe Enterprise" is not so easy. That is the reason why this certificate owns only 53 companies (in the Czech Republic), and indeed they are the best enterprises which ensure occupational health and safety and care for their employees. From a systemic approach to management at all levels of OSH is expected reduction of the current negative trend in employees' injuries and illnesses. Results and findings from the implementation of the program in companies fulfill these expectations.

In response to this program were developed other specific programs addressing other areas of the employment problem. These are the programs "Healthy Enterprise" and "Socially Responsible Business". These programs are by this time organized in form of competition, where the enterprise has to have specific intention according to defined rules and with this intention competes for the title "Healthy Enterprise" and "Socially Responsible Business". Interconnection of mentioned programs presents further development and possibilities for next improvement of safety culture in enterprises and maybe the better enforcement of the program abroad.

REFERENCES

State Labour Inspection Office (2009). *"Safe Enterprise": occupational health and safety management system.* Prague : State Labour Inspection Office, 2009. Available online also on www: < http://www.suip.cz/default/drvisapi.dll?LO=01000000d9c8b7a603000000030 00000a53f00008609b641000000000100ee0c0000000000000000000000000000 000000000000000000000000000000000000000000000000000000&type=appli cation/pdf>.

Chapter 102

Major Accident Prevention as a Tool to Increase the Level of Enterprise Safety Systems in the Czech Republic

Martina Prazakova, Stanislav Maly, Emanuel Dusek, Vilem Sluka

Occupational Safety Research Institute
116 52 Prague
Czech Republic

ABSTRACT

This contribution deals with the problems of safety systems, especially in the field of risk analysis and safety management systems in connection with the handling of dangerous chemical substances and set national major accident prevention system of the Czech Republic. Consistent approach of state authorities, education and other tools steadily and significantly increase overall safety of risk installations within major accident prevention system and also provide a significant positive progress in the level of enterprise safety systems. Permanent increase national authorities claims to ensure safety and to protect people and property in the vicinity of sources of risk and major accidents prevention related to dangerous chemical substances offers some specific problems.

Keywords: management systems, safety systems, prevention, chemical substances, dangerous substances, major accidents.

INTRODUCTION

Dangerous chemical substances and chemical preparations have one or more hazardous properties, from which most dangerous are explosiveness, flammability and toxicity. The danger arising from the properties of chemical substances and chemical preparations can take different forms, such as particularly an explosion, fire, toxic effects on organisms, contamination of the environment. When handling chemical substances and chemical preparations, there may be adverse events, which have different consequences and impacts to the human health and life, livestock, the environment and property in buildings and facilities and their surroundings. Everyone who handled with dangerous chemical substances, must necessary know their characteristics, and must also be able to analyze and evaluate the associated hazards.

LEGAL AND ADMINISTRATIVE FRAMEWORK

LEGAL FRAMEWORK

In the history, major accidents prevention related to dangerous chemical substances has been addressed in different countries in different ways. Major accidents have become usually have brought changes to legislation in this area.

In the EU Directive 82/501/EEC on the major accident hazards of certain industrial activities (Seveso I) was adopted in 1982 and was later replaced in 1996 by Directive 96/82/EC on the control of major accident hazards involving dangerous substances (Seveso II). Seveso II was amended by Directive 2003/105/EC.

Legal framework for the prevention of major accidents in the Czech Republic is the following: the first Major Accident Prevention Act was adopted in 1999. As a result of the amendment to the Seveso II Directive in 2003 the new Act was adopted in 2006 including related legal regulation, which provides details of the system. Legislation is supplemented by a number of detailed guidelines. Major accident prevention act designates the system for preventing major accidents for establishments and installations in which the selected dangerous chemical substance or chemical preparation is located with the aim of reducing the probability of incidence and the limitation of the consequence of major accidents to the human health and life, livestock, the environment and property in buildings and facilities and their surroundings. This Act regulates the obligations of legal persons and natural persons who own or use, or intend to use an establishment or an installation with the selected dangerous chemical substance and regulates the execution of the state administration.

SAFETY DOCUMENTATION AND PROCESS OF EVALUATION

In the Czech Republic nearly 200 establishments or installations were registered into two groups according Major Accident Prevention Act (Group A and Group B). An operator of an establishment or installations classified into Group A is obliged to prepare particularly a Safety Program as main safety document. An operator of an establishment or installations classified into Group B is obliged to prepare particularly a Safety Report. The Ministry of the Environment of the Czech Republic (MoE CR) lays down the scope and manner of preparing and updating the safety program and safety report in a legal regulation.

An operator prepares a draft safety program and safety report based on the results of the major accident risk assessment and includes within principles for prevention of major accidents, the structure and system of safety management to protect the human health and life, livestock, the environment and property.

On the basis of a Decision of the Regional Authority, an operator is obliged to include in the program and report preventative safety measures related to potential domino effects. An operator is obliged to submit a draft safety program or safety report or their up-dating for approval to the Regional Authority.

The Regional Authority sends the safety program or safety report or their updated version containing the prescribed requirements without delay for expression of a viewpoint to the Ministry, the affected bodies of the state administration and the affected municipalities also for the purpose of informing the public. The Ministry and the affected bodies of the state administration send their viewpoints on the safety documents to the Regional Authority. Within the same period of time, the affected municipalities send their viewpoints and the viewpoints of the public to the Regional Authority.

The Ministry of the Environment of the Czech Republic (MoE CR) utilizes for the evaluation of safety documentation the services of a professional workplace for the prevention of major accidents in the Occupational Safety Research Institute in Prague which has a developed an professional team which has become the center for specialists on individual problem of major accident prevention. In addition to assessing and evaluating the completeness and professional correctness of safety documents, this workplace participates in the professional preparation of employees of the state administration for the fulfillment of activities connected with the Act and also provides advisory activities in the application of the Major accident prevention act.

On the basis of the viewpoints of the Ministry, the affected bodies of the state administration, the affected municipalities and the public, the Regional Authority issue a decision within 90 days of receiving the safety program or safety report in which it approves the program or report or their updated version or requests that the operator remedy inadequacies found in the program or report. The Regional Authority sends a copy of its decision to the Ministry, affected bodies of the sate administration and municipalities for its information.

An operator of an establishment or installations classified into Group A or B is obliged to proceed according to the safety program or safety report for the protection of the human health and life, livestock, the environment and property.

An operator is obliged to inform the employees and other natural persons present in the establishment or at the installation of the risks of a major accident, of preventative safety measures and of desirable behavior in case of the occurrence of a major accident.

RISK ASSESSMENT INCLUDING HUMAN FACTOR IMPACT

OBLIGATIONS OF THE OPERATOR

An operator shall be obliged to assess the risk of a major accident in order to elaborate the safety program or safety report. The assessment of the risk of a major accident must include:

- identification of the source of the risk (hazard),
- description of the possible major accident scenarios and the conditions under which they can occur,
- estimation of the possible consequences of a major accident on the human health and life, livestock, the environment and property,
- estimation of the probability of the major accident scenarios,
- determination of the level of risk,
- assessment of the acceptability of the risk of a major accident.

The Ministry of the Environment shall lay down in a legal regulation the principles of assessment of the risk of a major accident.

GUIDELINES

Methodological guidelines are published in the Czech Republic which are intended for administrative offices but also help legal and natural persons and control bodies which perform inspections pursuant to the Act. Methodological guidelines for the analysis and assessment of risk have a recommendatory character and the operator and elaborator of the safety study must choose which methods are used. These methods must however present, characterize and substantiate. Methodological guidelines require that the risk analysis determines that all sources of risk of major accidents are identified and analyzed. The depth of the study should be proportionate to the risks from the handling of hazardous chemicals at the site.

Methodological guidelines for the analysis and assessment of risk are structured into following 17 chapters:

- overview of the establishment or installations specifying the type and volume of hazardous materials,
- overview of all hazardous substances or preparations in the establishment or installations, their classification and characteristics required for the analysis and assessment of risk,
- results of the evaluation and descriptions of hazardous chemical reactions caused by unwanted contact with chemical compounds in the establishment or installations or under unwanted operational conditions,
- results of the assessment and descriptions of the possible situation in the establishment or installation, which have the potential to cause damage to the human health and life, livestock, the environment and property,
- results of the evaluation and descriptions of possible situations beyond the establishment or installation which could cause a major accident,
- results of the identification and descriptions of the sources of risk of a major accident, relative evaluation of their importance and selection of risk sources for detailed risk analysis, including the indication of important sources of risk on a map of the enterprise,
- procedure and results of the identification of possible event scenarios and their causes which could create a major accident, and selection of representative scenarios of these events, including their description,
- procedures and results of the estimations performed on the consequences of representative major accident scenarios and their effects on the human health and life, livestock, the environment and property, including a graphic presentation of the most important results of the estimations,
- procedures and results of the estimation of probability of representative major accident scenarios,
- results and procedures for assessing the effect (reliability and errors) of human factors in relation to the relevant source of risk,
- presentation of the methods used in the risk analysis,
- detailed description of methods not accessible to the public,
- results of the specification of risk level of representative major accident scenarios,
- results of the assessment of the acceptable risk of major accidents,
- description of measures for unacceptable sources of risk, plan of their realization and control system for the fulfillment of this plan,
- description of the system for continuous monitoring the effectiveness of risk limitation measures,
- information on the performed assessment of the adequacy of safety and protection measures in relation to the existing risks.

RECOMMENDATIONS AND OBSERVATIONS

In relation to the analyses and risk assessment it is necessary to mention several recommendations and observations which come for decades of experience.

In essence the same structure and content of the analysis and assessment of risk is required for both groups of operators in the Czech Republic. There is the opinion in the Czech Republic that in reality there exists more or less risk operations in relation to the impact on the surroundings and thus there should be a single document for preparing the safety documentation with a greater or lesser extent taking into consideration the required information, procedures and steps. The depth of the analysis and assessment of risk is given the hazard posed by the chemicals and their handling and in particular the real threat to the surroundings of the site.

There is a separate approach described in a special methodological guideline for the method of assessing the cumulative, synergetic or domino effects.

It is recommended to use selection method elaborated for hazardous, toxic, flammable and explosive materials for identification of risk sources, relative evaluation of their importance and selection of risk sources in the framework of detailed analysis. This method is published in the manual "CPR 18E Guidelines for Quantitative Risk Assessment" (the Purple Book). For potentially harmful materials to the environment there are various methods and approaches e.g. the Czech methods ENVITech 03 and H&V index, the Swedish Environment-Accident-Index or the Dutch model PROTEUS.

Identification and description of the possible causes of a major accident represent key questions and greater emphasis must be placed on them. The attempt to use predefined scenarios can lead to important facts for the safety of the operation being overlooked at the given site and analyzed establishment or installation.

Modeling of the physical chemical processes and effects which appear during the event and the final conditions of the set scenarios is used for an estimation of the consequences of representative scenarios of major accidents and their impact on the human health and life, livestock, the environment and property. Usually it is necessary to select a conservative approach to modeling based on the assumption that for safety reasons it is necessary to consider input data on magnitudes which reflect the least favorable case. The actual risk activities (systems) should not be worse than results given by the conservative method. During the implementation of the resultant preventive measures the greatest attainable level of safety should be provided. Taking into consideration the uncertainty of data and the possibility of their variability and random error, the calculations have a certain level of uncertainty. This needs to be considered during risk management. Usually various computer programs are used but more simple procedures which use e.g. Dow index or Purple Book are sometimes sufficient.

Generic data is most commonly used to estimate the probability of a representative major accident scenarios, including fault tree analysis and event tree analysis methods, sometimes in combination with expert assessment.

The emphasis is placed on the assessment of human factor effects, both in safety systems and during the direct servicing of the installation, analyzed as a potential source of major accidents. The methodological guidelines of MoE are focused in particular on the recommendation of methods and extent of assessment of human factor effects at the site. This is mainly related to the assessment of reliability and human error during the performance of specific activities which have direct or indirect (system) relation to the potential occurrence of major accidents. Evaluation of the reliability of human factors makes up an inseparable part of the analysis and assessment of risk. The methodological approach in the Czech Republic for assessing the impact of the human factor attempts to narrow down the required extent to a minimal but sufficient level taking into account the safety of the site.

A principal approach – "social risk" is most commonly used for specifying the level of risk from representative major accident scenarios.

No specific values for acceptable risk are set in legislation for an evaluation of acceptable risk of the occurrence of a major accident so they are selected by the operator themselves although bodies of the state administration may be included in the decision making process. Risk reduction can be approached from the point of view of the principles of ALARA (As Low As Reasonably Achievable) or ALARP (As Low As Reasonable Practicable). Theoretic examples of criteria of acceptably for individual and social risk are stated in the methodological guideline which are taken from the manual "Land Use Planning Guidelines in the context of Article 12 of the Seveso II Directive 96/82/EC as amended by Directive 105/2003/EC. Christou, M.D., Struckl, M. and Biermann T. (Eds); JRC, September 2006". Should the risk become unacceptable then it must be effectively reduced.

In Czech legislation it is stated that: should the resulting risk value of a major accident appear to be unacceptable for the given source, a more detailed analysis is performed and if necessary organizational and technical measures are specified and elaborated for the reduction of this risk, and subsequently verified by repeating the risk analysis and assessment. The acceptability or unacceptability of risk for the given establishment or installation is given by a summary of the results of the performed risk analysis and assessment of other local conditions and factors, in particular social, economic, land use and others.

In the conclusion of the Czech safety documents there is information on the performance of suitable safety and protection measures in relation to the existing risks. The operator demonstrates and describes all essential measures to prevent and limit the consequences of a major accident on the basis of the results of the risk assessment in the document, where clarifying their benefit to the reduction of risk.

SAFETY MANAGEMENT SYSTEM

PRINCIPLES AND RULES FOR THE MANAGEMENT OF MAJOR ACCIDENT PREVENTION

Safety management in an establishment should not be limited to announcing policy from the side of the operator. The operator of the establishments in which there is a threat of a major accident should be able not only to formulate a safety policy but in particular to ensure its realization and fulfillment, i.e. to implement the appropriate preventive safety measures, systems and procedures, essential for reaching the set level of safety.

The operator must also be able to demonstrate that the accident prevention policy is:

- corresponding to the risk,
- sufficiently complex,
- materially up-to-date,
- constantly planned.

Safety management in the sense of major accident prevention can be divided into the following four phases.

The preparation phase assumes:

- knowledge/understanding of the risk sources and specification of risk activities (identification and assessment of risk),
- the performance of a quantitative risk analysis considering the possible impacts and consequences,
- clarification of the existing preventive measures and evaluation of their sufficiency and effectiveness.

The conception phase includes:

- establishment of policies and framework objectives,
- definition of methods and procedures for achieving the set objectives,
- definition of specific tasks (technical and organizational measures).

The performance phase envisages:

- elaboration of relevant organizational management documents or development on existing documents on elements of major accident prevention,
- implementation of major accident prevention management system,
- realization of specific technical measures.

The monitoring and evaluation phase includes:

- monitoring, assessment and evaluation of the benefit and effectiveness of the adopted measures,
- adoption of corrective measures if the measures adopted to-date are considered to be insufficient.

APPROACH OF THE OPERATOR TO MAJOR ACCIDENT PREVENTION

It is a very important, that the operator has met the main requirements and expected output which is a coherent and meaningful comprehension of the existing risk and the optimal and socially acceptable limits of operational safety in the establishment or installation which is safeguarded by adequate technical measures and a functional safety management system. Then the operator will be able to elaborate a safety document which will be defined by its logical correctness, completeness and sufficient detail.

REQUIREMENTS OF THE SAFETY DOCUMENTS

Operators of establishment classified into group A or B are obliged to elaborate the relevant safety documents.

The importance, extent and depth of elaboration of the required information are not the same for the individual items, or for establishment classified into various groups or for establishment classified into the same group.

In certain cases the maximum requirements can be partially moderated. The value, weight, importance and the expected level of detail of the information of the individual items differs case by case (establishment by establishment).

The relation between the specific establishment or installation must always be respected, in particular for:
- type and importance of the existing risk,
- character of risk activities,
- form of organization model and management used at the establishment,
- size of establishment,
- number employees,
- location of the establishment,
- situation in the surroundings of the establishment.

CONCLUSIONS

The main purpose of the Seveso directive and Major Accident Prevention Act is to encourage operators to search and evaluate the risks in the establishment, to perform an analysis of the existing prevention measures, to assess whether these prevention measures are sufficient in relation to the identified risks, to assess whether the existing safety management system is comprehensive, satisfactory and optimal.

Once the operator understands the sense and purpose of Seveso Directive and grasps the concept of the control of major accidents he/she will be able to elaborate a correct, complete and sufficiently detailed safety document.

To increase the level of overall safety of risk installations in the Czech Republic the MoE CR publishes some methodological guidelines on individual topics that are accessible on the Internet. There are also occupational health and safety web pages hosted on the special website and training at various levels for specific target groups, publications and lectures are provided. An expression of the will to disseminate these activities as much as possible is the launch of a new information portal on occupational health and safety and major accident prevention. Requirements on the increase in safety must be reasonable both in extent and intensity and must also be supported by an attempt to increase knowledge at all levels of the area in question.

Professional level of safety documentation is slowly but steadily increasing.

Can make some comments on the professional level of safety documentation. Problems of risk analysis underestimate the fact that the basis for operational safety management is a correctly performed analysis and evaluation of risk. Based on the results of analysis and risk assessment the safety management system must be set correctly and the corresponding level of risk posed by the given site determined. This is a very difficult task, which involves the subsequent demonstration of the fulfillment of the relevant requirements of the safety management system inspection bodies. The safety documents should be updated in order to complement the requirements taking into account the actual situation at the site, legislation and the effect of new findings and technical know-how on safety so the safety document is constantly up-to date.

It follows that the risk analysis and the safety management system are significant elements in terms of prevention and the mitigation of consequences of any major accident. Therefore, it is absolutely right that the issue is enshrined in Seveso Directive. Promoting the principles of Seveso Directive greatly increases overall safety of risk establishments and installations.

REFERENCES

OECD Guiding Principles for Chemical Accident Prevention, Preparedness and Response. Guidance for Industry (including Management and Labor), Public Authorities, Communities and other Stakeholders. OECD 2003.
Available on the www <URL:
English text: http://www.oecd.org/dataoecd/10/37/2789820.pdf ;
Czech text: http://www.oecd.org/dataoecd/15/63/34014622.pdf
OECD Guidance on Safety Performance Indicators. Guidance for Industry, Public Authorities and Communities for developing SPI Programmes related to Chemical Accident Prevention, Preparedness and Response. OECD 2003.
Available on the www URL: http://www2.oecd.org/safetyindicators/;
English text: http://www.oecd.org/dataoecd/60/39/21568440.pdf
http://www.env.cz/AIS/web.nsf/pages/havarie
http://www.env.cz/osv/edice.nsf/titletree
http://www.vubp.cz/oppzh.php

Chapter 103

Perceptions of Sport-Utility Vehicle (SUV) Safety by SUV Drivers and Non-Drivers

Christopher B. Mayhorn[1], Michael S. Wogalter[1],
Vincent C. Conzola[2]

[1]North Carolina State University
Raleigh, NC 27695-7650, USA

[2]IBM Corporation
Research Triangle Park, NC 27709, USA

ABSTRACT

The number of sport-utility vehicles (SUVs) on U.S. roads has grown substantially in recent years. Despite their popularity, SUVs have several potential disadvantages. The present study examined people's (N=370) perceptions of several SUV aspects (seeing above or around the SUV, collision involvement with smaller vehicles, headlights "blinding" other drivers, rollover, and low gas mileage) as a function of their SUV driving experience. Gas mileage was rated the most negative aspect of SUVs. Participants who had no SUV driving experience gave higher problem ratings to the SUV aspects than participants who drive an SUV or have some SUV driving experience. SUV drivers gave lower problem ratings than non-SUV drivers for aspects that could negatively affect non-SUV drivers (obscured line of sight, headlight glare, crash severity). Implications for driving safety and warnings are discussed.

Keywords: safety, beliefs, driver experience, risk communication.

INTRODUCTION

In recent years, sport-utility vehicles (SUVs) have become a popular vehicle on U.S. roadways. In 2001 over 17% of all light vehicles sold in the U.S. were SUVs (Wards 2002). One reason for their popularity is their perceived safety advantages compared to other vehicles (Pittle 2000). Previously, consumer buying decisions had been guided more by style attributes like color and vehicle model than safety (e.g., Pittle 2000). However, more recently, greater attention has been given to aspects that increase safety (e.g., anti-lock brakes, air bags). With that focus, some consumers apparently buy SUVs because they believe them to be safer than passenger vehicles (Pittle 2000). However, recent accident statistics have shown that SUVs are prone to a specific type of accident: rollovers (U.S. National Highway Transportation Safety Administration [NHTSA] 2006). According to the NHTSA, rollovers account for more than 25 percent of all vehicle-related deaths (Muller and Welch 2000, NHTSA 2006). Compared to SUVs, passenger cars and minivans have a much lower probability of rollover accidents (Stoller 2000). Two factors are primarily responsible for SUV's relative instability and propensity for rollover: (a) they have a higher center of gravity than automobiles, and (b) they are more likely to be overloaded. The higher center of gravity makes them more likely to overturn with rapid lateral changes in direction. Also, despite their large cargo area, the load capacity of most SUVs is "light-duty." The weight and position of the load vehicle can affect handling and stability and could promote loss of control (Lee 2002).

While rollover accidents primarily present a risk to SUV drivers and their passengers as many are single vehicle crashes, other aspects of these vehicles are a safety concern to other vehicles on the road. These include the following:

(a) size: many SUVs are so large that it is often difficult for automobile drivers to see around them in traffic.

(b) higher profile: the headlight beams of SUVs tend to project at a higher level from the ground, causing both direct and reflected glare, negatively affecting dark adaptation and, more colloquially, "blinding" the eyes of motorists in front of them.

(c) greater height and mass: in collisions, SUVs inflict more damage to smaller passenger vehicles than occurs in collisions involving two passenger vehicles of similar mass. Crashes between SUVs and vans and other vehicles account for the majority of fatalities in vehicle-to-vehicle collisions (Joksch 1998, Muller and Welch 2000, NHTSA 2006). Of all the fatalities that occurred in collisions between SUVs and cars, car occupants were 18.5 times more likely to be injured than the SUV occupants (Joksch, 1998, Muller and Welch 2000, NHTSA 2006).

Despite these accident data and the increased negative publicity surrounding SUVs, they are still a popular choice with many consumers. Given this frequent exposure to negative information regarding SUV safety, why do consumers continue to purchase these vehicles? One explanation is that SUV drivers are either not highly aware of the SUV safety problems or ignore them. It is likely that SUV drivers have been exposed to media presentations about SUV safety issues at least as much or more than non-SUV drivers. Indeed, one might reasonably expect SUV drivers to be more familiar and attuned with various aspects of their vehicles including safety issues because of their greater relevance due to ownership compared to non-SUV drivers. SUV drivers may be ignoring SUVs' negative aspects relative to persons who do not drive SUVs because they occur relatively infrequently (e.g., rollovers). Likewise SUV drivers may ignore other negative aspects because they are not directly affected by them (e.g., inability to see around an SUV or "blinding" headlights).

Non-SUV drivers' beliefs about SUVs may be affected in an opposite way because they are aware of the dangers of driving these vehicles and avoid their use. Non-SUV drivers' beliefs about SUVs might fit the availability heuristic somewhat better than SUV drivers. Non-SUV drivers may be more attuned because of the potential for greater negative consequences from SUVs to themselves (and in some cases to their passengers) relative to SUV drivers (e.g., a crash with less massive conventional passenger vehicle). Thus, for certain SUV aspects, non-SUV drivers may hold greater negative beliefs than SUV drivers due to their personal relevance in receiving negative effects to themselves.

The present study examined the responses of SUV and non-SUV drivers to determine whether they have different perceptions about the safety of SUVs. Participants were classified into one of two groups based on their responses to a question asking whether or not they had any experience driving SUVs.

Additionally, potential differences in demographic categories of gender, age, and college student vs. non-student were examined. Some research suggests that males tend to purchase and use SUVs more than females (Kweon and Kockelman 2003) and that younger males tend to be riskier drivers (e.g., being overrepresented as a group in crash statistics [Massie et al. 1997]). Age was examined because younger, less experienced drivers (teens and early twenties) tend be over represented in crashes compared to other groups (Massie et al. 1997). Older drivers may have age-related declines in perceptual, cognitive and motor abilities compared to younger adults and may be more adversely affected by some SUV attributes (Ball and Rebok 1994, Janke 1994). The third demographic categorization: college student vs. non-student is a classification that overlaps with age (college students tend to be younger than the non-students) but it was included to determine if the responses differ in comparison between them as a check on the potential generalizability of the student scores.

METHOD

SAMPLE

Three hundred seventy adults participated. Of these, 246 were undergraduate students (mean age = 21.2 years, SD = 3.9) from North Carolina State University who participated for research credit. The other 124 were non-student adults from the communities of the Research Triangle Region area of North Carolina (mean age = 34.1 years, SD = 14.3).

The overall sample included 228 males (mean age = 25.3 years, SD = 10.6) and 142 females (mean age = 26.0 years, SD = 10.9). Ninety-eight percent of the participants reported that they had a valid driver's license. Only 9 (2.4%) reported that they did not have access to a vehicle to drive. Participants reported driving an average of 13262 miles (SD = 9633) (mean = 21343 km, SD = 15503) in the prior 12 month period.

PROCEDURE

Participants completed a questionnaire that included several categories of items. Some concerned various kinds of automotive vehicle driving-related experiences including primary vehicle driven (year, make and model). One section of the questionnaire was titled with the heading "Sport Utility Vehicles." The first item asked, "Do you drive or have you ever driven a sport-utility vehicle (SUV)?" Respondents marked either "yes" (1) or "no" (0) to the question. The remaining five items in this section were responded to using a rating scale. They were statements concerning potential negative aspects of SUVs. Participants were asked to rate the statements according "To what extent do you think the following may be a problem with SUVs?" The specific items listed were: (a) Seeing above or around the SUV, (b) Involvement in collisions with smaller vehicles with less mass, (c) Headlights "blinding" the eyes of motorists in front of them in smaller vehicles,(d) Rollover, and (e) Low gas mileage.

Ratings were made using a 0- to 8-point Likert-type scale with the even numbers labeled with the following text anchors: (0) "Not a Problem At All;" (2) "Somewhat a Problem;" (4) "A Problem;" (6) "Very Much a Problem;" and (8) "Extremely a Problem."

RESULTS

Analyses comparing SUV versus non-SUV drivers were based on grouping participants according to whether they reported having had any experience driving SUVs. Additional exploratory analyses used the classifications: (a) whether their reported primary vehicle could be classified as an SUV, and (b) whether their reported primary vehicle could be classified as being in the light truck category (which includes SUVs). Also, demographic categories of gender, age group, and college student vs. non-student were examined.

DRIVERS WITH AND WITHOUT SUV DRIVING EXPERIENCE

Two hundred fifty three (68.4%) participants indicated they had experience driving SUVs, while the other 117 (31.6%) had never driven an SUV. Ratings of these two groups were examined with respect to each of the five problem statements. Table 1 (see Appendix) shows the mean ratings as a function of SUV driving experience.

A 2 (SUV driving experience) x 5 (SUV aspects) mixed-model analysis of variance (ANOVA) showed that both main effects were significant, $F(1, 368) = 16.63$, $MSe = 14.06$, $p < .0001$ and $F(4, 1472) = 13.97$, $MSe = 3.32$, $p < .0001$, for SUV driving experience and for SUV aspects, respectively. With respect to the first main effect, the means are shown on the bottom row of Table 1. Participants with no SUV driving experience gave higher problem ratings than participants with SUV driving experience. The means for the other main effect are shown on the right-most column of Table 1 (see Appendix). Comparisons among the SUV aspects' main effect means using Tukey's Honestly Significant Difference (HSD) test at $p < .05$ showed that participants were most concerned with low gas mileage and that this aspect was rated significantly higher compared to all other statements except for headlights "blinding" other drivers. The aspect of headlight "blinding" other drivers was rated significantly more of a problem than the remaining aspects except for collision with smaller vehicles. Collision with smaller vehicles was given significantly higher problem ratings than seeing above or around SUVs, but not significantly different from rollover. The latter two aspects (i.e., seeing above or around SUVs and rollover) did not differ.

The ANOVA also showed that the factors of SUV driving experience and SUV aspects significantly interacted, $F(4, 1472) = 10.28$, $MSe = 3.32$, $p< .0001$. The means are shown within the cells of (see Appendix) 1. Tests of simple effects showed that perceptions of three of the five SUV aspects significantly differed as a function of SUV experience. Experienced SUV drivers were less concerned than inexperienced SUV drivers about the aspects of: (1) seeing above or around the SUV, $F(1, 368) = 25.04$, $p < .0001$; (2) collisions involving smaller vehicles, $F(1, 368) = 19.83$, $p < .0001$; and (3) headlights "blinding" other drivers, $F(1, 368)=18.02$, $p < .0001$. The other two problem statements did not differ as a function of SUV driving experience.

PRIMARY SUV DRIVERS VERSUS DRIVERS OF NON-SUV VEHICLES

Participants were also asked what type of vehicle they drove most often. Fifty-three of the 370 participants indicated SUVs were their primary vehicle. Cell means for SUV driving experience and SUV aspects are shown in Table 2.

A 2 (SUV driving experience) x 5 (SUV aspects) mixed-model analysis of variance (ANOVA) showed that both main effects were significant, $F(1, 368) = 28.52$, $MSe = 13.64$, $p < .0001$, and $F(4, 1472) = 22.76$, $MSe = 12.02$, $p < .0001$, for SUV driving experience and SUV aspects, respectively. There was also a significant interaction, $F(4, 1472) = 3.56$, $MSe = 12.02$, $p < .01$. The pattern of means and simple effects analysis were similar to the preceding analysis, with one exception: there was also a significant difference in that participants who primarily drive SUVs rated SUV rollovers ($M = 3.87$) significantly less of a problem than those who primarily drive other vehicles ($M = 4.76$), $F(1, 1472) = 6.61$, $MSe = 5.43$, $p < .01$. One other comparison suggested a trend that did not attain the conventional criterion level for significance of $p < .05$. Participants who drive SUVs as their primary vehicle ($M = 5.00$) rated the problem of low SUV gas mileage somewhat less severely than primary drivers of other kinds of vehicles ($M = 5.63$), $F(1, 1472)$ = 3.36, $MSe = 5.43, p < .07$.

OTHER DEMOGRAPHICS

The relationship of three demographic factors (gender, age, and college student vs. non-college student status,) with respect to SUV problem perceptions was examined. Three separate mixed model ANOVAs were used. Each analysis utilized a single demographic category as a between-subjects (group) variable together with SUV aspects as the within-subjects (repeated measures) variable.

The ANOVA involving gender showed no main effect of gender or an interaction ($ps > .05$). The ANOVA showed the significant main effect of SUV aspects, $F(4, 1472) = 22.56$, $MSe = 3.41$, $p < .0001$, yielding the same pattern described earlier.

Age was also analyzed in a similar manner. A median split was used to divide participants into two, approximately equal groups of younger ($M = 19.9$, $SD = 1.2$) and older ($M = 31.6$, $SD = 12.8$) participants. The main effect of age group failed to reach the conventional criterion for significance but suggested that, in general, the older group ($M = 4.99$) perceived the SUV aspects to be more problematic than the younger age group ($M = 4.69$), $F(1, 368) = 2.99$, $MSe = 14.6$, $p = .08$. However, there was also a significant interaction, $F(4, 1472) = 3.46$, $MSe = 3.38$, $p < .01$. Simple effects analyses showed one significant difference: the older group ($M = 4.70$) rated seeing above and around the SUV as a greater problem than the younger group ($M = 3.94$), $F(1, 1472) = 9.45$, $MSe = 6.62, p < .01$.

The ANOVA with college students vs. non-students as the grouping factor showed that both main effects were significant, $F(1, 368) = 5.97$, $MSe = 14.46$, $p < .05$ and $F(4, 1472) = 22.72$, $MSe = 3.39$, $p < .0001$, for student vs. non-students and SUV aspects, respectively. In general, college students ($M = 4.68$) viewed the SUV aspects to be significantly less problematic than non-students ($M = 5.14$). There was also a significant interaction, $F(4, 1472) = 2.97$, $MSe = 3.39, p < .05$. Simple effects analysis showed that college students rated two SUV aspects significantly less problematic than non-students: (a) seeing above and around the SUV ($Ms = 3.99$ and 4.95, for college students and non-students, respectively), $F(1, 1132) = 1368$, $MSe = 5.60, p < .0001$, and (b) collisions involving smaller vehicles ($Ms = 4.55$ and 5.19, for college students and non-students, respectively), $F(1, 1132) = 5.89$, $MSe = 5.60, p < .02$.

DISCUSSION

SUVs are a popular vehicle type in the U.S. that offers both benefits and disadvantages to consumers. Benefits include greater ground clearance for off-road driving, higher seat height affording a better view of the road, and generally greater passenger and cargo space compared to conventional passenger cars. The disadvantages include poor gas mileage and rollover propensity, among others.

Perceptions of SUV disadvantages as a function of driving experience, as well as several demographic variables, were examined in the present study.

Based on the overall mean ratings and scale anchors, participants viewed all of the SUV aspects listed on the survey as being problematic. Because participants were required to assess how each of five safety aspects is a "problem," it is important to note that participants may have varied in their interpretation of this term. Previous safety research by Wogalter *et al.* (1991) suggests that participants would rate the items according to severity of injury rather than frequency (probability) of injury. However, of the items listed in this study, poor gas mileage was perceived by both SUV drivers and non-SUV drivers as the most problematic. The poor gas mileage issue was probably highly salient because of rising fuel prices in recent years which has drawn media attention to SUVs' relatively-high gas consumption per mile relative to conventional passenger cars. During much of the growth period for sales of SUVs, gasoline was relatively inexpensive and the difference in cost to fuel SUVs compared to cars was inconsequential. With rising gas prices and a greater awareness of the limits of oil reserves, and international conflicts among nations with larger supplies of oil, it is likely that drivers of both SUVs and non-SUVs are aware of poor gas mileage being a problem.

When asked to assess the aspects associated with safety, drivers of SUVs tended to have less negative perceptions of them. This was true in all three methods of categorizing SUV driving experience. The strongest effects noted were for those SUV aspects that could be classified as factors affecting other people, particularly non-SUV drivers and their passengers. In other words, from the perspective of the SUV driver, those aspects affecting other people (not themselves) were not as viewed as problematical relative to the viewpoint expressed by non-SUV drivers in their ratings. Statistically significant differences were consistently found between SUV and non-SUV drivers for three of the aspects: (1) seeing above or around the SUV, (2) collisions with smaller vehicles, and (3) headlights "blinding" other drivers.

Somewhat weaker differences were found for the two items that more directly affect SUV drivers and their passengers (but still could affect other people). SUV drivers had significantly lower problem ratings than non-SUV drivers for rollover for two of the three SUV driving experience classifications. Gas mileage showed the same trend but never reached conventional statistical significance.

The results suggest that SUV drivers are less critical about some SUV aspects than non-SUV drivers. A specific, definitive reason for the differences cannot be made at this point; further investigation is necessary to select among alternative explanations. However, several potential explanations can be offered. One is that SUV drivers might have had some or all of their beliefs regarding SUVs *before* they drove or purchased an SUV. That is, drivers may have already established their beliefs before using or purchasing an SUV or other vehicle types. Alternatively, their perceptions might have changed *after* they drove or purchased an SUV. Another, somewhat different, kind of explanation is that SUV drivers may have failed to pay attention to, or in some way ignored, negative information—before or after driving an SUV. Indeed one reasonable expectation expressed in the Introduction section is that SUV drivers probably know *more* about SUVs, in general, than non-SUV drivers. Although most SUV drivers have probably heard negative information, they may judge it to be less credible, perhaps in part due to other perceived benefits they derive from SUVs (e.g., generally greater passenger and cargo space than passenger cars). SUV drivers may focus more on the positive aspects of SUVs, believing that those aspects outweigh the negative aspects, than non-SUV drivers.

Non-SUV drivers have a different pattern of judgments than the SUV drivers with respect to some of the safety-related aspects because they are more negatively impacted by them than the SUV drivers are. For three of the five negatives listed, SUV drivers could have a safety advantage over non-SUV drivers. Obscured vision, greater risk in collisions with an SUV and headlights blinding are aspects that could detrimentally affect drivers of other (smaller) vehicles. SUV drivers may be less aware of or ignore these negative aspects because they are not adversely affected, their relevance is less, and/or they see them as advantages over other vehicle drivers. That is, some SUV drivers may realize that these aspects represent advantages of SUVs over other vehicles (non-SUVs), in which they are better off (i.e., less likely to be injured) in collisions involving smaller vehicles. Consequently they do not rate them as problematic as non-SUV drivers.

Several additional concepts can also be used to explain the belief differences between driver groups. For example, the perseverance effect (e.g., Anderson *et al.* 1980, Ross *et al.* 1975) says that beliefs, once formed, are relatively stable and resistant to change. SUV drivers may form positive attitudes and beliefs about the vehicles before or soon after they purchase an SUV. Therefore, despite the presence of evidence to the contrary, they maintain relatively positive perceptions of SUV safety. The third person effect (Perloff 1993) provides another explanation. One example of this effect was shown by Adams *et al.* (1998) who found that people often report that injury events are more likely to happen to other people than to them. Applying this to the findings in the present study, SUV drivers might believe that negative outcomes associated with SUVs are much more likely to occur to other drivers, not themselves. SUV drivers may be less concerned about safety-related problems simply because they are less directly affected by them compared to passenger car drivers, or in other words, they may be taking an egocentric perspective. This perspective, however, would not explain the concurrent failure to find significant differences between groups for low gas mileage, and to a lesser extent, rollover.

In addition, a few demographic differences were noted in the age and student/non-student analyses. The pattern of results between age and student/non-student analysis were relatively consistent such that the main finding indicated that college students and younger adults rated seeing around an SUV and collisions involving an SUV to be less problematic than non-students and somewhat older adults. Two explanations can be offered for these findings. They may partly reflect a mindset of invulnerability (Finn and Bragg 1986) that has been found in earlier research involving younger versus older adults. At the same time, older, non-students have more driving experience, and consequently, may be more aware of the problems than less-experienced, younger college students. Headlights blinding other drivers was rated more of a problem by older participants than the younger participants. This concurs with research search indicating that older adults have more problems with glare during night time driving than younger adults (Ball and Rebok 1994, Janke 1994).

Future research could explore several related issues. One is whether people are considering the economic and geopolitical context in their judgments of vehicle preferences and beliefs. Other issues of potential interest concern people's beliefs about safety-related aspects of their vehicles, the effects of educational efforts, and the influence of governmental regulations and industry standards on safety.

REFERENCES

ADAMS, A., BOCHNER, S. and BILIK, L., 1998, The effectiveness of warning signs in hazardous work places: cognitive and social determinants, *Applied Ergonomics,* **29,** 247-254.

ANDERSON, C.A., LEPPER, M.R. and ROSS, L., 1980, Perseverance of social theories: the role of explanation in the persistence of discredited information, *Journal of Personality and Social Psychology,* **39,** 1037-1049.

BALL, K. and REBOK, G., 1994, Evaluating the driving ability of older adults, *The Journal of Applied Gerontology,* **13,** 20-38.

FINN, P. and BRAGG, B.W.E., 1986, Perception of the risk of an accident by young and older drivers, *Accident Analysis & Prevention,* **18,** 289-298.

JANKE, M.K., 1994. *Age-related disabilities that may impair driving and their assessment: literature review.* Interim Report. DTNH 22-93-Y-5330, U.S. Department of Transportation, National Highway Traffic Safety Administration (Springfield, VA, National Technical Information Service).

JOKSCH, H., 1998, *Fatality risks in collisions between cars and light trucks.* Final Report, DOT-VNTSC-NHTSA-98-X, U.S. Department of Transportation, National Highway Traffic Safety Administration, (Springfield, VA, National Technical Information Service).

KWEON, Y.J. and KOCKELMAN, K.M., 2003, Overall injury risk to different drivers: Combing exposure, frequency, and severity models, *Accident Analysis and Prevention,* **35,** 441-450.

LEE, A.Y., 2002, Coordinated control of steering and ant-roll bars to alter vehicle rollover tendencies, *Journal of Dynamic Systems Measurement and Control-Transactions of the ASME,* **124** (1), 127-132.

MASSIE, D.L., GREEN, P.E. and CAMPBELL, K.L., 1997, Crash involvement rates by driver gender and the role of average annual mileage, *Accident Analysis and Prevention,* **29,** 675-685.

MULLER, J. and WELCH, D. 2000, Making safer SUVs: it's not rocket science, *Business Week,* October 16, 2000.

NATIONAL HIGHWAY TRANSPORTATION SAFETY ADMINISTRATION (NHTSA), 2006, *Motor vehicle traffic crash fatality counts and estimates of people injured for 2005,* Washington, DC: U.S. Department of Transportation.

PERLOFF, R., 1993, Third person effect research 1983-1992: a review and synthesis, *International Journal of Public Opinion Research,* **5,** 167-184.

PITTLE, R.D., 2000, SUV and stability control - a consumer view, Paper presented at the SAE Conference On Rollover and Stability Control, 16 May 2000, Troy, MI http://www.consumersunion.org/products/saseny500.htm (accessed 2 December 2006).

ROSS, L., LEPPER, M.R. and HUBBARD, M., 1975, Perseverance in self-perception and social perception: Biased attributional processes in the debriefing paradigm, *Journal of Personality and Social Psychology,* **32,** 880-892.

STOLLER, G. 2000, Formula predicts rollover risk, *USA Today*, July 17, 2000.

WARDS 2002 AUTOMOTIVE YEARBOOK, 2002, Alan K. Binder (Ed.) Southfield, MI: Wards Communications, pp. 244-245.

Table 1. Mean problem ratings of SUV aspects by drivers who have driven and who have not driven an SUV (SDs in parentheses).

SUV Aspects	SUV Driving Experience		
	No	Yes	Mean
Seeing above or around the SUV	5.27(2.27)*	3.87 (2.62)	4.31 (2.60)
Collision involvement with smaller vehicles	5.54(2.11)*	4.41 (2.34)	4.76 (2.33)
Headlights "blinding" other drivers	5.70(2.33)*	4.57 (2.42)	4.92 (2.45)
Rollover	4.80(2.20)	4.55 (2.18)	4.63 (2.19)
Low gas mileage	5.47(2.52)	5.58 (2.20)	5.54 (2.30)
Mean	5.36(2.28)*	4.59 (2.36)	
n	117	253	

Note. Higher scores indicate more negative perceptions of SUV problems.
* $p < .05$ between the two groups on SUV driving experience.

Table 2. Mean problem ratings of SUV aspects by participants who drive SUVs versus participants who primarily drive other kinds of vehicles (SD in parentheses).

SUV Aspects	SUV Driving Experience	
	Primary Vehicle is *not* SUV	Primary Vehicle is SUV
Seeing above or around the SUV	4.55 (2.56)*	2.91 (2.38)
Collision involvement with smaller vehicles	4.99 (2.48)*	3.42 (2.20)
Headlights "blinding" other drivers	5.18 (2.42)*	3.38 (2.04)
Rollover	4.76 (2.16)*	3.87 (2.23)
Low gas mileage	5.63 (2.28)	5.00 (2.41)
Mean	5.02 (2.34)*	3.71 (2.25)
n	317	53

Note. Higher scores indicate more negative perceptions of SUV problems.
* $p < .05$ between the two groups

CHAPTER 104

Sunscreen Labeling and Warnings: A Human Factors Analysis

Kelly A. Burke, Jonathan A. Dorris, Nathan T. Dorris

Dorris & Associates International, LLC
Atlanta, GA 30308, USA

ABSTRACT

Skin cancer is the most common form of cancer in the United States, affecting approximately one million Americans every year. An estimated 11,590 deaths from skin cancer occurred in 2009. Despite a plethora of compelling scientific evidence identifying the risks associated with excessive unprotected sun exposure, the efforts of the FDA and organizations such as the Skin Cancer Foundation, and a significant increase in public health promotion campaigns warning about these risks, the incidence of skin cancer continues to rise yearly. Clearly, an increase in the awareness of the risks of sun exposure on the skin is not fully translating into the adoption of appropriate sun protection behaviors. It is imperative that further research be conducted to better understand this lack of behavioral change. This paper attempted to investigate this phenomenon by examining the development of current labeling and warning messages relating to sunscreen products in light of the available behavioral literature on warning label design.

Keywords: Sunscreen, warnings, labeling

INTRODUCTION

Skin cancer is the most common form of cancer in the United States, affecting approximately one million Americans every year. An estimated 11,590 deaths from skin cancer occurred in 2009. Ultraviolet radiation (UV) from the sun is a significant risk factor for all three major types of skin cancer, and approximately 65% to 90% of malignant melanomas are caused by exposure to UV radiation (American Cancer Society (ACS), 2009). Unlike many other types of cancer, skin cancer is highly preventable by adhering to a complete program of sun protection and protecting the skin from intense sun exposure (ACS, 2009). Sunscreens are a key element of this program and have improved dramatically since their inception. The U.S. Food and Drug Administration (FDA) regulates sunscreens as over-the-counter (OTC) drugs, not as cosmetics. This designation is meaningful, as it generally requires higher standards for efficacy and safety testing. Additionally, sunscreen manufacturers are required to adhere to specific guidelines for package labeling.

In August 2007, the FDA proposed revisions to several aspects of the labeling on sunscreen products including: increase the maximum Sun Protection Factor (SPF) labeled value to 50; the use of symbols in conjunction with a descriptor word that will designate the level of UVA protection; and mandating very specific language on the label, warnings, indications and directions. While the FDA's 2007 Proposed Sunscreen Monograph has not been finalized, the Skin Cancer Foundation supports the FDA's position and is hopeful that if finalized, the new SPF and UVA protection rating systems will increase consumers awareness of the health hazards associated with excessive sun exposure and provide consumers with an easy way to ensure that they are getting a safe and effective sunscreen product, thereby decreasing the incidence of skin cancer.

However, despite a plethora of compelling scientific evidence identifying the risks associated with excessive unprotected sun exposure, the efforts of the FDA and organizations such as the Skin Cancer Foundation, and a significant increase in public health promotion campaigns warning about these risks, the incidence of skin cancer continues to rise yearly. Clearly, an increase in the awareness of the risks of sun exposure on the skin is not translating into the adoption of appropriate sun protection behaviors. It is imperative that further research be conducted to better understand this lack of behavioral change.

Skin cancer has long been recognized as a major public health problem. One million new cases are diagnosed in this country each year, the incidence is rising, and the cost of treatment runs in the billions (ACS, 2009). The incidence of malignant melanoma, the most dangerous form of the disease, is also escalating. Currently, the FDA standards require that sunscreen protect against UVB radiation, the portion of the UV spectrum that causes sunburn and contributes to the increasing incidence of cancer (The Skin Cancer Foundation, 2009). However, the FDA does not require protection against UVA radiation, which penetrates the skin more deeply than UVB rays and are a major cause of skin cancer and premature aging. The 2007 Proposed Sunscreen Monograph included a proposal for first-ever UVA standards, but has yet to finalize them to make them mandatory. The Center for Disease Control (CDC), National Cancer Institute (NCI), American Cancer Society (ACS), the American Academy of Dermatology (AAD), and the Skin Cancer Foundation recommend the use of sunscreen as part of a comprehensive sun protection regimen, but some consumers may be left wondering which of the hundreds of sunscreens on the market will best protect them and their families from the harmful effects of sun exposure.

INFORMATION ENVIRONMENT

The FDA, AAD, EPA, CDC, NCI, ACS, and the Skin Cancer Foundation have made significant efforts to radically increase public awareness via educational campaigns, which provide consumers with sources of information other than the label of sunscreen containers. Through the efforts of these organizations, skin cancer information and sun protection advice have been widely promulgated since the late 1980s. It is commonly believed that excessive sun exposure can increase the chances of developing skin cancer (Cokkinides, et al., 2001). Cokkinides et al. indicated that one-third of U.S. youth have received health care professional advice to protect skin from the sun, while one in two youths received information from friends and family. Through the continued efforts of public health, medical professionals and dermatology communities, skin cancer awareness and prevention information is widely disseminated to the public in a variety of ways including:

Public health campaigns – Public health campaigns across the country have attempted to educate the public about the dangers of UV radiation exposure. They can provide information through mass media (television, radio, magazines, etc.). Campaigns regarding sun protection have included activities to change the knowledge, attitude, beliefs, intentions, and sun protective behaviors of children and adults (Saraiya, et al., 2005). Several mass media campaigns were launched by the Skin Cancer Foundation, the ACS, and the Weather Channel. The CDC and the United States Department of Health and Human Services launched a five-year skin cancer prevention campaign, "Choose Your Cover" which emphasized that young people can have fun in the outdoors while protecting their skin by choosing five sun protection options: seeking shade, covering up, getting a hat, wearing sunglasses and applying sunscreen (see www.cdc.gov/cancer/skin/chooseyourcover).

Education in childcare facilities – As the number of preschool-aged children attending child care centers continues to grow, there has also been an increase in sun protection education programs at these facilities. These programs focus efforts to promote sun-protective behaviors among preschool-aged children. For example, the "Be Sunsafe" curriculum includes interactive classroom and take-home activities that promote covering up, finding shade, and asking for sunscreen (Saraiya, et al., 2004).

School based education programs – There are numerous education programs targeted at children of all ages and presented in the school environment. In 2002, the CDC published guidelines recommending that schools engage in skin cancer prevention activities. For example, the EPA's "SunWise" School Program strives to teach children and their caregivers how to protect themselves from overexposure to the sun (see www.epa.gov/sunwise/summary.html).

Healthcare provider counseling – Patient counseling and screening may prevent skin cancers or enable their early detection (Polster, et al., 1998). More than 95.1% of pediatricians agreed that cumulative sun exposure during childhood increases the risk of skin cancer in adulthood (Balk, et al., 2004). The American Academy of Pediatrics recommends that pediatricians incorporate sun protection counseling, including the correct use of sunscreen, into their practice. In a 2004 national survey of pediatricians, 22.3% reported counseling patients about sun protection behaviors and the most important sun protection recommendation named was sunscreen with an SPF ≥ 15 (Balk, et al., 2004).

However, despite these efforts, consumer behavior does not appear to be congruent with their increased awareness of the harmful effects of sun exposure (Nicol, et al., 2006). Previous studies (Robinson, et al., 1997; Santmyire, et al., 2001) have indicated that educational and primary skin cancer prevention campaigns, some initiated as early as 1985, have had little impact on the reported sun-protective behaviors of the U.S. adult population. Although educational campaigns have attempted to increase public knowledge of the nature of the risk, the harmful consequences of excessive sun exposure, and sun protection behaviors, these studies suggest that this increase in knowledge has not fully translated into increased sun-protective behavior, including the use of sunscreens.

SUNSCREEN PROTECTION BEHAVIOR

A review of the literature finds that numerous studies from a variety of different countries have attempted to examine and understand the factors associated with sun protection behaviors (Kasparian, et al., 2009). Sun protection behaviors have been studied in a wide range of participants including children, adolescents, young adults and adults. Research indicates that reported estimates of sunscreen use in the general population vary significantly, ranging from 7 to 90% (Kasparian, et al.). Research also shows that between 9 and 61% of study participants 'seldom' or 'never' use sunscreen when they are outdoors in the sun (Newman, et al., 1996), and that 75% of adolescent males discontinue using sunscreen after they become tanned. In a recent literature review Kasparian, et al. indicate that a large portion of individuals do not practice sun protection when outdoors in the sun. Furthermore, many studies have shown that a relatively large proportion (42%-76%) of young people and adults, primarily females, intentionally acquire a tan by either lying in the sun or using a tanning salon. A significant percentage of those seeking a tan have reported sunburns as a result of these activities (Boldeman, et al., 2001).

As the literature has clearly indicated (Branstrom, et al., 2001), there are many psychological factors associated with sun protection behavior including higher perceived risk of developing skin cancer, greater perceived benefits of sun protection, lower perceived barriers to sun protection, greater intentions to use sun protection, higher perceived value of health and appearance and increased skin cancer related anxiety (Kasparian, et al., 2009). Factors known in the Human Factors (HF) literature to impact the likelihood receivers will avail themselves of safety information or comply with admonishments include the following:

Cosmetic and social motives - The most widely cited motivational barrier to sun protection is the perceived physical attractiveness associated with having a tan (Mahler, et al., 2003). The research indicates that the primary motivation for tanning is to attain a tan, and the primary motive for tanning, especially with females, is perceived attractiveness. Participants report that they believe that they look and feel more attractive and healthy with a tan (e.g., Cokkinides, et al., 2001; Jones, et al., 2000; Mahler, et al.). In several studies, sun protection behaviors and in particular using sunscreen, are perceived as 'very important' by most participants in the sample. However, studies also indicate that the number of perceived benefits associated with sun exposure outweigh the harms (Jones, et al.). For example, Branstrom, et al. (2001) found that adolescents who perceived sunbathing as 'harmful' reported sunbathing just as often as those who believed sunbathing was 'healthy'.

Optimistic bias – One difficulty in attempting to effect health behavior change is that individuals are generally unrealistically optimistic about their risks for a variety of potential health problems and are resistant to feelings of personal susceptibility (Weinstein, 1983). Weinstein (1980) defined optimistic bias as perceiving oneself as less susceptible than others to unpleasant occurrences and reduces the worry about a range of health problems and decreases the likelihood that people will take preventative actions against these problems. Optimistic bias has been demonstrated in relation to a number of behaviorally related illnesses including lung cancer, asthma and smoking (Clarke, et al., 1997). Clarke, et al. argued that there is an optimistic bias in relation to the risk of getting skin cancer and in the mean number of years lost through skin cancer. Similarly, Penningroth (1996) found that participants showed optimistic bias for skin cancer risk and it was most strongly associated with higher perceived benefits from the sun and tanning.

Perceived consequences – Numerous studies suggest that health related decision making and behaviors are influenced by the perception of the probability of experiencing negative consequences of engaging, or not engaging, in the behavior (Turrisi, 1999). The use of sunscreen may have both perceived positive (prevents skin damage and skin cancer) and negative consequences (prevents tanning). Additionally, these consequences are associated with positive immediate (tanning) and negative long-term consequences (skin damage and cancer). It is possible that the perceptions of the negative consequences of excessive sun exposure, such as skin cancer, are considered too far off in the distant future to warrant an immediate behavior change (Branstrom et al., 2001). How individuals perceive these consequences will influence their decision making process about sunscreen use. As the literature has indicated, the perceived positive effects of sun exposure outweigh the negative long-term effect of increasing ones risk for skin cancer, especially among adolescents and young adults.

Risk acceptance – The perceived risk of a situation influences behavior, including the extent to which consumers will take preventative actions. Research suggests that risk perceptions are determined by two variables: the likelihood of injury and the severity of potential consequences (Slovic, 2000; Wogalter, et al., 1991). Slovic suggested that a number of factors influence the acceptability of risk across a range of potential hazards. As it relates to sun exposure, the potential danger has several characteristics that are associated with acceptance of greater levels of risk, including:

- the voluntary nature of this activity,
- consumers are familiar with the products and activities,
- the risk can be controlled by the individual,
- the risk is generally known or knowable, and
- the consequences are not immediate.

Label comprehension – There is a considerable amount of literature addressing the design of warnings and the behavioral responses to safety messages (e.g., Ayres, et al., 1998; DeJoy, 1989; Kalsher & Williams, 2006). For warning labels to evoke changes in behavior, it is necessary that the safety messages are understandable and provide the necessary information for consumers to use the product safely and effectively for its intended purposes. The FDA has used label comprehension testing results for many OTC drug products (Morris, et al., 1998) and has recently published a guide to provide recommendations to industry on conducting label comprehension studies. However, the rulemaking process for sunscreen labels is noticeably absent of any reference to published label comprehension studies or published human factors research concerning the effectiveness of the proposed label design.

Cognitive dissonance - The results of these studies are interesting because it suggests the possibility of cognitive dissonance; the unpleasant internal state that arises when individuals notice inconsistency between their attitudes and their behaviors. The literature suggests that especially with young people, the behavior of using sunscreen along with other sun protection behaviors is likely to be influenced by several competing beliefs, such as the desire to prevent skin cancer versus the desire to achieve a tan (Kasparian, et al., 2009). Robinson, et al., (1997) suggested that it is possible that these competing beliefs are reconciled through inadequate sunscreen use (using sunscreens with a low SPF, or failure to reapply sunscreen during extended sun exposure). This may lead young people to believe that they are complying with sun safety messages, while still achieving the desired tan.

When attempting to design effective warnings and safety messages that have the ability to change sun protection behavior, these psychological factors should be considered. As we will discuss in the next section, the FDA appears to have identified many of these issues and is attempting to address them in future sunscreen labeling.

SUNSCREEN REGULATIONS

The process by which the FDA promulgates regulations for OTC products involves a number of complicated steps that result in the publication of a series of notices or monographs in the Federal Register.

For sunscreens, this process started with the publication of the "Sunscreen Drug Products for Over-the-Counter Human Use: Establishment of a Monograph; Notice of Proposed Rulemaking" in 1978. In addition to describing the ingredients and concentrations that could be used in sunscreens, this initial monograph described the testing and labeling for efficacy, the sun protection factor (SPF) (FDA, 1978).

After 21 years, the FDA issued a final rule in 1999. Unfortunately, it has never gone into effect. The effective date was originally set as May 21, 2001, which was then extended, and has since been stayed indefinitely so that the FDA can address formulation, labeling and testing requirements for sunscreens providing UVA protections (see 21 CFR 352). Currently, sunscreen manufacturers are encouraged, but not required, to adhere to the regulations set forth in the 1999 final monograph. The industry appears to voluntarily comply with these guidelines, except the provisions regarding UVA issues, establishing the current custom and practice for manufacturers of sunscreen products sold in the United States.

HISTORY OF SUNSCREEN LABELING

The proposed rules of 1978 included the addition of one of two statements on the labeling that states that the product may reduce harmful effects of the sun including premature aging of the skin and skin cancer. The document makes note that there is insufficient data to make a claim that these products will prevent cancer and instead the language will read 'may reduce harmful effects of the sun' (FDA, 1978, pg. 38212). Since 1978, the FDA has developed a number of amendments addressing the evolution of new information about the stability and toxicity of various ingredients. However, the rules for sunscreen labels did not change significantly until the final rule was published in 1999.

Warnings - Since 1978 the warnings that were proposed for sunscreen labeling include specific safety messages regarding hazards that are intrinsic to the product itself (e.g. to avoid contact with the eyes) rather than the hazards of UV radiation and sun exposure. This falls in line with the FDA's role and approach in regulating OTC drug products, which includes sunscreen products. Prior to the final rule of 1999, there were admonitions permissible in the labeling of sunscreen products that indicated they were potentially an effective measure in the prevention of skin cancer but stopping short of saying that these products will in fact prevent skin cancer. The 2007 proposed rule included a prominent warning in the indications section of the labeling regarding skin cancer and skin damage in unambiguous language: "UV exposure from the sun increases the risk of skin cancer, premature skin aging, and other skin damage. It is important to decrease UV exposure by limiting time in the sun, wearing protective clothing, and using a sunscreen" (FDA, 2007, pg. 49113). This safety message communicates to readers that sunscreen use alone is not a sufficient protective action to prevent skin cancer.

Sun or Sunburn Protection Factor - The use of the term SPF on sunscreen product labeling originates in the 1978 proposed rules that adopt the use of the SPF rating already in use in Europe at that time as a guide for the consumer towards understanding the efficacy of the various sunscreen formulations (FDA, 1978, pg. 38213).

A substantial change to the SPF labeling was made in 2007, when the FDA proposed to change the words associated with the acronym to "sunburn protection factor" instead of "sun protection factor." Additionally, the term SPF will be associated with UVB, as in UVB SPF or UVB sunburn protection factor (FDA, 2007). This change is a shift in the representation of the uses and limitations of this class of product. This change is likely to go unnoticed by consumers as the acronym SPF is permissible on the label without the constituent words.

Substantivity – The effectiveness of a sunscreen product can be reduced by sweating and swimming. Resistance to removal by water and sweat is an important aspect of sunscreen performance. The ability of a sunscreen to bind to the skin and resist removal during swimming or sweating is known as substantivity (Agin, 2006). The 1978 proposed rule established the first labeling requirements for substantivity. A "water resistant" SPF claim meant that the SPF of a product "Retains its sun protection for at least 40 minutes in the water"; "waterproof" meant that the SPF of the product "Retains its sun protection for at least 80 minutes in the water"; and "sweat resistant" meant that the product "Retains is sun protection for at least 30 minutes of heavy sweating" (FDA, 1978). The 1999 final rule, which has been indefinitely stayed, mandated a change in the descriptors to "water resistant" and "very water resistant" and required the label to state "retains SPF after [40 or 80] minutes of" (select one or more of the following "activity in water," "sweating," or "perspiring"), no longer allowing the terms "waterproof" and "sweat resistant" to be used on the label.

UVA Labeling – In the 2007 Proposed Rule the FDA, for the first time, defined for manufacturers how to test and label for UVA protection. The most prominent change resulting from the new rule is a requirement for sunscreen product labels to include the UVA protection of the product. In addition to the UVB SPF number, the label will include a 1-4 star UVA rating indicating low to highest overall UVA protection and a UVA category descriptor (FDA, 2007).

Table 1. Star rating system for overall UVA protection (FDA, 2007).

Star category	Category descriptor
★☆☆☆	Low
★★☆☆	Medium
★★★☆	High
★★★★	Highest

LIMITATIONS OF LABELING

In 2000 the FDA stated that they are "concerned that an average sunscreen consumer may ascribe more to high SPF values than is clinically relevant and that such products may further encourage the use of sunscreens as a safe way to prolong sun exposure" (FDA, 2000, pg. 36322). The document goes on to say: "In addition, previously submitted labeling comprehension data, which were discussed at a public meeting (Ref. 14), indicated a fair amount of confusion concerning consumer comprehension of the SPF rating system" (pg. 36323).

If the most recent proposed rules in the process are initiated into the marketplace, then the result will be sunscreen products labeled with two separate ratings presented on two different scales; one providing a numeric value with a statement of low/medium/high/highest "UVB sunburn protection" while the other denoted in a quantity of stars with a statement of low/medium/high/highest "UVA protection." While it may be reasonable for the FDA to assume that a consumer may be able to appreciate the generalities of either scale (i.e., lower rating means less protection, higher rating means more protection), there is a lack of research to determine how this combination of scales will be interpreted by potential consumers. Questions abound: will consumers understand that one rating relates to a different spectrum of potential radiation protection? Will consumers be confused about which spectrum of radiation causes what harmful effects? Will consumers tailor their purchasing habits to one scale or the other? These questions and others remain unanswered.

The proposed rule of 2007 referenced several comments to the docket that purported to have data from labeling comprehension studies (the Division of Dockets Management informed the authors that they were not publicly available). These studies came to conflicting and contradictory conclusions about which manner of rating the efficacy of the UVA protection was most preferred and most comprehended by sunscreen users (FDA, 2007, pg. 49082-83). A proposed comprehension study on this proposed method of rating UVA protection by the Personal Care Products Council was offered to be conducted when they requested the FDA extend its period for commentary (FDA, 2007, C2462). The request for an extension was ultimately denied (C2463). It is not apparent from the rulemaking materials available that the FDA has conducted any similar comprehension studies to date.

The FDA, forced to make a determination from this conflicting body of evidence, proposed that UVA protection be denoted with a star rating and a category descriptor. Again this approach had previously been adopted in Europe, and these star ratings are a second attempt at harmonizing their labeling approach to international guidelines already in place (FDA, 2007, pg. 49083-84).

The rulemaking process indicates an approach that is focused on the effectiveness of the individual products marketed rather than a user-centered approach focused on consumer comprehension and perception of the risks involved. As discussed, the warnings on sunscreen labeling primarily identify the hazards of the sunscreen and not exposure to the sun. There is no research to indicate that consumers use the sunscreen container that they purchase as a primary source of information regarding their sun exposure habits and the associated risks; therefore any attempt to affect those sun exposure behaviors via the sunscreen labeling may be challenging. The FDA has identified several potential issues regarding the proposed labeling (e.g. comprehension, hazard communication) but noticeably absent from the rulemaking has been any overt reference to the published human factors literature regarding the effectiveness and design of warnings and safety communications.

As has been discussed, there is widespread knowledge about the risk of skin cancer therefore, labeling and warnings may provide limited benefit to those that understand and accept the risks of sun exposure. The labeling provided on sunscreen products will be most beneficial for those users that are ignorant to the potential hazards of sun exposure and the limitations of sunscreen effectiveness.

CONCLUSIONS

Skin cancer represents a major, growing public health problem in the United States, affecting approximately 1 million Americans each year (ACS, 2009). Future skin cancer incidence in today's young people can be reduced through the implementation of a widely recommended (ACS, CDC, FDA, etc.) sun protection regimen including the use of sunscreen. In light of the escalating rates of skin cancer, it is clearly important to develop effective means of communicating and presenting the warnings and other safety information on the sunscreen container are developed.

As the FDA is still finalizing the 2007 proposed rule, including the new label for UVA protection, it is important to determine whether consumers will be able to understand the information on the sunscreen container. The FDA has acknowledged the potential for consumers to be misled and stated that they want to provide information so that the consumer can make an informed decision about sunscreen use. However, under the proposed rule, the FDA will be requiring consumers to use a label with a specific scale designed to assist them make decisions about the effectiveness of the product, without empirical information concerning how the new scale will actually be interpreted.

Especially because the UVA protection terminology will be new, it is imperative that consumers perceive the terminology that would be consistent with UVA protection tests. It is important that consumers' expectations and perceptions regarding the level of sun protection afforded by sunscreen are consistent with the levels implied through the new rating system. One method that can offer invaluable information is label comprehension testing. In the context of this new labeling scheme, comprehension testing can offer the potential to improve the quality of the information offered to the consumers, possibly leading to increased, proper use of sunscreen as one of several sun protection behaviors.

REFERENCES

American Cancer Society (2009). *Cancer Facts & Figures 2009*. Atlanta: American Cancer Society; 2009.

Agin, P. (2006). Water resistance and extended wear sunscreens. *Dermatology Clinics, 24,* 75-79.

Ayres, T., Wood, C., Schmidt, R., Young, D. & Murray, J. (1998). Effectiveness of warning labels and signs: An update on compliance research. *Proceedings of the Silicon Valley Ergonomics Conference and Exposition,* 199-205.

Balk, S., O'Connor, K., Saraiya, M. (2004). Counseling parents and children on sun protection: A national survey of pediatricians. *Pediatrics, 11, 4*(4), 1056-1064).

Boldeman, C., Branstrom, R., Dal, H., Kristjansson, S., Rodvall, Y., Jansson, B. (2001). Sunbed use in a Swedish population age 13-50 years. *European Journal of Cancer, 37,* 2441-2448.

Branstrom, R., Brandberg, Y., Holm, L. (2001). Beliefs, knowledge and attitudes as predictors of sunbathing habits and use of sun protection among Swedish adolescents. *European Journal of Cancer Prevention, 10,* 337-345.

Clarke, V., Williams, T., Arthey, S. (1997). Skin type and optimistic bias in relation to the sun protection and suntanning behaviors of young adults. *Journal of Behavioral Medicine, 20*(2), 207-222.

Cokkinides, V., Johnston-Davis, K., Weinstock, M., O'Connell, M., Kalsbeek, W. (2001). Sun exposure and sun-protection behaviors and attitudes among U.S. youth, 11 to 18 years of age. *Preventive Medicine, 33,* 141-151.

DeJoy, D.M. (1989). Consumer product warnings: Review and analysis of effectiveness research. *Proceedings of the Human Factors Society 33rd Annual Meeting,* 936-940.

Food and Drug Administration. (1978). Sunscreen Drug Products for Over-the-Counter Human Use; Final Monograph. Docket No. 78N-0038. *Federal Register, 45*(98), 38208-38269.

Food and Drug Administration. (1999). Sunscreen Drug Products for Over-the-Counter Human Use; Final Monograph. Docket No. 78N-0038. *Federal Register, 64*(98), 27666-27693.

Food and Drug Administration. (2000). Sunscreen Drug Products for Over-the-Counter Human Use; Extention of Effective Date. Docket No. 78N-0038. *Federal Register, 65*(111), 36319-36324.

Food and Drug Administration. (2007). Sunscreen Drug Products for Over-the-Counter Human Use; Proposed Amendment of Final Monograph; Proposed Rule (2007). Docket No. 1978N-0038. *Federal Register, 72*(165), 49070-49122.

Jones, F., Harris, P., Chrispin, C. (2000). Catching the sun: An investigation of sun-exposure and skin protective behavior. *Psychology Health & Medicine, 5,* 131-141.

Kalsher, M.J. & Williams, K.J. (2006). Behavioral compliance: Theory, methodology, and results. In M.S. Wogalter (Ed.), *The handbook of warnings* (pp. 313-331). Mahwah, NJ.

Kasparian, N. & McLoone, J. (2009). Skin cancer-related prevention and screening behaviors: a review of the literature. *Journal of Behavioral Medicine, 32,* 406-428.

Mahler, H., Kulik, J., Gerrard, F., Harrell, J. (2003). Effects of appearance-based interventions on sun protection intentions and self-reported behaviors. *Health Psychology, 22*(22), 199-209.

Morris, L., Lechter, K., Weintraub, M., Bowen, D. (1998). Comprehension testing for OTV drug labels. *Journal of Policy & Marketing, 17*(1), 86-96.

Nicol, I., Gaudy, C., Gouvernet, J., Richard, M., Grob, J. (2006). Skin protection by sunscreens is improved by explicit labeling and providing free sunscreen. *Journal of Investigative Dermatology, 127,* 41-48.

Newman, W., Agro, A., Woodruff, S., Mayer, J. (1996). A survey of recreational sun exposure of residents of San Diego, California. *American Journal of Preventive Medicine, 12*, 186-194.

Polster, A., Lasek, R., Quinn, L., Chren, M. (1998). Reports by patients and dermatologists of skin cancer preventative services provided in dermatology office. *Archives of Dermatology, 134*, 1095-1098.

Robinson, J.K., Rigel, D.S., Amonette, R.A. (1997). Trends in sun exposure knowledge, attitudes and behaviors: 1986-1996. *Journal of the American Academy of Dermatology, 37*, 179-186.

Robinson J.K., Rigel D.S., Amonette R.A. (2000). Summertime sun protection used by adults for their children. *Journal of the American Academy of Dermatology, 42*, 746–753.

Santmyire, B.R, Feldman, S.R., Fleischer, Jr., A.B. (2001). Lifestyle high-risk behaviors and demographics may predict the level of participation in sun-protection behaviors and skin cancer primary prevention in the United States. *Cancer, 92*(5), 1315-1324.

Saraiya, M., Glanz, K. Briss, P.A. (2005). Interventions to prevent skin cancer by reducing exposure to ultraviolet radiation. *American Journal of Preventative Medicine, 27*(5).

Slovic, P. (Ed.). (2000). *The perception of risk.* London: Earthscan.

Thompson, S.C., Jolley, D., Marks, R. (1993). Reduction of solar keratoses by regular sunscreen use. *New England Journal of Medicine, 329*, 1147-51.

The Skin Cancer Foundation (2009). The Skin Cancer Foundation Website http://www.skincancer.org.

Turrisi, R., Hillhouse, J., Gerbert, C., Grimes, J. (1999). Examination of cognitive variables relevant to sunscreen use. *Journal of Behavioral Medicine, 22*(5).

Weinstein, N.D. (1980). Unrealistic optimism about future life events. *Journal of Personality and Social Psychology, 39*(5), 806-820.

Weinstein, N.D. (1983). Reducing unrealistic optimism about illness susceptibility. *Health Psychology, 2*(1), 11-20.

Wogalter, M.S., Brelsford, J., Desaulniers, D. (1991). Consumer product warnings: The role of hazard perception. *Journal of Safety Research, 22*, 71-82.

CHAPTER 105

The Safety Hierarchy and Its Role in Safety Decisions

Kenneth R. Laughery[1], Michael S. Wogalter[2]

[1]Psychology Department
Rice University
MS-25, P.O. Box 1892
Houston, Texas 77251-1892, USA

[2]Psychology Department
North Carolina State University
640 Poe Hall
Raleigh, NC 27695-7650, USA

ABSTRACT

The safety hierarchy, or hazard control hierarchy, is a priority scheme for dealing with product hazards. It is often referred to as the design, guard and warn sequence. In order of preference, alternative designs that eliminate or reduce the hazard should be given first consideration. Where alternative designs are not feasible, guarding is the next preferred approach. Guarding can be viewed as an effort to prevent contact between the product user and the hazard. But like alternative designs, guarding is not always a feasible solution. Warnings are the third line of defense. Warnings are intended to provide information needed to use the product safely. Several issues and/or questions are explored regarding the application of the hierarchy to product safety. Examples are presented as a context for exploring some of the issues.

Keywords: hazard, safety, hierarchy, warning, design, guarding, strategy, control

INTRODUCTION

There is a concept in safety, as well as in human factors, engineering and other disciplines, known as the safety hierarchy, or alternatively the hazard control hierarchy (National Safety Council, 1989; Sanders and McCormick, 1993). This concept concerns a priority scheme for dealing with hazards. In this article we explore some of the issues associated with the concept as it is applied to consumer products. The basic sequence of priorities in the hierarchy consists of three approaches: first to design it out, second to guard, and third to warn.

If a hazard exists with a product, the first step is to try to eliminate or reduce it through an alternative design. If a nonflammable propellant in a can of hair spray can be substituted for a flammable carrier and still adequately serve its function, then this alternative design would be preferred. Eliminating sharp edges on product parts or pinch points on industrial equipment are examples of eliminating hazards. But safe alternative designs are not always available.

The second approach to dealing with product hazards is guarding. The purpose of guarding is to prevent contact between people and the hazard. Guarding procedures can be divided into two categories: physical guards and procedural guards. Personal protective equipment such as rubber gloves and goggles, barricades on the highway, and bed rails on the side of an infant's crib are examples of physical guards. Designing a task so as to prevent people from coming into contact with a hazard is a procedural guard. An example would be the controls on a punch press that require the operator to simultaneously press two switches, one with each hand, ensures that fingers will not be under the piston when it strokes. Another example is a physician's prescription for a medication. Without it, the medication cannot be obtained. However, guarding, like alternative designs, are not always feasible solutions for dealing with hazards.

The third line of hazard defense is warnings. Warnings can be thought of as safety communications. One of the purposes of a warning is to provide to people the information needed to make informed decisions about how to use a product safely, including the choice on whether to use it at all. Warnings are third in the priority sequence because they are generally less reliable than design or guarding solutions. Even the best warnings are not likely to be 100% effective. People at risk may not see or hear a warning, or they may not understand it. Further, even warnings that are seen and/or heard and understood may not be successful in motivating compliance. It is these reasons that warnings are the third strategy in hazard control, behind design and guarding. Influencing human behavior is often difficult and seldom foolproof. A short comment related to these points makes sense to mention at this point. These concerns about reliability should not be regarded as a basis for not warning when appropriate to do so. Rather, warnings are one tool available to product manufacturers and designers for dealing with product safety, and they have an appropriate role in the safety hierarchy.

There are other approaches to dealing with product hazards, such as training (influencing how the product is used), personnel selection (influencing who uses it), and administrative controls (employer/supervisor sets and enforces rules). In the context of dealing with product hazards, these approaches are viewed as similar to warnings in that they mostly involve efforts intended to inform and influence behavior.

ISSUES ASSOCIATED WITH THE HIERARCHY

There are numerous questions or issues that may arise when applying the safety hierarchy. A starting point, of course, is to have a good understanding of the product hazards. While it is not within the scope of this short article to discuss the goals and methods of hazard analysis, there are two noteworthy points worth mentioning. The first point is that there are formal analytic procedures and/or tools for carrying out a product hazard analysis (Frantz, Rhoades, and Lehto, 1999). Examples of such procedures are fault-tree analysis and failure modes and effects analysis. Such procedures are widely recognized and practiced. A second point to note is that hazard analysis is, or should be, viewed as part of the design stage of product development. Hazard analysis ought to be carried out before it is made available to consumers. A product hazard that does not become recognized until the product has been in the marketplace can be costly both financially and with regard to safety outcomes. Recalls and retrofits are not a good substitute for timely and competent hazard analyses.

Once product hazards have been identified, whether through hazard analysis during design or through feedback after the product has been marketed (data about injury or health effects), the safety hierarchy comes into play in terms of decisions about how to address the hazards. In the following sections, we discuss some of the issues involved in such decisions.

Alternative Designs

The usually stated rule of thumb about when to implement an alternative design in dealing with a product hazard is "if a technologically and economically feasible alternative design is available, it should be implemented." Obviously the decision about whether to implement the alternative design is more complex than this phrase might imply. Clearly, alternatives must be technically possible, such as whether nonflammable carriers in hair sprays can be produced or whether there is a way to reduce automotive tire deterioration due to aging processes. But decisions about alternative designs must include consideration of issues such as the reliability and adequate function. If the alternative detracts from the effectiveness of the hair spray or causes the tire tread to wear faster, the alternative may not be an acceptable option, even though it addresses the hazard that led to its consideration.

It is also necessary to take into account economic feasibility is considering alternative designs. If the cost of eliminating a hazard with an alternative design is prohibitively expensive, it may not be an acceptable fix. Here again, however, the economically feasible decision may be considerably more complex than meets the eye. Such considerations also are not within the scope of this article, but one factor that is sometimes suggested or considered, rightly or wrongly, and that is the potential cost of defending lawsuits based on safety issues associated with the product.

One additional point related to decisions about alternative designs is not so much a technological or economic issue. This point concerns the situation where an alternative design that eliminates the hazard is feasible on both technical and economic dimensions, but its implementation creates another hazard. Perhaps an example would be a nonflammable carrier for hair spray that is extremely toxic if it gets into the eyes. Likewise, the harm could be to the environment that could indirectly affect of health of users and others. The carrier in hairsprays used to be chlorofluorocarbons (CFCs), but these were found to negatively influence the ozone layer and increase greenhouse gases, and it was banned from use in the U.S. and some other countries. Obviously, alternative designs that create as many or more hazards as they solve is not the intent of the safety hierarchy. The decision to ban CFCs was made to reduce a societal, environmental hazard but resulted in an increased personal use hazard.

Some Factors that Influence Decisions

In the above section on alternative designs, a few factors were described that influence decisions about how to address product hazards. Technological and economic feasibility and the potential creation of other hazards were noted. There are other factors that can play a role in deciding how to address hazards. One factor is what the consumer wants or will accept; or alternatively, what the manufacturer believes the consumer wants or will accept. An example of this issue in the context of a consumer product will help make the point. Most vehicles marketed in the U.S. have front seats that can be reclined to a nearly horizontal position. (Pickup trucks with bench seats are an example of an exception.) It is generally agreed that it is hazardous for a passenger to have the seat significantly reclined to where the shoulder belt is not in contact with the torso while the vehicle is moving. The problem is that when the occupant is in the reclined position, the restraint system loses its effectiveness. Virtually all manufacturers now warn in the vehicle owner's manual not to recline the seat while the vehicle is in motion. While the quality of such warnings varies, the warning approach has been chosen for addressing the hazard—the third line of defense in the safety hierarchy. Studies show that most people are unaware of this hazard, although when called to their attention, people do understand it (Leonard, 2006; Leonard & Karnes, 1998; Paige & Laughery, 2003; Rhoades & Wisniewski, 2004). Laughery and Wogalter (2008) have explored the use of warnings to address this hazard.

But an alternative approach exists for addressing the seat recline hazard. Apparently it would be technically and economically feasible to design the seat so that it cannot recline to an unsafe angle. In terms of the safety hierarchy, it would be a preferred solution compared to a warning approach. The point here is that vehicle manufacturers have taken into account at least two other factors in deciding to address the seat recline hazard with warnings. First, they considered a marketing factor based on the belief that customers want the seat recline feature. The second factor cited is that in circumstances where the driver is experiencing fatigue, it will be possible to rest by stopping and reclining the seat, a safety consideration.

There is also a guarding approach that has been proposed for addressing the seat recline hazard. This approach involves a classic "kill switch," a name that is unfortunate as it simply means to turn off the power to the product or equipment. The point is that, the vehicle cannot be driven from a stopped condition if the seat is reclined beyond some safe angle, and if the engine is running, the seat will not recline. Note that this guarding solution permits the fatigued driver to stop the vehicle, recline the seat, and rest. Like the above design alternative, it is likely to be more successful than warnings in dealing with the seat recline hazard. Note that there may be other design solutions, such as designing the restraint system so it also works while in a reclined belted position.

Warning Versus Alternative Design Versus Guarding

The above seat recline example illustrates a product where the hazard is understood and there are options as to how to deal with it. More specifically, there is a choice between a technologically and economically feasible alternative design, or guarding, or warnings. Note that successful the design and guarding options need to be fail-safe, unless of course there is some kind of structural failure or successful effort to override the kill switch. The effectiveness of a warning option depends on the communications successfully informing and motivating the occupant not to recline the seat in the moving vehicle. The differences in effectiveness, of course, illustrate the underlying value or purpose of the safety hierarchy.

Another example of a consumer product where the safety hierarchy could or should come into play is a turkey fryer. The base or stand for such a fryer, or cooker, is shown in Figure 1(a). A large aluminum pot sits on top of the propane-fueled base shown in the figure. A typical application or use of the product would be to put cooking oil such as peanut oil in the pot and cook turkey parts or other meat.

Figure 1. (a) Poultry Roaster, (b) Modified Poultry Roaster

A significant hazard associated with this product is that it is unstable and can tip over if intentionally or unintentionally bumped or moved. The resulting hot oil spill can result in severe or catastrophic burns. Such incidents have occurred in situations such as outdoor picnics or similar events where children or animals may be active in the vicinity of the cooker.

The cooker comes with an owner's manual. The manual contains a warning that includes a statement that the hot oil can cause severe burns and advising to keep children and pets away. Note that the instruction to keep children and pets away is an example of a warning recommending a guarding solution. Our concern here is not to evaluate the adequacy or inadequacy of the warning. Rather, the intent is to explore how the tip over hazard could or should be addressed from the perspective of the safety hierarchy.

There are several design aspects of the turkey fryer that contribute to its instability. Included among these characteristics are: the width of its base, the height of its center of gravity, and the fact that it has only three legs. In terms of alternatives, these are design features that can be improved in ways that result in a significant increase in stability. For example, adding a fourth leg, lowering the center of gravity by shortening the legs, or adding a ring at the base of the legs as shown in Figure 1(b) are examples of design alternatives that are readily achievable.

A FEW SUMMARY COMMENTS

The examples of the vehicle seat recline hazard and the turkey fryer tip over hazard were presented as a context for exploring some of the kinds of issues encountered in deciding how to address product hazards. The safety hierarchy provides some principles and/or guidelines based on what is likely to be most effective; that is, the design, guard and warn priority scheme. But, as indicated with the seat recline example, decisions about whether to seek solutions based on alternative design, guarding or warning may be complex. In addition to technological and economic feasibility, factors come into play such as secondary safety effects and customer preferences.

Sometimes, however, the decision may be relatively straightforward, as (we believe) is the case with the turkey fryer. Clearly it does not require a revision of Newton's laws of physics to come up with a more stable cooker at what would appear to be a modest, if any, increase in cost. Certainly in comparison to a warning that recommends a guarding solution (keep children and pets safely away), the design alternative that increases stability would appear to be more effective. The point, however, is not to suggest that children and pets need not be monitored around the fryer or that a warning spelling out the potential severe burn consequences of a tip over is not appropriate. Rather the point is that guarding and warnings should be viewed as a complement to better, safer design, not as a substitute for it.

We end this paper with a comment on the complimentary aspects of the design, guard, warn safety hierarchy. The hierarchy should not be viewed as a priority scheme consisting of three options from which a selection can/must be made. Rather, it defines a preference scheme based on what is likely to be most effective from a safety perspective. It is not meant to imply some sort of exclusion principle; for example, if you guard (such as putting up a fence around a power station), that there is no need to warn (hang a warning sign on the fence that emphasizes danger and not to enter). Instead, the matter may be better thought of as: even with a better design, it may still be appropriate to guard or warn, or both.

REFERENCES

Frantz, J.P., Rhoades, T.P. and Lehto, M.R. (1999) Practical considerations regarding the design evaluation of product warnings. In *Warnings and Risk Communication*, Wogalter, M.S., D D.M. and Laughery, K.R., eds. Taylor & Francis, Philadelphia, PA, 291-311.

Laughery, K.R. & Wogalter, M.S. (2008) On the symbiotic relationship between warnings resear forensics. *Human Factors, 50,* 3, 329-333.

Leonard, S.D. (2003) Who really knows about reclining the passenger seat? *Proceedings of the H Factors and Ergonomics Society 50th Annual Meeting (pp.855-859).* Santa Monica, CA: H Factors and Ergonomics Society.

Leonard, S.D. & Karnes, E.W. (1998) Perception of risk in automobiles: Is it accurate? *Proceedi the Human Factors and Ergonomics Society 42nd Annual Meeting (pp.1083-1087).* Santa Monica, CA: Human Factors and Ergonomics Society.

National Safety Council. (1989) *Accident Prevention Manual for Industrial Operation, 5th Editio* National Safety Council, Chicago, IL.

Rhoades, T.P. & Wisniewski, E.C. (2004) Judgments of risk associated with riding with a recline in an automobile. *Proceedings of the Human Factors and Ergonomics Society 48th Annual Meeting (pp. 1136-1139).* Santa Monica, CA: Human Factors and Ergonomics Society.

Sanders, M.S. & McCormick, E.J. (1993) *Human Factors in Engineering and Design, 7th Edition* McGraw-Hill, New York.

Paige, D.L. & Laughery, K.R. (2003) Risk perception: The effects of technical knowledge – or la it. *Proceedings of the XVth Triennial Congress of the International Ergonomics Associatio* Seoul, Korea: International Ergonomics Association.

Chapter 106

Reliance on Warnings as the Sole Remedy for Certain Hazards: Some Circumstances Where That Just Doesn't Work

Edward W. Karnes [1], David R. Lenorovitz [2], S. David Leonard [3]

[1] Edward W. Karnes, LLC
Morrison, CO, USA

[2] LENPRO Services, Inc.
Englewood, CO, USA

[3] University of Georgia
Athens, GA, USA

ABSTRACT

Some products manufactured and distributed for use by people contain hazards that may cause injuries. When such hazards are known to (or should have been known by) the manufacturers, they are obligated to design out, guard against, or, at the very least, provide users with an adequate warning of the associated danger. Warnings are at the lowest level in this hierarchy of remedial procedures principally because their success is dependent upon eliciting certain types of behaviors on the part of the product user. Some warnings developed for this purpose may be inadequate due to high costs of compliance, the requirement for users to perform behaviors beyond their capabilities, or because they require users to perform actions that are contrary to other user behaviors that may have become well entrenched through repetition or practice – i.e., ones having become conditioned responses or learned reflexes. This article describes some circumstances involving two different types of vehicles – a utility terrain vehicle and a personal watercraft – in which certain warnings were judged inadequate for just these types of reasons. The paper also examines actual or possible modifications to the products likely to correct the problems encountered.

Keywords: Warnings, Hazards, Hazard Control Hierarchy, Product Safety, Hazard Remedies, Reactive Motor Responses, Conditioned-Reflex Actions.

INTRODUCTION

It is widely acknowledged in engineering, technical, and legal disciplines that product manufacturers have a duty to exercise reasonable care in the design of products to provide safety for both planned uses and foreseeable misuses of the products that they manufacture. Reasonable care requires the manufacturer to provide remedies for identified hazards, especially serious hazards. A serious hazard is one that: may be unknown to or hidden from product users; has a reasonable likelihood of being encountered by users during foreseeable use (or likely misuse) of the product; and, most importantly, involves consequences of serious injury or death. The recognized, and well-accepted safety standard for providing remedies for hazards created by foreseeable use of products is the hazard control hierarchy (also referred to as the safety design hierarchy and/or systems safety engineering approach) originally developed in the military services during and shortly after World War II. The hazard control hierarchy serves as a guiding principle for proper design to provide safety for foreseeable uses and misuses of products.

The forensic human factors literature contains over thirty years of research addressing such topics as when and why to use warnings, how to best to configure warnings, and what factors or features are likely to make warnings more or less effective (Laughery, Wogalter & Young, 1994; Wogalter, DeJoy & Laughery, 1999; Wogalter, Young, & Laughery, 2001; Wogalter, 2006a). However, from a human factors and safety standpoint, relying on a warning – even an adequately composed and adequately located warning – as **the only** remedy for a given product hazard is often inadequate. This paper addresses two situations where warnings were initially provided as the sole remedy for identified product hazards. Before discussing those situations, however, it is necessary to examine the role of product warnings as remedies for product hazards.

References to the hazard control hierarchy and the systems safety principle have been addressed in the human factors, safety, and engineering literature for decades (Christensen, 1987, National Safety Council, 1989). The hazard control hierarchy requires a product manufacturer to provide hazard remedies by adopting the following priorities: (a) design out – i.e., eliminate the hazard (assuming that it is technologically and economically feasible to do so); (b) shield, protect, or guard the user from coming into contact with or being impacted by the hazard, and/or (at the very least); (c) adequately warn the user about the presence and consequences of the hazard, and recommend steps to avoid the hazard. The principle of systems safety requires that safety be designed into the product or system. Hazards should be eliminated or minimized in the design stage, and product designers should not rely on user's compliance with instructions and warnings as a substitute for designing a safe product in the first place.

While instructions and warnings are clearly less than perfect "solutions" for product hazards, they are nevertheless important aspects of product safety. Warnings, however, are no substitutes for safe design. Not even a very good warning can ever really make up for or overcome a poor (hazardous) design feature (Wogalter, 2006b).

The following two sections describe situations where a warning or a user's precautionary instruction was offered as the sole means of remedying an identified type of hazard (in lieu of redesigning the product or providing a means of shielding or guarding against the hazard), and discuss why such efforts were inadequate or ineffective in preventing or mitigating injuries associated with that hazard.

LIMB INJURIES IN UTILITY TERRAIN VEHICLE ROLL-OVER INCIDENTS

Utility terrain vehicles (UTVs) are off-road motorized vehicles that are designed for an operator and a passenger and provide a cargo bed. The vehicles are very popular for both recreational and utilitarian purposes, and provide safety accommodations in the form of roll bars and seat belts. However, numerous complaints, received by at least one manufacturer of UTVs, reported rollovers of the vehicles on flat, obstacle-free, open areas during turning maneuvers at low to moderate speeds, resulting in the occurrence of severe injuries to extremities – e.g., when an operator's or passenger's leg, foot, or arm came out of the vehicle during a rollover and was crushed by various portions of the vehicle's roll bar or other vehicle structure. The design of the UTV incorporated essentially open sides in the occupant compartment.

The manufacturer of the UTV had identified the hazards of vehicle rollovers and likely limb injuries prior to marketing the vehicle and had provided warnings as a remedy for these vehicle-instability and limb-injury hazards. Soon after the vehicle was introduced into the marketplace, the company became keenly aware of a continuing problem of limb injuries (mostly to the legs and feet) in rollover incidents. About three years after the UTV had been introduced into the marketplace, the company mailed significantly improved new warning labels to its UTV owners, warning riders to keep their arms and legs inside the vehicle in situations where the vehicle may be in danger of tipping over.

The UTV manufacturer, in response to legal complaints of injuries and deaths in

UTV rollover incidents, maintained that leg extensions were entirely voluntary responses performed with conscious deliberation and intention. The manufacturer claimed the injured persons negligently failed to comply with the adequate limb-injury warnings provided, and that the warnings would be effective in preventing limb injuries. However, the limb injuries continued unabated until the company (approximately four years after the vehicle had been introduced into the marketplace) instituted a design-change remedy (doors) that effectively constrained occupants' lower extremities within the occupant space of the vehicle.

The failure of using only warnings (to keep arms and legs inside the vehicle during a tip-over event) to remedy the limb-injury hazard rests with the fact that the warning could only be potentially useful or effective if an operator or passenger consciously, voluntarily, or intentionally extends a leg or arm in a protective manner during a roll over event. Warnings that instruct you "not to do something" – something that is typically done without conscious deliberation, awareness, or intention, but simply as a reactive response to a stimulus situation – cannot be effective (Leonard and Karnes, 2005a).

If one assumes that the operator or passenger riding in a UTV intentionally extends a leg or an arm during a rollover, a warning not to do that could conceivably have some benefit – if, and only if, there is sufficient time immediately preceding and during the rollover for the rider to make such a decision. However, given the fact that a vehicle rollover may occur in a second or less, the time available for decision-making would be critically limited. More importantly, the efficacy of limb-injury warnings is limited given that it is also likely that the limb extension can occur unintentionally, simply due to the physical forces associated with the rollover, or due to an individual's reactive response to the rollover event.

It is our opinion, that for some people, as opposed to all people, extending the leg outside of the occupant compartment in a UTV rollover event is an involuntary, reactive, learned protective motor response. While true human reflexes may range neurologically from simple monosynaptic reflexes (such as the "knee jerk" reflex) to polysynaptic reflexes (such as postural and righting reflexes) many learned reflex-like complex protective responses can become programmed and are executed rapidly without conscious deliberation or intention. Examples include extending the upper arms during the initiation of a fall event (the so-called parachute reflex); quickly stepping on the brake to avoid an auto collision; quickly moving out of the way when something falls; reaching out to catch an object unexpectedly thrown in our direction (or turning away or ducking to avoid the thrown object); and the frequently encountered "phantom braking reflex" (the response of a passenger in an automobile who perceives an imminent collision and quickly pushes his/her right leg to apply a nonexistent brake pedal). A defining characteristic of all these types of stimulus-driven reactive responses (including acquired reflexes and programmed motor responses) is that the person who executes such a response doesn't spend a lot of time deliberating or thinking about it — the person just reflexively does it. If a limb extension during a UTV rollover occurs involuntarily, without conscious thought or deliberation, then any warnings **not** to do it are essentially useless.

Human stimulus-driven reactive responses are programmed actions that require no conscious effort. Consequently, the response occurs without much delay. Relevant literature topic titles include conditioned reflexes, conditioned reactions, conditioned responses, condition avoidance responses, learned reactions, and acquired reflexes (reflexes gradually developed by training and association through the frequent repetition of a given stimulus-response situation). The behavioral science literature is replete with discussions of human reflexive-type responses (cf. Breese, 1917; Evarts, 1973; Sherwood, 2006; and Welchman, et al., 2010).

It has been commonly argued in defense of the efficacy of limb-injury warnings that the injured person consciously and deliberately extended the injured limb even though the injured person claimed that he/she did not. From a human factors standpoint, it is simply gross speculation for an observer (or anyone else attempting to interpret a person's rapid response to a stimulus situation) to opine that the person who executed the very rapid response did it volitionally, with conscious deliberation and intention. When the operator or passenger "extends" a leg or arm in a vehicle rollover event, and/or the limb is extended by physical forces accompanying the rollover, only the person involved knows whether the extension was done deliberately with conscious forethought. The most reasonable and logical way to assess whether a response was volitionally executed is to ask the person directly.

The issue of whether or not a rapidly executed response, such as a leg extension during a rollover, is a reactive response, a learned reflex, or something other than a "true" physiological reflex may make for an interesting academic debate. However, from a human factors and warning efficacy standpoint, it doesn't make an iota of difference whether the limb extension was a reactive response, a reflex, an acquired reflex, a programmed motor response, or whatever. What matters is whether or not the person who executed the warned-about response did it voluntarily with conscious intention and forethought. When the person who made the response says that it occurred without conscious deliberation or intention, it would be suspect for some other person to dispute that contention – given that an interpreter of a person's response to any situation cannot have direct access to the mind of the person who actually executed the response.

In the scores of UTV limb-injury incidents evaluated by the authors to date, not one person has testified that, during the rollover event, they had extended their limb with conscious intention or deliberation. To the contrary, they have testified that the response occurred rapidly and unintentionally. What is abundantly clear from a human factors standpoint is that any warning to not do something that is done without conscious deliberation or intention cannot be effective. The appropriate remedy for reactive stimulus-driven unintentional limb extensions is a design change that eliminates, or at a minimum, guards against exposure to the hazard.

Nearly four years after it entered the marketplace, this open-sided UTV was re-designed to have doors that did in fact accomplish that containment purpose. The company also offered free retrofit doors to owners of previously purchased vehicles.

ORIFICE INJURIES IN FALLS FROM PERSONAL WATERCRAFT

Personal watercraft (PWCs) are recreational watercraft on which a rider sits or stands. They have an inboard engine driving a pump jet with a screw-shaped impeller to create thrust for propulsion and steering. When PWCs were first introduced in the late 1980s they were single-rider, stand-up models. Soon after their introduction, major safety concerns arose due to the inability of operators to steer the PWCs when the throttle was released. In collision-avoidance situations, when operators reflexively released the throttle and attempted to steer away from a perceived collision, the vehicles could not be steered. The inability to steer without application of the throttle has been referred to as the off-throttle-steering (OTS) hazard. Manufacturers of PWCs attempted to remedy the OTS hazard by simply providing warnings until 2001. Then, engineering changes (throttle reapplication systems and deployable rudders) were introduced to provide steering capability, even when the operator was not applying thrust.

PWC models designed for two to three persons were subsequently introduced in the 1990s. Along with the introduction of these multi-person models came the occurrence of certain types of body orifice injuries. This type of hazard relates to the fact that lower-body orifice injuries (vaginal and/or anal) have occurred when persons riding as passengers on PWCs fall off to the rear of the watercraft with their legs abducted – directly in the path of the PWC's high-pressure water jet propulsion system. This has resulted in instances of severe physical trauma, and sometimes permanent damage to vulnerable body cavity tissues / organs. With few exceptions, the incidents have involved females riding as passengers. We are unaware of any orifice-injury complaints for PWC drivers. This is presumably due to the fact that when a driver falls, the engine shut-off lanyard (a form of "dead-man" switch) cuts power to the engine, thereby eliminating the high-pressure water jet stream.

Beginning in the mid-1990s, all PWC manufacturers, except one, provided some form of an orifice-injury warning. The warnings recommended that operators and passengers should wear wet suits, wet suit bottoms or other protective swimwear. They initially identified the injury hazard as resulting from impact with the water surface during a fall. In the later 1990s, the warnings were revised to include strong jet streams from the propulsion system as contributing to the hazard. Beginning in 2001, PWC manufacturers provided an industry-wide warning label that addressed numerous safety issues, including an orifice-injury hazard warning that stated:

> "WEAR PROTECTIVE CLOTHING. Severe internal injuries can occur if water is forced into body cavities as a result of falling into water or being near [the] jet thrust nozzle. Normal swimwear does not adequately protect against forceful water entry into body cavities. All riders must wear a wet suit bottom or clothing that provides equivalent protection (see

Owner's manual). Footwear, gloves, and goggles/glasses are recommended."

The 2001 industry-wide warning was tested for comprehension and was found to be adequately comprehended by focus-group participants. The warning, however, failed to clearly emphasize the fact that falling off to the rear of the PWC is especially dangerous for a person, especially a female, riding as a passenger because the strong stream of water that propels the vehicle can cause severe and permanent injuries due to the forceful injection of water into lower body cavities. The industry-wide warning provided essentially useless information that "being near [the] jet thrust nozzle" can cause severe internal injuries, given that a person who falls from the rear of a PWC has essentially no choice or control over where she/he falls or whether they are going to "be near" the jet thrust nozzle. The risk of severe internal injuries is simply caused by falling off to the rear of the PWC – which happens to be precisely where the powerful jet of the propulsion system is located.

Even with the provision of an adequately composed and properly located orifice-injury warning for passengers and operators, using warnings as **the only** remedy for the risk of injuries resulting from a rearward fall off a waterjet-propelled PWC is inadequate from a human factors and safety standpoint. The hazard control hierarchy dictates that the appropriate remedy for PWC orifice-injury hazards would involve design changes (assuming that they are technologically and economically feasible) that would decrease the likelihood of a passenger's falling off to the rear of the PWC in the first place. Examples would include providing a seat back support, or some form of a contoured seat with a back rest that would considerably reduce the risk of a passenger sliding off to the rear of the craft. Alternatively, lengthening the rear deck of the PWC enough to allow the jet's output pressure to have dissipated enough so that a fallen passenger would not be seriously injured by the flow; or configuring a contoured housing over the jet propulsion system that would direct or slide a fallen passenger far enough to the side or rear in order to reduce or eliminate this danger are approaches that could / should have been considered.

The inadequacy of using only a warning to remedy the hazard of forceful injection of water into body cavities of PWC passengers should have been apparent to PWC manufacturers. PWC manufacturers should also have realized that the likelihood of passenger compliance with a warning to wear a wet suit bottom would be low. Previous human factors research had identified various factors that can affect product warning compliance. First, the perception of risk for operating or riding on PWCs has been shown to be reliably lower than those associated with the use of other motorized recreational vehicles (cf. Karnes et al., 1988). In a study of risk perceptions for various activities involving possible safety concerns, (Leonard and Karnes, 2005a) the risk perception for riding as a passenger on a PWC was shown to be exceptionally low – not reliably different from other relatively low-risk activities (e.g., riding in a small airplane or using a phone during a thunderstorm). Low risk perception tends to result in decreased warning compliance.

Second, some warnings can be ineffective if there is a perceived high cost of compliance. A variety of studies have shown that cost of compliance affects the use of safety equipment (Kalsher & Williams, 2006). If an individual is invited to ride on a PWC, and doesn't see the warning until ready to mount the PWC, it is unlikely that she/he will just happen to have a wet suit or other protective garment readily available. Complying with such a warning would likely have been judged costly in terms of the time and effort needed to obtain and don a wet suit. When risk perception is low and cost of compliance is high, product warnings are highly unlikely to be effective.

Third, the tendency to emulate others when one is not very knowledgeable about hazards can affect compliance behaviors. Leonard and Karnes (1999) and Edworthy and Dale (2000) have noted that compliance is also reduced by observing others in the same situation who are not complying with available warnings, but do not appear to be suffering any negative consequences. When passengers are offered the opportunity to ride on a PWC, and observe that the driver and other users of PWCs are wearing only normal swimsuits, compliance with a warning to wear specialized protective garments is likely to be low.

PWC manufacturers have provided orifice-injury warnings advising the need to wear wet suits for over a decade. However, there is no evidence that any PWC manufacturer examined the ways in which users actually complied with those warnings. A study conducted by Smith and Perry (2003), and an ongoing study by Karnes, Burleson, and Leonard, have revealed that PWC passenger use of wet suit bottoms or neoprene shorts is practically nonexistent. Smith and Perry (2003) observed and surveyed 351 PWC users at 3 different lakes in Minnesota during the summer of 2003. They found 98.9% of these drivers or riders wore only swimsuits or cloth shorts. Questionnaires completed by 86 users showed that although respondents had some familiarity with the PWC hazard warning, they showed little understanding of the actual hazard posed by the propulsion system's water pressure, or knowledge of the apparel recommended to mitigate that hazard.

The on-going study by Karnes, Burleson, and Leonard has thus far involved observations of over 300 PWC users in seven states. Using data collected since the summer of 2004, it was found that wet suits, wet suit bottoms, or neoprene shorts were worn by only 4% of operators and by only 1% of passengers. Survey data collected for a study addressing failures of consumers to comply with warnings (Leonard and Karnes, 2005b) found that only 13 of 222 individuals who indicated they had been a passenger or both a driver and a passenger on a PWC, said they had ever worn wet suits. And 11 of those 13 said they sometimes wore only swimsuits. Only 3 of the persons who said they did wear wet suits reported that they only rode as passengers. Many people said the reason they didn't wear wetsuits was that none were available. These studies also have shown that riding as a passenger on a PWC is often an unplanned event – i.e., that the opportunity to ride often occurs serendipitously. Further, in all of the surveyed instances where riders rented PWCs, wetsuits or other protective clothing options were not available – and the only instructions provided concerned issues associated with operating the machine itself.

In summary, the failure of PWC users to comply with the warnings – assuming they had seen them – involved low perception of risk, likely lack of understanding of the specific hazard involved, along with the high cost of compliance of the stated remedy. The effects of these perceived costs were likely exacerbated by observing so many other riders who wore only swim suits (cf. Edworthy and Dale, 2000).

Manufacturers of PWCs had attempted to cure a design defect with the band-aid of a warning. Given the knowledge of the incidents, the manufacturers should have been aware that the likelihood of wearing wet suits or other forms of protective garments, especially in summer time, would be minimal. To design out and/or guard against the hazard by providing a back support (that would significantly reduce, if not eliminate the falling problem), or altering some of the other previously suggested PWC design features would be much more appropriate than only providing a warning – a warning that almost certainly would not be followed.

CONCLUSIONS

We have identified two types of products that embodied significant hazards to their users. The manufacturers of those products either knew about those hazards before the product was initially offered to the public, or became increasingly aware of the problems soon after product introduction. However, the manufacturers initially attempted to remedy the hazards simply by providing warnings. This paper discussed various reasons why these warnings were ineffective, and noted certain engineering design changes or types of guarding or shielding that were needed to more effectively remedy the situation.

The situations described here are examples of what has happened in a variety of cases where products contained design flaw hazards that were not abated by using warnings. One common feature of these cases is that manufacturers are likely to have made certain oversimplifying assumptions about their products – assumptions that nothing will go wrong and/or that the product will only be used in the manner originally conceived. When manufacturers do become aware of any problem(s), they may attempt to fix them by trying to modify users' behaviors (via instructions or warnings) so as to have them better conform to the designer's original product usage model. Incorporated in this approach are the ideas that usage of their products will be limited to the narrow range of cases they have considered, and/or that the users will follow all their instructions (no matter how difficult or unlikely that may be). Often overlooked are whether the users are likely to have the ability and/or inclination to comply with those cautions or instructions. Clearly, when it is difficult, cumbersome, or not possible for users to comply, compliance will be limited at best. It is incumbent upon designers and manufactures to produce items whose designs do not rely on user training or instructions to overcome existing behavioral propensities. At the same time, they should not expect users to procure and use protective items that are not readily available. Failure to recognize these possibilities may result in products that are unsafe for many of their potential users.

REFERENCES

Breese, B.B. (1917) *Psychology*, Scribner, New York.

Christensen, J.M. (1987) Comments on product safety. *Proceedings of the Human Factors Society, 31ˢᵗ Annual Meeting*. Human Factors Society, Santa Monica, CA, 1-12.

Edworthy, J. & Dale, S. (2000) Expanding knowledge of the effects of social influence in warnings compliance. *Proceedings of the IEA/HFES 2000 Congress*. Human Factors & Ergonomics Society, Santa Monica, CA, 4, 770-773.

Evarts, E.V. (1973) Motor reflexes associated with learned movement. *Science*, Vol. 179, 2/2/73, 501-503.

Kalsher, M.J. and Williams, K.J. (2006) Behavioral compliance: Theory, methodology, and results. In *Handbook of Warnings*, Wogalter, M.S., ed. Lawrence Erlbaum Associates, Mahwah, NJ, 313-331.

Karnes, E.W., Leonard, S.D., Schneider, T., Pedigo, W. and Krupa, D. (1988). Safety patterns for ATVs and other motorized recreational vehicles. In F. Aghazadch, ed. *Trends in Ergonomics/Human Factors V*, North Holland, New York, 647-655.

Laughery, K.R., Sr., Wogalter, M.S., & Young, S.L., eds. (1994) *Human factors perspectives on warnings*. Human Factors & Ergonomics Society, Santa Monica, CA.

Leonard, S.D. & Karnes, E.W. (1999) Socio-environmental effects on warnings. *International Journal of Industrial Ergonomics*, 25, 11-18.

Leonard, S.D. & Karnes, E.W. (2005a) Why some warnings don't work. *Proceedings of the XIX International Ergonomics and Safety Conference*, Las Vegas, NV, 6/26-29/2005, 15-19.

Leonard, S.D. & Karnes, E.W. (2005b) Risk perception: Has it changed? *Proceedings of the XIX International Ergonomics and Safety Conference*, Las Vegas, NV, 6/26-29/2005, 21-25.

National Safety Council. (1989) *Accident Prevention Manual for Industrial Operation, 5ᵗʰ Edition*. National Safety Council, Chicago, IL.

Sherwood, L. (2006) *Fundamentals of Physiology*. Thompson Brooks/Cole. Belmont, CA.

Smith, T.J. & Perry, R. (2003) A usability analysis of personal watercraft use. Report to Peter Riley, of Schwebel, Goetz, and Riley, Minneapolis, MN.

Welchman, A.E., Stanley, J., Schomers, M.R., Miall, R.C., Bulthoff, H.H. (2010) The quick and the dead: When reaction beats intention. *Proceedings of the Royal Society of Biological Sciences*. 02/03/2010.

Wogalter, M.S., DeJoy, D.M, & Laughery, K.R., eds. (1999) *Warnings and risk communication*. Taylor & Francis, Philadelphia, PA.

Wogalter, M.S., Young, S.L., & Laughery, K.R., Sr. (2001) *Human factors perspectives on warnings, Volume 2*. Human Factors and Ergonomics Society, Santa Monica, CA.

Wogalter, M.S., ed. (2006a) *Handbook of Warnings*. Lawrence Erlbaum Associates, Mahwah, NJ.

Wogalter, M.S. (2006b) Purposes and scope of warnings. In *Handbook of Warnings*, Wogalter, M.S., ed. Lawrence Erlbaum Associates, Mahwah, NJ, 3-9.

CHAPTER 107

The Effects of Background Color of Safety Symbols on Perception of the Symbols

Calvin K. L. Or[1], Alan H. S. Chan[2]

[1]Department of Industrial and Manufacturing Systems Engineering
The University of Hong Kong
Pokfulam, Hong Kong

[2]Department of Manufacturing Engineering and Engineering Management
The City University of Hong Kong
Kowloon Tong, Hong Kong

ABSTRACT

This present study investigated the effects of the background color of safety symbols on the way people perceive hazard and risk and consequent compliance to the symbols. Thirty-one Hong Kong Chinese rated perceived hazardousness, perceived severity of injury, immediacy of consequences, and likelihood of compliance for 21 safety symbols with various background colors and symbol types. A color was not tested if it was the same as the color of the symbol. This study showed that background color significantly influenced perception of the symbols. For hazard warning symbols (black background was not tested) and mandatory action symbols (blue background was not tested), red produced the highest levels of perceived hazard, injury severity, consequence immediacy, and compliance. For prohibition types of symbols (red background was not tested), hazard, severity, consequence immediacy, and compliance levels were perceived to be higher for black than for other background colors.

Keywords: Safety Symbol, Background Color, Hazard and Risk Perception, Behavioral Compliance

INTRODUCTION

Effective safety symbols should communicate critical safety information about a product or environment so as to improve the perceptions, decision-making, and safety behavior of the users of the symbols, and thereby prevent or reduce safety problems and accidents (Wogalter & Laughery, 2006). The literature shows that many design variables can affect how safety symbols convey information and how they impact on individual perceptions of safety. Color is one of the many obvious variables that has a large role in determining how people perceive hazards and risks, the attention that they pay to a safety symbol, and their consequent behavioral intention (Chan & Courtney, 2001; Braun & Silver, 1995; Leonard, 1999; Wogalter & Laughery, 2006). For example, Chan and Courtney (2001) examined the associations between various colors and the concepts conveyed by sixteen words, including those commonly used in warning signs, such as danger, caution and stop. In their study, strong color associations were found, such as the widespread association of red with danger and green with safety. Braun and Silver (1995) assessed the effect of color and signal words on hazard perceptions and compliance with warnings. Their study found that color was a significant main effect and that red resulted in the highest level of perceived hazard and behavioral compliance. In a similar study, Leonard (1999) paired various signal words with different background colors (red and green) to test their effects on risk perception. His study indicated that background color produced significant differences in perception.

While many previous studies have focused on the examination of the connoted hazard of signal words and colors and the impact of their interaction on perceptions, the purpose of the present study was to investigate the influence of various background colors for safety symbols on perceived hazard and risk and self-reported compliance to the symbols. This study consisted of two experiments which assessed safety perceptions for three types of safety symbols and eight different background colors. Experiment 1 was used to test individual comprehension of a set of safety symbols for each of the three symbol types in order to determine those with the highest comprehension rates (the most understood ones) for subsequent use in experiment 2. Experiment 2 was designed to investigate the effects of the background color of safety symbols on perceptions of various attributes of the symbols.

METHOD

Participants

For experiment 1, twenty-four Hong Kong Chinese participants (8 female and 16 male) from a local university were recruited. Their ages ranged from 16 to 25 years. For experiment 2, another group of participants were recruited; there were thirty-one Hong Kong Chinese (13 female and 18 male) whose ages ranged from 16 to 25 years. All participants selected for experiment 2 were tested and had normal color vision.

Design and procedure

In experiment 1, a set of 15 achromatic symbols (printed size: 20 x 20 mm each) of three types was presented to participants in a questionnaire to test comprehension of the symbols. The types of symbols used were: hazard warning (5 symbols), mandatory action (5 symbols), and prohibition (5 symbols). The participants were asked to write down the meaning of each of the 15 symbols in a blank space next to each symbol in the questionnaire. They were told to leave the space blank when they were unable to understand the symbol. Comprehension accuracy was then analyzed. For each symbol type, the one with the highest comprehension rate was selected to be used in experiment 2 to test the effects of background color on perception. Demographic data was collected at the end of experiment 1.

Prior to the beginning of experiment 2, participants were given the Pseudo-Isochromatic plate test for red green color deficiency. Participants who failed the test were excluded from the experiment. Based on the results of the comprehension evaluation in experiment 1, a symbol of each type was chosen for experiment 2. As shown in Figure 1, the symbols chosen were: the hazard warning symbol "risk of toxicity", the mandatory action symbol "head protection must be worn", and the prohibition symbol "mobile phones prohibited".

(a)   (b)   (c)

Figure 1. The three safety symbol types tested in experiment 2: (a) hazard warning, (b) mandatory action, and (c) prohibition.

These three symbols were paired with eight different background colors thus producing a total of 21 stimuli after exclusion of the three stimuli that had the same border and background color. The hazard warning symbol was black so a black background was not tested. Similarly a blue background was not tested for the blue mandatory action symbol and a red background was not tested for the red prohibition symbol. The eight background colors were red, green, blue, yellow, orange, grey, white, and black. The symbols were presented on a computer screen in a random order, and remained there until the participant responded. Participants were then asked to rate on a set of 9-point Likert-type scales, ranging from 0 (lowest level) to 8 (highest level), for the following four attributes for each symbol; 1) the perceived hazardousness, 2) the perceived severity of injury, 3) the immediacy of consequences, and 4) the likelihood of compliance. At the end of experiment 2, a questionnaire was used to collect demographic data of participants.

Data analysis

For experiment 1, the comprehension accuracy of the symbols was based on the correctness of the meaning given by the participants. For experiment 2, univariate analysis of variance (ANOVA) was performed to determine the effect of background color on the four dependent measures.

RESULTS

Experiment 1

Among the five symbols of each symbol type, the one that yielded the highest comprehension rate (the one that was most understood by the participants) was identified for each type: "risk of toxicity" (hazard warning type), "head protection must be worn" (mandatory action type), and "mobile phones prohibited" (prohibition type). Twenty-three of the 24 participants (96%) were able to correctly state the reference meaning of the "risk of toxicity" symbol. Twenty-four (100%) participants correctly described the reference meaning of the "head protection must be worn" symbol. Twenty-three (96%) participants were able to give the reference meaning of the "mobile phones prohibited" symbol.

Experiment 2

The combination of symbol type and background color yielded twenty-one stimuli after excluding the hazard warning symbol with black background, the mandatory action symbol with blue background, and the prohibition symbol with red background, since the border colors of the symbols were the same as their background colors. Means and standard deviations of the four dependent measures are shown in Table 1.

Table 1. Means and standard deviations of the four dependent measures

	Mean perceived hazardousness (SD)			Mean perceived severity of injury (SD)			Mean immediacy of consequences (SD)			Mean likelihood of compliance (SD)		
	HW	MA	P	HW	MA	P	HW	MA	P	HW	MA	P
Red	6.87 (1.34)	4.81 (1.60)	NA	6.77 (1.41)	4.48 (1.83)	NA	6.84 (1.49)	5.03 (2.35)	NA	6.84 (1.49)	4.77 (1.94)	NA
Green	5.26 (1.57)	3.39 (1.63)	2.16 (1.61)	5.23 (1.69)	3.19 (1.54)	2.29 (1.94)	5.35 (1.89)	3.77 (2.19)	2.29 (2.09)	5.55 (1.49)	3.65 (1.76)	3.10 (1.58)
Grey	5.52 (1.48)	3.48 (1.39)	2.81 (2.21)	5.52 (1.36)	3.81 (1.45)	2.97 (2.07)	5.68 (1.85)	4.23 (1.89)	3.19 (2.17)	5.84 (1.53)	3.97 (1.74)	4.26 (1.83)
Orange	5.32 (1.58)	4.00 (1.65)	2.29 (1.64)	5.45 (1.52)	4.00 (1.65)	2.42 (1.57)	5.61 (1.82)	4.52 (1.90)	3.10 (2.14)	5.58 (1.54)	4.10 (1.58)	3.71 (1.81)
Blue	5.13 (1.73)	NA	2.23 (1.96)	4.97 (1.76)	NA	2.26 (1.77)	5.19 (1.94)	NA	3.00 (2.37)	5.13 (1.69)	NA	3.65 (1.84)
Yellow	6.06 (1.34)	3.84 (1.27)	3.13 (1.63)	6.13 (1.28)	3.87 (1.50)	3.06 (1.81)	6.23 (1.73)	4.55 (1.91)	3.52 (2.17)	6.06 (1.48)	4.13 (1.59)	4.06 (1.41)
White	5.65 (1.50)	3.26 (1.61)	2.35 (1.64)	5.68 (1.38)	3.23 (1.38)	2.39 (1.76)	5.65 (1.94)	3.68 (2.18)	2.65 (2.29)	5.71 (1.55)	3.58 (1.93)	3.58 (2.06)
Black	NA	4.19 (1.91)	3.32 (2.27)	NA	4.26 (2.13)	3.29 (2.21)	NA	5.00 (2.39)	3.90 (2.48)	NA	4.39 (1.98)	5.16 (1.73)

HW=hazard warning symbol; MA=mandatory action symbol; P=prohibition symbol

Perceived hazardousness. The ANOVA showed that there was a significant main effect of color for all three symbol types: hazard warning, $F(6, 180)=10.93$, $p<0.001$; mandatory action, $F(6, 180)=6.56$, $p<0.001$; and prohibition, $F(6, 180)=4.96$, $p<0.001$. The hazard warning and mandatory action symbols with red backgrounds were perceived as most hazardous. The prohibition symbol with black background was rated as highest in hazard.

Perceived severity of injury. There was a significant main effect of color for all symbol types: hazard warning, $F(6, 180)=11.32$, $p<0.001$; mandatory action, $F(6, 180)=7.24$, $p<0.001$; and prohibition, $F(6, 180)=3.82$, $p<0.005$. Hazard warning and mandatory action symbols with red backgrounds and prohibition symbol with black background resulted in highest ratings of injury severity.

Immediacy of consequences. The effect of background color on perceived immediacy of consequences was significant for all symbol types: hazard warning, $F(6, 180)=9.57$, $p<0.001$; mandatory action, $F(6, 180)=4.55$, $p<0.001$; and prohibition, $F(6, 180)=5.28$, $p<0.001$. The hazard warning and mandatory action symbols with red background and the prohibition symbol with black background yielded highest perceived immediacy scores.

Likelihood of compliance. There was a significant main effect of color for all symbol types: hazard warning, $F(6, 180)=8.39$, $p<0.001$; mandatory action, $F(6, 180)=3.33$, $p<0.005$; and prohibition, $F(6, 180)=8.14$, $p<0.001$. Ratings of compliance likelihood were highest for hazard warning and mandatory action symbols with red background. For prohibition type, the symbol with black background yielded a highest compliance likelihood rating.

DISCUSSION AND CONCLUSIONS

This study tested perceptions of safety symbols for three types of symbols paired with various background colors. After excluding the three stimuli that had the same border and background color (i.e., the hazard warning symbol with black background, the mandatory action symbol with blue background, and the prohibition symbol with red background), 21 stimuli were tested. The study revealed that the background color of a safety symbol influenced perceptions of hazard, injury severity, immediacy of consequences, and likelihood of compliance with the warnings. For the hazard warning and mandatory action symbols, of those colors tested, it was found that a red background produced the highest levels of perceived hazard, injury severity, consequence immediacy, and behavioral compliance. This finding is consistent with some safety sign color coding standards and with previous research showing that red was perceived as having a significantly higher hazard connotation than other colors (Chan & Ng, 2009; Rodriguez, 1991; Wogalter & Laughery, 2006; Wogalter et al., 1995). Here, for the hazard warning symbol, yellow produced the second highest level of perceived hazard, injury severity, consequence immediacy, and compliance. However, for the mandatory action symbol type, the black background received the second highest mean ratings for the four dependent measures. For the prohibition symbol type, where a red background was not tested because the symbol was red, black conveyed a greater level of hazard, severity of injury, immediacy of consequences, and likelihood of compliance than all other background colors.

Although the use of color in safety symbols can be problematic for certain users, such as those with color deficiency, colored symbols are of great value in attracting attention more effectively than achromatic symbols (Wogalter and Laughery, 2006). Furthermore, previous studies have shown that color influenced perception of hazard as well as behavioral compliance with the symbols (Braun and Silver, 1995; Braun, Sansing, and Silver, 1994; Leonard, 1999). In the present study, the findings demonstrate that different background colors for safety symbols connoted different levels of perceived hazard, risk, and behavioral compliance. While color is one of the main design components of a safety symbol, obviously, color on its own cannot convey important safety information. Color can be used as an additional form of information dissemination, especially for gaining attention, in environments and on products where safety is an issue.

ACKNOWLEDGMENT

The work described in this paper was fully supported by a grant from the Research Grants Council of the Hong Kong Special Administrative Region, China [CityU 110508].

REFERENCES

Braun, C.C., Sansing, L., & Silver, N.C. (1994). The interaction of signal word and color on warning labels: differences in perceived hazard. Proceedings of the Human Factors and Ergonomics Society 38th Annual Meeting, 831-835.

Braun, C.C., & Silver, N.C. (1995). Interaction of signal word and colour on warning labels: differences in perceived hazard and behavioural compliance. Ergonomics, 38(11), 2207-2220.

Chan, A.H.S., & Courtney, A.J. (2001). Color associations for Hong Kong Chinese. International Journal of Industrial Ergonomics, 28(3-4), 165-170.

Chan, A.H.S., & Ng, A.W.Y. (2009). Perceptions of implied hazard for visual and auditory alerting signals. Safety Science, 47(3), 346-352.

Leonard, D.S. (1999). Does color of warnings affect risk perception? International Journal of Industrial Ergonomics, 23(5), 499-504.

Rodriguez, M.A. (1991). What makes a warning label salient? Proceedings of the Human Factors and Ergonomics Society 35th Annual Meeting, 1029-1033.

Wogalter, M.S., & Laughery, K.R. (2006). Warnings and hazard communications. In G. Salvendy (Ed.), Handbook of Human Factors and Ergonomics (3rd ed., pp. 889-911). New York: John Wiley & Sons.

Wogalter, M.S., Magurno, A.B., Carter, A.W., Swindell, J.A., Vigilante, W.J., & Daurity, J.G. (1995). Hazard associations of warning header components. Proceedings of the Human Factors and Ergonomics Society 39th Annual Meeting, 979-983.

Chapter 108

Hazard Perceptions of Consumer Products

William J. Vigilante, Jr.[1], Raymond W. Lim[2]

[1]Robson Forensic, Inc
Lancaster, PA 17603, USA

[2]Pierce College
Woodland Hills, CA, 91371, USA

ABSTRACT

Consumer's perceived level of hazard associated with usage scenarios of several common consumer products was examined. The research was conducted in conjunction with several forensic cases the lead author investigated. The cases involved residential fires caused by the malfunction of a candle, toaster, and battery charger. In each case the manufacturer-defendant argued the user was misusing the product at the time of the fire. This study sought to determine the perceived level of hazard associate with different uses of the incident products compared to other similar consumer products. Results indicate that although hazard perceptions varied, for the three consumer products of interest, the respondents did not perceive their use as dangerous but, on the contrary, relatively safe compared to similar types of products. Implications for consumer product manufacturers are discussed.

Keywords: Forensic Human Factors, Consumer product, Hazard / Risk perception, Warnings, Residential fire

INTRODUCTION

This research examined respondents' perceived level of hazard for several common consumer products. The research was conducted in conjunction with several forensic cases the lead author investigated. The cases involved residential fires caused by the malfunction of a candle, toaster, and battery charger.

In each case, the manufacturer-defendant argued that the user misused the product by leaving it unattended when the fire occurred. The defendants also argued that they had provided adequate warnings and instructions to inform the user of the fire potential associated with the product and the need to attend to the product while it was in use.

The first author's work in these cases showed that the warnings and instructions provided by the defense-manufacturers were not adequate. These opinions were expressed with particular consideration to the reality of how consumers use common household products and their beliefs and attitudes toward product safety. For example, the author identified a common underlying theme which was present in each of the cases and consistent with past warnings research: the product users were not thinking safety because they considered the products to be relatively safe and were comfortable with their use.

The introduction of this paper will begin with a description of the incidents, following by a discussion of the relevant warnings research, and finally the specific purpose of the study.

INCIDENT DESCRIPTIONS

In each of the cases, the paper's first author was retained by the home owner's insurance carrier to analyze the adequacy of the warnings and instructions that accompanied the products and to determine if the home owner's actions were reasonable and predictable. The insurance companies filed suit against the defense manufacturers in effort to recover the moneys they paid out to the home owners as a result of the fires. This particular type of legal action is referred to as subrogation and has become quite common in recent years. The premises of the subrogation claim is that the insurance company suffered a monetary loss caused by the negligence of the manufacturer of the defective product that caused the loss. Each case settled favorably for the plaintiff insurance company prior to trial.

The candle fire incident occurred at a single family residence on Christmas morning a few years ago and involved a scented jar candle. The family's holiday tradition included opening presents in the finished basement of the home after eating breakfast.

Upon awaking the morning of the fire, the home owner decided to light a scented jar candle in the basement. She had purchased the jar candle from a national discount retailer a few days prior to the fire. The candle was the same type of candle she had been purchasing and using for years.

Upon lighting the candle in the basement, the home owner placed it on an end table, waited a few minutes to ensure it was burning okay, and then proceeded back upstairs to the kitchen to prepare breakfast. About 15 minutes later, the home owner noticed smoke and discovered the basement was on fire. The fire was determined to be caused by the malfunction of the candle. Although the exact malfunction was not determined it was theorized that either a defect in the wax or glass resulted in the flame escaping from the confines of the glass jar.

The 3" tall glass jar candle possessed a peel-off front and back label as well as a label on the bottom of the jar. The front label presented the product and manufacturer's name and trademark as well as a short description of the product. The back label presented a longer description of the product and how the fragrance was made. The back label also presented a warning that included the instruction "Burn candle within sight." The bottom label repeated the warning text.

During her deposition, the home owner testified that although she did read the front of the jar label, she did not recall reading the back label, and was unaware of the label on the bottom of the jar. She also testified that although she knew candles could potentially start a fire, she believed that the contained flame of the jar candle made it relatively safe and that it was acceptable to light the candle and leave it alone in the room while doing chores in other rooms of the house.

The defense-manufacturer alleged that the candle warnings were adequate, that the home owner failed to comply with the warnings, and that by leaving the lit candle unattended the home owner had misused the product.

The toaster fire incident involved a toaster that malfunctioned and started a fire in the kitchen of a single family residence. The family had purchased the toaster two years prior to the fire from a large national retailer and had used it several times a week.

On the morning of the fire, the home owner placed waffles in the toaster for her son to eat before taking him to school, instructed her son to eat the waffles, and went up stairs to chair clothing. After dressing, the home owner came back down stairs and drove her son to school assuming her son had eaten the waffles. After returning home from dropping her son off at school, the home owner discovered her home on fire. The fire was determined to be caused by the malfunction of the toaster's carriage latch mechanism that failed to release resulting in the heating elements remaining energized, subsequently igniting the waffles, and spreading to the toaster housing and nearby combustibles.

During her deposition, the home owner testified that she knew that, like all heat producing appliances, there was a potential fire hazard associated with the toaster. However, she believed that the toaster was relatively safe and was not concerned with or thinking about the potential fire hazards when she toasted the waffles and left them for her son to eat. The home owner also testified that she did not read the instruction manual or warnings that accompanied the product because it was just a toaster and was simple to operate.

Similar to the prior case, the defense-manufacturer alleged that the toaster instructions and warnings, that stated never leave toaster unattended while toasting, were adequate, the home owner failed to comply with the warnings, and that by leaving the toaster unattended while toasting the home owner had misused the product.

The battery charger fire incident occurred at a single family residence and involved a battery pack for a remote control vehicle (RCV) that exploded while being charged. The home owner had purchased the battery charger in conjunction with a RCV truck and rechargeable battery packs a few years before the fire from a national hobby shop chain. The incident battery pack was purchased six months before the fire from the same hobby shop as a replacement for the original battery pack that had lost its charge holding ability. The home owner had used the battery charger and incident battery on numerous past occasions without incident.

The battery charger came with a mechanical timer to set the amount of time in which to "quick-charge" the battery packs. The timer had a 30 minute time limit after which the battery charger switched to a continuous "trickle-charge" unless the battery was removed or charger unplugged.

On the day of the fire, the home owner decided to charge the battery pack in his garage. After setting the mechanical timer, the home owner left the garage and entered the house. A few minutes later, the home owner exited the home through the garage and drove to his neighbor's house to pick up something and then returned home shortly thereafter. The home owner then went into his home office to work. A few hours later his wife discovered smoke coming from the garage. The fire was determined to have started when the charging battery exploded. The cause of the battery explosion was linked to its being overcharged by the battery charger.

During his deposition, the home owner testified that he read the battery charger instructions when he purchased the product a few years prior. However, he was unaware that:

- The battery could explode while being charged and burn down his home;
- The battery remained "trickle-charging" after the mechanical timer stopped.

The home owner also testified that he did not believe that he left the battery charging "unattended" because either he and/or his wife were at home at all times it was operating. Prior to the fire the home owner believed and expected the battery charger to be a safe product and not dangerous.

WARNINGS RESEARCH

As noted above, each of these cases shared a similar underlying theme: the home owners were not thinking safety because they considered the products to be relatively safe and were comfortable with their use. This fact was expected and is consisted with research on consumer product risk/hazard perception and its effect on consumers' behavior.

For example, research has shown that people do not think about unknown or unexpected hazards in familiar situations (Wagenaar, 1992; Weegles & Kanis, 2000; Woodson et al., 1992). The result is that people who do not expect a hazard to exist will not actively look for one. Furthermore, people who are not aware of a hazard do not realize that they are putting themselves at risk and do not knowingly take steps to avoid the unknown hazard (Wagenaar, 1992; Weegles & Kanis, 2000; Woodson et al., 1992).

Research has also shown that the perceived level of hazard associated with a product decreases as people become more familiar with it (Dejoy, 1999). As a person's perception of a hazard decreases so does the likelihood of him/her looking for or noticing a product warning (Wogalter et al, 1991). People are also unlikely to look for or read warnings on products they perceive as simple (Wright et al., 1982).

Also most people do not seek information when purchasing products they are highly familiar with (Dejoy, 1999; Godfrey & Laughery, 1984).

Research has also shown that peoples' knowledge of a product's hazards is frequently incomplete especially for non-open or non-obvious hazards (Leonard & Wogalter, 2000; Weegles & Kanis, 2000; Laughery, 1993). Furthermore, consumers may have knowledge of some hazards associated with a product but not others or they may require a cue (e.g., a warning) to recognize a particular hazard (Leonard & Wogalter, 2000; Weegles & Kanis, 2000; Laughery, 1993).

STUDY'S PURPOSE

The purpose of this study was to examine respondents' perceived level of hazard for different usage scenarios of common consumer products compared to the three products involved in the above mentioned fires. The first usage scenario involved leaving a product operating in one room of a house while the user was in another. The second scenario involved leaving home while product was operating.

METHODS

The questionnaire consisted of four sets of questions. The first set possessed the following demographic questions: gender, age, ethnic background, the highest level of schooling completed, and if English was the first language they learned.

The second set of questions was intended to capture the participants' ownership of the relevant consumer products. Participants were asked to answer "Yes" or "No" to the following questions:

- Do you own, or have you ever owned, any candles?
- Do you own, or have you ever owned, a toaster?
- Do you own any battery powered electronic devices (for example, cell phone, laptop computer, ipod, camera, etc.)
- Do you own any power tools or consumer products that have removable re-chargeable batteries (for example, cordless drill, flash light, dust-buster, etc.)?
- Do you own, or have you ever owned, a battery powered RCV (for example, a toy car, boat, airplane, etc.)?
- Do you own, or have you ever owned, a battery charger for the following types of products:
  o Re-chargeable batteries (for example, Duracell)?
  o Battery pack for remote control (toy) vehicles?
  o Car / Motorcycle / Boat batteries?

The third set of questions was intended to capture participants' product usage frequency. Respondents were asked to indicate how often they use the following products using a scale verbally anchored with: "never / yearly / several times a year

/ monthly / several times a month / weekly / several times a week / daily / several times a day:"

- Candles types: tapered, free standing, jar/glass, and scented;
- Toaster;
- Battery charger for: battery powered electronic devices, battery powered tools or consumer products, re-chargeable batteries (e.g., Duracell), Battery packs for RCV, and car/motorcycle/boat batteries.

The fourth set of questions was intended to capture the respondents' hazard perceptions toward the following consumer product and appliances:

- Candles: Free standing, Tapered stick, and Jar/glass;
- Home products and appliances: alarm clock, clothes dryer, computer, Crockpot, dishwasher, gas fireplace, humidifier, microwave oven, stereo/radio, stove-oven, stove-cook top/burner, toaster, TV, washer;
- Rechargeable batteries: cell phone, cordless drill, digital camera, flash light, laptop computer, motor vehicle battery, re-chargeable batteries, RCV battery pack.

Participants were asked to rate each product on the following two questions, customized for type of product, using an 11 point Likert-type scale with the following numerical and textual anchors: 1 = very dangerous,; 3 = dangerous; 5 = somewhat dangerous; 7 = somewhat safe; 9 = safe; 11 = very safe:

- Candles:
  - How SAFE is it to leave the following types of candles burning (lit) alone in one room while you are in another room of the house?
  - How SAFE is it to leave the following types of candles burning (lit) when no one is home?
- Home products and appliances:
  - How SAFE is it to leave the following products and appliances plugged-in and operating alone in a room while you are in another room of the house?
  - How SAFE is it to leave the following products and appliance plugged-in and operating when no one is home?
- Rechargeable batteries:
  - How SAFE is it to leave the following products plugged-in and charging (or re-charging) alone in a room while you are in another room of the house?
  - How SAFE is to leave the following products plugged-in and charging (or re-charging) when no one is home?

Pilot surveys were conducted with a dozen participants to ensure the questions were understandable. Questions and responses that were deemed inadequate or misunderstood by the pilot participants were re-worked or deleted.

The questionnaire was completed by 211 volunteers from the Los Angeles, CA area. The questionnaire was distributed by undergraduate students as part of a class assignment, while attending a senior level Psychology course at a local state university.

RESULTS

Respondents reported an average age of 34 years old (SD = 12 years) and 58% reported being female. Forty percent reported English as the first language they learned (all participants read and spoke English). Seven percent of the respondents had at least some high school education; 16% had a high school degree; 28% had some college level education or a technical school degree; 42% had a college degree, and 8% had at least some graduate level education.

Eighty-six percent of the respondents reporting owning or have owned candles; 94% toaster; 98% battery powered electronic devices; 75% power tools or consumer products that have removable re-chargeable batteries; 55% battery powered RCVs; and 66%, 50%, and 41% reported owning a battery charger for re-chargeable batteries, Battery pack for RCV, and car/motorcycle/boat batteries, respectively. Table 1 provides the participants' frequency of use responses for each of the products. Responses to the frequency of use questions were collapsed into the following four categories:

1. Never = Never
2. Seldom = Yearly / Several times a year
3. Occasionally = Monthly / Several times a month
4. Frequently = Weekly / Several times a week / Daily / Several times a day

Table 1 Product Usage Percentage

Product	Never	Seldom	Occasionally	Frequently
Tapered candle	.61	.31	.04	.04
Free Standing Candle	.49	.35	.07	.09
Jar/Glass candle	.23	.36	.21	.20
Scented candle	.20	.32	.21	.27
Toaster:	.06	.15	.20	.59
How often recharge:				
Electronic products:	.01	.02	.04	.93
Tools / consumer products:	.31	.24	.20	.26
Rechargeable batteries:	.40	.24	.25	.10
RCV batteries:	.50	.31	.12	.07
Car/MC/boat batteries:	.65	.25	.04	.06

Tables 2 presents the respondents' hazard perceptions presented as percentages. Respondent hazard perceptions were collapsed into the following five categories:

1. Dangerous = (1) very dangers to (3) dangerous
2. Somewhat Dangerous = (4) to (5) somewhat dangerous
3. Neutral = 6
4. Somewhat Safe = (7) somewhat safe to (8)
5. Safe = (9) safe to (11) very safe

Table 2 Percentage of respondents' hazard perceptions toward candles.

Product	dangerous	somewhat danger	neutral	somewhat safe	safe
Tapered candle:	.52 / .91	.21 / .05	.05 / .01	.14 / .02	.08 / .01
Free Standing Candle:	.47 / .88	.22 / .06	.07 / .03	.13 / .01	.11 / .02
Jar/Glass candle:	.14 / .70	.20 / .12	.04 / .05	.27 / .07	.36 / .07

* First percent is for the question: How Safe to leave lit alone in room?
** Second percent is for the question: How Safe to leave lit when no one is home?

Table 3 Percentage of respondents' hazard perceptions toward consumer products and appliances.

Product	dangerous	somewhat danger	neutral	somewhat safe	safe
Alarm clock:	.01 / .03	.02 / .01	.01 / .02	.05 / .07	.91 / .87
Clothes dryer:	.04 / .15	.06 / .11	.02 / .07	.16 / .19	.72 / .48
Computer:	.02 / .06	.03 / .06	.02 / .06	.09 / .17	.83 / .65
Crockpot:	.13 / .33	.12 / .24	.07 / .09	.30 / .12	.38 / .21
Dishwasher:	.05 / .16	.09 / .19	.04 / .10	.15 / .18	.68 / .37
Gas fireplace:	.39 / .77	.24 / .15	.07 / .02	.16 / .00	.14 / .06
Humidifier:	.14 / .40	.15 / .17	.06 / .05	.25 / .15	.39 / .23
Microwave oven:	.19 / .46	.11 / .18	.06 / .08	.20 / .07	.45 / .22
Stereo/radio:	.01 / .04	.02 / .08	.00 / .01	.09 / .10	.87 / .76
Stove-oven:	.28 / .69	.23 / .14	.09 / .03	.16 / .08	.25 / .06
Stove-cook top/burner:	.45 / .82	.21 / .09	.07 / .02	.15 / .01	.13 / .05
Toaster:	.24 / .62	.17 / .22	.08 / .02	.21 / .03	.30 / .10
TV:	.02 / .09	.04 / .07	.02 / .03	.11 / .14	.81 / .67
Washer:	.06 / .17	.04 / .15	.03 / .08	.20 / .21	.66 / .40

* First percent is for the question: How Safe to leave operating when in another room?

** Second percent is for the question: How Safe to leave operating when not at home?

Table 4 Percentage of respondents' hazard perceptions toward recharging batteries.

Product	dangerous	somewhat danger	neutral	somewhat safe	safe
Cell phone:	.01 / .06	.02 / .06	.02 / .01	.11 / .18	.84 / .69
Cordless drill:	.10 / .17	.10 / .12	.03 / .07	.17 / .19	.60 / .45
Digital camera:	.01 / .07	.02 / .05	.02 / .02	.11 / .18	.84 / .67
Flash light:	.01 / .09	.03 / .04	.04 / .03	.11 / .20	.81 / .64
Laptop computer:	.02 / .08	.04 / .09	.03 / .02	.11 / .19	.80 / .62
Motor vehicle battery:	.11 / .26	.15 / .17	.09 / .07	.20 / .26	.44 / .25
Re-chargeable batteries:	.05 / .16	.06 / .10	.05 / .04	.15 / .21	.69 / .48
RCV battery pack:	.06 / .20	.06 / .12	.04 / .04	.19 / .22	.64 / .42

* First percent is for the question: How Safe to leave charging alone in room?
** Second percent is for the question: How Safe to leave charging when no one is home?

DISCUSSION

The purpose of this study was to examine respondents' perceived level of hazard associated with usage scenarios of common consumer products and appliances. The usage scenarios included leaving the product operating in one room while the user was in another and leaving the product operating alone in the house. Of specific interest were the respondents' hazard perceptions concerning the use of candles, toasters, and battery chargers under two scenarios.

The first finding of interest involves the different levels of perceived hazard associated with the different types of candles. Although the majority of respondents believed it was at least somewhat dangerous to leave a tapered/stick or stand alone candle alone in a room, most of the respondents believed that it was not dangerous to leave a jar candle in a room by itself while the user was in a different room of the house. The respondents also believed that regardless of candle type, it was dangerous to leave one burning in the house while no one was home.

The findings for the toaster were not as clear cut compared to the candle. While the majority of the respondents believed it was not dangerous to leave the toaster toasting in the kitchen while the user was in a different room, some of the respondents did believe it was dangerous to do so. With respect to leaving the toaster alone in the house, most of the respondents believed it was not safe.

Similar to the jar candle and toaster, most of the respondents' believed that it was not dangerous to leave a RCV battery pack charging in one room of the house while the user was in another. The majority of the participants also believed that it was not dangerous to leave a RCV battery pack charging alone in the house. Although a minority of the respondents did perceive it as not safe.

With respect to the other products, the respondents' ratings indicated that cooking appliances tended to be perceived as more dangerous than non heat producing devices. The majority of participants did not perceive leaving a

Crockpot, microwave oven, gas fireplace, clothes dryer, or stove-oven operating in one room while the user was in another as dangerous. However, the majority of participants believed that it was dangerous to leave a stove-cook top operating while the user was in another room. Of course the stove-cook top involves an exposed flame and/or heating element.

The results also indicated that the respondents believed that it was safe to leave all of the battery powered products charging alone regardless of someone being home or not. These results were expected and seem to be consistent with the typical use of battery powered products. For example, cell phones, laptops, i-Pods, and digital cameras require charging on a daily basis. Often the charging period extends for hours. Other battery powered products such as dust-busters, baby monitors receivers, flashlights, and power tool batteries are intended to be left charging until needed.

The last result to note was also expected, the majority of respondents believed it was safe to leave non-heat producing electrical appliance plug-in and operating even when no one was home. These results are consistent with the ubiquitous use of consumer electronic products and their relative safety.

The results of this study are consistent with past research and indicate that consumers vary in their hazard perceptions and although they tend to relate to the actual risk of the product they do not always match. The implication of these tendencies indicates the need to provide adequate product warnings that explicitly communicate the specific hazard, its consequences, and how to avoid it. Furthermore, it cannot be assumed that consumers' beliefs and perceptions are uniform and/or correct. When adequate warnings are not provided consumers are deprived of the ability to make an informed decision with regards to their behavior.

REFERENCES

Dejoy (1999). "Chapter nine: Attitudes and Beliefs." In *Warnings and Risk Communications* (Eds. Wogalter, Dejoy, & Laughery). Pgs. 189 – 219. Taylor & Francis: Philadelphia, PA.

Frantz, P. (1994). "Effect of location and procedural explicitness on users processing of and compliance with product warnings." *Human Factors, 36 (3),* 532 – 546.

Godfrey, S.S. and Laughery, K.R. (1984). "The biasing effects of product familiarity on consumers' awareness of hazard." *In Proceedings of Human Factors and Ergonomics Society 28th Annual Meeting.* Pgs. 483 – 486. HFES: Santa Monica, CA.

Laughery, K.R. & Paige-Smith, D. (2006). "Chapter 31: Explicit information in warnings." In *Handbook of Warnings* (Ed. M. Wogalter). Pgs. 419 – 428. Lawrence Erlbaum Associates: Mahwah, NJ.

Laughery (1993). "Everybody Knows." *Ergonomics in Design. July,* 8 – 13.

Leonard & Wogalter (2000). "What you don't know can hurt you: household products and events." *Accident Analysis and Prevention, 32,* 383 – 388.

Peters, G. (1980). "15 cardinal principles to ensure effectiveness of warning systems." *Occupational Health and Safety, May,* 76 – 79.

Wagenaar, W.A. (1992). Chapter 9: Risk taking and accident causation. In *Risk Taking Behavior* (Ed. by J.F. Yates). Pgs. 257 – 280. John Wiley and Sons Ltd.: New York, NY.

1046

Weegles, W.F. & Kanis, H. (2000). "Risk perception in consumer product use." *Accident Analysis and Prevention, 32*, 365 – 370.

Wogalter, M.S., Brelsford, J.W., Desaulniers, D.R. & Laughery, K.R. (1991). "Consumer product warnings: the role of hazard perception." *Journal of Safety Research, 22*, 71 – 82.

Wogalter, M.S., Young, S.L., Brelsford, J.W., & Barlow, T. (1999). "The relative contributes of injury severity and likelihood information on hazard-risk judgments and warning compliance." *Journal of Safety Research, 30(3),*151 – 162.

Woodson, W., Tillman, P. & Tillman, B. (1992). *Human Factors Design Handbook* (2nd Ed.). McGraw Hill, Inc.: Columbus, OH.

Wright, P., Creighton, P., & Threlfall, S.M. (1982). "Some factors determining when instructions will be read." *Ergonomics, 25(3),* 225 – 237.

CHAPTER 109

Government, Warnings, and Safety Information: A Comparison of Inter-Agency Regulations and Guidance

Joseph B. Sala, Elizabeth A. Nichols, Rahmat Muhammad, Sunil D. Lakhiani, Robert Rauschenberger, and Christine T. Wood

Exponent® Failure Analysis Associates

ABSTRACT

Warning and safety information can be provided in an attempt to modify or encourage safe behavior with a given product or within a certain environment. Scientific investigations into the use and design of such warnings have yielded a vast literature and research base. However, manufacturers, employers, and practitioners must also satisfy and comply with governmental regulations related to the presentation of safety information for products, environments, and tasks. The specific regulatory requirements as well as non-mandatory guidelines put forward by a number of governmental agencies are reviewed, compared, and contrasted. The correspondence among such agency positions and a scientific understanding of the presentation of safety information is discussed.

Keywords: Government, regulations, warning

WARNINGS AND SAFETY INFORMATION

Warnings have generally been defined as information about possible negative consequences; "a message that something undesirable may occur to someone or something as a result of taking (or failing to take) some action" (Ayres et al., 1989). Over the past 30 years, a sizable literature has developed on behavioral responses to warnings. Significant reviews and annotated guides to the literature have been written elsewhere (e.g., Ayres et al., 1998; Miller & Lehto, 2001).

A variety of potential hazards can be identified for any number of products, environments, or tasks, and developers of warnings must make decisions as to which, if any, hazards to address. The scientific literature identifies factors to consider that include the likely effectiveness of providing a warning about a particular hazard (e.g., Ayres et al., 1989; Arndt et al., 1998; Dorris & Purswell, 1977), directives presented in relevant standards (e.g., Young et al., 2002, Wood et al., 2006), relationships between the number of warnings and attention and memory (e.g., Chen et al., 1997; Frantz et al., 1999), and selection based on hazards related to the highest impact in terms of injury count, risk analysis, or accident mode analysis (e.g., McCarthy et al., 1995; Ayres & Wood, 1995). The literature to which these studies belong allows for scientific investigation and findings to guide decisions as to content, format, and placement of warnings.

In addition to and often superseding warnings recommendations made in the literature, are the directives of regulatory agencies. The quantity and content of these regulated warnings and the presentation of safety information has largely increased over time (Diedrich et al., 2001). Furthermore, some of the governmental agencies have authored "guidelines" or made comments on the presentation of safety information to the industry that they regulate. In the present chapter, we characterize the involvement and directives of a number of governmental agencies with respect to the presentation of safety information. Our comparison highlights commonalities and differences in the requirements and recommendations across governmental agencies and relates these to relevant scientific research.

GOVERNMENTAL REGULATORY AGENCIES

CONSUMER PRODUCT SAFETY COMMISSION

The U.S. Consumer Product Safety Commission (CPSC), created in 1972, is charged with protecting "the public against unreasonable risks of injuries and deaths associated with consumer products."[1] The CPSC issues and enforces mandatory standards and works with industry to develop voluntary standards.

The CPSC has jurisdiction over about 15,000 types of consumer products and requires that specific warnings and safety information be included with some, but

---

[1] SEC. 2. [15 U.S.C. § 2051], b1

not all, of these products. Mandated safety information can be in the form of a warning label and/or instructions to be included on the product itself or in a product manual. For some product labels, the CPSC specifies that certain words or images be included while for others, the CPSC provides only general information about what must be addressed, without referencing specific wording. These requirements were established under various acts, some of which are further discussed.

The Consumer Product Safety Act (CPSA), enacted in 1972 and continually updated, established the CPSC and requires that warnings or safety information be included with a variety of products. For some products (e.g., certain self pressurized consumer products, walk-behind power lawn mowers, bunk beds, and CB base station antennas, TV antennas, and supporting structures), the CPSC requires that specific warning language or labels be included on the product, product container, and/or instruction manual. For other products (e.g., bicycle helmets) the CPSC requires that warnings, labels, or pictorials addressing general hazards be included on the product, product container, or instruction manual, although it does not specify the exact wording, label, or image. For some products, CPSC provides examples of warning labels. Example warnings labels for coal and wood burning appliances are included in 16 CFR 1406, and although the CPSC recommends one over the other, it does not require that one particular label be included.

Other acts require precautionary labeling on certain household products. For example, the Federal Hazardous Substances Act (FHSA) requires labeling if a product presents defined hazards (e.g., ingestion, inhallation). This Act specifies that, for these products, the label must include the appropriate signal word (e.g., Danger, Caution, or Warning), an affirmative statement of the principal hazard or hazards (e.g., Flammable, Harmful if swallowed), precautionary statements instructing users on self-protective measures, and the statement "Keep out of the reach of children." The FHSA provides guidelines on how to place the label in order to make it conspicuous. Also under the FHSA, toys with small parts intended for use by children ages 3 to 6 (e.g., marbles, small balls, and balloons) are required to have a specific warning regarding choking hazards. The Consumer Product Safety Improvement Act (CPSIA), issued in 2008, extends this by requiring the same choking warning statement in advertisements if the advertisement provides a direct means for purchase or order of the product (e.g., via catalogues and websites).

In addition to mandating warnings and safety information, the CPSC is also involved in the process of creating voluntary standards for consumer products. In 1981, Congress amended certain acts and authorized the CPSC to "give preference to voluntary standards over promulgating mandatory standards if [the Commission] determines that a voluntary standard will eliminate or adequately reduce an injury risk, and that there will be a likelihood of substantial compliance with the standard."[2] The amendments further require that the CPSC provide assistance to organizations engaged in developing voluntary standards.

The CPSC also provides general industry guidance to manufactures of consumer products in the form of published guidelines. For example, the

---

[2] 16 CFR Part 1031 (FR 71 p38755)

"Manufacturer's Guide to Developing Consumer Product Instructions" is a publication that illustrates "principles of good design that are generally applicable to all instructions associated with consumer products" and directs users to other sources of information. The CPSC also provides a publication titled "Age Determination Guidelines" which can be used to "help manufacturers accurately determine the appropriate age category for their toys" and to label, promote, and market those toys to that age group.

OCCUPATIONAL SAFETY AND HEALTH ADMINISTRATION

In 1970, U.S. Congress enacted the Occupational Safety and Health Act and formed the Occupational Safety and Health Administration (OSHA) under the Department of Labor "to assure safe and healthful working conditions for working men and women" by enforcing standards set forth in the Act and providing training in the field of occupational safety, among other actions.[3]

OSHA's regulations for the general, construction, agriculture, and maritime industry lay a wide array of requirements pertaining to warning signs, tags, and labels. OSHA requires warning signs and labels for informing employees about the hazards and safety precautions associated with hazardous materials (e.g., chemicals), environments (e.g., confined spaces, radiation hazards), and equipment (e.g., machine guarding) that they can encounter in the workplace. In some cases, OSHA mandates the content of precautionary labels and signs. Examples include: labeling associated with welding, cutting, and brazing operations, occupational exposures to asbestos, bloodborne pathogens, and electrical equipment use and installation. OSHA also establishes the criteria for the format and placement of precautionary signs. For example, with respect to manlifts, a top floor warning sign that reads "TOP FLOOR - GET OFF" in block letters no less than two inches in height must be located no more than two feet above the top terminal landing and within easy view of an ascending.[4]

For other work conditions OSHA provides examples of, but does not mandate, the specific language that would satisfy various requirements. For example, OSHA states that a sign reading "DANGER – PERMIT REQUIRED CONFINED SPACE, DO NOT ENTER" or similar language would satisfy the requirement that employers notify of a permit-required confined space.[5] With respect to machinery with parts that may rotate after power has been disengaged such as farming equipment, OSHA requires a warning sign that instructs employees to look and listen for evidence of rotation and not to remove the guard or access door until all components have stopped; however, no specific language is recommended.

In 1983, OSHA promulgated the Hazard Communication Standard (HCS), a generic and performance-oriented standard requiring that manufacturers, importers, or distributors of chemicals label the containers of hazardous chemicals to include

---

[3] Occupational Safety and Health Act of 1970 Public Law 91-596, 84 STAT. 1590
[4] 29 CFR 1910.68
[5] 29 CFR 1910.146(c)(2)

contact information, to identify the hazardous chemical(s), and to list "appropriate hazard warnings." However, this regulation does not specify the specific content, format, or placement of the labels. The act defines "hazard warning" as "any words, pictures, symbols, or combination thereof appearing on a label or other appropriate form of warning which convey the specific physical and health hazard(s)." It also requires manufacturers or importers to obtain or develop a material safety data sheet (MSDS) for each hazardous chemical they produce or import and provide it to distributors and employers. OSHA also requires that employers maintain copies of MSDSs for each hazardous chemical and ensure that they are readily available to employees in the work area.

Not all of OSHA's requirements list the specific elements of the warning signs and labels. However, they occasionally refer to other federal regulations or non-mandatory consensus standards such as those published by the American National Standards Institute (ANSI) or the American Society for Testing and Materials (ASTM). For example, OSHA's HCS has adopted definitions of the acute effects associated with hazardous materials from ANSI Z129.1.[6] OSHA also requires that precautionary labels associated with welding, cutting, and brazing to state "See ANSI Z49.1 – 1967 Safety in Welding and Cutting published by the American Welding Society." In many cases, OSHA refers to or requires compliance with specifications established by ASTM standards.

FOOD AND DRUG ADMINISTRATION

The U.S. Food and Drug Administration (FDA) is tasked, in part, with protecting the public health by assuring the safety, effectiveness, and security of medical products such as human drugs and medical devices. The agency's regulatory functions, first granted in 1906 with the passage of the Federal Food and Drug Act, have evolved; yet from the start and throughout its evolution, the FDA has regulated product labeling. The Food, Drug, and Cosmetic Act of 1938 expanded the FDA's regulatory functions to include cosmetics and medical devices, mandated pre-market approval of the safety of all new drugs, and required that drugs be labeled with adequate directions for safe use. To this day, the FDA relies on labeling to disseminate information necessary for safe and effective use of medical products including medical devices and drugs.

The Federal Food, Drug, and Cosmetic Act established that prescription drugs require professional supervision of a "practitioner licensed by the law to administer such drug." Therefore, a prescription drug's FDA approved labeling (also known as "package insert") is written in technical language for the healthcare practitioner as its principal audience and contains prescribing information as well as information for safe use. The labeling requirements for new and recently approved prescription drugs have changed over time; recent amendments have modified the content and format of package inserts.[7] While safety information can be found throughout the

---

[6] 48 FR 53295, dated 11/25/1983
[7] 71 FR 3922

package insert, the bulk of it is contained in the boxed warning, contraindications, warnings and precautions, and adverse reactions sections. Details of the content required in each section are provided in 21 CFR 201.57. The FDA has also released several guidance documents containing FDA recommendations intended to assist drug manufacturers in drafting the safety information sections. For example, 21 CFR 201.57 states that the boxed warning (if required by the FDA) is to provide information about special circumstances, particularly "those that may lead to death or serious injury." FDA's draft guidance on when to use a boxed warning expands on this and provides examples of circumstances that may require a boxed warning.

General labeling provisions for medical devices have been established in 21 CFR 801 and definitions of some terms, along with factors to consider and address in the label, are provided in the FDA's device labeling guidance.[8] Where applicable, the device labeling guidance paraphrases the labeling requirements for prescription drugs in order to maintain consistency between drug and device labeling content and to minimize misunderstandings by healthcare practitioners. As with prescription drugs, prescription medical device use is considered unsafe except under the supervision of a medical practitioner. Labeling on or within the package of the device must bear information for use as well as information about indications and "any relevant hazards, contraindications, side effects, and precautions under which practitioners licensed by law to administer the device can use the device safely and for the purpose for which it is intended." Exceptions can be made for "a device for which directions, hazards, warnings, and other information are commonly known" to licensed practitioners.[9]

The labeling content for non-prescription over-the-counter (OTC) drugs and medical devices must include safety information (e.g., warnings, contraindications, adverse reactions) and instructions for use.[10] This information is intended to be read and understood by the consumer in order to safely use the product for its intended purposes.[11] The FDA requires additional special warnings for specific medical devices. For example, denture reliners, pads, and cushions should include a conspicuous statement warning against longterm use.[12] OTC drug warnings include, as applicable, warnings about allergic reactions, Reye's syndrome, and alcohol, among others.[13] Non-prescription OTC drug and medical device labels must include instructions about actions to take if a user notices an adverse reaction or if they have any questions about the use of the product. In addition to dictating the contents of OTC drug and medical device labeling, the FDA also advises on the format and placement of this information. For medical devices, the principal display panel or "the part of a label that is most likely to be displayed, presented, shown, or examined" by the consumer, should accommodate all the mandatory label

---

[8] Device Labeling Guidance #G91-1
[9] 21 CFR 801.109
[10] 21 CFR 201.66; Device Labeling Guidance #G91-1
[11] 21 CFR 801.5; Guidance for Industry Labeling OTC Human Drug Products – Questions and Answers
[12] Guidance for Industry and FDA Staff - OTC Denture Cushions, Pads, Reliners, Repair Kits, and Partially Fabricated Denture Kits
[13] Guidance for Industry Labeling OTC Human Drug Products – Questions and Answers

information "with clarity and conspicuousness."[14] 21 CFR 201.66 details the placement of OTC drug information on product packaging. Information on OTC labels should appear as bulleted text format and FDA guidance document gives examples on how to convert a lengthy warning into the bulleted-text format.

NATIONAL HIGHWAY TRAFFIC SAFETY ADMINISTRATION

The National Highway Traffic Safety Administration (NHTSA), an agency of the U.S. Department of Transportation, was established by the Highway Safety Act of 1970. NHTSA directs the highway safety and consumer programs established by the National Traffic and Motor Vehicle Safety Act of 1966, the Highway Safety Act of 1966, the 1972 Motor Vehicle Information and Cost Savings Act, and succeeding amendments to these laws.

NHTSA requires that a warning label addressing hazards associated with air bags be affixed to car visors in positions that are equipped with air bags and that all utility vehicles have a warning label addressing rollover risk permanently affixed to the sun visor on the driver's side of the vehicle.[15] Samples of the required warning labels and pictograms are included in the final rules, and the manufacturer's label must conform to the sample. Required language for the air bag warning includes "The BACK SEAT is the SAFEST place for children," "NEVER put a rear-facing child seat in the front," and "Sit as far back as possible from the air bag." For both the air bag and rollover warnings, NHTSA provides specific instructions with respect to the color and size of the label, and requires that an alert label be present when the visor is in the up position if the warning is not visible when it is in that position. Further, for vehicles for which both air bag and rollover warning labels are required on the same sun visor, NHTSA mandates that a minimum of 3 cm separates the two warnings. NHSTA also requires that precautionary language addressing air bags and rollovers be included in the vehicle owner's manuals.

NHTSA also mandates warnings and safety information for child restraint systems used in motor vehicles and aircraft.[16] For example, child restraint systems that can be used in a rear facing position are required to have a permanently affixed warning label alerting users of the dangers of placing these restraint systems in front seats with air bags. NHTSA requires that the label conforms to the sample label included in the regulation.

DISCUSSION

Each of the reviewed governmental agencies, through distinct regulations, addresses warnings and the presentation of safety information. Each agency contends with unique circumstances and context in regulating the content, format, and how

---

[14] 21 CFR 801.60
[15] 49 CFR §571.208; 49 CFR §575.105
[16] 49 CFR 571.213

placement of such information is to be presented. It is beyond the scope of this chapter to detail an exhaustive history of the decisions that have shaped each agency's current position with respect to the presentation of safety information. However, commonalities and differences can be found among the policies and regulations of the different agencies reviewed and these have direct implications for the industry that each agency serves.

Across agencies, there are differences in the specificity and detail of regulations surrounding warnings and safety information. For example, both CPSC and OSHA regulations generally reference the need for warnings regarding hazards but are often silent about the content, format, and placement of such warnings. In only a few instances do these two agencies provide specific details regarding this information. In contrast, the FDA provides detailed information regarding required labeling of medical products in general and requires that labeling must have FDA approval or clearance.

Many agencies work with voluntary standards organizations and work groups to issue warning and safety information guidance (e.g., Collins, 1999). However, compliance with non-mandatory consensus standards often does not ensure regulatory compliance. For example, OSHA clarifies that compliance with national consensus standards is not a substitute for compliance with the provisions of the OSHA standard and specifically states that it "does not approve materials developed by the private sector" or provide any formal endorsement or approval.[17] Indeed, OSHA definitions of flammable liquids differ from that put forward by ANSI Z129.1. Other agencies have chosen to go against voluntary standards as well. For instance, in final rule makings NHTSA specifically contradicts recommendations found in warnings standards such as the ANSI Z535 series.[18] Elsewhere, the CPSC has officially endorsed a voluntary standard. With the passage of the CPSIA, the once voluntary ASTM F963 "Standard Consumer Safety Specification for Toy Safety" was made mandatory for all children's toys.

In addition to mandatory and voluntary standards, governmental agencies provide "guidance" documents to the members of the industry that they regulate. For example, both the CPSC and the FDA have authored manuals to assist manufacturers with the production of user-instruction manuals (Backinger & Kingsley, 1993; Singer, et. al., 2003). These guidance documents outline specific steps that manufacturers can follow to write instruction manuals that will be informative, clear, and understandable to users. These guidance documents recommend tasks that will help manufacturers identify necessary instructions and safety information, design and format them appropriately, and test user comprehension and opinion of the design. Many of the recommendations contained in these guidance documents are consistent with the ANSI Z535.6 standard for presenting safety information in such manuals.

These guidance documents also highlight a concern about "overwarning" - a commonly held concern in warnings research. Regarding warnings in user manuals,

---

[17] Standard Interpretation of 29 CFR 1910.1200 dated 02/09/94, titled, "Labeling provisions"
[18] 61 FR 60211

the CPSC instructs manufacturers to: "Avoid overwarning" and that "If you include too many safety messages for highly unlikely and trivial hazards, you weaken the effectiveness of the more significant messages" (Singer, et. al., 2003). In its guidance, the FDA notes: "Overwarning has the effect of not warning at all. The reader stops paying attention to excess warnings" (Backinger & Kingsley, 1993). While neither text presents explicit rules on the selection of warnings, they do discuss general issues and reference literature that can be consulted. Governmental agencies have indeed applied this scientific principle in deciding on regulations. For example, NHTSA has limited in-vehicle labeling and, at times, has expressly removed previously required information from in-vehicle labels[19] on the basis that overwarning in the interior of vehicles can reduce the impact of all labels.[20]

Another aspect of overwarning identified in the scientific literature is the inclusion of warnings about uncertain, unproven, or theoretical hazards. Should this practice become widespread, warnings will be viewed increasingly as false alarms and their impact will be reduced (Frantz et al., 1999). The FDA, in publishing its Final Rule regarding "Requirements on Content and Format of Labeling for Human Prescription Drug and Biological Products," states that "theoretical hazards not well-grounded in scientific evidence" can cause risk information to lose its significance, and subsequent overwarning can "have a negative effect on patient safety and public health." OSHA has required chemical labeling to specify health hazards including any target organ effects due to exposure to a chemical "for which there is statistically significant evidence based on at least one study conducted in accordance with established scientific principles that acute or chronic health effects may occur in exposed employees."[21]

All of the regulatory agencies reviewed here, CPSC, OSHA, FDA, and NHTSA, use warnings and safety information as a tool to assist in the safe interaction between an individual and a product, activity, or environment. They all provide mandated warnings well as non-mandatory guidance. Each agency addresses issues related to warning content, format and placement. In discussing these issues, the agencies recognize a number of principles reflected in the scientific literature and have acknowledged these principles in their rule-making.

REFERENCES

Arndt, S. R., Ayres, T. J., McCarthy, R. L., Schmidt, R. A., Wood, C. T., & Young, D. E. (1998). Warning labels and accident data. In *Proceedings of the Human Factors and Ergonomics Society 42nd annual meeting* (pp. 550-553). Santa Monica, CA: Human Factors and Ergonomics Society.

Ayres, T. J., Gross, M. M., Wood, C.T., Horst, D. P., Beyer, R. R., & Robinson, J. N. (1989). What is a warning and when will it work? In *Proceedings of the*

---

[19] e.g., 49 CFR 571.208; 68 FR 509
[20] 61 FR 60213
[21] 29 CFR 1910.1200

Human Factors Society 33rd annual meeting (pp. 426-430). Santa Monica, CA: Human Factors Society.

Ayres, T. J., & Wood, C. T. (1995). The warning label development process. In *Proceedings of the Silicon Valley Ergonomics Conference & Exposition, ErgoCon '95* (pp. 187-190).

Ayres, T. J., Wood, C. T., Schmidt, R. A., Young, D. E., & Murray, J. (1998). Effectiveness of warning labels and signs: An update on compliance research. In *Proceedings of the Silicon Valley Ergonomics Conference & Exposition, ErgoCon '98* (pp. 199-205).

Backinger, C.L., Kingsley, P.A. (1993) Write It Right: Recommendations for Developing User Instructions for Medical Devices Used in Home Health Care. HHS Publication FDA.

Chen, J. Y. C., Gilson, R. D., & Mouloua, M. (1997). Perceived risk dilution with multiple warnings. In *Proceedings of the Human Factors and Ergonomics Society 41st annual meeting* (pp. 831-835). Santa Monica, CA: Human Factors and Ergonomics Society.

Collins, B. (1999). Standards and Government Regulations in the USA. In M.S. Wogalter, D.M. DeJoy, & K.L. Laughery (Eds.), *Warnings and Risk Communication* (pp. 265-290). London: Taylor & Francis.

Diedrich, F. J., Wood, C. T., & Ayres, T. J. (2001). Trends in federally mandated warning labels for consumer products. In *Proceedings of the Human Factors and Ergonomics Society 45th annual meeting* (pp. 838-842). Santa Monica, CA: Human Factors and Ergonomics Society.

Dorris, A. L., & Purswell, J. L. (1977). Warnings and human behavior: Implications for the design of product warnings. *Journal of Products Liability, 1*, 255-263.

Frantz, J. P., Rhoades, T. P., Young, S. L., & Schiller, J. A. (1999). Potential problems associated with overusing warnings. In *Proceedings of the Human Factors and Ergonomics Society 43rd annual meeting* (pp. 916-920). Santa Monica, CA: Human Factors and Ergonomics Society.

McCarthy, R. L., Ayres, T. J., Wood, C. T., & Robinson, J. N. (1995). Risk and effectiveness criteria for using on-product warnings. *Ergonomics, 38*, 2164-2175.

Miller, J. M., & Lehto, M. R. (2001). *Warnings & safety instructions: Annotated and indexed* (4th ed.). Ann Arbor, MI: Fuller Technical Publications.

Singer, J.P., Balliro, G.M., and Lerner, N.D. (2003) Manufacturer's Guide to Developing Consumer Product Instructions. U.S. Consumer Product Safety Commission.

Therrell, J.A., Brown, P., Sutterby, J.A. and Thornton, C.D. (2002) Age Determination Guidelines: Relating Children's Ages to Toy Characteristics and Play Behavior. U.S. Consumer Product Safety Commission.

Wood, C. T., Sala, J. B., Sanders, K., & Cassidy, P. (2006). Trends in consumer product warnings found in voluntary standards. In *Proceedings of the Human Factors and Ergonomics Society 50th annual meeting* (pp. 1798-1802). Santa Monica, CA: Human Factors and Ergonomics Society.

Young, S. L., Frantz, J. P., Rhoades, T. P., & Darnell, K. R. (2002). Safety signs & labels: Does compliance with ANSI Z535 increase compliance with warnings? *Professional Safety, 47*(9), 18-23.

Chapter 110

Clarifying the Hierarchical Approach to Hazard Control

Steven M. Hall, Stephen L. Young, J. Paul Frantz, Timothy P. Rhoades,
Charles G. Burhans, Paul S. Adams

Applied Safety and Ergonomics, Inc.
Ann Arbor, MI 48108

ABSTRACT

Over the years, many different hazard-control methods have been categorized and rank-ordered into numerous hierarchies. These "safety hierarchies" are rules of thumb that can help guide consideration of options and strategies to achieve acceptable risk or conditions that are considered reasonably safe. Paradoxically, while options for hazard control and for hierarchies have proliferated, the concept of a hierarchical approach to hazard control has been grossly oversimplified, particularly in legal circles, to the point where the phrase "design, guard, warn" is commonly (and incorrectly) referred to as "*the* safety hierarchy" and presumed to be a rule that completely dictates safety decisions. This paper describes the concept and role of hierarchical approaches to hazard control, in light of common misconceptions held by legal professionals.

Keywords: Hazard control, safety hierarchy.

INTRODUCTION

Depending on the situation, hazards can be addressed in a wide variety of ways. The concept of grouping methods for addressing hazards and ordering them according to preference or priority originated many years ago in the field of safety management for industrial workers (cf. Heinrich, 1950; National Safety Council, 1955). This type of ordered list has been referred to as a "hierarchy of hazard controls" or "safety hierarchy." Since that time, the quantity and variety of strategies for hazard control have increased, as have the quantity and variety of safety hierarchies (cf. Barnett & Brickman, 1985). Paradoxically, while options for hazard control and hierarchies have proliferated, the concept of a hierarchical approach to hazard control has been grossly oversimplified, particularly in legal circles, to the point where the phrase "DESIGN, GUARD, WARN" is commonly (and incorrectly) referred to as "*the* safety hierarchy" and presumed to be a rule that completely dictates safety decisions. While this shorthand may be convenient for conversations between legal professionals and those involved in making hazard-control and safety decisions about environments, products, activities, and workplaces (referred to in this paper as "technical professionals"), the cost of this oversimplification seems to be widespread misunderstanding and misapplication of the concept of hazard-control hierarchies, particularly when technical professionals interact with legal professionals. The purpose of this paper is to:

1. briefly describe the concept and role of hierarchical approaches to hazard control from the perspective of technical professionals, and
2. identify and discuss several myths related to what is often referred to as "*the* safety hierarchy."

HIERARCHICAL HAZARD CONTROL

Generally speaking, a goal of technical professionals is to develop, maintain or manage environments, products, activities, or workplaces such that they are considered reasonably safe; recognizing that there can be no absolute safety and that the concept of safety reflects a freedom from risk that is considered unacceptable given the values of a society in a certain context at a certain time. Thus, it is expected that there will be a range of risk associated with different activities, products, workplaces, and environments that will be considered acceptable, given the current conventions of society and the interest of balancing risk considerations in light of the objectives, demands, goals, interests, and desires associated with a particular situation. In short, safety can be defined as freedom from unacceptable risk. Thus, a situation is considered reasonably safe when risk is in an acceptable range. If needed, this is accomplished through measures to affect the actual and/or acceptable level of risk.

Safety hierarchies can help *guide* consideration of options and strategies to achieve this goal, but they are simply *rules of thumb* rather than formulas or algorithms. According to Parker (2008), *rules of thumb* are easy-to-remember guides that fall somewhere between a mathematical formula and a "shot in the dark." Safety hierarchies, including "design, guard, warn," are *rules of thumb* in the sense that they are:

- concise and easy-to-remember simplifications used as shorthand to characterize the more complex considerations that go into actual decisions,

- general advice to be *considered*, but not algorithms that *dictate* decisions,
- not a statement of scientifically based fact or universally accurate rule, and
- not a set of criteria or a gauge for measuring or evaluating safety.

In our experience, there are many misconceptions regarding the nature and purpose of safety hierarchies that can lead to misapplication and/or overextension of this concept. Several of the more common misconceptions about safety hierarchies will be addressed as "myths" (i.e., widely held but false ideas) for the purposes of the following discussion.

MYTH 1—THERE IS ONLY ONE SAFETY HIERARCHY

It is not uncommon, especially in the context of litigation, to hear someone incorrectly state or imply that there is only one safety hierarchy with three levels of priority:

1. Design—eliminate a hazard through design
2. Guard—prevent contact with a hazard through guarding
3. Warn—provide warnings about hazards

This instance of a hierarchy is often referred to in shorthand using the phrase "DESIGN, GUARD, WARN." Moreover, it is often incorrectly described as "*the* safety hierarchy."

In reality, there are many different hierarchies. Barnett and Brickman (1985) identified 45 different published safety hierarchies that differed in terms of the number, types, and descriptions of hazard-control methods included. The authors concluded that "there is no such thing as **the** safety hierarchy; there are many hierarchies" (emphasis original). They proposed a safety hierarchy with five levels as an attempt to consolidate the many hierarchies that they observed. However, other hierarchies have been proposed that do not fit into their five-level scheme. For example, Hammer and Price (2001) proposed a complimentary pair of hierarchies with seven and five levels. Other hierarchy-like lists of hazard control methods include Haddon's unwanted energy release concept, which lists ten approaches (Haddon, 1970). Even a cursory review of the safety literature shows that it is simply untrue that there is only one hierarchy of hazard control or that one hierarchy is somehow superior to all the others in all or even most conditions.

MYTH 2—"*THE* SAFETY HIERARCHY" IS A SCIENTIFIC LAW THAT MUST BE FOLLOWED

In addition to the incorrect belief that there is only one safety hierarchy, there are those who treat "*the* safety hierarchy" as if it were a scientific law that must be followed. According to this thinking, every hazard must be "designed out" if it is "technologically and economically feasible," if it cannot be designed out it must be guarded, and if it can't be guarded it must be warned about. Other approaches to hazard control are generally ignored. Rigidly following this thinking would produce, in many instances, results inconsistent with our everyday life experience and widely considered unacceptable. Examples include:

- **Steaks**: The hazards posed by grilling steaks on a charcoal or gas grill (e.g., burns, fires, carbon monoxide) could be eliminated by cooking without gas or charcoal. It would be "technologically and economically feasible" to cook steaks on a stove or in an oven instead. However, these methods still pose risks of burns and fires. These hazards could be further reduced by cooking steaks only in a microwave.

- **Baseball:** It is possible to reduce the hazard of being hit by a pitch by replacing baseball with slow pitch softball, or eliminate it completely by playing tee-ball. This is "technologically and economically" feasible for all age groups and skill levels.

- **Bicycles**: Bicycles could be guarded from tipping using training wheels for all riders; training wheels are certainly "economically and technologically feasible." Alternately, bicycles could be replaced with tricycles or quadricycles.

- **Fire Hazards**: All clothing could be designed to be as fire-resistant as clothing worn by auto racers.

- **Stairs**: The Consumer Product Safety Commission estimates that more than 1.2 million injuries involving stairs or steps occur each year (NEISS database, 2008). The risks posed by stairs or steps could be "designed out" by eliminating them from homes (i.e., building only ranch-style houses) or by requiring elevators for multi-story homes.

Our everyday experiences as parents, teachers, coaches, citizens, etc. clearly do not reflect a rule that all hazards must be eliminated if possible, guarded if not eliminated, and warned about otherwise.

MYTH 3—"*THE* SAFETY HIERARCHY" MUST BE USED TO MINIMIZE AND ELIMINATE RISKS

Another misconception is that "*the* safety hierarchy" must be applied (and followed) to every hazard until risk is minimized or eliminated. However, such a position fails to recognize that the technical professionals' goal is to achieve reasonably safe conditions, as previously described.

Safety hierarchies do not distinguish between "acceptable" and "unacceptable" risk. In fact, they do not consider risk at all. They provide no guidance as to when additional efforts to reduce risk are no longer necessary or appropriate (i.e., when one has reached a level that is deemed "acceptable").

Determining the extent to which risks need to be mitigated or eliminated involves value judgments that fall outside the domain of safety hierarchies. The mere presence of a risk does not, in itself, require that it be eliminated or reduced. In fact, we as a society regularly and willingly accept and seek out many risks in return for various practical benefits (e.g., increases in efficiency, capability, quality, enjoyment, comfort, satisfaction, etc.). We also accept some degree of risk to the extent that the costs of reducing it (in terms of time, effort, resources, esteem, social standing, etc.) are viewed as disproportionate to the benefits gained by a reduction in the risk. Such value judgments (balancing these benefits and costs) can change over time and across situations, further complicating the application of safety hierarchies. For example, automobiles sold many years ago would generally not meet current standards for safety, and the acceptability of various driver behaviors have changed (e.g., the acceptable level of driver blood alcohol, the acceptability of non-driving activities concurrent with driving, and the acceptability of young drivers driving at night, with friends and/or after consuming any alcohol).

In addition, value judgments can be affected by macro socio-economic conditions (e.g., societies can vary in terms of their social norms, values, economic conditions, etc., which can influence tolerance for risks and the means to mitigate them). Context can also play an important role in rendering such value judgments. For example, the level of risk posed by logging or crab fishing would be considered too high (and in need of reduction) if observed in an office or clerical setting. Safety hierarchies do not consider these and other factors, and do not provide any guidance about what level of risk is acceptable.

MYTH 4—APPLYING "*THE* SAFETY HIERARCHY" WILL INVARIABLY LEAD TO A NET REDUCTION IN RISK

It is often assumed that higher-order controls in "*the* safety hierarchy" (e.g., designing out the hazard) reduce risk more than lower-order controls. This assumption, particularly with regard to hazard elimination, is premised on the simple notion that, if a hazard has been eliminated, it cannot cause harm. While this might be true logically, it does not follow that the elimination of a hazard will result in a net reduction in risk for the product or activity as a whole. In practice, hazard controls intended to mitigate or eliminate risk may not actually produce a net reduction in risk.

The elimination or mitigation of one hazard can give rise to other hazards. For example, the hazard of electric shock could be eliminated from a power drill by powering it instead with an internal combustion engine. While doing so would eliminate the electric shock hazard, it would introduce other risks that do not exist with the electric drill. In other cases, new hazards may be introduced that cannot be predicted based on the current state of scientific knowledge.

Another premise of safety hierarchies is that less preferred methods rely on human behavior and are thus expected to be less reliable in reducing risk. However, supposedly more preferred methods, including hazard elimination, do not completely eliminate reliance on human behavior. Research shows that people may adapt their behavior in response to safety interventions such that the expected benefit is reduced or eliminated, or such that overall risk is increased. People's altered response(s) to safety interventions is known in the literature as "behavioral adaptation." Young, Frantz and Rhoades (2002) concluded:

"...The literature related to behavioral adaptation indicates that more consequences may be lurking for those who blindly follow the safety hierarchy approach, particularly if one has a tendency to accept 'design' interventions as inevitably more effective than other interventions and if one completely separates design considerations from behavior (including unforeseen secondary behaviors). Even for those concerned with behavioral adaptation to changes in design, it is clear that predicting the nature and extent of behavioral change in design or environment can be difficult. As a result, it should not be surprising that unintended and/or unforeseen events may occur after such interventions." (p. 898)

Thus, hazard elimination (and guarding) will not always reduce the net risk associated with a product or activity. *"The* safety hierarchy" does not account for these kinds of scenarios and, therefore, it is inappropriate to assume that its application will invariably lead to a net reduction in risk.

MYTH 5—*"THE* SAFETY HIERARCHY" IS EQUALLY APPLICABLE TO ALL HAZARD-CONTROL DECISIONS

Historically, safety hierarchies originated in the field of safety management for workers in industrial settings (Heinrich, 1950; National Safety Council, 1955) and have since been adapted for use in other domains. However, because other domains have different characteristics, the safety hierarchy can be less useful for them. They are particularly ill-suited to domains where risk-taking and pushing the bounds of human performance are integral parts of the activity and are considered desirable (in contrast to, for example, working on an assembly line).

Many recreational and sporting activities fall in this category: riding bikes, playing on a jungle gym, skiing, swimming, surfing, stock car racing, football, baseball, ice skating, etc. These activities involve goals like thrill-seeking, adventure, entertainment, challenge, exploring, thrill-seeking, personal freedom, pushing the limits of abilities, and experiencing novelty—all goals that are not typically associated with industrial settings. Unlike in industry, the risks involved are often part of the appeal.

The safety hierarchy would suggest that measures should be taken to eliminate or reduce hazards, but such measures, while technologically and economically feasible, are not considered acceptable in the "real world," such as:

- limiting skiing to flat surfaces or small slopes (i.e., cross-country), or replacing skis with snowshoes,
- replacing stock car races with time trials to avoid multi-vehicle collisions, and replacing grandstands with wide-open pavement to prevent collisions with walls and protect spectators from flying debris,
- designing children's play structures to meet OSHA walking-working surfaces rules,
- limiting surfing to wave pools, thus avoiding hazards associated with the ocean, and
- requiring personal flotation devices (i.e., life jackets) for swimmers.

Instead, typical safety measures focus on use of personal protective equipment (e.g., helmets, pads) and matching difficulty level to skill level (e.g., difficulty graded ski slopes)—approaches that are not preferred according to safety hierarchies.

These examples illustrate the general concept that certain hazards are integral to the goals of a product or activity. Sometimes the connection between the hazard and what is widely considered a valid goal is obvious (e.g., a hypodermic needle must have a sharp point to perform its function, gasoline must be combustible to act as a fuel), but in other instances the utility associated with a hazard is less tangible or obvious, and the perceived validity of the goal may depend on an individual's values and preferences (e.g., the enjoyment riding a motorcycle rather than driving a car, the fashion value of wearing high-heeled shoes rather than sneakers that would reduce the risk of falls). In these instances, safety hierarchies can be a poor guide.

MYTH 6—"*THE* SAFETY HIERARCHY" IS AN APPROPRIATE GAUGE TO DETERMINE WHETHER SOMETHING IS SAFE

In the context of litigation, "*the* safety hierarchy" is sometimes incorrectly used to gauge whether something is reasonably safe. A typical argument along these lines would be: "If a hazard exists and if it could have been eliminated (or guarded), then the product or workplace is defective because the safety hierarchy states that hazards must be eliminated (or guarded)." This logic is flawed.

Safety hierarchies do not measure risk or determine whether it is acceptable, so they cannot determine whether something is reasonably safe. Even when a risk is considered unacceptable, it does not follow that it must be eliminated even if (or simply because) it is possible to do so. As discussed above, there are many reasons why hazard elimination (or guarding) may be a poor choice of hazard control. "*The* safety hierarchy" does not provide guidance about (much less dictate) how to weigh these factors, and it certainly is not a definitive test of the most appropriate method of addressing risks for a specific situation.

Another type of argument seen in litigation asserts that failure to follow "*the* safety hierarchy" as a procedure is evidence of a defective safety *process*, regardless of what hazard-control techniques may have been used. In essence, the misguided argument is that "*the* safety hierarchy" is *the* step-by-step process that must be used making such decisions. But "*the* safety hierarchy" does not account for numerous factors involved in hazard control selection, is not the only tool or advice that can or should be considered in making safety decisions, and as illustrated above, can produce absurd results when applied strictly. While "*the* safety hierarchy" is one rule of thumb that can be considered, there is no scientific basis for concluding it is a reliable way, let alone the one and only way to make judgments about what is considered reasonably safe.

Litigators argue that if "*the* safety hierarchy" had been properly followed, then different hazard controls would have been selected. But selecting hazard controls requires professional judgment; "*the* safety hierarchy" cannot replace this. Different technical professionals, faced with the same hazards, may legitimately select different approaches to hazard control, as there is more than one path to acceptable risk. Thus, using a safety hierarchy will not reliably produce the same decisions.

Setting aside previously discussed misconceptions, there is a unique issue in litigation in that "*the* safety hierarchy" is virtually always applied to a single hazard that is at issue without regard to other hazards, unintended consequences, or considerations other than simply minimizing risk. This narrow, retrospective view contrasts with the way safety is examined prospectively. Focusing only on one hazard may not produce the same conclusions as looking at safety more broadly, and generally wouldn't be expected to result in well-founded conclusions.

CONCLUSIONS

This paper discusses a number of problems associated with misconceptions of safety hierarchies. In doing so, we do not intend to suggest that the use of safety hierarchies is generally improper. The problems identified and discussed in this paper are the result of misunderstandings, misapplications, oversimplifications, and/or overextensions of the concept of safety hierarchies on the part of some individuals, especially in the context of litigation. The intent of this paper is to clarify the nature and purpose of safety hierarchies and limit the proliferation of common misconceptions.

REFERENCES

Barnett, L.B. and Brickman, D. (1985). Safety hierarchy. *Safety Brief. 3*, 2. Niles, IL: Triodyne, Inc.

Haddon, William J., Jr. (1970). On the escape of tigers: An ecological note. *Technology Review, 72*, 7.

Hammer, W. and Price, D. (2001). Occupational safety management and engineering. (5th ed.). Upper Saddle River, NJ: Prentice Hall.

Heinrich, H.W. (1950). Industrial accident prevention: A scientific approach. (3rd ed.). New York, NY: McGraw-Hill.

National Safety Council, Inc. (1955). Accident prevention manual for industrial operations. (3rd ed.). Itasca, IL: National Safety Council, Inc.

Parker, T. (2008). Rules of thumb: A life manual. New York, NY: Workman Publishing Company, Inc.

U.S. Consumer Product Safety Commission. (2008). National Electronic Injury Surveillance System (NEISS) On-line. January 1, 2008 – December 31, 2008. Product code 1842: Stairs, steps. https://www.cpsc.gov/cgibin/NEISSQuery/home.aspx. Washington, D.C.: U.S. Consumer Product Safety Commission.

Young, S.L., Frantz, J.P., and Rhoades, T.P. (2002). "Behavioral adaptation: Unintended consequences of safety interventions," Proceedings of the Human Factors and Ergonomics Society 46th Annual Meeting, pp. 895-899.

Chapter 111

Experimental Study on the Influence of Individual Cognitive Differences on Team Synergy Knowledge Integration Process

YI Shuping, SU Li,WANG Wei, HUANG Ran ,ZHANG Na

The State Key Laboratory of Mechanical Transmission
Mechanical Engineering College
Chongqing University, Chongqing 400030, China

ABSTRACT

Experimentally study on the process of designing remotely unlock device by the team composed of different major according to specified conditions. Analyze the team members' knowledge activity by Observer XT8.0 and analyze individual cognitive differences in the synergy design process of design -type R&D team using the deep interview about behavior and mentality on the basis of design contents. It is concluded from the analysis that knowledge foundation, psychology characteristics, cognitive style, cognitive ability and team cooperation willingness. Analyze the influence on team synergy knowledge integration process. And the results shows that positive influence of knowledge foundation, cognitive ability and team cooperation willingness and complex influence of psychology characteristics and cognitive style on the team synergy knowledge integration process. And then put forward approach for novice to improve team synergy design efficiency.

Key words: R&D Team, Individual Cognitive Differences, Team Synergy, Design, Experiment Study

INTRODUCTION

The characteristics of the design-type knowledge work showed outstandingly team synergy working mode mainly under the modern manufacturing environment. However, team synergy was composed of by different design workers. They form different thinking set and cognitive ability basing on different knowledge background, cultural background and work experience. There are individual differences in cognition. Thus barriers of knowledge exchange, knowledge conversion and knowledge sharing will be formed among its members. And even the serious deviations to understand the mission objectives appeared to reduce the efficiency of the team synergy work. The barriers of knowledge exchange, knowledge conversion and knowledge sharing caused by member's individual cognitive differences in the team synergy is due to the knowledge integration issues caused by the individual cognitive differences. At present, Research related to the paper focused on knowledge integration model, mechanism, implementations and so on.

Nonaka put forward SECI (Socialisation, Externalisation, Combination, Internalisation) model based on epistemology, including the integration between tacit knowledge, the integration from tacit knowledge to explicit knowledge, the integration from explicit knowledge to tacit knowledge, and the integration between explicit knowledge (Nonaka I et al., 2000). Shin put forward ITOI model based on ontology, including individual knowledge, team knowledge, organizational knowledge and inter-organizational knowledge in the same level and different levels of integration (Shin M et al., 2001). Berends put forward thinking along from the Perspective of cognitive and studied tacit knowledge integration mechanism between individuals (Berends J J, et al., 2004). Ni Yihua carried out the classification and description of ontology for manufacturers' knowledge and built the structural model of knowledge integration (Ni Yihua, et al., 2004). Qian Yadong discussed and studied the integration of knowledge and business on the integrated model of the auto companies' knowledge management, the processing model of the knowledge-intensive business, and business process and knowledge integration (Qian Yadong, et al., 2006). Lee proposed a system, object-based knowledge integration system (OBKIS) which supports the early stages of product development. XML is adopted to facilitate data exchange (Lee C K M, et al., 2005).

Influencing factors of team synergy working mode to knowledge integration are the current research gaps in the study of knowledge integration. The research of knowledge integration have focused mainly on models, mechanisms and other technical perspective, the study starting from the process perspective is not focused on the work process itself. Therefore, the study of integration starting from the knowledge activity perspective is the current research gap. How to make the team better achieve knowledge integration, improve the efficiency of team synergy design and explore those cognitive differences hidden in the design has become an urgent problem to solve.

EXPERIMENT STUDY

In order to uncover the black box that team synergy knowledge integration process and find out individual cognitive differences, the paper designs the product-design experiment oriented to team synergy. The experiment is used to observe and analyze team synergy knowledge integration process and dig out individual cognitive differences which influence knowledge integration.

EXPERIMENT CONTENT AND METHOD

The team composed of three postgraduates from different major designs remotely unlocks device and finishes the detailed design drawing by the way of team discussion. A's major is IE, B's major is management, and C's major is Mechanical Design. The team respectively finishes conceptual design, structure design and detail design of remotely unlock device in three discussions. And record the process of three discussions by monitoring system. Analyze the videos and compare individual cognitive differences in the action by Observer XT8.0. Analyze individual cognitive differences about behavior and mentality using the deep interview on the basis of design contents of the team synergy design process.

ANALYZE KNOWLEDGE ACTIVITY IN SYNERGY DESIGN PROCESS

(1) Conceptual Design Stage
The knowledge activity analysis, the total time and times of the three experimented members' knowledge activity in conceptual design stage are shown in Figure 1, Figure 2 and Figure 3.

Figure 1 shows the knowledge activity statistics of A, B and C in product conceptual design stage. Take time of team synergy work as the abscissa axis and the various knowledge activities in team synergy work process as the ordinate axis. The knowledge activities include preparation (preparing activities of team synergy work), speak (explaining the viewpoint and scheme), listen (attentive listening or listen with thinking), response (responding the action or the sound), question, answer, leaf data, write (noting or drawing) and irrelevant topics. The time length of various activities is massive in the figure.

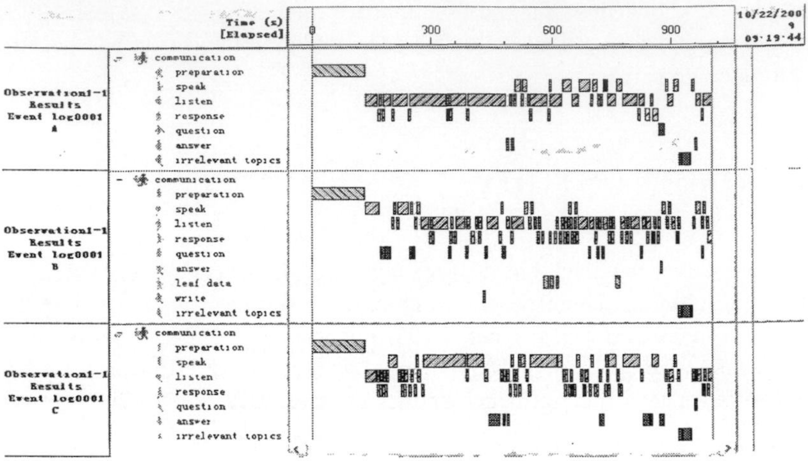

FIGURE 1. The knowledge activity analysis of the experimented members in product conceptual design stage

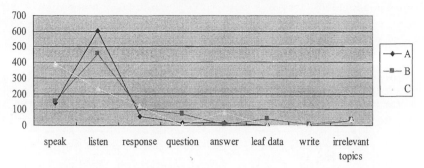

FIGURE 2. The total time statistics of the experimented members' knowledge activity in product conceptual design stage

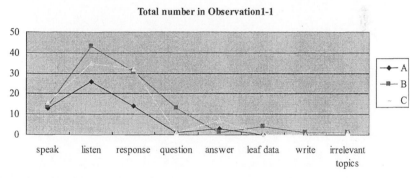

FIGURE 3 The total times statistics of the experimented members' knowledge activity in product conceptual design stage

In conceptual design stage, the speak time of C is longest, the total asking question time and times of C is shortest, the total answering time of C is longest, and C's listening time is shortest. It shows C's communication ability is good, and the design–type professional knowledge of C is most abundant. The total asking question time and times of B is most and the total answering time of B is shortest. It shows B has more intense team cooperation willingness, but his professional knowledge is worse. A's speak mainly centralize on the posterior discussion. And the total listening time of A is longest, the response times and total response time is shortest. It's because that A is familiar with the posterior discussion, and the conceptual design scheme which adopted by team is provided by A. It shows that A's creative ability is best.

(2) Structure Design Stage

Through the knowledge activity analysis, the total time and times of the three experiment members' knowledge activity in structure design stage are shown in Figure 4 and Figure 5.

Total Duration in Observation1-2

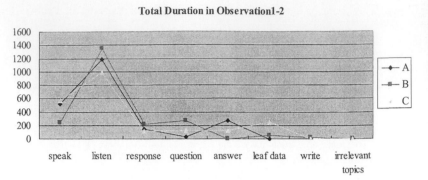

FIGURE 4 The total time statistics of the experimented members' knowledge activity in product structure design stage

Total number in Observation1-2

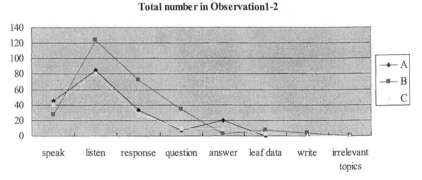

FIGURE 5 The total times statistics of the experimented members' knowledge activity in product structure design stage

In structure design stage, the experimented members' activities times are more than the first discussion. It shows the second discussion is more furious than the first discussion. The speaking time of C is longest. It shows C's communication ability is good, and his professional knowledge is abundant. The response times and total response time of B is most and the asking questions times and total asking time of B is most. It shows B has good cooperative ability and learning ability, and B is not familiar with the design–type knowledge and the experiment scheme. The answering times and total time of A is longest, while B's answering time is least. It shows A is familiar with the scheme, and B lacks the design-type knowledge. The leafing data time of C is longest, because he is most familiar with the difficult points in the scheme so that the problems are solved in the shortest time. The noting and writing time of B is longest, and it shows B has a serious attitude and higher learning ability.

(3) Detail Design Stage

Through the knowledge activity analysis, the total time and times of the three experimented members' knowledge activity in detail design stage are shown in Figure 6 and Figure 7.

Total Duration in Observation1-3

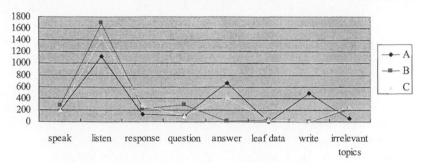

FIGURE 6 The total time statistics of the experimented members' knowledge activity in product detail design stage

Total number in Observation1-3

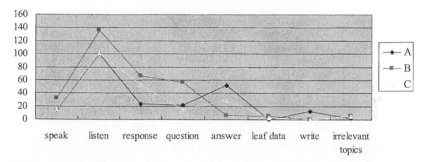

FIGURE7 The total times statistics of the experimented members' knowledge activity in product detail design stage

In detail design stage, the speaking time of the three members have little difference, while B is little more prominent. It shows they all have their own ideas about detail design. Lots of knowledge is involved in detail design besides the design-type knowledge. B has better cooperative ability and dare to speak out his idea. The third discussion is most furious. The listening time of the three members have little difference. The total response time of C is longest, while A's response time is shortest. But B's response times is most. It shows B's responses are action in majority, while the most C's responses are languages. The most questions come from B, and the most answers are from A. These show that their professional knowledge has difference. In order to confirm and check the scheme proposed by

A, C and B leaf data frequently. The writing time of A is longest. B and C' irrelevant topics are more than A. Because when A provided and drawn the scheme drawing, B and C had nothing to do.

According to the analysis of knowledge activity and statistical data, some conclusions are drawn as follows.

(1) Team synergy design process is knowledge activity process. According to the knowledge activity in different forms, all the tacit knowledge of individuals and team is exploited and converted into explicit knowledge gradually, thus integrating and forming new product and scheme finally. It is a dynamic process in which new knowledge system is constructed.

(2) The duration and times of the experimented members' knowledge activity in the three stages fully show the background diversity and qualitative difference of every team member in team synergy knowledge integration process.

(3) We can see that the duration and times of various knowledge activities in product structure design and detail design stage are more than in product conceptual design stage. This shows team cooperation in synergy design work process, and new product is designed in the communication and sharing of team members.

(4) According to the analysis of knowledge activity in the three stages of product synergy design, we can see that team members not only integrate their own knowledge, but also promote the emotion and stimulate inspiration each other, which help improve work efficiency and create new product.

(5) From the analysis of knowledge activity, the following individual cognitive differences which influence on team synergy design can be obtained: learning ability, creative ability, team cooperation, communication ability and initiative and so on.

COMPARATIVE ANALYSIS

Focused on the team members' behavior and cognitive in these three stages of the concrete design process, the paper analyzed and summed up the cognitive differences of each important topic by the deep interview. The case of conceptual design stage was shown in Table 1.

Individual cognitive differences that influence the design-type R&D team knowledge integration were obtained basing on the deep interview analysis.

(1) Knowledge foundation including Basic knowledge, professional knowledge and experience shortcuts. (2) Psychology characteristics including Personality, learning style and handling interpersonal conflict. (3) Cognitive style including Field independence and field dependence, impulsive style and reflective style.(4) Perception ability including Communication ability, perception ability, learning ability and creative ability.(5) Team cooperation including Team cooperation willingness.

THE METHOD OF IMPROVING SYNERGY WORK EFFICIENCY

Based on the above analysis, the following methods and measures can be adopted for the design-type R&D team to improve team synergy work efficiency.

(1) The influence of knowledge foundation on the team synergy knowledge integration process is positive. Team members should understand and communicate the relevant profession based on their expertise.

(2) The influence of psychology characteristics, cognitive style and cognitive ability on the team synergy knowledge integration process is complex. Team members need to pay more attention to develop their own personality, so as to merge into team and as soon as possible to adapt the team working environment.

(3) The influence of team cooperation willingness on the team synergy knowledge integration process is positive. In team synergy, enterprise should encourage team members from the material, spiritual and cultural aspects, mobilize team members' enthusiasm and creativity, and encourage the flow, transformation, sharing and innovation of tacit knowledge. In addition, enterprise should also establish sharing and meaningful team goals and create trusting, caring and supportive team atmosphere, and then guide team members to learn together.

Table 1 Analysis of the experimented members' cognitive differences in the conceptual design stage

Differences in team synergy		experimented members' behavior and cognitive		
Design Work Content	**Design Cognitiv Analysis Content**	**A**	**B**	**C**
The conceptual design stage	Cognitive style, Professional knowledge, Personality, Creative ability	Deep thinking, comparing with their scheme, not discussing.	Advocating infrared remote control, but not understanding it.	Not deep thinking, not comparing with their scheme, discussing.
		Clear concept, consistent with their programs, almost not discussing.	Unclear Concept, a lot of questions.	Understanding mechanical structure, denying the mechanical structure scheme and advocating electromagnetic principle.

		Advocating electromagnetic principle, and finding the circuit with similar function from the internet. Raised the fluid pressure scheme, but little knowledge about it. Determine the scheme: the infrared remote control switch plus electromagnetic principle	Not understanding A's scheme content, Agreeing with the A's views but having a lot of questions. Not understanding the fluid pressure, no opinion.	Agreeing with the A's views and understanding the content, not agreeing with fluid pressure.
Cognitive differences Analysis	In terms of professional knowledge, C is rich in the mechanical professional knowledge, and his explanation and expression during the discussion are very professional. The expertise of A and B are relatively weaker. But A has relatively well prepared for the data and could offset some of professional knowledge. At this stage B had a lot of doubt. On one hand, his professional knowledge was weak, on the other hand he has not well prepared. In terms of the personality, A is inward, and the enthusiasm of the discussion is not high, but he could be positive to think about someone else's view. B and C are outward and good at communication with others. In terms of the cognitive style, A prefers for field-independent, B and C prefer for the field-dependent. In terms of creative ability, A had a keen ability to solve problems, and stronger than the other two.			

CONCLUSIONS

The paper analyzes the cognitive differences in the process of team synergy design process, and summarizes the team members' differences in many aspects of the design process by the experiment analysis. On this basis, a number of approaches to improve the work efficiency of design-type R&D team synergy design are summed up, which is a very positive meaning to analyze team design behavior and improve team design capability.

REFERENCES

Berends J J, et al. (2004), *"Knowledge integration by thinking along"*, http://www.tm.test tue. nl/ecis/Working%20Papers/Eciswpl06.pdf

Nonaka, I., Toyama, R., Konno, N. (2000), "SECI, Ba and leadership: a unified model of dynamic knowledge creation", *Long Range Planning*, Vol. 33, No.1, pp. 5-34

Lee C K M, et al. (2005), "An object-based knowledge integration system for product development a case study", *Journal of Manufacturing Technology Management*, Vol. 16, No. 2, pp. 156-177

Ni, Y.H., Gu, X.J., Wu, Z.T. (2004), "Research on Knowledge Integration Ontology Theory and Framework", *China Mechanical Engineering*, Vol. 15, No.11, pp. 1954-1958

Qian, Y. D. , Gu, X.J., Wang, S.F., Li, X.S. (2006), "A Research on the Integration of Knowledge with Business in Knowledge Management of Automotive Enterprises", *Automotive Engineering*, Vol. 28, No. 6, pp. 598-601

Shin M et al. (2001), "From knowledge theory to management practice: towards an integrated approach", *Information Processing & Management*, Vol. 37, No. 2, pp. 335-355

Chapter 112

Empirically Evaluating and Developing Alarm Rate Standards for Liquid Pipeline Control Room Operators

Glen D. Uhack II, Craig M. Harvey

Louisiana State University

ABSTRACT

The liquid and gas pipeline community has recently been faced with the challenge of new governmental regulations set forth by Congress which are being implemented by PHMSA (an arm of the DOT). These new governmental regulations need to understand the role of the human-in-the-loop as part of alarm management systems. To investigate alarm rate standards a repeated measures design was developed that included a series of ten (10) operator scenarios utilizing high fidelity liquid pipeline simulation software (Stoner Pipeline Simulator). Participants completed two subsets of experiments, five were completed using an alarm display presenting alarms by time (chronological) and the remaining five experiments were completed using a categorical alarm display (by alarm priority).

Keywords: Alarm management, human-in-the-loop

INTRODUCTION

The liquid and gas pipeline community has recently been faced with the challenge of new governmental regulations set forth by Congress which are being implemented by PHMSA (an arm of the DOT). These new governmental regulations need to understand the role of the human-in-the-loop as part of alarm management systems. Past work in the area of alarm management concentrated on improving training programs, designing more efficient/effective Human Computer Interfaces (HCI), development of the alarm management lifecycle, development of industry consensus standards stating best practices, and reducing fatigue situations. The research and work performed here only begins to answer empirical questions with regards to alarm rate standards for liquid pipeline control room operators.

To investigate alarm rate standards a repeated measures design was developed that included a series of ten (10) operator scenarios utilizing high fidelity liquid pipeline simulation software (Stoner Pipeline Simulator). Participants completed two subsets of experiments, five were completed using an alarm display presenting alarms by time (chronological) and the remaining five experiments were completed using a categorical alarm display with alarm priority categories being high, medium, and low. For both, the chronological and categorical alarm display, experiments the alarm rates were derived from the Engineering Equipment Materials and Users Association (EEMUA) No. 191 average alarm rate standard and were randomly distributed between participants (EEMUA, 1999). The main operator performance metrics of interest were participant acknowledgement time, participant response time to take appropriate action to handle each alarm, accuracy of the response, and percentage of successful completion.

LITERATURE REVIEW

INTRODUCTION TO THE CHALLENGE

According to the ASM Consortium, it is estimated that poor alarm management practices cost the industry approximately $13 billion dollars each year (Dunn and Sands, 2005). An earlier finding found abnormal situations cost the industry over $20 billion each year (Nimmo, 1995). The purpose of quantifying alarm system performance is to ensure that it functions, thereby alerting the operator to events with safety and environmental consequences. Factors which contribute to the performance of the alarm system include human information processing characteristics, operator training, task load, workload, HMI, SCADA/DCS/alarm system components, the site's alarm management philosophy, alarm overloads, stale alarms, nuisance alarms, and bad alarm settings among others.

Human Factors Considerations

According to Ian Nimmo, studies conducted by the Abnormal Situation Management Consortium (ASM), American Institute of Chemical Engineers, American Petroleum Institute, American Chemistry Council, and similar organizations have concluded that about 80% of the root causes contributing to major accidents affecting safety, environment, and/or economics can be linked to human operator error (Nimmo, 2002). Human error can be caused by many variables, some of which are poor human-machine interface (HMI), poor situational awareness, operator experience, workload, communication, and shift-fatigue. There is little published research regarding specific best-practices for pipeline control room operators. However, the Engineering Equipment Materials and Users Association (EEMUA) No.191 authors imply that research conducted in the process industry or elsewhere in a control room setting, (e.g. a refinery or similar control room environments), can be used to improve human factors design and benchmark performance in the pipeline industry, as well as other industries (EEMUA, 1999).

EXPERIMENTAL METHOD

Experimental Design and Layout

The experimental model used for this study was a repeated measures design where all participants were subjected to the same conditions. Each participant was trained in an effort to orient participants to the type of situations they would experience during supervision and management of the simulated pipeline system. For the main experiment, participants completed ten separate experiments lasting ten minutes each. Five experiments were completed by each participant utilizing an alarm window that displays alarms in order of occurrence (chronological) and distinguishing the alarms by priority using color, i.e. three priority levels were used Red-High, Yellow-Caution, and White-Low. The remaining five experiments completed by participants were done using a different alarm window which displays alarms by grouping them categorically for each priority, although chronologically ordered within each category.

Since this study attempted to empirically determine an alarm rate standard the alarm rate/interval during each experiment was evenly distributed given the number of alarms. For example, for an experiment that would present 20 alarms in ten minutes to a participant, one alarm would display every 30 seconds. The alarm rates chosen for this study were derived from the Engineering Equipment Materials & Users' Association standard publication No. 191 (EEMUA, 1999).

To help reduce the chance that the study's experimental results could be biased toward a particular alarm display or alarm rate, the experimental scenarios completed by participants were randomized. Specifically, the order in which participants completed the five alarm rate experiments using the chronological alarm display and the five alarm rate experiments using the categorized display were randomized between participants. Also, the order in which participants completed the experiment starting with a particular alarm display was randomized between participants.

Participants

The participants in this study consisted of undergraduate and graduate students at Louisiana State University. The number of participants to complete the experiment was determined by a statistical power test. Participants completed the training, qualification testing/assessments, and the two experiments designed to measure specific human performance variables when supervising and operating a high fidelity pipeline simulator. After completing the experiment, participants were allowed to complete a subjective usability questionnaire.

To help reduce differences among participants attributed to other factors than those experimentally manipulated all participants completed a training presentation, demonstration/familiarization session, a multiple choice quiz, and training qualification assessment. This battery of pre-experiment training and testing was developed to help ensure only those participants who were able to successfully execute tasks representative to those during the actual experiment would be allowed to complete the experiment. Each participant was trained and oriented via a presentation presented by the experimenter.

Afterwards, any questions a participant had were then answered and the participants were shown a demonstration of the type of system they would supervise and manage to familiarize them with executing tasks representative of those during the actual experiment.

This training presentation provided each participant with a general overview of the actual pipeline they would operate upon successful completion of training and qualification and for this particular study it attempted to teach each participant the fundamental principles of abnormal situation assessment and responses which were required for each type of abnormal event simulated in this experiment (e.g. leak/pressure relief event). The principles emphasized were covered in the training presentation and any questions or ambiguities perceived by a participant were answered or clarified verbally and with illustrations/demonstration where appropriate.

A total of 39 participants volunteered for this study. Thirty-one participants were included in the data analysis. The first 4 participants were part of the pilot study which evaluated the experimental method; participants who were part of the pilot study were not included in the final data. All but one person made greater than 80% on the general knowledge quiz. Two participants failed the qualification assessment (passing was $\geq$ 50% and failing was < 50%), this is where each participant qualified using a scaled down version of the actual experiment (100% is 4/4). All participants passed the multiple choice quiz; passing was $\geq$ 4/6 and failing was < 4/6. Of the participants included in the final data analysis 16 participants started the experiment using the categorical alarm window first and 15 participants started the experiment using the chronological alarm window.

Equipment and Software

Germanischer Lloyd (GL)'s Advantica's Stoner Pipeline Simulator (SPS) software was used to develop a pipeline model that calculates the fluid hydraulics and transients occurring in the simulated pipeline; simply put SPS is the calculation engine for the simulated pipeline in this study. SPS is widely used in the pipeline community for engineering analysis. Schematic, GL's interface design module for SPS, in conjunction with Microsoft Visual Studio.NET 2008, was used to develop the graphical user interface (GUI) for the pipeline model that was developed with SPS. These screens were analogous to the SCADA screens used for online systems. SCADA systems are defined (PHMSA, 2008) as serving a principal function to alarm or notify a control room operator of abnormal process deviations, (e.g. pressure, flow, and temperature). A standalone Visual Basic program was developed to present alarms to participants. Our Visual Basic alarm presentation program allowed the manipulation of essential independent variables in this experiment (e.g. alarm category colors and different display methods). Sample pictures of the user interface are provided in Figures 1, 2, and 3.

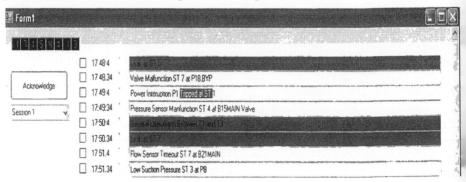

Figure 1 – Example Alarm List Display

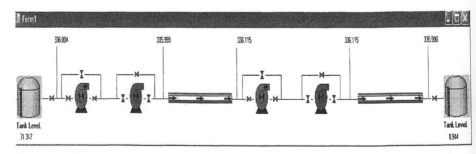

Figure 2 – User Interface Used for Qualifying Participants

The console for the study utilized four computer monitors, one 19-inch and three 17-inch monitors. The 19-inch monitor was used to display the participant qualifying and experiment overview displays. Two 17-inch monitors were dedicated to displaying detailed station displays of the simulated pipeline system. The remaining 17-inch monitor was used to display the alarm window and maintenance request form for the experiments. The additional screens were setup as an extension of the computer's desktop so one keyboard and mouse could control all system functions.

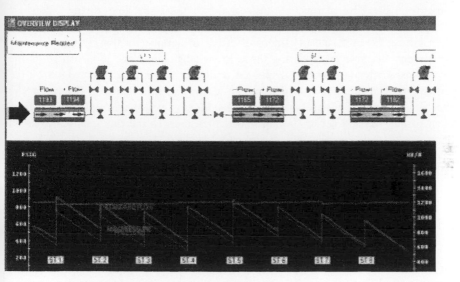

Figure 3 – Partial Snapshot of Overview Display Used During Actual Experiment

To capture each participant's observed performance and responses during each experimental scenario, TechSmith Corporation's Morae (video capture software) was installed on the same computer as that used by participants for their training and experiments. This data was analyzed after each participant finished all phases of the study to collect response times and scenario performance.

Independent and Dependent Variables and Their Measures

Two independent variables were manipulated in this experiment: alarm rates and alarm display

Alarm Rates – The alarm rates for this experiment were kept constant. Alarm rates were based on EEMUA No. 191 average alarm rate standard (EEMUA, 1999) and ISA average alarm rate standard (ISA, 2008). Table 1 for the alarm rates used for this experiment.

Alarm Displays – There were two types of alarm displays used for this experiment. One presented alarms chronologically (Figure 4) and the other presented them by category (Figure 5).

Each of the experimental dependent variables and their means of measurement are provided below.

Time Taken to Acknowledge Each Alarm – This is simply the time elapsed after an alarm is displayed until that alarm is acknowledged by the participant.

Response Time to Each Alarm – Response time can be broken down into three general sub-measures (1) Acknowledgement; (2) Analyze; and (3) Act. This framework is adapted from a study conducted by (Reising, 2004). Also, when an operator is handling alarms, by task analysis, there are four distinct steps an operator must complete to respond to any given alarm, these are (1) Acknowledge; (2) Analyze; (3) Act; and (4) Interpret effect of action (Reising, 2004). For this experiment, the time taken during the acknowledgement stage is measured as the time elapsed after an alarm is displayed until that alarm is acknowledged by the participant. The time taken to analyze an alarm and time taken to determine the effect of action was not analyzed separately for this experiment. Instead, the time taken to analyze and act on an alarm was measured as the time elapsed after an alarm was displayed until action was taken.

Fraction of Abnormal Situations Successfully Dealt With – A point system was defined to practically measure this variable. The point system was defined by 0 – Unsuccessful and 1 – Successful. "Successful" is defined here in a very literal sense as the participant executes the correct action for a given alarm. For clarity, the correct action did not have to be executed in the correct order to be awarded points for this measure.

Accuracy of Response – A point system was defined to practically measure this variable. The total number of points possible was calculated based on the minimum number of actions to successfully handle all abnormal situations for each alarm, for each unique experiment. Participants start the experiment with zero points and were awarded points when they took the appropriate action for each alarm. For example, one point could be awarded for submitting a maintenance request for a pump failure, one point could be awarded for shutting down the correct pump during a leak event, or one point could be awarded for opening/closing a block valve that malfunctioned. Every elemental defined action was worth one point. The maximum points a participant could achieve was 38 (20 Alarms – 10 Min.), 20 (10 Alarms – 10 Min.), 10 (5 Alarms – 10 Min.), 4 (2 Alarms – 10 Min.), and 2 (1 Alarms – 10 Min.). Unlike the Fraction of Abnormal Situations Successfully Dealt With measure, the Accuracy of Response measure evaluates a participant's ability to execute corrective action in the correct sequence.

Experimental Procedure

The experimental procedure begins with the participant completing a demographic survey (computer based). This survey included questions regarding generalized demographic information (e.g. major, student classification, GPA), a participants computer literacy, and a question asking if they have ever used Stoner's software (used to develop and complete experiments). Participants with any past experience using Stoner software were eliminated. Participants were then trained to navigate the user interface and become familiarized with standard operating procedures via a presentation and demonstration model. Participants were then administered a general knowledge quiz testing their ability to complete elemental tasks involved with the experiment.

Table 1 – Alarm Rates Used For Experiments

20 in 10 Minutes	Chronological and Categorized Alarm display
0 in 10 Minutes	Chronological and Categorized Alarm display
5 in 10 Minutes	Chronological and Categorized Alarm display
2 in 10 Minutes	Chronological and Categorized Alarm display
in 10 Minutes	Chronological and Categorized Alarm display

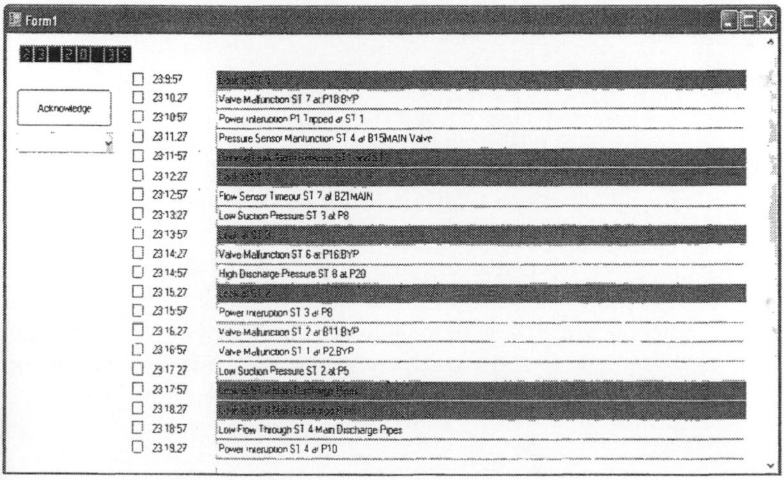

Figure 4 – Chronological Alarm Display

Upon passing the general knowledge quiz participants were then administered the qualification assessment (overview display for the qualification assessment can be found in Figure 2). The qualification assessment tested each participant's ability to correctly handle abnormal situations representative of those to be experienced during the experiment. Upon successful completion of the qualification assessment participants proceeded to complete the first part of the experiment. Finally, during day two a brief round of refresher training was provided to each participant to help ensure the skills taught during the first day were retained for the second portion of the experiment. After completing all experiments each participant was administered a subjective usability questionnaire evaluating their experiences during both days of

1084

the experiment.

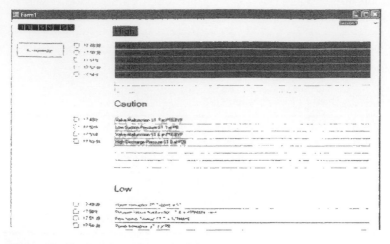

Figure 5 – Categorized Alarm Display

RESULTS

Figures 6 and 7 show how participant response time and acknowledgement time varied for the five different alarm rate experiments using both the Categorical and Chronological Alarm Displays, and by alarm type.

A total of 34 participants completed the subjective usability questionnaire. Of those surveyed, 79% (27/34) of participants preferred using the Categorical Alarm Display and 21% preferred using the Chronological Alarm Display. The average response (including all questions and participants surveyed) was 1.6 (1 - being strongly agree and 5 - being strongly disagree). Further results are forthcoming.

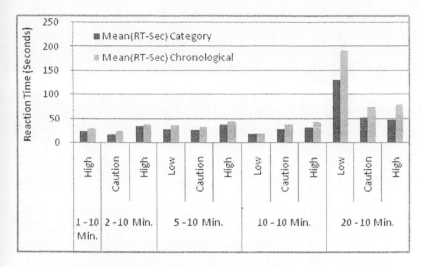

Figure 6 – Mean Response Times by Display Type

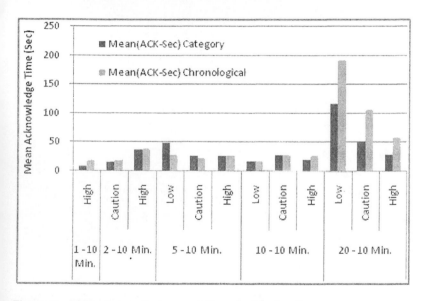

Figure 7 – Mean Acknowledgement Time by Display Type

REFERENCES

EEMUA (1999). Alarm systems: a guide to design, management and procurement. London, Engineering Equipment and Materials Users Association.

Dunn, D. G. and N. P. Sands (2005). ISA-SP18 - Alarm systems management and design guide, Chicago, IL, United States, ISA - Instrumentation, Systems, and Automation Society, Research Triangle Park, NC 27709, USA.

ISA (2008). ISA – 18.02 – 2008, Management of Alarm Systems for the Process Industries, ISA - Instrumentation, Systems, and Automation Society, Research Triangle Park, NC 27709, United States.

Nimmo, I. (1995). Abnormal situation management, New Orleans, LA, USA, Instrument Society of America, Research Triangle Park, NC, USA.

Nimmo, I. (2002). "It's time to consider human factors in alarm management." Chemical Engineering Progress 98(11): 30-38.

Pipeline and Hazardous Materials Safety Administration (PHMSA), D. O. T. (2008). Pipeline Safety: Control Room Management/Human Factors; Proposed Rule. Federal Register/Vol. 73, No. 178. 49 CFR Parts 192, 193, and 195.

Reising, D. V. C., J. L. Downs, et al. (2004). "Human Performance Models for Response to Alarm Notifications in the Process Industries: An Industrial Case Study." Human Factors and Ergonomics Society Annual Meeting Proceedings 48: 1189-1193.

Chapter 113

Case Study of Alarms Rates in Two Pipeline Companies vs. Alarm Rate Standards

Craig M. Harvey, Lisa D. Michelli, Glen D. Uhack, II

Louisiana State University

ABSTRACT

Industry controllers using alarm systems in the past have successfully handled pipeline failures, interruptions, leakages and miscellaneous alarms. Several recent reports and legislation have brought human factors to the forefront for operators of gas and hazardous liquid pipelines. In 2005, the National Transportation Safety Board (NTSB) released their Safety Study, "Supervisory Control and Data Acquisition (SCADA) in Liquid Pipelines." In that report, they reviewed the role of SCADA systems in 13 hazardous liquid line accidents from April 1992 to October 2004. In ten of the accidents, SCADA systems played some role. One key element of SCADA systems are alarms. This paper discusses data gathered from alarm reports of two oil companies show alarms occur at variable rates throughout various time intervals. Segregating alarms received into time-interval-lots allowed the authors to compare actual operating systems existing standards (e.g., EEMUA and ISA) as to what human operators were handling without discrepancy. The resulting analysis indicates a discrepancy between established standards and actual operator workloads. Further review of alarm rate standards and what constitutes an alarm requires evaluation.

Keywords: SCADA alarms, alarm rate standards, control room operations

INTRODUCTION

Several recent reports as well as legislation have brought human factors and human computer interaction to the forefront for operators of gas and hazardous liquid pipelines. In 2005, the National Transportation Safety Board (NTSB) released their Safety Study, "Supervisory Control and Data Acquisition (SCADA) in Liquid Pipelines." In that report, they reviewed the role of SCADA systems in the 13 hazardous liquid line accidents from April 1992 to October 2004. In ten of the accidents, SCADA systems played some role. This landmark study re-emphasized for the pipeline community the role of SCADA systems in pipeline safety. The report stated:

> *"The principal issue in the SCADA-related accidents investigated by the Board was the delay between controller's recognizing a leak and beginning efforts to reduce the effect of the leak. SCADA factors identified in these accidents include <u>alarms, display formats, the accuracy of SCADA screens, the controller's ability to accurately evaluate SCADA data during abnormal operating conditions</u>, the appropriateness of controller actions, the ability of the controller and supervisor to make appropriate decisions, and the effectiveness of training in preparing controllers to interpret the SCADA system and react to abnormal conditions." (p. vi) (underline added by author)*

Four items stand out as demanding the attention of the pipeline community from this statement: alarms, display formats, SCADA accuracy, and controller training. Prior to this report (2002), DOT implemented a non-prescriptive, performance-based regulation requiring operator qualification training for controllers. Later, in January 2007, the American Petroleum Institute released the API Recommended Practice 1165, Recommended Practice for Pipeline SCADA Displays (First Edition). Additionally API 1165 also briefly touches on accuracy of data and alarms. Last, the 109th Congress passed Senate Bill 3961, commonly referred to as the Pipeline Act of 2006. Late 2009, the Pipeline and Hazardous Materials Safety Administration (PHSMA), a division of the Department of Transportation, released 49 CFR Parts 192 and 195 entitled "Pipeline Safety: Control Room Management/Human Factors." This paper will address one of the areas, alarms, which received attention in this PHMSA's ruling and have received much attention by the industry itself as they have developed standards for alarms.

ALARMS

Alarm systems play an important role in maintaining a safe plant or pipeline. Poor performance of alarm systems and operator's response to alarms costs money and puts lives at risk. Bransby and Jenkinson (1999) surveyed 15 controlled sites and found eight specific events where alarms contributed to major plant damage, significant production loss, or environmental incidents. One resulted in a loss of $14 million and another in approximately $3.3 million. One can easily look back in history and see the impact alarms have had on control rooms. Three Mile Island was one accident that received much attention due to alarms. Many sources report operators were not able to think properly because of the loud noises generated within the control room by the alarms. In another incident, the Milford Haven refinery accident (Texaco) in 1994 resulted in 26 people receiving minor injuries. Alarm system shortcomings were a major contributor to the problem.

Even as recent as 2004, the failure of a pipeline controller to accurately evaluate data and promptly respond to a pipeline emergency was a major factor in a pipeline rupture that resulted in the release of 204,000 gallons of anhydrous ammonia in Kansas. Alarms once again were a major contributor. The operator stated, "His training did not specify which screens to use to analyze and evaluate the SCADA data." He further stated that, "from 11:15 a.m. to 11:48 a.m., an unusually high number of alarms and status events were displayed for the pipeline." NTSB found that during this 33-minute period, the SCADA system displayed 119 alarvms and status events. That is six alarms per minute. The company stated it would establish a goal to make the alarms more meaningful and significantly reduce the number of alarms that could distract controllers from more critical alarms. Questions remain as to what is the best practice for alarms. While the literature is abundant with many references to bad situations, little is available to guide control centers in what is the best design.

SO WHAT IS AN ALARM?

A review of the *API 1167-Recommended Practice for Alarm Management DRAFT Version 0.1* states, "An alarm is an audible or visible means of indication to the controller…or other condition **requiring a Controller's response**." While this standard is still in progress, API 1167 captures the essence of what most would conclude that an alarm entails. Now the question becomes, what is an appropriate alarm rate that an operator can handle? API 1167 attempts to answer this as well: "accordance with industry best practice guidelines" and "alarm rates … within the handling capability of the controller." So how are the best guidelines determined? Do we use the records of operators with safest records? Let us look at how alarms rates affect safety.

ALARM RATES

In a review of 96 refinery operators across 13 plants, Bransdy (2000) found that in normal operations the average operator received about one alarm every two minutes, day and night. Most of the alarms were of little value to the operator with around 50% repeat alarms. In addition, following a large plant disturbance there might be as many as 90 alarms in the first minute and 70 in the subsequent 10 minutes. Under this practice of alarms, there is very little doubt that alarms probably get very little review time and are most likely to overload controllers beyond their capacity.

So the question remains, how does one design an alarm system? Some have proposed that one best practice to adopt in the design of alarm systems is to vary the pitch, speed, and loudness of auditory alarms such that their urgency and response time requirement can be assessed (Haas and Edworthy, 1996). Haas and Edworthy did find operators were able to consistently perceive the urgency based through varying the pulse level and frequency. One has to be concerned with noise pollution hampering response to the alarms especially in alarm floods resulting from emergencies. Thus when to use sound and how much is something that has to be considered.

Another issue of concern is the reduction of "high" priority alarms. In the Milford accident, 87% of the 2040 alarms displayed were classified as "high" priority even though many of them were informational only (Wilkinson, J. and Lucas, D., 2002). Additionally, safety critical alarms were not clearly distinguishable.

Reactive or proactive? Plant management and controllers raise the question whether an operator is "system proactive" versus "alarm reactive". In general, managers and experienced operators agree that operators should be proactive. From their perspective, generally by the time the alarm has sounded, the operator is already behind the control curve. Designing better systems and alarm methods can possibly allow operators to stay in front of the control curve and ultimately prevent/reduce problems.

The Engineering Equipment and Materials User Association (EEMUA) 191- Alarm Systems (EEMUA, 1999) standard has put out a useful guide with high-level guidelines and training courses for industry (Table 1). However, their standard still relies primarily on only subjective evaluations. Their guidance states: (1) manageable steady state for an operator, 1 alarm every 10 minutes; or (2) average process rate, five per hour; or (3) flood state when an operator becomes overwhelmed, 10 in 10 minutes. This also assumes a distribution of 5%-high, 15-%-medium, and 80%-low alarms. Very little if any empirical research backs up their suggestions, but also very little empirical research exists for anyone to make such guidelines. ISA standards report acceptable alarm targets occur for one (1) alarm per any ten (10) minute time interval and have a maximum manageable rating at one (1) alarm per any five (5) minute interval (ISA, 2009) (Table 2).

Table 1 – EEMUA No. 191 Average Alarm Rate Standards

Long Term Average Alarm Rate in Steady Operation	Acceptability
>1 alarm per minute	Very likely to be unacceptable
1 alarm per two minute	Likely to be excessively demanding
1 alarm per five minutes	Manageable
<1 one alarm per ten minutes	Very likely to be acceptable

Table 2 - ISA 18.2 Average Alarm Rate Standards

Very Likely to be Acceptable	Maximum Manageable
~150 Alarms per day	~300 Alarms per day
~6 Alarms per hour (average)	~12 Alarms per hour (average)
~1 Alarms per 10 minutes (average)	~2 Alarms per 10 minutes (average)

ALARM RATES IN PRACTICE

Since the alarms set forth by industry are assumed acceptable, we decided to investigate alarms rates in practicing companies. Two companies were willing to share their alarm data with us to evaluate. The companies will remain nameless; however, we are sure this is representative of other companies in the industry. Company A only provided one day of data. Company B provided over 2 months of data. Figure 1 and Figure 2 below illustrate the average number of alarms per 2-minute time intervals for Company A and Company B respectively. Both figures provide the data by each console within the company. Company A had 10 consoles and Company B had four consoles. While we did not break down the alarms by priority, it is easy to see that both companies exceed EEMUA's manageable state of one alarm every 10 minutes. In fact, Company A's highest console rate was 6.56 alarms in 2 minutes and Company B's highest console rate was 29.31 alarms in 2 minutes during a single 2 minute period. Now it should be pointed out that during the time intervals for which this data was collected, there were no accidents or incidents aware of to the authors and yet both companies highly exceeded the EEMUA standards. So is the standard too low? Maybe these companies employ supermen and superwomen. Do the companies employ teams to address clusters of alarms? Have these companies simply been lucky? Alternatively, is an explanation definable?

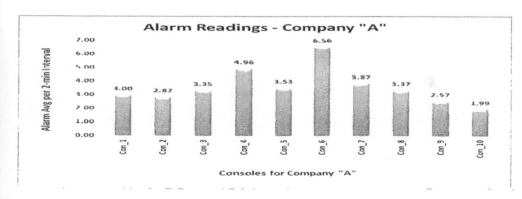

Figure 1 – Time Interval Average Comparisons for 2-minute segment as **Company A**

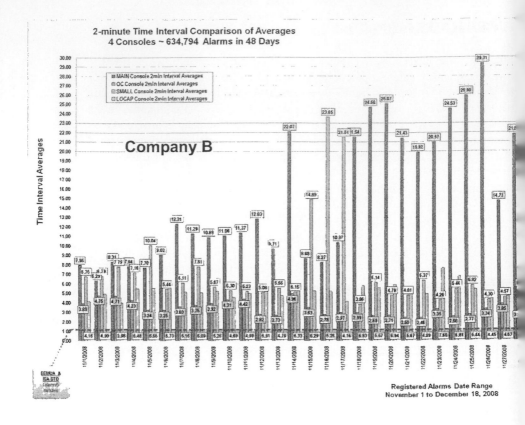

Figure 2 – Time Interval Average Comparisons for 2-minute segment as **Company B**

DISCUSSION

Alarms are not the only element that affects an operator's ability to handle alarms. Mix in one or two alarms every two minutes, with accompanying sound effects, frequent nuisance alarms, poor lighting, bad consoles, lack of ergonomically sound workstations, overtime, fatigue, and personal life stresses, and you could be asking for a disaster. Of course, one can also look to the amount of training, backup systems, and company emphasis on safety that also allows companies to prevent disasters. Case in point are Companies A and B. Both had sound management practices, as seen through an audit, highly trained operators, and a team support system in the control room. No operator operated in a vacuum. Thus while alarm rate standards are useful guidance, they are not the only thing that affects the effectiveness of a control room. Further work is needed to understand the impact of alarms on the operator. Work being done at Louisiana State University, funded by the Wright State University industry funded Center for Operator Performance, is focused on alarm management in SCADA systems.

CONCLUSION

There are no fail-proof systems. Even perfect systems cannot take into account, in a definable sense, the human factor. The topic of alarm management is of great importance to the pipeline industry, as well as many other industries that operate control rooms. There is also great diversity in regards to the practice of design principles, management philosophies, and equipment procurement. Organizations like EEMUA and API have dedicated themselves to the standardization of processes and technologies across industries that routinely transport/process large amounts of gas or hazardous liquids. Many successes have been realized when a site's management show a genuine commitment to an alarm management lifecycle to improve the performance of the alarm system. For example, NOVA Chemicals realized great success when an alarm assessment and rationalization was completed at one of their plants, and the company also expressed the most significant contributing factor to their success was consistently following the alarm management lifecycle (Errington, DeMaere et al. 2004). Xcel Energy exemplifies another success story when it reduced the alarm rate per shift from several hundred to only about eight, and during upsets a trip would typically produce no more than 20 alarms (Bass, 2007). It should be noted that Xcel Energy commissioned a team of experts who worked about 10 hours each week for a two-year period to obtain that level of performance. Good alarm system performance has been demonstrated to be achievable, however a company must be committed to a continually improving the system through a lifecycle approach.

REFERENCES

API (2007). API – RP 1165 – 2007, Recommended Practice for Pipeline SCADA Displays, API - American Petroleum Institute, API Publishing Services 1220 L Street, N.W., Washington, D.C. 20005, United States.

API (unpublished). API – RP 1167, Recommended Practice for Alarm Management Draft version 0.1, API - American Petroleum Institute, API Publishing Services 1220 L Street, N.W., Washington, D.C. 20005, United States.

Bass, J. (2007). Xcel Energy Implements an Alarm Management Strategy. Power Engineering 111(11): 196-204.

Bransby, M. L. (2000). Best Practice in Alarm Management. IEEE, 1-4.

Bransby, M., & Jenkinson, J. (1998). Alarming Performance. Computing & Control Engineering Journal, 61-67.

EEMUA (1999). Alarm systems: a guide to design, management and procurement. London, Engineering Equipment and Materials Users Association.

Errington, J., D. V. Reising, et al. (2006). ASM outperforms traditional interface. Chemical Processing 69(3): 55-58.

Haas, E. C., & Edworthy, J. (1996). Designing Urgency into Auditory Warnings using Pitch, Speed, and Loudness. Computing & Control

Engineering Journal, 193-198.

ISA (2008). ISA – 18.2 (2009). Management of Alarm Systems for the Process Industries, ISA - Instrumentation, Systems, and Automation Society, Research Triangle Park, NC 27709, United States.

NTSB. (2005). *Supervisory Control and Data Acquisition (SCADA) in Liquid Pipelines*. Washington, DC: National Transportation Safety Board (NTSB).

Pipeline and Hazardous Materials Safety Administration (PHMSA), D. O. T. (2009). Pipeline Safety: Control Room Management/Human Factors, Federal Register/Vol. 74, No. 231. 49 CFR Parts 192 and 195.

Wilkinson, J., & Lucas, D. (2002). Better alarm handling - a practical application of human factors. Measurement and Control, 35, 52-55.

Chapter 114

Understanding Accidents in the Great Outdoors: Testing a Risk Management Framework in the Led Outdoor Activity Domain

Paul M. Salmon*, Amy Williamson, Michael G. Lenné,
Eve Mitsopoulos-Rubens, Christina M. Rudin-Brown

Human Factors Group, Monash University Accident Research Centre
Building 70, Clayton Campus
Monash University
Victoria 3800, Australia

*Corresponding author, email: paul.salmon@muarc.monash.edu.au

ABSTRACT

In most safety critical domains, theoretically underpinned accident analysis is an accepted approach for identifying the causal factors involved in accidents, and for informing the development of strategies designed to improve safety. Such approaches are yet to be developed and applied within a led outdoor activity context, despite the industry recognizing the important role of accident analysis for enhancing safety. This paper presents an application of a popular risk management framework to the analysis of a high-profile led outdoor activity accident. This involved the development of an Accimap for the Lyme Bay sea canoeing incident, the aim being to test the usefulness of the approach for explaining how accidents occur in the outdoor activity sector. The outputs were used to test seven predictions regarding safety and accident causation made by the framework. In conclusion, the analysis provides supportive evidence for each prediction, and provides further support for the Accimap method as an accident analysis tool for complex sociotechnical systems.

Keywords: Led outdoor activity, accidents, accident analysis, Accimap

INTRODUCTION

Within the led outdoor activity domain (i.e. facilitated or instructed activities within outdoor education and recreation that have a learning goal associated with them), it is acknowledged that safety compromising accidents and incidents occur, and that they represent a key area for further research. A recent review commissioned by the Australian led outdoor industry sector found that, compared to other safety critical domains, relatively little is currently known about the causal factors involved in led outdoor activity accidents and incidents, and that the application of theoretically driven models and methods for accident analysis purposes has to date been limited (Salmon et al, 2009).

This paper presents an application of a systems-based risk management framework, widely used in other safety critical domains, to the analysis of a well known led outdoor activity accident. This involved the development of an Accimap (Rasmussen, 1997) for the Lyme Bay sea canoeing incident, the aim being to test the usefulness of the approach for explaining how accidents occur in the led outdoor activity sector. The outputs were then used to test seven predictions regarding safety and accident causation made by Rasmussen's risk management framework.

ACCIDENTS IN THE GREAT OUTDOORS

Compared to other domains in which safety compromising accidents and incidents represent a significant problem, relatively little is currently known regarding such events within the led outdoor activity domain. A number of accident causation models have been developed specifically for this area (e.g. Brackenreg, 1999; Davidson, 2007); however, these models appear to be limited in terms of scope, theoretical underpinning, and practical application (Salmon et al, 2009). In addition, theoretically underpinned accident analysis methods, the types of which are widely used in other safety critical domains, have not yet been developed for the led outdoor activity domain.

RASMUSSEN'S RISK MANAGEMENT FRAMEWORK

Rasmussen's risk management framework (Rasmussen, 1997) shows clear potential for accident analysis purposes in the led outdoor activity domain. The framework considers the various organizational levels involved in production and the control of safety and focuses on the mechanisms generating behavior within complex sociotechnical systems. The model views such systems as comprising a hierarchy of individuals and organizations (Cassano-Piche et al, 2009). Although the number of levels is not rigid and can vary according to the domain of application, the model typically focuses on the following levels: government, regulators, company, company management, staff, and work. Safety is viewed as an emergent property arising from the interactions between actors at each of the levels.

According to the framework, various levels are involved in safety management via the control of hazardous processes through laws, rules, and instructions. For systems to function safely, decisions made at high levels should promulgate down and be reflected in the decisions and actions occurring at lower levels of the system. Conversely, information at the lower levels regarding the system's status needs to transfer up the hierarchy to inform the decisions and actions occurring at the higher levels (Cassano-Piche et al, 2009). Without this so called 'vertical integration', systems can lose control of the processes that they control (Cassano-Piche et al, 2009). According to Rasmussen (1997), accidents are typically 'waiting for release', the stage being set by the routine work practices of various actors working within the system. Normal variation in behavior then serves to release accidents.

THE ACCIMAP METHOD

Accompanying Rasmussen's framework, the Accimap modeling approach (Rasmussen, 1997; Svedung & Rasmussen, 2002) is an accident analysis method that graphically represents the decision makers and decisions involved in producing the system in which an accident was permitted to occur. Following Rasmussen's framework, the Accimap method uses the following six organizational levels: government policy and budgeting; regulatory bodies and associations; local area government planning & budgeting, company management; technical and operational management; physical processes and actor activities; and equipment and surroundings. Failures at each of the levels are identified and linked between and across levels based on cause-effect relations.

Reflecting its utility and generic nature, the Accimap method has been used to describe a wide range of accidents occurring in the safety critical domains, ranging from police firearm mishaps (Jenkins, Salmon, Stanton & Walker, 2010) and public health incidents (Cassano-Piche et al, 2009; Vicente & Christoffersen, 2006) to loss of space vehicles (Johnson & de Almeida, 2008). One popular approach used to validate Rasmussen's risk management framework in new domains has been the use of Accimap outputs to test seven predictions regarding accident causation made by the framework (e.g. Cassano-Piche et al, 2009; Jenkins et al, 2010; Vicente & Christoffersen, 2006). These predictions are outlined in Table 1.1.

CASE STUDY APPLICATION: THE LYME BAY TRAGEDY

To evaluate the utility of the Accimap method for the analysis of led outdoor accidents and incidents, and to test the predictions made by Rasmussen's framework in the context of led outdoor activity accidents and incidents, an Accimap analysis was undertaken for a well known led outdoor activity incident, the Lyme Bay sea canoeing tragedy.

The incident involved the death of four students whilst on an outdoor education activity trip on March 22nd 1993. The activity involved a group of eight students, a school teacher, a junior instructor and a senior instructor engaging in an introductory open sea canoeing activity in Lyme Bay, Dorset, in the United Kingdom. After a series of initial capsizes, the junior instructor and eight students became separated from the senior activity instructor and their school teacher and were blown out to sea. Due to high wind and wave conditions, each canoe in the junior instructor/student group was subsequently swamped and capsized, and the junior instructor and students were left in the water with all canoes abandoned. Four students drowned as a result.

Methodology

One Human Factors researcher with significant experience in the application of the Accimap method, and other accident analysis approaches, performed the analysis based on the data contained in the official inquiry report (Jenkins & Jenkinson, 1993). In order to ensure reliability and validity, three Human Factors researchers independently reviewed the analysis and input data and any subsequent disagreements were resolved through further discussion and consensus. Following this, an expert panel of outdoor education and adventure activity providers reviewed and refined the analysis outputs (in consultation with all four Human Factors researchers).

RESULTS

The Accimap for the Lyme Bay incident is presented in Figure 1.1. A summary of the elements identified at each level is given below.

Equipment & surroundings
A failure by the activity centre and instructors to provide appropriate equipment for the canoeing activity was cited by the inquiry report as a key contributing factor to the accident. For example, spray decks, the device which creates a water tight seal at the point where the user sits in the canoe, were not provided in the students canoes, despite the students all being inexperienced in the operation of canoes and

also in sea canoeing in general (the instructors used spray decks). This was identified as a key causal factor in the ease in which the canoes were swamped and capsized. Further, the students' canoes also did not have supplementary buoyancy (which is recommended for sea canoeing) and were not fitted with deck lines, which meant that they could not be tied to each other or towed in the event of the group becoming separated. The students were also inadequately clothed for sea canoeing, and the lifejackets worn by the students were not inflated, on instruction from the instructors, and had no whistles attached to them.

Various other equipment that would normally be used on beginner sea canoeing trips was not provided. This included radios, distress flares, towlines and a survival bag. Also, the school teacher who capsized initially, an inexperienced sea canoeist, was given a Lazer three fifty canoe (Jenkins & Jenkinson, 1993), which was shorter and lighter than the other canoes and deemed inappropriate for use by a novice canoeist at sea. Finally, at this level the environmental conditions also played a role; although the wind speed level was not overly high, the offshore winds and high seas were key factors in sweeping the students out to sea.

Physical processes & actor activities

At the physical processes and actor activities level, most of the contributing factors identified relate to the instructors' and students' inability to respond to the unfolding situation, which was a function of their lack of experience and qualifications for such an activity and the absence of appropriate safety and emergency procedures. The trigger events were the continued capsizes of the teacher's canoe, and the collective inability of the senior instructor and teacher to successfully right the capsized canoe. Whilst the senior instructor attempted to right the capsized canoe, the junior instructor proceeded to raft together the students (i.e. organize tightly together in formation), although prior to this no advice had been given on group organization. Influenced by the offshore winds, the rafted group was quickly swept out to sea, and, due to the absence of distress flares and radios, the two groups (instructor/teacher group and rafted group, including eight students and one junior instructor) lost contact with each other. Due to the absence of tow or deck lines, the rafted group could not be tied together. As the students were swept further out to sea, the increasingly adverse conditions (e.g. higher waves), coupled with the lack of spray decks, led to further capsizes and the canoes becoming swamped with water until eventually all were submerged in the water with only one upturned canoe to hold on to. After a failed attempt to paddle ashore using the upturned canoe, the final canoe was abandoned. Whilst in the water, standard in-water procedures, designed to keep submerged entities warm and prevent hyperthermia, were not initiated.

The response to the unfolding situation is also important. The alarm was not raised on-shore until over three hours after the group should have returned. Further, initially the site manager did not immediately notify the coastguard; rather, he spent some time himself searching the shoreline for the missing canoeists. Finally, the coastguard was wrongly informed that the instructors were well qualified, over 18 years of age, and were well equipped for the activity, all of which was not the case.

1100

Once the rescue was underway, the coastguard themselves made a number of errors, although this is not the focus of the present analysis.

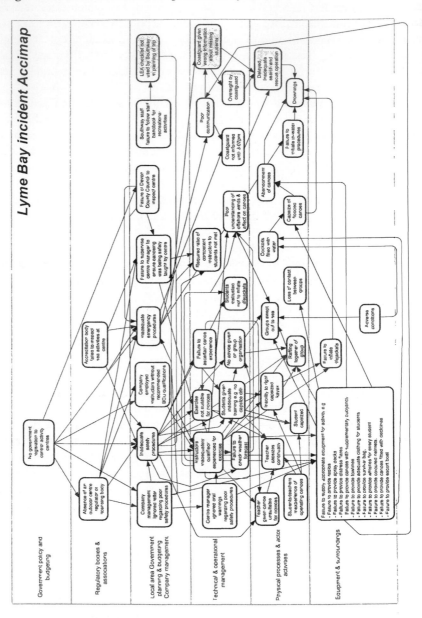

FIGURE 1.1 Lyme Bay Accimap.

Technical & operational management

At the technical and operational management level, decisions and actions both preceding the accident and on the day of the activity played a significant role in the failures at the lower levels. Before the incident, the centre's manager (and also the company management) failed to heed the content of a letter sent by two previous employees to management regarding unsafe practices, poor safety procedures, and inadequate equipment for the sea canoeing activities provided at the centre. The centre manager also failed to heed verbal warnings from the same two employees regarding the fact that canoeing was not being safely taught at the centre (Jenkins & Jenkinson, 1993). The instructors employed by the centre were also not sufficiently qualified for sea canoeing, which represents a failure on behalf of the manager to employ adequately qualified staff. The exercise itself was not suitable for novice canoeists; further, the inquiry report suggests that no attempt was made by staff to ascertain the experience levels of the students or teacher. As a result, the training given to the teacher and students prior to the trip in the centre's swimming pool was inadequate, involving no capsize drills whereby the procedure to right and re-board canoes are taught. This is particularly important with regard to the inability of the teacher to right the capsized canoe prior to the group becoming separated.

On the day of the activity, neither of the two instructors checked the weather forecast, and no attempt was made to ascertain the conditions out at sea. Students were also told not to inflate their lifejackets. Due to the instructors' lack of qualifications and the type of activity being undertaken, the required ratio of competent instructors to students was also not met.

Local area Government planning & budgeting, Company management

At the company management level, the company's management also failed to heed the content of the letter sent by two previous employees regarding the safety of canoeing activities at the centre in question. The managing director was charged with failing to devise and enforce safe procedures for executing sea canoeing activities, and the emergency procedures in place were also found to be inadequate. The employment of inadequately qualified instructors is also represented at this level, with the company management failing to procure employment of staff suitably qualified to provide safe canoeing activities. There was also a failure on behalf of the company management to supervise and ensure that safe activities were being provided at the centre in question. Finally, the inquiry report also concluded that the school involved had not planned the trip adequately by failing to completely follow their own staff handbook for organizing such trips, and a local education authority checklist for planning such activities was not used.

Regulatory bodies and associations

At the time of the incident, there was no regulatory body or licensing body for outdoor activity centers. This meant that unsafe practices and procedures could, to a large extent, continue unchecked without reprisals. Following the incident, the inquiry report recommended that a National independent system of registration and regulation of outdoor activity centers be developed as a matter of urgency, and also that no school or youth group should be permitted to use such centers until they had been approved for registration by an appropriate body. The absence of a regulator and legislation allowed the company management to continue with unsafe and inappropriate procedures, and to employ inadequately qualified staff. The report concluded that, if a regulatory body had been in place at the time, the concerns reported by previous employees would undoubtedly have been reported to the body, and appropriate action would have ensued.

In addition, although the centre in question was accredited by the British Activity Holidays Association, the inquiry report suggested that no examination of sea canoeing activities was undertaken, rather, only land and pool based activities were examined.

Government policy and budgeting

At the Government policy and budgeting level, the absence of legislation to control outdoor activity centers is the key failure involved. This meant that there was no regulating or licensing body overseeing outdoor activity providers, which enabled the centre to continue engaging in unsafe practices, despite these being identified and documented by previous employees of the centre.

DISCUSSION

The aim of this analysis was to test Rasmussen's risk management framework as a method for analyzing accidents and incidents in the led outdoor activity domain. In order to test its utility for explaining such incidents, the Accimap approach was used to ascertain whether the contributory events involved supported each of the predictions made by Rasmussen's framework. The predictions outlined for testing the risk management framework were all supported by the case study analysis. Table 1.1 (see Appendix) gives an overview of the predictions and examples of how the analysis supported them.

The results therefore provide further support for the risk management framework and Accimap approach as a valid and appropriate accident analysis methodology for use in the led outdoor activity domain. These findings also add to the considerable evidence presented in the literature regarding the utility of Rasmussen's framework (e.g. Cassano-Piche et al, 2009; Jenkins et al, 2010; Vicente & Christoffersen, 2006).

With participation on the increase, and safety compromising accidents and incidents continuing to occur, much further research is required. The analysis of accidents via Accimap represents one approach that can enhance safety through increasing activity provider's understanding of the nature of accidents and informing the development of appropriate strategies and countermeasures designed to prevent or mitigate future accidents and incidents. Other pertinent lines of work recommended are the development of standardized accident and incident reporting systems, the development of a universal database of outdoor activity accidents and incidents, and the development of taxonomies of system and human failures. In particular, the development of failure taxonomies for each of the levels used by the Accimap method may be one line of research which enhances the usability of the approach for analyzing accidents and incidents in the led outdoor activity domain.

Acknowledgement

This research was funded by the Department of Planning and Community Development (Sport and Recreation Victoria), along with contributions from various led outdoor activity providers. The authors also wish to acknowledge the members of the project steering committee, who provided initial data for the analysis and also validated and subsequently formed the expert panel for refining the analysis presented.

REFERENCES

Brackenreg, M. (1999). Learning from our mistakes – before it's too late. *Australian Journal of Outdoor Education*, 3(2), pp. 27-33.

Cassano-Piche, A. L., Vicente, K. J., & Jamieson, G. A. (2009). A test of Rasmussen's risk management framework in the food safety domain: BSE in the UK. *Theoretical Issues in Ergonomics Science*, 10:4, pp. 283-304.

Davidson, G. (2007). Towards understanding the root causes of outdoor education incidents. Presentation to the 15th National outdoor education conference, Ballarat, Victoria, 20-23 September 2007.

Jenkins, S. & Jenkinson, P. (1993). Report into the Lyme bay canoe tragedy. Devon County Council Report.

Jenkins, D. P., Salmon, P. M., Stanton, N. A. & Walker, G. H. (2010). A systemic approach to accident analysis: a case study of the Stockwell shooting. *Ergonomics*. 3:1. Pp 1 – 17.

Johnson, C.W. & Muniz de Almeida, I. (2008). Extending the borders of accident investigation: applying novel analysis techniques to the loss of the Brazilian space launch vehicle VLS-1 V03. *Safety Science*, 46:1. pp 38-53.

Rasmussen, J. (1997). Risk management in a dynamic society: A modelling problem. *Safety Science*, 27:2/3, pp. 183-213.

Salmon, P. M., Williamson, A., Mitsopoulos-Rubens, E., Rudin-Brown, C., & Lenné, M. G. (2009). The role of Human Factors in led outdoor activity incidents: literature review and exploratory analysis. Monash University Accident Research Centre Report.

Svedung, I., & Rasmussen, J. (2002). Graphic representation of accident scenarios: mapping system structure and the causation of accidents. Safety Science, 40, pp. 397-417.

Vicente, K. J. & Christoffersen, K. (2006) The Walkerton E. coli outbreak: a test of Rasmussen's framework for risk management in a dynamic society. *Theoretical Issues in Ergonomics Science*, 7, (2), pp. 93 – 112.

Table 1.1 Table showing the results of the testing of the risk management framework's predictions.

Prediction and case study evidence
1. Safety is an emergent property of a complex socio-technical system. It is impacted by the decisions of all of the actors—politicians, managers, safety officers and work planners— not just the front-line workers alone. *The Accimap demonstrates that actors at all levels of the system were involved in creating the conditions that led to the drowning incident. This is represented by the presence of failures at each of the levels specified by Rasmussen's framework.*
2. Threats to safety or accidents are usually caused by multiple contributing factors, not just a single catastrophic decision or action. *The Accimap demonstrates that no one factor identified was independently responsible, in isolation, for the incident. For example, failure of the students canoes to be fitted with spray decks can be attributed to a range of factors, including the company's employment of inexperienced, inadequately qualified instructors, the centre and company managements failure to heed verbal and written warnings from previous employees of the centre , and the absence of appropriate safety procedures at the centre in question.*
3. Threats to safety or accidents can result from a lack of vertical integration (i.e. mismatches) across levels of a complex socio-technical system, not just from deficiencies at any one level alone. *The Accimap demonstrates various mismatches between individuals and organizations across the different levels of the system responsible for delivering safe and efficient outdoor activities.*
4. The lack of vertical integration is caused, in part, by a lack of feedback across levels of a complex socio-technical system. Actors at each level cannot see how their decisions interact with those made by actors at other levels, so the threats to safety are far from obvious before an accident. *The Accimap demonstrates how decisions and actions made at the different levels were made without appreciation of, or feedback regarding, how they interacted with those made by actors at other levels. For example, the provision of sea canoeing activities at the centre was made without any acknowledgement or consideration of the other levels of the system. Feedback levels regarding the performance of such activities was not forthcoming from the higher levels, including company management and the accrediting body (who did not examine sea-based activities).*
5. Work practices in a complex socio-technical system are not static. They will migrate over time under the influence of a cost gradient driven by financial pressures in an aggressive competitive environment and under the influence of an effort gradient driven by the psychological pressure to follow the path of least resistance. *Economic and efficiency pressures played a part in the decisions made and actions taken by those involved, the result of which was the system migrating across the safe boundary of operation. Economic pressures were extant predominantly at the local area government and company management level, and efficiency pressures mainly affected the company management and technical and operational management levels.*
6. The migration of work practices can occur at multiple levels of a complex socio-technical system, not just one level alone. *In this case the migration of work practices occurred at four different levels: the local area government and company management level, technical and operational management level, the physical processes and actor activities level, and the equipment and surroundings level.*
**7. Migration of work practices causes the system's defences to degrade and erode gradually over time, not all at once. Accidents are released by a combination of this systematically- induced migration in work practices and a triggering event, not just by

an unusual action or an entirely new, one-time threat to safety.

Some of the practices that contributed to the accident had clearly been affecting the system for a significant portion of time prior to the accident occurring. The letter from previous employees regarding poor safety procedures and equipment in the context of sea canoeing activities, sent six months prior to the accident, is clear evidence that migration of work practices had been occurring for some time.

Chapter 115

Evacuation Flow Analysis for Aircraft Accident with Panic Passenger by Autonomous Agent and Multi-Agent Simulation

Tetsuya Miyoshi[1], Hidetoshi Nakayasu[2],
Yuka Ueno[2], Masaru Nakagawa[3]

[1]Toyohashi Sozo University
Toyohashi, Japan

[2] Konan University
Kobe, Japan

[3] Shiga University
Hikone, Japan

ABSTRACT

It is so critical for aircraft safety in case of emergency accident that the skill of flight crews and cabin internal arrangement must concentrate to save human lives, considering the possibility of the airplane being on fire. From these points of view, Federal Aviation Administration (FAA) makes its attention on these issues, therefore the new aircraft must satisfy several rules as Federal Aviation Regulation (FAR Part 25.803) where one of them is called "90 seconds rule." This rule means that the maximum seating capacity including the number of crewmembers can be evacuated from less than half number of emergency exits of the airplane to the ground under simulated emergency conditions within 90 seconds.

In this paper an internal model for autonomous agent and multi-agent was proposed to play a behavior of panic passenger in chaotic situation such as limited short time and closed space in an aircraft accident. The urgent evacuation simulation model of aircraft accident was also developed by the internal model as real as practical situation considering the possibility of the airplane being on fire. In order to evaluate the proposed models, the evacuation flow was evaluated entity similar to the actual aircraft DC-10-30. The simulation results suggest that the proposed model can represent the evacuation behavior of passengers on the limited situation. Especially, the behavior passenger with panic emotion tends to be selfish. Therefore, the passenger became to be non-adoptive to the other passengers and go ahead nevertheless the other passenger waiting. These tendencies include the mode, the jam of passenger appeared and it took more evacuation time.

Keywords: Aircraft evacuation, autonomous agent and multi-agent simulation, Psychological effect of passenger, panic passenger, non-adaptive behavior.

INTRODUCTION

It is well known that the development of effective evacuation system is the primary task for aircraft safety, while it is so critical for aircraft safety in case of emergency accident that the skill of flight crews and cabin internal arrangement must concentrate to save human lives considering the possibility of the airplane being on fire. From these points of view, Federal Aviation Administration (FAA) makes its attention on these issues, therefore the new aircraft must satisfy several rules as Federal Aviation Regulation (FAR) Part 25.803(Federal Aviation Regulation, 1990) where one of them is called "90 seconds rule." This rule means that the maximum seating capacity including the number of crewmembers can be evacuated from the airplane to the ground under simulated emergency conditions within 90 seconds. However, it is very difficult because the egress time necessary for evacuation is concerned with so many factors such as airframe (number, size and location of emergency exit, seat and aisle arrangement), passengers (age, health, sex, interrelationship and degree of panic) and flight crew (skill and training level).

The practical and traditional approach for evacuation experiments with participants are executed by airplane manufacturers (A380 evacuation test, 2006). However, these experiments are dangerous and expensive and not easily repeatable. On the other hand, a lot of simulation models on the evacuation from civil structures and transfer vehicles have been developed (National Transportation Safety Board, 2000, Santos et al, 2004, Galea et al, 2007). Especially, a dynamic model for aircraft evacuation system by Ceruti and Manzini (Ceruti et al, 2003) is proposed as a visual interactive simulation tool in order to develop and evaluate the several evacuation techniques. In the most of these works, a simulation tools used in manufacturing planning and industrial optimization is applied to the evacuation problems where passenger routings, interactions between people queuing and

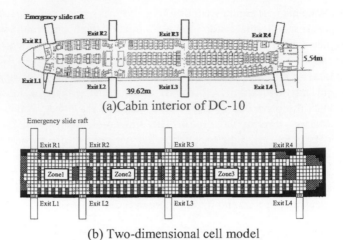

(a)Cabin interior of DC-10

(b) Two-dimensional cell model

Figure 1. Cabin interior and cell model.

operational time belong to a set of topics in industrial applications. However, most of the previous formulations of evacuation simulation have not been taken into account of psychological effect of passengers. It is very difficult to realize the evacuation behavior of passengers with remarkable high stress in chaotic situation such as limited 90 seconds and closed spece in an aircraft accident considering the possibility of the airplane being on fire.

In this paper an autonomous agent and multi-agent model about the behavior and psychological situation in the aircraft evacuation were proposed and the evacuation simulation systems were developed based on the passenger model. In this paper, it is tried to evaluate the flow of passengers to emergency exit in systems simulation based on an autonomous agent and multi-agent system. For the representation of behavior of panic passenger, an internal model of an agent as a passenger is introduced into the simulation model in order to reflect the unstable mind and selfish behavior of panic passenger who has remarkable high tension emotion when aircraft accident happens. By this internal model, an agent of panic passenger play a non-adaptive behavior that means do not go along the social rule to follow the queuing line to the emergency exit in the cabin.

SIMULATION MODEL BY AAMAS

MODEL AS ANALOGIES OF PASSENGERS AND EQUIPMENTS

In this paper an autonomous and multi-agents system (AAMAS) model is applied to construct the simulation model where a two-dimensional grid cell model shown in Figure1(b) is introduced to represent the passenger flow of the emergency evacuation in the aircraft. Figure 1(a) also represents a layout and allocation of

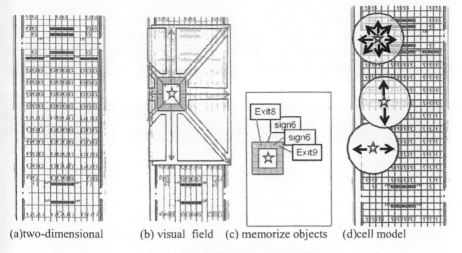

(a)two-dimensional   (b) visual field   (c) memorize objects   (d)cell model

Figure 2. Local rule about agent movements.

equipment in cabin of the DC-10-30, where it is seen that there are the three kinds of passenger cabins, zone1, zone2 and zone3. There are 24, 98 and 155 seats in each zone for the passenger capacities, respectively. Figure 1(b) shows the grid cell model to illustrate the movement of agent as the evacuation flow of passenger, where the cell means an analogy of the seat. In the model, the size of cell represents 0.43m square that means the area of the size of seat including the unit space around it of economy class in the cabin approximately. The equipments such as exit doors, exit signs, lavatories, galleries, counters, aisles and seats are also identified as approximately analogies in two- dimensional model so as the location and dimension of an aircraft. Passengers are represented as multi-agents that are initially placed in the cells and move around the grid cell by one's objectives whose behavior is controlled by the autonomous algorithm based on local rules previously specified. The generic passenger run speeds in evacuation were investigated by Galea (Galea et al. 2007), and the computerized evacuation simulation were performed using these evacuation speeds by Ceruti(Ceruti et al, 2003). From this study, the evacuation speed 0.98m/sec of old female that means the latest evacuation speed among various categories of passengers is selected to the agent's moving speed for evacuation since in order to estimate the upper bound of the estimated evacuation time as a guideline. The time step of computer simulation are derived as 0.43sec/step by the cell size (0.43m) and the evacuation speed of passenger (0.98m/sec). By the way, the additional duration time from exit to the ground by emergency escape slide is estimated as 2.2 sec by the aircraft evacuation practice(A380 Emergency Evacuation Test, 2006). It is also noted that the total egress time from the cabin to the ground is able to estimate the duration time to the emergency exit and the duration time from exit to the ground by emergency escape slide A time variable, T, is set to 0, and incremented by 1 every 0.43s. It is assumed in the previously specified rules that the multi-agents move to the exit each other in order to evacuate from an aircraft to the ground.   It is noted that though the individual behaviors of multi-agent are controlled by the local and autonomous algorithm, the final results of the movement of individuals by the autonomous and multi-agents will be seemed to yield a macro behavior controlled by group dynamics as an evacuation flow in an aircraft cabin.

ALGORITHM FOR AN AUTONOMOUS MOVEMENT OF MULTI-AGENT

The flow of passengers towards the exits is determined through four steps.

1) Each agent is then instructed to move towards the nearest exit which he could recognize. They gather the information on the location of emergency exits or exits signs at the first step (see Figure 2(b)). In Figure 2(b) the agent is denoted by star mark. He/she gathers the information about interior equipments within the visual field illustrated in Figure 2(b). The visual field of an agent is limited by the visual length d, which is predetermined as a simulation condition.

2) He/she memorizes the nearest direction into emergency exits or exits signs as an own action object if they could recognize the emergency exits or exit signs, otherwise they forecast the direction to the nearest exit route as a candidate object, which are illustrated as some samples in Figure 2(c). In this step the aim of agent is determined among the 16 directions as shown in Figure 2(b).

3) They determine the adjacent cell to move by considering both the memorized direction and current location itself, which is detected as a critical path so as to take the shorter time to go for exit subject to the physical limitation such as seats and equipments allocation (see Figure 2(d)). Adjacency is defined by the Moore neighborhood (i.e., the 8 surrounding patches). If the agent's current position is near the seats, the agents would move to the corridor. If the current position is on the corridor, he/she would move to the front space or back space (see Figure 2(d)).

4) Confirming whether the cell to move is occupied or not, they moved to the cell if it is not occupied, otherwise they must wait until the cell is empty. When a lot of agents would go for the same exit over the capacity of exits, the queuing of agents happened.

PSYCHOLOGICAL MODEL OF PASSENGERS

In the simulation model, the non-adaptive behavior of panic agent (Mintz, A., 1951) is represented by that of agent with internal model. This internal model shows various kinds of behavior depending on the level of accumulation of degrees of crisis factor such as remaining time for egress, queuing length to exit and difficulty of selecting exit. The agent acts non-adaptive behavior when the total level of crisis factor is larger than a threshold. On the other hand, the agent behaves adaptive action where he or she can follow the queuing line into emergency exit when the level of crisis factor is less than the threshold. In this paper, the former agent is called as a panic agent who does not obey the social rule and acts selfish and non-adaptive behavior. Thus this panic agent breaks into the queuing line and generates the traffic jam. Consequently this traffic jam is propagated into the accident crisis.

All movements are subject to the restriction that only one agent may occupy a cell during a unit time step. If the cell ahead is occupied by another agent, then it waits until the cell will be empty. In condition that the number of agents is relatively large, some agents do not respond in an orderly manner until a cell will be empty frequently, then the jams of passengers arise the representation of behavior of panic passenger, an internal psychological model that reflects the change of mind of passenger who has unstable emotion in aircraft accident is introduced into the system simulation. By this internal model, an agent of panic passenger plays an non-adaptive behavior that means do not adjust the social rule to follow the queuing line to the emergency exit in the cabin.

It is assumed that passengers would be panic whenever the degree of emergency and any other factors induced a panic emotion passes the level of threshold. The mental level L of agents that means the degree of psychological emotion is determined by the level for the remaining egress time (Lt), the level for queuing length to exit (Lq) and the level for difficulty of discovery (Ls) of exit as follows

$$L= Lt + Lq + Ls \qquad (1)$$

where Lt, Lq, Ls are called as crisis level and L is mental level of agent. Three crisis levels are determined by the linear functions of remaining egress time (t), queuing length to exit(n_q) and frequency of non-discovery of exit(n_s) up to t^*, n_q^*, n_s^*, respectively. t^*, n_q^*, n_s^*, are the parameters to control the degree of crisis factor for the situation of aircraft evacuation. The mental level L of each agent is calculated using this interior model, and if the agent' mental level is larger than the threshold L_0 and crisis level, the agent become a panic agent and plays the non-adaptive behavior.

SIMULATION RESULTS FOR AIRCRAFT EVACUATION

In this section the simulation results of aircraft evacuation are show in order to illustrate the performance of the proposed simulation system about the emergency evacuation, and discuss the factors for the evacuation efficiency in the aircraft evacuation. The evacuation simulations for DC10-30 shown in Figure 1 are performed in some conditions that the load factor is set to 100% or 50% and the initial allocation of the agents are determined by the patterns shown in Figure 3. The road factor means the ratio of passengers to the seat capacity. The visual length is assumed to be 20 cells, then it is assumed that the passengers are able to recognized the nearest emergent exit door. The evacuation flows and the completed evacuation time are shown as the simulation results in each simulation condition. In

1112

order to evaluate the pure multi-agents model without psychological model and the internal psychological model, two kinds of simulations are performed and illustrate the simulation results, separately.

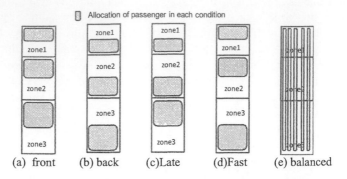

Figure 3. Allocation of passengers in each simulation condition.

SIMULATION RESULTS WITHOUT PSYCHOLOGICAL MODEL

Firstly, the results in no panic situation are shown in Figures 4, 5, 6. Figure 4 shows the evacuation flow of passenger in case that the load factor is 100%(277 passengers) and all of exit doors were opened. The agents moved by determining the next cell based on the collected equipments information and local rules.

The rates of remaining passengers within the aircraft are shown for the simulation time in Figure 5 in conditions that the all of emergent exit doors were opened or the 4 doors on the left hand side were opened. The crossing point to x-axis shows the complete evacuation time in the simulation condition. Comparing the complete times between the load factor 100% and 50%(136 passengers), then roughly speaking, the longer the evacuation time becomes, the larger load factors becomes in both condition about exit door. Even in the same load factor 50%, the total egress time is different according to the initial allocation of the agents. The total egress time is shortest in the condition "50%Fast" whose agents allocation is illustrated in Figure 3(d) and longest in the condition "50%Late" whose allocation is illustrated in Figure3(c). Considering the evacuation flow with these results, the efficiency of the evacuation depends on the amount of agents whose evacuation aim is set to the third emergent exit. The evacuation time became longer in case that some emergency exit doors are closed, since the number of passengers who went out were limited by closing the door (see Figure 6).

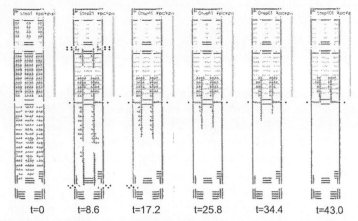

t=0   t=8.6   t=17.2   t=25.8   t=34.4   t=43.0

Simulation condition: L_0=1.0, 8 emergency exits were open. △: passenger

Figure 4.   Evacuation flow of passengers (no panic condition, 100% load factor)

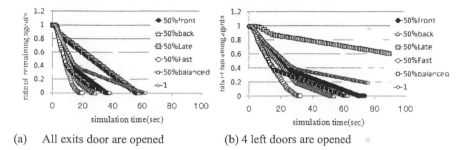

(a)   All exits door are opened   (b) 4 left doors are opened

Figure 5 Rate of remaining passengers in each simulation condition

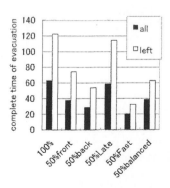

All exit door are opened

Figure 6 Rate of remaining passengers in each simulation condition

SIMULATION RESULTS WITH PSYCHOLOGICAL MODEL

Next, the simulation results considering the psychological model of passengers, in which the passenger psychological level determined by the three crisis factors, are illustrated. The parameters of three crisis factors, t^*, n_q^*, n_s^* are set to 90, 40, 20, respectively, and the threshold $L_0 = 0.8$ to judge whether the agent is panic or not. The conditions of load factor and allocation of passenger are set to the same ones with the first simulation results. The evacuation flow of passengers considering the psychological model is shown in Figure 7, in which the panic passenger raised at time 46.0 marked by black star ★ and jams marked by white star ☆raised, too. In order to illustrate the situation of aircraft evacuation, the rates of remaining passenger in the aircraft are plotted in Figure 8 in several conditions of load factor and allocation of passengers. The rates of remaining passenger, in case that 4 emergent exit doors were closed and 4 doors were opened, are plotted in Figure 8, too. Furthermore, each complete evacuation times in some conditions is illustrated in Figure 10, in which the complete evacuation time with all doors open condition and 4 doors open condition are represented. In case of 100% load factor and 50%Late with 4 door open condition, the evacuation time are not plotted since the simulations corresponding to the conditions were not completed, i.e., all of passengers could not get out within 100 seconds because of traffic jam.

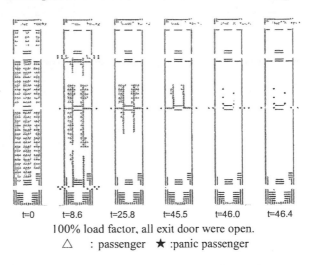

t=0   t=8.6   t=25.8   t=45.5   t=46.0   t=46.4

100% load factor, all exit door were open.

△ : passenger ★ :panic passenger

Figure 7   Evacuation flow of passengers including panic passengers

Comparing between Figures 6 and 10, the total egress times with agent's interior model are longer than that without interior model or equal to that in all cases except the 100% load factor and "50%Late" condition. In two conditions the simulation were not completed because of the simulation time up. Since the panic agents raised at 40 sec simulation time as shown in Figure 9, in the simulation whose evacuation time is shorter than 40sec was influenced little, but the ones whose evacuation time is long was influenced strongly. It may be the future work whether the interior model proposed in this paper is appropriate or not based on only these results, but the results suggest that the proposed model could represent the emotion of agents in several situations.

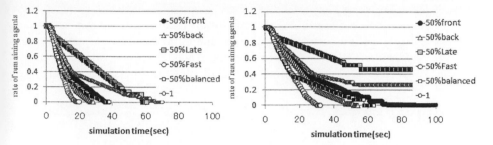

Figure 8 Rate of remaining passengers for simulation time

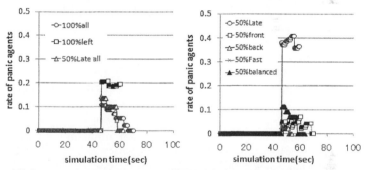

(a)   All doors were opened   (b) Half doors on the left hand side were opened

Figure 9   Rate of panic passenger

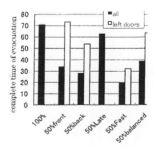

Figure 10. Complete evacuation time

CONCLUSIONS

In this paper the passenger model about the behavior and psychological situation in the aircraft evacuation were proposed and the evacuation simulation system were developed based on the passenger model for the actual aircraft DC-10-30.

Not only the factors such as load factor and allocation of agents to influenced the evacuation time but also the psychological affect were discussed through the considering the simulation results using the proposed method. Consequently, the simulation results suggested that the evacuation time became longer load factor became larger and the amount of the agent aiming to the third exit door became large. Furthermore, the proposed psychological model can partially represent the panic situation raised by passengers' emotion through the simulation results.

REFERENCES

National Transportation Safety Board(2000), "Safety Study : Emergency Evacuation of Commercial Airplanes."

Santos, G., Aguirre, B.E. (2004), " A Critical Review of Emergency Evacuation Simulation Models." *NIST Workshop on Building Occupant Movement during Fire Emergencies.*

Galea,E.R., Wang,Z., Togher,M., Jia,F. and Lawrence,P. (2007), "Predicting the Likely Impact of Aircraft Postcrash Fire on Aircraft Evacuation Using Fire and Evacuation Simulation." *International Fire & Cabin Safety Research Conference.*

Federal Aviation Regulation FAR Part 25.803 Emergency Evacuation, Part 25 Appendix J. Part 121.391 Flight assistants, Washington DC, USA. (1990), http : // www. Flightsimaviation .com/ data/FARS/part_25-803.html

Ceruti, A. and Manzini, R.(2003), "Aircraft evacuation dynamic model based on simulation tool." *Advanced Simulation Technologies Conference and Bussiness and Industry Symposium*, Modeling and Simulation International, 2003, http://www.scs.org /getDoc . cfm ?id =2166.

Galea, E.R.(1999), "Predicting the Evacuation Performance of Mass Transport Vehicles Using Computer Models." *Focus on Commercial Aviation Safety, 34*, pp.14-15

A380 Emergency Evacuation Test, http:/ /www. youtube. com/ watch ?v= XIaovi1JWy (2006)

Mintz, A(1951), "Non-adaptive group behavior", *Journal of Abnormal and Social Psychology, 46*, pp150-159.

CHAPTER 116

Risk Control Through the Use of Procedures – A Method for Evaluating the Change in Risk

Gregory Praino[1], Joseph Sharit[2]

[1]United Space Alliance, LLC
Cape Canaveral, FL 32920, USA

[2]Department of Industrial Engineering
University of Miami
Coral Gables, FL 33124, USA

ABSTRACT

Organizations use procedures to influence or control the behavior of their workers, but often have no basis for determining whether an additional rule, or procedural control will be beneficial. This paper outlines a proposed method for determining if the addition or removal of procedural controls will impact the occurrences of critical consequences.

The proposed method focuses on two aspects: how valuable the procedural control is, based on the inevitability of the consequence and the opportunity to intervene; and how likely the control is to fail, based on five procedural design elements that address how well the rule or control has been   Defined, Assigned, Trained, Organized and Monitored—referred to as the DATOM elements.

Keywords: Procedural control, critical consequences, control failure likelihood, control value

INTRODUCTION

Organizations frequently find themselves mired by rules that have questionable value. Often these rules are the result of a knee-jerk reaction to failures, near misses or even successes, with the organization layering on additional rules to address circumstances that are perceived as significant. Unfortunately, these organizations rarely reconsider these rules later, at best allowing unnecessary rules to clutter the policies and procedures that govern workers, or in the worst case, leaving rules that confuse workers and lead to undesired behavior.

After the Space Shuttle *Columbia* accident, NASA had reason to believe such clutter existed in the policies governing space shuttle ground processing work instructions. In response, a method was sought to systematically evaluate the rules in place—both to determine if some rules could be consolidated or eliminated, and also to ensure that there was no false sense of security where the abundance of rules masked uncontrolled risks.

The resulting method was structured to examine any critical process; in the case of the Space Shuttle Program, it was directed at activities where loss of life or of a space shuttle vehicle was possible. The general case addressed physical controls (i.e., barriers) as well as rules, or procedural controls. However, the scope of the current work is limited to the applications on procedural controls because the transactional nature of shuttle ground processing depends overwhelmingly on people performing the right task in the right way at the right time.

The Control Assessment method explicitly considers the risk associated with each rule in the process individually to determine if that particular rule reduces risk, increases it, or has no significant impact. The risk assessment is based on two main factors: how valuable the rule is at preventing a critical consequence and how likely the rule is to fail under the real-world conditions that exist when it is called on to function.

Control Value describes how necessary the function of that control is. Necessity is determined based on how inevitable the consequence is in the absence of any control, and if the control leaves sufficient opportunity to intervene once an initiating event has occurred.

Failure Likelihood depends on how well the control has been designed, which is based on how well it is defined, assigned, trained, organized and monitored. Each of these five elements, which can be remembered with the acronym *DATOM*, are necessary for the sustained performance of the control, with deficiencies in any area contributing to the likelihood that the control will fail.

BACKGROUND

During the investigation of the Space Shuttle *Columbia* accident, NASA began three distinct efforts. The debris recovery in west Texas and the reconstruction of the recovered hardware were the higher profile tasks because the proximate, physical causes would have been evident from an analysis of the debris. The National Transportation and Safety Board was consulted because the NTSB performs similar reconstructions of conventional aircraft mishaps to analyze the causes, and their methods and experience were expected to help speed the investigation.

The less visible task involved a complete review of the work instructions written during the prior two processing flows of *Columbia*, for the STS-109 and STS-107 missions. The review was intended to find any technical errors made by ground processing personnel that could have contributed to the accident. In its report, the Columbia Accident Investigation Board noted that in the roughly 16,500 work documents reviewed, there were no findings or observations that contributed to the accident. However, the board did note an accuracy rate of 99.75%, leaving a small number of work documents with "Technical Observations (technical concerns or process issues), and Documentation Observations (minor errors)" that revealed procedural issues (CAIB, 2003).

Interviews with the engineers who wrote the work documents revealed that many of the observations identified in the review were associated with rule interpretations. In some cases, rules still technically in-place could no longer be followed as-written or no longer provided the benefit intended because of process changes made since the rules were created. Other observations involved situations where a process improvement clearly implemented a better way of performing the function and the old rule was just never removed from the policy.

In light of the "Can-Do" culture in place at the Kennedy Space Center (Vaughan, 1996), it really comes as no surprise that technicians and inspectors on the floor would continue working when faced with some of these situations. It would be wasteful to stop working because a document that clearly described the task didn't comply with a formatting rule that no longer applied. There would be no value added by 'correcting' to comply with a rule intended for manually-typed instructions that had been phased out by the use of a computer-based authoring process several years before.

The real surprise is that errors like these weren't more common. Directions for engineers writing work instructions were distributed between 37 policy documents, so while one could argue the overwhelming compliance with obscure, redundant and ambiguous rules is somewhat wasteful, it is a testament to the thoroughness of those engineers that the accuracy was so high.

RISK DEFINITION

Typically, risk is defined in terms of failure likelihood and consequence severity of an outcome (Kumamoto and Henley, 1996), which in principle enables the expected loss or risk to be computed. Kaplan and Garrick's (1981) approach to understanding risk is based on obtaining answers to a triplet of questions: "What can go wrong?" "What are the consequences?" "What is the likelihood?" represents a more generalized approach to risk assessment that quantifies any hazard on an absolute scale. In the proposed approach to defining risk, which is in terms of control value and failure likelihood, the key points that differentiate it from the general case are that the analysis is limited to only consequences that the organization considers critical and that it provides an indication of relative risk between possible options. Thus this approach helps to effectively target the relevant factors when the scope of the assessment is limited.

By choosing 'critical' consequences as those which, if they occur, would threaten the existence of the organization, then any practical need for quantification is eliminated. In essence, if any critical hazard is realized, it could mean the end of the organization. In the case of the Space Shuttle Program, loss of another vehicle would almost definitely result in the immediate and permanent termination of operations (Block, 2008). While loss of a life would probably not result in the premature end of the program, it would impact the career of the decision maker who allowed the circumstances to exist. A parallel example from another industry would be a death resulting from surgical malpractice—the hospital may not choose to select this as a critical consequence because a single fatality would present a minimal threat to its existence, but if the Control Assessment was being performed by the surgeon it would almost definitely be critical.

Another important assumption regarding risk that must be addressed before proceeding is that the assessment does not attempt to provide an aggregate measure of risk, but instead addresses the differences between alternatives, asking the questions: "what is the benefit derived from the presence of this rule?", or "is rule A better than rule B?" Techniques like Probabilistic Risk Assessment attempt to account for all risks faced by the organization (Kumamoto and Henley, 1996), but the intent of the Control Assessment is to only consider those items that can be controlled. Consider the hazardous release of chemicals due to a railway accident. There is a small but real risk to a factory adjacent to the rail-line, but the manager of the factory wouldn't be expected to try controlling that risk. Such a risk could be controlled at the corporate level though, so a Control Assessment there might look at the task of selecting new facility locations.

CONTROL VALUE

The way Control Assessment considers the value of the control being examined is to evaluate the inevitability of the critical consequence in the absence of the control and the opportunity to intervene should something go wrong. A control to prevent a consequence that would only happen occasionally is less valuable than a control to avoid an inevitable consequence. Likewise, a control that leaves ample opportunity for an active response would be less valuable than one where the consequence would be immediate.

To illustrate the impact of inevitability on value, consider two circumstances for a control the Department of Transportation could put in place on Interstate Highways. The design standard could be changed to require runaway truck ramps on all interstates to address the hazard of failed brakes on trucks. Obviously, a ramp on steep section of road winding down through the mountains in Colorado would be far more valuable than on a straight and level section of road in the middle of Kansas farmland. On the steep road, a serious accident would be near inevitable because the lack of brakes would result in an increase in the speed of the truck, and impact with other vehicles that were under control. On the level road, the truck's speed would not increase and the vehicle could conceivably be allowed to come to a stop after running out of fuel, with no intervention necessary.

Opportunity to intervene can easily be seen in an example from space shuttle operations. Hypergolic rocket propellants are reactive enough that they will combust on contact, so no ignition system is required, allowing simpler and lighter thrusters to be used on the orbiter. On the other hand, the highly reactive chemicals pose a serious health risk to personnel; therefore one control in place at Kennedy Space Center restricts access to facilities when hypergols are being actively handled. A worker who entered the launch pad perimeter would have to cross an open field before getting close enough to operations to enter a dangerous concentration of propellant vapors. In the Orbiter Processing Facility, the hanger where maintenance and refurbishment of the orbiters is performed, a worker could be exposed to a hazardous concentration immediately upon entering.

In both cases, the control is necessary because a critical consequence would be inevitable during times that a hazardous concentration existed. However, the control value is much lower at the pad because there would be more opportunity to act once the worker enters the facility.

Whereas the Control Value can be used to explain where circumstances beyond the control of the organization are responsible for infrequent occurrences of critical consequences, the likelihood portion of the risk can be used to address the likelihood of the control failing. Rather than seeking an expected-value to describe risk, the procedural risk model focuses on whether the controls accomplish their intended function.

FAILURE LIKELIHOOD

Determining how often controls fail involves looking at the failure mechanisms of the controls. The most obvious case of a control that will not affect a worker's behavior is when the worker has a negative intent. Damage resulting from someone who intends to do harm by sabotage is outside the scope of control assessment because it is not the result of a control failure. Malicious compliance, on the other hand, is when an employee with a negative intent complies with a procedure they believe to be ineffective or counter-productive to the goals of the organization. This malicious compliance presents a procedural risk because the flawed procedure contains ineffective or failed controls—a worker who is aware that the procedure is not correct but who nonetheless follows the procedure would not be executing the actions desired by the organization but would be safe from reprisal.

Malicious compliance is a special case of the first way controls can fail: by not clearly agreeing with the organization's expectations. A control that is ambiguous or conflicts with expectations will leave a worker unaware of the correct action to perform, or in the case of a malicious worker, provide a plausible excuse for acting against the best interests of the organization.

The second way controls can fail is to instruct the worker to perform an action they are unable to, either by providing insufficient details or identifying actions that cannot be performed under the time or resource constraints. The classic example of this second control failure is the production vs. quality conflict. Turning out high volumes of a product increase profit, but the need for oversight or inspection to ensure delivery of satisfactory products often slows production.

The final way controls can fail is by calling for actions that are harmful to the worker. A worker who is aware of what harm may come will not proceed with the action. Usually, such a situation will also be in conflict with the goals of the organization because the costs associated with the organization's liability in such a case could harm the organization as well.

In each of these situations, the worker performs a different action than expected or refrains from performing any action. An unaware worker may happen to perform the correct action, but that is treated here as an incorrect action—it is not a desired mode of operating to count on happenstance to ensure that workers act correctly.

Although these three failure scenarios describe how a process fails, they are not practical for facilitating an analysis of procedural risk because the level of specification is too general; that is, failures are specified to be the result of badly selected or incompletely described controls. Further specification is necessary to describe the process in useful terms. To accomplish this objective, it is proposed that Control Assessment consider a set of characteristics to describe a process based on five of six basic questions: what? why? when? how? where? and who? 'Why' is excluded because it does not describe the process, but provides rationale for its existence. Providing this rationale can be helpful in motivating the workers who will be performing the task, but is not strictly necessary for successful task completion.

DATOM and Failure Likelihood

DATOM, the model that uses the five elements—define, assign, train, organize and monitor—to fully describe a process was based on the "5 Ws." The original intent behind use of the model was as a tool for process design, but it also had value as a means of spotting where incomplete processes could fail.

Whether designing a new process or examining an existing one, the first step in describing the process is to define the actions that are expected to take place. 'What' must be firmly established for the action to be part of a process. Without an overarching scheme, a worker will not reliably perform an action or sequence of actions to provide the needed output. Defining the 'what' involves deciding on the extent of the actions involved with the task, along with choosing or identifying the parameters that control the task actions.

The unique skills and limitations of the workers influence the 'how' 'when' and 'where' so 'who' must be addressed before progressing to the other remaining questions. Without clearly identifying 'who' will be assigned to the task, some level of confusion is inevitable because of the assumptions that must be made by the participants. However, problems persist even when there is an explicit assignment. An action may be consistently performed by the same worker under normal circumstances, but a substitution creates opportunities for misunderstanding. A substitute worker who is capable of performing the task may be unaware that a particular action needs to be performed, or may assume that the action is performed by another worker.

Once the task has been defined and a worker has been assigned to perform that task, 'how' the worker will perform the task becomes relevant. For the task to be effectively performed, the worker needs training in the process knowledge specific to the task and in the skills required to perform the expected actions.

'Where' and 'when' the task will be performed are linked together because both are limited by the defined process sequence. Some aspects of 'how' are similarly constrained, particularly in the context of tools, equipment, and other supporting resources. These three items together describe how the process is organized and determine the efficiency, quality, and safety of the process if a trained worker is assigned to the task.

The links between the 'when,' 'where' and 'how' demonstrate the shortcomings with simply using the five questions as the criteria for evaluating a process. In contrast, Control Assessment does not attempt to split the operational details of task performance, leaving the answers to those three questions together under the concept of how well the task is organized.

This restructuring of the five questions resulted in the rough approximation of what became DATOM. The answer to 'what' is equivalent to the Define element and 'who' provides the Assign information. 'How' is split between Train and Organize, with the remainder of Organize coming from 'when' and 'where'.

Failures within these four elements can cause failures of the process, but they do not provide feedback on whether the process actually produces the desired results. Without some form of check, the process will be vulnerable to changes in the inputs, the environment, or interpretations of the wording of the documented rules. Based on this need, Monitor is the necessary final element in the process evaluation criteria, even though it cannot itself cause a failure in a fully defined process.

The concept of monitoring includes activities that report on the 'health' of the process but are independent of the process itself. Inspection activities are similar but differ in a subtle and significant way from monitoring. Where inspections address quality during a specific instance of procedure execution, monitoring does not rely on acceptance criteria to determine if corrective action must be taken.

A nonconformance resulting from known and accepted process variation would need remedial action for that specific case, but no corrective action would be necessary as a process 'fix.' For example, consider a drilling operation where it's possible for the first hole to be drilled under-sized before the bit heats up from use. An undersized hole would be cause to reject the part, but it would not necessarily happen frequently enough to be worth the costs of changing the process. Monitoring, on the other hand, may catch a deficient process that is still producing conforming output. No short-term action would be needed but the process failure would eventually need to be corrected to prevent nonconformances. An example of this could be a machinist who makes a progressively larger, undocumented adjustment to a setting to compensate for a bad indicator on the machine—the output will conform, but the process is not sustainable.

SIGNIFICANCE

Since procedures are an organization's "mechanisms, techniques and processes that have been consciously and purposefully designed in order to try to control the organizational behavior" (Johnson and Gill, 1993), they are also the primary means for an organization to prevent the consequences that result from undesired action. The proposed method of considering procedures represents an attempt to understand not just if procedures are communicated effectively, but to see the procedure in context as an attempt at controlling workers' actions.

By using Control Value to understand if the organization's efforts have the potential to prevent or mitigate consequences and Failure Likelihood to determine if that potential is being realized, it should be possible for any organization to see if the actions taken by its workers to reduce a risk are succeeding. The technique won't provide the absolute measure of risk that a Probabilistic Risk Assessment would return, but it will allow the organization to see the relative impact on risk associated with the presence or absence of each control in its critical procedures.

The ultimate goal of utilizing the proposed technique is the development of a reliable tool that can be used to understand the relative risks when comparing alternative procedural controls. Using this tool, an organization could determine which rule among possible alternatives most effectively reduces risk. Similarly, the tool could be used when adding or removing a control to provide context to the change, particularly if the control under review is compared against existing controls intended to protect against the same hazard.

A series of validation exercises are under way within the workforce responsible for space shuttle ground processing and the initial results indicate that personnel with experience in a process can score a procedural control's elements— inevitability, opportunity for intervention, definition, assignment, training, organization and monitoring—consistently with the opinions of control value and failure likelihood provided by experts in risk assessment. The final configuration of the resulting tool is yet to be determined, but it appears that an assessment of procedural controls could be performed by a small group with expert knowledge of the process being assessed, facilitated by one who is familiar with the assessment technique, similar to the performance of a HAZOP analysis (AIChE, 1992).

REFERENCES

AIChE (American Institute of Chemical Engineers), (1992). *Guidelines for Hazard Evaluation Procedures: Second Edition with Worked Examples*. New York, NY: Center for Chemical Process Safety.

Block, R. (2008). NASA Chief: Odds grow for shuttle catastrophe. *Orlando Sentinel*. September 5, 2008

CAIB (Columbia Accident Investigation Board). (2003). *Columbia Accident Investigation Board Report*. Washington, DC: United States Government Printing Office

Johnson P. and Gill J. (1993). *Management Control and Organisational Behavior*. London: Paul Chapman Publishing Ltd.

Kaplan, S. and Garrick, B. J. (1981). On the Quantitative Definition of Risk. *Risk Analysis*, 1(1), 11-27.

Kumamoto, H. and Henley, E. J. (1996). *Probabilistic Risk Assessment and Management for Engineers and Scientists*. Piscattaway, NJ: IEEE Press

Vaughan, D. (1996). *The Challenger Launch Decision - Risky Technology, Culture and Deviance at NASA*. Chicago, IL: University of Chicago Press

A Human Factor Analysis to Mitigate Fall Risk Factors in an Aerospace Environment

Joylene Ware

Human Factors Integration
NASA/Kennedy Space Center
Kennedy Space Center, FL. 32899, USA

ABSTRACT

The objective of the research was to develop and validate a multifaceted model such as a fuzzy analytical hierarchy process (AHP) model that considers both qualitative and quantitative elements with relative significance in assessing the likelihood of falls and aid in the design of Ground Support Operations in aerospace environments. The model represented linguistic variables that quantified significant risk factor levels. Multiple risk factors that contribute to falls in NASA Ground Support Operations are task related, human/personal, environmental, and organizational. The subject matter experts were asked to participate in a voting system involving a survey where they judge risk factors using the fundamental pairwise comparison scale. The results were analyzed and synthesized using Expert Choice Software, which produced the relative weights for the risk factors. The following are relative weights for these risk factors: Task Related (0.314), Human/Personal (0.307), Environmental (0.248), and Organizational (0.130). The overall inconsistency ratio for all risk factors was 0.07, which indicates the model results were acceptable. The results show that task related risk factors are the highest cause for falls and the organizational risk are the lowest cause for falls in NASA Ground Support Operations. The rationale in this research is to justify using the priority vector to validate the weights by having two different sets of experts/decision makers create priority vectors separately and confirm the weights are similar. The fuzzy AHP model was validated by applying it to three scenarios in

NASA KSC Ground Support Operations regarding various case studies and historical data. The design of the experiment was a repeated measures analysis to evaluate three scenarios in NASAKSC Ground Support Operations. As a result, the predicted value was compared to the accepted value for each subject. The results from this model application confirmed that the predicted value and accepted value for the likelihood rating were similar. The three scenarios were Shuttle Landing Facility (SLF), Launch Complex Payloads (LCP), and Vehicle Assembly Building (VAB). The percentage error for the three scenarios was 0%, 33%, 0% respectively. The Kendall Coefficient of Concordance for assessment agreement between and within the subjects was significantly 1.00. Therefore, the appraisers are applying essentially the same standard when evaluating the scenarios. Multiple descriptive statistics for a 95% confidence interval and t-test are the following: coefficient of variation (21.36), variance (0.251), mean (2.34), and standard deviation (0.501). The results indicate there is minimal variability with fuzzy AHP modeling. As result, model evaluation and validation indicates that there is no difference between the current accepted NASA model and developed fuzzy AHP model. Future research includes developing fall protection guidelines.

Keywords: Human Factors Integration, Falls, Risk Factors, Analytical Hierarchy Process (AHP), fuzzy model, NASA/KSC Ground Support Operations

INTRODUCTION

NASA KSC environments such as facility maintenance, Space Shuttle operations, payloads, cranes, construction, and roofing are areas of concern for fall hazards. To address the issue NASA contracted with Gravitec Systems Inc., a fall-protection engineering firm, who assessed over 400 elevated work areas concerning fall hazards. Gravitec developed a hazard ranking system based on the assumption that multiple risk factors such as human factors, environmental factors, and working conditions have a uniform influence on falls. KSC environments such as facility maintenance, Space Shuttle operations, payloads, cranes, construction, and roofing are areas of concern for fall hazards. Under the umbrella of system safety, falls are events where an individual comes to rest unintentionally to a lower level and not by result of intrinsic event such as a stroke. An Ishikawa "Fishbone" Diagram (see Figure 1), was used as the conceptual model to represent the factors that lead to falls in NASA/KSC Ground Support Operations. The fishbone diagram identifies, sorts, and displays possible causes of a problem or quality characteristic. The cause and effect diagram displays the number of errors for the various risk factors that contribute to falls. There are extrinsic and intrinsic factors that contribute to falls. Extrinsic factors are characteristics from the outside. Intrinsic factors are original causes and characteristics within the human body. The extrinsic factors are organizational and environmental. The intrinsic factors are human/personal and task related.

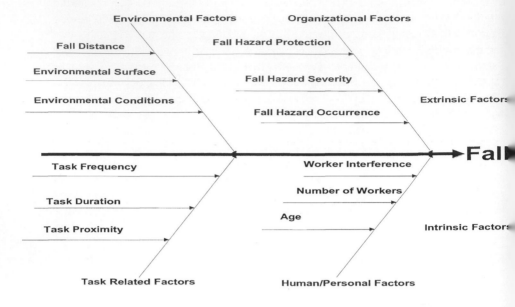

Figure 1. Conceptual Model

METHODOLOGY

The methodology (see Figure 3) in this study included twelve phases: knowledge acquisition, data collection, subject matter experts' interviews, Analytical Hierarchy Process (AHP), weight validation, fuzzification of variables, membership function development, fuzzy qualification using fuzzy set theory, fuzzy quantification using fuzzy set theory, model development, model usability, and model validation (Dagdeviren et al., 2008). The research hypotheses are: (1) the development of a conceptual model that characterizes risk factors can be useful in reducing the likelihood of falls in NASA Ground Support Operations; and (2) a fuzzy Analytical Hierarchy Process model can be developed and validated to predict the likelihood of falls in NASA Ground Support Operations. Multiple risk factors that contribute to falls in NASA Ground Support Operations are task related, human/personal, environmental, and organizational. The following Schematic Diagram represents the proposed model based on the fuzzy Analytical Hierarchy Process models.

RESULTS AND DISCUSSION

The Analytical Hierarchy Process (AHP) is a thorough mathematically process for prioritization and decision-making. Six subject matter experts were asked to participate in a voting system involving a survey where they judge risk factors using the fundamental AHP pairwise comparison scale. The results were analyzed and synthesized using Expert Choice Software, which produced the relative weights for the risk factors (see Figure 2). The following are relative weights for these risk factors: Task Related (0.314), Human/Personal (0.307), Environmental (0.248), and Organizational (0.130). The overall inconsistency ratio for all risk factors was 0.01, which indicates the model results were acceptable.

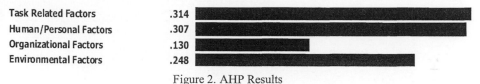

Figure 2. AHP Results

Weight Validation

The weights were validated by having two different teams of subject matter experts (SMEs) create priority vectors separately and confirm the weights are similar (see Table 1). Priority vectors are the average of the weights for each risk factor that lead to the falls. The validation is confirming that task related factors are a high priority at NASA due to the high-risk environment, schedule driven tasks, and the demanding task performance for safety.

Table 1: Weight Validation

Risk Factor	Priority Vector	Relative Weight	Rank
Task Related	0.304	0.314	1
Human/ Personal	0.302	0.307	2
Environmental	0.21	0.248	3
Organizational	0.17	0.13	4

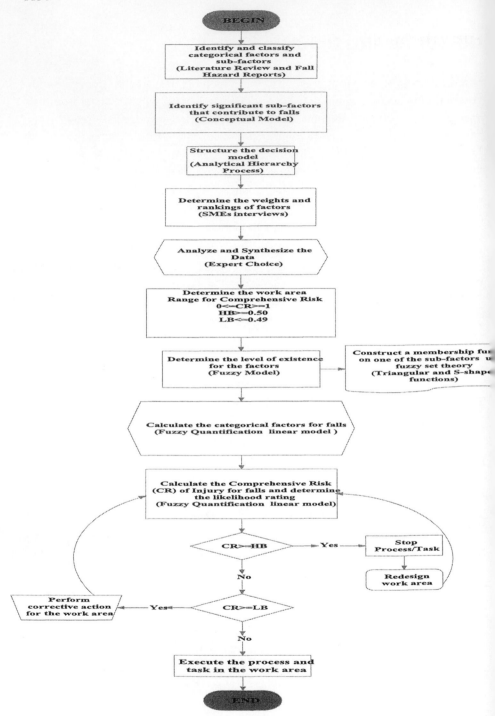

Figure 3. Schematic Diagram

Fuzzy Set Theory

Fuzzy Set Theory (FST) is a modeling technique frequently used where vague concepts and imprecise data are handled, and it is capable of managing both imprecision and uncertainty data (Bonissone, 1980). FST has been used for the development of the linguistic approach where any variable is treated as a linguistic variable (i.e. Low, Medium, and High). FST can be used to translate linguistic terms into numeric values to be used to get aggregate measures when given several inputs. FST characterizes the concept of approximation based on membership functions with a range between 0 and 1, inclusive, which provides the lower and upper approximations of a concept (Yao, 1992). Zimmermann identifies the necessity to use mathematical language to map several membership functions and develop FST models (Zimmermann, 1991).

Fuzzy AHP Model

The following fuzzy model uses the Fuzzy Quantification Linear equations (Bell-McCauley and Badiru, 1996).

Fuzzy Quantification Linear Models are shown below (Terano, 1992).

Task Related Risk:

$$X_1 = F(TR) = a_1 w_1 + a_2 w_2 + a_3 w_3 \qquad (1)$$

Human/Personal Risk:

$$X_2 = F(HP) = b_1 z_1 + b_2 z_2 + b_3 z_3 \qquad (2)$$

Organizational Risk:

$$X_3 = F(O) = c_1 u_1 + c_2 u_2 + c_3 u_3 \qquad (3)$$

Environmental Risk:

$$X_4 = F(E) = d_1 v_1 + d_2 v_2 + d_3 v_3 \qquad (4)$$

where risk sub-factors relative weights are

a = task related

b = human/personal

c = organizational risk

d = environmental risk

and risk sub-factors levels of existence are

w = task related

z = human/personal

u = organizational

v = environmental

The following equation was used to quantify the comprehensive risk of a fall as a result of all three categories:

$$Y = e_1 X_1 + e_2 X_2 + e_3 X_3 + e_4 X_4 \quad (5)$$

where Y   = comprehensive risk for the given condition

e_1 = weighting factor for the task related factors

e_2 = weighting factor for the human/personal factors

e_3 = weighting factor for the organizational factors

e_4 = weighting factor for the environmental factors

The weighting factors (e_1, e_2, e_3, e_4) represent the relative significance of the given risk factor category's contribution to the likelihood of injury.

The comprehensive risk is the aggregate value for the prediction of a (see Table 2), which is equal to the product of relative weight respective to the categorical risk factors.

Table 2: Aggregate Risk Value

Aggregate Risk Value	Risk Association	Likelihood Rating
0.00 - 0.20	Very Low risk	1
0.21 - 0.40	Low risk	2
0.41 - 0.60	Moderate risk	3
0.61 - 0.80	High risk	4
0.81 - 1.00	Very high risk	5

Model Usability and Statistical Analysis

The model usability involved a repeated measures analysis experimental design of fifteen subjects applying the fuzzy AHP model to three scenarios associated with falls in facilities or areas used in NASA Ground Support Operations. The three facilities or areas were the Shuttle Landing Facility (SLF), Launch Complex Payloads (LCP), and Vehicle Assembly Building (VAB). The predicted and accepted values for three scenarios (see Table 7). Then, a NASA safety manager compared the predicted value to the accepted value. The results from this model application confirmed that the predicted value and accepted value for the likelihood rating were similar. Equation 7 was used to calculate the percentage error for the three scenarios.

$$\%error = \frac{(predicted - accepted)}{accepted}$$ (6)

Table 3: Percentage Error

Scenario	Percentage Error
SLF	0%
LCP	33%
VAB	0%

The statistical analysis included tests for agreement of data and variability (Siegel, 1988). The test for data agreement was the Kendall Coefficient of Concordance (KC). The KC for assessment agreement between and within the subjects was significantly 1.00. Therefore, the appraisers are applying essentially the same standard when evaluating the scenarios. Multiple descriptive statistics for a 95% confidence interval and t-test are the following: coefficient of variation (21.36), variance (0.251), mean (2.34), and standard deviation (0.501). Therefore, there is minimal variability with fuzzy modeling. Table 6 shows the Statistical Analysis results. The statistical results show that the overall KC is 1.00, which indicates the outstanding high degree of agreement between and within the subjects. Because the p-values are greater than the alpha level (0.05) for all subjects, the null hypothesis is accepted. Agreement within the subject is due to a chance that the p-value provides the likelihood of obtaining a sample. However, the results for the subject indicate a slight disagreement for one of the three scenarios. The rationale for this outlier is that the subject probably did not understand the scenario. As a result, there is a relative agreement among the subjects in the likelihood of falls.

Model Validation

Model validation was the guarantee of agreement with the NASA standard. The model validation process was partitioned into three components: reliability, objectivity, and consistency. The model was validated by comparing the fuzzy AHP model to the NASA accepted model. The results indicate there was minimal variability with fuzzy AHP modeling (see Table 4). As a result, the fuzzy AHP model is confirmed valid.

Table 4: Model Validation

Scenario	Risk Value (Y)	Fuzzy AHP Model Rating Predicted	NASA/KSC Model Rating Accepted
SLF	0.404	2	2
LCP	0.351	2	3
VAB	0.451	3	3

CONCLUSIONS

In conclusion, the results indicate there is minimal variability with fuzzy modeling. The research focused on multiple risk factors that contribute to falls in an aerospace environment. The research emphasized the importance of system safety with respect to falls and concentrated on interdependence of falls. Ultimately, the result of the research was a systematic analysis for fall prevention that will lead to fall protection guidelines. The research hypotheses were validated by conceptual model, mathematical model, and the statistical analysis results. The research findings indicated that having cognizance of risk factors that lead to falls is beneficial and could prevent the likelihoods of falls in NASA Ground Support Operations. The results from the fuzzy AHP model were compared with the NASA accepted scale for the prediction of fall hazards. A fuzzy AHP model was developed and validated in the research.

REFERENCES

Bonissone, P. P. (1980), *"A fuzzy sets based linguistic approach: theory and applications,"* Proc. 12th Winter Simulation Conference, Orlando, Florida, 99-111.

Dagdeviren, M., Yuksel, I. (2008), "Developing a fuzzy analytical hierarchy process (AHP) model for behavior-based safety management," *Information Sciences*, 178, 1717-1733

McCauley-Bell, P., Badiru, A. (1996), "Fuzzy Modeling and Analytic Hierarchy Processing—Means to Quantify Risk Levels Associated with Occupational Injuries—Part II: The Development of a Fuzzy Rule-Based Model for the Prediction of Injury," *IEEE Transactions on Fuzzy Systems*, 4(2), 132-138.

Siegel, S., Castellan, N. (1988). Nonparametric Statistics for the Behavioral Sciences, 2nd ed., Columbus, OH: McGraw-Hill.

Terano, T., Asia, K. and Sugeno, M. (1992). *Fuzzy Systems Theory and Its Applications.* Boston, MA: Academic Publishers.

Yao ,Y. Y., Wong S., (1992), *"A decision theoretic framework for approximating concepts,"* International Journal of Man-machine studies, 37(6), 793-809.

Zimmermann, H. J., (1991). *Fuzzy Set Theory and its applications*, 2nd ed., Boston, MA: Academic Publishers

Chapter 118

Research on Astronaut Training Methods for Complicated Operation

Yijing Zhang, Xiang Zhang, Min Liu, Quanpeng Wang, Bin Wu*

State Key Laboratory of Space Medicine
Fundamental and Application
China Astronaut Research and Training Center
Beijing 100094, China

ABSTRACT

This paper is to explore suitable training methods for complicated operation in spaceflight. Three training method named Routine method, Compounding-feedback method and Integrated method were proposed from theory analysis and astronaut training experience. To validate these methods, two-factor experiment was designed. The two factors are training methods and operation complexity. Twenty-eight participants were divided into three method groups to finish three stages experiment. The average operation time, the average error, the subjective workload and the subjective survey of the first stage and the final stage were analyzed. Results show that different training methods impose varied effects on the two complexity levels at the three stages. With the increase of training method level (from Routine method to Compounding-feedback method and Integrated method), the average operation time of two complexity levels do not change significantly, neither do the average error of low-complexity level. However, compared with Routine method group, the average error of high-complexity level show a significant decrease in Compounding-feedback method group and Integrated method group (p=0.02, 0.039 for the first stage and the final stage respectively). The subjective workloads at both stages also show a degressive trend. Additionally, the subjective survey showed Compounding-feedback method and Integrated method are more effective on high-complexity operations than on low-complexity operations. The conclusion is that Compounding-feedback method and Integrated method are suitable to high-complexity operations.

Keywords: Astronaut training, Training method, Operation complexity, Spaceflight operation, Operation performance

INTRODUCTION

Astronaut training always serves as an essential part for preparing real spaceflight missions (Shayler, 2005; Morphew, 2001). With the development of future manned-spaceflight missions in China, such as long-term spaceflight, deep space exploration, the spaceflight operations become more difficult and complicated. Therefore, astronaut training is facing a new issue that is how to match the requirement of complex operations with limited resource. To gain high efficiency in astronaut training, it would be an effective way to apply different training methods to the operations with different complexity. This paper aims to explore suitable and efficient training methods for complicated operations in spaceflight.

TRAINING METHODS AND OPERATION COMPLEXITY MEASURES

TRAINING METHODS

Three training methods were proposed from astronaut training experience, theory analysis and Chinese astronaut surveys. They were named as Routine method, Compounding-feedback method and Integrated method.

Routine method is a general training method of astronaut training. It come from traditional education method and includes two parts: theory training and practice training. The two training part are usually used in astronaut training practice alternately.

Compounding-feedback method is based on feedback theory, feedforward theory and error management (Nalanagula et al., 2006; Laemlaksakul & Kaewkuekool, 2007; Thomas, 2003). This method includes two training form; one is feedforward and the other is feedback. Feedforward focus on error-prevention with giving operators pre-information about the common error and operation experience of Chinese astronauts. Feedback focus on error-correction with feeding them with their own performance and error analysis. Feedforward training will be applied to operators only in the first training stage and feedback training will be applied throughout the training procedure.

Integrated method is composed of compounding-feedback, mental imagery (Liao, 2008) and group discussion (Gall & Gall, 2003). According to the basic theory, the compounding-feedback can improve the operation reliability; mental imagery can help operators build up their psychological operation model; groups discuss help operators learn operations from different views and put them into a "mind storm".

OPERATION COMPLEXITY MEASURES

To assign different resources and training methods to different complicated operations, the complexity of spaceflight operations should be determined firstly. Therefore, two spaceflight operation complexity measures were developed. One is the operation complexity measure based on entropy (Zhang et al., 2009a); the other is the Weighted Halstead Measure (Zhang et al., 2009b).

The two complexity measures have been validated by correlation/regression analysis between complexity values and performance values with astronaut training data and the experiment data (Zhang et al., 2009a, 2009b).

The Operation Complexity Measure Based on Entropy

This operation complexity measure is proposed using a weighted Euclidean norm based on four factors: complexity of operation step size (COSS), complexity of operation logic structure (COLS), complexity of operation instrument information (COII), and complexity of space mission information (CSMI).

The development of the operation complexity measure followed four steps. First, four factors were identified to be reflected in the operation complexity measure for spaceflight. Second, the entropy theory for a graph was adopted to measure the four factors (Mowshowitz, 1967). Then, the weights of the four factors were determined based on a questionnaire survey of Chinese astronauts. Finally, the operation complexity values of spaceflight operations were determined by the weighted Euclidean norm of the four factors (Zhang et al., 2009a).

The Weighted Halstead Measure

This measure is based on Halstead method which indicated that one software program with more operators and operands has a more complicated procedure structure (Halstead, 1977). For the spaceflight operation, operator is defined as the action of astronauts; operand is defined as the instrument that is operated by astronauts. To reflect the complexity of spaceflight operations, the number of weighted operators and weighted operands were used in Halstead Volume formula. The operators were weighted based on the McCracken-Aldrich (1984) theory; the operands were weighted according to the complex level of operation interface. The measure can evaluate operation complexity from two aspects: the operation activity and the cognitive activity (Zhang et al., 2009b).

THE EXPERIMENT

DESIGN

In this experiment, the independent variable was training method and operation complexity. Training methods has three levels: Routine method, Compounding-feedback method and Integrated method; operation complexity have two levels: high-complexity and low-complexity.

Both objective and subjective indexes were used to evaluate the influence of training method and operation complexity. The objective indexes included: (1) *Operation time*: the time in which the participant succeeded in finishing the operations of one complexity level; (2) *Error count*: the total number of error happened in performing the operations of one complexity level. The subjective index was *Subjective Workload*, which was evaluated for the whole experiment operations by the NASA Task Load Index (NASA- TLX) questionnaire. *Subjective survey* was also used to evaluate the training outcome and effect.

This experiment had a within-participants design for operation complexity and a between-participants design for training method. Twenty-eight participants were divided into three training method groups. Every participant performed nine spaceflight operation units which were classified to two complexity levels.

PARTICIPANTS

Twenty-eight male volunteers from Tsinghua University were recruited as participants for this experiment. They all had college education or above in aeronautics and astronautics. These backgrounds were close to the standard for future astronaut selection. Their ages ranged from 21 to 26 years old (Mean =23.1, SD =1.57). They participated in the experiment based on signed consent forms.

EXPERIMENT TASK

The experiment tasks included nine spaceflight operation units with complexity values ranging from 0.9 to 1.7 (based on the measure of entropy) and 23.4 to 293(based on Weighted Halstead measure). One operation unit was the activity group for handling one malfunction of spacecraft system. Operation complexity values below 25 percentile and above 75 percentile were used as the criteria for the differentiation into "low-complexity" and "high-complexity" operations. Based on this value criterion, the two complexity levels were determined in this paper and each of them included three operation units. The nine operation units appeared in random sequence for each participant of one group. A participant finished a trial when he completed all the nine operation units.

FACILITIES

The experiment platform was a simulated spacecraft instrument system, which included the spacecraft instrument panel and the software modules (training support system). The training support can record important operation actions and the corresponding times. Another software program was developed for this experiment to record the time required to perform every operation unit. The whole experiment procedures were recorded by a video-recording system.

PROCEDURE

The experiment was carried out in China Astronaut Training Centre and had three phases. The first phase was a pilot experiment. The whole experiment procures were tested to make sure that the training method can be carried out smoothly and the facilities works in a reliable status.

The second phase was participant screening and pre-training. After explanation of the experiment, totally 33 male volunteers took cognitive ability tests and pre-training. Finally, 28 participants were admitted to participate in formal experiment. In this phase, the participants were asked to provide their personal information such as age and education background.

The third phase was the formal experiment. The whole experiment included three stages experiment, which corresponded to elementary training, advanced training, pre-mission training in astronaut training respectively. The three stages were conducted seven days apart. In every stage, training and practice were conducted alternately. The practice has two kinds of forms: one is an exercise with the help of operation manual of nine operation units; the other is a test without any manual. The fist stage included 4 trails of exercises. The second stage included 4 trails of exercises and 2 trails of test. The final stage included 3 trails of exercises and 3 trails of exam. After each stage of the experiment, they filled the NASA-TLI questionnaire and answered some questions related to their learning procedure.

DATA ANALYSIS

SPSS 15.0 was used for the data analysis in this study. The effects of training method at the two operation complexity levels at the first and final stages were analyzed in this section. The operation performance data did not pass the normality and homogeneity of variances tests, so multi-way ANOVA method was not applied to analyze the effects of all the two variables and their interactions together. The effects of training methods were analyzed under each of the two operation complexity levels separately. One-way ANOVA, nonparametric tests (Kruskal-Wallis Test), or paired-T test were applied according to the characteristics of the data samples. The results of subjective workload and subjective survey are also presented in this section.

AVERAGE OPERATION TIME

The average operation time of two complexity operations at the first stage and the final stage were analyzed in this section.

The first stage

At the first stage, the average operation time of high-complexity operations couldn't pass the normality and homogeneity of variances tests, and was explored by nonparametric tests (Kruskal-Wallis Test). The result were shown in Table 1 and showed that the average operation time did not have significant difference among three training method groups ($p=0.859$).

The average operation time of low-complexity operations passed the normality and homogeneity of variances tests, and was explored by ANOVA method. The result showed that the average operation time did not have significant difference among three training method groups ($p=0.940$).

Table 1. The analysis result of average operation time at the first stage

Variable	Routine method		Compounding-feedback method		Integrated method		F $(2, 25)$	p
	M	SD	M	SD	M	SD		
High-complexity*	407	99.6	386	62.3	398	44.4	$\chi^2=0.304$	0.859
Low-complexity	58.6	18.8	59.1	9.21	57.0	11.1	0.062	0.940

Note:* non-parameter analysis Kruskal-Wallis Test.

The final stage

At the final stage, the average operation time of high-complexity operations couldn't pass the normality and homogeneity of variances tests, and was explored by nonparametric tests (Kruskal-Wallis Test). The result showed that the average operation time did not have a significant difference among three training method groups ($p=0.559$) (see Table 2).

The average operation time of low-complexity operations passed the normality and homogeneity of variances tests, and was explored by ANOVA method. The result showed that the average operation time did not have significant difference among three training methods groups ($p=0.657$) (see Table 2).

AVERAGE ERROR COUNT

The average error count of two complexity operations at the first stage and the final stage were analyzed in this section.

Table 2. The analysis result of average operation time at the final stage

Variable	Routine method		Compounding-feedback method		Integrated method		F (2, 25)	p
	M	SD	M	SD	M	SD		
High-complexity*	205	31.3	198	25.5	213	31.6	χ^2=0.596	0.559
Low-complexity	26.7	3.98	29.0	6.87	28.4	5.04	0.427	0.657

Note:* non-parameter analysis Kruskal-Wallis Test.

The first stage

At the first stage, the average error count of high-complexity operations passed the normality and homogeneity of variances tests, and was analyzed by ANOVA (non-equal variances) test. The results showed that the training method imposed a significant effect on average error count of high-complexity operations (p=0.006) (see Table 3). Compared with that in Routine method group, the average error count showed a significant decrease in Compounding-feedback method group (p=0.004) and Integrated method group (p=0.013). The average error count did not show a significant difference between Compounding-feedback method group and Integrated method group (p=0.364).

Table 3. The analysis result of average error count at the first stage

Variable	Routine method		Compounding-feedback method		Integrated method		F (2, 25)	p
	M	SD	M	SD	M	SD		
High-complexity	3.44	2.33	0.90	0.636	1.17	0.650	10.39	0.006
Low-complexity*	0.639	0.435	0.250	0.167	0.389	0.486	χ^2=4.086	0.13

Note:* non-parameter analysis Kruskal-Wallis Test.

The average error count of low-complexity operations could not pass the normality and homogeneity of variances tests, and explored by nonparametric tests (Kruskal-Wallis Test). No significant difference for the average error count was found in the statistical result among three training method groups (p=0.13) (see Table 3).

The final stage

At the final stage, the average error count of high-complexity operations could not pass the normality and homogeneity of variances tests, and was explored by nonparametric tests (Kruskal-Wallis Test). The result (as shown in Table 4) showed that the training method imposed a significant effect on the average error count of high-complexity operations (p=0.034). To be specific, the average error count of Routine method group was significantly higher than that of Compounding-feedback method group (p=0.024), but it showed no significant difference with that of Integrated method group (p=0.505).

The average error count of low-complexity operations could not pass the normality and homogeneity of variances tests, and was explored by nonparametric tests (Kruskal-Wallis Test). No significant difference for the average error count was found in the statistical result among three training method groups (p=0.855) (see Table 4).

Table 4: the analysis result of average error count at the final stage

Variable	Routine method		Compounding-feedback method		Integrated method		χ^2	p
	M	SD	M	SD	M	SD		
High-complexity	2.04	1.49	0.700	0.554	1.52	1.11	6.765	0.034
Low-complexity	0.074	0.222	0.067	0.141	0.037	0.111	0.314	0.855

SUBJECTIVE RESULT

Subjective workload was evaluated for the whole experiment task and subjective survey was carried out after each stage of the experiment.

The average subjective workload at the first stage could not pass the normality and homogeneity of variances tests, and was explored by nonparametric tests (Kruskal-Wallis Test). The average subjective workload at the final stage passed the normality and homogeneity of variances tests, and was analyzed by one-way ANOVA. The results showed that the average subjective workload at the first stage do not change significantly among three method groups (χ^2=0.537, p=0.765), neither do the average subjective workload at the final stage (F=1.384, p=0.269). Though the statistical analysis did not show any difference among the three method groups at both the first and final stages, the average subjective workload showed a decline trend from Routine method group to Compounding-feedback method and Integrated method groups.

Additionally, the subjective surveys were carried out as a supplement for the data result. The participants in Integrated method group agreed with the decline trend of cognitive load with the help of the training method. And the subjective survey also showed that Compounding-feedback method and Integrated method were more effective on high-complexity operations than on low-complexity operations.

DISCUSSIONS

In this study, the average operation time, the average error count and the subjective workload of the first stage and the final stage were analyzed. Results show that different training methods impose varied effects on the two complexity levels at the two stages. With the increase of training method level (from Routine method to Compounding-feedback method and Integrated method), the average operation time of two complexity levels do not change significantly, neither do the average error count of low-complexity level. However, compared with Routine method group, the average error count of high-complexity level shows a significant decrease in Compounding-feedback method group and Integrated method group. The subjective workload at the first stage and the final stage also show a depressive trend.

In the first stage, the information of common error and operation experience of astronauts were given to the participants in Compounding-feedback method and Integrated method groups in the beginning of training. However, this feedforward training cannot help the participants improve their operation performance apparently but make them confused with the extra information. On the other hand, the feedback training part was considered as an efficient way to lower the error rate by feeding the participants with their own error information instantly according to the participants' view in both Compounding-feedback method group and Integrated method groups.

At the final stage, the compounding effect of the training methods and practice lead to the outcome that the operation time does not show any difference among three training methods. The participants in Integrated method group agreed with the decline trend in subjective workload. In addition, they thought that Integrated method could give them psychological support and keep them in a peaceful mood.

CONCLUSIONS

The conclusion is that Compounding-feedback method and Integrated method are more suitable to high-complexity operations than Routine method. Compared with Compounding-feedback method, Integrated method does not show a significant advantages on the operation performance. But, it is indicated by the workload evaluation and the subjective survey that Integrated method is helpful to maintain the operation performance and lower the workload. This point will be tested in further research.

REFERENCES

Gall, M. D. and Gall, J. P. (1976), "The discussion Method." In: N. L. Gage (Ed.), *The Psychology of Teaching Method,* pp. 166-216

Halstead, M. H. (1977), *Elements of software Science.* Amsterdam: Elsevier Press.

Laemlaksakul, V., and Kaewkuekool, S. (2007), *The effect of feedforward training to improve inspector performance.* IEEM 2007, pp. 423-426

Liao, J. L. (2008), "On the Imagery Training for Developing the Ability of Space Orientation for the Flight Student". *Journal of PLA Institute of Physical Education,* Vol. 27, No. 3, pp. 92-94

McCracken, J. H. and Aldrich, T. B. (1984), "Analyses of selected LHX mission functions implications for operator workload and system automation goals." *Human Factors Research in Aircrew Performance and Training.* MDA903-81-C-0504 ASI479-024-84

Morphew, M. E. "Psychological and Human Factors in Long Duration Spaceflight", McGill. *Journal of Medicine,* Vol. 6, No. 1, pp. 74-80

Mowshowitz A. (1967), Entropy and the complexity of graphs. University of Michigan, *Ph.D. Dissertation.* pp. 101-113

Nalanagula, D., Greenstein, J. S., and Granopadhye, A. K. (2006), "Evaluation of the effect of feedward training displays of search strategy on visual search performance." *International Journal of industrial Ergonomics,* Vol. 36, pp. 289-300

Shayler, D. J. (translated by Yuan, J. and Zhen, M.). (2005), *Disasters and Accidents in Manned Spaceflight.* Beijing: China Astronautics Press, pp. 265–282

Thomas, M. J. W. (2003), *Instructional use of error: the challenges facing effective error management in aviation training.* In Proceedings of the Sixth International Aviation Psychology Symposium, Sydney, Australia: Australian Aviation Psychology Association.

Zhang, Y. J., Wu, B., Wang, Y., et al. (2008), "Determination of index weights in the operation complexity model of spaceflight." *Space Medicine & Medical Engineering,* Vol. 21, No. 3, pp. 252-256

Zhang, Y. J., Li, Z. Z., Wu, B., and Wu, S. (2009a), "A Spaceflight Operation Complexity Measure and its Experimental Validation." *International Journal of Industrial Ergonomics,* Vol. 39, No. 5, pp. 756-765

Zhang, Y. J., Shi, L. J., Li, Z. Z., Wu, B., and Xu, S. (2009b), *Validation and Application of Spaceflight Operation Complexity Measure in Chinese Astronauts Training of Extra Vehicular Activity Mission.* The 60th International Astronautical Congress (IAC), Daejeon, South Korea, October 12-16, 2009.

* Corresponding author. E-mail address: **pubacc@126.com**

Chapter 119

The Preventive Strategies for the Third Industrial Accident in Korea

Youngsig Kang[1], Sunghwan Yang[2], Taegu Kim[3], Daysung Kim[4]*

[1]Dept. of Occupational Health & Safety Engineering, SEMYUNG University, Jecheon, Chungbuk, 390-711, South KOREA

[2]Dept. of Prosthetics & Orthotics, Korea National College of Rehabilitation and Welfare, Pyeongtaek, Gyeonggi, 459-070 South KOREA

[3*]Dep. of Occupational Health & Safety Engineering, College of Biomedical Science & Engineering, INJE University, Gimhae Gyeongnam, 621-749, South KOREA

[4]Occupational Safety & Health Research Institute, KOSHA, Incheon 403-711, South KOREA

Abstract

It's very important to evaluate the priority of prevention factor and strategies to minimize industrial accidents. It provides decisive information for prevention of industrial accidents and safety management. Therefore, this paper is proposing the evaluation of priority and concrete methods through statistical testing of prevention factors with a cause analysis in a cause and effects model and of existing strategies. This paper uses a priority matrix criterion to apply the rank and objectivity of questionnaire results. Also, this paper uses regression method (RA), exponential smoothing method (ESM), double exponential smoothing method (DESM), autoregressive integrated moving average (ARIMA) model and proposed analytical function method for trend analysis of accident data into the future. This paper standardizes questionnaire results of workers and managers in manufacturing and construction companies with less than 300 employees, located in the central Korean metropolitan areas where fatal accidents have occurred. Finally, this paper provides one method to construct safety management for industrial accident prevention and a forecasting method for industrial accident rates and fatality rates for occupational accidents per 10,000 persons.

Keywords: Prevention Factor and Strategies, Cause and Effects Model, Priority Matrix Criterion, Fatal Accidents, Safety Management, Forecasting Method

Introduction

Recently, the industrial accident rate is 0.71, which is a decrease of 1.4% compared with last year with 95,806 employment injury in Korea in 2008. The fatality rate of occupational accidents per 10,000 persons is 1.80 which is a decrease of 6.3% compared with last year with 2,422 persons dying. In addition, 1,448 persons died due to occupational diseases so the fatality rate of occupational related causes per 10,000 persons is 1.07, which is a decreased of 2.7% compared with last year. But because of inflation, the total cost including direct cost and indirect cost is 18.2 trillion won, which is an increase of 12.3% compared with last year.

Recording industrial accidents from the standpoint of number of occurrences, manufacturing companies are the highest with 37.4% or 35,819 persons and construction companies are second with 21.4% or 20,473 persons in 2008. Therefore, manufacturing companies and construction companies account for 58.8% of the industrial accidents. So, strategic method for industrial accident prevention in these two groups is required urgently. In particular 92% of occurrences and 83.7% of total deaths were in businesses with less than 300 employees.

Recording occupational accidents by type of occurrence, there were 18,527 persons involved in slips, 15,250 persons involved with stricture, 14,027 persons involved in falls, 8,670 persons involved in falling and flying, and 7,279 persons involved in collisions accounting for 74.1% of all occurrences in 2008. Falls, slips and strictures account for 55.5% of all occurrences.

Examining occupational diseases for 6 years from 2000 to 2005, work-related musculoskeletal disorders (WMSDs) accounted for 50%. So, the administration and laws for preventing WMSDs were strengthened and job analysis evaluation should be done systematically for work places with less than 50 persons.

Looking at the trend of industrial accident rates in Korea for 11 years from a general point of view, the industrial accident rate is static at 0.7.

Therefore, this paper intends to propose methods for minimizing industrial accidents by analysis and prioritization through statistical testing using questionnaires after developing prevention factors and strategies through cause analysis of industrial accidents for manufacturing companies and construction companies

Subjects and Method

Background

In Korea research reports exist covering the analysis of the causes of industrial accidents. A paper on the analysis of accident causes by type of occurrence and original object involved proposes that fatal accidents in construction companies can be cut by more than half if there is protective equipment, safety inspections before work begins and safety supervision (Park, 1997). In a paper classifying accidents in small businesses with less than 30 employees, more accidents were identified as occurring to workers in their 20's, to unskilled workers with less than 1 year experience, and on Saturdays.

Unsafe acts involving unsafe posture and motion were identified as the leading direct cause. This was followed by unsafe workplace conditions and environmental defects. The lack of safety management was the highest indirect cause (Kim, 1998). The study asserted that the solutions required safety education for young staff, improvement in posture, and development of systematic safety management programs (Kim, 1998).

An analysis of industrial accidents with fuzzy inference shows a lack of safety consciousness, indifference, conceit about safety, insufficient safety management and passive countermeasures for potential hazard as leading causes. The report concluded that the accident rate was low where the company and its staff had high safety awareness (Pae and Park, 2003).

Industrial accident aboard fishing vessels is caused by workers lack of qualifications and knowledge, insufficient safety management system and faults in the vessel itself (Lee and Jang, 2005). The solution proposed in the report was for ship owners to recognize the need for safety awareness by the workers and provide periodic education. A case study safety management text book should be developed with the contents made part of the test for the seaman engineer's license. The report also proposed that the safety management system and the emergency rescue system be developed as well as strengthening the vessel test procedures to remove accident causing factors (Lee and Jang, 2005).

A behavioral science viewpoint analyzed the intensity of stress factors through normal testing of several Korean life change unit (LCU) models developed from life stress factors adapted to actual life in Korea (Kang and Yang, 2006). A comparison of minor and major industrial accidents using 5 standard causes - work load, inadequate training, operating procedure, lack of knowledge and ignorance of danger - was made using statistical analysis. The results showed that the causes of industrial accidents in these two groups were not different, so there should be no differentiation in analyzing the two groups (Gyeke, 2003). In the analysis of cause and effect in the Bofal gas water leak, alarm absence, chemical potential hazard, insufficient system functionality, insufficient equipment, misinformation, and inadequate maintenance are causes of industrial accidents (Eckerman, 2005). The method for solving this to remove injury and risk is through a logical framework related to education, alarm, development of adequate maintenance systems, and emergency action plans (Eckerman, 2005).

In Korea, the prevention of fatal accidents in small and medium sized companies is being developed by using a Korean LCU model with life stress factors adapted to actual Korean life and proposing systematic management method for this (Kang, et al., 2008). In Korea, among the research for prevention of industrial accidents based on basic cause analysis, there is little research proposing synthesis of solutions for all companies. The methods for grasping the cause of industrial accident and preventing industrial accidents are mainly for specialized companies or characteristic facilities. So, in Korea, the development of models for all companies by type of occupational accident and disease has to be done in future.

Promoting method for priority of prevention factors and strategies

For minimizing industrial accidents, it's important to evaluate the priority of prevention factors and strategies which systematically provides decisive information for establishing the method of accident prevention and safety management. Also, forecasting the industrial accident rate provides very useful information for the strategy of prevention of industrial accidents in the long term. Therefore, promoting a system for preventing industrial accident by prevention factors and strategies is shown from Figure 1.

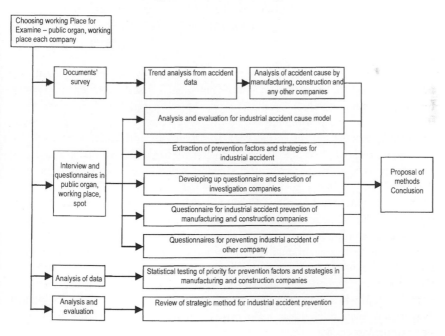

Figure 1. Industrial accident prevention method system based on prevention factors and strategies

Prevention and strategy by cause analysis of industrial accident

Recently, the government established a strategy to minimize accident deaths from traditional causes, such as falls, slips and strangulations, by 50%. Therefore, at the end of the third five-year plan for prevention of industrial accidents (2010-2014), a foundation for becoming an advanced country in occupational safety and health will be achieved with an industrial accident rate of 0.5 and occupational fatality rate of 0.76 per 10,000 persons.

Therefore, for preventing industrial accidents, this paper intends to draw prevention factors and strategies from existing cause and effect models with cause analysis and dramatically reduce industrial accidents by using priorities. So, the result of analyzing existing domestic/international cause and effect models and strategies, the number of prevention factors and strategies are set at 32.

Scope and Methodology

In Korea, 74% of the fatal accidents occur in the metropolitan and central areas. So, workers and managers in these areas are the subjects for the investigation. These areas include Seoul, Geyeonggi Province and Chungcheong Province. The main companies in the research are manufacturing companies and construction companies in places where staff is less than 300 persons. For developing the methods for prevention of industrial accidents systematically, the following procedure is for promoting priority evaluation of prevention factors and strategies.

First, forecasting of the industrial accident rate and occupational fatality rate per 10,000 persons has to be performed based on the existing data. Second, prevention factors and strategies of industrial accidents have to be developed from the cause analysis. Third, we create a questionnaire randomly arranging participating enterprise, prevention factors, and strategies. Fourth, the questionnaire is applied at enterprises in the metropolitan and central areas of Korea involved in manufacturing and construction companies with less than 300 persons. The subjects for the questionnaire are to be field workers and managers of safety and health. Fifth, under the subject of worker and manager of safety and health in middle area, questionnaire subjects have to be selected based on a simple random sampling method. Sixth, for securing objectivity of priorities from the questionnaire, priority matrix criterion is applied. Therefore, the sample response rate of prevention factors and strategies is measured from the principal diagonal based on the priority matrix. Then, the priority of each item is established. Seventh, an intensity analysis is performed to identify how much each prevention factor and strategy differs with the priority of prevention factors and strategies. For analyzing difference of intensity, normal testing has to be performed with a significance level of 0.05 (α=0.05) for sample response rate. Finally, based on the result of normal testing, the methods for minimizing industrial accidents are systematically established.

RESULT

In this paper, we used MFC (Microsoft Foundation Class) software for efficient forecasting of industrial accident rate and occupational fatality rate per 10,000 persons. Above all, with the industrial accident rate for 36 years, the forecasting for industrial accident rate for 10 years is shown in Figure 2. The Sum of Squared Errors (SSE) value of ARIMA Model in Figure 2 is 2.79. It's proved that ARIMA Model is the most desirable in case of forecasting the industrial accident rate in Korea. Also, in the case of maintaining the existing prevention strategy of industrial accidents, the industrial accident rate trend slowly becomes lower in Korea.

With existing material of occupational fatality rate per 10,000 persons for 18 years, the result of forecasting for 10 years in the future is shown in Figure 3. In case of forecasting occupational fatality rate per 10,000 persons, the SSE value obtained by doubling the exponential smoothing method is the lowest at 1.19. So, the double exponential smoothing method is proved to be most desirable.

Therefore, the death trend of occupational accidents appears to be stagnated for 10 years. Accordingly, epochal prevention strategies for industrial accidents have to be established and fulfilled in a sustainable direction.

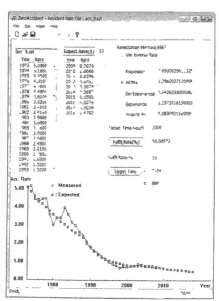

Figure 2. The forecasting of industrial accident rate for 10years

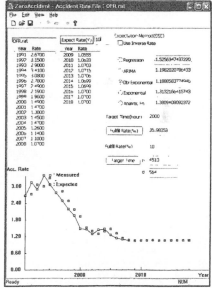

Figure 3. The forecasting of fatality rate for 10years

In this chapter, by making up 32 prevention factors and strategies for the questionnaire, the priorities were evaluated for workers and managers in the real field. The data for participating enterprises is the result of distributing questionnaires to 2,110 workers and managers in companies whose staff is less than 300 persons and collecting 1,545 questionnaires (collection rate : 73.2%).

1152

Table 1 is a priority evaluating model developed to match our actual life by evaluating prevention factors and strategies based on a priority matrix for dramatically minimizing industrial accidents for workers and managers of safety and health in middle regions of the country.

Therefore, with prevention factors and strategies, the priorities adapted to our actual life are analyzed leadership of company owner (1), prevention of stress (2), beginning of self-safety & health movement for conscious reform and reliable information from material safety data sheet (MSDS) (3), quantitative risk assessment (5), safety and health association (6), expansion of lockout/tagout for important tools (7), construction of OSHMS, enforcement of safety & health education, and hazard and risk prevention plan (8), small safety & health team (11) and to promote sufficient experts for prevention of musculoskeletal diseases (12). Among these strategies, leadership of company owner plays an important role to minimize accidents more than any prevention factor or any strategy.

Table 1. Priority model of prevention factors and strategies

rank	Prevention factors and strategies	frequency	Response rate
1	Leadership of company owner	307	0.199
2	Prevention of stress (job, working place, life stress, etc.)	124	0.080
3	Beginning of self-safety & health movement for conscious reform	86	0.056
3	Reliable information from material safety data sheet (MSDS)	86	0.056
5	Quantitative assessment of risk	85	0.055
6	Safety & health association	80	0.052
7	Expansion of lockout/tagout for important tools	73	0.047
8	Construction of occupational safety & health management system	66	0.043
8	Enforcement of safety & health education	66	0.043
8	Hazard and risk prevention plan	66	0.043
11	Small safety & health team	65	0.042
12	To promote sufficient experts for prevention of musculoskeletal Diseases	64	0.041
13	Evaluation of safety work plan daily and weekly	63	0.041
14	Industrial safety & health committee	62	0.040
15	Zero accident campaign	61	0.039
16	Enforcement of work study	60	0.039
17	High-five movement for fatal accident prevention	59	0.038
17	Process safety management (PSM) of chemical companies	59	0.030
19	Standardization of temporary structure construction	58	0.028
19	Safety management of aged workers	58	0.038
21	Enforcement of safety rules	56	0.036
22	Prevention of human error	55	0.024
23	Advanced technology using IT for accident prevention	53	0.034
24	Expansion of clean business with less than 50 workers	42	0.027
25	Increasing fines for a default	41	0.027
26	Education for children's safety & health	34	0.022
27	Safety education for foreign workers	33	0.021
28	Accident prevention to company picnic and athletic meetings	32	0.021
28	Expansion of safety & health research & development (R&D) cost	32	0.021
30	Enforcement of fall prevention education	28	0.018
31	Prevention of musculoskeletal diseases in service industries	26	0.017
32	Work-related diseases prevention of women workers	14	0.009

Especially, in Table 1 and 2, items with a significance level of 0.05 (α) are:
   (1) Leadership of company owner and prevention of stress(job, working place, life stress, etc)

(2) Prevention of stress and beginning of self-safety & health movement for conscious reform

(3) Beginning of self-safety & health movement for conscious reform and industrial safety & heal committee

(4) Industrial safety & heal committee and expansion of clean business with less than 50 workers

(5) Expansion of clean business with less than 50 workers and work-related diseases prevention of women workers

Table 2. The normal testing result by significance level (α =0.05)

Item of sample rate	Test statistic	Rejection region
$P_1 = P_2$	9.52	$Z \geq |\pm\ 1.96|$
$P_2 = P_3$	2.65	$Z \geq |\pm\ 1.96|$
$P_3 = P_{14}$	2.08	$Z \geq |\pm\ 1.96|$
$P_{14} = P_{24}$	2.01	$Z \geq |\pm\ 1.96|$
$P_{24} = P_{32}$	3.77	$Z \geq |\pm\ 1.96|$

The analysis of the difference in intensity of the sample response rate for prevention factors and strategies between these items was not significant.

Discussion and Conclusions

At present large companies have established OSHMS programs and have actively created systems for real time management of industrial accidents by recognizing the importance of safety and health in the work place. Also, small sized industries with less than 50 persons have actively performed clean business through government support.

From this standpoint, leadership of the company owner as ranking the first rank in the priority plays a decisive role in industrial accident prevention. So, for sustainable safety management from the top down, the number of managers studying safety and health needs to be enlarged as well as resurrection of laws and education of company owners with less than 50 staff must to be enlarged.

The main issue of advanced safety and health countries is stress. The results of this cause brain vein and heart disease, cancer, gastric ulcer, depression, tuberculosis, digestive trouble, and rheumatoid arthritis. Accordingly, more efficient methods for prevention of stress the second rank in the priority are required in national strategies. Also, our country has to search for fundamental solutions by expanding the focus to include the job, the workplace, and life stress for prevention of stress. Therefore, a Korean stress prevention model integrating job, workplace and life stress has to be developed and systematic prevention strategies have to be established for fundamental stress prevention.

The hazardous factors in the workplace are well known by workers in high risk work areas or processes. So, from the long-term standpoint, small safety & health team is an essential and core factor which has to be promoted for prevention of industrial accidents through beginning of self-safety & health movement for conscious reform in the workplace as the third rank in the priority. Our country's recognition and attitude for safety and health consciousness is insufficient in comparison with countries having advanced safety and health programs. Therefore, beginning of self-safety & health movement for conscious reform to minimize industrial accidents by establishment of safety culture should push consciousness education and improvement toward unconsciousness, indifference, disregard, ignorance, and recklessness. At the present, the reliable prevention information from harm from nano particles and safety of new materials by complex advanced factory automation are necessary. So, for reliable information from MSDS as the third rank in the priority, it is required definite proposal for existing/new MSDS as well as continuous management of a chemical material information system.

Advanced safety and health countries are introducing several incentives for quantitative assessment of risk as the fifth rank in the priority. Therefore, our country has to perform quantitative assessment of risk adapted to each workplace not only centered on huge companies, but also for workplaces with less than 50 workers.

For revitalization of the safety and health association as the sixth in the priority, it's gradually desirable to entrust related certification/verification toward association than government. Especially, the number of safety and health association has to be gradually increased with customized services to small and medium sized businesses for dramatically minimizing industrial accidents in the workplaces with less than 50 workers.

For expansion of lockout/tagout for important tools as the seventh in the priority, malfunction of equipment or non-existent safety equipment must be induced industrial accident in not only manufacturing companies but also in construction companies. Accordingly, installation of lockout/tagout for important tools program is urgently required for important equipment causing fatal accidents in workplaces with less than 50 workers.

At present, large companies have actively constructed OSHMS, but workplaces with less than 50 workers have limitations from a cost standpoint in establishing OSHMS. So, for construction of OSHMS as the eighth in the priority, workplaces with less than 50 workers must gradually construct OSHMS under government support to prevent industrial accidents. For enforcement of safety and health education as the eighth in the priority, it's essential to establish useful safety and health education adapted to each workplace and early safety education has to be introduced. In addition, in kindergarten and elementary/middle/high school, a model for safety education has to be established and operated systematically. Because advanced safety and health countries are performing safety and health education from childhood, the recognition of safety and health is deeply rooted. Therefore, they encounter accident prevention very easily because of a safety and health consciousness. But Korea performs very little safety and health education from childhood.

At present, our country ranks with advanced countries in the competition of quality for exporting. But in safety, small safety & health team as the eleventh in the priority has not activated. So, for minimizing industrial accidents by revitalization of small safety & health team, the prize rule or contest for task division of safety has to be promoted and the government has to positively support this so that it is maintained and developed from the bottom up.

Recently, the number of specialist in the prevention of musculoskeletal diseases as the twentieth in the priority is very insufficient and needs to be increased not only for repetition work and lifting tasks but also for systematic management method

Therefore, the main conclusions of this paper are:

First, it provides the very efficient information which can systematically establish strategies for preventing industrial accidents in the medium or short term by the optimal industrial accident rate and fatality rate for occupational accidents per 10,000 persons through evaluation of forecasting method.

Second, it standardizes positive prevention factors for the prevention of industrial accident and systematic safety management through the evaluation of priority matrix criterion which objectively and reasonably see prevention factors and strategies. Therefore, by unity of the safety norm, it can be reduced dramatically not only the fatal rate, but also the industrial accident rate.

Third, by industries, it actually provides very important information for planning of industrial accident prevention through concrete prevention method in-depth by proposed priority model and can easily establish safety management and improved direction. Therefore, it has a serious effect on workers' safety, health and welfare, and productivity by the safe work places.

Fourth, it can contribute much to improving the product and international competition through the level of work satisfaction by the safe work environment.

Finally, we will seek more harmony of labor and management because it is the best way to solve distrust and dissatisfaction at the work site.
A detailed study by correlation analysis for each prevention factor and strategy should be future research.

References

Bae, S.K., and Park, D.H. (2003), "An Evaluation Model for Human Attributes of Industrial Accidents", *Journal of Korean Society of Safety*, 18(4), 155-163.

Eckerman, I. (2005), "The Bhopal Gas Leak: Analysis of Cause and Consequences by Three Different Models", *Journal of Loss Prevention in the Process Industries*, 18, 213-217.

Gyekye, S.A. (2003), "Causal Attribution of Ghanaian Industrial Workers for Accident Occurrence Miners and Non-miners Perspective", *Journal of Safety Research*, 34, 533-538.

Kang, Y.S. (2008), *Safety Management System*, Hyoung Seol, 136.

Kang, Y.S., et al . (2006), *Ergonomics*, Shin Gwang, 28.

Kang, Y.S., Hahm, H. J. (2008), Yang, S. H., and Kim, T. G., "Application of the Life Change Unit Model for the Prevention of Accident Proneness among Small to Medium Sized Industries in Korea", *Industrial Health*, 46(5), 470-476.

Kim, Y.C. (1998), "The Characteristics of Occupational Injury in Small Manufacturing Factory", *Journal of Korean Society of Safety*, 13(2), 145-150.

KOSHA (1973-2008), *Statistics of the Industrial Accident (per years)*.

Lee, H.K., and Chang, S.R. (2005), "Cause Analysis and Prevention of Fishing Vessels Accident", Journal of Korean Society of Safety, 20(1), 153-157.

Park, J.K. (1997), "A Study on the Analysis Cause for Fatal Accident in construction Works", *Journal of Korean Society of Safety*, 12(4), 122-133.

Chapter 120

The Comparison Between OHSMS and PCMS for Business Area

Chan-O Kim, Hyoung-Jun Choi, Chung-Hwan Lee

Department of Safety Engineering
Seoul National University of Technology
Seoul, Republic of Korea

ABSTRACT

ISO/TC 223 (Societal Security) has published ISO/PAS 22399 (Guidelines for incident preparedness and operational continuity management) in 2007 and is providing a management system standard ISO 22301 (Preparedness and continuity management system - Requirements) so called PCMS. But the PCMS contains the important issues of OHSMS (occupational health and safety management system). Therefore, if this management system is aggressively pursued in business area, there will be occurred several severe conflicts or overrides with the current OHSMS, causing serious burden and chaos to the business area. This paper compared the PCMS and OHSMS and analyzed the influences of the occupational safety and health management with introducing the PCMS, and suggested a plan for improving the organizational system and function, and for enhancing the protection of workers in occupational health and safety management area.

Keywords: PCMS, BCP, OHSMS, IPOCM, safety, disaster, crisis, incident

INTRODUCTION

With rapidly changing the world-wide disaster situations, countries around the world are engaging in implementation of BCP (Business Continuity Program) in public and private areas. In particular ISO installed TC 223 in Societal Security area and published international standard ISO/PAS 22399 (Guidelines for incident preparedness and operational continuity management) in November 2007, which based on the concept of BCP. Furthermore, ISO/TC 223 is going on to provide a management system standard ISO 22301 (Preparedness and continuity management system - Requirements) so called PCMS and has a plan to publish the standard in 2011. This management system approaches with respect to safety management and security management that place weight on reducing risk of disaster or crisis. But this standard, in terms of safety management, includes the important details from occupational safety and health areas. Thus, if PCMS is aggressively pursued in business area, it will either run against or overlap with the current OHSMS. Consequently, it is safe to assume that there will be a heavy burden on business sector. This paper is going to compare the PCMS and OHSMS and analyze the influences of occupational safety and health management system with introducing PCMS, and suggest alternative approaches concerning company's system, performance and its internal structure and occupational safety and health management.

PCMS/IPOCM AND OHSMS

BCP AND ITS BASIC CONCEPTS

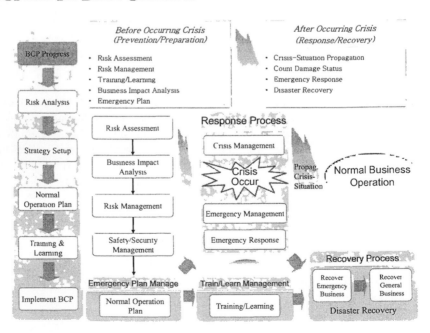

Figure 1. BCP Structural Overview[1]

In response to acting upon a situation whereby one or more critical function(s) of organization has/have been suspended due to disaster or crisis factor, BCP, an advanced disaster management standard, encompasses a system that identifies potential hazard from a disaster or crisis, evaluates risk involved and estimates impact of disaster, so that it may formulate and provide a workable emergency response plan, as well as a recovery strategy. BCP, also, signifies framework which acts to sustain and support such a system. Figure 1 shows BCP's Structural Overview.

INTERNATIONAL STANDARDIZATION OF DISASTER/CRISIS MANAGEMENT

International Cases on Disaster/Crisis Management Standards [1]-[6]

In order to respond efficiently to rapidly changing disaster situations, advanced countries around the world with better/best disaster management system, are putting forth disaster management standard and concurrent organization. After September 11, 2002, the United States of America passed a new act which creates the DHS (Department of Homeland Security). Simultaneously, it oversees FEMA (Federal Emergency Management Agency), which is responsible for disaster and crisis management. In addition, it is recommending to public and private for implementing BCP as an all-encompassing disaster and crisis management system. Furthermore, NFPA (National Fire Protection Association) established NFPA 1600: Standard on Disaster/Emergency Management and Business Continuity Programs (2004), as a BCP methodology. Since it received approval from DHS, FEMA and ANSI (American National Standards Institute), it has become a national standard of USA.

In order to prevent discontinuity of work and sustain work flow, the United Kingdom has been preparing implementation of BCP since 2003. Also, in 2006, BS 25999 (Code of Practice for Business Continuity Management) was established to make possible a-third-party certification approval, which is being maintained as the national standard with respect to BCM.

With SIS (Swedish Standards Institute), Sweden has actively participated in ISO/TC 223 and is greatly facilitating the cause for the establishment of international standard with regards to Societal Security and Social Responsibility.

The other concerned countries, including Australia (SA's HB 221), Israel (SII's HS 2-0142) and Japan (JISC), all have their own standard that pertain to BCP.

International Standardization of PCMS/IPOCM [2], [7]-[8], [10]

ISO/TC 223 was established in 2001 under the name Civil Defense. However, in 2004, conforming to performance result report by ISO/TMB AGS (Advisory Group on Security), the secretary country was changed from Russia to Sweden. In addition, in April 2006, at IWA (International Workshop Agreement) held in

Florence, Italy, TC's name was changed to Societal Security, as well as passing international standards that included range of application of international standardization, terminology clarification, common factors, shared variables, and the likes.

At this meeting, it has been agreed that the priority range of application should proceed from the perspective of Societal Security, which include EM (Emergency Management) and BCM (Business Continuity Management). In terms of content, such things as natural disaster, including earthquake, typhoon, tornado, tsunami, flood, and man-caused/made disaster, including waste chemical, radiation contamination, terror, labor strike, and technical crisis, including Y2K, telecommunication paralysis, are contained. With regards to clarification of terminologies, it has been deemed appropriate to clearly distinguish and compartmentalize EP/EM/BCM, so as to reflect the elements pertaining to danger recognition, warning, information guidance, mitigation strategy, impact analysis, communication between concerned parties, resource utilization, prevention strategy, recovery, training and assessment.

Establishing such international standards signifies the necessity of inclusion of The United States of America's NFPA 1600, The United Kingdom's BS 25999-1, Australia's HB 221, Israel's HS2-0142, as well as suggestions from Japan. With regards to technical details, the results that reflect the opinions of the aforementioned countries will be used. In response to this, the United States of America is strategically pursuing the promotion of its own BCP to IPOCM (incident preparedness and operational continuity management). As well, countries such as the United Kingdom, Australia, Israel and Japan are viewing the operation of international standard in light of their own national standard.

In November 2006, ISO/WD 22300 (Societal security - Fundamentals and vocabulary) was established. As well, in December 2007, ISO/PAS 22399 (Societal security - Guideline for incident preparedness and operational continuity management) was established. Furthermore, ISO/TC 223 is providing the management system standard ISO 22301 (Preparedness and continuity management system - Requirements) so called PCMS and has a plan to publish the standard in 2011.

Basic concept of PCMS/IPOCM [7], [9]

Figure 2 shows the basic concept of IPOCM, and PCMS has the same concept of IPOCM. Curve 2 in the Figure 2 signifies Incident which, in turn, represents outbreak or occurrence of disaster/crisis situation. The Curve 2 also shows complete malfunction of Operational Level and the lengthy time it takes to return to Normal Operational Level. PCMS/IPOCM aims to progressively improve, mitigate and prevent loss on both public and corporate level.

If PCMS/IPOCM is applied, via Prevention and Preparedness in response to Incident, it will be possible (as shown by Curve 1) to reduce the impact of Incident despite occurrence of disaster/crisis situation. As a result, core operation will be sustainable and the Emergency Response, which follows thereafter, will be able to shorten period of disruption. This will, in turn, minimize any loss incurred by operation dysfunction on national, corporate and public level. Lastly, such system, once in place, will allow operation of business to return to normal quickly.

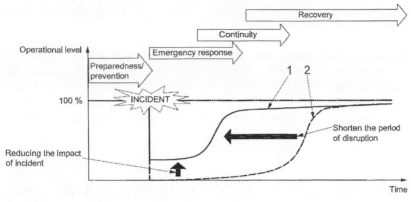

Figure 2. Basic concept of IPOCM

KOREA'S PREPARATION FOR DISASTER MANAGEMENT STANDARDIZATION

With Typhoon Rusa in August 2002 and Daegu Subway Fire Disaster in February 2003, it became apparent that Korea is also in dire need of an improved emergency response and emergency management system. As a result, in March 2004, legislation for disaster and safety management was passed and, in June 2004, on behalf of MOGAHA (Ministry of Government Administration and Home Affairs), NEMA (National Emergency Management Agency) was set up. In November 2005, based on NFPA 1600, an assessment of Local Authorities' disaster management capability was attempted.

In March 2008, with the inauguration of new government, MOPAS (Ministry of Public Administration and Security) underwent a series of changes and started overseeing NEMA. These two government bodies now oversee disaster and crisis management.

There has yet to be national standard on disaster and crisis management for Korea. However, starting in 2005, centering on (Private) Korea BCP Association and BCP Forum, and based on the NFPA 1600, government and corporate consultation is offered on business continuity management program.

As of February 29, 2008, foundation was laid to pave the way for government's control of operation of national disaster management standard and disaster mitigation activity on corporate level; it became possible after the passing of legislation concerning 'Support for Corporation's Disaster Mitigation Activities'.

MOPAS and NEMA have originally been pursuing implementation of disaster management standard based on BCP in Korea.

In December 2008, KATS (Korean Agency for Technology and Standards) published a new national standard KS A ISO/PAS 22399 (Societal security - Guideline for incident preparedness and operational continuity management) for private area. But MOPAS and NEMA are carefully reviewing establishment of national standards for public area.

Until now, disaster management primarily meant natural disaster. But, in the midst of considering broadening the scope of application for all kinds of disasters, based on 'the Principle Act of Disaster and Safety Management', it is expected that all realms of our society will be included in the government's disaster and crisis management effort.

COMPARATIVE ANALYSIS OF PCMS AND OHSMS

COMPARISON BETWEEN PCMS AND OHSMS

PCMS is based on the concept of IPOCM stemming from the basic concept of BCP. And KOSHA 18001, Korean occupational safety and health management system, based on the basic concepts of ILO standard, BS 8800 and OSHAS 18001. A comparison of IPOCM and KOSHA 18001 (OHSMS) has shown in Table 1.

Judging from the Table 1, it is evident that much of PCMS framework overlaps OHSMS. In addition, it is easy to see that Risk Assessment, OHSMS's essential issue, is also given priority in PCMS. As such, OHSMS and PCMS overlap, so as to reiterate each other.

Table 1. Comparison of PCMS(IPOCM) and OHSMS(KOSHA 18001)

Provision	PCMS(IPOCM)	OHSMS(KOSHA 18001)
Start	- define scope and boundaries for IPCOM - identify critical objectives, operation, functions, products and services - preliminary determination of likely risk scenarios and consequences	4.1 General principle 4.1.1 Self assessment of initial condition
Policy	5. Policy 5.3 Management leadership and commitment 5.4 Policy development	4.2 Safety and health policy

Plan	6. Planning
	6.2 Legal and other requirements
	6.3 Risk assessment and impact analysis
	6.4 IPOCM programs
	  6.4.1 Prevention and mitigation programs
	6.4.2 Response management programs
		4.3 Planning establishment
		  4.3.1 Risk assessment
		  4.3.2 Legal and other requirement
		  4.3.3 Objective
		  4.3.4 Safety and health manageme
		  planning

Implementation	7. Implementation and operation	4.4 Implementation and operation

Operation	7.1 Resources, Roles, responsibility and authority 7.2 Building and embedding IPOCM in the organization's culture 7.3 Competence, training and awareness 7.4 Communications and warning 7.5 Operational control 7.6 Finance and administration	4.4.1. Structure and responsibility 4.4.2. Education, training and qualification 4.4.3.Communications(Disclosure of information) 4.4.4. Documentation 4.4.5. Document management 4.4.6. Operation management 4.4.7. Emergency preparation and response
rformance sessment	8. Performance assessment 8.1 System evaluation 8.2 Performance measurement and monitoring 8.3 Corrective and preventative action 8.4 Maintenance 8.5 Internal audit and self assessments	4.5 Inspection and corrective action 4.5.1. Performance measurement and monitoring 4.5.2. Corrective and preventative action 4.5.3. Archiving 4.5.4. Self audit
anagement Review	9. Management Review	4.6 Manager Review

On a different note, whereas OHSMS considers subject Risk from the physical, chemical, biological, psychological and behavioral viewpoint, which embodies such things as mechanical failure, system condition, worker's human-error that are believed to give rise to accident, PCMS, on the other hand, not only includes non-intentional man induced accident, but it also include natural disaster, intentional man induced events. Therefore, it is easy to see that PCMS totally encompasses OHSMS. [8] - [9]

PROBLEMS ARISING FROM THE IMPLEMENTATION OF PCMS/IPOCM IN KOREA

PCMS(IPOCM) and Its Overlapping with the Existing Management Systems

The majority of domestic companies in Korea that are now operating globally employ OHSMS (Occupational Health and Safety Management System), in addition to ISO 9000 series (QMS, Quality Management System) and ISO 14000 series (EMS, Environment Management System). This is done in order to conform to the legal system and for employee protection. As such, there is much confusion, due to simultaneous practice of OHSMS, QMS and EMS. Therefore, recently, efforts are being made to uproot inefficiency arising as a result of integration of these three management systems. To that end, Korea has provided an alternative which would involve integration of the three management systems that tend to international standardization. There has actually been a case whereby a domestic company successfully implemented and employed SHE-Q (Safety, Health, Environment & Quality) System on-site.

However, if legally-binding PCMS is implemented in addition to the aforementioned three systems, a clash with OHSMS will be unavoidable. This concerns company's planning, corporate management, site operation and document management.

Mounting Opposition to Regulation Laxation and Corporate Burden as a result of Dualistic Corporate Governance

QMS and EMS are indispensable when it comes to obtaining international certification for product, system and service of companies that wish to put their name on world market. As well, with respect to employee protection that conforms to occupational safety and health regulation, OHSMS is unavoidable. Any employment or implementation of new management system that is legally-binding under such circumstance would surely yield nothing more than mounting burden and pressure for a company. However, it still does not sufficiently satisfy the system that is already in place which aims to provide corporate assistance. As well as running against corporate strengthening and policy of regulation laxation, there is a high probability of opposition from corporate community due to implementation of an additional management system.

Polarization across Industrial Spectrum Due to Discriminate Government Support and the Resulting Withering Effect on the Excluded Industries

As of February 29, 2008, according to legal stipulation for 'Support for Corporation's Disaster-Mitigation Activities', the companies that have performed successfully in terms of disaster mitigation will be given priority in government-sponsored business, insurance cut, tax support and financial aid. Considering disaster-induced loss, such support system may seem appropriate in encouraging corporate enthusiasm. However, concentration of such support may deter the government away from other areas of concern where some industries urgently find need for. Especially, OHSMS, having many similarities with PCMS, may present itself as an obstacle to the field of occupational safety and health.

OHSMS COUNTERMEASURE IN RESPONSE TO PCMS IMPLEMENTATION

Preservation of Independence for Occupational Safety and Health Domain

Because occupational safety and health domain in countries around the world conform to a similar format based on guidance from ILO (International Labor Organization), it is not easily assimilated into other management systems. Consequently, domain of occupational safety and health innately possess a sense of independence and exclusivity. Under such circumstance, if PCMS is implemented, there is no other option but for the two systems to coexist, creating much overlap. In order to solve this problem, there must first be a solution that recognize OHSMS and its distinct identity and find agreeable solution for overlapping plans, corporate management, on-site activity and document management before PCMS is introduced.

OHSMS and Its Change

It is predicted that future occupational safety and health issue cannot solely be focused on employment protection. Thus, change in occupational safety and health is unavoidable. Furthermore, moving on from merely protecting the employee through occupational facility, the future realm of occupational safety and health must include protection of employee from natural and intentional human induced disasters. This will facilitate a natural approach from occupational safety and health to disaster safety and enable occupational safety and health system to possess independence and, at the same time, become the leading influence on operation of disaster management system.

Domain IPOCM Operation that Includes OHSMS

Korea and the international community ought to take a closer look at overlapping of PCMS and OHSMS which requires a close examination. Because fundamental problems arise as a result of the implementation of PCMS which include Security and Safety Management, concerning Safety Management within a corporate community, prospective PCMS must first recognize the current OHSMS as a subsystem and foster an overall PCMS. In order to achieve this, the core conditions of PCMS must be reflected in OHSMS and can be recognized on external audit. In return, PCMS should explicitly acknowledge OHSMS and its core values as its subsystem, so that stability can be imbued all across a corporate community and its system.

Creation of Integrated Management System

Risk Assessment outlined by PCMS is an OHSMS derivative that is popularized all over the world. It is a safety technology employed by many in the international community. Those who are considered to be most knowledgeable in this field are those in precedent occupational safety and health field. These experts are actively involved in organizations in the field of occupational safety and health. Therefore, if enough manpower is supplemented in the field of disaster management in an OHSMS organization, PCMS operation becomes feasible within an organization. PCMS that partially employs some planning ideas from occupational safety field can ease planning in its early stage. Furthermore, for PCMS risk assessment, overlapping can be avoided and an integrated management system made possible by assessing all Incidents, including occupational accidents related to OHSMS.

Documentation should embody OHSMS as a sub-title and comprehensively compose PCMS document. Upon external audit, such a comprehensive documentation should enable execution of an integrated management system.

In order to enable this kind of an integrated management system, the concerned governments, MOL (Ministry of Labor) and MOPAS which are responsible for overseeing OHSMS and PCMS, ought to become aware of the differences in the two systems and acknowledge the need to create a mutual implementation between OHSMS and PCMS.

CONCLUSION

BCP/IPOCM-based international standard on ISO/TC 223 Societal Security is shaping up to be PCMS. In Korea, efforts are being made to disseminate disaster management standard to the corporate community in order to prepare for various disasters. However, it must be understood that overlapping in such areas as planning, corporate management, on-site activity and document management cannot be avoided with regards to integration of PCMS and OHSMS. In addition, risk assessment which is at the core of management system assumes the same process in the former and the latter. For this reason, without much planning beforehand, PCMS can only add to corporate burden and confusion and devastate other management systems, such as OHSMS, QMS and EMS.

Therefore, in order to avoid and prevent loss on corporate and national level and facilitate problem-free operation of IPOCM, the following must be considered:

(1) Occupational safety and health field ought to remain in its own realm.
(2) There needs to be an expansion of OHSMS.
(3) IPOCM system should include OHSMS as its subsystem.
(4) The requisite for each system is recognition where it is needed in both IPOCM and OHSMS, so that the need for an integrated operation of the two systems is served.

REFERENCES

[1] Jee-Hyup You (2006), "Management System of Social Facility and BCP", *13ᵗʰ Disaster Prevention Seminar,* Unpublished Manuscript.
[2] KISA (2006), "Report for National Strategy in 2nd International Conference of ISO/TC 223", KISA.
[3] NFPA (2004), "NFPA 1600: Standard on Disaster/Emergency Management and Business Continuity Programs (2004 Edition)" , NFPA.
[4] John Sharp (2008), "The Route Map to Business Continuity Management", BSI.
[5] BSI (2006), "BS 25999: Business Continuity Management - Part1: Code of practice" , BSI.

[6] BSI (2007), "BS 25999: Business Continuity Management - Part2: Specification", BSI.

[7] ISO (2007), "ISO/WD 22300: Societal security - Fundamentals and vocabulary", 2007-11-29, ISO.

[8] ISO (2007), "ISO/PAS 22399: Societal security - Guideline for incident preparedness and operational continuity management", 2007-12-10, ISO.

[9] KOSHA Code G-04-2003 (2003), "KOSHA 18001", KOSHA.

[10] Stefan Tangen (2009), "Enhancing societal resilience", *ISO Focus 2009*, 9-10, ISO.

Chapter 121

The Distribution of Pedestrian-Backing Vehicle Accidents by Backup Alarm Status and Vehicle Type

J.P. Purswell, Jerry L. Purswell

Purswell & Purswell
Engineering & Ergonomics, Inc.
Colorado Springs, CO 80920-1603, USA

ABSTRACT

The purpose of the current study was to update and expand upon an earlier study performed to review and categorize OSHA accident investigation records for pedestrian-backing vehicle accidents according to whether the backing vehicle had a backup alarm and whether the alarm was installed and functioning as intended. The current study includes an analysis of additional records as well as the business type (SIC code) of the employer. Incidents where a backup alarm was installed and operational, but possibly not audible above the background noise were also noted. The backing vehicle was also classified into one of three categories: industrial lift truck (forklift), some type of construction equipment, or a street vehicle such as a tractor trailer or straight truck. As with the previous analysis, OSHA accident investigation records provided the data for this analysis.

Keywords:Backup alarms, Backup alarm Effectiveness, Audibility

INTRODUCTION

OSHA regulations require that OSHA investigate incidents involving a worker fatality and one in which 3 or more employees are hospitalized (29 CFR 1904.39). OSHA regulations *for construction (29 CFR 1926)* require that vehicles with an obstructed rear view which are being backed up to have either an audible automatic alarm or an on the ground spotter directing the movement *(29 CFR 1926.601)*. To date, OSHA has not required industrial lift trucks or over-the-road trucks to have automatic backup alarms. In an earlier study, (Purswell & Purswell, 2001) reviewed the distribution of pedestrian-backing vehicle accidents contained in the OSHA Accident database (http://www.osha.gov/pls/imis/accidentsearch.html) identified by several specific search terms. More recently, OSHA has made available accident investigation summaries by keyword, one of which includes "backup alarms." From a review of these OSHA-categorized backup alarm-related cases, additional records involving pedestrian-backing vehicle accidents were identified and reviewed. In some of these instances, the investigation summary available from the OSHA website contained inadequate detail to discern what the status of the backup alarm (if any) of the backing vehicle was. Consequently, FOIA requests for the investigation files were filed with the particular OSHA offices that conducted the investigations. In some instances, FOIA responses clarified the backup alarm status of a backing vehicle. In a number of cases, the accident files were no longer available. In some cases, the FOIA response was still not determinative. Of the 348 accident files reviewed, 120 of them remained uncategorized as to the backup alarm status of the backing vehicle.

The current study expands upon the earlier study in several ways. First, due to a change in the OSHA website, more records of pedestrian-backing vehicle accidents were identified. In addition, records made available by OSHA since the earlier analysis have been incorporated. Finally, additional information has been extracted from each record reviewed. Specifically, the type of industry and the type of backing vehicle has been noted from the accident summaries. This information has been entered into a spreadsheet that contains a hyperlink to the accident summary, the SIC code, and the vehicle description from the accident summary. In instances involving multiple employers, the SIC code for the employer of the pedestrian was noted.

The records were categorized into five specific accident scenarios, and one additional category if the available records did not have enough information to otherwise classify the accident. The categories were as follows and the number of records corresponding to each category are shown below:

- No automatic audible backup alarm installed on the vehicle, (109).
- Automatic audible backup alarm installed, but no longer functioning, (85).
- Automatic audible backup alarm installed and functioning, (35).

- Automatic audible backup alarm installed and functioning, but (likely) inaudible above the background noise, (6).
- Inadequate information to determine the status of any automatic audible backup alarm, (121).

Due to space constraints, the 120 records of pedestrian-backing vehicle accidents which remained unclassified are not included here, but may be obtained by contacting the authors. The vehicle description used in the OSHA Accident Summary are adopted here. Vehicle descriptions which include a product manufacturer or brand name are included here as well. However, most records did not include such information.

ACCIDENT CATEGORIES

The following six tables contain the hyperlinked OSHA inspection numbers (so readers may easily access and review the records themselves) grouped by the four categories. The accident summaries may also be accessed by the inspection number at http://www.osha.gov/pls/imis/InspectionNr.html.

Table 1: No backup alarm installed on vehicle.

Inspection Number	SIC	Vehicle Type
119691715	1711	Truck
113172738	1794	Dump truck
306831793	1611	Dump truck
300782471	2511	tractor trailer
306096785	1731	tractor trailer
300779659	7699	tractor trailer
303688196	1381	winch truck
303647481	1623	aerial lift truck
125561712	0171	truck
119689966	1611	dump truck
303376925	4151	school bus
303933808	1541	telescopic forklift
304266299	4213	tractor trailer
304266216	4213	tractor trailer
304266208	2952	tractor trailer
304413438	1771	asphalt truck
303370944	3069	truck
126138692	4953	pickup truck
301043188	1794	dump truck
302297668	5181	electric forklift
302337977	5712	garbage truck

Table 1: No backup alarm installed on vehicle.

126793868	1611	pickup truck
102788908	1611	tractor trailer
301220026	2436	dump truck
122164585	1611	dump truck
126484088	7363	tractor trailer
126636158	3443	truck tractor
109651000	8711	dump truck
109168450	1622	dump truck
119743086	4731	delivery truck
107713133	1781	water truck
109295634	4215	tractor trailer
108801895	1611	bulldozer
123659039	1611	dump truck
103520557	4231	truck tractor
101390169	1611	scraper
102929528	4212	delivery truck
107302838	1611	transfer truck
109820753	1611	dump truck
100976489	1611	dump truck
109820746	1791	dump truck
18347518	1623	bulldozer
112203898	3315	forklift
112216254	7513	moving truck
103239950	1711	roll-off type container truck
111629754	1629	semi-tractor belly-dumping trucks
103299053	1622	backhoe
111822482	1794	scraper
111658415	1799	dump truck
111658407	1629	dump truck
111658373	1622	dump truck
111658399	1629	dump truck
102063294	8641	flatbed truck
106213465	1611	scraper
104382627	1611	truck
104694179	1629	earthmover
105598551	9199	front end loader
103282984	1611	front end loader
102829918	3271	truck
102552163	1531	dump truck
2435477	1389	oil field truck

Table 1: No backup alarm installed on vehicle.

104379425	7363	bulldozer
104379433	1629	bulldozer
105459945	1542	front end loader
105458822	9199	garbage truck
107039117	1542	scraper
101275626	1623	bucket truck
105879738	0179	flatbed truck
100938208	4491	forklift
105900476	0133	dump truck
102452968	1611	Road grader
101116861	1611	Scraper
104553342	3272	concrete truck
102217569	4131	dump truck
103426722	1623	Truck
101615169	1389	Truck
103099636	1611	dump truck
104703137	1389	frac truck
3328895	1611	asphalt truck
3328903	1611	asphalt truck
3328929	1611	asphalt truck
3328911	1611	asphalt truck
18749572	1541	dump truck
102040581	3089	asphalt truck
101718922	1611	dump truck
100355593	1611	grader
2877793	2421	forklift
100260959	3731	cherry picker truck
100314764	1622	scraper
2531655	1794	dump truck
14804256	4212	dump truck
1778034	1794	front end loader
1821818	1623	dump truck
1823459	1794	dump truck
15055890	1611	dump truck
14971048	1611	scraper
14971030	49199	scraper
1174028	1794	scraper
1701853	4212	dump truck
14478523	2085	dump truck
301993374	1611	dump truck

Table 1: No backup alarm installed on vehicle.

102755048	2656	tractor trailer
300510815	4953	dump truck
301994604	4491	front end loader
105237572	4911	flatbed truck
123746646	9621	dirt truck

Table 2: Backup alarm installed, but no longer functioning.

Inspection Number	SIC	Vehicle Type
306438755	4959	water truck
306309493	4212	front end loader
306309451	2679	front end loader
305569717	1771	bulldozer
305569725	1794	bulldozer
305997355	0783	front end loader
306073719	9199	garbage truck
305501793	1611	grader
301328142	1795	loader
303985568	0851	fertilizer loader
303985535	4213	fertilizer loader
103629242	1611	dump truck
303675144	4953	garbage truck
120278312	1629	grader
300611126	1622	sweeper truck
120278312	1629	grader
302535851	1751	dump truck
302535869	1751	dump truck
125674960	175	mechanical harvester
302090238	3272	front-end loader
125471110	1611	dump truck
301928909	1794	dump truck
301911590	1611	grader
119747491	8331	forklift
125311589	1611	excavator
100756436	3325	bulldozer
110098290	5093	front end loader
124765009	1794	scraper
119528321	1611	grader
109690560	1611	truck
106941578	1622	dump truck

Table 2: Backup alarm installed, but no longer functioning.

110092764	1611	dump truck
114848245	1629	semi trailer
100748730	1611	grader
106506918	1623	dump truck
109869214	4953	front end loader
123970527	1611	dump truck
116039025	4953	soil compactor
112160221	1623	loader
116039587	3316	forklift
100756436	3325	bulldozer
17874876	4953	bulldozer
106624240	4953	bulldozer
109018895	1794	dump truck
105933139	3399	W36 Articulator
107301897	1794	dump truck
107301889	1541	dump truck
102277415	1629	dump truck
102277407	1629	dump truck
111897625	4213	front-end loader
110390648	1629	scraper
17966078	4491	Ottawa yard hustlers
17966086	4491	Ottawa yard hustlers
103528477	1611	scraper
18061713	3271	dump truck
111510848	2411	loader
17853524	1623	dump truck
108499351	1542	scraper
104471412	1611	dump truck
110288958	1611	front end loader
106263791	1611	scraper
104393251	1611	pan
104399753	4212	front end loader
102538840	8711	dump truck
102538857	1611	dump truck
102538865	1611	dump truck
101030542	4214	compactor truck
104384151	1794	bulldozer
106317522	8711	dump truck
105686281	1794	Wabco pan
105686273	1794	Wabco pan
18391961	1611	dump truck

Table 2: Backup alarm installed, but no longer functioning.

2765733	1611	dump truck
18746685	1611	earthmover
105419626	1611	dump truck
105421275	1799	dump truck
18085522	1622	scraper
18085530	1791	scraper
17700873	8711	truck
17700899	1721	truck
101808483	1611	dump truck
2175305	1611	scraper
14989024	9190	grader
970970	1794	bucket loader
1976398	1623	gcraper
1657147	2061	front end loader
15247794	1611	dump truck
955518	7363	dump truck
116309592	1629	grader

Table 3: Backup alarm installed and audible.

Installed, Audible	SIC	Vehicle Type
307567727	1611	grader
300839925	1542	spray truck
305351330	1611	dump truck
303773824	2421	forklift
125917039	2657	forklift
304593858	1611	flatbed truck
304957020	1611	tack truck
303987101	1761	pickup truck
304161599	5093	front end loader
301273371	4491	forklift
302970884	1611	truck
125517425	1761	forklift
116508060	1611	truck
119992048	1623	asphalt truck
111910451	1611	slurry rig
105639744	1542	forklift
108012667	1611	grader

Table 3: Backup alarm installed and audible.

103038931	1542	asphalt truck
106105547	1611	dump truck
101278612	1611	scraper
3306966	1771	dump truck
105218614	9131	paver
2997625	1611	dump truck
2929461	1611	grader
2929453	1522	grader
970954	1611	asphalt truck
969568	4212	asphalt truck
969550	1611	asphalt truck
1151463	5093	wheel loader
1517903	3273	front end loader
301165304	4213	truck
115753774	1611	dump truck
303176622	1623	dump truck

Table 4: Backup alarm installed but inaudible.

Installed, inaudible	SIC	Vehicle Type
306377250	1623	backhoe
119820157	1771	dump truck
303092993	1611	dump truck
119802825	2033	forklift
122022536	3312	dump truck
102579984	1629	dump truck

ANALYSIS

The census of pedestrian-backing vehicle accidents developed from the OSHA Accident database reveals a number of interesting patterns. As might be expected, construction vehicles comprised a preponderance of the vehicle types and businesses employing persons injured in such accidents. Vehicles used in construction which may have an obstructed rear view are already required to have either a spotter or a spotter before backing (29 CFR 1926. 1926.601). To date, OSHA has not approved other methods such as cameras or ultrasonic sensors as alternate means to address the hazard of pedestrians being struck by backing vehicles. It should be noted that of the pedestrian-backing vehicle accidents in the OSHA database for which insufficient detail was available to classify the records according to one of the four categories in the above tables, 48 of 121 incidents involved workers employed by firms which fell under the Standard Industrial Classification Codes of construction. This compared with 72 of 109 construction employers involved in accidents where backup alarms were not installed, 61 of 85 instances where construction employers were involved in pedestrian-backing vehicle accidents where an installed reverse alarm was no longer functioning, and 26 of 35 instances where a backup alarm was reported to be audible at the time of the accident. As shown by Table 4, four of six instances involved construction employers in cases where a backing vehicle struck a pedestrian and the backup alarm was sounding, but likely not audible to the pedestrian above the background noise.

As ergonomics professionals, it must be a concern that so many of the back-up alarms were no longer functioning. Perhaps they were disabled because of their annoying properties, leaving a hazard in the workplace. It is also possible that the worker became habituated to the sound of the alarm. This study suggests that if a back-up alarm is to be relied on to prevent these type of accidents, then they must be regularly checked for proper functioning

REFERENCES

OSHA Accident Database, "Backup Alarm" category, OSHA Website: http://www.osha.gov/pls/imis/AccidentSearch.search?acc_keyword=%22Back-Up%20Alarm%22&keyword_list=on

Purswell, J.P., and Purswell, Jerry L. (2001), "The Effectiveness of Audible Backup Alarms as Indicated by OSHA Accident Investigation Records." *Advances in Occupational Ergonomics and Safety.* 444-450. IOS Press: 2001.

Chapter 122

Automated Analysis of Injury Narratives: Some Lessons Learned

Helen Marucci-Wellman[1], Mark Lehto[2], Helen Corns[1], Gary Sorock[3]

[1]Liberty Mutual Research, Institute for Safety
Hopkinton, Massachusetts, USA

[2]School of Industrial Engineering, Purdue University
West Lafayette, Indiana, USA

[3]Johns Hopkins Bloomberg School of Public Health
Glyndon, Maryland, USA

INTRODUCTION

The addition of narrative text information in electronic format to injury databases can be a useful adjunct to epidemiologic analyses and can suggest injury prevention strategies. Narrative fields provide many advantages if maintained in electronic format. Their ultimate use is to expand on coded injury data (e.g. exposure to noise) to better understand specific circumstances of an injury (Sorock et al., 1997). Grouping the data to satisfy investigative needs such as injury cause scenarios is an essential part of the analytic process. Researchers at the Liberty Mutual Research Institute and Purdue University have been involved for over a decade in developing methods for semi-automatic classification of injury narratives into classifications which describe and group the circumstances of injuries. These "injury narratives" can be used to understand both non- injury and injury-producing incidents as well as injuries sustained by people involved in these events. In this paper we summarize some of the major findings of this research.

KEYWORD SEARCHES FOR IDENTIFICATION AND CLASSIFICATION

Our earliest work began with "keyword" searches to identify and classify motor vehicle accidents in construction work zones from narrative text found in insurance claims (Lehto and Sorock, 1996; Sorock et al., 1996). Keyword mining techniques of injury narratives identified specific injury hazards that were not identifiable through the original coded data. Most crashes (26%) involved a stopped or slowing vehicle in the work zone and the most common crash (31%) was a rear-end collision (Table 1).

Table 1. Pre-crash activities of motor vehicles and crash types in construction work zones, U.S. 1990-1993

Pre-crash activities	No.	(%)	Crash type	No.	(%)
Stopping	961	26	Rear-end	1,142	31
Merging	332	9	Hit object (small)	413	11
Cutting off	203	6	Hit object (large)	257	7
Backing	177	5	Side impact	57	2
Driver error	120	3	Flip/overturn	16	1
Merging	332	9	Not specified	1,801	49
Not specified	1,887	51			
Total	3,686	100	Total	3,686	100

Adapted from Sorock, Ranney and Lehto, 1996

The injury narrative contained valuable information which allowed for subsequent classification of pre-crash actions and crash types that were not originally in the coded claims data and provided useful information on how to prevent future events which may lead to serious injuries for occupants and pedestrians.

BAYESIAN MODELING FOR ENHANCED MINING

A subsequent analyses enhanced the utility of this approach by integrating Bayesian tools to search out similar classifications of the accidents as was done earlier with text searches, based on the computer "learning" from the results of the earlier performed keyword searches (Lehto and Sorock, 1996). Additional narratives containing similar pre-crash and crash type information were mined accurately using this approach improving on the detection and accuracy achieved from keyword search alone (Table 2). The Bayesian learning approach appeared to offer a promising approach to accurately assign independent groupings of information contained in injury narratives beyond what was achievable from key word searching alone. Similar to the training of a human coder, who will subjectively infer the probability of a classification from the words in a narrative, conditional probabilities calculated by a Bayesian algorithm will uniformly and systematically identify similar intuitive predictors in a narrative.

Table 2 shows the overall accuracy of the predictions of pre-crash activity and crash type given by keyword search and the fuzzy Bayesian model (at three different probability thresholds), for initially classified and unclassified accident narratives. Accuracy is measured in terms of the probably of detecting the category assigned by the expert (P(d)), and the probability of false positives (P(fp)). For example, this table shows that in narratives in which the crash type was originally unclassifiable by an injury epidemiologist (right two columns) the model was able to correctly classify 20-30% with small percentages of false positive classification (3-8%). Separate training and evaluation datasets were used in this analysis (Lehto and Sorock, 1996).

Table 2. Comparison of Keyword Search and Bayesian Prediction

Keyword Search	Initially Classified (n=409)		Initially Unclassified (n=429)	
	Pre-crash Activity	Crash Type	Pre-crash Activity	Crash Type
P(d)	0.94	0.91	-	-
P(fp)	0.05	0.04	-	-
Bayesian Model				
p>/0.5				
P(d)	0.98	0.95	0.67	0.45
P(fp)	0.31	0.25	0.27	0.32
p>/0.7				
P(d)	0.96	0.93	0.2	0.28
P(fp)	0.07	0.07	0.05	0.08
p>/0.9				
P(d)	0.88	0.9	0.12	0.03
P(fp)	0.05	0.05	0.13	0.03

Adapted from Lehto and Sorock, 1996

BAYESIAN MODELS FOR CLASSIFICATION OF NARRATIVES INTO CAUSE OF INJURY GROUPS: NAÏVE AND FUZZY BAYES

This work led to several subsequent evaluations of both the classic Naive and Fuzzy Bayesian approaches for classifying narratives into pre-designated cause of injury classification structures. We have explored benefits and limitations to the utility of both Bayesian approaches for this application.

The Naïve Bayes model is a commonly applied method of text classification used for years in the field of information retrieval (Sebstiani, 2002). By making what is called the conditional independence assumption (Lewis, 1992), the probability of assigning a particular event code category can then be calculated using the expression:

$$P(E_i \mid n) = \prod_j \frac{P(n_j \mid E_i)P(E_i)}{P(n_j)}$$

(Eqn. 1)

where $P(E_i|n)$ is the probability of event code category E_i given the set of n words in the narrative. $P(n_j|E_i)$ is the probability of word n_j given category E_i. $P(E_i)$ is the probability of category E_i and $P(n_j)$ is the probability of word n_j in the entire keyword list.

The Naïve approach combines both positive and negative evidence using the Bayes's rule. However the independence assumption required by Naïve Bayes is clearly violated by the type of word dependencies found in injury narratives. For example the words "hit" and "struck". These two words are never found together. If they were independent words then they would by chance be found together some of the time.

The Fuzzy Bayes approach avoids making the conditional independence assumption by calculating $P(A_i \mid n)$ using the expression:

$$P(E_i \mid n) = MAX_j \frac{P(n_j \mid E_i)P(E_i)}{P(n_j)}$$

(Eqn. 2)

where each term i is assigned a value as explained in Eqn. 1 above.

The Fuzzy approach does not assume independence between words but is limited in that in its simplest form it only considers positive evidence when clearly negative evidence is helpful and intuitive for certain classifications. "He fell asleep at the wheel…" is an example of a narrative containing negative evidence that may not be considered by a Fuzzy Bayesian single word or paired word model. In this narrative the negative evidence is the word "asleep". The probability of the fall category given the word "fell" alone or "he fell" alone would be high (and not the transport category). However, the probability of the fall category considering all three words "he fell asleep" includes the negative evidence of asleep for the fall category and the probability of the fall category would be quite low while the probability of the transport category would be quite high.

THE FUZZY MODEL: IMPROVED PERFORMANCE WITH MULTIPLE WORD-PREDICTORS

In 2004 we investigated Fuzzy Bayesian methods for classifying injury narratives into external-cause-of-injury (e-code) categories using injury narratives and corresponding e-codes assigned by experts from the 1997 and 1998 United States National Health Interview Survey (Wellman et al, 2004). A Fuzzy Bayesian model was used to assign injury descriptions to 13 e-code categories. The computer program correctly classified 4695 (82.7%) of the 5677 injury narratives (which had been expertly assigned by coders with the National Center for Health Statistics) when multiple words were included as keywords in the model. We learned that the use of multiple word predictors compared with using single words alone in the fuzzy approach greatly improved both the sensitivity and specificity of the computer generated codes. This is illustrated using the "other transport" category. Only 43 of the 150 expertly coded narratives were correctly classified into the "other transport" category using the single word model, whereas 121 of the 160 were correctly classified in the multiple word model. The final results were promising. The overall sensitivity of classifying narratives at the category level was 0.82 with positive predictive values (PPV) ranging from a low of 0.83 in the foreign body eye category to a 1.0 of the other specified category. However, we concluded the results were likely overly optimistic because we used the same dataset for training and evaluation.

FUZZY VS. NAÏVE COMPARE AND CONTRAST: TWO-DIGIT CLASSIFICATIONS

More recently, using a larger dataset of 14,000 injury claims narratives, we were able to test the Bayesian methodology on separate training and prediction datasets (Lehto et al., 2009). We broadened the analyses to include both Fuzzy and Naïve Bayesian methods for classifying these narratives into Bureau of Labor Statistics Occupational Injury and Illness Classification System (OIICS) event categories. We also added word sequences and a smoothing constant to remove the effect of noise. Both models performed well and tended to predict one-digit BLS codes more accurately than two-digit codes. The overall sensitivity of the Fuzzy method was, respectively, .78 and .64 for one-digit and two-digit codes, specificity was .93 and .95, and positive predictive value (PPV) was .78 and .65. The Naive method showed slightly better accuracy: a sensitivity of .80 and .70, specificity of .96 and .97, and PPV of .80 and .70. Naïve and Fuzzy accuracy seemed to vary across categories, with different categories predicted better using one or the other model and improved sensitivity corresponded to lower positive predictive values. For example, although the mean sensitivity of the Naive model was higher for the fall category, the mean ppv of this category was lower for the Naïve model than for the Fuzzy model. The opposite was true for the bodily motion category where Fuzzy was more sensitive but had a much lower ppv than the Naïve model (Table 3).

Finally, the Fuzzy model contained more intuitive predictors (Table 4).

Table 3. Evaluation of Naïve and Fuzzy Bayes Model Prediction of One- and Two-Digit BLS OIICS1 Classifications

BLS OIICS	Description	Gold Standard		Naïve Bayes		Fuzzy Bayes	
		NGS2	%	Sen3	PPV5	Sen	PPV5
Contact – Group '0'		523	17.4	0.80	0.74	0.62	0.77
01	Struck against	145	4.8	0.46	0.65	0.28	0.66
02	Struck by	294	9.8	0.72	0.57	0.50	0.57
03	Caught/Compressed	73	2.4	0.60	0.58	0.51	0.61
Fall - Group '1'		521	17.4	0.85	0.70	0.83	0.76
11	Fall to lower level	185	6.2	0.70	0.62	0.62	0.64
13	Fall on same level	322	10.7	0.73	0.65	0.71	0.63
Bodily Motion -Group '2'		1013	33.8	0.76	0.90	0.87	0.77
21	Bodily reaction	124	4.1	0.52	0.63	0.37	0.74
22	Overexertion	535	17.8	0.85	0.79	0.92	0.57
23	Repetitive motion	76	2.5	0.86	0.71	0.80	0.75
Exposure to harmful substances or environment - Group '3'		303	10.1	0.88	0.80	0.86	0.82
31	Contact w electric current	28	0.9	0.75	0.81	0.68	0.63
32	Contact w temp extreme	93	3.1	0.88	0.78	0.89	0.75
34	Exposure caustic sub	110	3.7	0.76	0.84	0.75	0.78
35	Exposure to noise	37	1.2	0.89	0.87	0.97	0.90
37	Exposure to stress	33	1.1	0.85	0.61	0.88	0.58
Transportation - Group '4'		384	12.8	0.91	0.86	0.90	0.80
41	Highway accident	220	7.3	0.97	0.83	0.97	0.65
42	non highway accident	56	1.9	0.73	0.48	0.30	0.46
43	Ped struck by vehicle	104	3.5	0.63	0.78	0.44	0.61
Fire or Explosion - Group '5'		17	0.6	0.47	0.73	0.41	0.78
52	Explosion	11	0.4	0.45	0.83	0.55	0.67
Assaults & Violent Acts - Group '6'		87	2.9	0.76	0.68	0.54	0.77
61	Assaults	82	2.7	0.77	0.69	0.70	0.70
Non Classifiable - Group '9'							
99	Non classifiable	152	5.1	0.49	0.70	0.37	0.81
	General Unclassifiable7	52	1.7	0.04	0.50	0.08	0.67
	Other categories < 10	22	0.7	0.04	.	0.23	0.01
Total		3000	100.0	0.70	0.70	0.64	0.65

2Gold Standard Classifications

3Sensitivity

4Specificity

5Positive Predictive Value

6Negative Predictive Value

Adapted from Lehto et al., 2009

Table 4. Predictors Used in Fuzzy Bayes Model

	Frequency of Occurrence in Prediction Set	Frequency of Times Used to Predict	Prediction Strength
Fall to lower level			
FROM-LADDER	7	4	0.71
OFF-LADDER	7	4	0.68
FELL-OFF-LADDER	6	6	0.97
Fall on same level			
SLIPPED-FELL	24	21	0.79
SLIPPED-ON-ICE-FELL	12	12	0.93
SLIPPED-FELL-ON-ICE	9	9	0.95
Struck by			
CAME-DOWN	9	7	0.74
FELL-ON-FOOT	6	6	0.94
FELL-ONTO-FOOT	4	3	0.78
FELL-STRUCK	3	3	0.86
Highway			
REAR-ENDED	10	10	0.97
WAS-REAR-ENDED-BY	10	10	0.97
WAS-REAR-ENDED	10	10	0.96
Struck Against			
FINGER-ON	6	4	0.75
STRUCK-KNEE-ON	4	4	0.87
BRUISE&CAUSING	3	3	0.67
Overexertion			
FELT-PAIN-TO-LOWER	5	5	0.93
LIFTING-BOXES	4	4	0.94
PUSHING-PULLING	4	4	0.94
Bodily Reaction			
KNEELING	5	2	0.71
BENT-OVER	4	2	0.75
DOWN&KNEELING	3	3	0.75
Repetitive Motion			
TYPING	9	9	0.96
KEYBOARD	5	4	0.76
CARPAL-TUNNEL	3	3	0.93

Adapted from Lehto et al., 2009

FILTERING AND SEMI-AUTOMATIC APPROACHES

Both Bayesian models seemed to complement each other. The models assigned predictions with a degree of confidence that was strongly related to accuracy, supporting the possibility of filtering to improve accuracy further (Figure 1). The objective of our current research is to demonstrate that a computerized approach that classifies some narratives with high confidence and then intelligently selects out narratives for manual review is very effective for coding narratives from large administrative databases and for the purpose of public health surveillance.

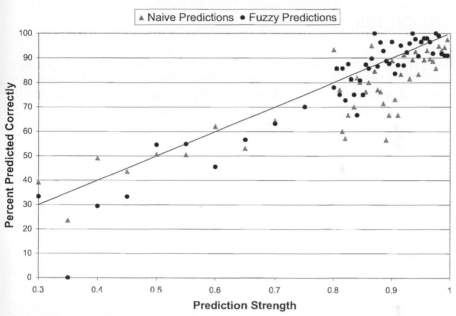

From Lehto et al., 2009
FIGURE 1. Calibration Curve for Naïve and Fuzzy Bayes Models

OTHER APPROACHES

Other ways in which the computer has been used to mine and code injury narratives include traditional data mining techniques where human-defined decision rules or predefined "text strings" were developed by experts in the field to mine a database (Kennedy et al., 2001; Williamson et al., 2001). However these require the development of lengthy decision rules and/or pre-defined text strings which require a human expert knowledgeable in the particular subject linguistics, acronyms and abbreviations. Newer, electronic approaches to text mining and data retrieval such as that of vector algorithms or neural networks may also be capable of performing classifications or groupings of narratives. However, there have been few evaluations on their utility for injury cause coding. The performance of neural networks and other multivariate methods using singular value decomposition and regression techniques have been compared to the Bayesian approach and found to be comparable in accuracy (Noorinaeini and Lehto, 2006; Chatterjee et al., 1995; Lehto et al., 1996). However, the Bayesian model required much less computation and the basis for predictions was more intuitive. One additional unique and useful quality of the Bayesian approach (which does not appear in other approaches) is the confidence parameter output for each prediction. The ability to set threshold levels based on the accuracy and resource requirements of the investigation significantly reduces the amount of manual coding required without sacrificing accuracy and allows the user to focus on the more difficult narratives.

CONCLUSION

Narrative text data can provide useful information for injury prevention (Sorock et al., 1997; Smith, 2001), case identification identifying pre-event circumstances (Lombardi et al., 2005; Lombardi et al., 2009), and developing hazard scenarios (Lincoln et al., 2004). Use of computer algorithms to intelligently select out those narratives that can be accurately coded by the computer has the potential of reducing the burden of reviewing and classifying large numbers of narratives and improving accuracy beyond what manual coders alone can achieve.

ACKNOWLEDGEMENTS

The authors would like to thank the many reviews, suggestions and support given over the years for this work by Theodore Courtney, Director of the Center for Injury Epidemiology at the Liberty Mutual Research Institute for Safety.

REFERENCES

Chatterjee, S., Sorock, G.S., and Lehto, M.R. (1995), "Classifying accident text descriptions using neural networks," presented at Institute for Operations Research and Management Science (INFORMS) Conference, New Orleans, LA, October 29-November 1.

Kennedy, B., Ramos-Santacruz, M., Sada, B., and Dodd, R. (2001), "Making effective use of aviation narratives: safety event coding using text mining." *Proceedings of the Federal Database Colloquium and Exposition*, San Diego, August 28-30, 357-373.

Lehto, M.R., Wellman, H.M., and Corns, H. (2009), "Bayesian methods: A useful tool for classifying injury narratives into cause groups." *Injury Prevention*, 15(4), 259-265.

Lehto, M.R., and Sorock, G. (1996), "Machine learning of motor vehicle accident categories from narrative data." *Methods of Information in Medicine*, 35,(4), 1-8.

Lehto, M.R., Sorock, G.S., and Chatterjee, S. (1996), "Classification of accidents and pre-crash activity from accident narratives." presented at 1st International Conference on Applied Ergonomics, Istanbul, Turkey, May 21-24.

Lewis, D.D. (1992), "Naïve Bayes at forty: The independence assumption in information retrieval." *Proceedings of the 15th Annual International ACM SIGIR Conference on Research and Development in Information Retrieval*, 37-50.

Lincoln, A.E., Sorock, G.S., Courtney, T.K., Wellman, H.M., Smith, G.S., and Amoroso, P.J. (2004), "Using narrative text and coded data to develop hazard scenarios for occupational injury interventions." *Injury Prevention*, 10(4), 249-254.

Lombardi, D.A., Matz, S., Brennan, M.J., Smith, G.S., Courtney, T.K. (2009), Etiology of work-related electrical injuries: A narrative analysis of workers' compensation claims." *Journal of Occupational and Environmental Hygiene*, 6(10), 612-23.

Lombardi, D.A., Pannala, R., Sorock, G.S., Wellman, H., Courtney, T.K., Verma, S., and Smith, G.S. (2005), "Welding related occupational eye injuries: A narrative analysis." *Injury Prevention*, 11(3), 174-179.

Noorinaeini, A., and Lehto, M.R. (2006), "Hybrid singular value decomposition: A model of text classification." *International Journal of Human Factors Modeling and Simulation*, 1(1), 95-118.

Sebstiani, F. (2002), "Machine learning in automated text categorization." *ACM Computing Surveys (CSUR)*, 34(1), 1-47.

Smith, G. (2001), "Public health approaches to occupational injury prevention: Do they work?" *Injury Prevention*, 7 (Suppl. I), i3-i10.

Sorock, G., Smith, G., Reeve, G., Dement, J., Stout, N., Layne, L., Pastula, S. (1997), "Three perspectives on work-related injury surveillance systems,." *American Journal of Industrial Medicine*, 32, 116-120.

Sorock, G., Ranney, T., and Lehto, M.R. (1996), "Motor vehicle crashes in roadway construction workzones: An analysis using narrative text from insurance claims," *Accident Analysis and Prevention*, 28(1), 131-138.

Wellman, H., Lehto, M.R., Sorock, G., and Smith, G. (2004), "Computerized coding of injury narrative data from the National Health Interview Survey," *Accident Analysis and Prevention*, 36(2), 165-171.

Williamson, A., Feyer, A-M., Stout, N. (2001), "Use of narrative analyses for comparison of the causes of fatal accidents in three countries: New Zealand, Australia and the United States." *Injury Prevention*, 7 (suppl I), i15-i26.

Chapter 123

Voluntary Guiding Principles for the Secure Handling of Nanomaterials

Wanda L. Greaves - Holmes

The University of Central Florida
Orlando, Florida 32816-2993, USA

ABSTRACT

In the absence of scientific clarity regarding the potential health effects of occupational exposure to nanoparticles, there is a need for guidance in making decisions about hazards, risks, and controls (Schulte & Salmanca-Buentello, 2007). Presently, no guiding principles have been universally accepted for personal protective equipment that is worn to prevent exposure to nanomaterials. The purpose of this article is to survey the literature and develop guiding principles in which all can utilize to minimize occupational exposure to nanomaterials.

INTRODUCTION

Although major trends may emerge later that show that nanomaterials are harmless to humans, I suggest precautions be established. Presently, there are no universally standardized, published guidelines or regulations for the safe handling of engineered nanomaterials. Research is inconclusive as to whether or not engineered nanoparticles may pose risk to human health due to their various compositions, sizes and ability to cross mammal's cell membranes. Engineered nanomaterials may exhibit higher toxicity due to their size compared to larger particles of the same composition. Current information about risks associated with nanoparticle exposure is limited. Until irrefutable evidence is available on the risks associated with nanomaterials voluntary precautions for the work place are highly recommended. Risk assessments and control strategies for nanotechnology research will be based on the most current toxicological data, exposure assessments, and exposure control information available from The National Institute for Occupational Safety and Health (NIOSH), Nanosafe (the United States Environmental Protection Agency (EPA), The Institut de recherche Robert-Sauvé en santé et en sécurité du travail (IRSST), the National Institutes of Health (NIH),Approaches to Safe Nanotechnology: A Informal Exchange with NIOSH, and Nanosafe were used to formulate these voluntary guidelines.   The following voluntary work place practices which may decrease the risk of human exposure to nanomaterials are suggested below.

Manufacturing Controls

Strict control of airborne nanoparticles can be accomplished by using conventional capture exhaust ventilation such as chemical fume hoods. Glove box containment is another effective method. Passing capture exhaust through a HEPA filter will provide protection against release of nanoparticles into the environment. Some actions that would require manufacturing controls include:

- Working with nanomaterial in a liquid media during pouring or agitation which could release aerosols size nanoparticles
- Fabricating nanoparticles
- Handling nanomaterials powder
- Maintenance or cleaning of equipment used to produce nanomaterials
- Utilize hand held personal environmental monitors like the TSI Aerotrak

Work Practices

Preventing inhalation, skin exposure, and ingestion are paramount if workers are to work safely with nanomaterials. This involves following standard procedures that should be followed for any particulate material with known or uncertain toxicity. Nanoparticle are so minute, these particles follow airstreams more easily than larger particles. Control of airborne exposure to nanoparticles will primarily be accomplished using engineering control similar to those used for general aerosols and vapors. These nanomaterials will be easily collected and retained in standard ventilated enclosures such as fume hoods. Additionally, nanoparticles are readily collected by HEPA filters. Respirators with HEPA filters will be sufficient protection for nanoparticles in case of immense spills. Many nanomaterials are synthesized in enclosed reactors or glove boxes. The enclosures are under vacuum or exhaust ventilation, which prevent exposure during the actual synthesis.

	Latex Gloves	Nitrile Gloves
Advantages:	• Cost effective • Satisfactory dexterity • Moderate quality • Moderate irritant & infectious protection	• Cost effective • Satisfactory dexterity • Superior quality
Disadvantages:	• Deterioration when used oil, greases & organic materials • Often imported • Frequent allergic reactions	• Deterioration when used with benzene, methlyene chloride, trichloroethylene, various ketones
Best utilize when using:	• Bases, alcohols, diluted water solutions	• Aliphatic chemicals, oils, grease, xylene, perchlorethylene & toluene

Table1.1 Comparison of Latex versus Nitrile gloves.

Inhalation exposure can occur during additional processing of materials removed from reactors; this processing should be done in fume hoods. In addition, maintenance on reactor parts that may release residual particles in the air should be done in fume hoods. Another process, the synthesis of particles using sol-gel chemistry, should be carried out in ventilated fume hoods or glove boxes. Good work practices will help minimize exposure to nanomaterials: These work practices are consistent with general good laboratory practice:

- Avoiding direct contact with nanomaterials especially when airborne or in liquid media (during a pouring and /or mixing processes with a high degree of agitation and process containment is not possible) using adequate personal protective equipment (PPE) such as:

Wearing FFP3 type masks or powered respirators incorporating helmets equipped with H14 high efficiency particulate air (HEPA) filters.

- Install and use efficient exhaust systems with particle filtration and ventilation systems filters to minimize free flowing airborne ultra fine particles.
- When working with dry, ultra fine particulate matter, a sturdy glove with good integrity should be used. Using two pair of disposable nitrile gloves are strongly recommended.
- Wear protective eye wear (e.g. safety goggles)
- Wear protective clothing and safety shoes
- Utilize ULPA filters (United Lightning Protection Association) to minimize combustion
- Storage or consumption of food or drink in areas where nanomaterials is handled is prohibited.
- Application of cosmetics, etc... in areas where nanomaterials is handled is prohibited
- Employees must wash hands before leaving the work area and after removing protective gloves
- Lab coats can easily become contaminated and should be removed before leaving the lab/ workplace.
- Personnel should avoid touching the face or other exposed skin after working with nanomaterials prior to hand washing.
- All containers containing nanomaterials must be labeled consistent with existing laboratory requirements.
- Cleaning of areas must be done with wet wiping or HEPA vacuuming. Dry sweeping or using compressed air is prohibited. Disposal of contaminated cleaning materials must comply with hazardous waste disposal policies.

Nanotechnology and nanosciences is a dynamic and rapidly growing field that offer the promise of technologically based innovations that will substantially improve the quality life for all human kind. The data currently available on some products reveal various information that, while preliminary, already allows us to conclude that engineered nanoparticles must be handled with care and that workers' exposure must be minimized, since these effects are extremely variable from one product to another. Therefore, a comprehensible, understanding of the possible drawbacks of nanotechnology is critical to realizing the significant benefits of nanotechnology. The majority of the initial nanomaterials research has focused on the probable hazards and risks of nanotechnology-based manufacturing. Although, toxicological research for nanotechnology is in its formative years, concerns about potential risks to the health and safety of workers, will require definitive answers. Questions will be focused on manufacturing practices, procedures and controls for the present and future uses of nanotechnology. Yet another area of interest is the environment. What is the fate of the environment when nanomaterials are disposed? What does "appropriate" disposal mean as it related to the field nanotechnology? What is obvious; however, is that the nanotechnology manufacturing industry must identify, develop and implement the optimum approach for protecting its employees, and the public at large. One promising option indicates that researchers may be able to "engineer out" unacceptable levels of toxicity in nanomaterials. If this undertaking comes to fruition, then the industry will be able to minimize the potentially negative implications to its worker and the environmental impact of nanomaterial-based manufacturing and products.

According to the documented toxic effects on living organisms as well as the unique physicochemical characteristics of nanomaterials validate immediate use of personal protective equipment ect., to limit exposure and protect the health of potentially exposed individuals. The introduction of strict universally standardized guidelines and procedures to prevent any risk of occupational disease in researchers, students or workers who synthesize, transform or use nanoparticles. A scientific approach to the identification, assessment, and mitigation of the risks posed by nanomaterial manufacturing and commercialization will protect the public, the environment and industry, thereby ensuring that the benefits of nanotechnology are shared by all.

REFERENCES

Greaves-Holmes, W.L.   (2010), *"A Voluntary Guide for Safe Handling of Engineered and Fabricated Nanomaterials." Journal of Technology Studies* Bowling Green, Ohio In Print

Schulte, P.A. and Salamana-Buentello, F. (2007), "Ethical and Scientific Issues of Nanotechnology in the Workplace." *Environmental Health Perspectives* , Volume 115, Number 1.

CHAPTER 124

Up-to-date Personal Protective Equipment

Katarzyna Majchrzycka

Central Institute for Labour Protection – National Research Institute
Czerniakowska 16,00-701
Warsaw, Poland

ABSTRACT

Personal protective equipment belongs to the group of equipment that protects the employees against dangers in the workplace. The decision to use such equipment must be preceded by all possible actions, both technical and organizational, aimed at eliminating or reducing the hazard to an admissible level. The basic development trends referring to particular types of personal protective equipment have been presented here. This especially concerns solutions in protective clothing, hand and foot protection, with particular emphasis put on producing clothing ensuring the sense of comfort and optima functioning for the user, applying new composites, intelligent textiles and nanotechnology. As for respiratory protective equipment and equipment protecting against falls from a height, directions in development of this equipment was discussed, emphasizing the use of new construction elements and electronic systems in order to inform the user of any changes in protective features and / or usage during the time of using those agents.

Keywords: Personal protective equipment, Smart an intelligent clothing, Polymer gloves, Equipment protecting against falls from a height, Active welding filter, Respiratory protective devices

INTRODUCTION

Personal protective equipment belongs to the group of equipment that protects the employees against dangers in the workplace. The decision to use such equipment must be preceded by all possible actions, both technical and organizational, aimed at eliminating or reducing hazard to an admissible level. Despite the above rule, the use of personal protective equipment is still very common in many workplaces. This includes, in particular, mining, construction, transportation and working in rooms with small capacity, for example containers, manholes, canals and refers to all types of emergency actions.

A common cause of accidents at work is the fact that employees fail to use personal protective equipment supplied by the employer. Another reason is lack of such equipment at the workplace or inappropriate selection of equipment. The employees' reluctance to use personal protective equipment may result from the fact that equipment is inappropriate for the conditions of a given workplace which provides only illusionary protection from danger. At the same time, it is worth mentioning that, new products are created with the use of modern technologies, technically advanced materials and electronic systems, which allow to introduce modern and competitive solutions already at the level of product design and production. This is possible due to the dynamic development of material engineering, information and communication technology that has been observed in the last few years.

PROTECTIVE CLOTHING

The dynamic development of technology resulting in the production of new materials, textile constituents and fibres provides the possibility to create new types of protective clothing, taking into consideration the aspects of ergonomy, significant for the user.

Current trends in developing protective clothing are aimed at producing clothing that is both comfortable and functional for the user. A new type of clothing has appeared that has captured the attention of the users – smart and intelligent textiles and clothing. Intelligent clothing receives impulses directly from the human body, as well as its surroundings, and reacts accordingly, introducing physical, chemical and biological changes, which are often reversible. The impulses that lead to an active reaction of textiles used in the design of protective clothing, include: stress, electromagnetic fields, temperature, humidity, UV, IR radiation, chemical substances. Intelligent textiles react by introducing an appropriate change, e.g. of their geometrical dimensions, focus level, stress positioning, light reflection coefficient. Skillful use of the properties of the new textiles and materials in the design of protective clothing may have outstanding effects resulting in the production of highly-functional clothing that cooperates with the human body and reacts to the changes of the external environment in a programmed way.

Intelligent textiles include textiles with phase change materials (PCM) that can change their state of matter, whenever heat is released from or absorbed by the body. In case of protective clothing, such materials may be used to change the thermal insulation of clothing.

The selected type of PCM in form of capsules may be incorporated into a textile product in various ways. Nowadays, there are produced fibres which include micro PCM capsules. They are totally surrounded by a polymer and permanently enclosed inside the fibre. The most popular is a polyacylonitrile one of trade name Outlast, with heat capacity from 4.2-8.4 J/g (Cox, 1998). The latest reports concerns viscose fibres with micro PCM capsules with heat capacity of 60 J/g.

At a conference devoted to innovation in chemical fibre industry that took place in Dornbirn in 2009, a new type of product and technology based on PCM (Vandendaele, 2009) was presented. Presented products were based on plant oil closed in microcapsules characterized by reactivity to cellulose, synthetic and protein fibres. This type of PCM is deposited on textile materials during conventional textile processes and their application does not require the use of any binding agent. The patented product is known under the name Thermic and is characterized by high heat capacity during the use and washing.

In many research centres, there are attempts to work out new products with PCM or new ranges of their use. Due to the fact that using protective clothing is connected with less comfort and heat load of the user, there have been attempts to apply PCM in protective clothing structures so as to cool the user's organism.

Research in PCM also concern underwear designed to be used under protective clothing [Pause, 2003; Pause 2006). Selected PCM microcapsules have been permanently connected to a polymer membrane that has later been placed between two layers of material. According to researchers, underwear made from this material is light, covers the whole body and is designed for multiple use and conservation. Tests on waistcoats with PCM, in which their positive effect in shaping favorable climate under clothing for surgeons has been confirmed, have been carried out at SNITEF (Health Research, Materials and Chemistry) in Norway (Reinerstein et al., 2008).

The design of intelligent clothing is possible due to bi-component fibres that change their shape depending on the surrounding temperature. They are flat at room temperature, but tend to expand irregularly and twist at a higher temperature. An example of modern high-technology that may be used in the design of clothing protecting from cold, includes fibres that are able to transform a wide solar light spectrum into thermal energy.

Polyurethanes with the memory of shape of glazing temperature above 55 °C, may be used in protective clothing and other means of individual protection, used i.e. in chemical, metallurgy or food industry. They are also applicable in intelligent waterproof and steam permeable membranes that are currently commonly used in clothing industry in textile and non-woven lamination as a waterproof layer in clothing protecting against bad weather. Together with the increase of temperature, exceeding activation temperature, there appears expansion of pores and increase in water steam permeability. This is beneficial from the point of view of thermoregulation of the user's organism as greater amount of sweat can be evaporated and taken outside.

Another group of active clothing, includes clothing equipped with integrated heating systems that guarantee precise temperature regulation. It may be used in workplaces situated in cold micro-climates, where the change of the employee's activity is followed by changes in the quantity of the released heat and the need to protect the body from the cold, and the use of clothing as passive protection does not provide the employee with thermal comfort. An appropriate level of protection and comfort may be achieved exclusively with the use of active protection that changes its insulation depending on the climate changes of the external environment and the level of heat released by the user.

Central Institute for Labour Protection – National Research Institute (CIOP-PIB) has created a model of protective clothing designed for work in cold environments, which reacts to the change of temperature and changes its thermal insulation in such a way so as to provide the user with maximum thermal comfort (Kurczewska and Leśnikowski, 2008). The active elements used in this type of clothing include six heating inserts made from steel fibres yarn. The heating inserts are controlled via a measuring and control system, which collects data from temperature micro-sensors situated at two points of the human body on the inside and outside of the clothing (see Figure 1).

a)

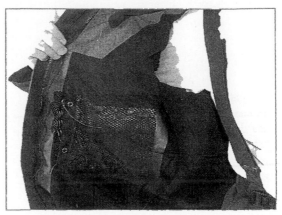

b)

Figure 1a Protective clothing for work in cold environment
1 b Active elements used in this type of clothing
(own source)

The new generation of protective clothing includes clothing with implemented sensors to monitor the physiological parameters and the external environment. Special clothing has been designed for emergency services, e.g. fire-fighting units, with implemented electronic micro-units that monitor the physiological parameters of the user of the clothing and the risk level he/she is exposed to. The aim of this type of clothing is to monitor the physiological condition of a fire-fighter, taking into consideration the external conditions and the strenuousness of the work (the body's energy expenditure and the resulting pressure on the human body).

HANDS AND LEGS PROTECTIVE EQUIPMENT

Changes in lifestyle and an increased professional activity of people using personal protective equipment have had significant effect on the requirements concerning equipment's functionality. The growing popularity of comfortable sports footwear has made users also look for comfort in case of other types of footwear, including protective footwear. The intensive development of textile production technology and advancements made in footwear design, which began with sports shoes, allowed for the introduction of new solutions in the production of protective equipment for legs and hands. Modern protective equipment intended for professional use, apart from its protective function, has to fulfil certain requirements concerning hygiene and the comfort of use.

As for protective gloves, a significant user's requirement is the provision of multifunctional protection, as well as the safety of the product, whose production involves the use of hand protective equipment, e.g. in food processing industry or electrical industry. Multifunctional protective gloves expand their scope of use. An example of new, multifunctional solutions used in the production of protective gloves, includes gloves produced with many layers of different polymers designed to protect hands against chemicals, mechanical injuries, burns resulting from touching hot objects, which have been approved for catering industry.

Another example includes polymer gloves that protect against harmful biological factors, including viruses, e.g. C and B-type jaundice, HIV virus and herpes virus, when the infection results from the use of infected needles. Between the external and internal layers of the glove made form thermoplastic elastomer, there is a solution for disinfection in form of dispersion. Modern polymer gloves designed by the Ansell company have a plate structure on the inside that releases a moisturizing substance when gloves are put on, which reduces friction. The above examples confirm the producers' concern to create products that are highly comfortable, but at the same time allow to eliminate factors qualified as harmful.

An interesting concept in construction of means of hand protection is working out the way of signalling protection time of activities, while these means are used for protection against chemical factors. A solution available on the market that enables to signal protective gloves leakage, is a product constructed by an American company CLI Colometric Laboratories. The solution based on colorimetric indices, relies on signaling contact with a certain group of chemical substances through the change of colours of self-adhesive pads. The principle is based on pasting pads directly on a hand, and signaling takes place after a chemical substance goes through the glove layer into the pad that is placed below it.

Another idea of signalling chemical leakage worked out by CIOP-PIB, includes the application of microcapsulation method, as a way of closing dyes that work selectively on selected chemical compound groups, i.e. acids, bases and organic solvents. Microcapsules that are created as a result of polymerization, together with dyes enclosed in them, are placed between the glove layers during the subsequent cycles of dipping method. A model cheme of multilayer structure of a glove with built-in system of sygnalling chemical leakage that uses plymer microcapsules with enclosed dye.

EQUIPMENT PROTECTING AGAINST FALLS FROM A HEIGHT

When considering the development of personal protective equipment that protects against falls from a height, it is worth noting that developments have been made both in the materials used, as well as in the design of such equipment.

In case of textile materials, such as technical straps or ropes, aramid fibres are used more and more often. They are also used in models which include polyamide or polyester fibres. These are used in case of textile materials used in the production of, e.g. safety harnesses, belts and ropes for positioning and safety ropes, which have increased resistance to the effect of hot agents, e.g. splattering of melting metal in welding.

Interesting new developments may be observed in case of light metal alloys, e.g. aluminium used in the production of catches. Due to that, the weight of the equipment is reduced, but its resistance is maintained and, its resistance to corrosion is increased, e.g. It is also worth noticing that modern plastic materials have been introduced in the production of, e.g. casing of self-locking equipment protecting against falls, thanks to which the weight of the equipment is reduced and its resistance to atmospheric conditions is increased.

With regards to the design of equipment that protects against falls from a height, the following changes have been observed around the world. With respect to harnesses, there is a tendency to reduce their weight and increase user's comfort. This is achieved by optimizing the position of the belts that transfer the weight and equipping them with special anti-pressure pads. More and more frequently, catches are produced in the self-locking and self-blocking version, which makes it easier for the user. Thanks to the common use of sewing machines, textile shock absorbers are produced mainly with the use of the sewing technique instead of the weaving technique, which reduces production costs. Moreover, shock absorbers produced in this way are less sensitive to changes of the atmospheric conditions.

A significant trend in the development of equipment that protects against falls from a height, in particular in the case of safety harnesses, is providing such equipment with devices that allow to notify about the accident, e.g. stopping the fall. Such devices may be started up manually by the equipment user or automatically, e.g. after exceeding the threshold level of the load, affecting the attachment device of the safety harness. In the past few years we have also observed the introduction of entirely new models of equipment designed for very specific use, e.g. anchorage subunits operating in form of a sucker designed to protect people working on wings and fuselage of airplanes when carrying out repairs and maintenance services.

EYES AND FACE PROTECTORS

Innovations in the design of protective visors and filters, mainly include the use of advanced material technology. This includes the use of modifications in spectral light transmission properties, which enables the adjustment of the light transmission coefficient to the values corresponding to specific light conditions, and therefore current user's requirements. The use of the photochromic effect allows to change the light transmission coefficient depending on the change in the light intensity of the external radiation, which is usually accompanied by ultraviolet radiation that causes the photochromic effect. In order to change the transmission of the optical radiation going through the filter (particularly in welder filters), the direction in the liquid crystal layer is changed as a result of applying the electromagnetic field, which can be generated by impulses of light.

Mass colouring or surface colouring allows to optimize the radiation spectrum that reaches the eye. The use of the light polarization effect and an appropriate positioning of the polarization surface with regard to the observed objects allows to eliminate the effects of adverse reflections of solar radiation (which may be blinding) against large reflexive surfaces, such as water or the surface of the road. In order to reduce the reflection of radiation, appropriate anti-glare coating can be put on the surface of protective visors and filters, as it reduces reflections resulting from light interference.

Another modification used to provide mechanical protection of the layers applied on the surface of the lens, is hardening done with the use of appropriate varnish. Additional layers that reduce misting may be applied on the internal surfaces. Apart from mechanical resistance and filter properties appropriate for specific use, resistance to ageing and the real mass of the material also play a significant role. The parameters described above have a significant effect on the quality and durability of the product and, as such, define the requirements concerning modern eye protection equipment.

An interesting solution prepared in CIOP-PIB is photochromic dye in arrangement of an active welding filter. Dangers related to exposure of workers to intensive ultraviolet, visible and infrared radiation occur during welding.

Because the visible radiation emitted by welding arches is very intensive, it is necessary to use very dark filters, with light transmittance coefficient below 0.061%. Using such dark filters makes it impossible to see the welded object before striking of the welding arch. Because of this, the worker is forced to 'peep' at the welding process when the welding arch strikes. One of the results of this momentary glare is the development of cataract – occupational diseases of welders. The use of active welding filters with variable shade number is an effective solution of this problem. The series of functionalized spirobenzopyranoindolins were synthesized and their photophysical and photochemical properties were investigated in solution using of absorption and emission spectroscopy. The model of an active welding filter with photochromic layer based on 1',3',3'-trimethyl-6-nitrospyro[2H-1-benzopyran-2,2'-indol] (6-nitroBPIS) have been developed. Conducted tests have shown that the developed model of filter meet the assumed requirements:

- the transmittance values were within the limits specified for protection level 5 in light state and levels 9 and 10 in dark state for welding with a covered electrode with current intensity between 100 A and 150 A,
- switching time of the filter 0.011 ms, guarantees effective eye protection against the harmful effects of light glare.

RESPIRATORY PROTECTIVE DEVICES

With regard to equipment that protects the respiratory tract, two main development trends should be mentioned. The first one is related to the development of technology concerning the production of construction elements used in the equipment that protects the respiratory tract, and the second one refers to the use of electronic devices that inform the user about the protective characteristics and/or functional properties of protection.

The development of technology related to the production of filter materials – filter non-woven fabric, which is the basic construction element of filters and filter face masks, is aimed at introducing different types of additional elements – modifiers introduced at the phase of fabric formation, which are designed to improve their capacity to retain dust particles, including nanoparticles, or provide them with new properties, e.g. bacteriostatic or antibacterial properties. A certain group of materials, treated as an additional element of the base material – polymer, is used to increase the capacity of providing the non-woven fabric with electrostatic load. As a result of maintaining the electrostatic load when using the filter non-woven fabric, the required level of protection is maintained for a longer time.

Bioactive filtering non-wovens have been developed by CIOP-PIB and introduced into production, which were designed to purify the air of pathogenic microorganisms and also bioactive filtrating half-masks and filters that were protecting respiratory system against biological factors (Brochocka and Majchrzycka, 2009). This equipment ensures effective keeping of microorganisms in the filtering layer with diameter of over 0,3 μm, at the same time inhibiting their growth. This ensures prolonged time of effective protection guaranteed by the filtrating material at the level of appropriate protective class against aerosol with stable and liquid dispersing phase, and also safe and multiple use of the device (see Figure 2).

Figure 2 Half - mask with bioactive filters
(own source)

Filtrating devices of respiratory system, in form of filtrating half-masks and filters completed with half-masks based on bioactive filtrating layers designed to be used in food industry, pharmaceutical industry, farming and in health care sector: hospitals, private surgeries, dentistry, larygology, prosectorium, etc.

In case of materials used to clean the air from harmful steam and gases, the development of technology related to the modification of such materials has been observed. The thermal, chemical and physical treatment using, e.g. low-temperature plasma, is aimed at increasing the adsorption capacity intended for specific steams of organic substances difficult to absorb.

Another trend in the development of equipment that protects the respiratory tract is the introduction of electronic elements. These usually include sensors that allow the users to monitor the protective parameters (entrance time) or functional parameters (breathing problems) of the protective equipment. Majority of studies have focused on the construction of a sensor indicating the exhaustion of the absorbent deposit, the use of which is related to the risk of breathing in harmful steam and gases. Electrical devices are becoming increasingly popular in the production of equipment that protects the respiratory tract as they allow for communication – voice communication modules in face masks, to monitor the parameters of the work environment (temperature, humidity) and, e.g. the amount of air in a pressurized air bottle.

CONCLUSIONS

Aspiration to assure comfortable living and working conditions for people, and greater care of health and life quality, creates favorable conditions as for the tendencies in developing intelligent means of personal protection. Apart from basic protective functions, these means also support physiological functions of an organism, monitor its condition, warn against external and internal danger. Modern materials used in constructing those means, fulfill this condition thanks to the advanced technologies that are applied in their production.

REFERENCES

Brochocka, A. and Majchrzycka, K. (2009), *Technology for the production of bioact melt-blown filtration materials applied to respiratory protective devices,* Fibres & Textiles in Eastern Europe 2009, Vol. 17, No 5(76)

Cox, R. (1998), *Synopsis of the new thermal regulating fiberOutlast.* Chemical Fibres International, No 6.

Kurczewska A. and Leśnikowski, J. (2008), *Variable-Thermoinsulation Garments Wi a Microprocessor Temperature Controller.* "JOSE", Vol. 14, No 1/2008

Pause, B. (2003), *Chemical protective garments with thermo-regulating properties.* Proceedings of 2nd ECPC, Switzerland

Pause, B. (2006), *New cooling undergarment for protective garment system.* 3rd European Conference on Protective Clothing (ECPC) and NOKOBETEF Conference, Gdynia, Poland

Reinertsen, R. E. et al. (2008), *Optimizing the Performance of Phase-Change Materia in Personal Protective Clothing Systems.* "Jose", Vol. 14, No 1/2008

Vandendaele, J. P. (2009), *Thermic^{TM}: Thermal Intelligent Comfort. A New Durable PCM Concept for fabrics.* Proceedings of 48th International Man-Made Fib Congress, Dornbirn, Austria

CHAPTER 125

Manufacturing Process Improvement Based on Reducing of Ergonomics Risks in Woodworking Enterprise

*Henrijs Kalkis[1,2], Irina Rezepina[1], Valerijs Praude[1],
Zenija Roja[3], Valdis Kalkis[3]*

[1]University of Latvia, Faculty of Economics and Management

[2]Riga Stradins University, Faculty of European Studies

[3]University of Latvia, The research centre of Ergonomics
Faculty of Chemistry

ABSTRACT

Many manufacturing processes in woodworking industry includes handling, lifting, monotonous and lasting arm movements, compulsory work positions, a.o. All of mentioned risk factors influences not only workers health, but also company's productivity. Woodworking is one of leading industries in Latvia. This work aimed to investigate how woodworking company's manufacturing process improvement reduces ergonomics risks at the workplaces. Results show that Ergonomics solutions improves woodworking manufacturing processes and employees work conditions in long-term. Besides complex approach of combining process management with ergonomics solutions can provide preventive activities for improving organization production and establishing safe and healthy work conditions.

Keywords: woodworking, manufacturing, ergonomics, improvement

INTRODUCTION

Woodworking is one of leading industries in Latvia. Its value added represents in almost 20% of total manufacturing. Industry is characterized with high export indicators, almost 2/3 of total manufacturing are export products (Economic Development of Latvia. Report, 2009). It is a complicated branch not only from the point of view of managing the technological processes of manufacturing, but also in terms of the organization of work and risk factors of the working environment, including the ergonomic ones. In the modern view process management allows for increasing the operation efficiency and production quality of the any organization and decreasing its costs (Asfahl, 2004).

Consequently, we can state that no business is possible without processes and their quality management. This is particularly important in the conditions of competition when the enterprise management must actively engage in the acquisition of the latest information and trends for continuous improvement of the existing and creation of new processes. Thus nowadays enterprise management must devote particular attention to the implementation of quality processes, process assessment, establishment of process management criteria as well as constant process control and supervision (Besterfield, 2004; Freivalds and Niebel, 2009). Many manufacturing processes in woodworking industry includes heavy load lifting, monotonous and lasting arm movements, compulsory work postures, a.o. All of mentioned risk factors influences not only workers health, but also company's productivity. Therefore ergonomics solutions are necessary for workers to choose appropriate work load and optimal work conditions in connection with manufacturing technologies, as well as improve work abilities and wellbeing at workplaces.

Ergonomic solutions are essential in quality process management and human wellbeing at the workplaces that results in increase of human resource workabilities and appropriate work methods and optimal working conditions associated with the production technologies.

The aim of the research was to analyze the improvement of the woodworking enterprise process management in manufacturing lines by applying the quality management and ergonomics investigation methods.

Medium size Latvian woodworking enterprise was chosen for the research. The study group consisted of 250 workers; all men aged $36.6 \pm SD$ 8.32. The main processes in the enterprise are sawing timber, its sorting, drying, packaging, assembling of the ready production and selling. The research was made in the period before and after the mechanization of the production lines and introduction of ergonomic solutions in the manufacturing processes from 2008 till 2009.

METHODS

QUESTIONNAIRES AND INTERVIEWS

Using in the ergonomics risk assessment questionnaire sheets and interviews it is possible to find out the opinion of the workers on the condition of existing workplace, as well as its compliance with his/her physiological and psychological abilities. It is possible to get know, what the worker complains about after the work, worker's opinion on necessary improvements, on psychological microclimate in the workplace etc.

WORKLOAD

Key indicator method (KIM) was used for assessment of the work hardness and ergonomic risks. By means of this method possible overloads lifting or moving heavy loads or performing other dynamic operations are identified (Buckle and Li, 1999). According to this method there are I – IV work risk degrees (Rd). When this method is applied, total workload (WL) is calculated as follows: WL = (M+S+C) L. Actions related with moving of heavy loads are assessed using points and taking into account following physical load identification criteria:
1) Weight of the object to move (kg) – M;
2) State of worker's body – S;
3) Conditions in which the work is performed – C;
4) Work intensity or length of work requiring physical load (hours) – L.

LOAD LIFTING AND MOVING

National Institute for Occupational Safety and Health (NIOSH) lifting equation is an assessment method for lifting and lowering tasks (Waters et al., 1993). The equation provides a recommended weight limit (RWL) based upon task parameters and the duration the task is performed. The RWL is obtained through the following equation (Dempsey, 2002):

$RWL = 23[25/H][1-(0.003|V-75|)] \times [0.82 + 4.5/D] [1-(0.0032 A)] FM \times CM$,

where H is the horizontal location in cm, V is the vertical location in cm, D is the distance in cm, FM is the frequency multiplier, A is the asymmetry angle in degrees, and CM is the coupling multiplier. The actual load lifted or lowered divided by the RWL provides the lifting index (LI). LI values greater than 1.0 are assumed to represent tasks posing risk to the worker population (Waters et al., 1994).

PROBLEM IDENTIFICATION

Failure Modes and Effects Analysis (FMEA) is a tool used in various industries to identify, prioritize, and eliminate known potential failures, problems, and errors from systems under design (Stamatis, 1995). Several industrial FMEA standards employ the Risk Priority Number (RPN) to measure risk and severity of failures. RPN is a product of three indices: Occurrence, Severity, and Detection. The analysis is organized around failure modes, which link the cause and effect of failures (Rhee and Ishii, 2003). FMEA has various applications, for example, it can be applied during the process design or in the enterprise risk (incl. economic, ergonomic, work environment risks) assessment procedures (McDermott et al., 1996). FMEA in research was used for identifying and prevention of the problems with products or processes before the errors take place and the possible unfavorable effects occur.

RESULTS AND DISCUSSION

Data gained from questionnaires indicated that before the plant mechanization workers exposed to physical strain, forced work postures, and during the work mainly pain appeared in back, legs, hands and arms. Workers mainly were subjected to dynamic work and intensity of load lifting varied from 10 to 500 times per shift according to job operations. This emphasizes the importance of the ergonomic risks in order to assure efficiency of the organization and pay greater attention to human resources or human capital as a key component in company's whole system.

The physical work load assessment was done in order to identify overload in various technological process cycles and discover necessary preventive ergonomics solutions. The assessment was carried out before ("0-process") and after the mechanization of lines with implemented ergonomics for optimal work process or techniques ("Ergo-process"). The first ergonomics risk assessment was done in 0-process applying the KIM method. The results are shown in Table 1.

Table 1. KIM method assessment scores and risk degree for professions in 0-process

PROFESSIONS	M	S	C	L	WL	Risk degree
	Scores					Rd = I – IV
Side log packaging operator	3	4	2	5	45	III
Side log sorting operator	2	3	2	2	14	II
Edging operator	2	2	1	1	5	I
Peeler operator	1	2	2	1	5	I
Central plank packaging operator	4	3	2	5	45	III
Log saw operator	4	3	2	3	27	III
Furniture packaging operator	7	8	2	4	68	IV

Based on the KIM method results the company established ergonomic solutions in several main manufacturing process cycles. Such action reduced physical work load, decreased probability of technological faults as well as assured the load lifting limits. KIM method results show that the workload reduced from III risk degree in

"0-process" (before the establishment of ergonomic solutions) to acceptable degree of risk severity II in "Ergo-process" (after the establishment of ergonomic solutions) for such professions: log packaging, log sorting, log sawing and product packaging operators. The results of reiterative ergonomics risk assessment in "Ergo-process" are shown in Table 2.

The highest risk degree IV was assessed for packing operators. According to survey data packing operators most frequently felt fatigue, marked pain in the back and during the work often made defective goods or delayed the work cycle due to heavy workload and inconvenient work conditions.

Table 2. KIM method assessment scores and risk degree for professions in Ergo-process

PROFESSIONS	M	S	C	L	WL	Risk degree
	Scores					Rd = I – IV
Side log packaging operator	3	4	2	5	45	II
Side log sorting operator	2	3	2	2	14	II
Edging operator	2	2	1	1	5	I
Peeler operator	1	2	2	1	5	I
Central plank packaging operator	4	3	2	5	45	II
Log saw operator	4	3	2	3	27	II
Furniture packaging operator	4	3	2	5	45	III

The ergonomics implementation involved training for safe and proper techniques of heavy load lifting and moving. For each profession the rest period was calculated and established. New mechanical and pneumatic supplementary aids were provided for the workers. Medical checkups were ensured for employees in connection with physical load at the workplaces. More attention was drawn to individual workers skills and abilities when arranging the workplace.

The establishment of new technologies for workplaces and ergonomic aids (pneumatic and automatic lifting, etc.) lifting and weight limit (RWL) recommended by NIOSH for all professions increased on average from 2.5 to 4 times, and the lifting index of lifting weight decreased an average of 3 to 6 times. The calculations of load lifting for various professions were done in "0-process" and "Ergo-process" and the results are shown in Table 3 and Table 4.

Table 3. Recommended weight limit (RWL), actual lifting and moving mass (M), lifting index (Li) in "0-process"

Professions	"0-process"		
	M	RWL	Li
Side plank packaging operator	8 ± 1 kg	2 kg	4,0
Side plank sorting operator	8 ± 1 kg	4 kg	2,0
Central plank packaging operator	9 ± 1 kg	5 kg	1,8
Small component packaging operator	2 ± 0,5 kg	0,5 kg	4,0
Furniture packaging operator	> 30 kg	10 kg	3,0

Table 4. Recommended weight limit (RWL), actual lifting and moving mass (M), lifting index (Li) in "Ergo-process"

Professions	"Ergo-process"		
	M	RWL	Li
Side plank packaging operator	8 ± 1 kg	12 kg	0,6
Side plank sorting operator	8 ± 1 kg	15 kg	0,5
Central plank packaging operator	10 ± 1 kg	18 kg	0,5
Small component packaging operator	1 ± 0,5 kg	5 kg	0,2
Furniture packaging operator	> 30 kg	23 kg	1,5

Human social and anthropometric indicators were taken into account in NIOSH calculations (worker's mean height 1.7 m, mean body mass 70 kg, sex: man, age: 18...30 years, physical preparedness: average).

The results show that worker may lift and move greater load during the work if reduced arm movements and the distance of moving the load or other NIOSH multipliers values. The exception in investigation was packing operator, when load exceeds 30 kg per one lifting time. For this profession the lifting index was not decreased although the ergonomics solutions were considered (e.g. mechanical lifting mechanisms). In this case were recommended longer rest periods and relaxation exercises for relieving muscle groups.

In order to work out recommendations for preventive measures of work environment improvement, the computer software ErgoEASER was applied. For example, if there is no possibilities to decrease lifting mass, recommendations are as follows: reduce the length of moving the load, body turning should be limited to 15°...20°, lift and move the mass (over 23 kg) 2 times per minute.

During the FMEA analysis the assessment was made in the main processes of the enterprise in the two manufacturing lines: raw material manufacturing (sawing line of timber) and the line of construction material manufacturing (furniture product line). After the assessment of the modes of errors, its consequences and causes as well as the existing and committed actions during the enterprise process optimization and improvement, the analysis was made of the equipment, ergonomic risks and production. In establishing the improvement actions the initially determined Risk Priority Number (RPN) = 9134 was taken into account. It is very high and suggested an urgent need for action both regarding the improvement of work organization as well as necessity of ergonomic measures.

After the implementation of ergonomic measures and technology modernization the RPN was decreased to 2444. It indicates that the implemented improvements have minimized the enterprise risk 3.73 times as the ergonomic risks and errors in the equipment operation were eliminated. As a result the production volume increased and its quality improved. The simplified results of FMEA are summarized in the Table 5.

Table 5. Management of risks and probabilities by FMEA

Process	Likely kind of fault	Before technological improvements				After technological improvements			
		Severity	Occurrence	Detection possibilities	RPS	Severity	Occurrence	Detection possibilities	RPS
Sawing line of timber (beam)	**Machinery:**								
	Damaged saw blade	7	8	9	504	7	3	5	105
	Damage in board	6	5	9	270	6	3	2	36
	Vibration	8	7	8	448	8	2	2	32
	Other faults*	6	6	6	216	6	3	3	54
	Ergonomics risks:								
	Uneven floor	8	10	9	720	8	4	5	160
	Heavy arm work	8	10	8	640	8	5	3	120
	Muscles fatigue	7	9	8	504	7	5	3	105
	Other faults*	6	6	6	216	6	6	3	108
	Production:								
	Dimension faults	8	8	8	512	8	5	4	160
	Design defects	8	7	7	392	8	4	5	160
	Other faults*	6	6	6	216	6	5	3	90
Furniture product line	**Machinery:**								
	Saw wear	8	7	9	720	8	4	5	160
	Hand protector lack	8	6	8	640	8	5	3	120
	Other faults*	6	6	6	504	7	5	3	105
	Ergonomics risks:								
	Muscles fatigue	8	8	9	576	8	6	5	160
	Heavy arm work	8	7	7	392	8	5	5	200
	Other faults*	6	6	6	216	6	6	3	106
	Production:								
	Dimension faults	8	10	9	720	8	4	5	160
	Design defects	8	8	8	512	8	6	4	192
	Other faults*	6	6	6	216	6	6	3	108
Total RPS before technological improvements („0- process")					9134	RPS in „Ergo-process"			2444

Worth to mention that implementation of quality and ergonomic improvements in manufacturing lines granted with economic efficiency calculations. Proved that the middle size woodworking company's total economic loss in "Ergo-process" compared to the "0-process", has decreased by 106200 EUR and the economic effect or benefit resulted in 154 000 EUR per year, taking into account and analyzing such factors: costs of illness, training, transfer costs, produced income, volume of production, costs avoided (anticipated future costs if no action is taken to avoid them). This case could be examined in further investigations as another cost-benefit success story of ergonomics implementation in process management of manufacturing enterprise.

This emphasizes the role of ergonomics for improving productivity and save costs resulting in total efficiency of company, not only considering it as issue of health and safety (Porter, 1998). From another point of view, this research illustrates good example where quality management improvements are integrated with ergonomics. Numerous studies showed that this is actual topic considering ergonomics as key factor that increases importance for core values in organizations such as productivity, quality and an inevitable change process (Klatte et al., 1997; Helander and Burri, 1995; Eklund, 1997) as well as essence of product design during production that influences the workload in relation with production costs and quality outcome (Helander and Nagamichi, 1992). Other investigations have focused on gaining positive effect on health outcomes procured by tool change, training in ergonomics of workers (McKenzie et al., 1985), and even identification of cost savings associated with human resources resulting in potential for a high return because recurring personnel costs, including training (Price, 1990) based on a one-time investment in human factors and ergonomics.

CONCLUSIONS

Complex approach of combining process management with ergonomics solutions can provide preventive activities for improving organization production and establishing safe and healthy work conditions. Ergonomics development can improve woodworking manufacturing processes and make better conditions of employees work as well as save and avoid costs in long-term based on this research results. In wood-working industry the FMEA is useful for using in the identification and prevention of problems in products and processes before the faults take place and the unfavorable consequences arise. FMEA, KIM and NIOSH are effective methods for ergonomics and quality analysis and improvement of main processes in woodworking enterprise. These methods would be suitable for complicated process improvement analysis in connection with ergonomics risks and solutions also for other economy sectors.

REFERENCES

Asfahl, C.R. (2004), *Industrial Safety and Health Management*. 5th ed. Upper Saddle River, NJ: Prentice Hall.

Besterfield D.H. (2004), *Quality Control*. New Jersey: Pearson Education.

BUCKLE, P., LI, G. (1999), "Current techniques for assessing physical exposure to work-related musculoskeletal risks, with emphasis on posture-based methods." *Journal of Ergonomics*, 42(5), 674–695.

Dempsey, P.G. (2002), Usability of revised NIOSH lifting equation. *Journal of Ergonomics*, 45 (12), 817–828.

Economic Development of Latvia. Report. (2009), Ministry of Economics, Republic of Latvia. Riga: Multineo.

Eklund, J. (1997), "Ergonomics, quality and continuous improvement - conceptual and empirical relationships in an industrial context." *Journal of Ergonomics*, 40, 982–1001.

Freivalds A., Niebel B. (2009), *Niebel's Methods, Standards, & Work Design*. 12th

Edition, McGraw-Hill.

Helander, M., Burri, G., (1995), "Cost effectiveness of ergonomics and quality improvements in electronics manufacturing", *International Journal of Industrial Ergonomics*, 15, 137–151.

Helander, M., Nagamichi, M. (Eds.), (1992), *Design for Manufacturability: a Systems Approach to Concurrent Engineering and Ergonomics*. Taylor & Francis, London.

Kalkis V. (2008), *Work Environment Risk Assessment Methods*. Riga: Fund of Latvian Education (in Latvian).

Klatte, T., Daetz, W., Laurig, W. (1997), Quality improvement through capable processes and ergonomic design. *International Journal of Industrial Ergonomics*, 20, 399–411.

McDermott, R., Mikulak, R., Beauregard, M. (1996), *The Basics of FMEA*. USA: Productivity, New York: Quality Resources.

McKenzie, F., Storment, J., Van Hook, P., Armstrong, T. (1985). A program for control of repetitive trauma disorders associated with hand tool operations in a telecommunications manufacturing facility. *American Industrial Hygiene Association*, 46, 674–678.

Porter, M. (1998), "Justifying the incorporation of ergonomics into organisational strategy—beyond single issue solving", *Advances in Occupational Ergonomics and Safety*, Vol. 2. IOS Press, Amsterdam, 119–122.

Price, H.E. (1990), "Conceptual system design and the human role. *MANPRINT: An Approach to Systems Integration.*" Van Nostrand Reinhold, New York.

Rhee, S., Ishii, K. (2003), "Using Cost based FMEA to Enhance Reliability and Serviceability", *Journal of Advanced Engineering Informatics*, vol.17, 179–188.

Stamatis DH. (1995) *Failure mode and effect analysis*. Milwaukee, WI: ASQ Quality Press.

Waters, T.R., Putz-Anderson, V., Garg, A., Fine, L.J. (1993), "Revised NIOSH equation for the design and evaluation of manual lifting tasks", *Journal of Ergonomics* 36 (7), 749-776.

Waters, T. R., Putz-Anderson, V., Garg, A. (1994), "Applications Manual for the Revised NIOSH Lifting Equation", *National Institute for Occupational Safety and Health, DHHS,* NIOSH Publication No. 94-110.

Chapter 126

Development of User Interfaces for STARtracker, a Web-Based Open Source Application For Tracking Oil and Gas Methane Emission Reductions

Gerald M. Knapp

Louisiana State University

ABSTRACT

The energy industry has recently been putting considerable efforts into reducing methane emissions in oil & gas exploration and production. This is motivated not only by the current environmental focus on greenhouse gas emissions, but by profit motive as well – when gas prices are high, emissions represent significant lost profits. The EPA GasSTAR program seeks to encourage and reward voluntary initiatives by companies aimed at methane emission reductions. A joint partnership between LSU, COMM Engineering, and Devon Energy led to the development of an open source web application for tracking emission reduction projects which was donated to the EPA and made available to industry. This paper explores usability analysis and issues addressed during development of the web interfaces for the application.

Keywords: Human Computer Interaction, Usability Analysis, Task Analysis

INTRODUCTION

The energy industry has recently been putting considerable efforts into reducing methane emissions and usage in oil & gas production. This is motivated not only by the current environmental focus on greenhouse gas emissions, but by profit motive as well – when gas prices are high, emissions represent significant lost profits. The EPA GasSTAR program seeks to encourage and reward voluntary initiatives by companies aimed at methane emission reductions. A joint partnership between LSU, COMM Engineering, and Devon Energy led to the development of an open source web application called STARtracker for tracking emission reduction projects. The software was donated to the EPA and made available to industry. Two iterations of the software have been developed, the first with basic management features, the second expanding capabilities based on user feedback and adding AJAX capabilities to improve feel. A third iteration is under way which is more formally addressing usability. This paper discusses usability issues that have been identified; solutions for these issues will be discussed during the presentation.

THE PRODUCTION PROCESS

During oil and gas exploration, the probable location of reservoirs and best location for extraction determined. A *well* is then drilled to extract the oil and gas. *Completion* is the starting point for oil & gas production, and is the process of capping a well and readying it for connection to a production *facility*. A production facility often serves multiple wells within a physical area. The purpose of the facility is to ready the natural gas and/or oil (often mixed together when coming out of a well). This involves:

- Separating gas, oil, and natural gas liquids (NGLs) into separate streams.
- Removing contaminants such as silica, which can corrode or sandblast pipelines when pressurized.
- Removing water, which causes pipeline corrosion.
- Pressurizing streams in order to bring them up to the line pressure of transport pipelines for distribution to processing facilities (process plants, storage facilities).

Figure 1 gives a basic idea of the process. A large production company may have tens of thousands of facilities and many more wells. These are typically organized into Countries ("US"), Divisions ("Southern"), fields ("Ft. Worth Basin"), facilities, and then individual wells.

Emission reduction opportunities in the production process generally fall into the following basic areas:

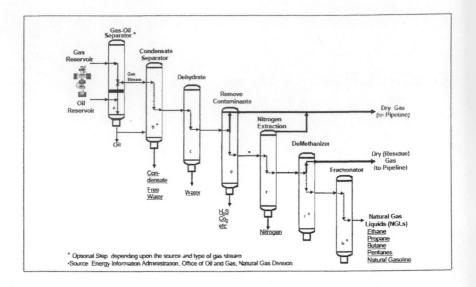

Figure 1 Generalized Natural Gas Processing Facility [Tobin et al. 2006].

- Completion-related. Large amounts of methane can be released during the capping process. This process can take from a couple of days to over a month (although emissions are not happening continuously during this time). "Green completions" use methods and technologies that reduce methane emissions to minimal values.
- Maintenance-related. During maintenance and repair activities, pipelines and equipment may need to be opened up or vented, releasing methane. These typically occur at a point in time or over a short duration of time.
- Facility operations. Seals and equipment can leak methane during normal operations. Also, because of the remoteness of most facilities, the only power source available is the natural gas or oil itself, and it is burned to power the facility equipment. Methane is typically emitted during the process, particularly if older and inefficient equipment is in use and no re-capture technology is in use. Another operational emission source is venting / flaring when there are over-pressures in the system. Emissions related to operations occur over the entire lifetime of the facility.

BEST MANAGEMENT PRACTICES

The EPA GasSTAR program had developed, and is looking to continuously add to, a set of "Best Management Practices", or BMPs, for methane emissions. BMP's include projects such as performing green completions, replacing seals with new types of seals that reduce emissions, replacing inefficient equipment with equipment that uses less fuel and emits less methane, and adding vapor recovery units (VRUs).

Each BMP has associated with it:
- Descriptive data on its implementation procedures and studies validating emission reduction effectiveness
- A set of calculations for calculating estimated emission reductions for each BMP implementation
- Optionally, other calculations such as economic value of the project (internal rate of return, present worth, etc).

GasSTAR industry partners track emission reduction projects, and report voluntary emissions reductions in a standardized format to the EPA GasSTAR program annually. The GasSTAR program provides awards and other recognition to its partners, and particularly to corporations which lead in emission reductions.

RATIONALE

Tracking large number of projects over thousands of facilities and many years presents a management challenge. The purpose of the web application was to:
- Centralizing both current and historical data, eliminating proliferation of spreadsheets.
- Providing access from anywhere, including field locations
- Simplifying reporting, allowing reporting anytime and flexible queries, as well as easy generation of GasSTAR annual reports.
- Simplifying overall management of corporate GasSTAR programs
- Providing a means of identifying opportunities for emissions reductions.

USABILITY ISSUES

USE CASES

Figure 2 summarizes the major users use case functions of the system. Details and usability issues of each are discussed in the following sections.

ADD, TRACK, AND MANAGE PROJECTS

Supervisory personnel enter new emission reduction projects into the application as they occur in the field. For point-in-time or short duration projects, the complete information for the project may be entered all at one time. For ongoing reductions (e.g., replacing old equipment with new), reductions may be accrued for years, projects are left "open". Thus there is a management task of coming back at a later date (possibly years later) to close the project. In addition, shut-ins (closing down of facilities for maintenance or because gas prices are too low) must be tracked as no operating emission reductions are accumulated during these times.

1216

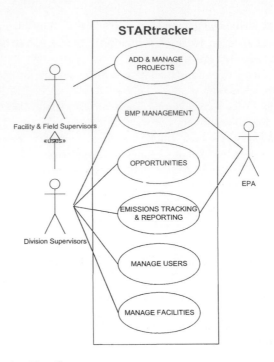

Figure 2 STARtracker Use Cases.

Figures 3 shows the interface from STARtracker version II for entering and editing reductions projects. Figure 4 shows the detail page for each project that was used in version I. In version II, BMP projects were generalized so new ones could be defined and existing ones extended over time, meaning the form fields can change. As a consequence, a technology decision was made to get rid of the detail page. From a usability standpoint, however, this turned out to be a mistake.

The following usability issues have been identified and are being studied for improvement in STARtracker version III:

- The reductions page revolves around BMPs. But user's – field personnel and supervisors - generally think and work in terms of facilities, not BMP's. The interface needs to be remapped to match the user's model.
- In a related vein, user's want to see all projects associated with a facility in one shot, in summary form (date, type, and reduction amount). Right now, they can only see all projects of a particular type at a facility.
- Once a facility is selected, user's want to be able to create any kind of BMP for it from that page (rather than switching BMP types). It turns out the users liked entering information through the detail page to enter a project. Although it presents a bit of a technology challenge (since fields must be generated on the fly). An icon-based toolbar may be added to provide easy access to create BMP types (with user's being able to define BMP icons).
- A better method is needed for selecting facilities. With large corporations having thousands of facilities and many levels in the hierarchy, the current

selection method can be confusing. A better search mechanism is needed.

- Mobile devices- field supervisors have requested a simple interface for reviewing and adding/entering reduction projects from mobile devices. The previous two bullets relate to this interface as well.
- A review / QC process workflow is needed for supervisors to check, approve, and possibly lock projects entered by personnel under them.
- Deleting (versus closing) a reductions project has potential implications for "changing reported history". Warnings and prompts have been added, but need to be improved as user's are sometimes confused by their meaning.
- Point-in-time and short duration reduction projects are specified by mcf "Total Reduction" (TR) in emissions, while ongoing projects are specified by an "Annual Reductions" (AR) rate. Some users have been confused by this distinction, and the interface must provide better clarification and distinction to the user between them.

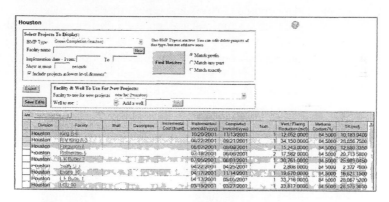

Figure 3 STARtracker Version II - Reduction Projects Page.

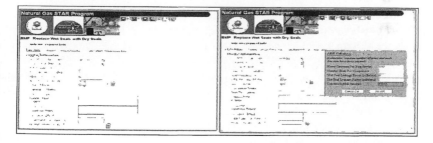

Figure 4 STARtracker Version I - Reduction Details Page.

DEFINE AND MANAGE BMPs

A significant feature of this application is the ability of companies to extend existing BMPs and add new BMPs over time. This required the creation of an internal calculation engine to handle user-defined formulas, as well as dynamic form creation, since BMP fields were no longer predefined. Figure 5 shows the screen used in STARtracker version II.

With this flexibility also comes complexity. Users must understand the distinction between the definition of a BMP (its fields, formulas, and help information), and instances of BMP projects (a green completion at a particular facility for instance). They must also understand the implications of changing an existing BMP definition (on history data), and understand concepts of formulas and formula execution order. Some new administrative users have struggled with the existing interfaces, and on at least a couple of occasions resulted in some major "woops" moments as a result of the confusion.

Identified usability issues under analysis for version III include:

- Provide additional information to users on how to choose the correct reduction type.
- Provide users with clearer information on impacts of new version creation, inactivating, and deleting on history and reporting.
- Provide additional ability to upload and manage information documents relating to the BMP.
- Provide BMP sharing and community capabilities, for instance:
  o The ability to share a BMP the company has developed with the EPA and other users of the STARtracker system
  o The ability to pull down (import) new BMPs from the EPA or other companies
  o RSS, blogging, and/or message & review boards pertaining to BMPs, to allow sharing of insights and information about BMPs with others in the community

OPPORTUNITIES

User feedback has been received that it would helpful if the application could guide users to where opportunities for reductions might lie. This capability is being considered for STARtracker version III. From an interface standpoint, it will involve:

- Additional data needing to be entered for each BMP type – such as what equipment types and factors (such as facility age) are relevant, and a means of estimating reductions and economic benefit.
- Additional data about facilities (equipment, ages) will need to be entered.
- An "intelligent" search mechanism for identifying opportunities.

EMISSIONS TRACKING AND REPORTING

The ultimate objective of the application is the generation of reports and charts summarizing emission reductions. These include being able to do so along geographic divisions (division, field, facility), by BMP type, and for different date ranges. Users also like to be able to export this data to spreadsheets for inclusion in things like monthly newsletters, which is provided currently in STARtracker. Users also want to be able to use the application to generate the annual report to the EPA, also a feature in STARtracker.

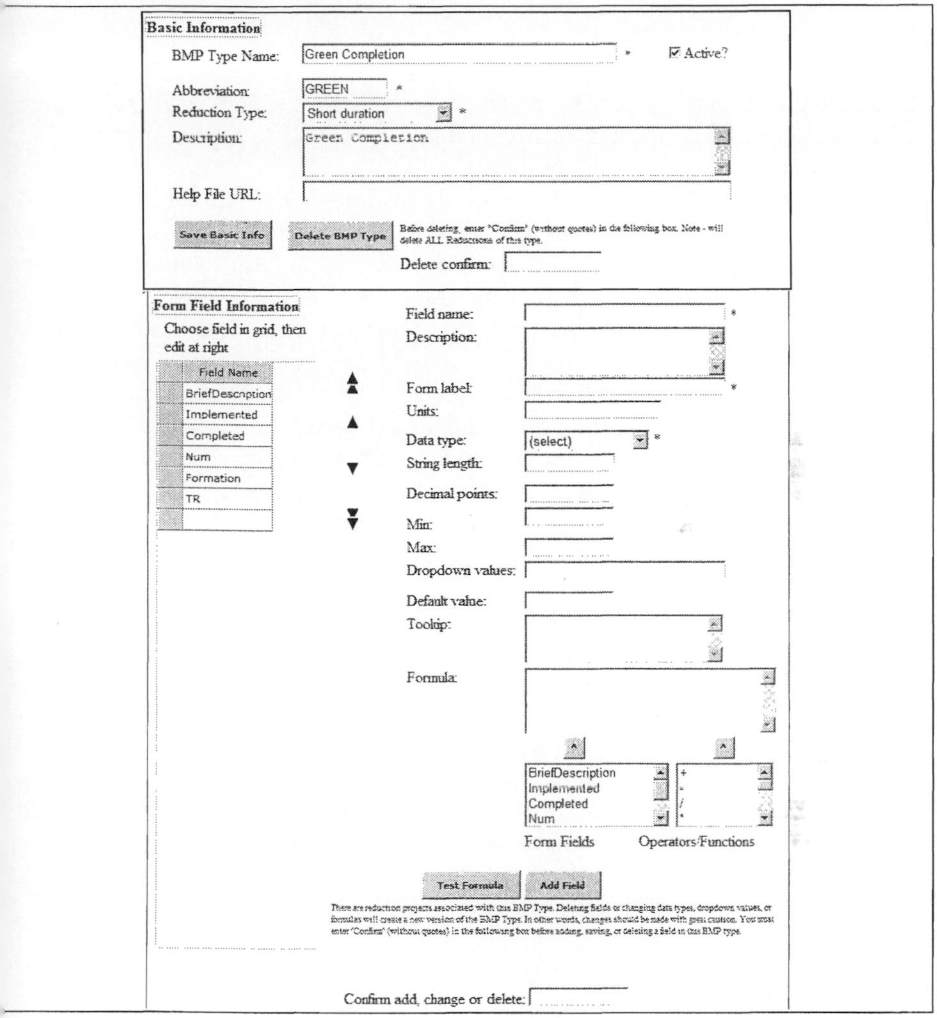

Figure 5 BMP Definition Form, STARtracker Version II.

Usability issues identified include the following:

- The current report query interface, while flexible, can be confusing to users. It has been suggested having some common defaults and moving off other options to an "Advanced" or "Extended Options" popup.
- Currently, reports only show total emission reductions. Users with smaller fields have indicated that this does not show relative progress versus the total amount of production in a field (reductions / production).

- User's want to be able to have direct email submission to EPA of the annual report right from the application, with capability to attach reports, and to automatically archive submitted reports and attachments.

FACILITIES MANAGEMENT

Production companies generally organize oil & gas facilities into geographic hierarchies. Reporting of emissions reduction corresponds to these hierarchies, so was important to retain in the application.

In version I of STARtracker, based on discussion with our prototype user pool we represented this through a fixed hierarchy of divisions and fields. Once the application went into the public, feedback came in that a more flexible representation was required as there were also country or global region levels, and some companies had levels between divisions and fields (or used different terminology for these). In version II, a more flexible representation was created allowing each company to define the number of layers in the hierarchy and the meaning to the organization. We also originally believed facilities would always be assigned to the lowest level in the hierarchy. As it turns out, some facilities fall outside recognized fields and therefore must be assigned higher up the hierarchy. Furthermore, we were surprised to find out how frequently fields and even divisions were reorganized. Reassigning facilities between fields or divisions was not possible within version I of STARtracker.

In version II, we improved the application by providing a tree interface for division management, which allows for any field, division, etc to be dragged and dropped to a different location within the hierarchy (see Figure 6). Adding and inactivating division nodes is also done right in the tree. And facilities are also now allowed to be assigned to non-leaf nodes in the tree.

A significant remaining usability issues are warnings and explanation on moving divisions. Moving divisions has implications for reduction history. Improvements are needed in making sure users understand the implications of such actions.

Figure 6 Divisions Management, STARtracker Version II.

MANAGE USERS

In discussions with users, it was determined that three access levels were sufficient:

- Administrator – manage users in their division; manage facilities and sub-divisions within assigned divisions; view/edit data in assigned divisions.
- Edit – could add and edit reduction projects within the assigned divisions, and view reports on these divisions.
- View – could view reduction reports for assigned divisions.

One set of users indicated they wanted the ability to assign each user different permissions at different levels in the hierarchy. For instance, a user might be administrator over a field but be assigned view permissions for an entire division so they could see how they compared to other areas. Version I provided this with a division / field capability. In version II, this was generalized to allow any level in an organizational hierarchy. No further usability issues have been identified in this use case.

CONCLUSION

The STARtracker application is being used by multiple companies with a variety of users of different technical capabilities. It is designed to be extensible (by defining and extending BMP project types), and has considerable complexities in scale (large number of facilities and projects), in project details (formulas and different project types), and history requirements. These characteristics present excellent opportunities for studying usability issues as pertains to the energy industry.

REFERENCES

EPA Natural Gas STAR program website (2010), http://www.epa.gov/gasstar/.

Tobin, J, Shambaugh, P, Mastrangelo, E. (2006), *Natural Gas Processing: The Crucial Link between Natural Gas Production and Its Transportation to Market, Natural Gas Production Data and Policy Papers,* Jan 2006 Special Report of the United States Energy Information Administration, Office of Oil and Gas.

CHAPTER 127

An Artificial Intelligent Approach for Evaluation of Teamwork Versus Health, Safety, Environment and Ergonomics (HSEE)

A. Azadeh, M. Rouzbahman, M. Saberi

Department of Industrial Engineering and Department of Engineering Optimization Research, Faculty of Engineering University of Tehran, Iran, P.O. Box 11365-4563

ABSTRACT

Researchers have been continuously trying to improve human performance with respect to HSE and ergonomics; hereafter referred to as HSEE. This study proposes a non-parametric efficiency frontier analysis method based on the adaptive neural network (ANN) technique for measuring job satisfaction among operators with respect to HSEE. In order to conduct this review, we have distributed questionnaires to four working shifts' operators in a gas refinery in which we used variety of questions related to HSE and Ergonomics. We have used ANN to predict the efficiency of HSE and Ergonomics and ranked the efficiencies. It should be noted that work groups have been used to satisfy the importance of teamwork in HSE and Ergonomics systems. By this approach, we assume some work teams according to some principles such as education, age and similarity in nature of works. The main objective of this article was Utilization of adaptive neural network for teamwork assessment versus HSE and Ergonomics. After defining system's efficiency against different shifts operators and ranking them, we began analyzing the results inside working groups. At first, we compared performances of individuals with teams and reviewed the average of these two by using T-test. We then reviewed the performances of different work groups (teams) by using ANOVA F-test.

Keywords: Neural network, Teamwork, HSE (Health, Safety, Environment), Ergonomics, ANOVA

INTRODUCTION

HSE-Ergonomics and Job Satisfaction

At the operational level, HSE attempts to decrease and eliminate dangerous accidents, contrary health influences and harmful events to the environment. In addition, by applying ergonomics in work system design we can achieve a balance between worker characteristics and task demands. By this means, we can enhance worker productivity and job satisfaction and provide more safe condition. Several studies have presented positive influences of using ergonomic principles to the workplace including machine, job and environmental design (Abou-Ali and Khamis , 2003; Shikdar and Sawaqed, 2004; Ayoub, 1990; Blanchard and Fabrychy, 1998; Azadeh et al., 2008a). There are close relationship between health, safety, environment and ergonomics factors. Improper design between operator and machine could lead to low level of safety. Moreover, improper design of system causes management error. Management error and work environment injurious factors could result to human error (Azadeh et al., 2008a). Essentially ergonomics is concerned with all those factors that can influence people and their behavior. HSE has defined human factors and ergonomics as the environmental, organizational and job factors, and human and individual characteristics which influence behavior at work. Careful consideration of human factors can improve health and safety by reducing the number of accidents and cases of ill-health at work.

Job satisfaction is defined as all the feelings that an individual has about his/her job (Spector, 1997). Researchers have tried to determine the various components of job satisfaction and determine the relative importance of each component of job satisfaction. They also want to know what influences these components have on operators' performance (Lu et al., 2005; Lu et al, 2007). A lot of factors are existed as significant factors in job satisfaction. Different researchers mentioned significant factors in job satisfaction.

Teamwork assessment with respect to Job satisfaction and HSEE

In this paper, the performance of HSE and Ergonomics against its operators has been reviewed and its efficiency has been ranked. In order to conduct this review, we have distributed questionnaires to different four working shifts' operators in a gas refinery in which we used variety of questions related to HSE and Ergonomics. We have used ANN to predict the efficiency of this integrated system by comparing ANN outputs with actual outputs and ranked these efficiencies. It should be noted that work groups have been used to satisfy the importance of team work in HSE and Ergonomics systems.

After organizing and analyzing the collected data from questionnaires, we divided the input indicators to four groups of Health, Safety, Environment and Ergonomics and defined job satisfaction as output indicator. In other words, we defined 4 groups of inputs (Health, Safety, Environment and Ergonomics) and a main output (Job Satisfaction) by bringing them out from our questionnaires. Later, we have analyzed the system efficiency against its operators by using ANN. The main aim of this note is to contribute to the use of neural networks in the efficiency measurement of HSE and Ergonomics system against 70 operators and 80 groups of operators in a gas refinery in Iran and Utilization of adaptive neural network for teamwork assessment versus HSSE. We have used some T- tests to compare teamwork with individual work and also we have used some ANOVA to compare different size of teams. To this end, for estimating output function, ANN method has been applied and for calculating the efficiency scores, a similar approach to econometric methods has been used. The Neural network efficiency will be determined using the predicted values obtained from the solution to the model. These efficiencies can be obtained by taking the ratio between the observed and predicted values for the output(s) of each operator.

ARTIFICIAL NEURAL NETWORK

Artificial neural networks (ANNs) are simply mathematical techniques and a consolidated technique in artificial and computational intelligence. ANN consists of interconnected neurons, and tries to simulate the behavior of biological neural networks. ANNs are generally used to model complex relationships and patterns (Azadeh et al., 2006; Azadeh et al., 2007; Azadeh et al., 2008b). Pattern recognition, data mining, identification, speech, vision, classification, forecasting and process modeling are some of main tasks of ANNs. Learning ability, generalization, parallel processing and error endurance are some of main features of ANNs in solving complex problems. Handling noise data and lack of data are other main features of neural networks. Neural networks can also be used to tackle problems that are difficult for human beings. Only one type of network, which is called the multi-layer perceptron (MLP) are discussed here. The data flows forward to the output continuously without any feedback n MLP. Error back propagation algorithm is the most applicable of learning rule in MLP (Azadeh et al., 2009b). Werbos proposed back propagation learning that is supervised learning (Werbos, 1997). Rumelhart and McClelland developed mentioned learning algorithm (Rumelhart and McClelland, 1986). The ANN uses a data set, which consists of input – desired output pattern pairs. There are three steps in working with ANN, (1) training, (2) testing and (3) implementation or recalling (Tang et al., 1991; Tang et al., 1993; Jhee and Lee, 1993; Hill et al., 1996; Kohzadi et al., 1996; Stern, 1996; Chiang et al., 1996; Indro et al., 1999; Brian, 2001; Azadeh et al., 2006; Azadeh et al.,2007; Azadeh et al., 2008b).

METHODOLOGY

For evaluation teamwork versus HSEE an integrated ANN algorithm is presented as follows (Azadeh et al., 2006; Azadeh et al., 2007):

1. Define inputs and output of our model and Collect data set S.
2. Divide S in to two subsets: training (S1) and test (S2) data.
3. Use ANN method to estimate relation between input(s) and output(s) by selecting architecture of ANN model, training data S1 and after that testing data S2. Select the best network architecture (ANN*) from the testing data error (MAPE) (Azadeh et al., 2006; Azadeh et al., 2007).
4. Run ANN* for all operators and calculate ANN model output (PANN*(i)) (Azadeh et al., 2006; Azadeh et al., 2007).
5. Estimate efficiency scores with real output (Preal(i)) and ANN model output (PANN*(i))(Azadeh et al., 2006; Azadeh et al., 2007).
6. Comparing individuals with work groups by using T-test.
7. Analyzing performances of different work groups (teams) by using ANOVA F-test.

THE EXPERIMENT

DATA COLLECTION

4 detailed questionnaires containing valuable information related to human factors, safety, environmental, management, teamwork and training are developed and presented to operators. Operators answer to the questions and we gave a score (weight) to each of their answers between 0 and 1. In order to conduct this review, we need defining inputs and outputs that are related with our system. After organizing the collected data from questionnaires, we divided the input indicators to four main categories which are Health, Safety, Environment, and Ergonomic and defined an output related to job satisfaction. The main objective of this article was Utilization of adaptive neural network for teamwork assessment versus HSSE. It should be noted that work teams have been used to satisfy the importance of team work in HSE and Ergonomics systems. We have used some T- tests to compare teamwork with individual work and also we have used some ANOVA to compare different size of teams.

Running the Proposed Algorithm

Step 1: As shown before, the first step is Determination of input(s) and output (P) variables of the model. We have chosen 4 main categories (Health, Safety, environment, Ergonomic) as input variables and analyze questions that were related with those categories. Then, for each category, we use the average score as final score for using in purposed ANN. In addition, we have chosen 1 main question among our questionnaires as output data. That was a question related to Job Satisfaction. 150 rows of data are collected from 70 operators, 34 mixtures of two operators, 22 mixtures of three operators, 13 mixtures of four operators and finally 11 mixtures of five operators.

Step 2: S1 is 135 rows of data for training and S2 is 15 rows of data for testing.

Step 3: In order to get the best ANN for this study, 25 ANN categories are tested to find the best model for the output (Job Satisfaction). We run each of these categories several times by changing their number of neurons from 1 to 100. After changing number of neurons from 1 to 100 for each category and analyzing their MAPE, we found min MAPE for each category. The architecture of the best ANN MLP model and its MAPE value is shown in Table 1.

Step 4: In order to found an ANN output for each operator, we tested our purposed model for all 150 operators.

Table 1: Architecture of the best ANN-MLP model and its associated relative error (MAPE)

Learning method	Number of neurons in first hidden layer	First transfer function	Second transfer function	MAPE
GDX	19	Tansig	Purelin	0.195696

Step 5: the results of this step is shown in Table 2. This table shows Estimation of efficiency scores for all operators by means of the proposed algorithm and rank operators according to their efficiencies.

Table 2: Estimation of efficiency scores for all operators by the proposed algorithm

tor m	Members of team	Preal (i)	PANN(i)	Fi	Rank	Operator /team	Members of team	Preal (i)	PANN(i)	Fi	Rank
	1	2	1.515565	1	1	51	51	1.5	1.580301	0.690053	104
	2	1.5	1.293303	0.86736	45	52	52	2	1.415656	1	1
	3	2	1.563409	1	1	53	53	2	1.570808	0.932774	14
	4	1.5	1.318483	0.812098	72	54	54	2	1.810075	0.842884	58
	5	2	1.55894	1	1	55	55	1.5	1.335435	0.794049	76
	6	2	1.720708	0.957713	7	56	56	2	1.333112	1	1
	7	2	1.212751	1	1	57	57	1.25	1.581329	0.558409	119
	8	2	1.449789	0.935155	13	58	58	2	1.744765	0.837292	62
	9	1.5	1.382058	0.751672	88	59	59	2	1.680461	0.86451	46
	10	1.5	1.604307	0.695611	102	60	60	1.75	1.661254	0.766089	85
	11	2	1.700764	0.908405	25	61	61	1.75	1.50455	0.82648	69
	12	2	1.486636	1	1	62	62	1.25	1.26193	0.669839	109
	13	1.5	1.378826	0.809228	73	63	63	1.75	1.521962	0.827284	68
	14	1.5	1.451203	0.792794	77	64	64	1.5	1.354388	0.773555	83
	15	2	1.379728	1	1	65	65	1.75	1.469171	0.856037	52
	16	1	0.948658	0.652564	110	66	66	1.25	1.50977	0.601735	115
	17	2	1.530915	0.962401	5	67	67	1.25	1.50977	0.604203	114
	18	2	0.97497	1	1	68	68	1.5	1.56926	0.707699	98
	19	2	1.887183	0.699714	100	69	69	2	1.434713	1	1
	20	2	1.439995	0.846523	55	70	70	1.5	1.381008	0.773455	84
	21	1	0.930028	0.55183	121	71	1,2	1.75	1.480387	0.861938	48
	22	1.5	1.471524	0.648845	111	72	3,4	1.5	1.483364	0.740304	90
	23	2	1.37517	0.918564	18	73	6,7	1.5	1.166982	0.881124	35
	24	1	1.01982	0.558168	120	74	8,9	1.5	1.210457	0.862759	47
	25	2	1.646134	0.838911	61	75	10,11	2	1.382533	1	1
	26	1	1.363389	0.481759	130	76	12,26	1	1.37397	0.503581	126
	27	2	1.507672	0.912879	21	77	14,16	1.5	1.502938	0.712397	95
	28	1	1.305658	0.508369	124	78	13,15	1.75	1.479081	0.844032	57
	29	1.5	1.567304	0.680396	107	79	20,21	1.75	1.306545	0.9243	17
	30	2	1.37694	1	1	80	22,23	1.5	1.323982	0.787809	78
	31	1	1.365834	0.507305	125	81	24,25	1.75	1.492279	0.847635	54
	32	1.5	1.596316	0.687581	105	82	28,29	1.5	1.350654	0.782662	80
	33	2	1.673137	0.893051	32	83	17,18	1.5	1.245979	0.831006	67
	34	2	1.610167	0.925966	15	84	27,30	2	1.309224	1	1
	35	1	1.179305	0.582929	116	85	5,19	1	1.567622	0.443846	131
	36	1.5	1.189847	0.877166	39	86	32,37	2	1.612383	0.874476	42
	37	2	1.373597	1	1	87	39,40	1	1.252654	0.52021	122
	38	1	1.446467	0.485713	129	88	31,38	2	1.545752	0.906984	26
	39	2	1.256227	1	1	89	35,36	1.75	1.634079	0.765296	86
	40	1	1.274507	0.499352	128	90	33,34	2	1.656947	0.868936	44
	41	1	1.283509	0.501557	127	91	41,43	2	1.69979	0.855639	53
	42	2	1.280896	1	1	92	46,48	1.5	1.523888	0.69577	101
	43	2	1.369837	0.965138	4	93	63,64	1.833	1.567828	0.836152	64
	44	2	1.382619	0.966612	3	94	51,52	1.5	1.29637	0.783323	79
	45	2	1.430623	0.951542	8	95	49,50	1.833	1.345032	0.936999	12
	46	1	1.701804	0.423535	133	96	42,47	1.5	1.597983	0.68069	106
	47	1	1.61045	0.443329	132	97	53,54	1.33	1.46077	0.645444	112
	48	2	1.565559	0.911252	22	98	56,57	1.5	1.187495	0.842325	60
	49	2	1.7573	0.842566	59	99	59,60	1.666	1.264664	0.899779	29
	50	2	1.677477	0.876591	40	100	61,62	1.666	1.458704	0.816772	71

Operator /team	Members of team	Preal (i)	PANN(i)	Fi	Rank
101	44,45	1.833	1.616053	0.836625	63
102	65,66	1.33	1.396824	0.676074	108
103	67,68	1.5	1.595826	0.694344	103
104	69,70	1.833	1.530852	0.877388	38
105	1,9,11	1.5	1.247658	0.83269	66
106	5,6,7	1.33	1.290837	0.722936	94
107	38,39,40	1.666	1.392223	0.860878	50
108	17,18,19	1.833	1.562283	0.872898	43
109	16,22,23	1.666	1.371919	0.874554	41
110	14,15,21	1.666	1.613278	0.777966	82
111	24,25,26	1.833	1.60396	0.861755	49
112	12,13,20	1.875	1.539732	0.911047	24
113	28,33,36	2	1.708	0.900294	28
114	30,31,32	1.625	1.273511	0.911242	23
115	38,39,40	1.625	1.668436	0.747548	89
116	34,35,37	1.5	1.545414	0.732906	92
117	41,42,43	1.75	1.29684	0.975863	2
118	2,3,4	1.75	1.483777	0.88562	33
119	44,45,49	1.5	1.307697	0.835052	65
120	50,51,52	1.375	1.403548	0.728162	93
121	53,54,55	1.625	1.489714	0.824879	70
122	56,57,65	1.875	1.711156	0.857335	51
123	46,47,48	1.875	1.743293	0.846394	56
124	59,60,61	1.875	1.617356	0.899047	30
125	62,63,64	1.5	1.643604	0.711321	96

Operator /team	Members of team	Preal (i)	PANN(i)	Fi
126	68,69,70	1.9	1.685547	0.885263
127	1,2,3,11	1.6	1.648347	0.759739
128	6,7,8,9	1.6	1.293676	0.915469
129	14,15,16,21	1.6	1.546717	0.801098
130	17,18,19,20	1.7	1.413847	0.913614
131	23,24,25,26	1.3	1.390179	0.708686
132	33,34,35,36	1.7	1.452278	0.898331
133	27,38,39,40	1.5	1.269097	0.879115
134	28,29,30,31	1.8	1.468946	0.946213
135	41,42,43,44	1.6	1.388852	0.879427
136	53,54,55	1.625	1.489714	0.824879
137	56,57,65	1.875	1.711156	0.857335
138	46,47,48	1.875	1.743293	0.846394
139	59,60,61	1.875	1.617356	0.899047
140	62,63,64	1.5	1.643604	0.711321
141	68,69,70	1.9	1.685547	0.885263
142	1,2,3,11	1.6	1.648347	0.759739
143	6,7,8,9	1.6	1.293676	0.915469
144	14,15,16,21	1.6	1.546717	0.801098
145	17,18,19,20	1.7	1.413847	0.913614
146	23,24,25,26	1.3	1.390179	0.708686
147	33,34,35,36	1.7	1.452278	0.898331
148	27,38,39,40	1.5	1.269097	0.879115
149	28,29,30,31	1.8	1.468946	0.946213
150	41,42,43,44	1.6	1.388852	0.879427

Step 6: After defining system's efficiency against different operators and ranking them, we began analyzing the results inside working groups. First, we compared performances of individuals with teams and reviewed the average of these two by using T-test. We assumed two samples of operators including 70 operators for individuals and 80 operators for teams. Table 3 is shown results of T- tests.

Table 3: Comparison of individuals with groups for job satisfaction

F-Test Two-Sample for Variances		
	Variable 1	Variable 2
Mean	0.805803	0.810756
Variance	0.03236	0.013027
Observations	70	80
df	69	79
F	2.484154889	
P(F<=f) one-tail	5.41992E-05	
F Critical one-tail	1.466502899	

T-Test: Two-Sample Assuming Unequal Variances		
	Variable 1	Variable 2
Mean	0.805803	0.810756
Variance	0.03236	0.013027
Observations	70	80
Hypothesized Mean Difference	0	
df	114	
t Stat	-0.198113	
P(T<=t) one-tail	0.421654	
t Critical one-tail	1.65832	
P(T<=t) two-tail	0.843309	
t Critical two-tail	1.980992234	

Step 7: We then reviewed the performances of different work groups (teams) by using ANOVA F-test (Single Factor). In Table 4 you can see the results of this test. Once again the equal average assumption was not rejected between different sizes of groups. From the above experiment we concluded that teamwork was not effective for job satisfaction among operators of this gas refinery.

Table4: ANOVA table for comparison of different size of teams for output (Job Satisfaction)

SUMMARY				
Groups	Count	Sum	Average	Variance
Column 1	34	26.75695	0.786969	0.020795
Column 2	22	17.74994	0.806815	0.007044
Column 3	13	11.05539	0.850415	0.005791
Column 4	11	9.298276	0.845298	0.007238

ANOVA						
Source of Variation	SS	df	MS	F	P-value	F crit
Between Groups	0.0531	3	0.0177	1.3795	0.255	2.724
Within Groups	0.9760	76	0.0128			
Total	1.0291	79				

CONCLUSION

A highly unique flexible ANN algorithm was proposed to measure and rank the HSEE's efficiency against operators because of nonlinearity of the neural networks in addition to its universal approximations of functions and its derivates, which makes them highly flexible. The proposed algorithm is composed of eight distinct steps. To show its applicability and superiority it was applied to a set of operators in a gas refinery. We have used ANN to predict the efficiency of HSEE against its operators and ranked the efficiency after comparing it with the actual data from system performance. For each operator we have a score between 0.42 and 1 that shows the performance of HSEE system against its operator is between 42 %(min) and 100 %(max). After defining system's efficiency against different shifts operators and ranking them, we began analyzing the results inside working groups. In the beginning we compared performances of individuals with teams and reviewed the average of these two by using T-test. The assumption of equal averages wasn't rejected for job satisfaction. We then reviewed the performances of different work groups (teams) by using ANOVA F-test. Once again the equal average assumption was not rejected for job satisfaction. From the above experiment we concluded that teamwork was not effective for job satisfaction among operators. Although, it is believed that ANNs can be a potential alternative for measuring technical efficiency, there is still a lack of both theoretical and empirical work in efficiency analysis and consequently optimization analysis.

REFERENCES

Abou-Ali, M.G., and Khamis, M. (2003), "An integrated intelligent defects diagnostic system for tire production and service." *Expert Systems with Applications*, 24, 247–259.

Ayoub, M.A. (1990), "Ergonomic deficiencies: I. Pain at work." *Journal of Occupational Medicine* 32 (1), 52–57.

Azadeh, A., Mohammad Fam, I., Khoshnoud, M., and Nikafrouz, M. (2008a), "Design and implementation of a fuzzy expert system for performance assessment of an integrated health, safety, environment (HSE) and ergonomics system: The case of a gas refinery." *Information Sciences*, 178, 4280–4300.

Azadeh, A., Ghaderi, S.F., Anvari, M., and Saberi, M. (2006), "Measuring performance electric power generations using artificial neural networks and fuzzy clustering." *In: Proceedings of the 32nd Annual Conference of the IEEE Industrial Electronics Society—IECON'06, Conservatoire National des Arts & Metiers,* Paris, France.

Azadeh, A., Ghaderi, S.F., Anvari, M., and Saberi, M. (2007), "Performance assessment of electric power generations using an adaptive neural network algorithm." Energy Policy, 35, 3155–3166.

Azadeh, A., Ghaderi, S.F., and Sohrabkhani, S. (2008b), "Annual electricity consumption forecasting by neural network in high energy consuming industrial sectors." *Energy Conversion and Management*, 49, 2272–2278.

Blanchard, W., and Fabrychy, J. (1998), "System Engineering and Analysis." *Prentice-Hall International*, Inc., USA, 112–123.

Brian Hwarng, H. (2001), "Insights into neural-network forecasting of time series corresponding to ARMA(p;q) structures." *Omega*, 29, 273–289.

Chiang, W.C., Urban, T.L., and Baldridge, G.W. (1996), "A neural network approach to mutual fund net asset value forecasting." Omega, *The International Journal of Management Science,* 24, 205–215.

Hill, T., O'Connor, M., and Remus, W. (1996), "Neural network models for time series forecasts. Management Science", 42 (7), 1082–1092.

Indro, D.C., Jiang, C.X., Patuwo, B.E., and Zhang, G.P. (1999), "Predicting mutual fund performance using artificial neural networks." *Omega, The International Journal of Management Science*, 27, 373–380.

Jhee, W.C., and Lee, J.K.(1993), "Performance of neural networks in managerial forecasting." *Intelligent Systems in Accounting Finance and Management*, 2, 55–71.

Kohzadi, N., Boyd, M.S., Kermanshahi, B., and Kaastra, I. (1996), "A comparison of artificial neural network and time series models for forecasting commodity prices." Neurocomputing, 10 (3), 169–181.

Lu, H., Alison E., While, K., and Barriball, L. (2005), "Job satisfaction among nurses: a literature review." *International Journal of Nursing Studies*, 42, 211–227.

Lu, H., Alison E., While, K., and Barriball, L. (2007), "Job satisfaction and its related factors: A questionnaire survey of hospital nurses in Mainland China." *International Journal of Nursing Studies,* 44, 574–588.

Shikdar, A.A., and Sawaqed, M.N. (2004), "Ergonomics, occupational health and safety in the oil industry: a managers' response." *Computers and Industrial Engineering*, 47, 223–232.

Spector, P.E. (1997), "Job Satisfaction: Application, Assessment, Causes, and consequences" *SAGE Publications*, London.

Stern, H.S. (1996), "Neural networks in applied statistic." Technometrics, 38 (3), 205–214.

Tang, Z., Almeida, C., and Fishwick, P. (1991), "A Time series forecasting using neural networks vs. Box–Jenkins methodology." *Simulation*, 57 (5), 303–310.

Tang, Z., and Fishwick, P.A. (1993), "A Back-propagation neural nets as models for time series forecasting." *ORSA Journal on Computing*, 5 (4), 374–385.

Werbos, P.I. (1974), "Beyond regression: new tools for prediction and analysis in the behavior sciences" (Ph.D. Thesis, Harvard University, Cambridge, MA).

Rumelhart, D.E., and McClelland, J.L. (1986), "Parallel Distributed Processing: Explorations in the Microstructure of Cognition." *Foundations*, 1. MIT Press, Cambridge, MA.

Printed and bound by CPI Group (UK) Ltd, Croydon, CR0 4YY

23/10/2024

01778268-0002